Springer Mineralogy

More information about this series at http://www.springer.com/series/13488

Nikita V. Chukanov • Marina F. Vigasina

Vibrational (Infrared and Raman) Spectra of Minerals and Related Compounds

Volume 1

Springer

Nikita V. Chukanov
Institute of Problems
of Chemical Physics
Russian Academy of Sciences
Chernogolovka, Russia

Marina F. Vigasina
Geological Faculty
Moscow State University
Moscow, Russia

Additional material to this book can be downloaded from http://extras.springer.com.

ISSN 2366-1585 ISSN 2366-1593 (electronic)
Springer Mineralogy
ISBN 978-3-030-26805-3 ISBN 978-3-030-26803-9 (eBook)
https://doi.org/10.1007/978-3-030-26803-9

This Springer imprint is published by the registered company Springer Nature Switzerland AG.
The registered company address is: Gewerbestrasse 11, 6330 Cham, Switzerland

Alexandr Dmitrievich Chervonnyi
1948–2017

This book is dedicated to the memory of the outstanding scientist, a specialist in the field of inorganic materials and chemistry of rare-earth elements Dr. Alexandr Dmitrievich Chervonnyi. Him belongs the idea to publish this book.

Preface

This volume is the third and final part of the series of reference books on vibrational spectra of minerals. Unlike the two previous parts (Chukanov 2014; Chukanov and Chervonnyi 2016), this book contains not only infrared (IR) spectra of minerals but also data on their Raman spectra.

In Chap. 1, numerous examples of the application of IR spectroscopy to the analysis of crystal-chemical features of minerals are considered. In particular, spectral bands that characterize different local situations around OH^- and BO_3^{3-} groups in vesuvianite-group minerals are revealed. The effect of symmetry on the parameters of IR spectra of vesuvianite-group minerals is discussed. By means of IR and Raman spectroscopic methods, it is shown that the clathrate mineral melanophlogite is not a single species but a mineral group including minerals with different combinations of small molecules (CO_2, CH_4, H_2S, N_2, H_2O, C_2H_6) entrapped in structural cages. Based on numerous IR spectra of nakauriite samples from different localities, it is demonstrated that this mineral does not contain sulfate groups, and its tentative simplified formula $(Mg_3Cu^{2+})(OH)_6(CO_3)\cdot 4H_2O$ is suggested. A close crystal chemical relationship between nepskoeite and shabynite is demonstrated based on their IR spectra, compositional, and X-ray diffraction data. Contrary to the formula $Mg_4Cl(OH)_7\cdot 6H_2O$ accepted for nepskoeite, this mineral is a borate with the tentative simplified formula $Mg_5(BO_3)(Cl,OH)_2(OH)_5\cdot nH_2O$ ($n > 4$). Consequently, shabynite may be a product of nepskoeite dehydration. Based on IR spectroscopic data, it is also shown that some nominally boron-free lead carbonate minerals (molybdophyllite, hydrocerussite, plumbonacrite, somersetite) often contain minor BO_3^{3-} admixture which is overlooked in structural and chemical analyses.

Chapter 2 contains IR spectra of 1024 minerals and related compounds which were not included in the preceding reference books of this series (Chukanov 2014; Chukanov and Chervonnyi 2016). Most spectra are accompanied by the information about the origin of reference samples, methods of their identification, and analytical data.

In Chap. 3, possibilities, advantages, and shortcomings of Raman spectroscopy as a method of investigation and identification of minerals are discussed. Numerous examples illustrate capabilities of Raman spectroscopy in identification of minerals and analysis of their crystal chemical features, orientation, and polarization effects, selection rules, as well as difficulties encountered in the study of microscopic inclusions in minerals and minerals that are unstable under laser beam.

Chapter 4 contains data on 2104 Raman spectra of more than 2000 mineral species taken from various periodicals. The data are accompanied by some experimental details and information on the reference samples used.

A supplementary chapter provides comments on published IR spectra which are erroneous, dubious, or of poor quality. This chapter is provided by a separate list of references.

This work was carried out with assistance of numerous colleagues. The working partnership with Prof. I.V. Pekov, Dr. A.D. Chervonnyi, and Dr. S.A. Vozchikova was the most important.

Reference samples and valuable analytical data were kindly granted by A.V. Kasatkin, S. Jančev, E. Jonssen, R. Hochleitner, E.V. Galuskin, S. Weiss, N.V. Sorokhtina, Ł. Kruszewski, and many other mineralogists, as well as mineral collectors, of which the contribution of R. Kristiansen, G. Möhn, W. Schüller, B. Ternes, G. Blass, B. Dünkel, S. Möckel, and C. Schäfer was the most important. Collaboration with the crystallographers N.V. Zubkova, R.K. Rastsvetaeva, S.M. Aksenov, D.I. Pushcharovsky, T.L. Panikorovskii, O.I. Siidra, S.N. Britvin, M.G. Krzhizhanovskaya, D.A. Ksenofontov, S.V. Krivovichev, and I. Grey, as well as with specialists in different areas of geosciences and analytical methods (J. Göttlicher, K.V. Van, D.A. Varlamov, V.N. Ermolaeva, D.I. Belakovskiy, Yu.S. Polekhovsky, P. Voudouris, A. Magganas, A. Katerinopoulos, N.V. Shchipalkina, V.O. Yapaskurt, L.A. Pautov, V.S. Rusakov, R. Scholz, A.R. Kampf, S. Encheva, P. Petrov, Ya.V. Bychkova, N.N. Koshlyakova, P. Yu. Plechov, C.L.A. de Oliveira, I.S. Lykova, and T.S. Larikova) was especially fruitful. All of them are kindly appreciated.

This work was partly supported by the Russian Foundation for Basic Research, grant no. 18-29-12007. A part of analytical data on reference samples was obtained in accordance with the Russian Government task, registration no. 0089-2016-0001.

Chernogolovka, Russia Nikita V. Chukanov
Moscow, Russia Marina F. Vigasina

Reference

Chukanov NV, Chervonnyi AD (2016) Infrared spectroscopy of minerals and related compounds. Springer, Cham. (1109 pp)

Contents

Contents for Volume 2

Some Examples of the Use of IR Spectroscopy in Mineralogical Studies 1

1.1 Characteristic Bands in IR Spectra of Vesuvianite-Group Minerals

Vesuvianite-group minerals (VGM) are widespread and occur in different geological formations including regional metamorphic rocks, skarns, rodingites, etc. Specific crystal-chemical features of these minerals reflect conditions of their crystallization. As a rule, high-temperature VGM have high-symmetry structures (space group *P*4/*nnc*), whereas low-temperature samples are characterized by the symmetry *P*4/*n* or *P*4*nc* (Allen and Burnham 1992). The simplified crystal-chemical formula of VGM is $X_{18}(X'Y1)Y2_4Y3_8T_{0-5}(SiO_4)_{10}(Si_2O_7)_4O_{1\text{-}2}W_9$ where *X*, X' = Ca, Na, K, Fe^{2+}, and *REE* (cations with coordination numbers from 7 to 9); *Y*1–*Y*3 = Al, Mg, Fe^{2+}, Fe^{3+}, Mn^{2+}, Mn^{3+}, Ti, Cr, Cu, Zn; *T* = B, Al, □; *W* = OH, F, O. The *Y*1 cations have tetragonal-pyramidal coordination, whereas the *Y*2 and *Y*3 cations occur in octahedra.

IR spectra of VGM are discussed in numerous publications (Paluszkiewicz and Żabiński 1992; Groat et al. 1995; Kurazhkovskaya and Borovikova 2003; Kurazhkovskaya et al. 2005; Borovikova and Kurazhkovskaya 2006); however, in most cases their interpretation is ambiguous. We have obtained IR spectra of 33 VGM samples from different kinds of localities which have been preliminarily investigated in detail using electron microprobe (including determination of boron), single-crystal X-ray structural analysis, DSC, ^{27}Al NMR, ICP-MS, and Mössbauer spectroscopy. As a result, characteristic IR bands corresponding to different local situations in the structures of VGM have been revealed. Data on crystal structures, crystal chemistry, and IR spectra of these samples are published by Britvin et al. (2003), Panikorovskii et al. (2016a–d, 2017a–d), and Aksenov et al. (2016). The most important results of this investigation are listed below in comparison with data published elsewhere.

1.1.1 O-H-Stretching Vibrations

The following empirical correlations between O–H stretching frequencies in IR spectra of minerals and O···O and H···O distances (from structural data) were established by E. Libowitzky (1999):

$$\nu\left(\text{cm}^{-1}\right) = 3592 - 304 \cdot 10^9 \cdot \exp\left[-d(\text{O}\cdots\text{O})/0.1321\right] \quad (1.1)$$

$$\nu\left(\text{cm}^{-1}\right) = 3632 - 1.79 \cdot 10^6 \cdot \exp\left[-d(\text{H}\cdots\text{O})/0.2146\right] \quad (1.2)$$

N. V. Chukanov, M. F. Vigasina, *Vibrational (Infrared and Raman) Spectra of Minerals and Related Compounds*, Springer Mineralogy, https://doi.org/10.1007/978-3-030-26803-9_1

Two decades ago this publication was of a great importance because it emphasized the existence of such correlations as a general trend. However, over time it became obvious that the Eqs. (1.1) and (1.2) are a very rough approximation and have a restricted applicability. First, it is to be noted that at high frequencies (above 3500 cm^{-1}) substantial deviations from the correlations (1.1) and (1.2) are common because O–H stretching frequencies depend not only on O···O and H···O distances, but also on the nature of cations coordinating O–H groups and H_2O molecules, as well as on the angle O–H···O, and the influence of these factors becomes most evident in case of weak hydrogen bonds. The Eqs. (1.1) and (1.2) predict that maximum possible values of O–H stretching frequencies for minerals are 3592 and 3632 cm^{-1} respectively. However, in many minerals including magnesium serpentines, brucite, kaolinite, amphiboles, etc. observed frequencies are much higher and even can exceed 3700 cm^{-1}.

In the IR spectra of VGM some absorption bands of O–H stretching vibrations are poorly resolved. In such cases, band component analysis is the most important source of errors and artifacts during data processing because of low correctness of inverse mathematical problems: small errors in experimental data lead to strong uncertainty of the final result. Additional uncertainty is connected with arbitrary choice of the band shape (Gauss, Lorentz, Voigt, or Lorentz-Gauss cross-product function), the number of components, and the acceptable values of the correlation coefficient R (e.g., 0.99, 0.995, or 0.999). This matter is discussed in detail by Chukanov and Chervonnyi (2016) (the section 1.1 "Sources of Errors and Artifacts in IR Spectroscopy of Minerals") where it is shown that different variants of band shape analysis may give a good and almost identical approximation accuracy (say, $R^2 \approx 0.9995$), but lead to totally different results.

For most VGM investigated by Chukanov et al. (2018) there are significant discrepancies between wavenumbers of observed O–H stretching bands and ν values calculated using correlations suggested by Libowitzky (1999). The above considerations explain why the attempts to apply Eqs. (1.1) and (1.2) to VGM failed.

Groat et al. (1995) distinguished 13 bands of O–H stretching vibrations in IR spectra of VGM, which have absorption maxima at the following wavenumbers (cm^{-1}): 3670 (A), 3635 (B), 3596 (C), 3567 (D), 3524 (E), 3487 (F), 3430 (G), 3383 (H), 3240 (I), 3210 (J), 3156 (K), 3120 (L), and 3054 (M). The polarization of these bands with respect to the fourfold c axis is as follows: $E \wedge c < 35°$ for the A–H bands, $E \perp c$ for the I band, and $E \| c$ for the J–M bands (Groat et al. 1995; Bellatreccia et al. 2005). Consequently, the bands A–H and J–M can be assigned to the vibrations of differently coordinated O11–H1 and O10–H2 groups, respectively. The I band was tentatively assigned to O–H stretching vibrations of silanol group (Chukanov et al. 2018).

Our data show that actually significant deviations of the A–M band positions from the "ideal" values indicated by Groat et al. (1995) take place. In particular, IR spectra of many VGM samples contain a band in the range 3440–3470 cm^{-1}, i.e., between F and G bands. Taking into account that in the group of 33 chemically and structurally investigated samples the intensity of this band shows distinct positive correlation with Ti content, it was assigned to vibrations of the O11–H1 group coordinated by Ti (Chukanov et al. 2018). Most Ti-rich samples are characterized by the space group *P4/nnc*. The only exception is a sample from the Ahkmatovskaya open pit, South Urals with 1.54 *apfu* Ti, space group *P4/n* showing bands at 3488 cm^{-1} (with a shoulder at 3460 cm^{-1}) and 3424 cm^{-1} instead of a single band in the range 3440–3470 cm^{-1} (Fig. 1.1).

Vesuvianite from the Ahkmatovskaya open pit is the only Ti-bearing VGM having space group *P4/n* among 33 samples investigated by Chukanov et al. (2018). The observed splitting of the band of TiO–H stretching vibrations is the result of distribution of Ti between the sites *Y*3A and *Y*3B, whereas in high-temperature VGM having the space group *P4/nnc* Ti is accumulated in the single *Y*3 site.

IR band in the range 3375–3380 cm^{-1} which is close to the H band by Groat et al. (1995) was observed by us only for two samples with the symmetry *P4/n* and high contents of Cu and

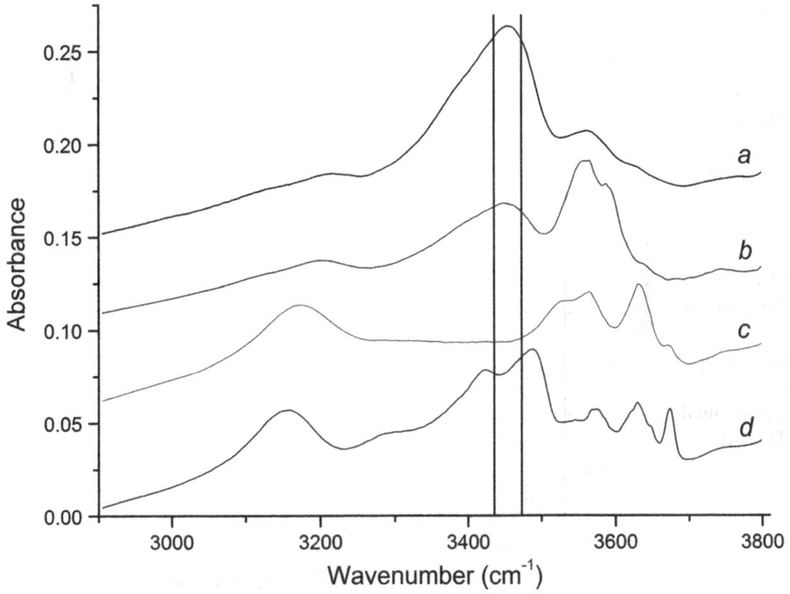

Fig. 1.1 IR spectra of vesuvianite-group minerals with different contents of Ti in the region of O–H-stretching vibrations: a sample from Alchuri, Shigar Valley, Pakistan (Aksenov et al. 2016) with $Ti_{2.21}$ (*a*); a sample from Hazlov, Karlovy Vary Region, Czech Republic with $Ti_{0.48}$ (*b*); a sample from Myrseter area, Drammen, Buskerud, Norway with $Ti_{0.00}$ (*c*); VGM from the Ahkmatovskaya open pit, South Urals, Russia with $Ti_{1.54}$ (anomalous Ti-rich sample, space group *P4/n*) (*d*). Two vertical lines outline the region of the band corresponding to the Ti···O11–H1 group in *P4/nnc* VGM

Mn^{3+}. This band is more intense in the IR spectrum of the sample from the N'Chwaning III mine, Kuruman, South Africa with a relatively higher content of Mn^{3+} (1.83 *apfu*). Based on these data, the band in the range 3375–3380 cm^{-1} can be assigned to the $^{Y3}Mn^{3+}$···O11–H1 group.

The nominal position of the D band is 3567 cm^{-1} (Groat et al. 1995), but in IR spectra of some samples this band is shifted towards lower wavenumbers (up to 3560 cm^{-1}). The D band is not observed in IR spectra of F-poor VGM and has the highest intensities in IR spectra of samples with most high contents of F (Britvin et al. 2003; Galuskin et al. 2003; Chukanov et al. 2018; see Fig. 1.2). Taking into account polarization $E \wedge c < 35°$ (Groat et al. 1995), the D band is to be assigned to the group O11–H1 in the situation when F occupies neighboring O11 site.

Galuskin et al. (2003) supposed that the J band corresponds to OH groups in the O10 site coordinating Fe in *Y*1 and forming hydrogen bond with F in the neighboring O10 site. However, this assumption was not confirmed by our investigations: IR spectra of most VGM, including F- and Fe-poor ones, contain distinct J band whose wavenumber varies from 3190 to 3225 cm^{-1}. These values correspond to strong hydrogen bonds, which is hardly possible in cases when F is the H-bond acceptor.

The weak B band (in the range from 3628 to 3632 cm^{-1}) is often observed in IR spectra of low-symmetry VGM. This band corresponds to very weak H-bonds formed by the groups O11–H1 with low values of the angle between O11–H1 and H1···O7 (see Lager et al. 1999).

1.1.2 B–O-Stretching Vibrations

In VGM boron can occupy sites with coordination numbers 3 or 4. IR spectra of most VGM samples contain shoulders in the range 1070–1170 cm^{-1} corresponding to stretching vibrations of $[BO_4]$ tetrahedra.

BO_3 groups are connected with $Y1O6_4O10$ polyhedra *via* O10 oxygen atom to form the cluster $T2Y1O_7$ (Fig. 1.3) where $T2 = B$ and $Y1 = Fe^{3+}$, Fe^{2+}, Mn^{3+}, Cu^{2+}, Al, or Mg. As a result, four

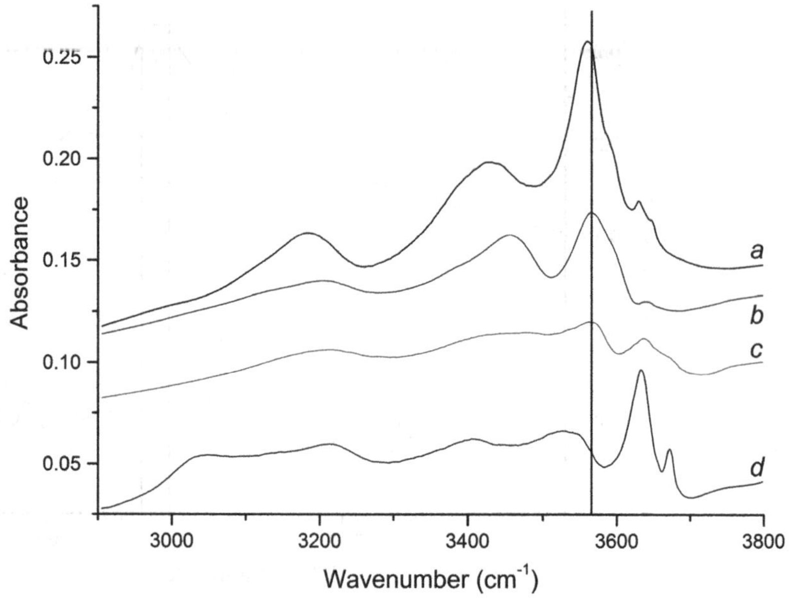

Fig. 1.2 IR spectra of vesuvianite-group minerals with different contents of F in the region of O–H-stretching vibrations: fluorvesuvianite holotype with 7.16 *apfu* F (*a*); VGM from Sakharyok massif, Keyvy Mts., Kola Peninsula with 3.06 *apfu* F (*b*); VGM from Gulshad, Kazakhstan with 0.24 *apfu* F (*c*); F-free alumovesuvianite holotype (*d*). Vertical line corresponds to the nominal position of the D band

degrees of freedom corresponding to the bond lengths *T*2–O12 (×2), *T*2–O10, and *Y*1–O10 are involved in stretching vibrations of BO_3^{3-}. This results in four nondegenerate modes and, consequently, the expected number of absorption bands in the region of stretching vibrations of BO_3 groups (i.e., 1200–1570 cm^{-1}) is equal to 4. However, in the IR spectrum of wiluite only three bands are observed in this region: the peaks at 1267 and 1373 cm^{-1} and the shoulder at 1415 cm^{-1} (Panikorovskii et al. 2017b; see Fig. 1.4). The fourth band corresponding to symmetric vibrations of BO_3^{3-} is forbidden for a regular BO_3 triangle and is weak in case of weak-distorted BO_3 triangle. The latter case takes place in wiluite: the bond lengths *T*2–O10 and *T*2–O12 are 1.39–1.40 and 1.32 Å, respectively (Panikorovskii et al. 2017b). Weak absorption between 1267 and 1373 cm^{-1} may correspond to the symmetric stretching mode of BO_3^{3-} groups (Fig. 1.4).

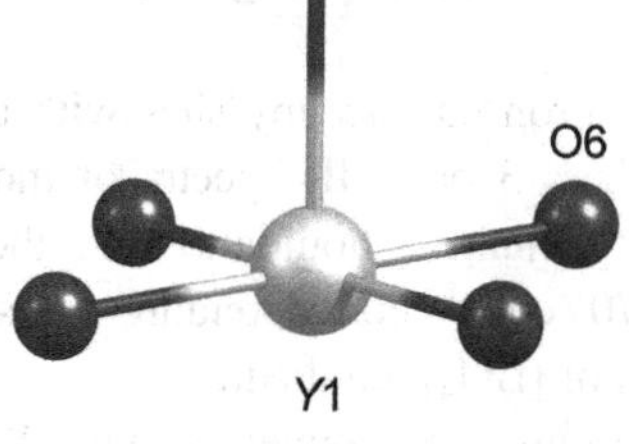

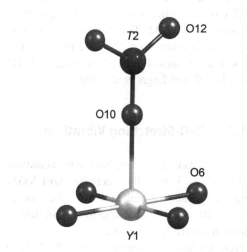

Fig. 1.3 Local environment of the *T*2 and *Y*1 sites in vesuvianite-group minerals

Unlike wiluite, most boron-rich VGM contain significantly distorted BO_3 triangle. For example, in a sample from Gulshad, Kazakhstan the bond lengths *T*2–O10 and *T*2–O12 are equal to 1.384 and 1.20 Å, respectively. As a result, four distinct IR bands (at 1557, 1467, 1419, and 1365 cm^{-1}) are observed in the range 1200–1570 cm^{-1} (curve *b* in Fig. 1.4). As compared to wiluite, these bands are substantially shifted towards high frequencies because of shorter B–O bonds and shortened *Y*1–O bond (2.044 Å for the mineral from Gulshad and 2.15 Å for wiluite: Chukanov et al. 2018; Panikorovskii et al. 2017b). These differences may be due to different predominant cations in the *Y*1 site: Mg in wiluite and Fe^{3+} in the sample from Gulshad.

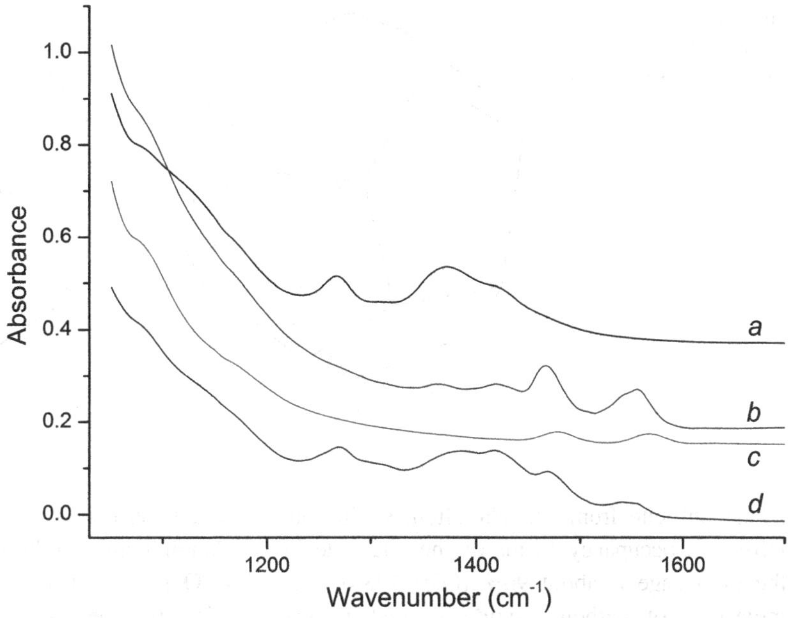

Fig. 1.4 IR spectra of vesuvianite-group minerals in the region of B–O-stretching vibrations: wiluite from its type locality (*a*), B-rich VGM from Gulshad (*b*), a typical B-bearing vesuvianite from Somma-Vesuvius complex, Italy (*c*), and anomalous B-rich VGM from Titivskoe (*d*)

An anomalous IR spectrum with six absorption bands in the range 1200–1570 cm^{-1} shows boron-rich VGM from Titivskoe boron deposit, Yakutia, Russia (curve *d* in Fig. 1.4). Structural investigation of this sample (Panikorovskii et al. 2016a) showed the presence of domains with different symmetry (*P*4/*nnc* and *P*4/*n*).

1.1.3 Stretching and Bending Vibrations of SiO_4^{4-} and $Si_2O_7^{6-}$ Groups

Based on the available data on IR spectra of a restricted set of VGM Kurazhkovskaya and Borovikova (2003) concluded that for low-symmetry samples the band of Si–O-stretching vibrations in the range from 960 to 990 cm^{-1} is shifted on 10–15 cm^{-1} towards lower frequencies as compared to high-symmetry VGM. Our data confirm this conclusion only partly. Indeed, among nine samples with the space group *P*4/*n*, eight samples show strong IR bands in the range 962–968 cm^{-1}, and in the IR spectrum of one more sample a band at 973 cm^{-1} is observed. Among 21 boron-poor samples with the space group *P*4/*nnc*, for 16 samples bands in the range 976–986 cm^{-1} are observed, but 5 samples show bands between 962 and 968 cm^{-1}.

Another specific feature of low-symmetry VGM indicated by Kurazhkovskaya and Borovikova (2003), as well as by Borovikova and Kurazhkovskaya (2006) is the doublet ~575 +615 cm^{-1} corresponding to O–Si–O bending vibrations. This regularity was confirmed by us as a general trend; however, among 21 boron-poor samples with the space group *P*4/*nnc*, 3 samples show doublets in the range ~575–615 cm^{-1} with components of approximately equal intensity.

1.2 Problem of Melanophlogite

Melanophlogite is a clathrate compound which contains guest molecules N_2, CO_2, and CH_4 entrapped within the cages of the 3D framework built by SiO_4 tetrahedra (Gies 1983; Nakagawa et al. 2001; Kolesov and Geiger 2003). The cubic unit cell of the structure of melanophlogite includes two $[5^{12}]$ cages and six $[5^{12}6^{2}]$ cages (Fig. 1.5). The former are considered to be occupied mainly by CH_4 molecules and the latter by N_2 and CO_2 (Gies et al. 1982; Gies 1983). In

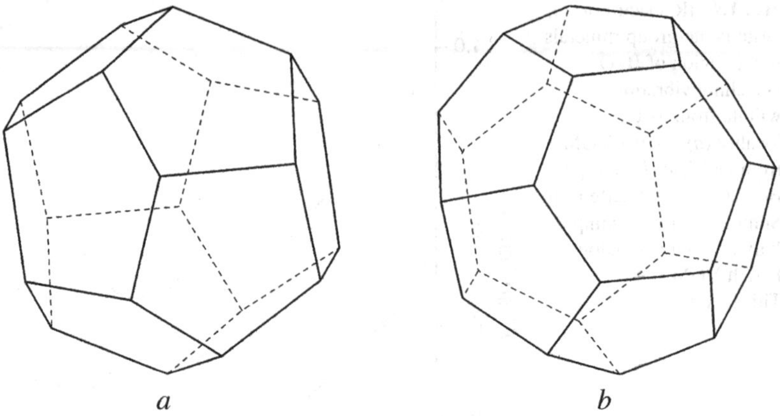

Fig. 1.5 $[5^{12}]$ cages (**a**) and $[5^{12}6^2]$ cages (**b**) in the structure of melanophlogite

melanophlogite from Mt. Hamilton, California, USA, the occupancy factor of the CH_4 site in the $[5^{12}]$ cage is about 90% (Gies 1983). The molecules of carbon dioxide located in the $[5^{12}6^2]$ cages can rotate and are statistically distributed between 12 possible equivalent orientations. Most part of N_2 and CO_2 occurs in the $[5^{12}6^2]$ cage, but minor part of these molecules can be present in the $[5^{12}]$ cage (Gies 1983; Kolesov and Geiger 2003). Based on the available structural data, the general formula of melanophlogite can be written as follows: $(CH_4,N_2,CO_2)_{2-x}(N_2,CO_2)_{6-y}(Si_{46}O_{92})$.

The IR spectrum of melanophlogite from Fortullino, Livorno province, Tuscany, Italy (cubic, with $a = 13.4051(13)$ Å, according to single-crystal X-ray diffraction data) contains a strong band at 2330–2336 cm^{-1} corresponding to antisymmetric vibrations of CO_2 molecules (Chukanov and Chervonnyi 2016; see Fig. 1.6).

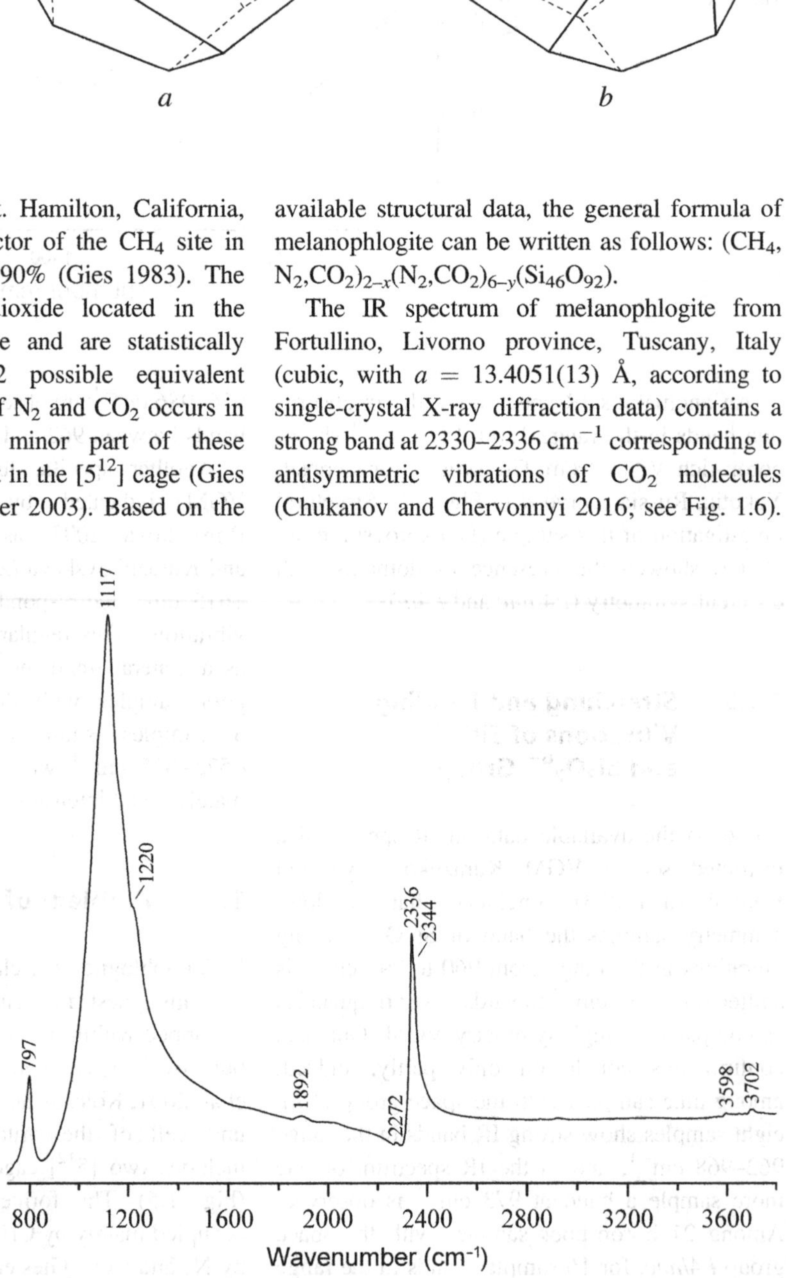

Fig. 1.6 Powder IR spectrum of melanophlogite from Fortullino, Italy

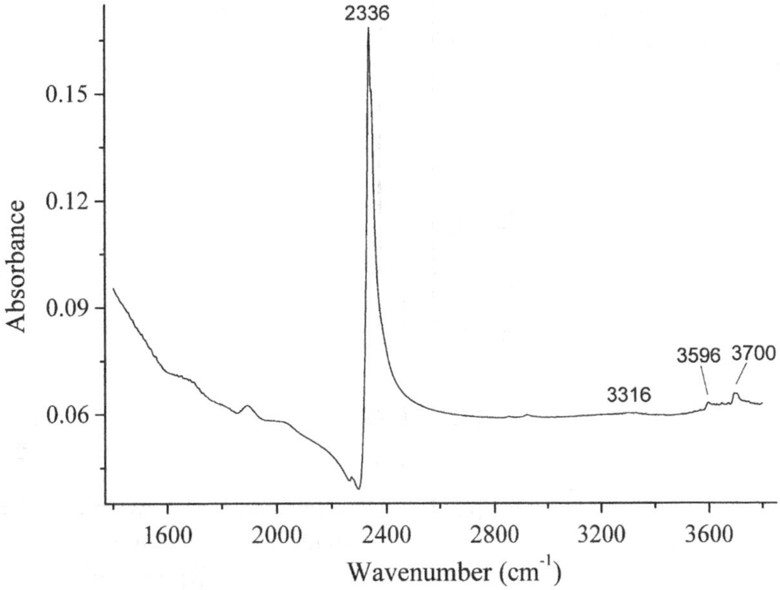

Fig. 1.7 IR spectrum of melanophlogite from Fortullino in the 1400–3800 cm^{-1} range. Very weak bands in the range from 2800 to 3000 cm^{-1} correspond to grease impurity

A weaker peak at ~2375 cm^{-1} present it IR spectra of some melanophlogite samples from this locality (Chukanov 2014) may be due to rotational splitting or correspond to a minor amount of CO_2 molecules in the [5^{12}] cages, which are predominantly occupied by CH_4. Weak bands at 3700 and 3596 cm^{-1} (Fig. 1.7) are, respectively, due to asymmetric and symmetric stretching vibrations of H_2O molecules that do not form hydrogen bonds. The band at 3316 cm^{-1} can be tentatively assigned to silanol groups or H-bonded H_2O molecules.

Two CO_2 modes at 1277 and 1378 cm^{-1} observed in the Raman spectrum of melanophlogite from Mt. Hamilton, California, USA correspond to the first overtone of the ν_2-bending mode and the symmetric ν_1-stretching mode, respectively, both bands being components of a vibrational system coupled via Fermi resonance (Kolesov and Geiger 2003).

Raman spectrum of melanophlogite from Mt. Hamilton has been investigated previously at 4 K (Kolesov and Geiger 2003). The bands at 2900 and 2909 cm^{-1} in the single-crystal Raman spectrum of melanophlogite from this locality have been assigned to asymmetric stretching modes of CH_4 located in the [5^{12}] and [$5^{12}6^2$] cages, respectively. Along with the main band at 1378.5 cm^{-1} assigned to CO_2 molecules in the [$5^{12}6^2$] cages, the shoulder at 1376 cm^{-1} was registered and was attributed to CO_2 in the [5^{12}] cages. Kolesov and Geiger (2003) also reported the presence of the band of symmetric N≡N-stretching vibrations located at 2321 cm^{-1} which corresponds to N_2 molecules. The IR forbidden band of N_2 in binary mixtures with other molecules has been observed at about 2328 cm^{-1} (Bernstein and Sandford 1999). It is to be noted that this value is close to the wavenumber of antisymmetric vibrations of CO_2 molecules, which is forbidden in Raman spectra. However, this band is not observed by us in Raman spectra of melanophlogite from different Italian localities (Figs. 1.8, 1.9 and 1.10).

Raman spectrum of the sample from Fortullino (Fig. 1.8) exhibits strong bands of CO_2 at 1277 and 1383 cm^{-1}. Two weak bands at 1257 and 1398 cm^{-1} accompanying the components of the Fermi-doublet are the so-called hot bands arising from the transitions from an excited vibrational level to the ground vibrational level (Wang et al. 2011). In accordance with Frezzotti et al. (2012) and Wang et al. (2011), carbon dioxide in this mineral is in the high density fluid state with $D \approx 1$ g/cm^3. This is confirmed by the down shift of the Fermi doublet frequencies from 1388

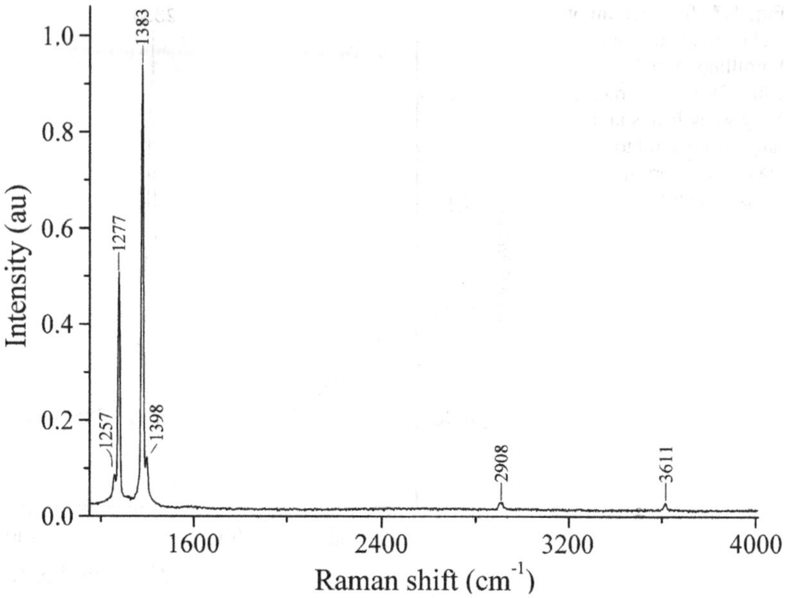

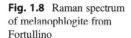

Fig. 1.8 Raman spectrum of melanophlogite from Fortullino

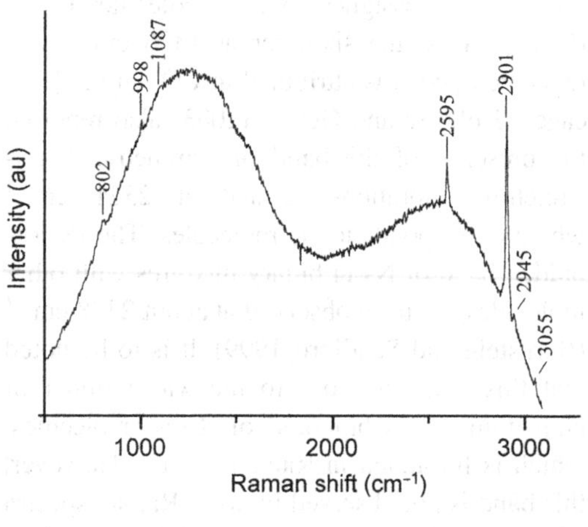

Fig. 1.9 Raman spectrum of melanophlogite from Racalmuto

and 1285 cm^{-1} ($\Delta = 103$ cm^{-1}) for CO_2 at normal conditions to the values 1383 and 1277 cm^{-1} ($\Delta = 106$ cm^{-1}); the increased value of Δ in the spectrum of melanoflogite from Fortullino (Fig. 1.8) also corresponds to a high-density CO_2 fluid. The very weak band at 2908 cm^{-1} which is attributed to CH_4 and the band at 3611 cm^{-1} attributed to OH-groups are in a good agreement with the IR spectrum of this sample (Chukanov and Chervonnyi 2016). Thus, nitrogen, which is a substantial component in melanophlogite from Mt. Hamilton, is absent in the sample from Rio Fortullino.

Melanophlogite from Racalmuto, Sicily (Fig. 1.9) is characterized by a higher content of CH_4 detected by the bands at 2901 and 3055 cm^{-1} and by a trace amount of C_2H_6 detected by the very weak bands at 998 and 2945 cm^{-1} (Kohlrausch 1943; Momma et al. 2011). The presence of H_2S molecules is detected by the presence in the Raman spectrum of the band at 2595 cm^{-1}. This sample does not contain N_2 and CO_2 molecules in detectable amounts. The sample of melanophlogite from Racalmuto contains a small amount of mineral impurities of calcite or aragonite (the band at 1087 cm^{-1}: Edwards et al. 2005) and celestine (998 cm^{-1}: Frezzotti et al. 2012). The weak band at 802 cm^{-1} and the bands at 363 and 276 cm^{-1} relate to the spectrum of the melanophlogite-type silicon-oxygen framework.

Melanophlogite from Miniera Giona, Sicily (Fig. 1.10) contains H_2S molecules (the band of H–S-stretching vibrations at 2593 cm^{-1}) and a low amount of CH_4 molecules (2904 cm^{-1}) whereas bands of N_2 and CO_2 are not observed in its Raman spectrum. This sample contains anhydrite admixture detected by the Raman bands at 1132 and 1004 cm^{-1} (Frezzotti et al.

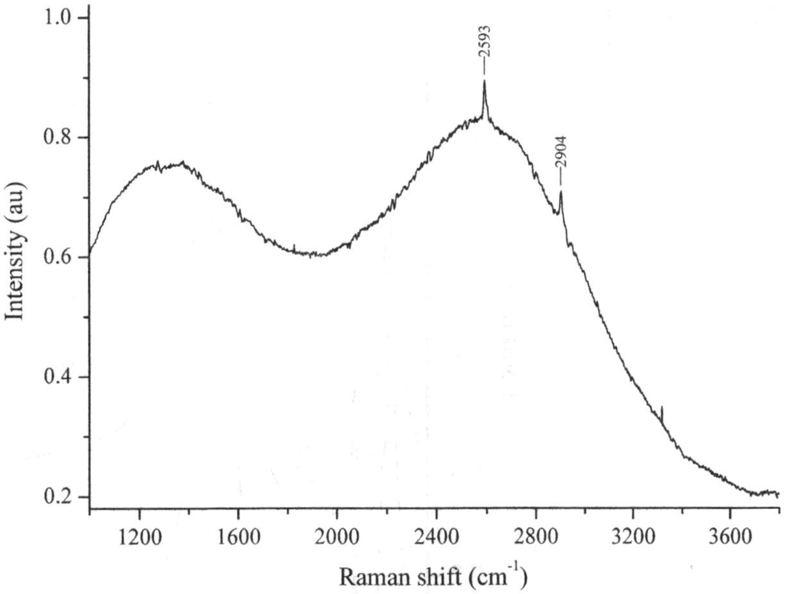

Fig. 1.10 Raman spectrum of melanophlogite from Miniera Giona

2012). Broad peaks in Figs. 1.9 and 1.10 are due to fluorescence.

Powder IR spectra of melanophlogite from some other localities do not show any presence of methane molecules. In the frequency range from 2800 to 3000 cm^{-1}, a sample from Chvaletice, Bohemia shows the presence of three overlapping, relatively broad bands indicative of the contamination by a polyatomic aliphatic hydrocarbon, most probably grease. Similar but much weaker bands are present in the IR spectrum of melanophlogite from Fortullino, Italy (Fig. 1.7), but no characteristic bands of methane are observed in this spectrum too.

The IR spectra of melanophlogite samples from Racalmuto and Miniera Giona (Fig. 1.11) show much lower contents of CO_2 (the bands at 2337–2338 cm^{-1}) and substantially higher contents of CH_4 (the bands at 3005–3008 cm^{-1}) as compared with the sample from Fortullino. Moreover, the bands in the range from 2850 to 2950 cm^{-1} in the IR spectrum of melanophlogite samples from Miniera Giona indicate the presence of hydrocarbons heavier than methane. IR spectrum of the sample from Racalmuto confirms the presence of H_2S (Fig. 1.12).

The band assignment and the distribution of different components between $[5^{12}]$ and $[5^{12}6^2]$ cages in melanophlogite samples from different localities are given in Table 1.1. These examples show that, in all probability, melanophlogite is not a single mineral species, but a mineral group including minerals with different combinations of small molecules (CO_2, CH_4, H_2S, N_2, H_2O, C_2H_6) entrapped in the $[5^{12}]$ and $[5^{12}6^2]$ cages. In particular, the mineral from Fortullino may be the CO_2-dominant analogue of melanophlogite, and in the samples from Racalmuto and Miniera Giona H_2S may be a species-defining component.

1.3 Problem of Nakauriite

Nakauriite was initially described as a new mineral with the general formula $(Mn,Ni,Cu)_8(SO_4)_4(CO_3)(OH)_6{\cdot}48H_2O$ (Suzuki et al. 1976). The mineral occurs in fissure-fillings in brucite-bearing serpentine at Nakauri, Aichi Prefecture, Japan, and is intimately intergrown with chrysotile. Most analytical data for nakauriite, including powder X-ray diffraction pattern, chemical composition and IR spectrum, have been obtained for a polymineral mixture, in

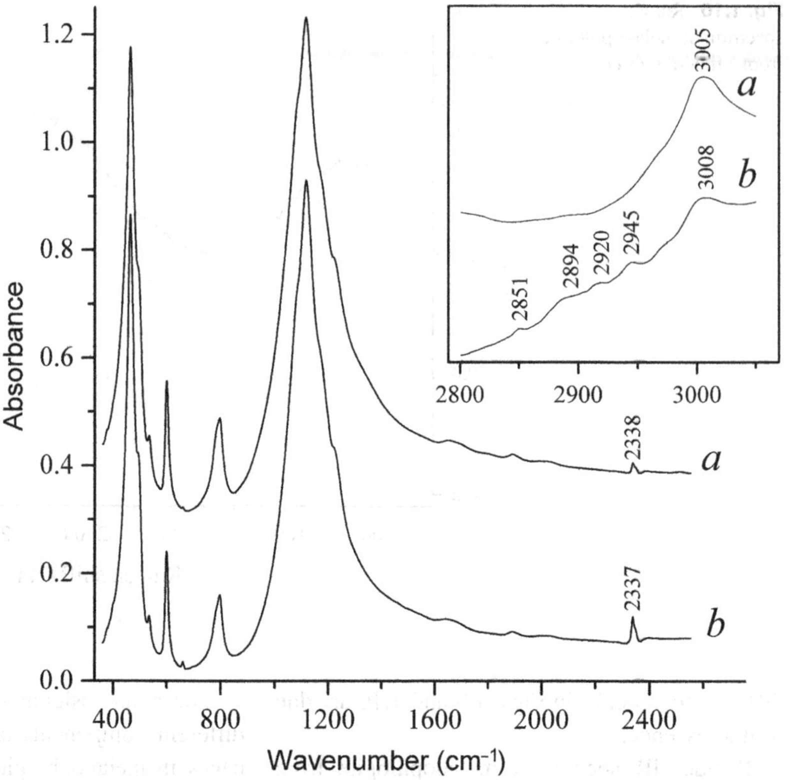

Fig. 1.11 Powder IR spectra of melanophlogite from Racalmuto (*a*) and Miniera Giona (*b*)

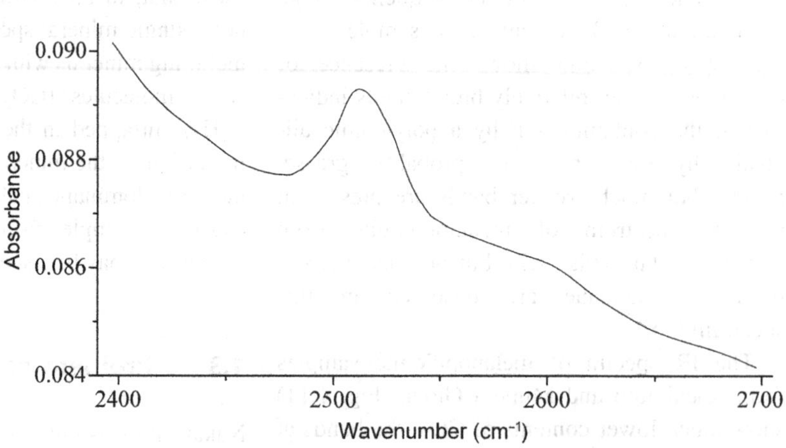

Fig. 1.12 IR spectrum of melanophlogite from Racalmuto in the region of S–H-stretching vibrations

which chrysotile is the main phase. The above formula does not conform to analytical data of Suzuki et al. (1976). Taking into account strong predominance of Cu over Mn and Ni, the simplified formula $Cu_8(SO_4)_4(CO_3)(OH)_6{\cdot}48H_2O$ is given for nakauriite in the IMA list of minerals. Published IR spectrum of nakauriite from its type locality contains strong bands of admixed chrysotile at 1075, 950 and 613 cm^{-1} (Suzuki et al. 1976; see curve *a* in Fig. 1.13), but characteristic bands of sulfate anions are not observed.

Later nakauriite was reported from several localities in Great Britain, USA, Austria, Italy, and Russia (Braithwaite and Pritchard 1983;

Table 1.1 The assignment of IR and Raman bands and the distribution of different enclathrated components between $[5^{12}]$ and $[5^{12}6^2]$ cages for melanophlogite samples from different localities

Locality	Wavenumber (cm^{-1})	Assignment
Mt. Hamilton[a]	3050w (R)	CH_4 (first overtone of the doubly degenerate ν_2-bending mode?)
	2909w (R)	CH_4 in $[5^{12}6^2]$ (symmetric stretching mode)
	2900 (R)	CH_4 in $[5^{12}]$ (symmetric stretching mode)
	2321 (R)	N_2 in $[5^{12}6^2]$
	1378.5 (R)	CO_2 in $[5^{12}6^2]$ (symmetric stretching mode)
	1376 (R)	CO_2 in $[5^{12}]$ (symmetric stretching mode)
	1277 (R)	CO_2: Fermi resonance between symmetric stretching mode and first overtone of the bending mode
Fortullino	3700 (IR)	H_2O (antisymmetric stretching mode)
	3611 (R)	H_2O (?) O–H-stretching mode
	3596 (IR)	H_2O (symmetric stretching mode)
	3316 (IR)	H_2O (?) O–H-stretching mode (H-bonded OH)
	2908w (R)	CH_4 in $[5^{12}6^2]$
	2344 (IR)	$^{12}CO_2$ in $[5^{12}]$ (antisymmetric stretching mode)
	2336 (IR)	$^{12}CO_2$ in $[5^{12}6^2]$ (antisymmetric stretching mode)
	2273w (IR)	$^{13}CO_2$ (antisymmetric stretching mode)
	1398w (R)	CO_2 ("hot band")
	1383 (R)	CO_2 in $[5^{12}6^2]$ (symmetric stretching mode)
	1377 (R)	CO_2 in $[5^{12}]$ (symmetric stretching mode)
	1277 (R)	CO_2: Fermi resonance between symmetric stretching mode and first overtone of the bending mode
	1257w (R)	CO_2 ("hot band")
Racalmuto	3055w (R)	Hydrocarbon other than methane
	3005w (IR)	CH_4 in $[5^{12}]$ (asymmetric stretching mode)
	2945w (R)	C_2H_6?
	2901 (R)	CH_4 in $[5^{12}]$ (symmetric stretching mode)
	2595 (R)	H_2S (symmetric stretching mode)
	2512 (IR)	H_2S (antisymmetric stretching mode)
	2338w (IR)	$^{12}CO_2$ in $[5^{12}6^2]$ (antisymmetric stretching mode)
Miniera Giona	3008w (IR)	CH_4 in $[5^{12}]$ (asymmetric stretching mode)
	2945w (IR)	Hydrocarbon other than methane
	2920w (IR)	Hydrocarbon other than methane
	2904w (R)	CH_4 in $[5^{12}]$ (symmetric stretching mode)
	2894w (IR)	Hydrocarbon other than methane
	2851w (IR)	Hydrocarbon other than methane
	2593 (R)	H_2S (symmetric stretching mode)
	2337 (IR)	$^{12}CO_2$ in $[5^{12}6^2]$ (antisymmetric stretching mode)

R Raman spectrum, *IR* infrared spectrum, *w* weak band
[a]Data from Kolesov and Geiger (2003)

Barnes 1986; Postl and Moser 1988; Palenzona and Martinelli 2007; Popov et al. 2016), as well as a secondary product in metallurgic slags. All available data indicate that nakauriite (1) does not contain sulfate anion and (2) usually forms fine intergrowths with magnesium serpentine-group minerals (see curve *b* in Fig. 1.13). In particular, electron microprobe analyses of nakauriite from Japan and from Nevada (Peacor et al. 1982) do not show substantial amounts of sulfur. Microchemical tests show only traces of SO_4^{2-} (Braithwaite and Pritchard 1983).

We managed to obtain an almost pure fraction of nakauriite from its type locality. Its IR

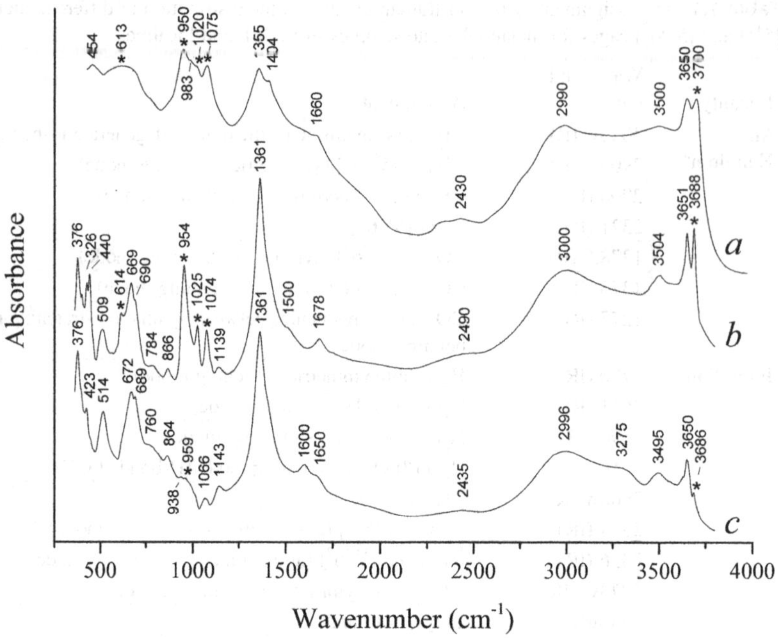

Fig. 1.13 IR spectra of nakauriite-bearing samples from Nakauri drawn using data from Suzuki et al. (1976) (*a*), and from Karkodin, South Urals, Russia (*b*) and IR spectrum of an almost pure nakauriite from Nakauri (*c*). Bands of admixed serpentine are marked by asterisk

spectrum is given in Fig. 1.13 (curve *c*). Specific features of the mineral are unusually low frequency of the non-split band of C–O-stretching vibrations (1361 cm^{-1}), as well as relatively low intensity and high width of the band of out-of-plane bending vibrations of CO_3^{2-} groups at 864 cm^{-1}. The bands at 1600 and 1650 cm^{-1} indicate the presence of H_2O molecules. The same features are inherent in carbonate members of the hydrotalcite group (Chukanov 2014) whose structures contain brucite-like layers and interlayer CO_3^{2-} anions and water molecules. Weak bands at 1066 and 1143 cm^{-1} may be due to the presence of trace amounts of SO_4^{2-} anions.

Our electron microprobe analyses of nakauriite samples from Nakauri (Japan), Karkodin, (Russia) and Monte Ramazzo mine (Italy) show that this mineral is Mg-dominant. Only three metal cations have been found, namely Mg, Cu^{2+}, and Ni^{2+}. The atomic ratio Mg:Cu:Ni in a sample from Nakauri is 0.75:0.23:0.02. In another sample from Nakauri Mg:Cu:Ni = 0.80:0.19:0.01. In the sample from Karkodin Mg:Cu:Ni = 0.75:0.23:0.02. Nakauriite from Monte Ramazzo does not contain Ni in detectable amounts; Mg:Cu = 0.80:0.20. In all analyzed samples the content of sulfur is below its detection limit. The predominance of Mg over Cu in nakauriite from Karkodin was noted also by Popov et al. (2016).

Thermal data for nakauriite from Suzuki et al. (1976) are nearly consistent with the general simplified formula $(Mg_3Cu^{2+})(OH)_6(CO_3)\cdot 4H_2O$. Hypothetically, nakauriite can be an interstratification of brucite and copper carbonate modules. Powder X-ray diffraction data of nakauriite from South Urals kindly provided by I.V. Pekov are given in Table 1.2.

Suzuki et al. (1976) reported the absence of the reflection near 7.8 Å in some samples from Nakauri. It was supposed that this reflection may be due to an impurity (Peacor et al. 1982; Braithwaite and Pritchard 1983). However, nakauriite from Karkodin shows a relatively strong reflection at 7.88 Å (Table 1.2). In our opinion, the reflections at 7.32 and 7.88 Å may correspond to nakauriite-type phases with different contents of interlayer water.

A further detailed investigation of nakauriite and revision of its chemical formula are required.

Table 1.2 Powder X-ray diffraction data of nakauriite from Karkodin, South Urals, Russia (MoKα radiation)

d, Å	*I*, %	*d*, Å	*I*, %	*d*, Å	*I*, %
7.88	19	2.845	4	1.5947	1
7.32	100	2.677	4	1.5762	1
5.11	12	2.612	8	1.5447	12
4.835	17	2.536	4	1.5185	5
4.629	12	2.365	43	1.5107	4
4.485	14	2.258	7	1.4569	5
4.217	4	2.221	4	1.4210	3
3.928	12	2.096	1	1.4035	3
3.642	13	2.037	2	1.3523	1
3.539	15	1.956	3	1.2985	4
3.335	6	1.910	24	1.2910	4
3.305	8	1.820	2	1.2404	1
3.105	4	1.782	3	1.2281	2
3.065	4	1.7206	4	1.2170	2
2.976	3	1.6951	2	1.2021	2
2.939	3	1.6657	2	1.1973	2
2.874	4	1.6439	5	1.1082	1

1.4 Relationship Between Nepskoeite and Shabynite

Nepskoeite was initially described as a new chloride-hydroxide mineral with the formula $Mg_4Cl(OH)_7{\cdot}6H_2O$ (Apollonov 1998). However, IR spectrum of nepskoeite published by Apollonov (1998) contains very strong band at 1297 cm^{-1} corresponding to stretching vibrations of orthoborate groups, as well as distinct band at cm^{-1} that may correspond to bending vibrations of orthoborate groups (Fig. 1.14).

Our reinvestigation of nepskoeite from the type material confirmed the presence of boron in this mineral. In particular, IR spectrum contains strong bands at 1301 and 734 cm^{-1} corresponding to stretching and bending vibrations of BO_3^{3-} anions, respectively (Chukanov 2014; see Fig. 1.15). Moreover, color reaction with quinalizarin shows a high content of boron in nepskoeite.

IR spectrum of nepskoeite shows similarity with that of shabynite $Mg_5(BO_3)(Cl,OH)_2(OH)_5{\cdot}4H_2O$ (Fig. 1.16). In particular, the IR spectrum of shabynite contains strong bands at 1302 and 732 cm^{-1}. However, substantial differences between these spectra are observed in the region of vibrations of H_2O molecules above 1500 cm^{-1}.

There is also close similarity between powder X-ray diffraction (PXRD) patterns of nepskoeite and shabynite (Table 1.3). However, nepskoeite shows strong reflections at 10.64 and 3.498 cm^{-1} that are absent in the PXRD pattern of shebynite. On the other hand, strongest reflections of different shabynite samples are observed at 9.62 and 3.191 cm^{-1}. Refraction indices of nepskoeite ($\alpha = 1.532$, $\beta \approx \gamma = 1.562$) are somewhat lower than those of shabynite ($\alpha = 1.543$, $\beta = 1.571$, $\gamma = 1.577$).

Based on these data, nepskoeite can be tentatively considered as a hydrous orthoborate chloride, a high-hydrated analogue of shabynite. Both minerals need further investigations, first of all, determination of unit-cell dimensions and H_2O content by means of direct methods.

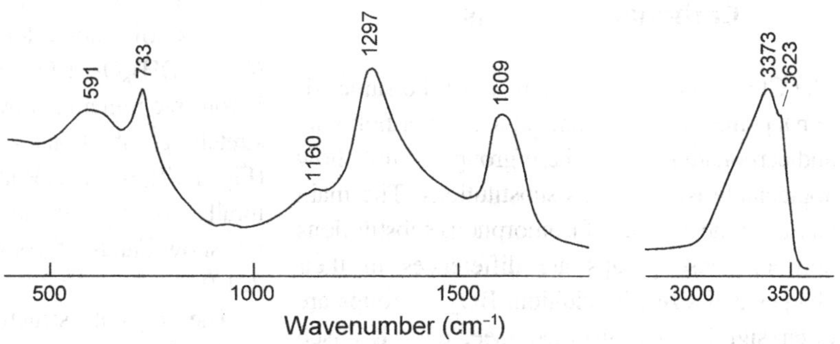

Fig. 1.14 IR spectrum of nepskoeite drawn using data from Apollonov (1998)

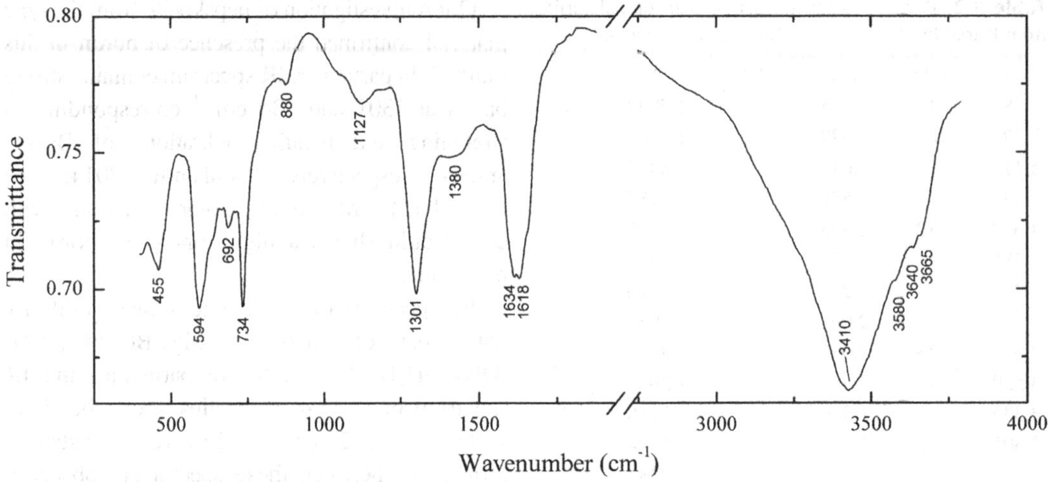

Fig. 1.15 IR spectrum of nepskoeite drawn using data from Chukanov (2014)

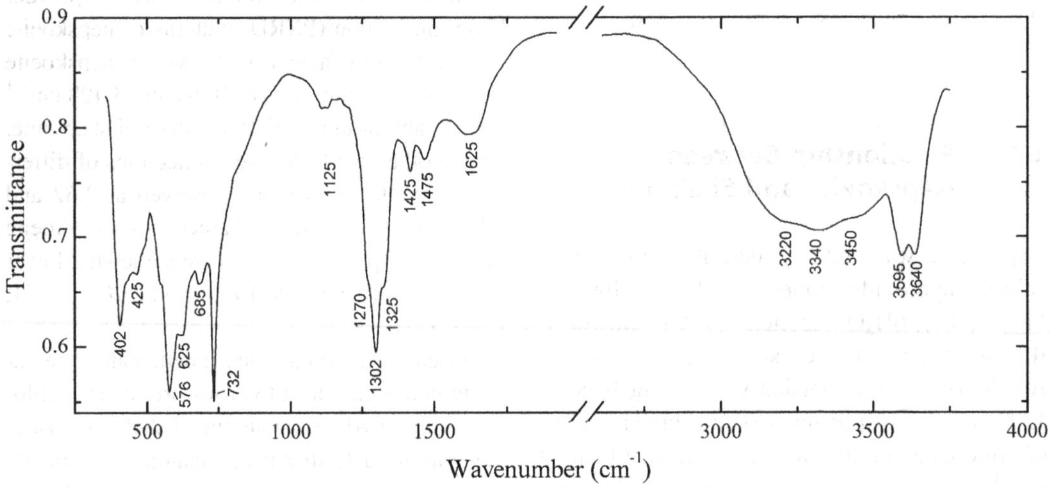

Fig. 1.16 IR spectrum of shabynite drawn using data from Chukanov (2014)

1.5 Orthoborate Groups in Lead Carbonate Minerals

The CO_3^{2-} and BO_3^{3-} anions have the same trigonal planar configuration, but in most orthoborate and carbonate minerals these groups do not show appreciable isomorphous substitutions. The main causes of the absence of isomorphous substitutions between these groups are differences in their charges and sizes. In addition, BO_3^{3-} groups are often significantly distorted (see, e.g., Kolitsch et al. 2012, as well as Sect. 1.1). However, some lead carbonate minerals are exceptions from this regularity. For example, IR spectra of some samples of molybdophyllite $Pb_8Mg_9(Si_{10}O_{28})(CO_3)_3(OH)_8O_2.H_2O$, which is nominally a boron-free mineral, show weak IR bands of B–O-stretching vibrations at 1170 and 1240 cm^{-1}) (Fig. 1.17), whereas Raman spectrum of the structurally investigated molybdophyllite sample does not show bands of borate groups (Kolitsch et al. 2012).

The crystal structure of molybdophyllite (Kolitsch et al. 2012) does not contain boron-

Table 1.3 Strongest lines (with $I \geq 10\%$) of the PXRD patterns of nepskoeite and shabynite

Nepskoeite (Apollonov 1998)		Shabynite, sample No. 3 (Pertsev et al. 1980)	
d, Å	*I*, %	*d*, Å	*I*, %
11.41	29	11.33	10
10.64	18	–[a]	–[a]
9.78	46	9.72	17
9.60	38	–	–
5.57	17	–	–
5.48	16	5.48 5.41	16 16
–	–	4.86	19
4.78	15	4.77	19
4.25	20	4.266 4.230	16 17
–	–	4.133	29
3.726	15	3.726	17
3.624	14	3.648	10
3.498	100	–	–
–	–	3.191	100
3.184	10	–	–
2.977	10	–	–
2.739	16	–	–
2.448	18	2.447	19
2.395	17	2.390	11
2.284	11	–	–
1.749	10	–	–

[a]The strongest reflection of the powder X-ray diffraction pattern of shabynite sample No. 2 (Pertsev et al. 1980) is observed at 9.62 Å

dominant sites and, consequently, BO_3^{3-} anions occur in CO_3^{2-}-dominant positions, unlike roymillerite $Pb_{24}Mg_9(Si_9AlO_{28})(SiO_4)(BO_3)(CO_3)_{10}(OH)_{14}O_4$ and britvinite $Pb_{15}Mg_9(Si_{10}O_{28})(CO_3)_2(BO_3)_4(OH)_{12}O_2$, in which orthoborate and carbonate groups are ordered in different sites (Yakubovich et al. 2008; Chukanov et al. 2017b; see Fig. 1.18).

The IR spectrum of hydrocerussite $Pb_3(OH)_2(CO_3)_2$ from Långban (curve *b* in Fig. 1.19) contains weak bands at 1230 (shoulder), 742, and 470 cm^{-1}. These bands may be assigned to stretching and bending vibrations of orthoborate anions partly substituting regular CO_3 triangles in the hydrocerussite structure. Indeed, in the IR spectrum of fluoborite containing regular BO_3 triangle strong bands of B–O-stretching and O–B–O bending vibrations are observed at 1241, 743, and 468 cm^{-1} (Chukanov 2014). It is important to note that bands of orthoborate groups are absent in the IR spectra of hydrocerussite from Merehead quarry, England (curve *a* in Fig. 1.19), some related minerals (curves *c* and *d* in Fig. 1.19), as well as synthetic analogues of hydrocerussite and plumbonacrite (Brooker et al. 1983). In all probability, the presence of borate groups in hydrocerussite from Långban is the result of high activity of boron that accompanied formation of this deposit where different borate minerals are common.

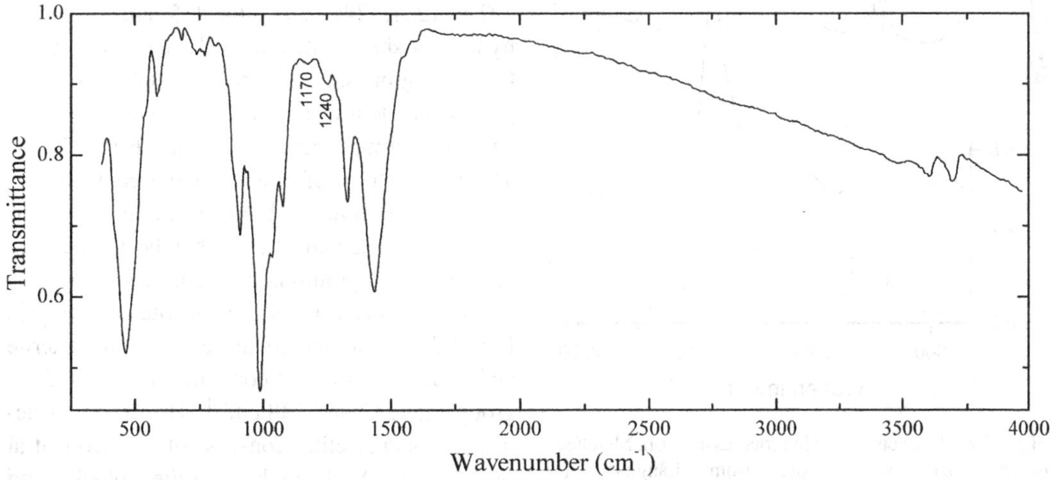

Fig. 1.17 IR spectrum of molybdophyllite from the Långban deposit, Bergslagen ore region, Filipstad district, Värmland, Sweden (Chukanov 2014, sample Sil247). The wavenumbers of BO_3^{3-} groups are indicated

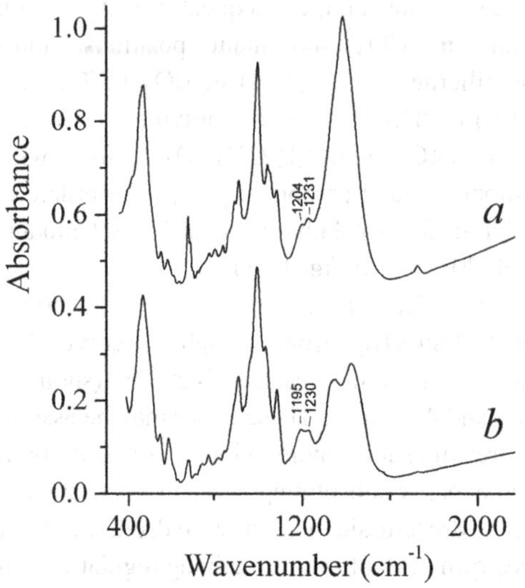

Fig. 1.18 IR spectra of (*a*) roymillerite and (*b*) britvinite from Långban, Värmland, Sweden. Bands of stretching vibrations of BO_3^{3-} groups are indicated

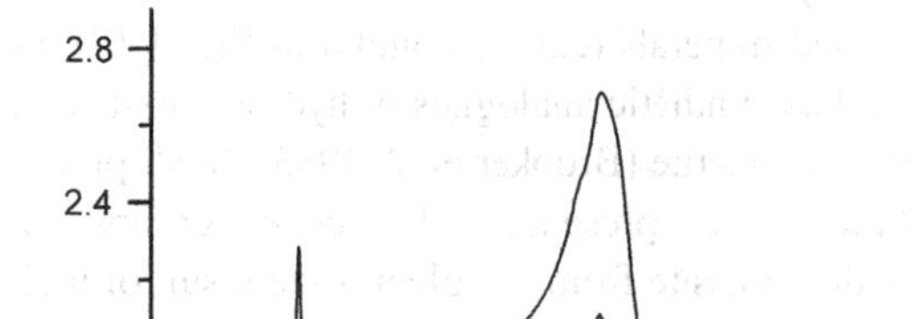

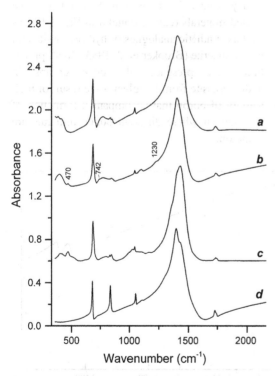

Fig. 1.19 IR spectra of (*a*) hydrocerussite from Merehead quarry, (*b*) hydrocerussite from Långban, (*c*) hydrocerussite-related phase $NaPb_5(CO_3)_4(OH)_3$ from Lavrion, and (*d*) cerussite from Merehead quarry. Bands of BO_3^{3-} groups are indicated

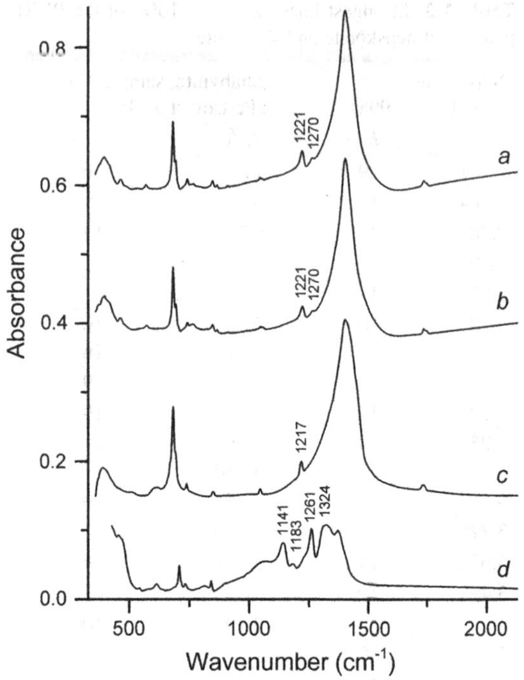

Fig. 1.20 IR spectra of (*a*) plumbonacrite from Merehead quarry, (*b*) plumbonacrite from Långban, (*c*) somersetite, and (*d*) mereheadite. Bands of stretching vibrations of BO_3^{3-} groups are indicated

The IR spectra of plumbonacrite $Pb_5O(OH)_2(CO_3)_3$ from Merehaad and Långban (Fig. 1.20) are similar and differ from the IR spectra of synthetic plumbonacrite analogue (Brooker et al. 1983) by additional bands of BO_3^{3-} groups at 1270, 1221, 742, and 464–465 cm^{-1}. Unlike hydrocerussite, a mineral with the only site of CO_3^{2-} groups, in the crystal structure, plumbonacrite is characterized by five positions of carbonate groups (Krivovichev and Burns 2000). The IR spectrum of plumbonacrite contains two bands of asymmetric B–O-stretching vibrations (at 1270 and 1221 cm^{-1}), which indicates the presence of BO_3^{3-} groups in different sites.

The IR spectrum of somersetite (curve *c* in Fig. 1.20) is similar to those of plumbonacrite and hydrocerussite and contains bands of BO_3^{3-} groups at 1217 and 738 cm^{-1}. The crystal structure of somersetite consists of electroneutral $[Pb_3(OH)_2(CO_3)_2]$ hydrocerussite block and electroneutral $[Pb_5O_2(CO_3)_3]$ block with the structure derivative from plumbonacrite. The

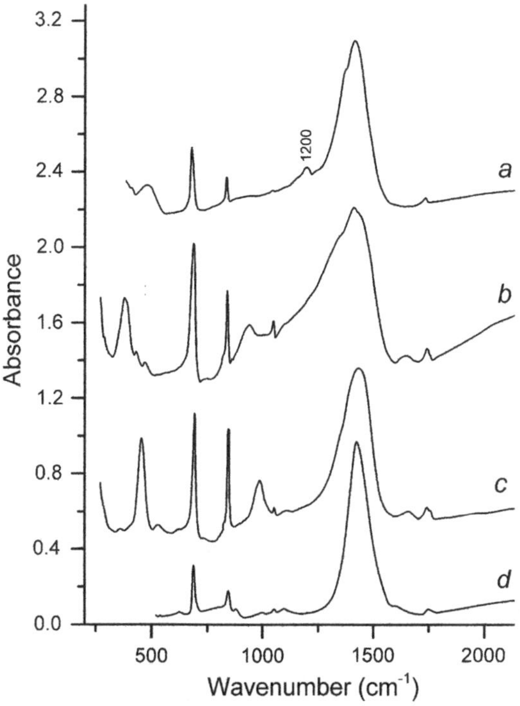

Fig. 1.21 IR spectra of (*a*) grootfonteinite holotype, (*b*) synthetic compound $KPb_2(CO_3)_2(OH)$ (Brooker et al. 1983), (*c*) synthetic compound $NaPb_2(CO_3)_2(OH)$ (Brooker et al. 1983), and (*d*) abellaite holotype (Ibáñez-Insa et al. 2017). Band of stretching vibrations of BO_3^{3-} groups is indicated

only band of B–O-stretching vibrations observed in the IR spectrum of somersetite indicates that all admixed BO_3^{3-} groups belong to the hydrocerussite block.

The relatively weak bands at 480 and 1200 cm^{-1} in the IR spectrum of grootfonteinite $Pb_3O(CO_3)_2$ indicate the presence of minor amounts of undistorted orthoborate groups, but these groups are absent in most samples of structurally related minerals and compounds (see Fig. 1.21).

In hydrocerussite, plumbonacrite, somersetite, and grootfonteinite CO_3/BO_3 are undistorted. As a result, bands of admixed B–O-stretching vibrations of BO_3^{3-} anions are observed in the narrow wavenumber range from 1200 to 1270 cm^{-1}. In contrast to these minerals, in mereheadite $Pb_{47}O_{24}(OH)_{13}Cl_{25}(BO_3)_2(CO_3)$ BO_3 triangles are significantly distorted with B–O distances varying from 1.23 to 1.32 Å (Krivovichev et al. 2009). This results in splitting of the band of B–O-stretching vibrations into several components (at 1141, 1183, and 1261 cm^{-1}); in addition, broad band at 1324 cm^{-1} corresponding to a mixed mode involving B–O- and C–O-stretching vibrations appears (curve *d* in Fig. 1.20).

2 IR Spectra of Minerals and Related Compounds, and Reference Samples Data

This chapter contains IR spectra of mineral species and varieties, most of which was not included in the preceding reference books (Chukanov 2014; Chukanov and Chervonnyi 2016). Along with spectra obtained by us, we provide most reliable new data on infrared spectra of 791 minerals and related synthetic compounds published elsewhere. Each spectrum is accompanied with analytical data on the reference sample, its occurrence and general appearance, associated minerals, as well as kind of sample preparation and/or method of registration of the spectrum.

The Sects. 2.1, 2.2, 2.3, etc. are arranged in ascending order of the atomic number Z_a of the main species-defining element for a given class of minerals: first for borate minerals (with $Z_a = 5$ for boron), then for carbon, carbides, carbonates, and organic substances (with $Z_a = 6$ for carbon), for nitrates (with $Z_a = 7$ for nitrogen), for oxides and hydroxides (with $Z_a = 8$ for oxygen), and so on.

A total of 174 spectra presented in this chapter have been obtained by one of the authors (NVC). In order to obtain absorption infrared spectra, powdered mineral samples have been mixed with anhydrous KBr, pelletized, and analyzed using an ALPHA FT IR spectrometer (Bruker Optics, Ettlingen, Germany) with a resolution of 4 cm^{-1} and 16 scans. IR spectrum of an analogous disc of pure KBr was used as a reference. It is important to note that reflectance mode IR spectra, IR spectra obtained without immersion medium (e.g., KBr), as well as IR spectra of single crystals, coarse-grained, or textured aggregates cannot be considered as stable and reliable diagnostic characteristics of mineral species due to specific effects induced by orientation, polarization, scattering, and reflection conditions. For example, in case of a single crystal, bands corresponding to normal vibrations with polarization vector parallel to the direction of propagation of IR radiation are absent in the spectrum. However these bands can be observed at another orientation of the crystal. In more detail these aspects were considered above (see Chukanov and Chervonnyi 2016, the section "Sources of Errors and Artifacts in IR Spectroscopy of Minerals"). For the above reasons, *only transmittance or absorbance IR spectrum of a pulverized sample dispersed in an immersion medium is a stable characteristic of a mineral and can be used as a diagnostic tool.*

Additional information includes general appearance, associated minerals, methods of the mineral species identification, and the list of wavenumbers of absorption bands with the indication of strong bands, weak bands, and shoulders. IR spectroscopy itself can be considered as an adequate identification method if IR spectrum is unique for a given mineral and coincides with IR spectrum of a well-investigated sample. For most synthetic samples the method of synthesis is shortly characterized.

For more than 100 samples (mainly holotypes of mineral species), a more detailed information is given including unit-cell dimensions, symmetry,

N. V. Chukanov, M. F. Vigasina, *Vibrational (Infrared and Raman) Spectra of Minerals and Related Compounds*, Springer Mineralogy, https://doi.org/10.1007/978-3-030-26803-9_2

strongest reflections of the powder X-ray diffraction pattern, empirical formula, optical data, density, etc.

The following *abbreviations* are used in this chapter:

Mt.	mountain
Co.	county
IR	infrared
D	density
D_{meas}	measured density
D_{calc}	calculated density
apfu	atoms per formula unit
Z	the number of formula units per unit cell
α, β, γ	refractive indices for biaxial minerals
ω, ε	refractive indices for uniaxial minerals
n	refractive index for isotropic minerals
$2V$	angle between optic axes
d	interplanar spacing
I	relative intensity of a line in the powder X-ray diffraction pattern
REE	rare-earth elements
Ln	lanthanides
s	strong band
w	weak band
sh	shoulder
□	vacancy

In most cases, the terms “strong band” and “weak band” mean band having transmittance minimum below and above any conventional values, respectively. As a rule, “shoulder” means inflection point of the spectral curve. For the convenience of visual perception, the positions of all peaks and shoulders in most figures are indicated by arrows.

For the numeration of samples, double letter-figure symbols are used. This numbering is a continuation of the numbering used in the previous books of this series (Chukanov 2014; Chukanov and Chervonnyi 2016). The meaning of letter parts of the symbols is explained in Table 2.1. It is to be noted that these designations are conventional and not unambiguous. For example, zirsilite-(Ce), $Na_{12-x}(Ce,Na)_3Ca_6Mn_3Zr_3NbSi_{25}O_{73}(OH)_3(CO_3)\cdot H_2O$, can be classified as cyclosilicate, as zirconosilicate or as carbonatosilicate.

Table 2.1 The meaning of letter symbols used in the numbering of reference samples

Symbol	Meaning of the symbol	Symbol	Meaning of the symbol
Bo	Borates with isolated orthogroups BO_3	PSi	Phosphato-silicates
B	Other borates	SSi	Sulfato-silicates
BC	Carbonatoborates	TiSi	Titanosilicates and related zircono-, niobo-, and stannosilicates
BAs	Arsenatoborates	AsSi	Arsenato-silicates
C	Carbon and carbonates	USi	Silicates with uranyl groups UO_2^{2+} (except nesosilicates)
Org	Organic compounds and salts of organic acids	P	Phosphides and phosphates
N	Nitrides and nitrates	S	Sulfates
O	Oxides and hydroxides	SC	Carbonato-sulfates
F	Fluorides	SP	Phosphato-sulfates
Sio	Nesosilicates (i.e., silicates with orthogroups SiO_4)	SMo	Sulfatomolybdates
Sid	Sorosilicates (i.e., silicates with diorthogroups Si_2O_7 or $SiAlO_7$)	Cl	Chlorides and hydroxychlorides
Siod	Silicates containing both orthogroups SiO_4 and diorthogroups Si_2O_7	V	Vanadates, V oxides, and hydroxides
Sit	Triorthosilicates with groups Si_3O_{10}	Cr	Chromates
Siot	Ortho-triorthosilicates	Ge	Germanates
Sir	Cyclosilicates (“r” means “ring”)	As	Arsenic, arsenides, arsenites, arsenates, and sulfato-arsenates
Sic	Inosilicates with chains formed by SiO_4 and AlO_4 tetrahedra	UAs	Uranyl arsenates

(continued)

Table 2.1 (continued)

Symbol	Meaning of the symbol	Symbol	Meaning of the symbol
Sib	Inosilicates with bands formed by SiO_4 and AlO_4 tetrahedra	AsS	Sulfato-arsenates
Sil	Phyllosilicates with layers formed by SiO_4 and AlO_4 tetrahedra	Se	Selenium, selenides, and selenites
Sif	Tectosilicates (aluminosilicates with 3d frameworks formed by SiO_4 and AlO_4 tetrahedra), except zeolites	Br	Bromides and bromates
Sif_Z	Zeolites	Mo	Molybdates and Mo-beariung oxides
Si	Silicon, silicides, and silicates with unknown or complex structures	Te	Tellurides, tellurites, and tellurates
Sia	Amorphous silicates	I	Iodides, iodites, and iodates
BeSi	Beryllosilicates	Xe	Xenates
BSi	Borosilicates and borato-silicates	W	Tungstates and W-bearing oxides
CSi	Carbonato-silicates		

2.1 Borates, Including Arsenatoborates and Carbonatoborates

Bo36 Barium strontium orthoborate fluoride $Ba_3Sr_4(BO_3)_3F_5$

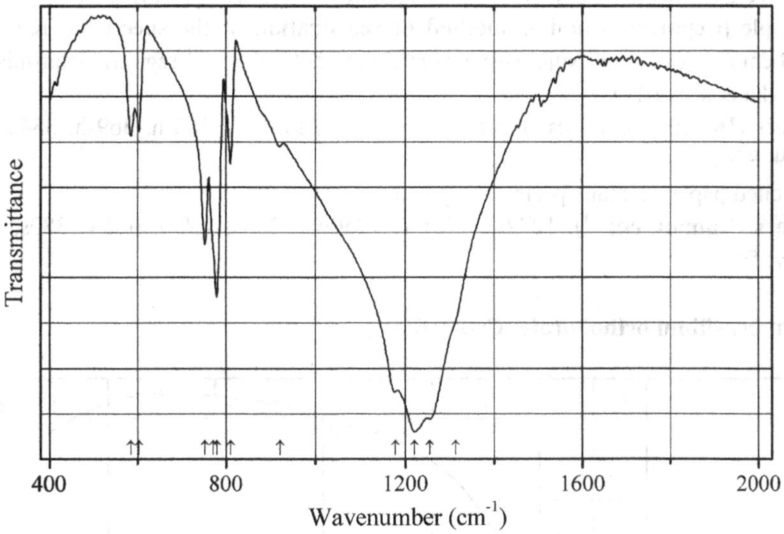

Origin: Synthetic.

Description: Prepared from BaF_2, $BaCO_3$, $SrCO_3$, and H_3BO_3 by using a high-temperature solid-state technique. The crystal structure is solved. Hexagonal, space group $P6_3mc$, a = 10.8953(16), c = 6.9381(15) Å, V = 713.3(2) Å^3, Z = 2. D_{calc} = 4.814 g/cm^3. Characterized by powder X-ray diffraction data.

Kind of sample preparation and/or method of registration of the spectrum: KBr disc. Transmission.

Source: Zhang et al. (2009a).

Wavenumbers (cm^{-1}): 1312sh, 1255s, 1221s, 1177s, 921w, 810, 779, 771sh, 753, 604, 585.

Note: The wavenumbers were determined by us based on spectral curve analysis of the published spectrum.

Bo37 Barium zirconium orthoborate $BaZr(BO_3)_2$

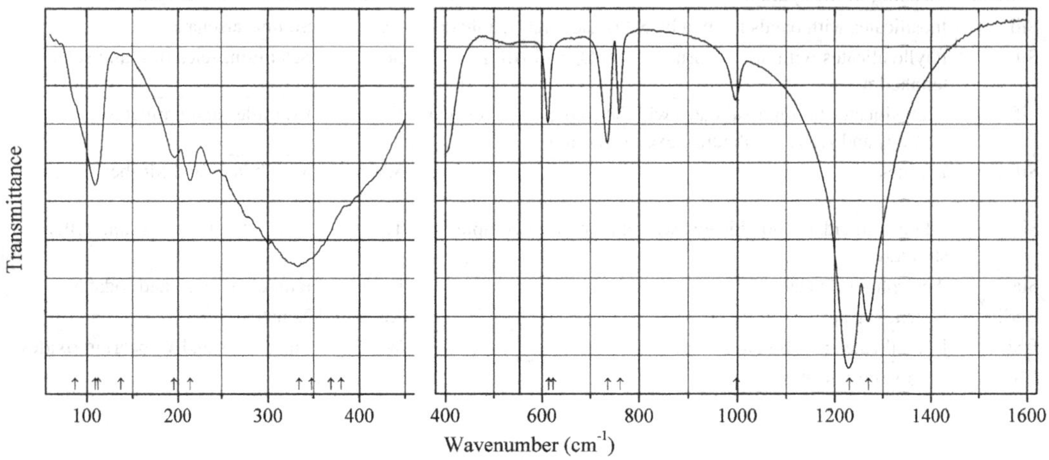

Origin: Synthetic.

Description: Powder obtained by means of standard solid-state reaction from the stoichiometric mixture of $BaCO_3$, ZrO_2, and B_2O_3 pressed into a pellet and heated first at 550 °C for 48 h and thereafter heated twice at 910 °C for 20 h. Trigonal, $a = 5.167$, $c = 33.913$ Å.

Kind of sample preparation and/or method of registration of the spectrum: KBr disc (in the 1500–400 cm^{-1} region) and Nujol suspension (in the 500–50 cm^{-1} region). Transmission.

Source: Mączka et al. (2015).

Wavenumbers (IR, cm^{-1}): 1270s, 1230s, 998w, 761, 736, 613, 380sh, 369sh, 334s, 214, 195w, 137w, 109, 87sh.

Note: In the cited paper, Raman spectrum is given.

Wavenumbers (Raman, cm^{-1}): 1272w, 1258w, 1250w, 1228s, 739w, 622w, 380s, 369s, 348w, 137, 112w, 59.

Bo38 Cesium beryllium orthoborate $CsBe_4(BO_3)_3$

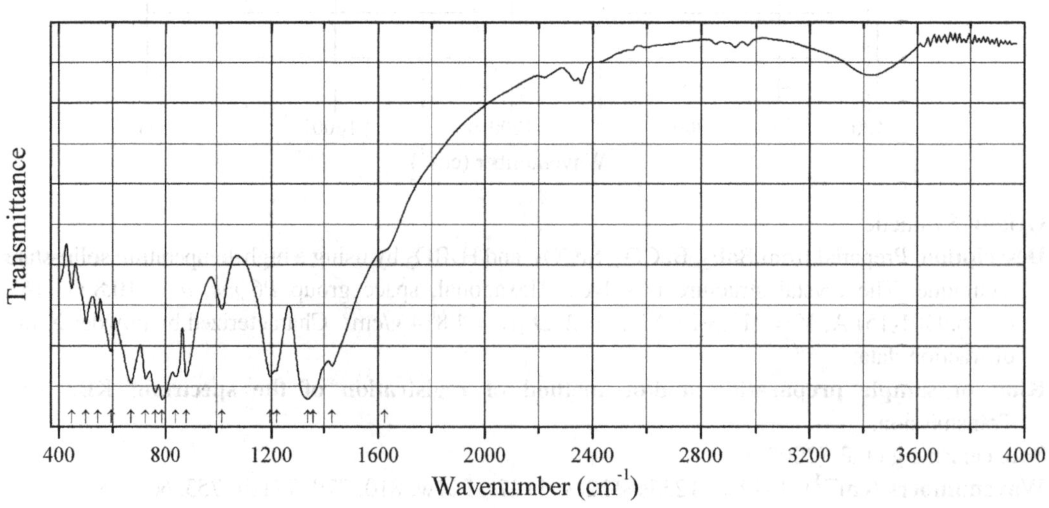

Origin: Synthetic.

Description: Prepared from Cs_2CO_3, BeO, and B_2O_3 by solid-state reaction at 800 °C for 48 h in air. The crystal structure is solved. Orthorhombic, space group *Pnma*, $a = 8.3914(5)$, $b = 13.3674(7)$, $c = 6.4391(3)$ Å, $V = 722.28(7)$ Å^3, $Z = 4$. $D_{calc} = 3.176$ g/cm^3. Characterized by powder X-ray diffraction data.

Kind of sample preparation and/or method of registration of the spectrum: KBr disc. Transmission.

Source: Huang et al. (2013a).

Wavenumbers (cm^{-1}): (3430), (1625sh), 1430, 1360sh, 1339s, 1220sh, 1197s, 1020, 881, 840, 785s, 760, 722, 669, 595, 545, 501, 448.

Note: The wavenumbers were partly determined by us based on spectral curve analysis of the published spectrum. The bands at 3430 and 1625 cm^{-1} may correspond to adsorbed water. Weak bands in the range from 2300 to 2400 cm^{-1} correspond to atmospheric CO_2.

Bo39 Calcium orthoborate fluoride $Ca_5(BO_3)_3F$

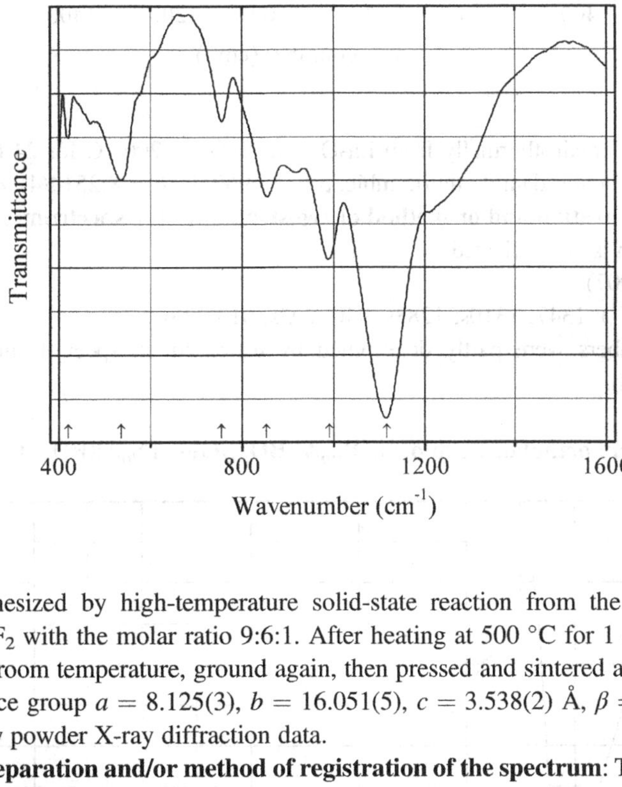

Origin: Synthetic.

Description: Synthesized by high-temperature solid-state reaction from the mixture of $CaCO_3$, H_3BO_3, and CaF_2 with the molar ratio 9:6:1. After heating at 500 °C for 1 day, the mixture was cooled down to room temperature, ground again, then pressed and sintered at 1000 °C for 2 days. Monoclinic, space group $a = 8.125(3)$, $b = 16.051(5)$, $c = 3.538(2)$ Å, $\beta = 100.90(4)°$, Z = 2. Characterized by powder X-ray diffraction data.

Kind of sample preparation and/or method of registration of the spectrum: Transmission. Kind of sample preparation is not indicated.

Source: Chen et al. (2006a).

Wavenumbers (cm^{-1}): 1117s, 990, 852, 755w, 536, 420w.

Note: The IR spectrum contains additional bands in the range from 1500 to 4000 cm^{-1} that correspond to adsorbed (?) water molecules.

Bo40 Lanthanum orthoborate La(BO_3)

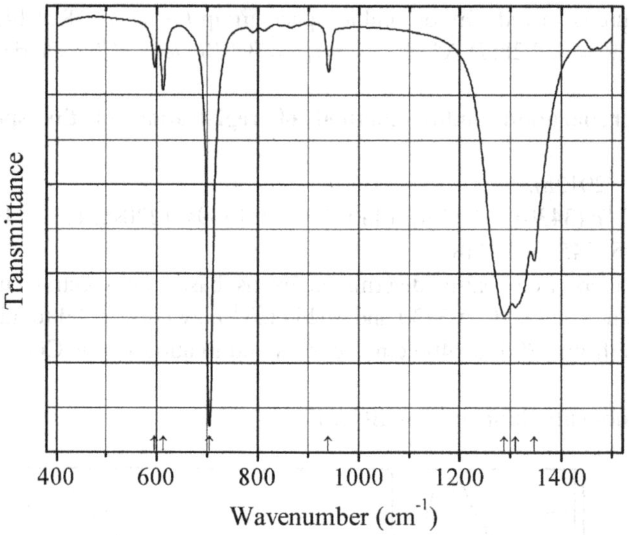

Origin: Synthetic.

Description: Prepared hydrothermally from La_2O_3 and B_2O_3 at 200 °C for 24 h. Characterized by powder X-ray diffraction data. Orthorhombic, $a = 5.0960(8)$, $b = 8.2514(4)$, $c = 5.8726(6)$ Å.

Kind of sample preparation and/or method of registration of the spectrum: Absorption. Kind of sample preparation is not indicated.

Source: Ma et al. (2007).

Wavenumbers (cm^{-1}): 1347, 1310s, 1288s, 940, 705s, 613, 596w.

Note: The wavenumbers were partly determined by us based on spectral curve analysis of the published spectrum.

Bo41 Lead aluminium orthoborate fluoride $Pb_6Al(BO_3)_2OF_7$ $Pb_6Al(BO_3)_2OF_7$

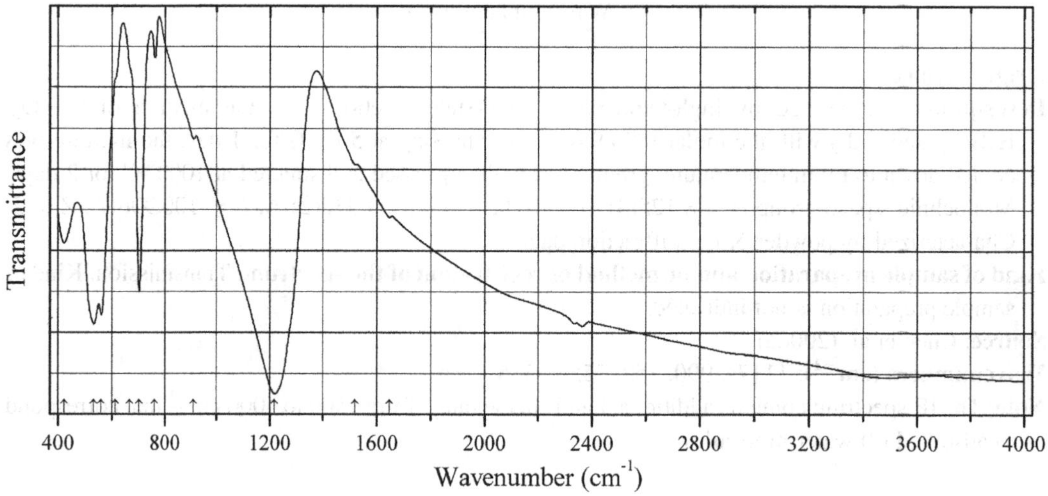

Origin: Synthetic.

Description: Prepared from a stoichiometric mixture of PbF_2, Al_2O_3, and H_3BO_3 at 430 °C with several intermediate grindings. Characterized by powder X-ray diffraction data. The crystal structure is solved. Orthorhombic, space group *Cmca*, $a = 11.649(7)$, $b = 18.300(11)$, $c = 6.394(4)$ Å, $V = 1363.1(15)$ Å^3, $Z = 4$. $D_{calc} = 7.488$ g/cm^3. In the structure, Al atoms coordinated by F (to form slightly distorted AlF_6 octahedra) are situated between the $[Pb_6BO_{11}F_{10}]$ layers.

Kind of sample preparation and/or method of registration of the spectrum: KBr disc. Transmission.

Source: Dong et al. (2012).

Wavenumbers (cm^{-1}): 1648w, 1520w, 1219s, 912w, 761w, 702s, 665sh, 612sh, 561s, 533s, 423.

Note: The wavenumbers were partly determined by us based on spectral curve analysis of the published spectrum.

Bo42 Lead bismuth orthoborate $PbBi(BO_3)O$

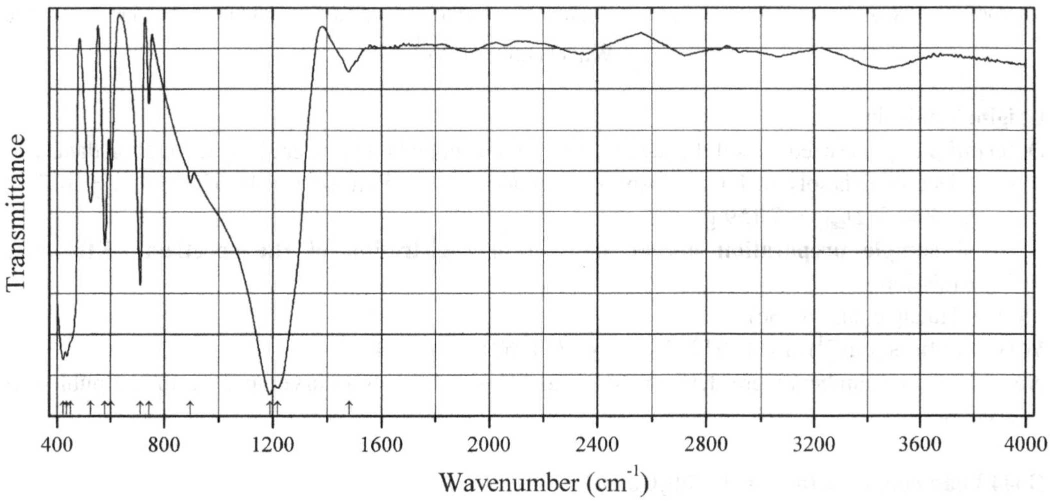

Origin: Synthetic.

Description: Synthesized by solid-state reaction of a stoichiometric mixture of PbO, Bi_2O_3, and H_3BO_3 powders. The crystal structure is solved. Orthorhombic, space group *Cmca*, a =10.782 (3), $b = 10.502(3)$, $c = 7.477(2)$ Å, $V = 846.7(4)$ Å^3, $Z = 8$. $D_{calc} = 7.704$ g/cm^3. Each Bi atom is coordinated to six O atoms. The BiO_6 octahedra are connected via common vertices and edges to form infinite $[BiO_4]$ layer.

Kind of sample preparation and/or method of registration of the spectrum: KBr disc. Transmission.

Source: Zhao et al. (2011).

Wavenumbers (cm^{-1}): 1480w, 1219s, 1189s, 897, 744, 711s, 603, 579, 526, 452sh, 437s, 422s.

Note: The wavenumbers were determined by us based on spectral curve analysis of the published spectrum.

Bo43 Lead cadmium orthoborate $Pb_8Cd(BO_3)_6$

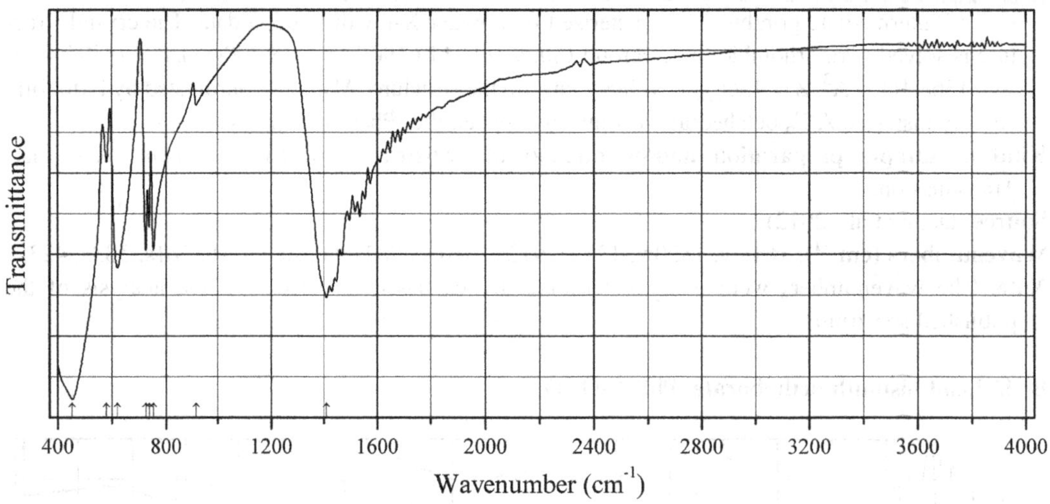

Origin: Synthetic.

Description: Synthesized via solid-state reaction. Characterized by powder X-ray diffraction data. The crystal structure is solved. Trigonal, space group *R*-3, $a = 9.5584(16)$, $c = 18.670(3)$ Å, $V = 1477.2(4)$ Å^3, $Z = 3$. $D_{calc} = 7.159$ g/cm^3.

Kind of sample preparation and/or method of registration of the spectrum: KBr disc. Transmission.

Source: Huang et al. (2013c).

Wavenumbers (cm^{-1}): 1411, 917, 752, 736, 724, 621, 581, 454.

Note: The wavenumbers were determined by us based on spectral curve analysis of the published spectrum.

Bo44 Lead copper orthoborate $Pb_2Cu(BO_3)_2$

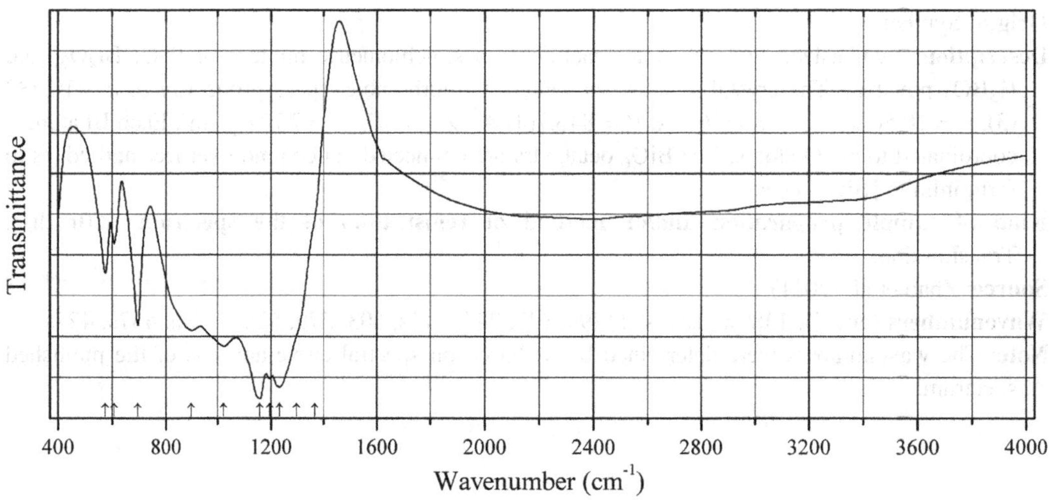

Origin: Synthetic.

Description: Prepared by a solid-state reaction method using PbO, CuO, and B_2O_3 as the starting components in the molar ratio 2:1:1. Characterized by powder X-ray diffraction data. The crystal structure is solved. Monoclinic, space group $P2_1/c$, $a = 5.6311(6)$, $b = 8.7628(9)$, $c = 6.2025(6)$ Å, $\beta = 115.7060(10)°$, $V = 275.77(5)$ Å^3, $Z = 2$. $D_{calc} = 7.172$ g/cm^3. Cu atoms have rectangular planar coordination.

Kind of sample preparation and/or method of registration of the spectrum: KBr disc. Transmission.

Source: Pan et al. (2006).

Wavenumbers (cm^{-1}): 1368sh, 1300sh, 1232s, 1195s, 1156s, 1024, 900, 694, 608, 575.

Note: The wavenumbers were determined by us based on spectral curve analysis of the published spectrum.

Bo45 Lead orthoborate tungstate $Pb_6(BO_3)_2(WO_4)O_2$

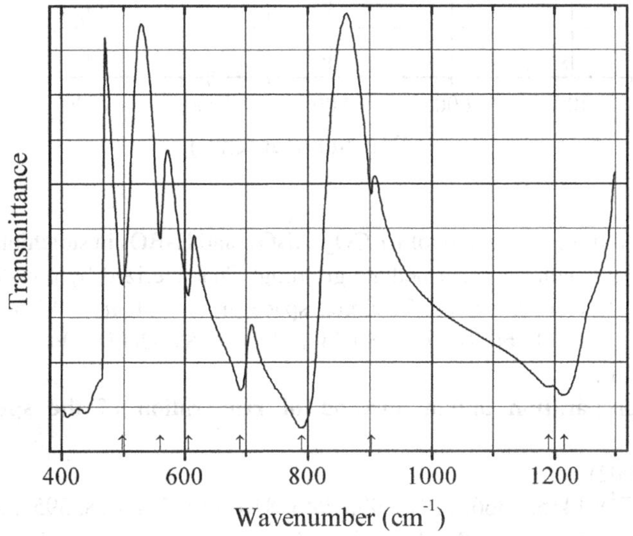

Origin: Synthetic.

Description: Crystals obtained from the melt of Bi_2O_3, PbO, WO_3, and H_3BO_3 with the molar ratio of 3:12:2:2. Characterized by powder X-ray diffraction data. The crystal structure is solved. Orthorhombic, space group *Cmcm*, $a = 18.4904(5)$, $b = 6.35980(10)$, $c = 11.6789(2)$ Å, $V = 1373.38$ (5) Å^3, $Z = 4$. $D_{calc} = 7.935$ g/cm^3.

Kind of sample preparation and/or method of registration of the spectrum: Transmission. Kind of sample preparation is not indicated.

Source: Li et al. (2011b).

Wavenumbers (cm^{-1}): 1215s, 1190, 902w, 790s, 690s, 606, 560, 498.

Note: The wavenumbers were partly determined by us based on spectral curve analysis of the published spectrum.

Bo46 Lithium aluminium orthoborate $Li_3Al(BO_3)_2$

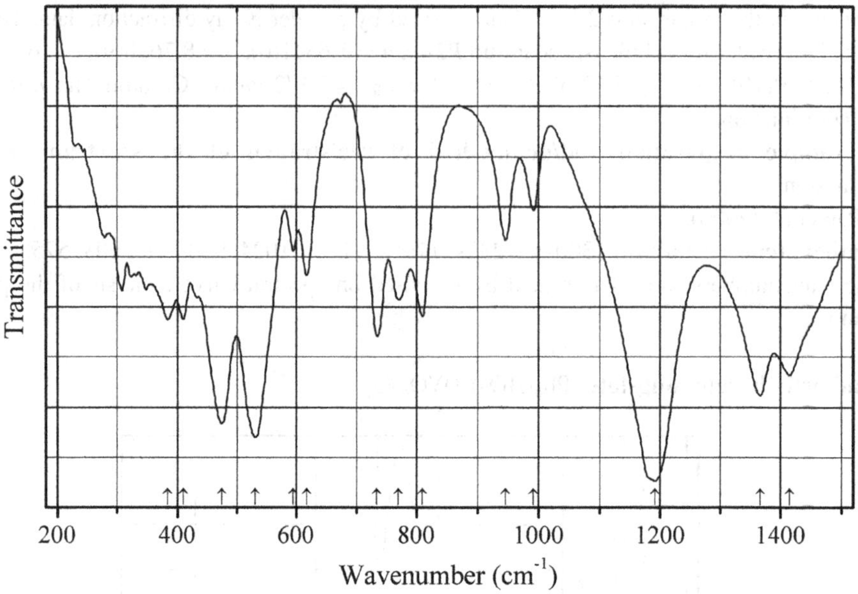

Origin: Synthetic.

Description: Prepared from the mixture of Li_2CO_3, Al_2O_3, and H_3BO_3 in stoichiometric proportion, at 690 °C for 1–2 days, with one intermediate grinding. Characterized by powder X-ray diffraction data. The crystal structure is solved. Triclinic, space group *P*-1, $a = 4.876(8)$, $b = 6.191(16)$, $c = 7.910(20)$ Å, $\alpha = 74.46(18)°$, $\beta = 89.44(17)°$, $\gamma = 89.52(18)°$, $V = 230.0(9)$ Å^3, $Z = 2$. $D_{calc} = 2.388$ g/cm^3.

Kind of sample preparation and/or method of registration of the spectrum: KBr disc. Transmission.

Source: He et al. (2002).

Wavenumbers (cm^{-1}): 1415, 1366, 1191s, 992, 946, 810, 770, 734, 618, 595, 531s, 475s, 410, 384.

Note: The wavenumbers were partly determined by us based on spectral curve analysis of the published spectrum.

Bo47 Lead orthoborate tungstate $Pb_6(BO_3)_2(WO_4)O_2$

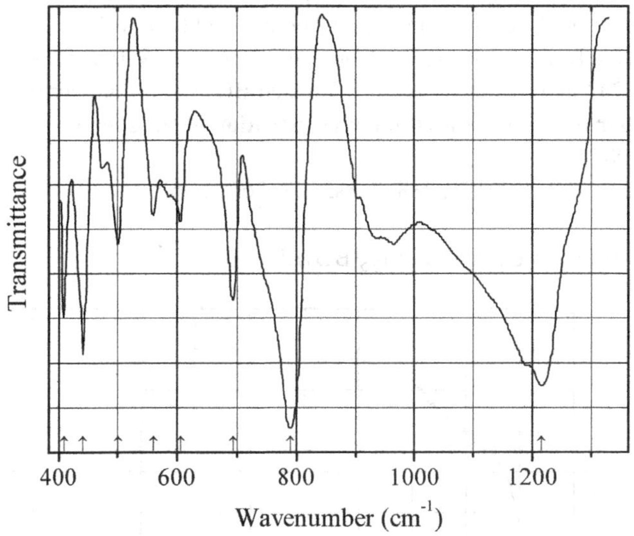

Origin: Synthetic.

Description: Obtained by a solid-state reaction method. The crystal structure is solved. Orthorhombic, space group *Cmcm*, $a = 18.480(4)$, $b = 6.3567(13)$, $c = 11.672(2)$ Å, $V = 1371.1(5)$ Å^3, $Z = 4$. $D_{calc} = 7.948$ g/cm^3.

Kind of sample preparation and/or method of registration of the spectrum: KBr disc. Transmission.

Source: Reshak et al. (2012b).

Wavenumbers (cm^{-1}): 1216s, 790s, 694, 606.5, 561, 500.5, 441, 409.

Bo48 Lithium strontium orthoborate $LiSr_4(BO_3)_3$

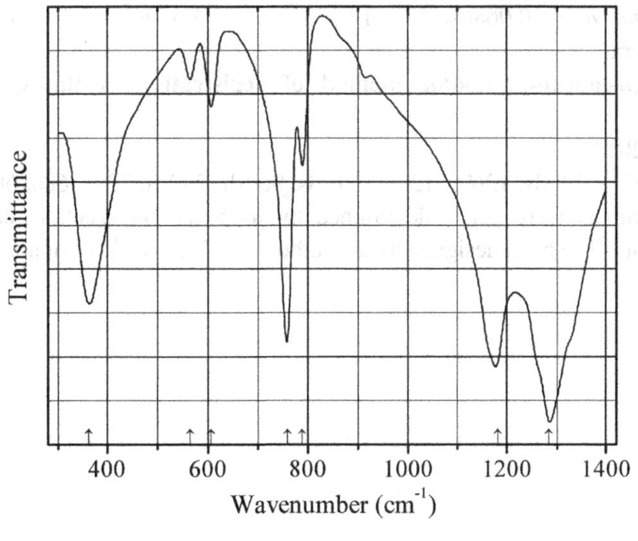

Origin: Synthetic.

Description: Obtained in the solid state reaction from the stoichiometric mixture of Li_2CO_3, $SrCO_3$, and H_3BO_3 at 750 °C, with several grindings. Characterized by powder X-ray diffraction data. The crystal structure is solved. Cubic, space group *Ia-3d*, $a = 14.95066(5)$ Å, $V = 3341.80(3)$ Å^3, $Z = 16$. $D_{calc} = 4.243$ g/cm^3. In the structure, isolated BO_3 groups are perpendicular to each other.

Kind of sample preparation and/or method of registration of the spectrum: KBr disc. Absorption.

Source: Wu et al. (2005).

Wavenumbers (cm^{-1}): 1284s, 1182s, 788, 758s, 607, 565w, 362.

Bo49 Magnesium orthoborate fluoride $Mg_5(BO_3)_3F$

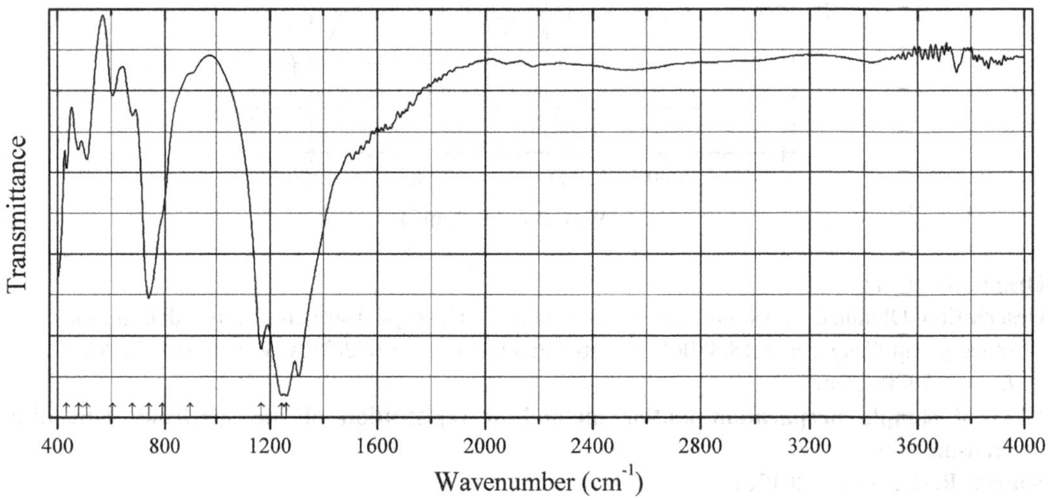

Origin: Synthetic.

Description: Crystals grown from the flux prepared from MgF_2, LiF, Na_2CO_3, and H_3BO_3. Characterized by powder X-ray diffraction data. The crystal structure is solved. Orthorhombic, space group *Pnma*, $a = 10.068(5)$, $b = 14.858(7)$, $c = 4.540(2)$ Å, $V = 679.2(6)$ Å^3, $Z = 4$. $D_{calc} = 3.100$ g/cm^3.

Kind of sample preparation and/or method of registration of the spectrum: KBr disc. Transmission.

Source: Bai et al. (2014).

Wavenumbers (cm^{-1}): 1308s, 1262s, 1243s, 1168s, 900sh, 790sh, 739s, 680, 606, 510, 478, 433.

Note: The wavenumbers were partly determined by us based on spectral curve analysis of the published spectrum. The band designed by the authors as 1275 cm^{-1} is a doublet 1262+1243 cm^{-1}.

Bo50 Potassium calcium orthoborate $KCa_4(BO_3)_3$

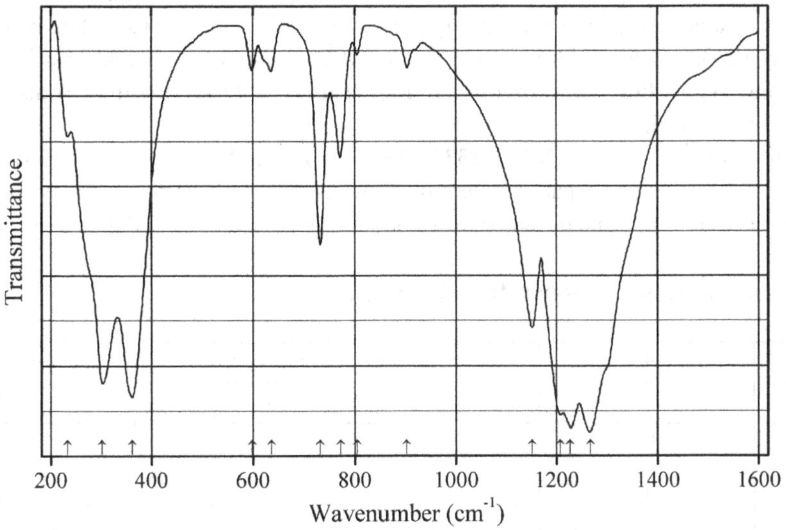

Origin: Synthetic.

Description: Obtained by heating stoichiometric mixture of K_2CO_3, $CaCO_3$, and H_3BO_3 first at 600 °C to decompose carbonates, and thereafter at 900 °C for 72 h. The crystal structure is solved. Orthorhombic, space group *Ama*2, a = 10.63455(10), b = 11.51705(11), c = 6.51942(6) Å, V = 798.49(2) Å^3, Z = 4. D_{calc} = 3.161 g/cm^3.

Kind of sample preparation and/or method of registration of the spectrum: KBr disc. Transmission.

Source: Wu et al. (2006a).

Wavenumbers (cm^{-1}): 1266s, 1226s, 1207s, 1150, 902, 804w, 772, 731, 636, 599, 362s, 301s, 232.

Note: The wavenumbers were partly determined by us based on spectral curve analysis of the published spectrum.

Bo51 Potassium magnesium orthoborate $KMg(BO_3)$

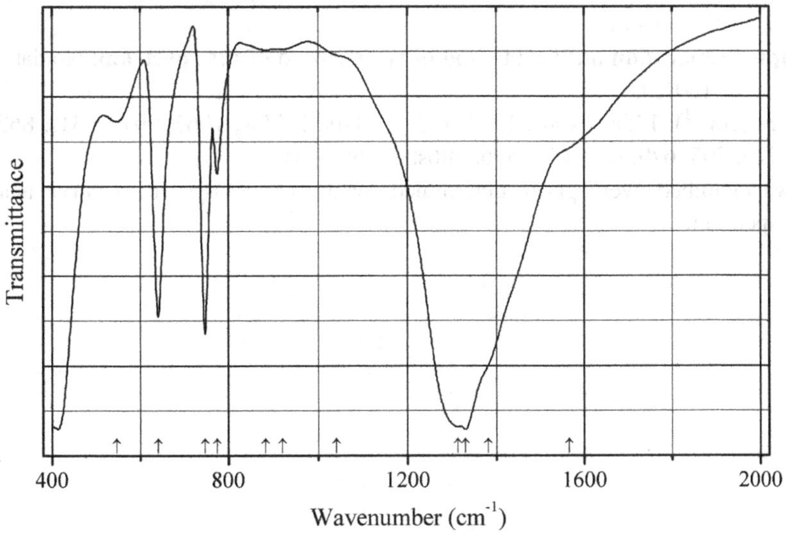

Origin: Synthetic.

Description: Prepared through solid-state reaction from the stoichiometric mixture of metal carbonates and H_3BO_3. Characterized by powder X-ray diffraction data. The crystal structure is solved. Cubic, space group $P2_13$, $a = 6.83443(4)$, $V = 319.23(1)$ Å^3, $Z = 4$. $D_{calc} = 2.543$ g/cm^3.

Kind of sample preparation and/or method of registration of the spectrum: KBr disc. Absorption.

Source: Wu et al. (2010a).

Wavenumbers (cm^{-1}): 1567sh, 1381sh, 1331s, 1314s, 1040sh, 920w, 882w, 774, 747s, 641s, 547.

Note: The wavenumbers were partly determined by us based on spectral curve analysis of the published spectrum. The band designed by the authors as 1318 cm^{-1} is a doublet 1331+1314 cm^{-1}.

Bo52 Potassium strontium orthoborate $KSr_4(BO_3)_3$

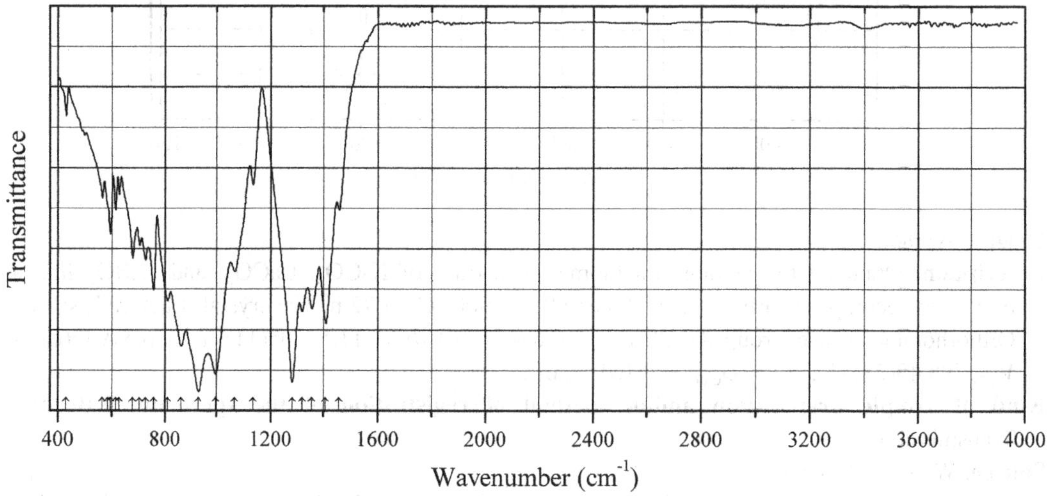

Origin: Synthetic.

Description: Synthesized by solid-state reaction of a stoichiometric mixture of K_2CO_3, $SrCO_3$, and H_3BO_3, heated first at 400 °C for 5 h and thereafter at 900 °C for 48 h with several intermediate grindings. Characterized by powder X-ray diffraction data. The crystal structure is solved. Orthorhombic, space group $Ama2$, $a = 11.025(10)$, $b = 11.977(10)$, $c = 6.872(6)$ Å, $V = 907.4(14)$ Å^3, $Z = 4$. $D_{calc} = 4.143$ g/cm^3.

Kind of sample preparation and/or method of registration of the spectrum: No data.

Source: Zhao et al. (2012).

Wavenumbers (cm^{-1}): 1458, 1408s, 1357, 1321, 1284s, 1135w, 1067, 997s, 931s, 862s, 930, 862, 809, 756, 726, 705, 679, 629, 615, 596, 586sh, 566w, 431w.

Note: The wavenumbers were partly determined by us based on spectral curve analysis of the published spectrum.

Bo53 Samarium orthoborate $Sm(BO_3)$

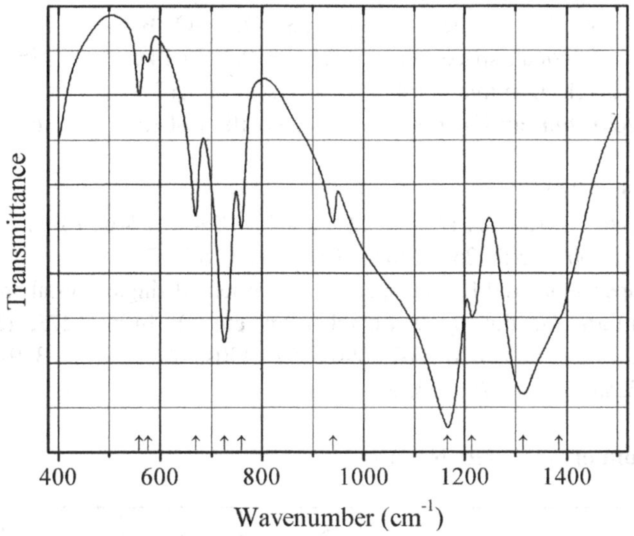

Origin: Synthetic.

Description: Synthesized by a solid-state method from stoichiometric amounts of H_3BO_3 and Sm_2O_3 first at 500 °C for 5 h, thereafter at 700 °C for 5 h, and finally at 900 °C for 5 h with intermediate grindings. Characterized by powder X-ray diffraction data. Triclinic (see JCPDS card No. 13-0489). Contains minor admixture of the vaterite-type polymorph.

Kind of sample preparation and/or method of registration of the spectrum: KBr disc. Transmission.

Source: Velchuri et al. (2011a).

Wavenumbers (cm^{-1}): 1384sh, 1315s, 1215, 1166s, 939, 759, 725, 669, 576w, 559w.

Bo54 Scandium lanthanum orthoborate $LaSc_3(BO_3)_4$

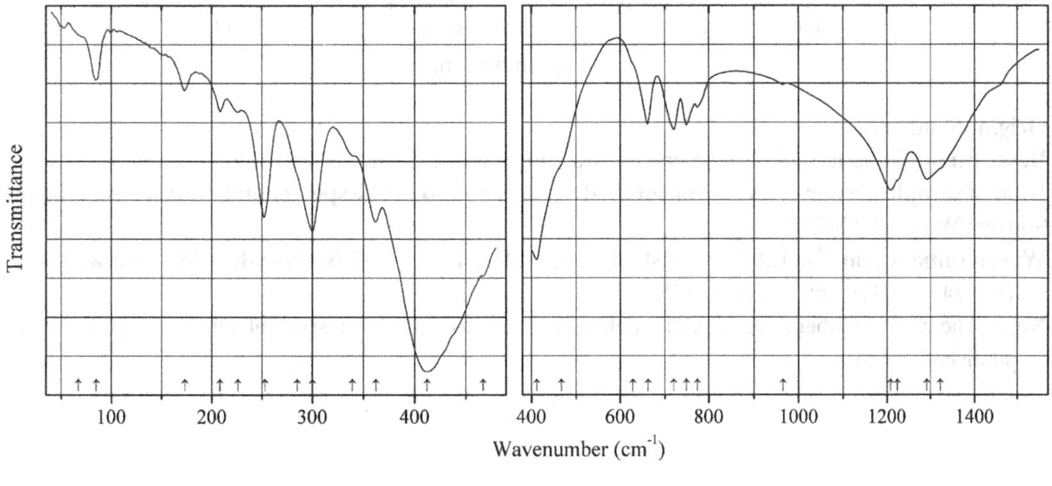

Origin: Synthetic.

Description: Bi-doped single crystals obtained by means of spontaneous crystallization from flux at starting molar composition of the melt La_2O_3:Sc_2O_3:Bi_2O_3:B_2O_3 = 1:1.5:13:13. The crystal structure is solved. Trigonal, space group $R32$, a = 9.8370(14), c = 7.9860(14) Å, Z = 3. The empirical formula is (electron microprobe): $Bi_{0.21}La_{0.91}Sc_{2.88}(BO_3)_4$.

Kind of sample preparation and/or method of registration of the spectrum: KBr disc and Nujol mull. Absorption.

Source: Mączka et al. (2010).

Wavenumbers (IR, cm^{-1}): 1322sh, 1292s, 1224sh, 1208s, 964w, 774w, 749, 720, 662, 628w, 467sh, 412s, 362w, 339w, 300, 285sh, 253, 226w, 208w, 173w, 85, 67w.

Note: In the cited paper, polarized Raman spectra of an oriented single crystal are given.

Wavenumbers (Raman, for the *x*(*yy*)*x* polarization, cm^{-1}): 1406, 1278, 1248sh, 1232, 1223, 983, 968w, 738, 712w, 663, 626, 607, 590, 457, 430s, 393sh, 384, 339s, 307sh, 298, 293, 248, 230, 227, 222sh, 207w, 176, 155, 90.

Bo55 Sodium calcium orthoborate $NaCa(BO_3)$

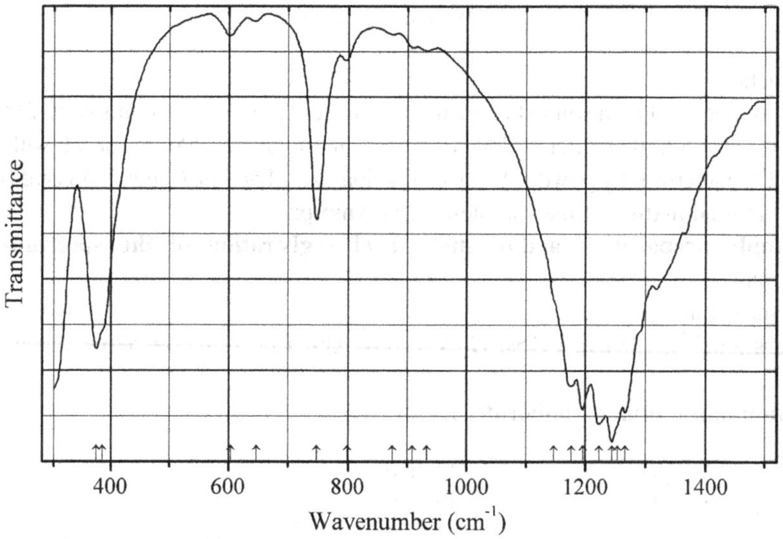

Origin: Synthetic.

Description: Orthorhombic. In the crystal structure, Na and Ca atoms are disordered.

Kind of sample preparation and/or method of registration of the spectrum: KBr disc. Absorption.

Source: Wu et al. (2006b).

Wavenumbers (cm^{-1}): 1267s, 1254sh, 1245s, 1223s, 1196s, 1177s, 1148sh, 933w, 909w, 876w, 798, 747w, 647w, 603, 385sh, 375.

Note: The wavenumbers were partly determined by us based on spectral curve analysis of the published spectrum.

Bo56 Sodium calcium orthoborate $NaCa_4(BO_3)_3$

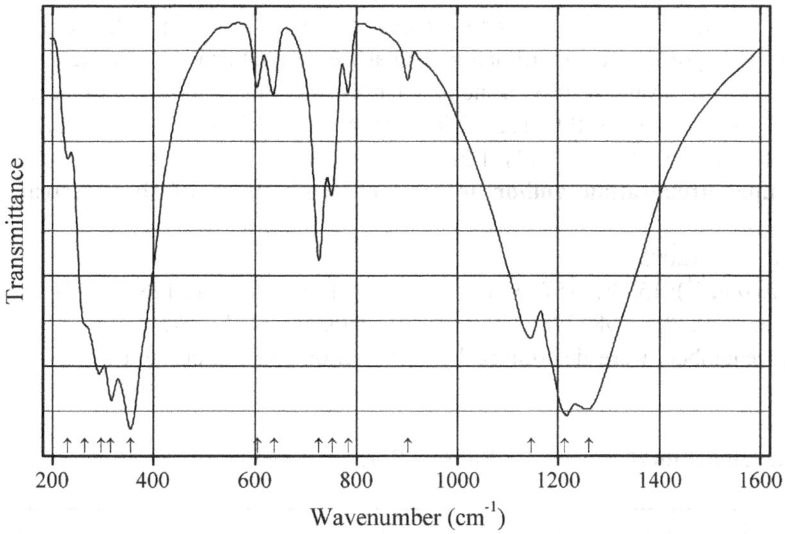

Origin: Synthetic.

Description: Polycrystalline sample prepared by sintering a stoichiometric mixture of Na_2CO_3, $CaCO_3$, and H_3BO_3, first at 600 °C and thereafter at 880 °C for 72 h with intermediate grinding. Characterized by powder X-ray diffraction data. The crystal structure is solved. Orthorhombic, space group *Ama*2, $a = 10.68004(11)$, $b = 11.28574(11)$, $c = 6.48521(6)$ Å, $V = 781.68(2)$ Å^3, $Z = 4$. $D_{calc} = 3.056$ g/cm^3.

Kind of sample preparation and/or method of registration of the spectrum: KBr disc. Absorption.

Source: Wu et al. (2006a).

Wavenumbers (cm^{-1}): 1261s, 1213s, 1146, 901w, 784w, 752, 726, 638w, 605w, 356s, 316s, 296s, 263sh, 229w.

Note: The wavenumbers were partly determined by us based on spectral curve analysis of the published spectrum.

Bo57 Sodium lanthanum orthoborate $Na_3La_9(BO_3)_8O_3$

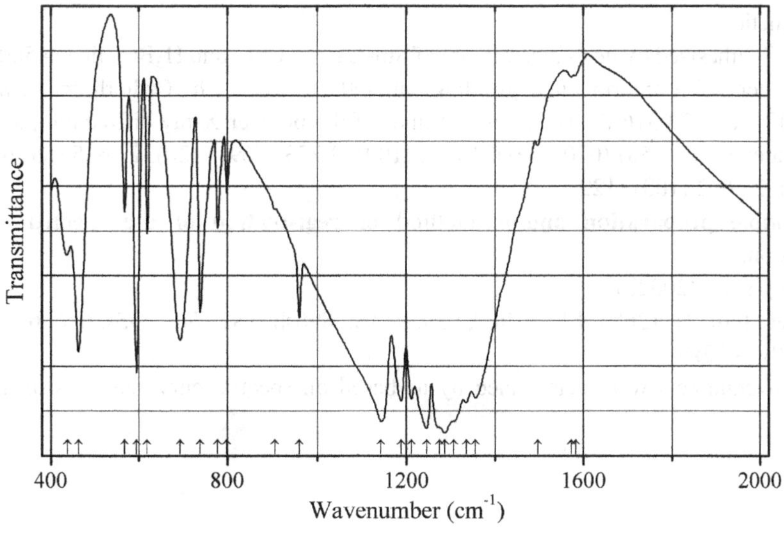

Origin: Synthetic.

Description: Synthesized from La_2O_3, Na_2CO_3, and H_3BO_3 by using high-temperature solid-state techniques, first at 600 °C for 10 h and thereafter at 1100 °C for 36 h with intermediate grinding. Characterized by powder X-ray diffraction data and elemental analysis. Hexagonal, $a = 78.9214$, $c = 8.7267$ Å. The strongest lines of the powder X-ray diffraction pattern [*d*, Å (*I*, %) (*hkl*)] are: 4.3378 (46) (002), 3.1032 (100) (112), 2.9038 (60) (210), 2.2140 (56) (302), 2.1724 (34) (004), 2.1682 (38) (221), 2.0724 (32) (311), 1.5661 (53) (412).

Kind of sample preparation and/or method of registration of the spectrum: KBr disc. Transmission.

Source: Zhang et al. (2005).

Wavenumbers (cm^{-1}): 1583w, 1574w, 1497w, 1356s, 1335s, 1307sh, 1287s, 1274sh, 1245s, 1212s, 1188s, 1143s, 959, 904, 798, 778, 739, 694, 620, 596, 568, 463, 437.

Note: The wavenumbers were determined by us based on spectral curve analysis of the published spectrum.

Bo58 Sodium samarium orthoborate $Na_3Sm_2(BO_3)_3$

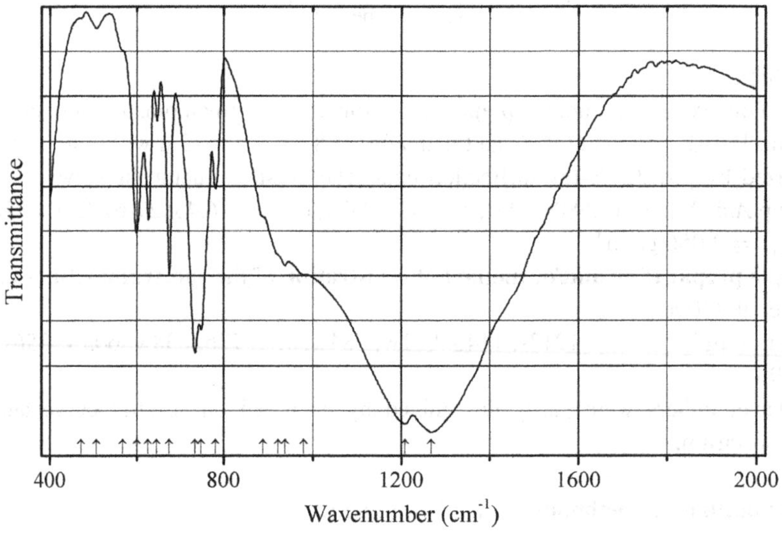

Origin: Synthetic.

Description: Synthesized by heating a mixture of Sm_2O_3, Na_2CO_3, and H_3BO_3 first at 500 °C for 10 h, and thereafter (after intermediate grinding) at 800 °C for 24 h. Orthorhombic, $a = 5.0585$, $b = 11.0421$, $c = 7.0316$ Å. The strongest lines of the powder X-ray diffraction pattern [*d*, Å (*I*, %) (*hkl*)] are: 5.5210 (58) (020), 5.0521 (80) (100), 3.7232 (69) (120), 2.9685 (65) (022), 2.8851 (81) (102), 2.5602 (100) (122).

Kind of sample preparation and/or method of registration of the spectrum: KBr disc. Transmission.

Source: Zhang et al. (2002b).

Wavenumbers (cm^{-1}): 1268s, 1208s, 980, 939, 923sh, 890sh, 781, 749, 735s, 676, 647w, 627, 600, 568sh, 509w, 473sh.

Note: The wavenumbers were determined by us based on spectral curve analysis of the published spectrum.

Bo59 Sodium strontium orthoborate $NaSr_4(BO_3)_3$

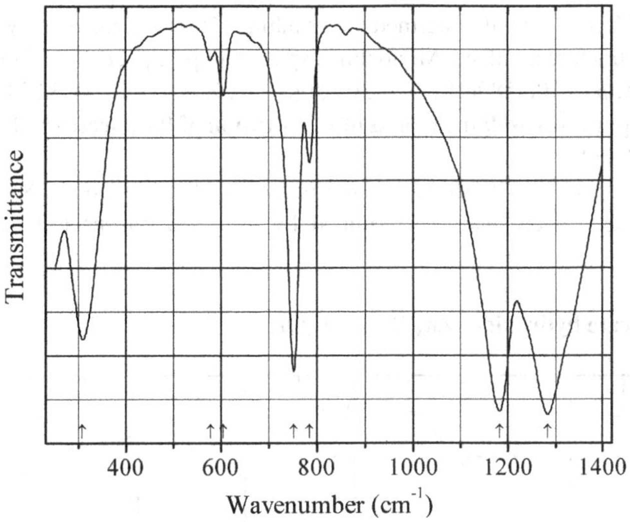

Origin: Synthetic.

Description: Obtained in the solid-state reaction from the stoichiometric mixture of Na_2CO_3, $SrCO_3$, and H_3BO_3 at 800 °C, with several grindings. Characterized by powder X-ray diffraction data. The crystal structure is solved. Cubic, space group *Ia-3d*, $a = 15.14629(6)$, $V = 3474.71(4)$ Å^3, $Z = 16$. $D_{calc} = 4.203$ g/cm^3. In the structure, isolated BO_3 groups are perpendicular to each other.

Kind of sample preparation and/or method of registration of the spectrum: KBr disc. Absorption.

Source: Wu et al. (2005).

Wavenumbers (cm^{-1}): 1283s, 1182s, 785, 753s, 605, 578w, 307s.

Bo60 Sodium strontium orthoborate $NaSr(BO_3)$

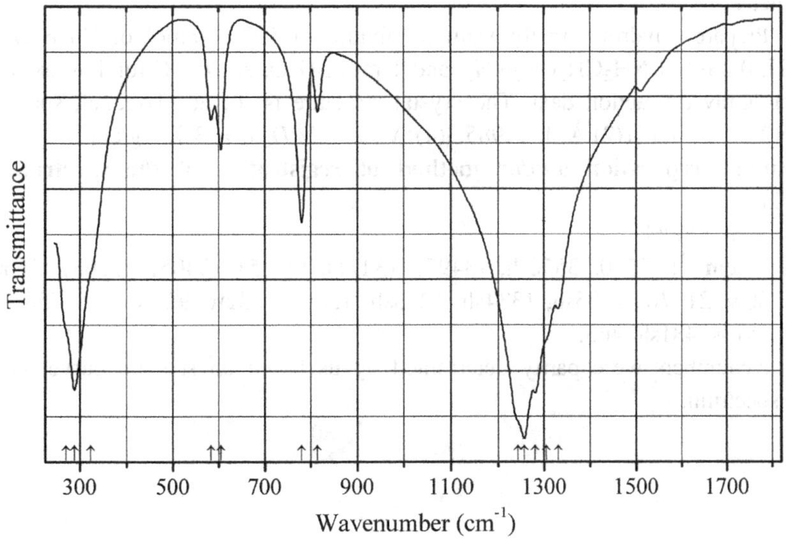

Origin: Synthetic.

Description: Prepared by heating a mixture of Na_2CO_3, $SrCO_3$, and H_3BO_3 first at 650 °C and thereafter at 850 °C for 72 h with intermediate grinding. Characterized by powder X-ray diffraction data. The crystal structure is solved. Monoclinic, space group $P2_1/c$, $a = 5.32446(7)$, $b = 9.2684(1)$, $c = 6.06683(8)$ Å, $\beta = 100.589(1)°$, $V = 294.30(8)$ Å^3, $Z = 4$. $D_{calc} = 3.824$ g/cm^3.

Kind of sample preparation and/or method of registration of the spectrum: KBr disc. Absorption.

Source: Wu et al. (2006b).

Wavenumbers (cm^{-1}): 1331, 1305sh, 1280s, 1257s, 1243sh, 815, 780, 605, 583, 323sh, 287s, 270sh.

Note: The wavenumbers were partly determined by us based on spectral curve analysis of the published spectrum.

Bo61 Zinc orthoborate hydroxide $Zn_8(BO_3)_3O_2(OH)_3$

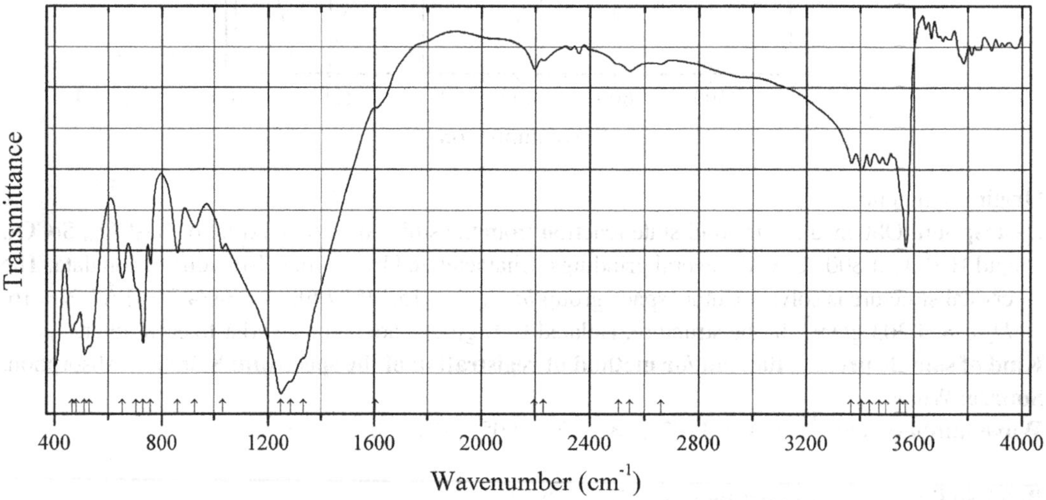

Origin: Synthetic.

Description: Prepared hydrothermally from a mixture of 0.637 mmol of $Zn_3B_2O_6$, 0.2 ml of CH_3COOH, 0.2 ml of $NH_2CH_2CH_2NH_2$, and 1 ml of H_2O, at 170 °C for 1 week. Characterized by powder X-ray diffraction data. The crystal structure is solved. Trigonal, space group $R32$, $a = 8.006(2)$, $c = 17.751(2)$ Å, $V = 985.3(4)$ Å^3, $Z = 3$. $D_{calc} = 3.956$ g/cm^3.

Kind of sample preparation and/or method of registration of the spectrum: KBr disc. Transmission.

Source: Chen et al. (2006b).

Wavenumbers (cm^{-1}): 3570, 3550sh, (3497), (3471), (3435), (3405), (3368), 2660w, 2546w, 2505sh, 2230w, 2197w, 1605sh, 1334sh, 1285sh, 1249s, 1032w, 927w, 861, 758, 730s, 705sh, 653, 530sh, 513s, 481sh, 465.

Note: The wavenumbers were partly determined by us based on spectral curve analysis of the published spectrum.

Bo62 Zinc orthoborate orthophosphate $Zn_3(BO_3)(PO_4)$

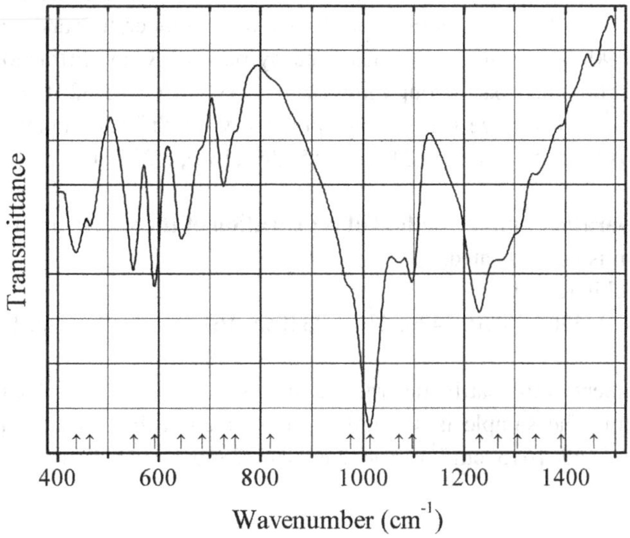

Origin: Synthetic.

Description: Prepared by heating a mixture of $ZnCO_3$, H_3BO_3, and $(NH_4)_2(HPO_4)$, taken in stoichiometric amounts, at 870 °C. Characterized by powder X-ray diffraction data. Hexagonal, $a = 8.435$ (4), $c = 13.032(6)$ Å. The strongest reflections are observed at 4.2113 and 2.5761 Å.

Kind of sample preparation and/or method of registration of the spectrum: KBr disc. Transmission.

Source: Wang et al. (2002).

Wavenumbers (cm^{-1}): 1458w, 1392w, 1342w, 1306sh, 1268, 1230s, 1098, 1071, 1013s, 975sh, 819sh, 751sh, 728, 685sh, 645, 592, 552, 464, 438.

Note: The wavenumbers were partly determined by us based on spectral curve analysis of the published spectrum.

Bo63 Cobalt dinickel orthoborate $CoNi_2(BO_3)_2$

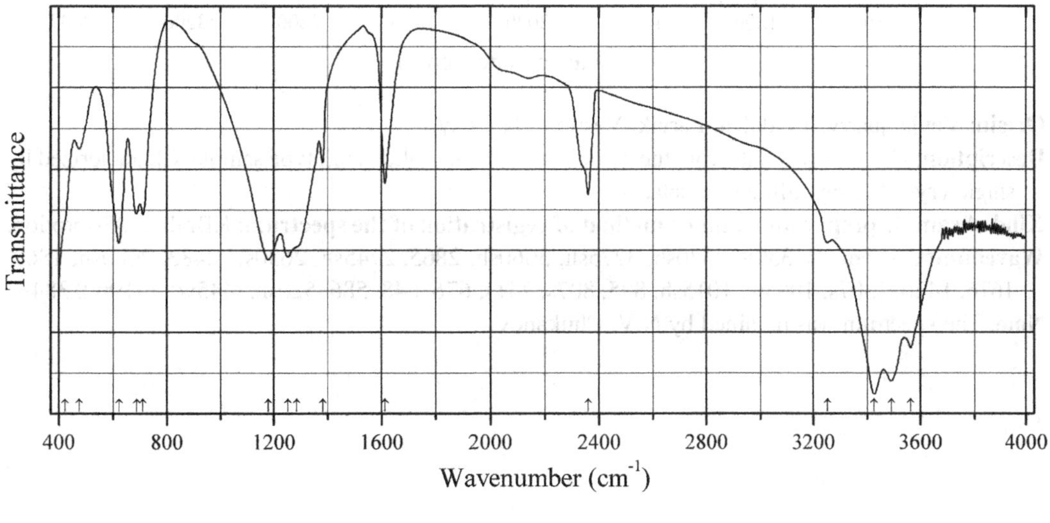

Origin: Synthetic.

Description: Synthesized by heating a stoichiometric mixture of cobalt nitrate, nickel nitrate, and boric acid, first at 450 °C for 4 h, thereafter (after cooling to room temperature and grinding) at 600 °C for 3 h, and finally at 900 °C for 48 h. Characterized by powder X-ray diffraction data. Isostructural with kotoite. Orthorhombic, space group *Pnmn*, $a = 5.419(9)$, $b = 8.352(0)$, $c = 4.478(8)$ Å, $Z = 2$. $D_{meas} = 4.48$ g/cm^3. The strongest lines of the powder X-ray diffraction pattern [*d*, Å (*I*, %) (*hkl*)] are: 3.9420 (38) (011), 2.6591 (100) (121), 2.4755 (36) (130), 2.2339 (55) (211), 1.7258 (30) (202), 1.6610 (33) (132).

Kind of sample preparation and/or method of registration of the spectrum: Transmission. Kind of sample preparation is not indicated.

Source: Güler and Tekin (2009).

Wavenumbers (cm^{-1}): 3564, 3490, 3426, 2360, 2340sh, 1613, 1382, 1285sh, 1253, 1180, 712, 688, 622, 476, 422sh.

Note: The wavenumbers were partly determined by us based on spectral curve analysis of the published spectrum. The sample is strongly contaminated with a hydrous phase (the bands at 3564, 3490, 3426, and 1613 cm^{-1}). The bands at 2360 and 2340 cm^{-1} may correspond to atmospheric CO_2.

Bo64 Berborite $Be_2(BO_3)(OH,F)\cdot H_2O$

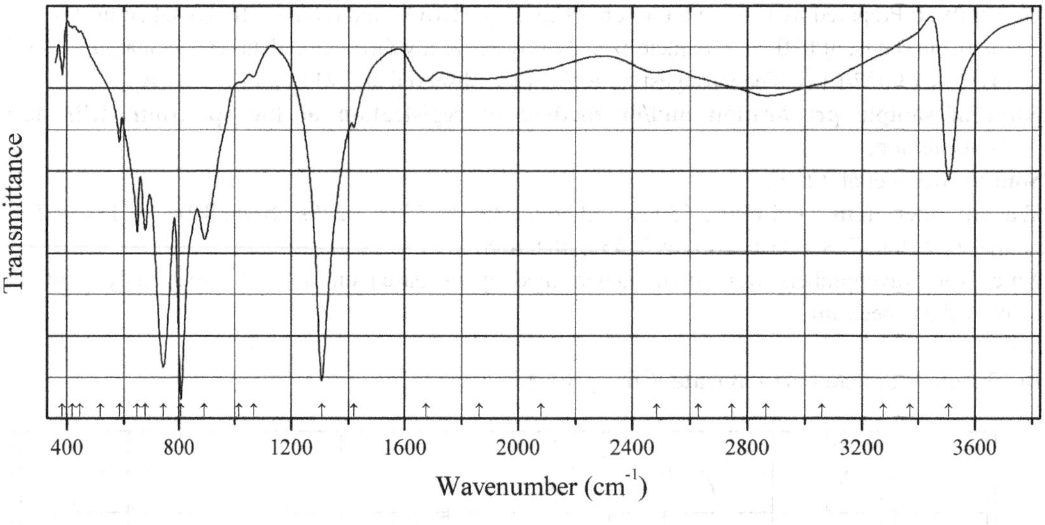

Origin: Vevja quarry, Tvedalen, Larvik, Vestfold, Norway.

Description: Colorless crystals from the association with natrolite. Holotype sample. Characterized by single-crystal X-ray diffraction data.

Kind of sample preparation and/or method of registration of the spectrum: KBr disc. Absorption.

Wavenumbers (cm^{-1}): 3508, 3370sh, 3275sh, 3060sh, 2865, 2745sh, 2630sh, 2483, 2080sh, 1862, 1676, 1420, 1307s, 1065w, 1015sh, 895, 807s, 741s, 676, 648, 586, 520sh, (445w), (419w), 384w.

Note: The spectrum was obtained by N.V. Chukanov.

B160 Lüneburgite $Mg_3[B_2(OH)_6(PO_4)_2]\cdot6H_2O$

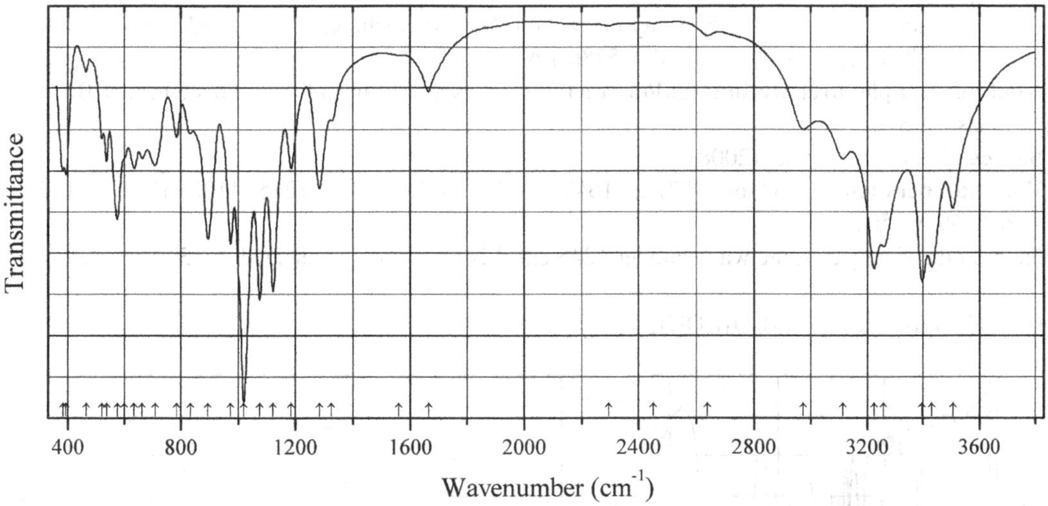

Origin: Morro Mejillones, Mejillones Peninsula, Mejillones, Antofagasta, II Region, Chile.

Description: Yellow nodule from clay. Investigated by I.V. Pekov. Identified by IR spectrum and qualitative electron microprobe analyses.

Kind of sample preparation and/or method of registration of the spectrum: KBr disc. Absorption.

Wavenumbers (cm^{-1}): 3505, 3431s, 3397s, 3258, 3225s, 3113, 2974, 2639w, 2450w, 2295w, 1665, (1560w), 1326, 1283, 1185, 1122s, 1075s, 1020s, 974, 895, 834, 785, 709, 665, 636, 600sh, 576, 538, 522, 465w, 393, 383.

Note: The spectrum was obtained by N.V. Chukanov.

B161 Ammonium pentaborate $(NH_4)B_5O_8$

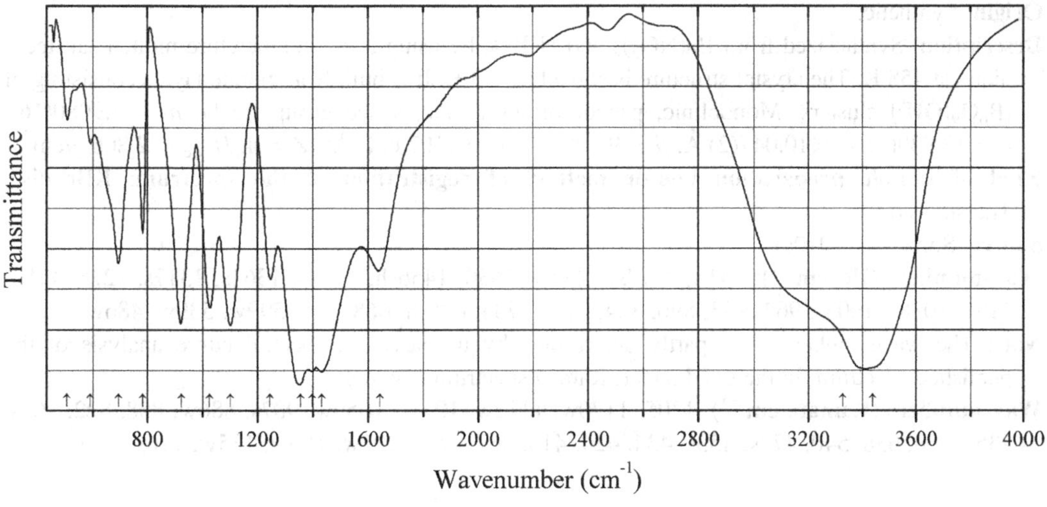

Origin: Synthetic.

Description: Crystals of commercially available ammonium pentaborate grown from aqueous solution. Characterized by powder X-ray diffraction data. Monoclinic, a = 7.189(5), b = 11.308(5), c = 7.217(6) Å, β = 100.12(7)°, V = 578(2) $Å^3$.

Kind of sample preparation and/or method of registration of the spectrum: KBr disc. Transmission.

Source: Balakrishnan et al. (2008).

Wavenumbers (cm^{-1}): 3436s, 3327sh, 1642, 1433s, 1397s, 1355s, 1245, 1102, 1024, 924, 782, 695, 591w, 504w.

Note: In the cited paper, the wavenumber 1245 cm^{-1} is erroneously indicated as 1345 cm^{-1}.

B162 Barium borate $BaB_8O_{11}(OH)_4$

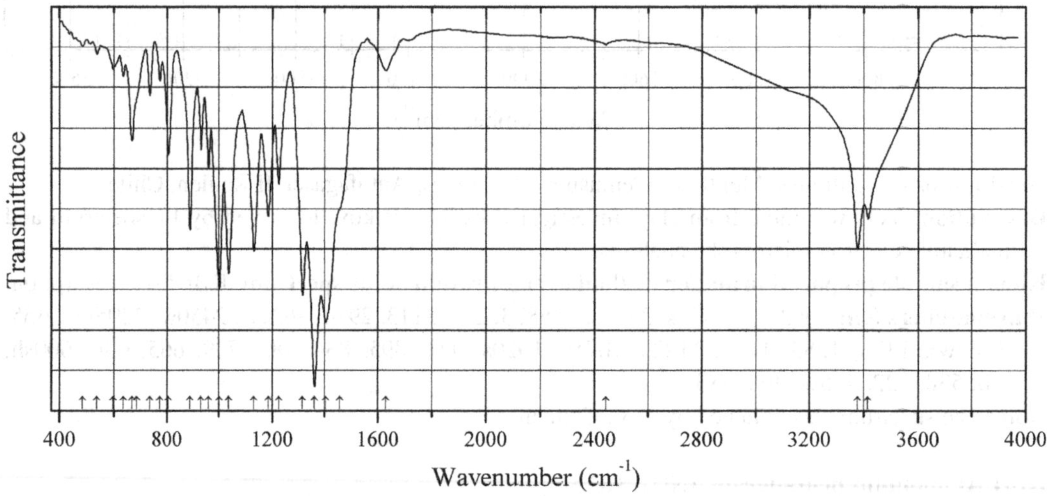

Origin: Synthetic.

Description: Synthesized from $Ba(NO_3)_2$ and H_3BO_3 by using a low-temperature molten salt technique at 458 K. The crystal structure is solved at 173 K. It is built from borate layers consisting of $[B_6O_9(OH)]$ clusters. Monoclinic, pseudo-orthorhombic, space group $P2_1/n$, a = 7.9080(16), b = 13.939(3), c = 10.047(2) Å, β = 90.00(3)°, V = 1107.6(4) $Å^3$, Z = 4. D_{calc} = 2.806 g/cm^3.

Kind of sample preparation and/or method of registration of the spectrum: KBr disc. Transmission.

Source: Sun et al. (2010a).

Wavenumbers (IR, cm^{-1}): 3416, 3377s, 2443w, 1630, 1456sh, 1404s, 1361s, 1317s, 1225, 1185, 1131, 1038s, 1003s, 962, 933, 890, 804, 771w, 734, 686sh, 668, 635, 599w, 539w, 486w.

Note: The wavenumbers were partly determined by us based on spectral curve analysis of the published spectrum. In the cited paper, Raman spectrum is given.

Wavenumbers (Raman, cm^{-1}): 1305, 1170w, 1112w, 1011s, 1050w, 987s, 886w, 858, 822, 761s, 736s, 658, 638, 540, 472s, 453s, 434, 424, 410, 367s, 243, 202w, 166w, 135w, 116s.

B163 Barium calcium diborate $BaCa(B_2O_5)$

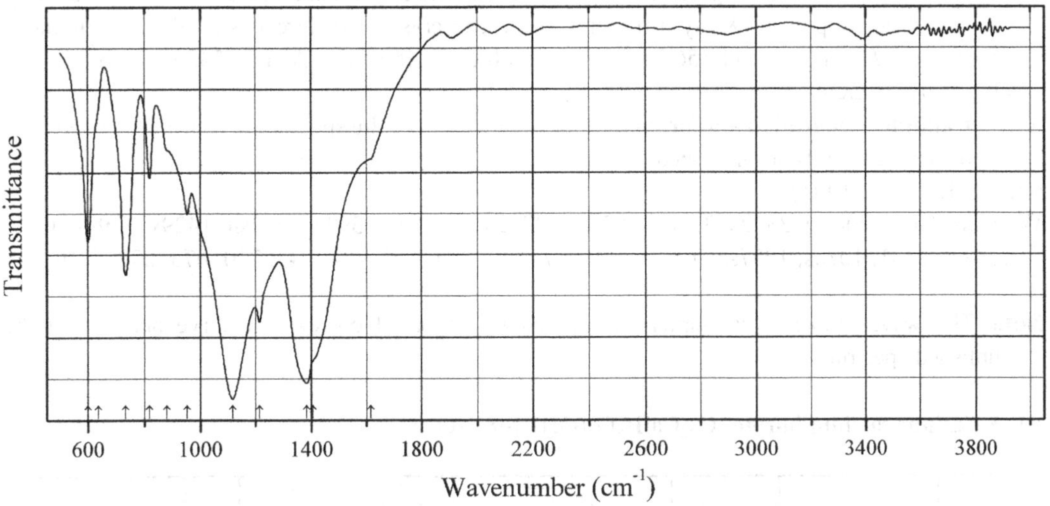

Origin: Synthetic.

Description: Crystals grown from the mixture of barium nitrate, calcium oxide, and boric acid first preheated at 500 °C for 7 h, then gradually heated to 900 °C, and kept at this temperature for 72 h with several intermediate grindings and mixing. The crystal structure is solved. Monoclinic, space group $P2_1/c$, $a = 6.568(2)$, $b = 20.545(7)$, $c = 8.201(2)$ Å, $\beta = 117.00(2)°$, $Z = 4$. $D_{calc} = 3.759$ g/cm^3.

Kind of sample preparation and/or method of registration of the spectrum: KBr disc. Transmission.

Source: Liu et al. (2015b).

Wavenumbers (cm^{-1}): 1615sh, 1405sh, 1383s, 1214, 1117s, 954, 881sh, 819, 735, 637sh, 600.

Note: The wavenumbers were partly determined by us based on spectral curve analysis of the published spectrum.

B164 Ammonium calcium borate $(NH_4)_2Ca[B_4O_5(OH)_4]_2{\cdot}8H_2O$

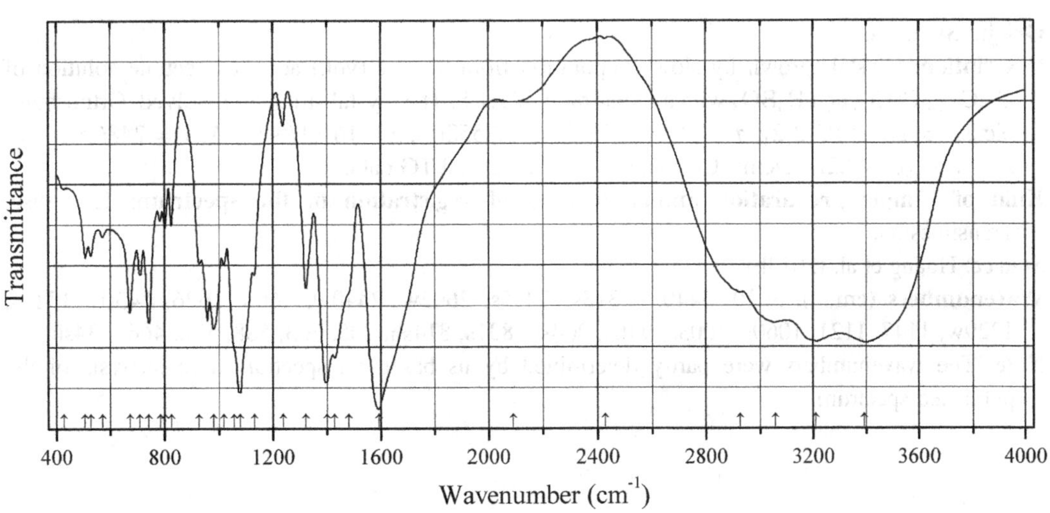

Origin: Synthetic.

Description: Synthesized through slow evaporation of a solution containing $CaCl_2$, H_3BO_3 and NH_3. Characterized by powder X-ray diffraction data. The crystal structure is solved. Orthorhombic, space group $P2_12_12_1$, a =11.556(7), b = 12.583(8), c = 16.679(8) Å, V = 2425.2(2) Å^3, Z = 4. D_{calc} = 1.651 g/cm^3.

Kind of sample preparation and/or method of registration of the spectrum: Transmission. Kind of sample preparation is not indicated.

Source: Li et al. (2011a).

Wavenumbers (cm^{-1}): 3392s, 3210s, 3055, 2927, 2427, 2090, 1592s, 1482sh, 1429s, 1399s, 1322, 1238w, 1131, 1078s, 1057sh, 1018, 980, 957, 928, 825, 802, 784, 743, 710, 673, 572, 529, 507, 428.

Note: The wavenumbers were partly determined by us based on spectral curve analysis of the published spectrum.

B165 Cesium calcium borate $Cs_2Ca[B_4O_5(OH)_4]_2 \cdot 8H_2O$

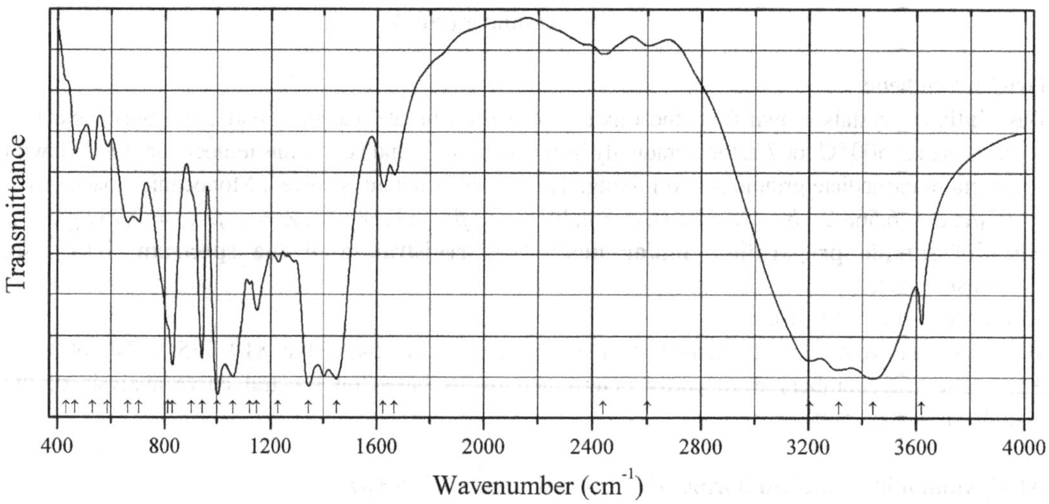

Origin: Synthetic.

Description: Crystals grown by slow evaporation from the 2:1 (vol.) aqueous acetone solution of Cs_2CO_3, $CaCl_2$, and H_3BO_3 with a molar ratio of 2:1:8. The crystal structure is solved. Orthorhombic, space group $P2_12_12_1$, a = 11.5158(7), b = 12.8558(7), c = 16.7976(10) Å, V = 2486.8(3) Å^3, Z = 4. D_{calc} = 2.224 g/cm^3. Characterized by DSC and TG data.

Kind of sample preparation and/or method of registration of the spectrum: KBr disc. Transmission.

Source: Huang et al. (2013b).

Wavenumbers (cm^{-1}): 3620, 3440s, 3313s, 3205s, 2602w, 2440w, 1668, 1626, 1450s, 1345s, 1229w, 1147, 1121, 1060s, 1003s, 946s, 904sh, 831s, 814sh, 705, 663, 588, 532, 466, 434sh.

Note: The wavenumbers were partly determined by us based on spectral curve analysis of the published spectrum.

B166 Calcium borate CaB_6O_{10}

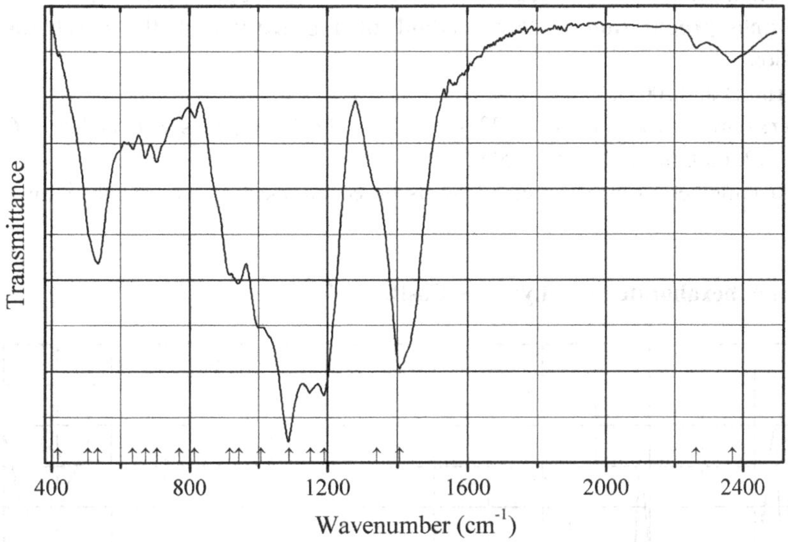

Origin: Synthetic.

Description: Colorless crystals prepared from corresponding oxides by solid-state reaction at 735 °C for 2 weeks. Characterized by powder X-ray diffraction data. The crystal structure is solved. Monoclinic, space group $P2_1/c$, a = 9.799(1), b = 8.705(1), c = 9.067(1) Å, β = 116.65(1)°, V = 691.23(13) $Å^3$, Z = 4. D_{calc} = 2.546 g/cm^3.

Kind of sample preparation and/or method of registration of the spectrum: KBr disc. Transmission.

Source: Chen et al. (2008b).

Wavenumbers (cm^{-1}): 2369w, 2264w, 1406s, 1342sh, 1150s, 1191s, 1088s, 1008sh, 942, 916, 815w, 769sh, 705w, 672w, 635w, 533, 504sh, 418w.

Note: The wavenumbers were partly determined by us based on spectral curve analysis of the published spectrum.

B167 Calcium tetraborate β-CaB_4O_7

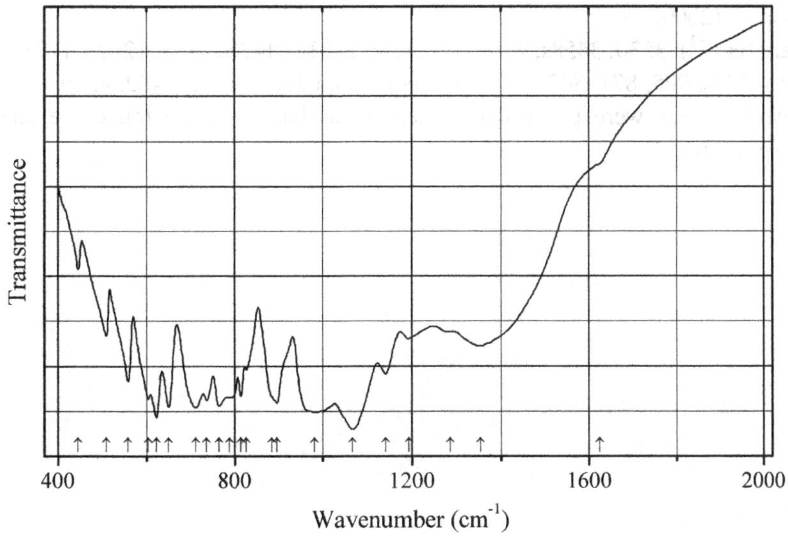

Origin: Synthetic.

Description: High-pressure β-modification isotypic with orthorhombic SnB_4O_7.

Kind of sample preparation and/or method of registration of the spectrum: KBr disc. Transmission.

Source: Knyrim et al. (2007).

Wavenumbers (cm^{-1}): 1623sh, 1353, 1285sh, 1193, 1142, 1067s, 982s, 896, 884sh, 826, 814, 788sh, 765, 736, 712, 650, 623s, 604, 558, 509, 446w.

Note: The wavenumbers were determined by us based on spectral curve analysis of the published spectrum.

B168 Dicalcium hexaborate monohydrate $Ca_2B_6O_{11}{\cdot}H_2O$

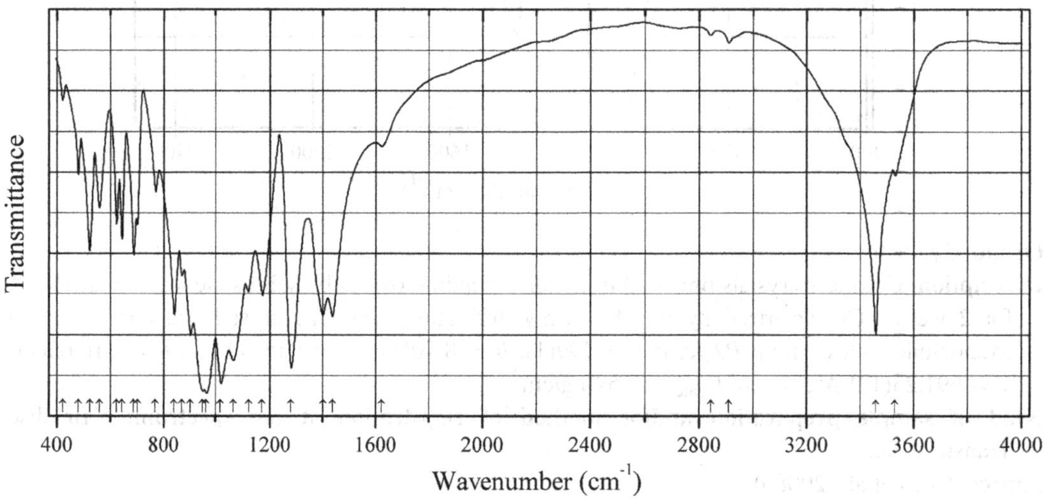

Origin: Synthetic.

Description: Synthesized hydrothermally from H_3BO_3 and CaO mixed with the mole ratio ranging from 1.5:1 to 2:1, at 234–300 °C for 48–96 h. Characterized by powder X-ray diffraction data. Orthorhombic, space group $Pbn2_1$.

Kind of sample preparation and/or method of registration of the spectrum: Transmission. Kind of sample preparation is not indicated.

Source: Guo et al. (2000).

Wavenumbers (cm^{-1}): 3530, 3458s, 2908w, 2842w, 1623w, 1438, 1403, 1282s, 1171, 1121, 1066s, 1019s, 966s, 953s, 905, 871, 842, 767, 701, 686, 643, 623, 559, 524, 482, 422w.

Note: The wavenumbers were partly determined by us based on spectral curve analysis of the published spectrum.

B169 Dipotassium sodium zinc pentaborate $K_2NaZnB_5O_{10}$

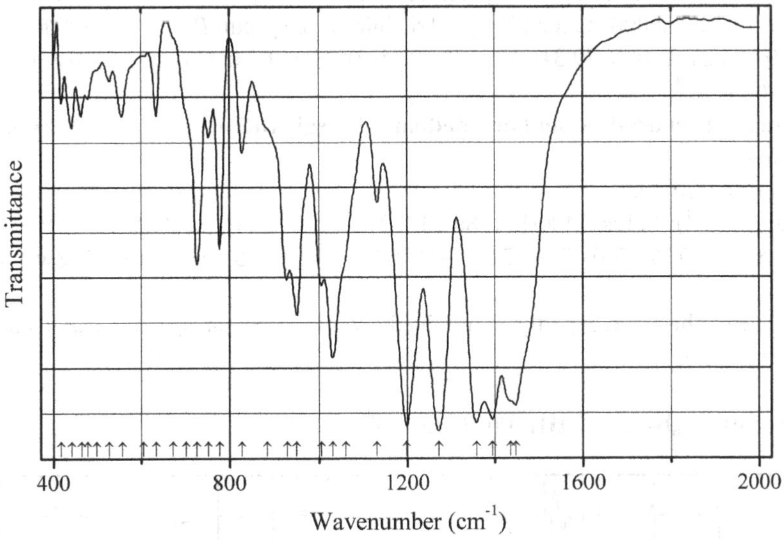

Origin: Synthetic.

Description: Synthesized by employing a high-temperature solution reaction method. The crystal structure is solved. Monoclinic, space group *C*2/*c*, $a = 7.9244(16)$, $b = 12.805(3)$, $c = 18.962(4)$ Å, $\beta = 99.39(3)°$, $V = 1898.4(7)$ Å^3, $Z = 8$. $D_{calc} = 2.663$ g/cm^3.

Kind of sample preparation and/or method of registration of the spectrum: KBr disc. Transmission.

Source: Chen et al. (2010).

Wavenumbers (cm^{-1}): 1447s, 1436sh, 1394s, 1358s, 1272s, 1199s, 1132, 1060sh, 1032, 1006, 949, 928, 883sh, 827, 777, 752, 726, 702sh, 673sh, 635w, 606sh, 557w, 527w, 498sh, 477w, 462w, 440w, 417w.

Note: The wavenumbers were partly determined by us based on spectral curve analysis of the published spectrum.

B170 Double-ring borate $(Na,K)_3Sr(B_5O_{10})$ $(Na,K)_3Sr(B_5O_{10})$

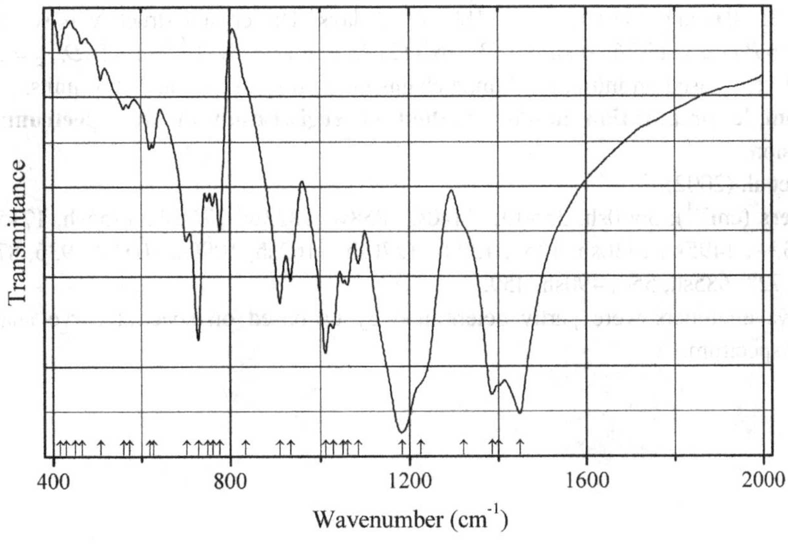

Origin: Synthetic.

Description: Synthesized from the melt prepared from Na_2CO_3, $SrCO_3$, H_3BO_3, and $K_2B_4O_7$ in the molar ratio 3:2:10:2 at 800 °C for 10 days. Triclinic, space group *P*-1, $a = 7.3900(15)$, $b = 7.6490(15)$, $c = 9.773(2)$ Å, $\alpha = 79.31(2)°$, $\beta = 70.85(2)°$, $\gamma = 62.09(1)°$, $V = 460.82(17)$ Å^3, $Z = 2$. $D_{calc} = 2.766$ g/cm^3.

Kind of sample preparation and/or method of registration of the spectrum: KBr disc. Transmission.

Source: Chen et al. (2011).

Wavenumbers (cm^{-1}): 1449s, 1400sh, 1386s, 1322sh, 1225sh, 1183s, 1085, 1061, 1051, 1029, 1011, 934, 910, 834sh, 776, 760, 748, 728, 702, 628w, 620w, 573w, 560w, 508w, 465w, 450sh, 430w, 416w.

Note: The wavenumbers were partly determined by us based on spectral curve analysis of the published spectrum.

B171 Lead borate $Pb_6B_{11}O_{18}(OH)_9$ $Pb_6B_{11}O_{18}(OH)_9$

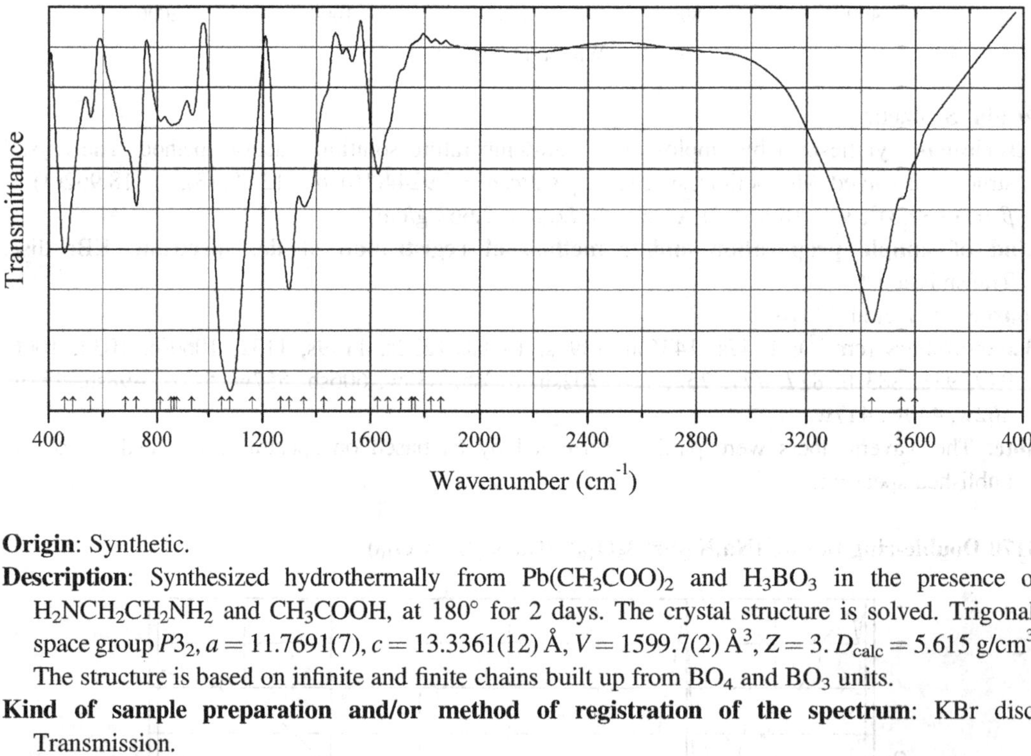

Origin: Synthetic.

Description: Synthesized hydrothermally from $Pb(CH_3COO)_2$ and H_3BO_3 in the presence of $H_2NCH_2CH_2NH_2$ and CH_3COOH, at 180° for 2 days. The crystal structure is solved. Trigonal, space group $P3_2$, $a = 11.7691(7)$, $c = 13.3361(12)$ Å, $V = 1599.7(2)$ Å^3, $Z = 3$. $D_{calc} = 5.615$ g/cm^3. The structure is based on infinite and finite chains built up from BO_4 and BO_3 units.

Kind of sample preparation and/or method of registration of the spectrum: KBr disc. Transmission.

Source: Yu et al. (2002).

Wavenumbers (cm^{-1}): 3600sh, 3550sh, 3440s, 1858w, 1823w, 1768sh, 1755sh, 1715sh, 1668sh, 1628, 1532w, 1495w, 1430sh, 1357, 1300s, 1270sh, 1162sh, 1080s, 1051sh, 936, 877sh, 867sh, 856, 814, 723, 685sh, 554, 490sh, 459.

Note: The wavenumbers were partly determined by us based on spectral curve analysis of the published spectrum.

B172 Lead borate PbB_4O_7 PbB_4O_7

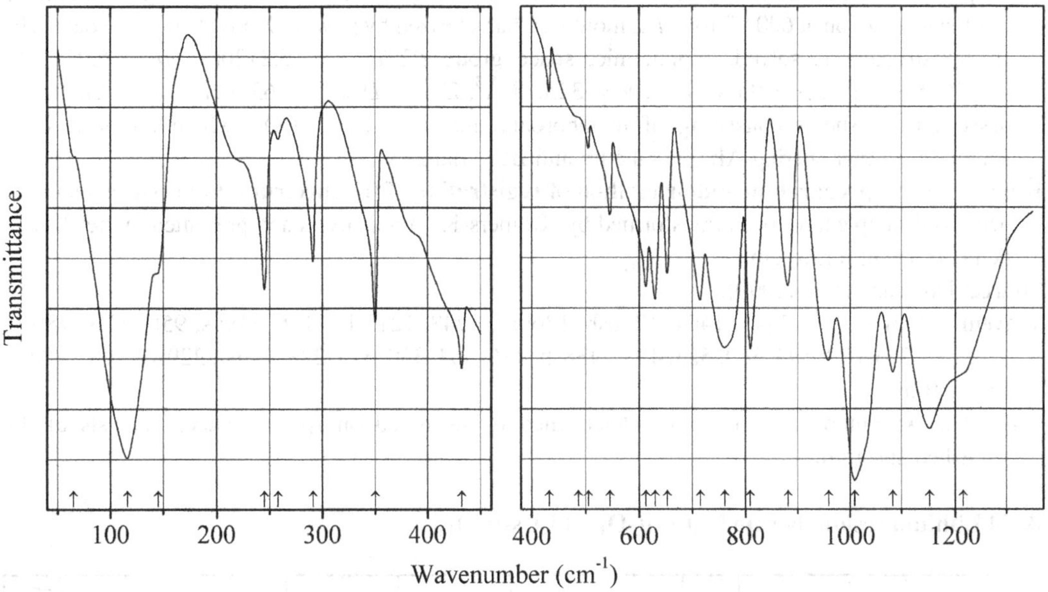

Origin: Synthetic.

Description: Obtained from stoichiometric melt of PbO and B_2O_3 heated to 1060 K. Orthorhombic *P*21*nm*, space group, *a* = 4.251, *b* = 4.463, *c* = 10.86 Å, *Z* = 2. Characterized by powder X-ray diffraction data.

Kind of sample preparation and/or method of registration of the spectrum: KBr disc (for the 1500–400 cm^{-1} region) and Nujol mull (for the 500–50 cm^{-1} region). Transmission.

Source: Hanuza et al. (2008b).

Wavenumbers (cm^{-1}): 1215sh, 1152s, 1083s, 1009s, 959s, 881, 808, 761, 715, 652, 629, 612, 545, 505w, 486sh, 432w, 350w, 291w, 258w, 245w, 145sh, 116, 65sh.

Note: In the cited paper, Raman spectra are given for different polarization and crystal orientation.

B173 Lithium aluminoborate Li_2AlBO_4

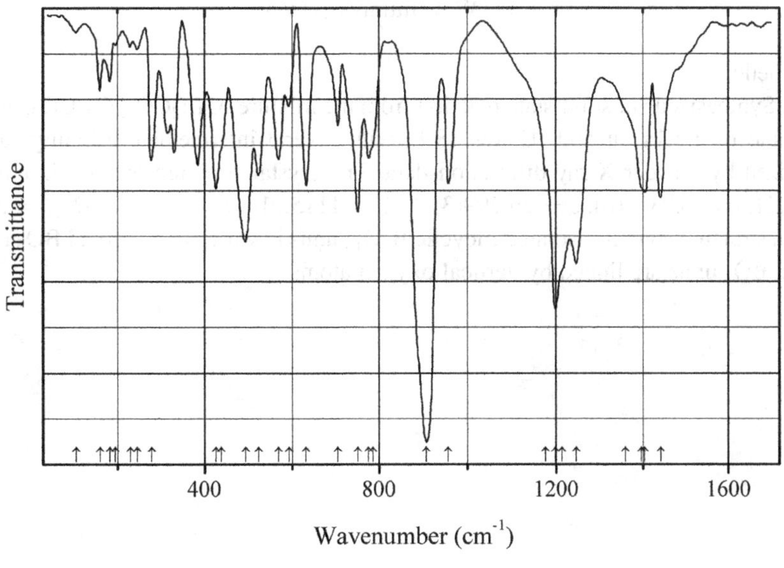

Origin: Synthetic.

Description: Prepared from a stoichiometric mixture of Li_2CO_3, B_2O_3, and Al_2O_3 (corundum) by solid-state reaction at 600 °C for *ca.* 2 months. Characterized by powder X-ray diffraction data. The crystal structure is solved. Monoclinic, space group $P2_1/c$, a = 6.2720(3), b = 5.0701(3), c = 10.2989(6) Å, β = 95.882(2)°, V = 325.78 Å^3, Z = 4. D_{calc} = 2.63 g/cm^3. The structure is based on the sheets consisting of metaboroaluminate rings, $B_2Al_2O_8$, which are formed of alternating corner-sharing AlO_4 tetrahedra and BO_3 triangles.

Kind of sample preparation and/or method of registration of the spectrum: Reflectance data for a pelletized sample have been transformed by Kramers-Kronig analysis and presented in the absorption coefficient formalism.

Source: Psycharis et al. (1999).

Wavenumbers (cm^{-1}): 1441, 1404, 1396sh, 1360sh, 1248, 1215sh, 1176, 1199s, 956, 907s, 786sh, 776, 751, 705, 632, 593, 569, 523, 493s, 438sh, 426, 384, 330, 315, 278, 245w, 229w, 195w, 182w, 159w, 105w.

Note: The wavenumbers were partly determined by us based on spectral curve analysis of the published spectrum.

B174 Lithium cesium borate $Li_4Cs_3B_7O_{14}$ $Li_4Cs_3B_7O_{14}$

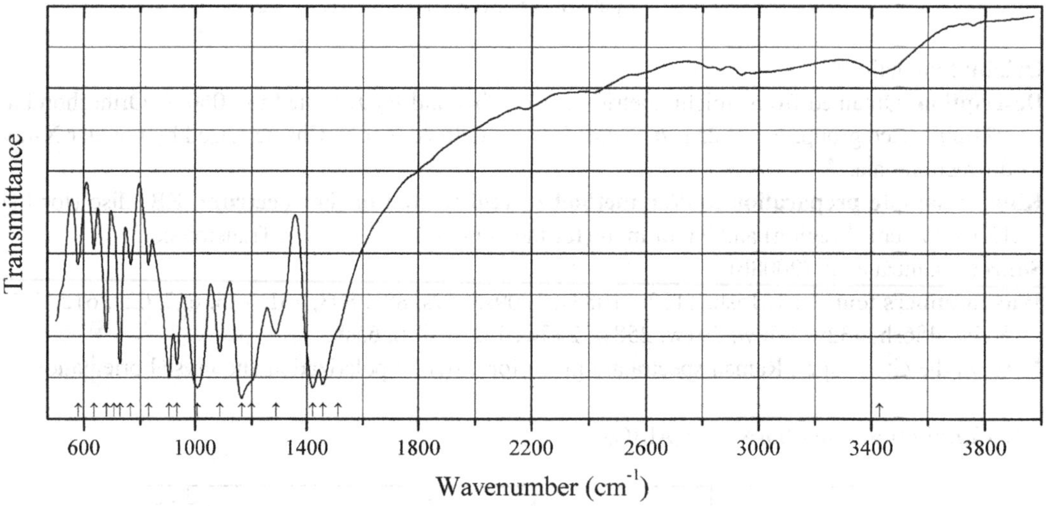

Origin: Synthetic.

Description: Synthesized *via* solid-state reaction from the mixture of Li_2CO_3, Cs_2CO_3, and H_3BO_3 in a molar ratio of 4:3:14, at 560 °C, for 48 h, with several intermediate grindings and mixings. Characterized by powder X-ray diffraction data. The crystal structure is solved. Trigonal, space group $P3_121$, a = 6.9313(6), c = 26.799(3) Å, V = 1115.01(19) Å^3, Z = 3. D_{calc} = 3.244 g/cm^3. The crystal structure contains isolated tricyclic B_7O_{14} units in which five trigonal BO_3 units and two tetrahedral BO_4 units are linked by vertical oxygen atoms.

Kind of sample preparation and/or method of registration of the spectrum: KBr disc. Transmission.

Source: Yang et al. (2011e).

Wavenumbers (cm^{-1}): (3428w), 1512sh, 1458s, 1421s, 1288, 1201sh, 1165s, 1090, 934, 905, 828, 766, 728, 707sh, 679, 635, 579.

Note: The wavenumbers were partly determined by us based on spectral curve analysis of the published spectrum. The wavenumber at 1129 cm^{-1} given by the authors is erroneous.

B175 Lithium sodium borate $LiNaB_4O_7$ $LiNaB_4O_7$

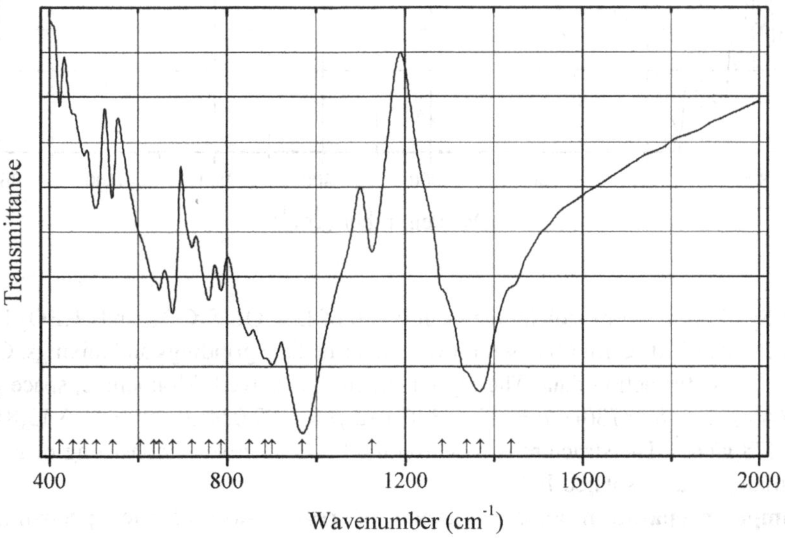

Origin: Synthetic.

Description: Synthesized from the mixture of Na_2CO_3, Ga_2O_3, H_3BO_3, and $Li_2B_4O_7$ in the molar ratio 2:1:8:1 by employing high temperature solution reaction method. Ga_2O_3 acted as a flux for the crystal growth. Characterized by powder X-ray diffraction data. The crystal structure is solved. Orthorhombic, space group *Fdd2*, a = 13.326(3), b = 14.072(3), c = 10.238(2) Å, V = 1919.9 (7) Å^3, Z = 16. D_{calc} = 2.563 g/cm^3. The basic structural unit is a bicyclic B_4O_9 group that consists of two vertex-sharing BO_4 tetrahedra and two bridging BO_3 triangles.

Kind of sample preparation and/or method of registration of the spectrum: KBr disc. Transmission.

Source: Reshak et al. (2012a).

Wavenumbers (cm^{-1}): 1440sh, 1369s, 1339sh, 1283sh, 1126, 968s, 900, 885sh, 848, 786, 759, 722, 678, 648, 637sh, 605sh, 543, 505, 479w, 453w, 423w.

Note: The wavenumbers were partly determined by us based on spectral curve analysis of the published spectrum.

B176 Lithium strontium borate $Li_2Sr_4B_{12}O_{23}$ $Li_2Sr_4B_{12}O_{23}$

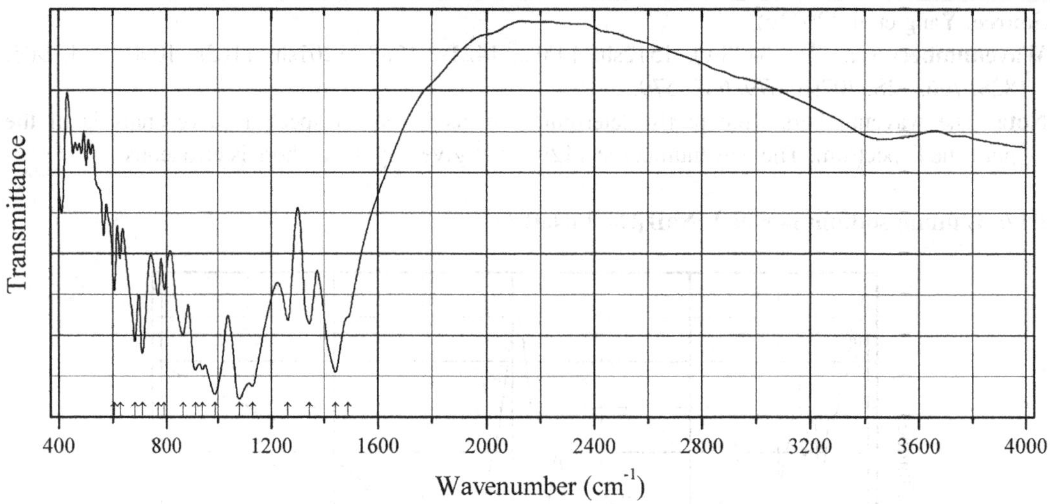

Origin: Synthetic.

Description: Synthesized from stoichiometric mixture of Li_2CO_3, $SrCO_3$, and H_3BO_3 by solid-state reaction method at 710 °C for 60 h, with several intermediate grindings and mixings. Characterized by powder X-ray diffraction data. The crystal structure is solved. Monoclinic, space group $P2_1/c$, a = 6.4664(4), b = 8.4878(4), c = 15.3337(8) Å, β = 102.024(3)°, V = 823.13(8) Å^3, Z = 2. D_{calc} = 3.478 g/cm^3. The structure is based on the $B_{10}O_{18}$ network, consisting of BO_4 tetrahedra and BO_3 triangles, and isolated B_2O_5 unit.

Kind of sample preparation and/or method of registration of the spectrum: KBr disc. Transmission.

Source: Zhang et al. (2012a).

Wavenumbers (cm^{-1}): 1488sh, 1442, 1345, 1264, 1129s, 1081s, 989s, 942s, 913s, 865, 791, 767, 708, 681, 627, 606, and a series of weak bands below 600 cm^{-1}.

Note: The wavenumbers were partly determined by us based on spectral curve analysis of the published spectrum.

B177 Lithium tetraborate $Li_2B_4O_7$

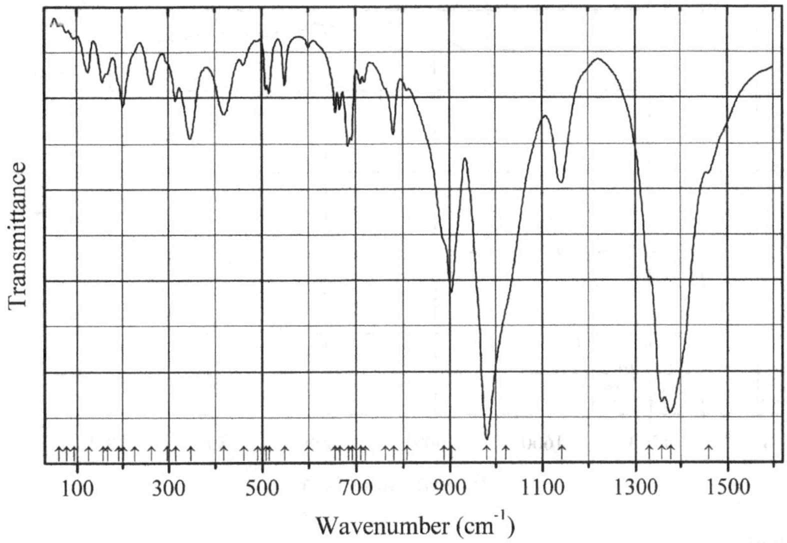

Origin: Synthetic.

Description: Single crystals grown by the Czochralski method from a stoichiometric melt. Characterized by powder X-ray diffraction data. The crystal structure is solved. Tetragonal, space group $I4_1cd$, $a = 9.477$, $c = 10.286$ Å, $Z = 8$. The structure contains pairs of BO_4 tetrahedra linked by a common O atom to form a B_2O_7 group. BO_3 triangles join these groups to yield a B_4O_7 network.

Kind of sample preparation and/or method of registration of the spectrum: KBr disc (for the region 500–1600 cm^{-1}), CsI disc (200–700 cm^{-1}), and polyethylene disc (50–300 cm^{-1}). Absorption.

Source: Zhigadlo et al. (2001).

Wavenumbers (cm^{-1}): 1458, 1378s, 1358s, 1330, 1142, 1021sh, 980s, 904s, 888sh, 809w, 780, 763sh, 719, 710, 691, 683, 666, 656, 600w, 549, 516, 509, 491w, 461w, 419, 348, 315, 297sh, 262, 226sh, 201, 191sh, 167, 156, 125, 94w, 77w, 61w.

Note: The wavenumbers were determined by us based on spectral curve analysis of the published spectrum.

B178 Magnesium strontium diorthoborate $MgSr(B_2O_5)$

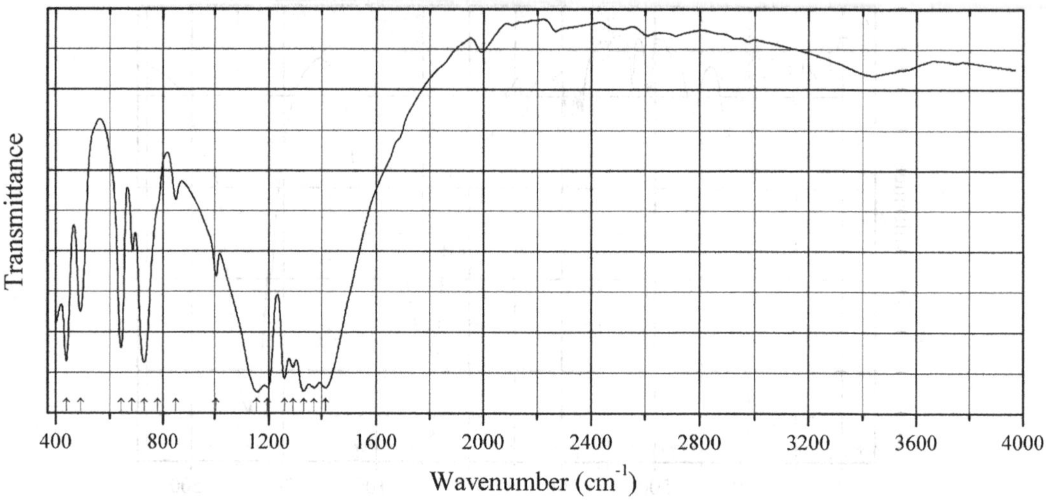

Origin: Synthetic.

Description: Synthesized by solid-state reaction from the stoichiometric mixture of MgO, $Sr(NO_3)_2$, and H_3BO_3 heated first at 500 °C for 24 h and then at 900 °C for 72 h with several intermediate grindings and mixings. Characterized by powder X-ray diffraction data. The crystal structure is solved. Monoclinic, space group $P2_1/c$, $a = 6.478(4)$, $b = 5.327(4)$, $c = 12.048(8)$ Å, $\beta = 102.805(8)°$, $V = 405.4(5)$ Å^3, $Z = 4$. $D_{calc} = 3.499$ g/cm^3.

Kind of sample preparation and/or method of registration of the spectrum: KBr disc. Transmission.

Source: Guo et al. (2014b).

Wavenumbers (cm^{-1}): 1415s, 1372s, 1333s, 1293s, 1260s, 1197s, 1157s, 1006, 849w, 780sh, 729, 685, 643, 492, 440.

Note: The wavenumbers were partly determined by us based on spectral curve analysis of the published spectrum.

B179 Potassium barium borate $KBaB_5O_9$ $KBaB_5O_9$

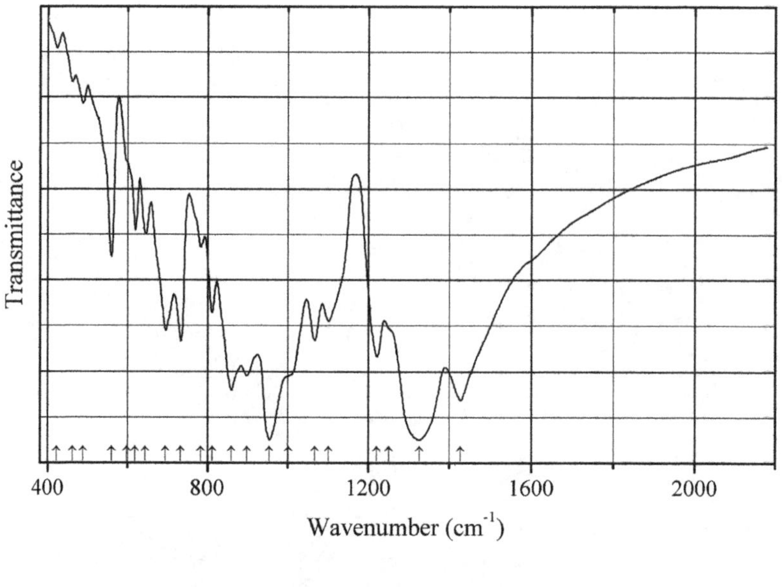

Origin: Synthetic.

Description: Prepared by the solid-state reaction of a stoichiometric mixture containing KNO_3, Ba $(NO_3)_2$, and H_3BO_3, first at 500 °C for 4 h, and thereafter (after regrinding) at 650 °C and for 48 h. Characterized by powder X-ray diffraction data. The crystal structure is solved. Monoclinic, space group $P2_1/c$, $a = 6.7168(11)$, $b = 8.2724(13)$, $c = 14.262(2)$ Å, $\beta = 92.724(2)°$, $V = 791.5(2)$ Å^3, $Z = 4$. $D_{calc} = 5.572$ g/cm^3.

Kind of sample preparation and/or method of registration of the spectrum: Transmission. Kind of sample preparation is not indicated.

Source: Yu et al. (2014).

Wavenumbers (cm^{-1}): 1428s, 1325s, 1251sh, 1221, 1100, 1065, 1002sh, 955s, 898s, 858s, 811, 784, 735, 697, 646, 620, 598sh, 560, 491w, 464w, 424w.

Note: The wavenumbers were partly determined by us based on spectral curve analysis of the published spectrum.

B180 Potassium borate $KB_3O_5 \cdot H_2O$ $KB_3O_5 \cdot H_2O$

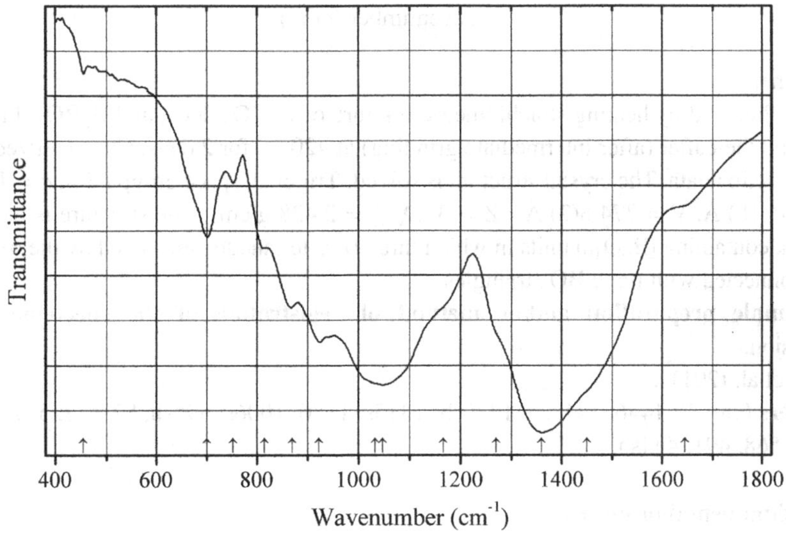

Origin: Synthetic.

Description: Prepared in the reaction between fine powders of $K_2B_4O_7 \cdot 4H_2O$ and $KB_5O_8 \cdot 4H_2O$ under exposure of water vapor, with subsequent heating to 110 °C. X-ray amorphous. Characterized by DTA.

Kind of sample preparation and/or method of registration of the spectrum: KBr disc. Transmission.

Source: Salentine (1987).

Wavenumbers (cm^{-1}): 1645sh, 1450sh, 1360s, 1270sh, 1165sh, 1048s, 1033sh, 923, 870, 816sh, 753, 701, 455w.

Note: The wavenumbers were partly determined by us based on spectral curve analysis of the published spectrum.

B181 Potassium chloride borate perovskite-related $K_3B_6O_{10}Cl$

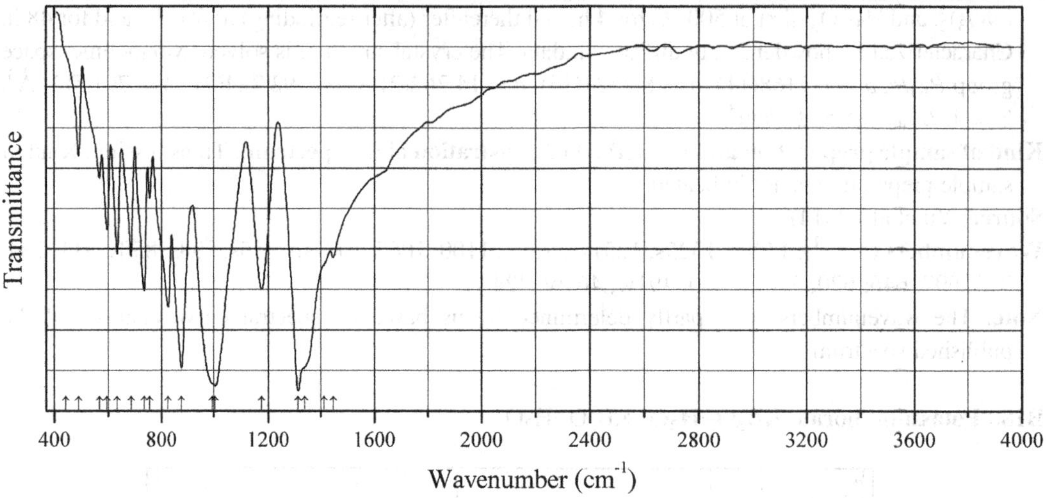

Origin: Synthetic.

Description: Prepared by heating stoichiometric mixture of K_2CO_3, KCl, and H_3BO_3, first at 500 °C for 10 h and thereafter (after intermediate grinding) at 720 °C for 2 days. Characterized by powder X-ray diffraction data. The crystal structure is solved. Trigonal, space group $R3m$, $a = 10.0624(14)$, $c = 8.8361(18)$ Å, $V = 774.8(2)$ Å^3, Z = 3. $D_{calc} = 2.428$ g/cm^3. The structure is based on a 3D framework containing $[B_6O_{10}]$ units in which three BO_4 tetrahedra are shared by the oxygen vertex and are connected with three BO_3 triangles.

Kind of sample preparation and/or method of registration of the spectrum: KBr disc. Transmission.

Source: Wu et al. (2011).

Wavenumbers (cm^{-1}): 1446, 1411sh, 1340sh, 1315s, 1176, 1006s, 996sh, 877s, 825, 755, 734, 686, 634, 597, 568, 491, 444sh.

B182 Potassium pentaborate KB_5O_8

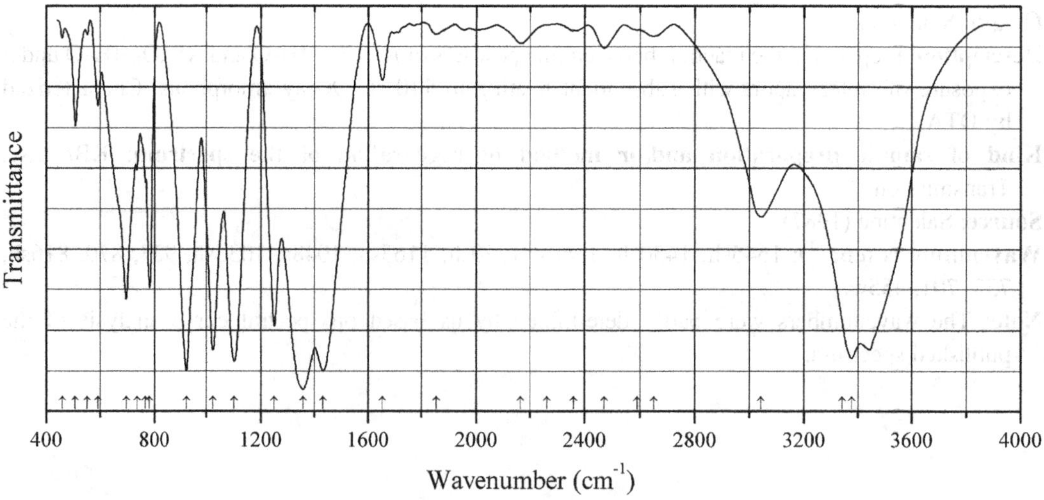

Origin: Synthetic.

Description: Crystals obtained by slow evaporation of aqueous solution containing potassium carbonate and boric acid in the stoichiometric ratio. Characterized by powder X-ray diffraction data. Orthorhombic, space group *Aba*2, $a = 11.065$, $b = 11.171$, $c = 9.054$ Å.

Kind of sample preparation and/or method of registration of the spectrum: KBr disc. Transmission. The procedure of baseline correction has been applied.

Source: Mary et al. (2008).

Wavenumbers (IR, cm^{-1}): 3443s, 3377s, 3042, 2650w, 2590sh, 2472w, 2360w, 2263sh, 2166w, 1854w, 1654, 1433s, 1358s, 1250, 1103s, 1025s, 925s, 782, 766, 735, 696, 591, 552w, 508, 459w.

Note: The wavenumbers were partly determined by us based on spectral curve analysis of the published spectrum. In the cited paper, Raman spectrum is given.

Wavenumbers (Raman, cm^{-1}): 917s, 786, 730, 559s, 456s, 373 (weak bands are not indicated).

B183 Potassium sodium zinc borate $K_2NaZnB_5O_{10}$ $K_2NaZnB_5O_{10}$

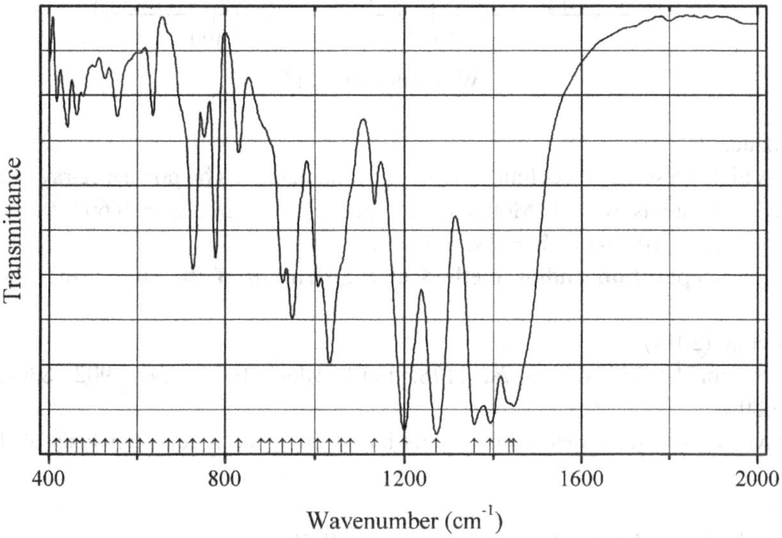

Origin: Synthetic.

Description: Colorless prismatic crystals obtained from K_2CO_3, ZnO, H_3BO_3, and $Na_2B_4O_7{\cdot}10H_2O$ in the molar ratio of 3:2:10:2 by employing a high-temperature solution reaction method. The crystal structure is solved. Monoclinic, space group *C*2/*c*, $a = 7.9244(16)$, $b = 12.805(3)$, $c = 18.962(4)$ Å, $\beta = 99.39$ (3)°, $V = 1898.4(7)$ Å^3, $Z = 8$. $D_{calc} = 2.663$ g/cm^3. The structure is based on the $[B_5O_{10}]^{5-}$ group that consists of one BO_4 tetrahedron and four BO_3 triangles condensed to a double ring via the common tetrahedron.

Kind of sample preparation and/or method of registration of the spectrum: KBr disc. Transmission.

Source: Chen et al. (2010).

Wavenumbers (cm^{-1}): 1447s, 1436sh, 1394s, 1358s, 1272s, 1199s, 1133, 1080sh, 1059sh, 1032s, 1006, 969sh, 949, 928, 900sh, 880sh, 830, 777, 752, 726, 698sh, 672sh, 636w, 607w, 583sh, 556w, 527w, 503w, 477w, 463w, 442w, 417w.

Note: The wavenumbers were partly determined by us based on spectral curve analysis of the published spectrum.

B184 Potassium triborate KB_3O_5

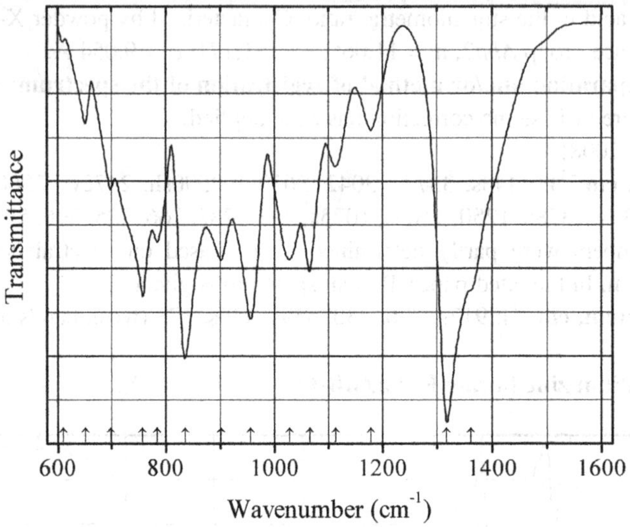

Origin: Synthetic.

Description: A high-pressure monoclinic polymorph. Characterized by powder X-ray diffraction data. The crystal structure is solved. Monoclinic, space group *C2/c*, a = 9.608(2), b = 8.770(2), c = 9.099(2) Å, β = 104.4(1)°, V = 742.8(3) Å^3, Z = 8.

Kind of sample preparation and/or method of registration of the spectrum: Attenuated total reflection (?).

Source: Sohr et al. (2014).

Wavenumbers (cm^{-1}): 1362sh, 1317s, 1178, 1113, 1066, 1028, 956s, 902, 836s, 783, 757s, 698, 651, 609w.

Note: The wavenumbers were determined by us based on spectral curve analysis of the published spectrum.

B185 Sodium aluminum borate $Na_2Al_2B_2O_7$ $Na_2Al_2B_2O_7$

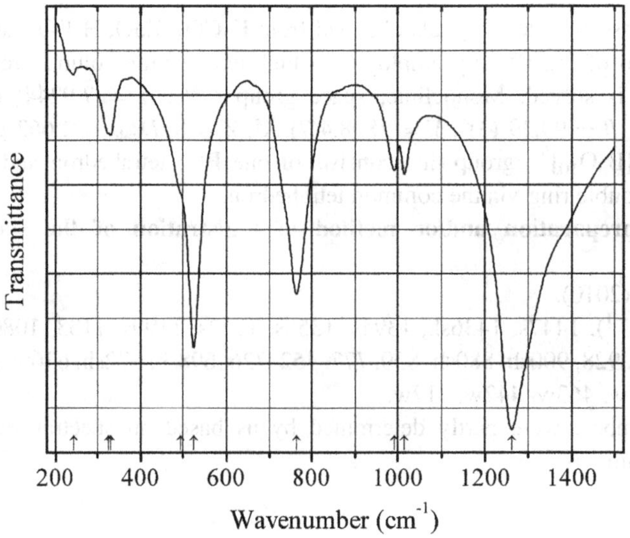

Origin: Synthetic.

Description: Prepared by heating stoichiometric mixture of $NaHCO_3$, Al_2O_3, and H_3BO_3, first at 400 °C for 10 h and thereafter at 950 °C for 2 days. The crystal structure is solved by the Rietveld technique. Trigonal, space group *P*-31*c*, $a = 4.8113(1)$, $c = 15.2781(3)$ Å, $V = 306.29(2)$ Å^3, $Z = 2$. $D_{calc} = 2.532$ g/cm^3. The structure contains infinite [$Al_2B_2O_7$] sheets. The strongest lines of the powder X-ray diffraction pattern [*d*, Å (*I*, %) (*hkl*)] are: 4.02 (59) (101), 3.820 (55) (004), 2.815 (100) (104), 2.406 (33) (110), 2.295 (36) (112).

Kind of sample preparation and/or method of registration of the spectrum: KBr disc. Absorption.

Source: He et al. (2001).

Wavenumbers (cm^{-1}): 1263s, 1015, 992, 765s, 525s, 495sh, 331, 325sh, 244w.

Note: The wavenumbers were partly determined by us based on spectral curve analysis of the published spectrum.

B186 Sodium barium borate $NaBaB_5O_9$ $NaBaB_5O_9$

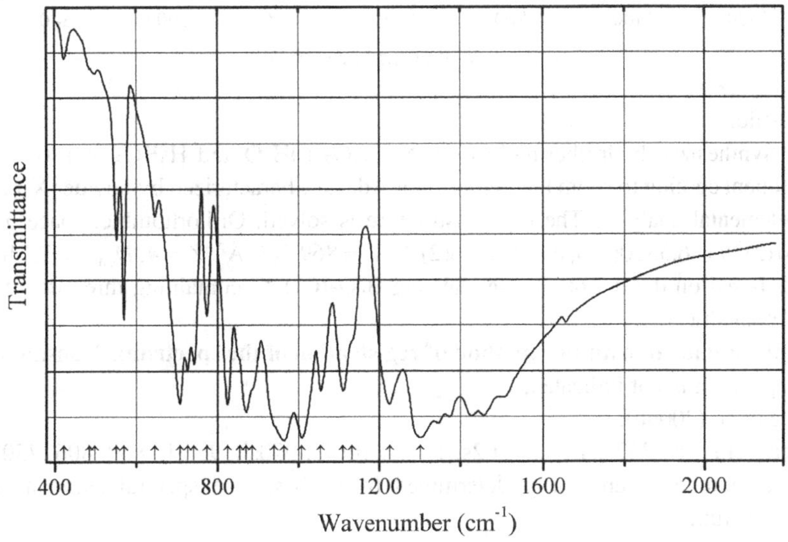

Origin: Synthetic.

Description: Prepared by solid-state reaction techniques from a stoichiometric ratio of $NaNO_3$, Ba $(NO_3)_2$, and H_3BO_3 preheated at 500 °C for 4 h and thereafter (after intermediate grinding) heated at 650 °C for 48 h. Characterized by powder X-ray diffraction data. The crystal structure is solved. Monoclinic, space group $P2_1/c$, $a = 6.5773(19)$, $b = 13.872(4)$, $c = 8.371(2)$ Å, $\beta = 105.393(3)$°, $V = 736.4(4)$ Å^3, $Z = 4$. $D_{calc} = 3.232$ g/cm^3. The structure is based on infinite corrugated layers containing $B_5O_9{}^{3-}$ double rings.

Kind of sample preparation and/or method of registration of the spectrum: Transmission. Kind of sample preparation is not indicated.

Source: Yu et al. (2014).

Wavenumbers (cm^{-1}): 1301s, 1225s, 1132sh, 1110, 1055, 1008s, 964s, 940sh, 885sh, 870s, 854sh, 824s, 773, 742, 726, 707s, 646, 570, 552, 498w, 486sh, 424w.

Note: The wavenumbers were partly determined by us based on spectral curve analysis of the published spectrum.

B187 Sodium borate $Na_2B_5O_8(OH){\cdot}2H_2O$ $Na_2B_5O_8(OH){\cdot}2H_2O$

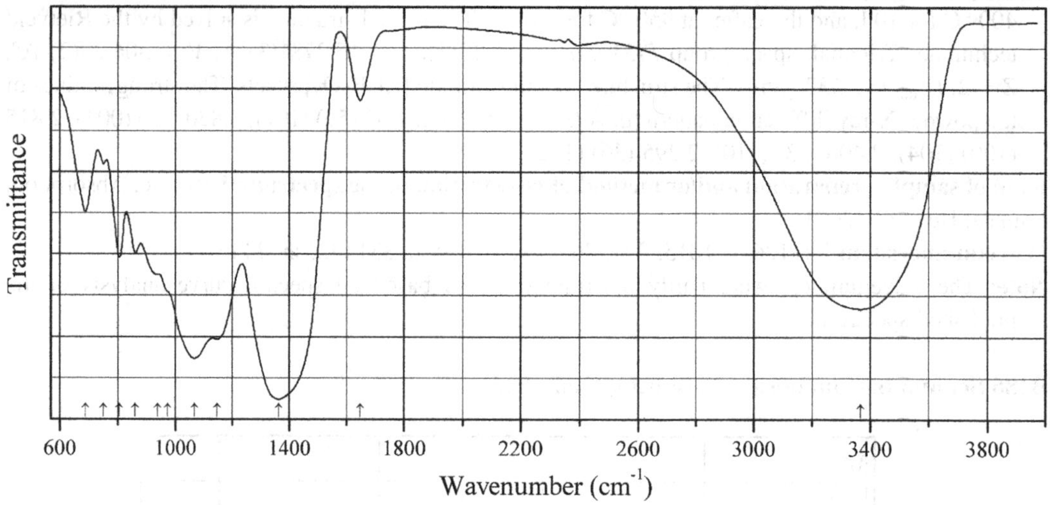

Origin: Synthetic.

Description: Synthesized hydrothermally from $Na_2B_4O_7{\cdot}10H_2O$ and H_3BO_3 at 180 °C for 3 days, with subsequent cooling to room temperature for 9 days. Characterized by powder X-ray diffraction data and elemental analysis. The crystal structure is solved. Orthorhombic, space group $Pna2_1$, $a = 11.967(2)$, $b = 6.5320(13)$, $c = 11.126(2)$ Å, $V = 869.7(3)$ Å^3, $Z = 4$. $D_{calc} = 2.146$ g/cm^3. The structure is based on the double hexagonal ring $B_5O_8(OH)^{2-}$ containing three BO_3 triangles and two BO_4 tetrahedra.

Kind of sample preparation and/or method of registration of the spectrum: Transmission. Kind of sample preparation is not indicated.

Source: Wang et al. (2009b).

Wavenumbers (cm^{-1}): 3367s, 1646, 1362s, 1149s, 1067s, 970sh, 936sh, 860, 806, 750, 688.

Note: The wavenumbers were partly determined by us based on spectral curve analysis of the published spectrum.

B188 Sodium borophosphate $Na_5(B_2P_3O_{13})$ $Na_5(B_2P_3O_{13})$

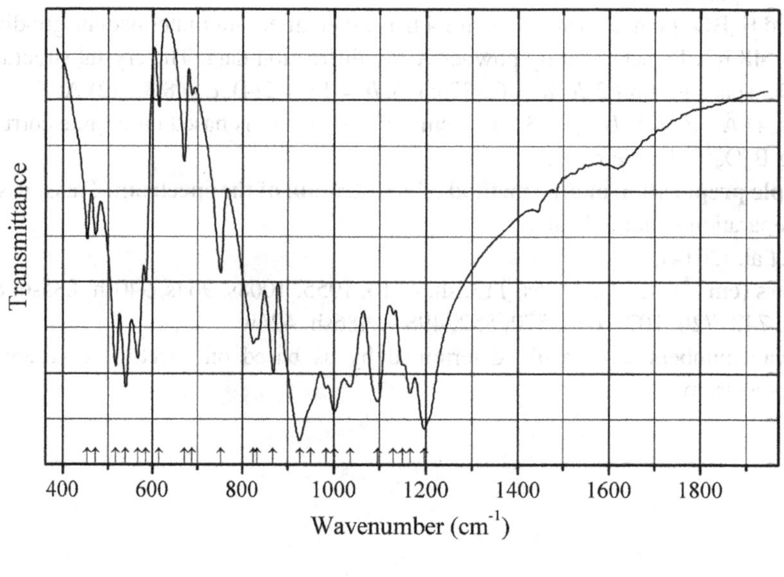

Origin: Synthetic.

Description: Prepared by solid-state reaction techniques, by heating the mixture of Na_2CO_3, H_3BO_3, and $(NH_4)(H_2PO_4)$ in the molar ratio 2.5:2:(3.01–3.05) first at 500 °C for 10 h and thereafter at 700 °C for 24 h with intermediate grinding. Monoclinic, space group *C*2. Characterized by powder X-ray diffraction data. $D_{meas} = 2.68$ g/cm^3.

Kind of sample preparation and/or method of registration of the spectrum: Transmission.

Source: Li et al. (2003).

Wavenumbers (cm^{-1}): 1197s, 1166, 1150sh, 1130, 1097s, 1036, 1002s, 983, 950sh, 926s, 869, 834, 825, 751, 688w, 671, 615w, 585, 566, 539, 517, 473, 455.

Note: The wavenumbers were determined by us based on spectral curve analysis of the published spectrum.

B189 Sodium borosulfate $Na_5[B(SO_4)_4]$ $Na_5[B(SO_4)_4]$

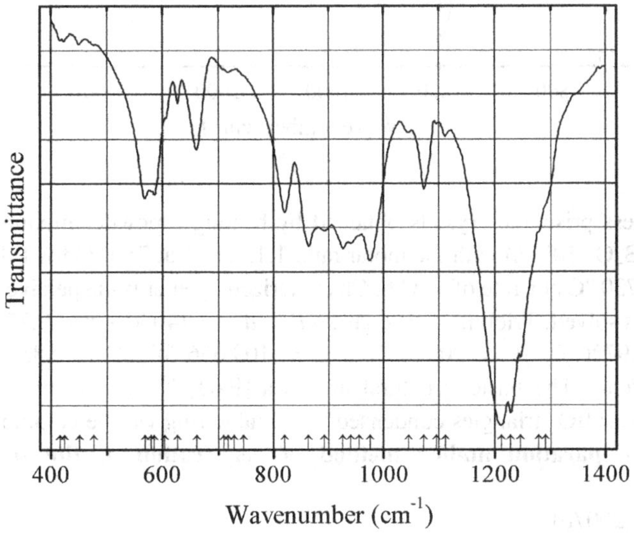

Origin: Synthetic.

Description: Colorless crystals prepared by solid-state reaction techniques, by heating the mixture of $NaHSO_4{\cdot}H_2O$ and $B(OH)_3$ at 673 K for 12 h. Characterized by powder X-ray diffraction data. The crystal structure is solved. Orthorhombic, space group *Pca*2$_1$, $a = 10.730(2)$, $b = 13.891(3)$, $c = 18.197(4)$ Å, Z = 8. $D_{calc} = 2.498$ g/cm^3. The structure contains open-branched pentameric anion $[B(SO_4)_4]^{5-}$ with the borate tetrahedron in the center.

Kind of sample preparation and/or method of registration of the spectrum: Attenuated total reflection.

Source: Daub et al. (2013).

Wavenumbers (IR, cm^{-1}): 1290sh, 1278sh, 1246s, 1229s, 1212s, 1111w, 1095, 1073, 1045, 975, 955sh, 941, 926, 892, 864, 821, 746sh, 729sh, 718w, 710sh, 661, 627, 604sh, 585, 580sh, 568, 477sh, 451w, 424w, 416w.

Note: The wavenumbers were determined by us based on spectral curve analysis of the published spectrum. In the cited paper, a figure of the Raman spectrum is given.

B190 Sodium calcium pentaborate $Na_3Ca(B_5O_{10})$ $Na_3Ca(B_5O_{10})$

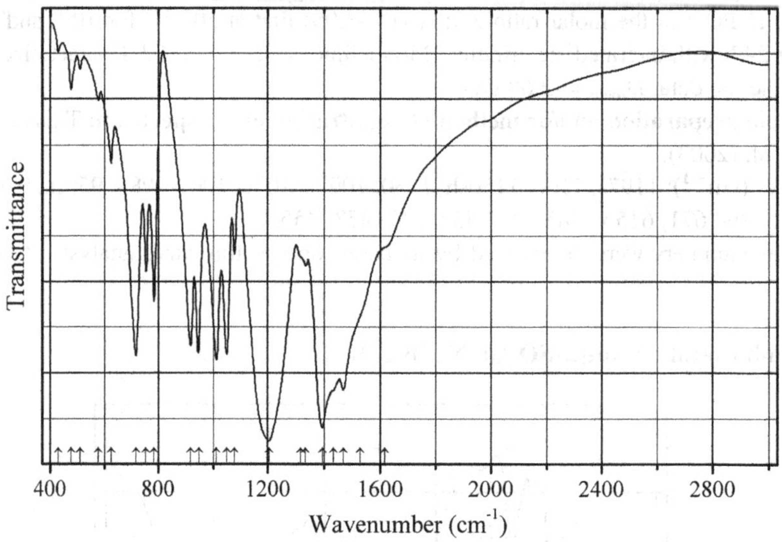

Origin: Synthetic.

Description: Colorless prismatic crystals obtained by heating a powder mixture of $CaCO_3$, Bi_2O_3, H_3BO_3, and $Na_2B_4O_7{\cdot}10H_2O$ with the molar ratio 1:1:2:4.63 at 750 °C for 1 day, with subsequent cooling down to 730 °C at a rate of 1 °C/h. Characterized by energy-dispersive X-ray analyses. The crystal structure is solved. Triclinic, space group *P*-1, $a = 7.4403(6)$, $b = 9.7530(10)$, $c = 12.9289(9)$ Å, $\alpha = 90.972(7)°$, $\beta = 90.073(7)°$, $\gamma = 109.656(6)°$, $V = 883.37(13)$ Å^3, $Z = 4$. $D_{calc} = 2.429$ g/cm^3. The basic structural unit is a $[B_5O_{10}]^{5-}$ group that consists of one BO_4 tetrahedron and four BO_3 triangles condensed to a double ring *via* the common tetrahedron.

Kind of sample preparation and/or method of registration of the spectrum: KBr disc. Transmission.

Source: Chen et al. (2007b).

Wavenumbers (cm^{-1}): 1617sh, 1530sh, 1468s, 1434sh, 1393, 1335, 1316sh, 1204s, 1078, 1050, 1012, 947, 918, 783, 754, 716, 625, 580w, 513w, 480w, 430w.

Note: The wavenumbers were partly determined by us based on spectral curve analysis of the published spectrum.

B191 Sodium magnesium pentaborate $Na_3MgB_5O_{10}$ $Na_3MgB_5O_{10}$

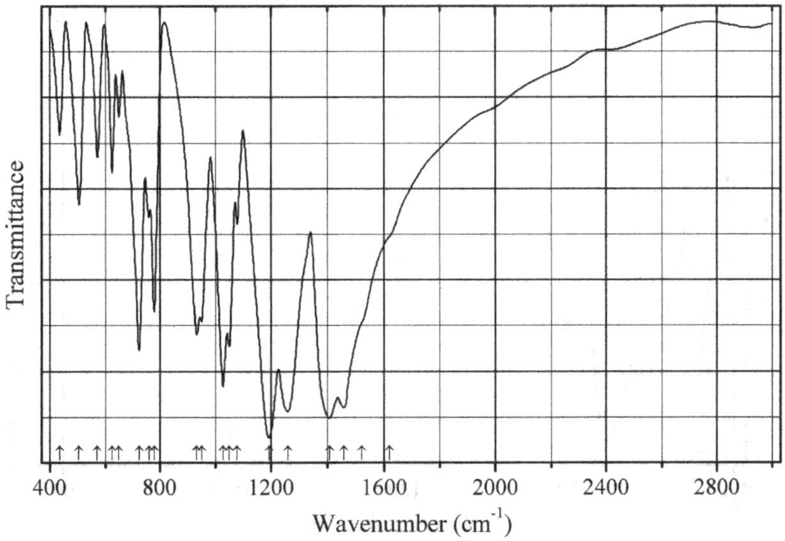

Origin: Synthetic.

Description: Colorless prismatic crystals obtained by heating a powder mixture of Na_2CO_3, Mg $(NO_3)_2$, H_3BO_3, and $Na_2B_4O_7{\cdot}10H_2O$ with the molar ratio 3:2:5:2 at 750 °C for 4 days, with subsequent cooling down to 700 °C at a rate of 1 °C/h and to 600 °C at a rate of 5 °C/h. Characterized by energy-dispersive X-ray analyses. The crystal structure is solved. Orthorhombic, space group *Pbca*, $a = 7.838(1)$, $b = 12.288(1)$, $c = 18.180(2)$ Å, $V = 1751.0(3)$ Å^3, $Z = 8$. $D_{calc} = 2.332$ g/cm^3. The basic structural unit is a $[B_5O_{10}]^{5-}$ group that consists of one BO_4 tetrahedron and four BO_3 triangles condensed to a double ring *via* the common tetrahedron.

Kind of sample preparation and/or method of registration of the spectrum: KBr disc. Transmission.

Source: Chen et al. (2007b).

Wavenumbers (cm^{-1}): 1620sh, 1522sh, 1458s, 1407s, 1260s, 1192s, 1079w, 1050, 1027s, 950, 932, 779, 760, 725, 651w, 626, 572, 506, 436.

Note: The wavenumbers were partly determined by us based on spectral curve analysis of the published spectrum.

B192 Sodium strontium aluminum borate $NaSr_7AlB_{18}O_{36}$ $NaSr_7AlB_{18}O_{36}$

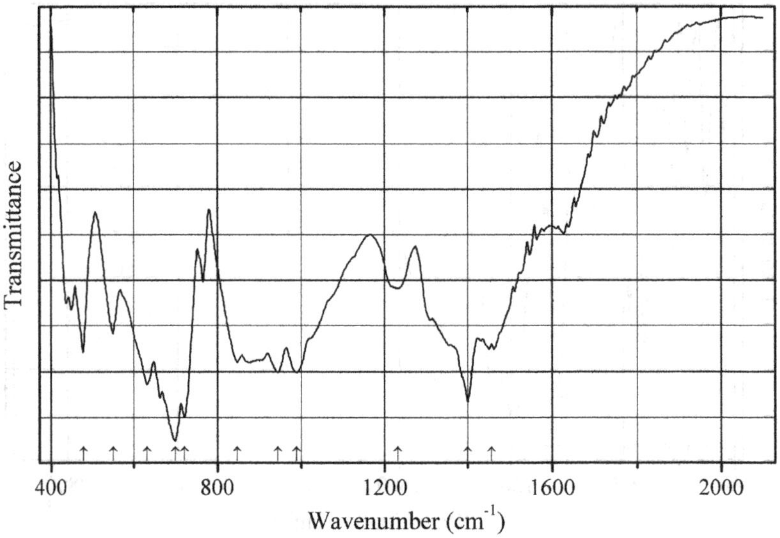

Origin: Synthetic.

Description: Crystals grown from the melt prepared from $SrCO_3$, $Al(OH)_3$, and $Na_2B_4O_7{\cdot}10H_2O$ (with the ratio 1:4:3) at 860 °C by cooling down to 700 °C at a rate of 2.0 °C/h, to 500 °C at 5.0 °C/h, and finally to room temperature at 20 °C/h. Characterized by powder X-ray diffraction data. The crystal structure is solved. Trigonal, space group *R*-3*c*, $a = 11.356(2)$, $c = 36.655(7)$ Å, $V = 4093.7(12)$ Å^3, $Z = 6$. $D_{calc} = 3.490$ g/cm^3. The crystal structure contains a polycyclic $B_{18}O_{36}$ building unit consisting of 12 BO_3 triangles and 6 BO_4 tetrahedra.

Kind of sample preparation and/or method of registration of the spectrum: KBr disc. Transmission.

Source: Chen et al. (2014a).

Wavenumbers (cm^{-1}): 1456, 1400s, 1232, 990s, 946s, 847, 765w, 721s, 700s, 663, 632s, 550, 479, 450, 437.

Note: The wavenumbers were partly determined by us based on spectral curve analysis of the published spectrum.

B193 Sodium vanadyl borate $Na_3(VO_2)B_6O_{11}$ $Na_3(VO_2)B_6O_{11}$

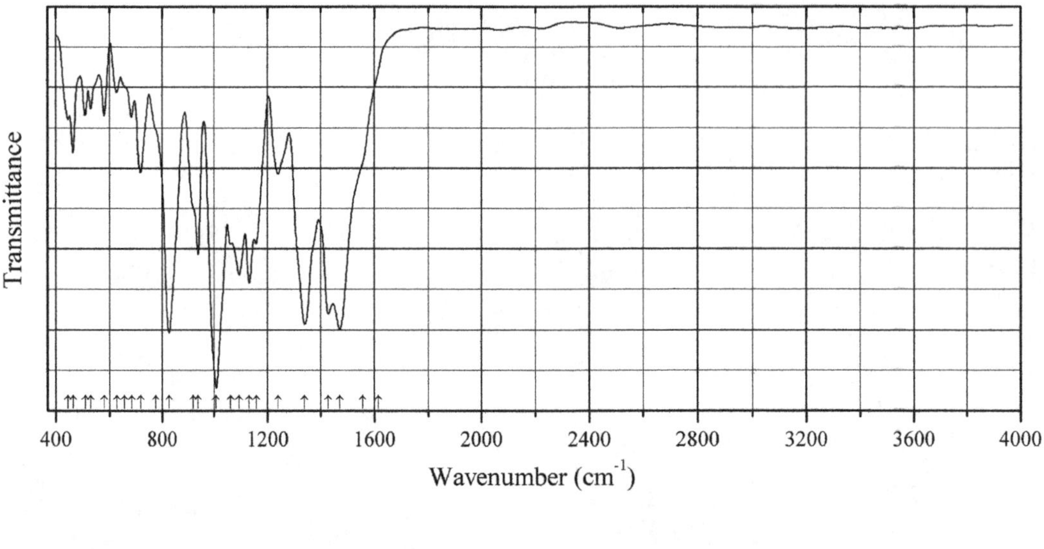

Origin: Synthetic.

Description: Synthesized by solid-state reaction between Na_2CO_3, V_2O_5, and H_3BO_3. A stoichiometric mixture of these reactants was heated first at 300 °C for 5 h, thereafter at 500 °C for 5 h, and finally at 600 °C for 2 days with several intermediate grindings. Orthorhombic, space group $P2_12_12_1$, a = 7.7359(9), b = 10.1884(12), c = 12.5697(15) Å, V = 990.7(2) Å^3, Z = 4. The strongest lines of the powder X-ray diffraction pattern [d, Å (I, %) (hkl)] are: 4.1325 (39) (121), 3.6679 (57) (103), 3.1309 (100) (004), 3.0821 (38) (220), 2.9814 (39) (032).

Kind of sample preparation and/or method of registration of the spectrum: KBr disc. Transmission.

Source: Fan et al. (2010).

Wavenumbers (cm^{-1}): 1615sh, 1555sh, 1469s, 1427s, 1340s, 1241, 1158, 1130, 1094, 1062, 1008s, 941, 921sh, 827s, 775sh, 716, 683w, 656sh, 628w, 581w, 531w, 510w, 465, 446.

Note: The wavenumbers were partly determined by us based on spectral curve analysis of the published spectrum.

B194 Sodium yttrium tellurate borate $Na_2Y_2(Te^{6+}B_2O_{10})$ $Na_2Y_2(Te^{6+}B_2O_{10})$

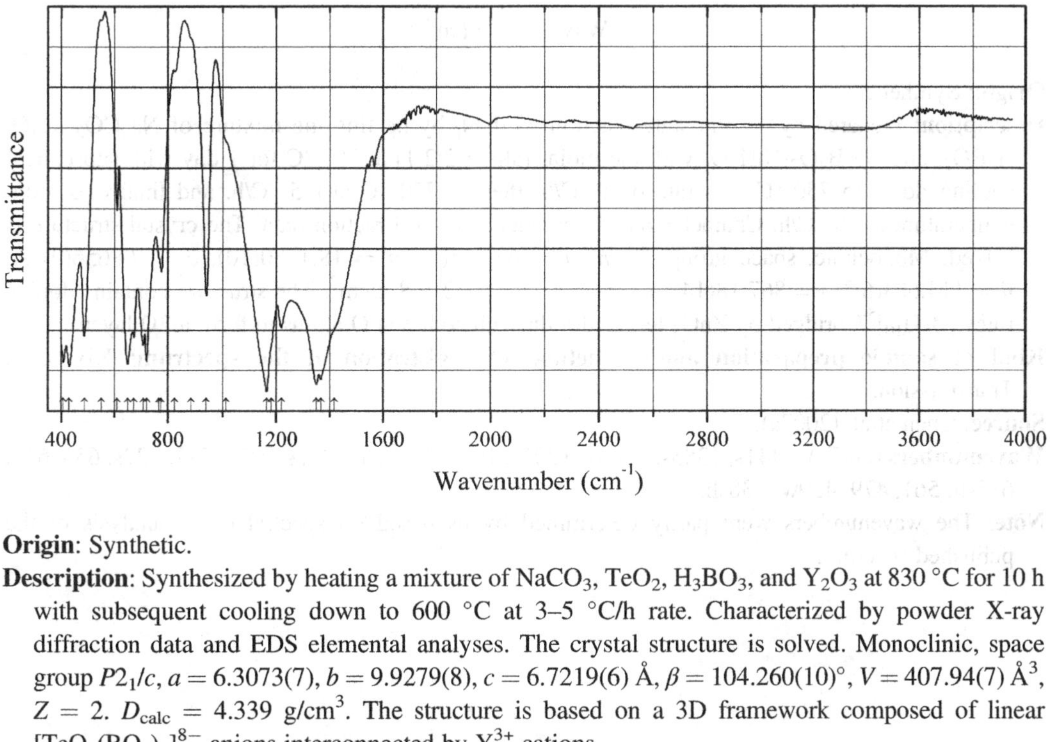

Origin: Synthetic.

Description: Synthesized by heating a mixture of $NaCO_3$, TeO_2, H_3BO_3, and Y_2O_3 at 830 °C for 10 h with subsequent cooling down to 600 °C at 3–5 °C/h rate. Characterized by powder X-ray diffraction data and EDS elemental analyses. The crystal structure is solved. Monoclinic, space group $P2_1/c$, a = 6.3073(7), b = 9.9279(8), c = 6.7219(6) Å, β = 104.260(10)°, V = 407.94(7) Å^3, Z = 2. D_{calc} = 4.339 g/cm^3. The structure is based on a 3D framework composed of linear $[TeO_4(BO_3)_2]^{8-}$ anions interconnected by Y^{3+} cations.

Kind of sample preparation and/or method of registration of the spectrum: Transmission? Kind of sample preparation is not indicated.

Source: Feng et al. (2015a).

Wavenumbers (cm^{-1}): 1418sh, 1366s, 1350s, 1220, 1183sh, 1165s, 1014sh, 941, 885sh, 825w, 775, 766sh, 720s, 708s, 674s, 650s, 610, 552, 488, 432s, 409s.

Note: The wavenumbers were partly determined by us based on spectral curve analysis of the published spectrum.

B195 Sodium zinc pentaborate $Na_3ZnB_5O_{10}$ $Na_3ZnB_5O_{10}$

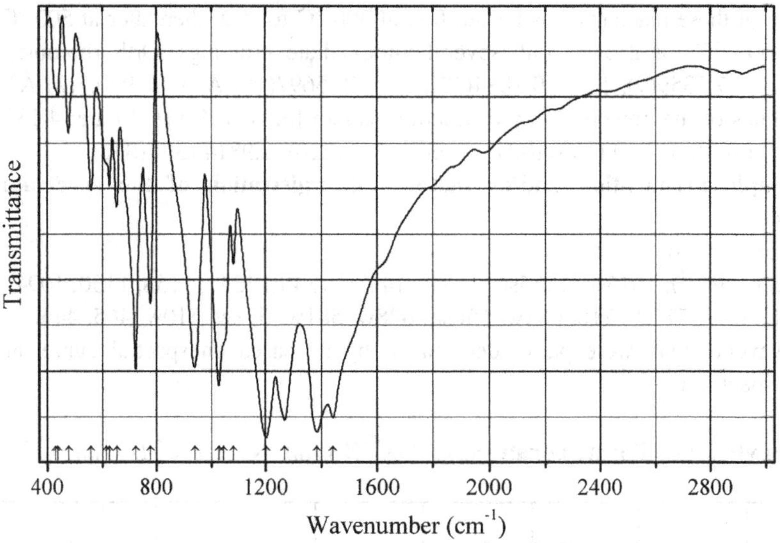

Origin: Synthetic.

Description: Prepared by a solid-state reaction method, by heating the mixture of Na_2CO_3, ZnO, H_3BO_3, and $Na_2B_4O_7{\cdot}10H_2O$ (with the molar ratio 1:2:2:1) at 750 °C for 1 day with subsequent cooling down to 730 °C at a rate of 1 °C/h, then to 720 °C at 0.5 °C/h, and finally to room temperature at 20 °C/h. Characterized by powder X-ray diffraction data. The crystal structure is solved. Monoclinic, space group $P2_1/n$, a = 6.6725(7), b = 18.1730(10), c = 7.8656(9) Å, β = 114.604(6)°, V= 867.18(14) Å^3, Z = 4. D_{calc} = 2.668 g/cm^3. The structure contains double rings $[B_5O_{10}]^{5-}$ bridged by ZnO_4 tetrahedra through common O atoms to form a 2D layer.

Kind of sample preparation and/or method of registration of the spectrum: KBr disc. Transmission.

Source: Chen et al. (2007a).

Wavenumbers (cm^{-1}): 1441s, 1383s, 1268s, 1201s, 1081, 1045sh, 1027s, 939s, 777, 722s, 653, 627, 615sh, 561, 479, 439w, 430sh.

Note: The wavenumbers were partly determined by us based on spectral curve analysis of the published spectrum.

B196 Strontium borate chloride $Sr_2B_5O_9Cl$ $Sr_2B_5O_9Cl$

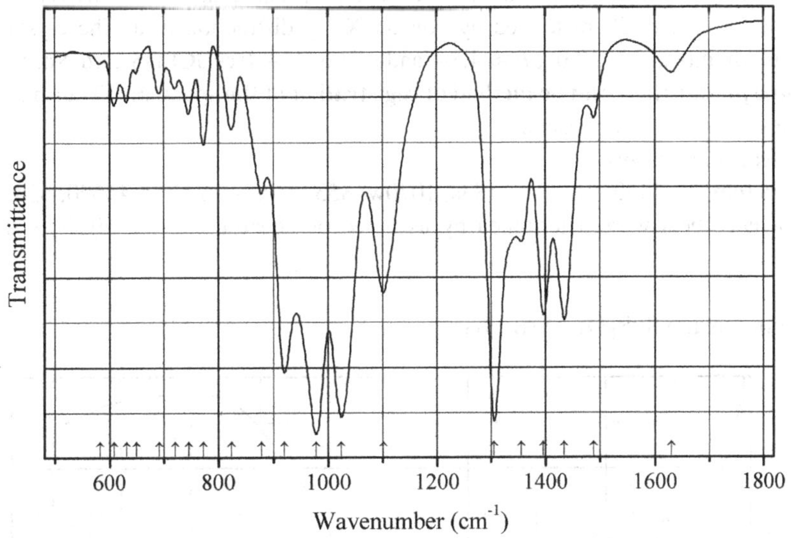

Origin: Synthetic.

Description: Obtained by stepwise heating a precipitate formed by adding $SrCl_2$ aqueous solution to $Na_2B_4O_7$ aqueous solution at 600, 700, and 800 °C for 8 h at each temperature. Characterized by powder X-ray diffraction data. Orthorhombic, a = 11.381, b = 11.319, and c = 6.498 Å (see JCPDS Card No. 27-0835).

Kind of sample preparation and/or method of registration of the spectrum: No data.

Source: Zhu et al. (2013).

Wavenumbers (cm^{-1}): 1630w, 1489w, 1435, 1396, 1355, 1306s, 1101, 1024s, 978s, 920s, 878, 824, 773, 721w, 746w, 692w, 650w, 632w, 609w, 582w.

Note: The wavenumbers were partly determined by us based on spectral curve analysis of the published spectrum.

B197 Strontium borate SrB_2O_4 SrB_2O_4

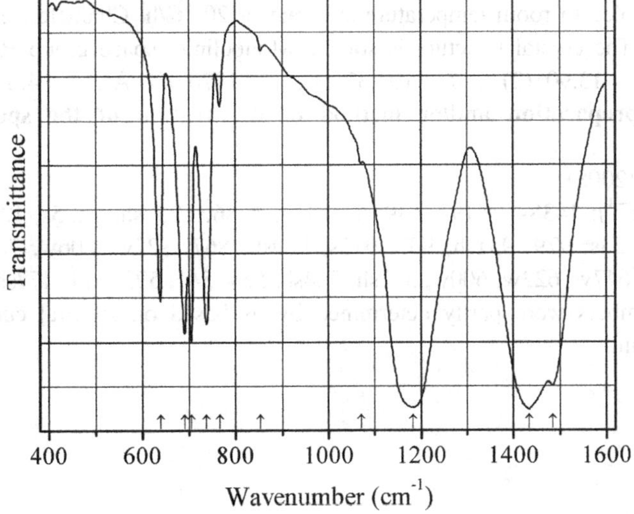

Origin: Synthetic.

Description: Synthesized by heating a mixture of appropriate amounts of $SrCO_3$ and H_3BO_3 at 1000 °C for 2 h in air. Characterized by powder X-ray diffraction data. The crystal structure is solved. Orthorhombic, $a \approx 12.01$, $b \approx 4.34$, and $c \approx 6.59$ Å (see JCPDS card No. 84-2175).

Kind of sample preparation and/or method of registration of the spectrum: Transmission. Kind of sample preparation is not indicated.

Source: Onodera et al. (1999).

Wavenumbers (cm^{-1}): 1483s, 1433s, 1183s, 1070w, 853w, 765w, 737, 704, 690, 638.

Note: The wavenumbers were determined by us based on spectral curve analysis of the published spectrum.

B198 Strontium borate SrB_8O_{13} SrB_8O_{13}

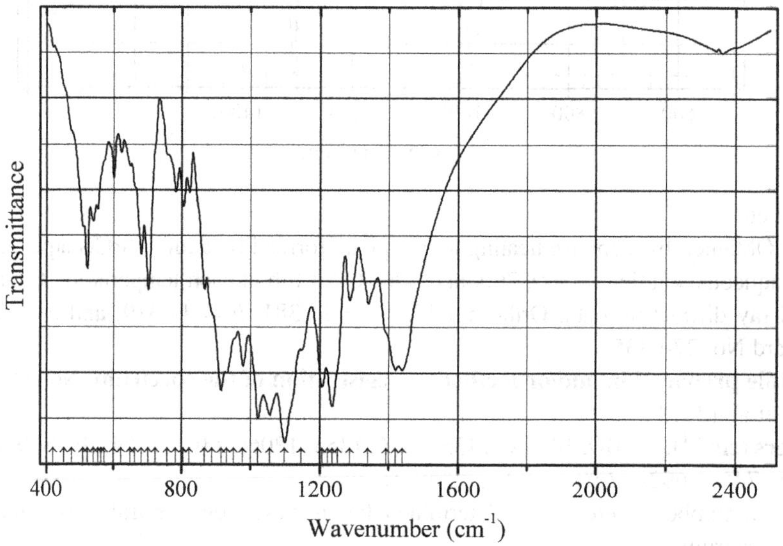

Origin: Synthetic.

Description: Synthesized by heating a powder mixture of Bi_2O_3, $SrCO_3$, and H_3BO_3 (with a molar ratio 3:3:26) at 735 °C for 2 weeks. The product was then cooled down to 500 °C at a rate of 5 °C/h and thereafter cooled to room temperature at a rate of 20 °C/h. Characterized by powder X-ray diffraction data. The crystal structure is solved. Monoclinic, space group $P2_1/c$, $a = 8.408(1)$, $b = 16.672(2)$, $c = 13.901(2)$ Å, $\beta = 106.33(1)°$, $V = 1870.0(4)$ Å^3, $Z = 4$. $D_{calc} = 2.714$ g/cm^3.

Kind of sample preparation and/or method of registration of the spectrum: KBr disc. Transmission.

Source: Tang et al. (2008).

Wavenumbers (cm^{-1}): 1438s, 1419s, 1392sh, 1340, 1286, 1250sh, 1235s, 1221sh, 1206s, 1147, 1101s, 1058s, 1023s, 979, 951sh, 932, 915s, 889sh, 867, 823w, 806w, 783w, 759sh, 724sh, 702, 681, 662sh, 647w, 622w, 600w, 572sh, 564sh, 554, 541, 523, 511, 477sh, 452sh, 420w.

Note: The wavenumbers were partly determined by us based on spectral curve analysis of the published spectrum.

B199 Strontium boroarsenate Sr(BAsO$_5$) Sr(BAsO$_5$)

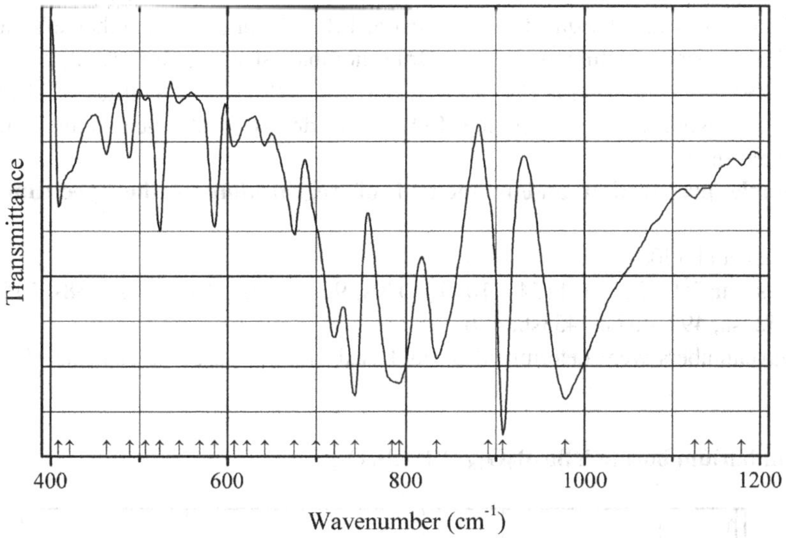

Origin: Synthetic.

Description: Synthesized by heating a mixture of $SrCO_3$, As_2O_5, and H_3BO_3 in a 2:1:2 molar ratio at 900 °C for 15 h. Characterized by powder X-ray diffraction data. The crystal structure is solved. Hexagonal, space group $P2_121$, $a = 7.056(3)$, $c = 6.898(1)$ Å, $V = 571.6(3)$ Å^3, $Z = 3$. In the infinite loop-branched [BAsO$_5$] chain, both B and As have fourfold coordination.

Kind of sample preparation and/or method of registration of the spectrum: KBr disc. Transmission.

Source: Birsöz and Baykal (2008).

Wavenumbers (IR, cm^{-1}): 1179w, 1141w, 1125w, 978s, 908s, 892sh, 834s, 792s, 785sh, 743s, 720s, 700sh, 675, 642w, 622sh, 607w, 585, 568sh, 545w, 523, 507w, 489, 463, 421sh, 409.

Note: The wavenumbers were determined by us based on spectral curve analysis of the published spectrum. In the cited paper, a figure of the Raman spectrum is given.

B200 Tin tetraborate β-SnB_4O_7

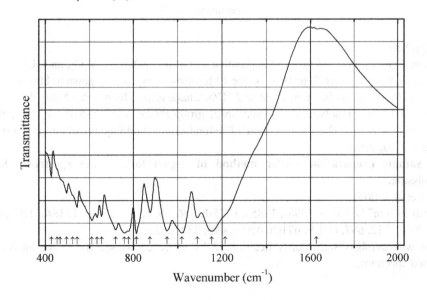

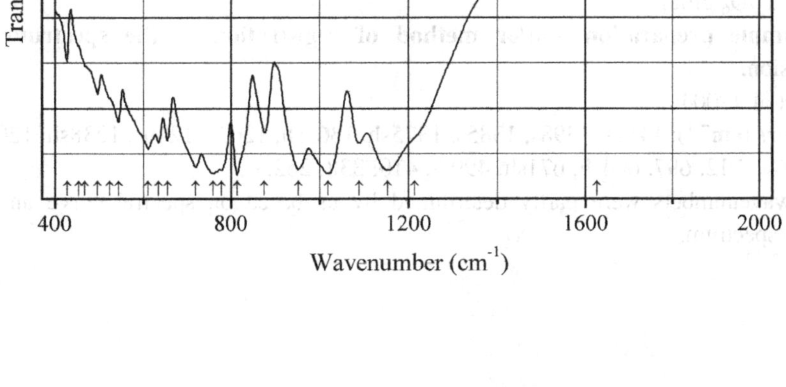

Origin: Synthetic.

Description: Prepared by compressing a mixture of SnO_2 and B_2O_3, taken in stoichiometric amounts, to 7.5 GPa for 3 h with subsequent heating first at 1100 °C for 5 min and thereafter at 750 °C for 15 min. The crystal structure is solved. Orthorhombic, space group $Pmn2_1$, a = 10.864(2), b = 4.4480(9), c = 4.2396(8) Å, V = 204.9(1) Å^3, Z = 2. D_{calc} = 4.44 g/cm^3. The structure is based on a network of corner-sharing BO_4 tetrahedra with channels built from four- and six-membered rings.

Kind of sample preparation and/or method of registration of the spectrum: KBr disc. Transmission.

Source: Knyrim et al. (2007).

Wavenumbers (cm^{-1}): 1213sh, 1154s, 1090, 1020s, 952s, 876, 813s, 777s, 758s, 719s, 655, 635, 611, 543, 523sh, 496, 467sh, 453sh, 426.

Note: The wavenumbers were determined by us based on spectral curve analysis of the published spectrum.

B201 Yttrium barium borate $YBa_3B_9O_{18}$ $YBa_3B_9O_{18}$

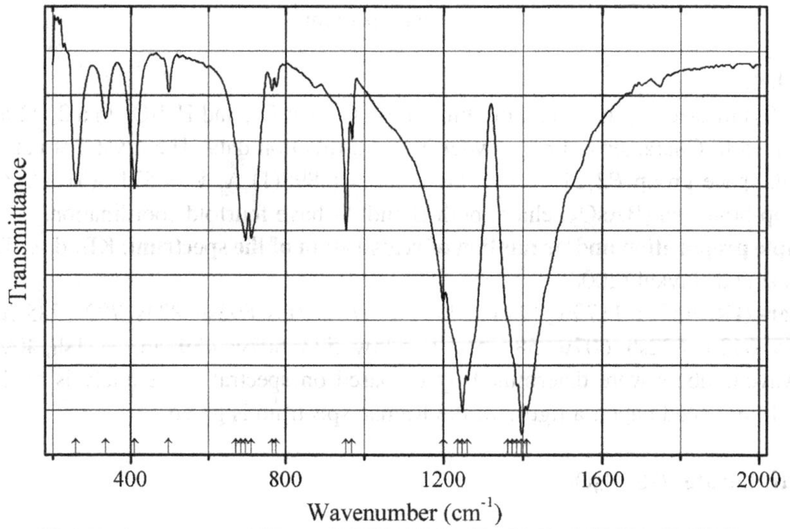

Origin: Synthetic.

Description: Crystals prepared by stepwise heating a mixture of $BaCO_3$, Y_2O_3, and H_3BO_3 (with the molar ratios Y:Ba:B = 1:1:9) to 1050 °C for 12 h followed by cooling down to 800 °C at a rate of 1 °C/h and from 800 to 600 °C at a rate of 2 °C/h. Characterized by powder X-ray diffraction data. The crystal structure is solved. Hexagonal, space group $P6_3/m$, a = 7.1761(6), c = 16.9657(6) Å, V = 756.1(1) Å^3, Z = 2. D_{calc} = 3.89 g/cm^3. The fundamental building unit of the crystal structure is the planar B_3O_6 group.

Kind of sample preparation and/or method of registration of the spectrum: KBr disc. Transmission.

Source: Li et al. (2004a).

Wavenumbers (cm^{-1}): 1411s, 1398s, 1385s, 1375sh, 1363sh, 1262s, 1249s, 1238sh, 1200, 969, 954, 777w, 766w, 712, 697, 684sh, 671sh, 499w, 410, 338, 262.

Note: The wavenumbers were partly determined by us based on spectral curve analysis of the published spectrum.

B202 Admontite $MgB_6O_{10}{\cdot}7H_2O$

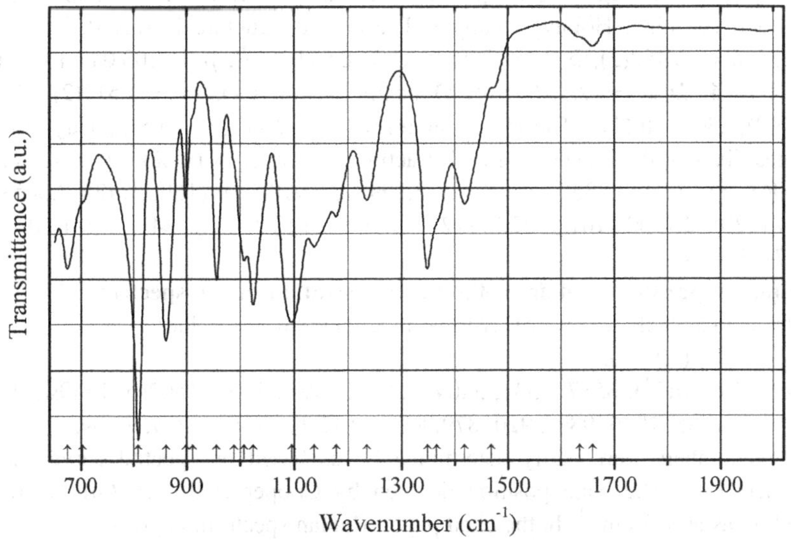

Origin: Synthetic.

Description: Prepared hydrothermally from MgO and H_3BO_3, taken in the molar ratio 1:6, at 100 °C for 120 min. Characterized by powder X-ray diffraction data.

Kind of sample preparation and/or method of registration of the spectrum: Attenuated total reflection of a powdered sample.

Source: Derun et al. (2015).

Wavenumbers (cm^{-1}): 1660w, 1636sh, 1468sh, 1419, 1365sh, 1348, 1236, 1178, 1137, 1095s, 1024s, 1007, 988sh, 956s, 911sh, 898, 861s, 809s, 703sh, 674.

Note: The wavenumbers were partly determined by us based on spectral curve analysis of the published spectrum.

B203 Fontarnauite $(Na,K)_2(Sr,Ca)(SO_4)[B_5O_8(OH)]{\cdot}2H_2O$

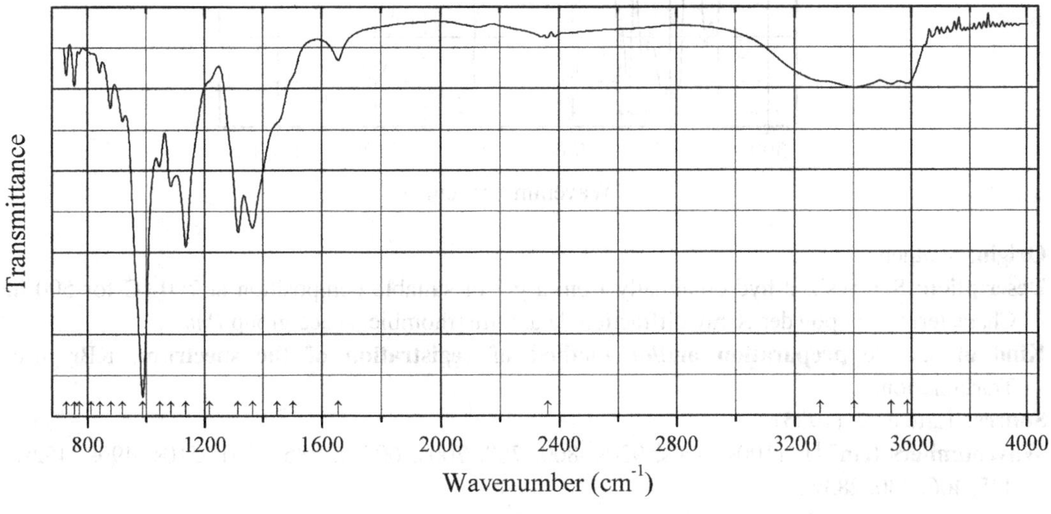

Origin: Village of Doğanlar, Kütahya Province, Western Anatolia, Turkey (type locality).

Description: Colorless to light-brown prismatic crystals from the association with probertite, glauberite, and celestine. Holotype sample. The crystal structure is solved. Monoclinic, space group $P2_1/c$, $a = 6.458(2)$, $b = 22.299(7)$, $c = 8.571(2)$ Å, $\beta = 103.047(13)°$, $V = 1202.5$ (10) Å^3, $Z = 4$. $D_{calc} = 2.533$g/cm^3. Optically biaxial (–), $\alpha = 1.517(2)$, $\beta = 1.517(2)$, $\gamma = 1.543(2)$, $2V = 46(1)°$. The empirical formula is $(Na_{1.84}K_{0.16})(Sr_{0.82}Ca_{0.18})S_{1.00}B_5H_5O_{15}$. The strongest lines of the powder X-ray diffraction pattern [*d*, Å (*I*, %) (*hkl*)] are: 11.1498 (100) (020), 3.3948 (8) (061), 3.3389 (20) (042), 3.1993+3.1990 (10) (160, –142), 3.0458 (10) (052), 3.0250 (7) (220), 2.7500 (10) (–222, 142), 2.3999 (8) (260), 2.2284 (7) (0.10.0, 222), 1.9237 +1.9237 (7) (311, –224).

Kind of sample preparation and/or method of registration of the spectrum: The spectrum was obtained from a small cleavage sheet crushed in a diamond-cell holder.

Source: Cooper et al. (2016b).

Wavenumbers (IR, cm^{-1}): 3587, 3531, 3404, 3288, 2359w, 1657, 1502sh, 1449sh, 1365s, 1315s, 1217sh, 1136s, 1085, 1046, 989s, 920, 879, 842w, 812sh, 773w, 756w, 729w.

Note: The wavenumbers were partly determined by us based on spectral curve analysis of the published spectrum. The band position denoted by Cooper et al. (2016b) as 862 cm^{-1} was determined by us at 842 cm^{-1}. In the cited paper, Raman spectrum is given.

Wavenumbers (Raman, cm^{-1}): 975s, 470, 430, 160, 129.

B204 Sinhalite $MgAl(BO_4)$

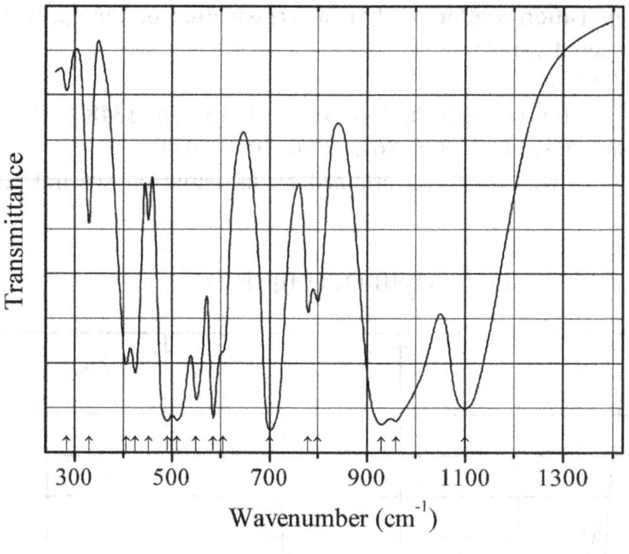

Origin: Synthetic.

Description: Synthesized hydrothermally from a gel of suitable composition at 700 °C for 500 h. Characterized by powder X-ray diffraction data. Orthorhombic, space group *Pnma*.

Kind of sample preparation and/or method of registration of the spectrum: KBr disc. Transmission.

Source: Tarte et al. (1985).

Wavenumbers (cm^{-1}): 1100s, 960s, 930s, 800, 780, 700s, 605sh, 586s, 551, 510s, 490s, 452w, 425, 406, 330, 283w.

B205 Sborgite $NaB_5O_6(OH)_4 \cdot 3H_2O$

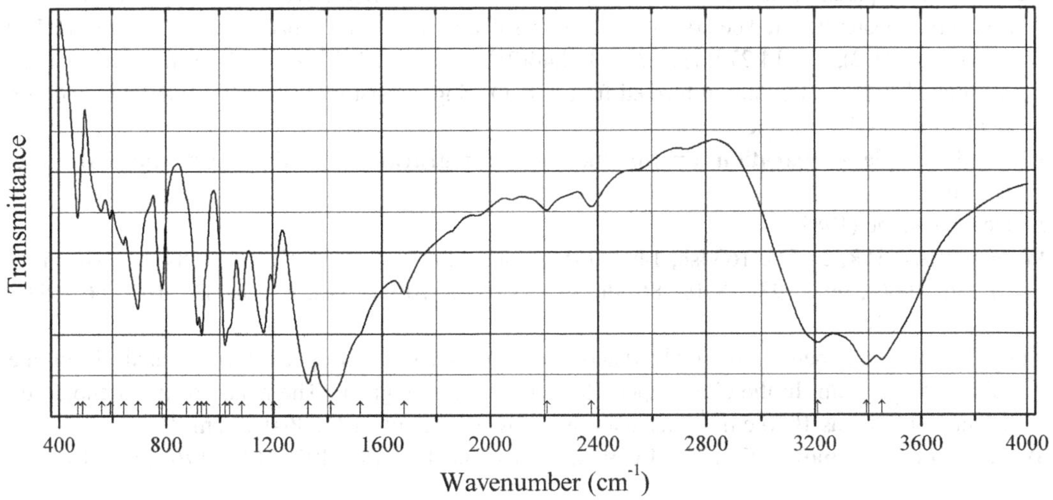

Origin: Synthetic.

Description: Synthesized from aqueous solutions of boric acid and borax. Characterized by powder X-ray diffraction data.

Kind of sample preparation and/or method of registration of the spectrum: KBr disc. Transmission.

Source: Chen and Pei (2016).

Wavenumbers (IR, cm^{-1}): 3450s, 3392s, 3210, 2375w, 2211w, 1681, 1518sh, 1413s, 1330s, 1205, 1166, 1084, 1038sh, 1021, 950sh, 933, 917, 875sh, 786, 775sh, 696, 643, 592, 559, 488, 471.

Note: The wavenumbers were partly determined by us based on spectral curve analysis of the published spectrum. In the cited paper, Raman spectrum is given.

Wavenumbers (Raman, cm^{-1}): 922, 856s, 774, 529s, 494, 468w, 386w.

B206 Potassium borate $KB_3O_3(OH)_4 \cdot H_2O$ $KB_3O_3(OH)_4 \cdot H_2O$

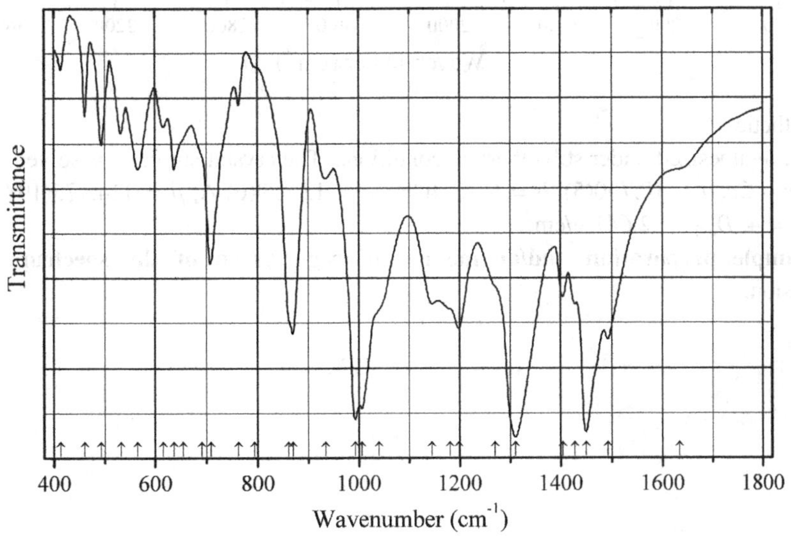

Origin: Synthetic.

Description: Prepared in the reaction between fine powders of $K_2B_4O_7{\cdot}4H_2O$ and $KB_5O_8{\cdot}4H_2O$ under exposure of water vapor. The crystal structure is solved. Monoclinic, space group *C2/c*, $a = 15.540$ (5), $b = 6.821(2)$, $c = 14.273(4)$ Å, $\beta = 104.44(2)°$, $V = 1465.1$ Å^3, $Z = 8$. The structure contains an isolated $B_3O_3(OH)_4{}^-$ anion formed from a B_3O_3 ring consisting of one $BO_2(OH)_2$ tetrahedron and two $BO_2(OH)$ triangles.

Kind of sample preparation and/or method of registration of the spectrum: KBr disc. Transmission.

Source: Salentine (1987).

Wavenumbers (IR, cm^{-1}): 1634sh, 1492, 1450s, 1428, 1404, 1310s, 1270sh, 1198s, 1181sh, 1145, 1041sh, 1006s, 992s, 934, 870s, 862sh, 795sh, 763w, 709, 691sh, 656sh, 638, 617, 567, 533, 494, 461, 413w.

Note: The wavenumbers were partly determined by us based on spectral curve analysis of the published spectrum. In the cited paper, Raman spectrum is given. The band position denoted by Salentine (1987) as 1000 cm^{-1} was determined by us as doublet (1006+992 cm^{-1}).

Wavenumbers (Raman, cm^{-1}): 1194w, 966, 753s, 624s, 488, 453, 407, 212w, 176, 137, 116.

B207 Tyretskite (monoclinic polytype) $Ca_2B_5O_9(OH){\cdot}H_2O$

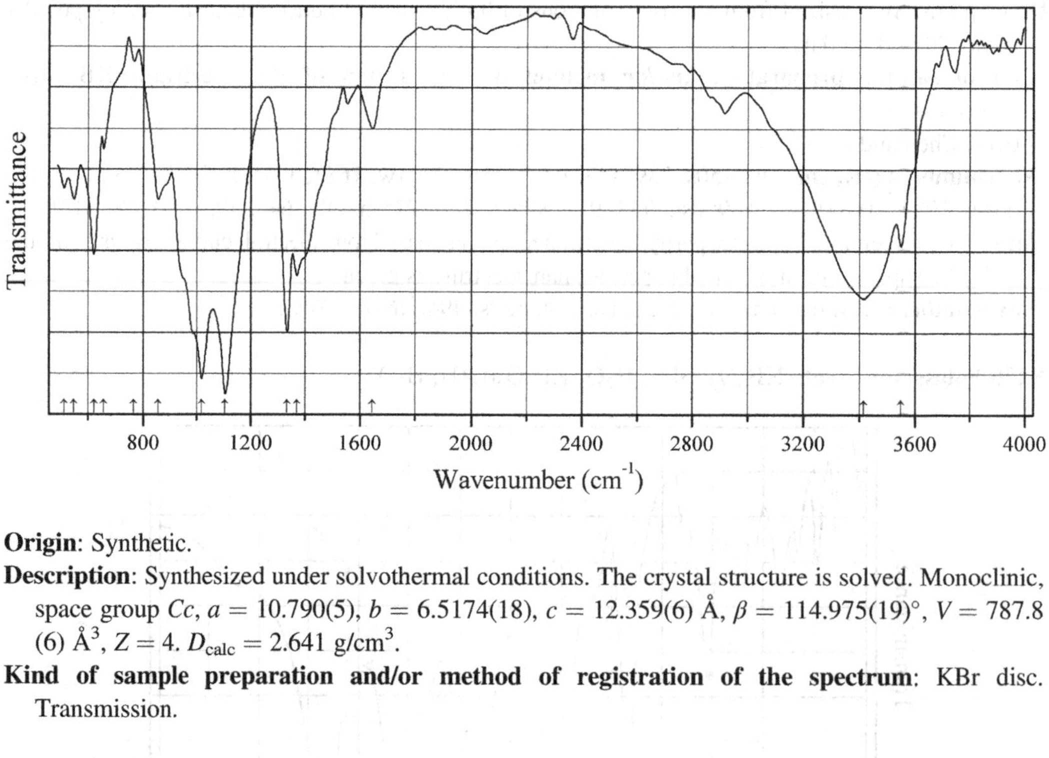

Origin: Synthetic.

Description: Synthesized under solvothermal conditions. The crystal structure is solved. Monoclinic, space group *Cc*, $a = 10.790(5)$, $b = 6.5174(18)$, $c = 12.359(6)$ Å, $\beta = 114.975(19)°$, $V = 787.8$ (6) Å^3, $Z = 4$. $D_{calc} = 2.641$ g/cm^3.

Kind of sample preparation and/or method of registration of the spectrum: KBr disc. Transmission.

Source: Wei et al. (2014).

Wavenumbers (cm^{-1}): 3549, 3415s, 1645, ~1371s, 1335s, 1108s, 1021s, 855, 767w, 657w, 622, 550, 514.

Note: The wavenumbers were partly determined by us based on spectral curve analysis of the published spectrum.

B208 Satimolite $KNa_2(Al_5Mg_2)[B_{12}O_{18}(OH)_{12}](OH)_6Cl_4·4H_2O$

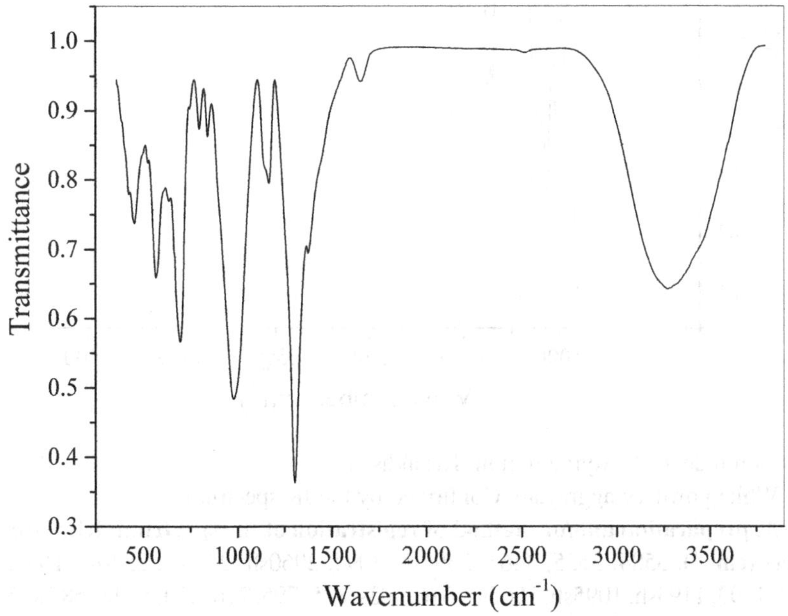

Origin: Chelkar salt dome, Aksai Valley, Aktobe (Aqtöbe) region, Kazakhstan.

Description: Isolated colorless crystals from the association with sylvite, halite, anhydrite, and boracite. Characterized by powder and single-crystal X-ray diffraction data. The crystal structure is solved. Trigonal, space group *R*-3*m*, $a = 15.1406(4)$, $c = 14.3794(9)$ Å, $V = 2854.7(2)$ Å^3. The structural formula is $(\square_{0.68}Na_{0.32})_6(Cl_{0.68}K_{0.22}\square h_{0.10})_6(Al_{0.66}Mg_{0.31}Fe^{3+}_{0.03})_7[B_{12}O_{18}(OH)_{12}](OH)_6·4H_2O$

Kind of sample preparation and/or method of registration of the spectrum: KBr disc. Absorption.

Wavenumbers (cm^{-1}): 3430, 3272s, 2518w, 1651, 1435sh, 1376, 1304s, 1166, 1150sh, 979s, 840, 797, 745w, 696s, 639, 568, 528, 455, 429.

Note: The spectrum was obtained by N.V. Chukanov.

B209 Priceite $Ca_4B_{10}O_{19}\cdot 7H_2O$

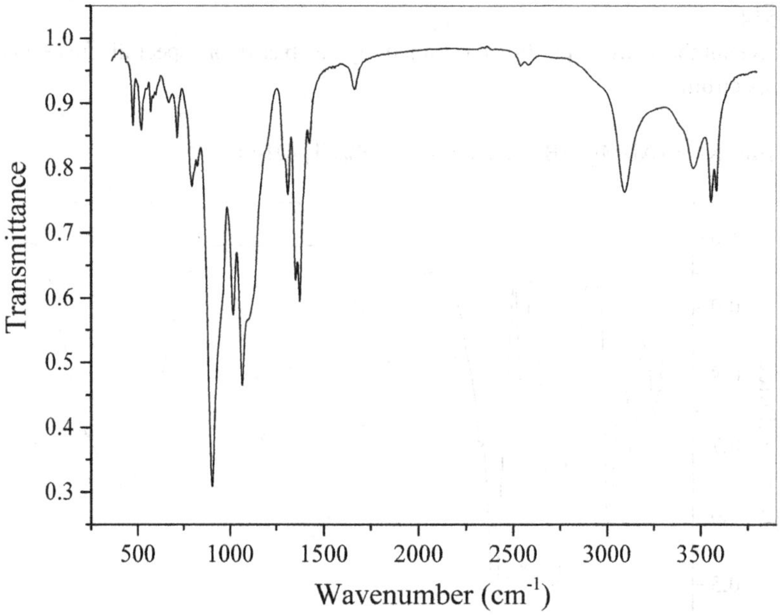

Origin: Inder boron deposit, Atyrau region, Kazakhstan.
Description: White powdery aggregate. Confirmed by the IR spectrum.
Kind of sample preparation and/or method of registration of the spectrum: KBr disc. Absorption.
Wavenumbers (cm^{-1}): 3584, 3555, 3461, 3390sh, 3094, 2960sh, 2580w, 2540w, 1663, 1424, 1373, 1352, 1311, 1293, 1195sh, 1095sh, 1065s, 1016, 903s, 825, 795, 716, 670, 599w, 587w, 572, 520, 474.
Note: The spectrum was obtained by N.V. Chukanov.

B210 Probertite $NaCaB_5O_7(OH)_4\cdot 3H_2O$

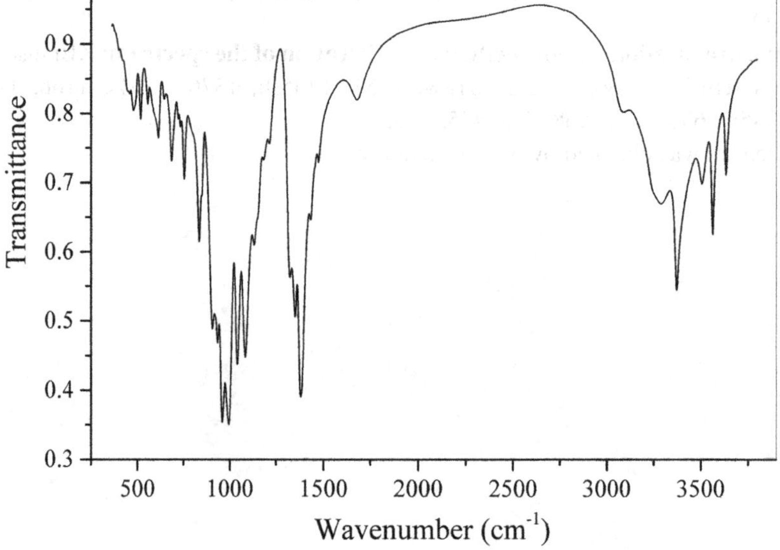

Origin: Banderma (Pandirma), Balikesir province, Turkey.

Description: Grey radial aggregate. Confirmed by the IR spectrum and qualitative electron microprobe analyses.

Kind of sample preparation and/or method of registration of the spectrum: KBr disc. Absorption.

Wavenumbers (cm^{-1}): 3634, 3564, 3506, 3372s, 3289, 3250sh, 3092, 1678, 1474, 1435, 1380s, 1351s, 1323, 1216, 1186, 1150sh, 1133, 1085s, 1041s, 994s, 959s, 934s, 906s, 850sh, 835, 805sh, 755, 737w, 726w, 686, 649w, 615, 600sh, 557, 517, 478, 456w, 447w, 424w, 400w.

Note: The spectrum was obtained by N.V. Chukanov.

BC10 Mereheadite $Pb_{47}Cl_{25}(OH)_{13}O_{24}(CO_3)(BO_3)_2$

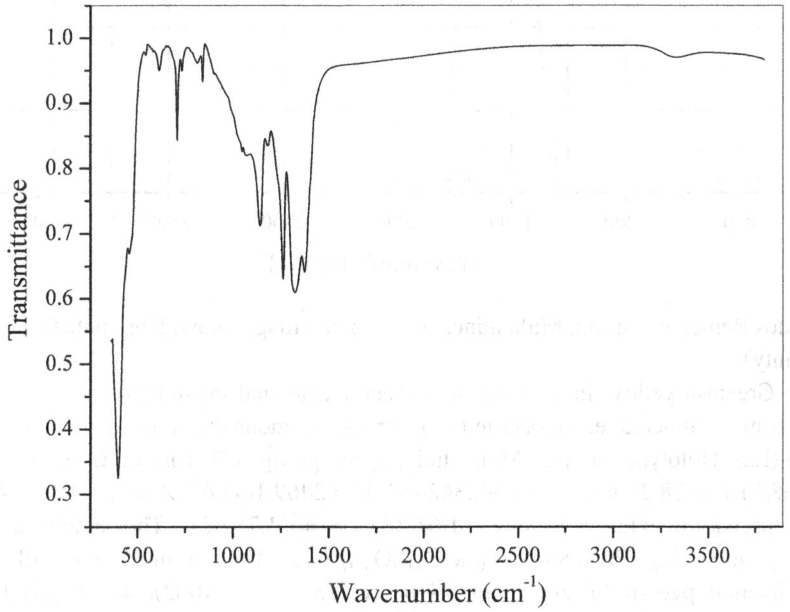

Origin: Merehead Quarry, Cranmore, Somerset, England, UK (type locality).

Description: Reddish-orange grains. The crystal structure is solved. Monoclinic, space group *Cm*, $a = 17.372(1)$, $b = 27.9419(19)$, $c = 10.6661(6)$ Å, $\beta = 93.152(5)°$, $V = 5169.6(5)$ Å^3, $Z = 2$. $D_{calc} = 7.236$ g/cm^3.

Kind of sample preparation and/or method of registration of the spectrum: KBr disc. Absorption.

Wavenumbers (cm^{-1}): 3331, 1373s, 1324s, 1261s, 1183, 1141s, 1071, 1050, 1000sh, 902w, 841, 813w, 734, 709, 615, 542w, 457s, 398s.

Note: The spectrum was obtained by N.V. Chukanov.

2.2 Carbonates

C342 Parisite-(La) $CaLa_2(CO_3)_3F_2$

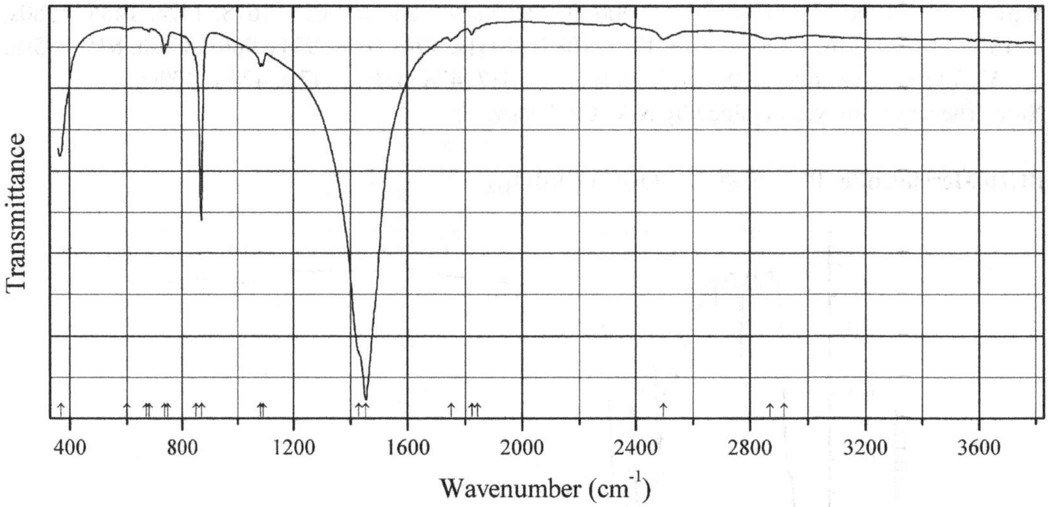

Origin: Rio dos Remédios Group, Mula mine, near Tapera village, Novo Horizonte Co., Bahia, Brazil (type locality).

Description: Greenish-yellow inner zone of a pseudohexagonal crystal from the association with hematite, rutile, almeidaite, fluocerite-(Ce), brockite, monazite-(La), rhabdophane-(La), and bastnäsite-(La). Holotype sample. Monoclinic, space group: *C*2, *Cm*, or *C*2/*m*, $a = 12.356(1)$, $b = 7.1368(7)$, $c = 28.299(3)$ Å, $\beta = 98.342(4)°$, $V = 2469.1(4)$ Å^3, $Z = 12$. $D_{calc} = 4.273$ g/cm^3. Optically pseudo-uniaxial (+), $\omega = 1.670(2)$, $\varepsilon = 1.782(5)$. The empirical formula is $Ca_{0.98}(La_{0.83}Nd_{0.51}Ce_{0.37}Pr_{0.16}Sm_{0.04}Y_{0.03})C_{3.03}O_{8.91}F_{2.09}$. The strongest lines of the powder X-ray diffraction pattern [*d*, Å (*I*, %) (*hkl*)] are: 13.95 (55) (002), 4.655 (37) (006), 3.555 (88) (020, –311), 2.827 (100) (026, 315, –317), 2.055 (58) (–331, –602), 1.950 (38) (0.2.12, 3.1.11, –3.1.13), 1.880 (36) (335, –337, 604, –608).

Kind of sample preparation and/or method of registration of the spectrum: KBr disc. Absorption.

Wavenumbers (cm^{-1}): 2918w, 2870w, 2497w, 1843w, 1823w, 1750w, 1454sm 1430sh, 1089, 1081, 871s, 850sh, 746w, 734, 679w, 670w, 602w, 368.

Note: The spectrum was obtained by N.V. Chukanov.

C343 Parisite-(Ce) $CaCe_2(CO_3)_3F_2$

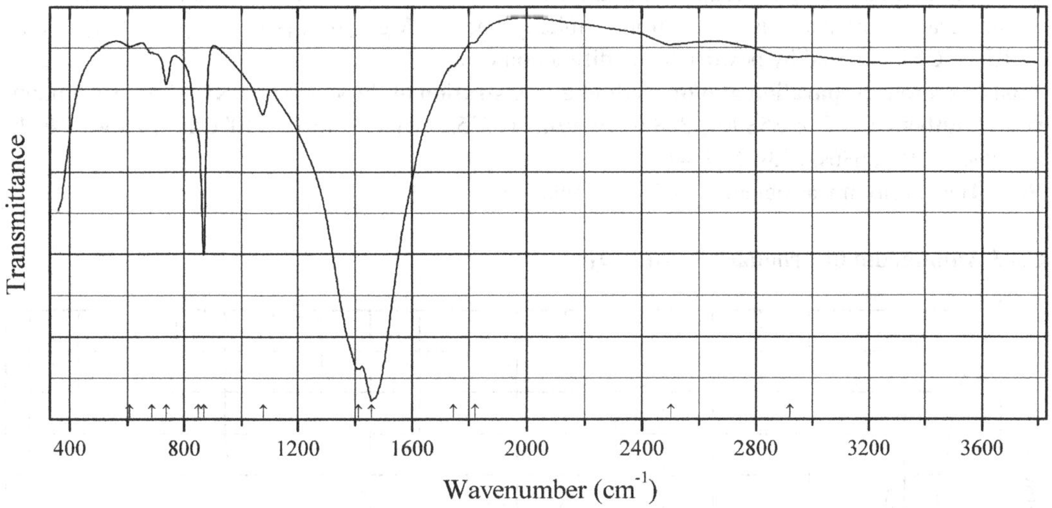

Origin: White Cloud Mine, Pyrites, Ravalli Co., Montana, USA.

Description: Beige crystal. The empirical formula is (electron microprobe): $Ca_{1.08}(Ce_{0.93}La_{0.47}Nd_{0.32}Pr_{0.06}Y_{0.08}Th_{0.06})(CO_3)_{3.00}F_{1.88}$.

Kind of sample preparation and/or method of registration of the spectrum: KBr disc. Absorption.

Wavenumbers (cm^{-1}): 2920w, 2500w, 1820w, 1746w, 1459s, 1411s, 1077, 869s, 850sh, 736, 685sh, 609w.

Note: The spectrum was obtained by N.V. Chukanov.

C344 Stichtite $Mg_6Cr_2(OH)_{16}(CO_3)\cdot 4H_2O$

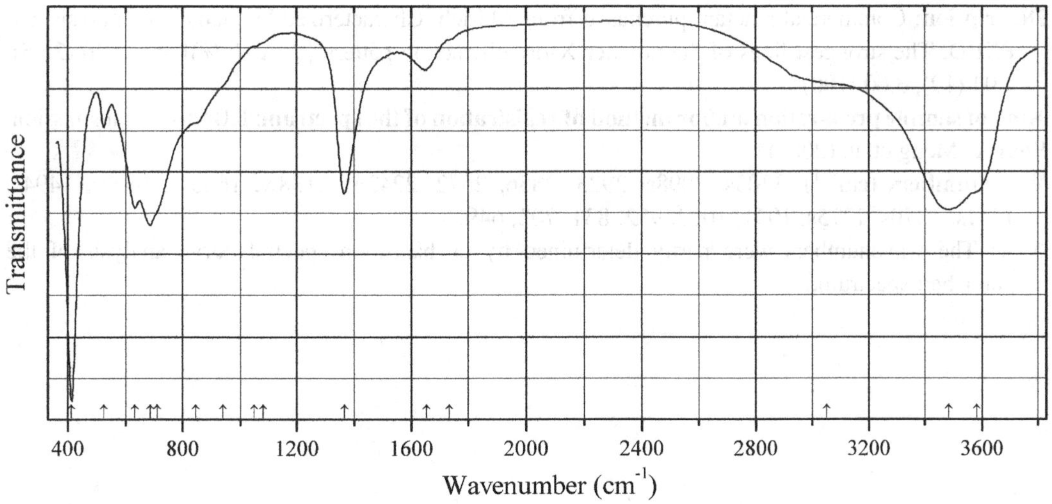

Origin: Kara-Uyuk stream, Terektinskiy ridge, Altai Mts., Siberia, Russia.

Description: Lilac scaly aggregate from the association with serpentine. An Al-rich variety. The empirical formula is (electron microprobe): $Mg_{5.98}(Cr_{1.13}Al_{0.73}Fe_{0.16})(OH)_{16}(CO_3)\cdot 4H_2O$. Characterized by powder X-ray diffraction data.

Kind of sample preparation and/or method of registration of the spectrum: KBr disc. Absorption.

Wavenumbers (cm^{-1}): 3580sh, 3482s, 3050sh, (1735sh), 1653, 1365s, 1080sh, 1050w, 940sh, 845sh, 710sh, 686s, 633s, 525, 413s.

Note: The spectrum was obtained by N.V. Chukanov.

C345 Ammonium bicarbonate NH_4HCO_3

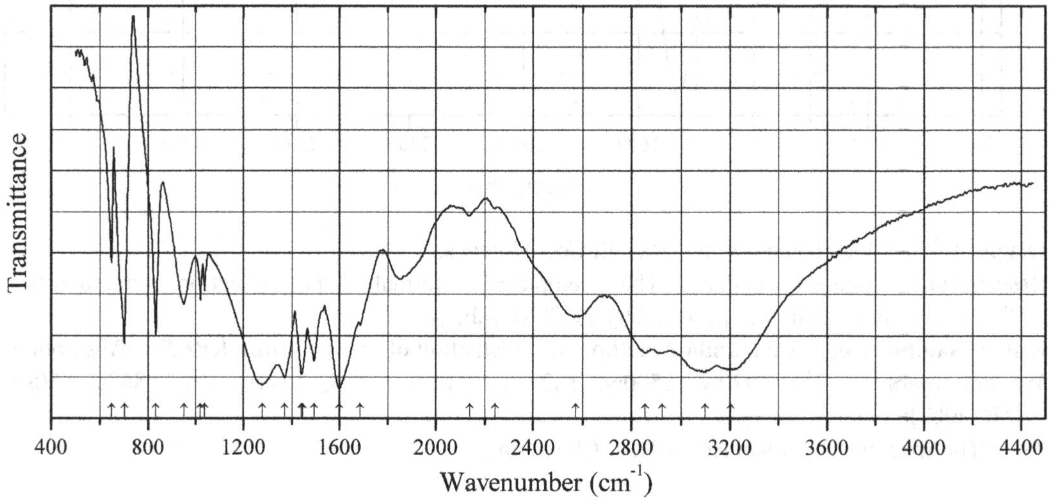

Origin: Synthetic.

Description: Commercial reactant purchased from Aldrich. Characterized by elemental analysis and PXRD. The strongest lines of the powder X-ray diffraction pattern [*d*, Å (*I*, %)] are: 5.36 (26.5), 4.04 (19), 3.00 (100).

Kind of sample preparation and/or method of registration of the spectrum: KBr disc. Transmission.

Source: Meng et al. (2005).

Wavenumbers (cm^{-1}): 3205s, 3098s, 2923, 2856, 2572, 2243w, 2138w, 1685w, 1600s, 1494s, 1441s, 1370s, 1275s, 1041, 1022, 952, 831, 702, 649.

Note: The wavenumbers were partly determined by us based on spectral curve analysis of the published spectrum.

C346 Copper(II) carbonate $CuCO_3$

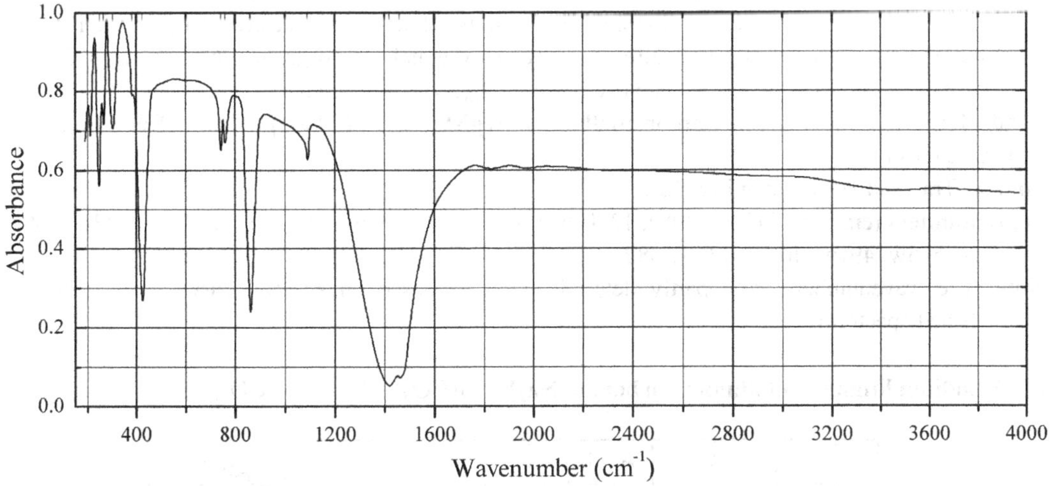

Origin: Synthetic.

Description: $CuCO_3$ can be prepared from azurite, malachite or CuO by reaction with CO_2 at a pressure of 20 kb and a temperature of 500 °C. Monoclinic, space group *Pa*, $a = 6.092$, $b = 4.493$, $c = 7.030$ Å, $\beta = 101.34°$, $V = 188.7$ Å^3, $Z = 4$. In the structure, Cu has the fivefold (square pyramid) coordination. $D_{meas} = 4.18$ g/cm^3, $D_{calc} = 4.35$ g/cm^3.

Kind of sample preparation and/or method of registration of the spectrum: KBr disc (above 190 cm^{-1}) and polyethylene mull (below 190 cm^{-1}). Transmission.

Source: Seidel et al. (1974).

Wavenumbers (cm^{-1}): 1460sh, 1420s, 1090w, 860s, 760, 743, 425, 383sh, 305, 268, 250, 212, 194, 166, 158, 151, 130, 113, 103, 97, 90, 83, 61, 55, 49, 44, 38, 31.

C347 Potassium lead carbonate fluoride $KPb_2(CO_3)_2F$

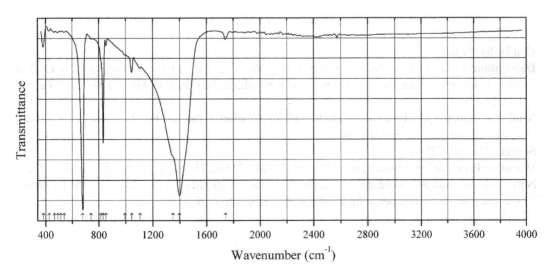

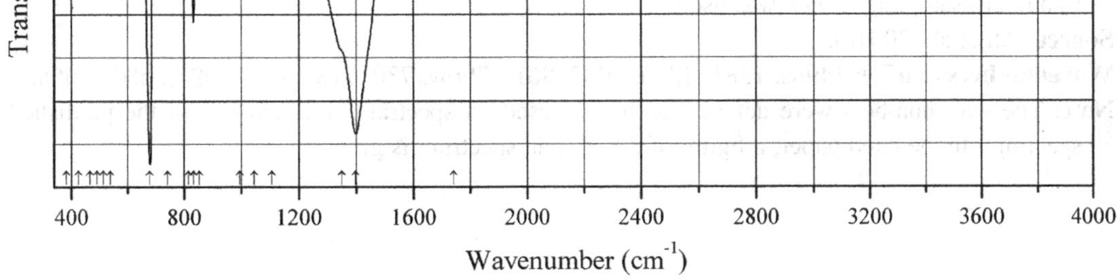

Origin: Synthetic.

Description: Synthesized by a conventional solid-state technique from the stoichiometric mixture of KF and $PbCO_3$, at 250 °C, in flowing CO_2 gas, for 2 days. Characterized by powder X-ray diffraction data. The crystal structure is solved. Hexagonal, space group *P*6$_3$/*mmc*, a = 5.3000 (2), c = 13.9302(8) Å, V = 338.88(3) Å^3, Z = 2. D_{calc} = 5.807 g/cm^3.

Kind of sample preparation and/or method of registration of the spectrum: Reflection of a powdered sample.

Source: Tran and Halasyamani (2013).

Wavenumbers (cm^{-1}): 1741w, 1398s, 1350sh, 1107w, 1045, 995w, 855w, 833, 815sh, 743w, 680s, 541w, 515w, 491w, 466w, 425w, 382.

Note: The wavenumbers were partly determined by us based on spectral curve analysis of the published spectrum.

C348 Sodium lithium gadolinium carbonate $Na_2LiGd(CO_3)_3$ $Na_2LiGd(CO_3)_3$

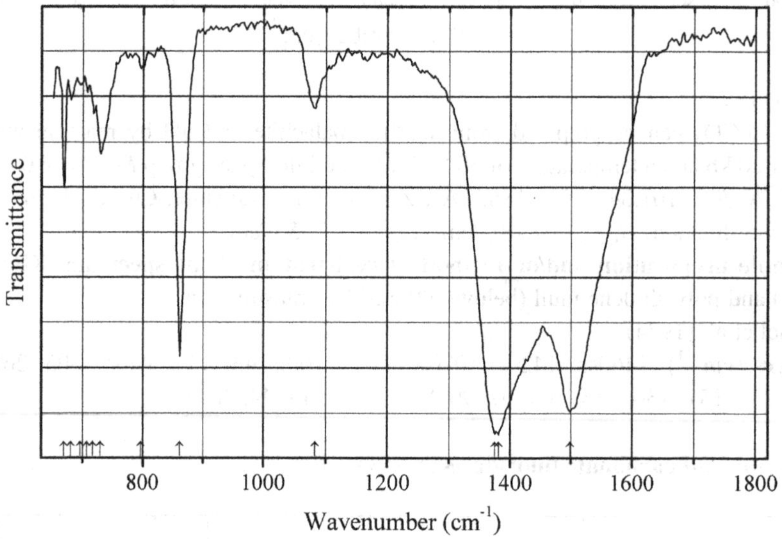

Origin: Synthetic.

Description: Obtained hydrothermally from GdF_3, Na_2CO_3, and Li_2CO_3 with molar ratio 1:6:2 at 230 °C for 48 h. Characterized by powder X-ray diffraction data. Cubic, space group *Fd*-3*m*, $a \approx$ 14.4 Å, Z = 8.

Kind of sample preparation and/or method of registration of the spectrum: A diamond-anvil cell as a micro-sampling device was used.

Source: Ali et al. (2004b).

Wavenumbers (cm^{-1}): 1498s, 1381+1375, 1083, 861s, 796w, 730, 718, (707), (696), 680w, 668.

Note: The wavenumbers were determined by us based on spectral curve analysis of the published spectrum. In the cited paper, a figure of the Raman spectrum is given.

C349 Sodium scandium carbonate $Na_5Sc(CO_3)_3{\cdot}2H_2O$ $Na_5Sc(CO_3)_3{\cdot}2H_2O$

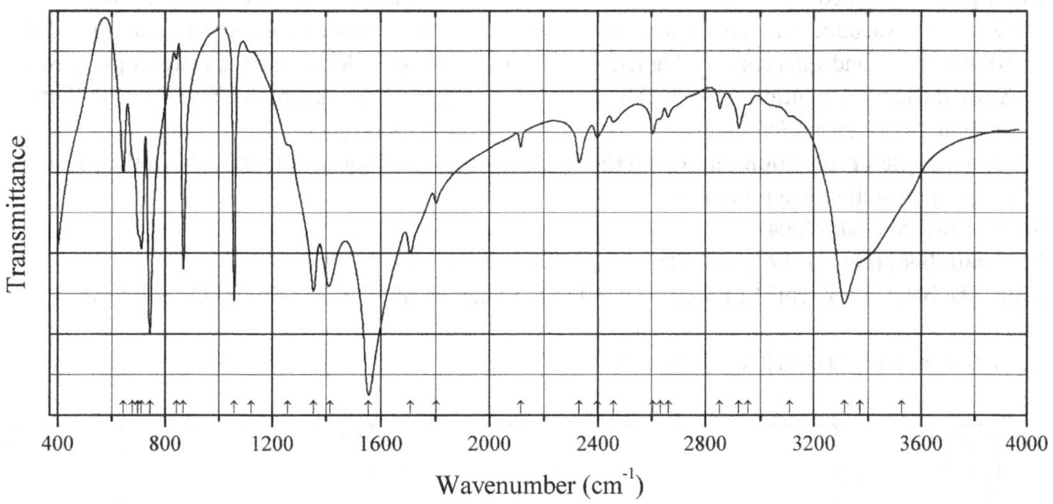

Origin: Synthetic.

Description: Obtained from aqueous solutions of scandium chloride and sodium carbonate. The crystal structure is solved. Tetragonal, space group $P\text{-}42_1c$, a = 7.4637(4), c = 11.570(2) Å, V = 644.55(13) Å^3, Z = 2. D_{meas} = 2.23 g/cm^3, D_{calc} = 2.246 g/cm^3.

Kind of sample preparation and/or method of registration of the spectrum: KBr disc and Nujol mull. Transmission.

Source: Dahm and Adam (2001).

Wavenumbers (IR, cm^{-1}): 3528sh, 3369sh, 3311s, 3108w, 2955sh, 2924, 2854w, 2662w, 2631sh, 2604, 2458w, 2399, 2330, 2114, 1804, 1708, 1556s, 1414s, 1354s, 1259sh, 1122sh, 1059s, 868s, 841w, 743s, 712s, 699sh, 680sh, 644, 359s, 261s, 230, 215, 195sh, 180, 165, 145, 122w, 101w, 93, 63w.

Note: In the cited paper, Raman spectrum is given.

Wavenumbers (Raman, cm^{-1}): 1739, 1702, 1601, 1575s, 1439, 1355, 1061s, 1039, 867, 764, 744s, 682, 668, 644, 354, 300, 277s, 245s, 231s, 216s, 187s, 165s, 132s.

C350 Strontium iron(III) oxycarbonate $Sr_4Fe_2O_6(CO_3)$

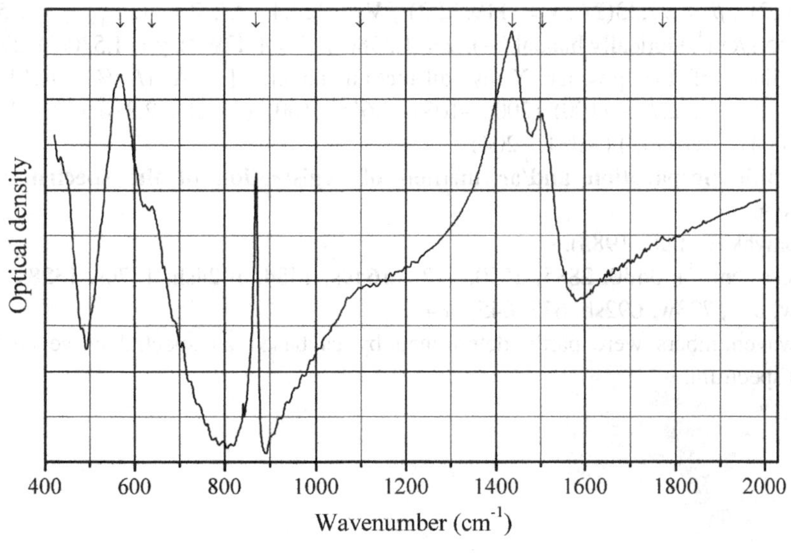

Origin: Synthetic.

Description: Prepared by heating an appropriate mixture of SrO, $SrCO_3$, and Fe_2O_3 at 1200 °C for 12 h under vacuum with subsequent cooling down to room temperature, annealing at 500 °C for 30 min in air and quenching. Characterized by EDS analysis, Mössbauer spectroscopy, powder X-ray diffraction, neutron powder diffraction, and single-crystal electron diffraction data. Orthorhombic, space group *I4/mmm*, $a = 3.88965(3)$, $c = 27.9906(1)$ Å.

Kind of sample preparation and/or method of registration of the spectrum: Absorption. Kind of sample preparation is not indicated.

Source: Bréard et al. (2004).

Wavenumbers (cm^{-1}): 1771w, 1506s, 1437s, 1098sh, 869, 640, 569s.

Note: The band at 569 cm^{-1} may correspond to polymerized $Fe^{3+}O_6$ and/or $Fe^{3+}O_5$ polyhedra.

C351 Barentsite $Na_7Al(HCO_3)_2(CO_3)_2F_4$

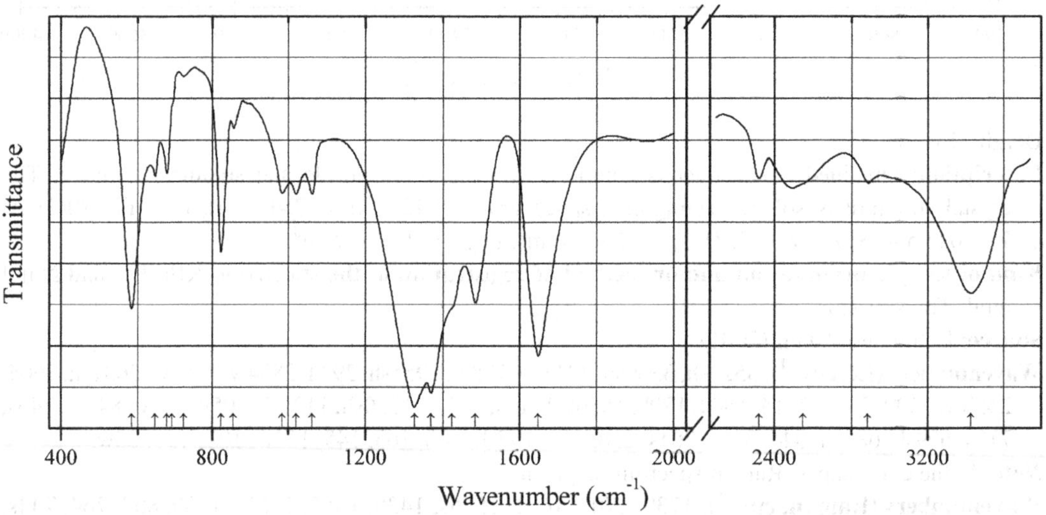

Origin: Restin'yun Mt., Khibiny massif, Kola Peninsula (type locality).

Description: Colorless anhedral grains from the association with shortite, albite, natrolite, trona, natrite, villiaumite, etc. Holotype sample. Triclinic, $a = 6.472(2)$, $b = 6.735(2)$, $c = 8.806(2)$ Å, $\alpha = 92.50(2)°$, $\beta = 97.33(2)°$, $\gamma = 119.32(2)°$, $V = 329.41$ Å^3, $Z = 1$. $D_{meas} = 2.56(2)$ g/cm^3, $D_{calc} = 2.55$ g/cm^3. Optically biaxial (−), $\alpha = 1.358(2)$, $\beta = 1.479(2)$, $\gamma = 1.530(2)$, $2V = 62°$. The strongest lines of the powder X-ray diffraction pattern [*d*, Å (*I*, %) (*hkl*)] are: 2.887 (84) (003, 2−11), 2.778 (100) (200, −103), 2.658 (100) (2−21), 2.316 (50) (2−22), 2.169 (70) 120, 004), 1.870 (42) (−331, −204).

Kind of sample preparation and/or method of registration of the spectrum: KBr disc. Transmission.

Source: Khomyakov et al. (1983).

Wavenumbers (cm^{-1}): 3430, 2885, 2550, 2322, 1649s, 1484, 1424sh, 1370s, 1328s, 1063, 1021, 984, 856w, 823, 723w, 692sh, 677, 645, 584s.

Note: The wavenumbers were partly determined by us based on spectral curve analysis of the published spectrum.

C352 Chukanovite $Fe_2(CO_3)(OH)_2$

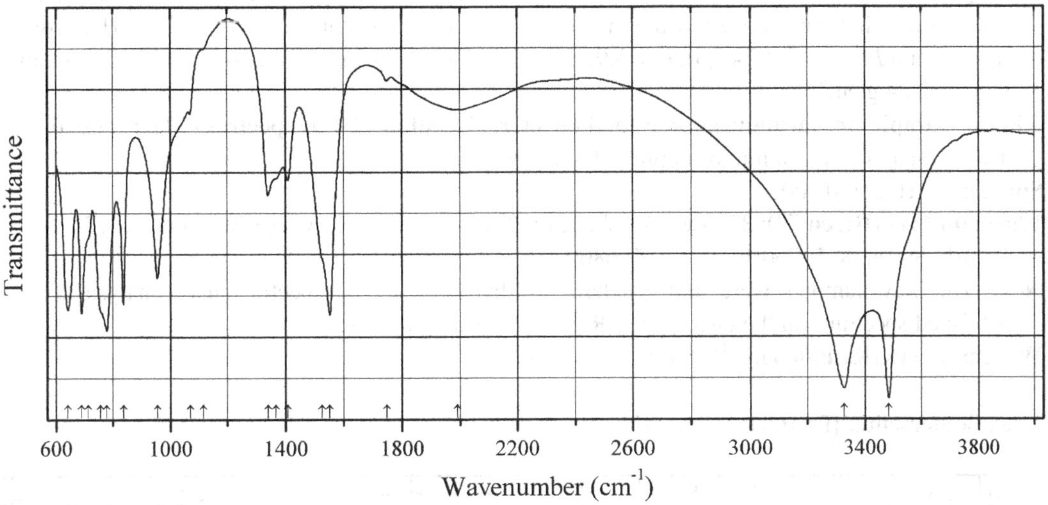

Origin: Synthetic.

Description: Obtained by heating a mixture of powdered claystone (a rock containing clay minerals, Ca-Mg-Fe carbonates and quartz as the main components) and iron powder, in the presence of iron plates and an aqueous solution containing NaCl and $CaCl_2$ at 90 °C for 6 months. The synthetic analogue of chukanovite was formed as randomly oriented powder on the iron plates. Characterized by powder X-ray diffraction and electron diffraction data. Monoclinic, space group $P2_1/a$, $a = 12.5$ (3), $b = 9.5(2)$, $c = 3.2(1)$ Å, $\beta = 97.6(5)°$, $V = 377(17)$ Å^3.

Kind of sample preparation and/or method of registration of the spectrum: Reflection from an iron plate covered by the synthetic analogue of chukanovite.

Source: Pignatelli et al. (2014).

Wavenumbers (cm^{-1}): 3485s, 3327s, 1992 (broad), 1748w, 1552s, 1525sh, 1408, 1368sh, 1340, 1116sh, 1069, 955, 837, 779, 757sh, 713sh, 690, 642.

Note: The wavenumbers were partly determined by us based on spectral curve analysis of the published spectrum.

C353 Lecoqite-(Y) $Na_3Y(CO_3)_3{\cdot}6H_2O$

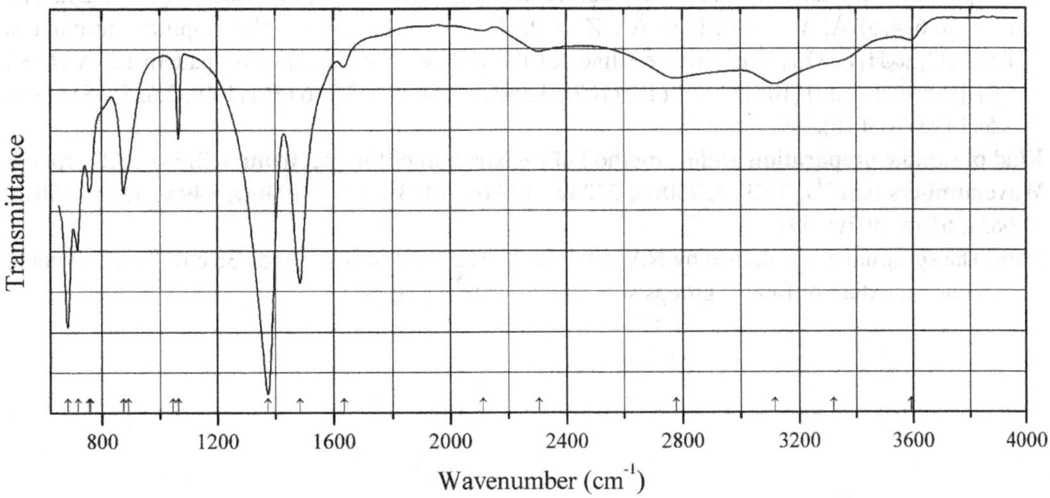

Origin: Synthetic.

Description: Prepared hydrothermally from Na_2CO_3, YF_3, and H_2O in the molar ratio 25:1:55 at 220 °C for 48 h. Characterized by thermoanalytical data. The crystal structure is solved. Hexagonal, space group $P6_3$, $a = 11.347(5)$, $c = 5.935(5)$ Å, $V = 661.8(5)$ Å^3, $Z = 2$. $D_{meas} = 2.25(5)$ g/cm^3, $D_{calc} = 2.24$ g/cm^3.

Kind of sample preparation and/or method of registration of the spectrum: Transmission. A diamond-anvil cell as a microsampling device was used.

Source: Ali et al. (2004a).

Wavenumbers (IR, cm^{-1}): 3593w, 3320sh, 3115, 2778, 2305w, 2112w, 1634w, 1485s, 1375s, 1065, 1045sh, 890sh, 875, 760sh, 755, 715, 680s.

Note: The wavenumbers were partly determined by us based on spectral curve analysis of the published spectrum. In the cited paper, Raman spectrum is given.

Wavenumbers (Raman, cm^{-1}): 1635, 715s, 680.

C354 Somersetite $[Pb_3(OH)_2(CO_3)_2][Pb_3(Pb_2O_2)(CO_3)_3]$

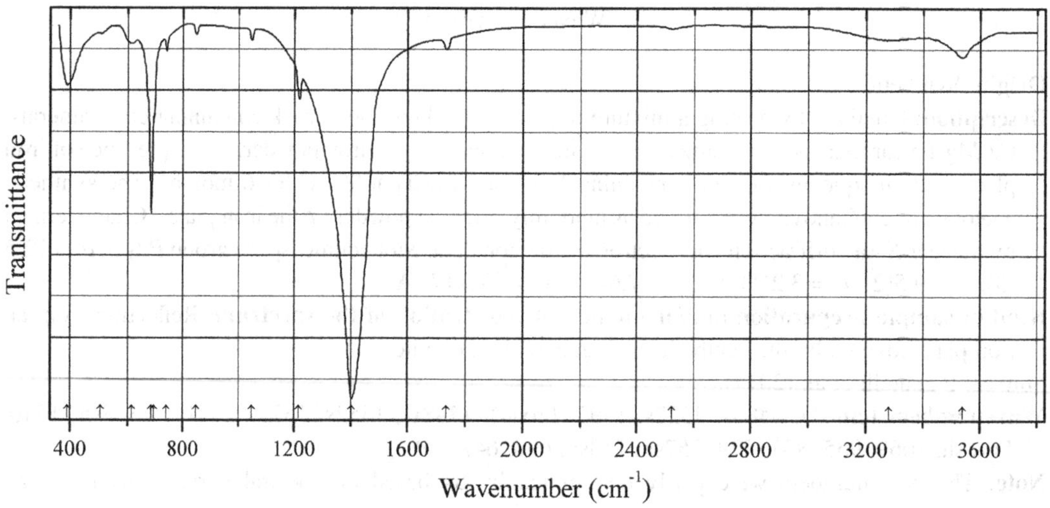

Origin: Torr Works ("Merehead Quarry"), Somerset, England, UK (type locality).

Description: Greenish grains from the association with calcite, aragonite, and quartz. Holotype sample. The crystal structure is solved. Hexagonal, space group $P6_3/mmc$, $a = 5.2427(7)$, $c = 40.624(6)$ Å, $V = 967.0(3)$ Å^3, $Z = 2$. $D_{calc} = 7.11$ g/cm^3. The empirical formula is $Pb_{8.004}C_{4.998}H_{1.998}O_{19}$. The strongest lines of the powder X-ray diffraction pattern [d, Å (I, %) (hkl)] are: 4.308 (33) (103), 3.581 (40) (107), 3.390 (100) (108), 3.206 (55) (109), 2.625 (78) (110), 2.544 (98) (0.0.16).

Kind of sample preparation and/or method of registration of the spectrum: KBr disc. Absorption.

Wavenumbers (cm^{-1}): 3539, (3280w), 2524w, 1734w, 1403s, 1217, 1046w, 849w, 738w, 690sh, 683s, 615w, 507w, 391.

Note: The spectrum was obtained by N.V. Chukanov. The bands at 1217 and 738 cm^{-1} correspond to a minor admixture of BO_3^{3-} groups substituting CO_3^{2-} groups.

C355 "Hydrucerussite-like mineral 9-40" Lead hydroxycarbonate

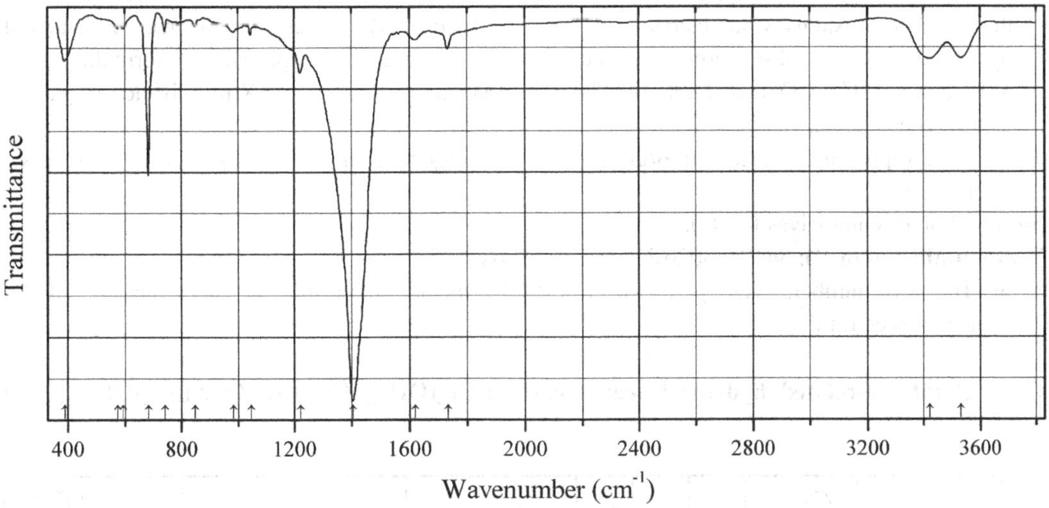

Origin: Torr Works ("Merehead Quarry"), Somerset, England, UK.

Description: Investigated by O.I. Siidra. Characterized by single-crystal X-ray diffraction data. Trigonal, $a = 9.0929(5)$, $c = 40.660(6)$ Å, $V = 2911.42(9)$ Å^3.

Kind of sample preparation and/or method of registration of the spectrum: KBr disc. Absorption.

Wavenumbers (cm^{-1}): 3531, (3422), 1733, (1619w), 1404s, 1220, 1046w, 984w, 848w, 741w, 683s, 595w, 575w, 391.

Note: The spectrum was obtained by N.V. Chukanov. The bands at 1220 and 741 cm^{-1} correspond to a minor admixture of BO_3^{3-} groups substituting CO_3^{2-} groups.

C356 Quintinite-related hydroxyde carbonate $Mg_4Cr_2(OH)_{12}(CO_3)·nH_2O$ $Mg_4Cr_2(OH)_{12}(CO_3)$ $·nH_2O$

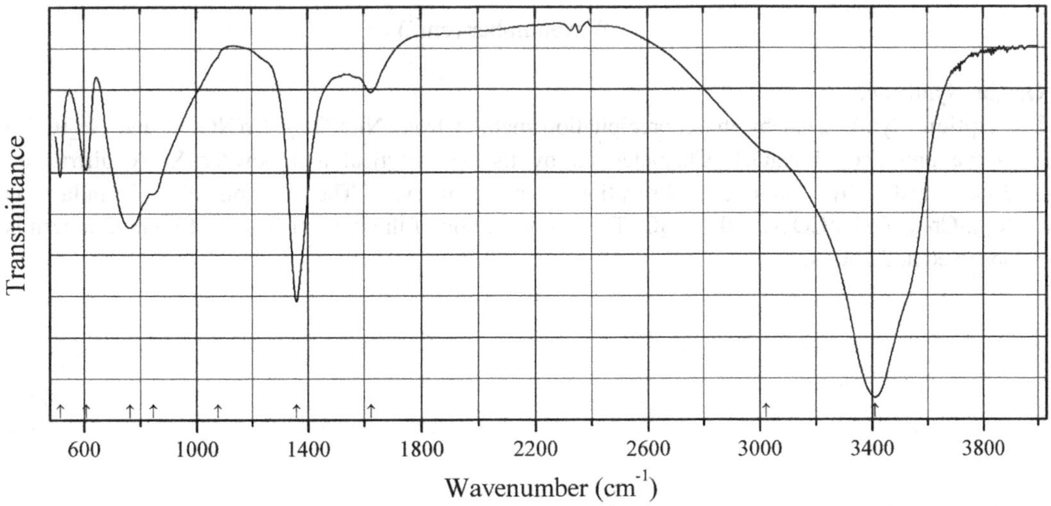

Origin: Synthetic.

Description: Synthesized by the coprecipitation method from $Mg(NO_3)_2$, $Cr(NO_3)_3$, and $Na_2(CO_3)$ in the presence of NaOH. Characterized by thermoanalytical and powder X-ray diffraction data, and by atomic absorption spectrometry. The empirical formula is $Mg_{0.68}Cr_{0.32}(OH)_2(CO_3)_{0.16}\cdot0.86H_2O$. The strongest line of the powder X-ray diffraction pattern is observed at 22.92 Å.

Kind of sample preparation and/or method of registration of the spectrum: KBr disc. Transmission.

Source: Labajos and Rives (1996).

Wavenumbers (cm^{-1}): 3410s, 3020sh, 1625w, 1360s, 1077sh, 845sh, 762, 608, 517.

Note: The wavenumbers were partly determined by us based on spectral curve analysis of the published spectrum.

C357 Quintinite-related hydroxyde carbonate $Ni_4Cr_2(OH)_{12}(CO_3)\cdot nH_2O$ $Ni_4Cr_2(OH)_{12}(CO_3)\cdot nH_2O$

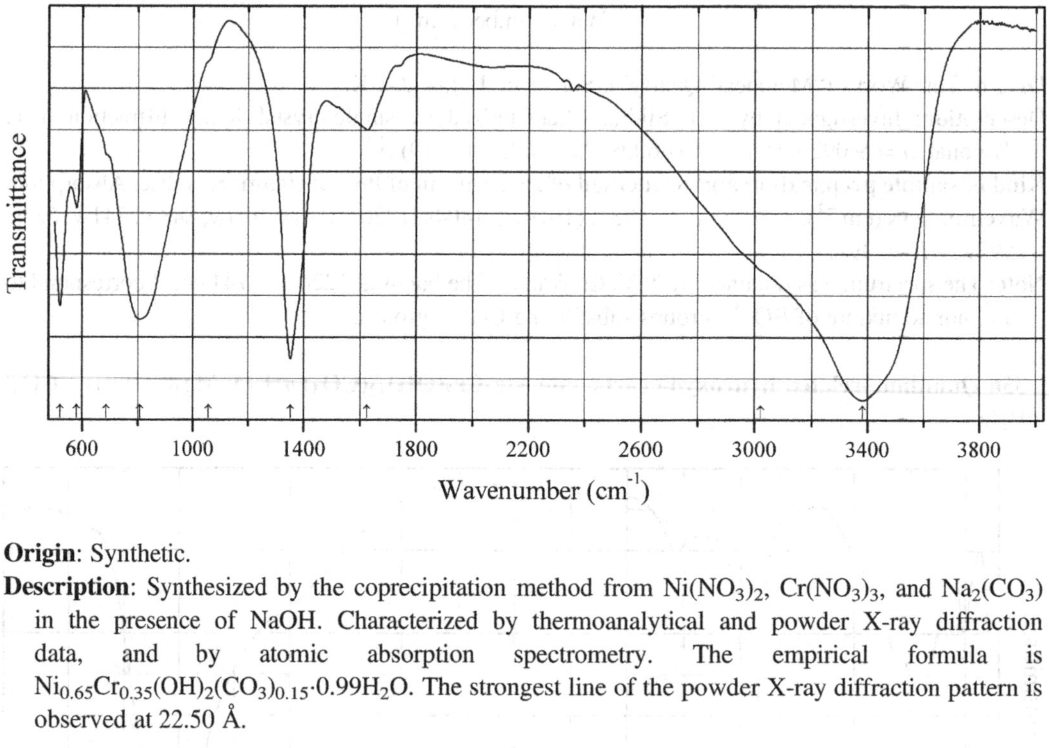

Origin: Synthetic.

Description: Synthesized by the coprecipitation method from $Ni(NO_3)_2$, $Cr(NO_3)_3$, and $Na_2(CO_3)$ in the presence of NaOH. Characterized by thermoanalytical and powder X-ray diffraction data, and by atomic absorption spectrometry. The empirical formula is $Ni_{0.65}Cr_{0.35}(OH)_2(CO_3)_{0.15}\cdot0.99H_2O$. The strongest line of the powder X-ray diffraction pattern is observed at 22.50 Å.

Kind of sample preparation and/or method of registration of the spectrum: KBr disc. Transmission.

Source: Labajos and Rives (1996).

Wavenumbers (cm^{-1}): 3382s, 3020sh, 1624w, 1354s, 1060sh, 806s, 687sh, 580, 521.

Note: The wavenumbers were partly determined by us based on spectral curve analysis of the published spectrum.

C358 Scarbroite $Al_5(CO_3)(OH)_{13}·5H_2O$

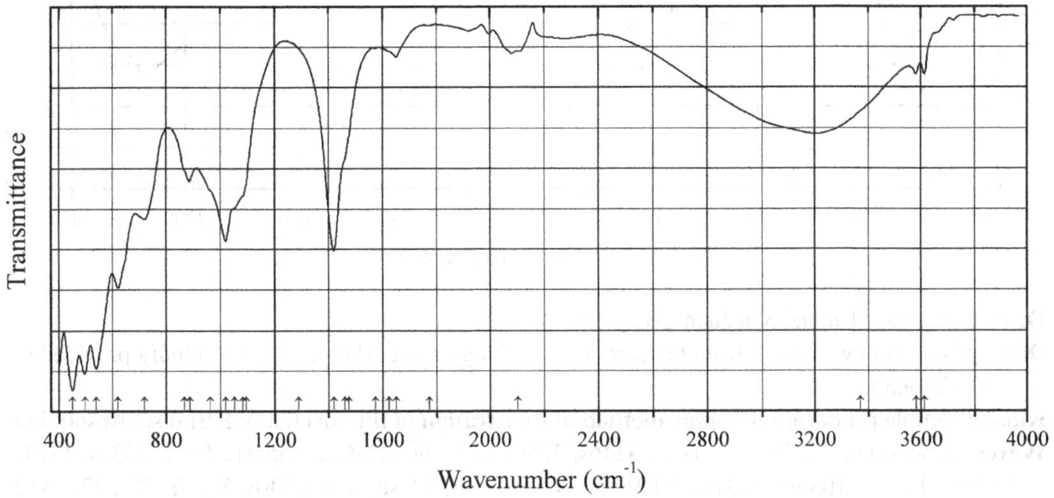

Origin: The former Fonte Civillina, the town of RecoaroTerme, Vicenza, NE Italy.

Description: Nests of microcrystalline aggregates from the association with quartz, baryte, galena, cerussite, etc. Characterized by powder X-ray diffraction data and semiquantitative electron microprobe analyses. Triclinic, $a = 9.892(1)$, $b = 14.934(2)$, $c = 26.321(4)$ Å, $\alpha = 98.89(1)°$, $\beta = 97.49(1)°$, $\gamma = 89.04(1)°$.

Kind of sample preparation and/or method of registration of the spectrum: KBr disc. Transmission.

Source: Boscardin et al. (2009).

Wavenumbers (IR, cm^{-1}): 3614w, 3583w, 3370 (broad), 2102w, (1773w), 1651w, 1622sh, (1573w), 1474sh, 1459sh, 1419s, 1288sh, 1096sh, 1084sh, 1052sh, 1020s, 963sh, 886, 868sh, 720, 622s, 540s, 497s, 451s.

Note: In the cited paper, the wavenumber 720 cm^{-1} is erroneously indicated as 750 cm^{-1}. The wavenumbers were partly determined by us based on spectral curve analysis of the published spectrum. In the cited paper, Raman spectrum is given.

Wavenumbers (Raman, cm^{-1}): 3668, 3614s, 3593, 2994s, 1421, 1342, 1107s, 983, 889, 696, 600s, 449, 376, 267.

C359 Bayleyite $Mg_2(UO_2)_2(CO_3)_3·18H_2O$

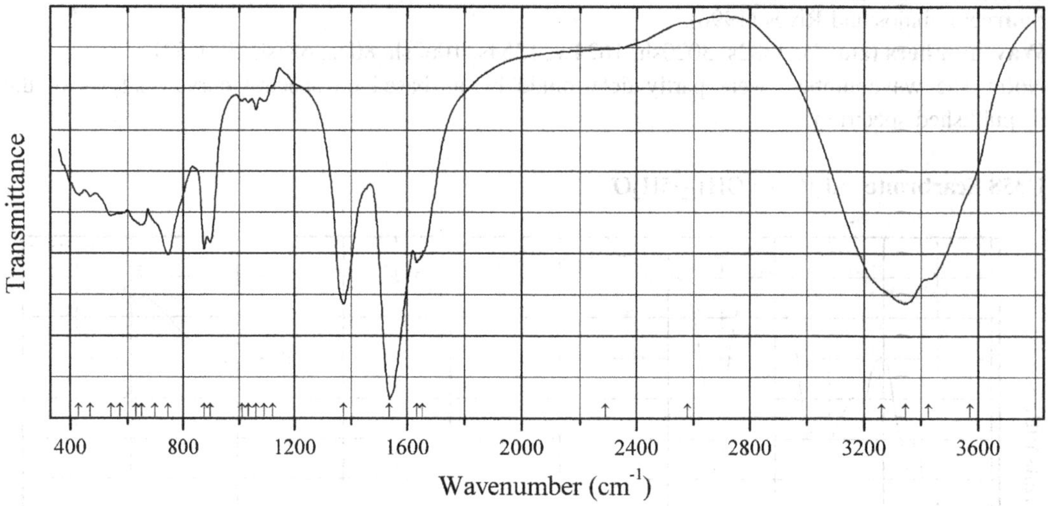

Origin: Hideout #1 mine, San Juan Co., Utah, USA.

Description: Yellow crystals from the association with gypsum. The sample was kindly provided by A.V. Kasatkin.

Kind of sample preparation and/or method of registration of the spectrum: KBr disc. Absorption.

Wavenumbers (cm^{-1}): 3570sh, 3425, 3346s, 3260sh, 2579w, 2290sh, 1655sh, 1635, 1539s, 1373s, 1120sh, 1088w, 1059w, 1031w, 1009w, 898, 876, 744, 695sh, 649, 630sh, 575sh, 542, 471, 430.

Note: The spectrum was obtained by N.V. Chukanov.

C360 Línekite $K_2Ca_3[(UO_2)(CO_3)_3]_2·8H_2O$

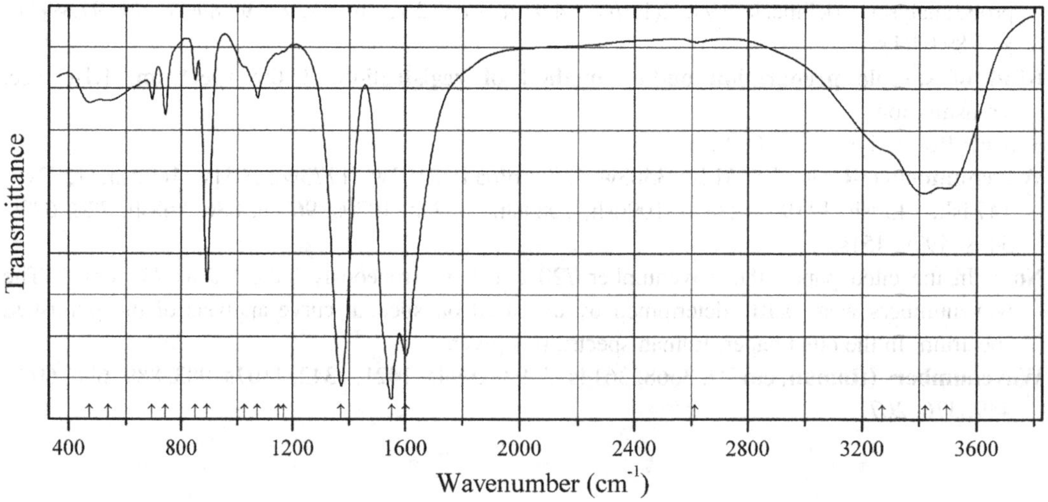

Origin: Geschieber vein, Svornost Mine, Jáchymov, Jáchymov District, Krušné Hory Mts, Karlovy Vary Region, Bohemia, Czech Republic (type locality).

Description: Greenish-yellow tabular crystals from the association with braunerite. Investigated by A.V. Kasatkin. The empirical formula is (electron microprobe): $K_{1.94}Ca_{3.03}(UO_2)_{2.00}(CO_3)_6 \cdot nH_2O$.

Kind of sample preparation and/or method of registration of the spectrum: KBr disc. Absorption.

Wavenumbers (cm^{-1}): 3496s, 3420s, 3270sh, 2612w, 1604s, 1555s, 1372s, 1168w, 1148w, 1071, 1025sh, 893s, 850, 741, 691, 538, 474.

Note: The spectrum was obtained by N.V. Chukanov.

C361 Wermlandite carbonate analogue $Mg_7Al_2(OH)_{18}[Ca(H_2O)_6](CO_3,SO_4)_2 \cdot 6H_2O$

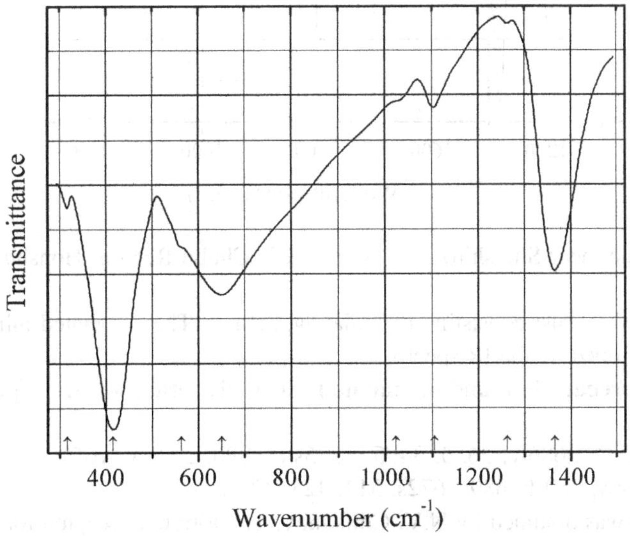

Origin: Långban deposit, Bergslagen ore region, Filipstad district, Värmland, Sweden (type locality).

Description: Pale greenish-gray hexagonal platelets on calcite crystals. The crystal structure is solved. Trigonal, space group $P\text{-}3c1$, $a = 9.303(3)$, $c = 22.57(1)$ Å, $V = 1692$ Å^3. $Z = 2$. $D_{meas} = 1.93$ g/cm^3, $D_{calc} = 1.96$ g/cm^3.

Kind of sample preparation and/or method of registration of the spectrum: KBr disc. Transmission.

Source: Rius and Allmann (1984).

Wavenumbers (cm^{-1}): 1365s, 1264, 1105w, 1024sh, 650s, 564sh, 416s, 317w.

Note: The wavenumbers were partly determined by us based on spectral curve analysis of the published spectrum. In the cited paper, the mineral is described as wermlandite *s.s.*, $Mg_7Al_2(OH)_{18}[Ca(H_2O)_6](SO_4)_2 \cdot 6H_2O$. However the intensities of the bands of asymmetric vibrations of carbonate and sulfate anions (at 1365 and 1105 cm^{-1}, respectively) indicate that it is a CO_3-dominant mineral. The value of measured density confirms this conclusion.

C362 Nakauriite $Cu_8(SO_4)_4(CO_3)(OH)_6{\cdot}48H_2O$

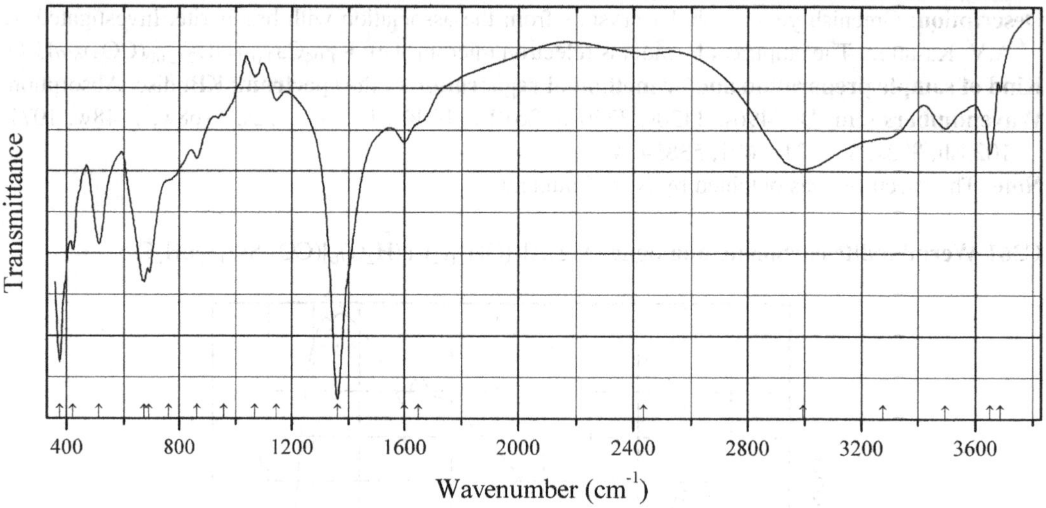

Origin: Nakauri mine, near Shinshiro city, Aichi pref., Chubu Region, Honshu Island, Japan (type locality).

Description: Light blue crust consisting of radial aggregates. The associated minerals are chrysotile and brucite. Confirmed by the IR spectrum.

Kind of sample preparation and/or method of registration of the spectrum: KBr disc. Transmission.

Wavenumbers (cm^{-1}): 3686w, 3650, 3495, 3275sh, 2996, 2435w, 1650sh, 1600, 1361s, 1143w, 1066w, 959w, 864w, 760sh, 689s, 672s, 514, 423, 376s.

Note: The spectrum was obtained by N.V. Chukanov. The formula accepted for nakauriite is wrong and is to be revised: actually, nakauriite does not contain sulfate groups. The weak bands at 959 and 3686 cm^{-1} correspond to chrysotile impurity.

C363 Nakauriite $Cu_8(SO_4)_4(CO_3)(OH)_6{\cdot}48H_2O$

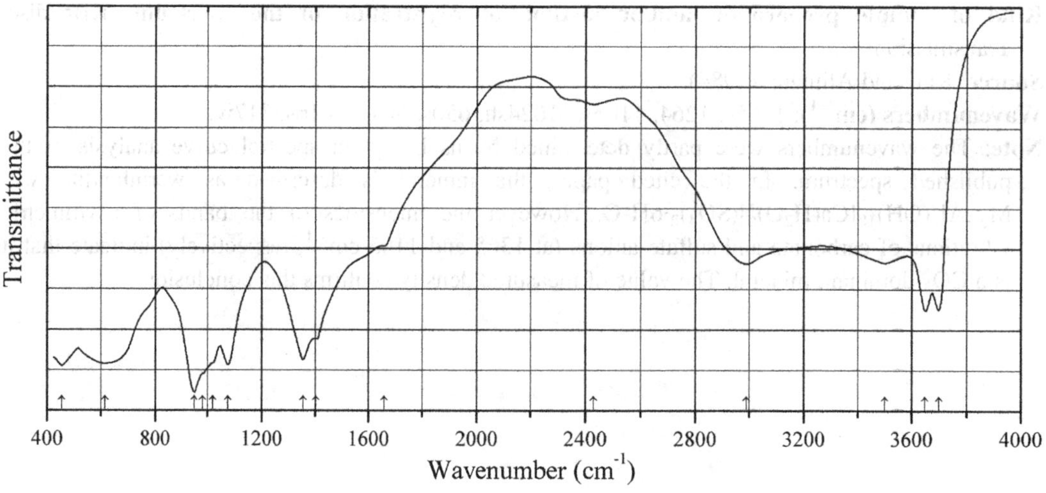

Origin: Nakauri mine, near Shinshiro city, Aichi pref., Chubu Region, Honshu Island, Japan (type locality).

Description: Holotype sample with impurities (see Sect. 1.3 in this book).

Kind of sample preparation and/or method of registration of the spectrum: KBr disc. Transmission.

Source: Suzuki et al. (1976).

Wavenumbers (cm^{-1}): 3700s, 3650s, 3500, 2990, 2430, 1660sh, 1404 (plateau), 1355s, 1075s, 1020sh, 983sh, 950s, 613s (broad), 454s.

Note: The strong bands at 3700, 1075, 950, 613, and 454 cm^{-1} correspond to chrysotile impurity.

C364 Nakauriite $Cu_8(SO_4)_4(CO_3)(OH)_6{\cdot}48H_2O$

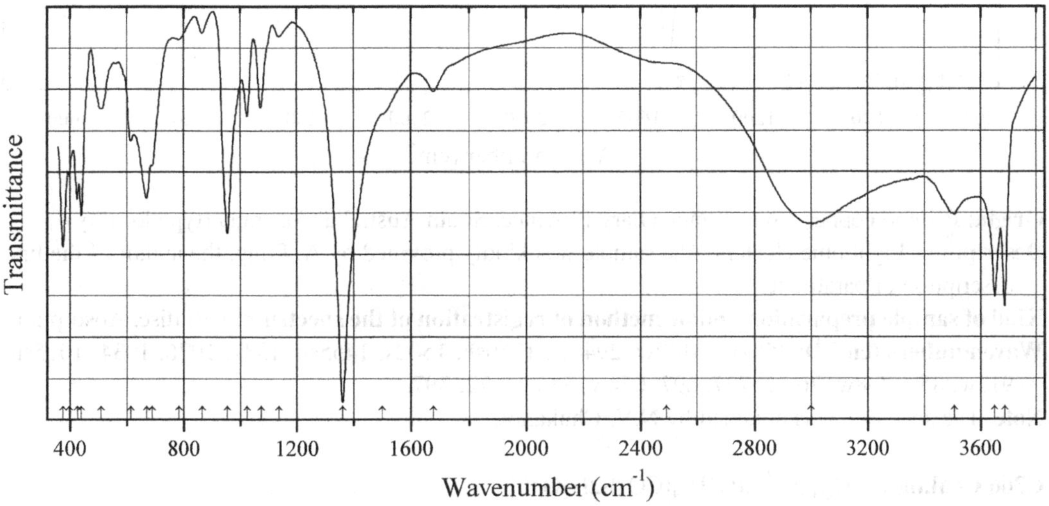

Origin: Chromite quarry near Karkodin railway station, Chelyabinsk region, South Urals.

Description: Blue crystalline crust on serpentine. Investigated by I.V. Pekov and N.V. Chukanov. For the description see Sect. 1.3 in this book.

Kind of sample preparation and/or method of registration of the spectrum: KBr disc. Transmission.

Wavenumbers (cm^{-1}): 3688s, 3651s, 3504, 3000s, 2490sh, 1678, 1500sh, 1361s, 1139w, 1074, 1025, 954s, 866w, 784w, 690sh, 669s, 614, 509, 440s, 426s, 398, 376s.

Note: The spectrum was obtained by N.V. Chukanov. The formula accepted for nakauriite is wrong and is to be revised: actually, nakauriite does not contain sulfate groups. The bands at 3688, 1074, 1025, 954, 614, and 376 cm^{-1} correspond to serpentine.

C365 Paratooite-(La) $(La,Ca,Na,Sr)_{12}Cu_2(CO_3)_{16}$

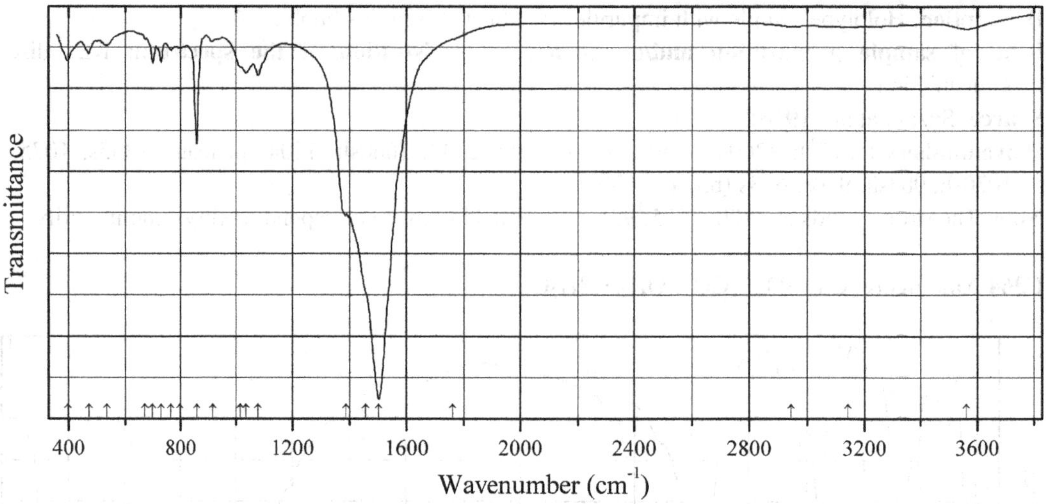

Origin: Paratoo copper mine, Yunta, Olary Province, South Australia, Australia (type locality).

Description: Light blue clusters. The sample was kindly provided by A. Pring, the author of the first description of paratooite.

Kind of sample preparation and/or method of registration of the spectrum: KBr disc. Absorption.

Wavenumbers (cm^{-1}): 3561w, 3142w, 2944w, 1760sh, 1502s, 1455sh, 1387, 1076, 1034, 1015sh, 917w, 859, 798w, 764w, 727, 697, 674w, 536w, 472, 397.

Note: The spectrum was obtained by N.V. Chukanov.

C366 Coalingite $Mg_{10}Fe^{3+}{}_2(OH)_{24}(CO_3)\cdot 2H_2O$

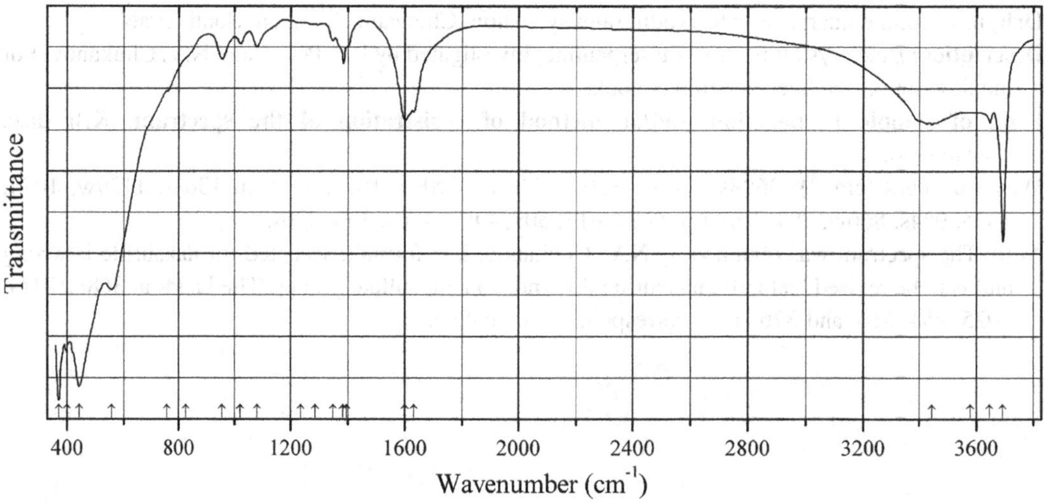

Origin: Union Carbide Asbestos pit, New Idria district, Diablo Range, Fresno Co., California, USA (type locality).

Description: Brown crust on serpentine. Identified by the IR spectrum.

Kind of sample preparation and/or method of registration of the spectrum: KBr disc. Absorption.

Wavenumbers (cm^{-1}): 3693s, 3647, 3445, 1632, 1601, 1395sh, 1384, 1349w, 1285w, 1235w, 1077, 1021w, 958, 825sh, 755sh, 558s, 443s, 399s, 371s.

Note: The spectrum was obtained by N.V. Chukanov.

C369 Ikaite $Ca(CO_3)·6H_2O$

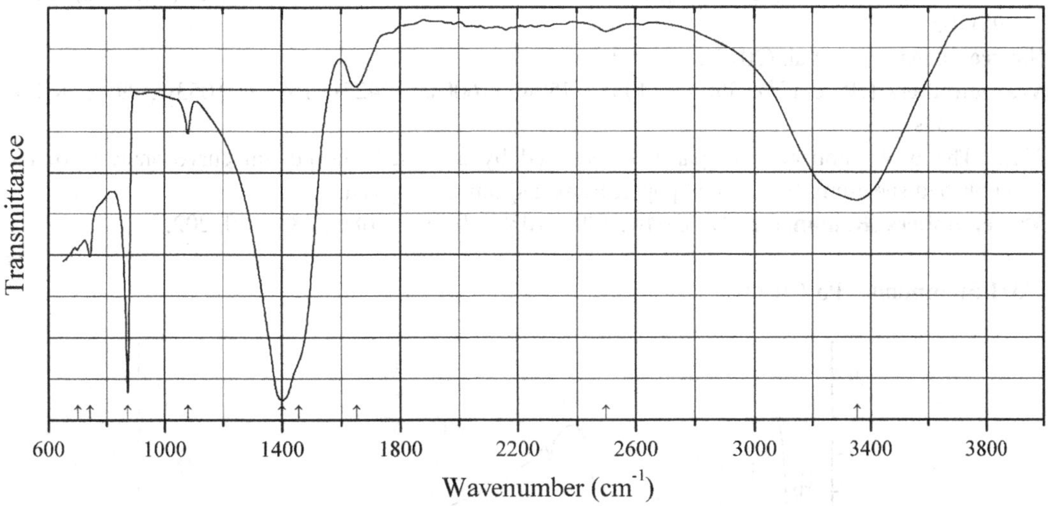

Origin: Artificial river bed in an alpine valley situated in the eastern part of Austria.

Description: Beige precipitate collected in February, 2014. Characterized by powder X-ray diffraction data.

Kind of sample preparation and/or method of registration of the spectrum: Attenuated total reflection of powdered mineral.

Source: Boch et al. (2015).

Wavenumbers (cm^{-1}): 3350s, 2000w, 1650, 1455sh, 1400s, 1080, 873s, 743, 700w.

Note: The wavenumbers were partly determined by us based on spectral curve analysis of the published spectrum.

C370 Abellaite $NaPb_2(CO_3)_2(OH)$

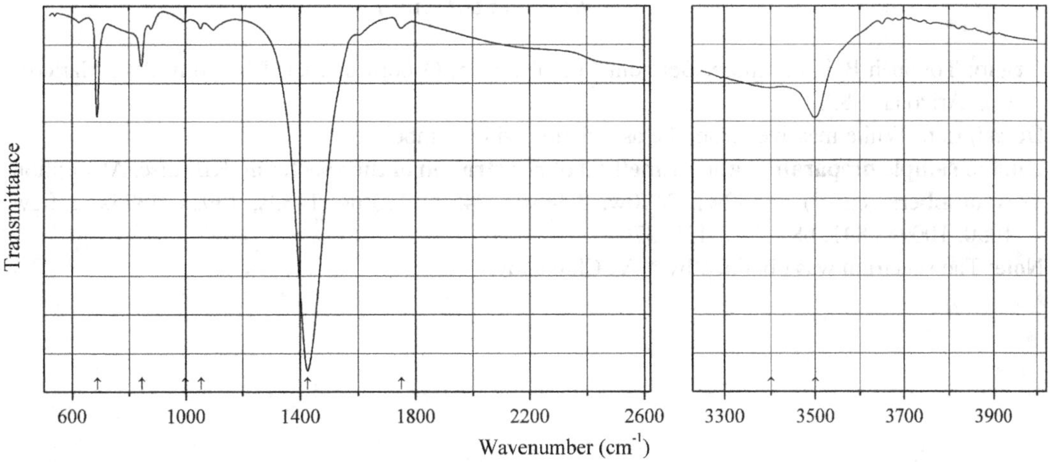

Origin: Eureka mine, southern Pyrenees, Lleida province, Catalonia, Spain (type locality).

Description: White coating on the surface of the aggregate of primary minerals (roscoelite, pyrite, uraninite, sulfides, etc.). Holotype sample. Trigonal, space group *P*63*mc*, $a = 5.254(2)$, $c = 13.450(5)$ Å, $V = 321.5(2)$ Å^3, $Z = 2$. $D_{calc} = 5.93$ g/cm^3. The empirical formula is $Na_{0.96}Ca_{0.04}Pb_{1.98}(CO_3)_2(OH)$. The strongest lines of the powder X-ray diffraction pattern [*d*, Å (*I*, %) (*hkl*)] are: 3.193 (100) (013), 2.627 (84) (110), 2.275 (29) (020), 2.242 (65) (021, 006), 2.029 (95) (023).

Source: Ibáñez-Insa et al. (2017).

Wavenumbers (IR, cm^{-1}): 3500, (3400), 1750w, (1600sh), 1425s, 1098w, 1053w, 998w, 878w, 844, 688s.

Note: The wavenumbers were partly determined by us based on spectral curve analysis of the published spectrum. In the cited paper, Raman spectrum is given.

Wavenumbers (Raman, cm^{-1}): 3504w, 1391, 1058s, 1038w, 868w, 683, 280, 202.

C371 Shannonite $Pb_2O(CO_3)$

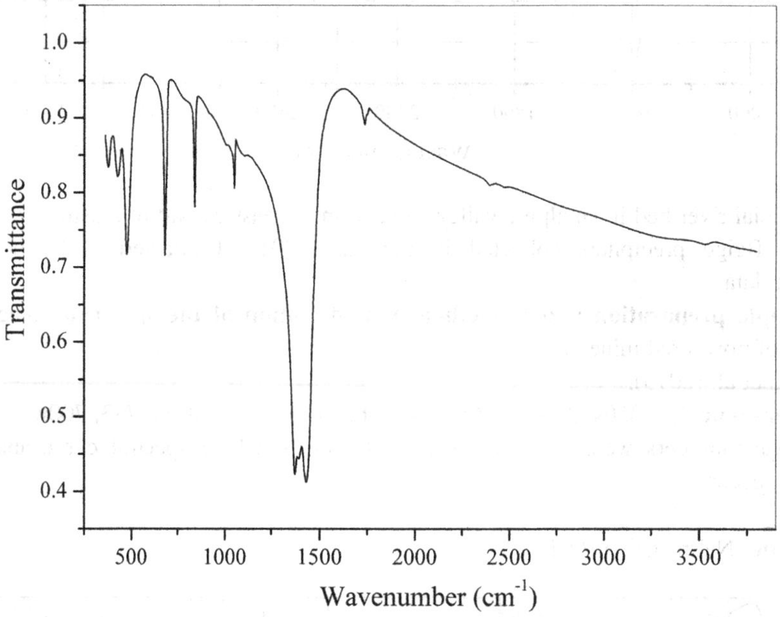

Origin: Tonopah-Belmont mine, Belmont Mt., Tonopah, Osborn district, Big Horn Mts., Maricopa Co., Arizona, USA.

Description: White massives from the association with plumbojarosite.

Kind of sample preparation and/or method of registration of the spectrum: KBr disc. Absorption.

Wavenumbers (cm^{-1}): (3532w), 2470w, 2396w, 1745sh, 1735w, 1433s, 1392s, 1374s, 1102w, 1050, 1009w, 841, 682, 473, 425, 376.

Note: The spectrum was obtained by N.V. Chukanov.

CSi30 Roymillerite $Pb_{24}Mg_9(Si_9AlO_{28})(SiO_4)(BO_3)(CO_3)_{10}(OH)_{14}O_4$

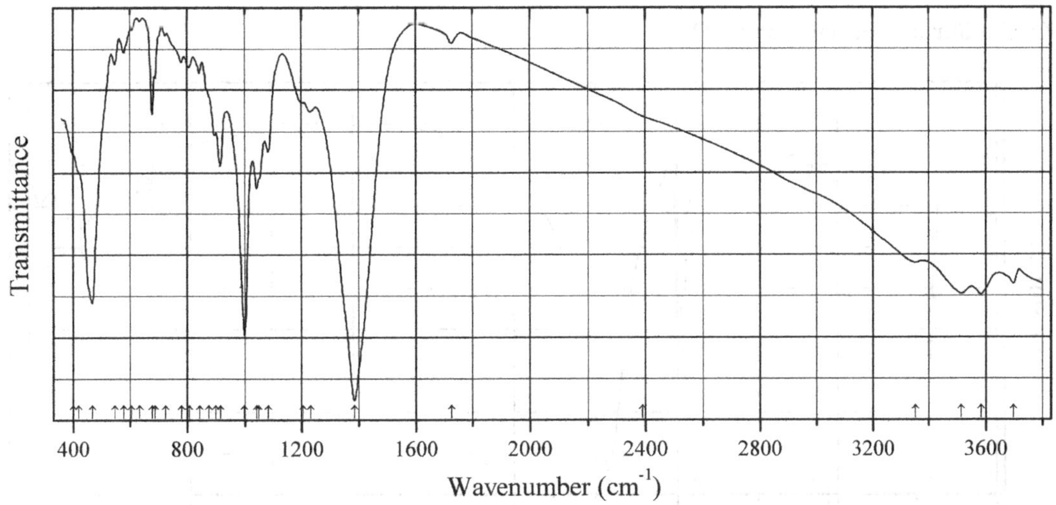

Origin: Kombat Mine, Grootfontein district, Otjozondjupa region, Namibia (type locality).

Description: Colorless platy single-crystal grains from the association with jacobsite, cerussite, hausmannite, sahlinite, rhodochrosite, baryte, grootfonteinite, Mn-Fe-oxides, and melanotekite. Holotype sample. The crystal structure is solved. Triclinic, space group *P*-1, a = 9.315(1), b = 9.316(1), c = 26.463(4) Å, α = 83.295(3)°, β = 83.308(3)°, γ = 60.023(2)°, V = 1971.2 (6) $Å^3$, Z = 1. D_{calc} = 5.973 g/cm^3. Optically biaxial (–), α = 1.86(1), $\beta \approx \gamma$ = 1.94(1), $2V$ = 5(5)°. The empirical formula is $Pb_{24.12}Mg_{8.74}Mn_{1.25}Fe_{0.94}B_{1.03}Al_{1.04}C_{9.46}Si_{9.39}H_{14.27}O_{83}$. The strongest lines of the powder X-ray diffraction pattern [*d*, Å (*I*, %) (*hkl*)] are: 25.9 (100) (001), 13.1 (11) (002), 3.480 (12) (017, 107, –115, 1–15), 3.378 (14) (126, 216), 3.282 (16) (–2–15, –1–25), 3.185 (12) (–116, 1–16), 2.684 (16) (031, 301, 030, 300, 332, –109, 0–19, 1–18), 2.382 (11) (0.0.–11).

Kind of sample preparation and/or method of registration of the spectrum: KBr disc. Absorption.

Wavenumbers (cm^{-1}): 3700, 3583, 3513, 3352, 1726w, 1385s, 1231, 1204, 1083, 1050sh, 1042, 999s, 915, 898, 875sh, 842w, 806w, 780w, 725w, 688, 679, 635w, 605w, 580w, 548w, 467s, 420sh, 400sh.

Note: The spectrum was obtained by N.V. Chukanov.

2.3 Organic Compounds and Salts of Organic Acids

Org70 Barium formate $Ba(HCO_2)_2$

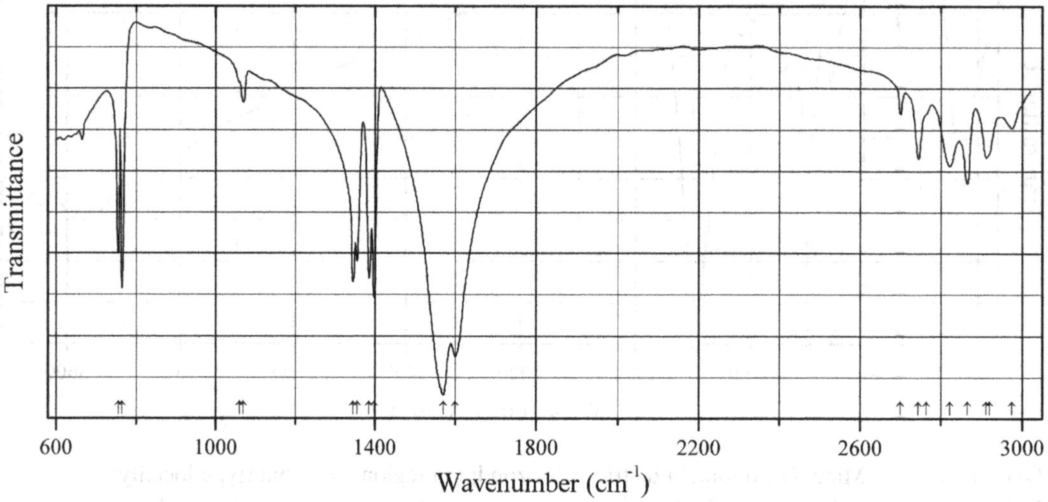

Origin: Synthetic.

Description: Prepared from formic acid and barium carbonate. Orthorhombic, space group $P2_12_12_1$, $a = 6.81$, $b = 8.91$, $c = 7.67$ Å, $Z = 4$.

Kind of sample preparation and/or method of registration of the spectrum: KBr disc. Absorption.

Source: Harvey et al. (1963).

Wavenumbers (cm^{-1}): 2975w, 2920sh, 2912, 2865, 2822, 2765sh, 2745, 2700w, 1600s, 1570s, 1398, 1385, 1355, 1345, 1069w, 1060sh, 765, 756.

Note: For the IR spectrum of barium formate see also Liu et al. (2001).

Org71 Cadmium formatedihydrate $Cd(HCOO)_2 \cdot 2H_2O$

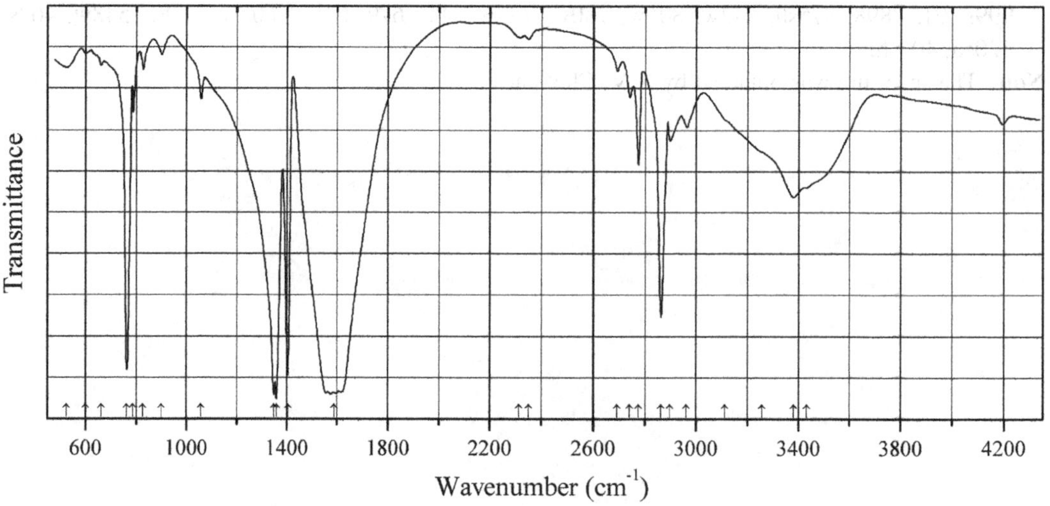

Origin: Synthetic.

Description: Monolinic, space group $P2_1/c$, $Z = 4$.

Kind of sample preparation and/or method of registration of the spectrum: KBr and polyethylene discs. Transmission.

Source: Abraham and Aruldhas (1994).

Wavenumbers (cm^{-1}): 3432sh, 3382, 3257sh, 3112sh, 2962w, 2896, 2862s, 2774, 2740w, 2690w, (2352w), (2314w), 1588s, 1405s, 1358s, 1348s, 1059w, 901w, 826w, 788w, 764s, 662w, 600w, 525w, 361, 299, 264, 231, 195, 154, 138, 121, 113, 103, 84, 68.

Note: Weak bands in the range from 2300 to 2400 cm^{-1} may correspond to atmospheric CO_2.

Org72 Cadmium oxalate trihydrate $Cd(C_2O_4)\cdot 3H_2O$

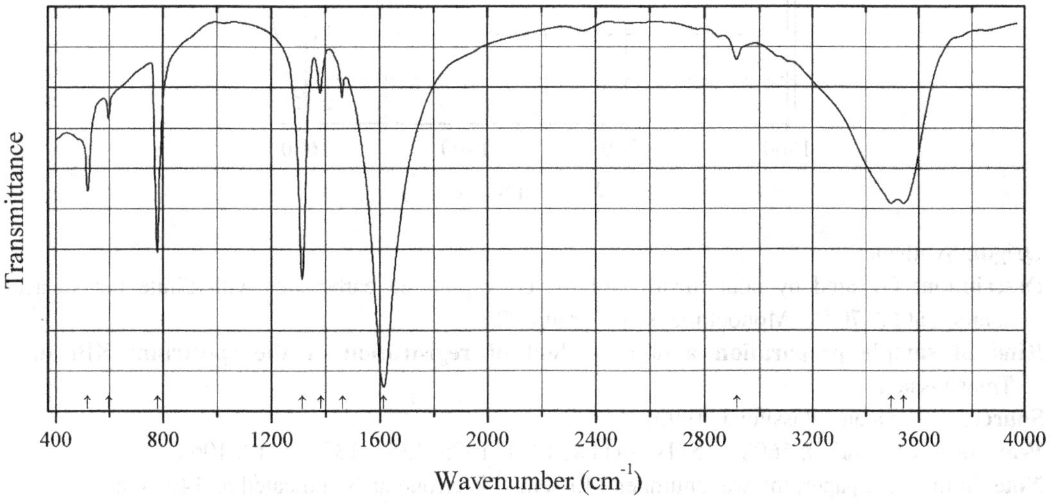

Origin: Synthetic.

Description: Colorless crystals grown at room temperature in silica gel, in the presence of Cd^{2+} ions impregnated with oxalic acid. Triclinic, $a = 6.0059$, $b = 6.66$, $c = 8.49$ Å, $\alpha = 105.71°$, $\beta = 98.99°$, $\gamma = 74.66°$.

Kind of sample preparation and/or method of registration of the spectrum: KBr disc. Transmission.

Source: Raj et al. (2008).

Wavenumbers (cm^{-1}): 3542s, 3496s, 2919w, 1613s, 1461, 1381, 1314s, 778s, 599, 519.

Org73 Copper strontium formate $CuSr(HCOO)_4$

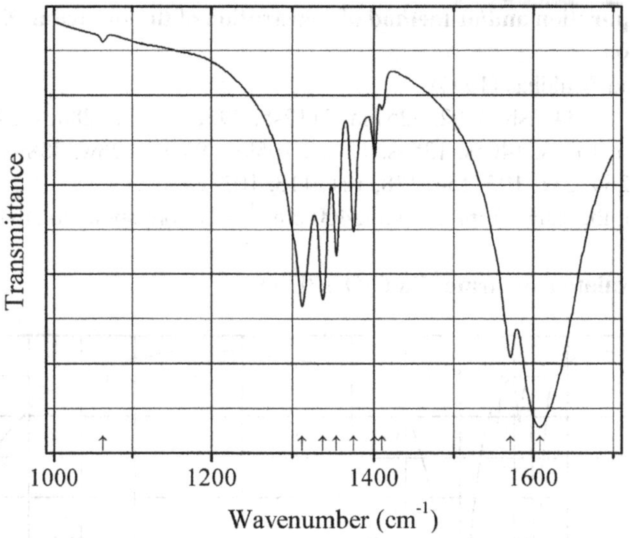

Origin: Synthetic.

Description: Prepared by neutralization of the corresponding carbonates with dilute formic acid solution at 60–70 °C. Monoclinic, space group *P2/c*.

Kind of sample preparation and/or method of registration of the spectrum: KBr disc. Transmission.

Source: Stoilova and Vassileva (1999).

Wavenumbers (cm^{-1}): 1608s, 1571s, 1411w, 1401, 1375, 1354, 1337, 1312, 1063.

Note: In the cited paper, the wavenumber 1401 cm^{-1} is erroneously indicated as 1407 cm^{-1}.

Org74 Lead(II) oxalate $Pb(C_2O_4)$

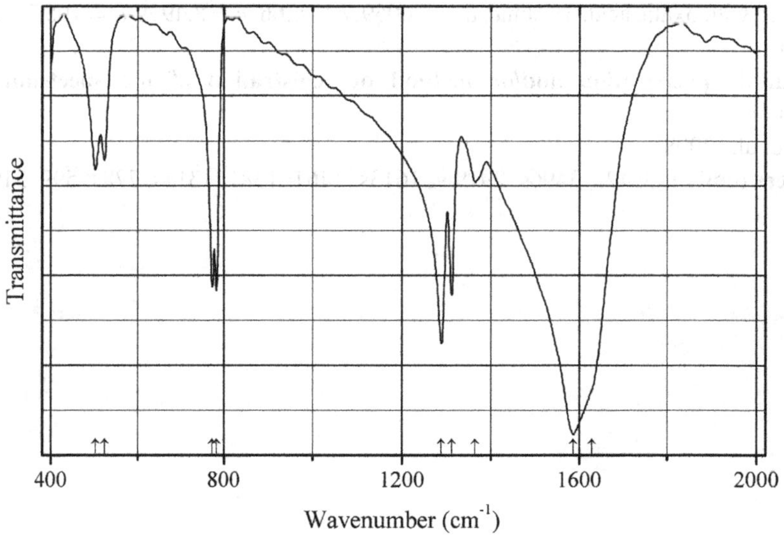

Origin: Synthetic.

Description: Prepared by reacting equimolecular amounts of 0.2 M solutions of lead nitrate and $(NH_4)(HC_2O_4)$. Characterized by powder X-ray diffraction data. Triclinic, space group *P*-1, $Z = 2$ (see JCPDF 14–0803).

Kind of sample preparation and/or method of registration of the spectrum: KBr disc. Transmission.

Source: Mancilla et al. (2009b).

Wavenumbers (IR, cm^{-1}): 1630sh, 1587s, 1365w, 1312, 1289s, 782, 773, 524, 504.

Note: In the cited paper, Raman spectrum is given.

Wavenumbers (Raman, cm^{-1}): 1707w, 1589s, 1476s, 1436s, 1400w, 1366w, 911w, 891, 854, 572w, 497s, 484s.

Org75 Neptunium(IV) oxalate hexahydrate $Np(C_2O_4)_2{\cdot}6H_2O$

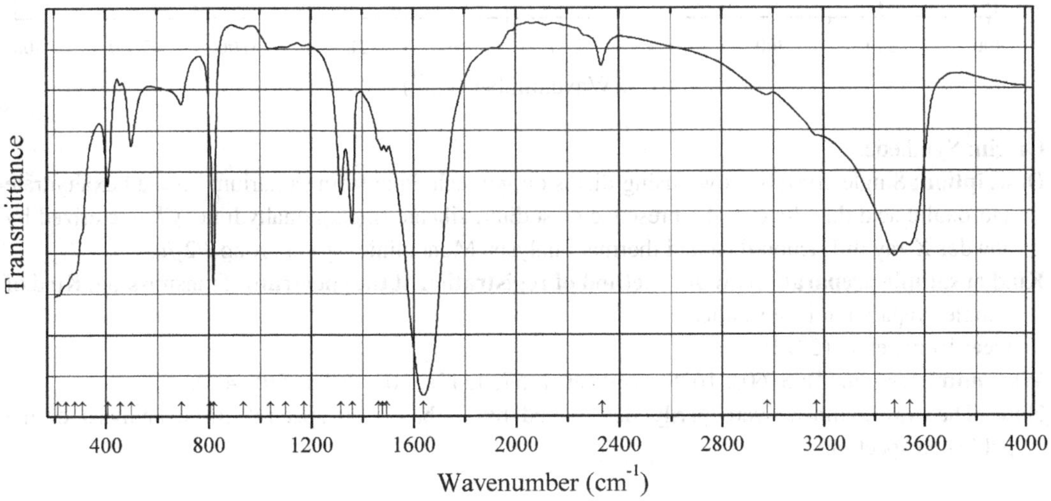

Origin: Synthetic.

Description: Obtained by precipitation from aqueous solution. The strongest lines of the powder X-ray diffraction pattern [*d*, Å (*I*, %)] are: 7.88 (70), 6.36 (100), 5.04 (15), 4.91 (20), 3.93 (90), 3.18 (15).

Kind of sample preparation and/or method of registration of the spectrum: KBr disc. Transmission.

Source: Lindsay et al. (1970).

Wavenumbers (cm^{-1}): 3542, 3480s, 3170sh, 2974w, 2332w, 1638s, 1495w, 1477w, 1461sh, 1360s, 1316, 1172w, 1098w, 1039w, 934w, 818s, 805sh, 694, 499, 456, 407, 309sh, 279sh, 245sh, 212sh.

Note: The wavenumbers were determined by us based on spectral curve analysis of the published spectrum.

Org76 Samarium oxalate decahydrate $Sm_2(C_2O_4)_3{\cdot}10H_2O$

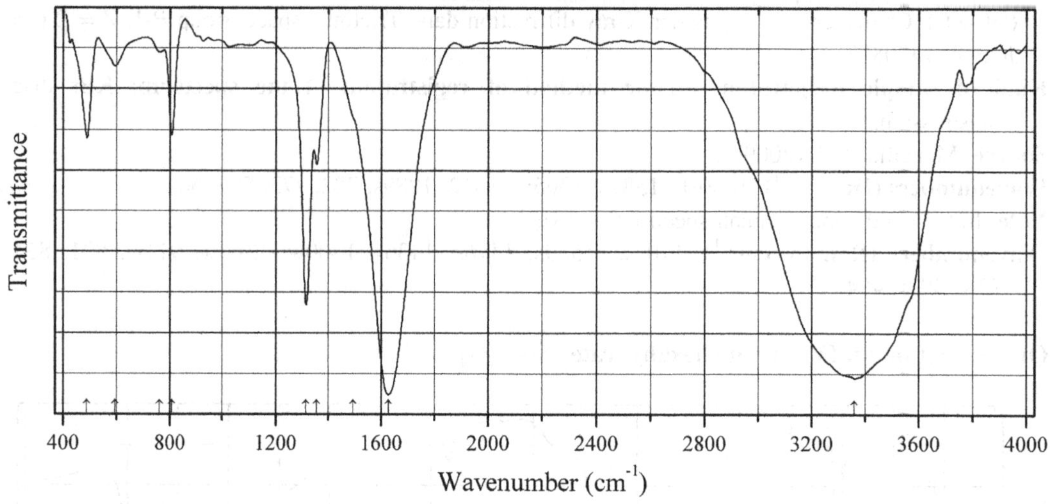

Origin: Synthetic.

Description: Single crystals grown using diffusion gel technique from samarium nitrate hexahydrate and oxalic acid dihydrate in the presence of sodium silicate (meta)nonahydrate. Characterized by powder X-ray diffraction data and thermal analysis. Monoclinic, space group $P2_1/c$.

Kind of sample preparation and/or method of registration of the spectrum: Transmission. Kind of sample preparation is not indicated.

Source: Vimal et al. (2014).

Wavenumbers (cm^{-1}): 3360s, 1628s, 1495sh, 1357, 1317s, 807, 761w, 595, 490.

Note: The wavenumbers were partly determined by us based on spectral curve analysis of the published spectrum.

Org77 Uranium(IV) oxalate fluoride hydrate $U_2(C_2O_4)F_6{\cdot}2H_2O$

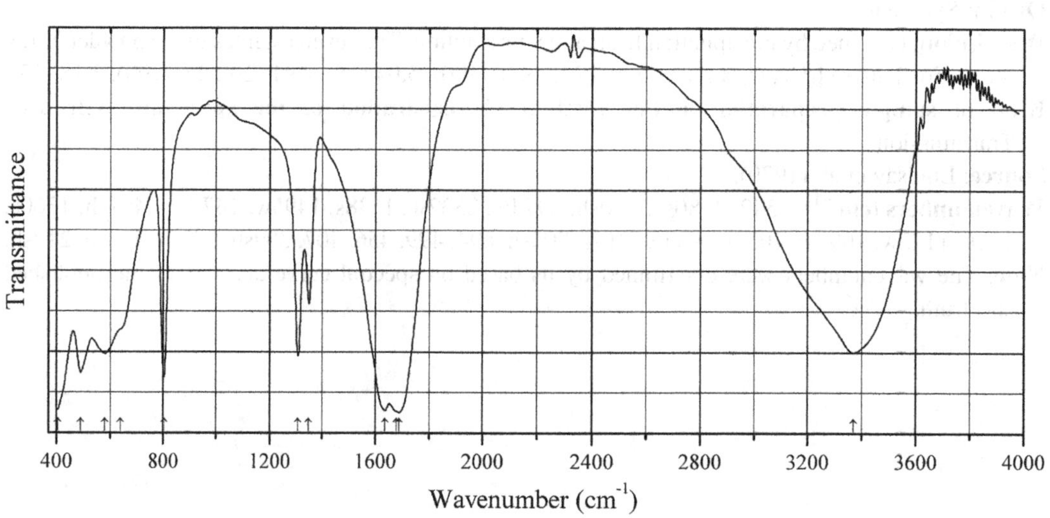

Origin: Synthetic.

Description: Prepared hydrothermally from UO_2, HF, and oxalic acid dihydrate at 120 °C for 3 days. Characterized by powder X-ray diffraction data. The crystal structure is solved. Monoclinic, space group *C2/c*, $a = 17.246(3)$, $b = 6.088(1)$ Å, $c = 8.589(2)$ Å, $\beta = 95.43(3)°$, $Z = 8$.

Kind of sample preparation and/or method of registration of the spectrum: KBr disc. Transmission.

Source: Wang et al. (2006b).

Wavenumbers (cm^{-1}): 3370 (broad), 1691s, 1680sh, 1638s, 1350, 1310, 804s, 638sh, 582s, 491, 405s.

Note: The wavenumbers were partly determined by us based on spectral curve analysis of the published spectrum.

Org78 Zirconium basic oxalate $Zr(C_2O_4)(OH)_2{\cdot}0.5H_2O$

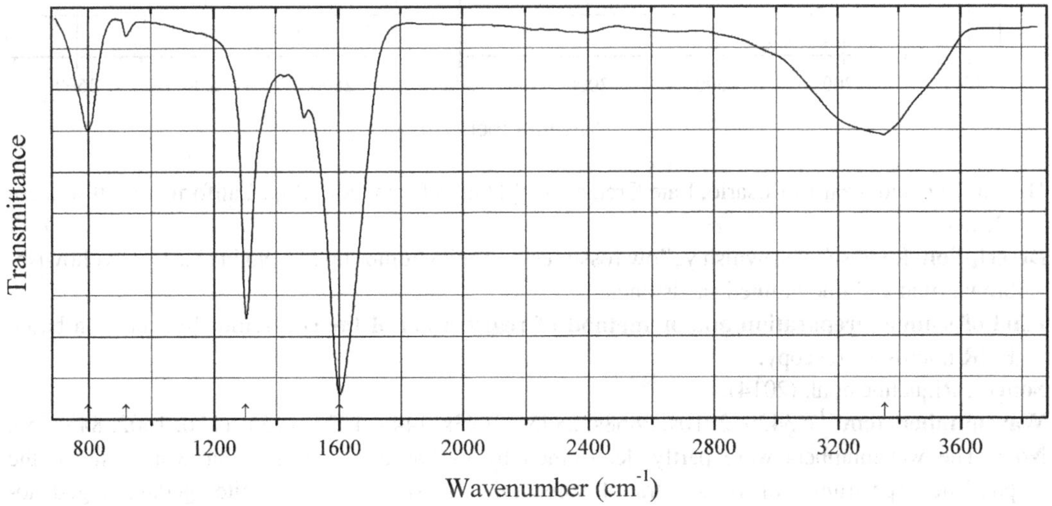

Origin: Synthetic.

Description: Obtained by interdiffusion of oxalic acid and $ZrO(NO_3)_2$ in silicate gel. Characterized by TG analysis. The crystal structure is solved. Tetragonal, space group *I4/m*, $a = 12.799(5)$, $c = 7.527(5)$ Å, $V = 1233.0(1)$ Å^3, $Z = 8$. $D_{calc} = 2.35$ g/cm^3.

Kind of sample preparation and/or method of registration of the spectrum: Transmission. Kind of sample preparation is not indicated.

Source: Hamdouni et al. (2013).

Wavenumbers (cm^{-1}): 3350, 1600s, 1300s, 920w, 798.

Org79 Bacalite

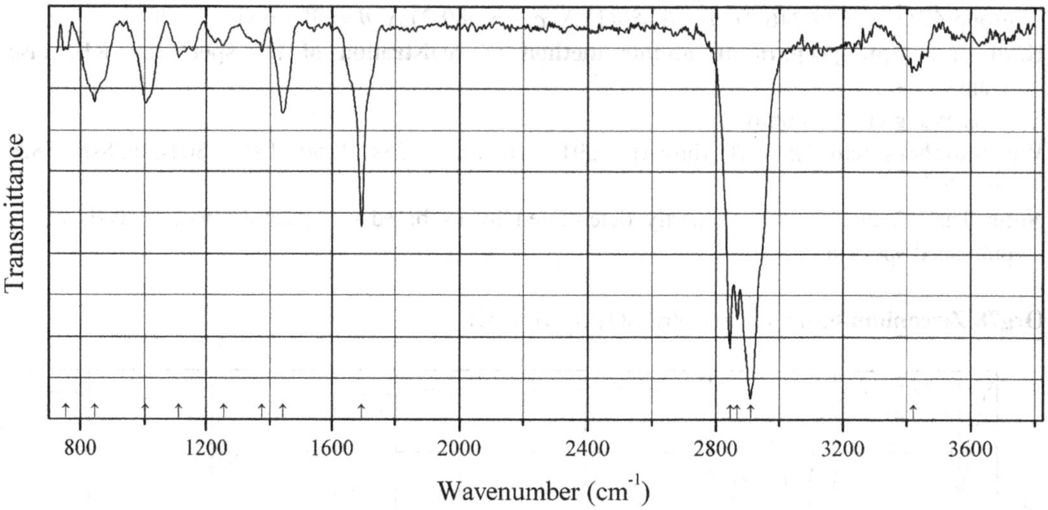

Origin: El Gallo, near El Rosario, Late Cretaceous El Gallo Formation, Baja California, northwestern Mexico.

Description: Reddish to brownish yellow fossil resin forming lumps up to 5 cm in size from yellowish-brown mud and fine-grained sandstones.

Kind of sample preparation and/or method of registration of the spectrum: Synchrotron-based FTIR microspectroscopy.

Source: Riquelme et al. (2014).

Wavenumbers (cm^{-1}): 3420, 2910s, 2868s, 2845s, 1693s, 1444, 1375, 1255, 1110, 1002, 845, 755.

Note: The wavenumbers were partly determined by us based on spectral curve analysis of the published spectrum. For IR spectra of some other fossil resins (succinite, gedanite, gedano-succinite, rumanite, and retinite) see Golubev and Martirosyan (2012).

Org80 Caoxite $Ca(C_2O_4){\cdot}3H_2O$

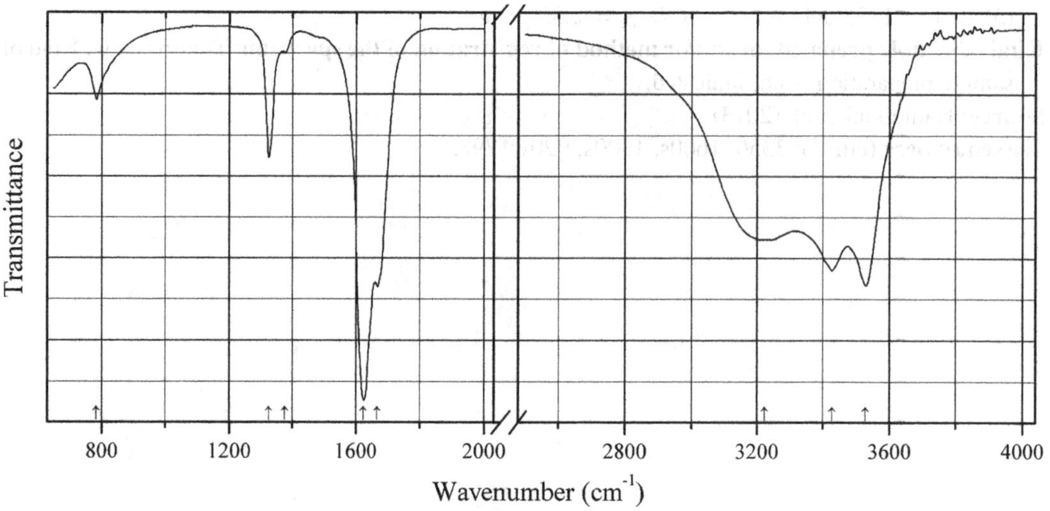

Origin: Synthetic.

Description: Obtained by the reaction between an aqueous solution of diethyl oxalate and calcite crystals. Characterized by powder X-ray diffraction data. Triclinic, space group *P*-1, $a = 6.1097$ (13), $b = 7.1642(10)$, $c = 8.4422(17)$ Å, $\alpha = 76.43(1)°$, $\beta = 70.19(2)°$, $\gamma = 70.91(2)°$, $V = 325.3$ (1) Å^3, $Z = 2$.

Kind of sample preparation and/or method of registration of the spectrum: A single crystal pressed in a diamond anvil cell. Transmission.

Source: Conti et al. (2015).

Wavenumbers (IR, cm^{-1}): 3528s, 3427s, 3222, 1668s, 1624s, 1377w, 1327, 783.

Note: The wavenumbers were partly determined by us based on spectral curve analysis of the published spectrum. In the cited paper, Raman spectrum is given. Raman bands at 2941 and 2882 cm^{-1} may correspond to a compound with C–H bonds. Raman shifts above 3000 cm^{-1} are not given in the cited paper.

Wavenumbers (Raman, cm^{-1}): 2941, 2882, 1472s, 912s, 507, 156.

Org81 Caoxite $Ca(C_2O_4)\cdot 3H_2O$

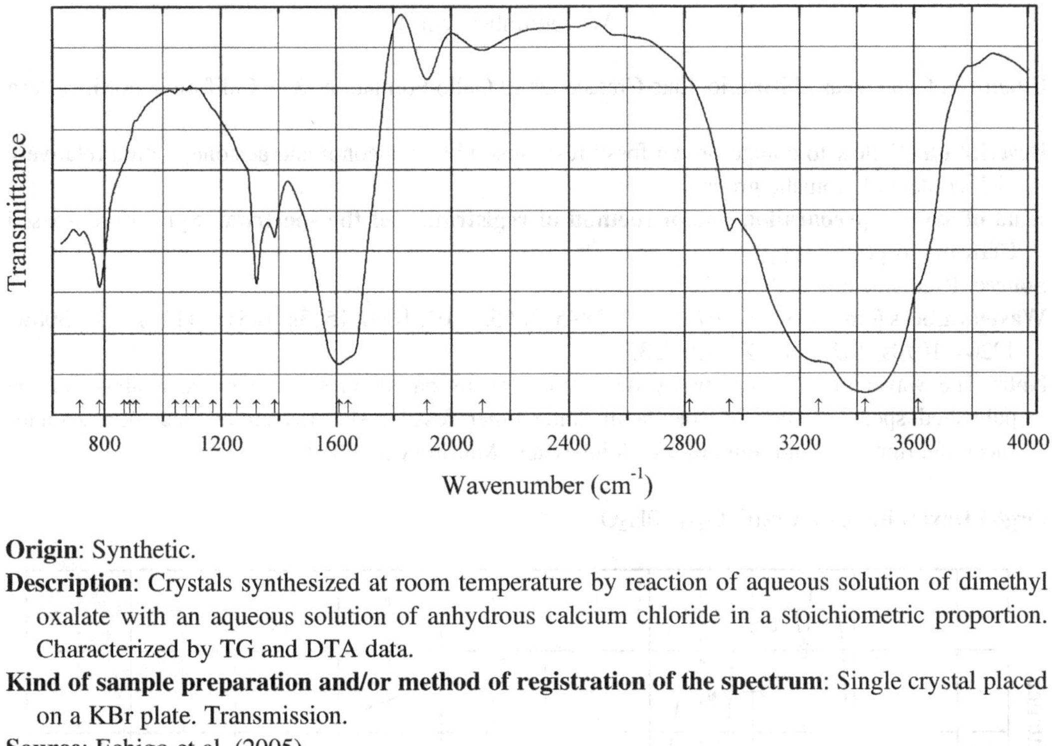

Origin: Synthetic.

Description: Crystals synthesized at room temperature by reaction of aqueous solution of dimethyl oxalate with an aqueous solution of anhydrous calcium chloride in a stoichiometric proportion. Characterized by TG and DTA data.

Kind of sample preparation and/or method of registration of the spectrum: Single crystal placed on a KBr plate. Transmission.

Source: Echigo et al. (2005).

Wavenumbers (cm^{-1}): 3615sh, 3429s, 3265sh, 2955, 2815sh, 2105, 1915, 1640sh, 1610s, 1386, 1323, 1254sh, 1173sh, 1114sh, 1080, 1041, 908sh, 887sh, 867sh, 783, 714.

Note: The wavenumbers were partly determined by us based on spectral curve analysis of the published spectrum. The bands in the range from 2800 to 3000 cm^{-1} correspond to the admixture of an organic substance with C–H bonds.

Org82 Coahuilite

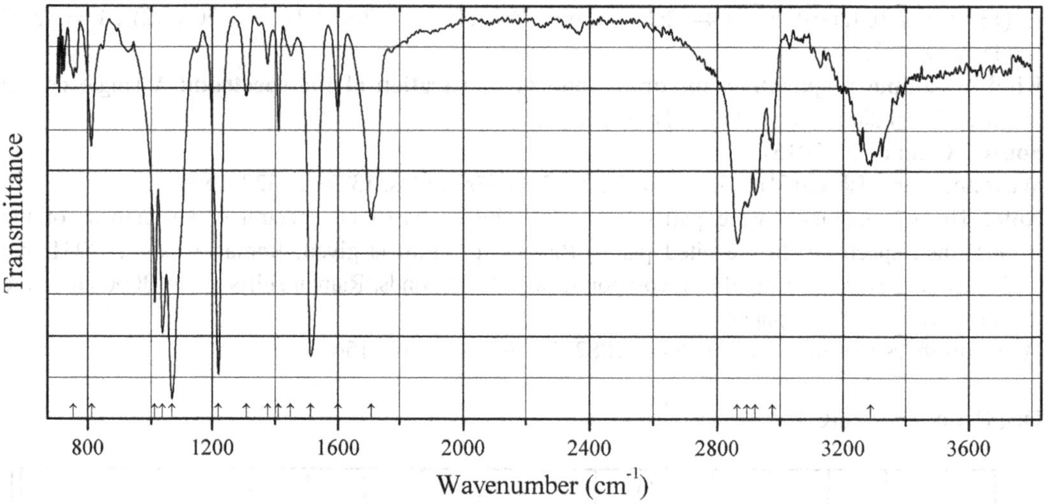

Origin: El Gallo, near El Rosario, Late Cretaceous El Gallo Formation, Baja California, northwestern Mexico.

Description: Yellow to orange-brown fossil resin insoluble in alcohol and acetone, with a relatively high content of aromatic groups.

Kind of sample preparation and/or method of registration of the spectrum: Synchrotron-based FTIR microspectroscopy.

Source: Riquelme et al. (2014).

Wavenumbers (cm^{-1}): 3287, 2975, 2921, 2895, 2865, 1707, 1600, 1513s, 1451, 1411, 1375, 1308w, 1220s, 1070s, 1038s, 1012, 812, 753.

Note: The wavenumbers were partly determined by us based on spectral curve analysis of the published spectrum. For IR spectra of some other fossil resins (succinite, gedanite, gedano-succinite, rumanite, and retinite) see Golubev and Martirosyan (2012).

Org83 Deveroite-(Ce) $Ce_2(C_2O_4)_3 \cdot 10H_2O$

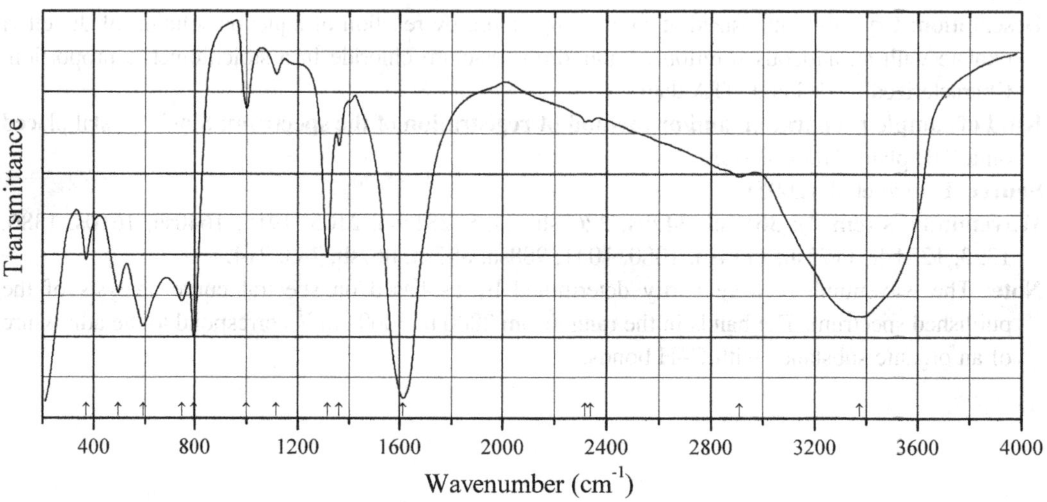

Origin: Synthetic.

Description: Synthesized by mixing aqueous solutions of stoichiometric amounts of $CeCl_3 \cdot 7H_2O$ and oxalic acid. Characterized by powder X-ray diffraction data, DTA–TG–DTG, and elemental analyses.

Kind of sample preparation and/or method of registration of the spectrum: KBr disc. Transmission.

Source: Gabal et al. (2012).

Wavenumbers (cm^{-1}): 3375, 2910w, 1615s, 1361, 1316, 1118w, 999, 797s, 749, 591s, 495, 371.

Note: The wavenumbers were partly determined by us based on spectral curve analysis of the published spectrum.

Org84 Refikite $C_{20}H_{32}O_2$

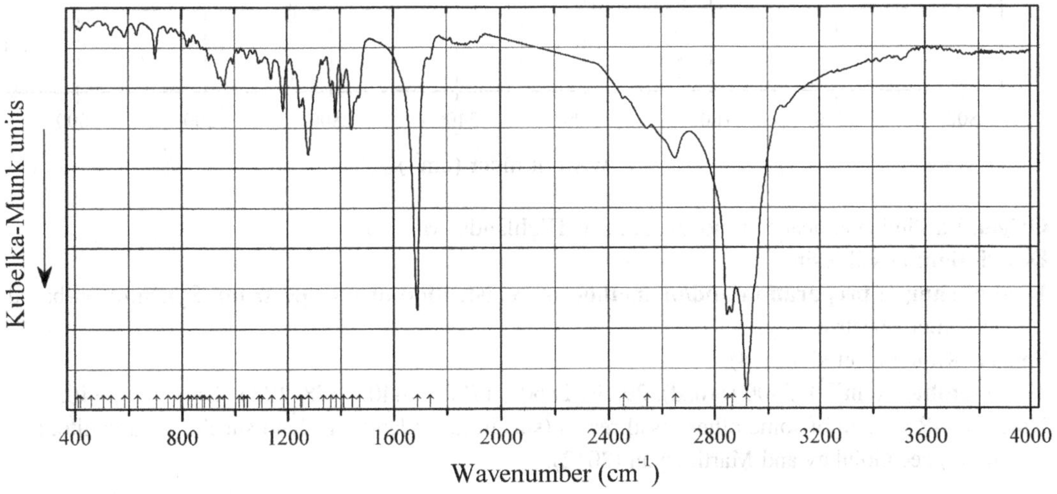

Origin: Krásno, Slavkovský Les Mts., western Bohemia, Czech Republic.

Description: Polycrystalline crusts on pinetree bark and wood. The crystal structure is solved. Orthorhombic, space group $P2_12_12$, a = 22.6520(7), b = 10.3328(3), c = 7.6711(2) Å, V = 1795.49(9) Å^3, Z = 4. D_{calc} = 1.1334 g/cm^3. The empirical formula is $C_{19}H_{33}COOH$.

Kind of sample preparation and/or method of registration of the spectrum: Attenuated total reflection.

Source: Pažout et al. (2015).

Wavenumbers (IR, cm^{-1}): 2919s, 2865s, 2847s, 2651, 2545, 2456, 1737w, 1689s, 1472, 1442, 1410, 1383, 1363, 1334, 1279, 1254, 1248, 1225w, 1185, 1139, 1101w, 1092w, 1046w, 1032w, 1015w, 960, 940, 904w, 885w, 875w, 856w, 836w, 823w, 794w, 770w, 747w, 703, 633w, 586w, 535w, 507w, 461w, 420w, 412w.

Note: The wavenumbers were partly determined by us based on spectral curve analysis of the published spectrum. In the cited paper, Raman spectrum is given.

Wavenumbers (Raman, cm^{-1}): 3013, 2952s, 2935s, 2890s, 2844s, 1474s, 1452s, 1383, 1362, 1249, 1202, 739, 725s.

Org85 Simojovelite

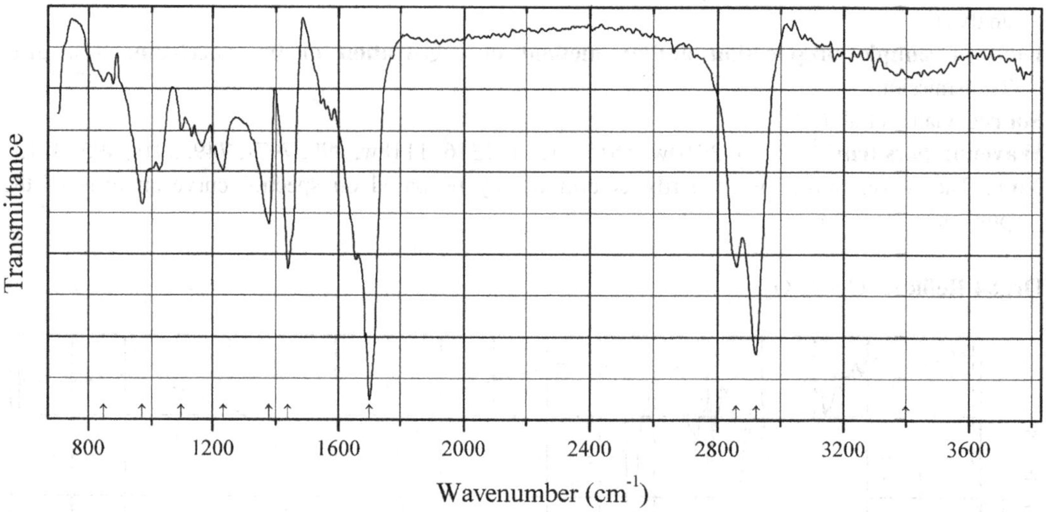

Origin: La Pimienta, near Simojovel, Chiapas Highlands, Mexico.

Description: Fossil resin.

Kind of sample preparation and/or method of registration of the spectrum: Synchrotron-based FTIR microspectroscopy.

Source: Riquelme et al. (2014).

Wavenumbers (cm^{-1}): 3400 (broad), 2923s, 2860s, 1700s, 1440, 1378, 1235, 1098, 970, 846.

Note: For IR spectra of some other fossil resins (succinite, gedanite, gedano-succinite, rumanite, and retinite) see Golubev and Martirosyan (2012).

Org86 Stepanovite $NaMgFe^{3+}(C_2O_4)_3 \cdot 8–9H_2O$

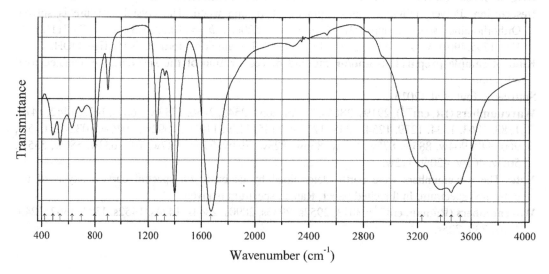

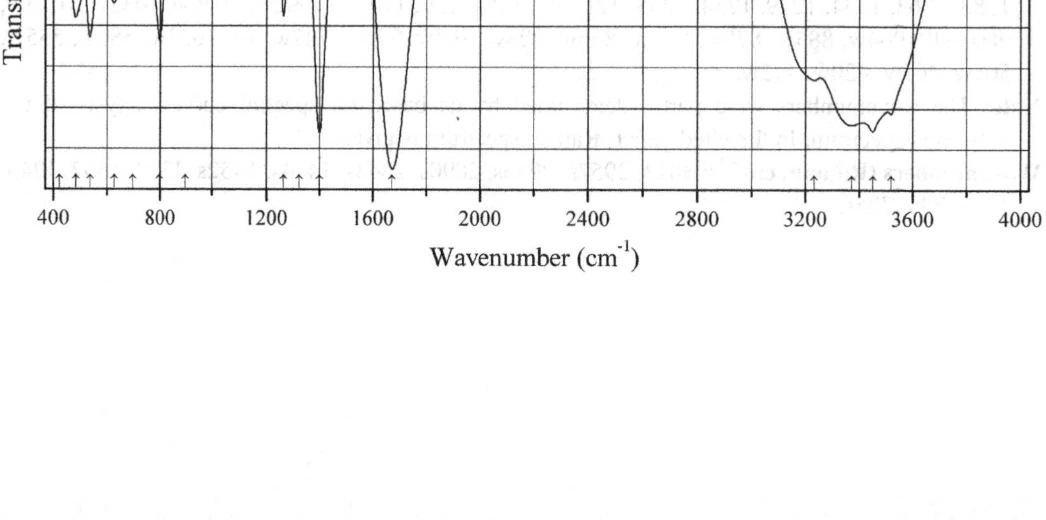

Origin: Synthetic.

Description: Obtained by the reaction of an aqueous solution of $Na_3[Fe(C_2O_4)_3]\cdot 5H_2O$ (prepared by reaction of a suspension of freshly precipitated $Fe(OH)_3$ with an aqueous solution of $NaHC_2O_4$) with a great excess of $MgCl_2$. The crystal structure is solved. Trigonal, space group *P*3*c*1, $a = 17.0483(4)$, $c = 12.4218(4)$ Å, $V = 3126.7(1)$ Å^3, $Z = 6$. $D_{calc} = 1.687$ g/cm^3.

Kind of sample preparation and/or method of registration of the spectrum: KBr disc. Transmission.

Source: Piro et al. (2016).

Wavenumbers (IR, cm^{-1}): 3518s, 3450s, 3371s, 3233, 1674s, 1400s, 1324w, 1265, 900, 798, 696, 627, 536, 483, 423.

Note: In the cited paper, Raman spectrum is given.

Wavenumbers (Raman, cm^{-1}): 3450w, 3350w, 3275w, 1728s, 1666, 1478, 1460, 1523, 1398, 1267, 903, 599, 537, 480s.

Org87 Zhemchuzhnikovite $NaMgAl(C_2O_4)_3\cdot 8H_2O$

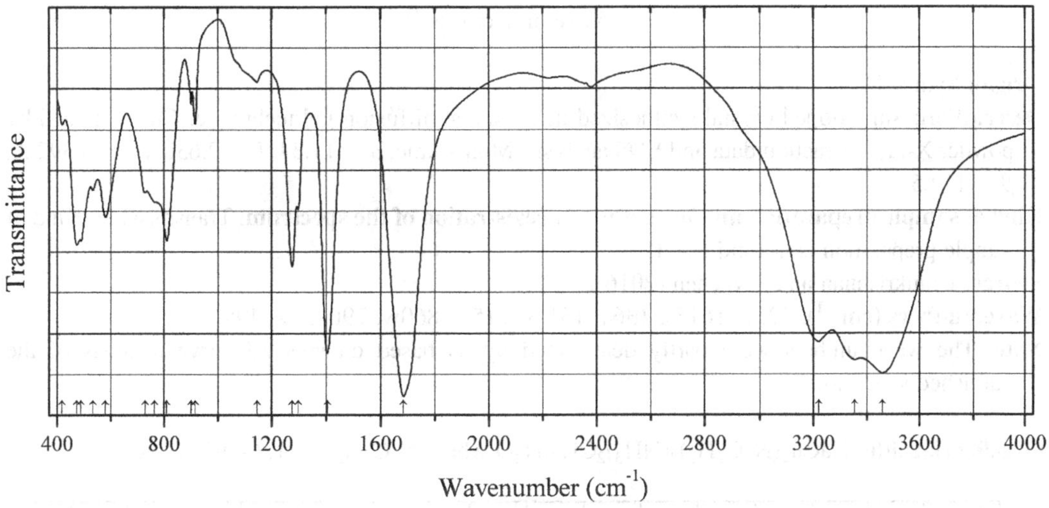

Origin: Synthetic.

Description: Fe-rich variety, $NaMg(Al_{0.55}Fe^{3+}_{0.45})(C_2O_4)_3\cdot nH_2O$, obtained by the mixing aqueous solutions of $NaMg[Fe(C_2O_4)_3]\cdot 9H_2O$ and $NaMg[Al(C_2O_4)_3]\cdot 9H_2O$. The crystal structure is solved. Trigonal, space group *P*3*c*1, $a = 16.8852(5)$, $c = 12.5368(5)$ Å, $V = 3095.5(2)$ Å^3, $Z = 6$. $D_{calc} = 1.652$ g/cm^3.

Kind of sample preparation and/or method of registration of the spectrum: KBr disc. Transmission.

Source: Piro et al. (2016).

Wavenumbers (IR, cm^{-1}): 3458s, 3355s, 3219s, 1684s, 1404s, 1298, 1276, 1146w, 916, 901, 811, 765sh, 729, 581, 534, 490, 475, 419w.

Note: In the cited paper, Raman spectrum is given.

Wavenumbers (Raman, cm^{-1}): 3467, 3222, 1788s, 1688, 1520w, 1479, 1440s, 1266, 991, 923, 856, 565, 533, 479s.

Org88 Deveroite-(Ce) $Ce_2(C_2O_4)_3 \cdot 10H_2O$

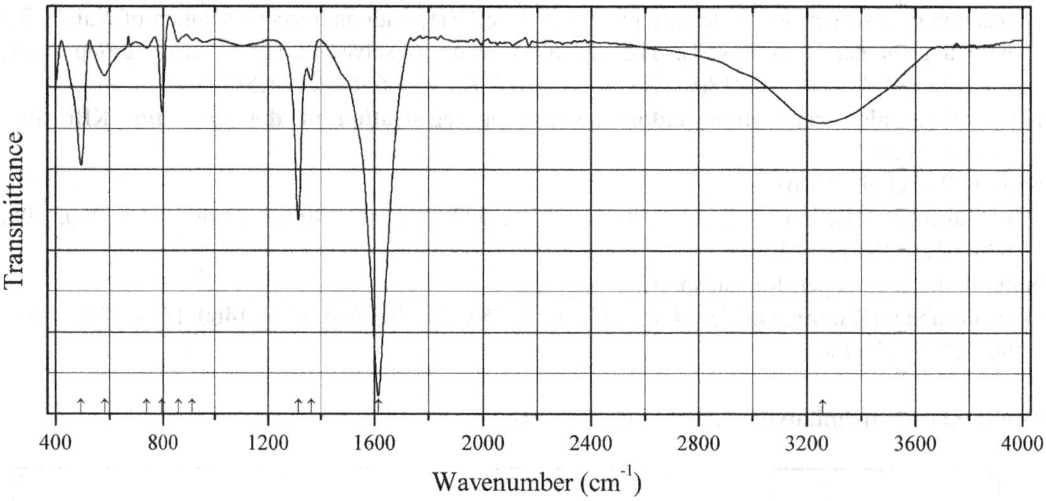

Origin: Synthetic.

Description: Sm^{3+} doped crystal synthesized using single diffusion gel technique. Characterized by powder X-ray diffraction data and EDS analysis. Monoclinic, $a = 11.34$, $b = 9.630$, $c = 10.392$ Å, $\beta = 114.5°$.

Kind of sample preparation and/or method of registration of the spectrum: Transmission. Kind of sample preparation is not indicated.

Source: Unnikrishnan and Ittyachen (2016).

Wavenumbers (cm^{-1}): 3257, 1615s, 1364, 1316s, 915w, 860w, 796, 582, 495.

Note: The wavenumbers were partly determined by us based on spectral curve analysis of the published spectrum.

Org89 Triazolite $NaCu_2(N_3C_2H_2)_2(NH_3)_2Cl_3 \cdot 4H_2O$ where $N_3C_2H_2^-$ is 1,2,4-triazolate anion

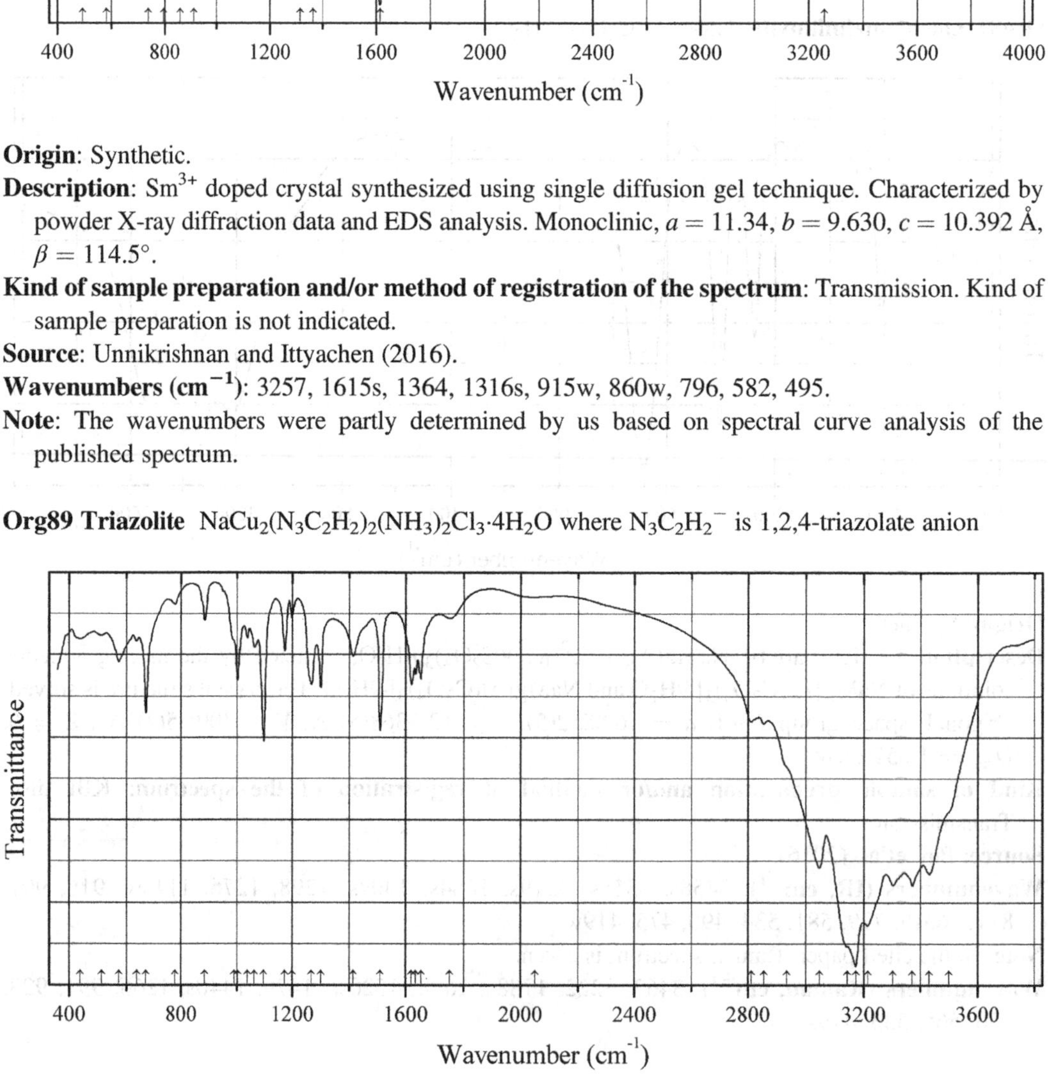

Origin: Pabellón de Pica Mountain, 1.5 km south of Chanabaya village, Iquique Province, Tarapacá Region, Chile (type locality).

Description: Clusters and radiated aggregates of prismatic crystals from the association with salammoniac, halite, ammineite, joanneumite, chanabayaite, nitratine, natroxalate, and möhnite. Holotype sample. The crystal structure is solved. Orthorhombic, space group $P2_12_12_1$, $a = 19.3575(5)$, $b = 7.15718(19)$, $c = 12.5020(4)$ Å, $V = 1732.09(8)$ Å^3, $Z = 4$. $D_{calc} = 2.028$ g/cm^3. Optically biaxial (–), $\alpha = 1.582(4)$, $\beta = 1.625(3)$, $\gamma = 1.625(3)$, $2V = 5(3)°$. The empirical formula is $Na_{1.14}(Cu_{1.86}Fe_{0.14})(Cl_{2.99}S_{0.23})N_{9.23}C_{3.43}H_{23.34}O_{4.29}$. The strongest lines of the powder X-ray diffraction pattern [*d*, Å (*I*, %) (*hkl*)] are: 10.22 (97) (101), 6.135 (40) (011), 5.696 (17) (301), 5.182 (59) (202), 5.119 (100) (211), 4.854 (19) (400), 3.752 (16) (312, 501), 3.294 (18) (221), 2.644 (17) (404), 2.202 (18) (324, 713).

Kind of sample preparation and/or method of registration of the spectrum: KBr disc. Absorption.

Wavenumbers (cm^{-1}): 3500sh, 3431s, 3371s, 3302s, 3215s, 3173s, 3145sh, 3047s, 2935sh, 2855, 2812, 2050w, 1751, 1653, 1635, 1621, 1510, 1414, 1300, 1267, 1198w, 1172, 1095, 1062, 1036w, 1002, 990sh, 887, 778w, 669, 638w, 575, 513w, 440w.

Note: The spectrum was obtained by N.V. Chukanov.

2.4 Nitrides and Nitrates

N20 Hexaamminenickel(II) nitrate $[Ni(NH_3)_6](NO_3)_2$

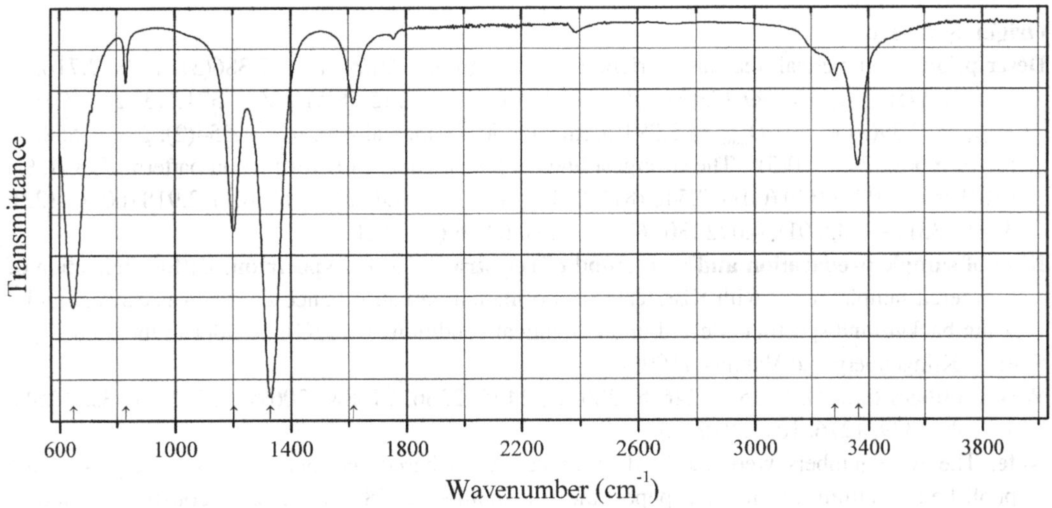

Origin: Synthetic.

Description: Prepared in the reaction between nickel nitrate hexahydrate and gaseous ammonia in the presence of silica gel. The crystal structure is solved. Cubic, space group *Fm*-3*m*, $a = 10.8738(6)$ Å, $V = 1285.73(7)$ Å^3, $Z = 4$. $D_{calc} = 1.471(1)$ g/cm^3.

Kind of sample preparation and/or method of registration of the spectrum: Attenuated total reflection of a powdered sample.

Source: Breternitz et al. (2015).

Wavenumbers (cm^{-1}): 3364, 3282w, 1616w, 1329, 1202s, 832, 648s.

Note: In the cited paper, the wavenumber 832 cm^{-1} is erroneously indicated as 823 cm^{-1}.

N21 Lantanum nitrate hexahydrate $La(NO_3)_3{\cdot}6H_2O$

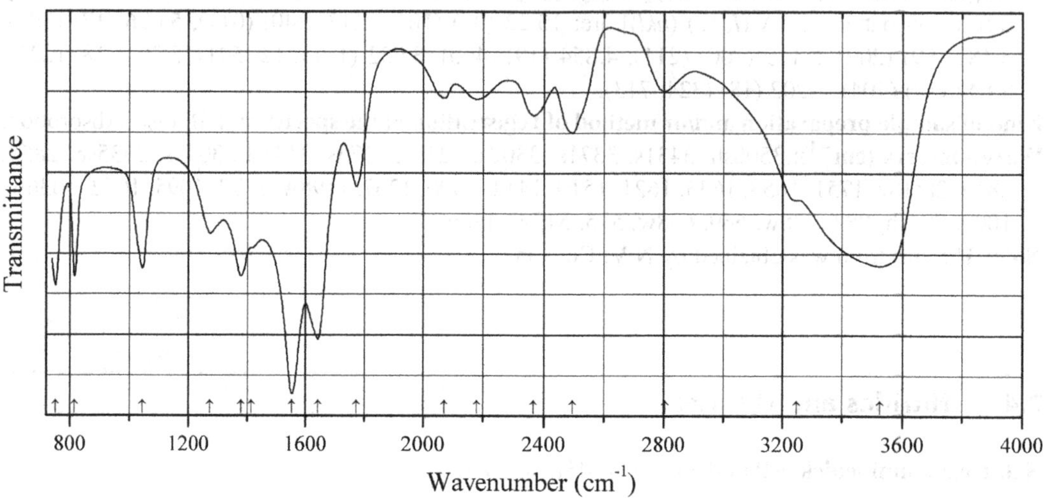

Origin: Synthetic.

Description: Commercial reactant. Triclinic, space group $P2_1/c$, a = 7.386(3), b = 7.716(3), c = 11.345(4) Å, α = 99.773(5)°, β = 91.141(6)°, γ = 115.58(5)°, V = 571.6(3) Å^3, Z = 2. D_{meas} = 2.39(3) g/cm^3, D_{calc} = 2.391 g/cm^3. Optically biaxial (–), α = 1.554(2), β = 1.558(2), γ = 1.566(2), 2V = 70(5)°. The strongest lines of the powder X-ray diffraction pattern [d, Å (I, %) (hkl)] are: 11.089 (100) (001), 3.540 (81) (0–13, –1–12), 5.484 (79) (002, 101), 2.918 (60) (–122), 3.089 (33) (–113, 201), 4.022 (30) (102, –112), 6.826 (23) (010).

Kind of sample preparation and/or method of registration of the spectrum: Diffuse reflection of powdered sample mixed with KBr. The transformation into absorbance spectra was carried out by using background spectra collected under identical conditions with KBr powder in the holder.

Source: Klingenberg and Vannice (1996).

Wavenumbers (cm^{-1}): 3525s, 3246sh, 2804w, 2495, 2366, 2108w, 2069w, 1772, 1643s, 1554s, 1415sh, 1379, 1276, 1042, 815, 752.

Note: The wavenumbers were partly determined by us based on spectral curve analysis of the published spectrum. In the cited paper, the wavenumber 2178 cm^{-1} is erroneously indicated as 2108 cm^{-1}.

N22 Uranyl nitrate hexahydrate $UO_2(NO_3)_2{\cdot}6H_2O$

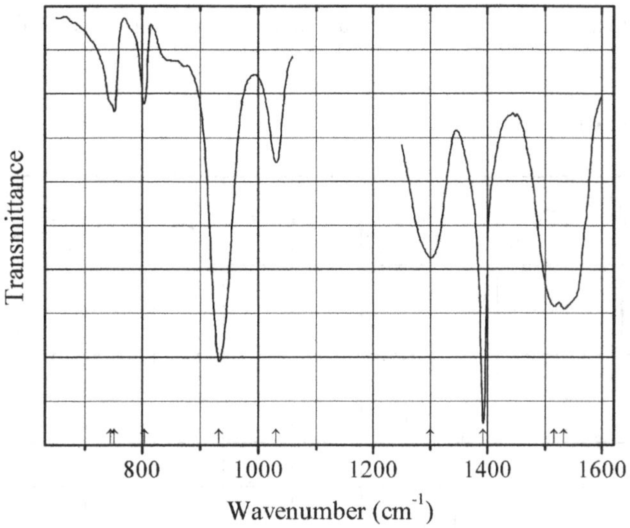

Origin: Synthetic.
Description: Commercial reactant.
Kind of sample preparation and/or method of registration of the spectrum: A mixture with KBr. Transmission.
Source: Caldow et al. (1960).
Wavenumbers (cm^{-1}): 1531s, 1515s, 1392s, 1300, 1032, 933s, 804, 752, 745sh.
Note: The band at 1392 cm^{-1} may correspond to KNO_3 formed in the reaction between $UO_2(NO_3)_2{\cdot}6H_2O$ and KBr. Consequently, the presence of uranyl bromide in the sample is not excluded.

N23 Nierite β-Si_3N_4

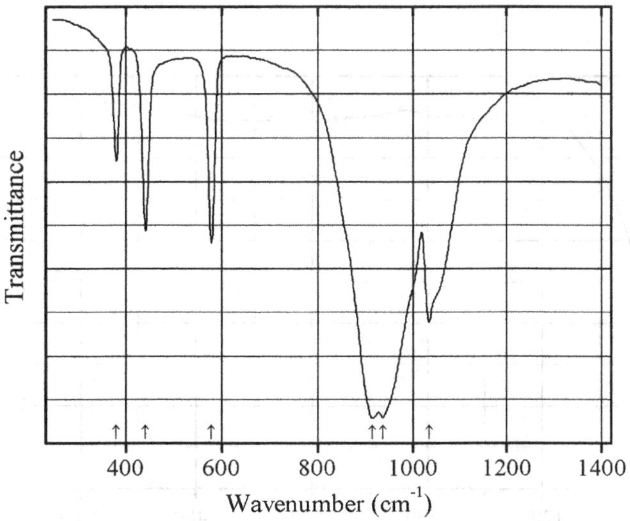

Origin: Synthetic.
Description: Prepared by reacting silicon powder with nitrogen at 1350 °C for 2 h followed by heating at 1500 °C for 16 h. Hexagonal, space group $P6_3/m$, $Z = 2$.
Kind of sample preparation and/or method of registration of the spectrum: CsI disc. Transmission.

Source: Wild et al. (1978).
Wavenumbers (cm^{-1}): 1035, 938s, 915s, 579, 441, 380.
Note: The wavenumbers were determined by us based on spectral curve analysis of the published spectrum.

N24 Nitratine Na(NO$_3$)

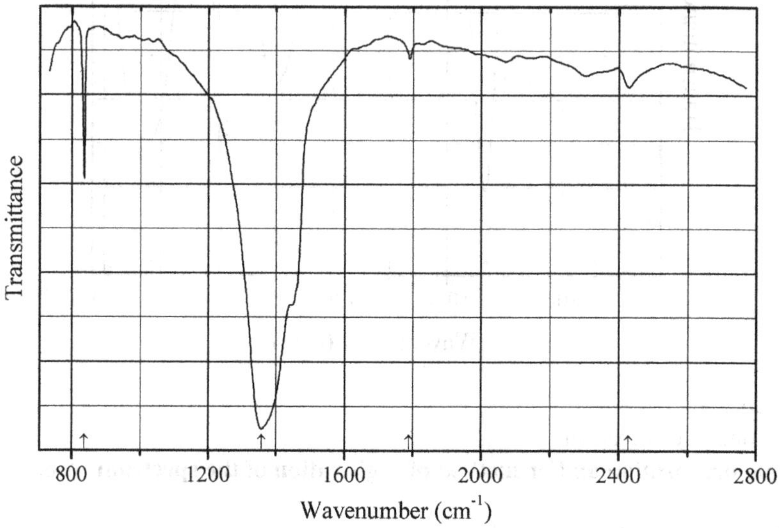

Origin: Synthetic.
Kind of sample preparation and/or method of registration of the spectrum: Nujol mull. Transmission.
Source: Miller and Wilkins (1952).
Wavenumbers (cm^{-1}): 2428w, 1790w, 1358s, 836.
Note: A shoulder near 1447 cm^{-1} corresponds to Nujol.

N25 Osbornite TiN

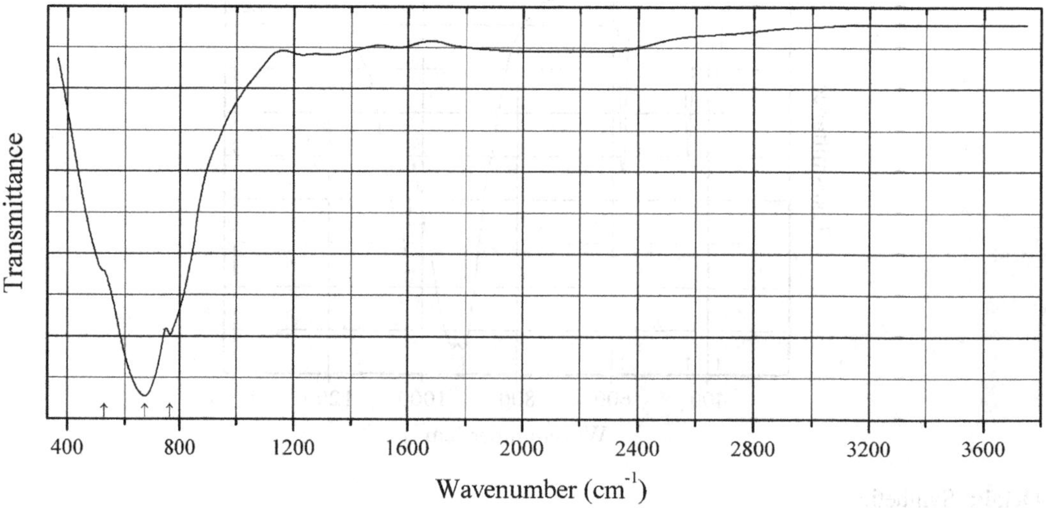

Origin: Synthetic.
Description: A layer deposited using sequential additions of $TiCl_4$ and NH_3 on fumed silica powder.
Kind of sample preparation and/or method of registration of the spectrum: Diffuse reflection of a mixture with KBr powder.
Source: Snyder et al. (2006).
Wavenumbers (cm^{-1}): 765, 670s, 530sh.

N26 Qingsongite BN

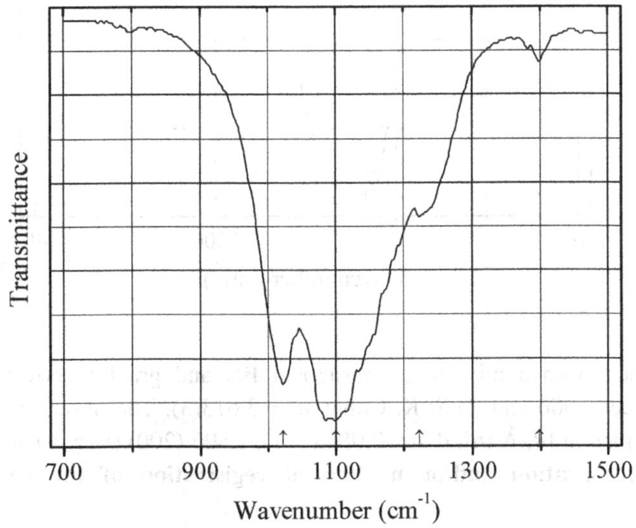

Origin: Synthetic.
Description: Produced from hexagonal BN by spontaneous high pressure (5.5 GPa) and high temperature (1800–1900 K) nucleation using Mg as a solvent-catalyst.
Kind of sample preparation and/or method of registration of the spectrum: KBr disc. Transmission.
Source: Kutsay et al. (2010).
Wavenumbers (IR, cm^{-1}): 1398, 1223, 1100, 1022.
Note: The wavenumbers were determined by us based on spectral curve analysis of the published spectrum. The weak band at 1398 cm^{-1} corresponds to the admixture of hexagonal BN. In the cited paper, a figure of qingsongite Raman spectrum is given.

N27 Qingsongite (C-bearing) $C_{0.3}(BN)_{0.7}$

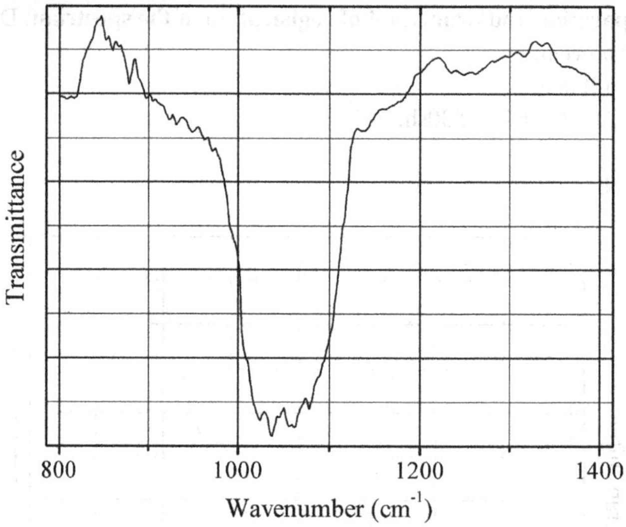

Origin: Synthetic.

Description: Obtained from a mixture of hexagonal BN and graphite powders at 30 GPa and temperature between 2000 and 2500 K. Cubic, $a = 3.613(3)$. The observed lines of the powder X-ray diffraction pattern [*d*, Å (*hkl*)] are: 2.086 (111), 1.806 (200) (very weak), 1.276 (220).

Kind of sample preparation and/or method of registration of the spectrum: KBr disc. Transmission.

Source: Knittle et al. (1995).

Wavenumbers (IR, cm^{-1}): ~1045.

Note: In the cited paper, Raman spectrum is given.

Wavenumbers (Raman, cm^{-1}): 1323.

N28 Sinoite Si_2N_2O

Origin: Synthetic.

Description: Orthorhombic, space group *Cmc*2_1.

Kind of sample preparation and/or method of registration of the spectrum: No data.

Source: Mirgorodsky et al. (1989).

Wavenumbers (cm^{-1}): 1130, 1070sh, 1030sh, 990, 953s, 906s, 730sh, 679, 648w, 542, 496, 448, 327, 252.

2.5 Oxides and Hydroxides

O495 Ferricoronadite $Pb(Mn^{4+}{}_6Fe^{3+}{}_2)O_{16}$

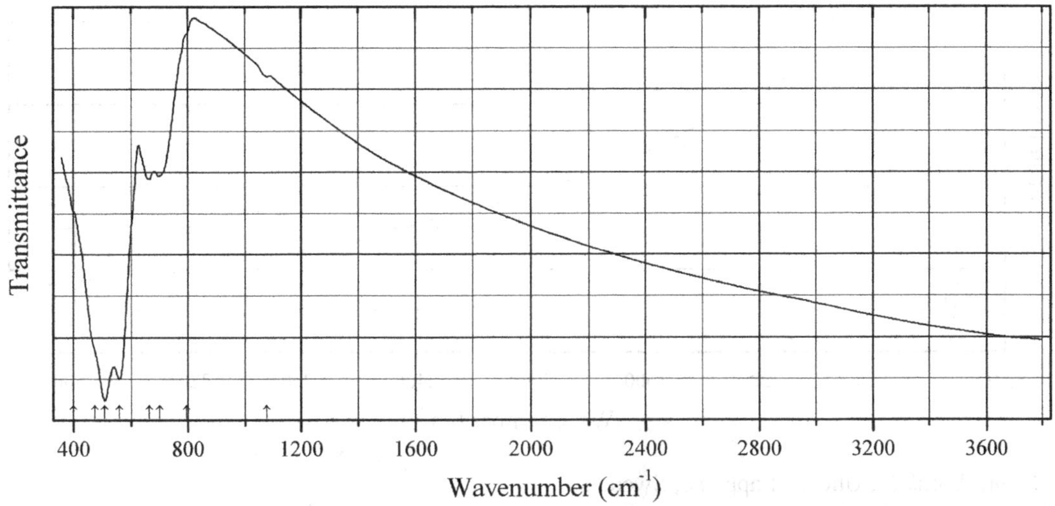

Origin: "Mixed Series" metamorphic complex near the Nežilovo village, Pelagonian massif, Republic of Macedonia (type locality).

Description: Veinlets in granular aggregate consisting of franklinite, gahnite, hetaerolite, roméite, almeidaite, Mn-analogue of plumboferrite, högbomite-group minerals, Zn-bearing talc, baryte, quartz, etc. Holotype sample. The crystal structure is solved. Tetragonal, space group *I*4/*m*, $a = 9.9043(7)$, $c = 2.8986(9)$ Å, $V = 284.34(9)$ Å^3, $Z = 1$. $D_{calc} = 5.538$ g/cm^3. The empirical formula is (electron microprobe): $Pb_{1.03}Ba_{0.32}(Mn^{4+}{}_{4.85}Fe^{3+}{}_{1.35}Mn^{3+}{}_{1.18}Ti_{0.49}Al_{0.09}Zn_{0.04})O_{16}$. According to the Mössbauer spectrum, all iron is trivalent. The Mn *K*-edge XANES spectroscopy shows that Mn is predominantly tetravalent, with subordinate Mn^{3+}. The strongest lines of the powder X-ray diffraction pattern [*d*, Å (*I*, %) (*hkl*)] are: 3.497 (33) (220), 3.128 (100) (−130, 130), 2.424 (27) (−121, 121), 2.214 (23) (240, –240), 2.178 (17) (031), 1.850 (15) (141, –141), 1.651 (16) (060), 1.554 (18) (−251, 251).

Kind of sample preparation and/or method of registration of the spectrum: KBr disc. Absorption.

Wavenumbers (cm^{-1}): 1078w, 795sh, 700, 665, 560s, 510s, 475sh, 400sh.

Note: The spectrum was obtained by N.V. Chukanov.

0496 Ferrihollandite $Ba(Mn^{4+}_6Fe^{3+}_2)O_{16}$

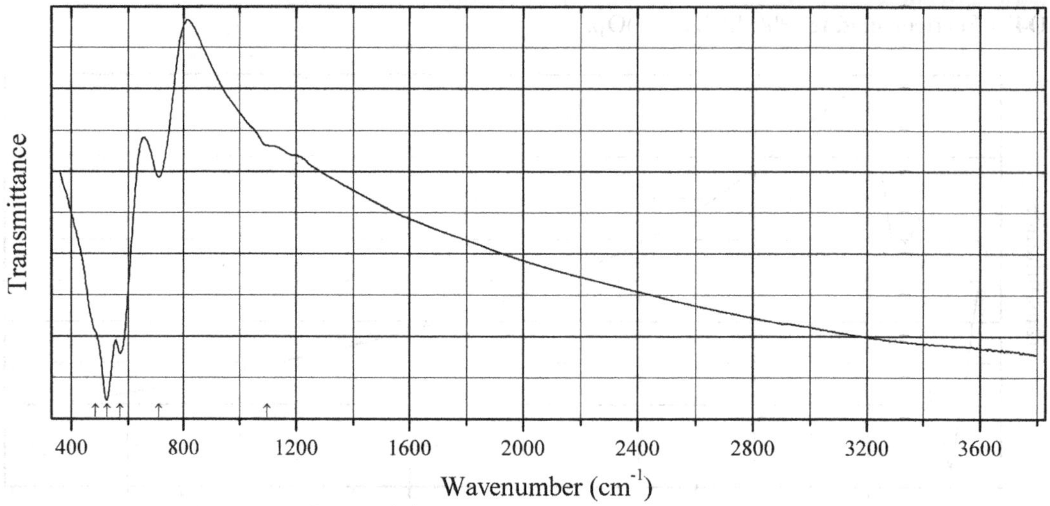

Origin: Sörhårås, Ultevis, Lappland, Sweden.

Description: Black prismatic crystals. The empirical formula is (electron microprobe): $(Ba_{0.79}K_{0.17}Pb_{0.11}Sr_{0.07}Na_{0.07})(Mn_{6.22}Fe_{0.93}Al_{0.43}Ti_{0.32}Mg_{0.10}Zn_{0.02})O_8$.

Kind of sample preparation and/or method of registration of the spectrum: KBr disc. Absorption.

Wavenumbers (cm^{-1}): 1095w, 708, 572s, 525s, 485sh.

Note: The spectrum was obtained by N.V. Chukanov.

0497 Cesàrolite $PbMn^{4+}_3O_6(OH)_2$

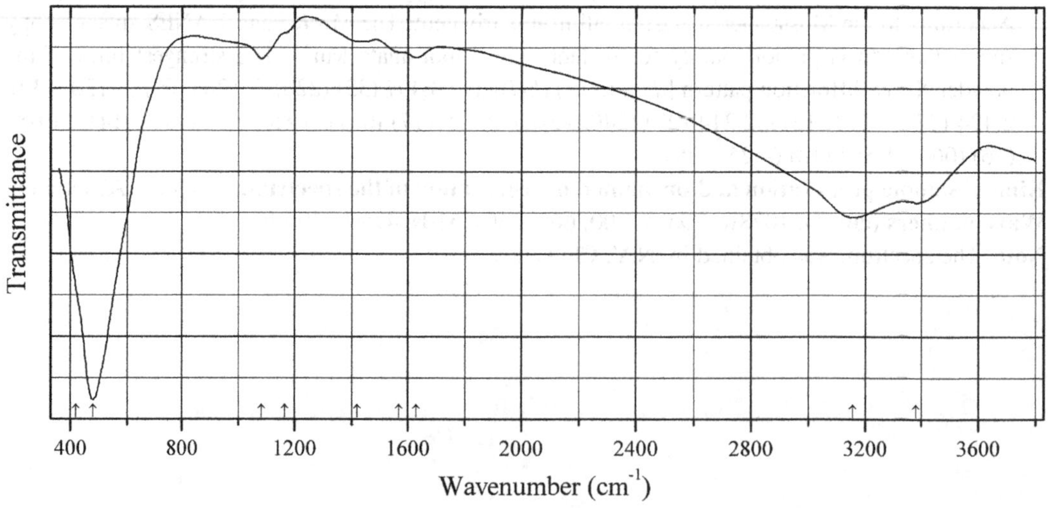

Origin: Belorechenskoe deposit, Adygea (Adygeya) Republic, Northern Caucasus, Russia.

Description: Black massive, with brown streak, from the association with baryte, dolomite, fluorite, galena, and gypsum. Investigated by A.V. Kasatkin. The empirical formula is (electron microprobe): $Pb_{0.75}Cu_{0.2}Zn_{0.1}Mn_{3.0}(O,OH)_8$. The strongest lines of the powder X-ray diffraction pattern are observed at 3.42, 3.13, 2.39, 2.21, 2.11, 1.88, 1.77, 1.69, 1.57, 1.48, and 1.41 Å.

Kind of sample preparation and/or method of registration of the spectrum: KBr disc. Absorption.

Wavenumbers (cm^{-1}): 3381, 3158s, 1630, 1570, 1420w, 1165sh, 1080, 480, (420sh).

Note: The band at 1630 cm^{-1} may correspond to adsorbed water; the weak band at 1420 cm^{-1} may be due to dolomite impurity.

Note: The spectrum was obtained by N.V. Chukanov.

O498 Sodalite Ca-Al-Mo-W-analogue $Ca_8(Al_{12}O_{24})[(MoO_4)_{1.5}(WO_4)_{0.5}]$

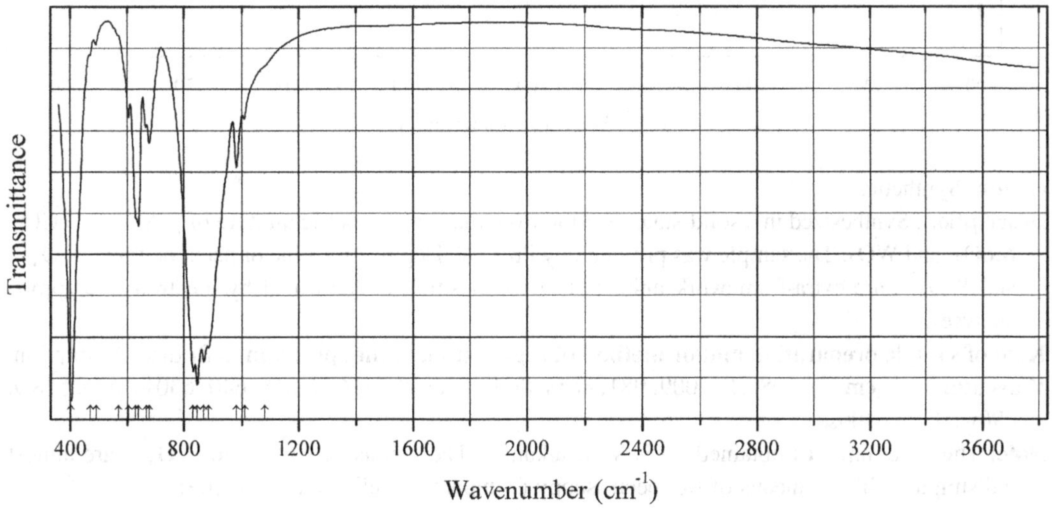

Origin: Synthetic.

Description: Synthesized in a solid-state reaction from the stoichiometric mixture of γ-Al_2O_3, $CaCO_3$, MoO_3, and WO_3. The sample was provided by Prof. W. Depmeier. Cubic or pseudocubic. MoO_4^{2-} and WO_4^{2-} are extra-framework anions. The composition is confirmed by electron microprobe analyses.

Kind of sample preparation and/or method of registration of the spectrum: KBr disc. Absorption.

Wavenumbers (cm^{-1}): 1080sh, 1009, 982, 885s, 868s, 846s, 832s, 679, 668, 640, 630sh, 606, 570sh, 491w, 470w, 403s.

Note: The spectrum was obtained by N.V. Chukanov. The anions MoO_4^{2-} and WO_4^{2-} are almost indistinguishable by means of IR spectroscopy (compare powellite and scheelite).

0499 Sodalite Ca-Al-Mo-W-analogue $Ca_8(Al_{12}O_{24})[(MoO_4)(WO_4)]$

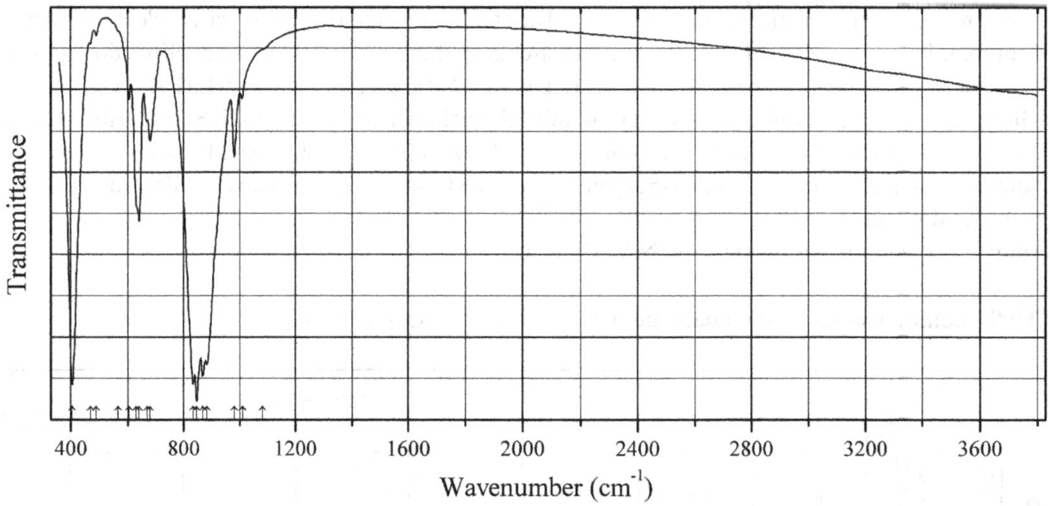

Origin: Synthetic.

Description: Synthesized in a solid-state reaction from the stoichiometric mixture of γ-Al_2O_3, $CaCO_3$, MoO_3, and WO_3. The sample was provided by Prof. W. Depmeier. Cubic or pseudocubic. MoO_4^{2-} and WO_4^{2-} are extra-framework anions. The composition is confirmed by electron microprobe analyses.

Kind of sample preparation and/or method of registration of the spectrum: KBr disc. Absorption.

Wavenumbers (cm^{-1}): 1080sh, 1009, 981, 885s, 869s, 846s, 833s, 679, 668, 640, 630sh, 606, 568w, 490w, 470w, 406s.

Note: The spectrum was obtained by N.V. Chukanov. The anions MoO_4^{2-} and WO_4^{2-} are almost indistinguishable by means of IR spectroscopy (compare powellite and scheelite).

0500 Sodalite Ca-Al-Mo-analogue $Ca_8(Al_{12}O_{24})[(MoO_4)(WO_4)]$

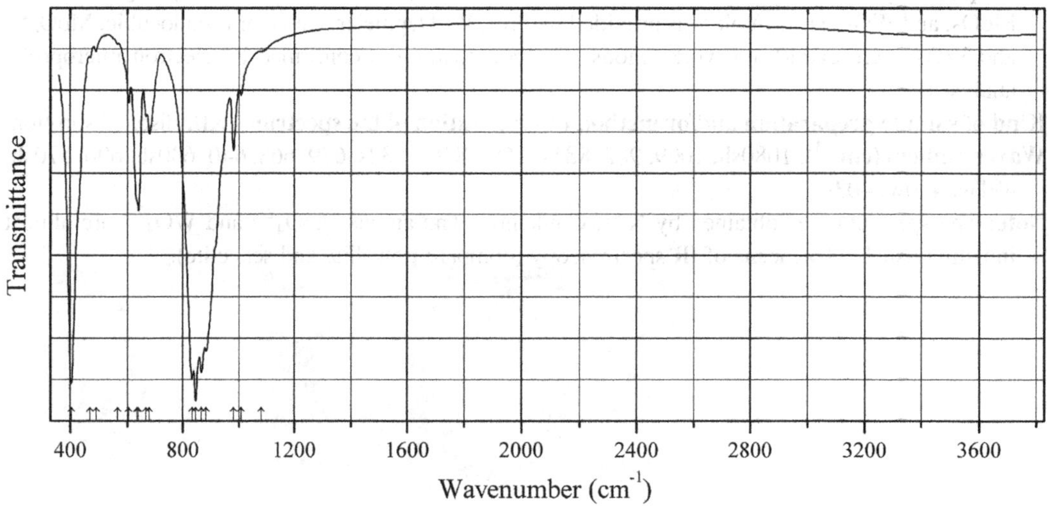

Origin: Synthetic.

Description: Synthesized in a solid-state reaction from the stoichiometric mixture of γ-Al_2O_3, $CaCO_3$, and MoO_3. The sample was provided by Prof. W. Depmeier. Cubic or pseudocubic. $MoO_4{}^{2-}$ is extra-framework anion. The composition is confirmed by electron microprobe analyses.

Kind of sample preparation and/or method of registration of the spectrum: KBr disc. Absorption.

Wavenumbers (cm^{-1}): 1080, 1008, 981, 885s, 868s, 846s, 833s, 679, 667, 641, 635sh, 606, 568w, 491w, 470w, 404s.

Note: The spectrum was obtained by N.V. Chukanov.

O501 Lesukite Cu-bearing variety $(Al,Cu)_2(OH)_{5-x}Cl{\cdot}nH_2O$ ($n \approx 2$)

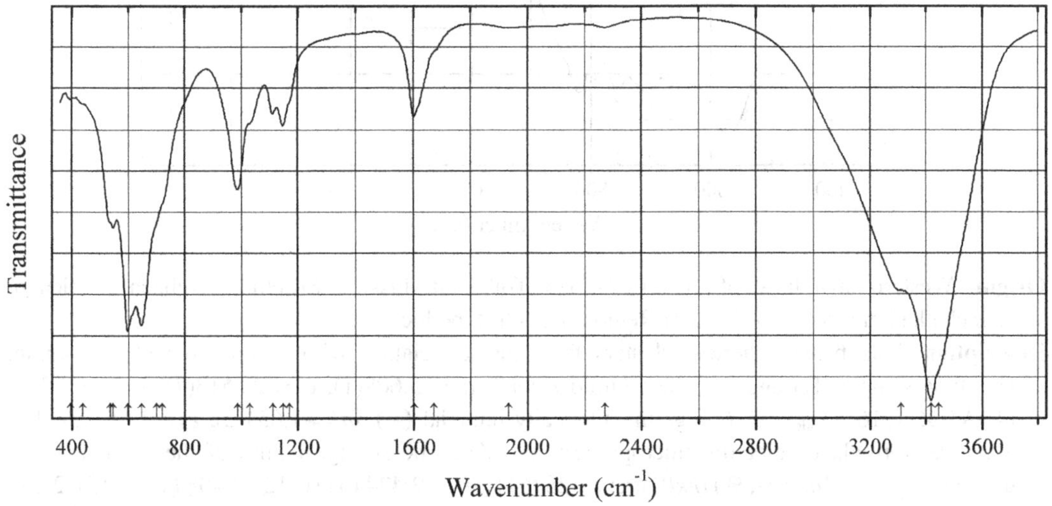

Origin: Cerro Mejillones, Mejillones Peninsula, Mejillones, Antofagasta, II Region, Chile.

Description: Lemon-yellow powdery aggregate consisting of microscopic cubic crystals from the association with gypsum, atacamite, and goethite. Investigated by I.V. Pekov. Characterized by powder X-ray diffraction data. The empirical formula is (electron microprobe): $(Al_{1.85}Cu_{0.15})(OH)_{4.85}Cl_{1.00}{\cdot}nH_2O$.

Kind of sample preparation and/or method of registration of the spectrum: KBr disc. Absorption.

Wavenumbers (cm^{-1}): 3445sh, 3418s, 3310s, 2271w, 1933w, 1670sh, 1602, 1170sh, 1148, 1111, 1030sh, 986, 720sh, 700sh, 647s, 598s, 546, 535sh, 440w, 395w.

Note: The spectrum was obtained by N.V. Chukanov.

O502 Deltalumite δ-Al_2O_3

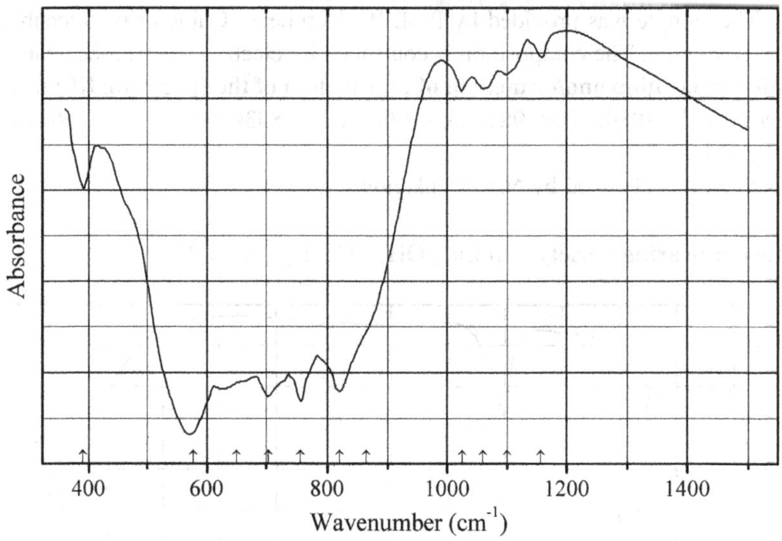

Origin: Western lava flow of the 2012–2013 Tolbachik Fissure Eruption, Tolbachik volcano, Kamchatka Peninsula, Far-Eastern Region, Russia (type locality).

Description: Pale beige spherical clusters from the association with corundum and moissanite. Holotype sample. Tetragonal, space group *P*-4*m*2, $a = 5.608(1)$, $c = 23.513(7)$ Å, $V = 739.4$ (4) $Å^3$, $Z = 16$. $D_{calc} = 3.663$ g/cm^3. Optically uniaxial (–), $\omega = 1.654(2)$, $\varepsilon = 1.653(2)$. The empirical formula is (electron microprobe): $Al_{2.00}O_3$. The strongest lines of the powder X-ray diffraction pattern [*d*, Å (*I*, %) (*hkl*)] are: 2.728 (61) (202), 2.424 (51) (212), 2.408 (49) (213), 2.281 (42) (206), 1.993 (81) (1.0.11, 220, 221), 1.954 (48) (0.0.12), 1.396 (100) (327, 3.0.11, 400, 401, 2.1.14, 2.2.12).

Kind of sample preparation and/or method of registration of the spectrum: KBr disc. Absorption.

Wavenumbers (cm^{-1}): 1157w, 1100w, 1063w, 1024w, 865sh, 820s, 755s, 703s, 628, 571s, 391.

Note: The spectrum was obtained by N.V. Chukanov.

O503 Magnesiohögbomite-2*N*3*S* $(Mg,Fe,Zn,Ti)_4(Al,Fe)_{10}O_{19}(OH)$

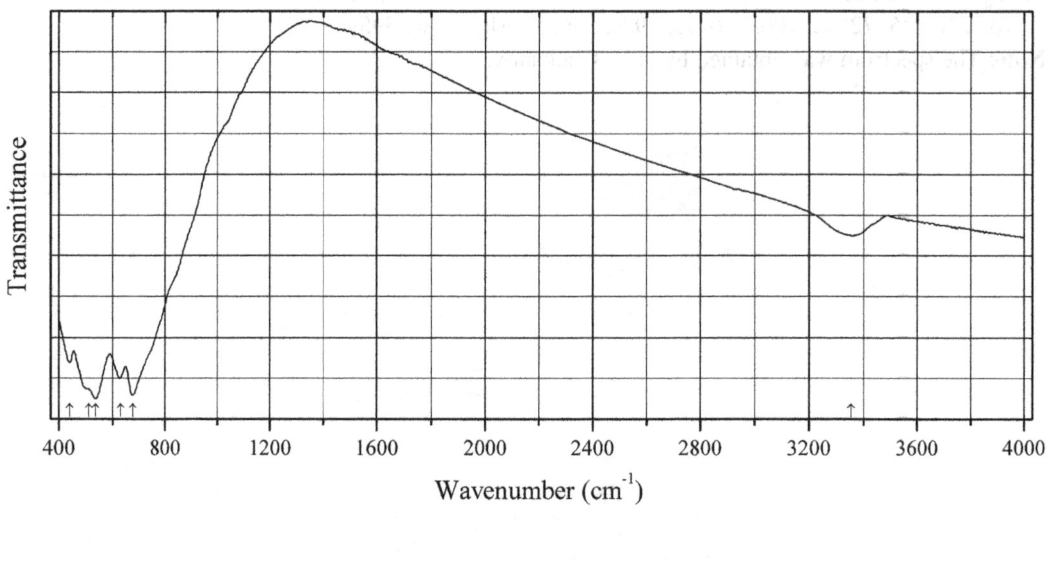

Origin: Sadok Lake, Chelyabinsk region, South Urals.

Description: Dark brown platy grains from clinopyroxenite. Characterized by powder X-ray diffraction data and Mössbauer spectroscopy. Hexagonal, $a = 5.715(5)$, $c = 23.931(2)$ Å, $V = 677.01(4)$ Å^3. The empirical formula is $(Mg_{5.4–5.7}Fe^{3+}{}_{1.4–1.7}Fe^{2+}{}_{0.8–0.9})(Al_{18.0–18.6}Ti_{1.0–1.1}Fe^{3+}{}_{0.4–0.9}Cr_{0–0.1})O_{38}(OH)_2$.

Kind of sample preparation and/or method of registration of the spectrum: KBr disc. Transmission.

Source: Korinevsky et al. (2016).

Wavenumbers (cm^{-1}): 3355, 675s, 630s, 536s, 511sh, 440.

O504 Woodallite $Mg_6Cr_2(OH)_{16}Cl_2·4H_2O$

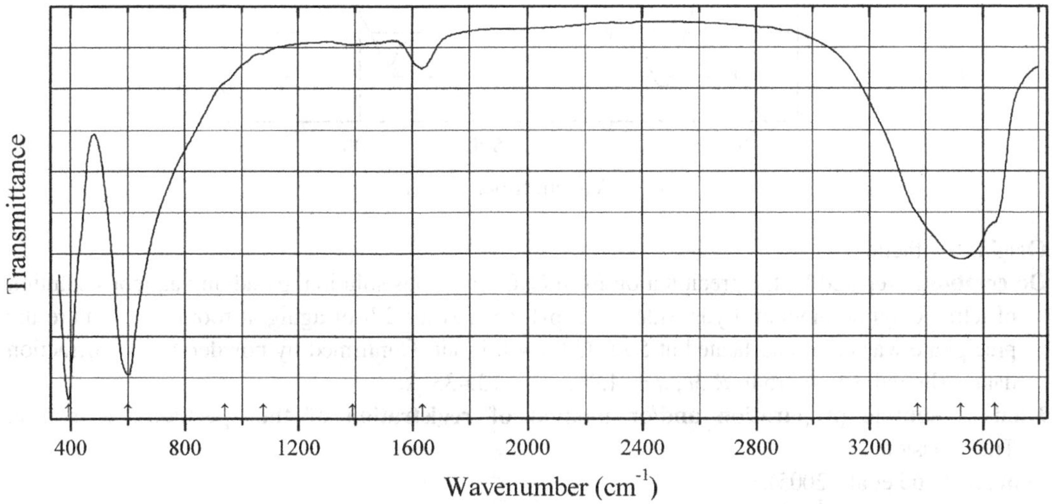

Origin: Kyzyl-Uyuk stream, Terektin ridge, Altai Mts., Siberia, Russia.

Description: Lilac crust on serpentine. Investigated by I.V. Pekov. The empirical formula is (electron microprobe): $(Mg_{5.90}Fe_{0.10})(Cr_{0.94}Fe_{0.89}Al_{0.17})Cl_{1.92}(SO_4)_{0.02}(CO_3)_x(OH)_{16}·4H_2O$ ($x \ll 1$). The sample contains zones with Fe:Cr ≈ 1:1 in atomic units.

Kind of sample preparation and/or method of registration of the spectrum: KBr disc. Absorption.

Wavenumbers (cm^{-1}): 3640sh, 3520s, 3370sh, 1635, 1390w, 1076w, 940sh, 600s, 392s.

Note: The spectrum was obtained by N.V. Chukanov.

0505 Gallium(III) oxide α-Ga_2O_3

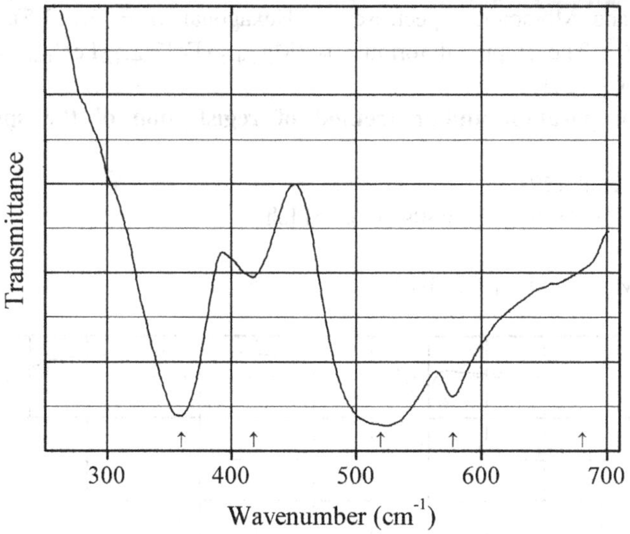

Origin: Synthetic.

Description: Prepared by the precipitation from $GaCl_3$ aqueous solution, by adding aqueous solution of tetramethyl ammonium hydroxide up to pH 7.82. After 2 h of aging at room temperature the precipitate was dried and heated at 500 °C for 4 h in air. Confirmed by powder X-ray diffraction data. Trigonal, space group *R*-3*c*, $a = 4.982$, $c = 13.433$ Å.

Kind of sample preparation and/or method of registration of the spectrum: KBr disc. Transmission.

Source: Ristić et al. (2005).

Wavenumbers (cm^{-1}): 680sh, 577, 520s, 418, 360s.

0506 Gallium(III) oxyhydroxide α-GaOOH

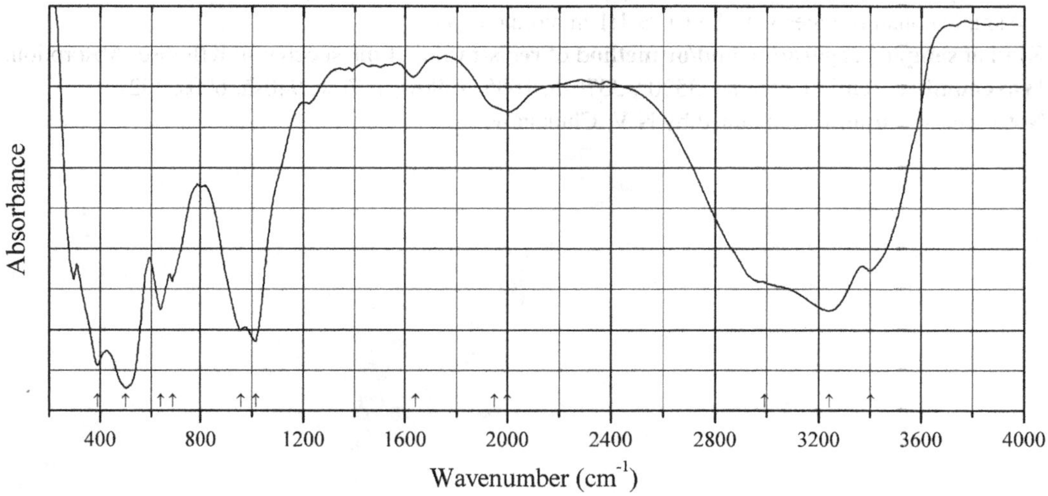

Origin: Synthetic.

Description: Prepared by the precipitation from $GaCl_3$ aqueous solution, by adding aqueous solution of tetramethyl ammonium hydroxide up to pH 7.82. Confirmed by powder X-ray diffraction data. Isostructural with goethite. Orthorhombic, space group *Pbnm*, $a = 4.58$, $b = 9.80$, $c = 2.97$ Å.

Kind of sample preparation and/or method of registration of the spectrum: KBr disc. Transmission.

Source: Ristić et al. (2005).

Wavenumbers (cm^{-1}): 3403, 3243s, 2990, 2000, 1950, 1642w, 1221w, 1015s, 958s, 688, 640, 500s, 388s, 295.

Note: The wavenumbers were partly determined by us based on spectral curve analysis of the published spectrum.

O507 Aluminium niobate $AlNbO_4$

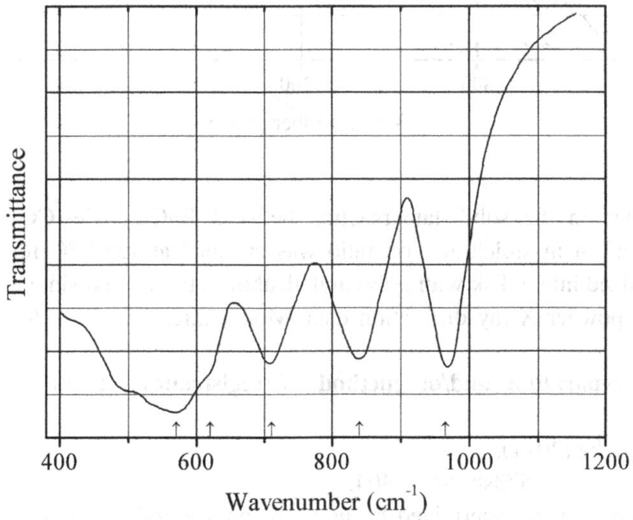

Origin: Synthetic.

Description: Prepared by firing intimate mixture of $Al(OH)_3$ and Nb_2O_5 in air at 1350 °C. Monoclinic. Characterized by powder X-ray diffraction data.

Kind of sample preparation and/or method of registration of the spectrum: KBr disc. Transmission.

Source: Blasse and 'T Lam (1978).

Wavenumbers (IR, cm^{-1}): 965, 840, 710, 620sh, 570.

Note: In the cited paper, Raman spectrum is given.

Wavenumbers (Raman, cm^{-1}): 940sh, 800, 730, 690, 600, 400.

O508 Barium cerium tantalite Ba_2CeTaO_6

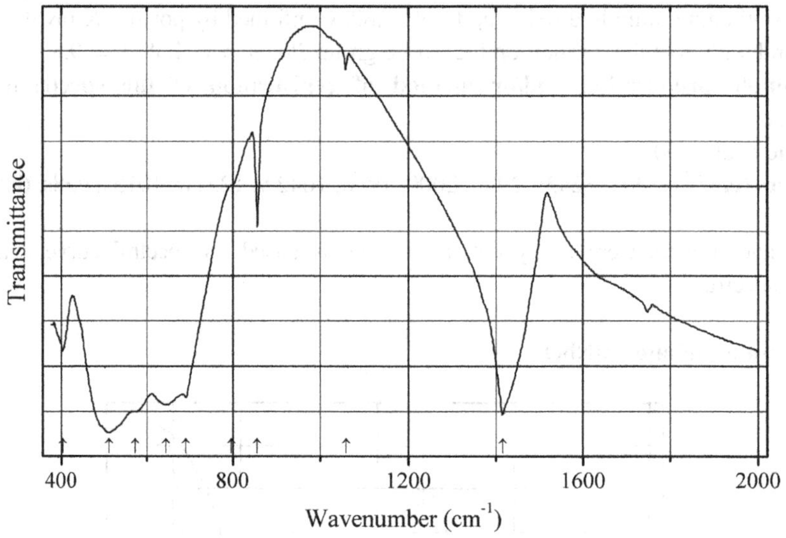

Origin: Synthetic.

Description: Obtained in the solid-state reaction between $BaCO_3$, $Ce_2(CO_3)_3$, and Ta_2O_5. The reactant mixture taken in stoichiometric ratio was calcined at 1350 °C for 15 h. The calcined sample was palletized into a disk with polyvinyl alcohol as binder and sintered at 1370 °C for 5 h. Characterized by powder X-ray diffraction data. Monoclinic, $a = 9.78$, $b = 9.02$, $c = 4.27$ Å, $\beta = 93.8°$.

Kind of sample preparation and/or method of registration of the spectrum: KBr disc. Transmission.

Source: Bharti and Sinha (2011).

Wavenumbers (cm^{-1}): 645, 573sh, 512s, 404.

Note: The wavenumbers were determined by us based on spectral curve analysis of the published spectrum. The bands at 1747, 1415, 1058, 857, 796(sh), and 691 cm^{-1} correspond to the admixture of a carbonate. In the cited paper, the absorptions in the ranges 1700–1800 and 1400–1500 cm^{-1} have been erroneously assigned to the presence of adsorbed moisture in KBr and symmetric stretching vibrations of TaO_6 octahedra, respectively.

O509 Barium cobalt antimonate $Ba_3CoSb_2O_9$

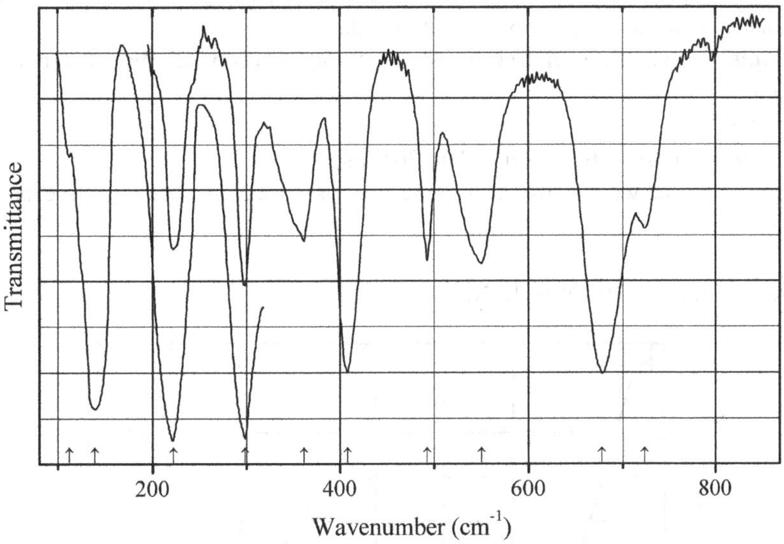

Origin: Synthetic.

Description: A compound with ordered hexagonal perovskite-type structure.

Kind of sample preparation and/or method of registration of the spectrum: KBr and polyethylene discs. Transmission.

Source: Liegeois-Duyckaerts (1985).

Wavenumbers (cm^{-1}): 723, 678s, 551, 493, 408s, 361, 298, 222, 139, 112.

O510 Barium cobaltate Ba_2CoO_4

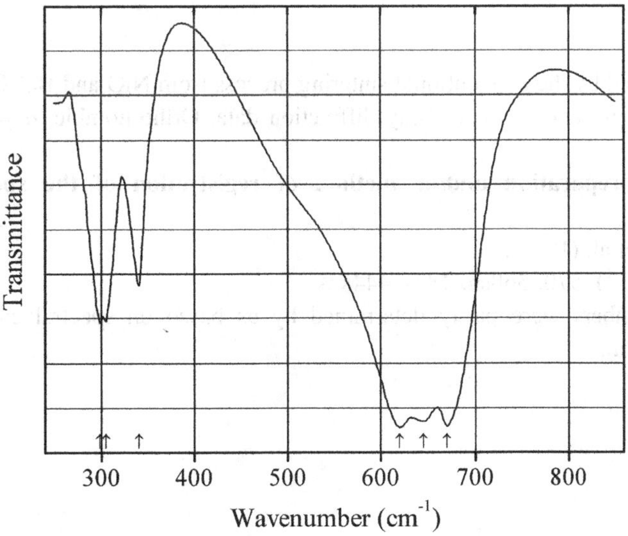

Origin: Synthetic.

Description: Synthesized from the mixture of barium and cobalt carbonates at 950 °C. Confirmed by chemical analyses and powder X-ray diffraction data.

Kind of sample preparation and/or method of registration of the spectrum: KBr disc. Transmission.

Source: Baran (1973).

Wavenumbers (cm^{-1}): 670s, 645s, 620s, 340, 305, 299.

Note: The wavenumbers were partly determined by us based on spectral curve analysis of the published spectrum.

O511 Barium nickel oxide $BaNiO_2$ $BaNiO_2$

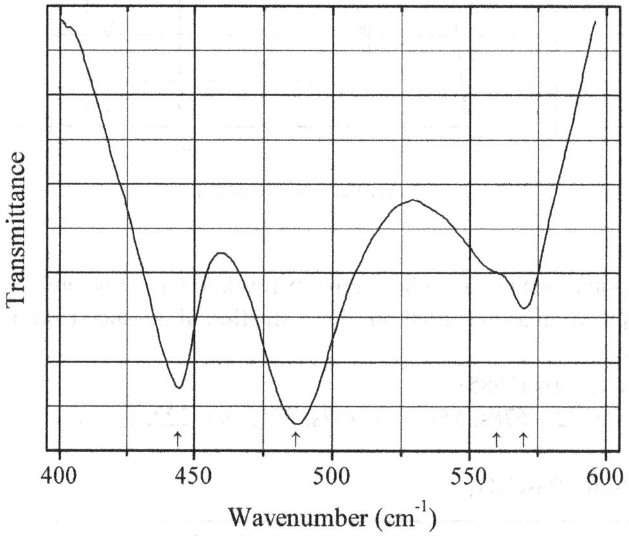

Origin: Synthetic.

Description: Prepared by the conventional sintering process from NiO and $BaCO_3$. Characterized by Mössbauer spectrum and powder X-ray diffraction data. Orthorhombic, $a = 5.737$, $b = 9.190$, $c = 4.760$ Å.

Kind of sample preparation and/or method of registration of the spectrum: KBr disc. Transmission.

Source: Gottschall et al. (1998).

Wavenumbers (cm^{-1}): 570, 560sh, 487s, 444s.

Note: The wavenumbers were partly determined by us based on spectral curve analysis of the published spectrum.

O512 Barium nickel oxide BaNiO$_3$ BaNiO$_3$

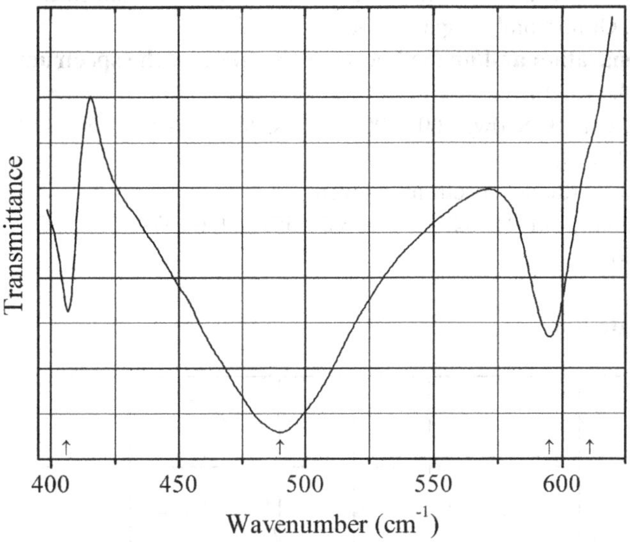

Origin: Synthetic.

Description: Prepared by the conventional sintering process from NiO and BaCO$_3$. Characterized by Mössbauer spectrum and powder X-ray diffraction data. Hexagonal, $a = 5.635$, $c = 4.8041$ Å. Hypothetically, Ni is trivalent and the formula is $Ba^{2+}Ni^{3+}O^{2-}{}_2(O^{\cdot})^-$.

Kind of sample preparation and/or method of registration of the spectrum: KBr disc. Transmission.

Source: Gottschall et al. (1998).

Wavenumbers (cm^{-1}): 611sh, 595, 490s, 406.

Note: The wavenumbers were partly determined by us based on spectral curve analysis of the published spectrum.

O513 Barium niobate BaNb$_2$O$_6$

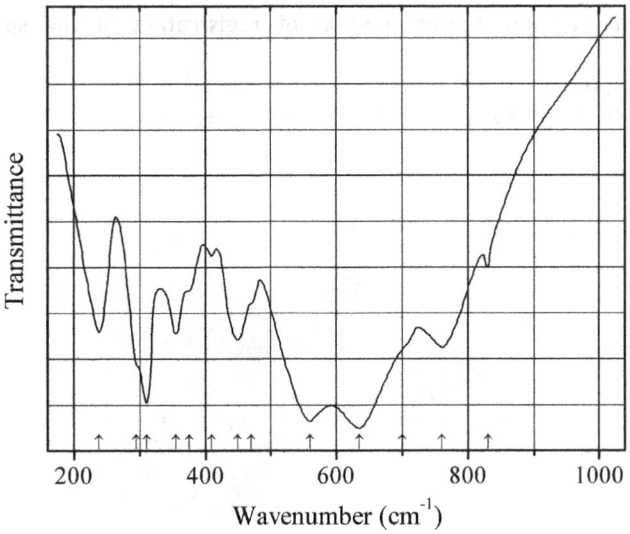

Origin: Synthetic.
Description: Prepared by heating stoichiometric mixture of Nb_2O_5 and $BaCO_3$ pressed in a pellet at 1200 °C for 60 h. Orthorhombic, space group *Pbmm*, $Z = 2$.
Kind of sample preparation and/or method of registration of the spectrum: Transmission.
Source: Repelin et al. (1979).
Wavenumbers (IR, cm^{-1}): 830w, 760, 700sh, 635s, 560s, 470sh, 450, 410w, 375sh, 355, 310s, 295sh, 238, 170, 151, 87.
Note: In the cited paper, Raman spectrum is given.
Wavenumbers (Raman, cm^{-1}): 847, 712s, 633, 557s, 496, 379, 366, 306, 280, 230, 200, 190, 141, 120, 112, 100.

O514 Barium titanate Ba_2TiO_4

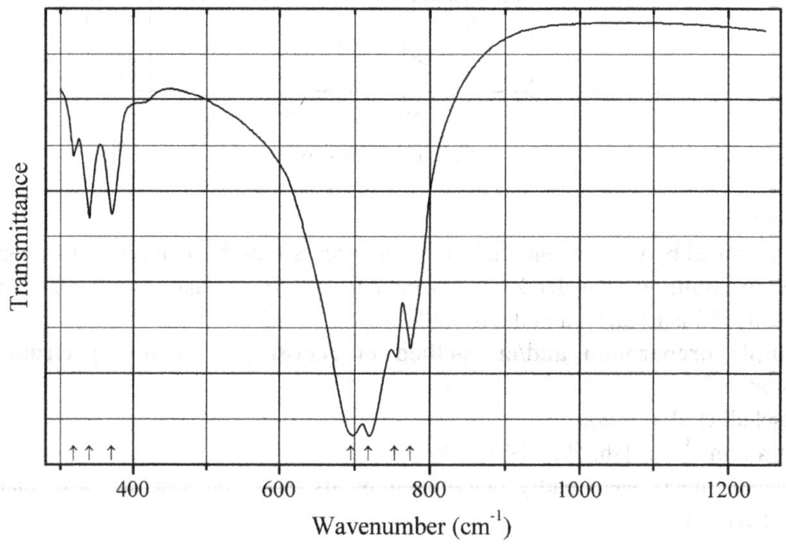

Origin: Synthetic.
Description: Obtained by heating a mixture of TiO_2 and $BaCO_3$ at 1200–1300 °C for 1–3 days.
Kind of sample preparation and/or method of registration of the spectrum: KBr disc. Transmission.
Source: Wijzen et al. (1994).
Wavenumbers (cm^{-1}): 774, 753, 719s, 695s, 370, 340, 319w.

O515 β-Gallium(III)-oxide β-Ga_2O_3

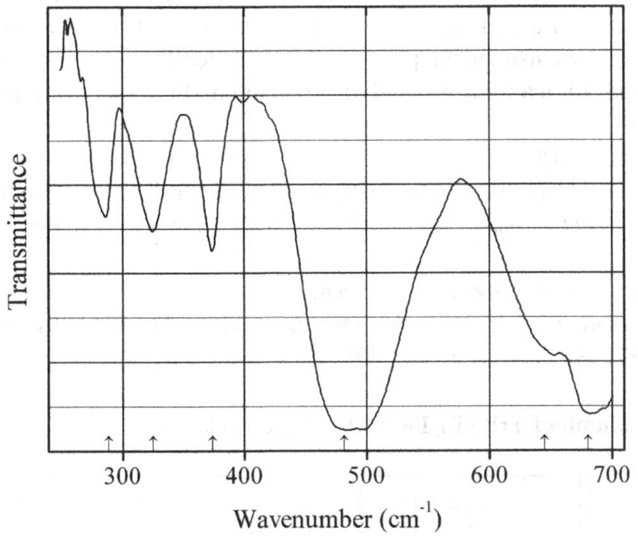

Origin: Synthetic.
Description: Monoclinic, space group *C2/m*, $a = 12.21$, $b = 3.037$, $c = 5.798$ Å, $\beta = 103.838°$.
Kind of sample preparation and/or method of registration of the spectrum: KBr disc. Transmission.
Source: Ristić et al. (2005).
Wavenumbers (cm^{-1}): 680, 644, 482s, 374, 325, 289.

O516 Bismuth(III) aluminate $Bi_2Al_4O_9$ $Bi_2Al_4O_9$

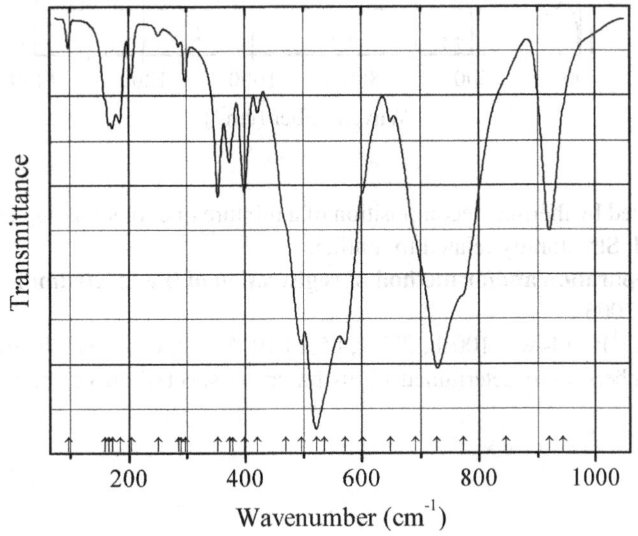

Origin: Synthetic.

Description: Synthesized from a stoichiometric mixture of $Bi(NO_3)_3 \cdot 5H_2O$ and $Al(NO_3)_3 \cdot 9H_2O$ together with 10 wt.% of glycerine. The mixture was heated first at 353 K, then at 473 K for 2 h, and finally (after homogenization the powder) at 1210 K for 48 h. Structurally related to mullite.

Kind of sample preparation and/or method of registration of the spectrum: KBr and polyethylene discs. Absorption.

Source: Murshed et al. (2015).

Wavenumbers (IR, cm^{-1}): (943sh), 920, 846sh, 772sh, 728s, 691sh, 649, 601sh, 572s, 536sh, 522s, 497s, 470sh, 421w, 399, 378sh, 373, 353, 297w, 289sh, 286w, 251w, 204w, 185, 172, 166, 159, 97w.

Note: In the cited paper, Raman spectrum is given.

Wavenumbers (Raman, cm^{-1}): 840, 763, 691, 633, 615, 571, 509, 491, 478, 441, 406, 385, 373, 345, 322, 312, 281, 274s, 251, 184, 138, 120s, 103.

O517 Bismuth(III) aluminoferrite $Bi_2Fe_3AlO_9$ $Bi_2Fe_3AlO_9$

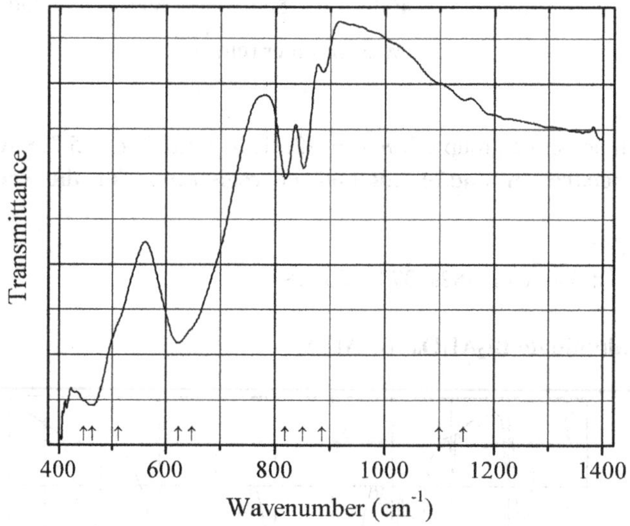

Origin: Synthetic.

Description: Produced by thermal decomposition of a mixture of corresponding metal nitrates using a glycerine method. Structurally related to mullite.

Kind of sample preparation and/or method of registration of the spectrum: KBr disc. Absorption.

Source: Voll et al. (2006).

Wavenumbers (cm^{-1}): 1144w, 1100sh, 887w, 852, 819, 650sh, 624s, 512sh, 464s, 448sh.

Note: The wavenumbers were determined by us based on spectral curve analysis of the published spectrum.

O518 Bismuth(III) stannate pyrochlore-type $Bi_2Sn_2O_7$

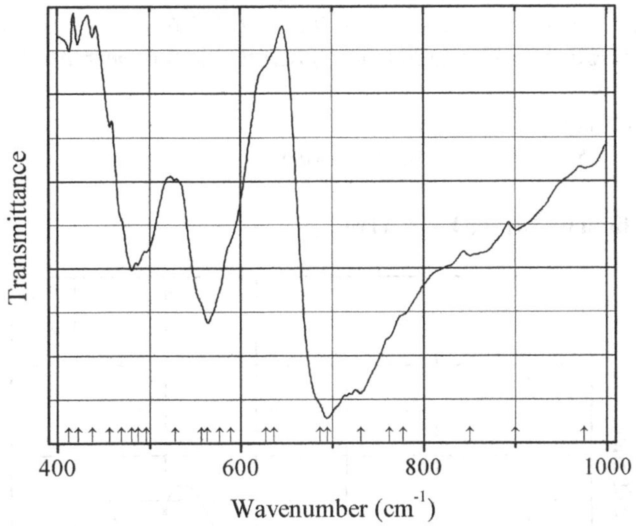

Origin: Synthetic.

Description: Obtained by sintering at 1173 K of pelletized precipitate formed after adding ammonia solution to the 0.01 M solution containing bismuth chloride and stannous oxy chloride in stoichiometric ratio. Tetragonal, $a = 21.328$, $c = 21.4$ Å.

Kind of sample preparation and/or method of registration of the spectrum: Transmission. Kind of sample preparation is not indicated.

Source: Ravi et al. (1999).

Wavenumbers (cm^{-1}): 731s, 694s, 686sh, 636sh, 627sh, 588sh, 576sh, 563, 556sh, 496sh, 487, 480, 469sh.

Note: The wavenumbers were determined by us based on spectral curve analysis of the published spectrum.

O519 Bismuth(III) tantalate $Bi_7Ta_3O_{18}$ $Bi_7Ta_3O_{18}$

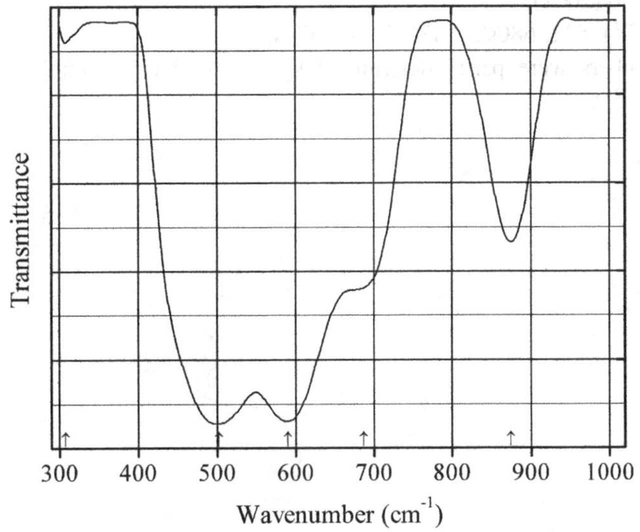

Origin: Synthetic.

Description: Obtained from Bi_2O_3 and Ta_2O_5 by solid-state method at 950 °C for 18 h. Monoclinic, space group *C*2/*m*, *a* = 34.060(3), *b* = 7.618 (9), *c* = 6.647(6) Å, *β* = 109.210 (7)°, *Z* = 4.

Kind of sample preparation and/or method of registration of the spectrum: Transmission. Kind of sample preparation is not indicated.

Source: Chon et al. (2014).

Wavenumbers (cm^{-1}): 874, 687sh, 590s, 502s, 308w.

O520 Bismuth(III) titanate $Bi_4Ti_3O_{12}$ $Bi_4Ti_3O_{12}$

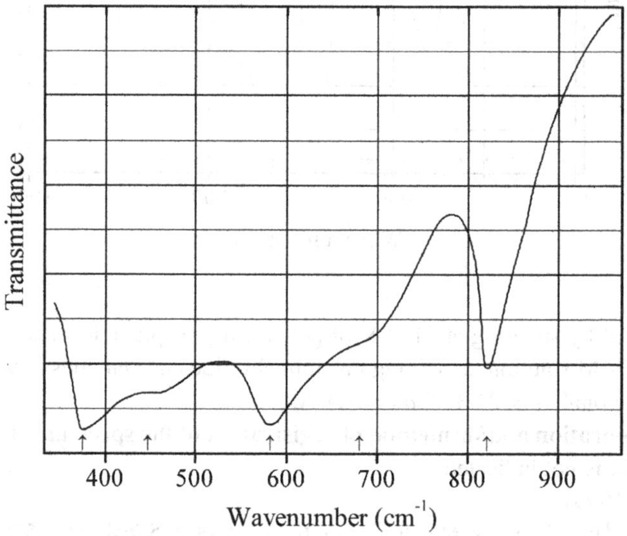

Origin: Synthetic.

Description: Prepared hydrothermally. Structurally related to perovskite. The strongest lines of the powder X-ray diffraction pattern are observed at 3.81, 2.95, 2.72, 2.26, 1.92, and 1.61 Å.

Kind of sample preparation and/or method of registration of the spectrum: KBr disc. Transmission.

Source: Chen and Jiao (2001).

Wavenumbers (cm^{-1}): 822, 680sh, 582s, 447sh, 374s.

Note: The wavenumbers were partly determined by us based on spectral curve analysis of the published spectrum.

O521 Bismuth ferrite $BiFeO_3$

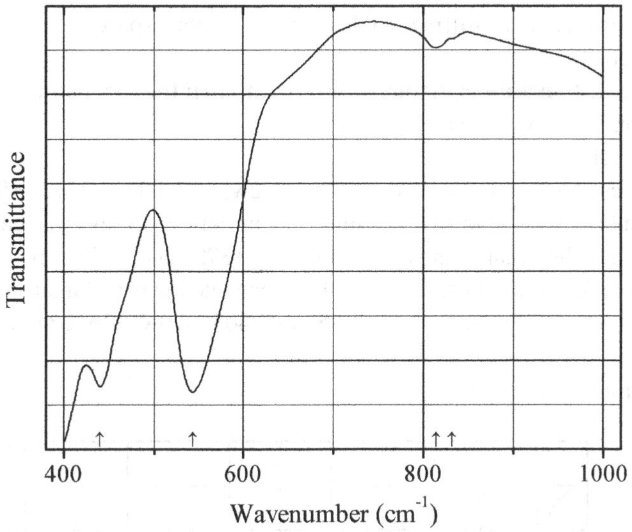

Origin: Synthetic.

Description: Obtained by two-stage solid phase synthesis from Bi_2O_3 and Fe_2O_3, first at 923 K for 1 h and thereafter (after re-grounding and pressing) at 1123 K for 2 h. The compound has rhombohedrally distorted perovskite structure, space group *R*3*c*.

Kind of sample preparation and/or method of registration of the spectrum: KBr disc. Transmission.

Source: Bujakiewicz-Korońska et al. (2011).

Wavenumbers (cm^{-1}): 832sh, 814w, 543s, 440s.

Note: The wavenumbers were partly determined by us based on spectral curve analysis of the published spectrum. In the cited paper, also FIR spectrum for a sample suspended in Apiezon N grease is given.

O522 Calcium indium oxide Ca_2InO_4 Ca_2InO_4

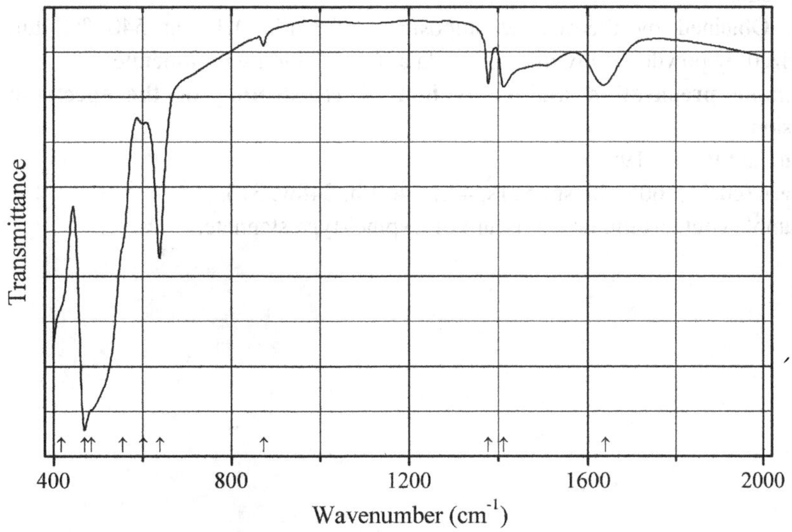

Origin: Synthetic.

Description: Prepared from indium and calcium nitrates, in solid-state reaction at 1173 K for 2 h. Characterized by powder X-ray diffraction data. Orthorhombic, space group $Pca2_1$ or $Pbcm$, $Z = 4$ (see JCPDS 017-0643).

Kind of sample preparation and/or method of registration of the spectrum: Transmission. Kind of sample preparation is not indicated.

Source: Zheng et al. (2012).

Wavenumbers (IR, cm^{-1}): 1639w, 1412w, 1378w, 872w, 639, 602, 556sh, 485sh, 470s, 417sh.

Note: The wavenumbers were partly determined by us based on spectral curve analysis of the published spectrum. The weak bands at 1639, 1412+872, and 1378 cm^{-1} may correspond to H_2O, CO_3^{2-}, and NO_3^- impurities, respectively. In the cited paper, Raman spectrum is given.

Wavenumbers (Raman, cm^{-1}): 648w, 543s, 495, 455, 403, 370w, 336, 284w, 258, 199, 113s.

O523 Cadmium stannate $CdSnO_3$

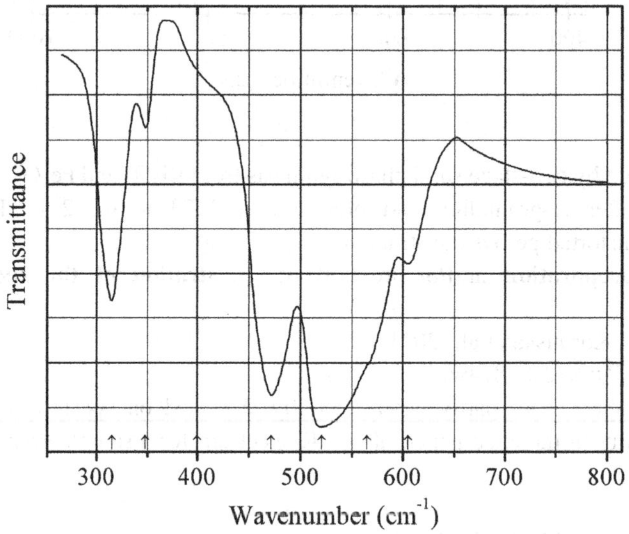

Origin: Synthetic.

Description: Obtained by thermal decomposition of $CdSn(OH)_6$ at 540 °C during 15 min. Characterized by powder X-ray diffraction data. Isostructural with ilmenite.

Kind of sample preparation and/or method of registration of the spectrum: CsIr disc. Transmission.

Source: Botto and Baran (1980).

Wavenumbers (cm^{-1}): 605, 565sh, 521s, 472s, 400sh, 348w, 315.

Note: The sample contains minor admixture of a spinel-type stannate.

O524 Cesium uranyl niobate $Cs_2(UO_2)_2(Nb_2O_8)$ $Cs_2(UO_2)_2(Nb_2O_8)$, or $CsUNbO_6$

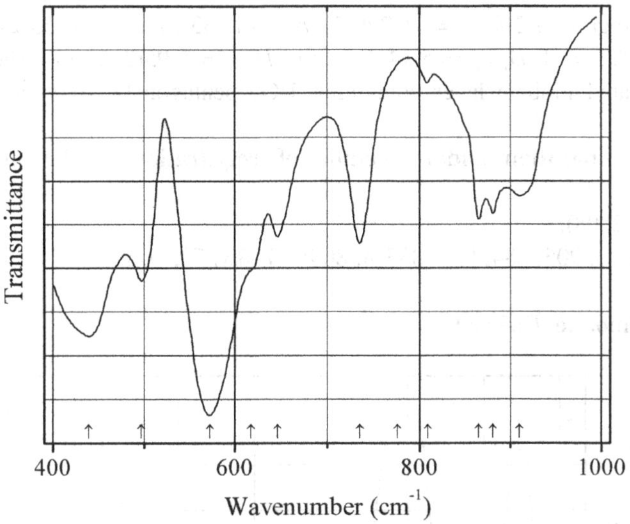

Origin: Synthetic.

Description: Carnotite-type niobate with $UNbO_6$ layers.

Kind of sample preparation and/or method of registration of the spectrum: KBr disc. Transmission.

Source: Saad et al. (2008).

Wavenumbers (cm^{-1}): 910, 881, 865, 809w, 776sh, 735, 646, 617sh, 572s, 497, 439s.

O525 Cesium uranyl niobate $Cs_9[(UO_2)_8O_4(NbO_5)(Nb_2O_8)_2]$ $Cs_9[(UO_2)_8O_4(NbO_5)(Nb_2O_8)_2]$

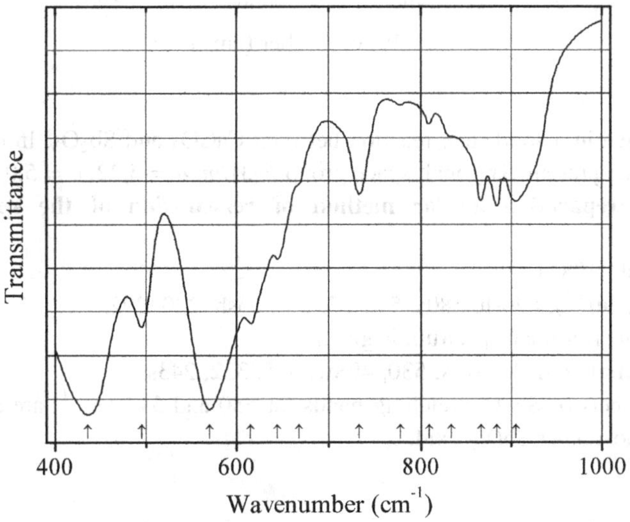

Origin: Synthetic.

Description: Prepared from $CsNO_3$, U_3O_8, and Nb_2O_5 by solid-state reaction at 1000 °C in air. Monoclinic, space group $P2_1/c$, $a = 16.729(2)$, $b = 14.933(2)$, $c = 20.155(2)$ Å, $\beta = 110.59(1)°$, $V = 4713.5(1)$ Å^3, $Z = 4$. $D_{meas} = 5.94(2)$ g/cm^3, $D_{calc} = 5.95(3)$ g/cm^3. The crystal structure is based on the uranyl niobate layer containing UO_7 pentagonal bipyramids and NbO_5 square pyramids.

Kind of sample preparation and/or method of registration of the spectrum: KBr disc. Transmission.

Source: Saad et al. (2008).

Wavenumbers (cm^{-1}): 905, 884, 866, 833sh, 809w, 778w, 733, 668sh, 644, 615, 570s, 495, 436s.

O526 Calcium antimonite $CaSb_2O_6$

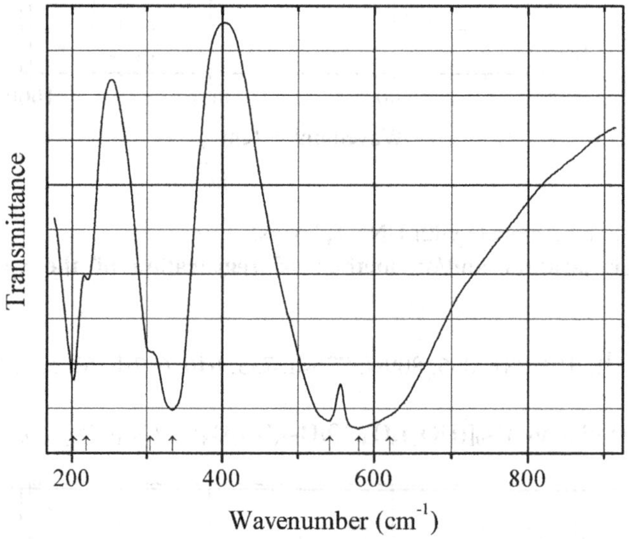

Origin: Synthetic.

Description: Obtained in a solid-state reaction between $CaCO_3$ and Sb_2O_3. In the crystal structure, SbO_6 octahedra are present. Trigonal, space group P-$31/m$, $a = 5.22$, $c = 5.01$ Å.

Kind of sample preparation and/or method of registration of the spectrum: CsI disc. Transmission.

Source: Husson et al. (1984).

Wavenumbers (IR, cm^{-1}): 620sh, 580s, 542s, 335s, 305sh, 220, 202.

Note: In the cited paper, Raman spectrum is given.

Wavenumbers (Raman, cm^{-1}): 678s, 530, 498w, 345, 332, 243s.

Note: The wavenumbers of Sb–O stretching bands (at 580 and 542 cm^{-1}) are anomalously low as compared with most other Sb(V) oxides.

O527 Calcium copper titanate $CaCu_3Ti_4O_{12}$ $CaCu_3Ti_4O_{12}$

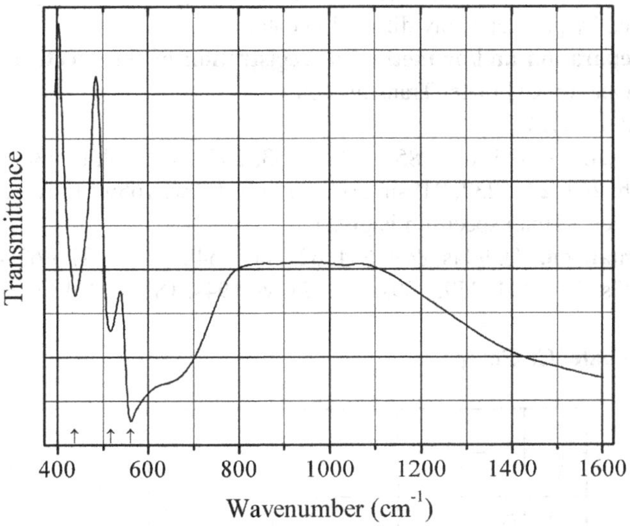

Origin: Synthetic.

Description: Nano-sized powder synthesized by a polymerization-based complex method and calcined at 800 °C in air for 8 h. A perovskite-type compound. Characterized by powder X-ray diffraction data. Cubic, $a = 7.398(2)$ Å.

Kind of sample preparation and/or method of registration of the spectrum: Transmission. Kind of sample preparation is not indicated.

Source: Masingboon et al. (2008).

Wavenumbers (cm^{-1}): 561s, 516, 437.

Note: The sample exhibits a giant dielectric constant (Masingboon et al. 2009)

O528 Calcium niobate columbite-type $CaNb_2O_6$

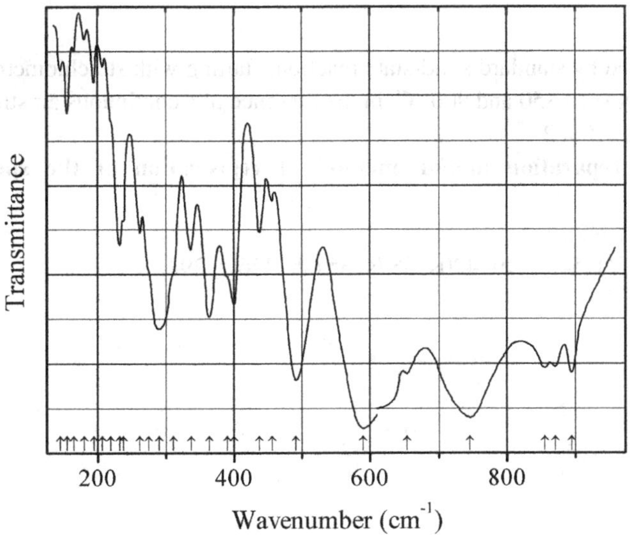

Origin: Synthetic.

Description: White solid prepared from the mixture of Nb_2O_5 and $CaCO_3$ powders at 1300 °C for 48 h. Characterized by powder X-ray diffraction data.

Kind of sample preparation and/or method of registration of the spectrum: Powder spread on polyethylene film and on CsI plate. Transmission.

Source: Husson et al. (1977a).

Wavenumbers (IR, cm^{-1}): 895, 870, 855, 745s, 653, 590s, 492s, 456, 438, 400, 390, 364, 337, 311sh, 290, 275sh, 262, 237, 232, 219sh, 207w, 194w, 180w, 165w, 155w, 145w.

Note: In the cited paper, Raman spectrum is given.

Wavenumbers (Raman, cm^{-1}): 904s, 849, 664, 627, 600, 540s, 495, 487, 462, 430, 385s, 379, 369, 344, 340, 314, 293s, 286, 264, 259, 239s, 223, 213w, 194s, 186, 162, 136s, 127, 108, 84, 63.

O529 Calcium plumbate Ca_2PbO_4

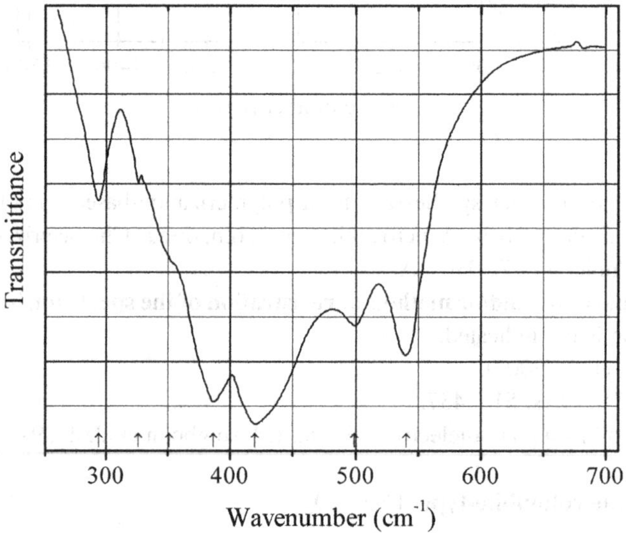

Origin: Synthetic.

Description: Obtained by standard solid-state reaction, starting with stoichiometric mixture of PbO_2/$CaCO_3$, heated between 850 and 900 °C, in the presence of a continuous air stream. Orthorhombic, space group *Pbam*, $Z = 2$.

Kind of sample preparation and/or method of registration of the spectrum: KBr disc. Transmission.

Source: Diez et al. (1995).

Wavenumbers (cm^{-1}): 540, 499, 420s, 387s, 351sh, 326w, 294.

O530 Chromium uranium oxide Cr_2UO_6 Cr_2UO_6

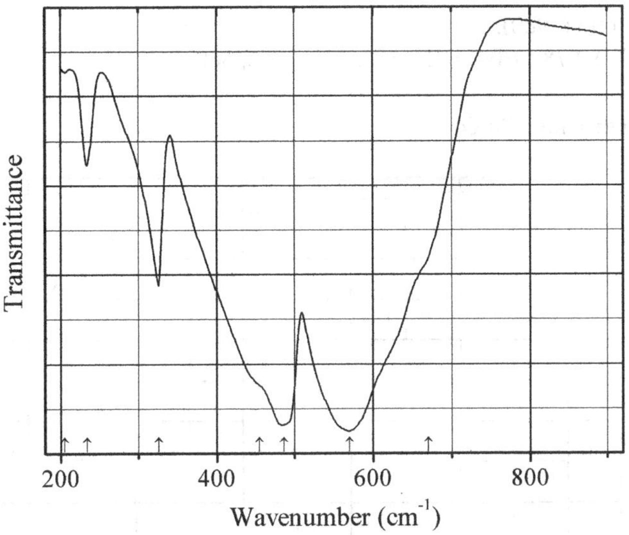

Origin: Synthetic.

Description: Synthesized hydrothermally from $Cr(NO_3)_3{\cdot}9H_2O$ and γ-UO_3 at 325–425 °C. Characterized by powder X-ray diffraction data.

Hexagonal, $a = 4.988(1)$, $c = 4.620(1)$ Å. $D_{calc} = 7.31$ g/cm^3.

Kind of sample preparation and/or method of registration of the spectrum: KBr disc and Nujol mull. Absorption.

Source: Hoekstra and Siegel (1971).

Wavenumbers (cm^{-1}): 670sh, 570s, 487s, 455sh, 326, 234, 205w.

O531 Cobalt zinc tellurium oxide $Co_3Zn_2TeO_8$

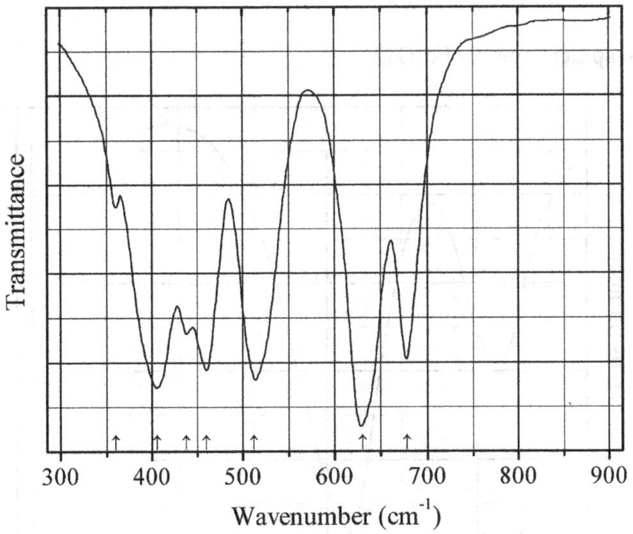

Origin: Synthetic.

Description: Spinel-type compound obtained in the solid-state reaction between TeO_2, $CoCO_3$, and $ZnCO_3$ at 1050 °C.

Kind of sample preparation and/or method of registration of the spectrum: KBr disc. Transmission.
Source: Baran and Botto (1980).
Wavenumbers (cm^{-1}): 678, 630s, 512s, 460, 438, 406s, 360w.

O532 Copper(II) hydroxide $Cu(OH)_2$

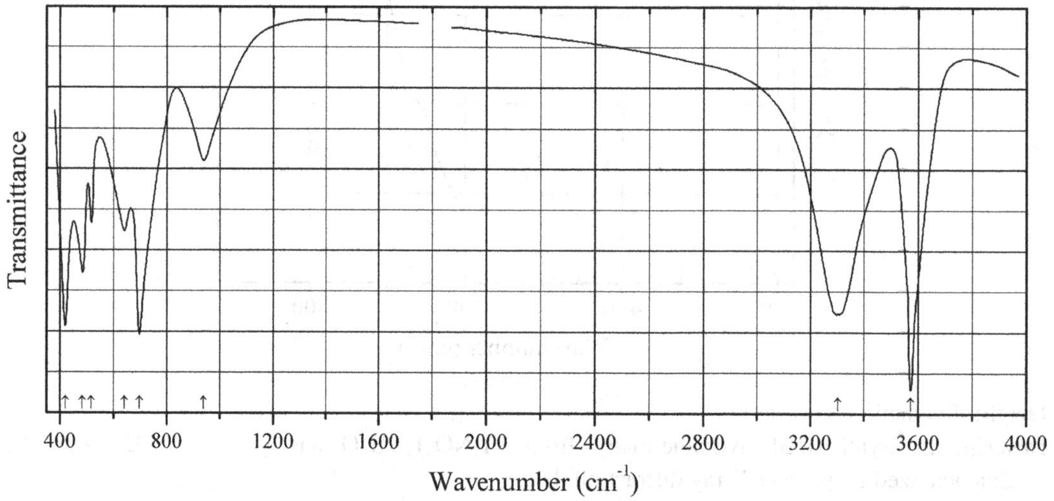

Locality: Synthetic.
Description: Characterized by powder X-ray diffraction data. Orthorhombic, space group *Cmcm*, $a = 2.936(5)$, $b = 10.54(1)$, $c = 5.238(8)$ Å.
Kind of sample preparation and/or method of registration of the spectrum: KBr disc. Transmission.
Source: Schönenberger et al. (1971).
Wavenumbers (cm^{-1}): 3574s, 3304s, 940, 695s, 640, 517, 485, 420.

O533 Cobalt ferrite spinel-type $CoFe_2O_4$

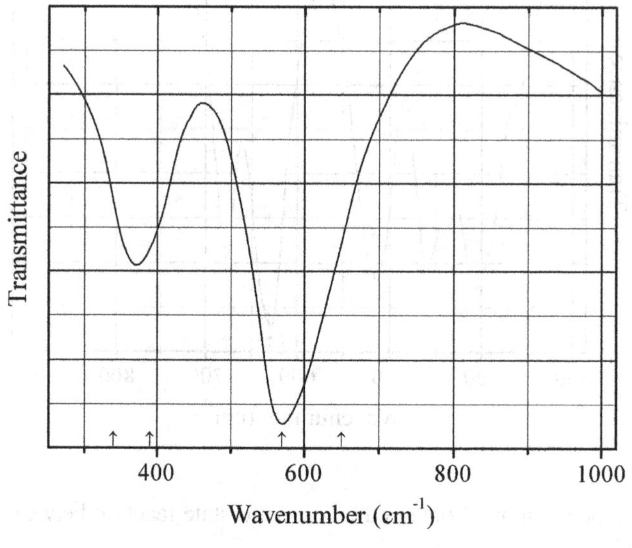

Origin: Synthetic.

Description: Prepared using a conventional ceramic technique. The powder X-ray diffraction showed a single phase and a spinel structure.

Kind of sample preparation and/or method of registration of the spectrum: KBr disc. Transmission.

Source: Srinivasan et al. (1984).

Wavenumbers (cm^{-1}): 650, 570, 390, 340.

O534 Indium oxide In_2O_3

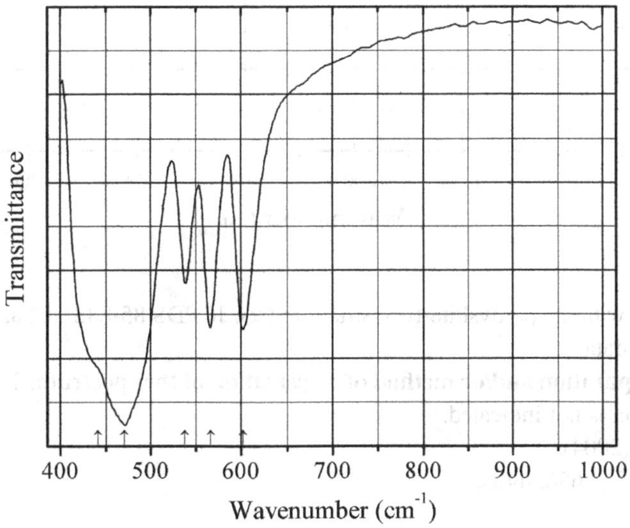

Origin: Synthetic.

Description: Commercial reactant.

Kind of sample preparation and/or method of registration of the spectrum: Diffuse reflection of a powdery sample pressed into pellet with KBr. The absorption spectrum was calculated from reflection spectrum by using the Kubelka-Munk function.

Source: Jiang et al. (2011).

Wavenumbers (cm^{-1}): 602, 566, 538, 471s, 441sh.

Note: The wavenumbers were determined by us based on spectral curve analysis of the published spectrum.

O535 Lanthanum aluminum oxide $LaAlO_3$

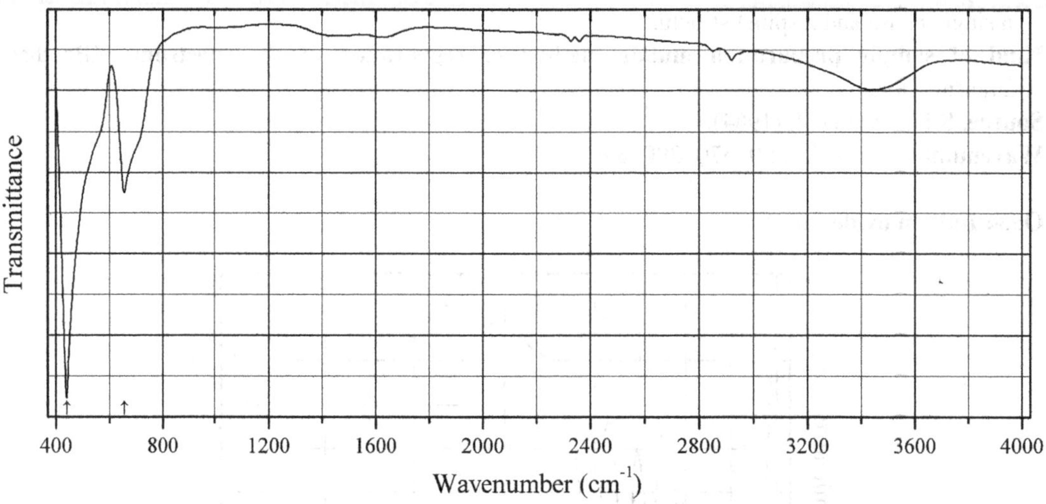

Origin: Synthetic.

Description: Cubic, with the perovskite-type structure (see JCPDS 85-848). Characterized by powder X-ray diffraction data.

Kind of sample preparation and/or method of registration of the spectrum: Transmission. Kind of sample preparation is not indicated.

Source: Zhou et al. (2004).

Wavenumbers (cm^{-1}): 656, 440s.

O536 Lanthanum iron(III) oxide $LaFeO_3$

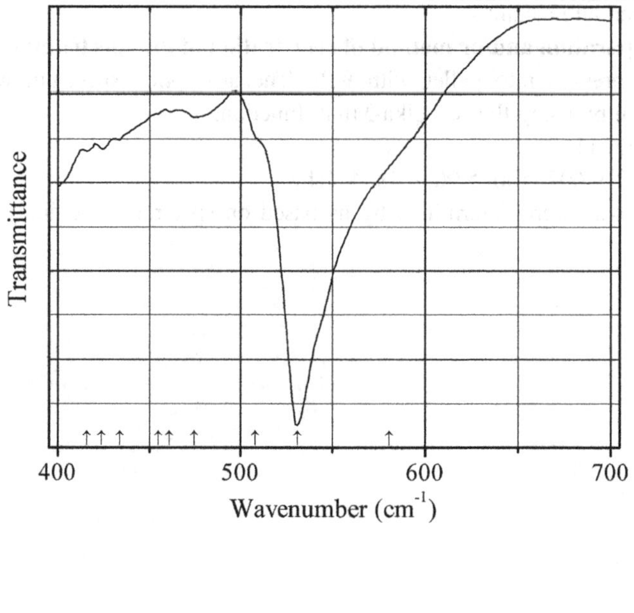

Origin: Synthetic.

Description: Prepared from stoichiometric mixture of La_2O_3 and Fe_2O_3 in the presence of excess of the euthectic mixture of NaCl and KCl at 900 °C for 6 h. Characterized by powder X-ray diffraction data. Orthorhombic, space group *Pnma*, $a = 5.5676(2)$, $b = 7.8608(3)$, $c = 5.5596(2)$ Å.

Kind of sample preparation and/or method of registration of the spectrum: Attenuated total reflection of a powdered sample.

Source: Romero et al. (2014).

Wavenumbers (IR, cm^{-1}): 581sh, 531s, 508sh, 475w, 461sh, 455w, 434w, 424w, 416w.

Note: In the cited paper, Raman spectrum is given. The Raman bands at 1310 and 1143 cm^{-1} have been assigned to second-order excitations.

Wavenumbers (Raman, cm^{-1}): 1310s, 1143, 650s, 500, 486, 433s, 431, 411, 288, 264, 173, 151, 101.

O537 Lead(II) stannate Pb_2SnO_4 Pb_2SnO_4

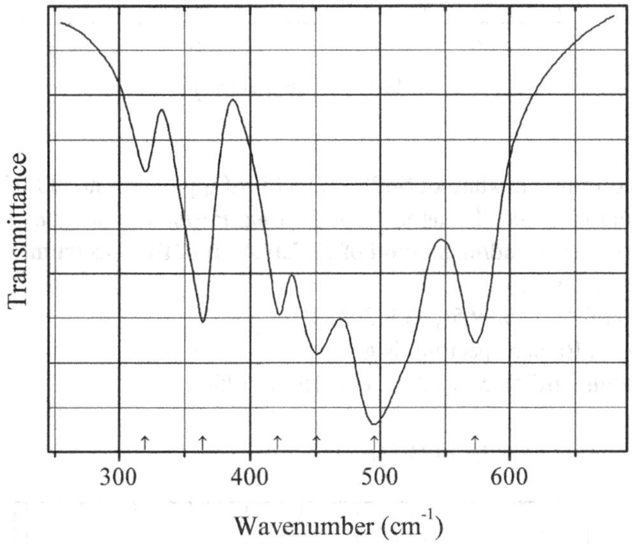

Origin: Synthetic.

Description: Prepared from the stoichiometric mixture of PbO and SnO_2 at 700 °C for 25 h in air. Tetragonal, $a = 8.74$, $c = 6.30$ Å (see JCPDS No. 24-0589).

Kind of sample preparation and/or method of registration of the spectrum: CsI disc. Transmission.

Source: Vigouroux et al. (1982).

Wavenumbers (IR, cm^{-1}): 573, 495s, 451, 421, 364, 320w.

Note: In the cited paper, Raman spectrum is given.

Wavenumbers (Raman, cm^{-1}): 613w, 540, 457s, 379, 292s, 275, 196, 129s, 80, 35.

0538 Lead tin oxide $Pb^{2+}{}_4Pb^{4+}Sn^{4+}O_8$ $Pb^{2+}{}_4Pb^{4+}Sn^{4+}O_8$

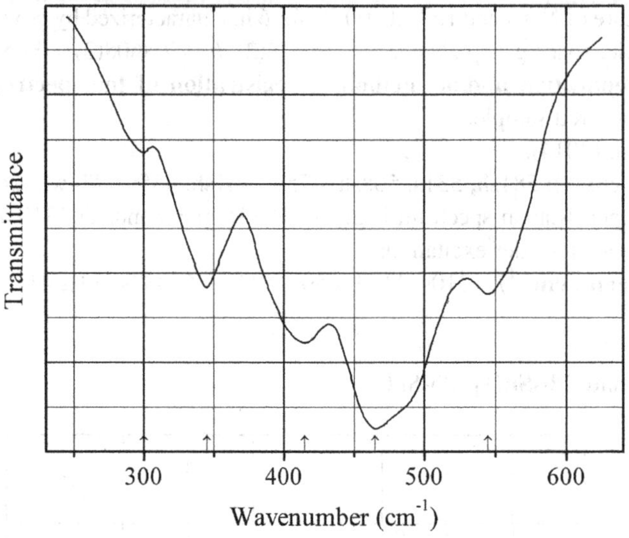

Origin: Synthetic.

Description: Prepared from a mixture of $SnPb_2O_4$ and Pb_3O_4 powders at 580 °C for several months. The crystal structure is solved. Tetragonal, space group *P*4₂/*m*, *a* = 8.77, *c* = 6.43 Å.

Kind of sample preparation and/or method of registration of the spectrum: CsI disc. Absorption.

Source: Vigouroux et al. (1982).

Wavenumbers (IR, cm^{-1}): 545, 465s, 415, 345, 300w.

Note: In the cited paper, Raman spectrum is given.

Wavenumbers (Raman, cm^{-1}): 550s, 260, 195, 152w, 125s.

0539 Lithium aluminate $LiAl_5O_8$ $LiAl_5O_8$

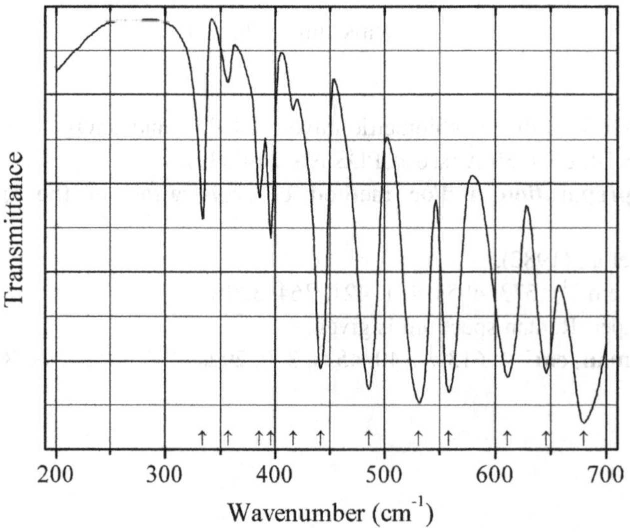

Origin: Synthetic.

Description: Prepared by sintering a mixture of Li_2CO_3 and Al_2O_3 at 1300 °C. A compound with ordered spinel-type structure. Characterized by powder X-ray diffraction data.

Kind of sample preparation and/or method of registration of the spectrum: KBr disc. Transmission.

Source: Brabers (1976).

Wavenumbers (cm^{-1}): 680s, 646s, 611s, 558s, 531s, 486s, 442s, 417w, 396, 385, 357w, 334.

Note: The wavenumbers were determined by us based on spectral curve analysis of the published spectrum.

O540 Lithium aluminate $LiAlO_2$-beta β-$LiAlO_2$

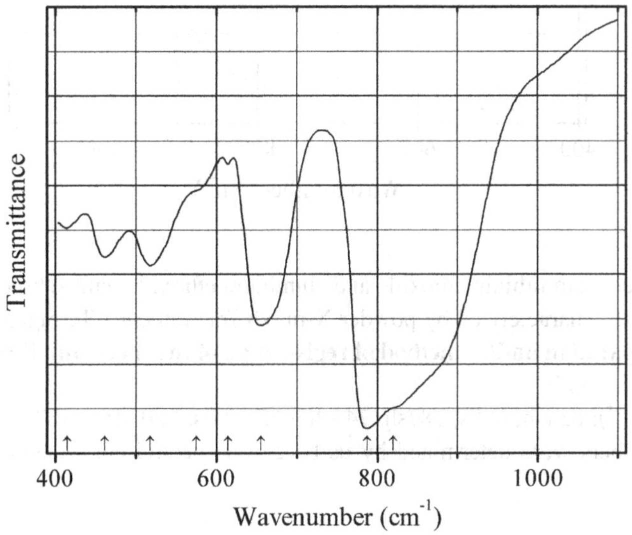

Origin: Synthetic.

Description: Prepared from lithium ethoxide and aluminu methoxide with subsequent hydrolysis and heating to 600 °C. Characterized by powder X-ray diffraction data. Monoclinic.

Kind of sample preparation and/or method of registration of the spectrum: KBr disc. Transmission.

Source: Hirano et al. (1987).

Wavenumbers (cm^{-1}): 820sh, 787s, 655, 614w, 575sh, 518, 462, 415w.

Note: The wavenumbers were determined by us based on spectral curve analysis of the published spectrum.

O541 Lithium aluminate $LiAlO_2$-gamma γ-$LiAlO_2$

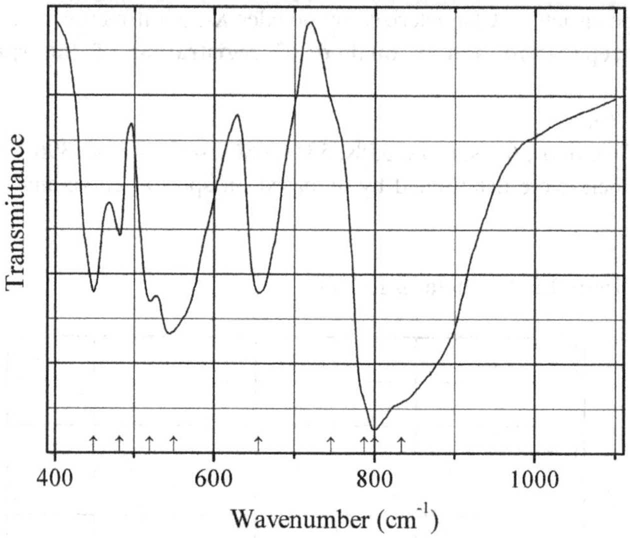

Origin: Synthetic.

Description: Prepared from lithium ethoxide and aluminum ethoxide with subsequent hydrolysis and heating to 1000 °C. Characterized by powder X-ray diffraction data. Tetragonal.

Kind of sample preparation and/or method of registration of the spectrum: KBr disc. Transmission.

Source: Hirano et al. (1987).

Wavenumbers (cm^{-1}): 834sh, 800s, 787sh, 745sh, 655, 550s, 520, 482, 450.

Note: The wavenumbers were determined by us based on spectral curve analysis of the published spectrum.

O542 Lithium aluminium oxide-alpha α-$LiAlO_2$

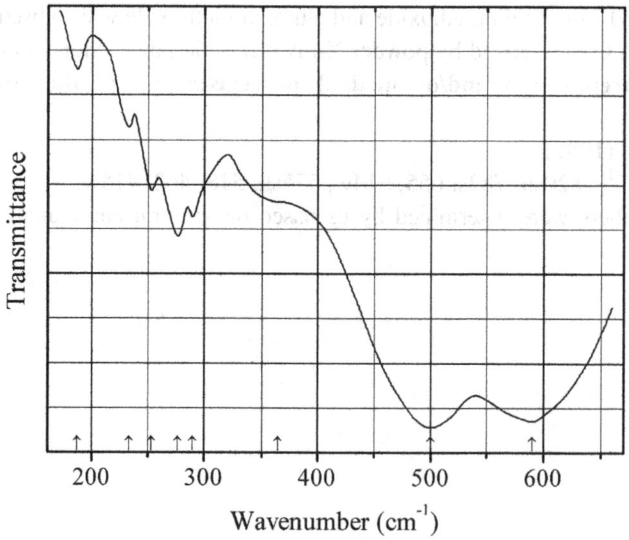

Origin: Synthetic.

Description: The polymorph with trigonally distorted NaCl-type structure synthesized in a solid-state reaction from Al oxide and Li carbonate.

Kind of sample preparation and/or method of registration of the spectrum: KBr disc. Transmission.
Source: Moore and White (1970).
Wavenumbers (cm^{-1}): 590s, 500s, 365sh, 290, 277, 253, 233w, 187w

O543 Lithium aluminium oxide-gamma γ-$LiAlO_2$

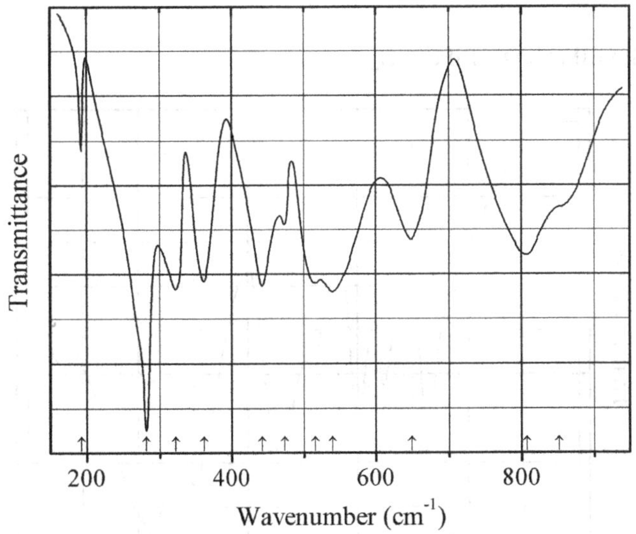

Origin: Synthetic.
Description: Synthesized in a solid-state reaction from Al oxide and Li carbonate. The structure consists of corner-linked tetrahedra in which both Li and Al are four-coordinated. The space group is $P4_12_12$, $Z = 4$.
Kind of sample preparation and/or method of registration of the spectrum: KBr disc. Transmission.
Source: Moore and White (1970).
Wavenumbers (cm^{-1}): 850sh, 807, 649, 540, 516, 473w, 442, 361, 322, 282s, 192w.

O544 Lithium cobalt(III) iron(III) oxide delafossite-type $LiCo_{0.5}Fe_{0.5}O_2$

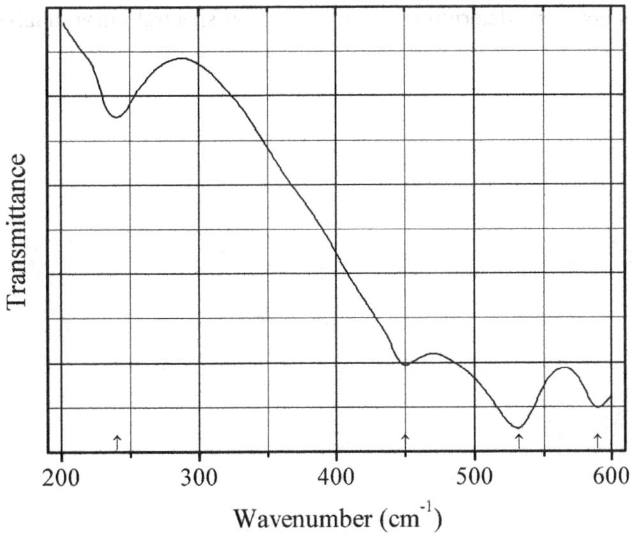

Origin: Synthetic.

Description: Prepared by heating up to 600 °C a gel obtained from the stoichiometric mixture of Li, Co, and Fe nitrates and aqueous solution of maleic acid. Characterized by powder X-ray diffraction data. Trigonal, space group *R*3*m*.

Kind of sample preparation and/or method of registration of the spectrum: KBr disc. Absorption.

Source: Khosravi et al. (2013).

Wavenumbers (cm^{-1}): 590s, 532s, 450, 240w.

O545 Lithium ferrite $LiFe^{3+}{}_5O_8$ $LiFe^{3+}{}_5O_8$

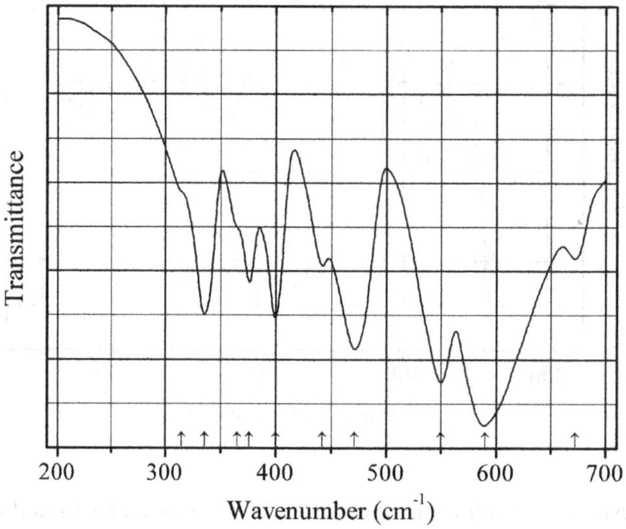

Origin: Synthetic.

Description: Prepared by sintering a mixture of Li_2CO_3 and Fe_2O_3 at 1300 °C. A compound with ordered spinel-type structure. Characterized by powder X-ray diffraction data.

Kind of sample preparation and/or method of registration of the spectrum: KBr disc. Transmission.

Source: Brabers (1976).

Wavenumbers (cm^{-1}): 672w, 590s, 550s, 471, 442, 400, 376, 365sh, 336, 315sh.

Note: The wavenumbers were determined by us based on spectral curve analysis of the published spectrum.

O546 Lithium iron(III) oxide γ-$LiFeO_2$

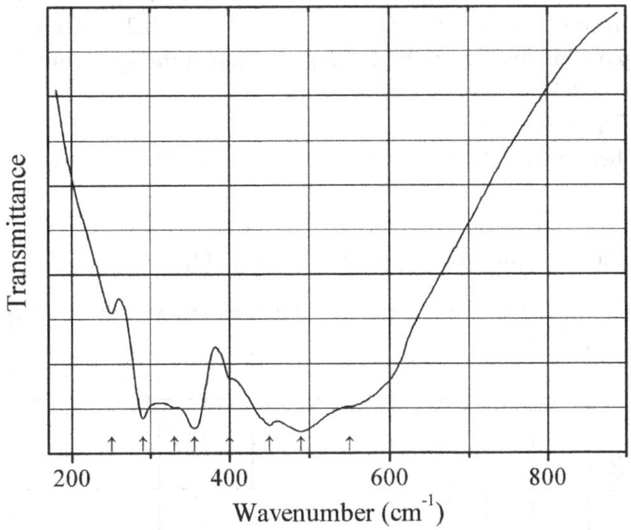

Origin: Synthetic.

Description: Synthesized in a solid-state reaction from Al oxide and Li carbonate. The structure consists of corner-linked tetrahedra in which both Li and Al are four-coordinated. The space group is $P4_12_12$, $Z = 4$.

Kind of sample preparation and/or method of registration of the spectrum: KBr disc. Transmission.

Source: Moore and White (1970).

Wavenumbers (cm^{-1}): 550sh, 490s, 450s, 400sh, 355s, 330sh, 290s, 250.

O547 Lithium magnesium manganese(IV) oxide spinel-type $Li_2MgMn_3O_8$

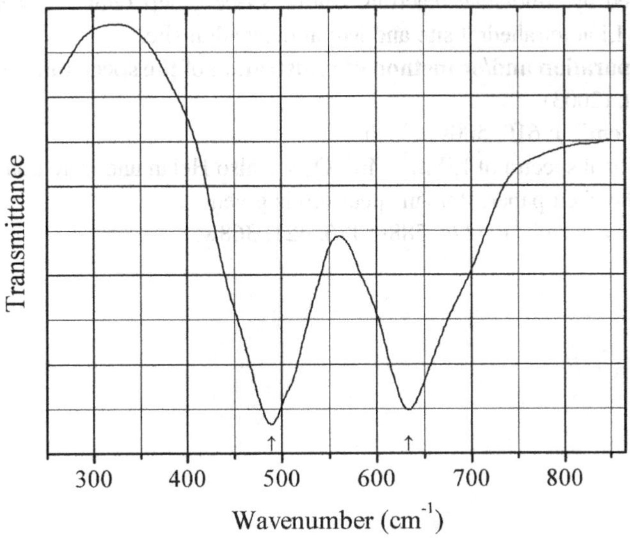

Origin: Synthetic.
Description: Prepared by solid-state reaction using Li_2CO_3, MnO_2, and MgO. Characterized by powder X-ray diffraction data. Cubic, space group *Fd*3*m*, *a* = 8.2794(2) Å.
Kind of sample preparation and/or method of registration of the spectrum: TlBr disc. Absorption.
Source: Strobel et al. (2003).
Wavenumbers (cm^{-1}): 633s, 489s.
Note: The wavenumbers were determined by us based on spectral curve analysis of the published spectrum.

O548 Lithium manganese oxide spinel-type $LiMn^{3+}Mn^{4+}O_4$

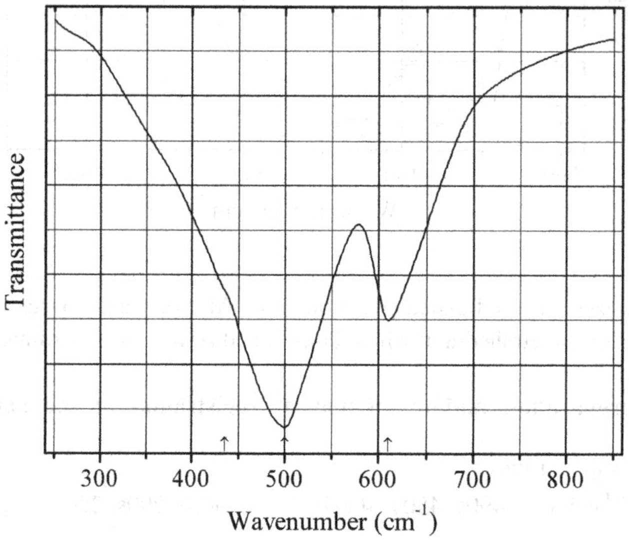

Origin: Synthetic.
Description: Prepared by solid-state reaction. Cubic, space group *Fd*3*m*, *a* = 8.1967 Å. A normal spinel containing Li at tetrahedral site and Mn at octahedral site.
Kind of sample preparation and/or method of registration of the spectrum: TlBr disc. Absorption.
Source: Strobel et al. (2003).
Wavenumbers (IR, cm^{-1}): 610, 500s, 435sh.
Note: For the vibrational spectra of $LiMn^{3+}Mn^{4+}O_4$ see also Helan and Berchmans (2011) and Julien et al. (1998). In the cited paper, Raman spectrum is given.
Wavenumbers (Raman, cm^{-1}): 627s, 588s, 486, 421, 368w.

O549 Lithium niobateilmenite-type $LiNbO_3$

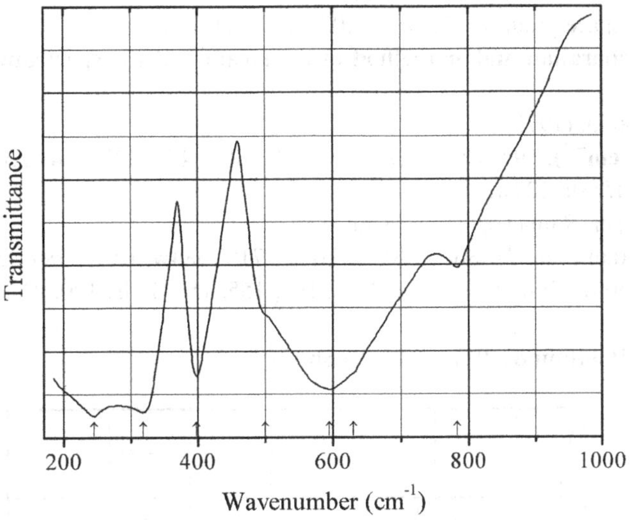

Origin: Synthetic.

Description: Obtained by hydrothermal synthesis. Metastable modification isostructural with ilmenite.

Kind of sample preparation and/or method of registration of the spectrum: KBr disc. Transmission.

Source: Baran et al. (1986).

Wavenumbers (IR, cm^{-1}): 783, 630sh, 595s, 500sh, 398, 318s, 245s.

Note: In the cited paper, Raman spectrum is given.

Wavenumbers (Raman, cm^{-1}): 735s, 677w, 470, 381w, 291, 275, 214, 173w.

O550 Lithium zinc niobium oxide spinel-type $LiZnNbO_4$

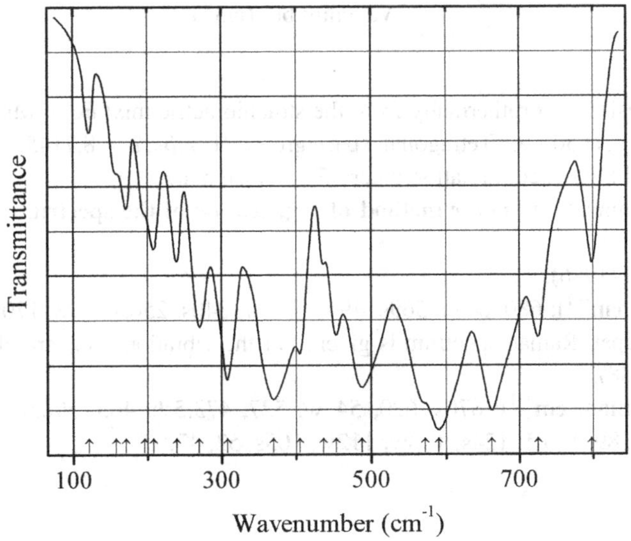

Origin: Synthetic.

Description: Synthesized in the solid-state reaction between Li_2CO_3, ZnO, and Nb_2O_5 at 1000 °C for 10 h. Tetragonal, space group $P4_122$, $a = 6.079$, $c = 8.401$ Å.

Kind of sample preparation and/or method of registration of the spectrum: Polyethylene disc. Transmission.

Source: Keramidas et al. (1975).

Wavenumbers (IR, cm^{-1}): 798, 725, 662s, 590s, 572, 488s, 454, 437w, 405, 370s, 308s, 270, 240, 208, 196sh, 171, 158sh, 122w.

Note: In the cited paper, Raman spectrum is given.

Wavenumbers (Raman, cm^{-1}): 868w, 819s, 762w, 718, 684w, 616s, 589s, 582, 496, 460, 435, 366w, 335, 323, 300w, 266, 250s, 237s, 222s, 193, 155, 152, 134s, 120, 94s.

O551 Manganese(II) antimony(III) oxide $MnSb_2O_4$

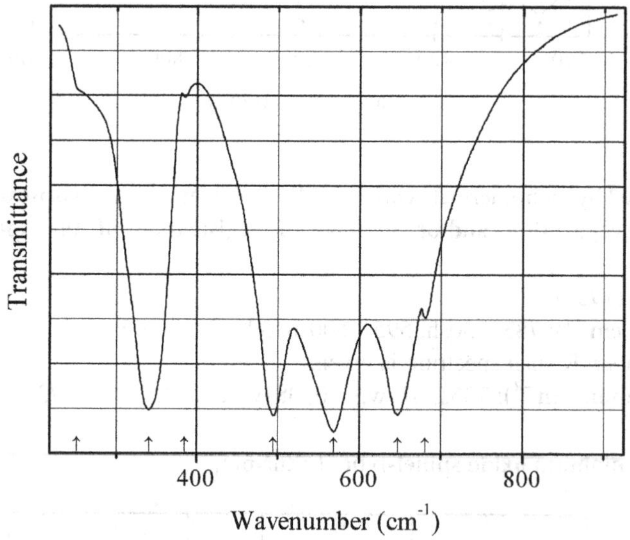

Origin: Synthetic.

Description: Synthesized hydrothermally from the stoichiometric mixture of MnO and Sb_2O_3 in the presence of 5% HF, at 500 °C. Tetragonal, space group $P4_2/mbc$, $a = 8.7145$, $c = 6.0011$ Å, $Z = 4$. Sb has tetrahedral SbO_3E coordination where E is a lone pair.

Kind of sample preparation and/or method of registration of the spectrum: KBr and CsI discs. Transmission.

Source: Chater et al. (1986).

Wavenumbers (IR, cm^{-1}): 680, 647s, 569s, 495s, 385w, 340s, 250sh, 198, 170, 134, 96.5, 79.

Note: In the cited paper, Raman spectrum is given. For the vibrational spectra of $MnSb_2O_4$ see also Gavarri et al. (1988).

Wavenumbers (Raman, cm^{-1}): 670s, 620, 547w, 527, 473.5sh, 465, 398.5w, 350, 345sh, 292s, 254.5, 221, 215, 189, 155.5, 124s, 118w, 112w, 105s, 52, 47.

O552 Nickel manganese(IV) oxide Ni_6MnO_8

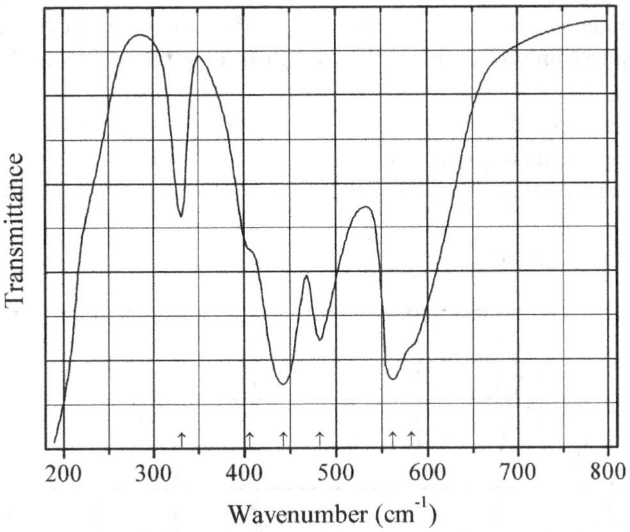

Origin: Synthetic.

Description: Prepared by addition an excess of oxalic acid to boiled solution of Ni(II) and Mn (II) acetates in acetic acid (25%) with subsequent drying and calcination at 873 K for 3 h. Characterized by powder X-ray diffraction data. Cubic, space group *Fm3m*, *a* = 8.306(3).

Kind of sample preparation and/or method of registration of the spectrum: KBr disc. Transmission.

Source: Porta et al. (1991).

Wavenumbers (cm^{-1}): 582sh, 562s, 482, 443s, 406sh, 331.

O553 Potassium diuranate $K_2U_2O_7$

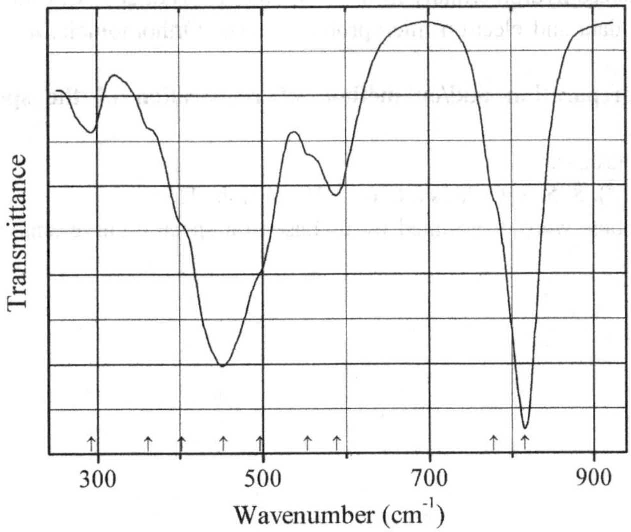

Origin: Synthetic.

Description: Prepared by heating stoichiometric mixture of U_3O_8 and K_2CO_3 in air. Characterized by powder X-ray diffraction data. Orthorhombic, $a = 6.95(2)$, $b = 7.97(2)$, $c = 22.16(2)$ Å.

Kind of sample preparation and/or method of registration of the spectrum: KBr disc. Absorption.

Source: Volkovich et al. (1998).

Wavenumbers (IR, cm^{-1}): 816s, 778sh, 588, 554sh, 497sh, 452s, 401sh, 361sh, 292.

Note: In the cited paper, Raman spectrum is given.

Wavenumbers (Raman, cm^{-1}): 778s, 562w, 491w, 434, 336, 287, 267w, 245w, 150sh, 133, 100w.

O554 Potassium niobate $KNbO_3$

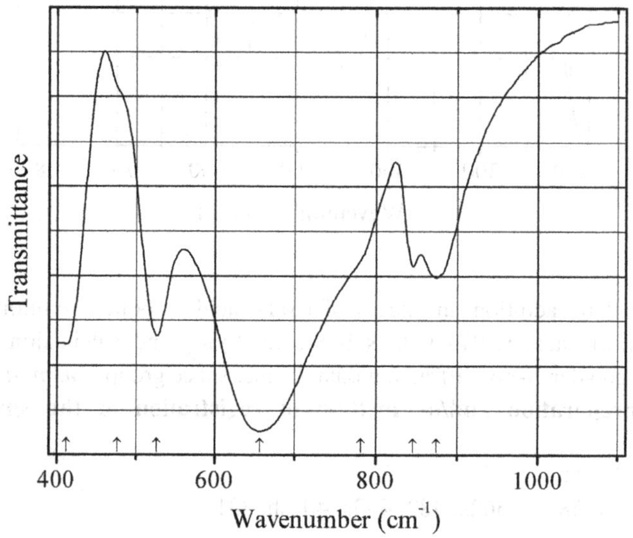

Origin: Synthetic.

Description: Synthesized hydrothermally from Nb_2O_5 and KOH at 200 °C. Characterized by powder X-ray diffraction data and electron microprobe analysis. Orthorhombic, $a = 5.697$, $b = 3.971$, $c = 5.721$ Å.

Kind of sample preparation and/or method of registration of the spectrum: KBr disc. Transmission.

Source: Wang et al. (2007).

Wavenumbers (cm^{-1}): 875, 846, 782sh, 656s, 526s, 477sh, 412s.

Note: The wavenumbers were determined by us based on spectral curve analysis of the published spectrum.

O555 Potassium niobate KNb_7O_{18} KNb_7O_{18}

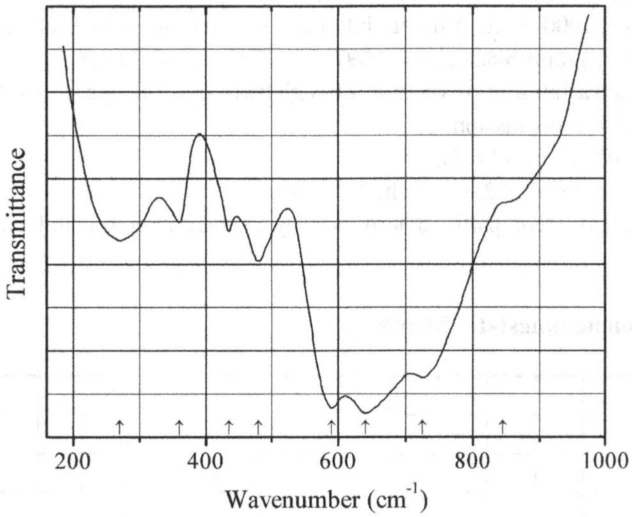

Origin: Synthetic.

Description: Structurally related to $TlNb_7O_{18}$ (tetragonal, space group *P4/mbm*, $a = 27.50$, $c = 3.94$ Å, $Z = 8$).

Kind of sample preparation and/or method of registration of the spectrum: CsI disc. Transmission.

Source: Bhide et al. (1980).

Wavenumbers (cm^{-1}): 845sh, 725s, 640s, 590s, 480, 435, 360, 270.

O556 Potassium niobate perovskite-type $KNbO_3$

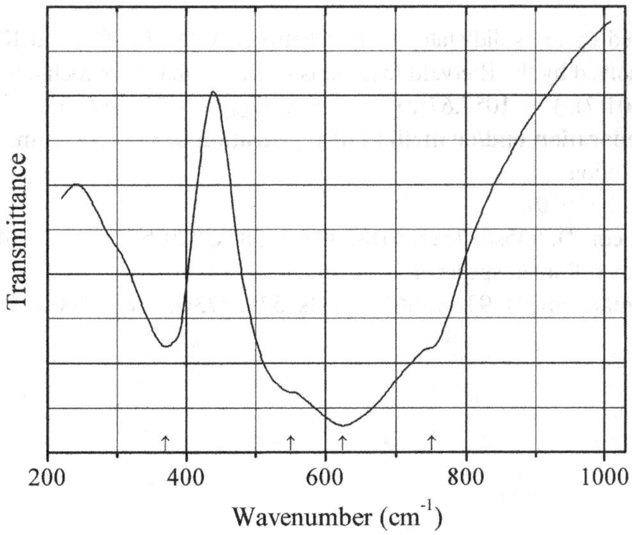

Origin: Synthetic.

Description: Prepared in a solid-state reaction, by double-ply heating the stoichiometric mixture of Nb_2O_5 and KNO_3 at 1000 °C for 1 h with intermediate grinding. Structurally related to perovskite. Orthorhombic, space group *Bmm*2, $a = 5.697$, $b = 3.971$, $c = 5.721$ Å.

Kind of sample preparation and/or method of registration of the spectrum: KBr or CsI disc and polyethylene matrix. Transmission.

Source: Rocchiccioli-Deltcheff (1973).

Wavenumbers (cm^{-1}): 750sh, 625s, 550sh, 370, 180.

Note: The wavenumbers were partly determined by us based on spectral curve analysis of the published spectrum.

O557 Potassium niobate tungstate $KNbWO_6$

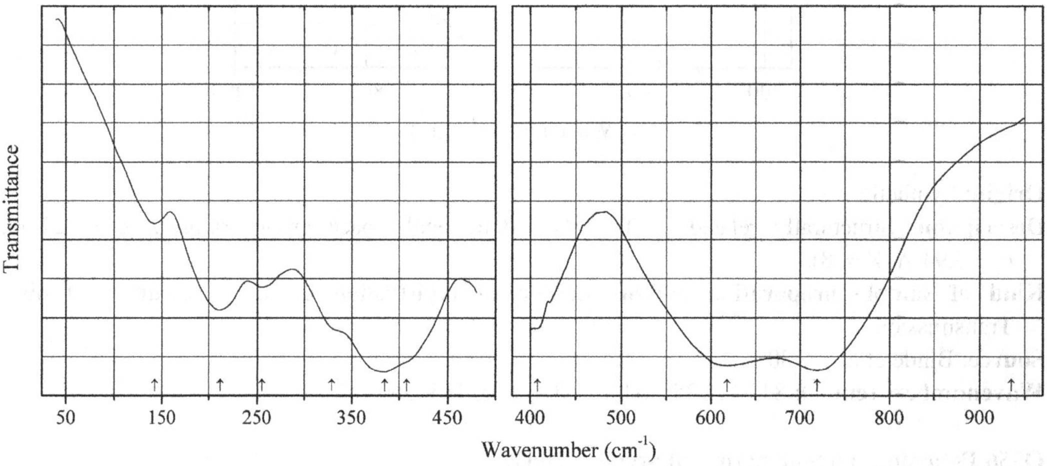

Origin: Synthetic.

Description: Prepared by the solid-state reaction between WO_3, Nb_2O_5, and KNO_3 at 973 K. The crystal structure solved by the Rietveld method is related to that of pyrochlore. Cubic, space group *Fd*3*m*, $a = 10.5001(7)$, $V = 1057.67(2)$ Å^3, $Z = 8$. $D_{calc} = 4.7529$ g/cm^3.

Kind of sample preparation and/or method of registration of the spectrum: KBr disc and Nujol suspension. Absorption.

Source: Knyazev et al. (2010).

Wavenumbers (IR, cm^{-1}): 935sh, 719s, 618s, 408sh, 385s, 329sh, 255, 212, 142.

Note: In the cited paper, Raman spectrum is given.

Wavenumbers (Raman, cm^{-1}): 934w, 861w, 664s, 576, 473w, 438w, 360w, 246sh, 196s, 152s.

O558 Potassium tantalite perovskite-type $KTaO_3$

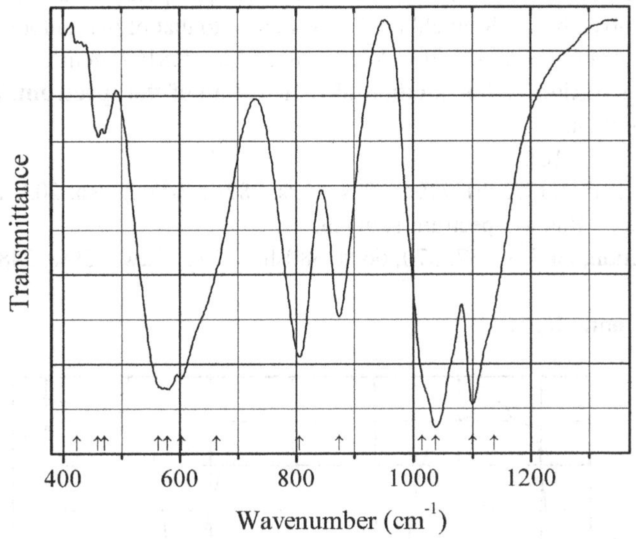

Origin: Synthetic.

Description: Prepared in a solid-state reaction, by double-ply heating the stoichiometric mixture of Nb_2O_5 and KNO_3 at 1230 °C for 4 h with intermediate grinding. Structurally related to perovskite. Cubic, $a = 3.989$ Å.

Kind of sample preparation and/or method of registration of the spectrum: KBr or CsI disc and polyethylene matrix. Transmission.

Source: Rocchiccioli-Deltcheff (1973).

Wavenumbers (cm^{-1}): 755sh, 605s, 340.

Note: The wavenumbers were determined by us based on spectral curve analysis of the published spectrum.

O559 Potassium tantalate tungstate $KTaWO_6$

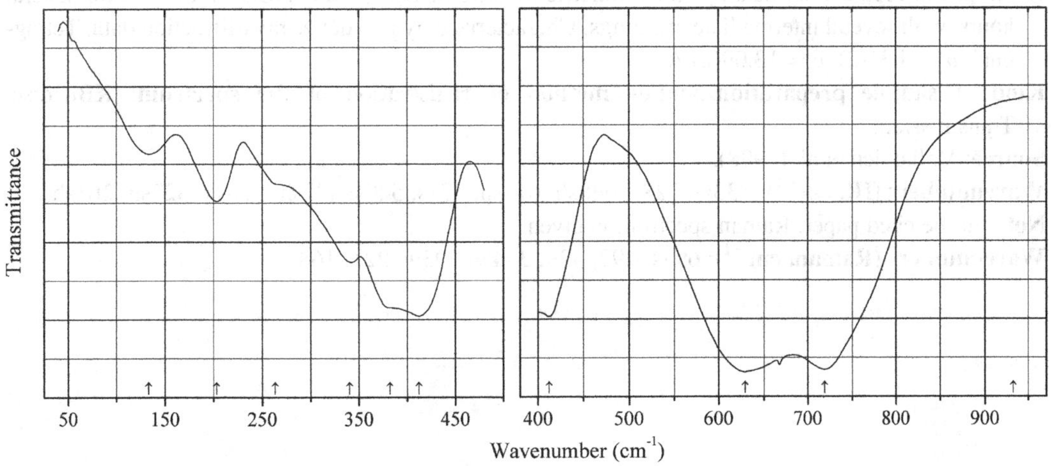

Origin: Synthetic.

Description: Prepared by the solid-state reaction between WO_3, Nb_2O_5, and KNO_3 at 973 K. The crystal structure solved by the Rietveld method is related to that of pyrochlore. Cubic, space group *Fd*3*m*, $a = 10.4695(1)$, $V = 1147.57(3)$ Å^3, $Z = 8$. $D_{calc} = 5.8197$ g/cm^3.

Kind of sample preparation and/or method of registration of the spectrum: KBr disc and Nujol suspension. Absorption.

Source: Knyazev et al. (2010).

Wavenumbers (IR, cm^{-1}): (933sh), 719s, 629s, 412s, 382sh, 340, 263sh, 203, 133.

Note: In the cited paper, Raman spectrum is given.

Wavenumbers (Raman, cm^{-1}): 937, 870, 664s, 580sh, 481w, 452w, 359w, 248s, 191s, 145s.

O560 Potassium urinate K_2UO_4

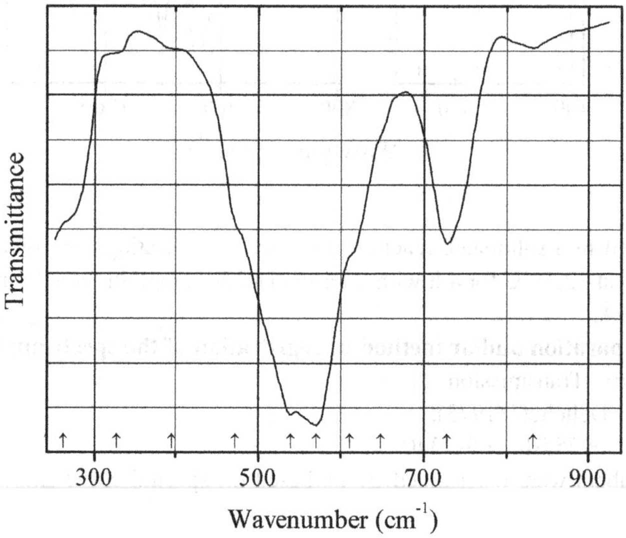

Origin: Synthetic.

Description: Prepared by heating stoichiometric mixture of U_3O_8 with K_2CO_3 at 800 °C for several hours, with several intermediate grindings. Characterized by powder X-ray diffraction data. Tetragonal, $a = 4.31(2)$, $c = 13.09(2)$ Å.

Kind of sample preparation and/or method of registration of the spectrum: KBr disc. Transmission.

Source: Volkovich et al. (1998).

Wavenumbers (IR, cm^{-1}): 833w, 728s, 648sh, 611sh, 571s, 540s, 472sh, 395sh, 327sh, 261sh.

Note: In the cited paper, Raman spectrum is given.

Wavenumbers (Raman, cm^{-1}): 694s, 492, 439, 344w, 293w, 221, 168.

O561 Sodium diuranate $Na_2U_2O_7$

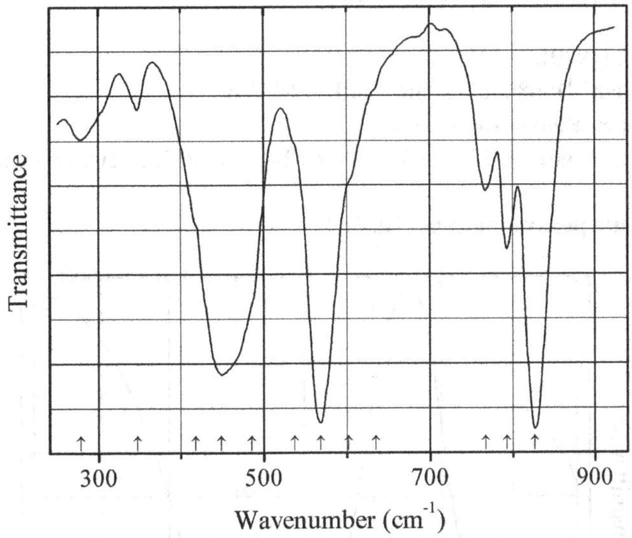

Origin: Synthetic.

Description: Prepared by heating stoichiometric mixture of U_3O_8 and Na_2CO_3 in air. Characterized by powder X-ray diffraction data. Orthorhombic, $a = 6.77(1)$, $b = 7.97(1)$, $c = 18.32(1)$ Å.

Kind of sample preparation and/or method of registration of the spectrum: KBr disc. Absorption.

Source: Volkovich et al. (1998).

Wavenumbers (IR, cm^{-1}): 827s, 793, 767, 634sh, 602sh, 568s, 537sh, 486sh, 449s, 418sh, 347w, 278.

Note: In the cited paper, Raman spectrum is given.

Wavenumbers (Raman, cm^{-1}): 826w, 788s, 779s,752, 599sh, 584, 536w, 420s, 357w, 313s, 274, 233, 202, 146, 117s, 100w.

O562 Sodium stannate Na_4SnO_4

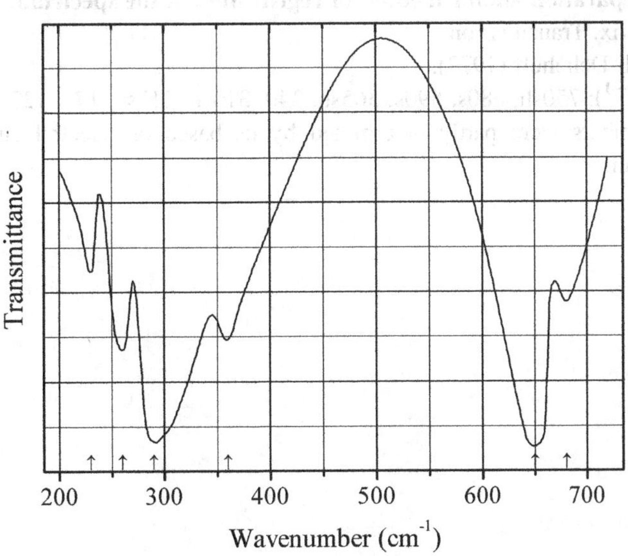

Origin: Synthetic.

Kind of sample preparation and/or method of registration of the spectrum: CsI disc. Transmission.

Source: Kessler et al. (1979).

Wavenumbers (IR, cm^{-1}): 680, 650s, 360, 290s, 260, 230.

Note: In the cited paper, Raman spectrum is given.

Wavenumbers (Raman, cm^{-1}): 679, 664s, 638, 612w, 310, 235, 212w, 180, 160, 145, 100.

O563 Sodium tantalite perovskite-type $NaTaO_3$

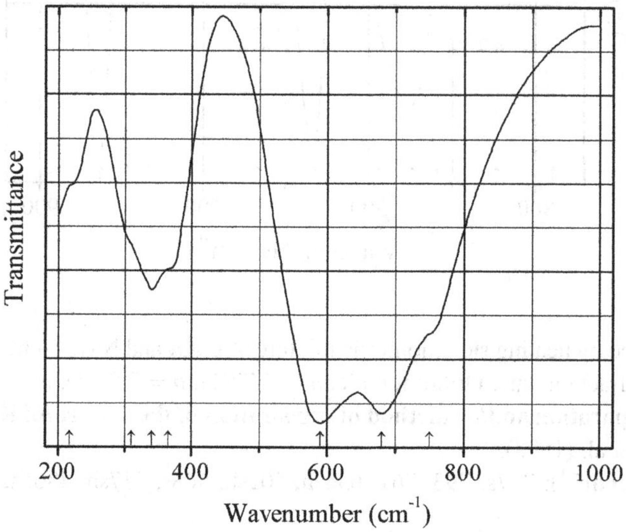

Origin: Synthetic.

Description: Prepared in a solid-state reaction, by double-ply heating the stoichiometric mixture of Ta_2O_5 and $NaNO_3$ at 1200 °C for 6 h and for 12 h with intermediate grinding. Structurally related to perovskite. Orthorhombic, space group $Pc2_1n$, $a \approx 5.51$–5.52, $b \approx 7.75$–7.79, $c \approx 5.48$–5.50 Å.

Kind of sample preparation and/or method of registration of the spectrum: KBr or CsI disc and polyethylene matrix. Transmission.

Source: Rocchiccioli-Deltcheff (1973).

Wavenumbers (cm^{-1}): 750sh, 680s, 590s, 365sh, 340, 310sh, 217sh, 173, 127.

Note: The wavenumbers were partly determined by us based on spectral curve analysis of the published spectrum.

O564 Sodium uranate Na_2UO_4

Origin: Synthetic.

Description: Prepared by heating stoichiometric mixture of U_3O_8 and Na_2CO_3 at 800 °C for several hours, with several intermediate grindings. Characterized by powder X-ray diffraction data. Orthorhombic, $a = 6.77(1)$, $b = 7.97(1)$, $c = 18.32(1)$ Å.

Kind of sample preparation and/or method of registration of the spectrum: KBr disc. Transmission.

Source: Volkovich et al. (1998).

Wavenumbers (IR, cm^{-1}): 809w, 774s, 732s, 706w, 518s, 488, 453s, 310, 257s.

Note: The IR data are taken from Volkovich et al. (1998). There are strong discrepancies between these data and the figure of the IR spectrum of Na_2UO_4 given in this paper. In the cited paper, Raman spectrum is given.

Wavenumbers (Raman, cm^{-1}): 736w, 712s, 547w, 506w, 442w, 362w, 329w, 238s, 177w, 140w.

O565 Sodium yttrium titanate $NaYTiO_4$

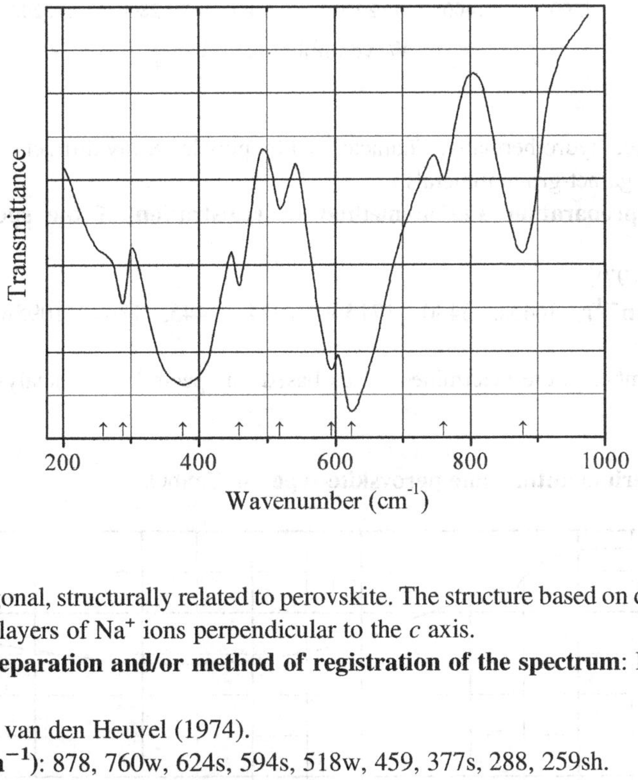

Origin: Synthetic.

Description: Tetragonal, structurally related to perovskite. The structure based on double layers of Y^{3+} ions and double layers of Na^{+} ions perpendicular to the *c* axis.

Kind of sample preparation and/or method of registration of the spectrum: No data in the cited paper.

Source: Blasse and van den Heuvel (1974).

Wavenumbers (cm^{-1}): 878, 760w, 624s, 594s, 518w, 459, 377s, 288, 259sh.

Note: The wavenumbers were determined by us based on spectral curve analysis of the published spectrum. The band at 878 cm^{-1} is ascribed by the authors to stretching vibrations of the Ti–O bond directed towards the Na-layers. In the cited paper, a figure of the Raman spectrum is given.

O566 Strontium aluminum hydroxide $Sr_3Al_2(OH)_{12}$

Hydrogarnet $Sr_3Al_2(OH)_{12}$ $Sr_3Al_2(OH)_{12}$

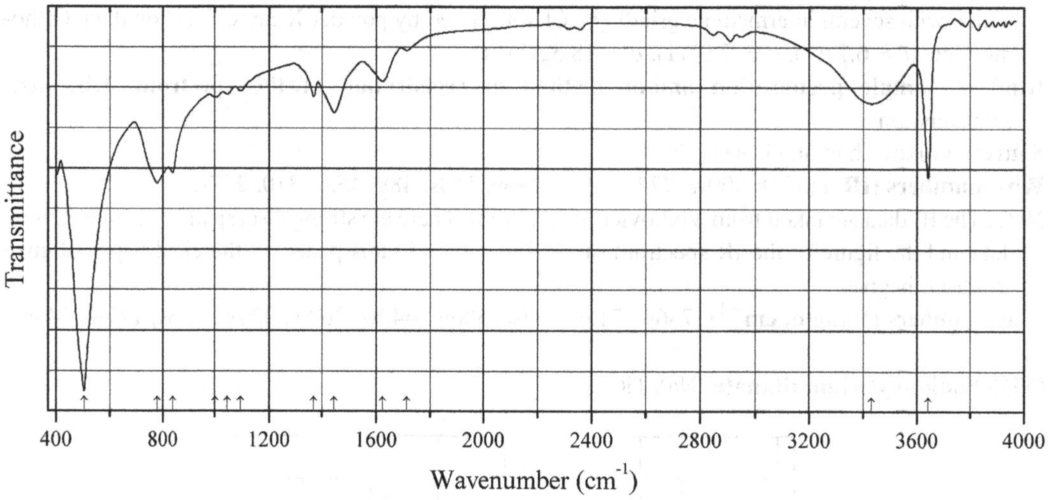

Origin: Synthetic.

Description: Prepared hydrothermally. Characterized by powder X-ray diffraction data. Cubic, structurally related to garnet-group minerals.

Kind of sample preparation and/or method of registration of the spectrum: KBr disc. Transmission.

Source: Li et al. (1997).

Wavenumbers (cm^{-1}): 3643s, 3430, 1715w, 1624, 1445, 1369, 1095w, 1048w, 1002w, 839, 776, 505s.

Note: The wavenumbers were determined by us based on spectral curve analysis of the published spectrum.

O567 Strontium cerium antimonate perovskite-type Sr_2CeSbO_6

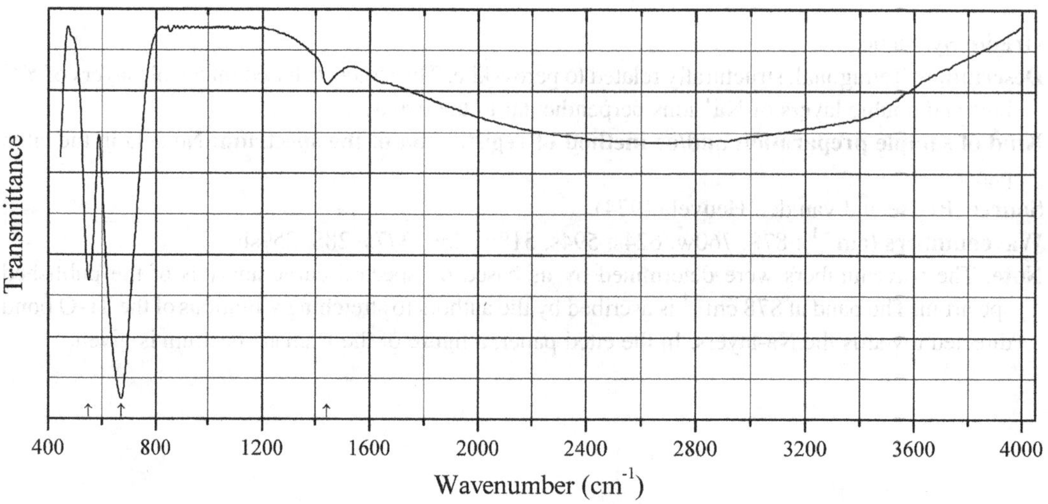

Origin: Synthetic.

Description: A compound with ordered perovskite-type structure prepared by heating a mixture of $SrCO_3$, $Ce_2(CO_3)_3$, and Sb_2O_5, taken in stoichiometric ratio, at 1350 °C for 15 h. Characterized by powder X-ray diffraction data and energy dispersive X-ray spectrum. Orthorhombic, $a = 8.84$, $b = 6.22$, $c = 5.83$ Å.

Kind of sample preparation and/or method of registration of the spectrum: KBr disc. Transmission.

Source: Bharti, and Sinha (2010).

Wavenumbers (cm^{-1}): 1443w, 670s, 550.

Note: The wavenumbers were partly determined by us based on spectral curve analysis of the published spectrum. The weak band at 1443 cm^{-1} may correspond to the admixture of a carbonate.

O568 Strontium magnesium niobate $Sr_3MgNb_2O_9$

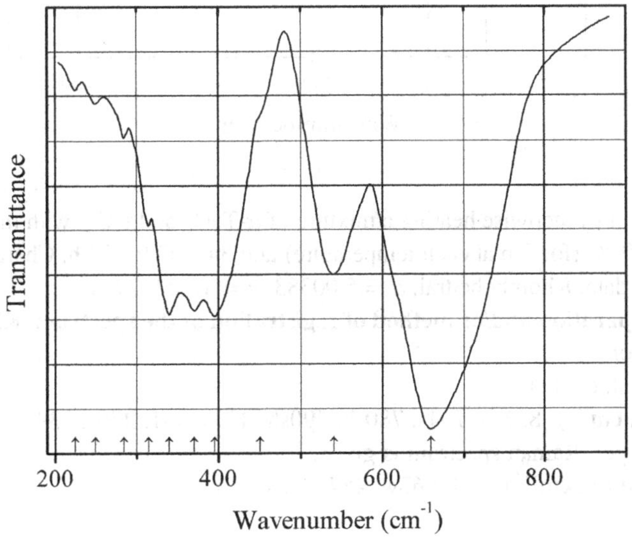

Origin: Synthetic.

Description: A compound with ordered perovskite-type structure prepared by a solid-state reaction technique. Characterized by powder X-ray diffraction data.

Kind of sample preparation and/or method of registration of the spectrum: CsI disc (?). Transmission.

Source: Blasse, and Corsmit (1974).

Wavenumbers (IR, cm^{-1}): 660s, 540, 450sh, 395s, 370s, 340s, 315, 285w, 250w, 225w.

Note: In the cited paper, Raman spectrum is given.

Wavenumbers (Raman, cm^{-1}): 830s, 535w, 455, 400, 310w, 240w.

O569 Tellurite rhombohedral polymorph TeO_3

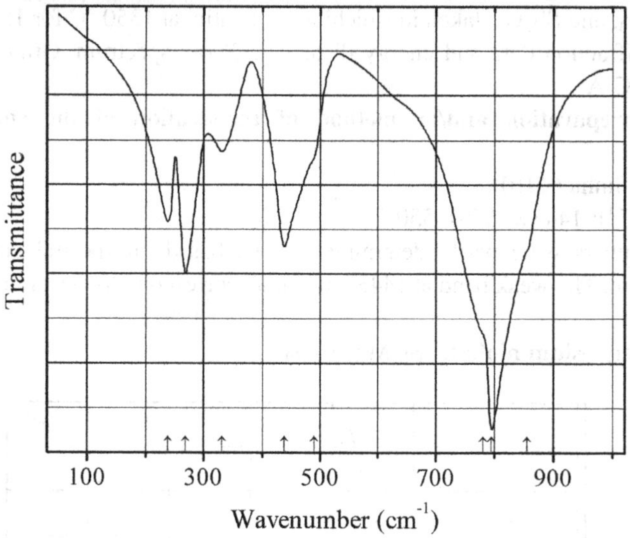

Origin: Synthetic.

Description: Prepared by stepwise heating a mixture of α-TeO_2 and I_2O_5 (with the molar ratio 1:2) at 250, 280, and 310 °C (for 2 h at each temperature) and 340 °C for 24 h. Characterized by powder X-ray diffraction data. Rhombohedral, $a = 5.00383$, $c = 13.22429$ Å.

Kind of sample preparation and/or method of registration of the spectrum: KBr and polyethylene discs. Transmission.

Source: Cornette et al. (2011).

Wavenumbers (IR, cm^{-1}): 855sh, 794s, 780sh, 490sh, 439s, 331, 270s, 239.

Note: In the cited paper, Raman spectrum is given.

Wavenumbers (Raman, cm^{-1}): 844, 663w, 487, 333s, 261.

O570 Tellurium(IV) tin oxide Te_3SnO_8 Te_3SnO_8

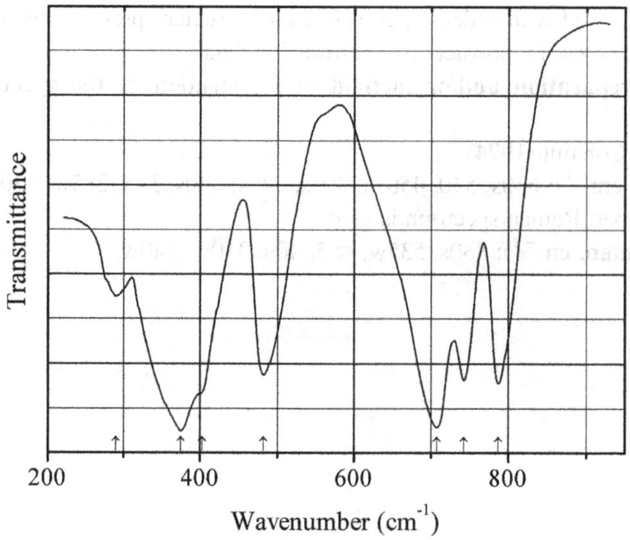

Origin: Synthetic.

Description: Prepared by heating a mixture of TeO_2 and SnO_2 taken in stoichiometric amounts, at 700–750 °C for 12–15 h with several intermediate grindings. Cubic, space group *Ia*3, Z = 8.

Kind of sample preparation and/or method of registration of the spectrum: KBr disc. Transmission.

Source: Botto and Baran (1981).

Wavenumbers (cm^{-1}): 788, 743, 708s, 482, 403sh, 375s, 290w.

O571 Tellurium(IV) titanium oxide Te_3TiO_8 Te_3TiO_8

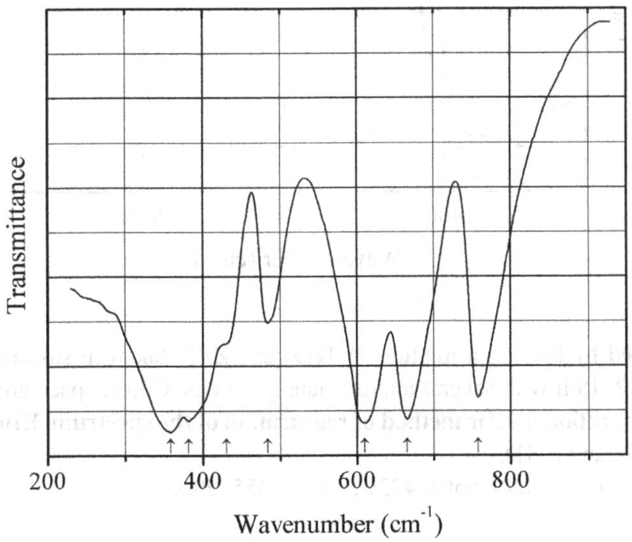

Origin: Synthetic.

Description: Prepared by heating a mixture of TeO_2 and TiO_2 taken in stoichiometric amounts, at 700–750 °C for 12–15 h with several intermediate grindings. Cubic, space group *Ia*3, Z = 8.

Kind of sample preparation and/or method of registration of the spectrum: KBr disc. Transmission.

Source: Botto and Baran (1981).

Wavenumbers (cm^{-1}): 758, 665s, 610s, 484, 430sh, 382sh, 358s.

O572 Tellurium(IV) zirconium oxide Te_3ZrO_8 Te_3ZrO_8

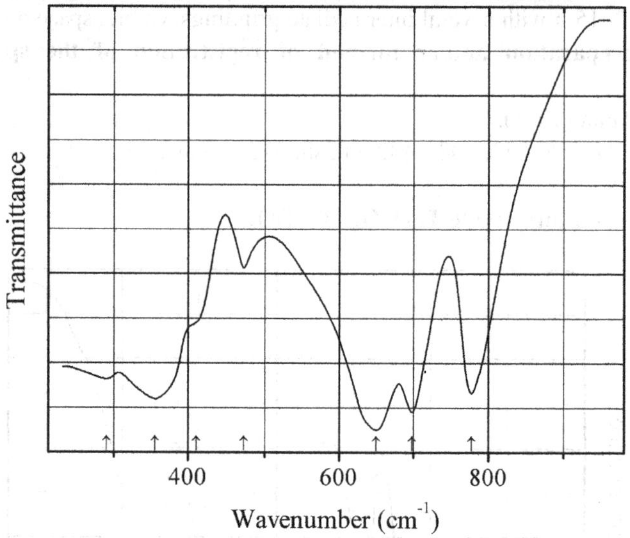

Origin: Synthetic.

Description: Prepared by heating a mixture of TeO_2 and ZrO_2 taken in stoichiometric amounts, at 700–750 °C for 12–15 h with several intermediate grindings. Cubic, space group *Ia*3, *Z* = 8.

Kind of sample preparation and/or method of registration of the spectrum: KBr disc. Transmission.

Source: Botto and Baran (1981).

Wavenumbers (cm^{-1}): 778, 698, 650s, 472w, 410sh, 355, 290.

O573 Tin(IV) hydroxide $Sn(OH)_4$

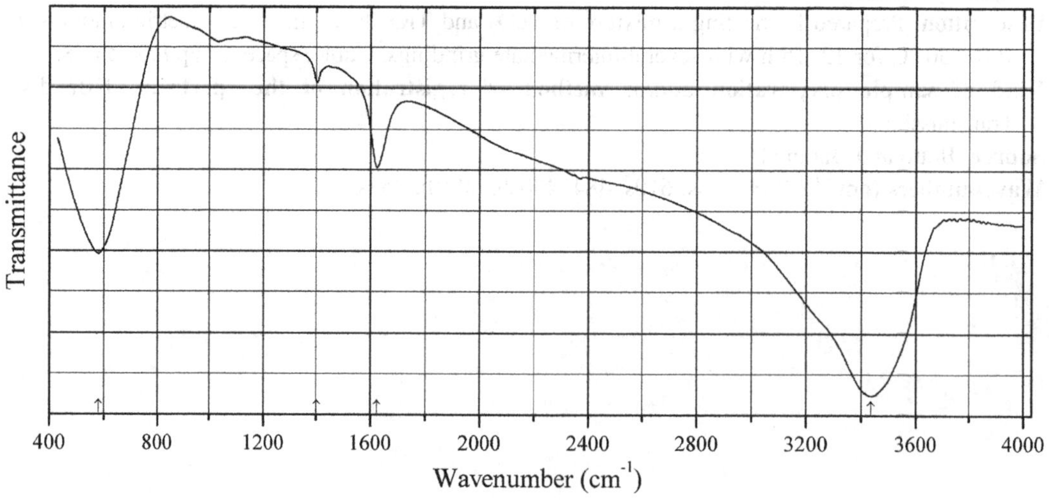

Origin: Synthetic.

Description: Synthesized by dissolving tin metal in concentrated HCl followed by addition of concentrated NH_4OH.

Kind of sample preparation and/or method of registration of the spectrum: Transmission. Kind of sample preparation is not indicated.

Source: Prodjosantoso et al. (2015).

Wavenumbers (cm^{-1}): 3436s, 1625, 1402w, 580s.

O574 Tungsten trioxide monoclinic WO_3

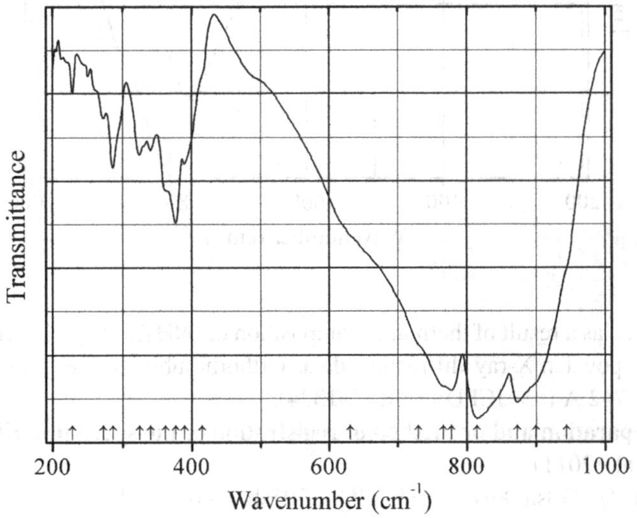

Origin: Synthetic.

Description: Obtained as a result of thermal decomposition of $(NH_4)_{10}W_{12}O_{41}{\cdot}nH_2O$ at 1223 K for 5 h. Characterized by powder X-ray diffraction data. Monoclinic, space group $P2_1/n$, $a = 7.319$, $b = 7.556$, $c = 7.722$ Å, $\beta = 90.48°$ (see JCPDS card 43-1035).

Kind of sample preparation and/or method of registration of the spectrum: KBr disc. Absorption.

Source: Kustova et al. (2011).

Wavenumbers (cm^{-1}): 943sh, 900sh, 873s, 815s, 777s, 766sh, 730sh, 631sh, 415sh, 390, 377, 362sh, 341, 325, 287, 273w, 228w.

0575 Tungsten trioxide orthorhombic WO_3

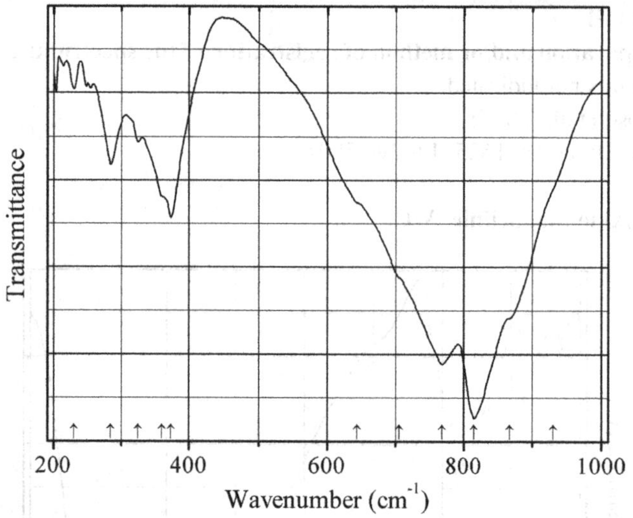

Origin: Synthetic.

Description: Obtained as a result of thermal decomposition of $(NH_4)_{10}W_{12}O_{41}{\cdot}nH_2O$ at 873 K for 6 h. Characterized by powder X-ray diffraction data. Orthorhombic, space group *Pcnb*, $a = 7.339$, $b = 7.574$, $c = 7.742$ Å (see JCPDS card 20-1324).

Kind of sample preparation and/or method of registration of the spectrum: KBr disc. Absorption.

Source: Kustova et al. (2011).

Wavenumbers (cm^{-1}): 930sh, 867sh, 815s, 768s, 705sh, 644sh, 374, 360sh, 325, 285, 230w.

0576 Tungsten trioxide triclinic WO_3

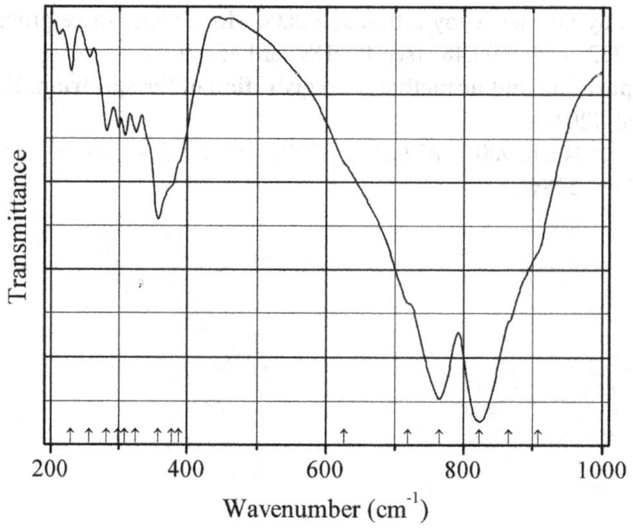

Origin: Synthetic.

Description: Obtained by mechanical treatment (grinding and pressing) of monoclinic WO_3. Characterized by powder X-ray diffraction data. Orthorhombic, space group *P*-1, $a = 7.309$, $b = 7.522$, $c = 7.671$ Å, $\alpha = 88.8°$, $\beta = 90.93°$, $\gamma = 90.93°$ (see JCPDS card 32-1395).

Kind of sample preparation and/or method of registration of the spectrum: KBr disc. Absorption.

Source: Kustova et al. (2011).

Wavenumbers (cm^{-1}): 908sh, 866sh, 823s, 765s, 719sh, 627sh, 389sh, 378sh, 358, 325, 309, 299, 282w, 257, 229w.

O577 Vanadium oxide bariandite-type $V_{10}O_{24}{\cdot}9H_2O$

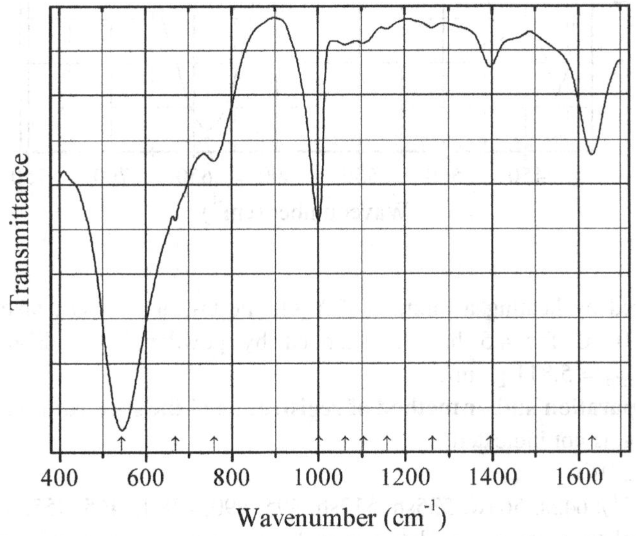

Origin: Synthetic.

Description: Mixed valence vanadium(IV)/(V) nanostructured oxide with a bariandite-like structure prepared by sol-gel processing of themolecular vanadium(IV) alkoxide $[V_2(OPr^i)_8]$ (OPr^i = isopropoxide).

Kind of sample preparation and/or method of registration of the spectrum: KBr disc. Transmission.

Source: Menezes et al. (2009).

Wavenumbers (IR, cm^{-1}): 1632, 1396, 1261w, 1158w, 1104w, 1062w, 1001, 758, 669, 544s.

Note: The wavenumbers were partly determined by us based on spectral curve analysis of the published spectrum. In the cited paper, Raman spectrum is given.

Wavenumbers (Raman, cm^{-1}): 1022, 908s, 518s, 429w, 409w, 270s.

O578 Yttrium iron antimony(V) oxide pyrochlore-type Y_2FeSbO_7

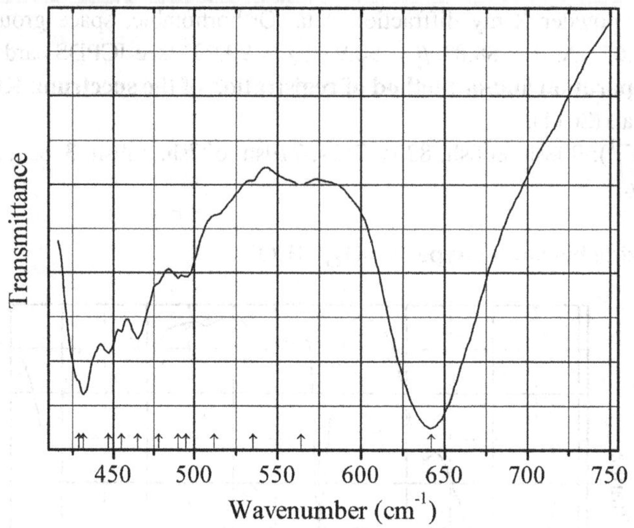

Origin: Synthetic.

Description: Prepared by heating a mixture of Y_2O_3, Fe_2O_3, and Sb_2O_3, taken in stoichiometric amounts, at 1150 °C for 4.5 h. Characterized by powder X-ray diffraction data. Cubic, $a = 10.223$ Å. $D_{calc} = 5.811$ g/cm^3.

Kind of sample preparation and/or method of registration of the spectrum: Transmission. Kind of sample preparation is not indicated.

Source: Jana et al. (2016).

Wavenumbers (cm^{-1}): 642s, 564w, 535sh, 512sh, 495, 490, 478sh, 465, 455, 447, 432s, 429sh.

Note: The wavenumbers were partly determined by us based on spectral curve analysis of the published spectrum.

O579 Yttrium oxide Y_2O_3

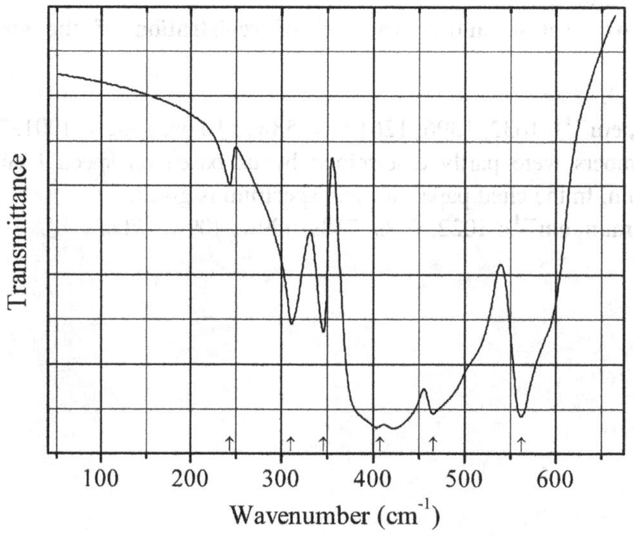

Origin: Synthetic.

Description: Commercial reactant. Characterized by powder X-ray diffraction data. Cubic, space group *Ia*3.

Kind of sample preparation and/or method of registration of the spectrum: Thin powdery film on polyethylene sheet. Transmission.

Source: White and Keramidas (1972).

Wavenumbers (IR, cm^{-1}): 562s, 465s, 408s, 346, 311, 243.

Note: In the cited paper, Raman spectrum is given.

Wavenumbers (Raman, cm^{-1}): 603, 576w, 480s, 440, 389s, 337, 325, 236w, 162.

O580 Franklinite $ZnFe_2O_4$

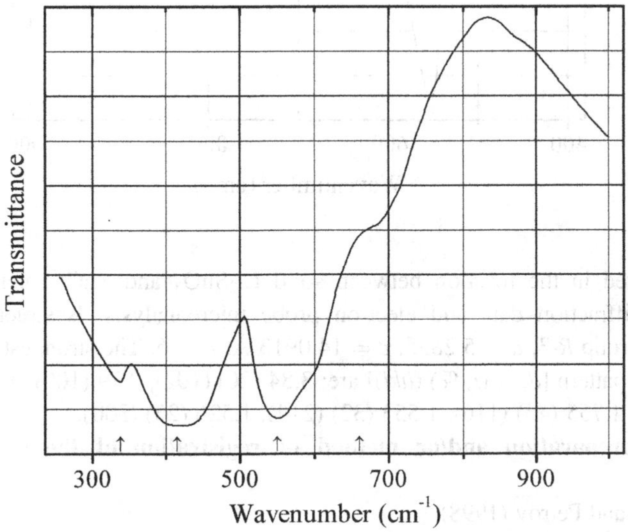

Origin: Synthetic.

Description: Prepared using a conventional ceramic technique. Characterized by powder X-ray diffraction data.

Kind of sample preparation and/or method of registration of the spectrum: KBr disc. Transmission.

Source: Srinivasan et al. (1984).

Wavenumbers (cm^{-1}): 660sh, 550s, 415s, 338.

0581 Zinc stannate $ZnSnO_3$

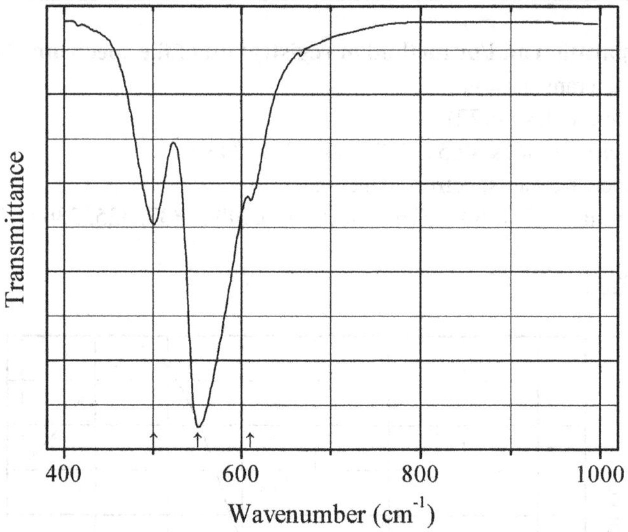

Origin: Synthetic.

Description: Prepared in the reaction between solid Li_2SnO_3 and $ZnCl_2$ melt. Characterized by powder X-ray diffraction data and electron probe microanalysis. Isostructural with ilmenite. Trigonal, space group *R*-3, $a = 5.2835$, $c = 14.0913$ Å, $Z = 6$. The strongest lines of the powder X-ray diffraction pattern [*d*, Å (*I*, %) (*hkl*)] are: 3.84 (20) (102), 2.79 (100) (104), 2.64 (88) (110), 1.918 (26) (204), 1.755 (45) (116), 1.553 (32) (214), 1.526 (25) (300).

Kind of sample preparation and/or method of registration of the spectrum: KBr disc. Transmission.

Source: Kovacheva and Petrov (1998).

Wavenumbers (cm^{-1}): 610, 552s, 501.

0583 Bismite α-Bi_2O_3

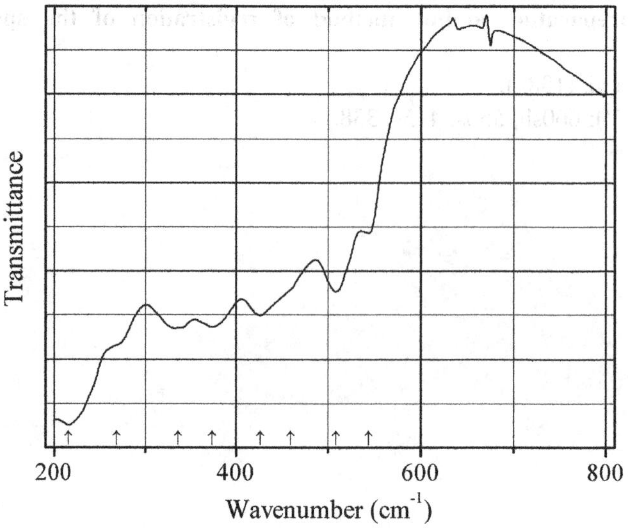

Origin: Synthetic.

Kind of sample preparation and/or method of registration of the spectrum: KBr disc. Absorption.

Source: Cahen et al. (1980).

Wavenumbers (cm^{-1}): 544w, 509, 460sh, 426, 374, 337, 269sh, 216s.

Note: The wavenumbers were determined by us based on spectral curve analysis of the published spectrum.

O584 Bismutocolumbite $BiNbO_4$

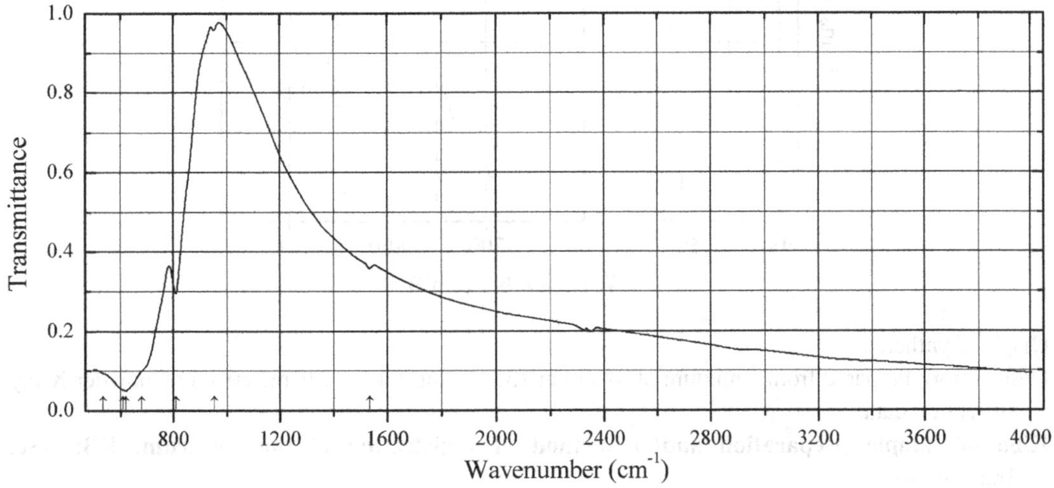

Origin: Synthetic.

Description: Prepared by stepwise heating of stoichiometric mixture of Bi_2O_3 and Nb_2O_5 at 700, 800, and 900 °C for 6 h at each temperature. Characterized by powder X-ray diffraction data. Orthorhombic.

Kind of sample preparation and/or method of registration of the spectrum: KBr disc. Transmission.

Source: Rao and Buddhudu (2010).

Wavenumbers (IR, cm^{-1}): 1534w, 952w, 809, 680sh, 620s, 610sh, 538sh.

Note: The wavenumbers were partly determined by us based on spectral curve analysis of the published spectrum. In the cited paper, Raman spectrum is given.

Wavenumbers (Raman, cm^{-1}): 984w, 840s, 746w, 621, 540, 474w, 422, 242s, 196s.

O585 Bismutotantalite triclinic dimorph $BiTaO_4$

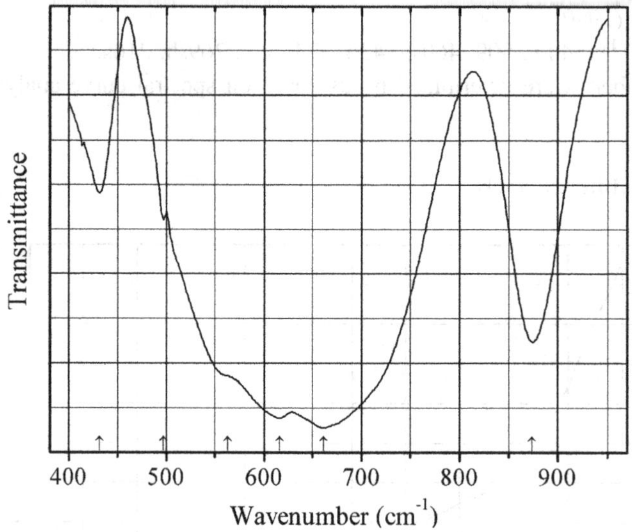

Origin: Synthetic.

Description: Prepared from a mixture of oxides at 1373 K for 2 days. Characterized by powder X-ray diffraction data.

Kind of sample preparation and/or method of registration of the spectrum: KBr disc. Transmission.

Source: Zhang et al. (2009b).

Wavenumbers (cm^{-1}): 874, 661s, 616s, 563sh, 497w, 432.

Note: The wavenumbers were determined by us based on spectral curve analysis of the published spectrum.

O586 Braunite $Mn^{2+}Mn^{3+}{}_6O_8(SiO_4)$

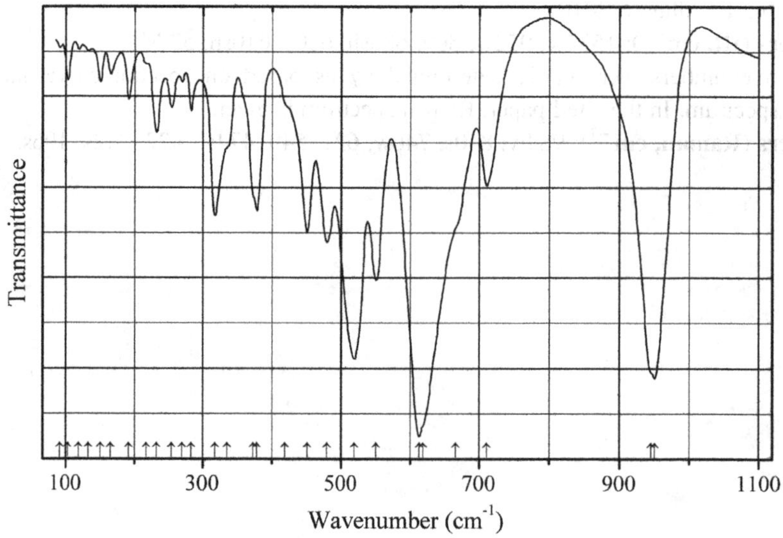

Origin: Synthetic.

Description: Prepared from MnO_2, $MnCl_2$, and SiO_2 above 615 °C (i.e. above the $MnCl_2$ melting point). Characterized by powder X-ray diffraction data. The crystal structure is solved. Tetragonal, space group $I4_1/acd$, $a = 9.371(2)$, $c = 18.847(8)$ Å, $V = 571.6(3)$ Å^3, $Z = 8$.

Kind of sample preparation and/or method of registration of the spectrum: Nujol mull between polyethylene plates. Absorption.

Source: Palvadeau et al. (1991).

Wavenumbers (cm^{-1}): 951s, 945sh, 710, 666sh, 619sh, 613s, 551, 519s, 479, 450, 419sh, 378, 373sh, 335sh, 317, 283w, 270w, 255w, 233, 218sh, 192w, 166w, 151w, 134w, 120w, 103w, 91w.

Note: The wavenumbers were partly determined by us based on spectral curve analysis of the published spectrum.

O587 Brizzite polymorph $Na_2Sb_2O_6$

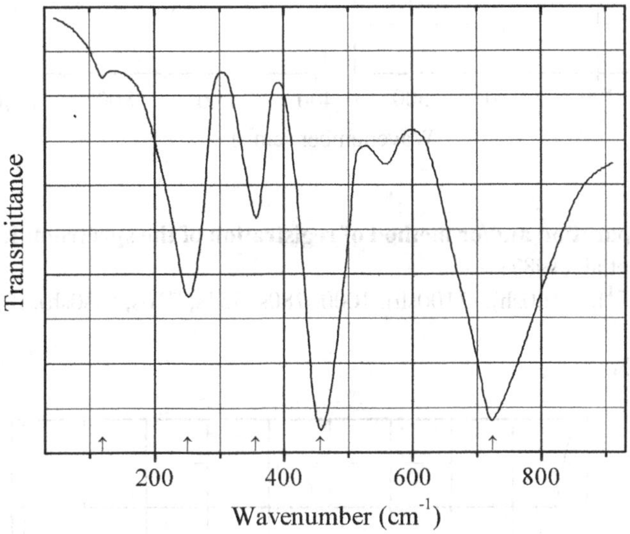

Origin: Synthetic.

Description: Structurally related to pyrochlore.

Source: Vandenborre et al. (1982).

Wavenumbers (cm^{-1}): 725s, 455s, 355, 250, 120w.

Note: A weak band between 500 and 600 cm^{-1} corresponds to H_2O impurity.

O588 Bromellite BeO

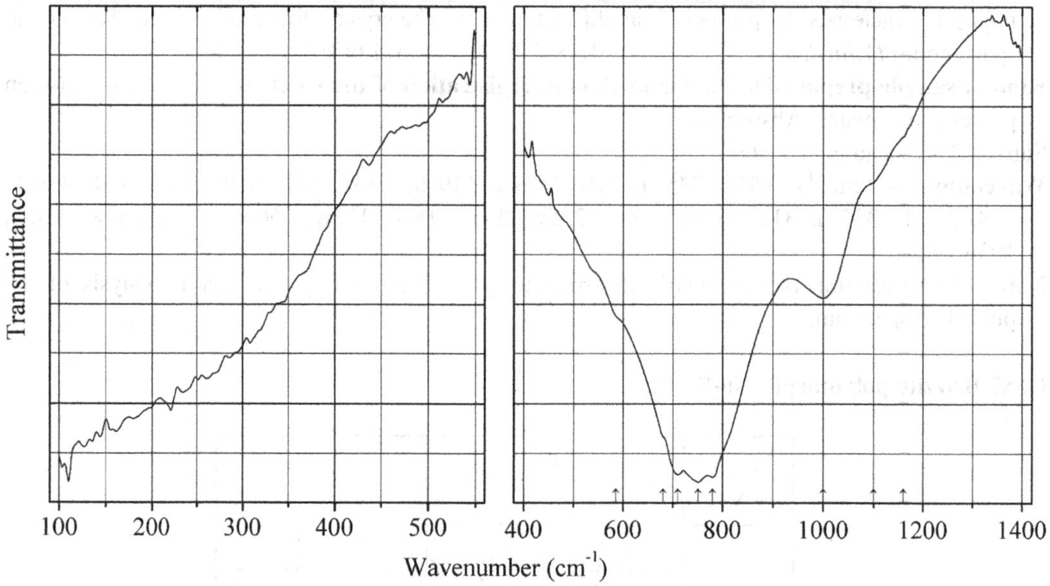

Origin: Synthetic.
Kind of sample preparation and/or method of registration of the spectrum: KBr disc. Absorption.
Source: Hofmeister et al. (1987).
Wavenumbers (cm^{-1}): (1160sh), (1100sh), 1000, 780s, 751s, 710s, (680sh), (585sh).

O589 Brookite TiO_2

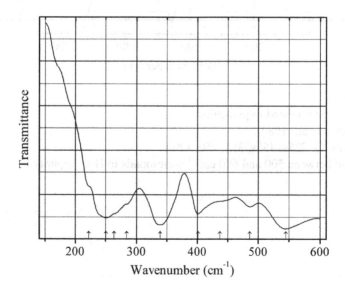

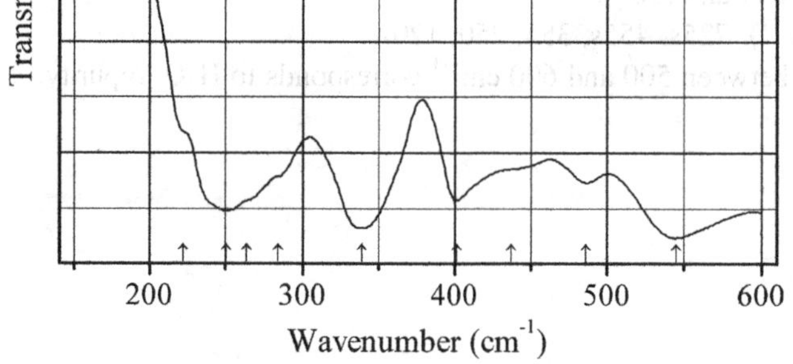

Origin: Synthetic.

Description: Nanocrystals prepared hydrothermally. Characterized by powder X-ray diffraction data.

Kind of sample preparation and/or method of registration of the spectrum: KBr disc. Transmission.

Source: Yanqing et al. (2000).

Wavenumbers (cm^{-1}): 545s, 486, 437sh, 401s, 339s, 284sh, 264sh, 251s, 222sh.

Note: The wavenumbers were partly determined by us based on spectral curve analysis of the published spectrum.

O590 Brucite Co-analogue *β*-$Co(OH)_2$

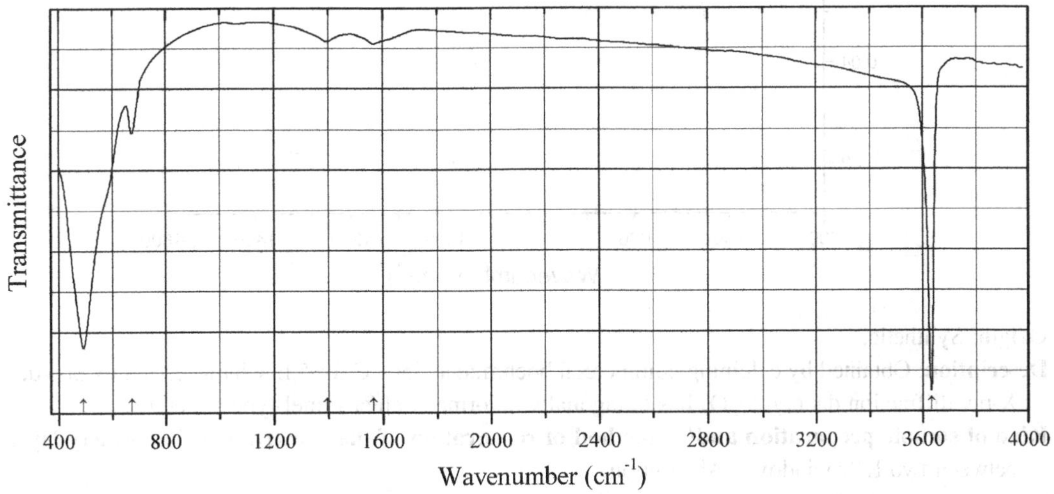

Origin: Synthetic.

Description: Nanoplates prepared hydrothermally from $CoCl_2$ and NaOH at 120 °C for 3 h. Characterized by powder X-ray diffraction data.

Kind of sample preparation and/or method of registration of the spectrum: Transmission. Kind of sample preparation is not indicated.

Source: Zhan (2009).

Wavenumbers (cm^{-1}): 3633s, 1566w, 1394w, 675, 490s.

Note: The wavenumbers were partly determined by us based on spectral curve analysis of the published spectrum. The bands in the range of 600–1600 cm^{-1} may correspond to impurities.

O591 Gamma-alumina γ-Al_2O_3

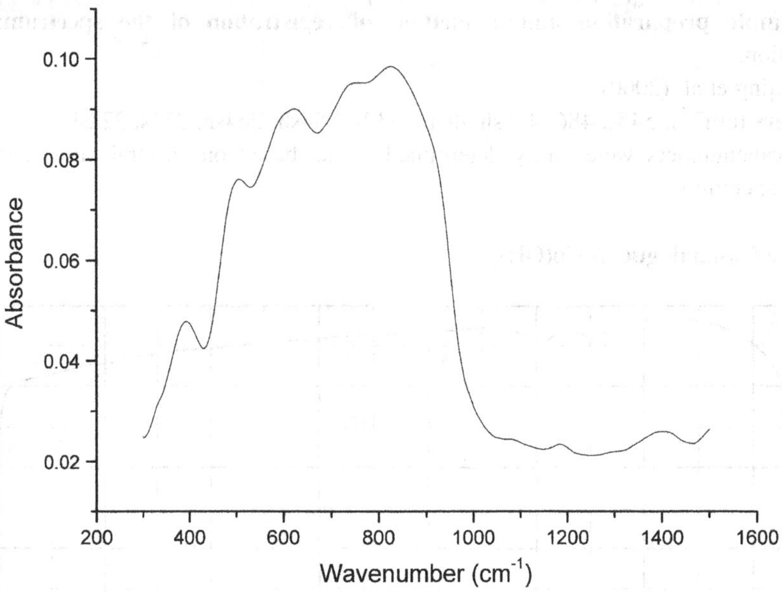

Origin: Synthetic.

Description: Obtained by calcining commercial boehmite at 550 °C for 5 h. Characterized by powder X-ray diffraction data. γ-Al_2O_3 has tetragonally deformed defect spinel-type structure.

Kind of sample preparation and/or method of registration of the spectrum: Thin powdery layer between two KBr windows. Absorption.

Source: Saniger (1995).

Wavenumbers (cm^{-1}): 1403w, 1184w, 1085w, 826s, 764sh, 634s, 506, 393.

Note: The wavenumbers were partly determined by us based on spectral curve analysis of the published spectrum.

O592 Chlormayenite $Ca_{12}Al_{14}O_{32}(\square_4Cl_2)$

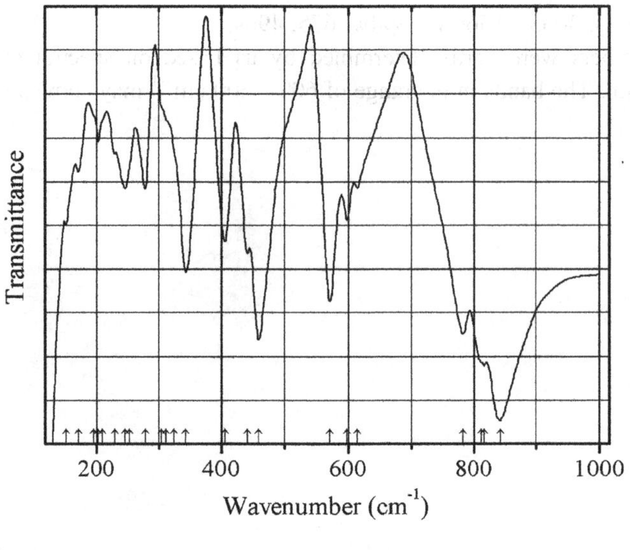

Origin: Synthetic.

Description: Synthesized by heating of a stoichiometric mixture of $CaCO_3$, Al_2O_3, and $CaCl_2$ first at 1323 K for 16 h and thereafter at 1473 K for 24 h with intermediate grinding. Characterized by powder X-ray diffraction data, neutron powder diffraction, and EDX analysis. The crystal structure is solved. Cubic.

Kind of sample preparation and/or method of registration of the spectrum: Transmission. Kind of sample preparation is not indicated.

Source: Schmidt et al. (2014).

Wavenumbers (cm^{-1}): 842s, 817s, 812sh, 783s, 614, 598, 571, 458s, 440, 405, 344s, 325sh, 312sh, 304sh, 278, 253sh, 246, 229, 209sh, 203w, 195sh, 171w, 150.

Note: The wavenumbers were partly determined by us based on spectral curve analysis of the published spectrum. In the cited paper, the wavenumber 783 cm^{-1} is erroneously indicated as 738 cm^{-1}.

O593 Chrysoberyl $BeAl_2O_4$

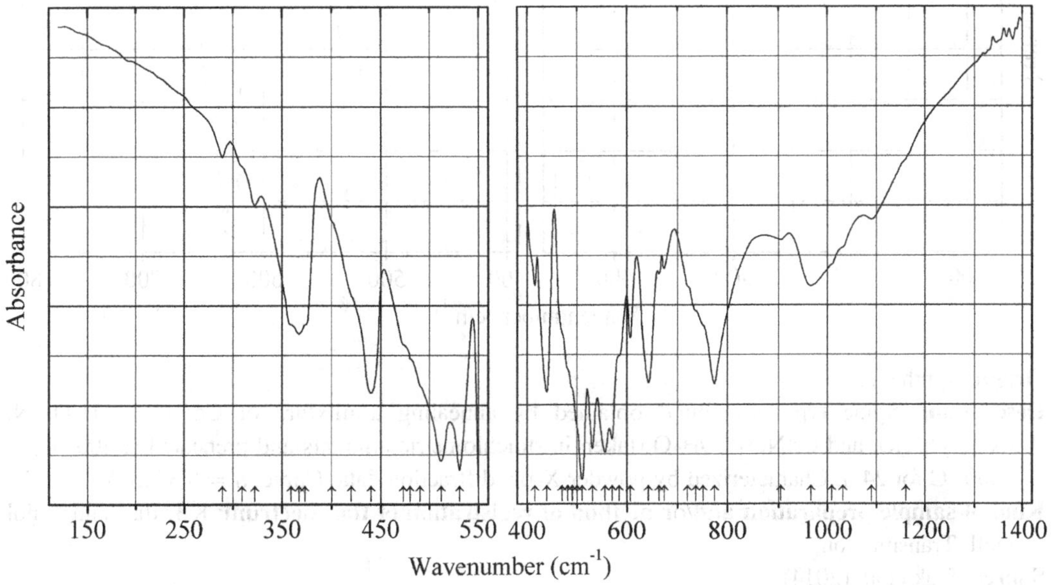

Origin: Colatine, Espirito, Santo, Brazil.

Description: The sample contains 3.1 wt% Fe_2O_3, which corresponds to 2.5 mol% $BeFe_2O_4$.

Kind of sample preparation and/or method of registration of the spectrum: KBr disc and powder dispersed in petroleum jelley. Absorption.

Source: Hofmeister et al. (1987).

Wavenumbers (cm^{-1}): 1161sh, 1090, 1032sh, 1009sh, 967, 904, 774, 751sh, 736sh, 719sh, 673, 663, 642, 606, 584, 570s, 556s, 531s, 510s, 500sh, 490sh, 481sh, 470sh, 439, 415 (for a sample pressed in a disc with KBr); 531s, 512s, 490sh, 480sh, 473, 440, 420sh, 400sh, 373sh, 367, 359sh, 322, 309w, 289w.

Note: In the cited paper, the wavenumber 531 cm^{-1} is erroneously indicated as 538 cm^{-1}.

O594 Cochromite Ni-bearing $Co_{0.9}Ni_{0.1}Cr_2O_4$

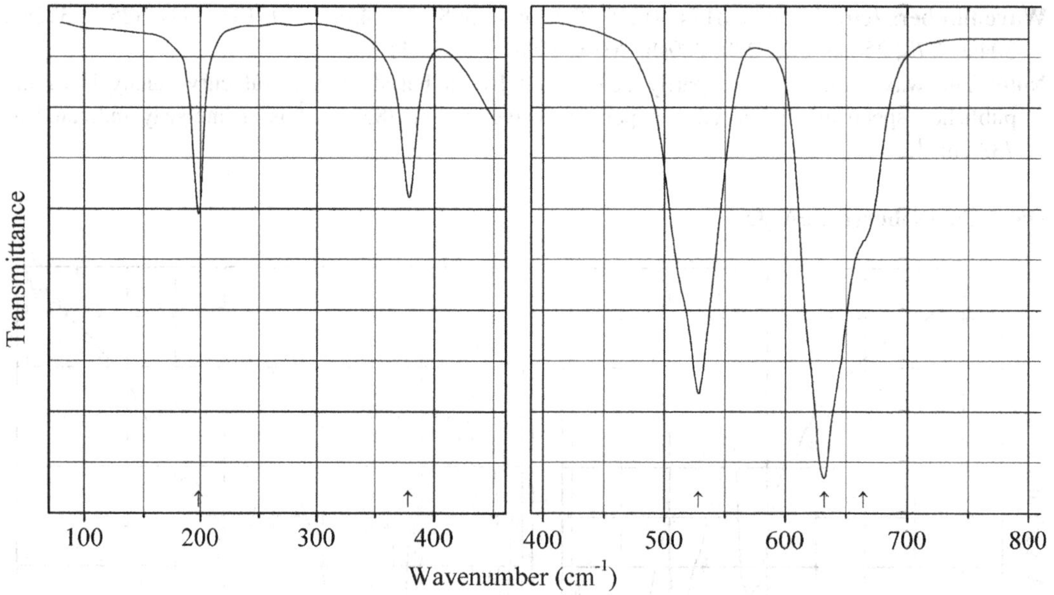

Origin: Synthetic.

Description: Spinel-type compound obtained by annealing a mixture of $Co(NO_3)_2 \cdot 6H_2O$, $Ni(NO_3)_2 \cdot 6H_2O$, and $Cr(NO_3)_3 \cdot 9H_2O$ (taken in stoichiometric amounts and preheated at 400 °C) at 1000 °C for 24 h. Characterized by powder X-ray diffraction data. Cubic, a = 8.3323 Å.

Kind of sample preparation and/or method of registration of the spectrum: KBr disc and Nujol mull. Transmission.

Source: Ptak et al. (2014).

Wavenumbers (IR, cm^{-1}): 664sh, 632s, 528s, 378, 198.

Note: The wavenumbers were partly determined by us based on spectral curve analysis of the published spectrum. For the vibration spectra of synthetic cochromite analogue see also Mączka et al. (2013). In the cited paper, Raman spectrum is given.

Wavenumbers (Raman, cm^{-1}): 683, 601w, 514, 450, 195s.

O595 Columbite-(Mn) $Mn^{2+}Nb_2O_6$

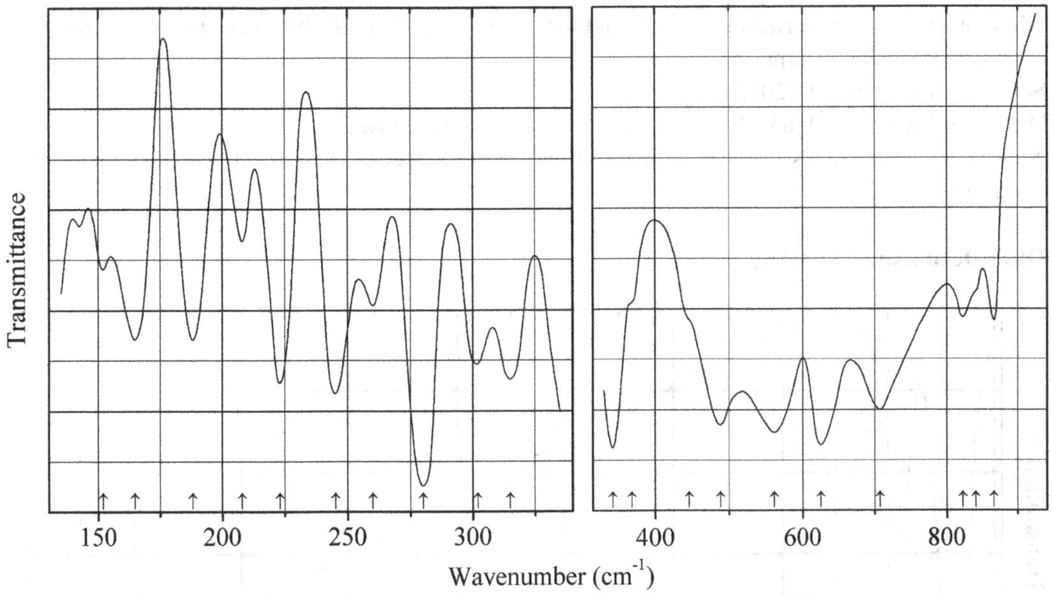

Origin: Synthetic.

Description: Prepared using conventional solid-state reaction techniques. Characterized by powder X-ray diffraction data. Orthorhombic, $a = 14.413(17)$, $b = 5.759(5)$, $c = 5.083(7)$ Å.

Kind of sample preparation and/or method of registration of the spectrum: CsI disc (above 340 cm^{-1}) and polyethylene film (below 340 cm^{-1}). Transmission.

Source: Husson et al. (1977a, b).

Wavenumbers (IR, cm^{-1}): 865, 840sh, 822, 707s, 625s, 560s, 488s, 445sh, 368sh, 342s, 315, 302, 280s, 260, 245, 223, 208w, 188, 165, 152w.

Note: In the cited paper, Raman spectrum is given.

Wavenumbers (Raman, cm^{-1}): 877s, 823, 707w, 634w, 624w, 606, 531s, 487, 440w, 399, 386, 361, 315s, 298w, 288, 275, 264w, 245s, 215, 207, 179, 160w, 140s, 127, 113, 89s.

O596 Corundum α-Al_2O_3

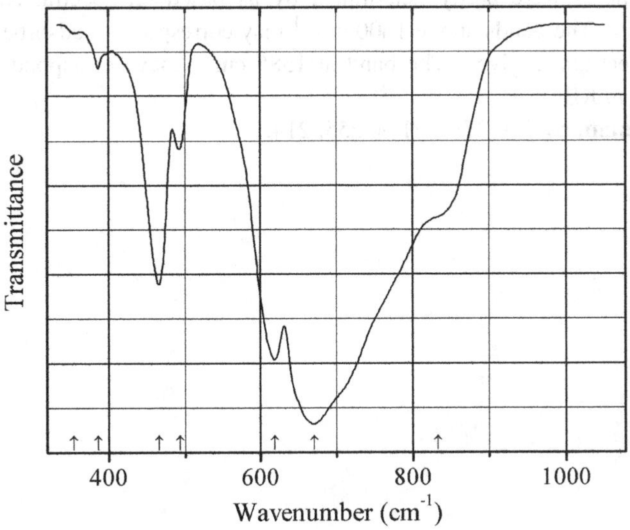

Origin: Synthetic.

Description: Irregular grains from 1 to 3 μm across.

Kind of sample preparation and/or method of registration of the spectrum: Free particles dispersed in air. Absorption.

Source: Mutschke et al. (2013).

Wavenumbers (cm^{-1}): 833sh, 670s, 618s, 493, 466, 386w, 354w.

Note: The wavenumbers were determined by us based on spectral curve analysis of the published spectrum.

0597 Delafossite Al analogue $Cu^{1+}AlO_2$

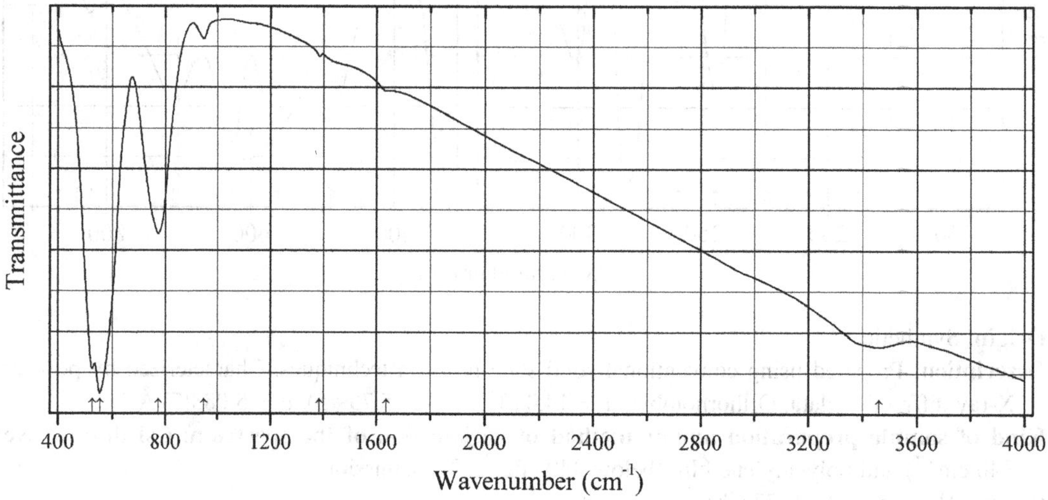

Origin: Synthetic.

Description: Synthesized from CuO and $Al(NO_3)_3·9H_2O$ by a sol-gel method using ethylene glycol as solvent. Characterized by powder X-ray diffraction data. Trigonal, space group *R*-3*m*, *a* = 2.852, *c* = 16.830 Å.

Kind of sample preparation and/or method of registration of the spectrum: KBr disc. Absorption.

Source: Benreguia et al. (2015).

Wavenumbers (IR, cm^{-1}): (3460), 1637sh, 1385w, 949w, 772, 554s, 526s.

Note: The wavenumbers were partly determined by us based on spectral curve analysis of the published spectrum. The bands above 1600 cm^{-1} may correspond to adsorbed water. In the cited paper, Raman spectrum is given. The band at 1385 cm^{-1} may correspond to the admixture of potassium nitrate in KBr.

Wavenumbers (Raman, cm^{-1}): 773s, 417s, 255, 214.

O598 Fluorcalciomicrolite (Ca,Na,□)$_2$Ta$_2$O$_6$F

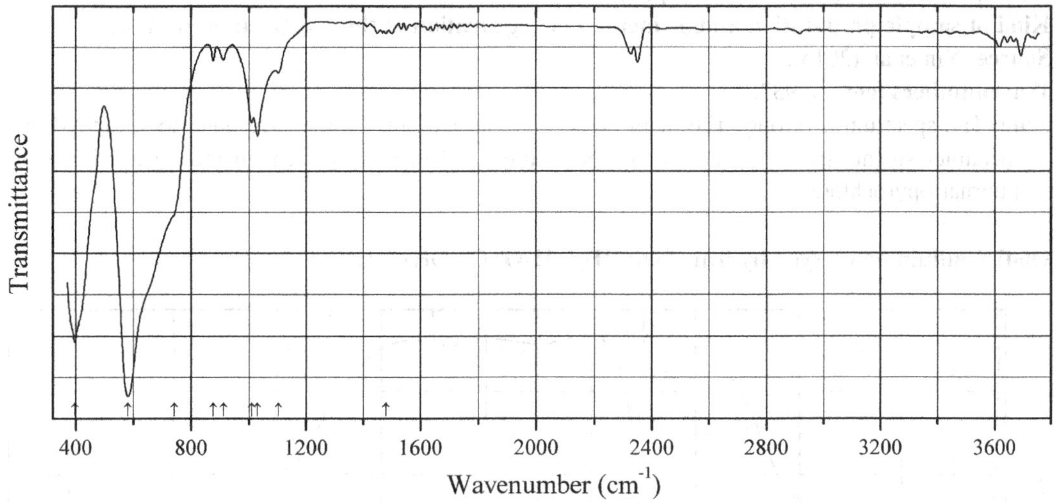

Origin: No data.

Description: The empirical formula is (electron microprobe): (Ca$_{1.23}$Na$_{0.745}$*REE*$_{0.01}$Sr$_{0.01}$)(Ta$_{1.78}$Nb$_{0.08}$Ti$_{0.08}$Si$_{0.06}$)O$_6$[F$_{0.57}$(OH,O,□)$_{0.43}$].

Kind of sample preparation and/or method of registration of the spectrum: KBr disc. Transmission.

Source: Geisler et al. (2004).

Wavenumbers (IR, cm^{-1}): (1480w), 1105w, 1031, 1010, 912w, 876w, 740sh, 580s, 396s.

Note: Weak bands in the range from 2300 to 2400 cm^{-1} correspond to atmospheric CO_2. The weak band at 1480 cm^{-1} may correspond to the admixture of a carbonate. In the cited paper, Raman spectrum is given.

Wavenumbers (Raman, cm^{-1}): 835, 690s, 504s, 341, 295, 155.

O599 Fluornatropyrochlore (Na,Pb,Ca,*REE*,U)$_2$Nb$_2$O$_6$F

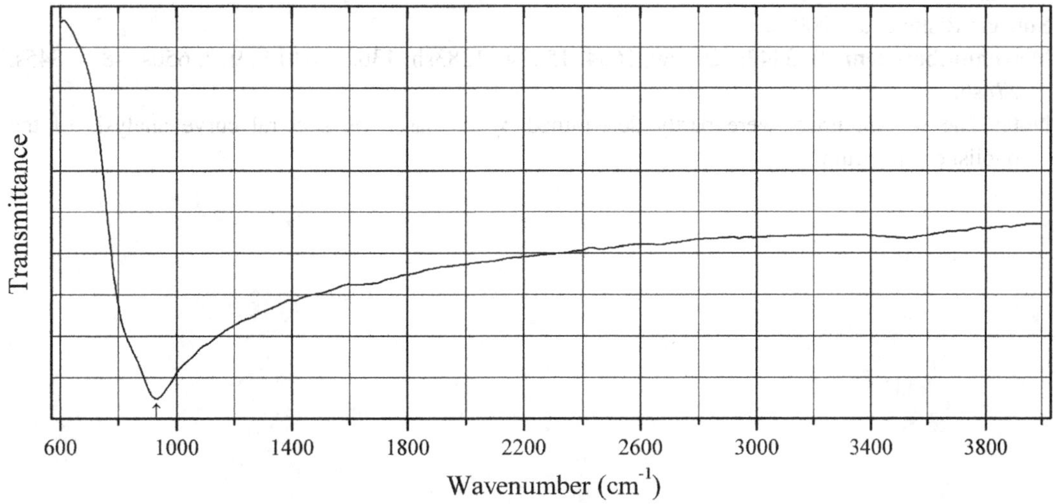

Origin: Boziguoer *REE* deposit, Baicheng County, Akesu, Xinjiang, China (type locality).
Description: A grain from the holotype specimen.
Kind of sample preparation and/or method of registration of the spectrum: Reflection.
Source: Yin et al. (2015).
Wavenumbers (cm^{-1}): 932.
Note: The spectrum is wrong. The main broad band at 932 cm^{-1} may correspond to an anhydrous metamict silicate [possibly, thorite or chevkinite-(Ce)] that are present in the association with fluornatropyrochlore.

O601 Gallium hydroxyde hydrate $Ga(OH)_3{\cdot}nH_2O$ $Ga(OH)_3{\cdot}nH_2O$

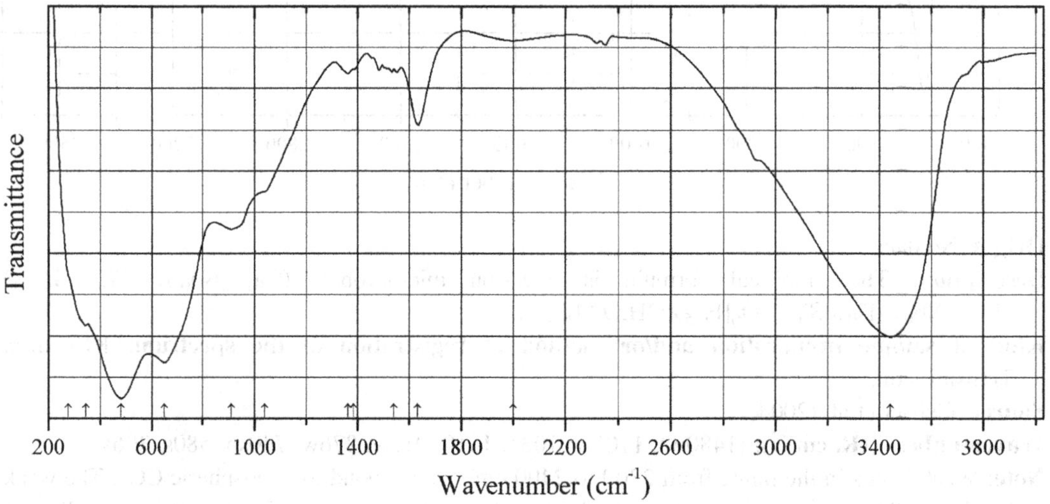

Origin: Synthetic.
Description: Obtained by addition of hot water and tetramethylammonium hydroxide solution to the solution of gallium(III)-isopropoxide dissolved in 2-propanol. X-ray amorphous.
Kind of sample preparation and/or method of registration of the spectrum: KBr disc. Transmission.
Source: Ristić et al. (2005).
Wavenumbers (cm^{-1}): 3443s, 2000w, 1634, 1542w, 1383sh, 1362, 1040sh, 912, 650s, 482s, 345s, 278sh.
Note: The wavenumbers were partly determined by us based on spectral curve analysis of the published spectrum.

O602 Harmunite cubic polymorph $CaFe_2O_4$

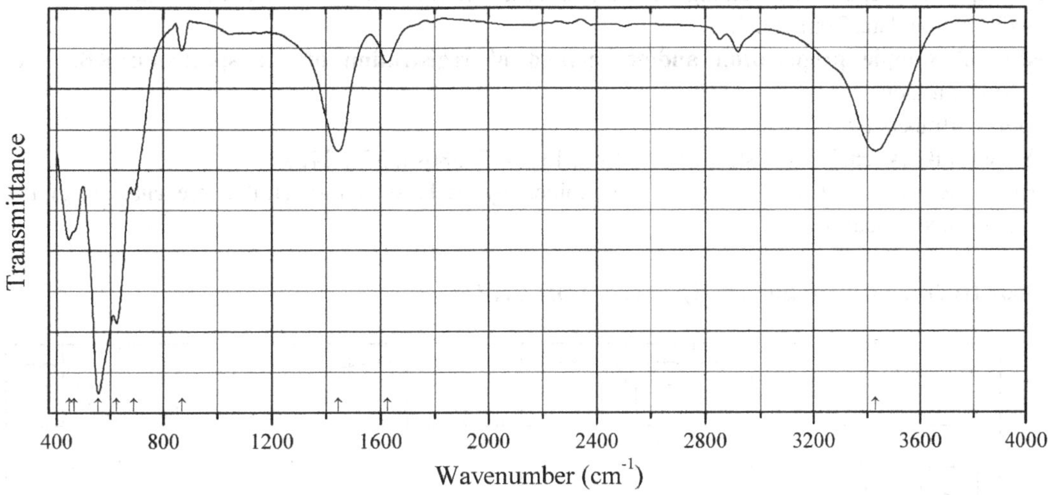

Origin: Synthetic.

Description: Nanoparticles synthesized from a stoichiometric mixture of calcium chloride and iron (III) nitrate by a sol-gel technique with subsequent annealing. Characterized by powder X-ray diffraction data.

Kind of sample preparation and/or method of registration of the spectrum: Transmission. Kind of sample preparation is not indicated.

Source: An et al. (2015).

Wavenumbers (cm^{-1}): 3431, 1625w, 1445, 869w, 690, 624s, 557s, 466sh, 447.

Note: The wavenumbers were determined by us based on spectral curve analysis of the published spectrum. The bands at 3431 and 1625 cm^{-1} may correspond to water molecules adsorbed on the surface of the nanoparticles. The bands at 1445 and 869 cm^{-1} may correspond to the admixture of a carbonate. The assignment of the band at 1445 cm^{-1} given in the cited paper is erroneous.

O603 Hausmannite $Mn^{2+}Mn^{3+}{}_2O_4$

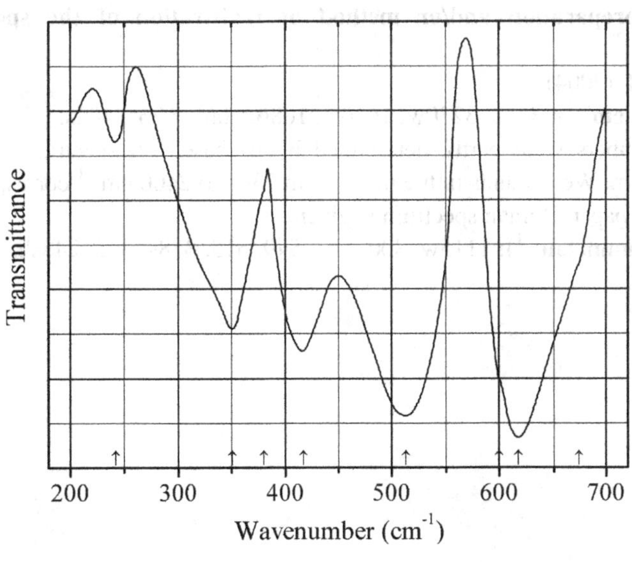

Origin: Synthetic.

Description: Prepared by heating $MnCO_3$ at 1185 °C for 5 h. Characterized by powder X ray diffraction data. Tetragonal.

Kind of sample preparation and/or method of registration of the spectrum: KBr disc. Transmission.

Source: Brabers (1969).

Wavenumbers (cm^{-1}): 674sh, 617s, 600sh, 513s, 417, 380sh, 351, 242w

Note: The wavenumbers were partly determined by us based on spectral curve analysis of the published spectrum.

O604 Hydrokenomicrolite $(\square,H_2O)_2Ta_2(O,OH)_6(H_2O)$

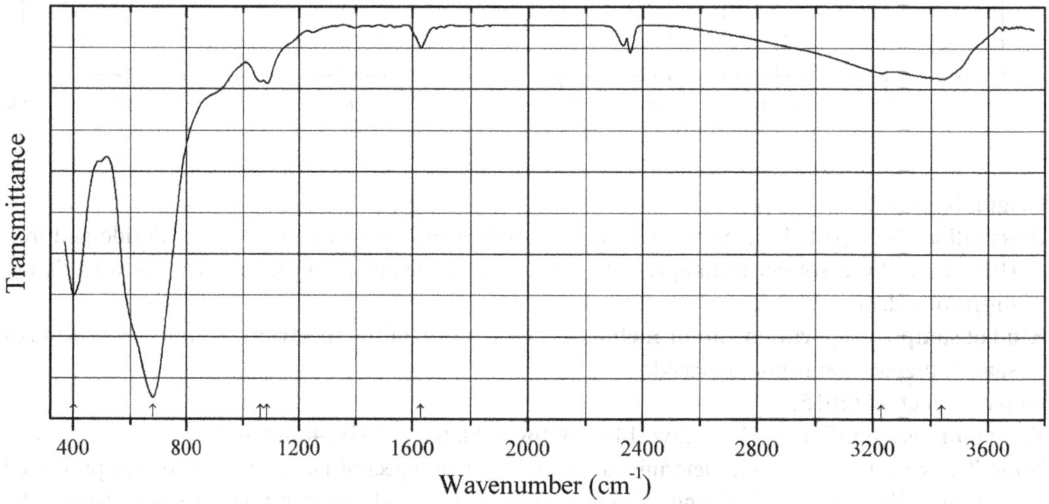

Origin: Artificial.

Description: Prepared by treating natural fluorcalciomicrolite at 175 °C in a 1 M HCl-$CaCl_2$ solution for 14 days. The grains are heterogeneous, with fluorcalciomicrolite core and hydrokenomicrolite outer zones.

Kind of sample preparation and/or method of registration of the spectrum: KBr disc. Transmission.

Source: Geisler et al. (2004).

Wavenumbers (IR, cm^{-1}): 3440, 3230w, 1630w, 1086, 1061, 676s, 402s.

Note: The wavenumbers were partly determined by us based on spectral curve analysis of the published spectrum. Weak bands in the range from 2300 to 2400 cm^{-1} correspond to atmospheric CO_2. In the cited paper, Raman spectrum is given.

Wavenumbers (Raman, cm^{-1}): 1115w, 800, 637, 549, 515, 338s, 302, 248sh, 159, 138, 120.

O605 Ilsemannite $Mo_3O_8{\cdot}nH_2O$ (?)

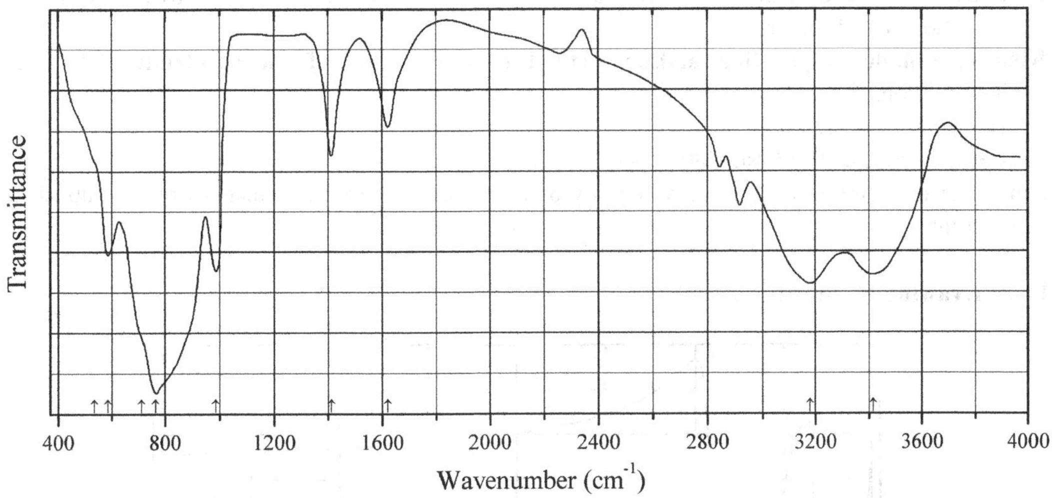

Origin: Synthetic.

Description: Blue amorphous product of ultrasound irradiation of a slurry of molybdenum hexacarbonyl. Characterized by DSC and TG data.

Kind of sample preparation and/or method of registration of the spectrum: KBr disc. Absorption.

Source: Dhas and Gedanken (1997).

Wavenumbers (cm^{-1}): 3415, 3180, 1621, 1412, 986, 765s, 712sh, 587, 535sh.

Note: The wavenumbers were determined by us based on spectral curve analysis of the published spectrum. The assignment of the band at 1412 cm^{-1} made by the authors of the cited paper is questionable. Weak bands in the range from 2800 to 3000 cm^{-1} correspond to the admixture of an organic substance.

O606 Iwakiite-hausmannite intermediate member $Mn^{2+}(Mn^{3+}Fe^{3+})O_4$

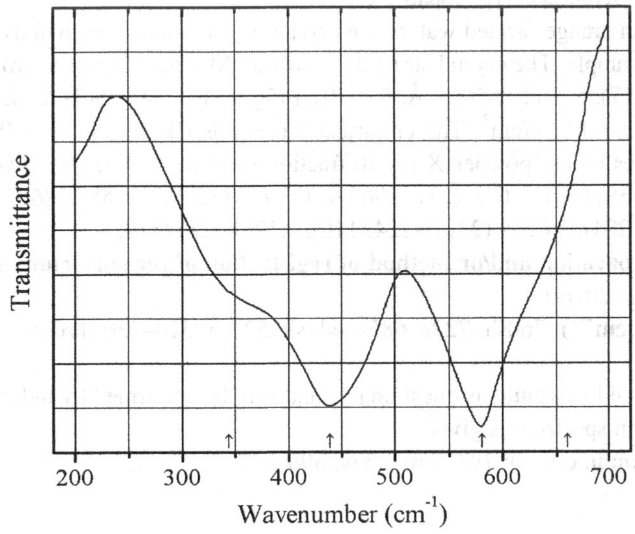

Origin: Synthetic.

Description: Prepared by heating $MnCO_3$ at 1360 °C for 5 h in air. Characterized by powder X-ray diffraction data. Tetragonal.

Kind of sample preparation and/or method of registration of the spectrum: KBr disc. Transmission.

Source: Brabers (1969).

Wavenumbers (cm^{-1}): 660sh, 580s, 439s, 344sh.

Note: The wavenumbers were determined by us based on spectral curve analysis of the published spectrum.

O607 Kyawthuite $Bi^{3+}Sb^{5+}O_4$

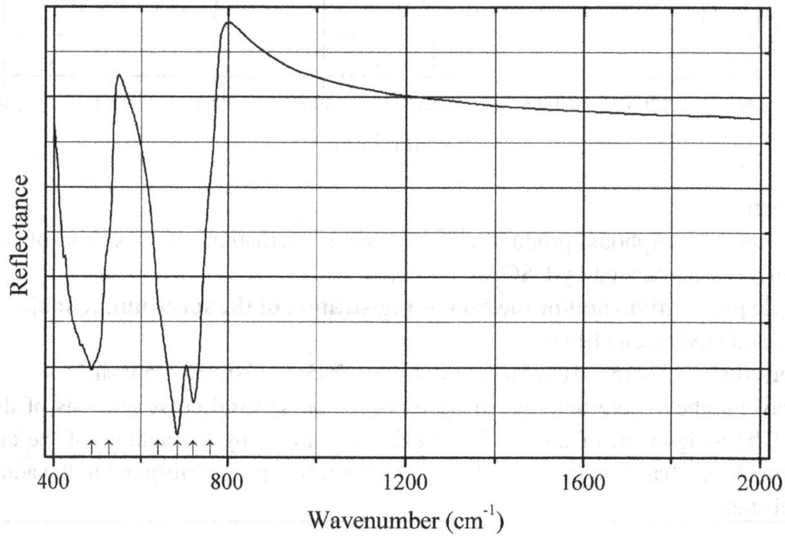

Origin: Chaung-gyi-ah-le-ywa, Chaung-gyi valley, 5 km north-northeast of the town of Mogok, Pyin-Oo-Lwin district, Myanmar (type locality).

Description: Reddish orange faceted waterworn crystal of gem quality from alluvium. Holotype (and the only known) sample. The crystal structure is solved. Monoclinic, space group *I2/c*, $a = 5.4624$ (4), $b = 4.88519(17)$, $c = 11.8520(8)$ Å, $\beta = 101.195(7)°$, $V = 310.25(3)$ Å^3, $Z = 4$. $D_{meas} = 8.256$ (5) g/cm^3, $D_{calc} = 8.127$ g/cm^3. The empirical formula is $(Bi^{3+}_{0.83}Sb^{3+}_{0.18})(Sb^{5+}_{0.99}Ta^{5+}_{0.01})O_4$. The strongest lines of the powder X-ray diffraction pattern [*d*, Å (*I*, %) (*hkl*)] are: 3.266 (100) (−112), 2.900 (66) (112), 2.678 (24) (200), 2.437 (22) (020, –114), 1.8663 (21) (024), 1.8026 (43) (−116, 220, 204), 1.6264 (23) (−224, 116), 1.5288 (28) (312, –132).

Kind of sample preparation and/or method of registration of the spectrum: Specular reflection.

Source: Kampf et al. (2016i).

Wavenumbers (IR, cm^{-1}): 760sh, 722s, 685s, 641sh, 527sh, 510–430 (broad, with an extremum at 488 cm^{-1}).

Note: Natural origin of kyawthuite is questionable and is to be confirmed by independent finds. In the cited paper, Raman spectrum is given.

Wavenumbers (Raman, cm^{-1}): 793, 736, 453s, 396s, 322, 258, 173s.

O608 Layered perovskite $BaBi_2Ta_2O_9$ $BaBi_2Ta_2O_9$

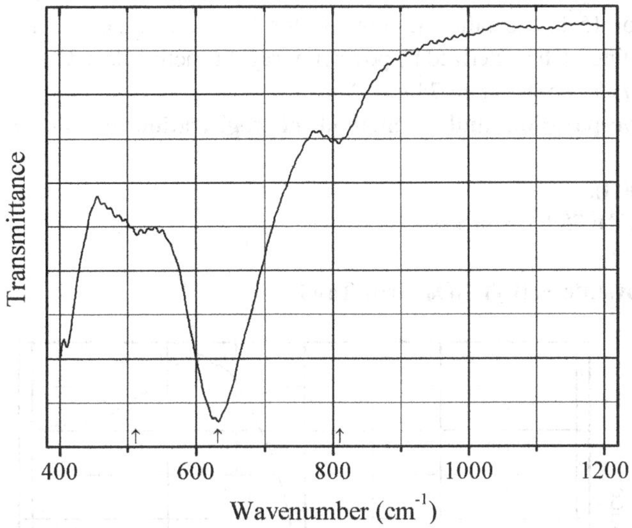

Origin: Synthetic.

Description: Prepared by solid-state reaction of $BaCO_3$, Bi_2O_3, and Ta_2O_5 at 1000 °C for 72 h with intermediate grindings. Characterized by powder X-ray diffraction data. Tetragonal, space group *I*4/*mmm*, $a = 3.954$, $c = 25.487$ Å.

Kind of sample preparation and/or method of registration of the spectrum: KBr disc. Transmission.

Source: Li et al. (2008).

Wavenumbers (cm^{-1}): 809.5, 631s, 512.

O609 Layered perovskite $CaBi_2Ta_2O_9$ $CaBi_2Ta_2O_9$

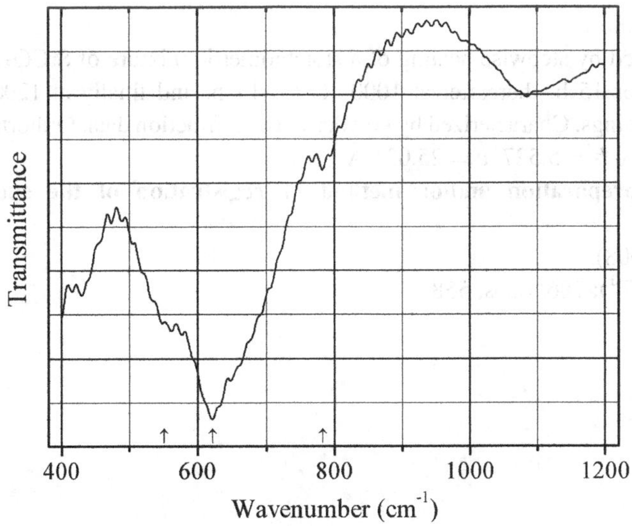

Origin: Synthetic.

Description: Prepared by stepwise heating of a stoichiometric mixture of $CaCO_3$, Bi_2O_3, and Ta_2O_5, first at 900 °C for 15 h, thereafter at 1000 °C for 15 h, and finally at 1200 °C for 24 h with intermediate grindings. Characterized by powder X-ray diffraction data. Orthorhombic, space group $A2_1am$, $a = 5.467$, $b = 5.427$, $c = 24.931$ Å.

Kind of sample preparation and/or method of registration of the spectrum: KBr disc. Transmission.

Source: Li et al. (2008).

Wavenumbers (cm^{-1}): 782, 622s, 551.

O610 Layered perovskite $SrBi_2Ta_2O_9$ $SrBi_2Ta_2O_9$

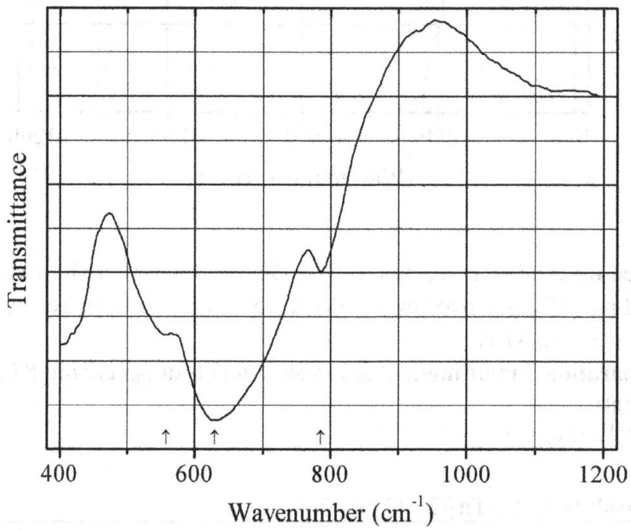

Origin: Synthetic.

Description: Prepared by stepwise heating of a stoichiometric mixture of $SrCO_3$, Bi_2O_3, and Ta_2O_5, first at 900 °C for 15 h, thereafter at 1000 °C for 15 h, and finally at 1200 °C for 24 h with intermediate grindings. Characterized by powder X-ray diffraction data. Orthorhombic, space group $A2_1am$, $a = 5.473$, $b = 5.527$, $c = 25.031$ Å.

Kind of sample preparation and/or method of registration of the spectrum: KBr disc. Transmission.

Source: Li et al. (2008).

Wavenumbers (cm^{-1}): 786, 629s, 558.

O611 Lueshite $NaNbO_3$

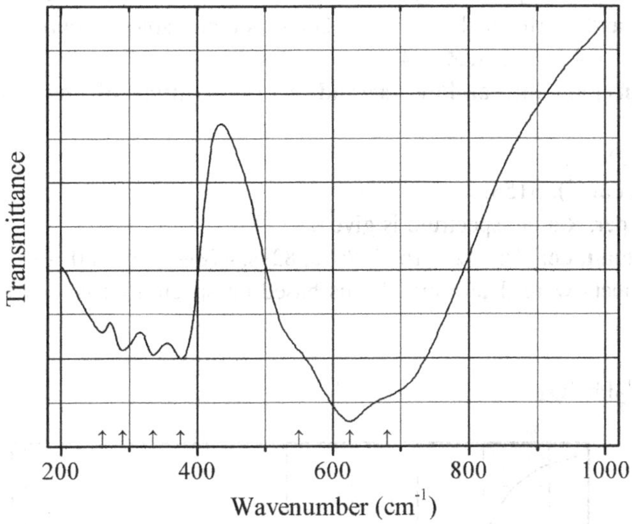

Origin: Synthetic.

Description: Prepared by heating a stoichiometric mixture of $NaNO_3$ and Nb_2O_5 at 1000 °C for 1 h. Characterized by powder X-ray diffraction data. Orthorhombic, space group *Pbma*, $a = 5.666$, $b = 15.520$, $c = 5.506$ Å.

Kind of sample preparation and/or method of registration of the spectrum: CsI disc. Transmission.

Source: Rocchiccioli-Deltcheff (1973).

Wavenumbers (cm^{-1}): 680sh, 625s, 550sh, 375, 335, 290, 260.

Note: The wavenumbers were partly determined by us based on spectral curve analysis of the published spectrum.

O612 Metastudtite $UO_3 \cdot 2H_2O$

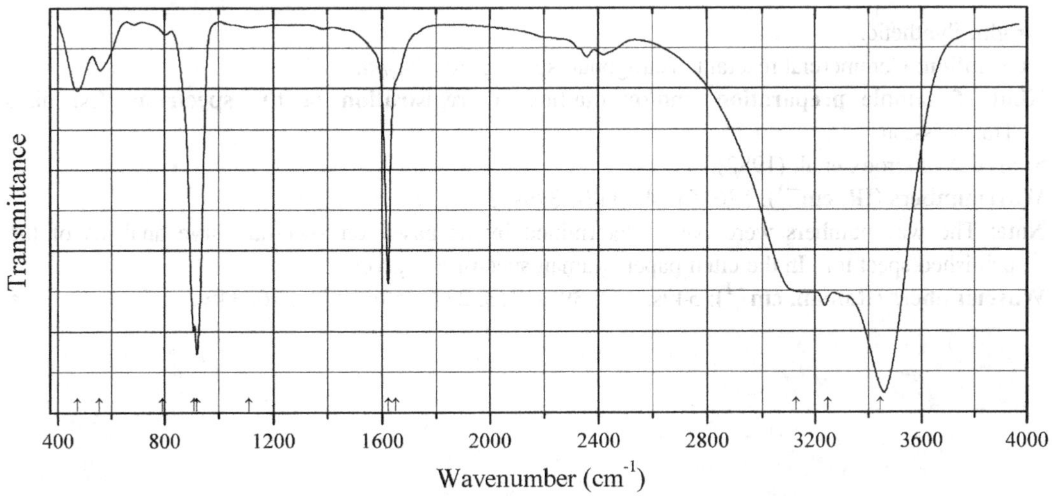

Origin: Synthetic.

Description: Obtained by dehydrating synthetic studtite at 90 °C for 48 h. Characterized by powder X-ray diffraction and thermoanalytical data. Orthorhombic, space group *Pnma*, $a = 8.4184(4)$, $b = 8.7671(4)$, $c = 6.4943(3)$ Å, $Z = 4$.

Kind of sample preparation and/or method of registration of the spectrum: KBr disc. Transmission.

Source: Guo et al. (2014a).

Wavenumbers (IR, cm^{-1}): 3153.

Note: In the cited paper, Raman spectrum is given.

Wavenumbers (Raman, cm^{-1}): 1725, 1652, 869s, 829s. 478w, 356, 280, 190, 156.

Note: The wavenumbers were determined by us based on spectral curve analysis of the published spectrum.

O613 Minium $Pb^{2+}{}_2Pb^{4+}O_4$

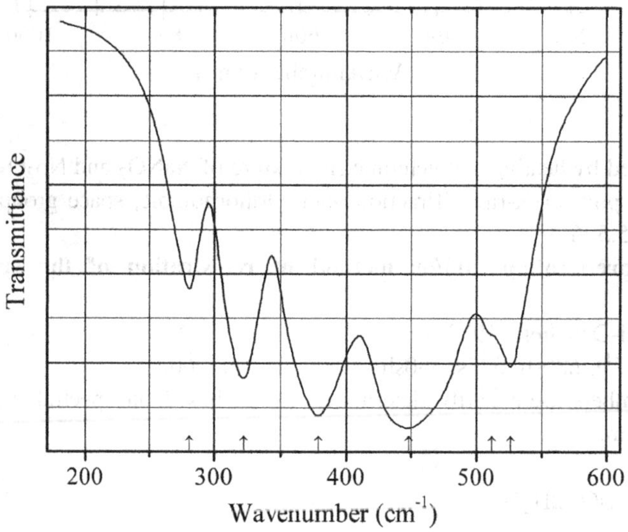

Origin: Synthetic.

Description: Commercial reactant. Tetragonal, space group $P4_2/m$.

Kind of sample preparation and/or method of registration of the spectrum: CsI disc. Transmission.

Source: Vigouroux et al. (1982).

Wavenumbers (IR, cm^{-1}): 526, 512sh, 448s, 379s, 322s, 281.

Note: The wavenumbers were partly determined by us based on spectral curve analysis of the published spectrum. In the cited paper, Raman spectrum is given.

Wavenumbers (Raman, cm^{-1}): 549s, 477, 391s, 313, 232, 152w, 121s, 76, 34w.

O614 Mopungite $NaSb(OH)_6$

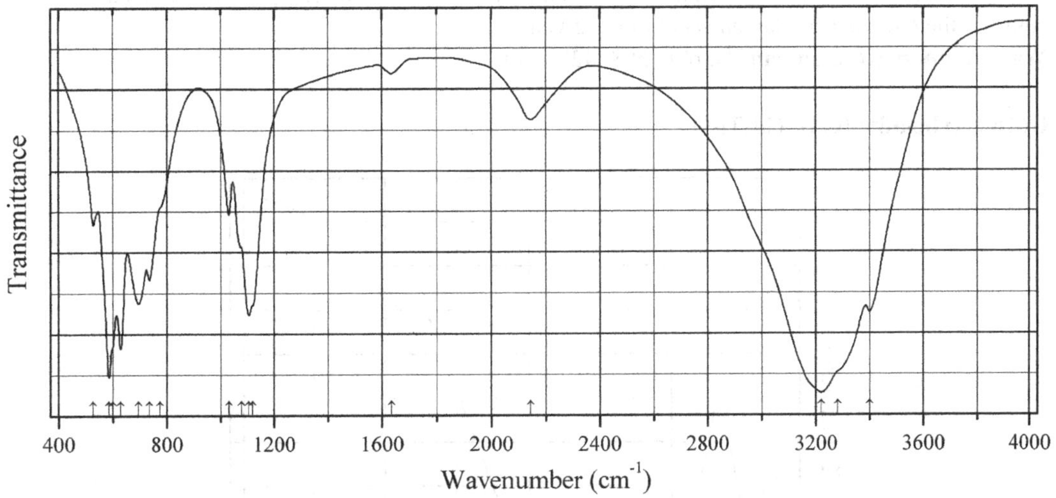

Origin: Synthetic.
Description: Obtained by boiling aqueous solution of NaCl with potassium antimonate.
Kind of sample preparation and/or method of registration of the spectrum: KBr disc. Absorption.
Source: Siebert (1959).
Wavenumbers (cm^{-1}): 3400, 3280sh, 3220s, 2145, 1635w, 1120sh, 1105s, 1075sh, 1030, 775, 735, 695, 628s, 600sh, 586s, 528.

O615 Nichromite Ni_2CrO_4

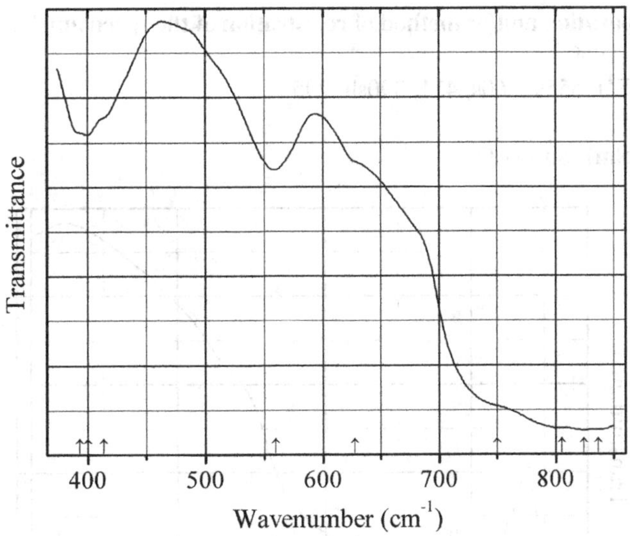

Origin: Synthetic.
Description: Synthesized by heating a mixture of corresponding nitrates. Characterized by powder X-ray diffraction data. Cubic. Contains ~25% admixture of tetragonal (pseudocubic), phase, space group *I*4_1/*amd*. For the main phase, $a = 8.3186(2)$.
Kind of sample preparation and/or method of registration of the spectrum: A spectrometer fitted with a photoacoustic detector was used.

Source: Hosterman et al. (2013).
Wavenumbers (IR, cm^{-1}): No quantitative data (only a figure of the IR spectrum is given).
Note: In the cited paper, Raman spectrum is given.
Wavenumbers (Raman, cm^{-1}): 678, 508, 427, 190.

O616 Oxybetafite-(Gd) $Gd_2Ti_2O_7$

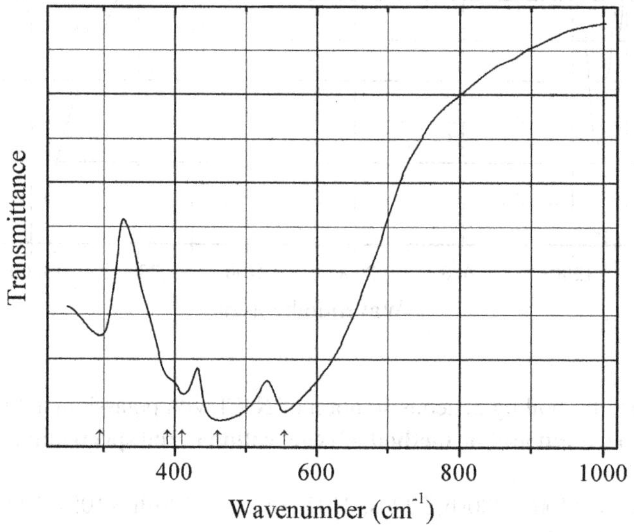

Origin: Synthetic.
Description: Obtained using a solid-state reaction technique. Characterized by powder X-ray diffraction data. Cubic.
Kind of sample preparation and/or method of registration of the spectrum: KBr disc. Transmission.
Source: Knop et al. (1969).
Wavenumbers (cm^{-1}): 555s, 460s, 411, 390sh, 295.

O617 Oxybetafite-(Sm) $Sm_2Ti_2O_7$

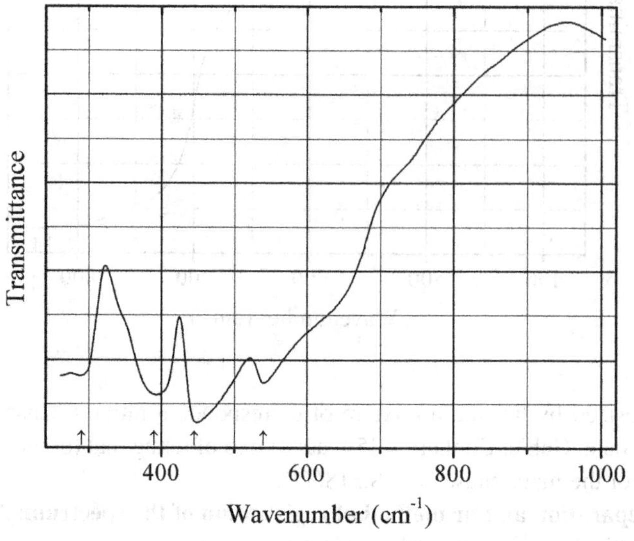

Origin: Synthetic.

Description: Obtained using a solid-state reaction technique. Characterized by powder X-ray diffraction data. Cubic.

Kind of sample preparation and/or method of registration of the spectrum: KBr disc. Transmission.

Source: Knop et al. (1969).

Wavenumbers (cm^{-1}): 539, 445s, 390s, 290.

O618 Oxybismuthobetafite $Bi_2Ti_2O_7$

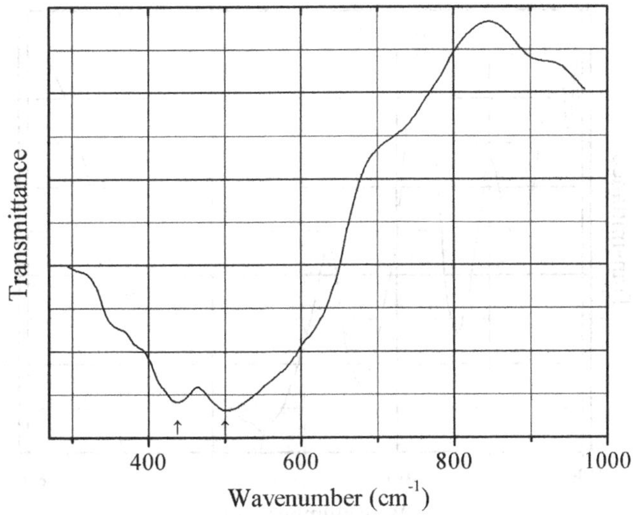

Origin: Synthetic.

Description: Obtained using a solid-state reaction technique. Characterized by powder X-ray diffraction data. The symmetry is not cubic.

Kind of sample preparation and/or method of registration of the spectrum: KBr disc. Transmission.

Source: Knop et al. (1969).

Wavenumbers (cm^{-1}): 500s, 437s.

O619 Paramelaconite $Cu^{1+}{}_2Cu^{2+}{}_2O_3$

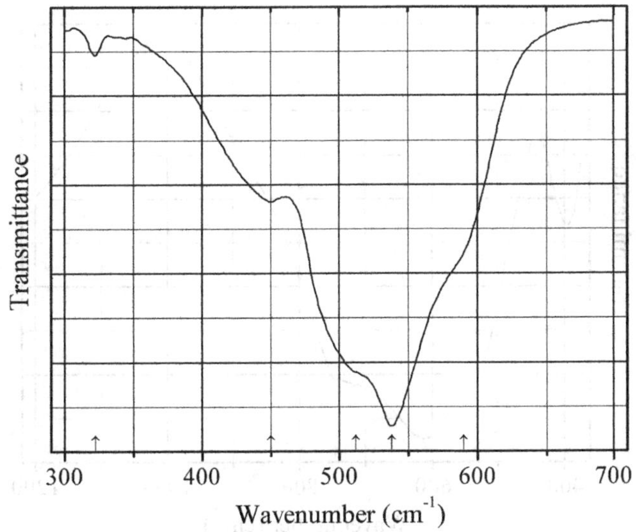

Origin: Synthetic.
Description: Prepared by oxidation of Cu_2O or reduction of CuO. Trtragonal, $a \approx 5.84$, $c \approx 9.93$ Å.
Kind of sample preparation and/or method of registration of the spectrum: Absorption. Kind of sample preparation is not indicated.
Source: Djurek et al. (2015).
Wavenumbers (cm^{-1}): 590sh, 538s, 512sh, 450, 322w.

O620 Pyrophanite $MnTiO_3$

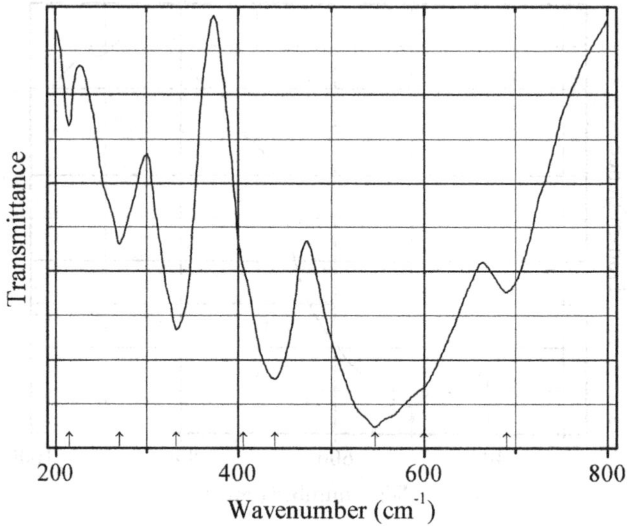

Origin: Synthetic.
Description: Prepared using solid-state reaction techniques. Characterized by powder X-ray diffraction data.
Kind of sample preparation and/or method of registration of the spectrum: CsI disc. Transmission.
Source: Baran and Botto (1978).
Wavenumbers (cm^{-1}): 690, 600sh, 547s, 440s, 405, 333, 270, 215w.

O621 Rutile TiO_2

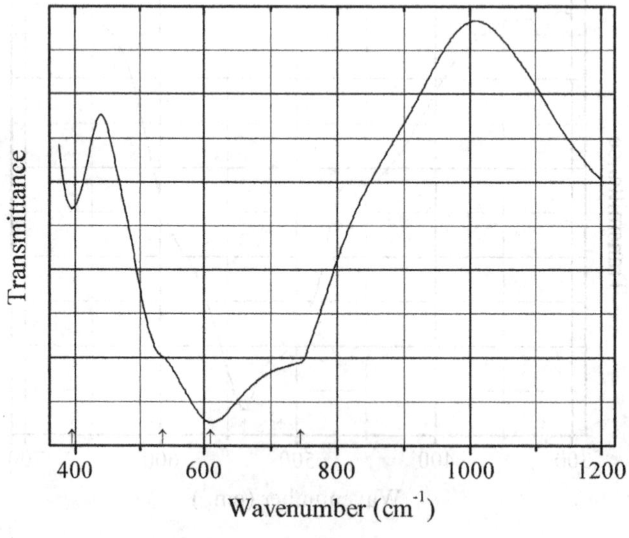

Origin: Unknown.
Description: Natural fibrous crystals included in quartz.
Kind of sample preparation and/or method of registration of the spectrum: KBr disc. Absorption.
Source: Peng et al. (1995).
Wavenumbers (cm^{-1}): 745sh, 610s, 535sh, 395.

O622 Schoenfliesite $MgSn(OH)_6$

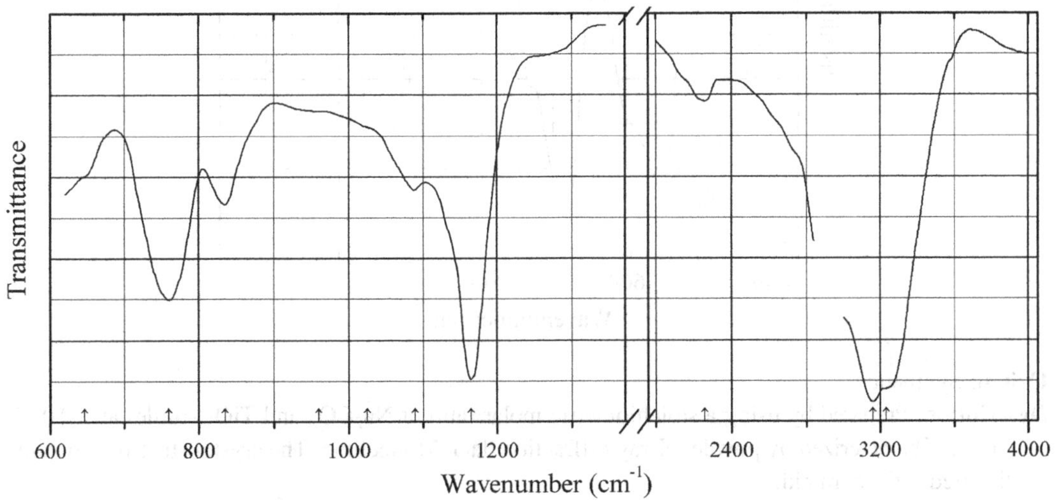

Origin: Pitkäranta, Karelia, Russia.
Description: Fibrous affregate from the association with serpentine, chlorite, chondrodite, diopside, luorite, calcite, dolomite, magnetite, and cassiterite. Cubic, $a = 7.77(1)$. $D_{meas} = 3.32$ g/cm^3, $D_{calc} = 3.49$ g/cm^3. The empirical formula is $(Mg_{0.94}Mn_{0.13})Sn_{0.97}(OH)_{6.00}$. The strongest lines of the powder X-ray diffraction pattern [*d*, Å (*I*, %) (*hkl*)] are: 4.495 (60) (111), 3.898 (100) (200), 2.758 (60) (220), 2.349 (40) (311), 1.741 (50) (420).
Kind of sample preparation and/or method of registration of the spectrum: Nujol mull. Transmission.
Source: Nefedov et al. (1977).
Wavenumbers (cm^{-1}): 3230sh, 3160s, 2260w, 1165s, 1089, 1025sh, 960sh, 835, 760s, 647sh.
Note: The wavenumbers were partly determined by us based on spectral curve analysis of the published spectrum.

O623 Sodium titanate $Na_2Ti_3O_7$ $Na_2Ti_3O_7$

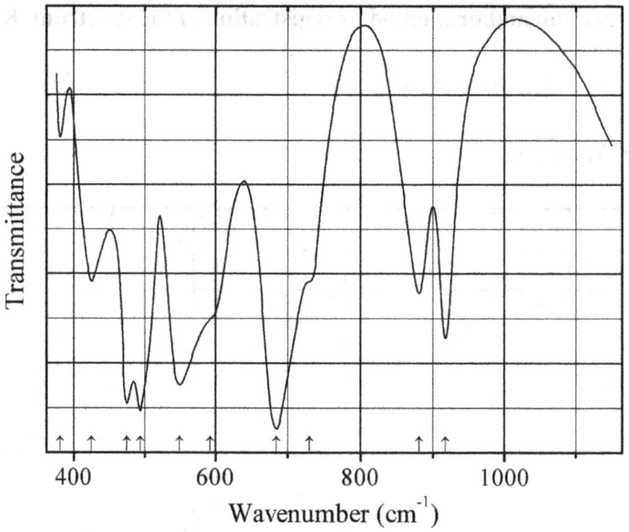

Origin: Synthetic.

Description: Prepared by using a stoichiometric molar ratio of Na_2CO_3 and TiO_2 (rutile) at 1250 °C for 4 h. Characterized by powder X-ray diffraction data. Monoclinic. The crystal structure contains distorted TiO_5 pyramid.

Kind of sample preparation and/or method of registration of the spectrum: KBr disc. Absorption.

Source: Peng et al. (1995).

Wavenumbers (cm^{-1}): 918, 881, 730sh, 684s, 593sh, 550s, 493s, 475s, 425, 380w.

O624 Lepidocrocite $Fe^{3+}O(OH)$

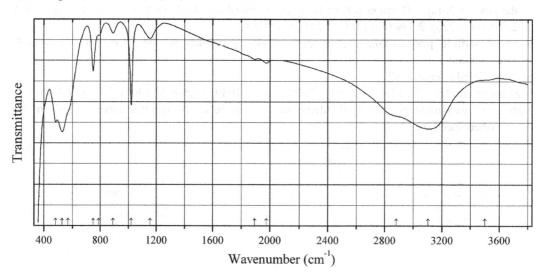

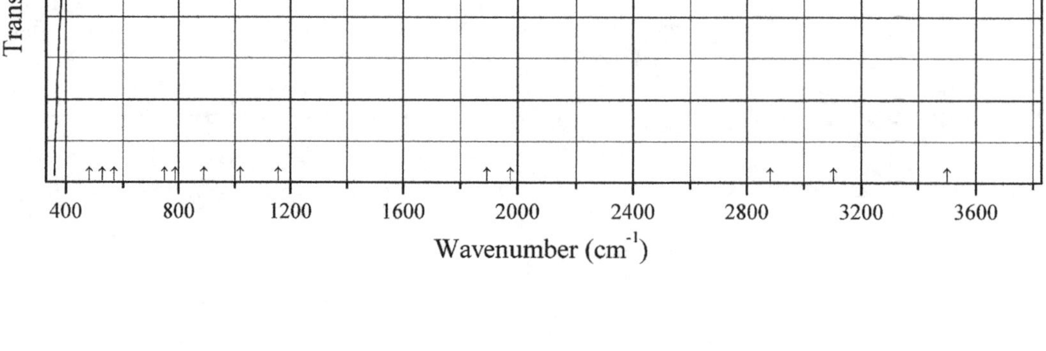

Origin: Hilarion mine, Agios Konstantinos, Lavrion mining District, Attikí (Attika, Attica) Prefecture, Greece.

Description: Yellowish-brown columnar aggregate from the association with goethote and hematite. Pseudomorph after rail. The sample was kindly granted by I.V. Pekov. Confirmed by the IR spectrum.

Kind of sample preparation and/or method of registration of the spectrum: KBr disc. Absorption.

Wavenumbers (cm^{-1}): (3500sh), 3105s, 2880sh, 1976w, 1894w, 1155, 1020s, 892, 788, 748, 570sh, 528s, 483s.

Note: The spectrum was obtained by N.V. Chukanov.

O625 Spertiniite $Cu(OH)_2$

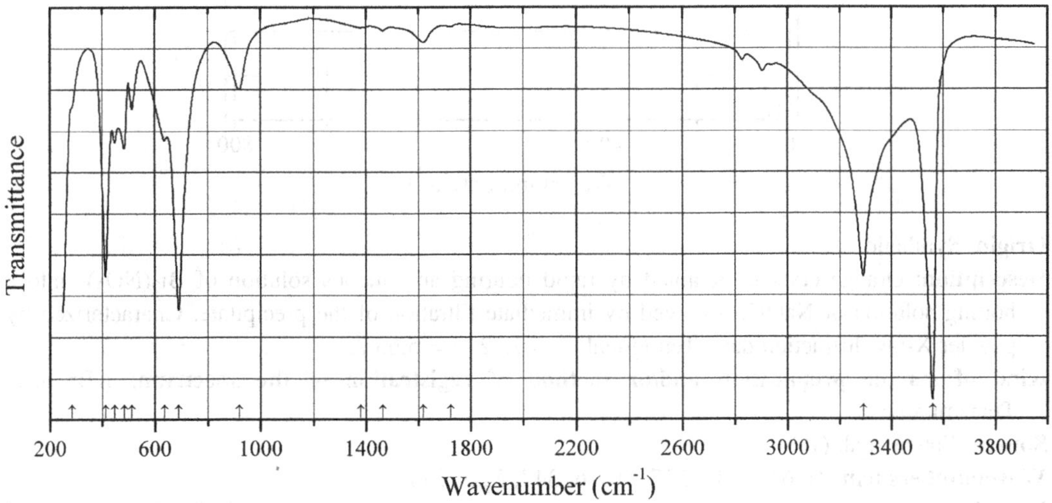

Origin: Synthetic.

Description: Acicular crystals. Confirmed by powder X-ray diffraction data and electron microdiffraction pattern.

Kind of sample preparation and/or method of registration of the spectrum: Absorption. Kind of sample preparation is not indicated.

Source: Rodríguez-Clemente et al. (1994).

Wavenumbers (cm^{-1}): 3560s, 3292s, 1725, 1620w, 1465w, 1380w, 917, 689s, 636, 514, 484, 448, 412s, 284sh.

Note: The wavenumbers were partly determined by us based on spectral curve analysis of the published spectrum. The band at 1620 cm^{-1} corresponds to adsorbed water. Weak bands in the range from 2800 to 3000 cm^{-1} correspond to the admixture of an organic substance.

O626 Sphaerobismoite β-Bi_2O_3

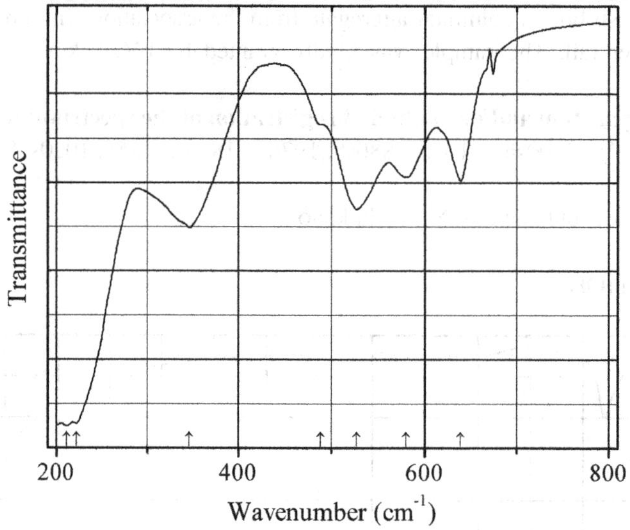

Origin: Synthetic.

Description: Orange crystals obtained by rapid pouring an aqueous solution of $Bi_2(NO_3)_3$ into a boiling solution of NaOH, followed by immediate filtration of the precipitate. Characterized by powder X-ray diffraction data. Tetragonal, $a = 7.72$, $c = 5.63$ Å.

Kind of sample preparation and/or method of registration of the spectrum: KBr disc. Transmission.

Source: Cahen et al. (1980).

Wavenumbers (cm^{-1}): 639, 580, 527, 489sh, 347, 222s, 211s.

Note: The wavenumbers were determined by us based on spectral curve analysis of the published spectrum.

O627 Tantite Ta_2O_5

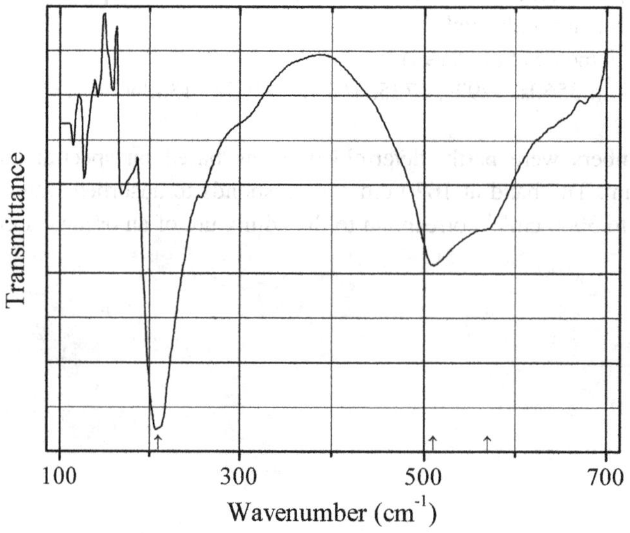

Origin: Synthetic.

Description: A crystalline film deposited on Si substrate by using CVD at 430 °C and annealed at 800 °C for 10 min. Surface SiO_2 was removed by using a diluted HF solution.

Kind of sample preparation and/or method of registration of the spectrum: A film 100 nm thick. Absorption.

Source: Ono et al. (2001).

Wavenumbers (cm^{-1}): 570sh, 510, 210s.

O628 Tistarite Ti_2O_3

Origin: Synthetic.

Kind of sample preparation and/or method of registration of the spectrum: Reflectance of a single crystal.

Source: Lucovsky et al. (1977).

Wavenumbers (cm^{-1}): 448s, 343 (for $E \parallel c$); 511, 451s, 376s, 280w (for $E \perp c$).

O629 Trevorite Co-analogue $CoFe_2O_4$

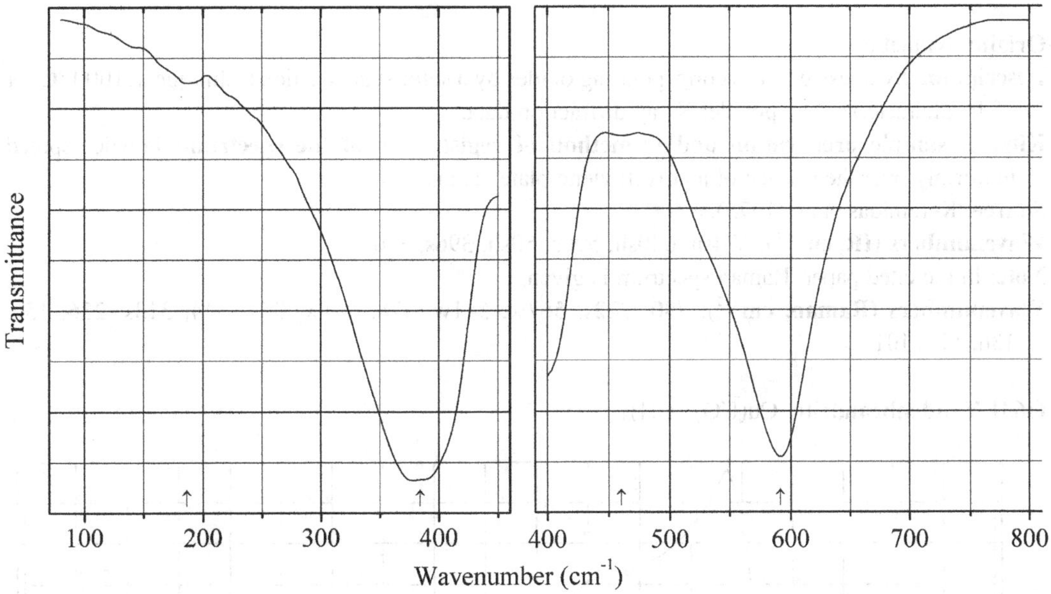

Origin: Synthetic.

Description: Synthesized from corresponding nitrates by a solid-state reaction technique at 1000 °C for 24 h. Characterized by powder X-ray diffraction data. Cubic, space group *Fd*-3*m*.

Kind of sample preparation and/or method of registration of the spectrum: KBr disc and Nujol mull. Absorption.

Source: Ptak et al. (2014).

Wavenumbers (IR, cm^{-1}): 591, 460w, 384, 186w.

Note: The wavenumbers were partly determined by us based on spectral curve analysis of the published spectrum. In the cited paper, Raman spectrum is given.

Wavenumbers (Raman, cm^{-1}): 470.

0630 Ulvospinel Zn-analogue $TiZn_2O_4$

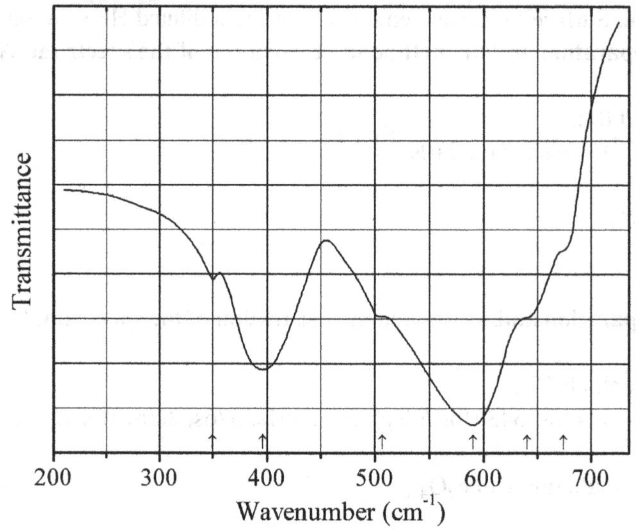

Origin: Synthetic.

Description: Synthesized from corresponding oxides by a solid-state reaction technique at 1000 °C for 25 h. Characterized by powder X-ray diffraction data.

Kind of sample preparation and/or method of registration of the spectrum: Powder spread uniformly over the surface of a polyethylene plate. Transmission.

Source: Keramidas et al. (1975).

Wavenumbers (IR, cm^{-1}): 674sh, 640sh, 590s, 50sh, 396s, 350.

Note: In the cited paper, Raman spectrum is given.

Wavenumbers (Raman, cm^{-1}): 780, 722s, 567w, 541w, 478, 441w, 351, 344, 313s, 256, 156, 136, 117, 101.

0631 Vandenbrandeite $Cu(UO_2)(OH)_4$

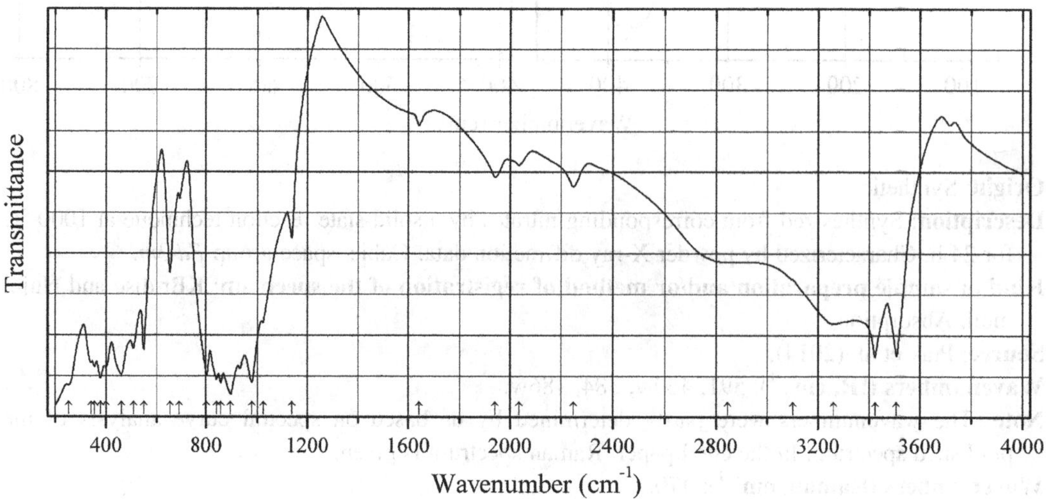

Origin: Kalongwe deposit, Shaba, province, Zaire (type locality).

Description: Green tabular crystals from the association with kasolite. Characterized by powder X-ray diffraction data and electron microprobe analyses.

Kind of sample preparation and/or method of registration of the spectrum: KBr disc. Transmission.

Source: Botto et al. (2002).

Wavenumbers (IR, cm^{-1}): 3508s, 3423s, 3262, 2845, 3100sh, 2247, 2035, 1943, 1643w, 1139, 1024, 978s, 897s, 859s, 842s, 803s,694, 655, 548s, 508s, 460s, 400s, 377s, 354s, 340s, 252s.

Note: The wavenumbers were partly determined by us based on spectral curve analysis of the published spectrum. In the cited paper, Raman spectrum is given.

Wavenumbers (Raman, cm^{-1}): 2233, 2018, 1946, 862, 805s, 474, 186.

O632 Vandenbrandeite hydrogen-free analogue $CuUO_4$

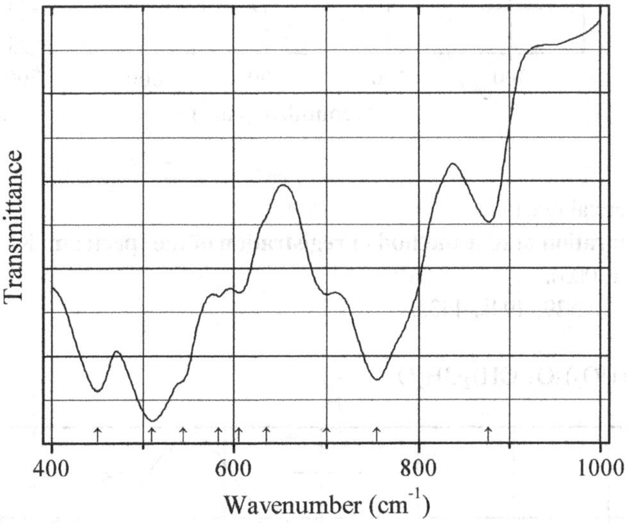

Origin: Synthetic.

Description: Prepared in a solid-state reaction between CuO and UO_3. The strongest lines of the powder X-ray diffraction pattern [*d*, Å (*I*, %)] are: 5.08 (50), 4.15 (90), 3.76 (50), 3.44 (100), 3.26 (50), 2.71 (90).

Kind of sample preparation and/or method of registration of the spectrum: Nujol mull. Transmission.

Source: Jakeš et al. (1968).

Wavenumbers (cm^{-1}): 877, 755s, ~700, 635sh, ~605, 583, ~544sh, ~510s, ~450s.

Note: The wavenumbers were partly determined by us based on spectral curve analysis of the published spectrum. In the cited paper, also IR spectra of $MnUO_4$, MnU_3O_{10}, $CoUO_4$, CoU_3O_{10}, $NiUO_4$, NiU_3O_{10}, CuU_3O_{10}, $AgUO_4$, and $HgUO_4$ are given.

O633 Zincite ZnO

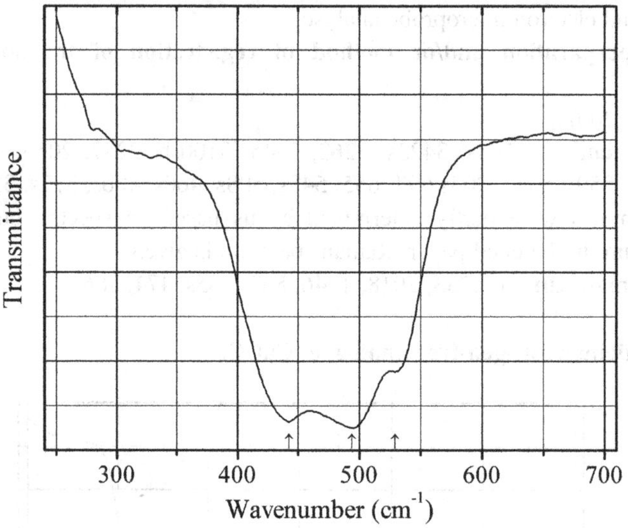

Origin: Synthetic.
Description: Commercial reactant.
Kind of sample preparation and/or method of registration of the spectrum: KBr disc. Transmission.
Source: Musić et al. (2003).
Wavenumbers (cm^{-1}): 529, 494s, 443s.

O634 Masuyite $Pb(UO_2)_3O_3(OH)_2 \cdot 3H_2O$

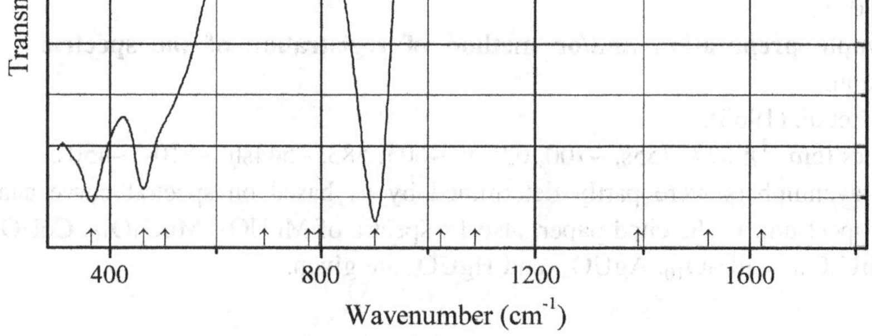

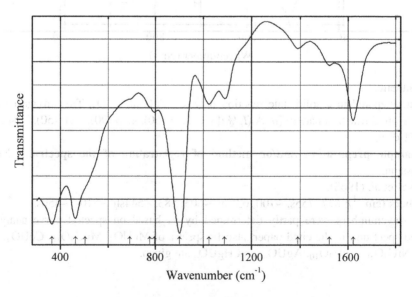

Origin: Rum Jungle (?), northern Australia.
Description: Specimen No. 26494 from the collections of the Department of Geology and Mineralogy, University of Queensland. Confirmed by powder X-ray diffraction data.
Kind of sample preparation and/or method of registration of the spectrum: Transmission.
Source: Wilkins (1971).
Wavenumbers (cm^{-1}): 1622, 1522w, 1391w, 1089, 1024, 900s, 796w, 778sh, 692sh, 503sh, 462s, 364s.

O635 Sodium titanate $Na_2Ti_6O_{13}$ $Na_2Ti_6O_{13}$

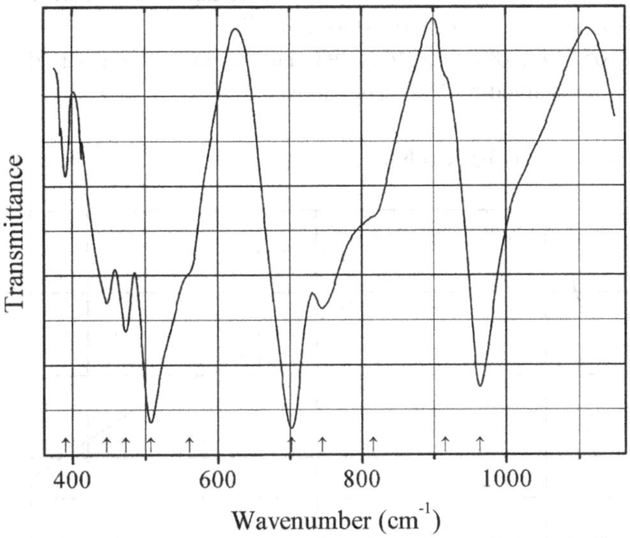

Origin: Synthetic.

Description: Prepared by using a stoichiometric mixture of Na_2CO_3 and TiO_2 (rutile) at 1250 °C for 4 h. Characterized by powder X-ray diffraction data. Monoclinic, with distorted TiO_6 octahedra.

Kind of sample preparation and/or method of registration of the spectrum: KBr disc. Transmission.

Source: Peng et al. (1995).

Wavenumbers (cm^{-1}): 964s, 916sh, 816sh, 745, 702s, 560sh, 507s, 473, 447, 390w.

O640 Chibaite $SiO_2{\cdot}n(CH_4,C_2H_6,C_3H_8,C_4H_{10})$ (n_{max} = 3/17)

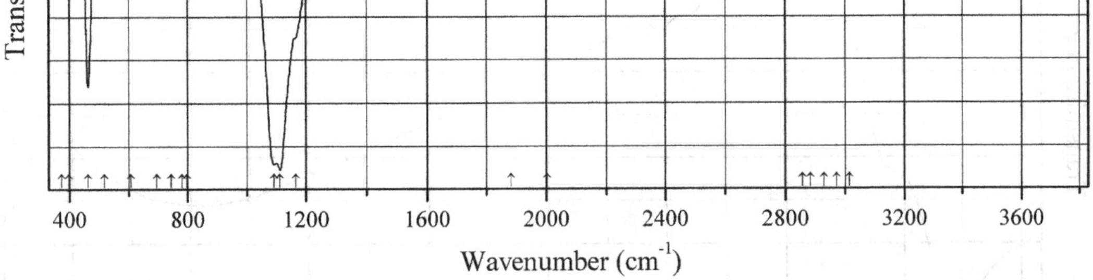

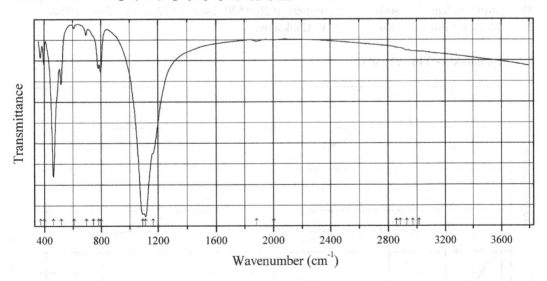

Origin: Arakawa, Minamiboso city, Chiba prefecture, Kanto region, Honshu Island, Japan (type locality).

Description: White semitransparent crystals. Intergrowths with quartz.

Kind of sample preparation and/or method of registration of the spectrum: KBr disc. Absorption.
Wavenumbers (cm^{-1}): 3014w, 2970w, 2928w, 2880w, 2854w, 2003w, 1882w, 1165sh, 1113s, 1093s, 797, 780, 745sh, 695, 609w, 519, 464s, 398, 374.
Note: The bands at 1165, 797, 780, 695, 464s 398, and 374 cm^{-1} are partly due to quartz.
Note: The spectrum was obtained by N.V. Chukanov.

O641 Samarskite-(Y) (Y,Ce,U,Fe,Nb)(Nb,Ta,Ti)O_4

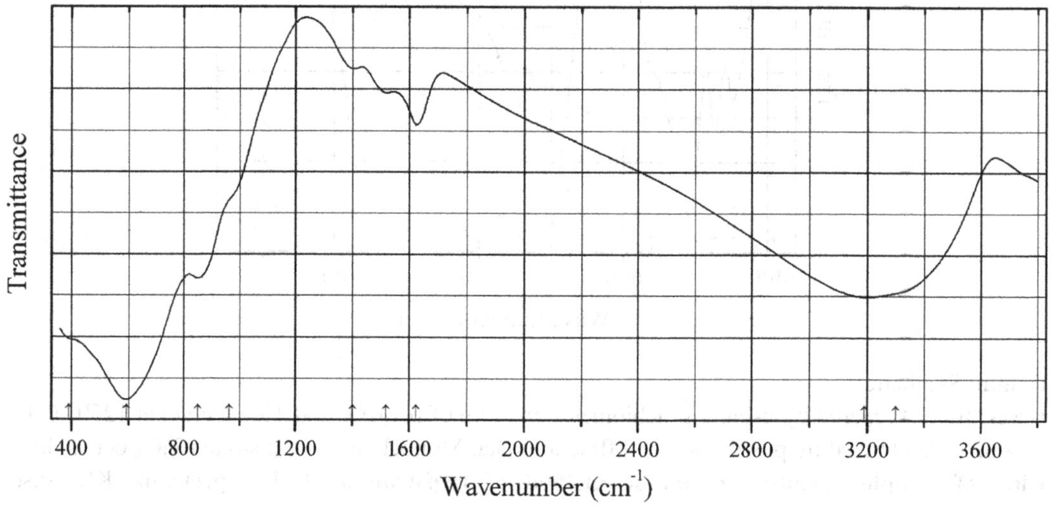

Origin: Herrebøkasa, Aspedammen, Idd, Halden, Østfold, Norway.
Description: Black grains. X-ray amorphous, metamict. The empirical formula is (electron microprobe): $(Y_{0.25}U_{0.2}Ca_{0.2}Fe_{0.2}Ln_{0.1})(Nb_{0.85}Ti_{0.1}Ta_{0.05})O_9 \cdot nH_2O$.
Kind of sample preparation and/or method of registration of the spectrum: KBr disc. Absorption.
Wavenumbers (cm^{-1}): 3300sh, 3192s (broad), 1626, 1520, 1403w, 965sh, 851, 593s, 390sh.
Note: The spectrum was obtained by N.V. Chukanov.

O642 Samarskite-(Yb) $YbNbO_4$

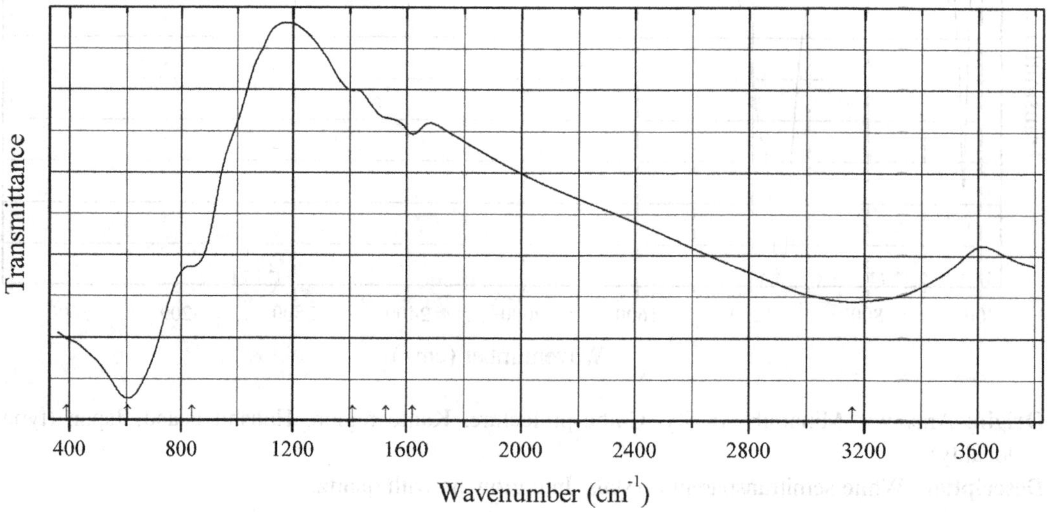

Origin: Little Patsy pegmatite (Patsy pegmatite), South Platte Pegmatite District, Jefferson Co., Colorado, USA (type locality).

Description: Black grains in pegmatite. X-ray amorphous, metamict. The empirical formula is (electron microprobe): $(Yb_{0.13}Dy_{0.07}Er_{0.04}Y_{0.01}La_{0.01})Ca_{0.24}U_{0.12}Th_{0.08}Fe_{0.18}Mn_{0.02}(Nb_{0.93}Ta_{0.11}Ti_{0.01})\cdot nH_2O$.

Kind of sample preparation and/or method of registration of the spectrum: KBr disc. Absorption.

Wavenumbers (cm^{-1}): 3160s (broad), 1623, 1530sh, 1412w, 838, 607s, (390sh).

Note: The spectrum was obtained by N.V. Chukanov.

O643 Gauthierite $KPb[(UO_2)_7O_5(OH)_7]\cdot 8H_2O$

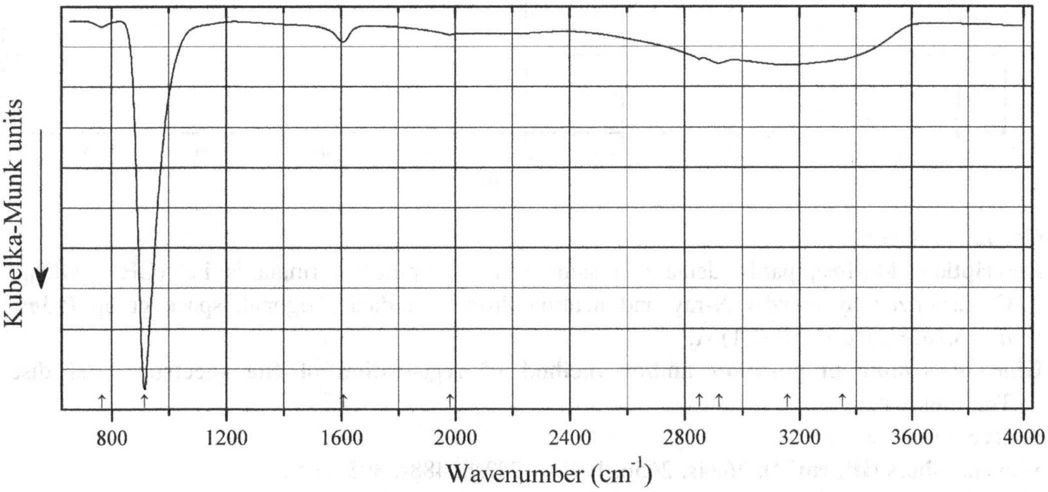

Origin: Shinkolobwe mine, Shinkolobwe, Katanga province, Democratic Republic of Congo (type locality).

Description: Yellowish orange crystals from the association with soddyite and a metazeunerite–metatorbernite series mineral. Holotype sample. The crystal structure is solved. Monoclinic, space group $P2_1/c$, $a = 29.844(2)$, $b = 14.5368(8)$ Å, $c = 14.0406(7)$ Å, $\beta = 103.708(6)°$, $V = 5917.8(6)$ Å^3, $Z = 8$. $D_{calc} = 5.437$ g/cm^3. Optically biaxial (–), $\alpha = 1.780(5)$, $\beta = 1.815(5)$, $\gamma = 1.825(5)$, $2V = 70(5)°$. The empirical formula is $K_{0.67}Pb_{0.78}U_7O_{34}H_{23.77}$. The strongest lines of the powder X-ray diffraction pattern [*d*, Å (*I*, %) (*hkl*)] are: 7.28 (49) (020, 400), 3.566 (67) (040, –802, –204), 3.192 (100) (622, –224), 2.541 (18) (–842, –244), 2.043 (14) (406), 2.001 (23) (662,–264, 14.2.0), 1.962 (14) (426, –146), 1.783 (17) (12.0.4, –10.4.6).

Kind of sample preparation and/or method of registration of the spectrum: Attenuated total reflection of a polycrystalline sample.

Source: Olds et al. (2016a).

Wavenumbers (IR, cm^{-1}): 3350sh, 3154 (broad), 2919, 2852, 1980w, 1607, 915s, 764w.

Note: The bands in the range from 2800 to 3000 cm^{-1} correspond to the admixture of an organic substance. In the cited paper, Raman spectrum is given.

Wavenumbers (Raman, cm^{-1}): 833s, 821s, 696, 558, 539, 464, 454, 403, 355, 328, 260, 204, 160, 128.

O645 Amakinite $(Fe^{2+},Mg)(OH)_2$

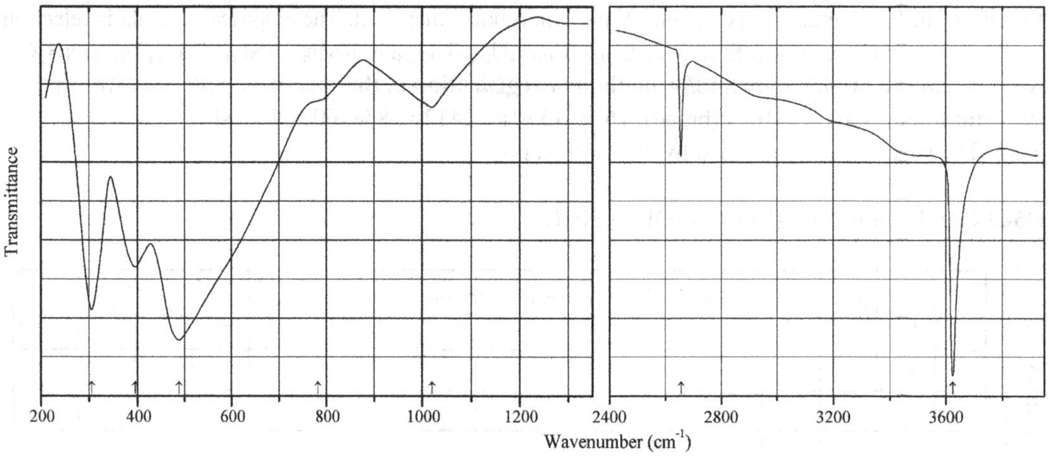

Origin: Synthetic.

Description: Mg-free, partly deuterated sample. The empirical formula is $Fe^{2+}(OH)_{1.7}(OD)_{0.3}$. Characterized by powder X-ray and neutron diffraction data. Trigonal, space group *P*-3*m*1, $a = 3.2628(1)$, $c = 4.604(1)$ Å.

Kind of sample preparation and/or method of registration of the spectrum: CsI disc. Transmission.

Source: Lutz et al. (1994).

Wavenumbers (IR, cm^{-1}): 3624s, 2656, 1020w, 782sh, 488s, 395, 305s.

Note: The wavenumbers were partly determined by us based on spectral curve analysis of the published spectrum. The band at 2656 cm^{-1} corresponds to D–O-stretching vibrations. In the cited paper, Raman spectrum is given.

Wavenumbers (Raman, cm^{-1}): 3573s, 407, 260.

O646 Clarkeite $Na(UO_2)O(OH)\cdot nH_2O$

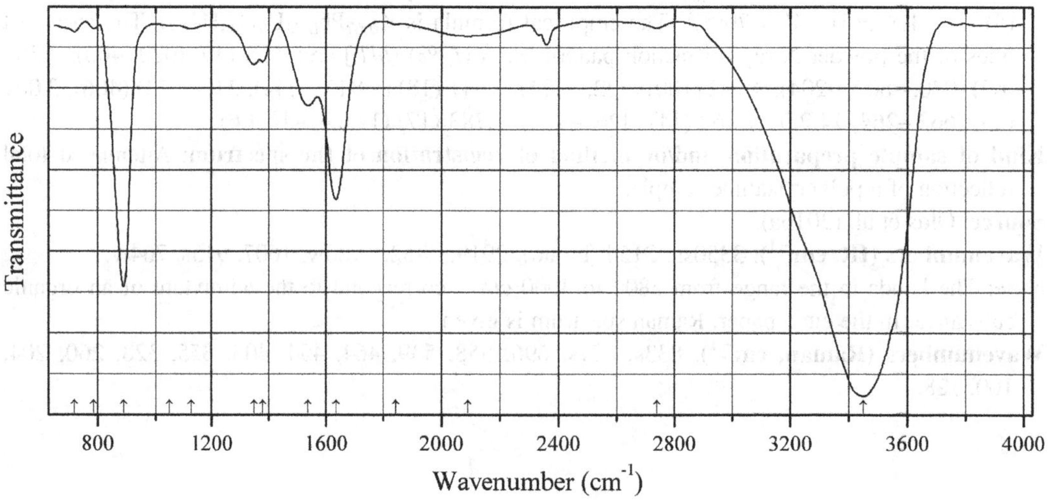

Origin: Synthetic.

Description: Obtained by precipitation generated in a uranyl peroxycarbonato complex solution at pH 14 controlled by NaOH, with subsequent drying in vacuum at 100 °C for 3 h.

Kind of sample preparation and/or method of registration of the spectrum: Transmission. Kind of sample preparation is not indicated.

Source: Kim et al. (2009).

Wavenumbers (cm^{-1}): 3450s (broad), 2738w, 2090w (broad), 1845, 1636s, 1537, 1378, 1347, 1126w, 1048sh, 888s, 786w, 720w.

Note: The wavenumbers were determined by us based on spectral curve analysis of the published spectrum.

O647 Duttonite $V^{4+}O(OH)_2$

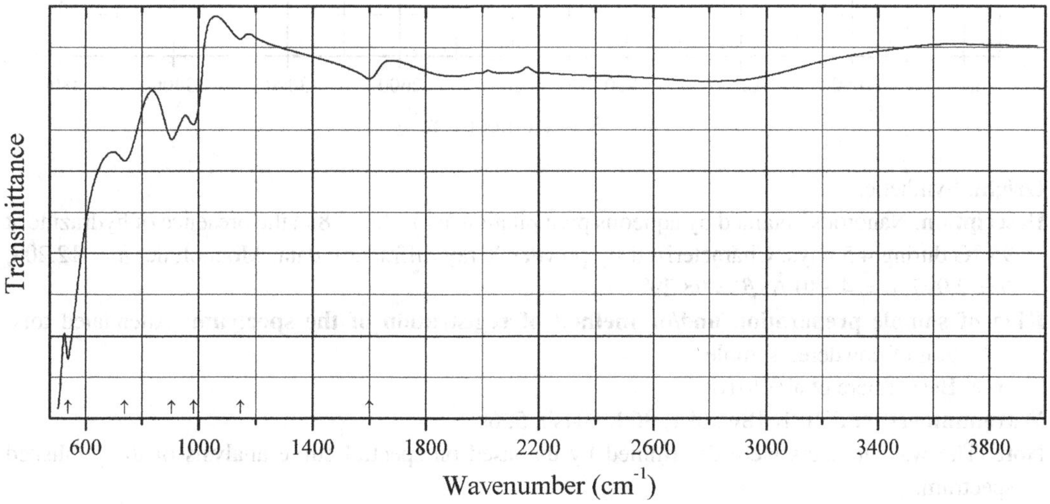

Origin: Synthetic.

Description: Nanorods obtained by aqueous precipitation at pH 4.0 in the presence of hydrazine at 95 °C during 4.5 days. Characterized by powder X-ray diffraction data.

Kind of sample preparation and/or method of registration of the spectrum: Attenuated total reflection of a powdered sample.

Source: Besnardiere et al. (2016).

Wavenumbers (cm^{-1}): 1600, 1145w, 982, 905, 738, 536s, and broad bands near 2800 and 1900 cm^{-1}).

Note: The wavenumbers were determined by us based on spectral curve analysis of the published spectrum.

O648 Häggite $V^{3+}V^{4+}O_2(OH)_3$

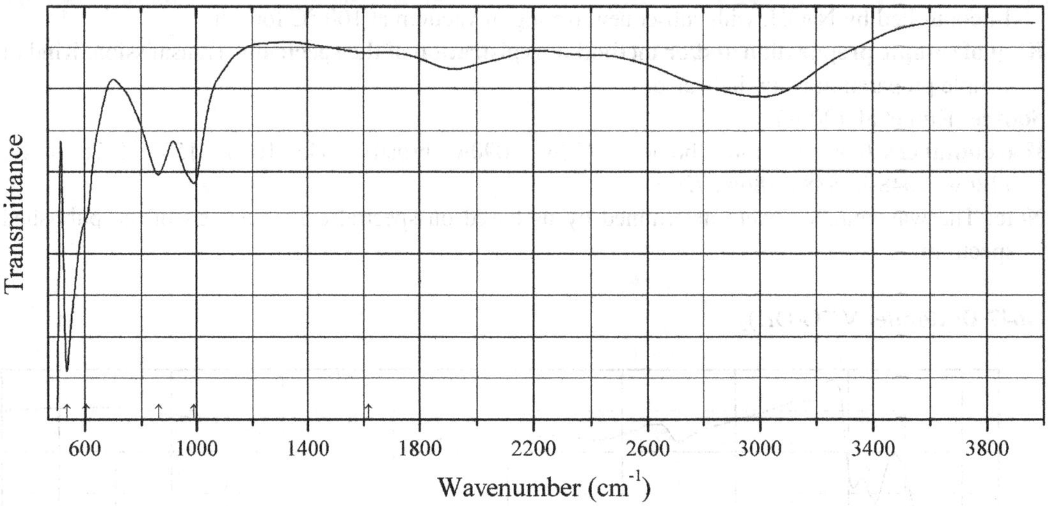

Origin: Synthetic.

Description: Nanorods obtained by aqueous precipitation at pH 3.6–3.8 in the presence of hydrazine at 95 °C during 4.5 days. Characterized by powder X-ray diffraction data. Monoclinic, $a = 12.208$, $b = 2.997$, $c = 4.840$ Å, $\beta = 98.29°$

Kind of sample preparation and/or method of registration of the spectrum: Attenuated total reflection of powdered sample.

Source: Besnardiere et al. (2016).

Wavenumbers (cm^{-1}): 1618w, 991, 860, 604sh, 536s.

Note: The wavenumbers were determined by us based on spectral curve analysis of the published spectrum.

O649 Jianshuiite $MgMn^{4+}{}_3O_7{\cdot}3H_2O$

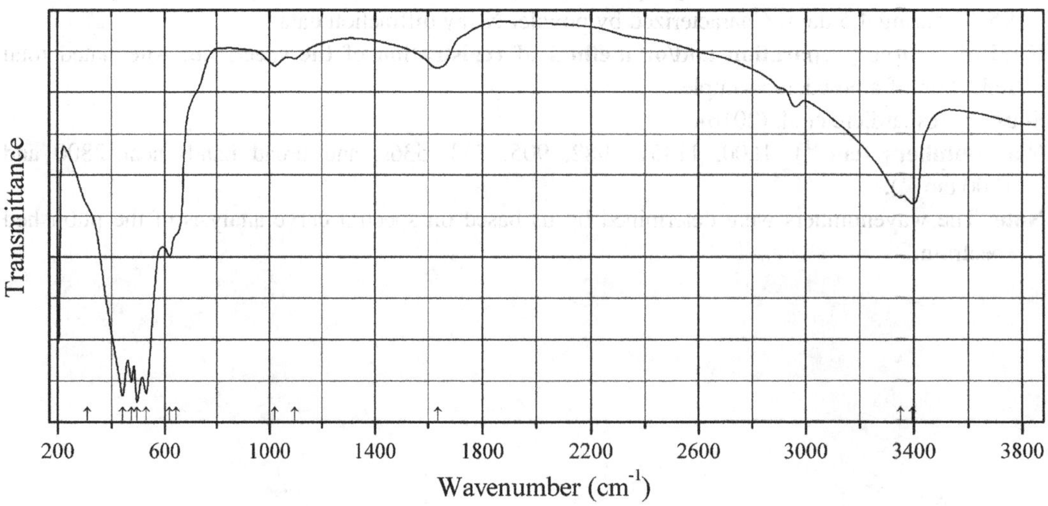

Origin: Luzhai Mn deposit, Jianshui Co., Honghe Autonomous Prefecture, Yunnan, China (type locality).

Description: Aggregate of dark brown microcrystals. Holotype sample. Characterized by powder X-ray diffraction data. Triclinic, space group *P*-1, $a = 7.534(4)$, $b = 7.525(6)$, $c = 8.204(8)$ Å, $\alpha = 89.753(8)°$, $\beta = 117.375(6)°$, $\gamma = 120.000(6)°$. $D_{meas} = 3.50–3.60$ g/cm^3. The empirical formula is $(Mg_{0.85}Mn^{2+}_{0.05})Mn^{4+}_{3.15}O_{7.20}·2.80H_2O$.

Kind of sample preparation and/or method of registration of the spectrum: Transmission.

Source: Yan et al. (1992).

Wavenumbers (cm^{-1}): 3396, 3352, 1635, 1096w, 1020w, 645sh, 620, 530s, 496s, 475s, 442s, 312sh.

Note: The wavenumbers were partly determined by us based on spectral curve analysis of the published spectrum. Weak bands in the range from 2800 to 3000 cm^{-1} correspond to the admixture of an organic substance.

O650 Mushistonite $Cu^{2+}Sn^{4+}(OH)_6$

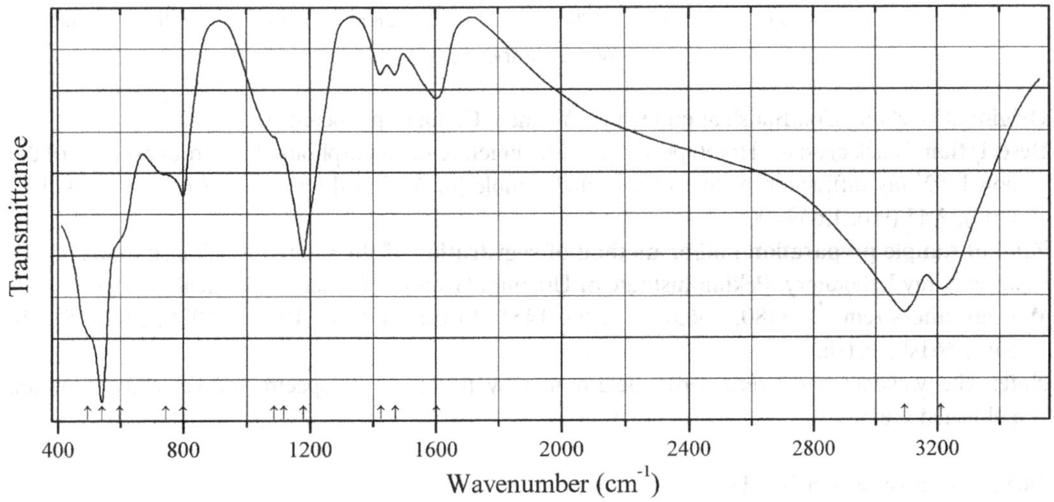

Origin: Mushiston Sn deposit, Kaznok valley, Penjikent, Zeravshan range, Tajikistan (type locality).

Description: Pseudomorphs after stannite. Holotype sample. Cubic, $a = 7.735$ Å. The strongest lines of the powder X-ray diffraction pattern [*d*, Å (*I*, %) (*hkl*)] are: 3.88 (100) (111), 2,740 (50) (220), 2.230 (20) (222), 1.932 (16) (400), 1.729 (35) (420), 1.578 (23) (422).

Kind of sample preparation and/or method of registration of the spectrum: KBr disc. Transmission.

Source: Marshukova et al. (1984).

Wavenumbers (cm^{-1}): 3211s, 3093s, 1603, 1471, 1424, 1178s, 1117sh, 1087sh, 800, 745sh, 597sh, 540s, 495sh.

Note: The wavenumbers were determined by us based on spectral curve analysis of the published spectrum.

0651 Orthobrannerite $U^{4+}U^{6+}Ti_4O_{12}(OH)_2$

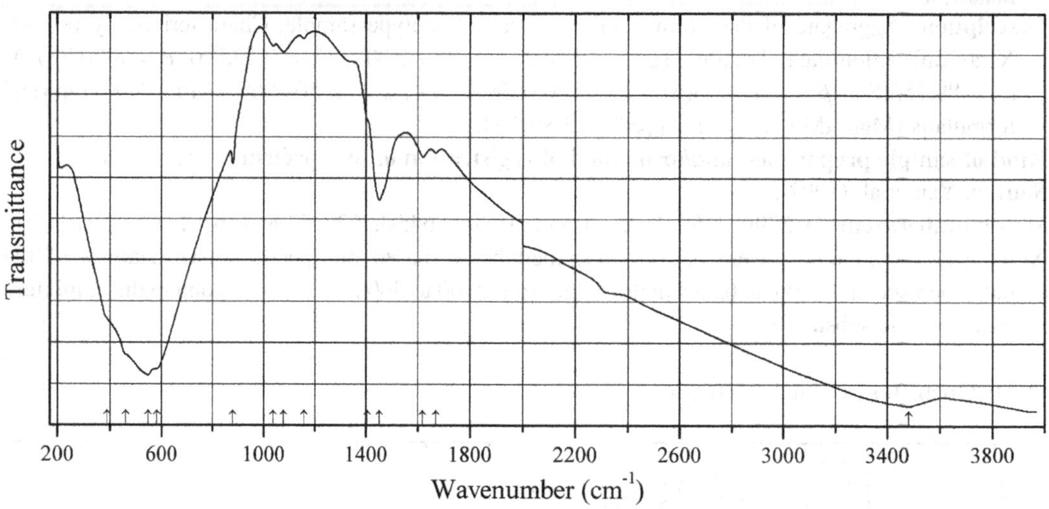

Origin: Dengzhong Co., Baoshan prefecture, Yunnan, China (type locality).

Description: Black crystals. Holotype sample. Metamict, X-ray amorphous. The strongest lines of the powder X-ray diffraction pattern of a heated sample [*d*, Å (*I*, %)] are: 4.87 (70), 3.89 (80), 3.17 (100), 2.45 (90), 1.659 (90).

Kind of sample preparation and/or method of registration of the spectrum: Transmission.

Source: X-ray Laboratory, Peking Institute of Uranium Geology, Wuhan Geological College (1978).

Wavenumbers (cm^{-1}): 3480, 1669w, 1620w, 1451, 1405sh, 1157w, 1078w, 1036w, 880, 581sh, 546s, 461sh, 391sh.

Note: The wavenumbers were partly determined by us based on spectral curve analysis of the published spectrum.

0652 Pyrochroite $Mn^{2+}(OH)_2$

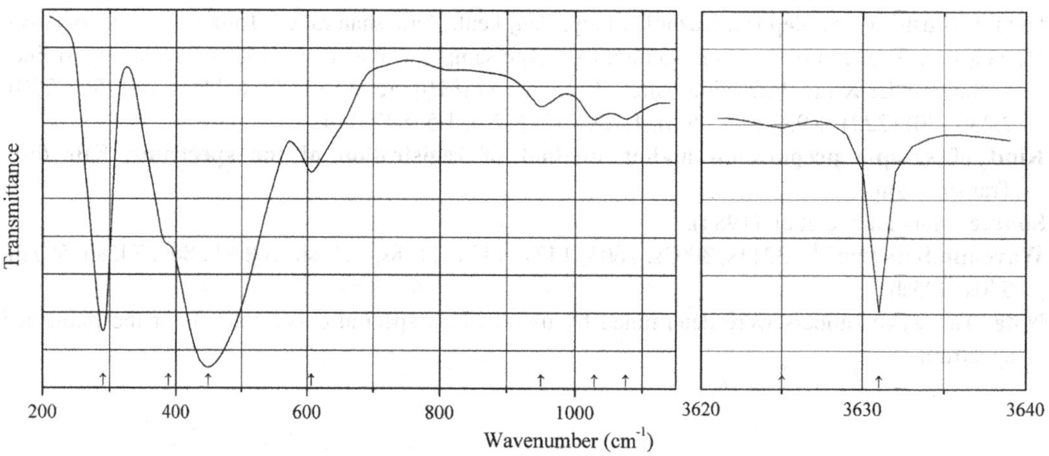

Origin: Synthetic.

Kind of sample preparation and/or method of registration of the spectrum: CsI disc. Transmission.

Source: Lutz et al. (1994).

Wavenumbers (IR, cm^{-1}): 3631s, 1075w, 1029w, 950w, 606, 450s, 388sh, 290s.
Note: In the cited paper, Raman spectrum is given.
Wavenumbers (Raman, cm^{-1}): 3578s, 401, 234.

O653 Fluorcalciopyrochlore $(Ca,Na)_2(Nb,Ti)_2O_6F$

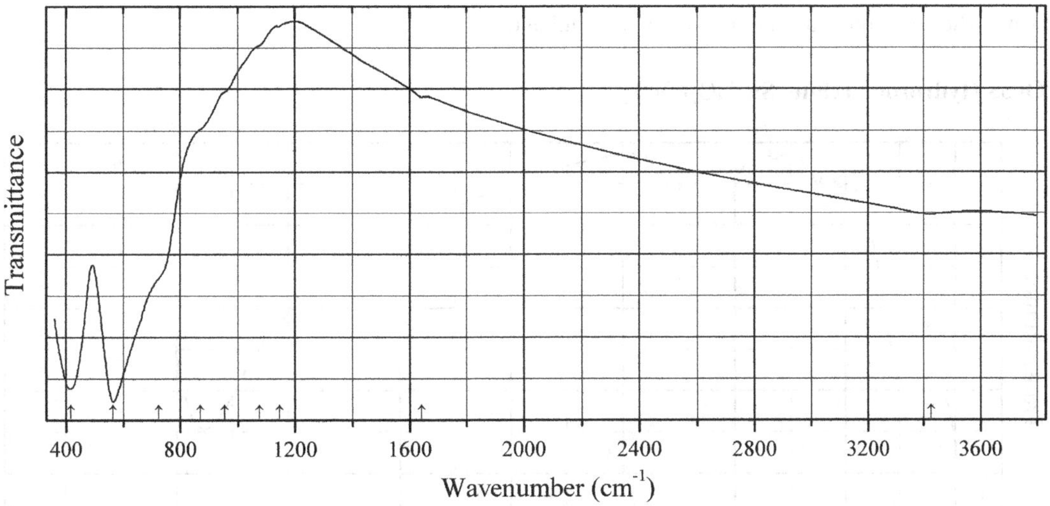

Origin: Tatarka River, Krasnoyarskiy Kray, Siberia, Russia.
Description: Yellow-brown octahedral crystal. The empirical formula is (electron microprobe): $(Ca_{0.85}Na_{0.19}REE_{0.02})(Nb_{1.91}Ti_{0.04}Fe_{0.03}Re_{0.02})(O,OH)_{6.36}F_{0.64}\cdot nH_2O$ ($n \ll 1$).
Kind of sample preparation and/or method of registration of the spectrum: KBr disc. Absorption.
Wavenumbers (cm^{-1}): 3426w, 1643w, 1145w, 1075sh, 955sh, 870sh, 725sh, 565s, 416s.
Note: The spectrum was obtained by N.V. Chukanov.

O654 Ferronigerite-2*N*1*S* $(Al,Fe,Zn)_2(Al,Sn)_6O_{11}(OH)$

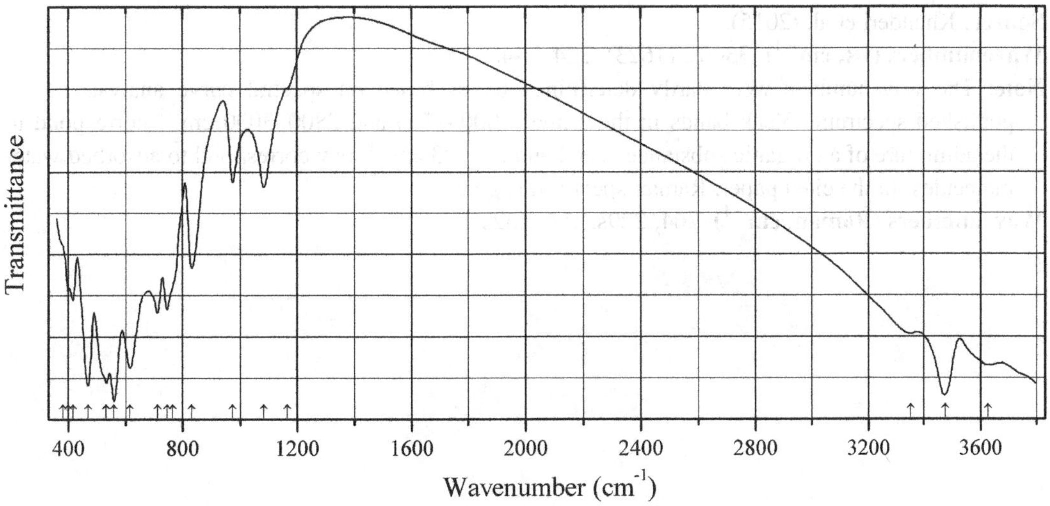

Origin: Three Aloes Mine, Uis, Damaraland District, Kunene Region, Namibia.

Description: Orange-brown tabular crystal. The empirical formula is (electron microprobe): $Al_{5.97}Fe_{0.79}Zn_{0.15}Mg_{0.07}Cr_{0.02}(Sn_{0.93}Ti_{0.07})O_{11}(OH)$.

Kind of sample preparation and/or method of registration of the spectrum: KBr disc. Absorption.

Wavenumbers (cm^{-1}): 3627w, 3473, 3352w, 1165sh, 1083, 975, 832, 765sh, 746, 712, 616s, 561s, 534s, 470s, 417, 399, 380sh.

Note: The spectrum was obtained by N.V. Chukanov.

O655 Hydroromarchite $Sn^{2+}{}_3O_2(OH)_2$

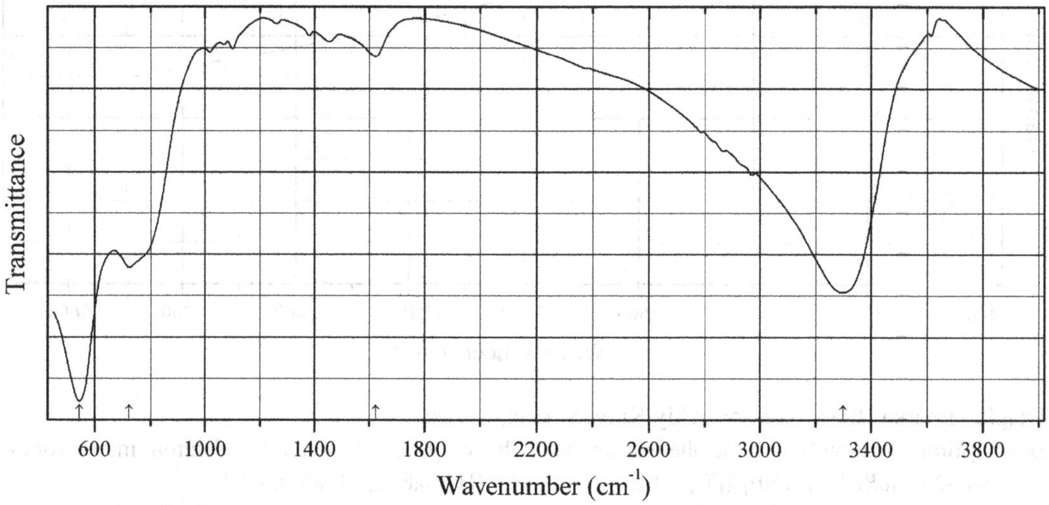

Origin: Synthetic.

Description: Yellow precipitate obtained by addition of water to diethylether solution of $Sn(NMe_2)_2$. Characterized by TG and powder X-ray diffraction data.

Kind of sample preparation and/or method of registration of the spectrum: Transmission. Kind of sample preparation is not indicated.

Source: Khanderi et al. (2015).

Wavenumbers (IR, cm^{-1}): 3300s, (1623), 724, 544s.

Note: The wavenumbers were partly determined by us based on spectral curve analysis of the published spectrum. Weak bands in the ranges 1000–1500 and 2800–3000 cm^{-1} correspond to the admixture of an organic substance. The band at 1623 cm^{-1} may correspond to adsorbed water molecules. In the cited paper, Raman spectrum is given.

Wavenumbers (Raman, cm^{-1}): 264, 229s, 186, 132s.

O656 Kusachiite $Cu^{2+}Bi^{3+}{}_2O_4$

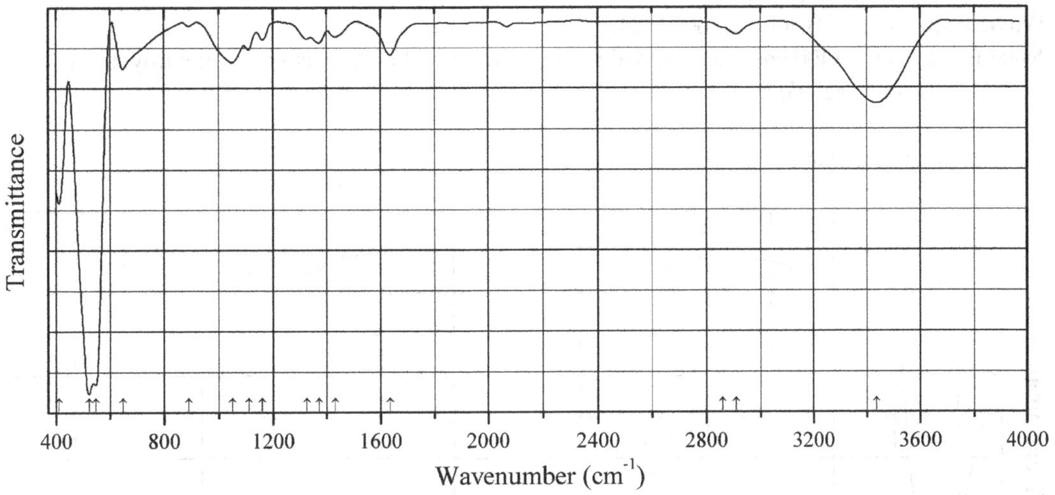

Origin: Synthetic.

Description: Prepared hydrothermally from bismuth acetate and copper nitrate in the presence of NaOH at 140 °C for 12 h. Characterized by powder X-ray diffraction data. Tetragonal, space group *P4/ncc*, $a = 8.567$, $c = 5.791$ Å, $V = 425.17$ Å^3.

Kind of sample preparation and/or method of registration of the spectrum: Transmission. Kind of sample preparation is not indicated.

Source: Yuvaraj et al. (2016).

Wavenumbers (IR, cm^{-1}): 890w, 648, 547s, 522s, 410.

Note: The wavenumbers were partly determined by us based on spectral curve analysis of the published spectrum. Bands above 1000 cm^{-1} correspond to impurities. In the cited paper, Raman spectrum is given.

Wavenumbers (Raman, cm^{-1}): 575w, 399, 257s, 187w, 123s, 84w.

O657 Montroydite HgO

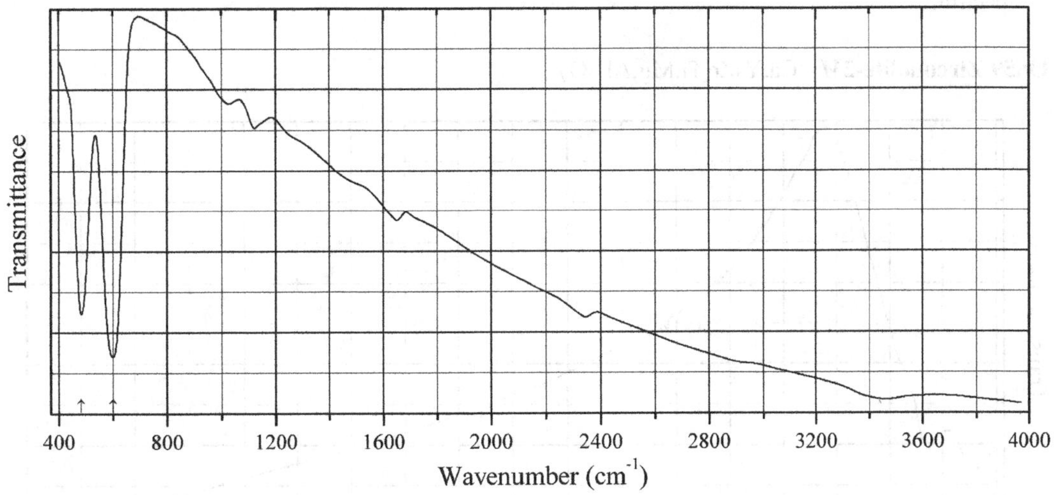

Origin: Synthetic.

Description: Commercial reactant. Characterized by powder X-ray diffraction data.

Kind of sample preparation and/or method of registration of the spectrum: No data.
Source: Xhaxhiu et al. (2013).
Wavenumbers (cm^{-1}): 1122w, 600, 484.
Note: Bands above 1400 cm^{-1} correspond to impurities. For the IR spectrum of montroydite see also Godelitsas et al. (2003).

O658 Romarchite SnO

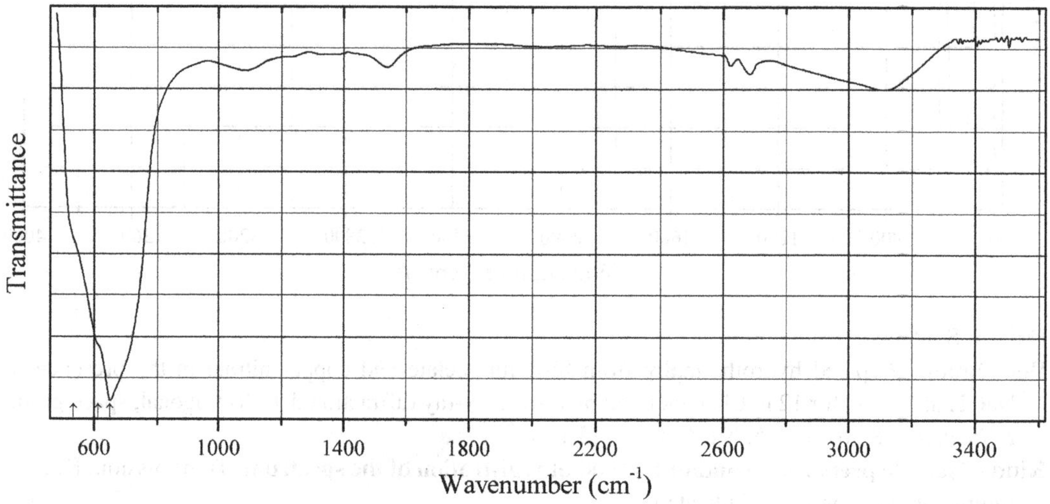

Origin: Synthetic.
Description: Nanoparticles. Characterized by powder X-ray diffraction data.
Kind of sample preparation and/or method of registration of the spectrum: Transmission. Kind of sample preparation is not indicated.
Source: Krishnakumar et al. (2008).
Wavenumbers (cm^{-1}): 651s, 613sh, 535sh.
Note: The wavenumbers were determined by us based on spectral curve analysis of the published spectrum.

O659 Zirconolite-2*M* (Ca,Y)Zr(Ti,Mg,Al)$_2$O$_7$

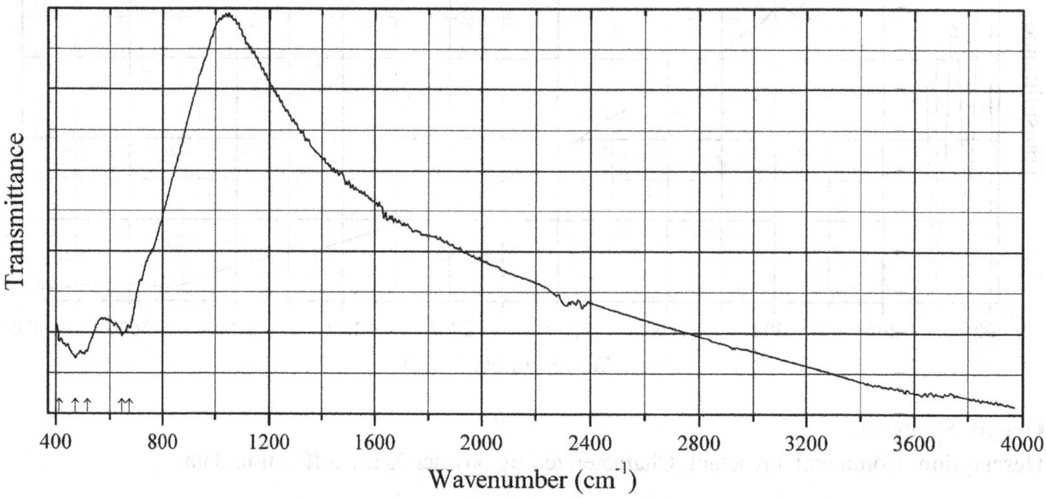

Origin: Synthetic.

Description: Synthesized by a solid-state reaction method at 1400 °C for 16 h. Characterized by powder X-ray diffraction data. Monoclinic, *a* = 12.441, *b* = 7.239, *c* = 11.341 Å, *β* = 100.694°. The empirical formula is $Ca_{0,83}Ce_{0,17}ZrTi_{1,66}Al_{0.34}O_7$.

Kind of sample preparation and/or method of registration of the spectrum: KBr disc. Transmission.

Source: Souag et al. (2015).

Wavenumbers (cm^{-1}): 685, 646, 517s, 473s, 413sh.

Note: The band position denoted by Souag et al. (2015) as 685 cm^{-1} was determined by us at 675 cm^{-1}.

O660 Magnesiohögbomite-2*N*3*S* $(Mg,Fe,Zn,Ti)_4(Al,Fe)_{10}O_{19}(OH)$

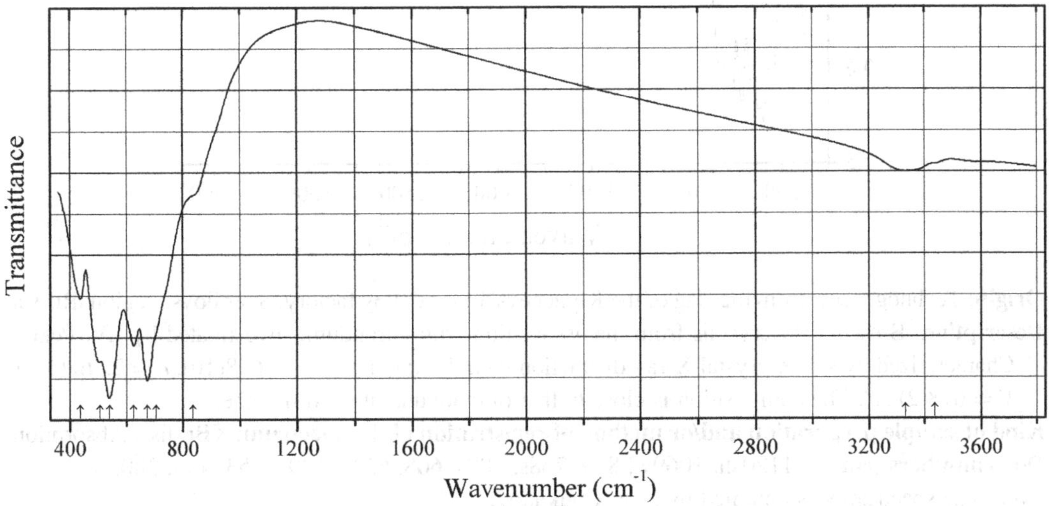

Origin: Zelentsovskaya pit, near Zlatoust, Chelyabinsk region, Southern Urals, Russia.

Description: Crystals from the association with clinochlore, magnetite, and spinel. Investigated by I.V. Pekov. Characterized by single-crystal X-ray diffraction data. Hexagonal, *a* = 5.743(3), *c* = 23.10(2) Å, *V* = 659.7(8) Å^3. The empirical formula is (electron microprobe): $(Mg_{2.04}Fe_{1.89}Zn_{0.06}Mn_{0.02})(Al_{9.52}Ti_{0.77})O_{19}(OH)$.

Kind of sample preparation and/or method of registration of the spectrum: KBr disc. Absorption.

Wavenumbers (cm^{-1}): 3435sh, 3332, 835sh, 705sh, 674s, 626s, 540s, 507s, 437.

Note: The spectrum was obtained by N.V. Chukanov.

O661 Diaoyudaoite $NaAl_{11}O_{17}$

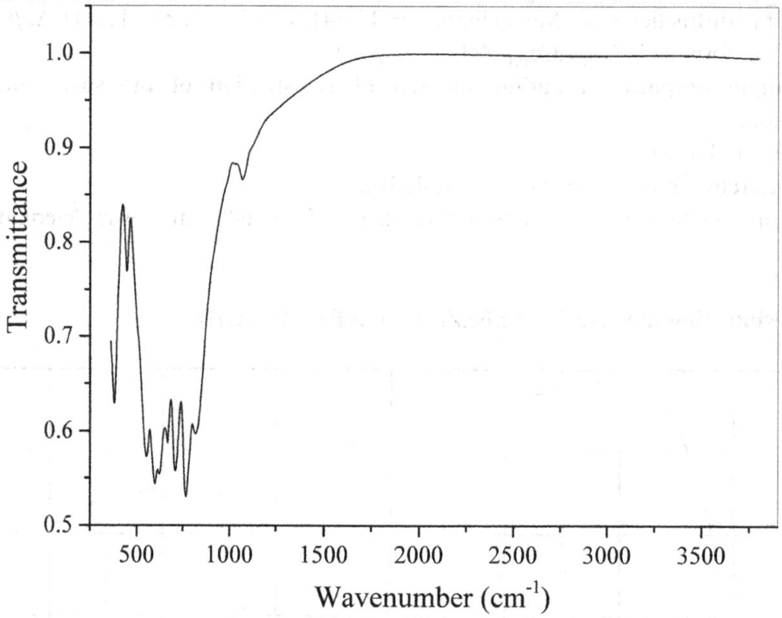

Origin: Technogenetic, from the slag of the Klyuchevskoi ferroalloy factory, Sverdlovsk region, Russia.
Description: Brown platy crystals from the association with corundum. Investigated by I.V. Pekov. Characterized by single-crystal X-ray diffraction data. Hexagonal, $a = 5.618(10)$, $c = 22.62(3)$ Å, $V = 618(2)$ Å^3. The composition is close to that of diaoyudaoite end-member.
Kind of sample preparation and/or method of registration of the spectrum: KBr disc. Absorption.
Wavenumbers (cm^{-1}): 1120sh, 1069w, 819, 768s, 709s, 668, 624s, 599s, 553, 445, 380.
Note: The spectrum was obtained by N.V. Chukanov.

O662 Zincovelesite-6*N*6*S* $Zn_3(Fe^{3+},Mn^{3+},Al,Ti)_8O_{15}(OH)$

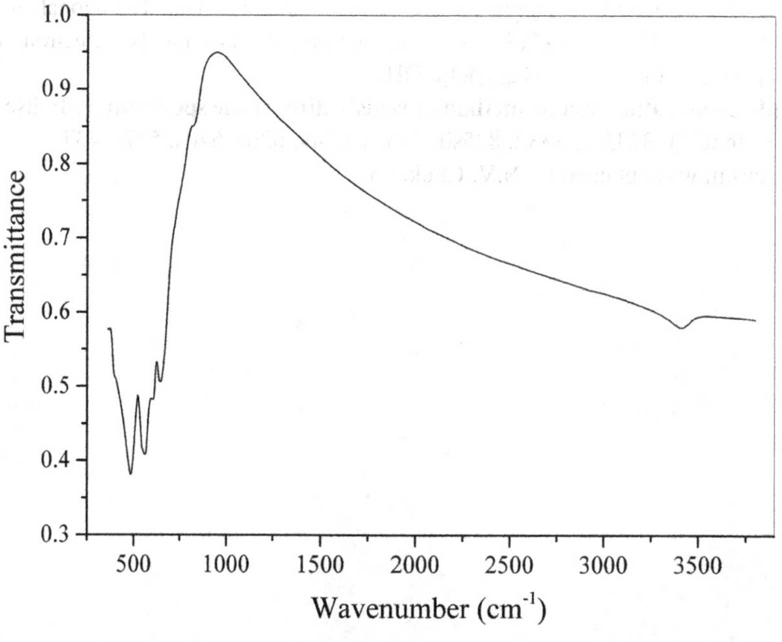

Origin: "Mixed Series" metamorphic complex near the Nežilovo village, Jacupica Mountains, Pelagonia mountain range, Macedonia (type locality).

Description: Black lenticular aggregates from the association with franklinite, gahnite, hetaerolite, zincochromite, ferricoronadite, baryte, As-rich fluorapatite, dolomite, Zn-bearing talc, almeidaite, etc. Holotype sample. The crystal structure is refined by the Rietveld technique. Trigonal, probable space group $P\text{-}3m1$, $a = 5.902(2)$ Å, $c = 55.86(1)$ Å, $V = 1684.8(9)$ Å^3, $Z = 6$. $D_{calc} = 5.158$ g/cm^3. The empirical formula is $H_{1.05}Zn_{3.26}Mg_{0.21}Cu_{0.05}Fe^{3+}_{3.18}Mn^{3+}_{2.32}Al_{1.38}Ti_{0.57}Sb_{0.20}O_{16}$. The strongest lines of the powder X-ray diffraction pattern [*d*, Å (*I*, %) (*hkl*)] are: 2.952 (62) (110), 2.881 (61) (1.0.16), 2.515 (100) (204), 2.493 (88) (1.1.12), 2.451 (39) (1.0.20), 1.690 (19) (304, 2.1.16), 1.572 (19) (2.0.28), 1.475 (29) (221).

Kind of sample preparation and/or method of registration of the spectrum: KBr disc. Absorption.

Wavenumbers (cm^{-1}): 3407w, 817sh, 647, 605, 563s, 550sh, 484s, 400sh.

Note: The spectrum was obtained by N.V. Chukanov.

O663 Wölsendorfite $Pb_7(UO_2)_{14}O_{19}(OH)_4 \cdot 12H_2O$

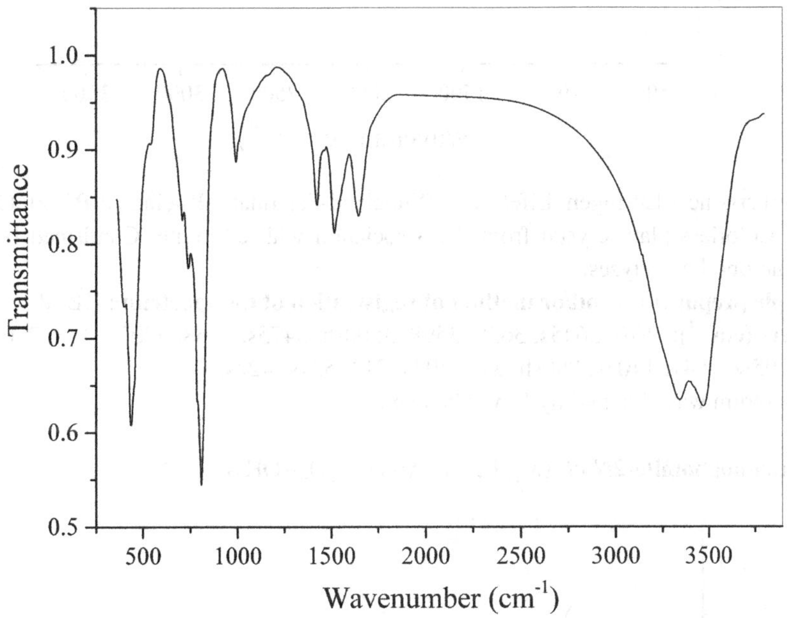

Origin: Shinkolobwe, Katanga (Shaba), Democratic Republic of Congo.

Description: Orange-red grains. Investigated by A.V. Kasatkin. The empirical formula based on semiquantitative electron microprobe analysis is $Pb_{5.3}Ca_{1.0}As_{0.4}(UO_2)_{14.35}(O,OH)_x \cdot nH_2O$.

Kind of sample preparation and/or method of registration of the spectrum: KBr disc. Absorption.

Wavenumbers (cm^{-1}): 3470s, 3344s, 1645, 1515, 1423, 992, 808s, 739, 707, 542w, 437s, 402.

Note: The spectrum was obtained by N.V. Chukanov.

O664 Hydrocalumite $Ca_2Al(OH)_6(Cl,OH)\cdot 3H_2O$

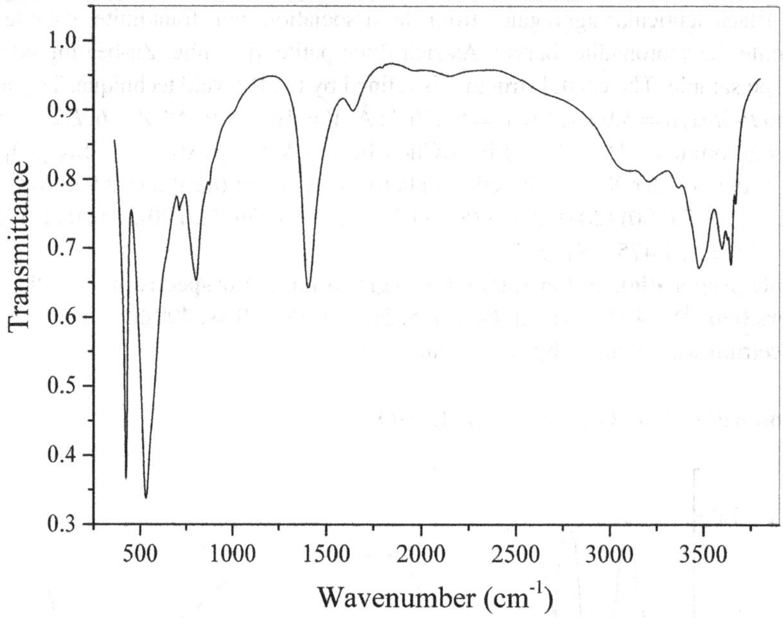

Origin: Bellerberg, near Ettringen, Eifel Mts., Rhineland-Palatinate (Rheinland-Pfalz), Germany.
Description: Colorless platy crystal from the association with ettringite. Confirmed by qualitative electron microprobe analyses.
Kind of sample preparation and/or method of registration of the spectrum: KBr disc. Absorption.
Wavenumbers (cm^{-1}): 3670, 3645s, 3629, 3598, 3500sh, 3475s, 3368, 3207, 3108, 2700sh, 2149w, 1973w, 1795w, 1641, 1404s, 990sh, 874, 804s, 713, 531s, 422s.
Note: The spectrum was obtained by N.V. Chukanov.

O665 Magnesiohögbomite-2*N*4*S* $(Mg,Fe^{2+})_{10}Al_{22}Ti^{4+}{}_2O_{46}(OH)_2$

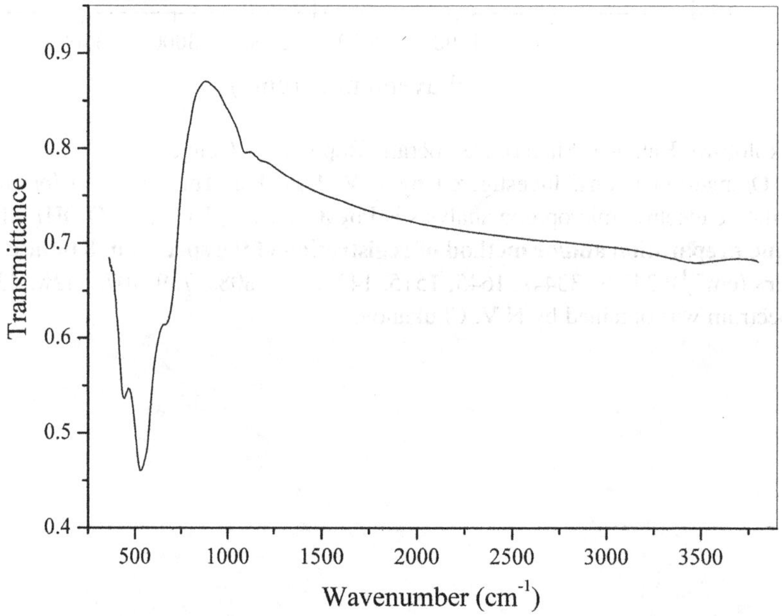

Origin: Kastor, Kirkkjokk, Sweden.

Description: Black grains. The empirical formula is (electron microprobe): $(Mg_{5.64}Fe_{4.26}Mn_{0.12})(Al_{23.2}Fe_{0.80})(Ti_{1.87}Fe_{0.13})O_{44}(OH)_2$.

Kind of sample preparation and/or method of registration of the spectrum: KBr disc. Absorption.

Wavenumbers (cm^{-1}): 3490sh, 3475, 1185sh, 1100sh, 667, 530s, 442, 370sh.

Note: The spectrum was obtained by N.V. Chukanov.

O667 Hyalite $SiO_2{\cdot}H_2O$

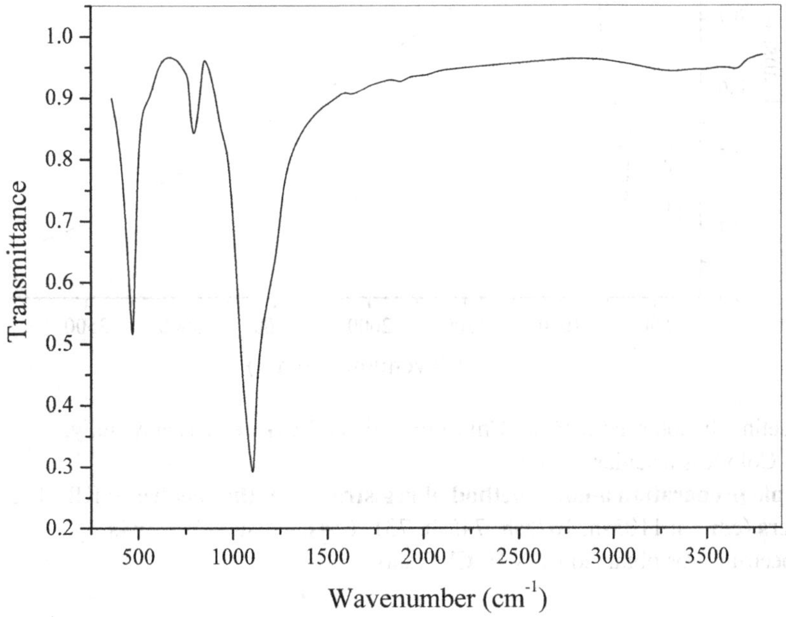

Origin: Tarcal, Borsod-Abaúj-Zemplén Hungary.

Description: Colorless sinter aggregate.

Kind of sample preparation and/or method of registration of the spectrum: KBr disc. Absorption.

Wavenumbers (cm^{-1}): 3652w, 3471sh, 3324w, 2000w, 1879w, 1625w, 1185sh, 1104s, 797, 540sh, 470s.

Note: The spectrum was obtained by N.V. Chukanov. The spectrum is very close to that of quartz glass but contains weak bands of H_2O molecules.

O668 Clinocervantite $Sb^{3+}Sb^{5+}O_4$

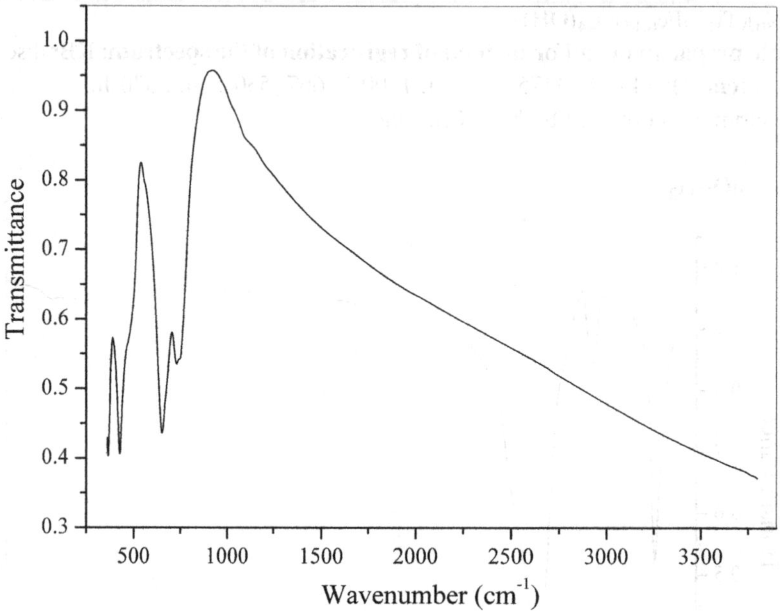

Origin: Le Cetine di Cotorniano Mine, Chiusdino, Siena Province, Tuscany, Italy.
Description: Colorless acicular crystals.
Kind of sample preparation and/or method of registration of the spectrum: KBr disc. Absorption.
Wavenumbers (cm^{-1}): 1105sh, 1025sh, 745sh, 731, 654s, 470sh, 425s, 364s.
Note: The spectrum was obtained by N.V. Chukanov.

O669 Ishikawaite $U_{1-x}FeNb_2O_8$

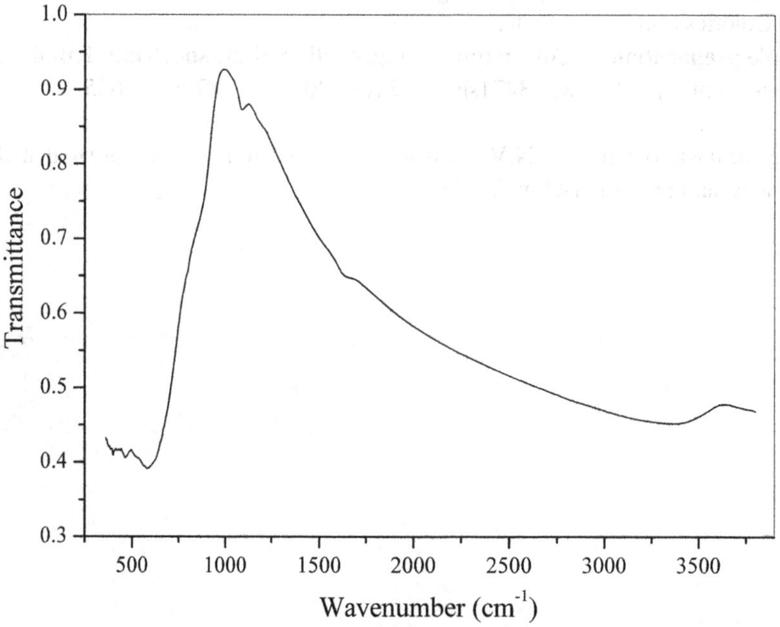

Origin: Pit No. 298, Ilmenskiy Natural Reserve, Ilmeny Mts., Chelyabinsk region, Southern Urals, Russia.

Description: Black prismatic crystal from the association with corundum. X-ray amorphous, metamict. The empirical formula is (electron microprobe): $U_{0.42}Th_{0.18}\mathit{REE}_{0.11}Fe_{0.88}Nb_{1.91}Ti_{0.09}O_8 \cdot nH_2O$.

Kind of sample preparation and/or method of registration of the spectrum: KBr disc. Absorption.

Wavenumbers (cm^{-1}): 3364 (broad), 1645w, 1088w, 825sh, 583s (broad), 463s, 400s.

Note: The spectrum was obtained by N.V. Chukanov.

O670 Samarskite-(Y) $YFe^{3+}Nb_2O_8$

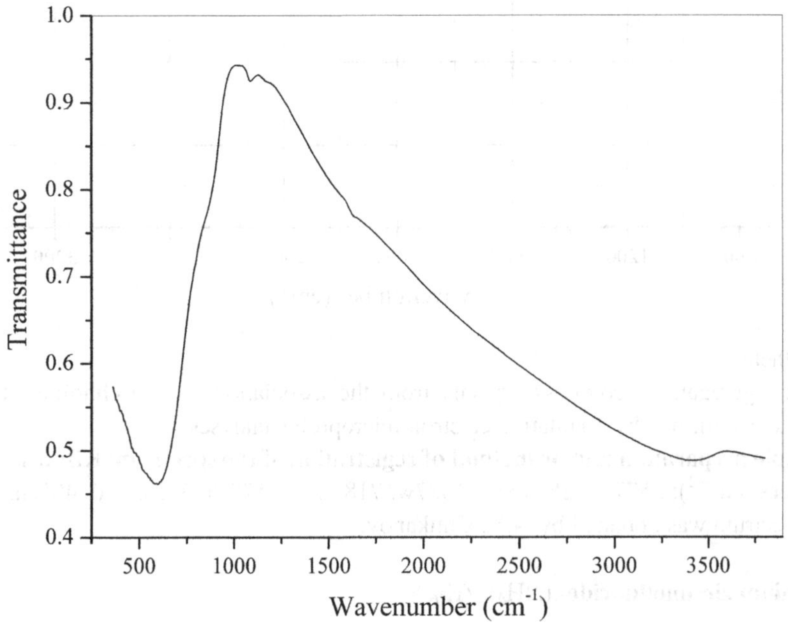

Origin: Blyumovskaya Pit (Pit No. 50), Ilmenskiy Natural Reserve, Ilmeny Mts., Chelyabinsk region, Southern Urals, Russia (type locality).

Description: Black prismatic crystal. X-ray amorphous, metamict. Confirmed by qualitative electron microprobe analyses.

Kind of sample preparation and/or method of registration of the spectrum: KBr disc. Absorption.

Wavenumbers (cm^{-1}): 3418 (broad), 1655w, 1088w, 850sh, 599s, 530sh, 475sh.

Note: The spectrum was obtained by N.V. Chukanov.

2.6 Fluorides and Fluorochlorides

F76 Carlhintzeite $Ca_2AlF_7 \cdot H_2O$

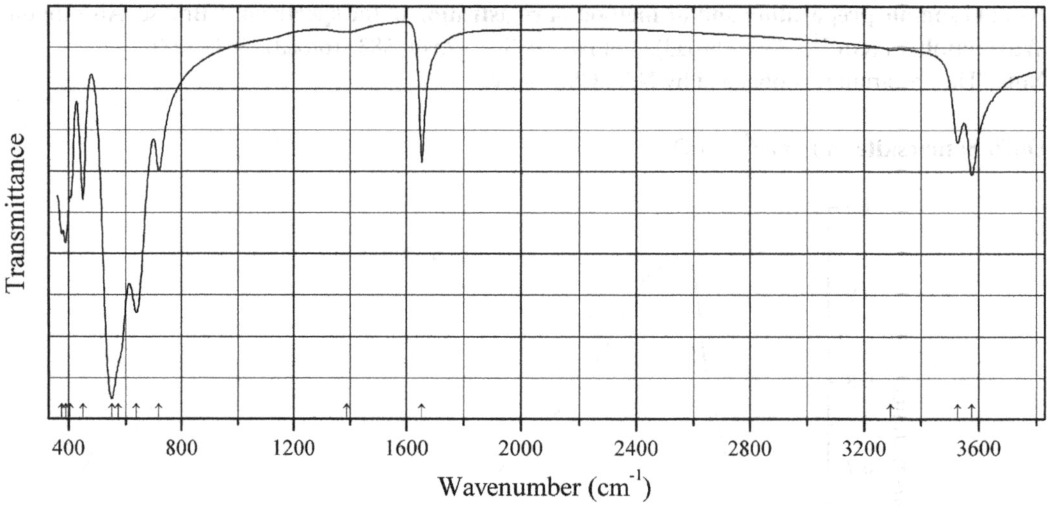

Origin: Synthetic.

Description: Aggregate of colorless crystals from the association with pachnolite, strengite, and tobermorite. Confirmed by qualitative electron microprobe analyses.

Kind of sample preparation and/or method of registration of the spectrum: KBr disc. Absorption.

Wavenumbers (cm^{-1}): 3577, 3528, 1655, 1387w, 718, 638s, 575sh, 552s, 450, 405sh, 390, 380.

Note: The spectrum was obtained by N.V. Chukanov.

F77 Ammonium zirconofluoride $(NH_4)_2ZrF_6$

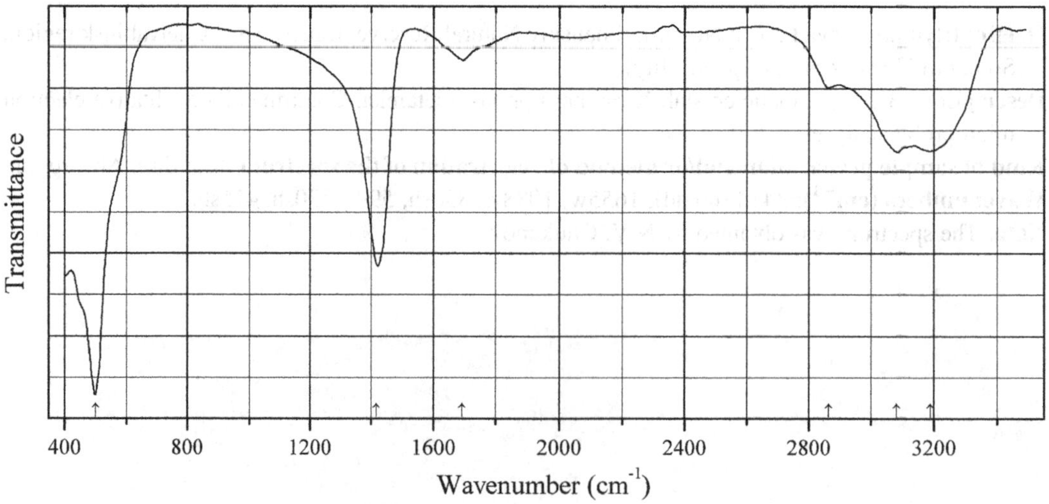

Origin: Synthetic.

Description: Prepared by dissolving stoichiometric quantities of zirconium chloride and ammonium fluoride in concentrated hydrofluoric acid. Orthorhombic, space group $Pca2_1$, $Z = 8$.

Kind of sample preparation and/or method of registration of the spectrum: KBr disc. Transmission.

Source: Kruger and Heyns (1997).
Wavenumbers (IR, cm^{-1}): 3190, 3080, 2864w, 1690w, 1417s, 500s.
Note: The band at 1690 cm^{-1} indicates the presence of H_2O. In the cited paper, Raman spectrum is given.
Wavenumbers (Raman, cm^{-1}): 3146s, 1698, 1412w, 536s, 475w, 377w.

F78 Barium magnesium fluoride $BaMgF_4$

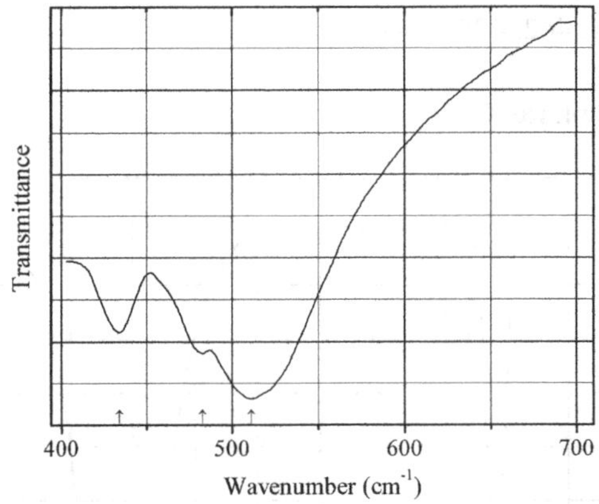

Origin: Synthetic.
Description: Synthesized hydrothermally from BaF_2 and MgF_2 in the presence of CF_3COOH, at 230 °C for 24 h. Orthorhombic, space group $Cmc2_1$. Characterized by DSC and powder X-ray diffraction data.
Kind of sample preparation and/or method of registration of the spectrum: Transmission. Kind of sample preparation is not indicated.
Source: Kim et al. (2010b).
Wavenumbers (cm^{-1}): 511s, 483, 434.

F79 Barium manganese fluoride $BaMnF_4$

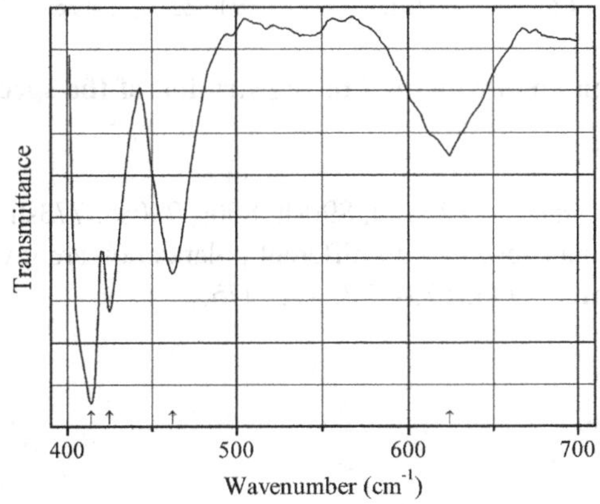

Origin: Synthetic.

Description: Synthesized hydrothermally from BaF_2 and MnF_2 in the presence of CF_3COOH, at 230 °C for 24 h. Orthorhombic, space group *Cmc*2_1. Characterized by DSC and powder X-ray diffraction data.

Kind of sample preparation and/or method of registration of the spectrum: Transmission. Kind of sample preparation is not indicated.

Source: Kim et al. (2010b).

Wavenumbers (cm^{-1}): 624, 462, 425, 414s.

F80 Cesium hexafluorphosphate $CsPF_6$

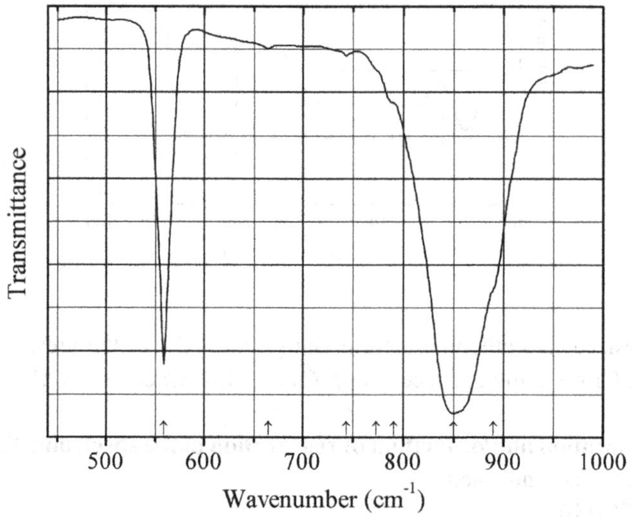

Origin: Synthetic.

Description: Prepared by crystallization from aqueous solution containing equimolar quantities of HPF_6 and Cs_2CO_3. Cubic, $a = 8.218$ Å. The strongest lines of the powder X-ray diffraction pattern [*d*, Å (*I*, %) (*hkl*)] are: 4.113 (100) (200), 2.9051 (29) (220), 2.4787 (29) (311), 2.3726 (8) (222), 1.8381 (9) (420).

Kind of sample preparation and/or method of registration of the spectrum: No data in the cited paper.

Source: Heyns et al. (1981).

Wavenumbers (IR, cm^{-1}): 1410w, (1260w), 890sh, 850s, 790sh, 773sh, 743w, 665w, 559s, 76s.

Note: In the cited paper, Raman spectra with different polarization are given.

Wavenumbers (Raman, cm^{-1}): 744, 575–577, 472–475.

F81 Cesium hexafluorphosphate $CsPF_6$

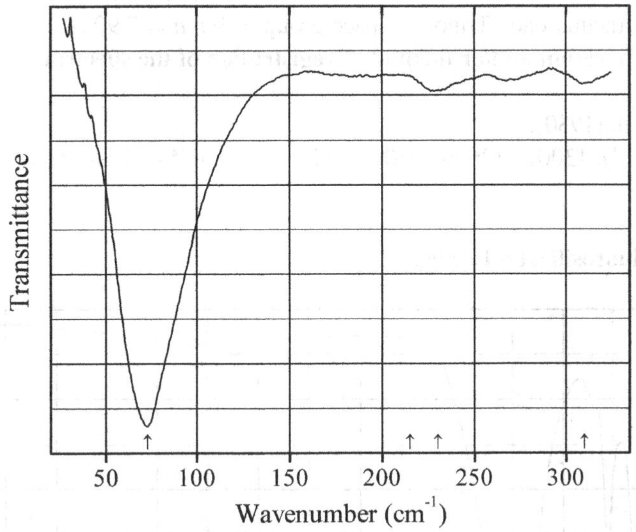

Origin: Synthetic.

Description: Cubic, space group *Fm*3*m*, $a = 8.228(5)$ Å, $V = 557(1)$ Å^3, $Z = 4$.

Kind of sample preparation and/or method of registration of the spectrum: KBr and polyethylene discs. Transmission.

Source: English and Heyns (1984).

Wavenumbers (IR, cm^{-1}): 1405w, 890sh, 846s, 790w, 743w, 559s, 470s, 310w, 230w, 215w, 79.

Note: In the cited paper, Raman spectrum is given.

Wavenumbers (Raman, cm^{-1}): 744, 578. 476.

F82 Cesium stibiofluoride $CsSbF_6$

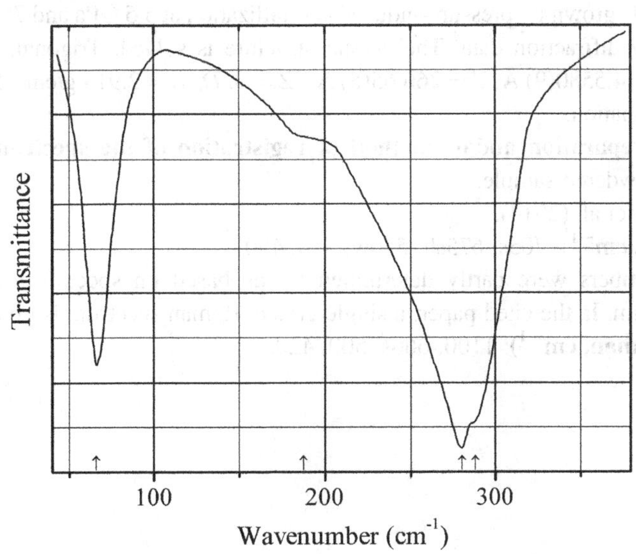

Origin: Synthetic.

Description: Prepared from Cs_2CO_3, Sb_2O_5, and an excess of hydrofluoric acid. Characterized by powder X-ray diffraction data. Trigonal, space group *R*-3*m*, *a* = 7.9037, *c* = 8.2543 Å.

Kind of sample preparation and/or method of registration of the spectrum: KBr pellet and Nujol mull. Transmission.

Source: De Beer et al. (1980).

Wavenumbers (cm^{-1}): 1300w, 1260w, 950w, 845w, 668sh, 655s, 635sh, 560w, 450w, 288sh, 280s, 180–195sh, 66s.

F83 Lithium hexafluorosilicate Li_2SiF_6

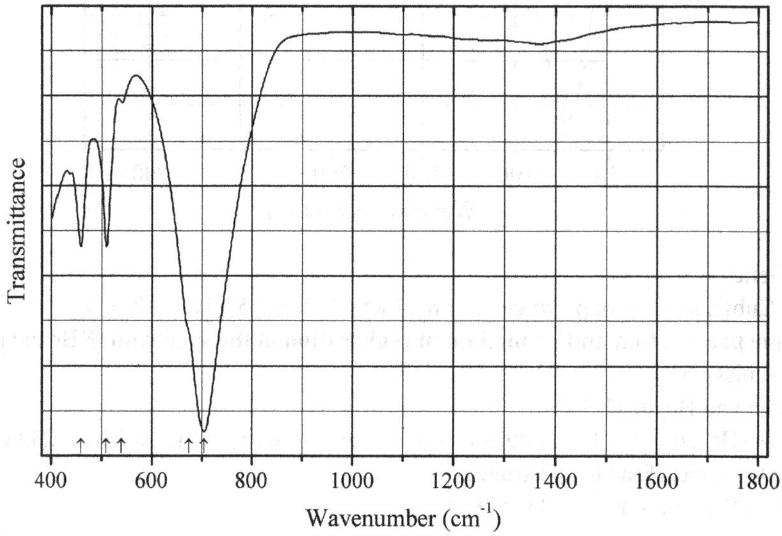

Origin: Synthetic.

Description: Crystals grown by pressure-induced crystallization at 5.5 GPa and 750 °C. Characterized by powder X-ray diffraction data. The crystal structure is solved. Trigonal, space group *P*321, $a = 8.219(2)$, $c = 4.5580(9)$ Å, $V = 266.65(8)$ Å^3, $Z = 3$. $D_{calc} = 2.914$ g/cm^3. Both Li and Si have octahedral coordination.

Kind of sample preparation and/or method of registration of the spectrum: Attenuated total reflection of a powdered sample.

Source: Hinteregger et al. (2014).

Wavenumbers (IR, cm^{-1}): 705s, 675sh, 540w, 510, 460.

Note: The wavenumbers were partly determined by us based on spectral curve analysis of the published spectrum. In the cited paper, a single-crystal Raman spectrum is given.

Wavenumbers (Raman, cm^{-1}): 1100, 660s, 500, 420.

F84 Nickel antimonate fluoride $Ni_3Sb_4O_6F_6$

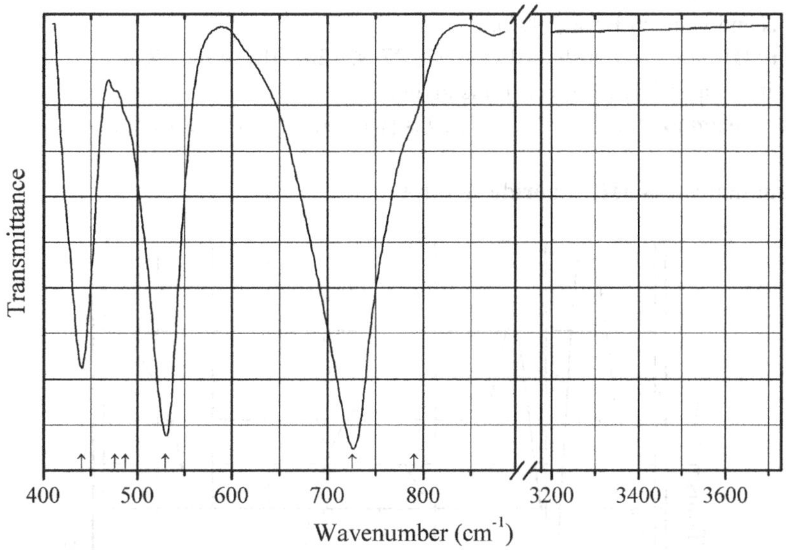

Origin: Synthetic.

Description: Green crystals obtained hydrothermally from NiF_2 and Sb_2O_3 at 230 °C for 4 days. The crystal structure is solved. Cubic, space group *I*-43*m*, $a = 8.0778(1)$, $V = 527.08(1)$ Å^3, $Z = 2$. $D_{calc} = 5.501$ g/cm^3.

Kind of sample preparation and/or method of registration of the spectrum: KBr disc. Reflection.

Source: Hu et al. (2014).

Wavenumbers (cm^{-1}): 790sh, 726s, 530s, 487sh, 476w, 440s.

Note: The wavenumbers were determined by us based on spectral curve analysis of the published spectrum.

F85 Potassium antimony fluoride $KSbF_6$

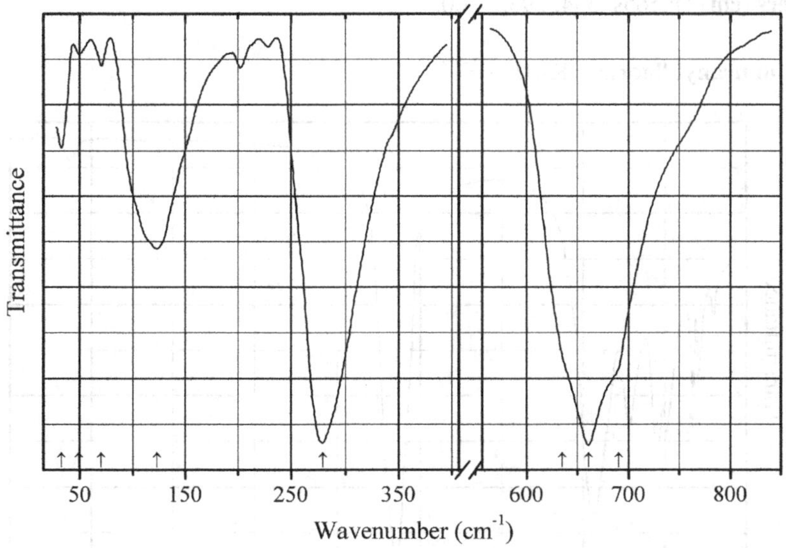

Origin: Synthetic.

Description: Tetragonal, space group *P*-42*m*, $a = 5.16(1)$, $c = 10.07(2)$ Å.

Kind of sample preparation and/or method of registration of the spectrum: KBr and polyethylene discs. Transmission.
Source: Heyns and van den Berg (1995).
Wavenumbers (IR, cm^{-1}): 690sh, 661s, 635sh, 279s, 123, 70w, 49w, 32.
Note: In the cited paper, Raman spectrum is given.
Wavenumbers (Raman, cm^{-1}): 661s, 575, 292s [with the *z*(*yy*)*x* polarization].

F86 Potassium manganese(III) fluoride K_3MnF_6

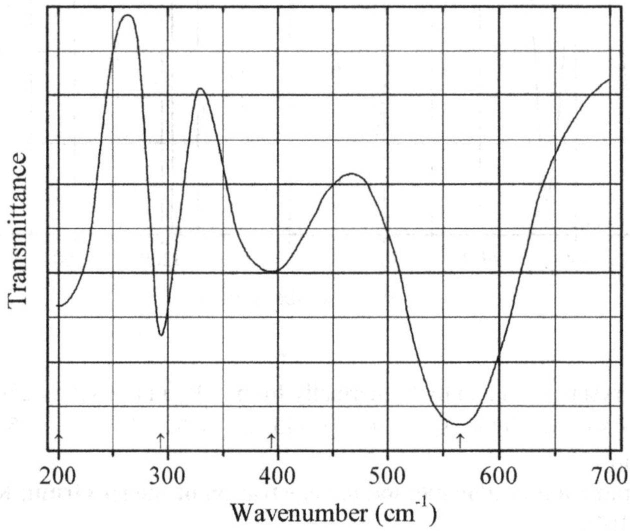

Origin: Synthetic.
Description: The crystal structure contains distorted MnF_6^{3-} octahedron.
Kind of sample preparation and/or method of registration of the spectrum: KBr disc. Absorption.
Source: Wieghardt and Siebert (1971).
Wavenumbers (cm^{-1}): 565s, 394, 293, ~200.

F87 Potassium uranyl fluoride $K_3(UO_2)F_5$

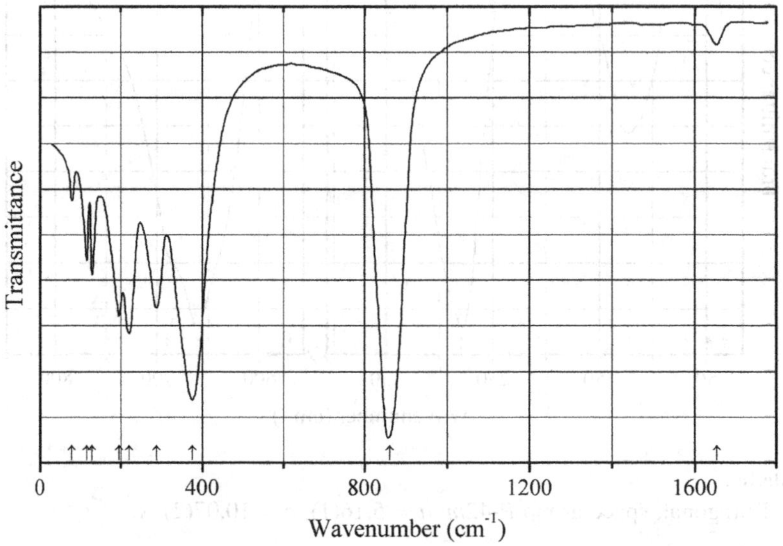

Origin: Synthetic.

Description: Yellow crystals obtained by heating an aqueous solution containing stoichiometric amounts of uranyl and potassium fluorides to 80 °C with subsequent cooling and adding ethanol. Tetragonal, *a* = 9.160, *c* = 18.167 Å, *Z* = 8. Characterized by powder X-ray diffraction data.

Kind of sample preparation and/or method of registration of the spectrum: Nujol mull. Transmission.

Source: Ohwada et al. (1972).

Wavenumbers (cm^{-1}): 1655w, 860s, 376s, 289, 220, 195, 131, 118, 81w.

F88 Tetrammine zinc borofluoride $[Zn(NH_3)_4](BF_4)_2$

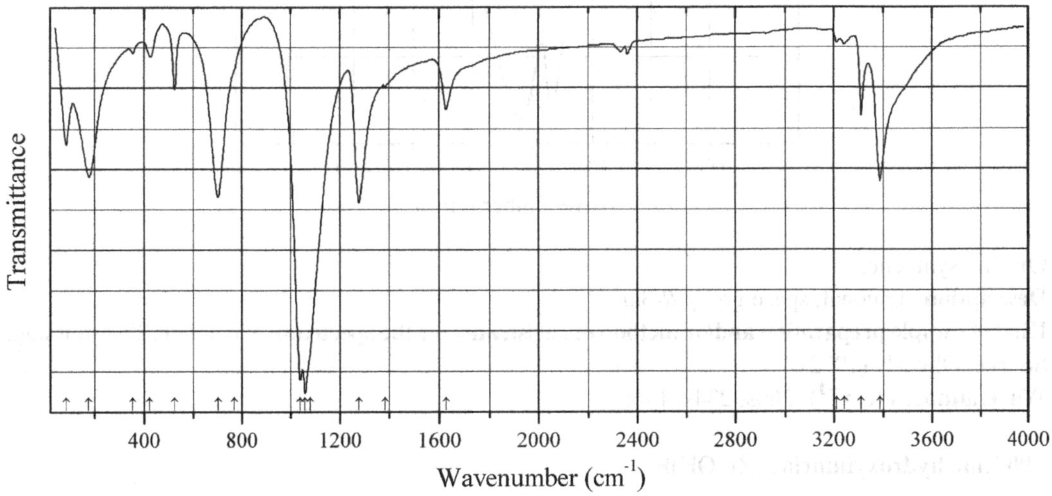

Origin: Synthetic.

Description: Orthorhombic, space group *Pnma*, *a* = 10.523, *b* = 7.892, *c* = 13.354 Å.

Kind of sample preparation and/or method of registration of the spectrum: KBr disc and Nujol mull. Absorption.

Source: Mikuli et al. (2007).

Wavenumbers (IR, cm^{-1}): 3385s, 3310, 3240w, 3211w, 1628, 1383sh, 1277s, 1080sh, 1057s, 1039s, 769w, 702s, 523, 423, 354w, 175s, 83.

Note: In the cited paper, Raman spectrum is given.

Wavenumbers (Raman, cm^{-1}): 3381, 3314s, 3210, 1624, 1085w, 1045sh, 770s, 704w, 525, 436s, 355, 172s.

F89 Uranyl fluoride UO_2F_2

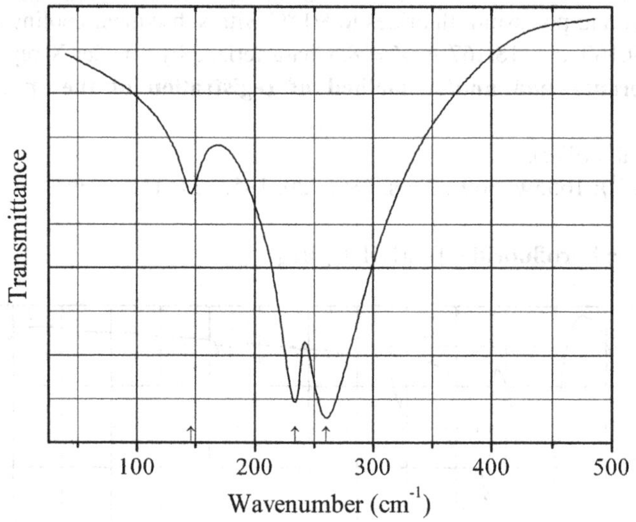

Origin: Synthetic.
Description: Trigonal, space group *R*-3*m*.
Kind of sample preparation and/or method of registration of the spectrum: Nujol mull. Transmission.
Source: Ohwada (1972).
Wavenumbers (cm^{-1}): 260s, 234s, 146.

F90 Zinc hydroxyfluoride Zn(OH)F

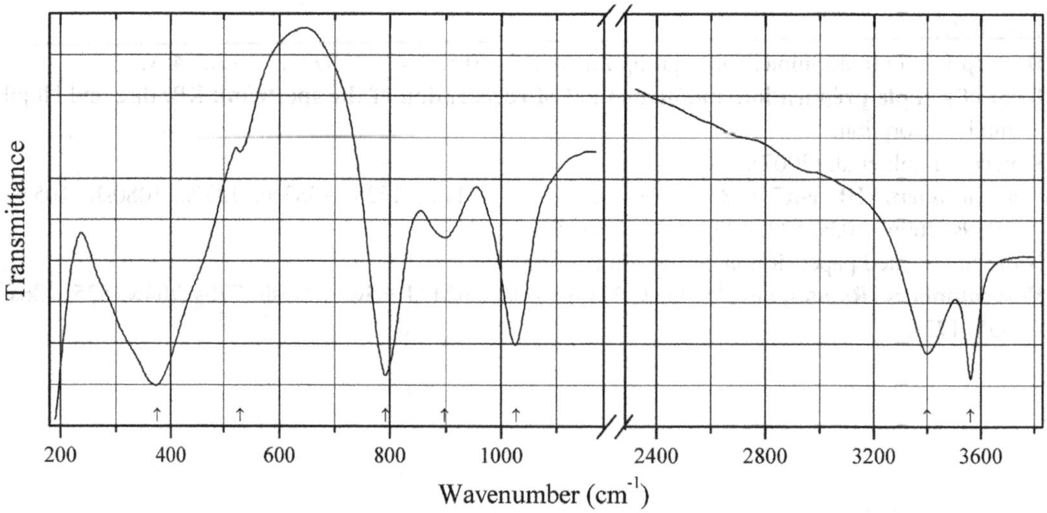

Origin: Synthetic.
Description: Obtained by boiling an aqueous solution of ZnF_2. Orthorhombic, space group *Pna*2_1, $Z = 4$.
Kind of sample preparation and/or method of registration of the spectrum: KBr disc and Nujol mull. Transmission.
Source: Lutz et al. (1993).
Wavenumbers (cm^{-1}): 3562s, 3401s, 1028s, 898, 792s, 530w, 375s.

F91 Gananite BiF_3

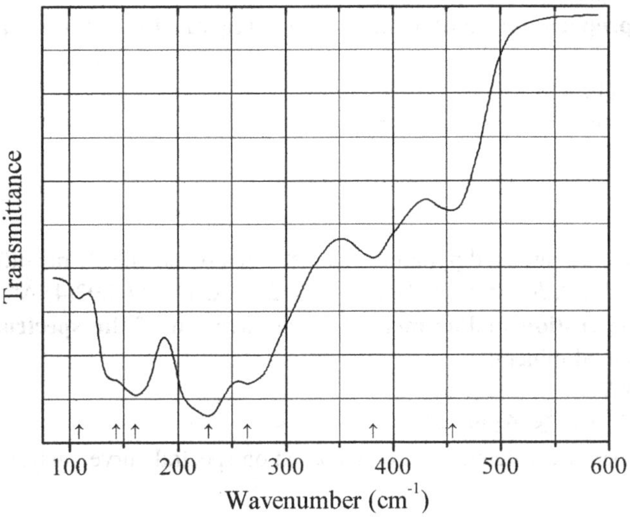

Origin: Synthetic.

Description: Characterized by powder X-ray diffraction data. The crystal structure is solved.

Kind of sample preparation and/or method of registration of the spectrum: Nujol mull or polyethylene disc.

Source: Ignat'eva et al. (2006).

Wavenumbers (cm^{-1}): 455w, 381w, 264s, 228s, 161s, 143sh, 109w.

Note: The wavenumbers were partly determined by us based on spectral curve analysis of the published spectrum.

F92 Waimirite-(Yb) YbF_3

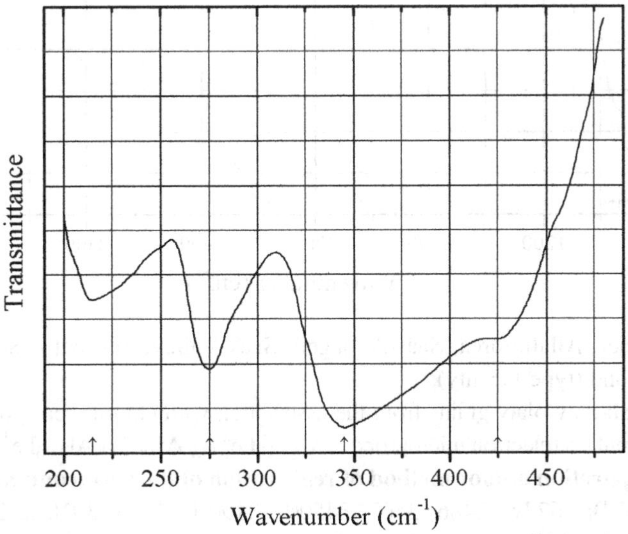

Origin: Synthetic.

Description: Reagent grade commercial product. Orthorhombic.

Kind of sample preparation and/or method of registration of the spectrum: CsI disc. Transmission.

Source: Taylor et al. (1972).

Wavenumbers (cm^{-1}): 425, 345s, 275, 215.

F93 Heklaite

Origin: Synthetic.

Description: Colorless hexagonal dipyramid. Characterized by powder X-ray diffraction data. Orthorhombic, $a = 9.3375(5)$, $b = 5.5009(3)$, $c = 9.7912(7)$ Å, $V = 502.92(4)$ Å^3, $Z = 2$.

Kind of sample preparation and/or method of registration of the spectrum: Attenuated total reflection of powdered mineral.

Source: RRUFF (2007).

Wavenumbers (cm^{-1}): 712s, 663w, 479.

Note: The wavenumbers were determined by us based on spectral curve analysis.

2.7 Silicates

Sio160 Magnesiochloritoid $MgAl_2O(SiO_4)(OH)_2$

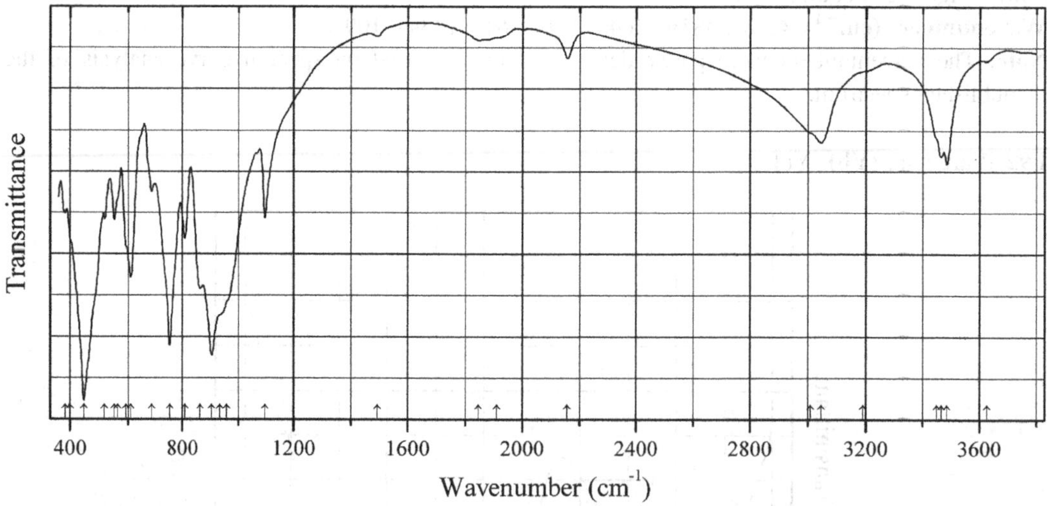

Origin: Allalin glacier, Allalin area, Saas-Almagell, Saas Valley, Zermatt – Saas Fee area, Wallis (Valais), Switzerland (type locality).

Description: Greenish-gray platy grains from the association with chlorite, paragonite, and amphibole. The empirical formula is (electron microprobe): $(Mg_{1.22}Fe^{2+}_{0.77}Mn_{0.01})(Al_{3.96}Fe^{3+}_{0.04})Si_{2.00}O_{10}(OH)_4$.

Kind of sample preparation and/or method of registration of the spectrum: KBr disc. Absorption.

Wavenumbers (cm^{-1}): 3627w, 3486, 3467, 3450sh, 3190sh, 3045, 3010sh, 2158, 1910w, 1845, 1495w, 1095, 960sh, 935sh, 906s, 866, 809, 753s, 685, 611, 595sh, 565sh, 553, 520, 448s, 400sh, 383.

Note: The spectrum was obtained by N.V. Chukanov.

Sio161 Morimotoite $Ca_3(TiFe^{2+})(SiO_4)_3$

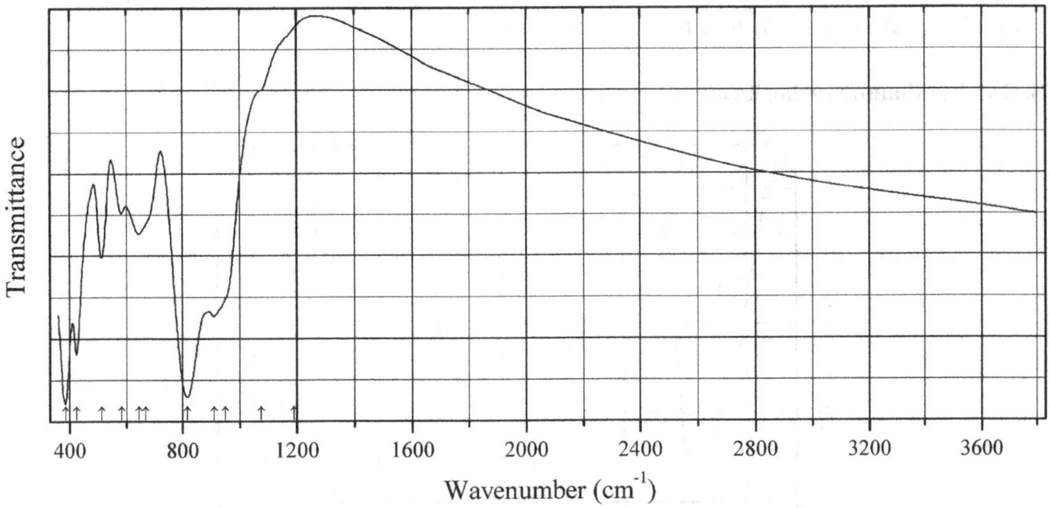

Origin: Odihkincha alkaline massif, Taimyr district, Krasnoyarsk Krai, Siberia, Russia.

Description: Black grains from the association with diopside. Investigated by I.V. Pekov. The empirical formula is (electron microprobe): $(Ca_{2.75}Mg_{0.17}Na_{0.05}Mn_{0.03})(Ti_{0.89}Fe^{2+}_{0.79}Fe^{3+}_{0.38}Zr_{0.02})(Si_{2.69}Fe^{3+}_{0.22}Al_{0.09})O_{12}$.

Kind of sample preparation and/or method of registration of the spectrum: KBr disc. Absorption.

Wavenumbers (cm^{-1}): 1190sh, 1075sh, 950sh, 910s, 816s, 670sh, 645, 584, 513, 424s, 385s.

Note: The spectrum was obtained by N.V. Chukanov.

Sio162 Wadalite $Ca_6Al_5Si_2O_{16}Cl_3$

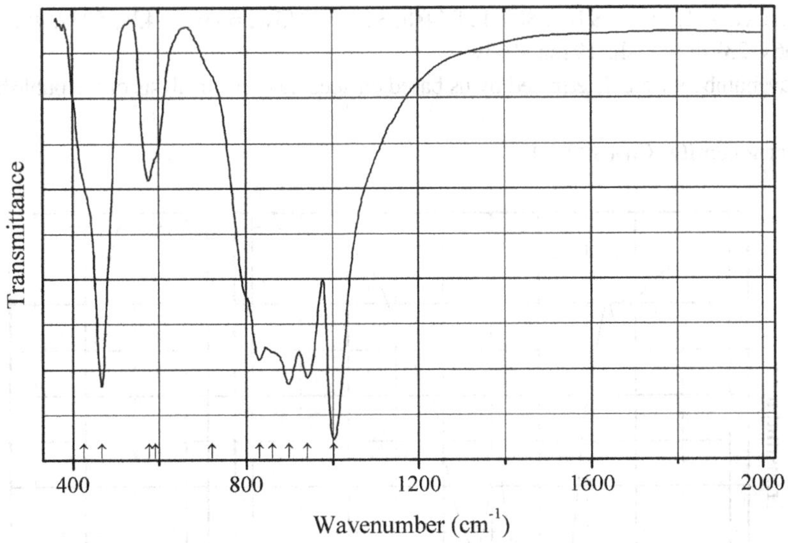

Origin: Bellerberg, near Ettringen, Eifel Mts., Rhineland-Palatinate (Rheinland-Pfalz), Germany.

Description: Yellow crystals from the association with calcite and gypsum. The empirical formula is (electron microprobe): $(Ca_{5.30}Mg_{0.70})(Al_{2.62}Fe_{0.35}Ti_{0.03})(Si_{2.04}Al_{1.96})Cl_{3.25}O_x$.

Kind of sample preparation and/or method of registration of the spectrum: KBr disc. Absorption.
Wavenumbers (cm^{-1}): 1003s, 941s, 898s, 860sh, 830s, 720sh, 590sh, 576, 467s, 425sh.
Note: The spectrum was obtained by N.V. Chukanov.

Sio163 Lanthanum orthosilicate $La_{9.33}(SiO_4)_6O_2$

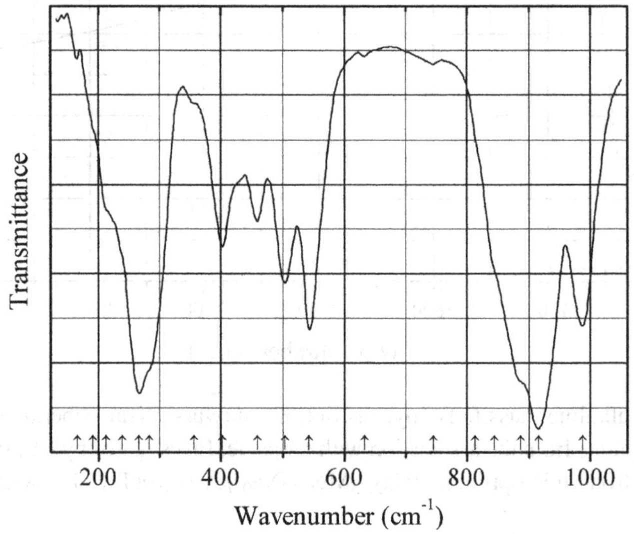

Origin: Synthetic.
Description: Apatite-type compound.
Kind of sample preparation and/or method of registration of the spectrum: Absorption. Kind of sample preparation is not indicated.
Source: Smirnov et al. (2010).
Wavenumbers (cm^{-1}): 988s, 916s, 887sh, 844sh, 813sh, 745w, 633w, 543s, 503s, 458, 403, 358sh, 283sh, 266s, 239sh, 213sh, 191sh, 165w.
Note: The wavenumbers were determined by us based on spectral curve analysis of the published spectrum.

Sio164 Fluorchegemite $Ca_7(SiO_4)_3F_2$

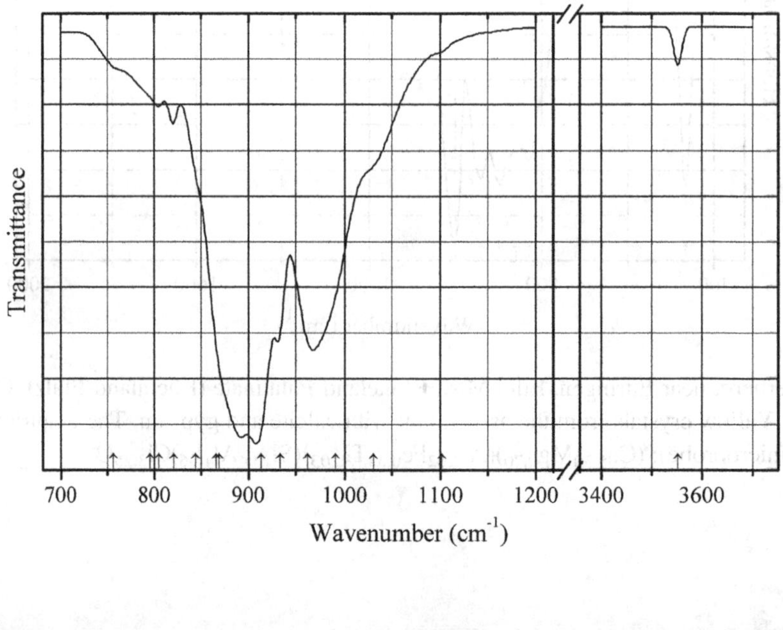

Origin: Lakargi Mt., Upper Chegem caldera, Kabardino-Balkarian Republic, Northern Caucasus, Russia (type locality).

Description: Lens-shaped aggregate. The associated minerals arelarnite, edgrewite, wadalite, eltyubyuite, rondorfite, lakargiite, Th-rich kerimasite, as well as their alteration products. Holotype sample. Orthorhombic, space group *Pbnm*, $a = 5.0620(1)$, $b = 11.3917(2)$, $c = 23.5180(3)$ Å, $V = 1356.16(4)$ Å^3, $Z = 4$. $D_{calc} = 2.91$ g/cm^3. Optically biaxial (–), $\alpha = 1.610(2)$, $\beta = 1.615(2)$, $\gamma = 1.619(2)$, $2V = 80(8)°$. The empirical formula is (electron microprobe, OH calculated): $Ca_{7.01}Mg_{0.01}Ti_{0.01}Si_{2.98}O_{12}F_{1.40}(OH)_{0.60}$. The experimental powder X-ray diffraction pattern was not obtained.

Kind of sample preparation and/or method of registration of the spectrum: Reflection from a polished grain.

Source: Galuskina et al. (2015).

Wavenumbers (IR, cm^{-1}): 3552w, 1102sh, 1031sh, 991sh, 962s, 934, 914s, 889s, 866sh, 844sh, 821w, 805w, 795sh, 756sh.

Note: The wavenumbers are indicated only for the maxima of individual bands obtained by Galuskina et al. (2015) as a result of the spectral curve analysis. In the cited paper, Raman spectrum is given.

Wavenumbers (Raman, cm^{-1}): 3552, 3548s, 3539, 992s, 843, 817s, 560, 442, 410, 297, 258w.

Sio165 Hatrurite triclinic polymorph $Ca_3(SiO_4)O$

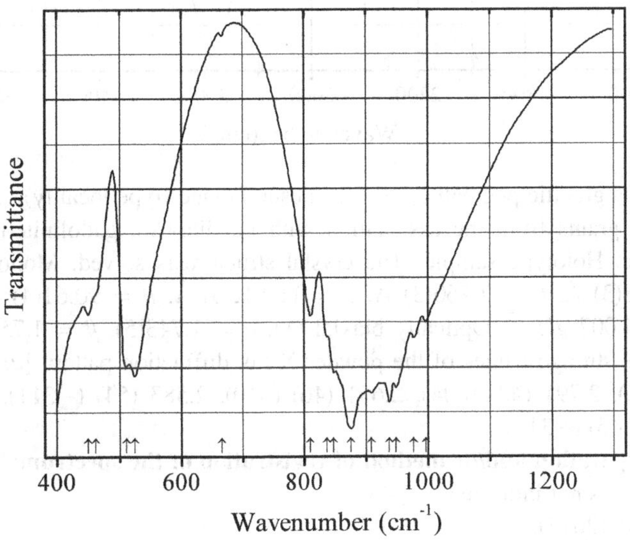

Origin: Synthetic.

Description: Prepared by heating a mixture of calcium carbonate and silica gel (with the $CaO:SiO_2$ molar ratio of 3:1) pressed into a pellet, at 1450 °C. Characterized by powder X-ray diffraction data. Triclinic, $a = 11.630$, $b = 14.216$, $c = 13.690$ Å, $\alpha = 105.345°$, $\beta = 94.558°$, $\gamma = 89.845°$.

Kind of sample preparation and/or method of registration of the spectrum: KBr disc. Absorption.

Source: Del Bosque et al. (2014).
Wavenumbers (cm^{-1}): 996, 977, 949, 938s, 909, 875s, 848, 837sh, 810, (666), 525, 511, 462sh, 450.
Note: The wavenumbers were partly determined by us based on spectral curve analysis of the published spectrum. The band at 666 cm^{-1} corresponds to atmospheric CO_2. In the cited paper, the wavenumber 949 cm^{-1} is erroneously indicated as 959 cm^{-1}.

Sio166 Pilawite-(Y) $Ca_2Y_2Al_4(SiO_4)_4O_2(OH)_2$

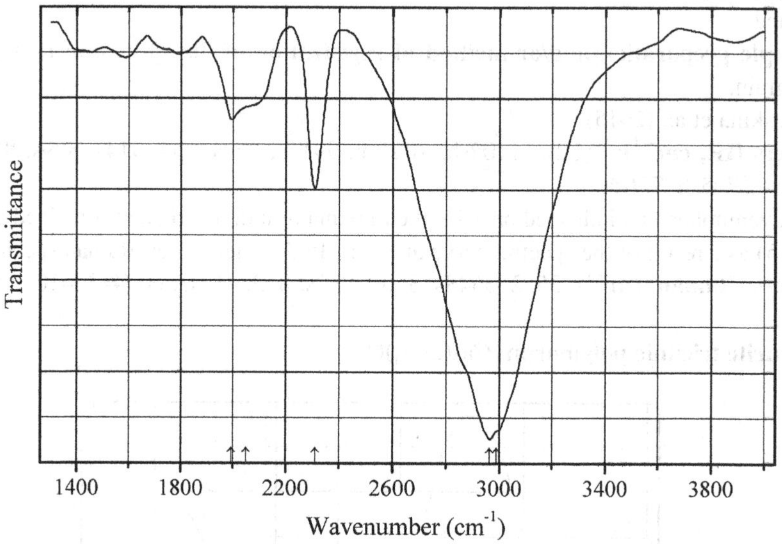

Origin: Piława Górna granitic pegmatite, Lower Silesia, Poland (type locality).
Description: White grains from the association with keiviite-(Y), gadolinite-(Y), hingganite-(Y), xenotime-(Y), etc. Holotype sample. The crystal structure is solved. Monoclinic, space group $P2_1/c$, a = 8.558(3) Å, b = 7.260(3) Å, c = 11.182(6) Å, β = 90.61(4)°, V = 694.7(4) Å^3, Z = 2. D_{calc} = 4.007 g/cm^3. Optically biaxial (+), α = 1.743(5), β = 1.754(5), γ = 1.779(5), $2V$ = 65(2)°. The strongest lines of the powder X-ray diffraction pattern [d, Å (I, %) (hkl)] are: 3.044 (100) (022), 2.791 (43) (004), 2.651 (46) (310), 2.583 (54) (−311), 2.485 (62) (−222, 114, 123), 2.408 (45) (−312).
Kind of sample preparation and/or method of registration of the spectrum: Absorption. Kind of sample preparation is not indicated.
Source: Pieczka et al. (2015).
Wavenumbers (IR, cm^{-1}): 2990sh, 2965s, 2309, 2050sh, 1995.
Note: The wavenumbers were partly determined by us based on spectral curve analysis of the published spectrum. The narrow band at 2309 cm^{-1} may correspond to CO_2 molecules present in cavities of the heteropolyhedral framework. In the cited paper, a figure of the Raman spectrum is given.

Sio167 Spessartine Ca-rich $(Mn_{1.31}Ca_{1.02}Fe^{2+}_{0.52}Mg_{0.10})(Al_{1.59}Fe^{3+}_{0.46}Ti_{0.06})Si_{2.90}O_{12}$

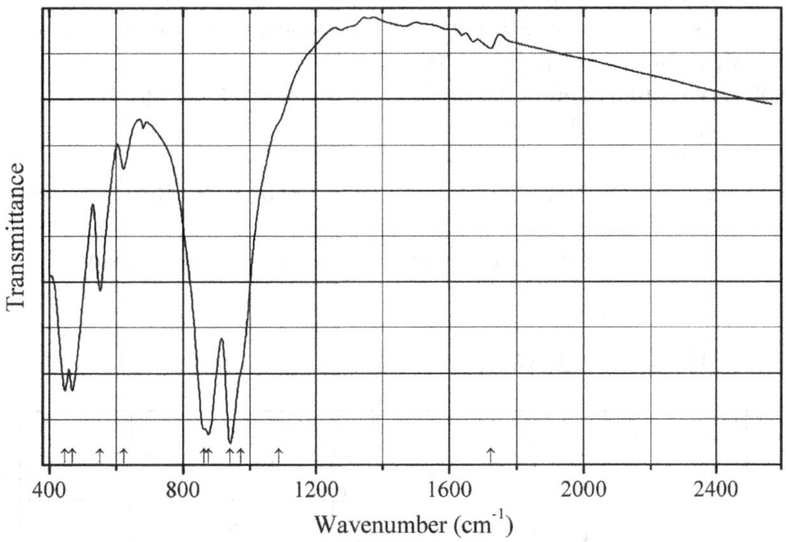

Origin: Pit no. 287, Ilmeny Mts., Chelyabinsk region, South Urals, Russia.

Description: Brown-red grains from the association with diopside, quartz, and scapolite. Cubic, a = 11.736 Å. Mössbauer spectroscopy indicates that 46% of iron is trivalent. The strongest lines of the powder X-ray diffraction pattern [d, Å (I, %)] are: 2.935 (38), 2.624 (100), 2.396 (26), 1.904 (24), 1.627 (23), 1.569 (33).

Kind of sample preparation and/or method of registration of the spectrum: KBr disc. Transmission.

Source: Korinevsky (2015).

Wavenumbers (IR, cm^{-1}): 1724w, 1087sh, 972sh, 941s, 876s, 863sh, 623, 553, 469s, 447s.

Note: The wavenumbers were partly determined by us based on spectral curve analysis of the published spectrum. In the cited paper, Raman spectrum is given.

Wavenumbers (Raman, cm^{-1}): 1009, 891s, 834, 623w, 536, 354s, 218, 154.

Sio168 Kirschsteinite $CaFe^{2+}(SiO_4)$

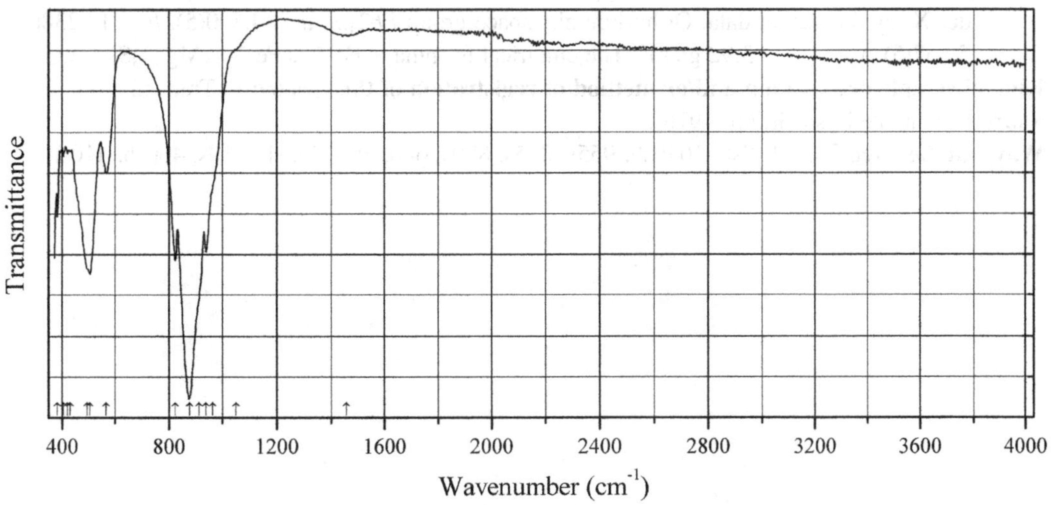

Origin: Dolores mine, Pastrana, Mazarrón, Murcia, Spain.

Description: Orthorhombic, $a = 4.8613(3)$, $b = 11.0995(5)$, $c = 6.3989(8)$ Å, $V = 345.28(4)$ Å^3, $D_{meas} = 2.39(3)$ g/cm^3, $D_{calc} = 2.391$ g/cm^3. The empirical formula is (electron microprobe): $(Ca_{0.95}Mn_{0.02}Mg_{0.02}Na_{0.01})(Fe^{2+}{}_{0.83}Mg_{0.16}Fe^{3+}{}_{0.01})Si_{1.00}O_4$.

Kind of sample preparation and/or method of registration of the spectrum: Attenuated total reflection of powdered mineral.

Source: RRUFF (2007).

Wavenumbers (cm^{-1}): 1456, 1049, 964sh, 939, 912sh, 876s, 823, 567, 504s, 496sh, 432, 420, 405, 384.

Note: The wavenumbers were determined by us based on spectral curve analysis of the published spectrum.

Sio169 Laihunite $(Fe^{3+},Fe^{2+},\square)_2(SiO_4)$

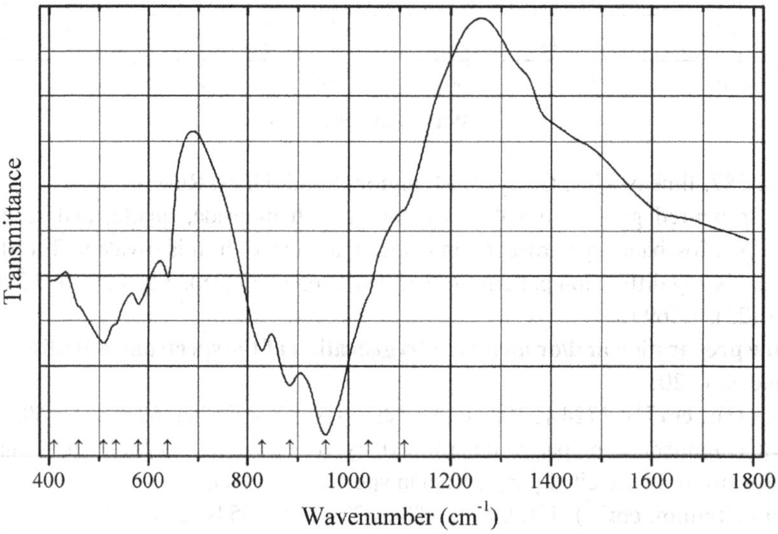

Origin: Laihe Fe deposit, Qianshan District, Liaoning Province, China (type locality).

Description: Black tabular crystals. Holotype sample. Characterized by Mössbauer spectroscopy and powder X-ray diffraction data. Orthorhombic, space group $Pb2_1m$, $a = 4.800(5)$, $b = 10.238(5)$, $c = 5.857(5)$ Å. $D_{meas} = 3.92$ g/cm^3. The empirical formula is $Fe^{3+}{}_{1.50}Fe^{2+}{}_{0.58}Mg_{0.03}Si_{0.96}O_4$.

Kind of sample preparation and/or method of registration of the spectrum: Transmission.

Source: Laihunite Resh Group (1976).

Wavenumbers (cm^{-1}): 1110sh, 1040sh, 955s, 885s, 830s, 640, 580, 535sh, 510s, 460sh, 410sh.

Sio170 Oxybritholite thorium analogue $Th_2Ca_3(SiO_4)_3O$

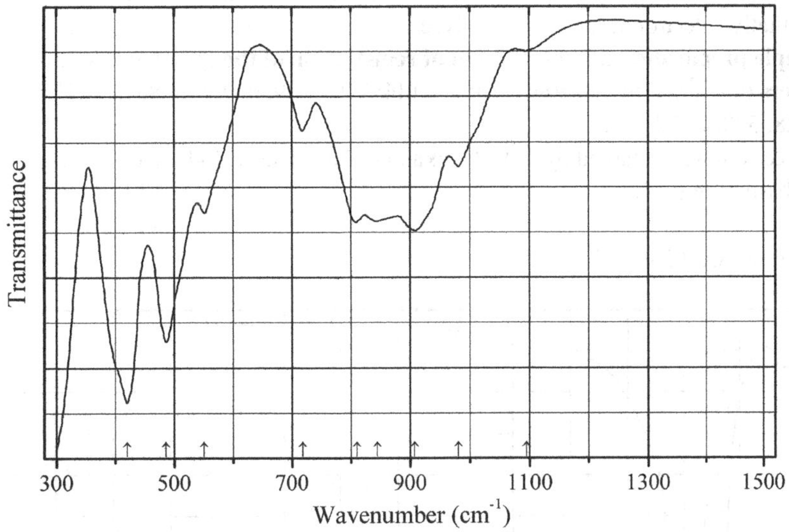

Origin: Synthetic.

Description: Synthesized by solid-state reaction between thorium nitrate, calcium nitrate, and silicon oxide at 1553 K for 30 h with intermediate grindings every 2 h. The crystal structure is solved. Hexagonal, space group $P6_3/m$, $a = 9.50172(9)$, $c = 6.98302(8)$ Å, $V = 545.98(1)$ Å^3, $Z = 2$. $D_{calc} = 4.966$ g/cm^3. The crystal-chemical formula is $(Ca_{3.84}Th_{0.16})(Th_{3.21}Ca_{2.79})(SiO_4)_6O_2$.

Kind of sample preparation and/or method of registration of the spectrum: Attenuated total reflection of a powdered sample.

Source: Bulanov et al. (2015).

Wavenumbers (cm^{-1}): 1095w, 981, 908s, 845, 810, 718, 550, 485s, 420s.

Sio171 Ulfanderssonite-(Ce) $(Ce_{15}Ca)Mg_2(SiO_4)_{10}(SiO_3OH)(OH,F)_5Cl_3$

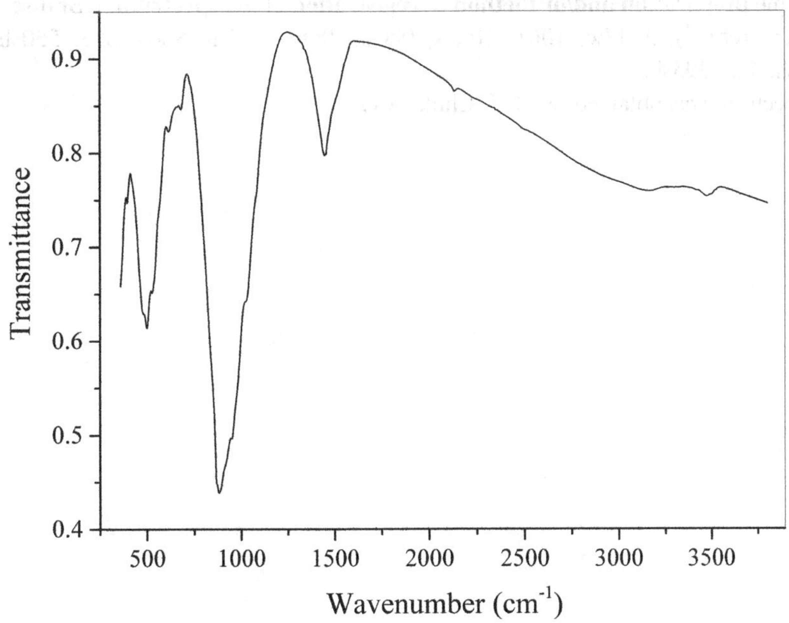

Origin: Malmkärra Mine, Norberg, Västmanland, Sweden (type locality).

Description: Grey grains from the association with cerite-(Ce) and bastnäsite-(Ce). Confirmed by semiquantitative electron microprobe analyses.

Kind of sample preparation and/or method of registration of the spectrum: KBr disc. Absorption.

Wavenumbers (cm^{-1}): 475w, 3160w, 2135w, (1448), 1025, 950, 915sh, 881s, (875sh), 682w, 616w, 565sh, 528, 500, 485sh, 395w.

Note: The spectrum was obtained by N.V. Chukanov. The bands at 1448 and 857 cm^{-1} correspond to admixed bastnäsite-(Ce).

Sid47 Cuspidine $Ca_8(Si_2O_7)_2F_4$

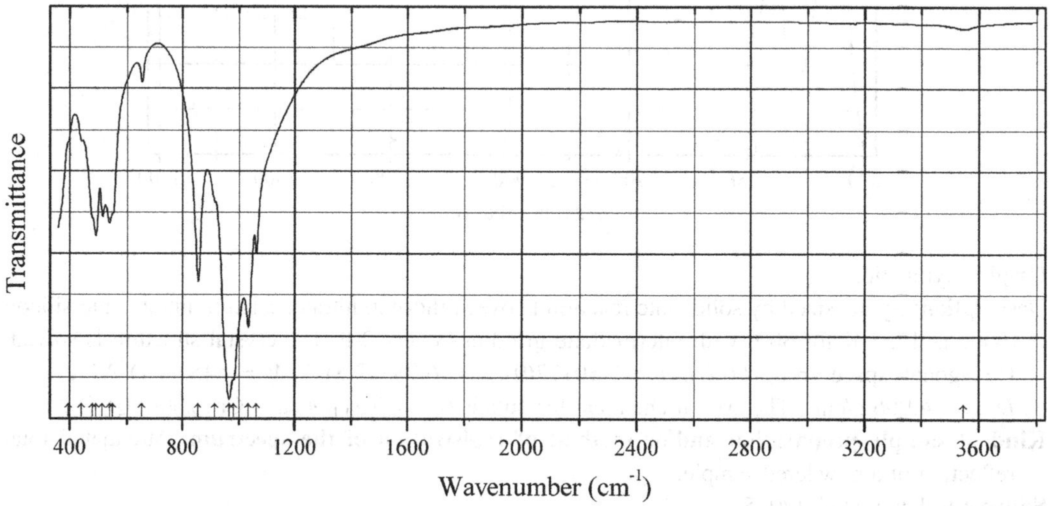

Origin: Bellerberg, near Ettringen, Eifel Mts., Rhineland-Palatinate (Rheinland-Pfalz), Germany.

Description: Colorless crystals from the association with dorrite, clinopyroxene, spinel, and gypsum. Confirmed by the IR spectrum.

Kind of sample preparation and/or method of registration of the spectrum: KBr disc. Absorption.

Wavenumbers (cm^{-1}): 3547w, 1061, 1032s, 980sh, 966s, 915sh, 856s, 653, 550sh, 540, 515, 491, 480sh, 442, 395sh.

Note: The spectrum was obtained by N.V. Chukanov.

Sid48 Nasonite $Ca_4Pb_6(Si_2O_7)_3Cl_2$

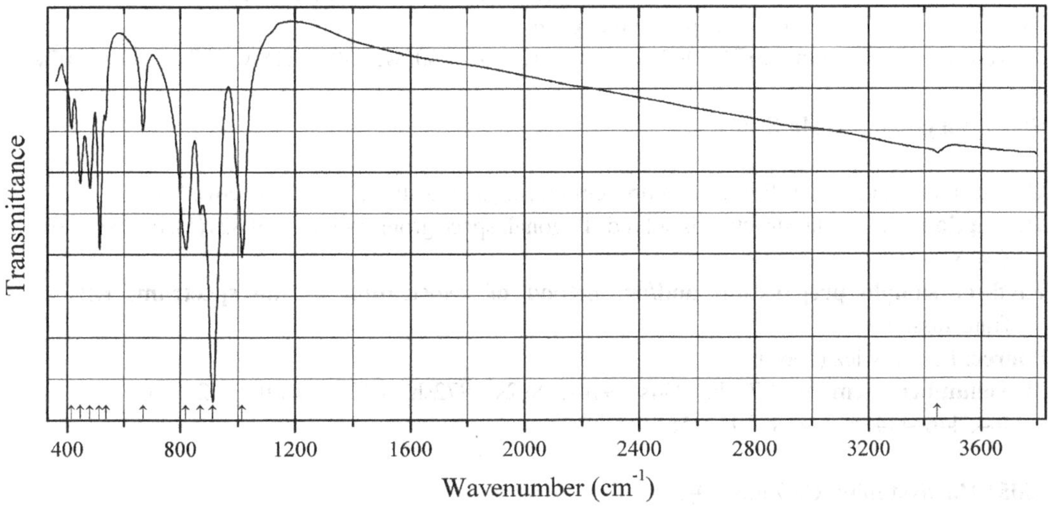

Origin: Långban deposit, Bergslagen ore region, Filipstad district, Värmland, Sweden.

Description: Lemon yellow grains from the association with barysilite and jacobsite. The empirical formula is (electron microprobe): $(Pb_{5.78}Ca_{0.22})(Ca_{3.98}Mn_x)Si_{6.00}O_{21}[Cl_{1.75}(OH)_{0.25}]$ ($x \ll 1$).

Kind of sample preparation and/or method of registration of the spectrum: KBr disc. Absorption.

Wavenumbers (cm^{-1}): 3446w, 1015s, 913s, 869, 820s, 668, 537, 515s, 481, 446, 415.

Note: The spectrum was obtained by N.V. Chukanov.

Sid49 Åkermanite $Ca_2Mg(Si_2O_7)$

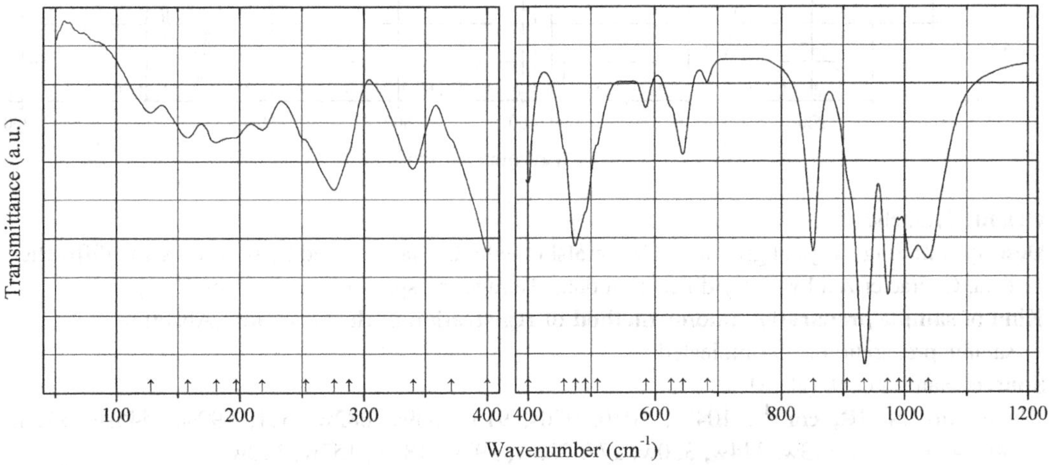

Origin: Synthetic.

Description: Single crystal grown by Czochralski method. Characterized by X-ray diffraction data. Tetragonal, space group $P\text{-}42_1m$, $Z = 2$.

Kind of sample preparation and/or method of registration of the spectrum: Absorption. Kind of sample preparation is not indicated.

Source: Hanuza et al. (2012).

Wavenumbers (IR, cm^{-1}): 1038s, 1009s, 991, 972s, 934s, 906sh, 852s, 683w, 644w, 625sh, 586w, 510sh, 491sh, 475s, 457sh, 400, 371sh, 340w, 288sh, 276w, 253sh, 218w, 197, 181, 158, 128.

Note: In the cited paper, Raman spectrum is given.

Wavenumbers (Raman, cm^{-1}): 992w, 941w, 910s, 666, 605w, 450w, 318w, 227w, 211w, 107w.

Sid50 Barysilite $Pb_8Mn(Si_2O_7)_3$

Origin: Långban deposit, Bergslagen ore region, Filipstad district, Värmland, Sweden.

Description: The crystal structure is solved. Trigonal, space group *R*-3*c*, *a* = 9.821(5), *c* = 38.38(6) Å, Z = 6.

Kind of sample preparation and/or method of registration of the spectrum: KBr disc. Transmission.

Source: Lajzérowicz (1966).

Wavenumbers (cm^{-1}): 970sh, 934s, 914s, 892s, 872sh, 832s, (780), 702, 553, 528, 479, 464, 446–444, 423–418, 393, 258.

Sid51 Hardystonite $Ca_2Zn(Si_2O_7)$

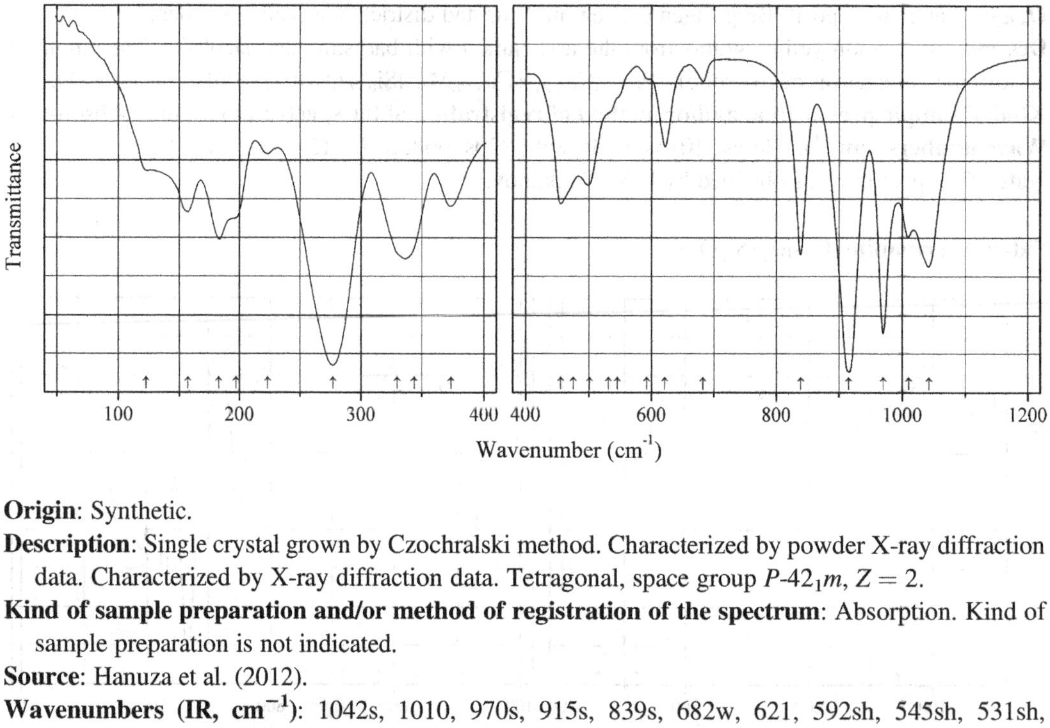

Origin: Synthetic.

Description: Single crystal grown by Czochralski method. Characterized by powder X-ray diffraction data. Characterized by X-ray diffraction data. Tetragonal, space group *P*-42$_1$*m*, Z = 2.

Kind of sample preparation and/or method of registration of the spectrum: Absorption. Kind of sample preparation is not indicated.

Source: Hanuza et al. (2012).

Wavenumbers (IR, cm^{-1}): 1042s, 1010, 970s, 915s, 839s, 682w, 621, 592sh, 545sh, 531sh, 501, 475sh, 455, 373w, 344w, 330w, 277, 223w, 197w, 183w, 157w, 122w.

Note: In the cited paper, Raman spectrum is given.

Wavenumbers (Raman, cm^{-1}): 1019w, 1004w, 994w, 939w, 908s, 664s, 615w, 551w, 480w, 445w, 315w, 280w, 204w, 147w, 100w.

Sid52 Keiviite-(Yb) β-$Yb_2Si_2O_7$

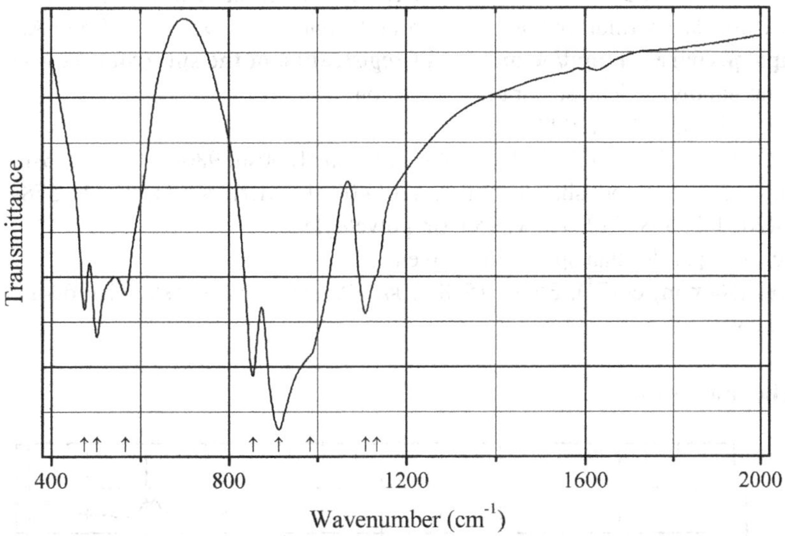

Origin: Synthetic.

Description: Synthesized from $Yb(NO_3)_3$ and $Si(OC_2H_5)_4$ by a sol-gel method with subsequent calcination at 1200 °C for 2 h. Characterized by powder X-ray diffraction data. Monoclinic, space group *C2/m*, $a = 6.80053$, $b = 8.87508$, $c = 4.70740$ Å, $\beta = 101.984$.

Kind of sample preparation and/or method of registration of the spectrum: Transmission. Kind of sample preparation is not indicated.

Source: Zhao et al. (2013a).

Wavenumbers (cm^{-1}): 1133sh, 1108, 983sh, 912s, 853s, 567, 502, 474.

Sid53 Lawsonite $CaAl_2(Si_2O_7)(OH)_2{\cdot}H_2O$

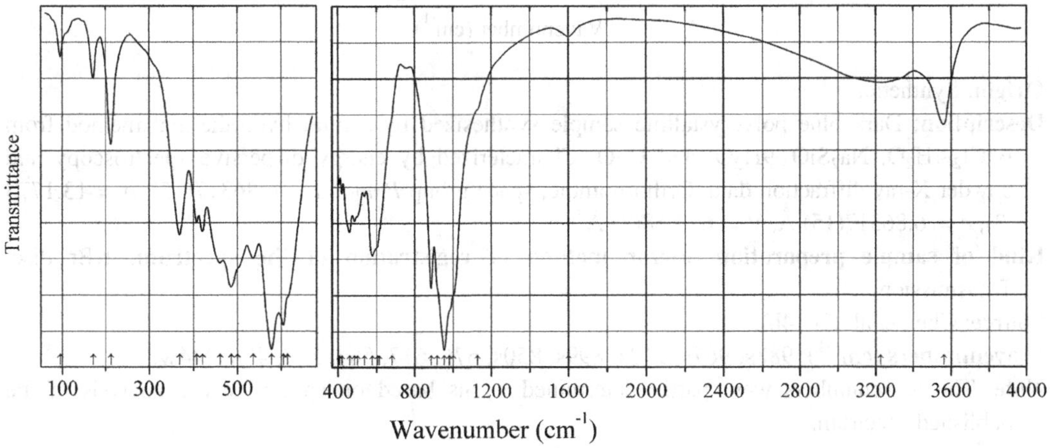

Origin: Tiburon Peninsula, California, USA.

Description: Equant to tabular crystals. Characterized by electron microprobe analyses and powder X-ray diffraction data. Orthorhombic, space group *Ccmm*, $a = 8.795$ Å, $b = 5.847$ Å, $c = 13.142$ Å.

Kind of sample preparation and/or method of registration of the spectrum: Powder dispersed in KBr disc and in polyethylene substrate. Absorption.

Source: Le Cléac'h and Gillet (1990).

Wavenumbers (IR, cm^{-1}): 3560, 3225, 1600w, 1125sh, 1000sh, 980sh, 953s, 915sh, 885s, 680sh, 614sh, 605sh, 578, 536, 500sh, 486s, 460, 420, 407 (in KBr); 614sh, 605sh, 578s, 536, 500sh, 486, 460, 420, 407, 368, 212, 171w, 95w (in polyethylene).

Note: In the cited paper, Raman spectrum is given.

Wavenumbers (Raman, cm^{-1}): 3540, 1578, 1047, 955sh, 935s, 915sh, 810, 694s, 455w, 400w, 372w, 328, 280s.

Sid54 Scottyite $BaCu_2Si_2O_7$

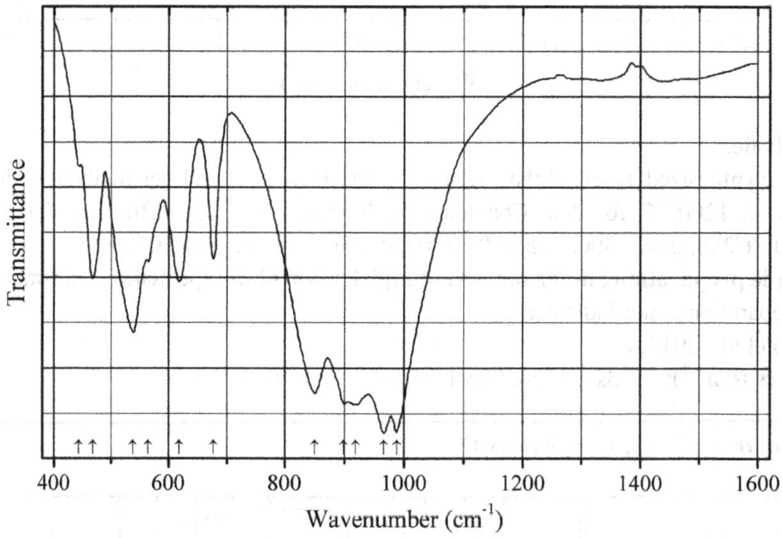

Origin: Synthetic.

Description: Dark blue polycrystalline sample synthesized by a mild hydrothermal method from $BaCl_2{\cdot}4H_2O$, $Na_2SiO_3{\cdot}9H_2O$, and CuO. Characterized by energy dispersive spectroscopy and powder X-ray diffraction data. Orthorhombic, space group *Pnma*, $a = 6.86317(15)$, $b = 13.1773(3)$, $c = 6.86317(15)$ Å, $V = 623.68(2)$ Å^3.

Kind of sample preparation and/or method of registration of the spectrum: KBr disc. Transmission.

Source: Chen et al. (2014b).

Wavenumbers (cm^{-1}): 988s, 967s, 919s, 899s, 850s, 676, 617, 564, 537, 468, 444w.

Note: The wavenumbers were partly determined by us based on spectral curve analysis of the published spectrum.

Sid57 Junitoite $CaZn_2Si_2O_7 \cdot H_2O$

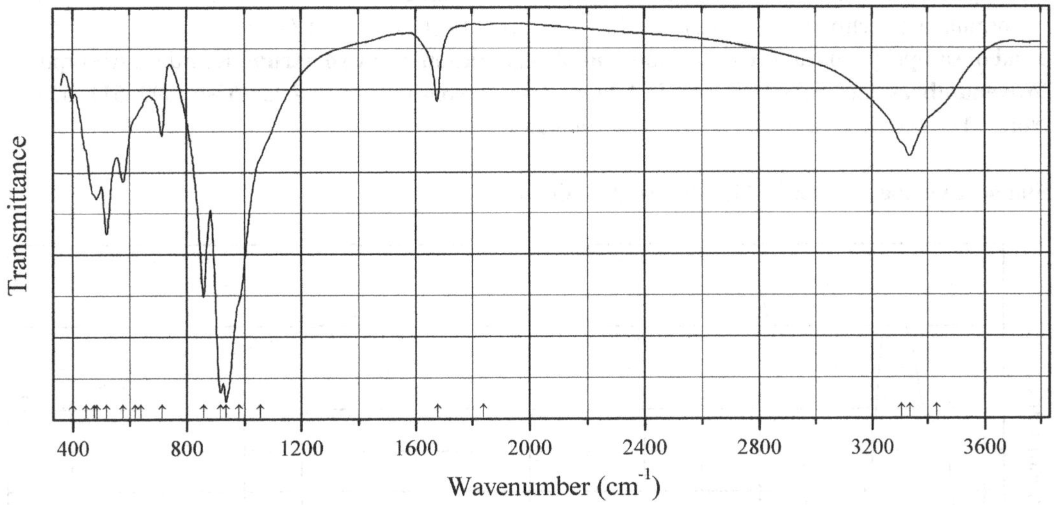

Origin: Christmas Mine, Christmas area, Banner District, Dripping Spring Mts, Gila Co., Arizona, USA (type locality).

Description: Aggregate of colorless platelets. The empirical formula is (electron microprobe): $Ca_{1.05}Zn_{2.07}Si_{1.85}Al_{0.02}Fe_{0.01}O_7 \cdot H_2O$.

Kind of sample preparation and/or method of registration of the spectrum: KBr disc. Absorption.

Wavenumbers (cm^{-1}): 3430sh, 3336, 3305sh, 1838w, 1677, 1055sh, 980sh, 935s, 915s, 857s, 715, 640sh, 620sh, 578, 520, 484, 475sh, 445sh, 399w.

Note: The spectrum was obtained by N.V. Chukanov.

Sid58 Rowlandite-like mineral $REE_4FeSi_4(O,F,OH)_{16} \cdot nH_2O$

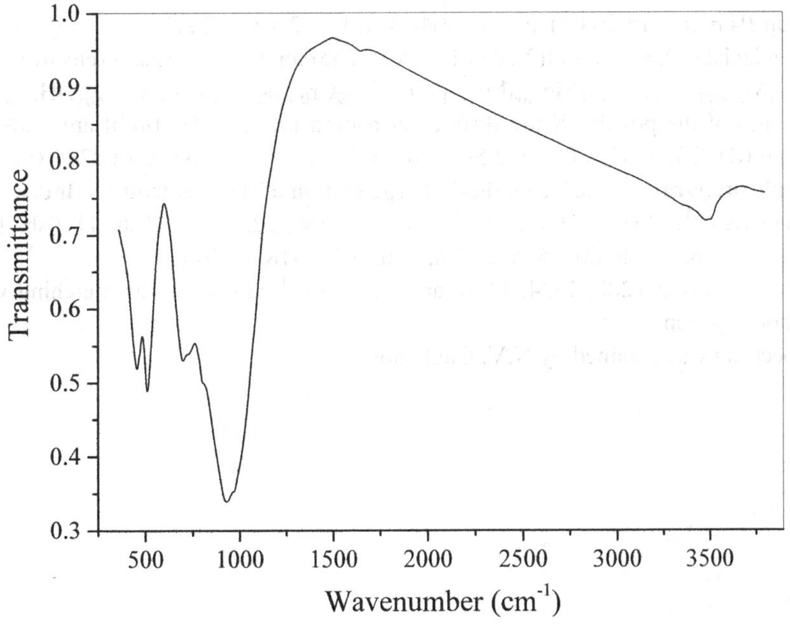

Origin: Heftetjern pegmatite, Tørdal, Telemark, Norway.

Description: Anhedral grains from the association with fluorite. Amorphous, metamict. The empirical formula is (electron microprobe): $Y_{2.5}Ln_{1.6}Fe_{0.8}Si_{4.05}O_{14}(F,OH)_2 \cdot nH_2O$.

Kind of sample preparation and/or method of registration of the spectrum: KBr disc. Absorption.

Wavenumbers (cm^{-1}): 3500sh, 3483. 3380sh, 1649w, 965sh, 931s, 807sh, 765sh, 699, 511, 457.

Note: The spectrum was obtained by N.V. Chukanov.

Siod66 Cyprine $Ca_{19}Cu^{2+}(Al_{10}Mg_2)Si_{18}O_{68}(OH)_{10}$

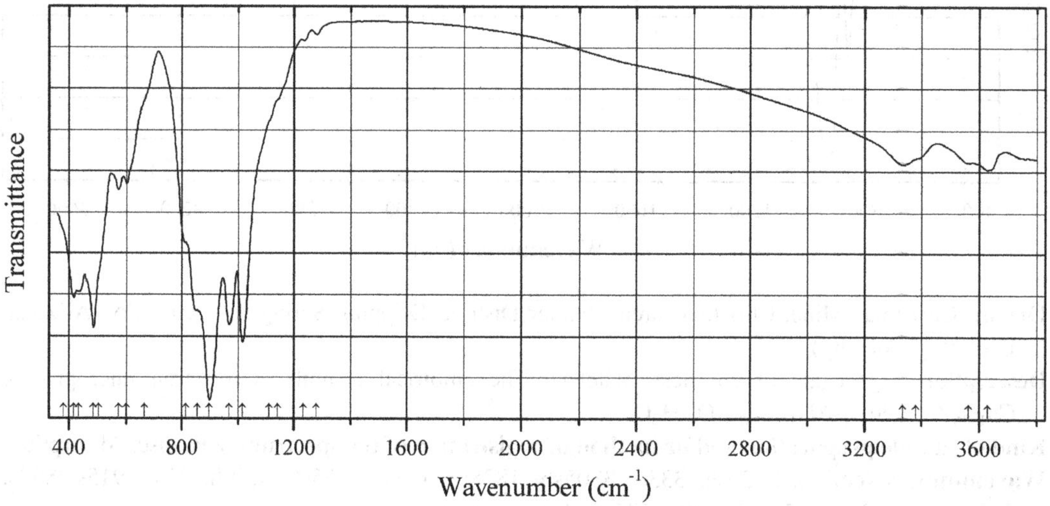

Origin: Wessels mine, near Hotazel, Kalahari Manganese Field, North Cape province, South Africa (type locality).

Description: Dark red prismatic crystals from the association with calcite, apatite, andradite, henritermierite, and rhodochrosite. Holotype sample. The crystal structure is solved. Tetragonal, space group *P4/n*, $a = 15.569(1)$, $c = 11.804(1)$ Å, $V = 2861.6(2)$ Å^3, Z = 4. $D_{calc} = 2.65$ g/cm^3. Optically uniaxial (–), $\omega = 1.744(2)$, $\varepsilon = 1.732(2)$. The empirical formula is (electron microprobe): $Ca_{19}(Cu_{0.91}Mg_{0.09})_{\Sigma 1.00}(Al_{8.38}Mg_{1.64}Mn^{3+}_{1.87}Fe^{3+}_{0.29}Cr_{0.10})_{\Sigma 12.28}Si_{17.86}O_{67.86}(OH_{9.28}O_{0.72})$. The strongest lines of the powder X-ray diffraction pattern [*d*, Å (*I*, %) (*hkl*)] are: 5.89 (12) (002), 2.950 (47) (004), 2.752 (100) (432), 2.594 (76) (522), 2.459 (35) (620), 1.622 (28) (672).

Kind of sample preparation and/or method of registration of the spectrum: KBr disc. Absorption.

Wavenumbers (cm^{-1}): 3630, 3563, 3380sh, 3335, 1280w, 1234w, 1140sh, 1110sh, 1015s, 969s, 900s, 855sh, 815sh, 665sh, 602, 574, 505sh, 486s, 434, 418, (380sh).

Note: The weak bands at 1280, 1234, 1140, and 1110 cm^{-1} correspond to stretching vibrations of admixed borate groups.

Note: The spectrum was obtained by N.V. Chukanov.

Siod67 Alumovesuvianite $Ca_{19}Al(Al_{10}Mg_2)Si_{18}O_{69}(OH)_9$

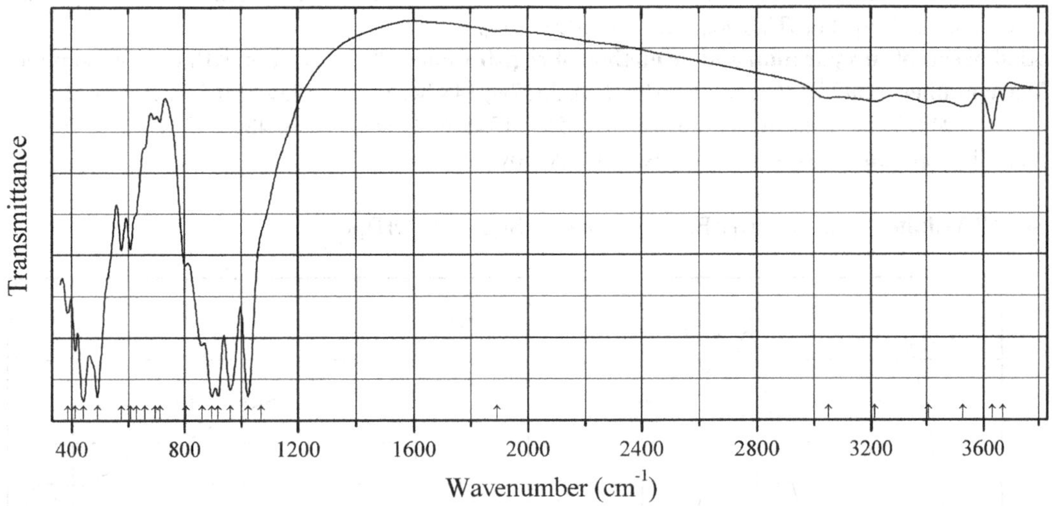

Origin: Jeffrey mine, Asbestos, Estrie Region, Québec, Canada (type locality).

Description: Pink prismatic tetragonal crystals from the association with diopside and prehnite. Holotype sample. Characterized by powder MAS NMR data. The crystal structure is solved. Tetragonal, space group *P4/n*, unit-cell parameters refined from the powder data are a = 15.5603(5) Å, c = 11.8467(4) Å, V = 2868.3(4) Å^3, Z = 2. D_{meas} = 3.31(1) g/cm^3, D_{calc} = 3.36 g/cm^3. Optically uniaxial (–), ω = 1.725(2), ε = 1.722(2). The empirical formula is $Ca_{19.00}(Al_{0.92}Fe^{3+}_{0.08})_{\Sigma 1.00}(Al_{9.83}Mg_{1.80}Mn^{3+}_{0.25})_{\Sigma 11.88}Si_{17.98}O_{69.16}(OH)_{8.44}$. The strongest lines of the powder X-ray diffraction pattern [*d*, Å (*I*, %) (*hkl*)] are: 2.96 (22) (004), 2.761 (100) (432), 2.612 (61) (224), 2.593 (25) (600), 1.7658 (20) (831), 1.6672 (20) (734), 1.6247 (21) (912), 1.3443 (22) (880).

Kind of sample preparation and/or method of registration of the spectrum: KBr disc. Absorption.

Wavenumbers (cm^{-1}): 3671w, 3632, 3527, 3407, 3212, 3051, 1890w, 1070sh, 1024s, 962s, 919s, 897s, 863, 804, 713w, 695w, 660sh, 630sh, 609, 577, 491s, 442s, 412, 386.

Note: The spectrum was obtained by N.V. Chukanov.

Siod68 "Ferrovesuvianite" $Ca_{19}Fe^{2+}(Al,Mg)_{12}Si_{18}O_{69}(OH)_9$

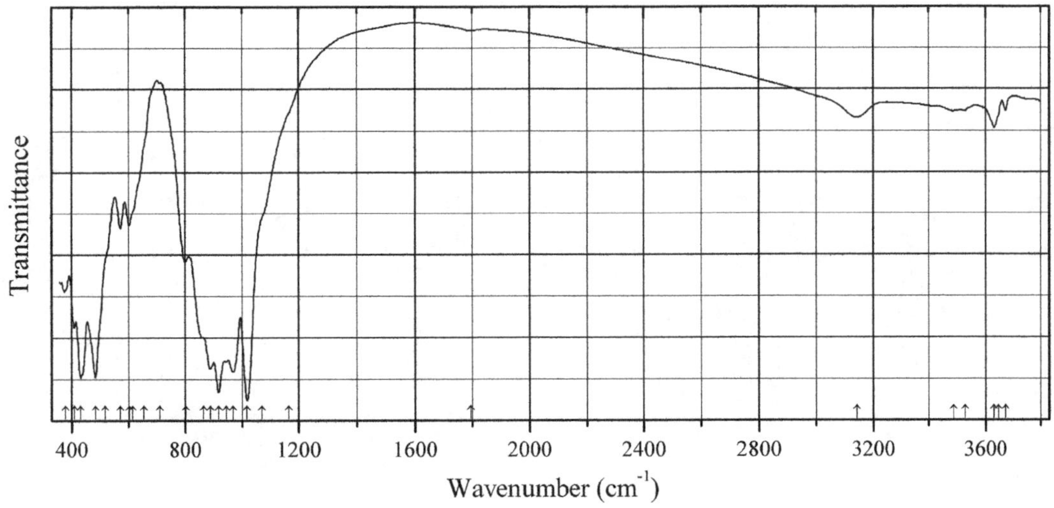

Origin: Valle d'Aosta (Aosta valley), Italy.
Description: Dark brownish-green crystals. The empirical formula is (electron microprobe): $Ca_{19.0}(Al_{9.3}Mg_{1.9}Fe_{1.2}Ti_{0.4}Mn_{0.1}Cr_{0.1})Si_{18}(O,OH)_{78}$.
Kind of sample preparation and/or method of registration of the spectrum: KBr disc. Absorption.
Wavenumbers (cm^{-1}): 3673w, 3645sh, 3630, 3527w, 3487w, 3143, 1795w, 1165sh, 1070sh, 1018s, 969s, 944, 917s, 888s, 865sh, 802, 711w, 655, 615sh, 603, 573, 520sh, 484s, 432s, 409, 379.
Note: The spectrum was obtained by N.V. Chukanov.

Siod69 Wiluite $Ca_{19}(Al,Mg)_{13}(B,\square,Al)_5(SiO_4)_{10}(Si_2O_7)_4(O,OH)_{10}$

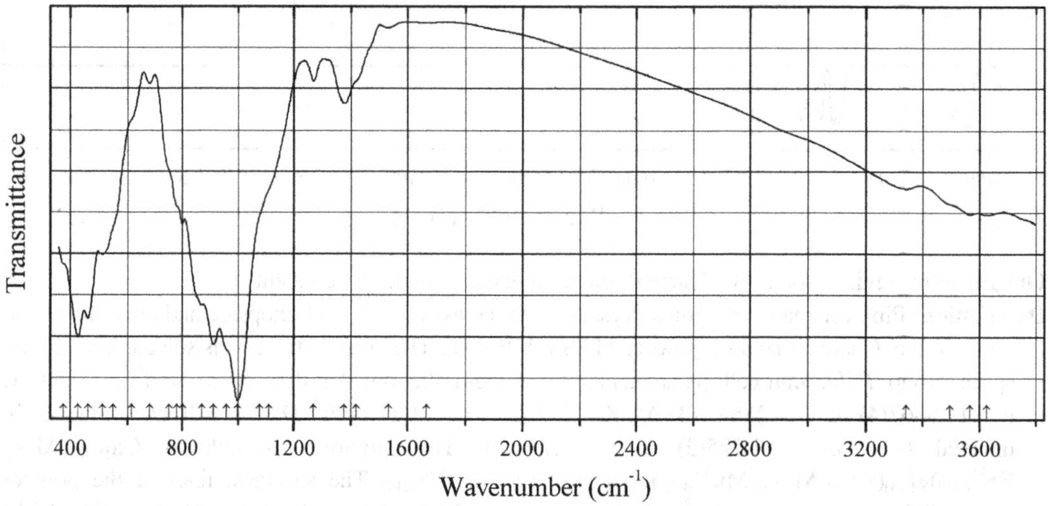

Origin: Siki-Yadunskiy fault, Siki River basin, Evenki Autonomous Area, Siberia, Russia.
Description: Greenish-gray short-prismatic crystals from the association with grossular. The empirical formula is (electron microprobe): $(Ca_{18.78}Na_{0.17})(Al_{5.77}Mg_{5.16}Fe_{1.37}Ti_{0.58}Mn_{0.06}Cr_{0.06})(B_xAl_{1.03}\square_y)Si_{18}O_{68}(O,OH)_{10}$.
Kind of sample preparation and/or method of registration of the spectrum: KBr disc. Absorption.
Wavenumbers (cm^{-1}): 3626w, 3566, 3495sh, 3340, 1667w, 1535w, 1420sh, 1376, 1269, 1110sh, 1075sh, 998s, 960sh, 913s, 871, 801, 780sh, 750sh, 679w, 615sh, 550sh, 514, 463, 427s, 377sh.
Note: The spectrum was obtained by N.V. Chukanov.

Siod70 Magnesiovesuvianite $Ca_{19}Mg(Al_{10}Mg_2)Si_{18}O_{68}(OH)_{10}$

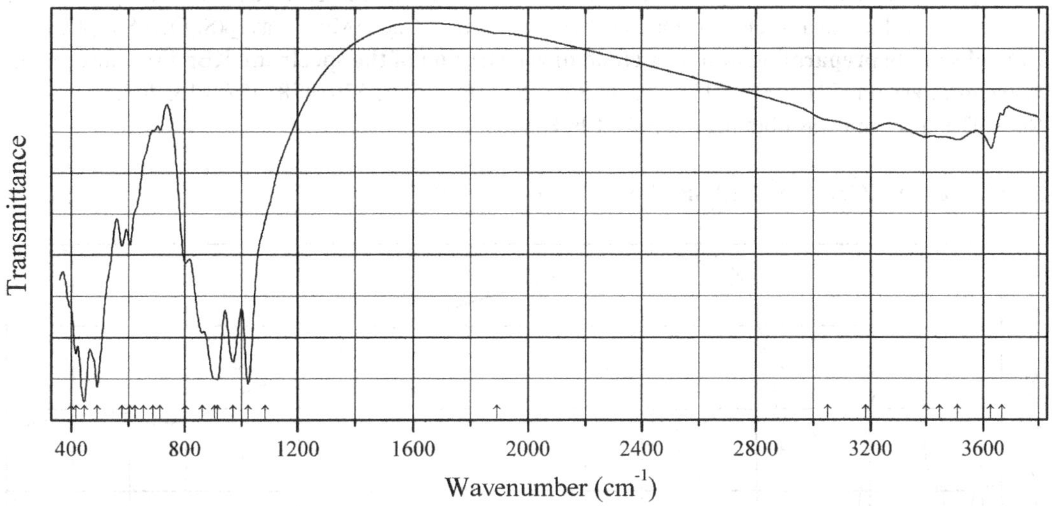

Origin: Tuydo combe, near Lojane, Republic of Macedonia (type locality).

Description: Light pink acicular tetragonal crystals from the association with calcite, garnet of the grossular-andradite series, and clinochlore. Holotype sample. The crystal structure is solved. Tetragonal, space group *P*4/*n*, $a = 15.5026(3)$, $c = 11.7858(5)$ Å, $V = 2832.4(2)$ Å^3, $Z = 2$. $D_{meas} = 3.30(3)$ g/cm^3, $D_{calc} = 3.35$ g/cm^3. Optically uniaxial (−), $\omega = 1.725(2)$, $\varepsilon = 1.721(2)$. The empirical formula is (electron microprobe): $(Ca_{18.99}Na_{0.01})_{\Sigma19.00}(Mg_{0.60}Al_{0.40})_{\Sigma1.00}(Al_{11.05}Mg_{0.70}Mn_{0.07}Fe_{0.02})_{\Sigma11.84}Si_{17.84}O_{68.72}(OH)_9$. The strongest lines of the powder X-ray diffraction pattern [*d*, Å (*I*, %) (*hkl*)] are: 10.96 (23) (110), 3.46 (22) (240), 3.038 (33) (510), 2.740 (100) (432), 2.583 (21) (522), 2.365 (94) (620), 2.192 (19) (710), 1.6165 (25) (672).

Kind of sample preparation and/or method of registration of the spectrum: KBr disc. Absorption.

Wavenumbers (cm^{-1}): 3668w, 3627, 3508, (3445), 3398, 3183, 3050sh, 1890w, (1085sh), 1024s, 971s, 915s, 905s, 862, 803, 713w, 689w, 655sh, 625sh, 607, 579, 491s, 445s, 416, 400sh.

Note: The spectrum was obtained by N.V. Chukanov.

Siod71 Ganomalite $Pb_9Ca_6(Si_2O_7)_4(SiO_4)O$

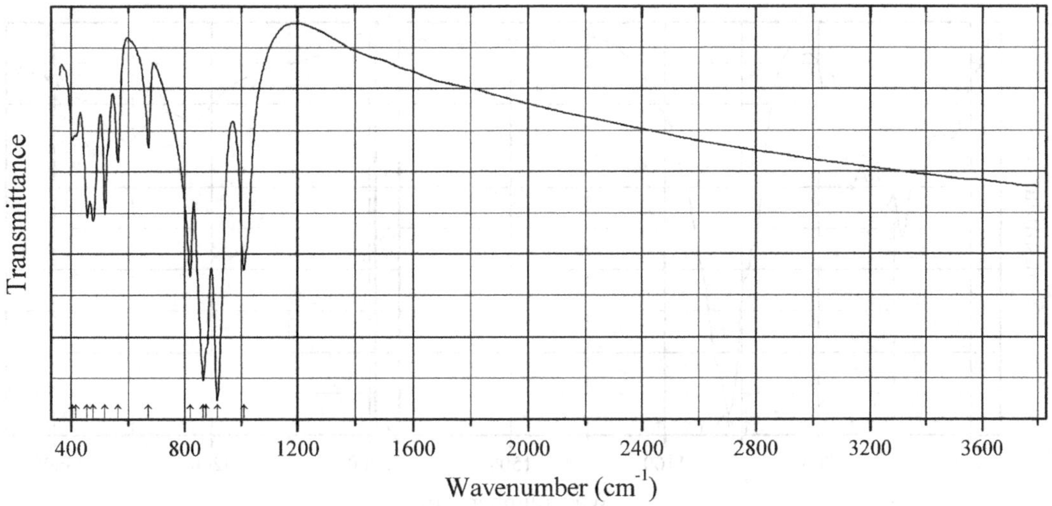

Origin: Långban deposit, Bergslagen ore region, Filipstad district, Värmland, Sweden (type locality).
Description: Colorless anhedral grains from the association with native lead, baryte, calcite, and pyrochroite. The empirical formula is (electron microprobe): $(Pb_{8.58}Ca_{0.42})Ca_{5.00}(Mn_{0.68}Ca_{0.32})(Si_2O_7)_4(SiO_4)O$.
Kind of sample preparation and/or method of registration of the spectrum: KBr disc. Absorption.
Wavenumbers (cm^{-1}): 1009, 915s, 875sh, 866s, 819, 672, 565, 520, 478, 457, 416, 405.
Note: The spectrum was obtained by N.V. Chukanov.

Siod72 Uedaite-(Ce) $Mn^{2+}CeAl_2Fe^{2+}(Si_2O_7)(SiO_4)O(OH)$

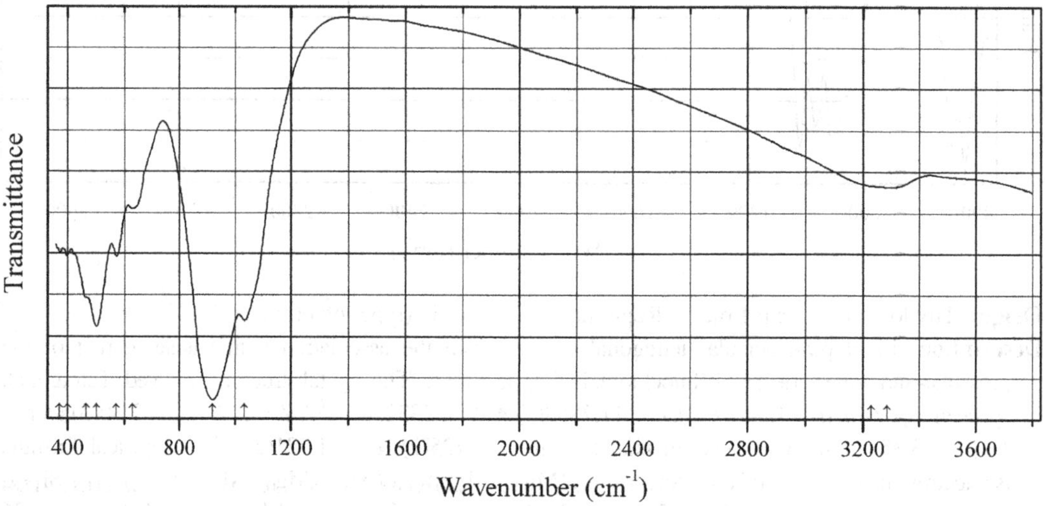

Origin: Heftetjern amazonite pegmatite, Tørdal, Telemark, Norway.
Description: Black zone in an allanite-(Ce) crystal. The empirical formula is (electron microprobe): $(Ce_{0.59}La_{0.27}Nd_{0.12})(Mn_{0.61}Ca_{0.39})(Al_{1.54}Fe_{1.35}Ti_{0.04}Mg_{0.04}Mn_{0.04})(Si_{2.94}Al_{0.06})(O,OH)_{13}$.
Kind of sample preparation and/or method of registration of the spectrum: KBr disc. Absorption.
Wavenumbers (cm^{-1}): 3285, 3230sh, 1033s, 921s, 630, 573, 502s, 465sh, 400, 372.
Note: The spectrum was obtained by N.V. Chukanov.

Siod73 Okhotskite $Ca_2(Mn,Mg)(Mn^{3+},Al,Fe^{3+})_2(Si_2O_7)(SiO_4)(OH)_2{\cdot}H_2O$

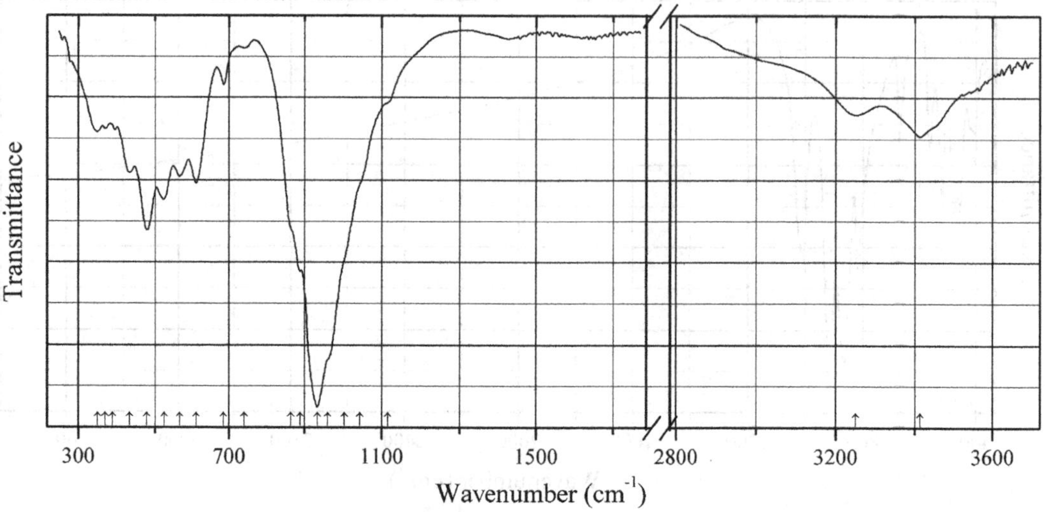

Origin: South Minusa Intermontane Trough, Siberia, Russia.

Description: Main component of the okhotskite-braunite ore. Characterized by powder X-ray diffraction data and electron microprobe analyses. Contains Al-enriched zones corresponding to pumpellyite-(Mn^{2+}).

Kind of sample preparation and/or method of registration of the spectrum: Absorption. Kind of sample preparation is not indicated.

Source: Kassandrov and Mazurov (2009).

Wavenumbers (cm^{-1}): 3414, 3250, 1115sh, 1042sh, 1002sh, 960sh, 933s, 888s, 864sh, 741, 687, 613, 568, 525, 480s, 435, 390, 371, 350.

Note: The wavenumbers were determined by us based on spectral curve analysis of the published spectrum.

Siod74 Wiluite $Ca_{19}(Al,Mg)_{13}(B,\square,Al)_5(SiO_4)_{10}(Si_2O_7)_4(O,OH)_{10}$

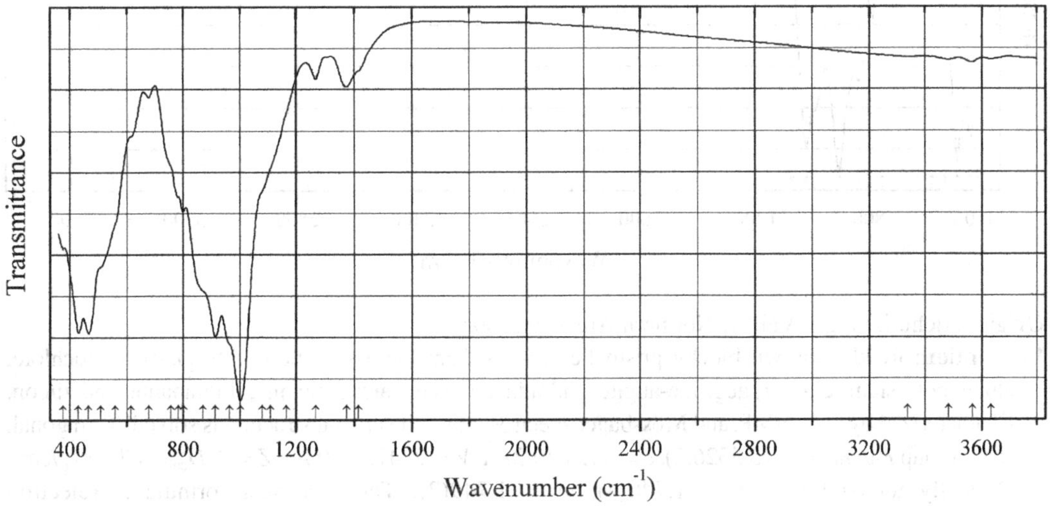

Origin: Wiluy River, Yakutia, Russia (type locality).

Description: Dark green crystal from the association with a chlorite-group mineral, Al-bearing diopside, fluorapatite, goethite, pyrite, grossular, apatite, wollastonite, and perovskite. Characterized by Mösbauer spectroscopy. The crystal structure is solved. Tetragonal, space group *P4/nnc*, a = 15.7027(3), c = 11.7008(3) Å, V = 2885.1(1) Å^3. The crystal-chemical formula is $^{X1}(Ca)_{2.00}{}^{X2}(Ca)_{8.00}{}^{X3}(Ca)_{8.00}{}^{X4}(Ca)_{1.00}{}^{Y1}(Mg_{0.56}Fe^{2+}{}_{0.27}Fe^{3+}{}_{0.17})_{\Sigma 1.00}{}^{Y2}(Al_{3.90}Fe^{2+}{}_{0.10})_{\Sigma 4.00}{}^{Y3}(Al_{3.82}Mg_{3.14}Ti_{0.63}Fe^{3+}{}_{0.21}Fe^{2+}{}_{0.16}Mn_{0.04})_{\Sigma 8.00}{}^{Z1}(Si)_{2.00}{}^{Z2}(Si)_{8.00}{}^{Z3}(Si)_{8.00}(O)_{68.00}{}^{T1+T2}(B_{3.04}Al_{0.72}\square_{1.24})_{\Sigma 5.00}{}^{W}(O_{8.32}OH_{0.96})_{\Sigma 9.28}{}^{12}O_{1.52}$.

Kind of sample preparation and/or method of registration of the spectrum: KBr disc. Absorption.
Wavenumbers (cm^{-1}): 3634w, 3566w, 3481w, 3338w, 1415sh, 1373, 1267, 1165sh, 1110sh, 1080sh, 1002s, 965sh, 914s, 870sh, 803, 782, 755sh, 679w, 615sh, 560sh, 510sh, 466s, 430s, 376.
Note: The spectrum was obtained by N.V. Chukanov.

Siod75 Ferrovesuvianite $Ca_{19}Fe^{2+}(Al,Fe,Ti)_{12}(SiO_4)_{10}(Si_2O_7)_4(OH,O)_{10}$

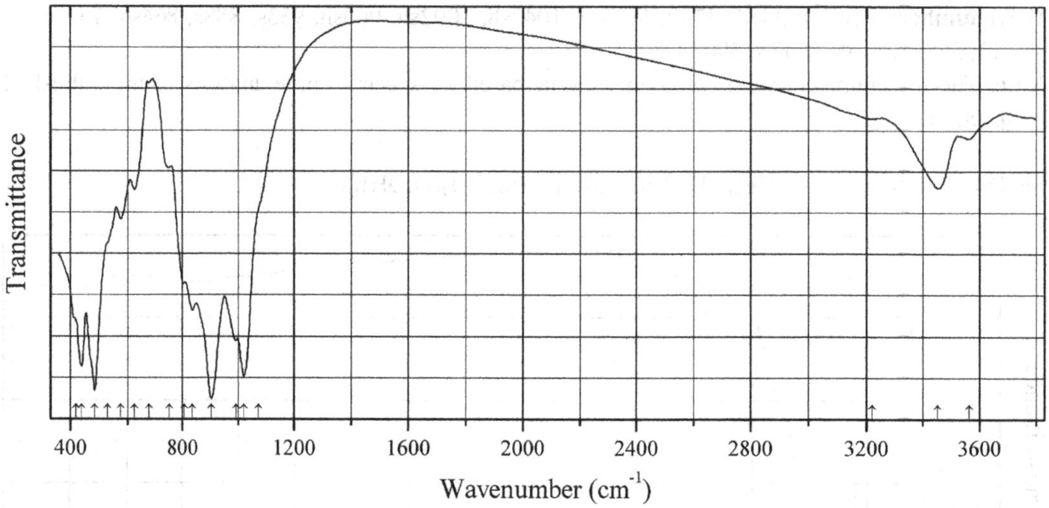

Origin: Alchuri, Shigar Valley, Northern Areas, Pakistan.
Description: Reddish-brown blocky prismatic crystals from the association with quartz, clinochlore, albite, potassium feldspar, aegirine-augite, andradite, zoisite, calcite, titanite, fluorapatite, and zircon. Characterized by ^{27}Al NMR and Mössbauer spectroscopy. The crystal structure is solved. Tetragonal, space group *P4/nnc*, $a = 15.5326(2)$, $c = 11.8040(2)$ Å, $V = 2847.87(8)$ Å^3, $Z = 2$. $D_{calc} = 3.460$ g/cm^3. Optically uniaxial (−), $\varepsilon = 1.740(4)$, $\omega = 1.749(2)$. The empirical formula is (electron microprobe): $(Ca_{18.11}Na_{0.885})(Mg_{0.63}Fe^{2+}_{0.79}Fe^{3+}_{1.765}Al_{7.99}Ti_{2.21})Si_{17.62}O_{69.92}(OH)_{7.37}F_{1.33}$. The crystal-chemical formula is $^{[8-9]}(Ca_{17.1}Na_{0.9})$ $^{[8]}Ca_{1.0}$ $^{[5]}(Fe^{2+}_{0.44}Fe^{3+}_{0.34}Mg_{0.22})$ $^{[6]}(Al_{3.59}Mg_{0.41})$ $^{[6]}(Al_{4.03}Ti_{2.20}Fe^{3+}_{1.37}Fe^{2+}_{0.40})$ $Si_{18}O_{68})$ $[(OH)_{5,84}O_{2.83}F_{1.33}]$.
Kind of sample preparation and/or method of registration of the spectrum: KBr disc. Absorption.
Wavenumbers (cm^{-1}): 3563w, 3453, 3220w, 1070sh, 1021s, 993s, 906s, 836s, 806, 750, 677w, 626, 578, 530sh, 485s, 439s, 420sh.
Note: The spectrum was obtained by N.V. Chukanov.

Siod76 Fluorvesuvianite $A_3O_4H_2O_3P_4O_3(H){\cdot}H_2O$

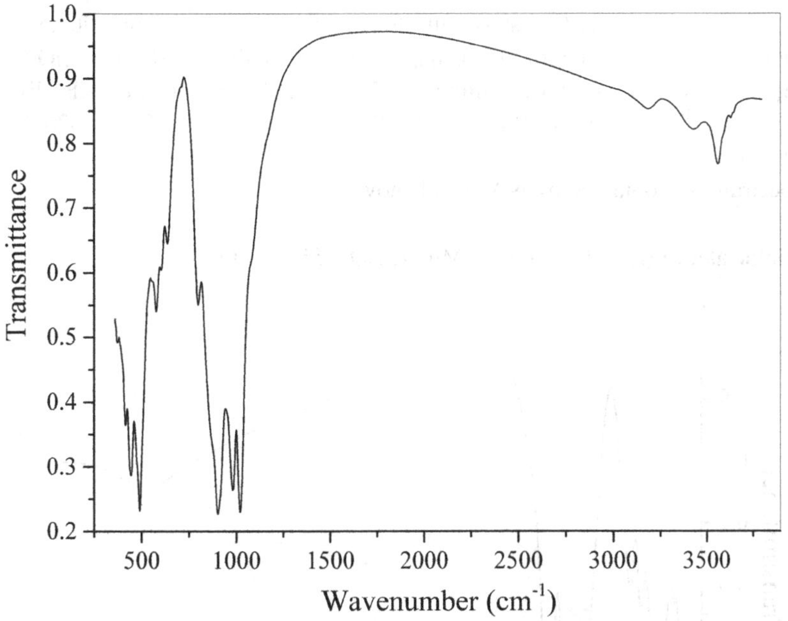

Origin: Abandoned Lupikko iron mine, Pitkäranta, Karelia, Russia (type locality).

Description: Colorless acicular crystals from the association with sphalerite and clinochlore. Confirmed by the IR spectrum.

Kind of sample preparation and/or method of registration of the spectrum: KBr disc. Absorption.

Wavenumbers (cm^{-1}): 3645sh, 3630w, 3560, 3431, 3183w, 1160sh, 1075sh, 1020s, 982s, 903s, 875sh, 799, 710w, 638, 605, 578, 560sh, 491s, 445s, 416, 376.

Note: The spectrum was obtained by N.V. Chukanov.

Siod77 Epidote-(Sr) $CaSr(Al_2Fe^{3+})(Si_2O_7)(SiO_4)O(OH)$

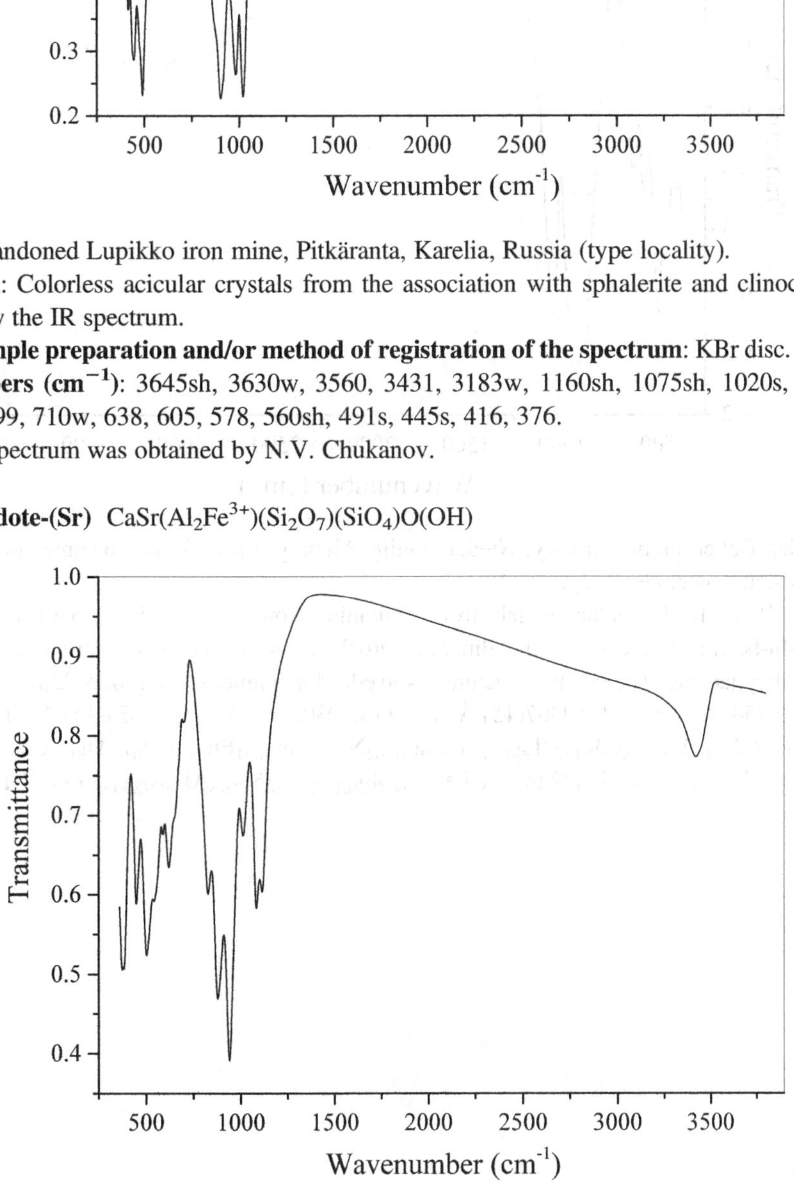

Origin: N'Chwaning Mine, Kuruman, Kalahari manganese fields, Northern Cape province, South Africa.

Description: Clusters of dark red crystals with thin zones of manganipiemontite-(Sr) . The typical composition corresponds to the formula $Ca_{1.0}Sr_{1.0}(Al_{1.8}Fe_{0.9}Mn_{0.3}(Si_2O_7)(SiO_4)O(OH)$.

Kind of sample preparation and/or method of registration of the spectrum: KBr disc. Absorption.

Wavenumbers (cm^{-1}): 3420, 1114, 1083, 1012, 940s, 878s, 827, 669w, 650sh, 620, 589, 543, 503s, 449, 394s, 376s.

Note: The spectrum was obtained by N.V. Chukanov.

Siod78 Ferriakasakaite-(La) $CaLaFe^{3+}AlMn^{2+}(Si_2O_7)(SiO_4)O(OH)$

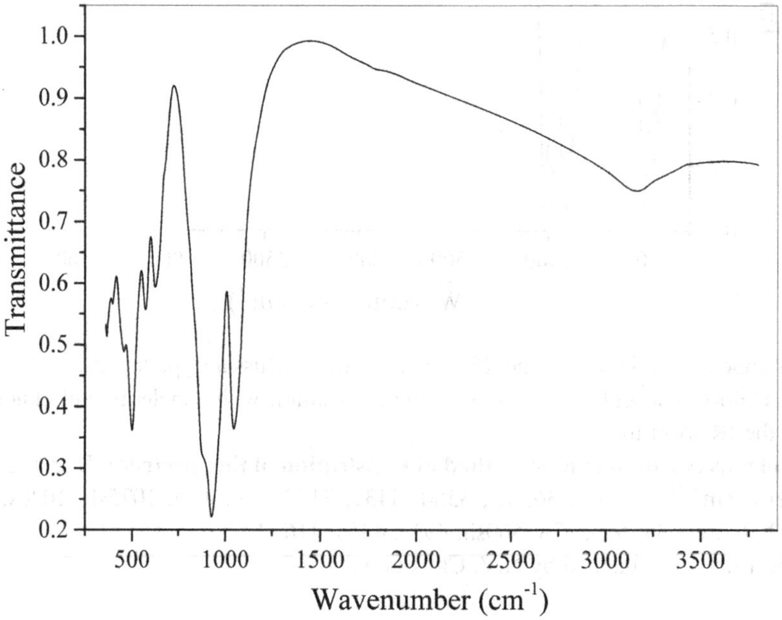

Origin: In den Dellen pumice quarry, Niedermendig, Mendig, Laach Lake volcanic complex, Eifel, Rhineland-Palatinate, Germany.

Description: Black thick-tabular crystals from sanidinite, from the association with nosean and/or haüyne, Mn-bearing biotite, magnetite, ilmenite-pyrophanite series members, Mn-bearing zirconolite, and secondary jarosite. The crystal structure is solved. Monoclinic, space group $P2_1/m$, $a = 8.90540(13)$, $b = 5.75454(7)$, $c = 10.10367(15)$ Å, $\beta = 114.1030(18)°$, $V = 472.634(11)$ Å^3. The empirical formula is $(Ca_{0.68}Mn^{2+}_{0.32})_{\Sigma1.00}(La_{0.49}Ce_{0.39}Pr_{0.02}Nd_{0.02}Sm_{0.01}Eu_{0.01}Gd_{0.01}Th_{0.01}Ca_{0.04})$ $(Fe^{3+}_{0.52}Fe^{2+}_{0.04}Al_{0.34}Ti^{4+}_{0.10})_{\Sigma1.00}Al_{1.00}(Mn^{2+}_{0.53}Fe^{2+}_{0.34}Mg_{0.13})_{\Sigma1.00}(Si_{2.98}Al_{0.02})_{\Sigma3.00}O_{12.00}(OH)$.

Kind of sample preparation and/or method of registration of the spectrum: KBr disc. Absorption.
Wavenumbers (cm^{-1}): (3300sh), 3164, 1045s, 928s, 885sh, 675sh, 626, 573, 499s, 455, 394, 366.
Note: The spectrum was obtained by N.V. Chukanov.

Siod79 Vesuvianite $Ca_{19}Fe^{3+}[Al_{10}(Fe^{2+},Mn^{2+},Mg)_2](Si_2O_7)_4(SiO_4)_{10}O(OH)_9$,

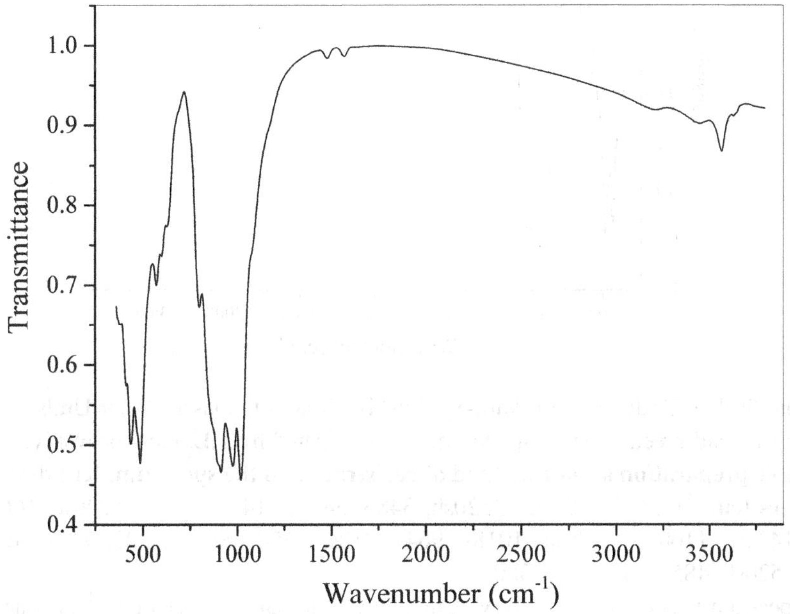

Origin: Somma-Vesuvius volcanic complex, Campania region, Italy (type locality).
Description: Greenish brown crystals from skarn xenolith. Neotype sample used for vesuvianite formula revision. Characterized by Mössbauer spectroscopy, ^{27}Al MAS NMR, powder X-tay diffraction and thermal analysis. The crystal structure is solved. Tetragonal, space group *P4/nnc*; *a* = 15.5720(3), *c* = 11.8158(5). The crystal-chemical formula is $^{X1}(Ca)_{2.00}{}^{X2}(Ca)_{8.00}{}^{X3}(Ca)_{8.00}{}^{X4}(Ca_{0.97}Na_{0.03})_{1.00}{}^{Y1}(Fe^{3+}_{0.50}Mg_{0.28}Fe^{2+}_{0.22}){}^{Y2}(Al_{3.85}Fe^{2+}_{0.15}){}^{Y3}(Al_{5.26}Mg_{1.83}Fe^{3+}_{0.54}Fe^{2+}_{0.26}Mn_{0.11}){}^{Z1}(Si)_{2.00}{}^{Z2}(Si)_{8.00}{}^{Z3}(Si)_{8.00}(O)_{68.00}{}^{T1+T2}(Al_{0.44}B_{0.25}\square_{4.31}){}^{W}(OH_{5.65}F_{2.00}O_{1.30}Cl_{0.05})$.
Kind of sample preparation and/or method of registration of the spectrum: KBr disc. Absorption.
Wavenumbers (cm^{-1}): 3565, 3450w, 3629w, 3210w, 1568w, 1478w, 1170sh, 1075 sh, 1018s, 976s, 914s, 900sh, 870sh, 799, 629, 599, 572, 486s, 436s, 415, 384.
Note: The spectrum was obtained by N.V. Chukanov.

Siod80 Vesuvianite Cr-bearing $Ca_{19}Fe^{3+}[(Al,Cr)_{10}(Fe^{2+},Mn^{2+},Mg)_2](Si_2O_7)_4(SiO_4)_{10}O(OH)_9$

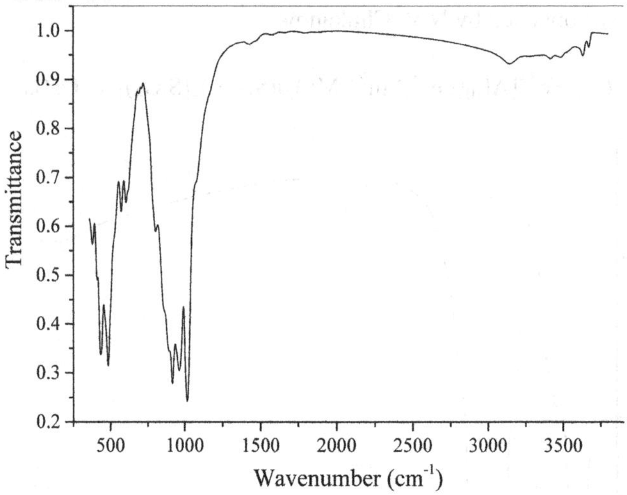

Origin: Lekhoilinskoe Cr deposit, Voikaro-Syn'inskiy ultrabasite massif, Polar Urals.

Description: Emerald-green crystals on chromite. Investigated by T.L. Panikorovskiy.

Kind of sample preparation and/or method of registration of the spectrum: KBr disc. Absorption.

Wavenumbers (cm^{-1}): 3672w, 3633, 3520sh, 3483, 3416, 3146, 1880w, 1796w, 1661w, 1577w, 1465sh, 1424w, 1160sh, 1075sh, 1018s, 964s, 918s, 895sh, 865sh, 803, 710w, 690w, 615sh, 605, 573, 520sh, 485s, 434s, 412, 380.

Note: The spectrum was obtained by N.V. Chukanov. The band at 3416 cm^{-1} is characteristic for Cr-bearing vesuvianite.

Siod81 Vesuvianite S-bearing $Ca_{19}Fe^{3+}[Al_{10}(Fe^{2+},Mn^{2+},Mg)_2](Si_2O_7)_4(SiO_4)_{10}S_x(OH,O)_{10}$ (?)

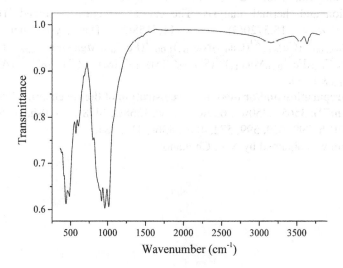

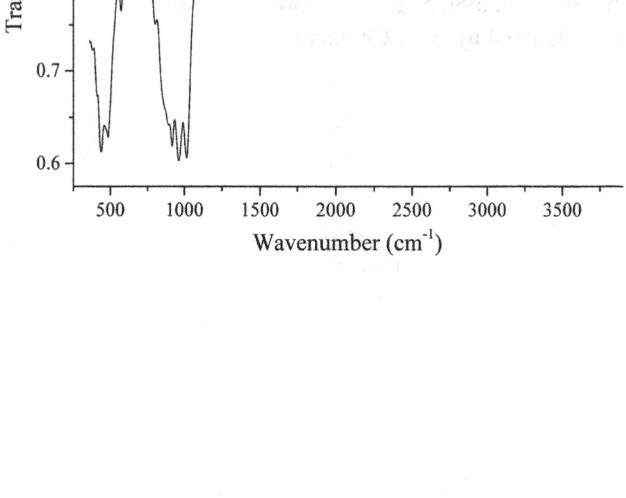

Origin: Monzoni Mts., Fassa valley, Trento Province, Trentino-Alto Adige (Trentino-Südtirol), Italy.
Description: Yellow crystal. Investigated by T.L. Panikorovskiy.
Kind of sample preparation and/or method of registration of the spectrum: KBr disc. Absorption.
Wavenumbers (cm^{-1}): 3665w, 3634, 3535, 3146, 1588sh, 1560w, 1463w, 1412sh, 1355sh, 1165sh, 1075sh, 1017s, 963s, 918s, 894, 870sh, 803, 705sh, 683w, 606, 573, 485s, 440s, 415, 384.
Note: The spectrum was obtained by N.V. Chukanov.

Siod82 "Hydrovesuvianite" $Ca_{19}Fe^{3+}[Al_{10}(Fe^{2+},Mn^{2+},Mg)_2](Si_2O_7)_4[SiO_4,(OH)_4]_{10}O(OH)_9$

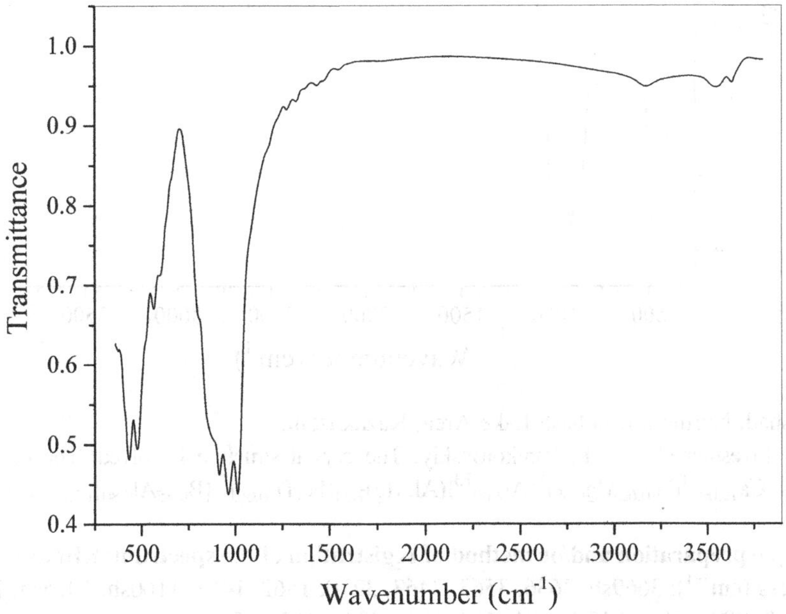

Origin: Vilyui River Basin (Wilui River Basin), Sakha Republic (Yakutia), Eastern-Siberian Region, Russia.
Description: Epitaxy on wiluite crystals from the association with grossular. Investigated by T.L. Panikorovskiy.
Kind of sample preparation and/or method of registration of the spectrum: KBr disc. Absorption.
Wavenumbers (cm^{-1}): 3631, 3546, 3160, 1547w, 1465sh, 1435w, 1363sh, 1323w, 1275w, 1165sh, 1013s, 963s, 914s, 880sh, 655sh, 595, 566, 520sh, 477s, 433s, 376.
Note: The spectrum was obtained by N.V. Chukanov. The band at 3546 cm^{-1} may correspond to $(OH)_4$ tetrahedra.

Siod83 Vesuvianite B-bearing

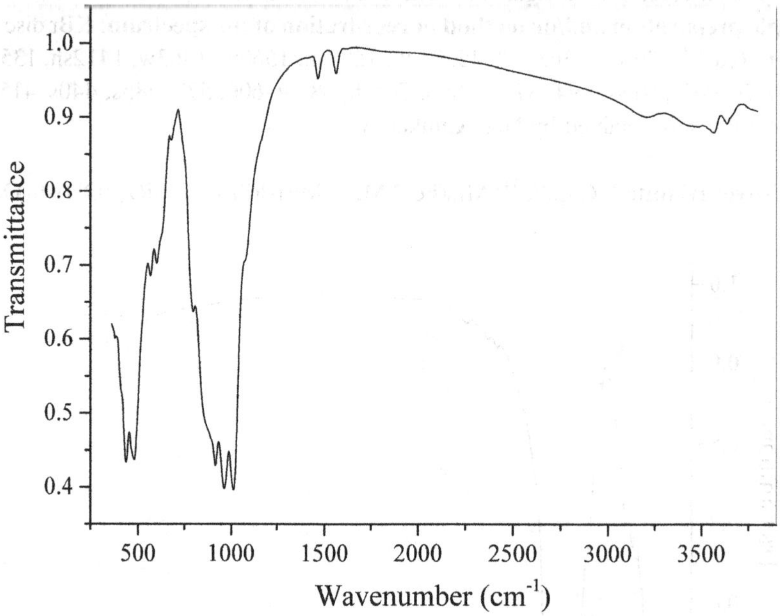

Origin: Gulshad, Northern Balkhash Lake Area, Kazakhstan.

Description: Investigated by T.L. Panikorovskiy. The crystal structure is solved. The crystal-chemical formula is $Ca_{19.00}{}^{Y1}(Fe_{0.62}Mg_{0.38})^{Y2}Al_{4.00}{}^{Y3}[(Al,Mg)_{7.34}(Fe,Ti)_{0.66}]^{T1}[B_{0.45}Al_{0.80}]_{1.25}{}^{T2}B_{0.50}Si_{18}O_{68}(OH,O)_{10}$.

Kind of sample preparation and/or method of registration of the spectrum: KBr disc. Absorption.

Wavenumbers (cm^{-1}): 3669sh, 3636, 3563, 3463, 3210, 1562, 1464, 1160sh, 1075sh, 1012s, 962s, 915s, 875sh, 800, 680w, 625sh, 601, 568, 479s, 434s, 415sh, 380.

Note: The spectrum was obtained by N.V. Chukanov. The bands at 1562 and 1464 cm^{-1} correspond to BO_3 triangles with shortened (as compared to wiluite) B–O bonds.

Siod84 Vesuvianite B-bearing

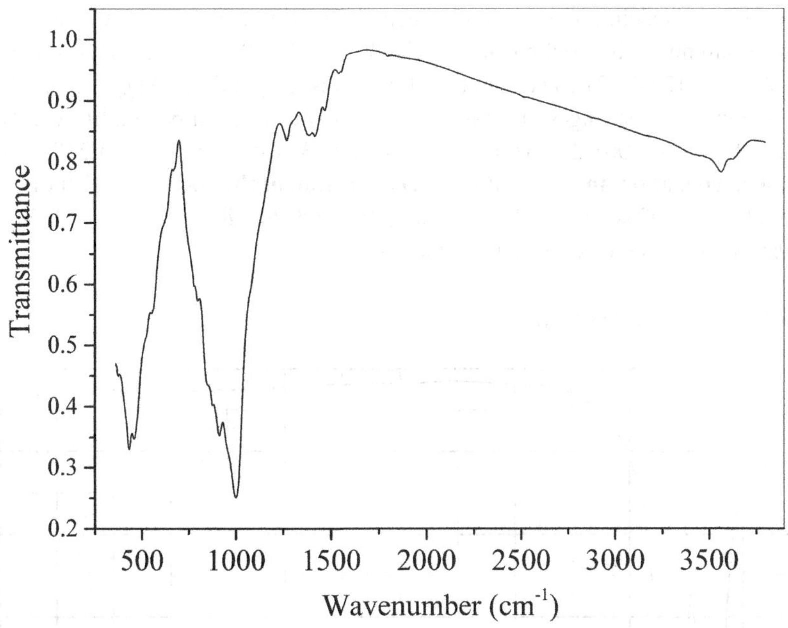

Origin: Gulshad, Northern Balkhash Lake Area, Kazakhstan.

Description: Investigated by T.L. Panikorovskiy. The crystal structure is solved. The crystal-chemical formula is $Ca_{19.00}{}^{Y1}(Fe_{0.87}Mg_{0.13})^{Y2}Al_{4.00}{}^{Y3}[(Al,Mg)_{6.45}(Fe,Mn)_{1.55}]^{T1}B_{2.06}{}^{T2}B_{1.00}Si_{18}O_{68}(OH,O)_{12}$.

Kind of sample preparation and/or method of registration of the spectrum: KBr disc. Absorption.

Wavenumbers (cm^{-1}): 3623w, 3565, 3450sh, 1800w, 1543w, 1469, 1417, 1388, 1300sh, 1269, 1130sh, 1070sh, 1000s, 913s, 882, 855sh, 802, 782, 670w, 620sh, 552, 520sh, 460s, 434s, 376.

Note: The spectrum was obtained by N.V. Chukanov. The bands at 1543 and 1469 cm^{-1} correspond to BO_3 triangles with shortened (as compared to wiluite) B–O bonds.

Sic27 Khesinite $Ca_4(Mg_3Fe^{3+}{}_9)O_4(Fe^{3+}{}_9Si_3)O_{36}$

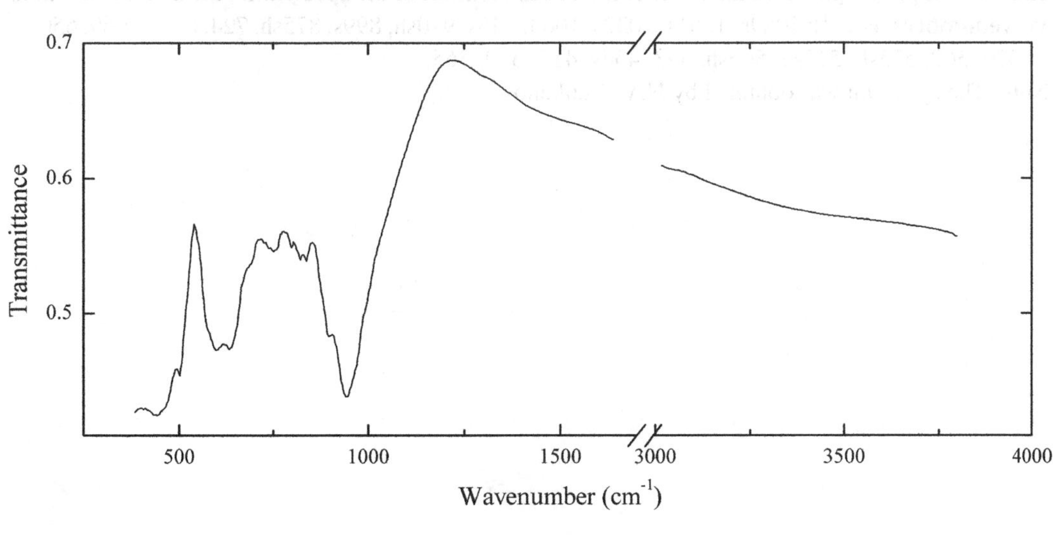

Origin: Burned dump of the Korkinskiy quarry, Chelyabinsk coal basin, Kopeisk, South Urals, Russia.
Description: Black tabular crystals from the association with melilite, pyroxene, amphibole, wollastonite, anorthite, and calcium ferrites. Technogenetic. Investigated by B.V. Chesnokov. Related to aenigmatite-group minerals. Triclinic, $a = 10.58(3)$, $b = 10.90(3)$, $c = 9.10(4)$ Å, $\alpha = 107.08(2)°$, $\beta = 95.02(2)°$, $\gamma = 124.45(2)°$. The empirical formula is $Ca_{1.16}Fe^{3+}_{4.16}Mg_{0.32}Ti_{0.02}Al_{0.64}Si_{0.65}O_{10}$. $D_{calc} = 4.09$ g/cm^3. The strongest lines of powder X-ray diffraction pattern [d, Å (I, %)] are 2.993 (70), 2.721 (80), 2.587 (100), 2.526 (90), 2.473 (40), 2.132 (55), 1.626 (52), 1.517 (70), 1.506 (50).
Kind of sample preparation and/or method of registration of the spectrum: KBr disc. Absorption.
Wavenumbers (cm^{-1}): 953s, 902s, 838, 824, 801, 750, 690sh, 640s, 596s, 500s, 463s.
Note: The spectrum was obtained by N.V. Chukanov.

Sic105 Fowlerite $CaMn_3Zn(Si_5O_{15})$

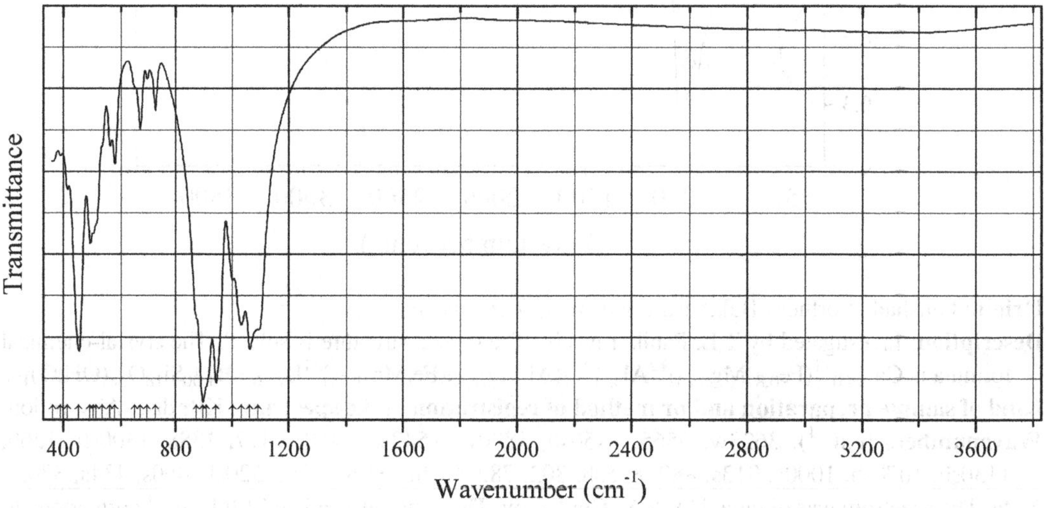

Origin: Franklin, Ogdensburg, Sussex Co., New Jersey, USA.
Description: Pinkish-brown grains. The empirical formula is $Ca_{0.75}Mn_{3.16}Zn_{0.56}Fe_{0.31}Mg_{0.25}(Si_5O_{15})$.
Kind of sample preparation and/or method of registration of the spectrum: KBr disc. Absorption.
Wavenumbers (cm^{-1}): 1089s, 1061s, 1032s, 1003, 945s, 910sh, 899s, 875sh, 724, 693w, 669, 650sh, 579, 562, 535sh, 515sh, 505sh, 493, 454s, 415, 390, 365.
Note: The spectrum was obtained by N.V. Chukanov.

Sic106 Ferrorhodonite $CaMn_3Fe(Si_5O_{15})$

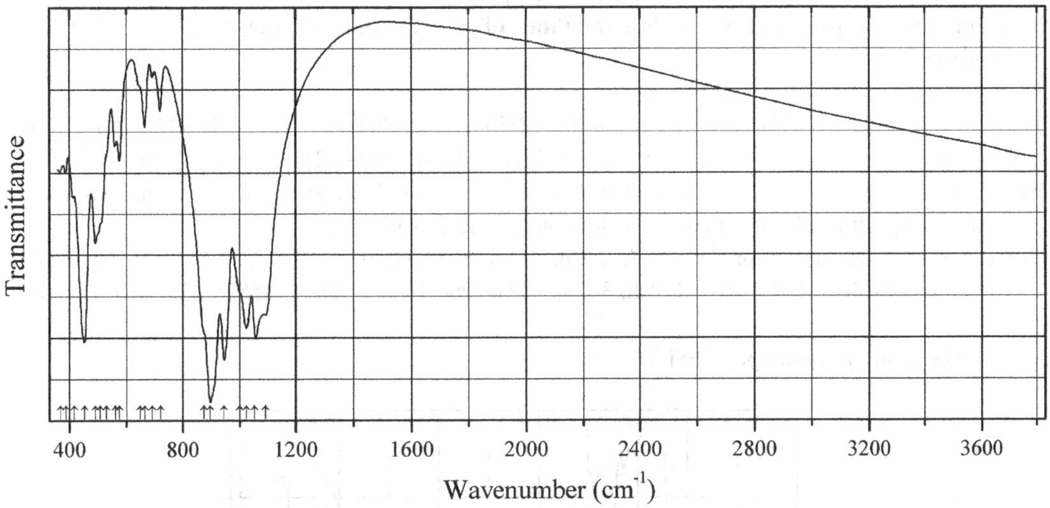

Origin: Broken Hill Pb-Zn deposit, Yancowinna Co., New South Wales, Australia (type locality).

Description: Brownish red coarse crystalline aggregates in the association with galena, chalcopyrite, spessartine, and quartz. Holotype sample. The crystal structure is solved. Triclinic, space group *P*-1, $a = 6.6766(5)$, $b = 7.6754(6)$, $c = 11.803(1)$ Å, $\alpha = 105.501(1)°$, $\beta = 92.275(1)°$, $\gamma = 93.919(1)°$, $V = 580.44(1)$ Å^3, $Z = 2$. $D_{meas} = 3.71(2)$ g/cm^3, $D_{calc} = 3.701$ g/cm^3. Optically biaxial (+), $\alpha = 1.731(4)$, $\beta = 1.736(4)$, $\gamma = 1.745(5)$, $2V = 80(10)°$. The crystal-chemical formula is $(Ca_{0.81}Mn_{0.19})(Mn_{2.52}Fe_{0.48})(Fe^{2+}{}_{0.81}Mn_{0.12}Mg_{0.04}Zn_{0.03})(Si_5O_{15})$. The strongest lines of the powder X-ray diffraction pattern [*d*, Å (*I*, %) (*hkl*)] are: 3.337 (32) (−1−13), 3.132 (54) (−210), 3.091 (41) (0−23), 2.968 (100) (−2−11), 2.770 (91) (022), 2.223 (34) (−204), 2.173 (30) (−310).

Kind of sample preparation and/or method of registration of the spectrum: KBr disc. Absorption.

Wavenumbers (cm^{-1}**)**: 1092s, 1053s, 1025, 1000sh, 946, 897, 875sh, 721, 694, 667, 650sh, 577, 563sh, 530sh, 510sh, 492, 453s, 416sh, 388, 368.

Note: The spectrum was obtained by N.V. Chukanov.

Sic107 Lithium metasilicate Li_2SiO_3

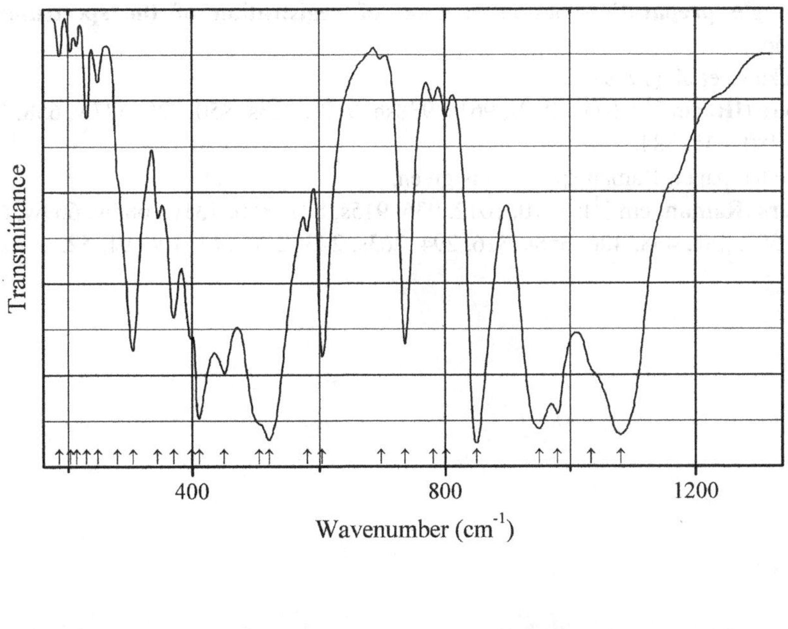

Origin: Synthetic.

Description: Commercial reactant. Orthorhombic, space group $Cmc2_1$, $Z = 4$.

Kind of sample preparation and/or method of registration of the spectrum: CsI disc. Transmission.

Source: Devarajan and Shurvell (1977).

Wavenumbers (IR, cm^{-1}): 1080s, 1034sh, 980, 950s, 850s, 801w, 781w, 735, 697w, 604, 580w, 520s, 505sh, 450, 410, 398, 370, 345w, 305, 280sh, 248w, 230w, 214w, 204w, 196w.

Note: The wavenumbers were partly determined by us based on spectral curve analysis of the published spectrum. In the cited paper, Raman spectrum is given.

Wavenumbers (Raman, cm^{-1}): 1087, 1034, 1001sh, 983s, 945, 852, 731, 645, 610s, 587sh, 567, 520, 496, 465, 450w, 410, 398sh, 345w, 325w, 291sh, 297, 273w, 258w, 234, 210, 186, 141.

Sic108 Alamosite polymorph $PbSiO_3$

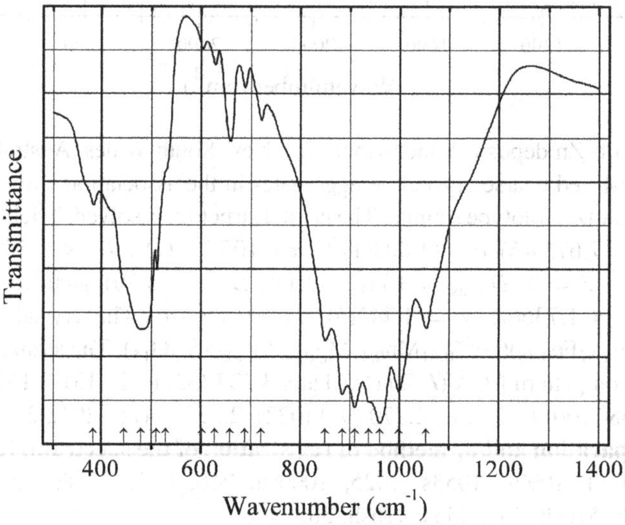

Origin: Synthetic.

Description: Prepared by crystallization at 650 °C from the undercooled melt. Characterized by powder X-ray diffraction data. Hexagonal.

Kind of sample preparation and/or method of registration of the spectrum: KBr disc. Transmission.

Source: Furukawa et al. (1979).

Wavenumbers (IR, cm^{-1}): 1054, 999s, 961s, 938sh, 910s, 885s, 850, 720, 687w, 658, 629w, 604w, 530, 511, 480, 445, 384.

Note: In the cited paper, Raman spectrum is given.

Wavenumbers (Raman, cm^{-1}): 1070, 1012, 936, 915s, 871, 850, 731w, 685w, 669w, 628w, 615w, 545, 513, 503, 484, 438, 406, 358s, 316, 294, 263s, 243, 231, 144, 106, 91, 52.

Sic109 Dorrite $Ca_4[Mg_3Fe^{3+}{}_9]O_4[Si_3Al_8Fe^{3+}O_{36}]$

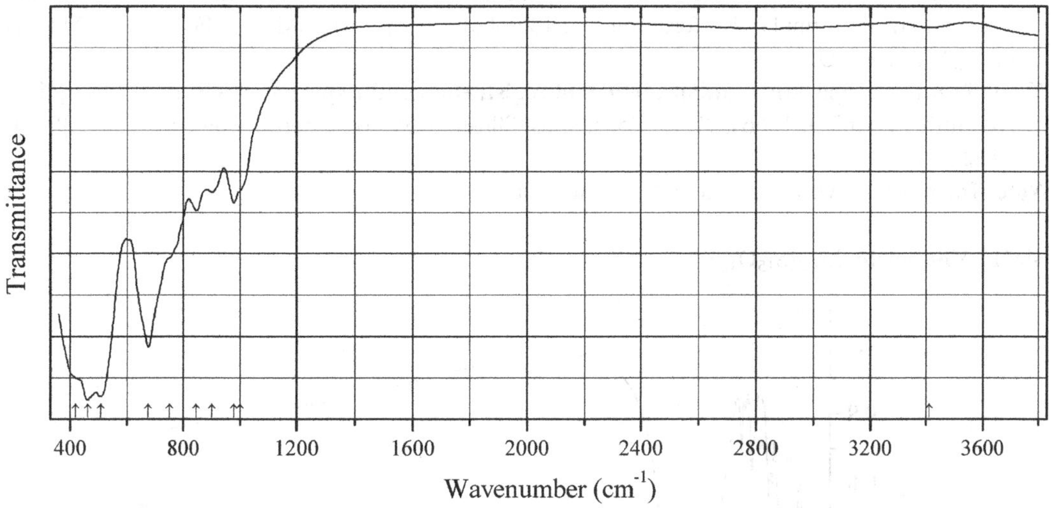

Origin: Bellerberg, near Ettringen, Eifel Mts., Rhineland-Palatinate (Rheinland-Pfalz), Germany.

Description: Brown equant crystals from the association with cuspidine, clinopyroxene, spinel, and gypsum. The empirical formula is (electron microprobe): $Ca_{4.1}(Mg_{3.6}Mn_{0.2}Fe_{7.4}Al_{0.4}Ti_{0.3})(Si_{3.8}Al_{8.2})O_{40}$.

Kind of sample preparation and/or method of registration of the spectrum: KBr disc. Absorption.

Wavenumbers (cm^{-1}): (3412w), 1000sh, 979, 902, 846, 750sh, 677s, 510s, 463s, 420sh.

Note: The spectrum was obtained by N.V. Chukanov.

Sic110 Aegirine-augite $(Ca,Na)(Fe^{3+},Mg,Fe^{2+})Si_2O_6$

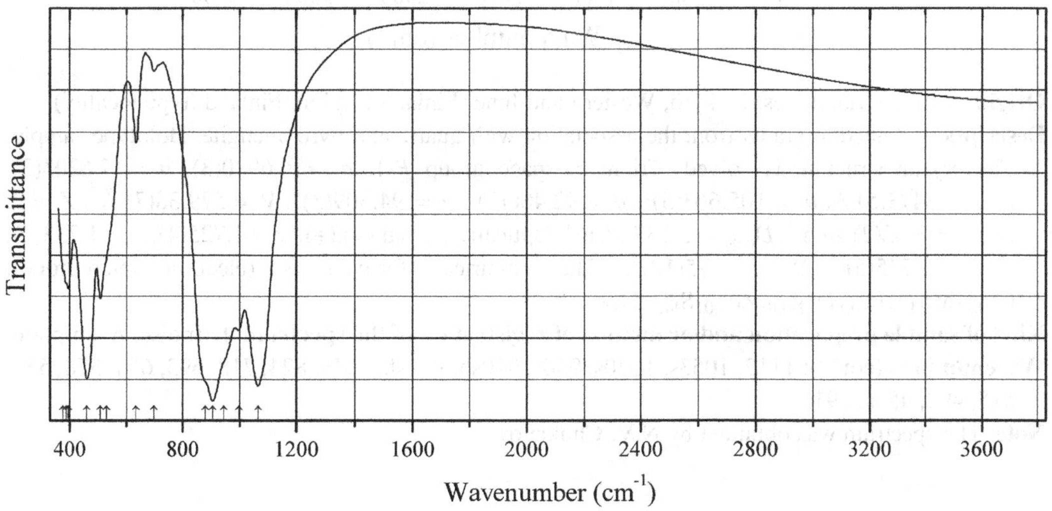

Origin: Harstigen Mine, Pajsberg, Persberg district, Filipstad, Värmland, Sweden.
Description: Olive-green anhedral grains from the association with julgoldite-(Fe^{3+}) and calcite. The empirical formula is (electron microprobe): $(Ca_{0.49}Na_{0.42}Mn_{0.09})(Fe_{0.52}Mg_{0.44}Mn_{0.04})(Si_{1.96}Al_{0.03}Fe_{0.01}O_6)$.
Kind of sample preparation and/or method of registration of the spectrum: KBr disc. Absorption.
Wavenumbers (cm^{-1}): 1066s, 997s, 945sh, 905s, 880sh, 698w, 636, 530sh, 509, 461s, 395, 387sh, 377sh.
Note: The spectrum was obtained by N.V. Chukanov.

Sic111 Vittingeite $Mn_5(Si_5O_{15})$

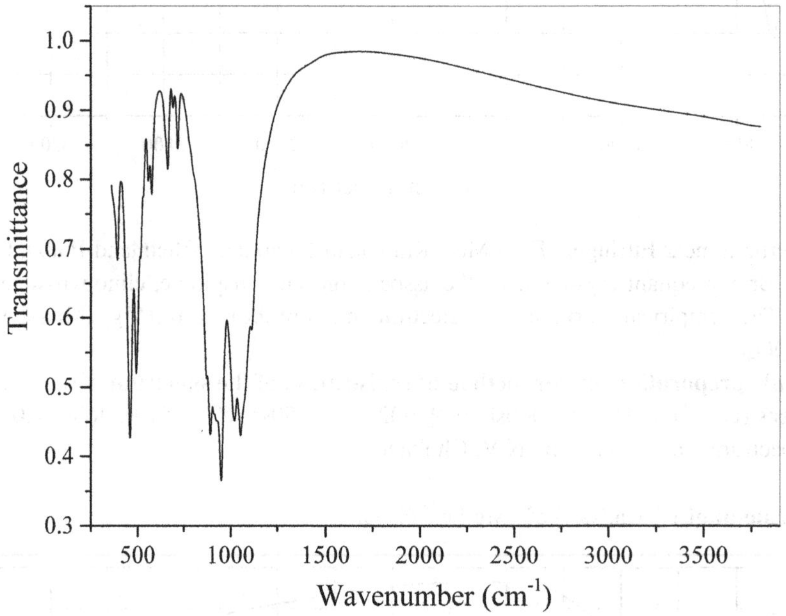

Origin: Vittinge iron mines, Isokyrö, Western and Inner Finland Region, Finland (type locality).
Description: Anhedral grains from the association with quartz and pyroxmangite. Holotype sample. The crystal structure is solved. Triclinic, space group *P*-1, $a = 6.6980(3)$, $b = 7.6203(3)$, $c = 11.8473(5)$ Å, $\alpha = 105.663(3)°$, $\beta = 92.400(3)°$, $\gamma = 94.309(3)°$, $V = 579.38(7)$ Å^3, $Z = 2$. $D_{meas} = 3/62(2)$ g/cm^3, $D_{calc} = 3.737$ g/cm^3. Optically biaxial (+),(+), $\alpha = 1.725(4)$, $\beta = 1.733(4)$, $\gamma = 1.745(5)$, $2V = 75(10)°$. The empirical formula is (electron microprobe): $Ca_{0.11}Mn_{4.71}Fe_{0.11}Mg_{0.08}Zn_{0.01}Si_{4.99}O_{15}$.
Kind of sample preparation and/or method of registration of the spectrum: KBr disc. Absorption.
Wavenumbers (cm^{-1}): 1112, 1053s, 1020s, 950s, 930sh, 915sh, 891s, 824, 718, 693, 664, 578, 559, 515, 492, 458s, 391.
Note: The spectrum was obtained by N.V. Chukanov.

Sic112 Haradaite $Sr(VO)(Si_2O_6)$

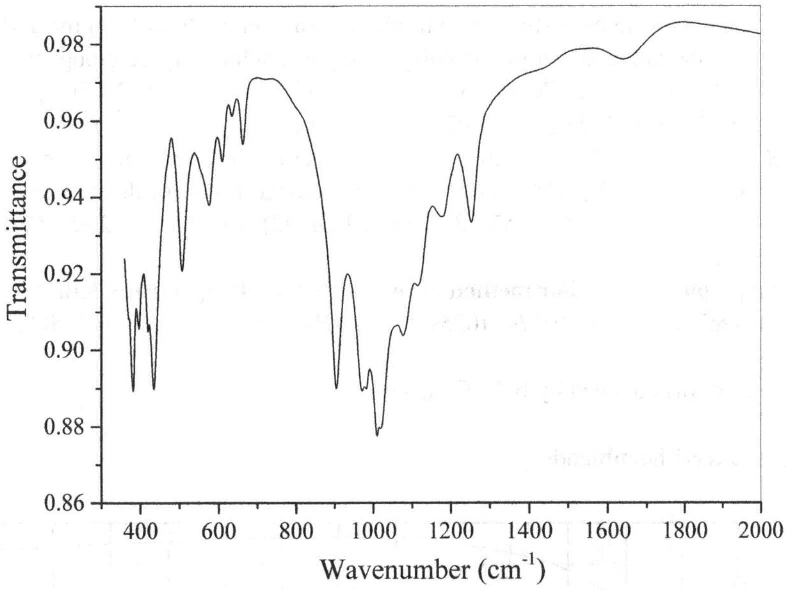

Origin: Yamato mine, Amami-Oshima Island, Kagoshima Prefecture, Nansei Archipelago, Kyushu region, Japan (type locality).
Description: Bright green grains.
Kind of sample preparation and/or method of registration of the spectrum: KBr disc. Absorption.
Wavenumbers (cm^{-1}): 1254, 1177, 1115, 1076s, 1020s, 1010s, 971s, 904s, 663, 635w, 610, 576, 507, 434s, 394, 380s.
Note: The spectrum was obtained by N.V. Chukanov.

Sic113 Dalnegorskite $Ca_5Mn(Si_3O_9)_2$

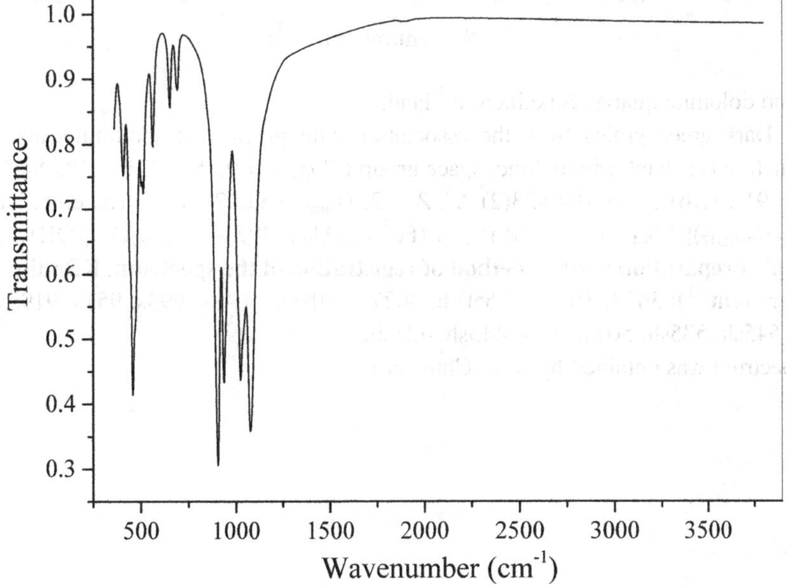

Origin: Dalnegorskoe boron deposit, town of Dalnegorsk, Primorskiy Kray, Russian Far East, Russia (type locality).

Description: Tight aggregate consisting of split thin fiber-like individuals from the association with Mn-bearing hedenbergite and datolite. Holotype sample. Triclinic, Space group: *P*-1, $a = 7.2588(11)$, $b = 7.8574(15)$, $c = 7.8765(6)$ Å, $\alpha = 88.550(15)°$, $\beta = 62.582(15)°$, $\gamma = 76.621(6)°$, $V = 386.23(11)$ Å^3, $Z = 1$. $D_{meas} = 3.02(2)$ g/cm^3, $D_{calc} = 3.062$ g/cm^3. Optically biaxial (−), $\alpha = 1.640(3)$, $\beta = 1.647(3)$, $\gamma = 1.650(3)$, $2V = 75(10)°$. The empirical formula is (electron microprobe): $Ca_{5.03}Mn_{0.50}Fe_{0.36}Mg_{0.04}Si_{6.03}O_{18}$. The strongest lines of the powder X-ray diffraction pattern [*d*, Å (*I*, %)] are: 3.80 (57), 3.48 (57), 3.28 (42), 2.952 (100), 2.951 (66), 1.815 (34), 1.708 (34), 1.703 (34).

Kind of sample preparation and/or method of registration of the spectrum: KBr disc. Absorption.

Wavenumbers (cm^{-1}): 1892w, 1077s, 1025s, 937s, 905s, 693, 653, 563, 513, 505, 470sh, 458s, 445sh, 407.

Note: The spectrum was obtained by N.V. Chukanov.

Sib152 Magnesio-ferri-hornblende

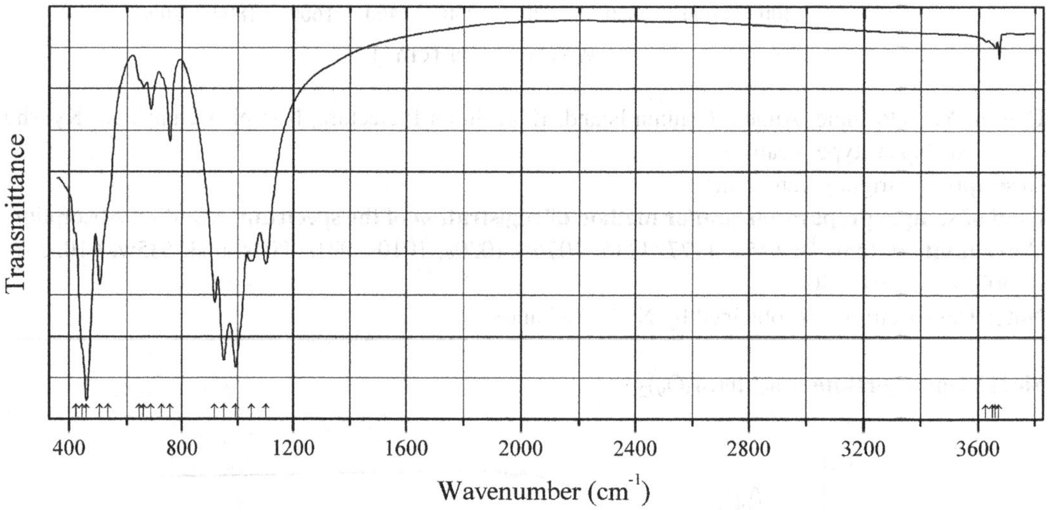

Origin: Otamo dolomite quarry, Siikainen, Finland.

Description: Dark green grains from the association with plagioclase, dolomite, and calcite. The crystal structure is solved. Monoclinic, space group *C*2/*m*, $a = 9.855(1)$, $b = 18.084(1)$, $c = 5.289(1)$ Å, $\beta = 91.141(6)°$, $V = 104.853(2)$ Å^3, $Z = 2$. $D_{calc} = 3.057$ g/cm^3. The empirical formula is $K_{0.03}(Ca_{1.92}Na_{0.07})[(Mg_{4.01}Fe^{2+}_{0.33}Mn^{2+}_{0.03})(Fe^{3+}_{0.48}Al_{0.15})][(Si_{7.43}Al_{0.57}O_{22}](OH)_2$.

Kind of sample preparation and/or method of registration of the spectrum: KBr disc. Absorption.

Wavenumbers (cm^{-1}): 3673, 3660w, 3650sh, 3627w, 1099s, 1049s, 993s, 951s, 919s, 755, 725sh, 686, 660, 645sh, 535sh, 507s, 461s, 445sh, 425sh.

Note: The spectrum was obtained by N.V. Chukanov.

Sib153 Ferro-ferri-katophorite $Na(NaCa)(Fe^{2+}{}_4Fe^{3+})(Si_7Al)O_{22}(OH)_2$

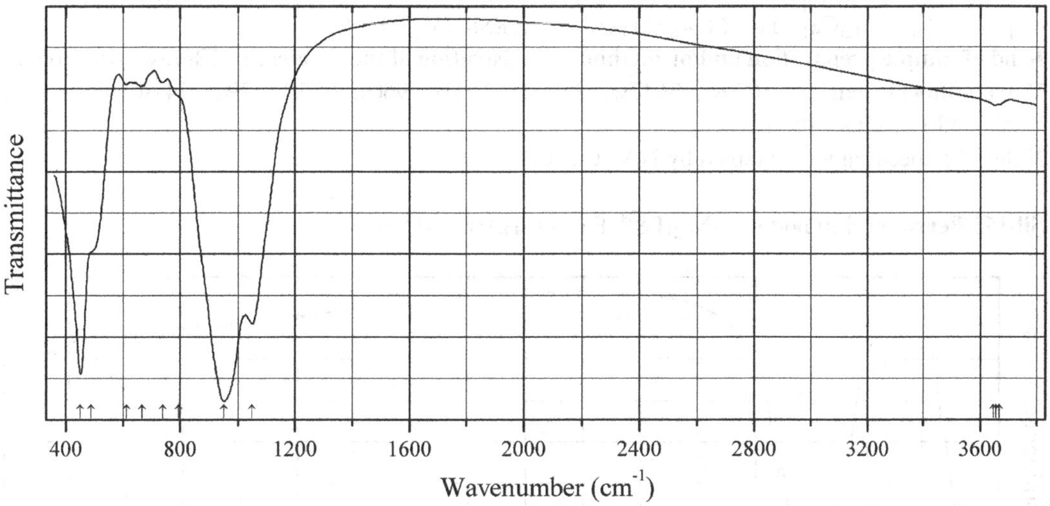

Origin: In den Dellen (Zieglowski) pumice quarry, 1.5 km NE of Mendig, Laacher See volcano, Eifel region, Rhineland-Palatinate (Rheinland-Pfalz), Germany.

Description: Black crystals on sanidinite. Characterized by Mössbauer spectrum. The empirical formula is (electron microprobe): $(Na_{0.68}K_{0.32})(Ca_{1.31}Na_{0.69})(Mg_{1.17}Fe^{2+}{}_{1.79}Mn_{0.66}Fe^{3+}{}_{1.19}Ti_{0.19})(Si_{6.20}Al_{1.80}O_{22})(OH)_{1.92}O_{0.08}$.

Kind of sample preparation and/or method of registration of the spectrum: KBr disc. Absorption.

Wavenumbers (cm^{-1}): 3667w, 3655sh, 3647w, 1050s, 953s, 795sh, 738, 667, 613, 490sh, 452s.

Note: The spectrum was obtained by N.V. Chukanov.

Sib154 Ferro-pargasite $NaCa_2(Fe^{2+}{}_4Al)(Si_6Al_2)O_{22}(OH)_2$

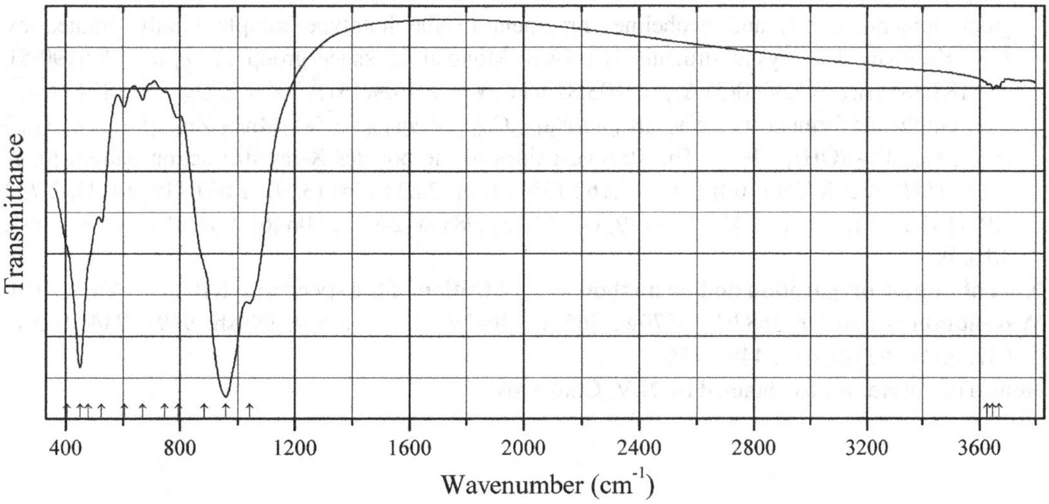

Origin: Ilmeny (Il'menskie) Mts., South Urals, Russia.

Description: Black grains with blue streak from fenite. The empirical formula is (electron microprobe): $K_{0.4}Na_{1.5}Ca_{1.1}(Fe_{3.2}Mg_{0.8}Mn_{0.3}Al_{0.6}Ti_{0.1})(Si_{6.3}Al_{1.7}O_{22})(OH)_2$.

Kind of sample preparation and/or method of registration of the spectrum: KBr disc. Absorption.

Wavenumbers (cm^{-1}): 3670w, 3647w, 3627w, 1041s, 960s, 885sh, 794, 747w, 669, 605, 526, 480sh, 450s, 405sh.

Note: The spectrum was obtained by N.V. Chukanov.

Sib155 Ferro-ferri-nybøite $NaNa_2(Fe^{2+}{}_3Fe^{3+}{}_2)(Si_7Al)O_{22}(OH)_2$

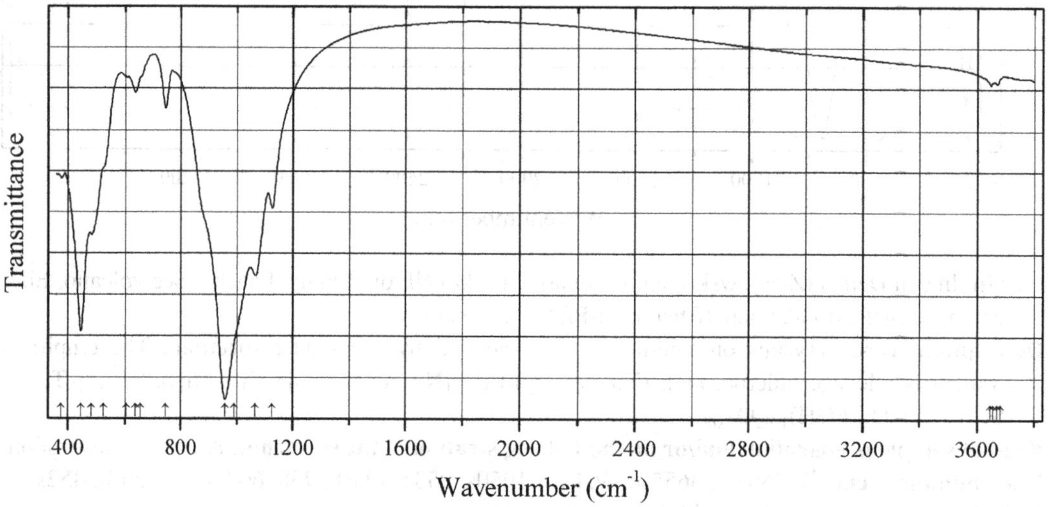

Origin: Poudrette quarry, Mont Saint-Hilaire, La Vallée-du-Richelieu RCM, Montérégie (Rouville) Co., Québec, Canada (type locality).

Description: Black crystals from the association with a eudialyte-group mineral, an astrophyllite-group mineral, albite, and nepheline. Fragment of the holotype sample kindly granted by A.V. Kasatkin. The crystal structure is solved. Monoclinic, space group *C2/m*, $a = 9.9190(5)$, $b = 18.0885(8)$, $c = 5.3440(3)$ Å, $\beta = 103.813(1)°$, $V = 931.09(13)$ Å^3, $Z = 2$. $D_{calc} = 3.424$ g/cm^3. The empirical formula is $(Na_{0.68}K_{0.27})(Na_{1.83}Ca_{0.17})(Mg_{0.06}Fe^{2+}{}_{3.17}Mn_{0.31}Zn_{0.01}Fe^{3+}{}_{1.36}Ti_{0.06})$ $(Si_{7.41}Al_{0.59}O_{22})(OH)_{1.58}F_{0.42}$. The strongest lines of the powder X-ray diffraction pattern [*d*, Å (*I*, %) (*hkl*)] are: 8.520 (100) (110), 3.162 (55) (310), 2.834 (24) (330), 1.671 (19) (461), 2.732 (10) (151), 2.552 (10) (−202), 2.344 (9) (−351), 3.298 (7) (240), 2.606 (6) (061), 1.446 (6) (−661, 4.10.0).

Kind of sample preparation and/or method of registration of the spectrum: KBr disc. Absorption.

Wavenumbers (cm^{-1}): 3683sh, 3670w, 3655w, 3647w, 1124, 1064s, 990sh, 959s, 744, 655sh, 637, 607w, 525sh, 483, 446s, 376.

Note: The spectrum was obtained by N.V. Chukanov.

Sib156 Potassic-magnesio-fluoro-arfvedsonite $KNa_2(Mg_4Fe^{3+})Si_8O_{22}F_2$

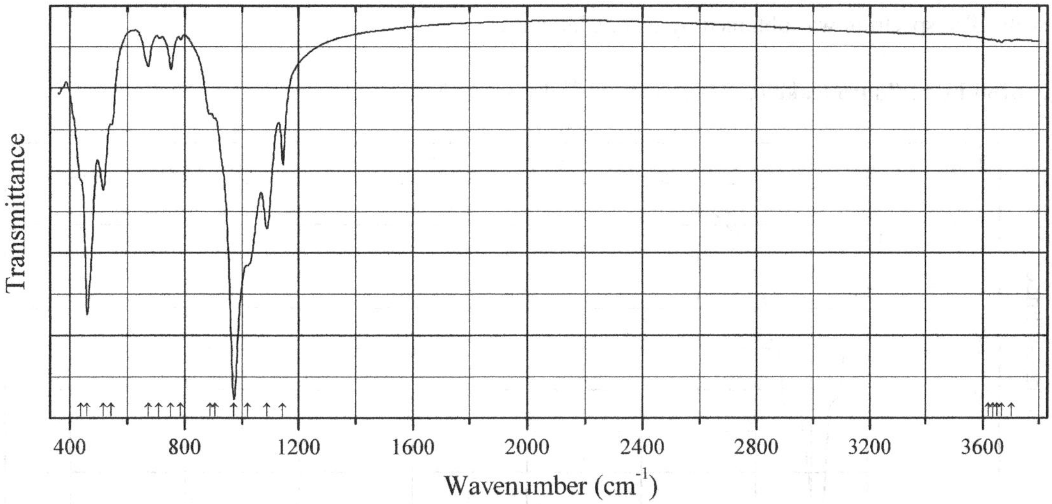

Origin: Highway 366 road cut, Val-des-Monts, Québec, Canada (type locality).

Description: Fragment of the holotype sample kindly granted by A.V. Kasatkin.

Kind of sample preparation and/or method of registration of the spectrum: KBr disc. Absorption.

Wavenumbers (cm^{-1}): 3704w, 3667w, 3651w, (3636w), (3619w), 1144, 1088s, 1020s, 972s, 905sh, 890, 785w, 752, 711w, 673, 546, 518s, 461s, 440sh.

Note: The spectrum was obtained by N.V. Chukanov.

Sib157 Ferro-ferri-fluoro-leakeite $NaNa_2(Fe^{2+}{}_2Fe^{3+}{}_2Li)Si_8O_{22}F_2$

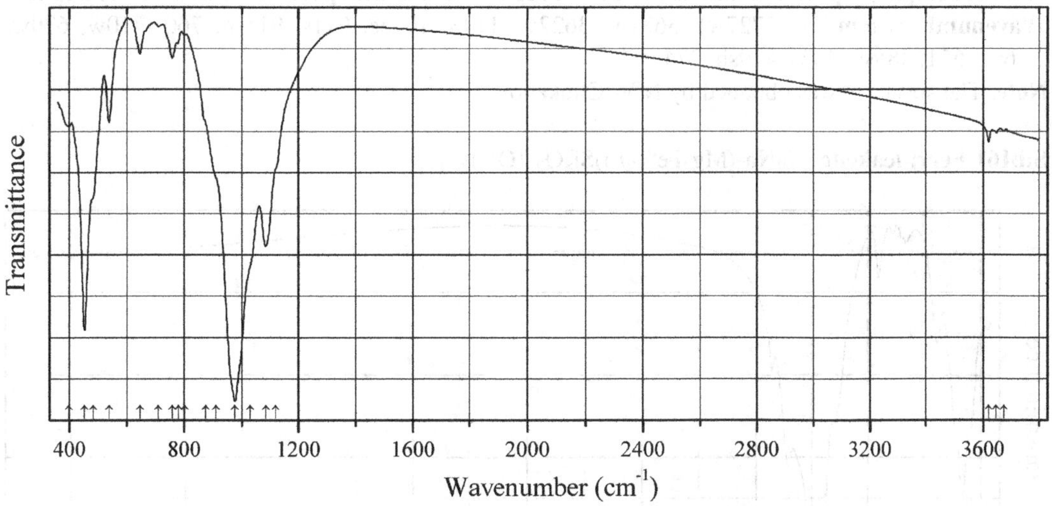

Origin: Aryskan *REE* deposit, Tyva Republic, Russia.

Description: Black prismatic crystals from the association with aegirine, polylithionite, quartz, and albite. Investigated by A.V. Kasatkin. The empirical formula is (electron microprobe, Li calculated): $(Na_{0.46}K_{0.32})Na_{2.00}(Mg_{0.09}Fe^{2+}{}_{1.99}Li_{0.80}Mn_{0.10}Zn_{0.06}Fe^{3+}{}_{1.73}Al_{0.16}Ti_{0.06})(Si_{8.00}O_{22})F_{1.42}(OH)_{0.58}$.

Kind of sample preparation and/or method of registration of the spectrum: KBr disc. Absorption.

Wavenumbers (**cm^{-1}**): 3674w, 3646w, 3620w, 1120sh, 1084s, 1030sh, 977s, 910sh, 875sh, (801w), (779w), 758, 710w, 646, 540, 485sh, 454s, 397.

Note: The spectrum was obtained by N.V. Chukanov.

Sib160 Ferri-fluoro-leakeite $NaNa_2(Mg_2Fe^{3+}{}_2Li)Si_8O_{22}F_2$

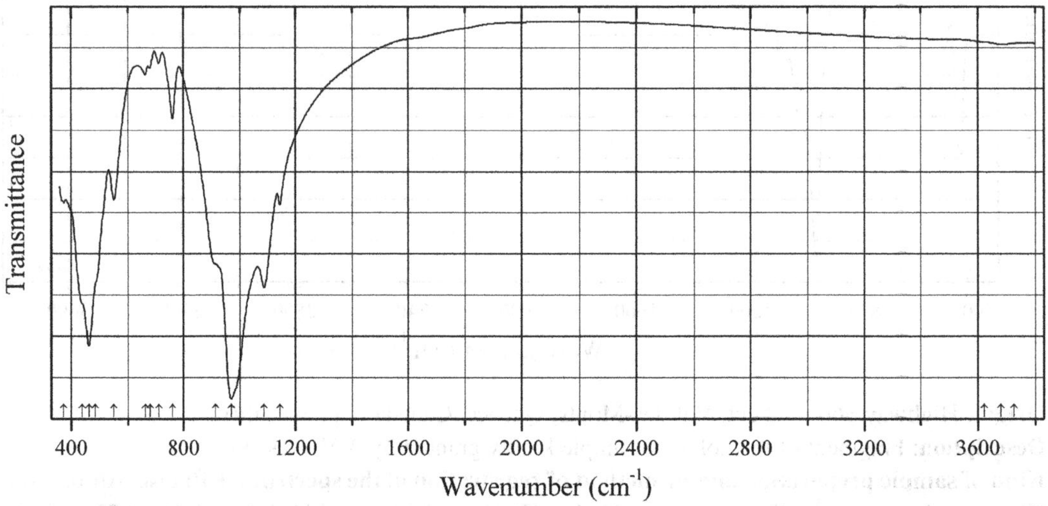

Origin: Norra Kärr, Gränna, Jönköping, Småland, Sweden.

Description: Dark green prismatic crystals from the association with aegirine and albite. The empirical formula is (electron microprobe; ICP MS analysis for Li): $(Na_{0.55}K_{0.44})(Na_{1.97}Ca_{0.02}Mn_{0.01})(Mg_{1.78}Mn_{0.09}Zn_{0.06})Li_{1.05}(Fe_{1.44}Al_{0.51}Ti_{0.08})(Si_{7.83}Al_{0.17}O_{22})F_{1.18}(OH)_{0.82}$.

Kind of sample preparation and/or method of registration of the spectrum: KBr disc. Absorption.

Wavenumbers (**cm^{-1}**): (3727w), 3681w, 3622w, 1143, 1085s, 971s, 915sh, 760, 710w, 679w, 661, 551, 485sh, 462s, 439sh, 374.

Note: The spectrum was obtained by N.V. Chukanov.

Sib161 Ferri-leakeite $NaNa_2(Mg_2Fe^{3+}{}_2Li)Si_8O_{22}(OH)_2$

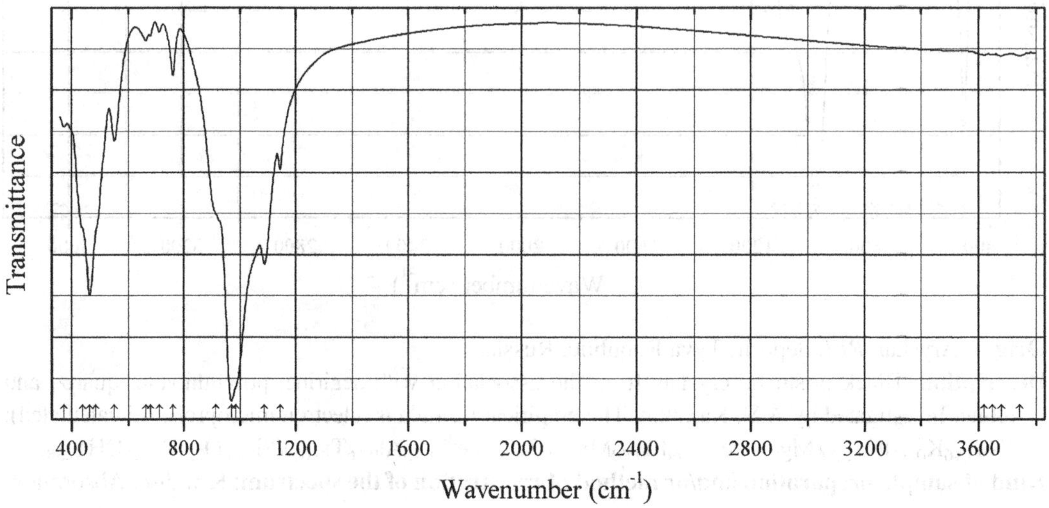

Origin: Norra Kärr, Gränna, Jönköping, Småland, Sweden.

Description: Dark green prismatic crystals from the association with aegirine and albite. Characterized by Mössbauer spectroscopy. The empirical formula is (electron microprobe; ICP MS analysis for Li): $(Na_{0.56}K_{0.44})(Na_{1.90}Mn_{0.08}Ca_{0.02})(Mg_{1.88}Fe^{2+}_{0.10}Mn_{0.02})Li_{1.12}(Fe^{3+}_{1.08}Al_{0.50}Fe^{2+}_{0.24}Ti_{0.06})(Si_{7.97}Al_{0.03}O_{22})(OH)_{1.11}F_{0.89}$.

Kind of sample preparation and/or method of registration of the spectrum: KBr disc. Absorption.

Wavenumbers (cm^{-1}): (3742w), 3680w, 3648w, 3620w, 1143, 1086s, 985sh, 971s, 915sh, 761, 710w, 679w, 661, 551, 485sh, 464s, 440sh, 374.

Note: The spectrum was obtained by N.V. Chukanov.

Sib162 Potassic-ferri-leakeite $KNa_2(Mg_2Fe^{3+}{}_2Li)Si_8O_{22}(OH)_2$

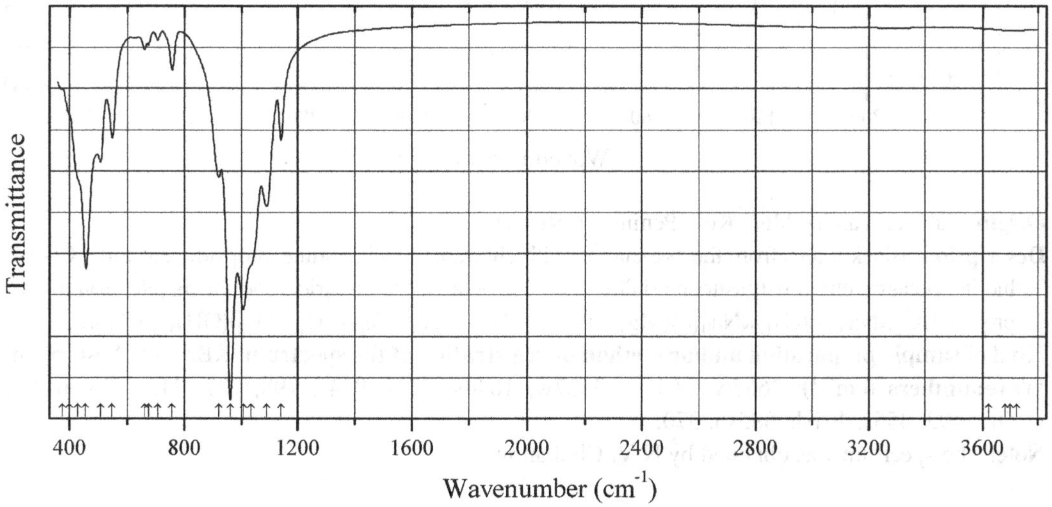

Origin: Kedykverpakhk Mt., Lovozero alkaline complex, Kola peninsula, Murmansk region, Russia.

Description: Greenish-gray fibrous aggregate from the association with natrolite and ussingite. Investigated by I.V. Pekov.

Kind of sample preparation and/or method of registration of the spectrum: KBr disc. Absorption.

Wavenumbers (cm^{-1}): 3727w, 3701w, 3683w, 3622w, 1141, 1089s, 1035sh, 1007s, 961s, 920, 757, 707w, 675w, 661, 549, 508, 457s, 430sh, 400sh, 375.

Note: Actually, this sample may be the F-dominant analogue of potassic-ferri-leakeite. The spectrum was obtained by N.V. Chukanov.

Sib163 Potassic-ferro-pargasite $KCa_2(Fe^{2+}{}_4Al)(Si_6Al_2)O_{22}(OH)_2$

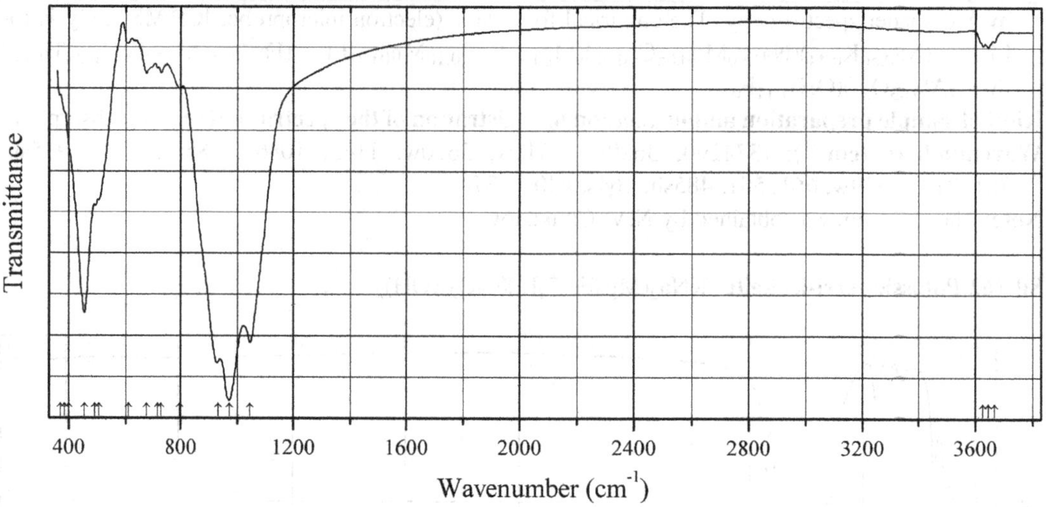

Origin: Sal'nye Tundry Mts., Kola Peninsula, Russia.

Description: Black grains from the association with chlorapatite, almandine, diopside, enstatite, Cl-rich biotite, potassic-chloropargasite, marialite, and plagioclase. The empirical formula is (electron microprobe): $(K_{0.55}Na_{0.42})(Ca_{1.98}Na_{0.02})(Mg_{1.98}Fe^{2+}{}_{2.11}Al_{0.65}Ti_{0.26})(Si_{6.03}Al_{1.97}O_{22})(OH)_{1.62}Cl_{0.38}$.

Kind of sample preparation and/or method of registration of the spectrum: KBr disc. Absorption.

Wavenumbers (cm^{-1}): 3667w, 3645w, 3627w, 1044s, 973s, 934s, 798, 727, 715, 674, 611w, 507, 493, 455s, 401sh, 385sh, 370.

Note: The spectrum was obtained by N.V. Chukanov.

Sib164 Oxo-magnesio-hastingsite $NaCa_2(Mg_2Fe^{3+}{}_3)(Si_6Al_2)O_{22}O_2$

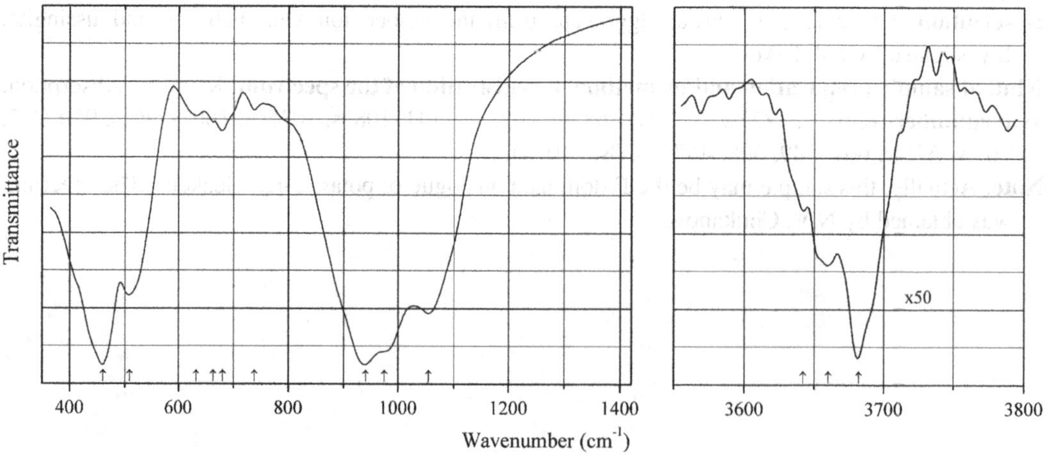

Origin: Deeti volcanic cone, Gregory rift, northern Tanzania (type locality).

Description: Brown megacryst from volcanic tuff. Holotype sample. The crystal structure is solved. Monoclinic, space group *C*2/*m*, $a = 9.8837(3)$, $b = 18.0662(6)$, $c = 5.3107(2)$ Å, $\beta = 105.278(1)°$, $V = 914.77(5)$ Å^3, $Z = 2$. $D_{meas} = 3.19(1)$ g/cm^3. Optically biaxial (–), $\alpha = 1.706(2)$, $\beta = 1.715(2)$, $\gamma = 1.720(2)$. The empirical formula is $(Na_{0.67}K_{0.33})(Ca_{1.87}Ma_{0.14}Mn_{0.01})(Mg_{3.27}Fe^{3+}{}_{1.25}Ti_{0.44}Al_{0.08})(Si_{6.20}Al_{1.80}O_{22})[O_{1.40}(OH)_{0.60}]$. The strongest lines of the powder X–ray diffraction pattern [*d*, Å (*I*, %) (*hkl*)] are: 3.383 (62) (131), 2.708 (97) (151), 2.555 (100) (–202), 2.349 (29) (–351), 2.162 (36) (261).

Kind of sample preparation and/or method of registration of the spectrum: KBr disc. Transmission.

Source: Zaitsev et al. (2013).

Wavenumbers (cm^{-1}): 3682w, 3660w, 3642w, 1055s, 975sh, 940s, 737w, 681, 664, 633, 508s, 460s.

Note: The band positions denoted by Zaitsev et al. (2013) as 3662, 3652, and 3645 cm^{-1} were determined by us at 3682, 3660, and 3642 cm^{-1}, respectively.

Sib165 Ferri-fluoro-katophorite $Na(CaNa)(Mg_4Fe^{3+})(AlSi_7O_{22})F_2$

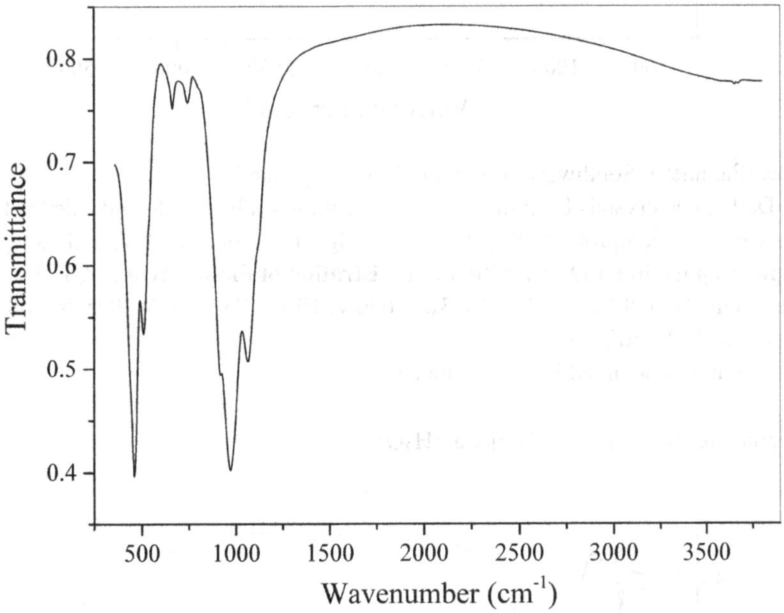

Origin: Bear Lake diggings, Monmouth township, Haliburton Co., Ontario, Canada (type locality).

Description: Black crystal. Fragment of holotype.

Kind of sample preparation and/or method of registration of the spectrum: KBr disc. Absorption.

Wavenumbers (cm^{-1}): 3671w, 3652w, 1115sh, 1065s, 971s, 922s, 805sh, 745, 664, 510s, 460s.

Note: The spectrum was obtained by N.V. Chukanov.

Sib166 Fluoro-pargasite $NaCa_2Mg_4Al(Al_2Si_6O_{22})(F,OH)_2$

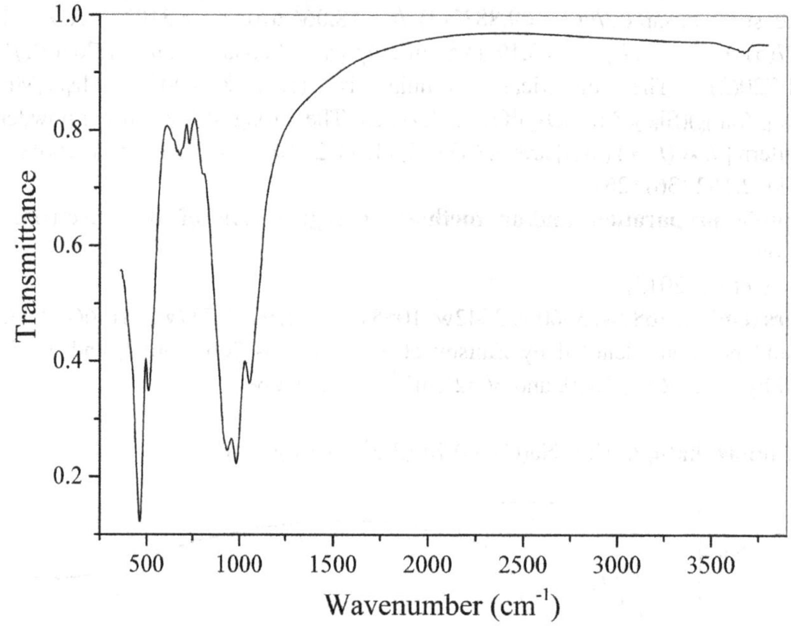

Origin: Pargas (Parainen), Southwestern Finland Region, Finland.

Description: Dark green crystals from the association with fluorphlogopite and calcite. The empirical formula is (electron microprobe): $(Na_{0.7}K_{0.3})Ca_{2.0}(Mg_{3.6}Fe_{0.7}Al_{0.6}Ti_{0.1})(Si_{6.4}Al_{1.6}O_{22})F_{1.7}(OH)_{0.3}$.

Kind of sample preparation and/or method of registration of the spectrum: KBr disc. Absorption.

Wavenumbers (cm^{-1}): 3688w, 3674w, 3653w, 3630w, 1056s, 984s, 935s, 925sh, 808, 734, 695sh, 681, 667, 650sh, 510s, 462s.

Note: The spectrum was obtained by N.V. Chukanov.

Sib167 Tobermorite $[Ca_4Si_6O_{17}\cdot 2H_2O](Ca\cdot 3H_2O)$

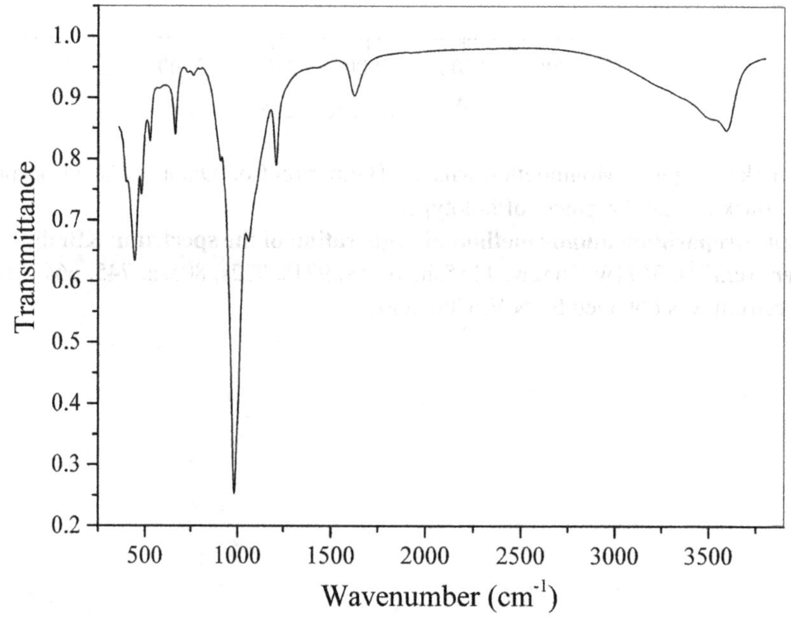

Origin: Pervomaiskiy quarry, Crimea, Russia.

Description: White fibrous aggregate from the association with prehnite and laumontite. Investigated by I.S. Lykova. The empirical formula is (electron microprobe): $Ca_{4.68}Si_6O_{15}(O,OH)_2 \cdot nH_2O$.

Kind of sample preparation and/or method of registration of the spectrum: KBr disc. Absorption.

Wavenumbers (cm^{-1}): 3594, 3525sh, 3355sh, 3240sh, 1624, 1420w, 1211, 1061, 983s, 910, 798w, 764w, 732w, 665, 527, 480, 444s, 402.

Note: The spectrum was obtained by N.V. Chukanov.

Sir199 Colinowensite $BaCuSi_2O_6$

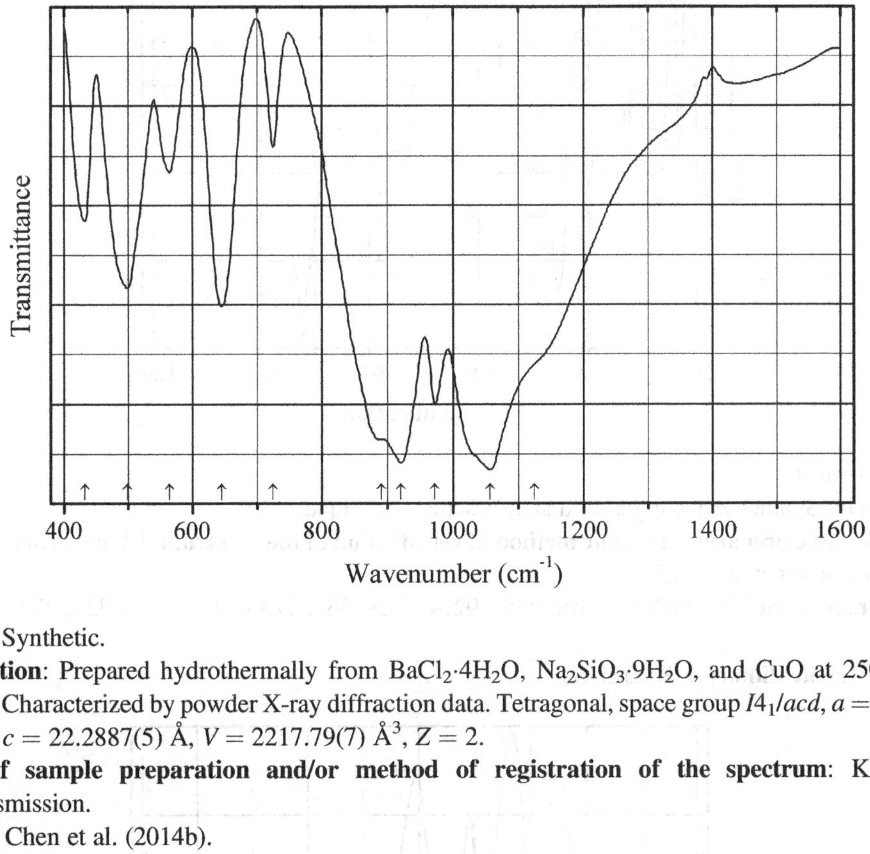

Origin: Synthetic.

Description: Prepared hydrothermally from $BaCl_2 \cdot 4H_2O$, $Na_2SiO_3 \cdot 9H_2O$, and CuO at 250 °C for 48 h. Characterized by powder X-ray diffraction data. Tetragonal, space group $I4_1/acd$, $a = 9.97511$ (17), $c = 22.2887(5)$ Å, $V = 2217.79(7)$ Å^3, $Z = 2$.

Kind of sample preparation and/or method of registration of the spectrum: KBr disc. Transmission.

Source: Chen et al. (2014b).

Wavenumbers (cm^{-1}): 1125sh, 1057s, 972, 920s, 891sh, 724, 644, 564, 498, 432.

Note: The wavenumbers were partly determined by us based on spectral curve analysis of the published spectrum.

Sir200 Gerenite-(Y) $(Ca,Na)_2Y_3Si_6O_{18} \cdot 2H_2O$

Origin: Strange Lake peralkaline complex, Quebec-Labrador boundary, Canada (type locality).

Description: Creamy aggregate. Holotype sample. Triclinic, $a = 9.245(5)$, $b = 9.684(6)$, $c = 5.510$ (3) Å, $\alpha = 97.44(6)°$, $\beta = 100.40(6)°$, $\gamma = 116.70(6)°$. $D_{calc} = 3.46$ g/cm^3. Optically biaxial (–), $\alpha = 1.602(l)$, $\beta = 1.607(1)$, $\gamma = 1.611(1)$, $2V = 73(3)°$.

Kind of sample preparation and/or method of registration of the spectrum: No data.
Source: Jambor et al. (1998).
Wavenumbers (cm^{-1}): 3480, 1655, 943, 668, 430, 340, 328, 306.
Note: Some other bands overlap with strong bands of admixed quartz.

Sir201 Pabstite $BaSnSi_3O_9$

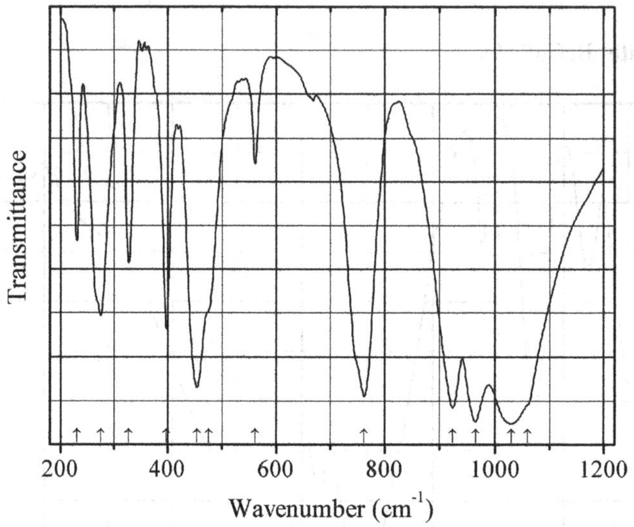

Origin: Synthetic.
Description: Synthesized using a solid-state reaction technique.
Kind of sample preparation and/or method of registration of the spectrum: KI disc. Transmission.
Source: Choisnet et al. (1975).
Wavenumbers (cm^{-1}): 1060sh, 1030s, 965s, 923s, 762s, 561, 476sh, 455s, 397, 327, 275, 231.

Sir202 Wadeite dimorph $K_2ZrSi_3O_9$

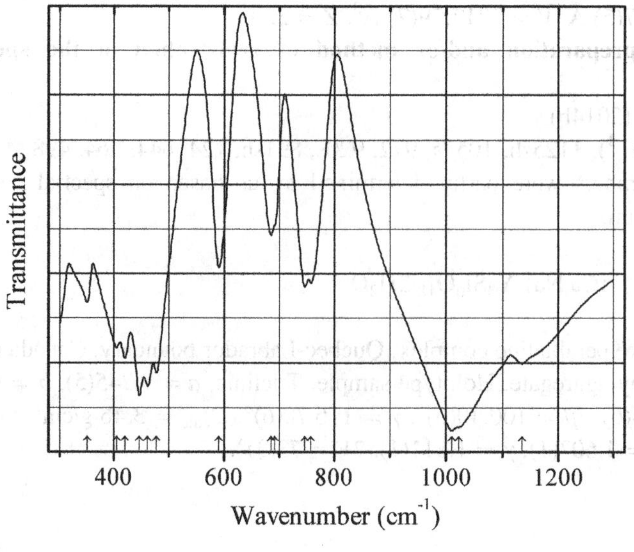

Origin: Synthetic.

Description: Synthesized from a glass having wadeite composition by repeatedly crushing and heating at 973–993 K. Characterized by powder X-ray diffraction data. The structure contains four-membered rings of SiO_4 tetrahedra.

Kind of sample preparation and/or method of registration of the spectrum: KBr disc. Transmission.

Source: Geisinger et al. (1987).

Wavenumbers (IR, cm^{-1}): 1135, 1022sh, 1010s, 758, 747, 692, 685, 591, 477, 461, 446s, 420, 405, 351.

Note: The wavenumbers were partly determined by us based on spectral curve analysis of the published spectrum. In the cited paper, Raman spectrum is given.

Wavenumbers (Raman, cm^{-1}): 1154, 1074, 1014w, 800, 744w, 539, 524, 511s, 419s, 400, 339, 311, 304, 215, 179, 149, 111, 90, 67, 52.

Sir203 Wadeite Rb analogue $Rb_2TiSi_3O_9$

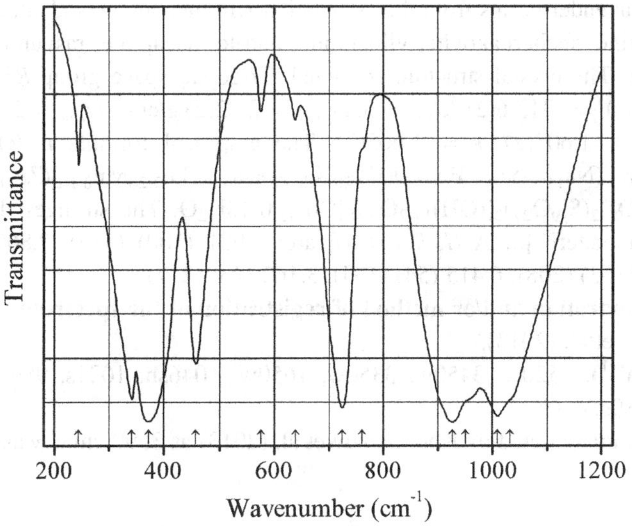

Origin: Synthetic.

Description: Prepared from Rb_2CO_3, TiO_2, and SiO_2 using a solid-state reaction technique.

Kind of sample preparation and/or method of registration of the spectrum: KI disc. Transmission.

Source: Choisnet et al. (1975).

Wavenumbers (cm^{-1}): 1033sh, 1010s, 950sh. 926s, 760sh, 724s, 639w, 576, 458, 372s, 342s, 245.

Note: The wavenumbers were partly determined by us based on spectral curve analysis of the published spectrum.

Sir204 Davinciite $Na_{12}K_3Ca_6Fe^{2+}{}_3Zr_3(Si_{26}O_{73}OH)Cl_2$

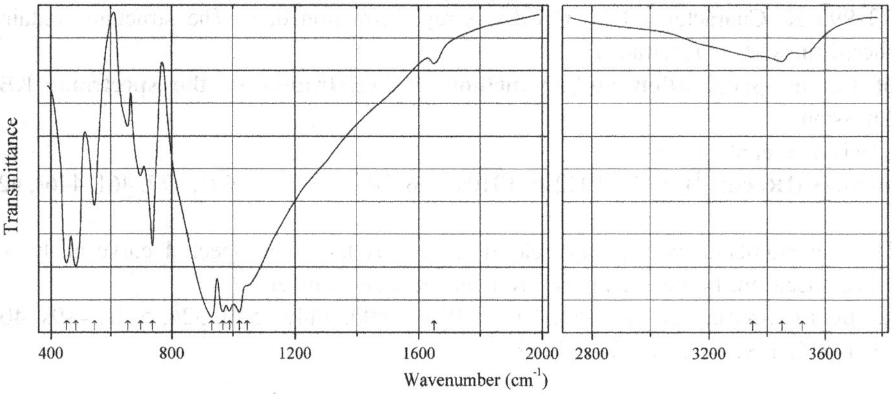

Origin: Rasvumchorr Mt., Khibiny alkaline complex, Kola peninsula, Murmansk region, Russia (type locality).

Description: Dark lavender grains from the association with nepheline, sodalite, potassium feldspar, delhayelite, aegirine, shcherbakovite, villiaumite, natrite, nacaphite, rasvumite, and djerfisherite. Holotype sample. The crystal structure is solved. Trigonal, space group *R*3*m*, $a = 14.2956(2)$, $c = 30.0228(5)$ Å, $V = 5313.6(2)$ Å^3, $Z = 3$. $D_{meas} = 2.82(2)$ g/cm^3, $D_{calc} = 2.848$ g/cm^3. Optically uniaxial (+), $\omega = 1.603(2)$, $\varepsilon = 1.605(2)$. The empirical formula is (electron microprobe, H_2O calculated): $(Na_{11.75}Sr_{0.29}Ba_{0.03})(K_{2.28}Na_{0.72})Ca_{5.99}(Fe_{2.26}Mn_{0.16})(Zr_{2.80}Ti_{0.15}Hf_{0.03}Nb_{0.02})(Si_{1.96}Al_{0.04})(Si_3O_9)_2(Si_9O_{27})_2[(OH)_{1.42}O_{0.58}]Cl_{1.62}{\cdot}0.48H_2O$. The strongest lines of the powder X-ray diffraction pattern [*d*, Å (*I*, %) (*hkl*)] are: 2.981 (100) (315), 2.860 (96) (404), 4.309 (66) (205), 3.207 (63) (208), 6.415 (54) (104), 3.162 (43) (217).

Kind of sample preparation and/or method of registration of the spectrum: No data.

Source: Khomyakov et al. (2013).

Wavenumbers (cm^{-1}): 3520w, 3450w, 3350w, 1650w, 1046sh, 1021s, 989s, 967s, 930s, 738s, 699, 657, 544, 480s, 450s.

Note: The band position denoted by Khomyakov et al. (2013) as 3590 cm^{-1} was determined by us at 3520 cm^{-1}.

Sir205 Rippite $K_2(Nb,Ti)_2(Si_4O_{12})O(O,F)$

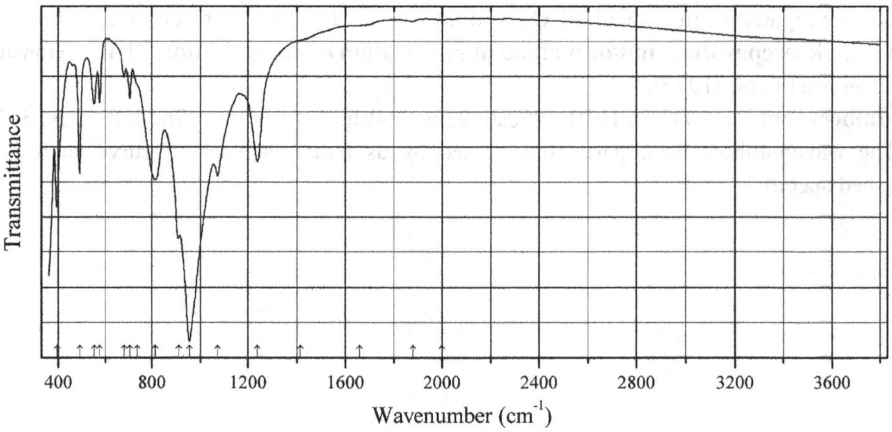

Origin: Chuktukon carbonatite massif, Chadobets upland, southern Siberian craton, Krasnoyarsky Kray, Russia (type locality).

Description: Prismatic crystals from the association with pyrochlore-supergroup minerals, quartz, goethite, baryte, monazite-(Ce), K-feldspar, fluorite, fluorapatite, Ca-*REE*-fluorcarbonates, Nb-rich rutile, olekminskite, aegirine, etc. Holotype sample. The crystal structure is solved. Tetragonal, space group *P4bm*, $a = 8.7388(2)$, $c = 8.1277(2)$ Å, $V = 620.69(2)$ Å^3, $Z = 2$. $D_{meas} = 3.17(2)$ g/cm^3, $D_{calc} = 3.198$ g/cm^3. Optically uniaxial (+), $\omega = 1.738(2)$, $\varepsilon = 1.747(2)$. The empirical formula is (electron microprobe): $K_{2.00}(Nb_{1.88}Ti_{0.10}Zr_{0.02})(Si_{4.00}O_{12})O(O_{0.88}F_{0.12})$. The strongest lines of the powder X-ray diffraction pattern [*d*, Å (*I*, %) (*hkl*)] are: 6.205 (100) (001), 4.383 (83) (020), 4.082 (90) (002), 3.530 (87) (121), 2.985 (81) (022), 2.822 (70) (122), 2.768 (99) (130).

Kind of sample preparation and/or method of registration of the spectrum: KBr disc. Absorption.

Wavenumbers (cm^{-1}): 1999w, 1878w, 1660w, 1415sh, 1238s, 1073s, 954s, 910s, 813s, 740sh, 707, 683, 577, 552, 489s, 396s.

Note: The spectrum was obtained by N.V. Chukanov.

Sir206 Dualite $Na_{30}(Ca,Na,Ce,Sr)_{12}(Na,Mn,Fe,Ti)_6Zr_3Ti_3MnSi_{51}O_{144}(OH,H_2O,Cl)_9$

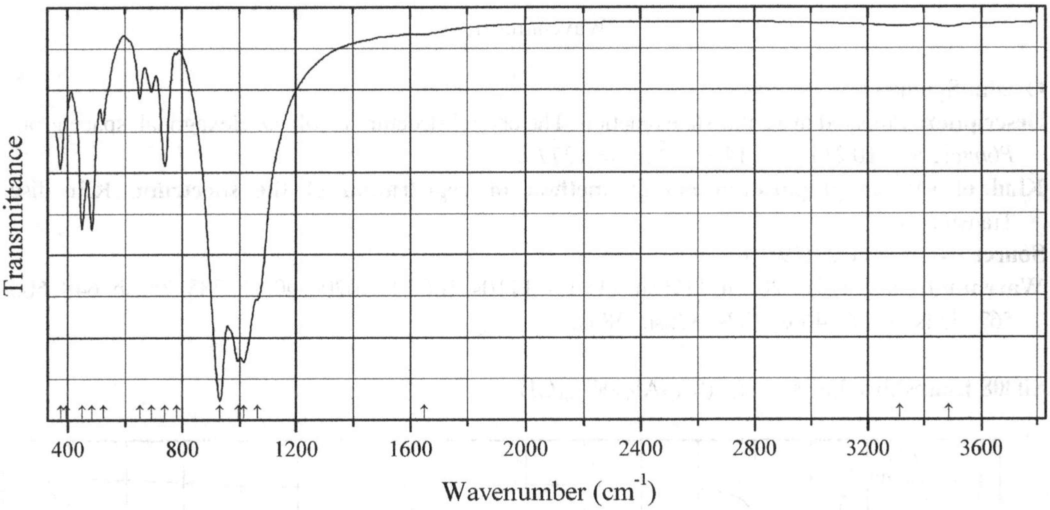

Origin: Alluaiv Mt., Lovozero alkaline complex, Kola peninsula, Murmansk region, Russia (type locality).

Description: Yellow anhedral grains from the association with K-Na feldspar, nepheline, sodalite, cancrinite, aegirine, alkaline amphibole, eudialyte, lovozerite, lomonosovite, vuonnemite, lamprophyllite, sphalerite, and villiaumite. Holotype sample. The crystal structure is solved. Trigonal, space group *R3m*, $a = 14.153(9)$, $c = 60.72(5)$ Å, $V = 10{,}533(22)$ Å^3, $Z = 3$. $D_{meas} = 2.84$ (3) g/cm^3, $D_{calc} = 2.814$ g/cm^3. Optically uniaxial (+), $\omega = 1.610(1)$, $\varepsilon = 1.613(1)$. The crystal-chemical formula is $(Na_{29.79}Ba_{0.1}K_{0.10})_{\Sigma 30}(Ca_{8.55}Na_{1.39}REE_{1.27}Sr_{0.79})(Na_{3.01}Mn_{1.35}Fe^{2+}_{0.87}Ti_{0.77})$ $(Zr_{2.61}Nb_{0.39})(Ti_{2.52}Nb_{0.48})(Mn_{0.82}Si_{0.18})(Si_{50.77}Al_{0.23})O_{144}[(OH)_{6.54}(H_2O)_{1.34}\cdot Cl_{0.98}]$. The strongest lines of the powder X-ray diffraction pattern [*d*, Å (*I*, %) (*hkl*)] are: 7.11 (40) (110), 4.31 (50) (0.2.10), 2.964 (100) (1.3.10), 2.839 (90) (048), 2.159 (60) (2.4.10, 0.4.20), 1.770 (60) (2.4.22, 4.0.28, 440), 1362 (50) (5.5.12, 3.0.42).

Kind of sample preparation and/or method of registration of the spectrum: KBr disc. Absorption.

Wavenumbers (cm^{-1}): 3485w, 3314w, 1650w, 1063s, 1015s, 995s, 931s, 780w, 740, 694, 652, 525, 485, 450, 395sh, 375.

Note: The spectrum was obtained by N.V. Chukanov.

Sir207 Roedderite Na-free analogue $K_2Mg_2(Mg_3Si_{12})O_{30}$

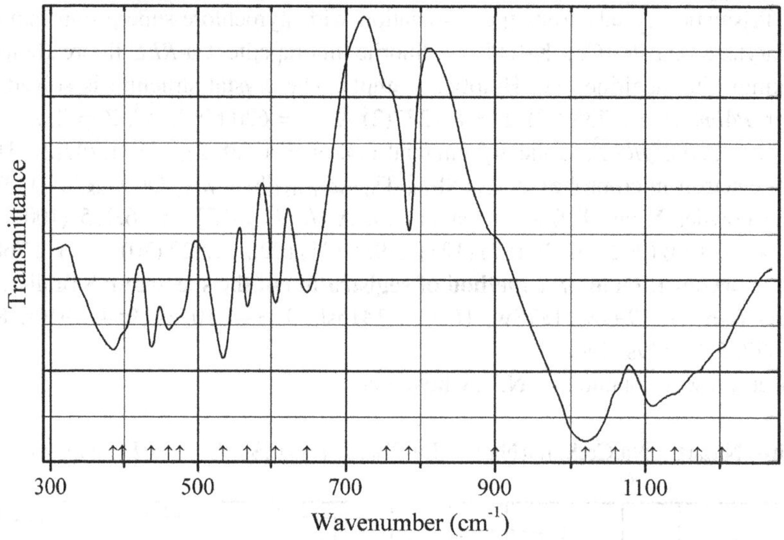

Origin: Synthetic.

Description: Prepared in a solid-state reaction. The crystal structure is solved. Hexagonal, space group *P6/mcc*, $a = 10.211$, $c = 14.152$ Å, $V = 1277.8$ Å^3.

Kind of sample preparation and/or method of registration of the spectrum: KBr disc. Transmission.

Source: Nguyen et al. (1980).

Wavenumbers (cm^{-1}): 1203sh, 1170sh, 1140sh, 1110s, 1060sh, 1020s, 900sh, 785, 755sh, 648, 606, 567, 534s, 476sh, 460s, 437s, 398sh, 385s.

Sil308 Luanshiweiite $KLiAl_{1.5}(Si_{3.5}Al_{0.5})O_{10}(OH)_2$

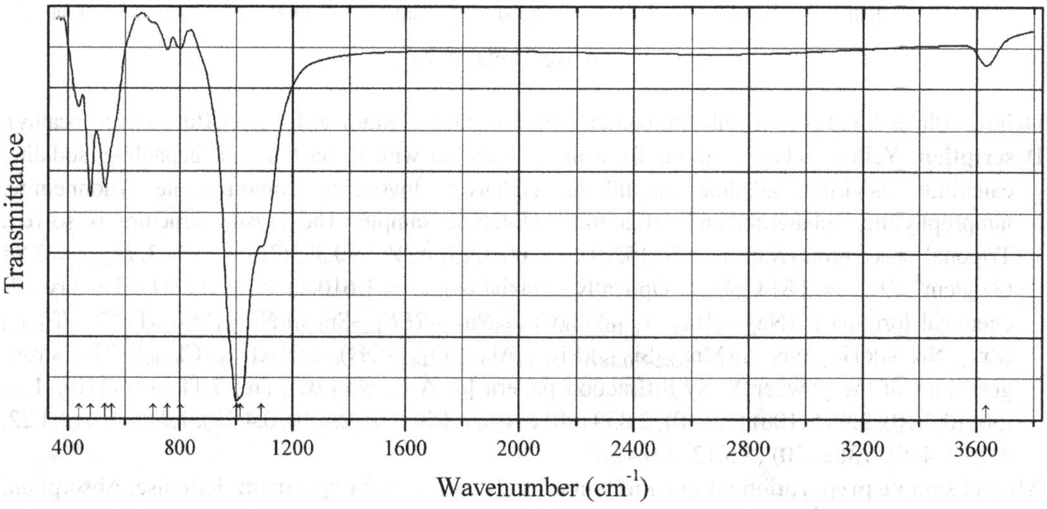

Origin: Ognyovskoe Ta deposit, Ognyovka-Bakennoe pegmatite field, Kalba ridge, Kazakhstan.

Description: Lilac grains from the association with albite, microcline, and quartz. Investigated by A.V. Kasatkin. The empirical formula is $(K_{0.86}Na_{0.05}Cs_{0.02})(Li_{1.10}Al_{1.51}Mn_{0.09})(Si_{3.26}Al_{0.74}O_{10})(OH)_{1.53}F_{0.47}$.

Kind of sample preparation and/or method of registration of the spectrum: KBr disc. Absorption.

Wavenumbers (cm^{-1}): 3634, 1087s, 1015sh, 1002s, 801, 753, 700sh, 555sh, 531s, 479s, 438.

Note: The spectrum was obtained by N.V. Chukanov.

Sil309 Sodium lithium aluminosilicate $Na_3Li_2(AlSi_2O_8)$ $Na_3Li_2(AlSi_2O_8)$

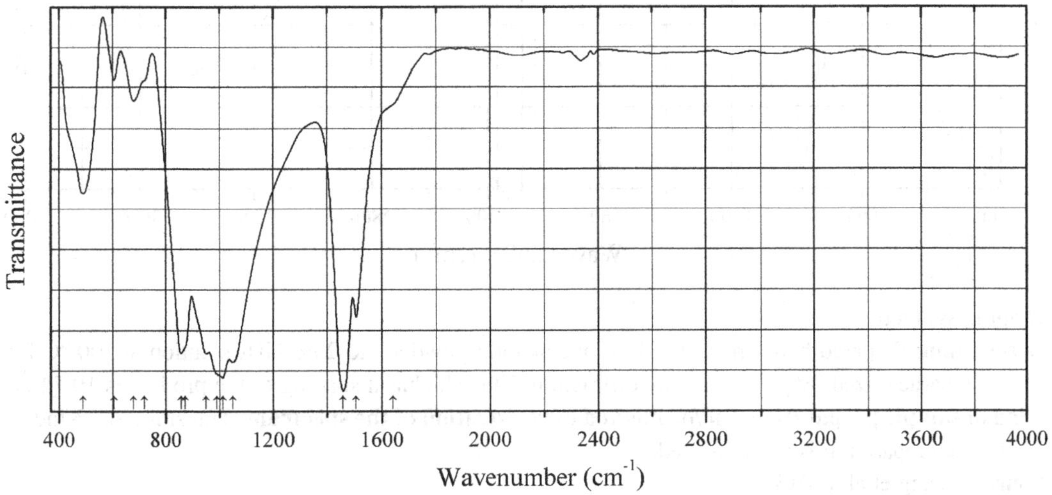

Origin: Synthetic.

Description: Synthesized by heating a stoichiometric mixture of Li_2CO_3, Na_2CO_3, Al_2O_3, and SiO_2 at 630 °C for 48 h with several intermediate grindings and mixings. Characterized by powder X-ray diffraction data. The crystal structure is solved. Orthorhombic, space group *Cmca*, $a = 14.1045(19)$ Å, $b = 14.7054(19)$ Å, $c = 7.0635(9)$ Å, $V = 1465.1(3)$ Å^3, $Z = 8$. $D_{calc} = 2.666$ g/cm^3. The structure is based on a 2D layer, which is composed of $[Al_2Si_2O_{12}]$ rings and SiO_4 tetrahedra.

Kind of sample preparation and/or method of registration of the spectrum: KBr disc. Transmission.

Source: Han et al. (2013).

Wavenumbers (cm^{-1}): 1640sh, 1505, 1458s, 1051s, 1012s, 990sh, 951s, 872sh, 858s, 720sh, 680, 607w, 491.

Note: The wavenumbers were partly determined by us based on spectral curve analysis of the published spectrum. The band at 1458 cm^{-1} (erroneously assigned by the authors to the Si–O–Al bridges) indicates that the sample is contaminated by a carbonate.

Sil310 Tobelite hydrated variety $(NH_4,H_2O)Al_2[(Si,Al)_4O_{10}](OH)_2{\cdot}nH_2O$ (?)
"Ammonium illite"

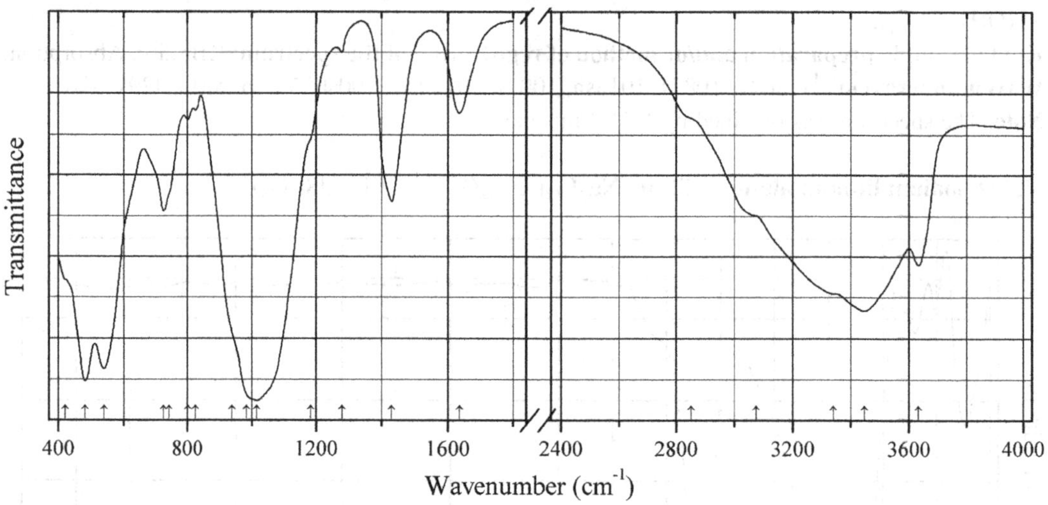

Origin: Synthetic.

Description: Prepared hydrothermally from metakaolin powder and 25% NH_3 solution at 300 °C for 1 h. Characterized by powder X-ray diffraction data. The basal spacing of the product is 10.74 Å.

Kind of sample preparation and/or method of registration of the spectrum: Transmission. Kind of sample preparation is not indicated.

Source: Wang et al. (2013).

Wavenumbers (cm^{-1}): 3634, 3446, 3340sh, 3075sh, 2850sh, 1635, 1432, 1278w, 1182sh, 1014s, 982sh, 940sh, 825w, 801w, 741sh, 723, 538s, 481s, 421sh.

Note: The wavenumbers were partly determined by us based on spectral curve analysis of the published spectrum.

Sil311 Falcondoite $Ni_4Si_6O_{15}(OH)_2{\cdot}6H_2O$

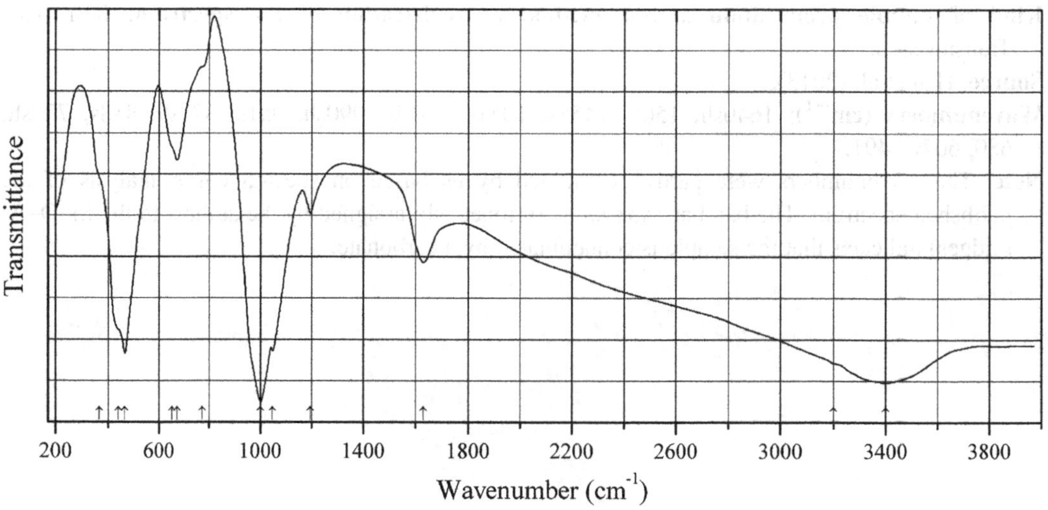

Origin: Falcondo Mine, Bonao, La Vega Province, Dominican Republic (type locality).

Description: Green sample confirmed by powder X-ray diffraction data.

Kind of sample preparation and/or method of registration of the spectrum: Nujol mull. Transmission.

Source: Reddy et al. (1987).

Wavenumbers (cm^{-1}): 3400, 3200, 1630, 1195, 1048, 1000s, 775sh, 674, 651sh, 465s, 440sh, 365sh.

Note: The wavenumbers were partly determined by us based on spectral curve analysis of the published spectrum.

Sil312 Imogolite $Al_2SiO_3(OH)_4$

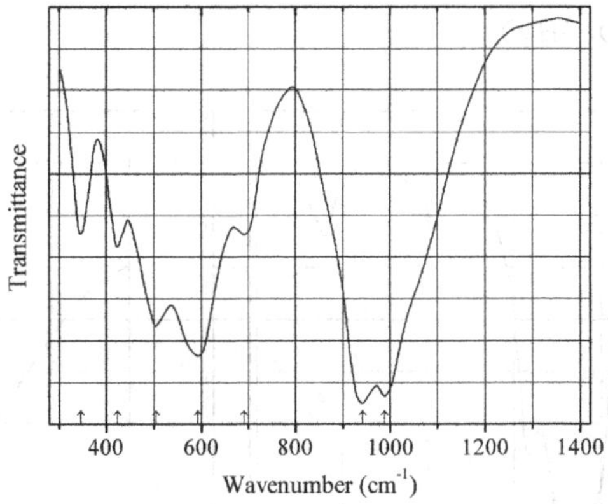

Origin: Natural sample; the locality is not indicated.

Kind of sample preparation and/or method of registration of the spectrum: KBr disc. Transmission.

Source: Farmer et al. (1979).

Wavenumbers (cm^{-1}): 989s, 941s, 691, 593s, 504, 423, 346.

Note: The wavenumbers were determined by us based on spectral curve analysis of the published spectrum.

Sil313 Imogolite $Al_2SiO_3(OH)_4$

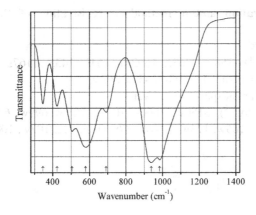

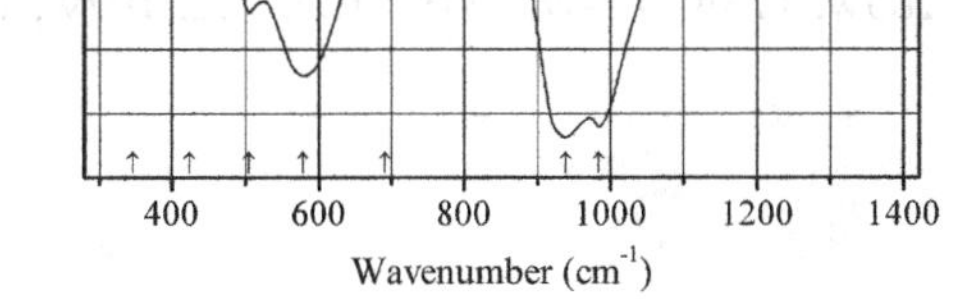

Origin: Synthetic.

Description: Amorphous product of interaction between hydroxyaluminium and orthosilicic acid in dilute aqueous solutions of pH < 5.

Kind of sample preparation and/or method of registration of the spectrum: KBr disc. Transmission.

Source: Farmer et al. (1979).

Wavenumbers (cm^{-1}): 985s, 938s, 691, 579s, 505, 423, 346.

Note: The wavenumbers were determined by us based on spectral curve analysis of the published spectrum.

Sil314 Kanemite $HNaSi_2O_5{\cdot}3H_2O$

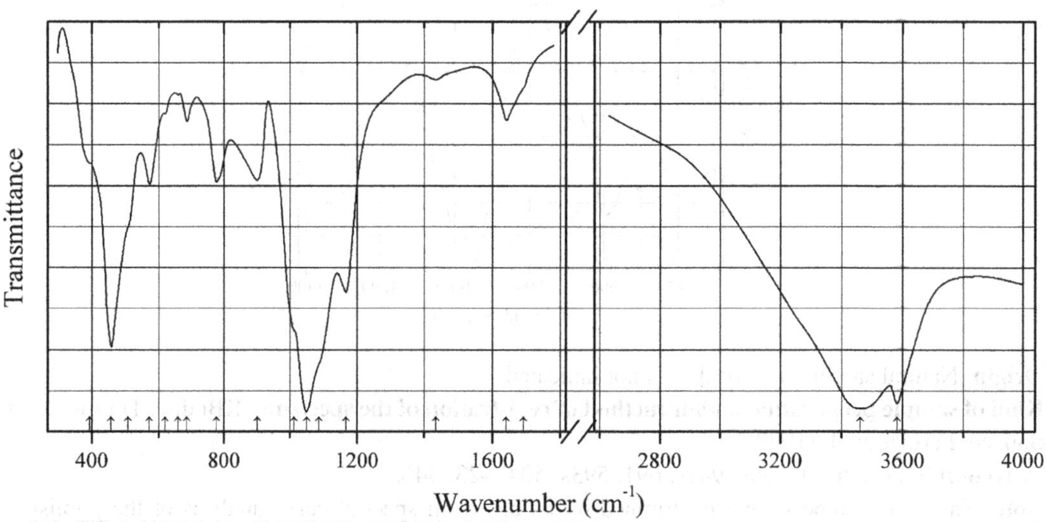

Origin: Synthetic.

Description: Synthesized according to known methods. Characterized by powder X-ray diffraction data.

Kind of sample preparation and/or method of registration of the spectrum: KBr disc. Transmission.

Source: Huang et al. (1998).

Wavenumbers (IR, cm^{-1}): 3582, 3463s, 1692sh, 1642, 1433, 1167s, 1087sh, 1049s, 1012sh, 899, 777, 687w, 661w, 619sh, 571, 508sh, 457s, 393sh.

Note: In the cited paper, Raman spectrum is given.

Wavenumbers (Raman, cm^{-1}): 3150w, 1060s, 1015s, 788w, 699w, 646w, 503w, 489w, 465s, 419, 372, 285, 261, 237, 185w, 173w, 154w, 137w, 129, 122, 107w, 100w.

Sil315 Margarite $CaAl_2Si_2Al_2O_{10}(OH)_2$

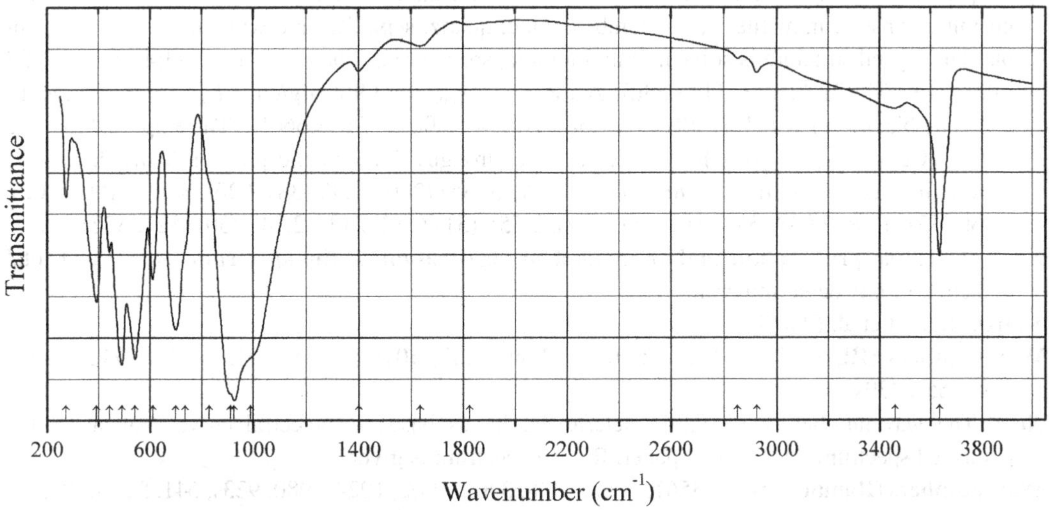

Origin: Enontekiö, northern Finland.

Description:A Li- and Be-poor variety.

Kind of sample preparation and/or method of registration of the spectrum: KBr disc. Transmission.

Source: Lahti and Saikkonen (1985).

Wavenumbers (cm^{-1}): 3634, 3461w, 1825w, 1634w, 1400w, 987sh, 925s, 910sh, 828sh, 734sh, 698s, 609, 541s, 490s, 442, 392, 273.

Note: The wavenumbers were partly determined by us based on spectral curve analysis of the published spectrum. Weak bands in the range from 2800 to 3000 cm^{-1} correspond to the admixture of an organic substance.

Sil316 Plumbophyllite $Pb_2Si_4O_{10}{\cdot}H_2O$

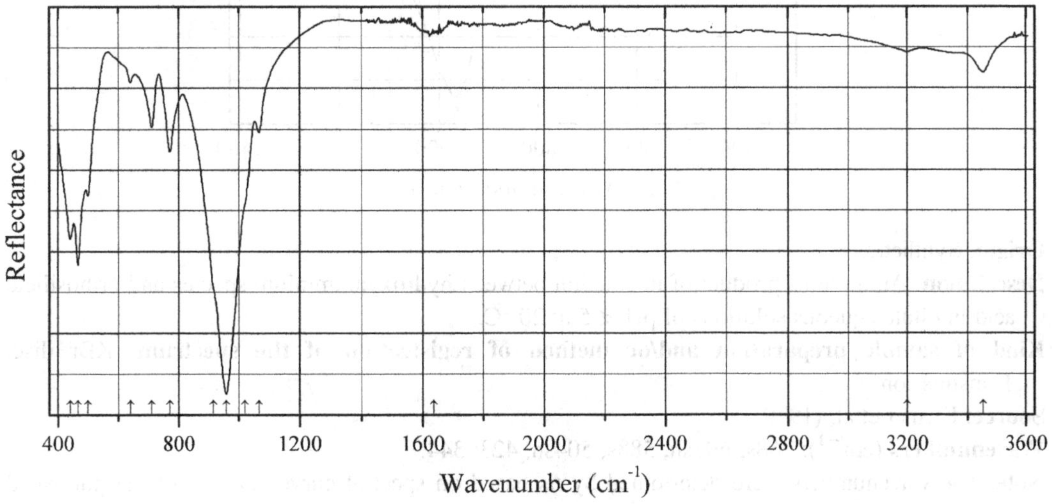

Origin: Blue Bell claims, near Baker, San Bernardino Co., California, USA (type locality).

Description: Pale blue prismatic crystals from the association with cerussite, chrysocolla, fluorite, goethite, gypsum, mimetite, opal, plumbotsumite, quartz, sepiolite, and wulfenite. Holotype sample. The crystal structure is solved. Orthorhombic, space group *Pbcn*, $a = 13.2083(4)$, $b = 9.7832(3)$, $c = 8.6545(2)$ Å, $V = 1118.33(5)$ Å^3, $Z = 4$. $D_{meas} = 3.96(5)$ g/cm^3, $D_{calc} = 3.940$ g/cm^3. Optically biaxial (+), $\alpha = 1.674(2)$, $\beta = 1.684(2)$, $\gamma = 1.708(2)$, $2V = 66(2)°$. The empirical formula is $Pb_{1.79}Cu_{0.02}Si_{4.00}O_{9.62}(OH)_{0.38}·1.02H_2O$. The strongest lines of the powder X-ray diffraction pattern [*d*, Å (*I*, %) (*hkl*)] are: 7.88 (97) (110), 6.63 (35) (200), 4.90 (38) (020), 3.623 (100) (202), 3.166 (45) (130), 2.938 (57) (312, 411, 222), 2.555 (51) (132, 213), 2.243 (50) (521, 332).

Kind of sample preparation and/or method of registration of the spectrum: Attenuated total reflection of powdered mineral.

Source: Kampf et al. (2009b).

Wavenumbers (IR, cm^{-1}): 3452, 3200w, 1636w, 1067, 1019sh, 958s, 915sh, 771, 711, 640w, 499, 465s, 439s.

Note: The wavenumbers were partly determined by us based on spectral curve analysis of the published spectrum. In the cited paper, Raman spectrum is given.

Wavenumbers (Raman, cm^{-1}): 3561, 3468, 3338, 3209, 1062, 1024s, 980, 923s, 641, 506, 480, 347, 329, 251, 209, 146, 94s.

Sil317 Protoimogolite

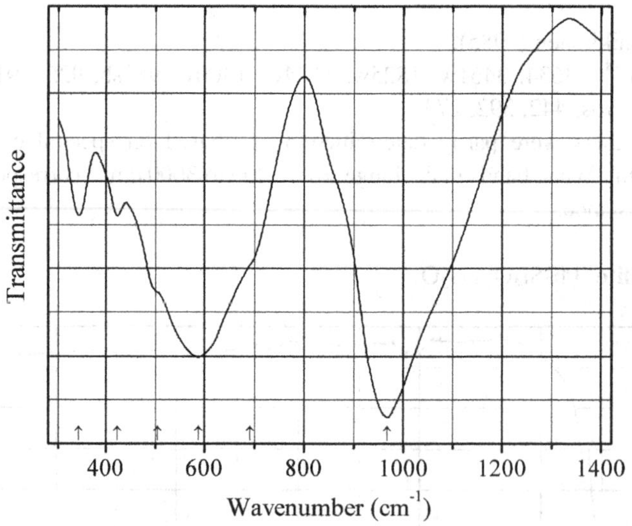

Origin: Synthetic.

Description: Amorphous product of interaction between hydroxyaluminium species and orthosilicic acid in dilute aqueous solutions of pH < 5 at 20 °C.

Kind of sample preparation and/or method of registration of the spectrum: KBr disc. Transmission.

Source: Farmer et al. (1979).

Wavenumbers (cm^{-1}): 968s, 691sh, 588s, 504sh, 423, 344.

Note: The wavenumbers were determined by us based on spectral curve analysis of the published spectrum.

Sil318 Pyrophyllite $Al_2Si_4O_{10}(OH)_2$

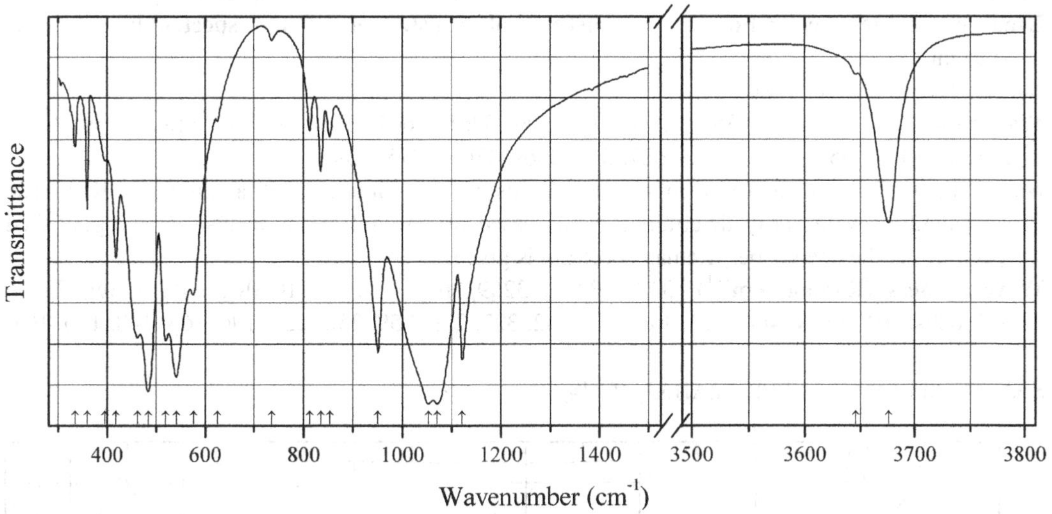

Origin: Nakamuraguchi, Yano-Shokozan area, Hiroshima Prefecture, Japan.

Description: Fine-grained aggregate. Characterized by powder X-ray diffraction data and electron microprobe analyses. Triclinic, $a = 5.16(1)$, $b = 8.96(1)$, $c = 9.37(2)$ Å, $\alpha = 90.8(2)°$, $\beta = 101.0(2)°$, $\gamma = 89.8(2)°$. The empirical formula is close to that of the end-member.

Kind of sample preparation and/or method of registration of the spectrum: KBr disc. Transmission.

Source: Wiewióra and Hida (1996).

Wavenumbers (cm^{-1}): 3676, 3646, 1121s, 1070s, 1052s, 950s, 853, 835, 813, 737w, 625w, 576, 541s, 520, 484s, 462, 418, 396w, 359, 334w.

Note: The wavenumbers were partly determined by us based on spectral curve analysis of the published spectrum.

Sil319 Silinaite $NaLiSi_2O_5 \cdot 2H_2O$

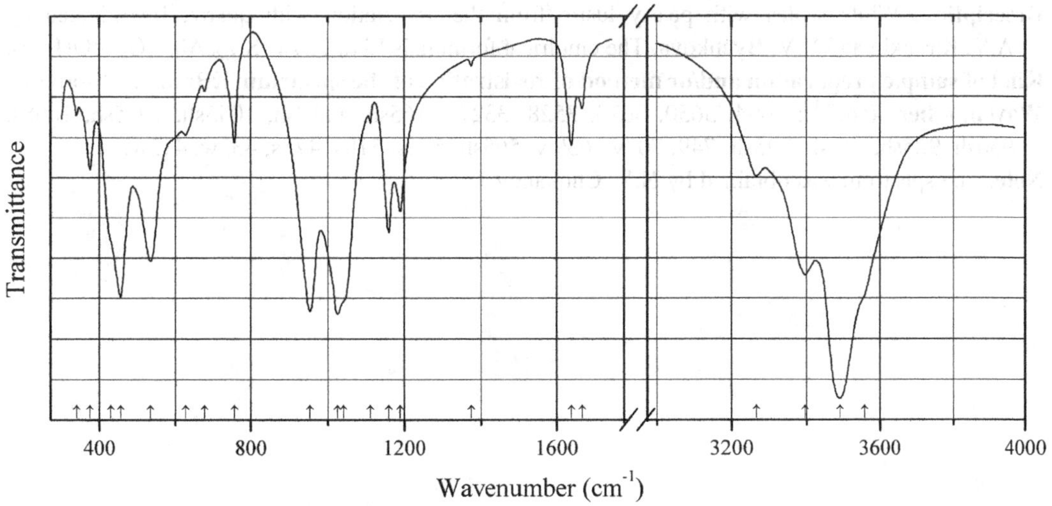

Origin: Synthetic.

Description: Characterized by powder X-ray diffraction data. Monoclinic, space group *A*2/*n*, $Z = 4$.

Kind of sample preparation and/or method of registration of the spectrum: KBr disc. Transmission.

Source: Huang et al. (1999a).

Wavenumbers (IR, cm^{-1}): 3558sh, 3490s, 3396, 3265, 1667, 1639, 1376w, 1189, 1159, 1111, 1042sh, 1026s, 955s, 757, 677w, 627, 535, 456s, 430sh, 375, 339w.

Note: The weak band at 1376 cm^{-1} corresponds to the admixture of a nitrate in KBr. The wavenumbers were partly determined by us based on spectral curve analysis of the published spectrum. In the cited paper, Raman spectrum is given.

Wavenumbers (Raman, cm^{-1}): 3472, 3389, 3259, 1652, 1068s, 1000w, 956, 773w, 734w, 635, 609s, 501, 490w, 466, 423, 408, 371, 342, 327, 279, 259, 250, 227, 196, 163, 148, 126, 107.

Sil320 Cookeite $(Al,Li)_3Al_2(Si,Al)_4O_{10}(OH)_8$

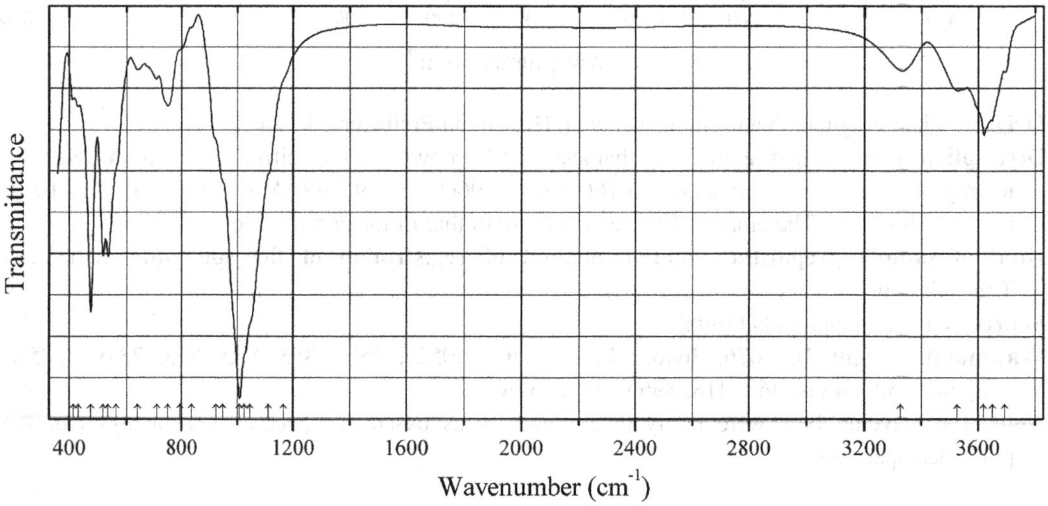

Origin: Nagol'nyi Kryazh, Lugansk region, Ukraine (type locality).

Description: White scales with pearly lustre from the association with quartz. Investigated by A.V. Kasatkin and Y.V. Bychkova. The empirical formula is $Li_{0.92}Al_{3.97}(Si_{3.29}Al_{0.71}O_{10})(OH,O)_8$.

Kind of sample preparation and/or method of registration of the spectrum: KBr disc. Absorption.

Wavenumbers (cm^{-1}): 3694, 3650, 3620, 3528, 3327, 1165sh, 1110sh, 1045sh, 1025sh, 1008s, 950sh, 925sh, 835sh, 795sh, 749, 707w, 639w, 565sh, 536s, 518s, 475s, 432w, 415w.

Note: The spectrum was obtained by N.V. Chukanov.

Sil321 Windhoekite Na-bearing variety $(Ca,Na)_2Fe^{3+}{}_{3+x}[(Si,Al)_8O_{20}](OH)_4 \cdot nH_2O$

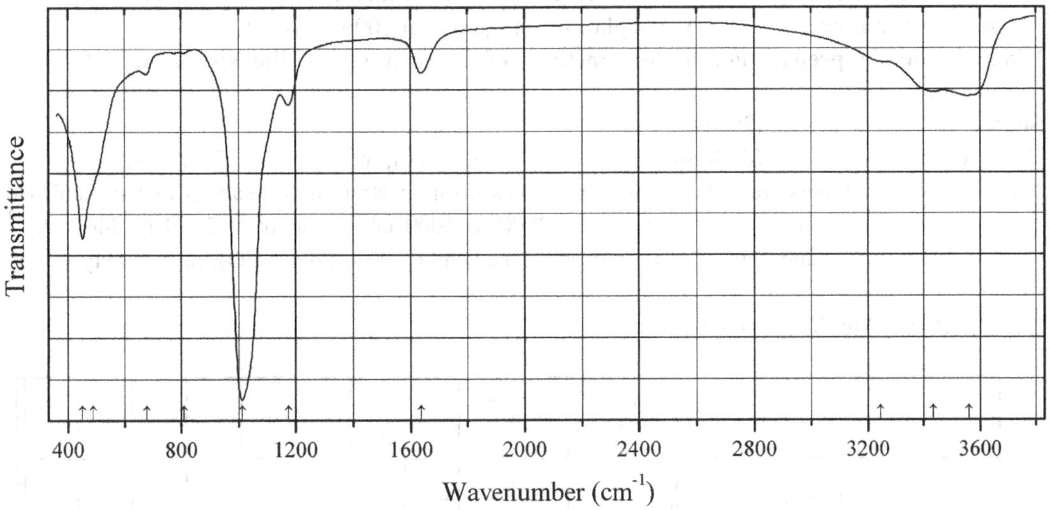

Origin: Ariskop Quarry, Aris, near Windhoek, Windhoek district, Khomas Region, Namibia (type locality).

Description: Clusters of brown acicular crystals from the association with fluorapofillite, aegirine, and microcline. The crystal structure is solved. Monoclinic, space group *C*2/*m*, a = 14.0626(3), b = 17.9007(8), c = 5.2527(2) Å, β = 104.4(6)°, V = 1280.3(2) Å^3, Z = 2. A specific feature of Na-rich windhoekite is the presence of the fivefold coordination M(4)-site occupied by Ca and Na in the atomic ratio 1:1, which is attached to a ribbon consisting of Fe^{3+}- and Ca-centered octahedra and sandwiched between two opposing tetrahedral ribbons. The empirical formula is (electron microprobe): $K_{0.08}Na_{0.42}Ca_{1.15}Fe^{3+}{}_{3.52}Mn_{0.41}Cr_{0.04}Ti_{0.10}(Si_{7.44}Al_{0.56})O_{20}(OH)_x \cdot nH_2O$.

Kind of sample preparation and/or method of registration of the spectrum: KBr disc. Absorption.

Wavenumbers (cm^{-1}): 3558, 3434, 3245w, 1636, 1175, 1012s, 810w, 678w, 490sh, 452s.

Note: The spectrum was obtained by N.V. Chukanov.

Sil322 Stevensite $(Ca,Na)_xMg_{3-y}Si_4O_{10}(OH)_2$

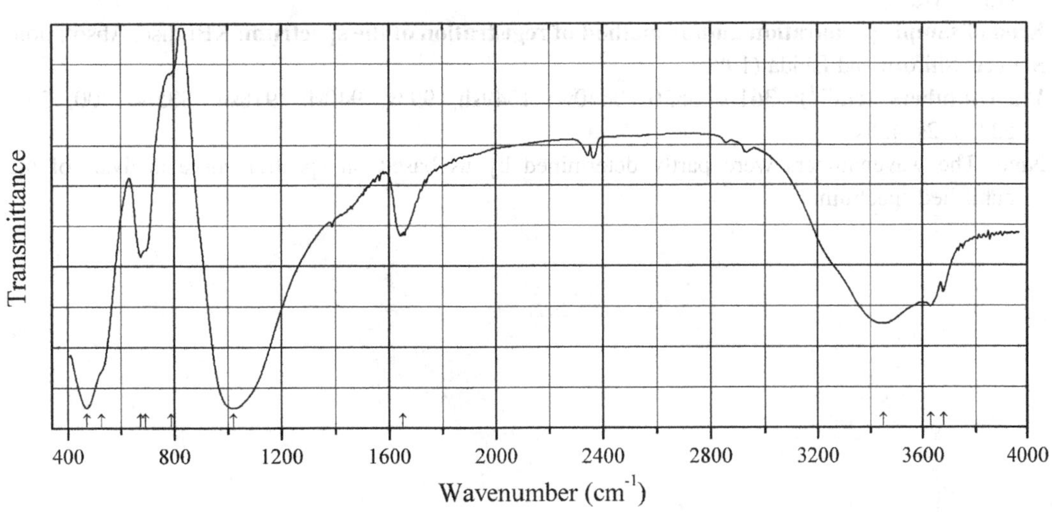

Origin: Ghassoul locality, Atlas Mts., Morocco.

Description: Characterized by powder X-ray diffraction data, thermal and chemical analyses. Basal spacing value is equal to 13.5 Å. Weight loss on ignition at 1000 °C is 7.19%.

Kind of sample preparation and/or method of registration of the spectrum: KBr disc. Transmission.

Source: Benhammou et al. (2011).

Wavenumbers (cm^{-1}): 3678, 3630, 3447, 1650, 1022s, 787sh, 692sh, 675, 527sh, 471s.

Note: The wavenumbers were determined by us based on spectral curve analysis of the published spectrum. Weak bands in the ranges from 2800 to 3000 cm^{-1} and from 2300 to 2400 cm^{-1} correspond to the admixture of an organic substance and to atmospheric CO_2, respectively.

Sil323 Sudoite $Mg_2Al_3(Si_3Al)O_{10}(OH)_8$

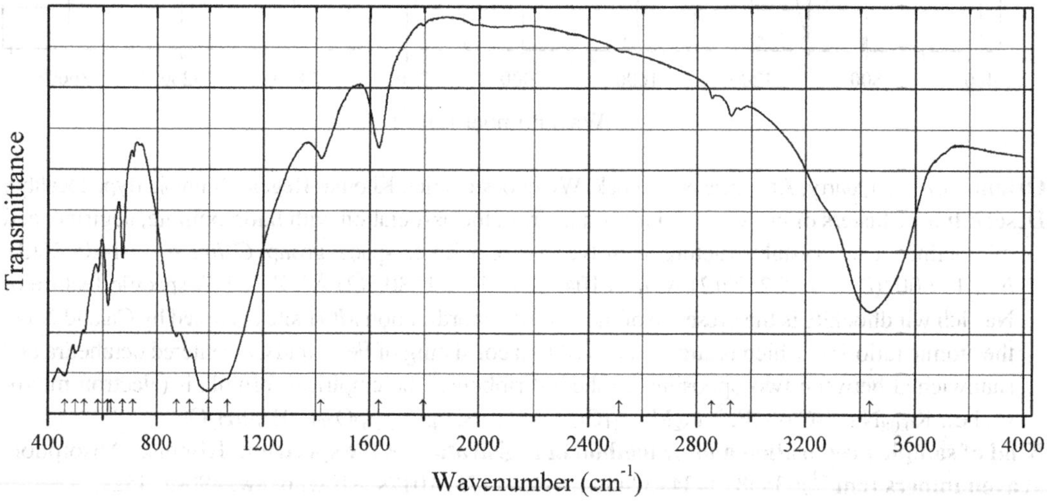

Origin: Kamikita, Aomori prefecture, Japan.

Description: The empirical formula is $(K_{0.01}Na_{0.115}Ca_{0.005}Mg_{1.385}Fe^{3+}{}_{0.035}Al_{3.29})(Si_{3.125}Al_{0.875})O_{10}(OH)_8$.

Kind of sample preparation and/or method of registration of the spectrum: KBr disc. Absorption.

Source: Shirozu and Ishida (1982).

Wavenumbers (cm^{-1}): 3610, 3530, 3340w, 1049sh, 993s, 940sh, 918sh, 835w, 700, 557, 527, 472s, 455s.

Note: The wavenumbers were partly determined by us based on spectral curve analysis of the published spectrum.

Sil324 "Tetraferrinontronite" $Na_{0.75}Fe_2(Si_{3.25}Fe_{0.75}O_{10})(OH)_2$

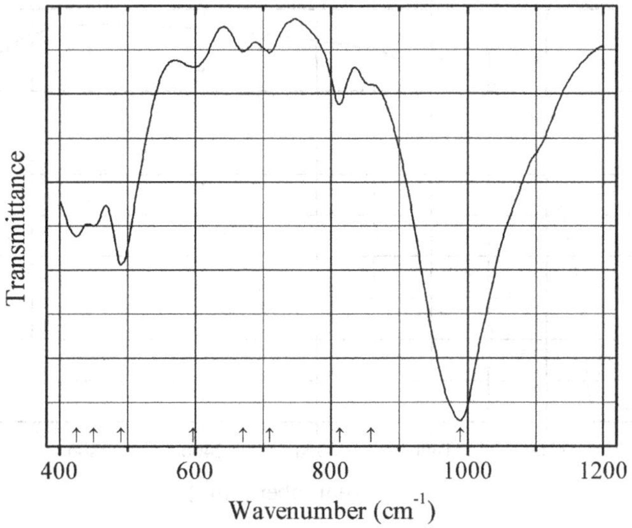

Origin: Synthetic.

Description: Synthesized hydrothermally. Characterized by powder X-ray diffraction data; *b* ≈ 9.25 Å.

Kind of sample preparation and/or method of registration of the spectrum: KBr disc. Absorpion.

Source: Petit et al. (2015).

Wavenumbers (cm^{-1}): 989, 859sh, 813, 710, 671, 597, 490, 450, 425.

Note: The wavenumbers were determined by us based on spectral curve analysis of the published spectrum. In the cited paper, the following wavenumbers are given: 3562, 993s, 851sh, 812, 710w, 669w, 598w, 490s, 449, 420.

Note: The compound is described as a smectite, but compositional and PXRD data correspond to a dioctahedral mica.

Sil325 Yangzhumingite $KMg_{2.5}Si_4O_{10}F_2$

Origin: A lamproitic dyke at the Kvaløya Island, North Norway.

Description: Grains from the association with low-Al phlogopite, apatite, Fe-bearing potassium feldspar, quartz, and alkali amphibole. The crystal structure is solved. Monoclinic, space group *C*2/*m*, $a = 5.2677(3)$, $b = 9.1208(5)$, $c = 10.1652(6)$ Å, $\beta = 100.010(4)°$. The empirical formula is $(K_{0.98}Na_{0.03})(Mg_{2.35}Fe_{0.23}Cr_{0.01}Ti_{0.02}Ni_{0.01}\square_{0.38})(Si_{3.66}Al_{0.34}O_{10})F_{1.16}(OH)_{0.84}$.

Kind of sample preparation and/or method of registration of the spectrum: Micro-FTIR measurement on a crystal mounted on glass capillary.

Source: Schingaro et al. (2014).

Wavenumbers (cm^{-1}): 3605w, 3586, 3550, 3537, 1638w.

Note: The wavenumbers were partly determined by us based on spectral curve analysis of the published spectrum.

Sil326 Montmorillonite $(Na,Ca)_{0.3}(Al,Mg)_2(Si_4O_{10})(OH)_2{\cdot}nH_2O$

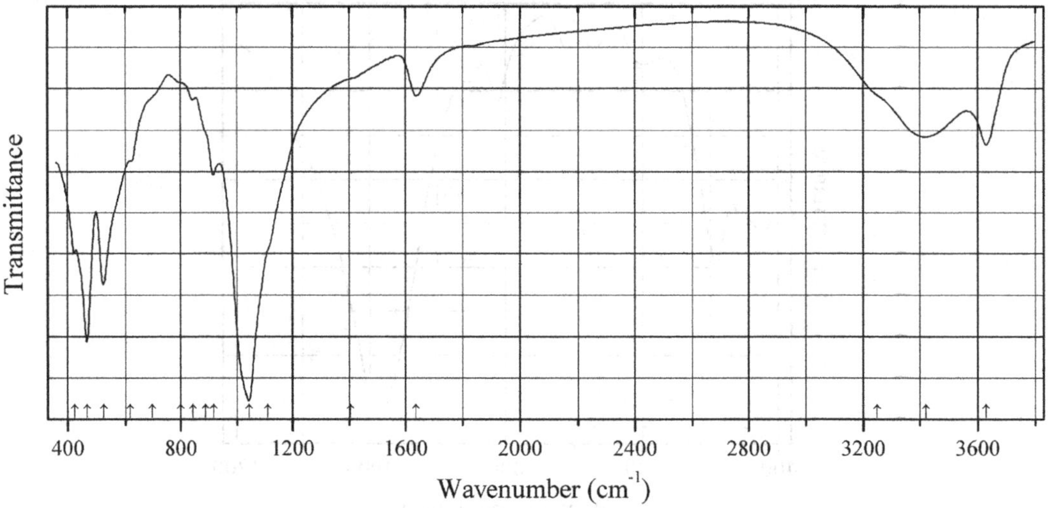

Origin: Voudia bentonite quarry, Milos Island, Greece.

Description: Gray nodule from the association with natroalunite. The empirical formula is (electron microprobe): $(Ca_{0.12}K_{0.06}Na_{0.04})(Al_{1.40}Mg_{0.35}Fe_{0.20}Ti_{0.04}Cr_{0.01})(Si_{3.83}Al_{0.17}O_{10}){\cdot}nH_2O$.

Kind of sample preparation and/or method of registration of the spectrum: KBr disc. Absorption.

Wavenumbers ($\mathbf{cm^{-1}}$): 3630, 3420, 3250sh, 1638, 1405sh, 1110sh, 1041s, 917, 890sh, 842w, 800sh, 695sh, 619, 525s, 467s, 424.

Note: The spectrum was obtained by N.V. Chukanov.

Sil328 Hydrobiotite $K(Mg,Fe^{2+})_6(Si,Al)_8O_{20}(OH)_4{\cdot}nH_2O$

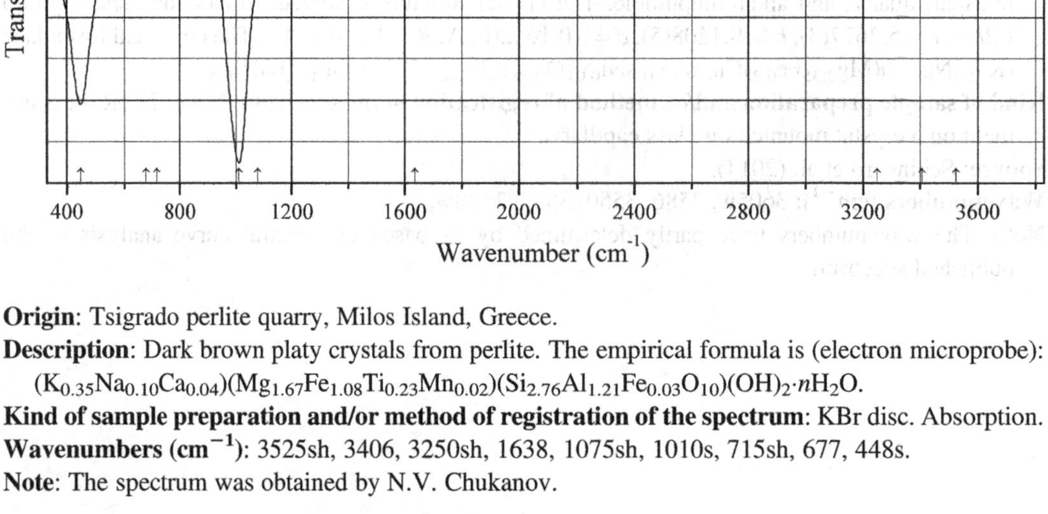

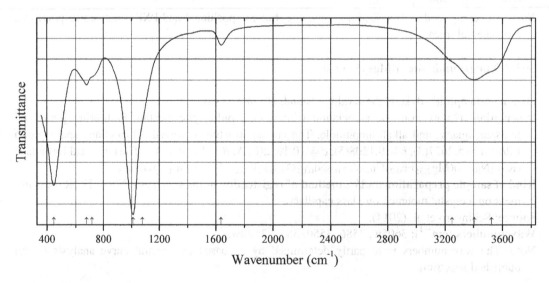

Origin: Tsigrado perlite quarry, Milos Island, Greece.

Description: Dark brown platy crystals from perlite. The empirical formula is (electron microprobe): $(K_{0.35}Na_{0.10}Ca_{0.04})(Mg_{1.67}Fe_{1.08}Ti_{0.23}Mn_{0.02})(Si_{2.76}Al_{1.21}Fe_{0.03}O_{10})(OH)_2{\cdot}nH_2O$.

Kind of sample preparation and/or method of registration of the spectrum: KBr disc. Absorption.

Wavenumbers ($\mathbf{cm^{-1}}$): 3525sh, 3406, 3250sh, 1638, 1075sh, 1010s, 715sh, 677, 448s.

Note: The spectrum was obtained by N.V. Chukanov.

Sil329 Cookeite $(Al,Li)_3Al_2(Si,Al)_4O_{10}(OH)_8$

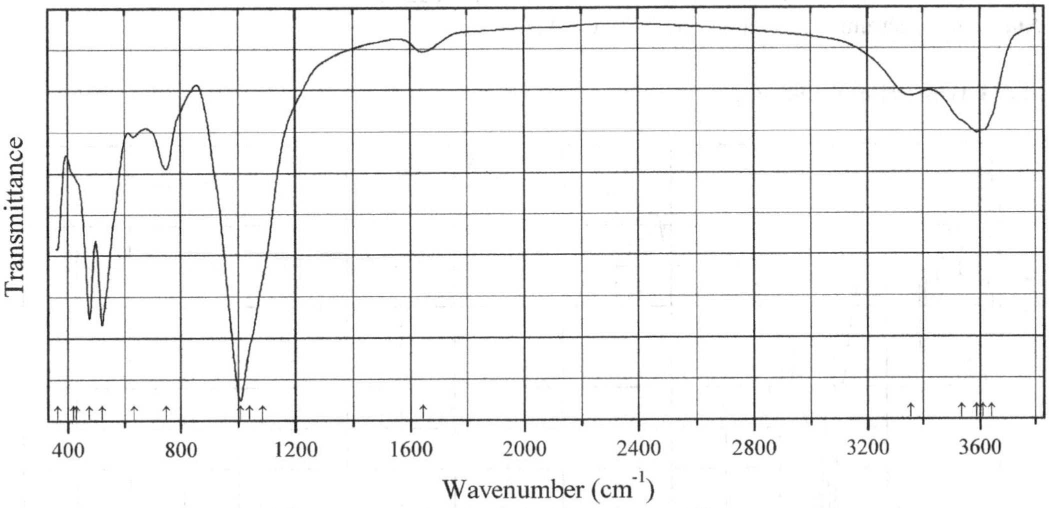

Origin: Coronel Murta, Minas Gerais, Brazil.

Description: Yellow split platelets from the association with F-rich muscovite and elbaite. A hydrated sample. The empirical formula is $(Li_{1.07}Al_{3.82}Fe_{0.11})(Si_{3.14}Al_{0.86}O_{10})(OH)_{7.8}F_{0.2}\cdot nH_2O$.

Kind of sample preparation and/or method of registration of the spectrum: KBr disc. Absorption.

Wavenumbers (cm^{-1}): 3640sh, 3610sh, 3588, 3535sh, 3355, 1644w, 1085sh, 1040sh, 1008s, 749, 635w, 521s, 476s, 430sh, 420sh, (363).

Note: The spectrum was obtained by N.V. Chukanov.

Sil330 Ferrisepiolite $(Fe^{3+},Fe^{2+},Mg)_4[(Si,Fe^{3+})_6O_{15}](O,OH)_2\cdot 6H_2O$

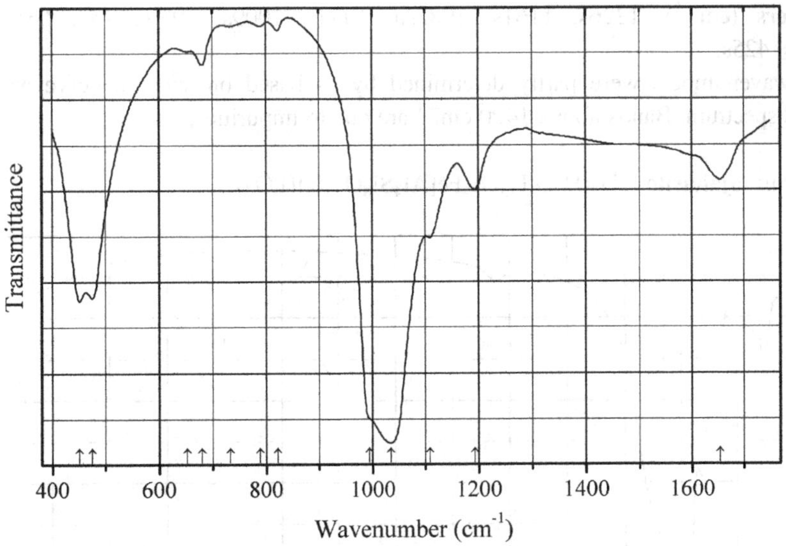

Origin: Flora (Selsurt) Mt., Lovozero alkaline complex, Kola Peninsula, Murmansk region, Russia.

Description: Beige fibrous aggregate from the association with yofortierite and narsarsukite. Characterized by semiquantitative electron microprobe analyses. Confirmed by the IR spectrum.

Kind of sample preparation and/or method of registration of the spectrum: KBr disc. Absorption.
Wavenumbers (cm^{-1}): 1653, 1193, 1108, 1034s, 995sh, 822w, 789w, 732w, 680, 652, 475s, 451s.
Note: The spectrum was obtained by N.V. Chukanov.

Sil331 Wesselsite $SrCuSi_4O_{10}$

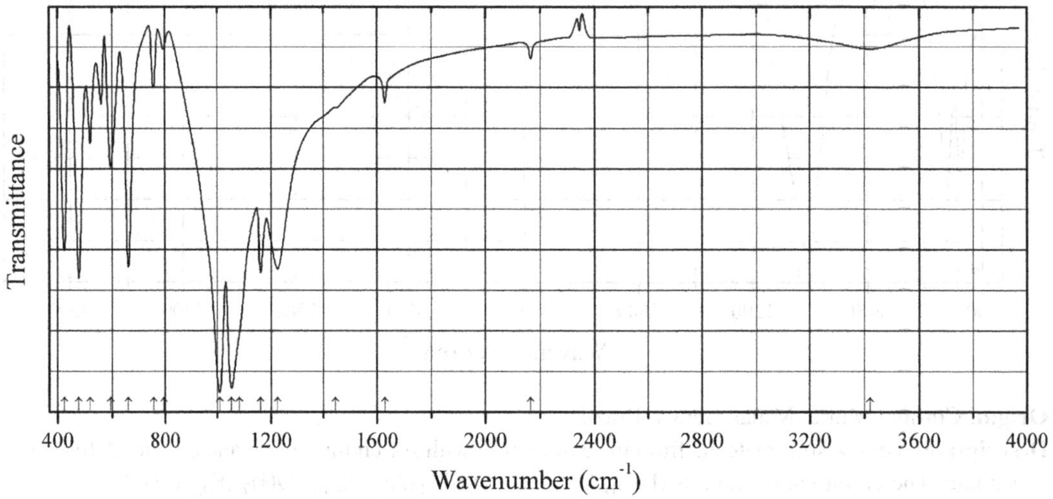

Origin: Synthetic.
Description: Synthesized via sol-gel method and calcined at 900 °C. Characterized by powder X-ray diffraction data and EDS analyses. Tetragonal, $a = 7.366$, $c = 15.574$ Å.
Kind of sample preparation and/or method of registration of the spectrum: Transmission. Kind of sample preparation is not indicated.
Source: Zhang et al. (2016b).
Wavenumbers (cm^{-1}): 1226s, 1161s, 1082sh, 1054s, 1009s, 793w, 753, 661s, 596, 562, 521, 479s, 425s.
Note: The wavenumbers were partly determined by us based on spectral curve analysis of the published spectrum. Bands above 1400 cm^{-1} are due to impurities.

Sil332 Hydronaujakasite $Na_2(H_2O,H_3O)_4Fe[Al_4Si_8O_{22}(OH,O)_4]$

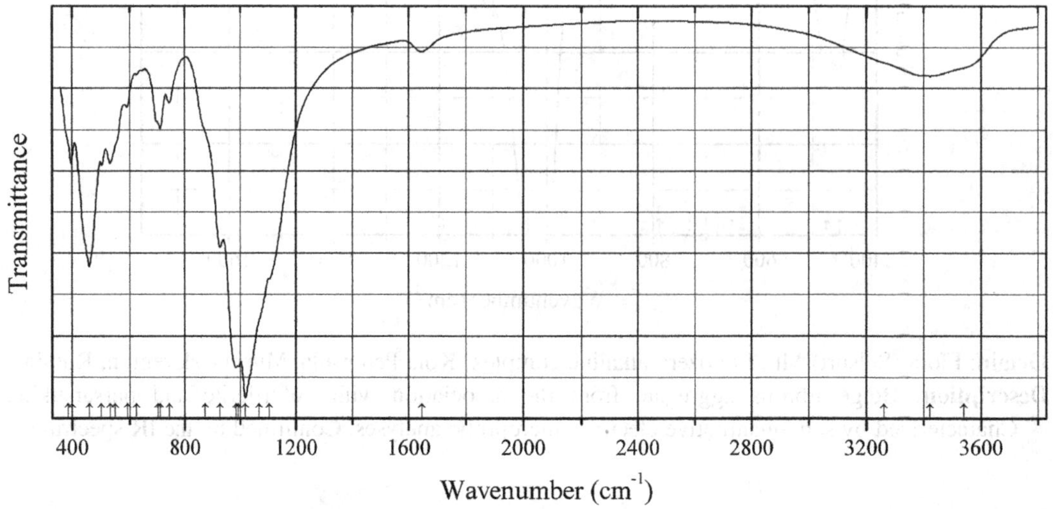

Origin: Tuperssuatsiait bay, southern part of the Ilímaussaq alkaline complex, Narsaq, Kujalleq, South Greenland (type locality).

Description: Peripheral zones (replacement rims) up to 1 mm thick of naujakasite crystals. The crystal structure is solved. Monoclinic, space group *C2/m*, $a = 14.983(8)$, $b = 7.998(4)$, $c = 10.403(6)$ Å, $\beta = 113.874(8)°$, $V = 1140.0(11)$ Å^3, $Z = 2$. $D_{meas} = 2.66(1)$ g/cm^3, $D_{calc} = 2.673$ g/cm^3. Optically biaxial (+), $\alpha = 1.525(2)$, $\beta = 1.530(2)$, $\gamma = 1.545$. The empirical formula is $H_{10.78}Na_{1.83}Ca_{0.09}Fe^{2+}_{0.90}Mn_{0.20}Al_{3.95}Si_{8.05}O_{29.52}$.

Kind of sample preparation and/or method of registration of the spectrum: KBr disc. Absorption.

Wavenumbers (cm^{-1}): 3540sh, 3420, 3260sh, 1646w, 1103s, 1065sh, 1018s, 994s, 985s, 928s, 875sh, 742, 708, 700sh, 625w, 593, 550sh, 533, 502, 460s, 399, 385sh.

Note: The spectrum was obtained by N.V. Chukanov.

Sil333 "Hydrochamosite-1*M*" $(Fe,Al,Mg)_6(Si,Al)_4O_{10}(OH)_8 \cdot nH_2O$

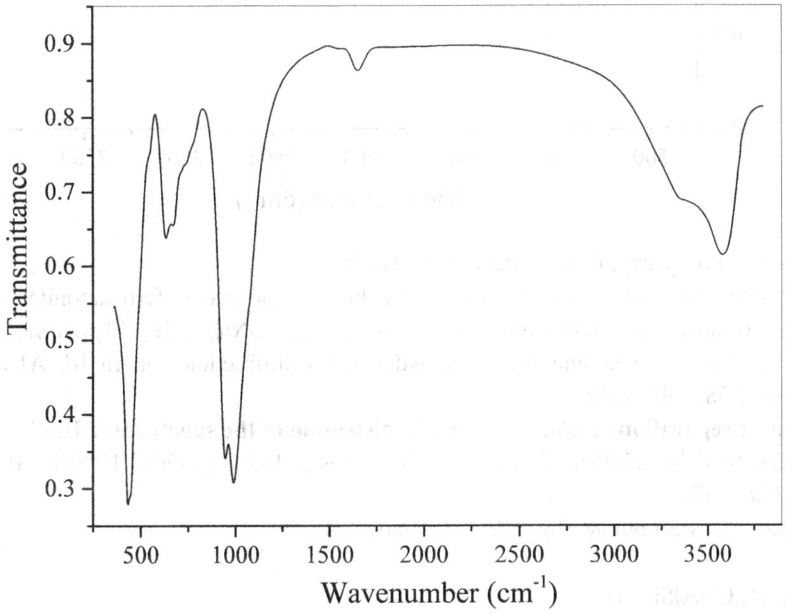

Origin: Karagach ridge, Karadag Mts., Crimea Peninsula, Russia.

Description: Olive-green, powdery. Investigated by A.V. Kasatkin. The observed lines of the powder X-ray diffraction pattern [*d*, Å] are: 14.50s, 7.18s, 4.80w, 4.61, 3.57, 2.85, 2.65, 2.49s, 2.09s, 1.55w, 1.52, 1.47w, 1.42w, 1.34w, 1.30, 1.23w, 1.18w.

Kind of sample preparation and/or method of registration of the spectrum: KBr disc. Absorption.

Wavenumbers (cm^{-1}): 3579s, 3380sh, 1652, 1075sh, 991s, 947s, 770sh, 730sh, 671, 634, 545sh, 445sh, 433s.

Note: The spectrum was obtained by N.V. Chukanov.

Sil334 "Ferrisaponite" $Ca_x(Fe^{3+},Mg)_{3-y}[(Si,Al)_4O_{10}](O,OH)_2 \cdot nH_2O$

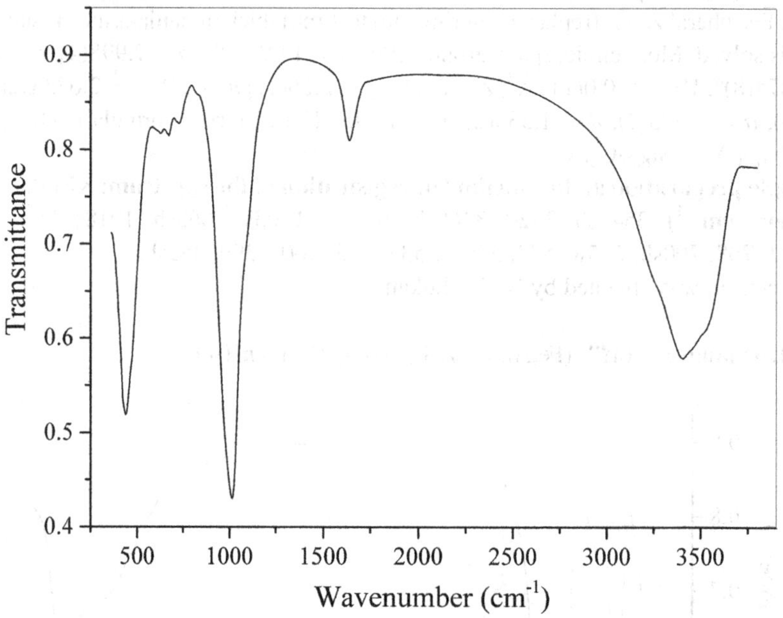

Origin: Pervomaiskiy quarry, Crimea Peninsula, Russia.

Description: Brownish-black grains with submetallic lustre. A product of ferrosaponite oxidation. The empirical formula is (electron microprobe): $Ca_{0.32}Na_{0.10}(Fe_{1.7}Mg_{1.1})(Si_{3.2}Al_{0.8}O_{10})(O,OH)_2 \cdot nH_2O$. The observed lines of the powder X-ray diffraction pattern [*d*, Å] are: 14.82w, 4.58s, 2.64s, 2.58, 2.42, 2.30, 1.54s.

Kind of sample preparation and/or method of registration of the spectrum: KBr disc. Absorption.

Wavenumbers (cm^{-1}): 3510sh, 3440sh, 3406, 3250sh, 1635, (1420), 1090sh, 1014s, 830sh, 727, 673, 628, 437s.

Note: The spectrum was obtained by N.V. Chukanov.

Sif148 Glass $(K,Cs)AlSi_{5-6}O_x$

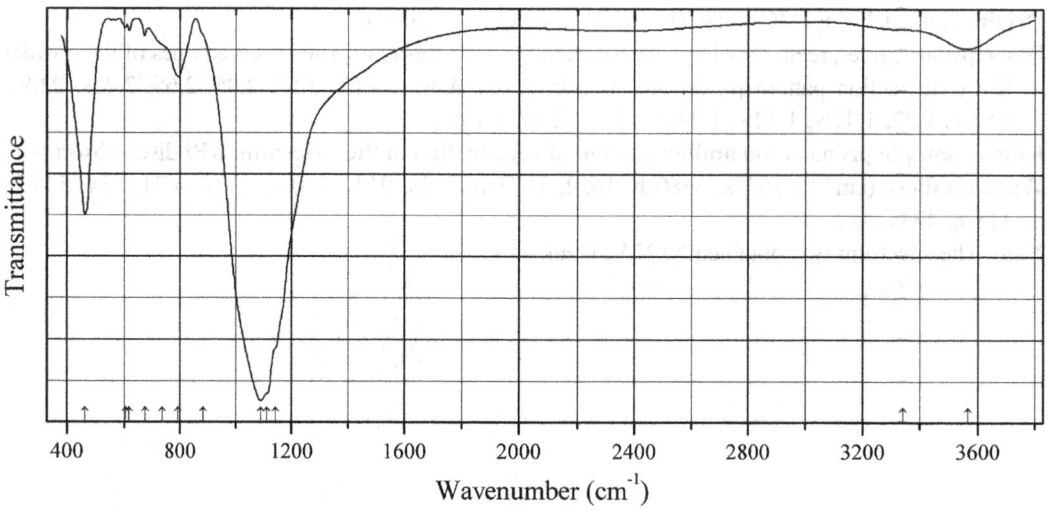

Origin: Arsenatnaya fumarole, Second scoria cone of the Northern Breakthrough of the Great Tolbachik Fissure Eruption, Tolbachik volcano, Kamchatka peninsula, Russia.

Description: Greenish with conchoidal fracture. Investigated by I.V. Pekov. Characterized by qualitative electron microprobe analyses.

Kind of sample preparation and/or method of registration of the spectrum: KBr disc. Absorption.

Wavenumbers (cm^{-1}): 3565, (3340w), 1140sh, 1110sh, 1087s, 885sh, 793, 735sh, 673, 619w, 606w, 463s.

Note: The spectrum was obtained by N.V. Chukanov.

Sif149 Hydroxycancrinite (?) $Na_{8-x}(Si_6Al_6O_{24})(CO_3)_{<1}\cdot nH_2O$
Cancrinite CO_3-deficient

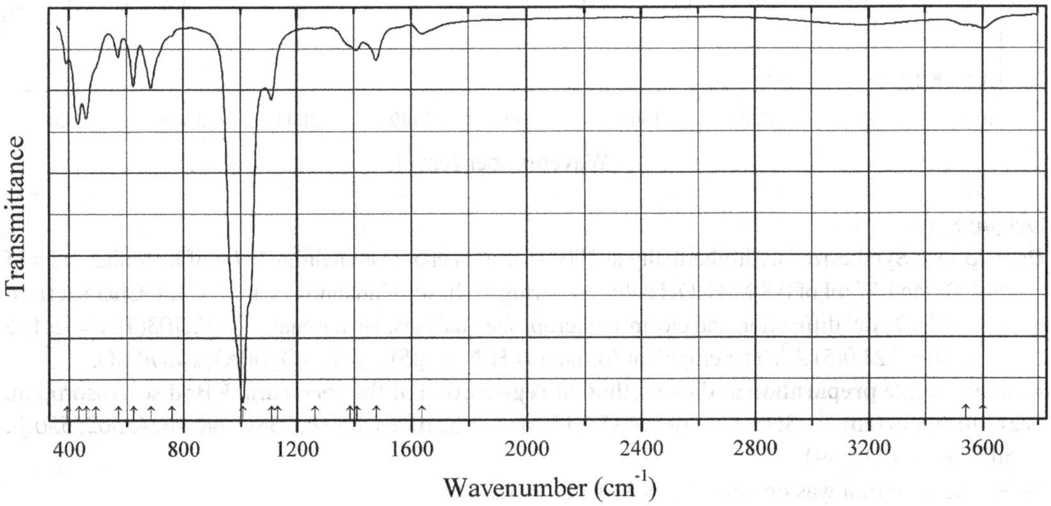

Origin: Synthetic.

Description: Synthesized hydrothermally at 200 °C from a charge containing 0.7 g of kaolinite, 0.425 g of Na_2CO_3, and 17 ml of 0.8 M NaOH solution during 22 h, see Chukanov et al. (2012a). Characterized by powder X-ray diffraction and semiquantitative electron microprobe analyses. Hexagonal, a = 12.703(1), c = 5.181(1) Å, V = 723.9(3) Å^3. The empirical formula is $Na_{6-7}(Si_{6.1}Al_{5.9}O_{24})(CO_3)_x\cdot nH_2O$ ($x \approx 0.45$).

Kind of sample preparation and/or method of registration of the spectrum: KBr disc. Absorption.

Wavenumbers (cm^{-1}): 3600, 3540w, 1636, 1477, 1409w, 1385sh, 1262w, 1110, 1130sh, 1007s, 758w, 686, 626, 572, 495sh, 461s, 433s, 394.

Note: The spectrum was obtained by N.V. Chukanov.

Sif150 Depmeierite $Na_8[Al_6Si_6O_{24}](PO_4,CO_3)_{1-x}{\cdot}3H_2O$ ($x < 0.5$)

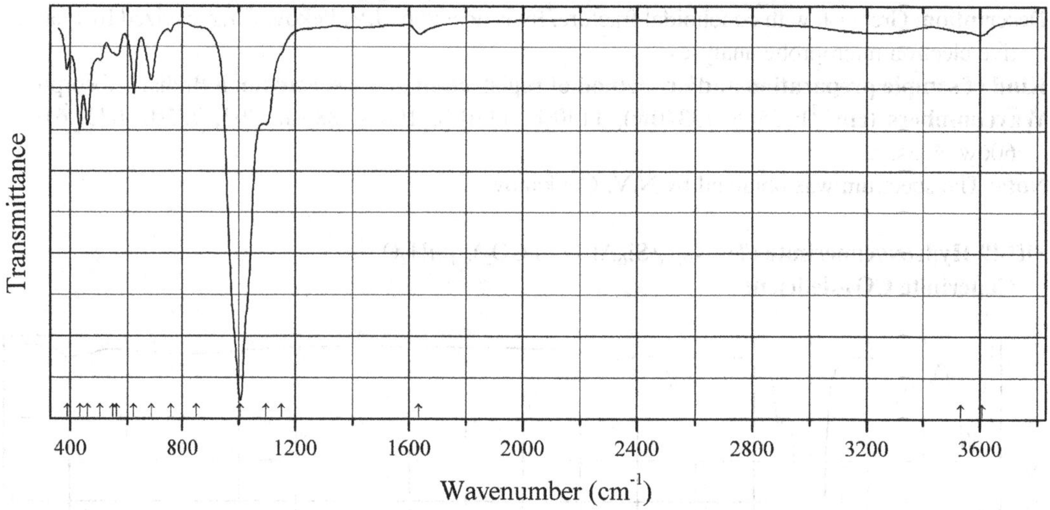

Origin: Synthetic.

Description: Synthesized hydrothermally at 200 °C from a charge containing 0.7 g of kaolinite, 1.7 g of Na_3PO_4, and 17 ml of 0.8 M NaOH solution during 10 h, *see* Chukanov et al. (2012a). Characterized by powder X-ray diffraction and electron microprobe analyses. Hexagonal, $a = 12.703(4)$, $c = 5.166(2)$ Å, $V = 722.0(5)$ Å^3. The empirical formula is $H_xNa_{7.15}(Si_{6.22}Al_{5.78}O_{24})(PO_4)_{0.59}{\cdot}nH_2O$.

Kind of sample preparation and/or method of registration of the spectrum: KBr disc. Absorption.

Wavenumbers (cm^{-1}): 3607, 3533w, 1635, 1150sh, 1095, 1006s, 850w, 758w, 685, 624, 562, 550sh, 505, 460s, 433s, 391.

Note: The spectrum was obtained by N.V. Chukanov.

Sif151 Vishnevite potassium analogue $K_2Na_6[Al_6Si_6O_{24}](SO_4){\cdot}nH_2O$

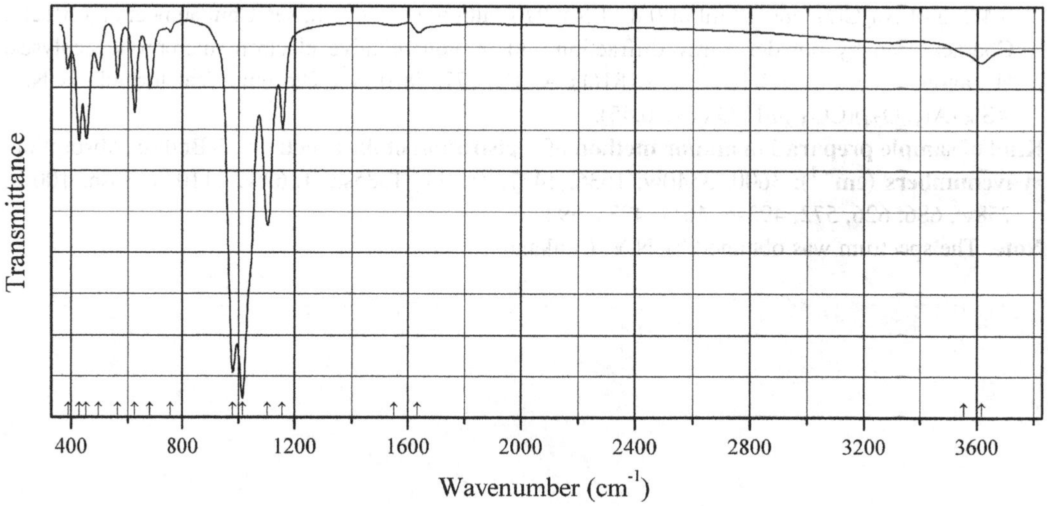

Origin: Synthetic.

Description: Synthesized hydrothermally at 200 °C from a charge containing 0.7 g of kaolinite, 1.7 g of K_2SO_4, and 17 ml of 0.8 M NaOH solution during 10 h, *see* Chukanov et al. (2012a). Characterized by powder X-ray diffraction and electron microprobe analyses. Hexagonal, a = 12.800(1), c = 5.246 (1) Å, V = 744.3(2) Å^3. The empirical formula is $K_{2.02}Na_{6.17}(Si_{6.01}Al_{5.99}O_{24})(SO_4)_{0.94}(OH)_x \cdot nH_2O$.

Kind of sample preparation and/or method of registration of the spectrum: KBr disc. Absorption.

Wavenumbers (cm^{-1}): 3617, 3555sh, 1635, (1552w), 1156, 1103s, 1014s, 981s, 756w, 679, 625, 564, 497, 454s, 429s, 391.

Note: The spectrum was obtained by N.V. Chukanov.

Sif152 Vishnevite $Na_8(Si_6Al_6O_{24})(SO_4) \cdot 2H_2O$

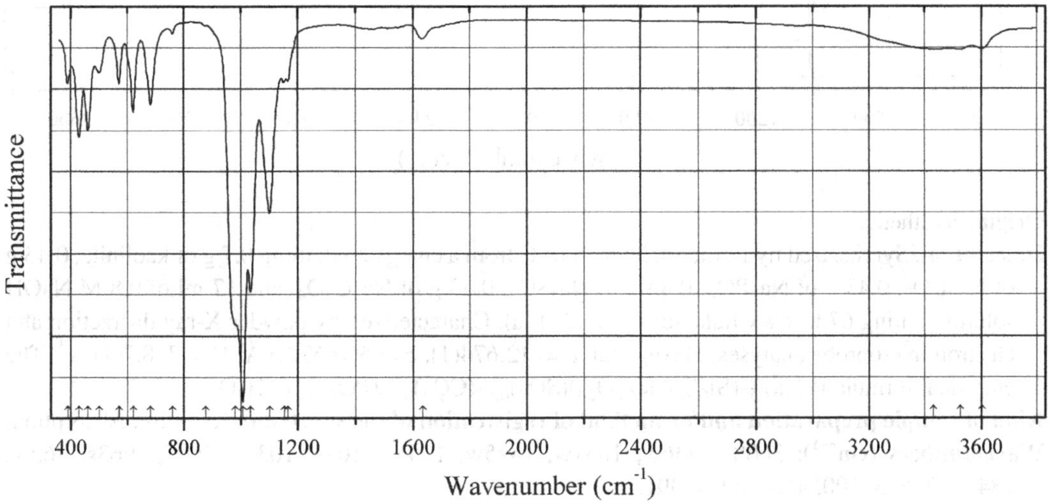

Origin: Synthetic.

Description: Synthesized hydrothermally at 200 °C from a charge containing 0.7 g of kaolinite, 1.7 g of Na_2SO_4, and 17 ml of 0.8 M NaOH solution during 10 h, *see* Chukanov et al. (2012a). Characterized by powder X-ray diffraction and electron microprobe analyses. Hexagonal, a = 12.689(1), c = 5.180 (1) Å, V = 722.2(2) Å^3. The empirical formula is $Na_{8.03}(Si_{6.01}Al_{5.99}O_{24})(SO_4)_{0.96}(CO_3,OH)_x \cdot nH_2O$.

Kind of sample preparation and/or method of registration of the spectrum: KBr disc. Absorption.

Wavenumbers (cm^{-1}): 3603, 3526w, 3430, 1635, 1164, 1152, 1101s, 1035s, 1007s, 980sh, 876w, 762w, 684, 622, 572, 503, 461, 430, 391.

Note: The spectrum was obtained by N.V. Chukanov.

Sif153 Vishnevite CO$_3$-bearing $Na_8(Si_6Al_6O_{24})(SO_4,CO_3)\cdot 2H_2O$

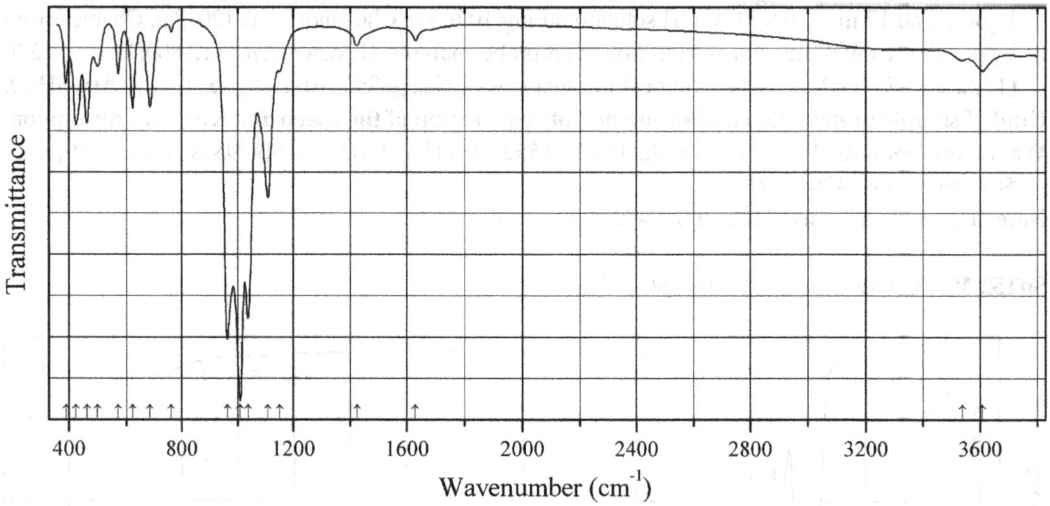

Origin: Synthetic.

Description: Synthesized hydrothermally at 200 °C from a charge containing 0.7 g of kaolinite, 0.43 g of Na_2CO_3, 0.43 g of Na_3PO_4, 0.43 g of Na_3SO_4, 0.43 g of $Na_2C_2O_4$, and 17 ml of 0.8 M NaOH solution during 67 h, *see* Chukanov et al. (2012a). Characterized by powder X-ray diffraction and electron microprobe analyses. Hexagonal, $a = 12.674(1)$, $c = 5.1667(2)$ Å, $V = 718.7(4)$ Å^3. The empirical formula is $Na_{7.52}(Si_{6.18}Al_{5.82}O_{24})(SO_4)_{0.62}(CO_3)_{0.19}(PO_4)_{0.03}\cdot nH_2O$.

Kind of sample preparation and/or method of registration of the spectrum: KBr disc. Absorption.

Wavenumbers (cm^{-1}): 3608, 3536w, 1631w, 1425w, 1151, 1107s, 1036s, 1008s, 963s, 762w, 684, 623, 573, 500, 463s, 425s, 391.

Note: The spectrum was obtained by N.V. Chukanov.

Sif154 Cancrinite NO$_3$-analogue $Na_8(Si_6Al_6O_{24})(NO_3,CO_3)_{2-x}\cdot 3H_2O$

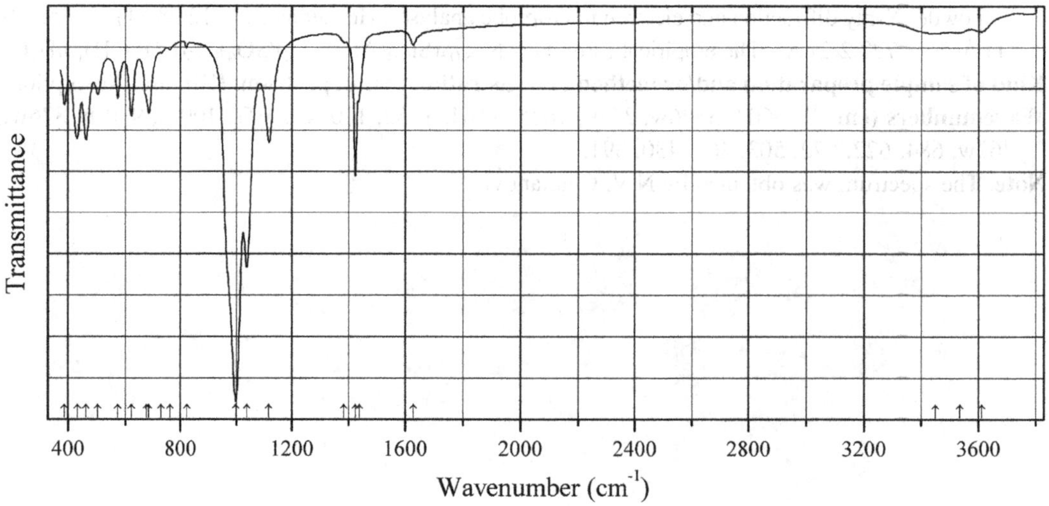

Origin: Synthetic.

Description: Synthesized hydrothermally at 160 °C from a charge containing 1 g of kaolinite, 4 g of $NaNO_3$, and 45 ml of 8 M NaOH solution during 120 h (see Chukanov et al. 2011, 2012a). Characterized by powder X-ray diffraction and electron microprobe analyses and gas chromatography of annealing products. The crystal structure is solved. Hexagonal, space group $P6_3$, $a = 12.6743(2)$, $c = 5.18289(13)$ Å, $V = 721.02(2)$ Å^3. The empirical formula is $Na_{7.8}(Si_{6.05}Al_{5.95}O_{24})(NO_3)_{1.32}(CO_3)_{0.27}\cdot 3.3H_2O$.

Kind of sample preparation and/or method of registration of the spectrum: KBr disc. Absorption.

Wavenumbers (cm^{-1}): 3611, 3535, 3449, 1629, 1438sh, 1423s, 1384w, 1117, 1036s, 998s, 823w, 763w, 730w, 684, 675sh, 622, 575, 504, 464s, 433s, 389.

Note: The spectrum was obtained by N.V. Chukanov.

Sif155 Cancrinite NO$_3$-analogue low-hydrous $Na_8(Si_6Al_6O_{24})(NO_3,CO_3)_{2-x}\cdot H_2O$

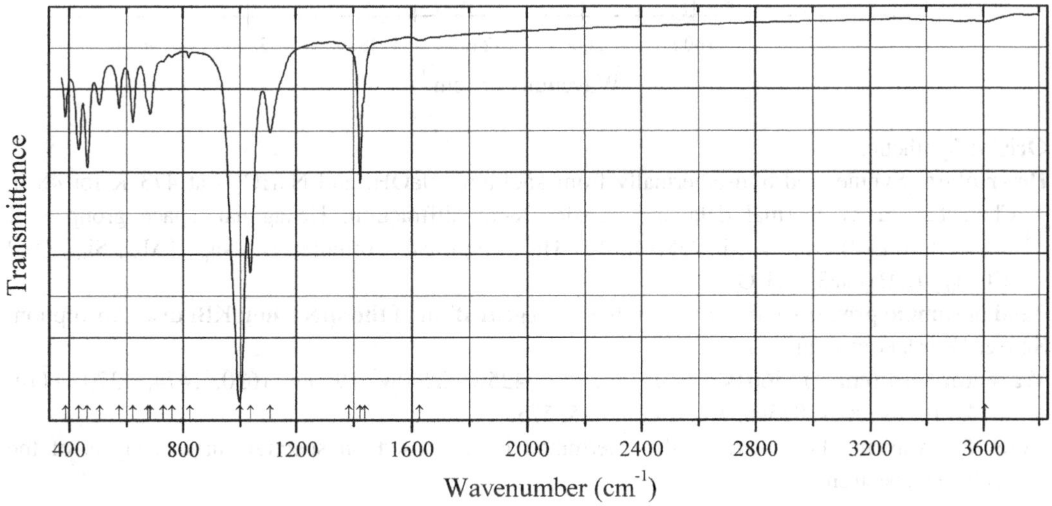

Origin: Synthetic.

Description: Product of partial dehydration of the NO$_3$-analogue of cancrinite Sif154 at 300 °C (see Chukanov et al. 2011, 2012a). The empirical formula is $Na_{7.8}(Si_{6.05}Al_{5.95}O_{24})(NO_3)_{1.32}(CO_3)_{0.27}\cdot nH_2O$ ($n \approx 1$).

Kind of sample preparation and/or method of registration of the spectrum: KBr disc. Absorption.

Wavenumbers (cm^{-1}): 3604w, 1628w, 1438sh, 1422s, 1384w, 1108, 1036s, 997s, 823w, 762w, 730w, 684, 675sh, 624, 575, 507, 464s, 434s, 387.

Note: The spectrum was obtained by N.V. Chukanov.

Sif156 Cancrinite Ca-free analogue $Na_8[Al_6Si_6O_{24}](CO_3)\cdot 4H_2O$

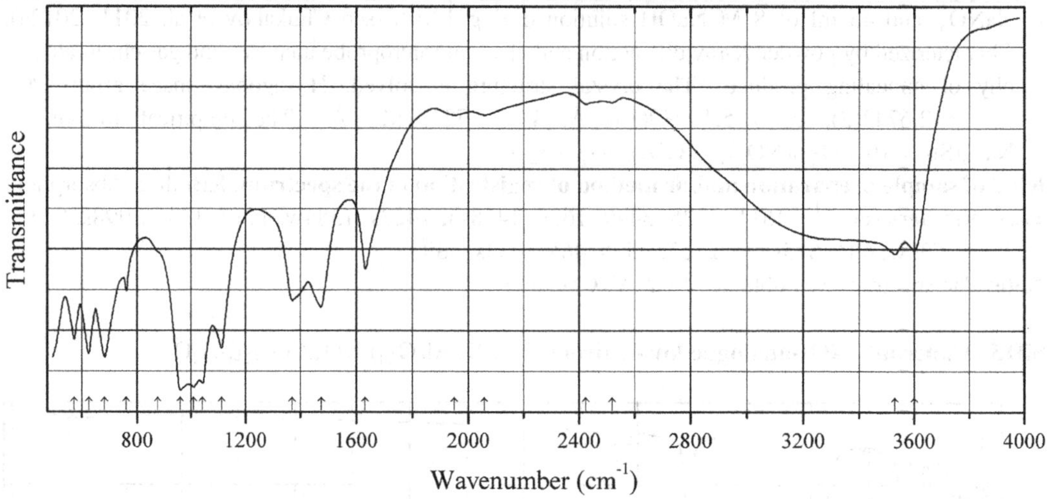

Origin: Synthetic.

Description: Synthesized hydrothermally from kaolinite, NaOH, and $NaHCO_3$ at 473 K for 48 h. Characterized by thermal data and powder X-ray diffraction. Hexagonal, space group $P6_3$, a = 12.663(2), c = 5.1738(9) Å. The empirical formula is $Na_{8.28}[Al_{5.93}Si_{6.07}O_{24}](CO_3)_{0.93}(OH)_{0.49}\cdot 3.64H_2O$.

Kind of sample preparation and/or method of registration of the spectrum: KBr disc. Absorption.

Source: Kurdakova et al. (2014).

Wavenumbers (cm^{-1}): 3601s, 3530s, 2520w, 2425w, 2060w, 1950w, 1630, 1474, 1370, 1114, 1044s, 1010s, 960s, 875sh, 761w, 683, 625, 573.

Note: The wavenumbers were partly determined by us based on spectral curve analysis of the published spectrum.

Sif157 Carnegieite (high) $Na(AlSiO_4)$

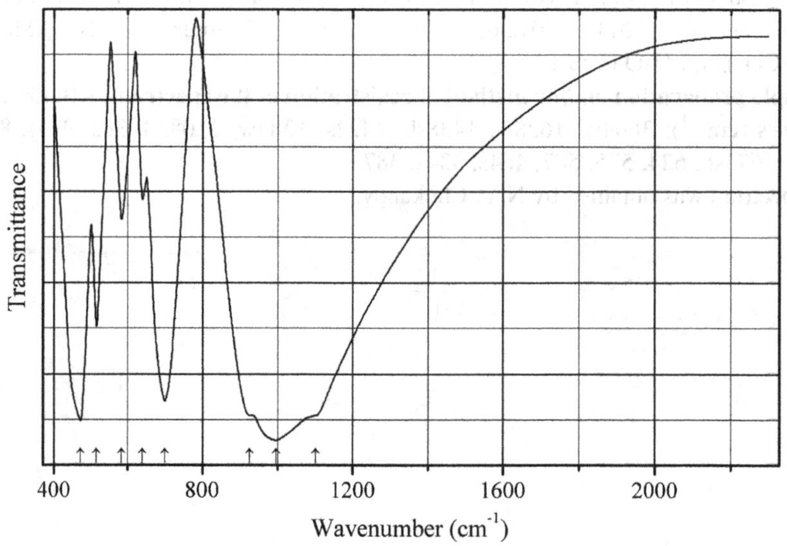

Origin: Synthetic phase polymorphous with nepheline.

Description: Prepared from hydrated alumina gel, NaOH, and highly reactive aerosol silica at 800° with subsequent annealing at 1300 °C. Characterized by powder X-ray diffraction data. Cubic, space group $P2_13$.

Kind of sample preparation and/or method of registration of the spectrum: KBr disc. Transmission.

Source: Nayak and Kutty (1998).

Wavenumbers (cm^{-1}): 1100sh, 996s, 925sh, 700s, 640, 583, 517, 474s.

Sif158 Carnegieite (low) $Na(AlSiO_4)$

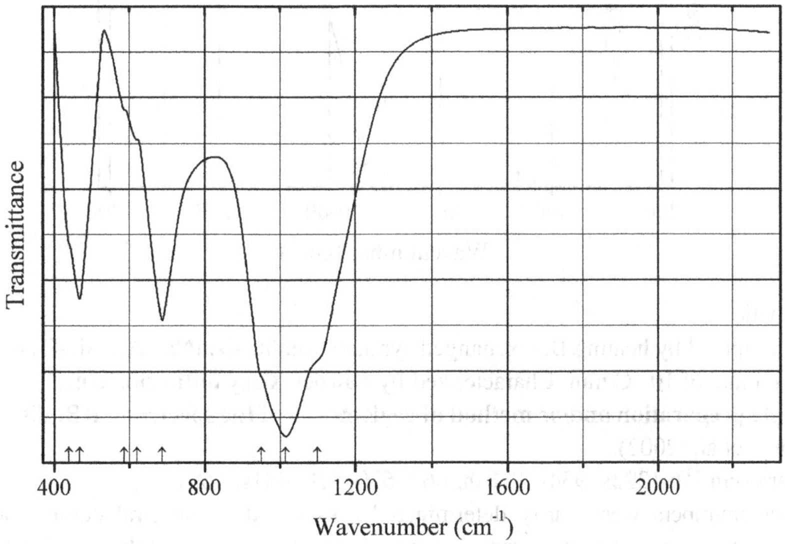

Origin: Synthetic phase polymorphous with nepheline.

Description: Prepared from hydrated alumina gel, NaOH, and highly reactive aerosol silica at 800°. Characterized by powder X-ray diffraction data. Orthorombic, space group $Pb2_1a$.

Kind of sample preparation and/or method of registration of the spectrum: KBr disc. Transmission.

Source: Nayak and Kutty (1998).

Wavenumbers (cm^{-1}): 1100sh, 1016s, 950sh, 687, 620sh, 586sh, 468, 439sh.

Sif159 Hexacelsian $Ba(Al_2Si_2O_8)$

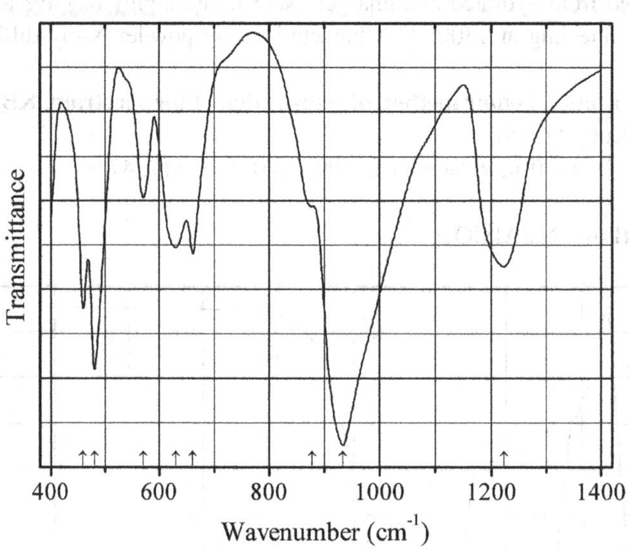

Origin: Synthetic.

Description: Prepared by heating Ba-exchanged synthetic zeolite 4A ($Na_{12}Al_{12}Si_{12}O_{48}\cdot27H_2O$) up to 1300 °C at a rate of 10 °C/min. Characterized by powder X-ray diffraction data.

Kind of sample preparation and/or method of registration of the spectrum: KBr disc. Absorption.

Source: Aronne et al. (2002).

Wavenumbers (cm^{-1}): 1223s, 934s, 878sh, 662, 630, 570, 481s, 460s.

Note: The wavenumbers were partly determined by us based on spectral curve analysis of the published spectrum. For the IR spectra of hexacelsian and its polymorphs see also Dondur et al. (2005) and Colomban et al. (2000).

Sif160 Rubicline $Rb(AlSi_3O_8)$

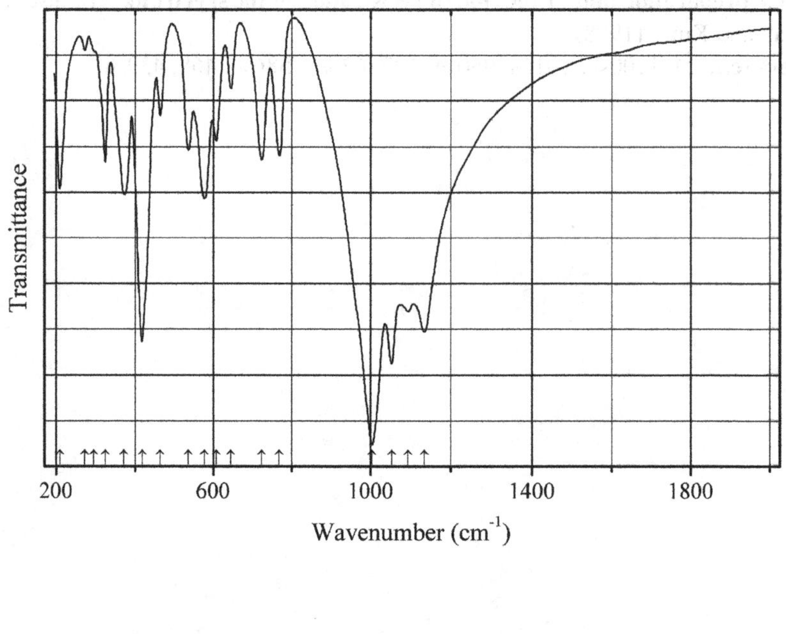

Origin: Synthetic.

Kind of sample preparation and/or method of registration of the spectrum: KBr disc. Transmission.

Source: Roy (1987).

Wavenumbers (cm^{-1}): 1134s, 1093s, 1053s, 1004s, 770, 725, 645, 607, 578, 537, 465, 419s, 372, 325, 297sh, 274w, 210.

Note: The wavenumbers were partly determined by us based on spectral curve analysis of the published spectrum.

Sif161 Sodalite nitrite analogue $Na_8[AlSiO_4]_6(NO_2)_2 \cdot nH_2O$

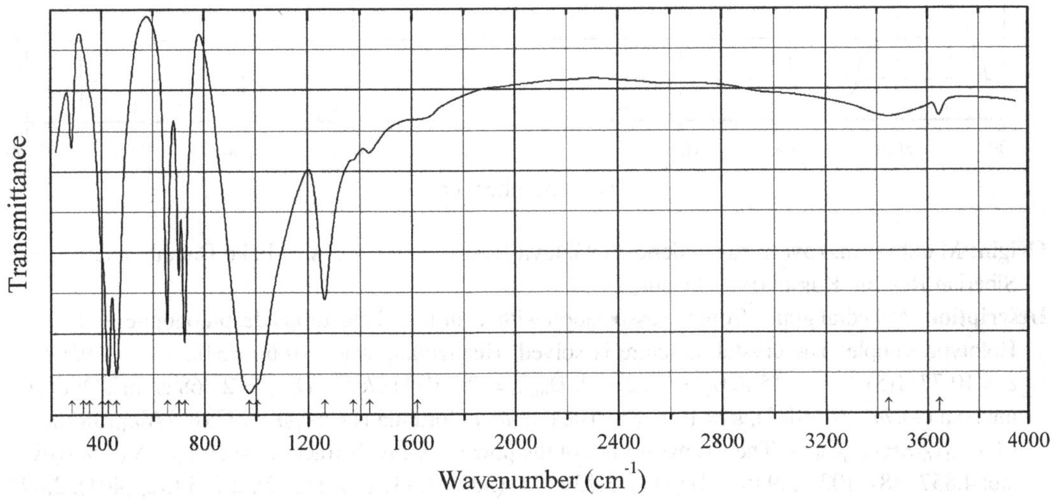

Origin: Synthetic.

Description: Prepared hydrothermally from kaolinite, in the presence of $NaNO_2$. Characterized by powder X-ray diffraction data. Cubic, $a = 8.931(1)$ Å.

Kind of sample preparation and/or method of registration of the spectrum: KBr disc. Transmission.

Source: Buhl (1991).

Wavenumbers (cm^{-1}): 3645w, 3445w, 1620sh, 1435, 1372sh, 1264, 1005sh, 972s, 722s, 699, 654s, 455s, 424s, 400sh, 350sh, 326sh, 282.

Note: The wavenumbers were determined by us based on spectral curve analysis of the published spectrum. The bands at 1264 and 1372 cm^{-1} correspond to stretching vibrations of NO_2^- and NO_3^-, respectively. The band at 1435 cm^{-1} indicates the presence of CO_3^{2-}.

Sif162 Sulfhydrylbystrite $Na_5K_2Ca[Al_6Si_6O_{24}](S_5)^{2-}(SH)^-$

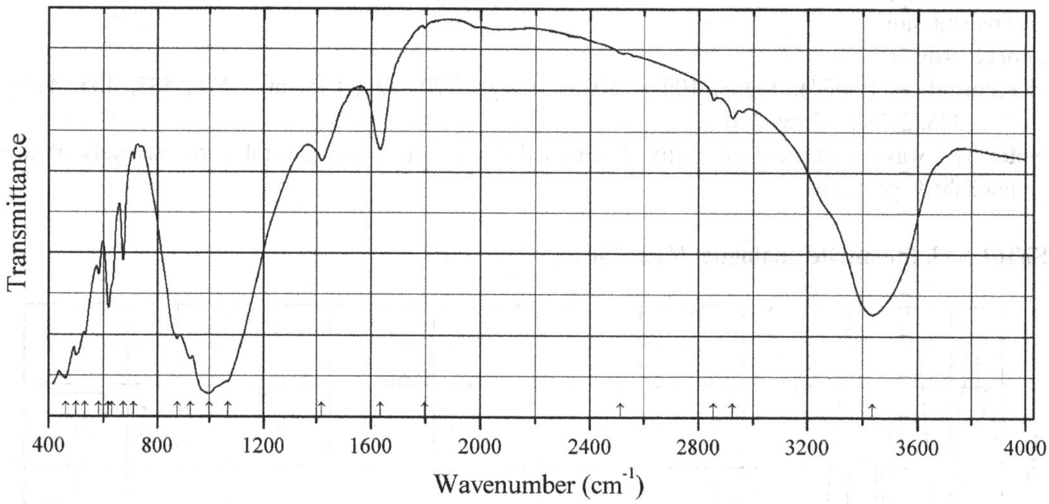

Origin: Malobystrinskoye lazurite deposit, Malaya Bystraya River basin, Lake Baikal area, Eastern Siberian Region, Russia (type locality).

Description: Anhedral grains from the association with lazurite, calcite, diopside, phlogopite, and pyrite. Holotype sample. The crystal structure is solved. Hexagonal, space group *P*31*c*, $a = 12.9567(6)$, $c = 10.7711(5)$ Å, $V = 1566.0(1)$ Å^3, $Z = 2$. $D_{meas} = 2.391(1)$ g/cm^3, $D_{calc} = 2.368$ g/cm^3. Optically uniaxial (+), $\omega = 1.661(2)$, $\varepsilon = 1.584(2)$. The empirical formula is $Na_{5.17}K_{1.87}Ca_{0.99}(Al_{6.01}Si_{5.99}O_{24})(S_5)^{2-}{}_{0.86}(SH)_{0.86}Cl_{0.07}$. The strongest lines of the powder X-ray diffraction pattern [*d*, Å (*I*, %) (*hkl*)] are: 4.857 (48) (102), 3.948 (38) (211), 3.739 (94) (300), 3.331 (100) (212), 2.715 (32) (401), 2.692 (56) (004).

Kind of sample preparation and/or method of registration of the spectrum: KBr disc. Transmission.

Source: Sapozhnikov et al. (2016).

Wavenumbers (cm^{-1}): 3436, 2926, 2855, 2514, 1798, 1634, 1071, 1000, 926, 876, 712, 673, 631sh, 619, 583, 534, 501, 462.

Note: The bands at 3436 and 1634 cm^{-1} correspond to H_2O molecules that are not indicated in the chemical formula of sulfhydrylbystrite. Weak bands in the range from 2800 to 3000 cm^{-1} correspond to the admixture of an organic substance. The assignment of the very weak band at 2514 cm^{-1} to S–H-stretching vibrations made by the authors is ambiguous and questionable.

Sif163 Thallium feldspar $TlAlSi_3O_8$

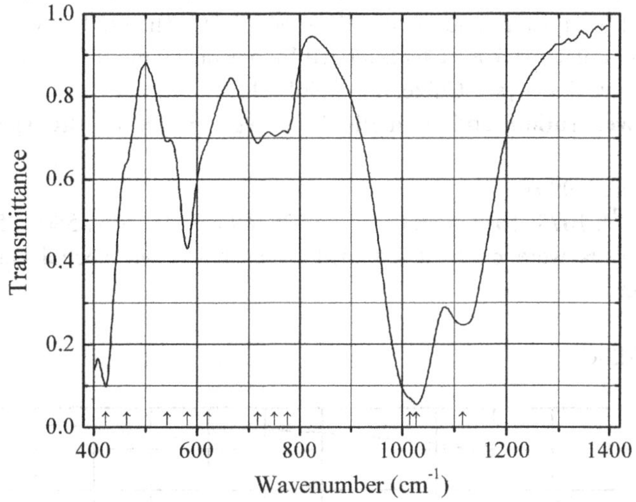

Origin: Synthetic.

Description: Synthesized from a powdered sample of natural low albite and $TlNO_3$ in a 1:1 weight ratio under hydrothermal conditions, at 550 °C for 5 days. The crystal structure is solved. Monoclinic, space group *C*2/*m*, *a* = 8.882(3), *b* = 13.048(2), *c* = 7.202(2) Å, *β* = 116.88(1)°, *V* = 744.5(4) Å^3, *Z* = 4. D_{calc} = 3.958 g/cm^3. The strongest lines of the powder X-ray diffraction pattern [*d*, Å (*I*, %) (*hkl*)] are: 3.94 (64) (200, 111), 3.63 (70) (13–1), 3.62 (80) (22–1), 3.45 (92) (11–2), 3.372 (100) (220), 3.197 (54) (002), 2.992 (78) (131).

Kind of sample preparation and/or method of registration of the spectrum: KBr disc. Transmission.

Source: Kyono and Kimata (2001).

Wavenumbers (cm^{-1}): 1116s, 1027s, 1014sh, 776, 751, 718, 620sh, 581, 542, 463sh, 423s.

Note: The wavenumbers were determined by us based on spectral curve analysis of the published spectrum.

Sif164 Thallium sodalite $Tl_6(Al_6Si_6O_{24})$

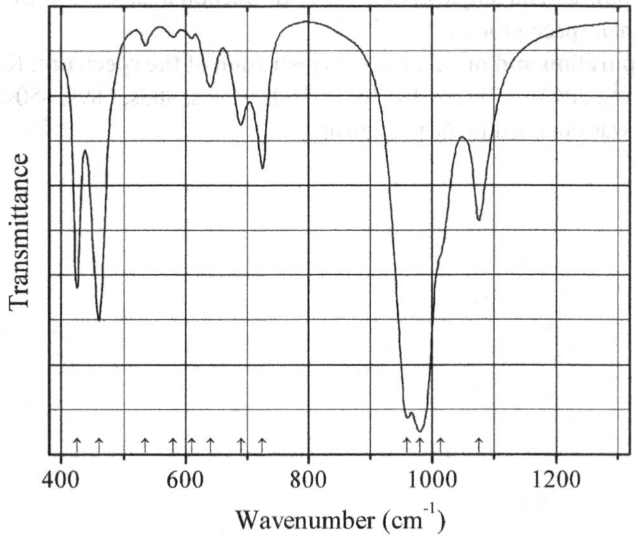

Origin: Synthetic.

Description: Obtained in the ion-exchange reaction between hydroxysodalite $Na_8(Al_6Si_6O_{24})(OH)_2{\cdot}2H_2O$ and 1 M aqueous solution of $TlNO_3$ at 100 °C. The product was dried at 425 °C for 5 h under a vacuum of 10^{-5} Torr. Characterized by powder X-ray diffraction data. The structure was refined by Rietveld analysis. Cubic, a = 8.9653(1) Å.

Kind of sample preparation and/or method of registration of the spectrum: KBr disc. Transmission.

Source: Latturner et al. (1999).

Wavenumbers (cm^{-1}): 1075, 1013sh, 980s, 960s, 725, 690, 640, 610w, 580w, 535w, 460s, 425s.

Note: The wavenumbers were partly determined by us based on spectral curve analysis of the published spectrum.

Sif165 Kalsilite $KAlSiO_4$

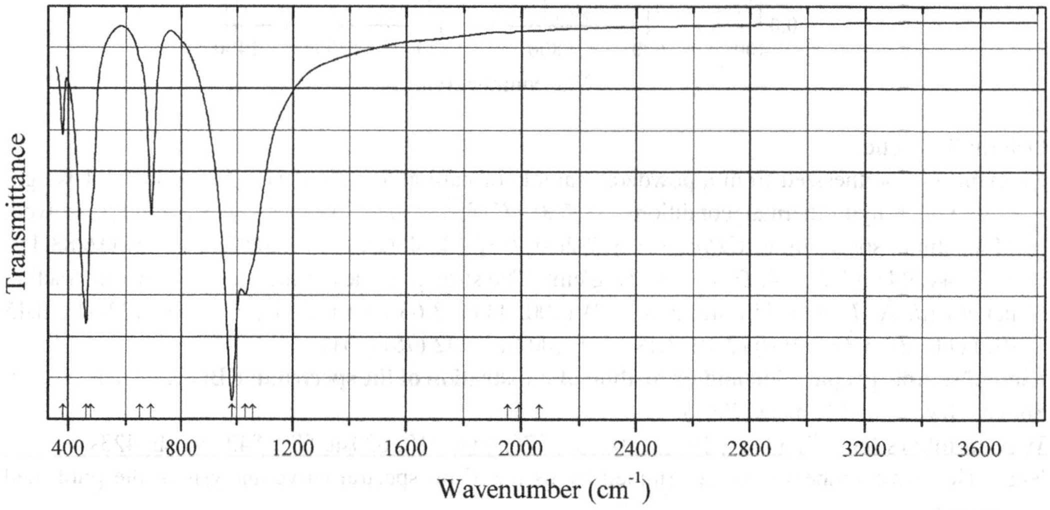

Origin: Koashva Mt., Khibiny alkaline complex, Kola peninsula, Murmansk region, Russia.

Description: Crystals from the association with carbobystrite. Investigated by I.V. Pekov. Characterized by powder and single-crystal X-ray diffraction data, as well as electron microprobe analyses. Hexagonal, space group $P6_3$.

Kind of sample preparation and/or method of registration of the spectrum: KBr disc. Absorption.

Wavenumbers (cm^{-1}): 2063w, 1992w, 1953w, 1055sh, 1030s, 983s, 689s, 650sh, 480sh, 462s, 383.

Note: The spectrum was obtained by N.V. Chukanov.

Sif166 Dmisteinbergite $Ca(Al_2Si_2O_8)$

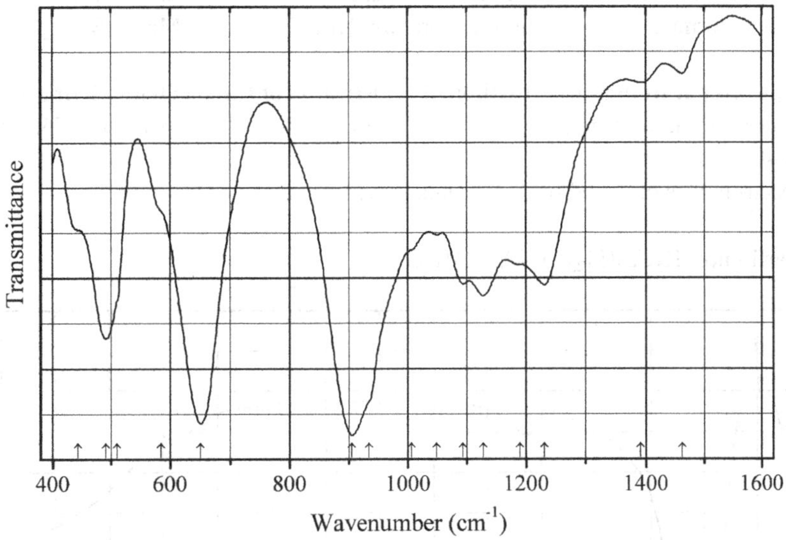

Origin: Burned dump of the Chelyabinsk coal basin, Kopeisk, South Urals, Russia (type locality).

Description: Hexagonal platelets from the association with anorthite, svyatoslavite, troilite, and cohenite. The empirical formula is (electron microprobe): $Ca_{1.00}Al_{2.01}Si_{2.03}O_{8.07}$.

Kind of sample preparation and/or method of registration of the spectrum: KBr disc. Absorption.

Source: Simakin et al. (2010).

Wavenumbers (IR, cm^{-1}): (1463), (1392), 1231, 1190sh, 1128, 1094, 1049, 1007sh, 935sh, 905s, 650s, 584sh, 510sh, 490s, 443sh.

Note: The wavenumbers were partly determined by us based on spectral curve analysis of the published spectrum. Weak bands in the range 1300–1500 cm^{-1} correspond to the admixture of a carbonate. In the cited paper, Raman spectrum is given.

Wavenumbers (Raman, cm^{-1}): 903s, 809, 651, 485s, 432s.

Sif_Z127 Erionite-K $K_{10}[Si_{26}Al_{10}O_{72}]·30H_2O$

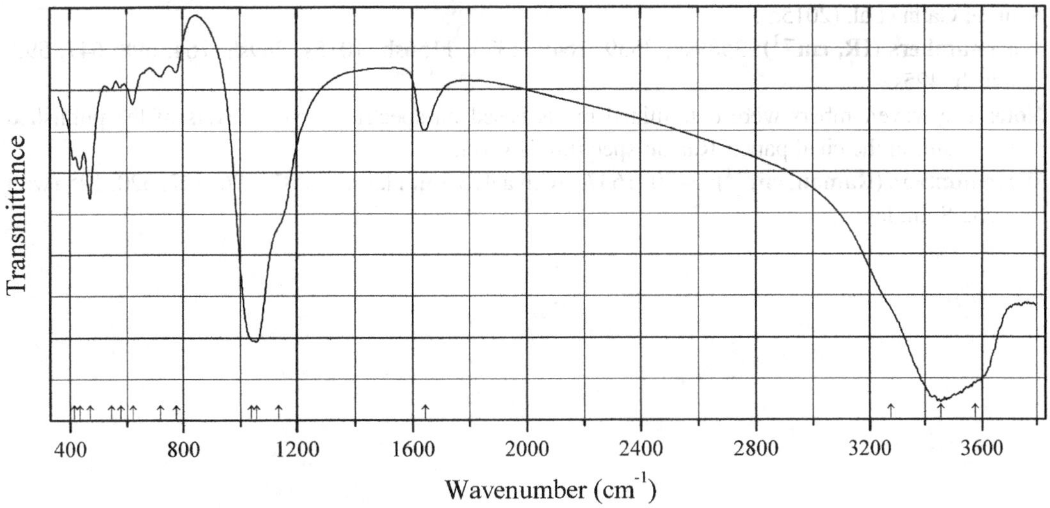

Origin: Karadag Mts., Crimea Peninsula, Russia.

Description: Light green crystals of erionite-K with zones of erionite-Na. Investigated by A.V. Kasatkin. The empirical formula is (electron microprobe): $K_{1.82}Ca_{1.78}Na_{1.70}Mg_{0.57}(Si_{27.42}Al_{8.27}Fe_{0.30}O_{72})\cdot nH_2O$.

Kind of sample preparation and/or method of registration of the spectrum: KBr disc. Absorption.

Wavenumbers (cm^{-1}): 3575sh, 3455s, 3275sh, 1645, 1135sh, 1058s, 1040sh, 775, 720, 623, 579, 545, 470s, 435, 415.

Note: The spectrum was obtained by N.V. Chukanov.

Sif_Z128 Merlinoite $K_5Ca_2(Si_{23}Al_9)O_{64}\cdot 24H_2O$

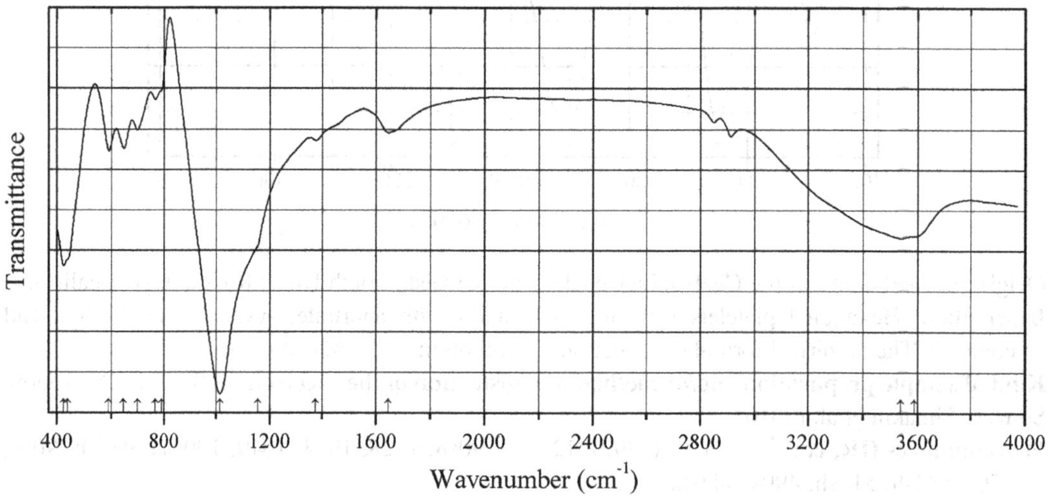

Origin: Fosso Attici, Sacrofano, Italy.

Description: Yellowish prismatic crystals from the association with phillipsite. The crystal structure is solved. Orthorhombic, space group *Immm*, a = 14.066(5), b = 14.111(5), c = 9.943(3) Å (at 100 K). D_{calc} = 2.177 g/cm^3. The empirical formula is $(K_{5.69}Na_{0.37})(Ca_{1.93}Ba_{0.40}Mg_{0.01})(Si_{21.38}Al_{10.55}Fe^{3+}_{0.02})O_{64}\cdot 19.6H_2O$.

Kind of sample preparation and/or method of registration of the spectrum: KBr disc. Absorption.

Source: Gatta et al. (2015a).

Wavenumbers (IR, cm^{-1}): 3589sh, 3539, 1649, 1375, 1156sh, 1015s, 789sh, 764, 698, 647, 592, 440sh, 425s.

Note: The wavenumbers were determined by us based on spectral curve analysis of the published spectrum. In the cited paper, Raman spectrum is given.

Wavenumbers (Raman, cm^{-1}): 3470, 1637, (with a 473.1 nm laser); 1087, 496, 422, 320, 125 (with a 632.8 nm laser).

Sif_Z129 Phillipsite-NH$_4$ $(NH_4,Na)_9(Al_9Si_{27}O_{72}]\cdot 24H_2O$

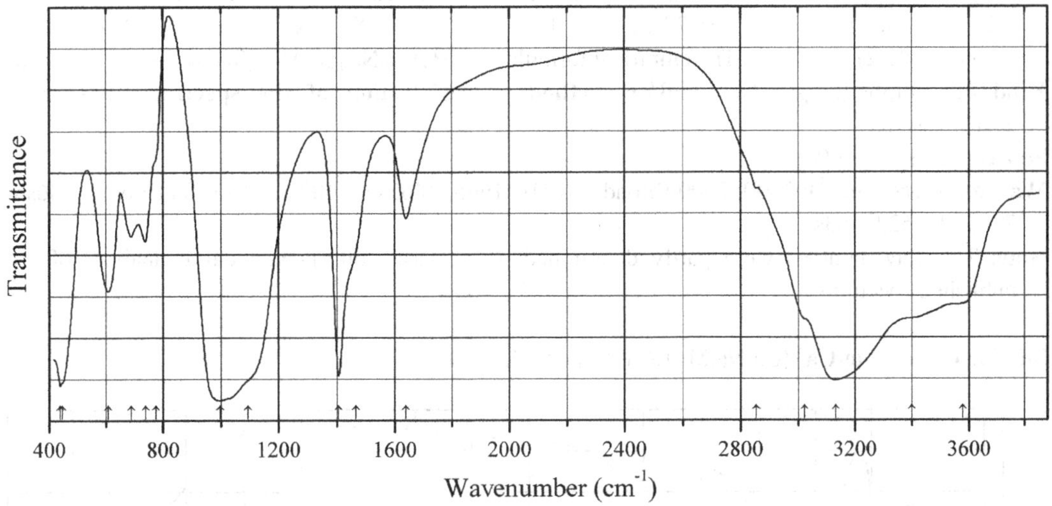

Origin: Artificial.

Description: NH_4^+-exchanged Si-poor phillipsite from Vallerano, Rome, Italy. The crystal structure is solved. Monoclinic, a = 10.0507(5), b = 14.2016(8), c = 8.7281(8) Å, β = 125.123(5)°, V = 1019.0(9) Å^3, Z = 4. The empirical formula is $(NH_4)_{11.41}Na_{1.36}(Al_{13.36}Si_{22.64}O_{72.45})\cdot 20.6H_2O$.

Kind of sample preparation and/or method of registration of the spectrum: KBr disc. Transmission.

Source: Gualtieri (2000).

Wavenumbers (cm^{-1}): 3580, 3402sh, 3131s, 3022sh, 1640, 1465sh, 1406s, 1092sh, 996s, 775sh, 739, 689, 607, 446sh, 438s.

Note: The wavenumbers were partly determined by us based on spectral curve analysis of the published spectrum.

Sif_Z130 Phillipsite-NH$_4$ $(NH_4,Na)_9(Al_9Si_{27}O_{72}]\cdot 24H_2O$

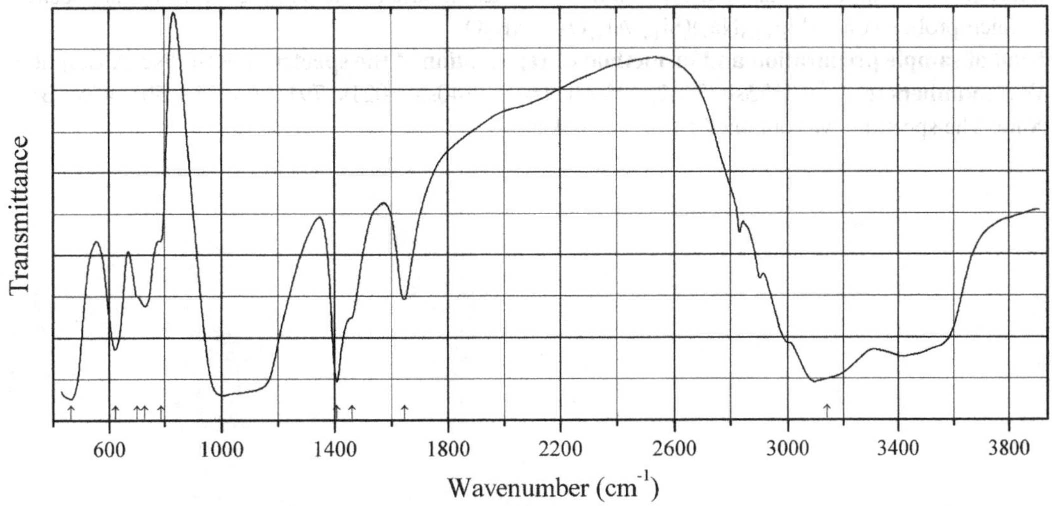

Origin: Artificial.

Description: NH_4^+-exchanged phillipsite from Perrier, Puy du D?me, France. The crystal structure is solved. Monoclinic, a = 10.0122(8), b = 14.1943(12), c = 8.7284(17) Å, β = 125.024(11)°, V = 1015.81(2) Å^3, Z = 4. The empirical formula is $(NH_4)_{9.89}Na_{0.45}(Al_{9.92}Si_{26.08}O_{71.96})\cdot 18.2H_2O$.

Kind of sample preparation and/or method of registration of the spectrum: KBr disc. Transmission.

Source: Gualtieri (2000).

Wavenumbers (cm^{-1}): 3400–3600 (broad), 3141s, 1645, 1460sh, 1407s, 980–1180 (broad), 786sh, 729, 700, 622, 463s.

Note: The wavenumbers were partly determined by us based on spectral curve analysis of the published spectrum.

Sif_Z131 Faujasite-Ca $(Ca,Na,Mg)_2(Si,Al)_{12}O_{24}\cdot 15H_2O$

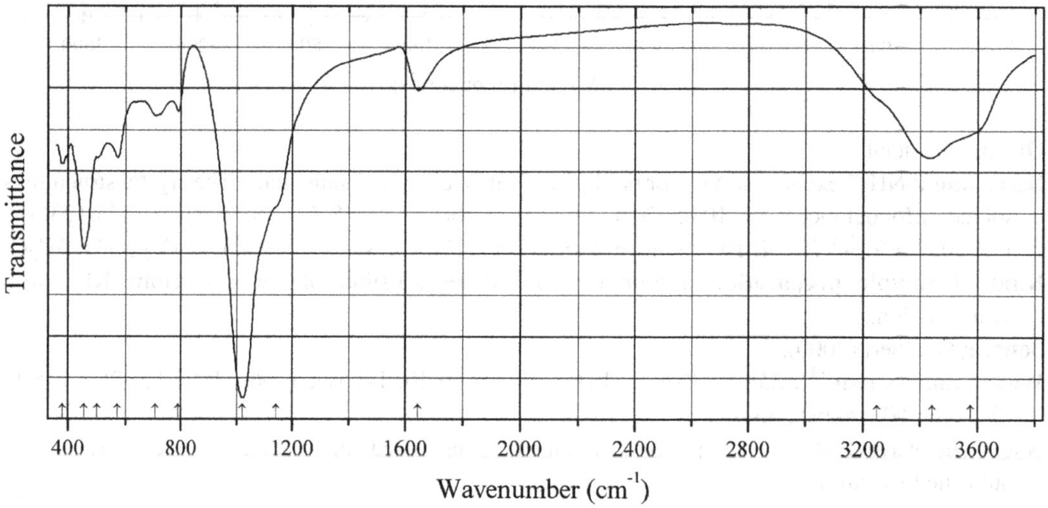

Origin: Quarry No. 1, Limberg, Sasbach, Germany.

Description: Colorless octahedral crystals from cavities in basalt. The empirical formula is (electron microprobe): $(Ca_{11.5}Mg_{10.5}Na_7)(Si_{141}Al_{51}O_{384})\cdot nH_2O$.

Kind of sample preparation and/or method of registration of the spectrum: KBr disc. Absorption.

Wavenumbers (cm^{-1}): 3575sh, 3442, 3250sh, 1645, 1140sh, 1021s, 791, 706, 573, 503, 456s, 381.

Note: The spectrum was obtained by N.V. Chukanov.

Sif_Z132 Tschernichite $CaAl_2Si_6O_{16}·8H_2O$

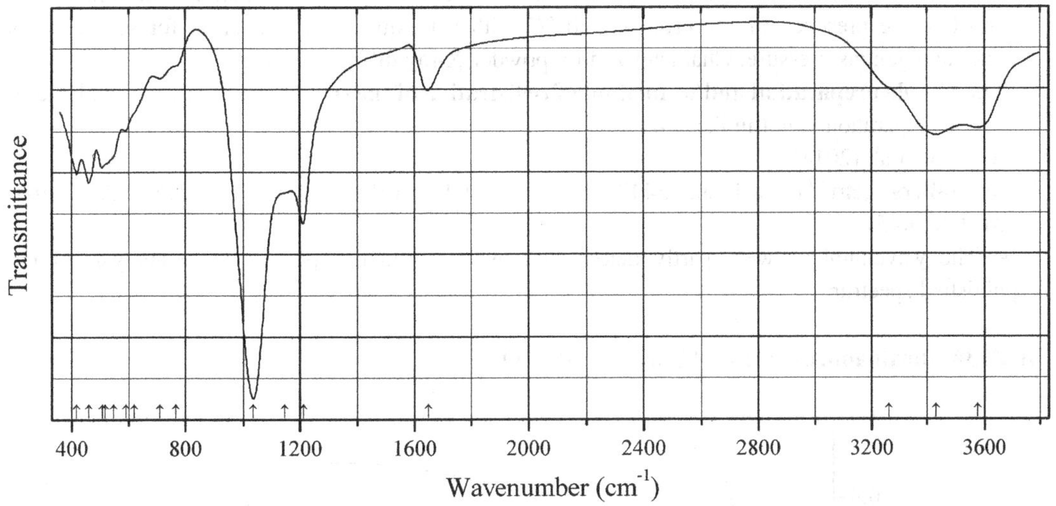

Origin: Markaz, Lis-Kas-Kő, Hungary.

Description: Colorless crystals. Identified by morphological features and qualitative electron microprobe analyses. Confirmed by the IR spectrum.

Kind of sample preparation and/or method of registration of the spectrum: KBr disc. Absorption.

Wavenumbers (cm^{-1}): 3577, 3430, 3260sh, 1648, 1210s, 1145, 1035s, 765sh, 711, 620sh, 591, 549sh, 520sh, 508, 460s, 418.

Note: The spectrum was obtained by N.V. Chukanov.

Sif_Z133 Mazzite-Na $Na_8(Si_{28}Al_8)O_{72}·30H_2O$

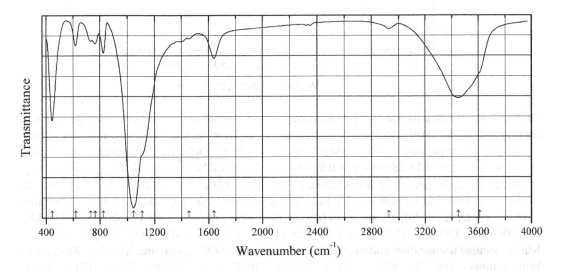

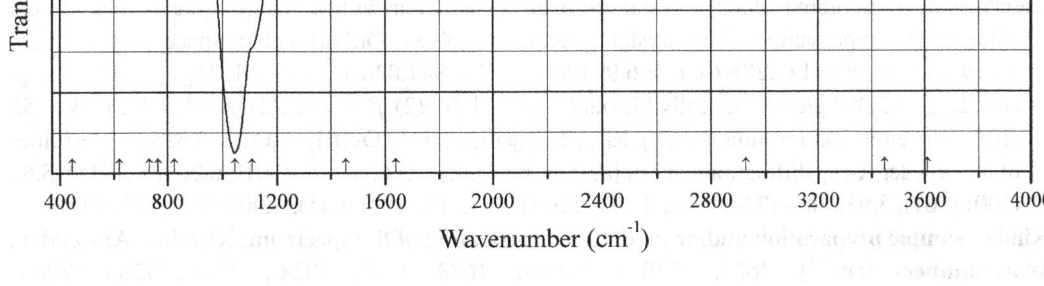

Origin: Synthetic.

Description: Prepared hydrothermally in the reaction between magadiite, sodium aluminate, and NaOH in the presence of glycerol, at 120 °C with subsequent crystallization for several days under autogenous pressure. Characterized by powder X-ray diffraction data.

Kind of sample preparation and/or method of registration of the spectrum: Transmission. Kind of sample preparation is not indicated.

Source: Cui et al. (2014).

Wavenumbers (cm^{-1}): 3610sh, 3447s, 2927w, 1641, 1454w, 1107sh, 1046s, 825, 764, 731, 620, 445s.

Note: The wavenumbers were partly determined by us based on spectral curve analysis of the published spectrum.

Sif_Z134 Martinandresite $Ba_2(Al_4Si_{12}O_{32})\cdot 10H_2O$

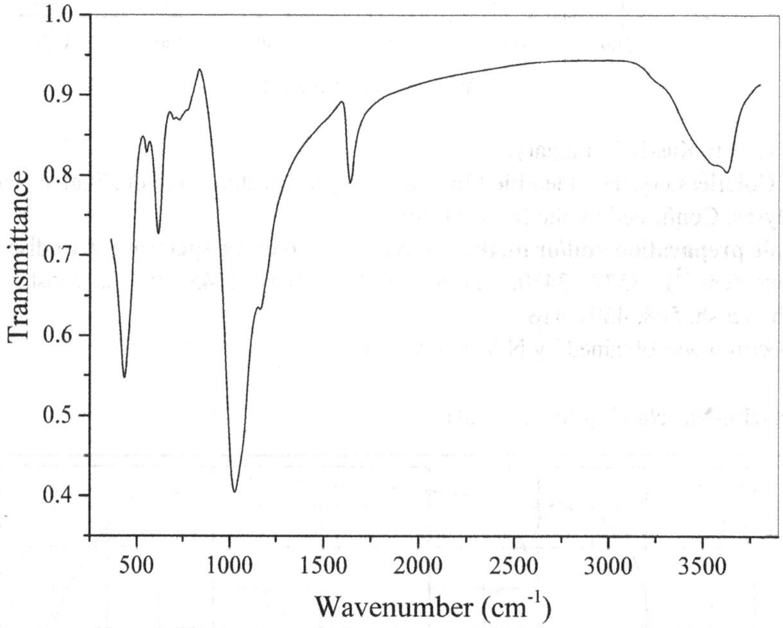

Origin: Wasenalp, near the Isenwegg peak, Ganter valley, Simplon region, Switzerland (type locality).

Description: Tan-colored blocky crystal from the association with armenite, quartz, dickite, and chlorite. Holotype sample. The crystal structure is solved. Orthorhombic, space group *Pmmn*, $a = 9.4640(5)$, $b = 14.2288(6)$, $c = 6.9940(4)$ Å, $V = 941.82(8)$ Å^3, $Z = 1$. $D_{meas} = 2.482(5)$ g/cm^3, $D_{calc} = 2.495$ g/cm^3. Optically biaxial (–), $\alpha = 1.500(2)$, $\beta = 1.512(2)$, $\gamma = 1.515(2)$, $2V = 55(10)°$. The empirical formula is $Na_{0.17}K_{0.04}Ba_{2.00}(Al_{4.19}Si_{11.81}O_{32})H_{19.85}O_{9.93}$. The strongest lines of the powder X-ray diffraction pattern [*d*, Å (*I*, %) (*hkl*)] are: 6.98 (74) (001), 6.26 (83) (011), 5.61 (100) (101), 3.933 (60) (220, 031), 3.191 (50) (112), 3.170 (62) (041), 3.005 (79) (231, 141).

Kind of sample preparation and/or method of registration of the spectrum: KBr disc. Absorption.

Wavenumbers (cm^{-1}): 3625, 3570sh, 3260sh, 1638, 1167, 1028s, 774w, 728w, 700w, 616, 551, 432s.

Note: The spectrum was obtained by N.V. Chukanov.

Sif_Z135 Rongibbsite $Pb_2(Si_4Al)O_{11}(OH)$

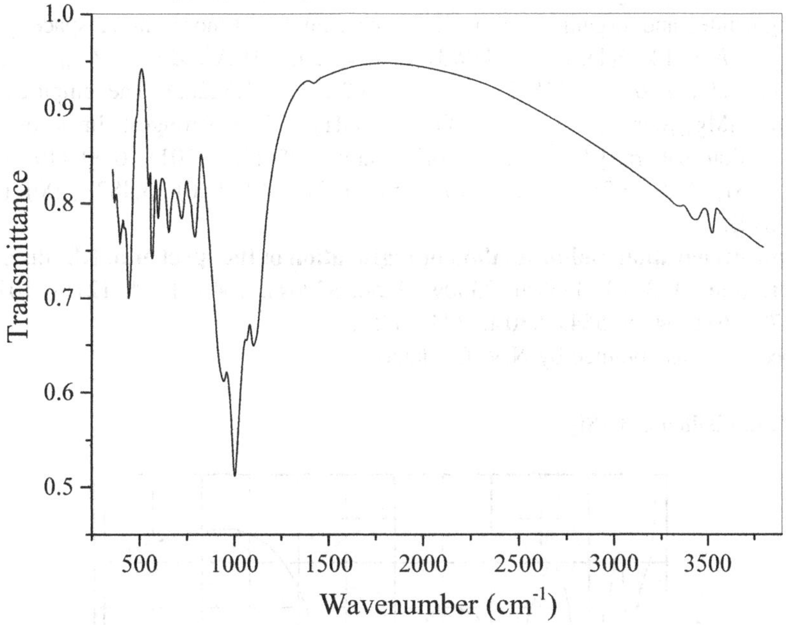

Origin: Big Horn Mts, Maricopa Co., Arizona, USA (type locality).

Description: Colorless prismatic crystals. The empirical formula is (electron microprobe): $Pb_{2.0}Si_{3.75}Al_{1.1}Mg_{0.05}(O,OH)_{12}$.

Kind of sample preparation and/or method of registration of the spectrum: KBr disc. Absorption.

Wavenumbers (cm^{-1}): 3524, 3436, (3347), (1425w), 1103s, 1065s, 1004s, 946s, 794, 724, 656, 600, 569, 549, 485sh, 444s, 425sh, 400, 373.

Note: The spectrum was obtained by N.V. Chukanov.

Sif_Z136 Ferrierite-NH$_4$ $(NH_4,Mg_{0.5})_5(Al_5Si_{31}O_{72})\cdot 22H_2O$

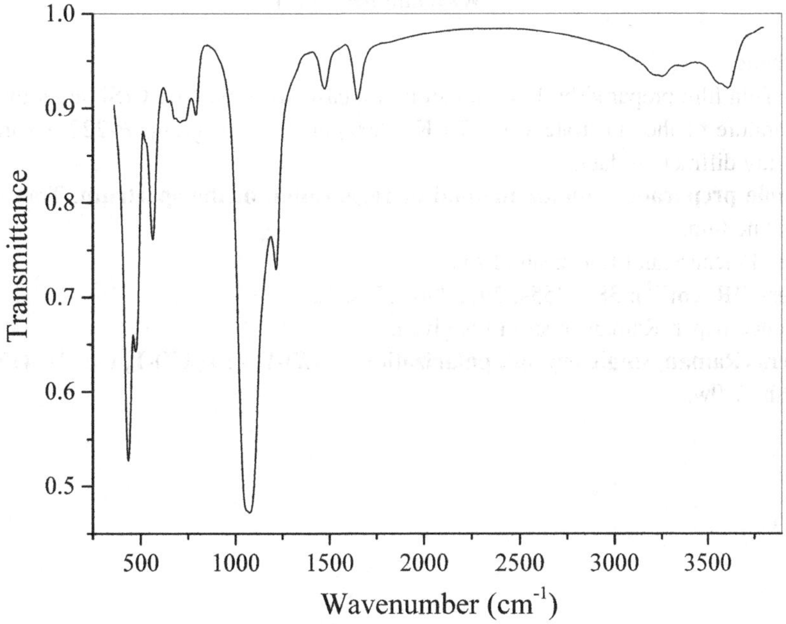

Origin: Libous lignite quarry, near Chomutov, Ústí Region, Bohemia, Czech Republic (type locality).

Description: Radiated aggregates consisting of fibrous crystals from the association with siderite, opal, kaolinite, goethite, and organic matter. Holotype sample. Orthorhombic, space group *Immm*, $a = 19.10(1)$, $b = 14.15(1)$, $c = 7.489(3)$ Å, $V = 2024(3)$ Å^3, $Z = 1$. $D_{calc} = 2.154$ g/cm^3. Optically biaxial (+), $\alpha = 1.518(2)$, $\beta = 1.520(2)$, $\gamma = 1.522(2)$. The empirical formula is $H_{0.35}[(NH_4)_{2.74}Mg_{1.07}Na_{0.21}](Al_{5.44}Si_{30.56}O_{72})\cdot 21.55H_2O$. The strongest lines of the powder X-ray diffraction pattern [*d*, Å (*I*, %) (*hkl*)] are: 6.95 (28) (101), 6.60 (19) (011), 3.988 (61) (321, 031, 420), 3.784 (19) (330), 3.547 (73) (112, 040), 3.482 (100) (202), 3.143 (37) (141, 312).

Kind of sample preparation and/or method of registration of the spectrum: KBr disc. Absorption.

Wavenumbers (cm^{-1}): 3610, 3565sh, 3360w, 3250, 3220sh, 1646, 1474, 1216s, 1076s, 1060sh, 791, 730, 707, 681, 647w, 564s, 530sh, 474s, 435s.

Note: The spectrum was obtained by N.V. Chukanov.

Si49 Chromium disilicide $CrSi_2$

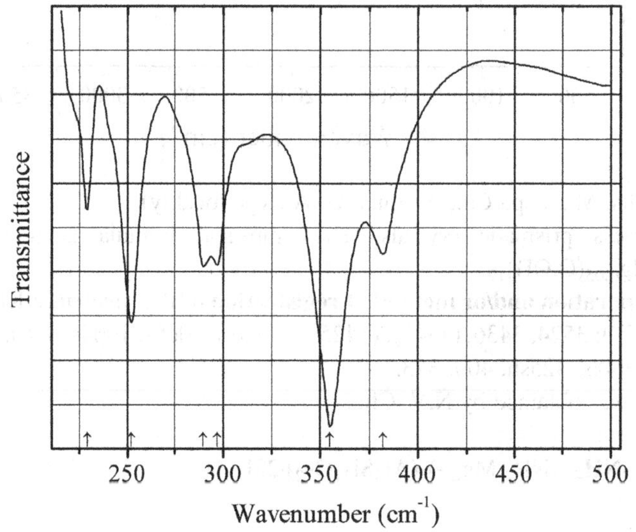

Origin: Synthetic.

Description: Thin film prepared by laser ablation of a cast stoichiometric CrSi target under vacuum. The temperature of the substrate was 773 K. Hexagonal, space group $P6_222$. Characterized by powder X-ray diffraction data.

Kind of sample preparation and/or method of registration of the spectrum: Transmission of a polycrystalline film.

Source: Chaix-Pluchery and Lucazeau (1998).

Wavenumbers (IR, cm^{-1}): 382, 355s, 297, 290, 252s, 229.

Note: In the cited paper, Raman spectrum is given.

Wavenumbers (Raman, single crystal, polarization *Y(XX)-Y* + *εY(XZ)-Y*, cm^{-1}): 412s, 397, 354, 305s, 300sh, 290w.

Si50 Bridgmanite $MgSiO_3$

Magnesium silicon oxide perovskite-type

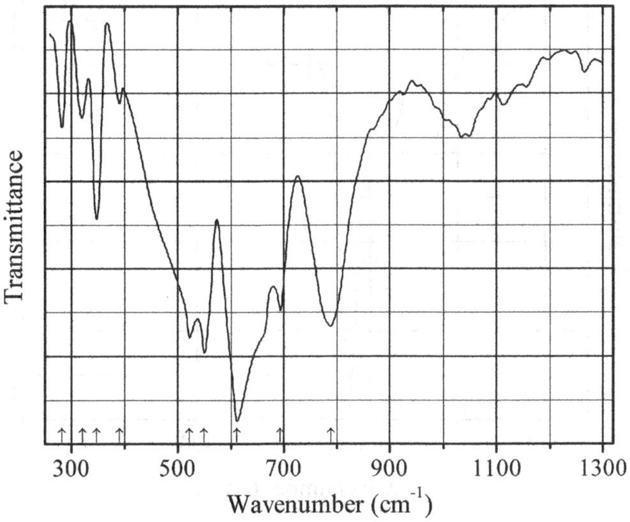

Origin: Synthetic.

Description: Obtained from the $MgSiO_3$ glass in a laser heated diamond anvil cell at 500 kbar.

Kind of sample preparation and/or method of registration of the spectrum: CsI or KBr disc. Absorption.

Source: Madon and Price (1989).

Wavenumbers (cm^{-1}): 789s, 694, 612s, 550s, 522s, 390w, 347, 320, 282.

Si51 Akimotoite $MgSiO_3$

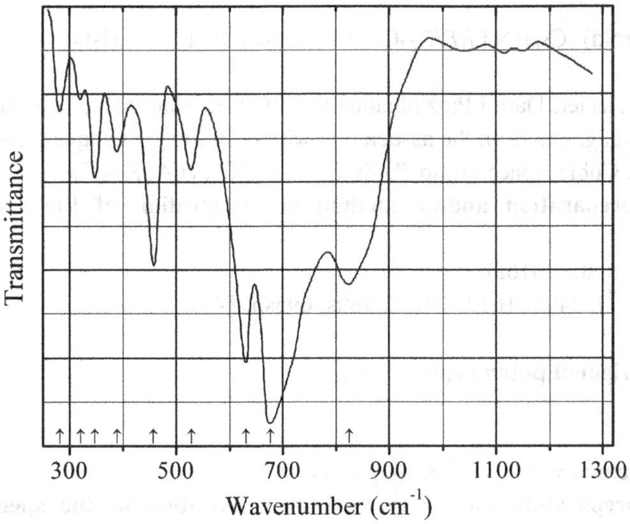

Origin: Synthetic.

Description: Synthesized from $MgSiO_3$ glass in a laser heated diamond anvil cell at 500 kbar.

Kind of sample preparation and/or method of registration of the spectrum: KBr or CsI disc. Absorption.

Source: Madon and Price (1989).

Wavenumbers (cm^{-1}): 825s, 677s, 631s, 528, 457, 388, 347, 320w, 282.

Si52 Luobusaite *β*-$FeSi_2$

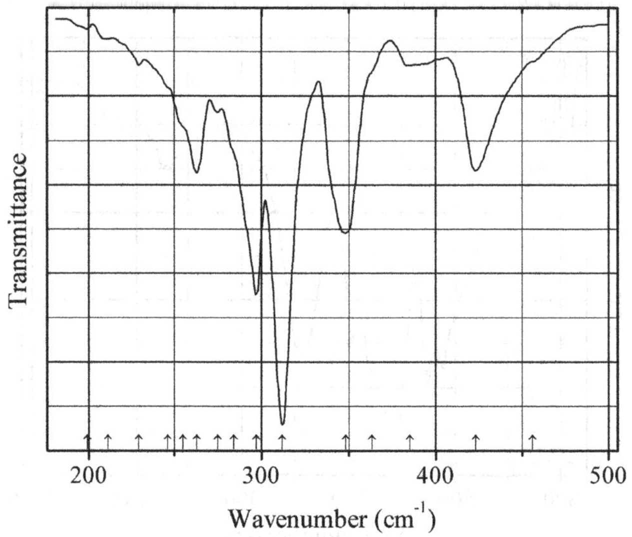

Origin: Synthetic.

Kind of sample preparation and/or method of registration of the spectrum: Thin film on single crystal Si substrate. Absorption.

Source: Fenske et al. (1996).

Wavenumbers (cm^{-1}): 456sh, 423, 385w, 363sh, 348s, 312s, 297s, 284sh, 275w, 263, 255sh, 246sh, 229w, 211sh, 199w.

Note: The wavenumbers were determined by us based on spectral curve analysis of the published spectrum.

Si53 Mendeleevite-(Nd) $Cs_6[(Nd,REE)_{23}Ca_7)(Si_{70}O_{175})(OH,F)_{19}\cdot 16H_2O$

Origin: Dara i Pioz glacier, Dara i Pioz alkaline massif, Tien Shan Mts., Tajikistan (type locality).

Description: Anhedral grains from the association with pectolite grains, quartz, aegirine, fluorite, etc. Holotype sample. Cubic, space group *Pm*-3, $a = 21.9106(4)$ Å, $Z = 2$.

Kind of sample preparation and/or method of registration of the spectrum: KBr disc. Transmission.

Source: Agakhanov et al. (2016a).

Wavenumbers (cm^{-1}): 3408, 1612, 1011s, 980s, 695sh, 547sh.

Si54 Bridgmanite trigonal polymorph $MgSiO_3$

Origin: Synthetic.

Description: Isostructural with ilmenite. Single crystal.

Kind of sample preparation and/or method of registration of the spectrum: Unpolarized reflection.

Source: Hofmeister and Ito (1992).

Wavenumbers (cm^{-1}): 951, 665, 619, 536w, 448, 377.

Si55 Sapphirine $Mg_4(Mg_3Al_9)O_4[Si_3Al_9O_{36}]$

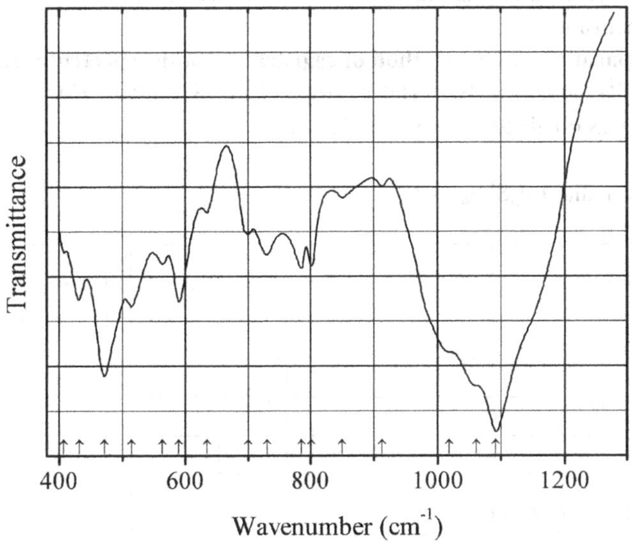

Origin: Betroka, Anosy Region, Tuléar (Toliara) province, Madagascar.

Description: Metacrystal from metamorphic schist.

Kind of sample preparation and/or method of registration of the spectrum: KBr disc. Transmission.

Source: Povarennykh (1970).

Wavenumbers (cm^{-1}): 1092s, 1060sh, 1018sh, 913w, 850w, 801, 785, 730, 700, 635w, 590, 564, 515, 472s, 432, 408w.

Sia24 Pearlite $(Na,K)_x(Si_{1-x}Al_xO_2)\cdot nH_2O$ ($x \ll 1$)

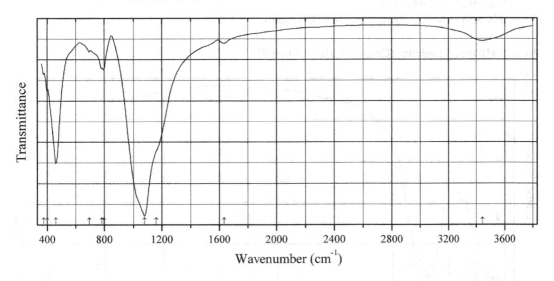

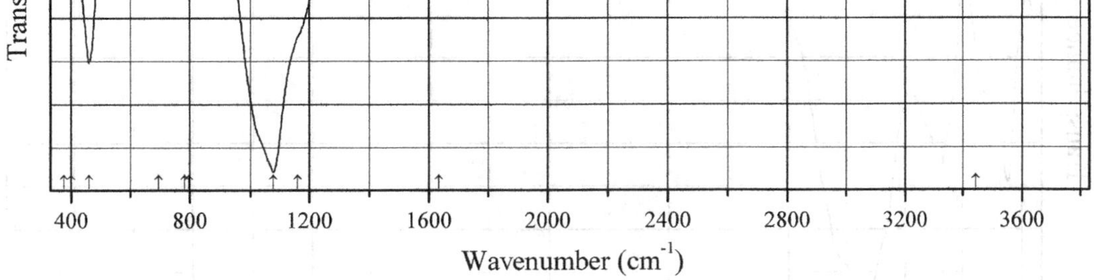

Origin: Tsigrado quarry, Milos Island, Greece.

Description: Light gray, semitransparent, massive, from the association with hydrobiotite. The sample contains quartz inclusions.

Kind of sample preparation and/or method of registration of the spectrum: KBr disc. Absorption.

Wavenumbers (cm^{-1}): 3442, 1634w, 1160sh, 1078s, 796, 783, 695w, 461s, 400, 375.

Note: The spectrum was obtained by N.V. Chukanov.

BeSi74 Sphaerobertrandite $Be_3SiO_4(OH)_2$

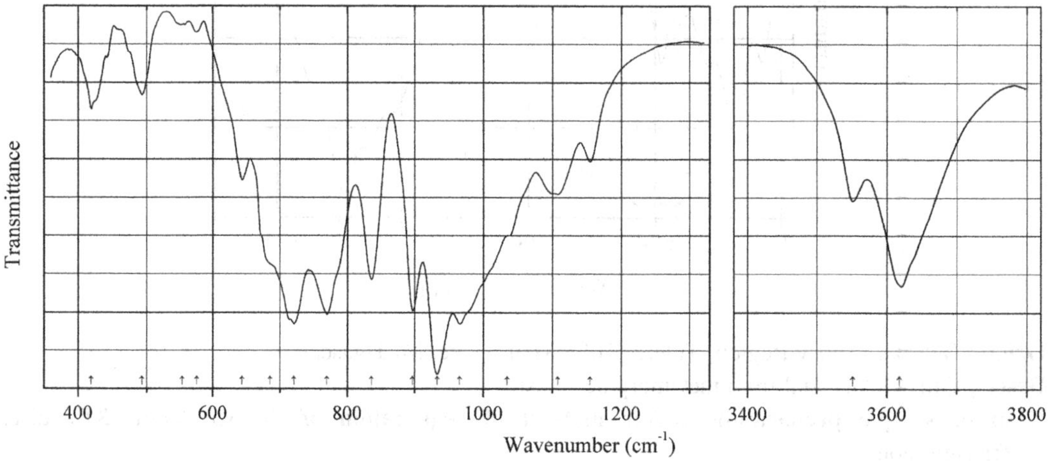

Origin: Sagåsen, Tvedalen, Larvik, S. Norway.

Description: Yellow spherulites from the association with diaspor. Identified by the IR spectrum.

Kind of sample preparation and/or method of registration of the spectrum: KBr disc. Absorption.

Wavenumbers (cm^{-1}): 3618s, 3551s, 1155, 1108, 1035sh, 965s, 933s, 897s, 836s, 770s, 721s, 685sh, 643, 576w, 554w, 493, 419.

Note: The spectrum was obtained by N.V. Chukanov.

BeSi75 Hydroxylgugiaite $(Ca_3\square)(Si_{3.5}Be_{2.5})O_{11}(OH)_3$

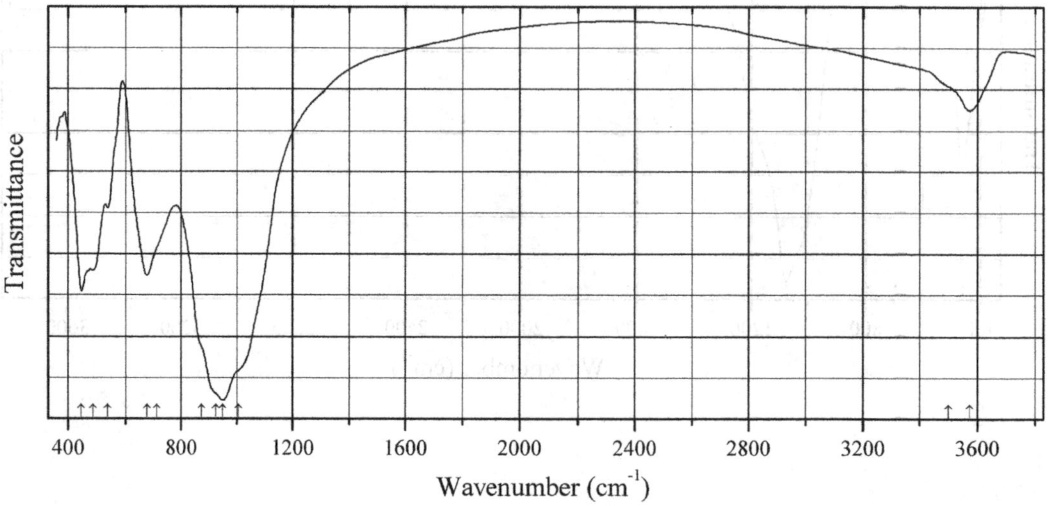

Origin: Larvik plutonic complex, Porsgrunn, Telemark, Norway (type locality).

Description: Pale grey crystals from the association with chlorite and calcite. Holotype sample. The crystal structure is solved. Tetragonal, space group $P\text{-}42_1/m$, a = 7.4151(2), c = 4.9652(1) Å, V = 272.9(1) Å^3, Z =12. Optically uniaxial (+), ω = 1.622(2), ε = 1.632(2). The empirical formula is $(Ca_{2.76}Na_{0.31}Mn_{0.05}Fe_{0.01})(Si_{3.45}Be_{2.53}Al_{0.07})O_{11}[(OH)_{2.57}F_{0.43}]$.

Kind of sample preparation and/or method of registration of the spectrum: KBr disc. Absorption.

Wavenumbers (cm^{-1}): 3574, 3500sh, 1105sh, 950s, 925sh, 875sh, 710sh, 676, 539, 487, 447.

Note: The spectrum was obtained by N.V. Chukanov.

BeSi76 "Hydroxylgadolinite-(Y)" $(Y,Ca)_2(Fe,\square)Be_2Si_2O_8(OH,O)_2$

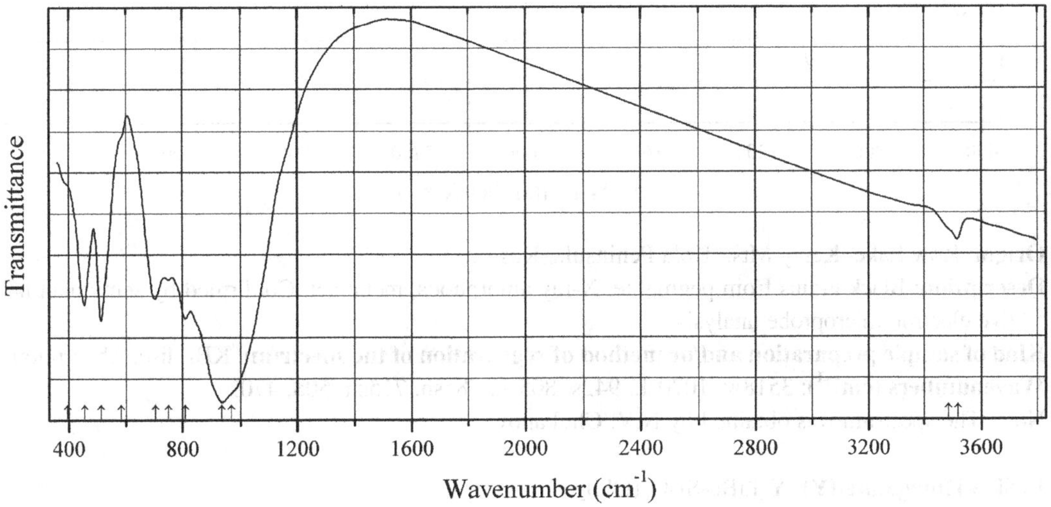

Origin: Heftetjern granitic pegmatite, Southern Norway.

Description: Bottle-green grains and short-prismatic crystals from the association with feldspar, quartz, and intermediate members of the gadolinite-(Y)–hingganite-(Y) solid-solution series. The crystal structure is solved. Monoclinic, space group $P2_1/c$, a = 4.7514(10), b = 7.5719(16), c = 9.9414(2) Å, β = 90.015(4)°, V = 357.663(3) Å^3, Z = 2. D_{calc} = 3.967 g/cm^3. Optically biaxial (+), α = 1.760(4), β = 1.770(4), γ = 1.785(4), $2V$ = 80(10)°. The empirical formula is $(Y_{1.285}Ca_{0.55}Ce_{0.07}La_{0.04}Nd_{0.01})Fe^{2+}{}_{0.57}Be_{2.02}Si_{1.995}O_{8.48}(OH)_{1.52}$. The strongest lines of the powder X-ray diffraction pattern [d, Å (I, %) (hkl)] are: 4.761 (48) (100), 3.554 (30) (021), 3.452 (30) (10–2, 102), 3.138 (81) (11–2, 112), 2.972 (39) (120), 2.849 (100) (12–1, 121), 2.570 (59) (11–3, 113, 12–2), and 2.215 (27) (211, 12–3, 123).

Kind of sample preparation and/or method of registration of the spectrum: KBr disc. Absorption.

Wavenumbers (cm^{-1}): 3517, 3485sh, 970sh, 936s, 809, 752, 704, 585sh, 514, 455, 395sh.

Note: The spectrum was obtained by N.V. Chukanov.

BeSi77 Gadolinite-(Y) $Y_2Fe^{2+}Be_2Si_2O_{10}$

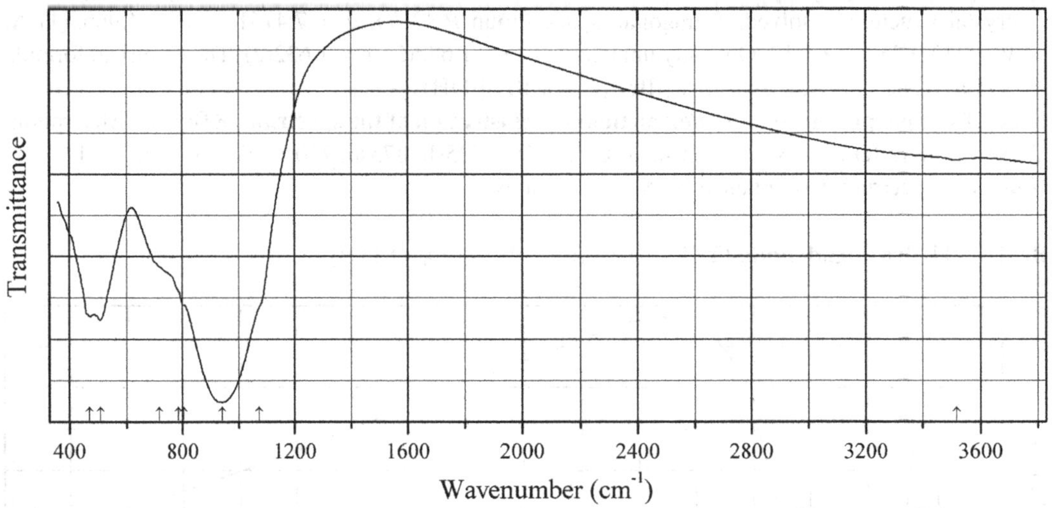

Origin: Row Lake, Keivy Mts., Kola Peninsula, Russia.
Description: Black grains from pegmatite. X-ray amorphous, metamict. Confirmed by semiquantitative electron microprobe analysis.
Kind of sample preparation and/or method of registration of the spectrum: KBr disc. Absorption.
Wavenumbers (cm^{-1}): 3518w, 1070sh, 942s, 805sh, 785sh, 715sh, 508, 470.
Note: The spectrum was obtained by N.V. Chukanov.

BeSi78 Hingganite-(Y) $Y_2\square Be_2Si_2O_8(OH)_2$

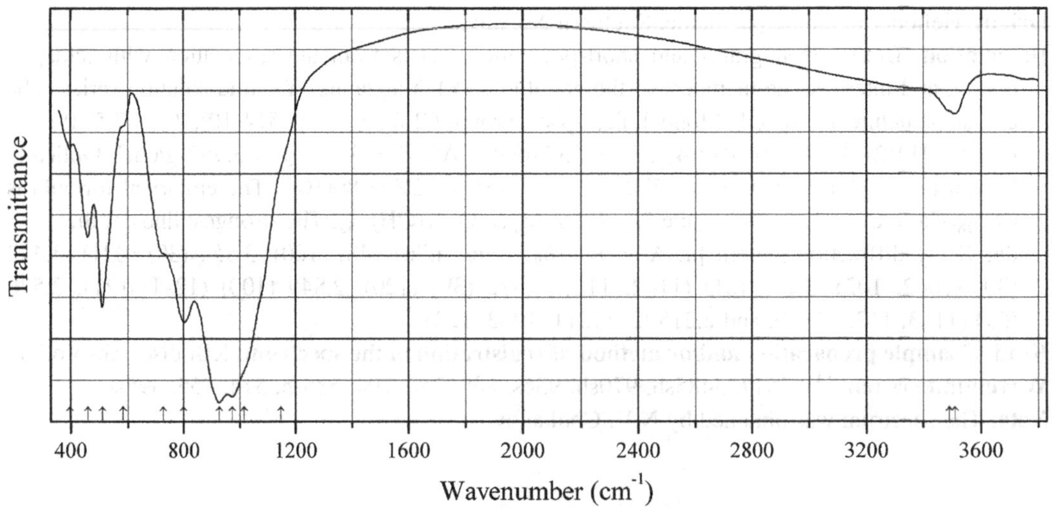

Origin: Heftetjern pegmatite, Tørdal, Telemark, Norway.
Description: Olive-green crystals from the association with *REE*-bearing epidote. A Ca- and Fe-rich variety. The empirical formula is (electron microprobe): ~$(Y_{1.1}Ca_{0.8}Ln_{0.1})(Fe_{0.35}Al_{0.05})Be_2Si_2O_{8.05}(OH)_{1.95}$.
Kind of sample preparation and/or method of registration of the spectrum: KBr disc. Absorption.
Wavenumbers (cm^{-1}): 3510, 3490sh, 1145sh, 1015sh, 973s, 927s, 800, 725sh, 585sh, 511, 460, 398w.
Note: The spectrum was obtained by N.V. Chukanov.

BeSi79 Bityite $CaLiAl_2(Si_2BeAl)O_{10}(OH)_2$

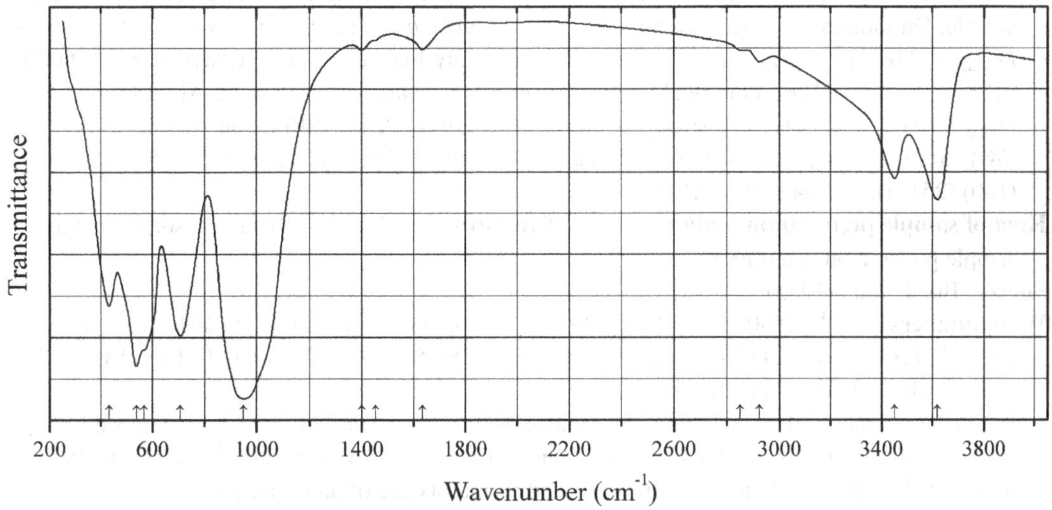

Origin: Maantienvarsi pegmatite, Eräjärvi area, Orivesi, southern Finland.

Description: Fine-scaled white to yellowish mass with a pearly lustre from the association with beryl, bertrandite, fluorite, and fluorapatite. Characterized by powder X-ray diffraction data. Monoclinic, space group *C2/c* or *Cc*, $a = 4.99$, $b = 8.68$, $c = 19.04$ Å, $\beta = 95.17°$, $V = 821.33$ Å^3. $D_{meas} = 3.05$ g/cm^3, $D_{calc} = 3.12$ g/cm^3. Optically biaxial (–), $\alpha = 1.650$, $\beta = 1.658$, $\gamma = 1.660$, $2V = 52.9°$. The empirical formula is (wet chemical analysis, $Z = 2$): $(Ca_{1.93}K_{0.03}Na_{0.02})(Li_{1.19}Al_{3.68}Mg_{0.35}Fe_{0.13})(Si_{4.26}Be_{2.21}Al_{1.53})O_{19.30}(OH)_{4.54}F_{0.16}$.

Kind of sample preparation and/or method of registration of the spectrum: KBr disc. Transmission.

Source: Lahti and Saikkonen (1985).

Wavenumbers (cm^{-1}): 3620, 3451, (2924), (2853sh), 1634w, 1453sh, 1400w, 949s, 707s, 568sh, 537s, 431.

Note: The wavenumbers were partly determined by us based on spectral curve analysis of the published spectrum.

BeSi80 Chiavennite $CaMn^{2+}(BeOH)_2Si_5O_{13}{\cdot}2H_2O$

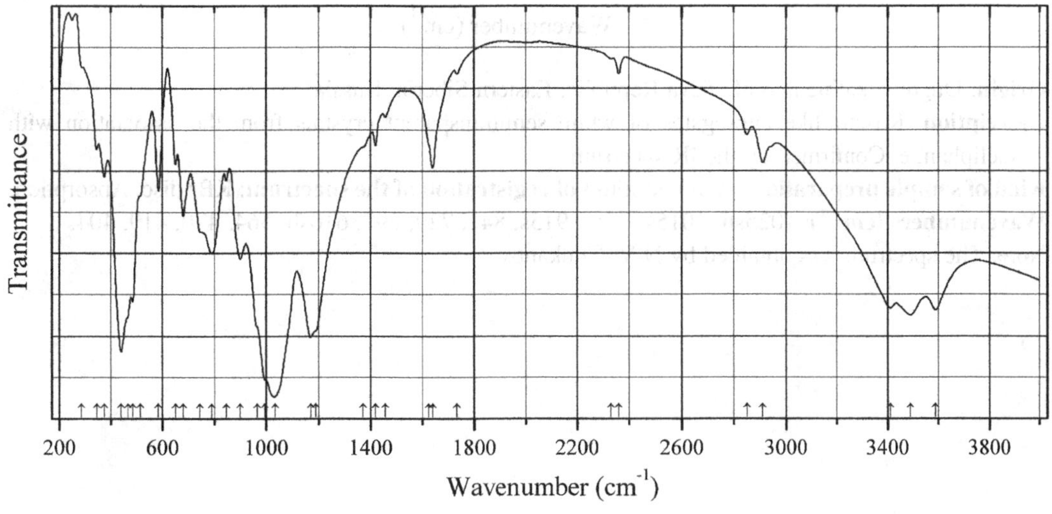

Origin: Chiavenna, Valchiavenna, Sondrio Province, Lombardy, Italy (type locality).

Description: Euhedral orange grains and crusts from the association with beryl and bavenite. Holotype sample. Orthorhombic, space group $P2_1ab$, $a = 8.729(5)$, $b = 31.326(11)$, $c = 4.903(2)$ Å, $Z = 4$. $D_{meas} = 2.64(1)$ g/cm^3, $D_{calc} = 2.657$ g/cm^3. Optically biaxial (−), $\alpha = 1.581(1)$, $\gamma = 1.600(1)$, $2V = 70(5)°$. The empirical formula is $(Ca_{0.97}Na_{0.05})Mn_{0.97}(Be_{1.98}Al_{0.03})(Si_{4.65}Al_{0.35})O_{12.63}(OH)_{2.37}{\cdot}2.16H_2O$. The strongest lines of the powder X-ray diffraction pattern [*d*, Å (*I*, %) (*hkl*)] are: 15.7 (100) (020), 4.15 (30) (041), 3.93 (30) (080), 3.82 (30) (240), 3.28 (75) (201), 2.903 (100) (251, 181), 1.944 (30) (3.12.0).

Kind of sample preparation and/or method of registration of the spectrum: Absorption. Kind of sample preparation is not indicated.

Source: Bondi et al. (1983).

Wavenumbers (cm^{-1}): 3590, 3490, 3410, (2910), (2850), (2360), (2330), 1735, 1640, 1625sh, 1458, 1420, 1372sh, 1190sh, 1170, 1035, 995sh, 965sh, 900, 845, 790, 745sh, 680, 650, 585, 515sh, 485, 465sh, 440, 375, 345, 285sh.

Note: The wavenumbers were partly determined by us based on spectral curve analysis of the published spectrum. Weak bands in the ranges from 2300 to 2400 cm^{-1} and from 2800 to 3000 cm^{-1} correspond to atmospheric CO_2 and the admixture of an organic substance.

BeSi81 Gugiaite $Ca_2BeSi_2O_7$

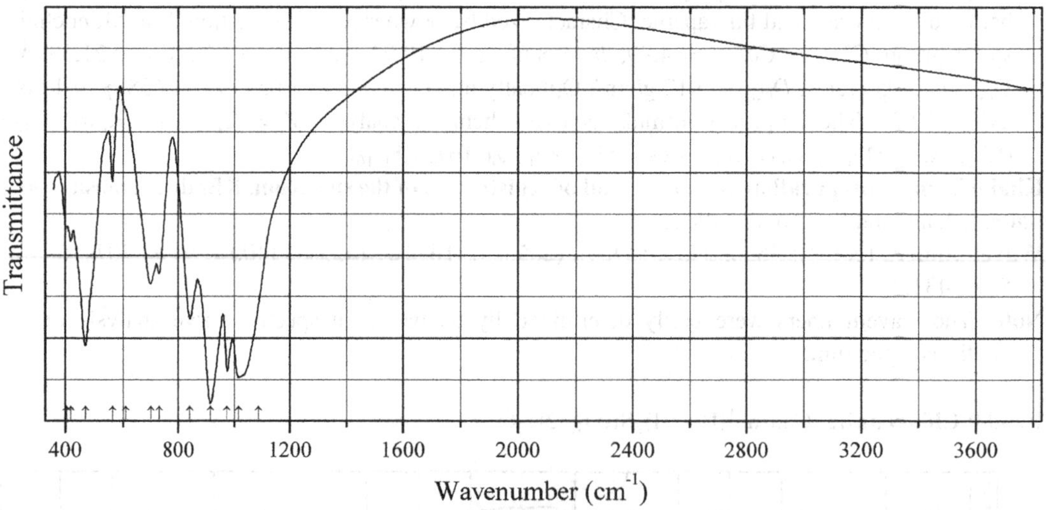

Origin: Dugdu alkaline massif, Tuva Republic, Eastern Siberia, Russia.

Description: Rosette-like aggregates of white semitransparent crystals from the association with meliphanite. Confirmed by the IR spectrum.

Kind of sample preparation and/or method of registration of the spectrum: KBr disc. Absorption.

Wavenumbers (cm^{-1}): 1025sh, 1015s, 975s, 915s, 841, 729, 697, 610sh, 564, 470, 419, 404.

Note: The spectrum was obtained by N.V. Chukanov.

BeSi82 Leifite $NaNa_6Be_2Al_3Si_{15}O_{39}F_2$

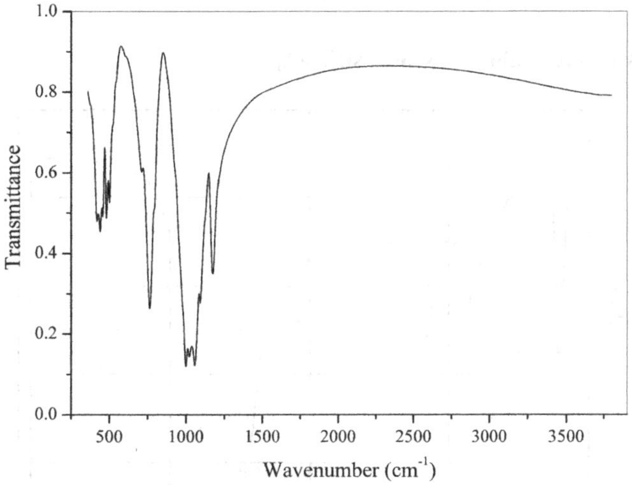

Origin: Poudrette (Demix) quarry, Mont Saint-Hilaire, Rouville RCM (Rouville Co.), Montérégie, Québec, Canada.

Description: White radiated aggregate of prismatic crystals. Investigated by A.V. Voloshin. Confirmed by the IR spectrum.

Kind of sample preparation and/or method of registration of the spectrum: KBr disc. Absorption.

Wavenumbers (cm^{-1}): 1175s, 1094s, 1058s, 1023s, 1000s, 935sh, 795sh, 764s, 712, 610sh, 545sh, 515sh, 502, 481, 457, 440s, 419.

Note: The spectrum was obtained by N.V. Chukanov.

BeSi83 Danalite $Fe^{2+}{}_4Be_3(SiO_4)_3S$

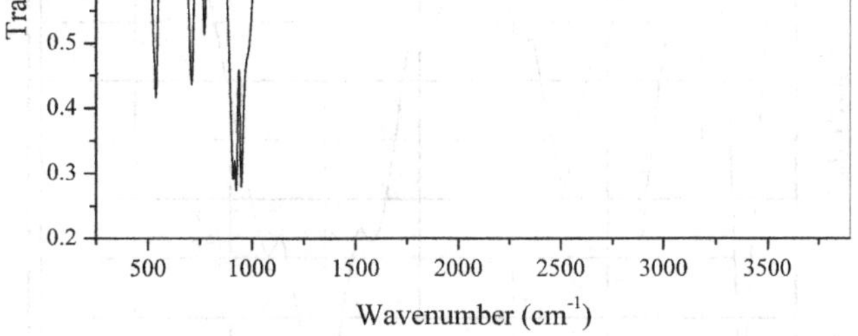

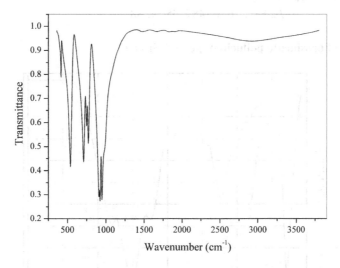

Origin: Lupikko deposit, near Pitkäranta, Ladoga lake, Karelia, Russia.

Description: Dark red grain in skarn, in the association with clinochlore, calcite, and fluorite. The empirical formula is (electron microprobe): $H_xFe_{2.6}Mn_{0.8}Zn_{0.6}Be_3Si_3O_{12}S_{0.9}$.

Kind of sample preparation and/or method of registration of the spectrum: KBr disc. Absorption.

Wavenumbers (cm^{-1}): 2922 (broad), 1907w, 1830w, 1667w, 1473w, 975sh, 948s, 924s, 910s, 771, 747, 710, 538, 416.

Note: The spectrum was obtained by N.V. Chukanov. The broad band at 2922 cm^{-1} indicates the presence of OH groups forming strong hydrogen bonds.

BSi99 Cesium borosilicate pollucite-type $CsBSi_2O_6$

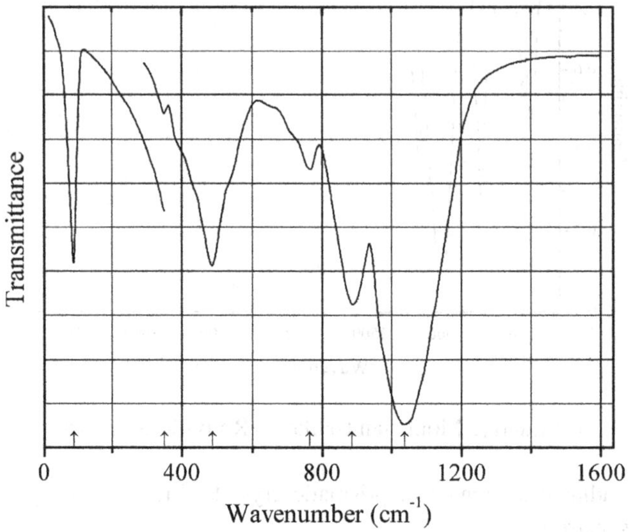

Origin: Synthetic.
Description: Synthesized from the stoichiometric mixture of Cs_2CO_3, H_3BO_3, and SiO_2 by a conventional solid-state reaction technique at 800 °C. Structurally related to pollucite.
Kind of sample preparation and/or method of registration of the spectrum: KBr and polyethylene discs. Transmission.
Source: Rulmont and Tarte (1987).
Wavenumbers (cm^{-1}): 1037s, 886, 765w, 487, 350w, 88.

BSi100 Potassium borosilicate pollucite-type $K(BSi_2O_6)$

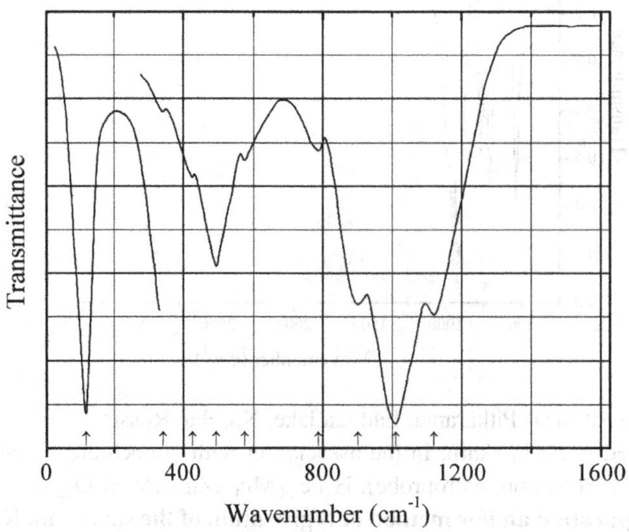

Origin: Synthetic.

Description: Obtained by heating the stoichiometric mixture of K_2CO_3, H_3BO_3, and SiO_2 to 800 °C. Characterized by powder X-ray diffraction data.

Kind of sample preparation and/or method of registration of the spectrum: KBr and polyethylene discs. Transmission.

Source: Rulmont and Tarte (1987).

Wavenumbers (cm^{-1}): 1120s, 1010s, 900s, 788w, 575w, 495, 427, 340w, 117s.

BSi101 Nolzeite $NaMn_2(Si_3BO_9)(OH)_2·2H_2O$

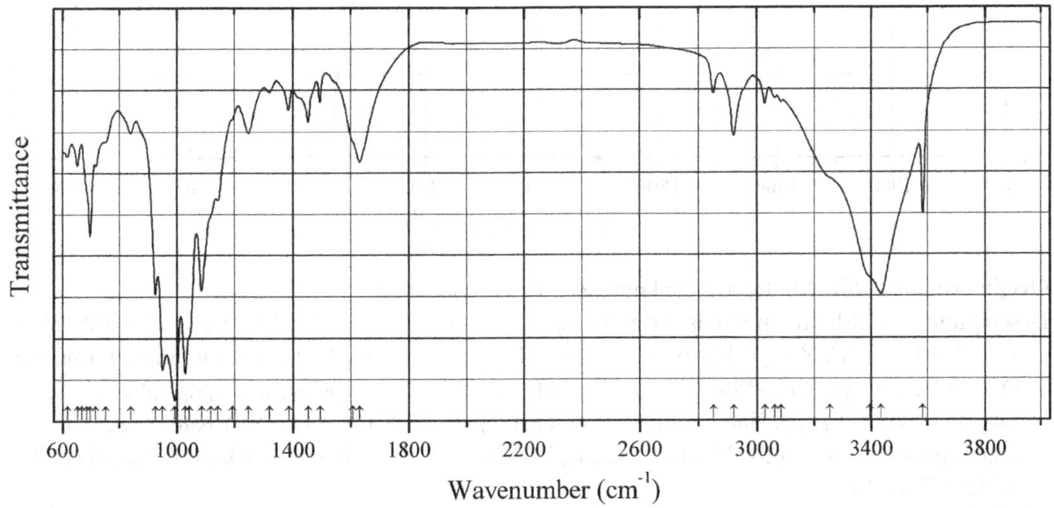

Origin: Poudrette quarry, La Vallée-du-Richelieu, Montérégie (formerly Rouville County), Québec, Canada (type locality).

Description: Pale green acicular crystals from the association with aegirine, nepheline, sodalite, eudialyte-group minerals, analcime, natron, pyrrhotite, catapleiite, and steedeite. Holotype sample. The crystal structure is solved. Boron has fourfold coordination. Triclinic, space group *P*-1, $a = 6.894$ (1), $b = 7.632(2)$, $c = 11.017(2)$ Å, $\alpha = 108.39(3)°$, $\beta = 99.03(3)°$, $\gamma = 103.05(3)°$, $V = 519.27$ Å^3, $Z = 2$. $D_{calc} = 2.79$ g/cm^3. Optically biaxial, $n_{min} = 1.616(2)$, $n_{max} = 1.636(2)$. The empirical formula is $Na_{1.04}(Mn_{1.69}\square_{0.24}Fe_{0.05}Ca_{0.02})(Si_{2.96}S_{0.04})(B_{0.70}Si_{0.30})O_9(OH)_2·2H_2O$. The strongest lines of the powder X-ray diffraction pattern [*d*, Å (*I*, %) (*hkl*)] are: 10.113 (100) (00−1), 6.911 (16) (0−10), 3.593 (13) (0−13), 3.026 (15) (0−23), 2.808 (50) (211, 2−20), 2.675 (12) (0−1−3).

Kind of sample preparation and/or method of registration of the spectrum: Transmission. Kind of sample preparation is not indicated.

Source: Haring and McDonald (2016).

Wavenumbers (IR, cm^{-1}): 3583, 3434s, 3395sh, 3254sh, 3083w, 3062w, 3028w, 2922, 2852w, 1631, 1607sh, 1493w, 1453w, 1385w, 1320w, 1247, 1192sh, 1143s, 1117sh, 1086s, 1044sh, 1030, 993s, 951, 926, 840, 752sh, 717, 697, 685sh, 653w, 618w.

Note: The wavenumbers were partly determined by us based on spectral curve analysis of the published spectrum. Weak bands in the range from 2800 to 3100 cm^{-1} may correspond to the admixture of an organic substance. Weak bands in the range from 1240 to 1500 cm^{-1} may correspond to the admixture of an organic substance and/or (partly) carbonate groups. In the cited paper, Raman spectrum is given.

Wavenumbers (Raman, cm^{-1}): 3548, 3399, 1588, 1290, 1009, 842, 626, 553, 390, 341, 268, 223, 167.

BSi102 Proshchenkoite-(Y) (Y,*REE*,Ca,Na,Mn)$_{15}$Fe^{2+}Ca(P,Si)Si$_6$B$_3$(O,F)$_{48}$

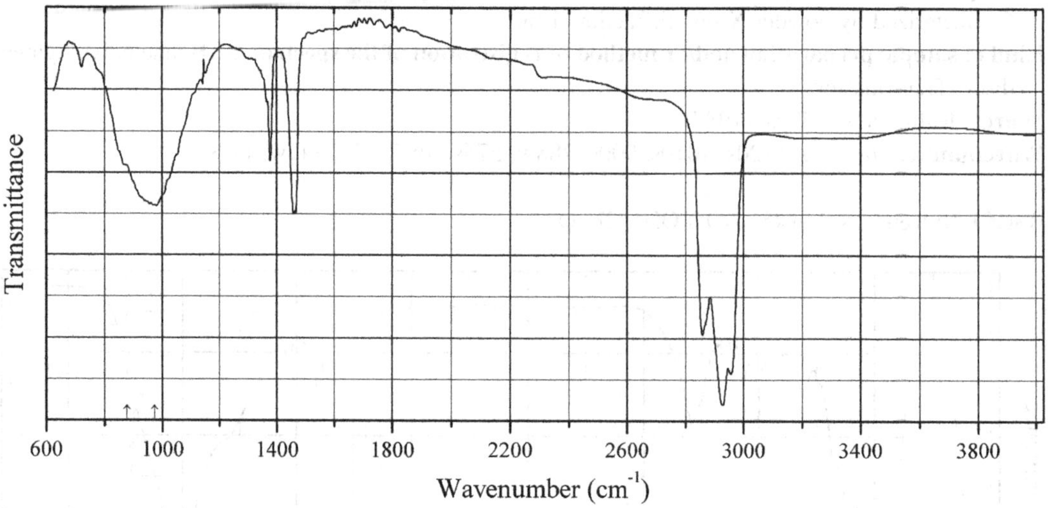

Origin: Tommot *REE*-Nb deposit, Yakutia, Russia (type locality).

Description: Reddish-brown grains. Trigonal, space group *R*3*m*, *a* = 10.7527(7), *c* = 27.4002(18) Å, *V* = 2743.6(6) Å^3, *Z* = 3. Uniaxial (−), ω = 1.734(2), ε = 1.728(2). The empirical formula is (Y$_{3.70}$*REE*$_{7.54}$Ca$_{1.55}$Na$_{1.16}$Mn$_{0.77}$Th$_{0.10}$Pb$_{0.01}$)(Fe$^{2+}_{0.83}$Mn$_{0.15}$Ti$_{0.02}$)Ca$_{1.00}$(P$_{0.70}$Si$_{0.26}$As$_{0.04}$) Si$_{6.05}$B$_{3.20}$(O$_{34.55}$F$_{13.45}$). The strongest lines of the powder X-ray diffraction pattern [*d*, Å (*I*, %) (*hkl*)] are: 4.441 (36) (202), 3.144 (77) (214), 3.028 (45) (009), 2.968 (100) (027), 1.782 (32) (330), 1.713 (32) (1.2.14).

Kind of sample preparation and/or method of registration of the spectrum: Nujol mull. Transmission.

Source: Kristiansen (2016); for the sample description see Raade et al. (2008).

Wavenumbers (cm^{-1}): 972s, 871sh.

Note: Strong bands above 1200 cm^{-1} correspond to Nujol.

BSi103 Darrellhenryite Na(LiAl$_2$)Al$_6$(BO$_3$)$_3$Si$_6$O$_{18}$(OH)$_3$O

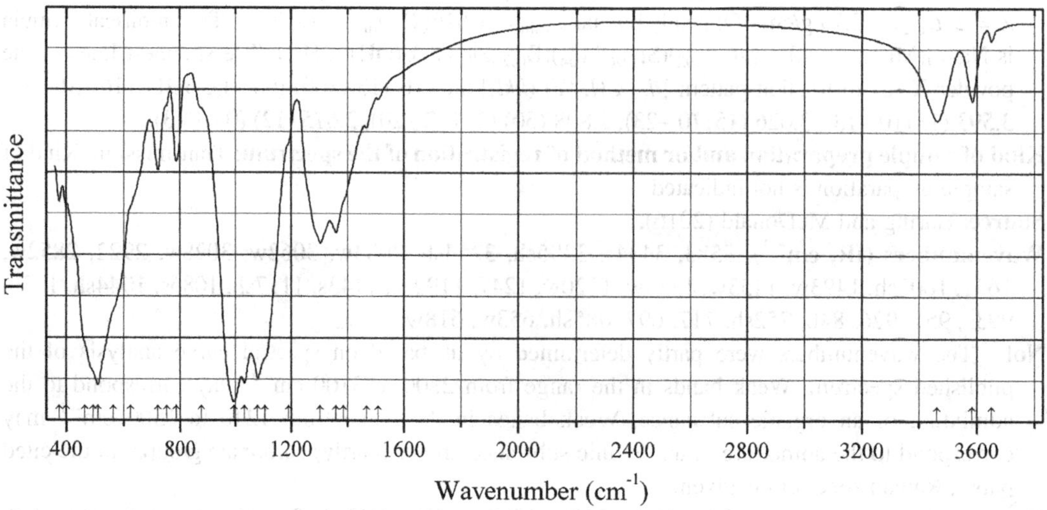

Origin: Nová Ves, Český Krumlov, South Bohemia Region, Czech Republic (type locality).
Description: Pink grains. Fragments of the cotype sample. Characterized by electron microprobe analyses.
Kind of sample preparation and/or method of registration of the spectrum: KBr disc. Absorption.
Wavenumbers (cm^{-1}): 3650w, 3584, 3463, 1509w, 1467w, 1385sh, 1359, 1304, 1190sh, 1105sh, 1079s, 1032s, 996s, 880sh, 786, 750sh, 720, 625sh, 565sh, 511s, 495sh, 465sh, 400w, 376w.
Note: The spectrum was obtained by N.V. Chukanov.

BSi104 Darrellhenryite $Na(LiAl_2)Al_6(BO_3)_3Si_6O_{18}(OH)_3O$

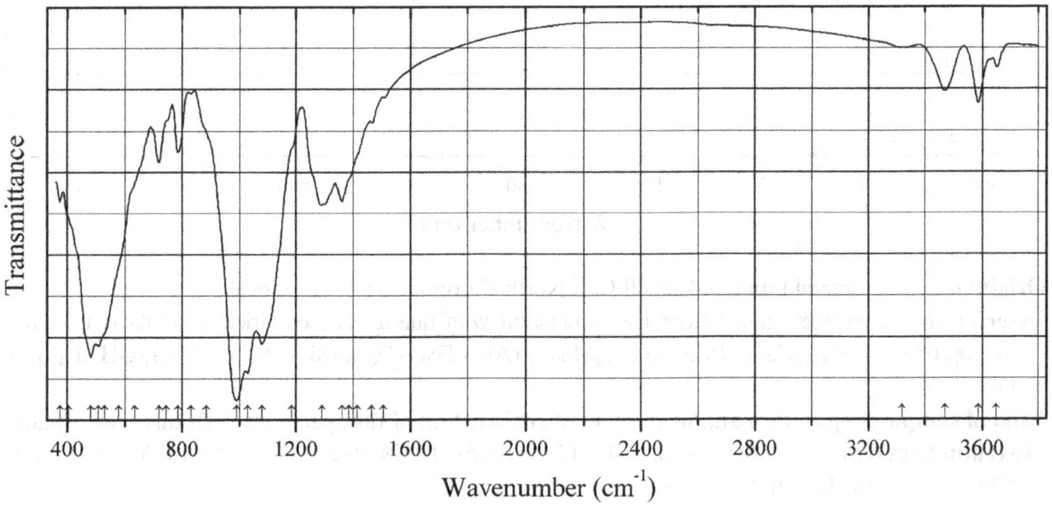

Origin: Aleksandrovskoe Ta deposit, Irkutsk region, Siberia, Russia.
Description: Colorless grains from pegmatite. Investigated by A.V. Kasatkin. The empirical formula is (electron microprobe, Li calculated): $Na_{0.59}LiAl_{7.98}Ti_{0.02}(BO_3)_3(Si_6O_{18})(OH)_3(O,OH)$.
Kind of sample preparation and/or method of registration of the spectrum: KBr disc. Absorption.
Wavenumbers (cm^{-1}): 3649w, 3585, 3469, 3320w, 1507w, 1465w, 1415sh, 1385sh, 1360, 1280, 1185sh, 1080s, 1029s, 990s, 885sh, 830w, 785, 745sh, 719, 635sh, 580sh, 530sh, 510s, 483s, 405sh, 375.
Note: The spectrum was obtained by N.V. Chukanov.

BSi105 Fluor-dravite $NaMg_3Al_6(Si_6O_{18})(BO_3)_3(OH)_3F$

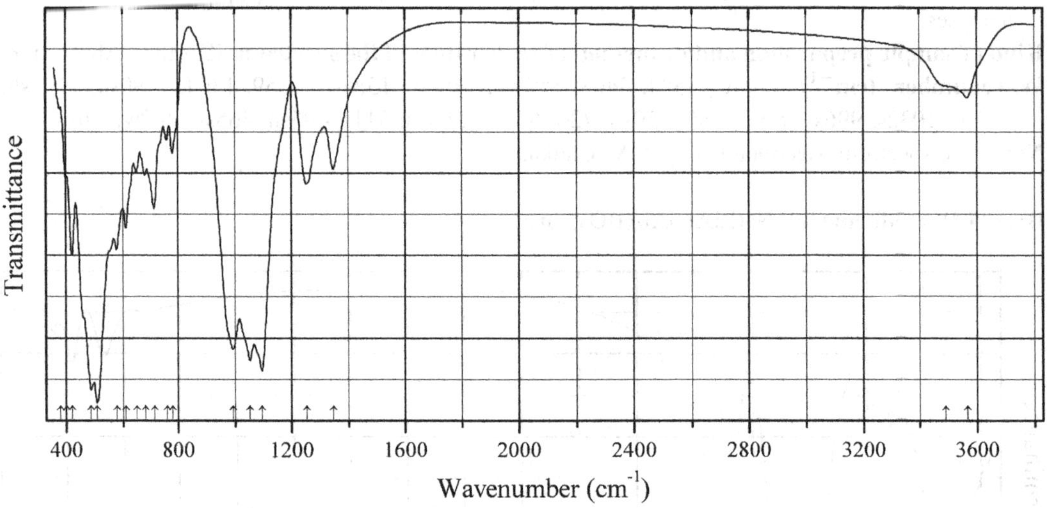

Origin: Crabtree Emerald mine, Mitchell Co., North Carolina, USA (type locality).

Description: Olive-green grains from the association with quartz. The empirical formula is (electron microprobe): $(Na_{0.78}Ca_{0.05})(Mg_{1.75}Fe_{1.14}Mn_{0.11})(Al_{5.78}Fe_{0.17}Ti_{0.05})(Si_{5.91}Al_{0.09})(BO_3)_3(OH)_3F_{0.68}(O,OH)_{0.32}$.

Kind of sample preparation and/or method of registration of the spectrum: KBr disc. Absorption.

Wavenumbers (cm^{-1}): 3566, 3490sh, 1347, 1253, 1095s, 1052s, 992s, 780, 760, 715, 681, 649, 611, 579, 508s, 488s, 421, 402w, 381w.

Note: The spectrum was obtained by N.V. Chukanov.

BSi106 Luinaite-(OH) $NaFe^{2+}{}_3Al_6(Si_6O_{18})(BO_3)_3(OH)_3(OH)$

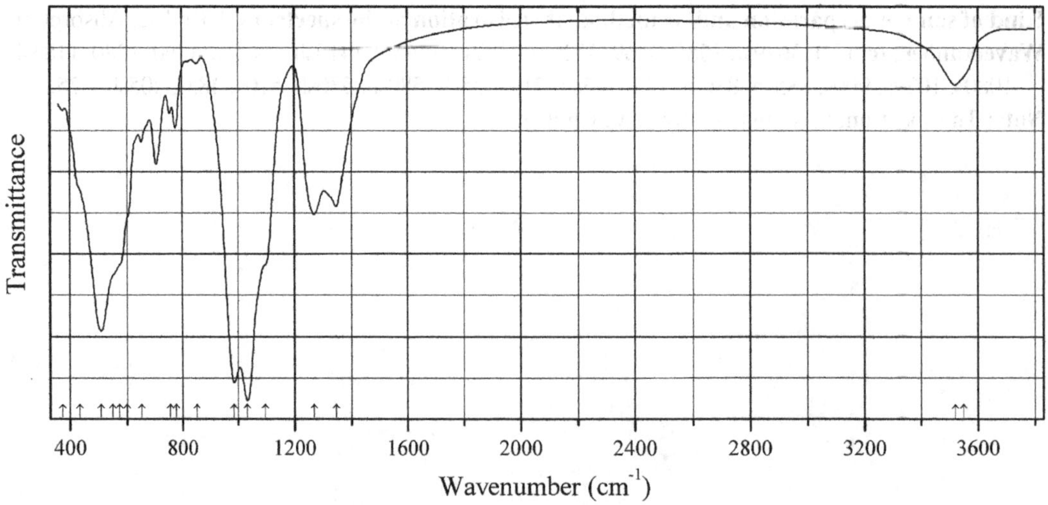

Origin: Cleavland Tin mine, Luina, Waratah, Tasmania, Australia (type locality).

Description: Aggregate of light green acicular crystals.

Kind of sample preparation and/or method of registration of the spectrum: KBr disc. Absorption.

Wavenumbers (cm^{-1}): 3550sh, 3521, 1346, 1268, 1093sh, 1033s, 986s, 854w, 775, 753, 704, 652, 600sh, 575sh, 550sh, 509s, 435sh, 374w.

Note: The spectrum was obtained by N.V. Chukanov.

SSi14 Chlorellestadite $Ca_{10}[(SiO_4)_3(SO_4)_3]Cl_2$

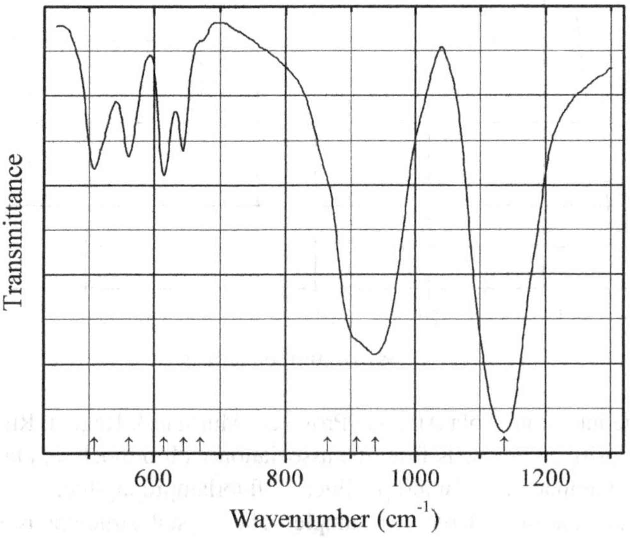

Origin: Synthetic.

Description: Synthesized by heating of an appropriate mixture of CaO, $CaSO_4$, SiO_2, $CaCl_2$, and CaF_2 first at 900 °C for 5 h and then at 950 °C for 9 h with intermediate grinding. Characterized by powder X-ray diffraction data. The crystal structure is solved. Hexagonal, space group $P6_3/m$, $a = 9.6239(3)$, $c = 6.87749(3)$ Å, $V = 551.64(2)$ Å^3. The formula is $Ca_{10}(SiO_4)_3(SO_4)_3Cl_{1.6}F_{0.4}$.

Kind of sample preparation and/or method of registration of the spectrum: KBr disc. Absorption.

Source: Fang et al. (2011).

Wavenumbers (cm^{-1}): 1138s, 938s, 910sh, 865sh, 670sh, 644, 615, 561, 508.

Note: The wavenumbers were partly determined by us based on spectral curve analysis of the published spectrum. In the cited paper, the wavenumber 670 cm^{-1} is erroneously indicated as 660 cm^{-1}.

TiSi318 Fluorbarytolamprophyllite $(Ba,Sr)_2[(Na,Fe^{2+})_3(Ti,Mg)F_2][Ti_2(Si_2O_7)_2O_2]$

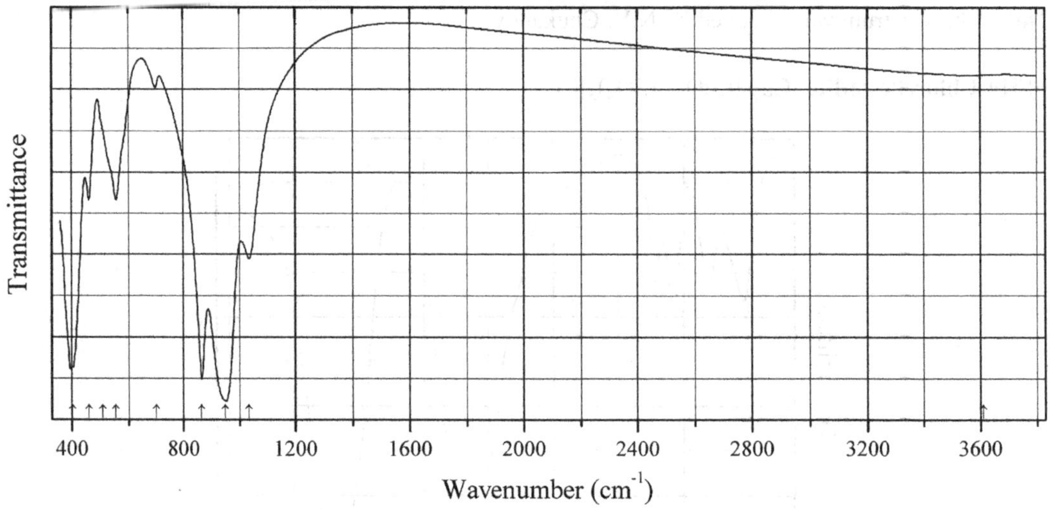

Origin: Niva alkaline intrusion, Kola Alkaline Province, Murmansk Region, Russia (type locality).

Description: Brown prismatic crystals from the association with orthoclase, titanian aegirine-augite, arfvedsonite, aenigmatite, lamprophyllite, fluorlamprophyllite, barytolamprophyllite, shcherbakovite, and natrolite. Holotype sample. The crystal structure is solved. Monoclinic, space group: *C2/m*, $a = 19.538(1)$, $b = 7.092(1)$, $c = 5.391(2)$ Å, $\beta = 96.704(8)°$, $V = 741.8(3)$ Å^3, $Z = 2$. $D_{calc} = 3.662$ g/cm^3. Optically biaxial (+), $\alpha = 1.738$ (3), $\beta = 1.745(4)$, $\gamma = 1.777$ (4), $2V = 55(5)°$. The empirical formula is $(Ba_{0.865}Sr_{0.44}K_{0.46}Na_{0.26})(Na_{2.38}Ca_{0.09}Fe_{0.47}Mn_{0.06})(Ti_{2.79}Mg_{0.09}Fe_{0.035}Nb_{0.06}Zr_{0.015}Ta_{0.01})(Si_{3.99}Al_{0.01})O_{16}[F_{1.04}O_{0.72}(OH)_{0.24}]$. The strongest lines of the powder X-ray diffraction pattern [*d*, Å (*I*, %) (*hkl*)] are: 9.692 (40) (200), 3.726 (59) (−311), 3.414 (67) (311, 510, 401), 3.230 (96) (600), 3.013 (53) (−5−11), 2.780 (100) (221).

Kind of sample preparation and/or method of registration of the spectrum: KBr disc. Absorption.

Wavenumbers (cm^{-1}): (3610w), 1031, 950s, 866s, 701w, 556, 510sh, 461, 402s.

Note: The spectrum was obtained by N.V. Chukanov.

TiSi319 Polyakovite-(Ce) $(Ce,Ca)_4MgCr_2(Ti,Nb)_2Si_4O_{22}$

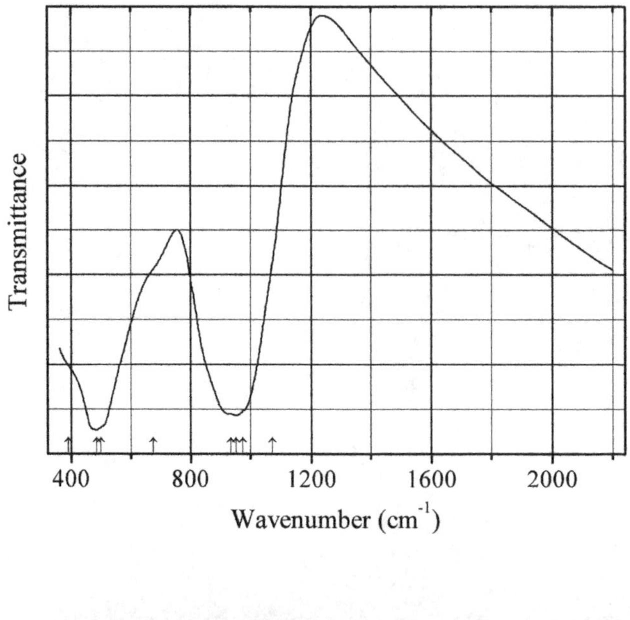

Origin: Pit No. 97, Ilmen (Il'menskie) Mts, Chelyabinsk region, Southern Urals, Russia (type locality).

Description: Black grain from the association with calcite, dolomite, fluororichterite, phlogopite, forsterite, monazite-(Ce), clinohumite, chromite, and davidite-(Ce). X-ray amorphous, metamict. The empirical formula is (electron microprobe): $(Ce_{1.8}La_{1.2}Nd_{0.4}Pr_{0.2}Ca_{0.3}Th_{1.15})(Cr_{1.3}Fe_{0.8})Mg_{0.7}Ti_{1.6}Nb_{0.4}Si_4O_{22}$.

Kind of sample preparation and/or method of registration of the spectrum: KBr disc. Absorption.

Wavenumbers (cm^{-1}): 1070sh, 975sh, 952s, 935sh, 675sh, 500sh, 483s, 390sh.

Note: The spectrum was obtained by N.V. Chukanov.

TiSi320 Yoshimuraite $Ba_2Mn^{2+}{}_2Ti(Si_2O_7)(PO_4)O(OH)$

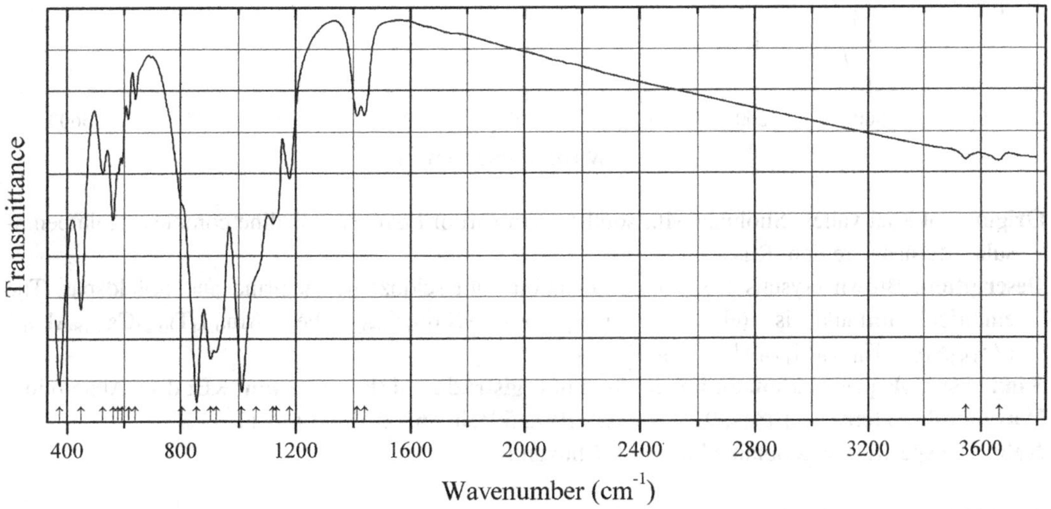

Origin: Tanohata mine, Tanohata-mura, Shimohei-gun, Iwate Prefecture, Tohoku Region, Honshu Island, Japan.

Description: Brown platelets. The empirical formula is (electron microprobe): $(Ba_{3.6}Sr_{0.3}Na_{0.1})(Mn_{3.5}Fe_{0.3}Mg_{0.1})(Ti_{1.8}Fe_{0.2})(Si_2O_7)_{2.05}[(PO_4)_{1.1}(SO_4)_{0.7}(CO_3)_x]O_2(OH)_{1.1}F_{0.9}$.

Kind of sample preparation and/or method of registration of the spectrum: KBr disc. Absorption.

Wavenumbers (cm^{-1}): 3665w, 3546w, 1438, 1413, 1177, 1130sh, 1120, 1060sh, 1011s, 922s, 904s, 856s, 805sh, 641w, 617w, 594, 580sh, 562, 526, 448, 374s.

Note: The spectrum was obtained by N.V. Chukanov.

TiSi321 Lavenite Fe-analogue (Na,Ca)$_2$(Fe,Mn)(Zr,Ti,Nb)(Si_2O_7)(O,F)$_2$

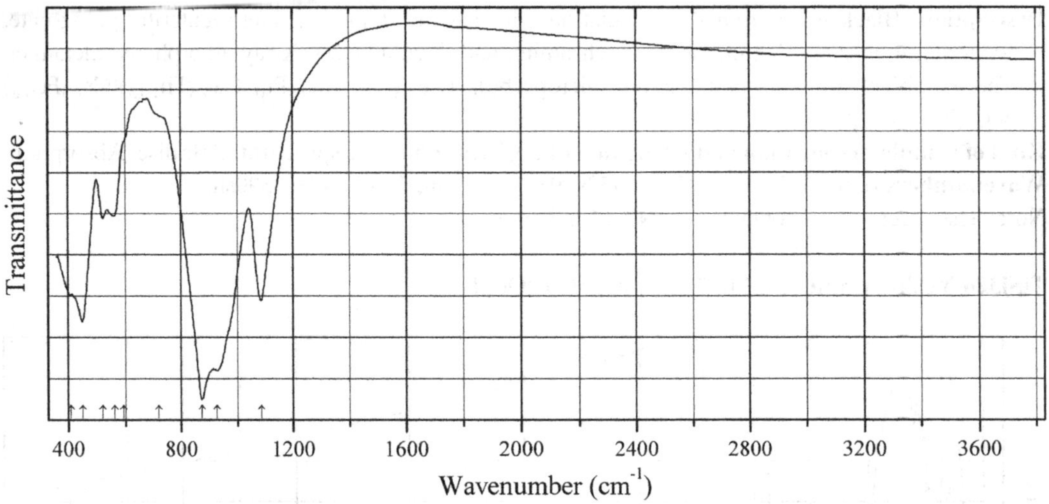

Origin: Ankisuai valley, Suoluaiv Mt., southeastern part of Lovozero alkaline complex, Kola peninsula, Murnansk region, Russia.

Description: Brown crystals from the association with seidozerite, aegirine and K-feldspar. The empirical formula is (electron microprobe): ($Na_{1.66}Ca_{0.34}$)($Fe_{0.27}Mn_{0.24}Ti_{0.24}Ca_{0.18}Al_{0.05}$)($Zr_{0.84}Nb_{0.15}Ti_{0.01}$)($Si_{1.99}Al_{0.01}$)$O_7(OH,F)_2$.

Kind of sample preparation and/or method of registration of the spectrum: KBr disc. Absorption.

Wavenumbers (cm^{-1}): 1084s, 928s, 875s, 720sh, 595sh, 562, 522, 450s, 411.

Note: The spectrum was obtained by N.V. Chukanov.

TiSi322 Potassium titanium silicate $K_2TiSi_3O_9 \cdot H_2O$ $K_2TiSi_3O_9 \cdot H_2O$

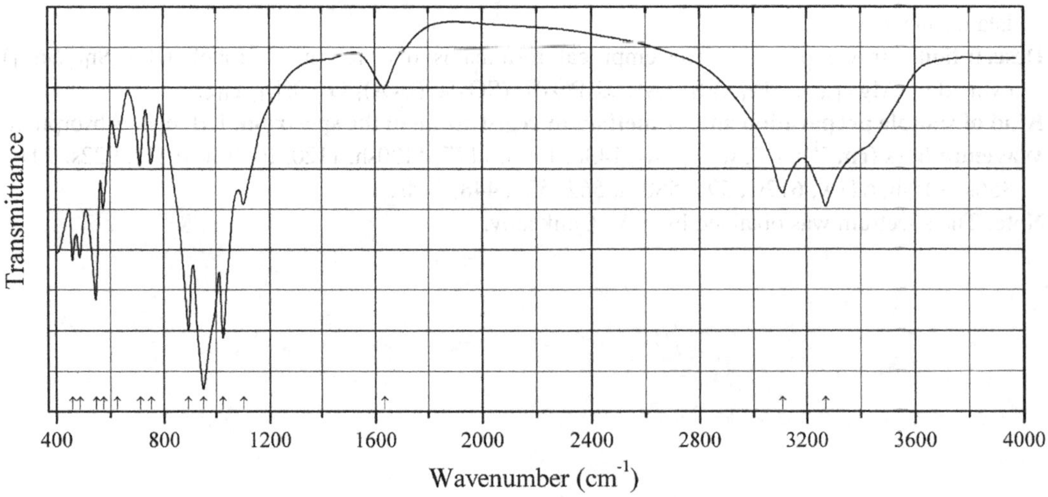

Origin: Synthetic.

Description: Prepared hydrothermally from a solution containing $TiCl_4$, H_2O_2, SiO_2, KOH, and NaOH, at 180 °C for 7 days. Characterized by powder X-ray diffraction data and elemental analysis. The crystal structure is solved. Orthorhombic, space group $P2_12_12_1$, $a = 9.9081(4)$, $b = 12.9445(5)$, $c = 7.1384(3)$ Å, $V = 915.5$ Å^3, $Z = 4$. $D_{calc} = 2.701$ g/cm^3. The microporous structure is based on a heteropolyhedral framework containing Si_3O_9 chains.

Kind of sample preparation and/or method of registration of the spectrum: KBr disc. Transmission.

Source: Bortun et al. (2000).

Wavenumbers (cm^{-1}): 3270, 3110, 1635, 1103, 1026s, 952s, 893s, 752, 711, 626w, 575, 547, 487, 461.

TiSi323 Sodium titanium silicate $Na_2TiSi_2O_7{\cdot}2H_2O$ $Na_2TiSi_2O_7{\cdot}2H_2O$

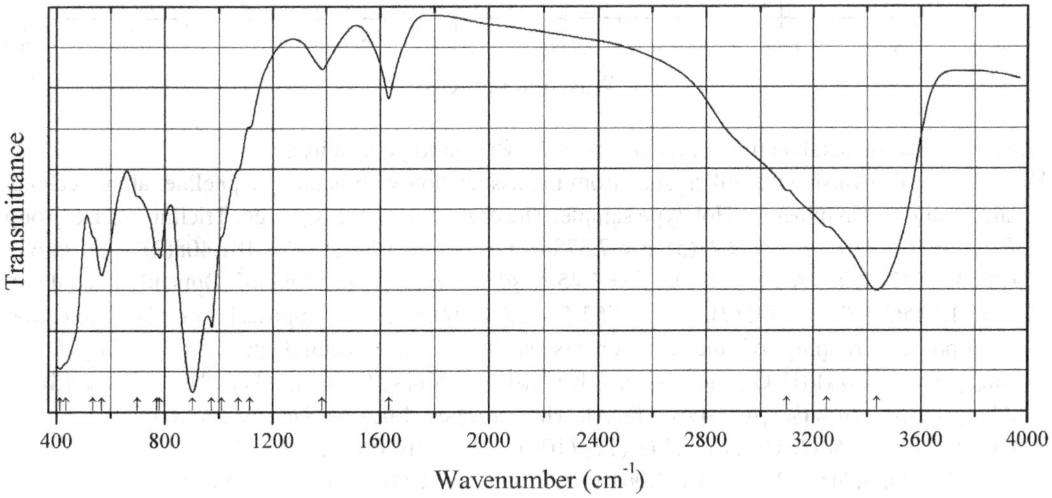

Origin: Synthetic.

Description: White powder. A compound with layered structure synthesized hydrothermally from $TiCl_4$, silicic acid, NaOH, and H_2O_2 at 200 °C for 7 days. Characterized by powder X-ray diffraction, TG analysis, MAS ^{29}Si, and ^{23}Na NMR. The strongest reflection is observed at 14.97 Å.

Kind of sample preparation and/or method of registration of the spectrum: KBr disc. Transmission.

Source: Clearfield et al. (1997).

Wavenumbers (cm^{-1}): 3436s, (3250), (3100), 1632, 1385w, 1115sh, 1073sh, 1011sh, 973s, 902s, 779, 768sh, 698sh, 568, 536sh, 435sh, 412s.

Note: The wavenumbers were partly determined by us based on spectral curve analysis of the published spectrum.

TiSi324 Batievaite-(Y) $Y_2Ca_2Ti(Si_2O_7)_2(OH)_2{\cdot}4H_2O$

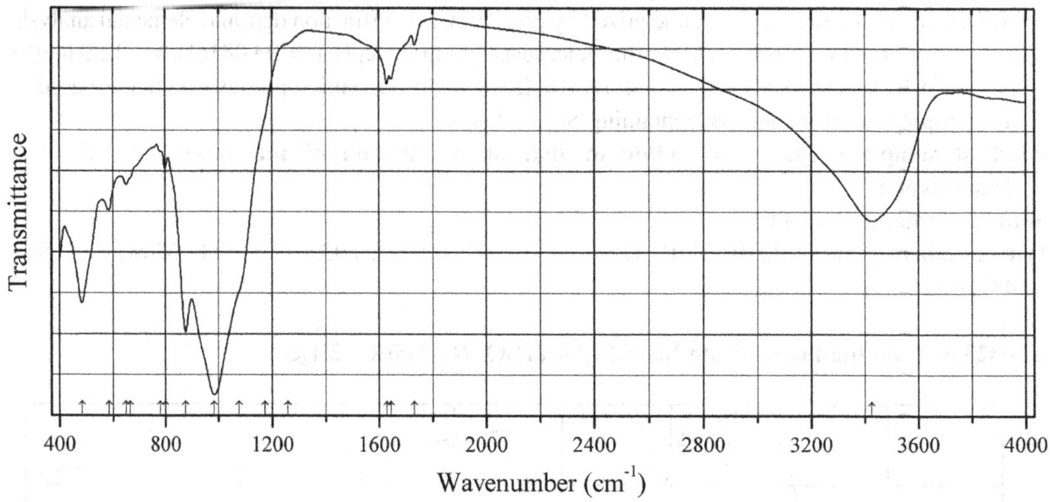

Origin: Sakharjok alkaline massif, Kola Peninsula, Russia (type locality).

Description: Brownish euhedral crystals from the association with hainite, nepheline, albite, calcite, and zeolite-group minerals. Holotype sample. The crystal structure is solved. Triclinic, space group *P*-1, $a = 9.4024(8)$, $b = 5.5623(5)$, $c = 7.3784(6)$ Å, $\alpha = 89.919(2)°$, $\beta = 101.408(2)°$, $\gamma = 96.621(2)°$, $V = 375.65(6)$ Å^3, Z = 1. $D_{meas} = 3.45(5)$ g/cm^3, $D_{calc} = 3.357$ g/cm^3. Optically biaxial (+), $\alpha = 1.745(5)$, $\beta = 1.745(7)$, $\gamma = 1.752(5)$, $2V = 60(5)°$. The empirical formula is (electron microprobe; grouping of the components is based on structural data): $(Y_{0.81}Ca_{0.65}Ln_{0.23}Mn_{0.15}Zr_{0.12}Fe_{0.04})[(H_2O)_{0.75}Ca_{0.70}\square_{0.55}]Ca_{2.00}[\square_{0.61}Na_{0.25}(H_2O)_{0.14}](Ti_{0.76}Nb_{0.15}Zr_{0.09})[Si_{3.91}Al_{0.09}O_{14}][(OH)_{1.56}F_{0.44}][(H_2O)_{1.27}F_{0.73}]$. The strongest lines of the powder X-ray diffraction pattern [*d*, Å (*I*, %) (*hkl*)] are: 9.145 (17) (100), 7.238 (36) (00−1), 4.350 (23) (0−1−1), 4.042 (16) (11−1), 3.745 (13) (2−10), 3.061 (30) (300), 2.991 (100) (11−2), 2.819 (16) (3−10).

Kind of sample preparation and/or method of registration of the spectrum: KBr disc. Transmission.

Source: Lyalina et al. (2016).

Wavenumbers (cm^{-1}): 3426, 1732w, 1646, 1630, 1258sh, 1172sh, 1077sh, 985s, 877, 800w, 780sh, 664sh, 649, 584, 493s.

Note: The wavenumbers were partly determined by us based on spectral curve analysis of the published spectrum. The band at 1732 cm^{-1} indicates possible presence of H_3O^+ or $H_5O_2^+$ groups.

TiSi325 Bazirite $BaZr(Si_3O_9)$

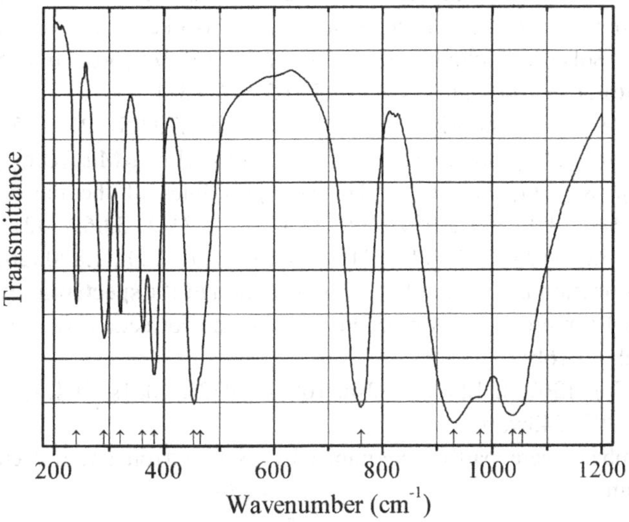

Origin: Synthetic.

Description: Synthesized using solid-state reaction thchniques from the stoichiometric mixture of $BaCO_3$, ZrO_2, and SiO_2.

Kind of sample preparation and/or method of registration of the spectrum: KI disc. Transmission.

Source: Choisnet et al. (1975).

Wavenumbers (cm^{-1}): 1055sh, 1038s, 979sh, 930s, 760s, 465sh, 454s, 382, 361, 320, 291, 240.

TiSi326 Fogoite-(Y) $Na_3Ca_2Y_2Ti(Si_2O_7)_2OF_3$

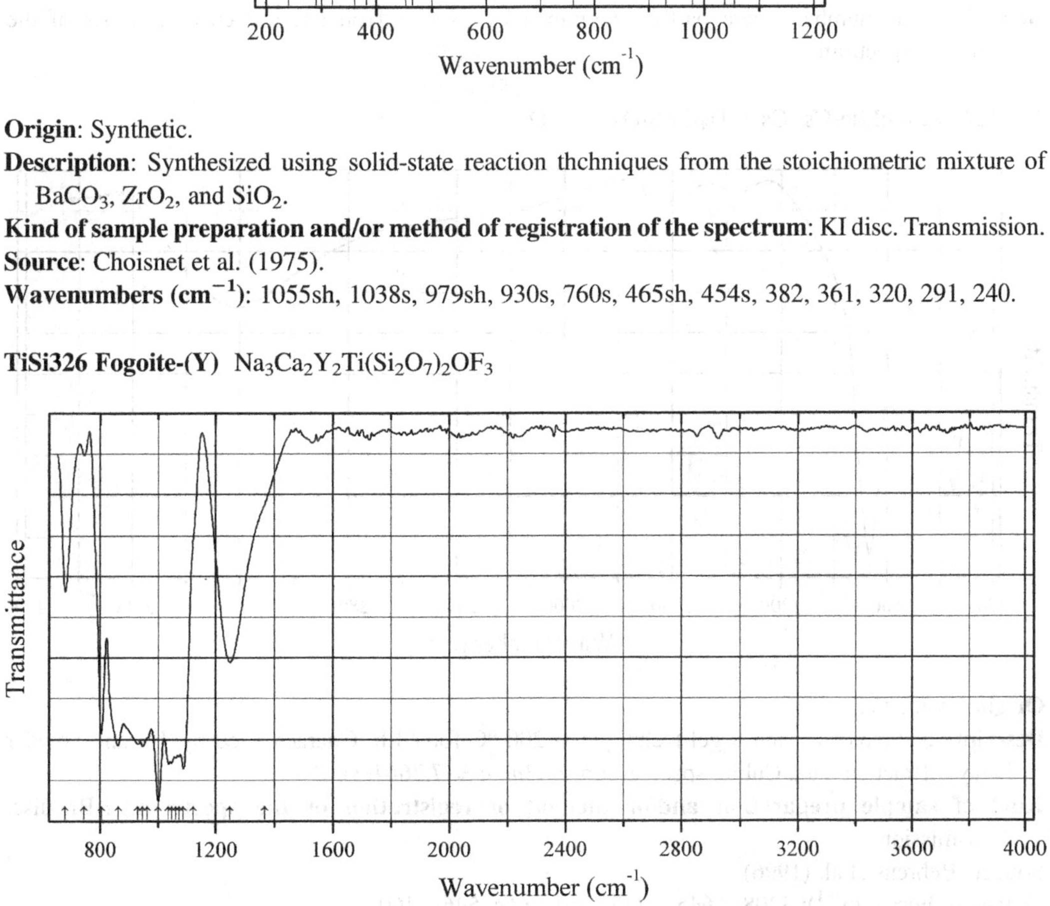

Origin: Lagoa do Fogo, the São Miguel Island, the Azores, Portugal (type locality).

Description: Colorless long-prismatic crystals from the association with sanidine, astrophyllite, fluornatropyrochlore, ferrokentbrooksite, quartz, and ferro-katophorite. Holotype sample. The crystal structure is solved. Triclinic, space group *P*-1, $a = 9.575(6)$, $b = 5.685(4)$, $c = 7.279(5)$ Å, $\alpha = 89.985(6)°$, $\beta = 100.933(4)°$, $\gamma = 101.300(5)°$, $V = 381.2(7)$ Å^3, $Z = 1$. $D_{calc} = 3.523$ g/cm^3. Optically biaxial (+), $\alpha = 1.686(2)$, $\beta = 1.690(2)$, $\gamma = 1.702(5)$, $2V = 57(1)°$. The empirical formula is (electron microprobe): $(Na_{2.74}Mn_{0.15})Ca_2(Y_{1.21}Ln_{0.35}Mn_{0.16}Zr_{0.11}Nb_{0.09}Fe_{0.07}Ca_{0.01})(Ti_{0.76}Nb_{0.23}Ta_{0.01})(Si_{4.03}O_{14})O_{1.12}F_{2.88}$. The strongest lines of the powder X-ray diffraction pattern [*d*, Å (*I*, %) (*hkl*)] are: 2.954 (100) (−1−12, −310), 3.069 (42) (300, 0−12), 2.486 (24) (310, 2−12), 3.960 (23) (−1−11, −210), 2.626 (21) (−220), 1.820 (20) (−104).

Kind of sample preparation and/or method of registration of the spectrum: Absorption of a crystal fragment using an IR microscope. A procedure of baseline correction was applied.

Source: Cámara et al. (2016b).

Wavenumbers (cm^{-1}): 1367sh, 1250, 1122sh, 1088s, 1072s, 1061s, 1046s, 1036s, 1000s, 963sh, 946s, 930s, 862s, 803, 743w, 677.

Note: The wavenumbers were partly determined by us based on spectral curve analysis of the published spectrum.

TiSi327 Ivanyukite-Cs $Cs_3HTi_4O_4(SiO_4)_3·4H_2O$

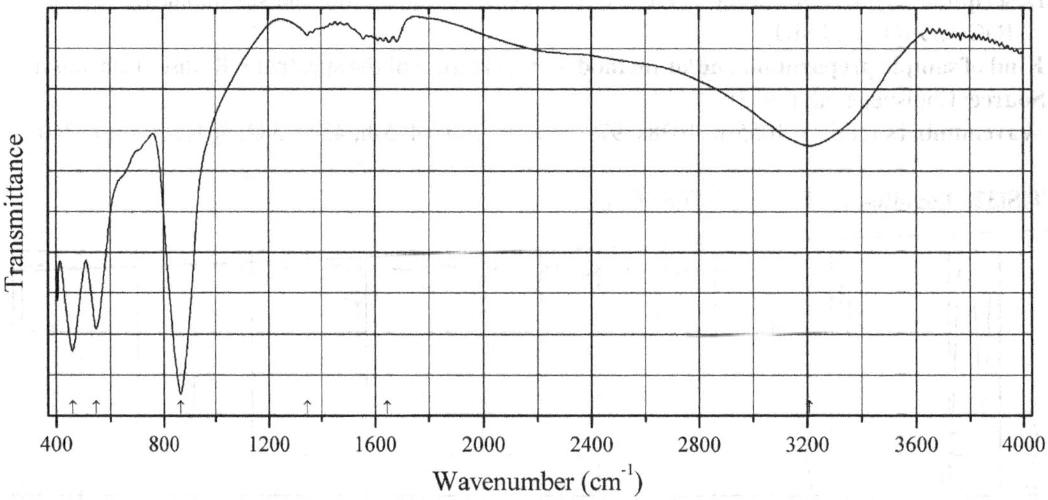

Origin: Synthetic.

Description: Prepared using a gel technique at 200 °C for 48 h. Characterized by TG and powder X-ray diffraction data. Cubic, space group *P*-43*m*, $a = 7.7644(3)$, $Z = 1$.

Kind of sample preparation and/or method of registration of the spectrum: KBr disc. Transmission.

Source: Behrens et al. (1996).

Wavenumbers (cm^{-1}): 3208, 1645w, (1347w), 866s, 546s, 460s.

Note: The wavenumbers were partly determined by us based on spectral curve analysis of the published spectrum.

TiSi328 Schülerite-type mineral $Ba_2Na_2Mg_2Ti_2(Si_2O_7)_2O_2F_2$

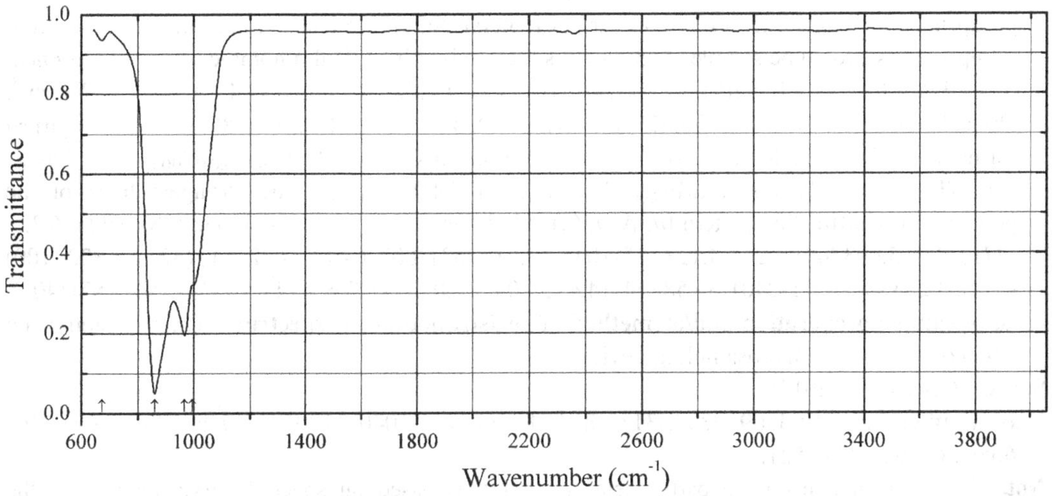

Origin: Eifel volcanic region, Germany.

Description: Pale yellow zones of the two grains obtained from an American mineral collector. Erroneously described as "schülerite." The crystal structure is solved. Triclinic, space group *P*-1, $a = 5.396(1)$, $b = 7.071(1)$, $c = 10.226(2)$ Å, $\alpha = 99.73(3)°$, $\beta = 99.55(3)°$, $\gamma = 90.09(3)°$, $V = 379.1(2)$ Å^3, $Z = 1$. $D_{calc} = 3.879$ g/cm^3. Structurally related to schülerite. The empirical formula is $(Ba_{1.57}Sr_{0.14}K_{0.14})(Na_{1.10}Ca_{0.43}Mn_{0.30}Fe_{0.17})(Fe_{0.88}Mg_{0.79}Na_{0.33})(Ti_{1.67}Fe_{0.21}Nb_{0.09}Zr_{0.02}Al_{0.01})Si_{3.95}O_{15.93}F_{2.07}$. The Mössbauer spectrum given by Sokolova et al. (2013) cannot be used for precise determination of the Fe^{2+}:Fe^{3+} ratio because of a strong scatter of experimental points. Consequently, the existence of a Mg-dominant site is questionable.

Kind of sample preparation and/or method of registration of the spectrum: Transmission using an IR microscope and a diamond micro compression cell.

Source: Sokolova et al. (2013).

Wavenumbers (cm^{-1}): 995sh, 967s, 860s, 675w.

TiSi329 Hydroterskite $Na_2ZrSi_6O_{12}(OH)_6$

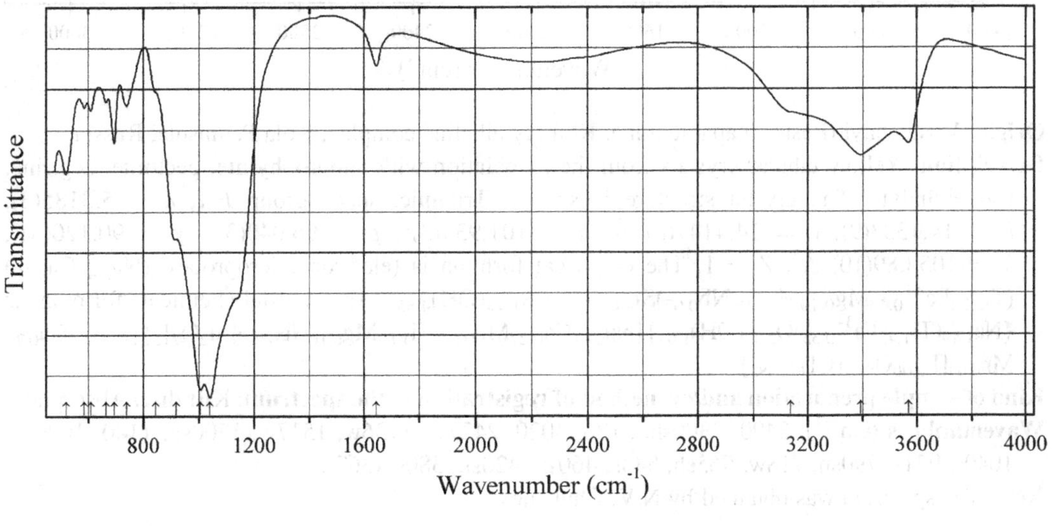

Origin: Saint-Amable sill, Demix-Varennes quarry, near Varennes, Québec, Canada (type locality).

Description: Short prismatic crystals from the association with aegirine, analcime, an astrophyllite-group mineral, catapleiite, a eudialyte-group mineral, fluorite, monazite, natrolite, and a rinkite-group species. Holotype sample. The crystal structure is solved. Orthorhombic, space group *Pnca*, $a = 13.956(6)$, $b = 14.894(7)$, $c = 7.441(4)$ Å, $V = 1546.8(20)$ Å^3, $Z = 4$. $D_{calc} = 2.57$ g/cm^3. Optically biaxial (−), $\alpha = 1.562(2)$, $\beta = 1.567(2)$, $\gamma = 1.571(2)$, $2V = 86(3)°$. The empirical formula (after excluding of trace components) is $(Na_{1.54}K_{0.01}Ca_{0.07}Ce_{0.01}La_{0.05})(Zr_{0.74}Ti_{0.09}Nb_{0.05}Th_{0.005}Fe_{0.08}Mn_{0.06}Al_{0.01})Si_{6.09}O_{12}(OH)_{5.96}F_{0.035}$. The strongest lines of the powder X-ray diffraction pattern [*d*, Å (*I*, %) (*hkl*)] are: 7.427 (56) (020), 6.638 (48) (011), 6.327 (47) (210), 5.093 (49) (220), 4.123 (55) (031), 3.716 (53) (002, 040), 3.482 (51) (321), 3.322 (100) (022), 3.283 (80) (202, 240), 3.158 (54) (420), 3.091 (50) (411), 2.625 (48) (042), 2.544 (57) (402).

Kind of sample preparation and/or method of registration of the spectrum: Transmission using adiamond-anvil cell microsampling device.

Source: Grice et al. (2015).

Wavenumbers (cm^{-1}): 3569, 3393, 3138sh, 1644, 1139sh, 1040s, 1005s, 917sh, 843sh, 740, 695, 665w, 610w, 587w, 521.

Note: The wavenumbers were partly determined by us based on spectral curve analysis of the published spectrum.

TiSi330 Betalomonosovite $Na_2\square_4Na_2Ti_2Na_2Ti_2(Si_2O_7)_2[PO_3(OH)][PO_2(OH)_2]O_2(OF)$

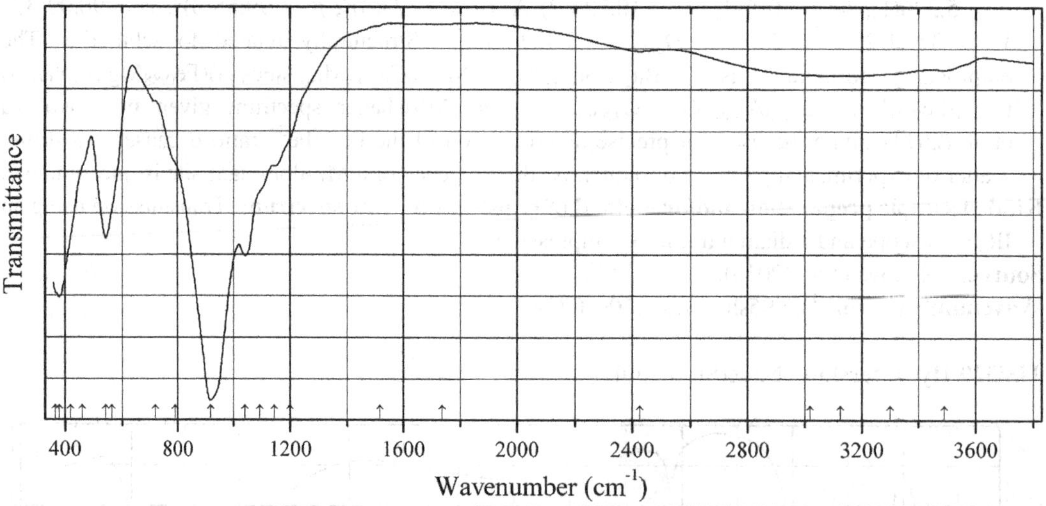

Origin: Vostochnyi (Eastern) apatite mine, Khibiny alkaline complex, Kola Peninsula, Russia.

Description: Yellow tabular crystals from the association with lamprophyllite, pectolite, aegirine, and eudialyte. The crystal structure is solved. Triclinic, space group *P*-1, $a = 5.3185(3)$, $b = 14.1333(9)$, $c = 14.4147(8)$ Å, $\alpha = 101.934(3)°$, $\beta = 96.040(3)°$, $\gamma = 90.120(3)°$, $V = 1053.89(10)$ Å^3, $Z = 1$. The empirical formula is (electron microprobe): $(Na_{5.40}Ca_{0.50})(Ti_{3.22}Fe^{3+}_{0.43}Mg_{0.10}Mn_{0.09}Nb_{0.07})Si_{4.00}P_{1.97}O_{21.15}(OH)_{4.85}$. The crystal-chemical formula is $\{Na_{1.49}(Ti_{1.45}Fe^{3+}_{0.55})O_{2.47}(OH)_{1.63}\}\{Na_{0.81}Ca_{0.27}Mn_{0.06}(Ti_{1.84}Mg_{0.10}Nb_{0.06})[Si_2O_7]_2\}\{Na_{3.22}Ca_{0.22}Mn_{0.03}[P_{2.00}O_{5.62}(OH)_{2.38}]\}$.

Kind of sample preparation and/or method of registration of the spectrum: KBr disc. Absorption.

Wavenumbers (cm^{-1}): 3490, 3300sh, 3126, 3020, 2427w, 1736w, 1517w, 1200sh, 1140, 1090sh, 1040s, 921s, 790sh, 718w, 565sh, 543s, 460sh, 420sh, 380s, (367).

Note: The spectrum was obtained by N.V. Chukanov.

TiSi331 Betalomonosovite $Na_2\square_4Na_2Ti_2Na_2Ti_2(Si_2O_7)_2[PO_3(OH)][PO_2(OH)_2]O_2(OF)$

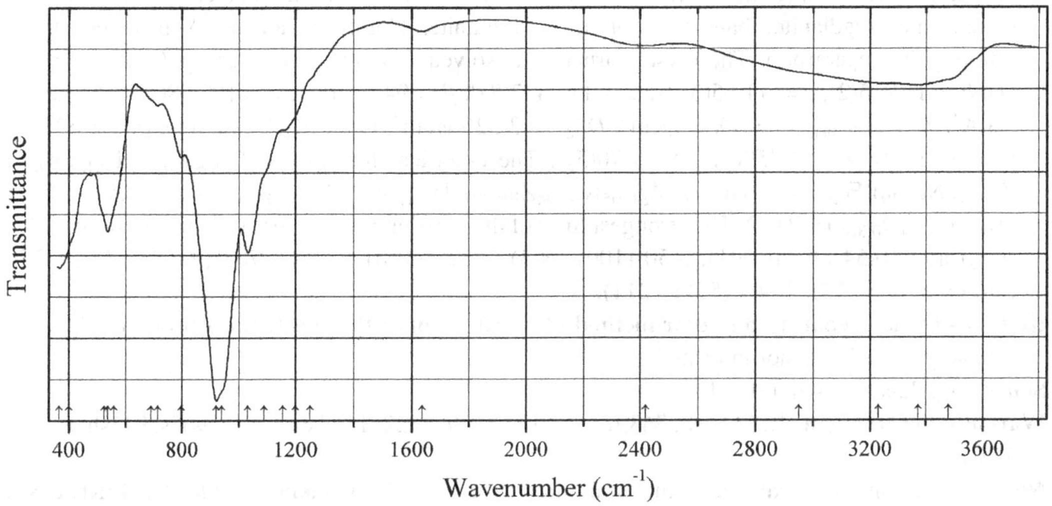

Origin: Olenii Ruchei (Reindeer Creek) open pit of the Olenii Ruchei apatite mine, Niorkpakhk Mt., Khibiny alkaline complex, Kola Peninsula, Russia.

Description: Beige lamellae crystals from the association with microcline, aegirine, and lamprophyllite. The crystal structure is solved. Triclinic, space group *P*-1, $a = 5.30090(18)$, $b = 14.1460(4)$, $c = 14.4435(4)$ Å, $\alpha = 103.3862(15)°$, $\beta = 90.4128(17)°$, $\gamma = 90.4128(17)°$, $V = 1046.21(6)$ Å^3, $Z = 1$. The empirical formula is (electron microprobe): $(Na_{4.40}Ca_{0.52}K_{0.01})(Ti_{3.14}Fe^{3+}_{0.41}Mg_{0.12}Nb_{0.11}Mn_{0.10})(Si_{3.98}Al_{0.02})P_{1.94}O_{19.91}(OH)_{6.06}$. The crystal-chemical formula is $\{Na_{1.04}(Ti_{1.42}Fe^{3+}_{0.46}Mn_{0.12})O_{1.60}(OH)_{2.40}\}\{Na_{1.03}Ca_{0.22}(Ti_{1.74}Mg_{0.16}Nb_{0.10})[Si_2O_7]_2\}\{Na_{2.33}Ca_{0.24}[P_{2.00}O_{4.80}(OH)_{3.20}]\}$.

Kind of sample preparation and/or method of registration of the spectrum: KBr disc. Absorption.

Wavenumbers (cm^{-1}): 3480sh, 3373, 3230sh, 2950sh, 2420w, 1636w, 1250sh, 1200sh, 1156, 1090sh, 1033s, 940sh, 921s, 798, 714w, 690sh, 560sh, 538s, 525sh, 400sh, 367s.

Note: The spectrum was obtained by N.V. Chukanov.

TiSi332 Bulgakite $Li_2(Ca,Na)Fe^{2+}_7Ti_2(Si_4O_{12})_2O_2(OH)_4(F,O)(H_2O)_2$

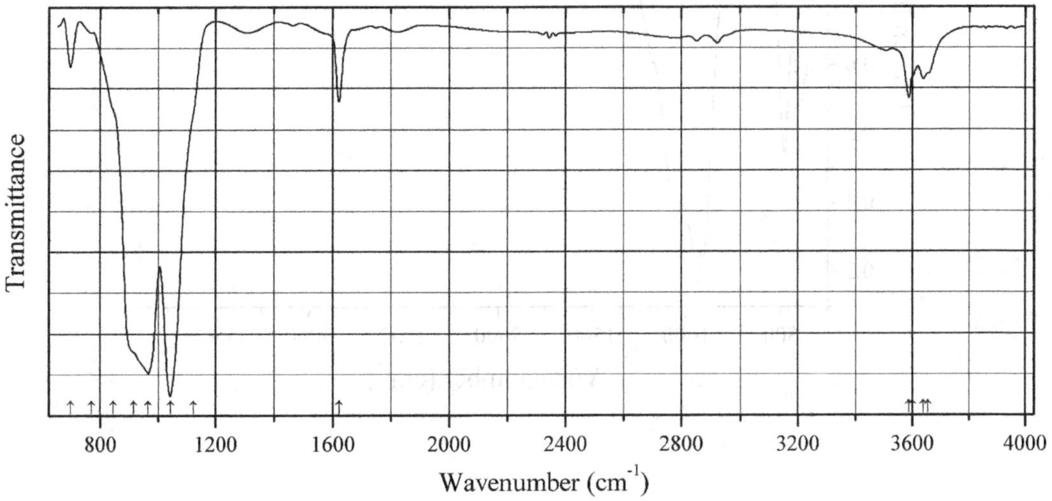

Origin: Dara-i Pioz glacier, Dara-i Pioz alkaline massif, Tien Shan Mts., Tajikistan (type locality).

Description: Brownish orange grains from the association with amphibole, quartz, feldspar, brannockite, sogdianite, bafertisite, albite, and titanite. Holotype sample. A member of the astrophyllite supergroup. The crystal structure is solved. Triclinic, space group *P*-1, $a = 5.374$ (1), $b = 11.965(2)$, $c = 11.65(3)$ Å, $\alpha = 113.457(8)°$, $\beta = 94.533(8)°$, $\gamma = 103.08(1)°$, $V = 657.5$ (8) Å^3, $Z = 1$. $D_{meas} = 3.30(2)$ g/cm^3, $D_{calc} = 3.326$ g/cm^3. Optically biaxial (+), $\alpha = 1.695(3)$, $\beta = 1.711(2)$, $\gamma = 1.750(3)$, $2V = 70(5)°$. The empirical formula is $(Li_{0.94}K_{0.91}Rb_{0.12}Cs_{0.03})(Ca_{0.60}Na_{0.40})(Fe_{5.34}Mn_{1.32}Li_{0.25}Mg_{0.05}Na_{0.04}Zn_{0.02})(Ti_{1.82}Sn_{0.10}Nb_{0.05}Zr_{0.04})[(Si_{7.78}Al_{0.24})O_{24}]O_{2.30}(OH)_4O_{0.30}{\cdot}0.94H_2O$. The strongest lines of the powder X-ray diffraction pattern [*d*, Å (*I*, %) (*hkl*)] are: 10.54 (100) (001), 3.50 (100) (003), 2.578 (100) (130), 2.783 (90) (1−42), 1.576 (68) (3–51, –3−22), 2.647 (55) (−211).

Kind of sample preparation and/or method of registration of the spectrum: Absorption. Kind of sample preparation is not indicated.

Source: Agakhanov et al. (2016b).

Wavenumbers (IR, cm^{-1}): 3655sh, 3639w, 3600sh, 3589, 1622, 1122sh, 1041s, 965s, 915sh, 844sh, 768sh, 695.

Note: Based on chemical data, the simplified formula of bulgakite should be $LiK(Ca,Na)Fe^{2+}{}_7Ti_2(Si_4O_{12})_2O_2(OH)_4(F,O)(\square,H_2O)_2$. The wavenumbers were partly determined by us based on spectral curve analysis of the published IR spectrum. The band position denoted by Agakhanov et al. (2016b) as 940 cm^{-1} actually corresponds to a strong peak with absorption maximum at 965 cm^{-1} and a shoulder at 915 cm^{-1}. The IR band at 695 cm^{-1} is erroneously assigned to Si–O-stretching vibrations. In the cited paper, Raman spectrum is given.

Wavenumbers (Raman, cm^{-1}): 1041, 910s, 785sh, 733s, 660s, 569s, 420, 395, 367, 258, 233, 170, 133.

TiSi333 Catapleiite heating product $Na_6Zr_3[Si_9O_{27}]$

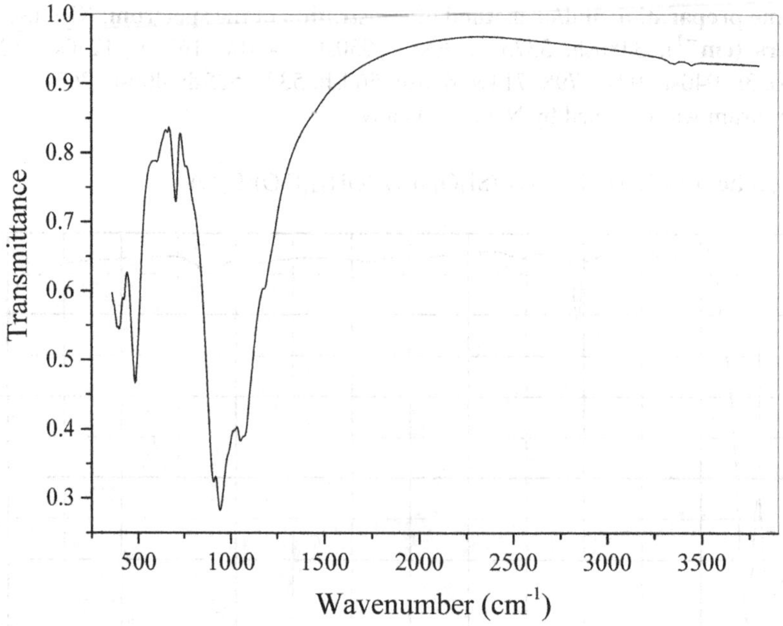

Origin: Artificial.

Description: Product of heating (from room temperature to 950 °C at a rate of 5 °C/min) of a catapleiite crystal from Aikuaivenchorr Mt., Khibiny alkaline complex, Kola peninsula, Murmansk region, Russia. The crystal structure is solved. Hexagonal, space group $P6_3/mcm$, $a = 11.5901(9)$, $c = 9.9546(9)$ Å, $V = 1158.05(16)$ Å^3. The structure is based on the heteropolyhedral framework which principally differs from that of catapleiite and is built by isolated [ZrO_6] octahedra connected with each other by nine-membered rings [Si_9O_{27}] formed by SiO_4 tetrahedra.

Kind of sample preparation and/or method of registration of the spectrum: KBr disc. Absorption.

Wavenumbers (cm^{-1}): 3446w, 3356w, 3345sh, 1181, 1070sh, 1051s, 1019, 942s, 912s, 760w, 707, 660w, 599w, 483, 424, 400, 390.

Note: The spectrum was obtained by N.V. Chukanov. The mands in the range from 3300 to 3600 cm^{-1} may correspond to adsorbed water.

PSi11 Calcium orthophosphate orthosilicate $Ca_5(PO_4)_2(SiO_4)$

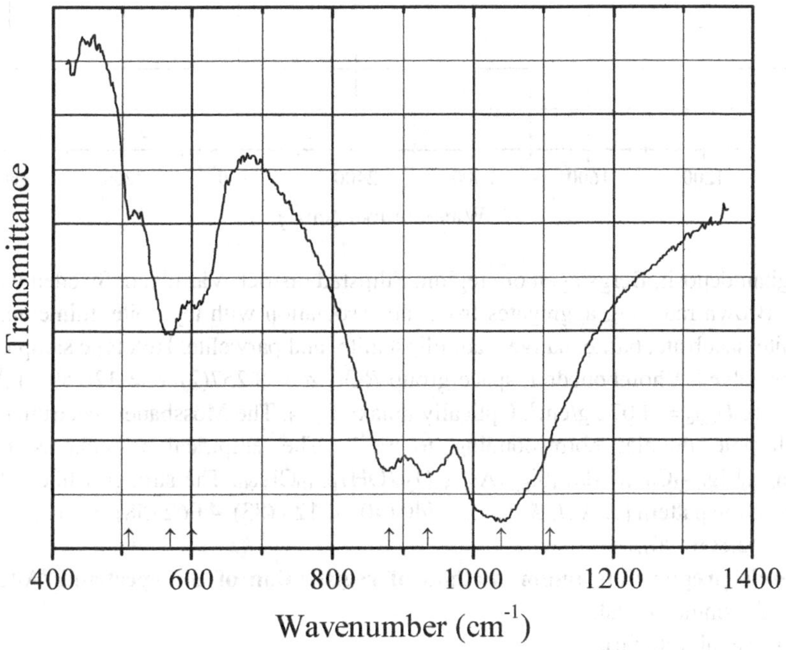

Origin: Synthetic.

Description: Obtained by sintering of compacted mixture of calcium hydrogen phosphate, calciumcarbonate, and silicon oxide. Characterized by powder X-ray diffraction data. Hexagonal.

Kind of sample preparation and/or method of registration of the spectrum: KBr disc. Transmission.

Source: Lugo et al. (2015).

Wavenumbers (IR, cm^{-1}): 1110sh, 1040s, 935s, 880s, 601, 570, 510.
Note: The wavenumber 935 cm^{-1} is erroneously indicated by Lugo et al. (2015) as 960 cm^{-1}. In the cited paper, Raman spectrum is given.
Wavenumbers (Raman, cm^{-1}): 1084, 1058, 963s, 857s, 642, 587, 439, 435, 402, 297w, 218.

AsSi14 Wiklundite $Pb_2(Mn^{2+},Zn)_3(Fe^{3+},Mn^{2+})_2(Mn^{2+},Mg)_{19}(As^{3+}O_3)_2[(Si,As^{5+})O_4]_6(OH)_{18}Cl_6$

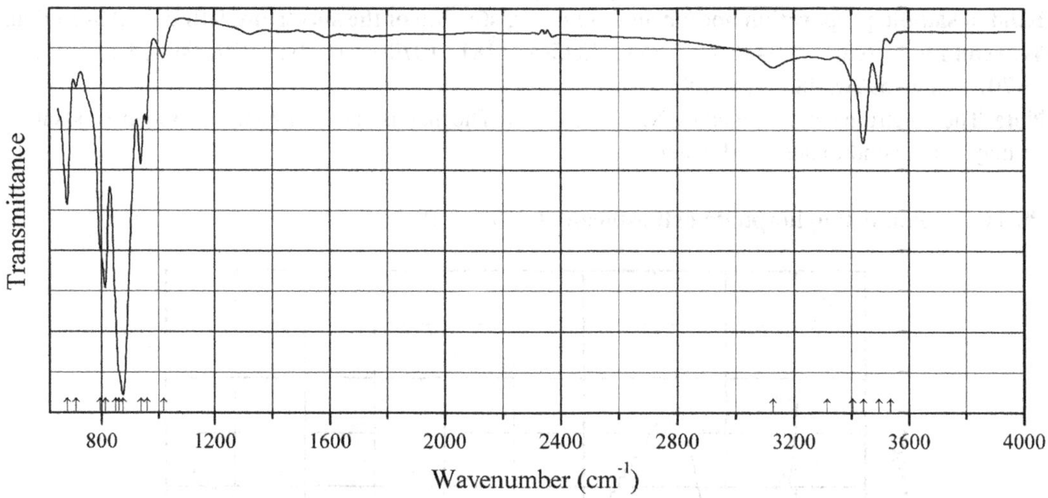

Origin: Långban deposit, Bergslagen ore region, Filipstad district, Värmland, Sweden (type locality).
Description: Brown radiating aggregates from the association with tephroite, mimetite, turneaurite, johnbaumite, jacobsite, barite, native lead, filipstadite, and parwelite. Holotype sample. The crystal structure is solved. Rhombohedral, space group *R*-3*c*, *a* = 8.257(2), *c* = 126.59(4) Å, *V* = 7474 (6) Å^3, *Z* = 6. D_{calc} = 4.072 g/cm^3. Optically uniaxial (–). The Mössbauer spectrum contains only one quadrupole doublet corresponding to Fe^{3+}. The empirical formula is $Pb_{2.04}Mn_{21.23}Fe^{3+}_{1.76}Zn_{0.30}Mg_{0.23}Ca_{0.05}Al_{0.04}Si_{5.85}As_{2.37}O_{30}(OH)_{18.10}Cl_{5.90}$. The strongest lines of the powder X-ray diffraction pattern [*d*, Å (*I*, %)] are: 4.740 (40), 4.128 (83), 4.062 (58), 3.561 (40), 3.098 (81), 2.882 (100), 2.806 (90).
Kind of sample preparation and/or method of registration of the spectrum: Attenuated total reflection of a single crystal.
Source: Cooper et al. (2016c).
Wavenumbers (cm^{-1}): 3536w, 3496, 3441, 3404sh, 3316w, 3128w, 1020w, 959, 938, 875s, 861sh, 848sh, 813s, 795sh, 712w, 681.
Note: The presence of Mn^{2+} in the Fe^{3+}-dominant site *M*(1) with the mean *M*(1)–O distance of 2.06 Å is questionable.

USi11 Potassium uranium(V) sorosilicate $K_3(U_3O_6)(Si_2O_7)$

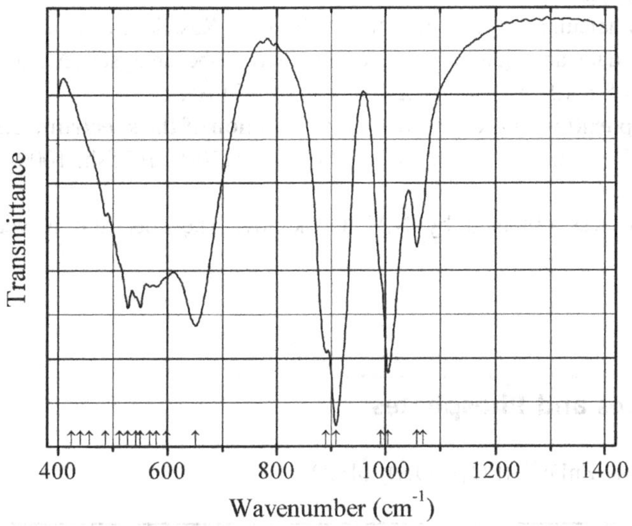

Origin: Synthetic.

Description: Dark red needle crystals synthesized hydrothermally KOH, KF, UO_3, and SiO_2 (in the molar ratio K:U:Si:F = 15:1:2:10) at 600 °C for 5 days. Characterized by powder X-ray diffraction data. The crystal structure contains an uranate column formed by corner-sharing UO_6 octahedra.

Kind of sample preparation and/or method of registration of the spectrum: KBr disc. Transmission.

Source: Lin et al. (2008).

Wavenumbers (IR, cm^{-1}): 1068sh, 1057, 1005s, 991sh, 909s, 891s, 651s, 598sh, 579, 567, 550, 541, 527, 512sh, 487, 457sh, 441sh, 425sh.

Note: The wavenumbers were partly determined by us based on spectral curve analysis of the published spectrum. In the cited paper, Raman spectrum is given.

Wavenumbers (for Raman bands indicated by the authors, cm^{-1}): 972, 924, 888, 770, 570, 361, 232.

USi12 Swamboite-(Nd) $Nd_{0.333}[(UO_2)(SiO_3OH)]\cdot 2.5H_2O$

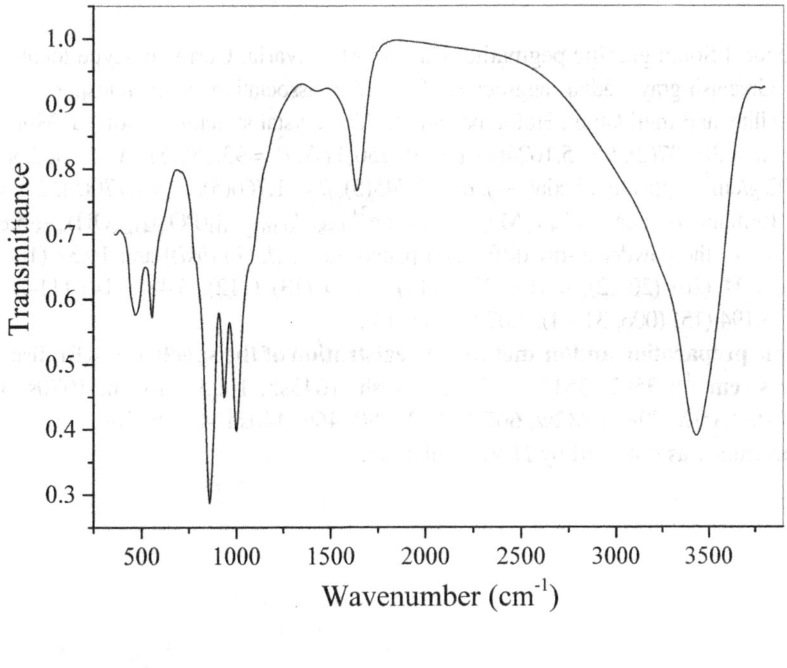

Origin: Swambo Hill (Swambo Mine), Kambove District, Katanga (Shaba), Democratic Republic of Congo (type locality).

Description: Yellow acicular crystals. Investigated by A.V. Kasatkin. Characterized by single-crystal X-ray diffraction data and qualitative electron microprobe analyses. Monoclinic, a = 6.70(4), b = 7.010(7), c = 8.86(2) Å, β = 102.2(3)°, V = 407(3) Å^3.

Kind of sample preparation and/or method of registration of the spectrum: KBr disc. Absorption.

Wavenumbers (cm^{-1}): 3433s, 3250sh, 1638, (1427w), 1150sh, 1075sh, 1000s, 936s, 859s, 785sh, 610sh, 556, 471.

Note: The spectrum was obtained by N.V. Chukanov. The spectrum is very close to that of uranophane.

2.8 Phosphides and Phosphates

P652 Zincoberaunite $ZnFe^{3+}{}_5(PO_4)_4(OH)_5 \cdot 6H_2O$

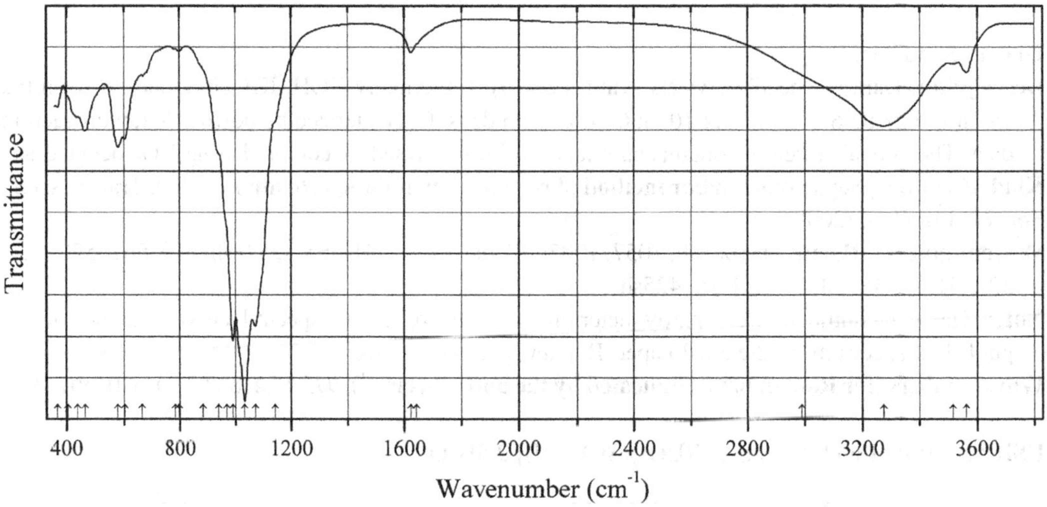

Origin: Hagendorf South granitic pegmatite, Hagendorf, Bavaria, Germany (type locality).

Description: Greenish-gray radial aggregates from the association with feldspar, quartz, jungite, phosphophyllite, and mitridatite . Holotype sample. The crystal structure is solved. Monoclinic, space group *C2/c*, a = 20.837(2), b = 5.1624(4), c = 19.250(1) Å, β = 93.252(5)°, V = 2067.3(3) Å^3, Z = 4. D_{calc} = 2.92 g/cm^3. Optically biaxial (–), α = 1.745(5), β = 1.760(5), γ = 1.770(5), $2V$ = 80(5)°. The empirical formula is $(Zn_{0.83}Ca_{0.08}Mg_{0.06})_{\Sigma 0.97}(Fe^{3+}{}_{4.88}Al_{0.16})_{\Sigma 5.04}(PO_4)_{4.09}(OH)_{4.78} \cdot 5.86H_2O$. The strongest lines of the powder X-ray diffraction pattern [d, Å (I, %) (hkl)] are: 10.37 (100) (200), 9.58 (32) (002), 7.24 (26) (20–2), 4.817 (22) (111), 4.409 (13) (112), 3.483 (14) (11–4, 600), 3.431 (14) (404), 3.194 (15) (006, 31–4), 3.079 (33) (314).

Kind of sample preparation and/or method of registration of the spectrum: KBr disc. Absorption.

Wavenumbers (cm^{-1}): 3562, 3515w, 3276, 2990sh, 1645sh, 1625, 1140sh, 1070s, 1032s, 990s, 970sh, 940sh, 885sh, 798w, 782w, 665sh, 603, 580, 466, 440sh, 405sh, 368.

Note: The spectrum was obtained by N.V. Chukanov.

P653 Beraunite $Fe^{2+}Fe^{3+}_5(PO_4)_4(OH)_5·6H_2O$

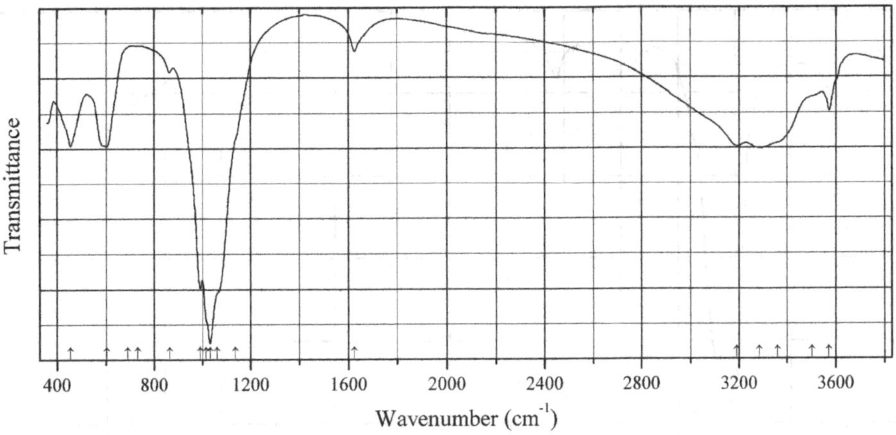

Origin: Levaäniemi mine, Svappavaara, Kiruna district, Lappland, Sweden.

Description: Black spherulites from the association with cacoxenite. Al-bearing variety. Investigated by A.V. Kasatkin. The empirical formula is (electron microprobe): $Fe_{5.26}Al_{0.74}(PO_4)_{4.00}(OH)_5·6H_2O$.

Kind of sample preparation and/or method of registration of the spectrum: KBr disc. Absorption.

Wavenumbers (cm^{-1}): 3572, 3500sh, 3360sh, 3286, 3192, 1626, 1135sh, 1060sh, 1031s, 1015sh, 991s, 866, (733w), 604, 690sh, 456.

Note: The spectrum was obtained by N.V. Chukanov.

P654 Natrodufrénite $NaFe^{2+}Fe^{3+}_5(PO_4)_4(OH)_6·2H_2O$

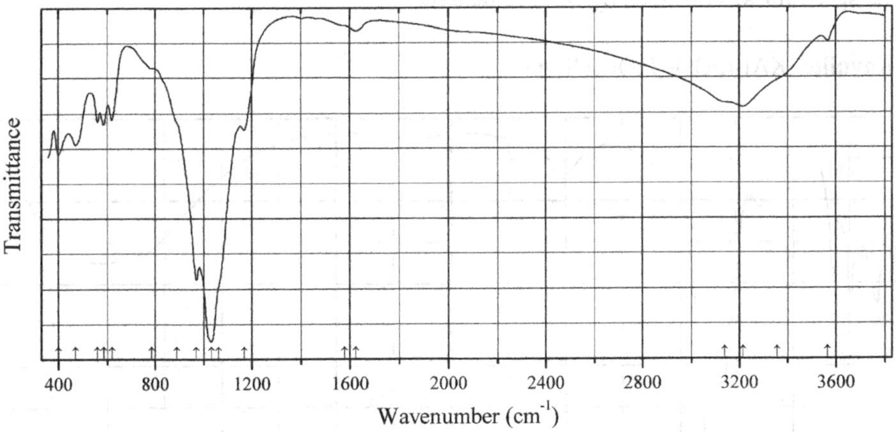

Origin: Chino open pit, near Santa Rita, New Mexico, USA.

Description: Black spherulites with greenish-blue streak. Investigated by A.V. Kasatkin. The empirical formula is (electron microprobe): $Na_{0.93}Ca_{0.15}Fe_{5.50}Al_{0.42}(PO_4)_{4.00}(OH)_6·2H_2O$.

Kind of sample preparation and/or method of registration of the spectrum: KBr disc. Absorption.

Wavenumbers (cm^{-1}): 3567w, 3355sh, 3215, 3140, 1625w, 1580sh, 1167s, 1060sh, 1031s, 970s, 890sh, 786w, 620, 586, 561, 471s, 401s.

Note: The spectrum was obtained by N.V. Chukanov.

P655 Tvrdýite $Fe^{2+}Fe^{3+}{}_2Al_3(PO_4)_4(OH)_5(H_2O)_4·2H_2O$

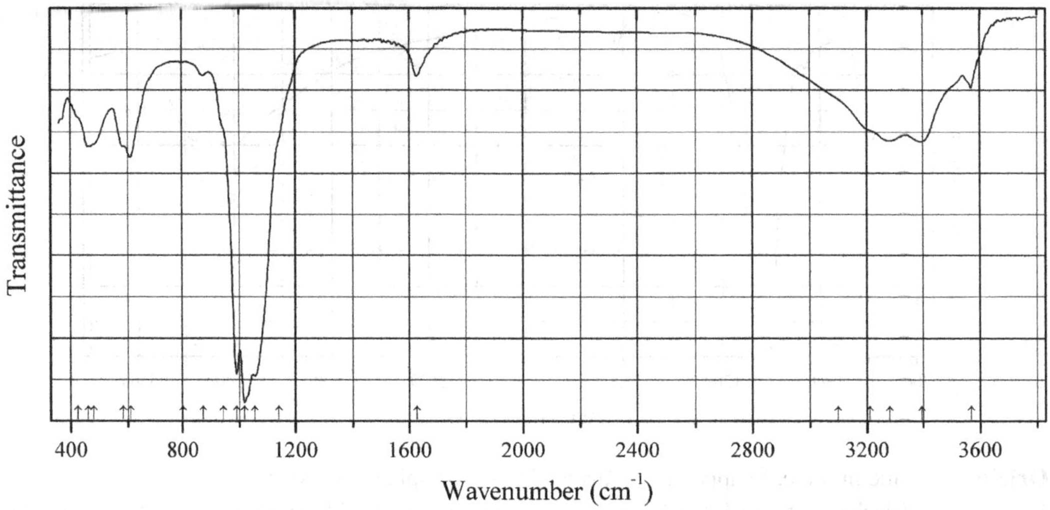

Origin: Hagendorf South pegmatite, Cornelia mine, Hagendorf, Waidhaus, Upper Palatinate, Bavaria, Germany.
Description: Greenish-gray radial-fibrous aggregates. Al-deficient variety. The empirical formula is (electron microprobe): $Ca_{0.05-0.08}Zn_{0.34-0.50}Mn_{0.04-0.07}Mg_{0-0.05}Fe_{3.85-3.99}Al_{1.44-1.66}Cr_{0.06}(PO_4)_4(OH)_5(H_2O)_4·2H_2O$.
Kind of sample preparation and/or method of registration of the spectrum: KBr disc. Absorption.
Wavenumbers (cm^{-1}): 3569, 3394, 3280, 3210sh, 3100sh, 1627, 1140sh, 1054s, 1018s, 990s, 945sh, 874, 802w, 610, 585sh, 480sh, 460, 425sh.
Note: The spectrum was obtained by N.V. Chukanov.

P656 Minyulite $KAl_2(PO_4)_2(F,OH)·4H_2O$

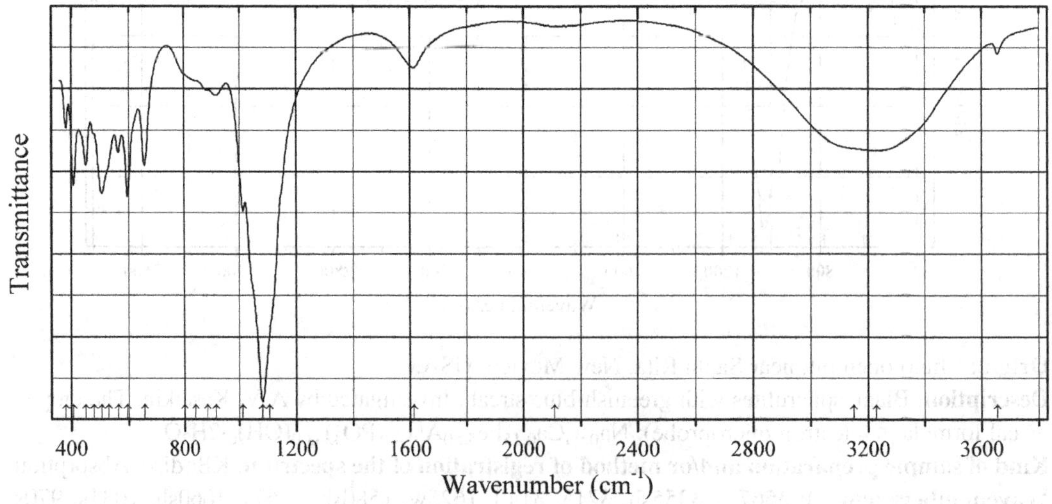

Origin: Cerro Mejillones, Mejillones Peninsula, Mejillones, Antofagasta, II Region, Chile.
Description: A F-rich sample. Investigated by I.V. Pekov.
Kind of sample preparation and/or method of registration of the spectrum: KBr disc. Absorption.
Wavenumbers (cm^{-1}): 3656w, 3229 (broad), 3150sh, 2110w (broad), 1615, 1105sh, 1080s, 1013s, 917, 886, 840sh, 800sh, 656s, 597s, 564, 530sh, 507, 480sh, 451s, 408s, 381.
Note: The spectrum was obtained by N.V. Chukanov.

P657 Variscite-4*O* $Al(PO_4)\cdot 2H_2O$

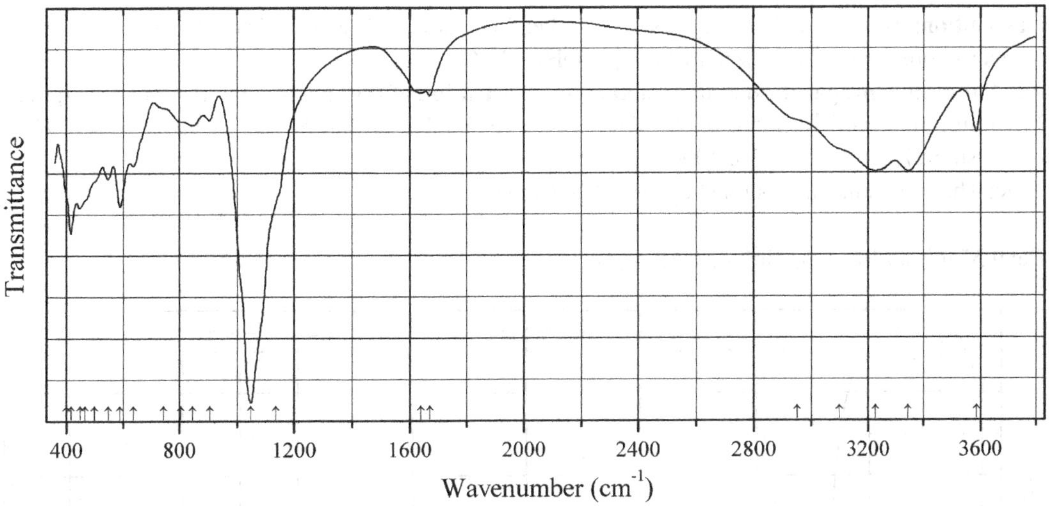

Origin: Cerro Mejillones, Mejillones Peninsula, Mejillones, Antofagasta, II Region, Chile.

Description: Colorless crystals from the association with gypsum and minyulite. Investigated by I.V. Pekov. Characterized by single-crystal X-ray diffraction data. Orthorhombic, $a = 9.675(4)$, $b = 9.893(4)$, $c = 17.203(9)$ Å, $V = 1647(1)$ Å^3. The empirical formula is (electron microprobe): $Al_{1.00}(PO_4)_{1.00}\cdot 2H_2O$

Kind of sample preparation and/or method of registration of the spectrum: KBr disc. Absorption.

Wavenumbers (cm^{-1}): 3585, 3344s, 3228s, 3100sh, 2950sh, 1670, 1639, 1135sh, 1049s, 905, 846, 804, 745sh, 637, 590s, 547, 500sh, 465sh, 448s, 416s, 405sh.

Note: The spectrum was obtained by N.V. Chukanov.

P658 Beraunite $Fe^{2+}Fe^{3+}{}_5(PO_4)_4(OH)_5\cdot 6H_2O$

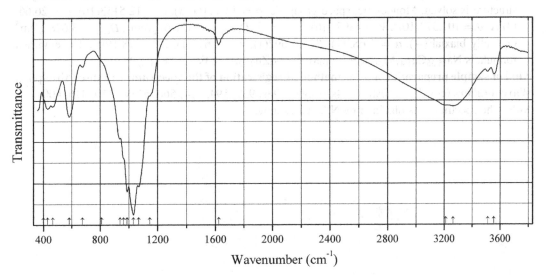

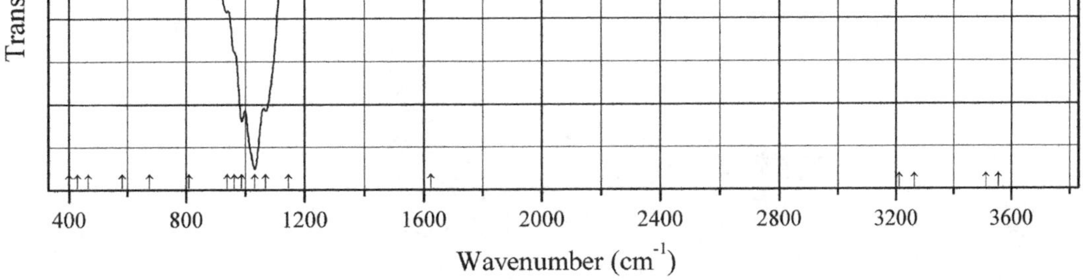

Origin: Hagendorf South pegmatite, Cornelia mine, Hagendorf, Waidhaus, Upper Palatinate, Bavaria, Germany.

Description: Black radial aggregates. A Mn-bearing variety. The empirical formula is (electron microprobe): $Fe_{5.44}Mn_{0.49}Zn_{0.07}(PO_4)_{4.00}(OH)_5 \cdot 6H_2O$.

Kind of sample preparation and/or method of registration of the spectrum: KBr disc. Absorption.

Wavenumbers (cm^{-1}): 3554, 3510, 3265, 3213, 1625, 1145sh, 1069s, 1032s, 989s, 965sh, 940s, 810sh, 677, 584s, 465, 430, 400sh.

Note: The spectrum was obtained by N.V. Chukanov.

P659 Manitobaite $Na_{16}Mn^{2+}{}_{25}Al_8(PO_4)_{30}$

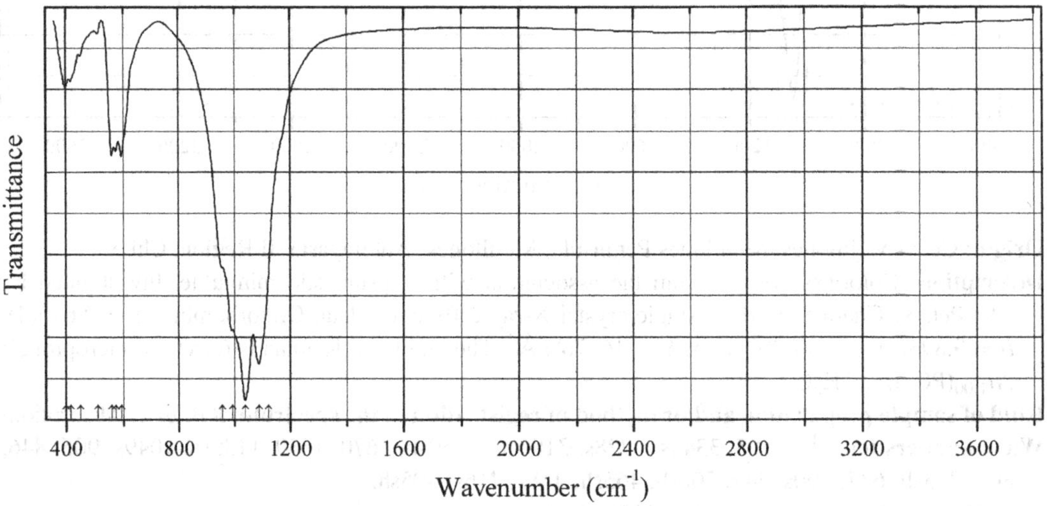

Origin: Cross Lake, Manitoba, Canada (type locality).

Description: Green grain. A fragment of holotype kindly granted by A.V. Kasatkin. The crystal structure is solved. Monoclinic, space group *Pc*, $a = 13.4516(15)$, $b = 12.5153(16)$, $c = 26.661(3)$ Å, $\beta = 101.579(10)°$, $V = 4397.1(6)$ Å^3, $Z = 2$. $D_{meas} = 3.621(6)$ g/cm^3, $D_{calc} = 3.628$ g/cm^3. Optically biaxial (−), $\alpha = 1.682(1)$, $\beta = 1.692(1)$, $\gamma = 1.697(1)$, $2V = 78.1(6)°$. The empirical formula is $Na_{15.55}Ca_{1.47}Mg_{0.88}Fe^{2+}{}_{4.19}Mn^{2+}{}_{18.78}Zn_{0.32}Al_{6.54}Fe^{3+}{}_{1.05}P_{30.08}O_{120}$.

Kind of sample preparation and/or method of registration of the spectrum: KBr disc. Absorption.

Wavenumbers (cm^{-1}): 1120sh, 1084s, 1036s, 993s, 957, 594, 574, 560, 511w, 450, 416, 400.

Note: The spectrum was obtained by N.V. Chukanov.

P660 Ammonium vanadyl pyrophosphate α-$(NH_4)_2(VO)_3(P_2O_7)_2$

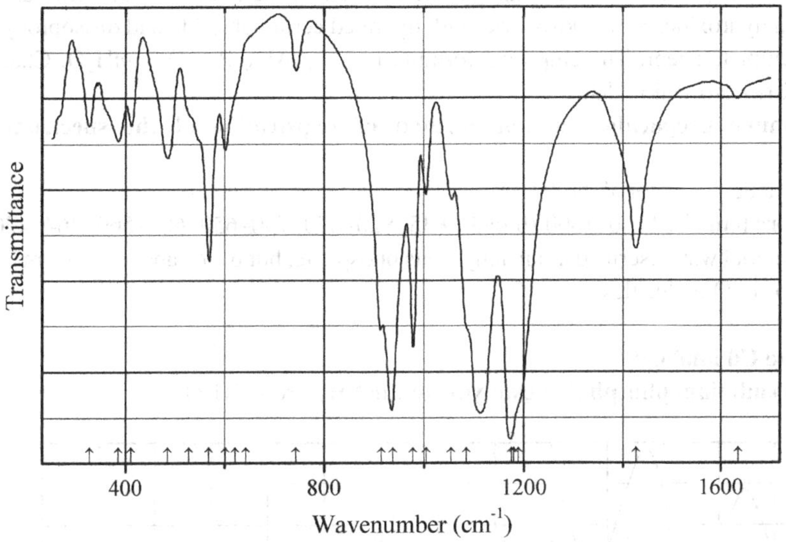

Origin: Synthetic.

Description: Prepared from the mixture of V_2O_5 and $(NH_4)_2(HPO_4)$ which was firstly heated up to 200 °C, homogenized and further heated at 325 °C for 2 h in air. Orthorhombic, space group *Pnma*, $Z = 4$.

Kind of sample preparation and/or method of registration of the spectrum: KBr disc. Transmission.

Source: Baran and Rabe (1999).

Wavenumbers (cm^{-1}): 3262 (broad), 1635w, 1427, 1189sh, 1175s, 1180s, 1085sh, 1054w, 1114s, 1004, 978s, 938s, 915w, 744, 643, 622sh, 601, 568s, 548sh, 527sh, 485, 411w, 385, 326.

Note: The wavenumbers were partly determined by us based on spectral curve analysis of the published spectrum.

P661 Aluminium phosphate hydrate $Al(PO_4){\cdot}nH_2O$

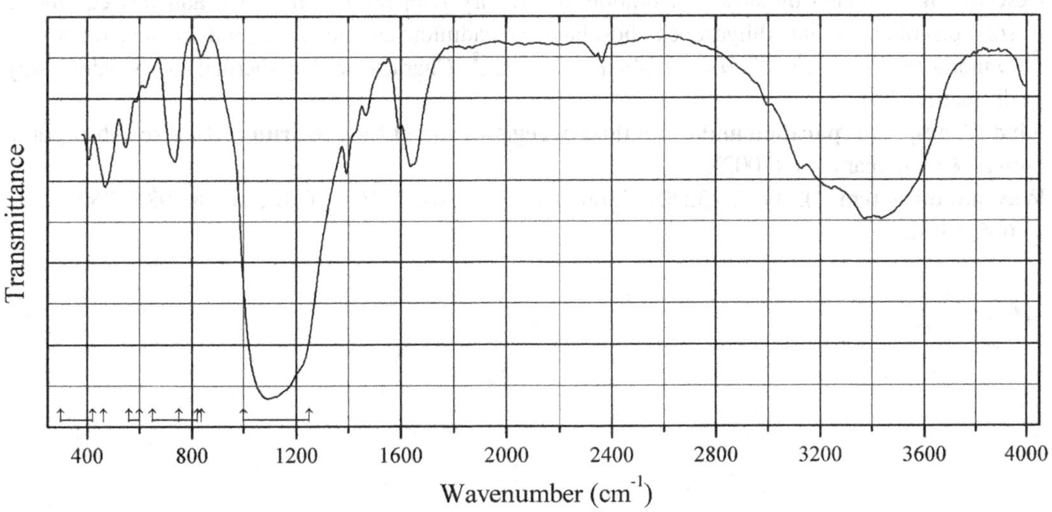

Origin: Synthetic.

Description: Ce-doped mesoporous material synthesized by the hydrothermal method, starting from aluminium hydroxide, 85% phosphoric acid, hydrated ceriumchloride and di-isopropylamine as an organic template agent. The empirical formula is $Ce_{0.04}Al_{0.97}P_{1.01}O_{4.04}\cdot nH_2O$. Characterized by powder X-ray diffraction data.

Kind of sample preparation and/or method of registration of the spectrum: KBr disc. Transmission.

Source: Souza de Araujo et al. (1997).

Wavenumbers (cm^{-1}): 1250–1000 (broad), 835, 820–650, 750–650, 600–560, 464, 420–300.

Note: The material was described as an anhydrous phosphate, but bands above 1550 cm^{-1} indicate the presence of H_2O molecules.

P662 Struvite Cd analogue
Ammonium cadmium phosphate hexahydrate $Cd(NH_4)(PO_4)\cdot 6H_2O$

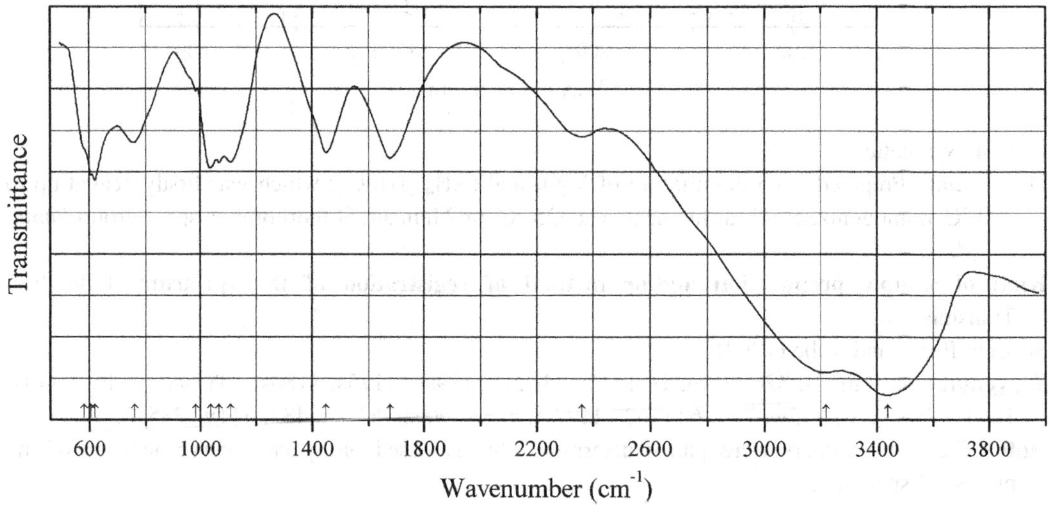

Origin: Synthetic.

Description: Obtained by slow evaporation at ordinary temperatures from the aqueous equimolar solutions of ammonium dihydrogen phosphate and cadmium sulfate. Structurally related to struvite. Orthorhombic, $a = 13.882$, $b = 12.249$, $c = 11.395$ Å. Characterized by thermal and powder X-ray diffraction data.

Kind of sample preparation and/or method of registration of the spectrum: KBr disc. Absorption.

Source: Ravikumar et al. (2002).

Wavenumbers (cm^{-1}): 3440s, 3220s, 2360, 1680s, 1450s, 1110s, 1070s, 1040s, 985, 760, 620s, 605, 580sh.

P663 Ammonium iron(II) phosphate hydrate $(NH_4)Fe(PO_4){\cdot}H_2O$

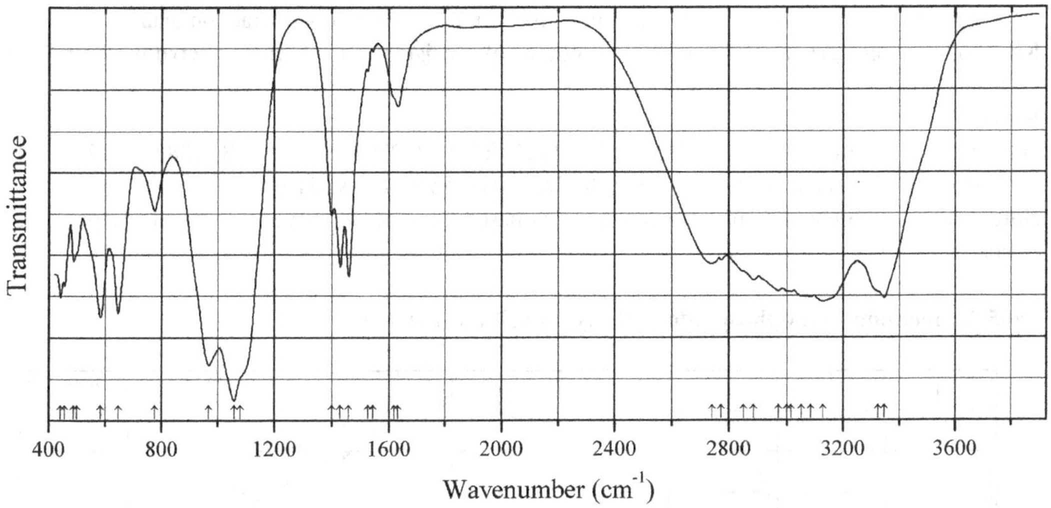

Origin: Synthetic.

Description: White crystals obtained from the equimolar mixture of $(NH_4)_3(PO_4){\cdot}3H_2O$ and $FeSO_4{\cdot}7H_2O$ ground in the presence of the surfactant PEG-400 and heated at 40 °C for 48 h. Characterized by powder X-ray diffraction data. Orthorhombic, space group $Pmm2_1$, $a = 5.660$, $b = 8.825$, $c = 4.826$ Å, $V = 241$ Å^3.

Kind of sample preparation and/or method of registration of the spectrum: KBr disc. Transmission.

Source: Yuan et al. (2008).

Wavenumbers (cm^{-1}): 3347, 3325sh, 3130, (3085), (3052), (3014), (3000), (2970), (2886), (2850sh), (2773), 2740, 1634, 1620sh, 1545, 1528, 1460, 1431, 1400, 1081sh, 1058s, 970s, 775, 646, 584, 500sh, 488, 457, 442.

Note: The wavenumbers were partly determined by us based on spectral curve analysis of the published spectrum.

P664 Ammonium magnesium phosphate $(NH_4)Mg(PO_4){\cdot}H_2O$

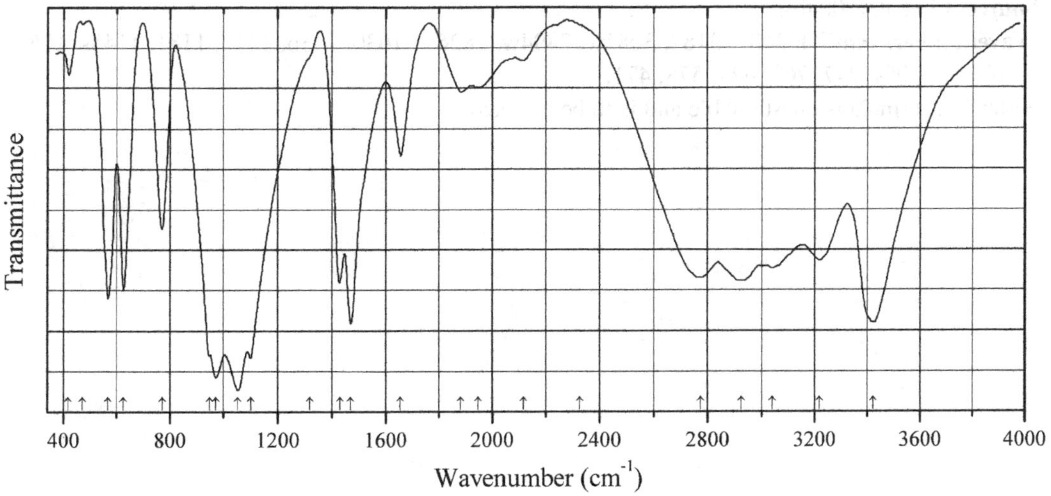

Origin: Synthetic.
Description: Obtained by adding 0.5 M solution of $MgCl_2 \cdot 6H_2O$ to an excess of saturated $(NH_4)_2H(PO_4)$ solution. Characterized by thermal analysis and powder X-ray diffraction data.
Kind of sample preparation and/or method of registration of the spectrum: KBr disc. Transmission.
Source: Sronsri et al. (2014).
Wavenumbers (cm^{-1}): 3424s, 3218, 3040, 2924, 2775, 2327w, 2118, 1948, 1882, 1657, 1470s, 1430, 1319sh, 1102s, 1055s, 974s, 949sh, 772, 627s, 568s, 472w, 419.
Note: The wavenumbers were determined by us based on spectral curve analysis of the published spectrum.

P665 Ammonium titanophosphate $(NH_4)_2Ti_2O(HPO_4)(PO_4)_2$?

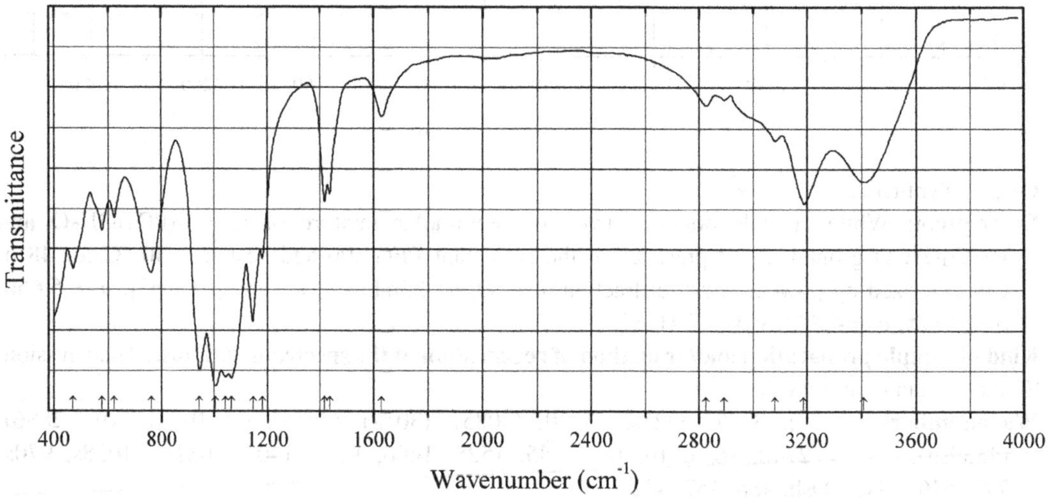

Origin: Synthetic.
Description: Synthesized hydrothermally from $Ti(SO_4)_2$, $H_3(PO_4)_3$, and ammonia solution at 200 °C for several hours. Characterized by powder X-ray diffraction data and thermal analysis.
Kind of sample preparation and/or method of registration of the spectrum: No data.
Source: Li et al. (2004b).
Wavenumbers (cm^{-1}): 3407, 3187, 3081w, 2894w, 2826w, 1630, 1436, 1417, 1181, 1145s, 1067s, 1044s, 1006s, 947, 760, 624, 578, 471.
Note: The formula is questionable and is to be checked.

P666 Antimony(III) phosphate $SbPO_4$

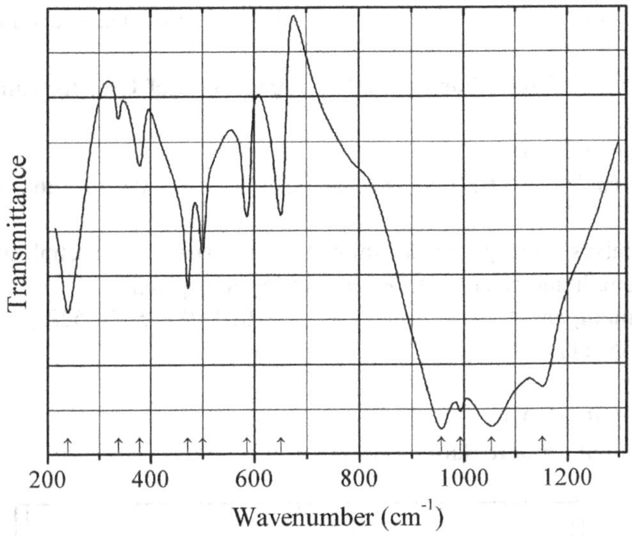

Origin: Synthetic.

Description: Prepared in the reaction of antimony with freshly prepared metaphosphoric acid at high temperatures. Characterized by powder X-ray diffraction data. Monoclinic, space group $P2_1/m$, $Z = 2$.

Kind of sample preparation and/or method of registration of the spectrum: KBr and polyethylene discs. Transmission.

Source: Brockner, and Hoyer (2002).

Wavenumbers (IR, cm^{-1}): 1152s, 1054s, 993s, 957s, 651, 586, 500, 472, 378, 337w, 239.

Note: In the cited paper, Raman spectrum is given.

Wavenumbers (Raman, cm^{-1}): 1054, 977, 937w, 623, 584, 548w, 478, 356s.

P667 Antimony(V) oxophosphate $SbO(PO_4)$

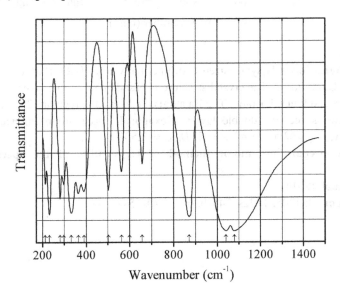

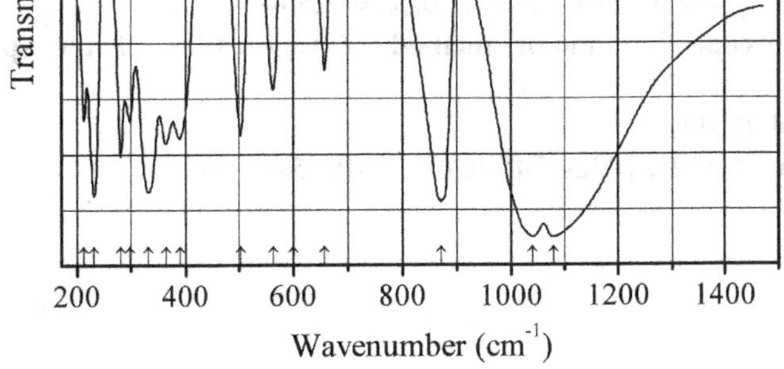

Origin: Synthetic.

Description: Monoclinic, space group *C*2*c*, *a* = 6.791(1), *b* = 8.033(1), *c* = 7.046(1) Å, β = 115.90 (1)°, Z = 4. The crystal structure consists of chains of corner shared distorted octahedra linked together by PO_4 tetrahedra.

Kind of sample preparation and/or method of registration of the spectrum: See Husson et al. (1988b).

Source: Husson et al. (1988a).

Wavenumbers (IR, cm^{-1}): 1080s, 1040s, 872s, 658, 600w, 564, 504, 390, 364, 332s, 297, 280, 231s, 212.

Note: The wavenumbers were partly determined by us based on spectral curve analysis of the published spectrum. In the cited paper, Raman spectrum is given.

Wavenumbers (Raman, cm^{-1}): 1122, 1060w, 1050, 1010, 788, 585s, 528s, 460w, 417, 375, 312s, 278, 197, 152, 135, 117w.

P668 Minjiangite $BaBe_2(PO_4)_2$
Barium beryllium phosphate $BaBe_2(PO_4)_2$

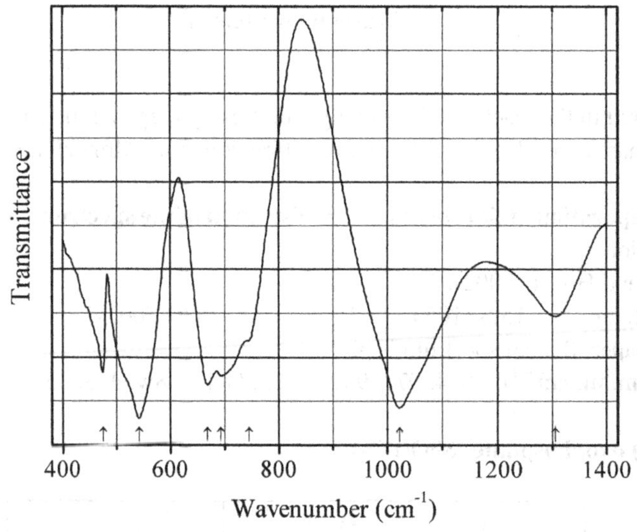

Origin: Synthetic.

Description: Colorless crystals synthesized hydrothermally from BeO, $Ba(OH)_2{\cdot}8H_2O$, and H_3PO_4 (85%) at 200 °C for 7 days. The crystal structure is based on of double layers of tetrahedra, which contain both Be and P in a 1:1 ratio. The Ba atoms are located in regular 12-coordinated polyhedra and connect two successive double layers. Hexagonal, space group *P*6/*mmm*, *a* = 5.028(1), *c* = 7.466(1) Å, *V* = 162.51(1) Å^3, Z = 1. D_{calc} = 3.507 g/cm^3.

Kind of sample preparation and/or method of registration of the spectrum: KBr disc. Transmission.

Source: Dal Bo et al. (2014).

Wavenumbers (cm^{-1}): 1305, 1022s, 745sh, 694, 670s, 541s, 475.

P669 Barium chromium pyrophosphate $BaCr_2(P_2O_7)_2$

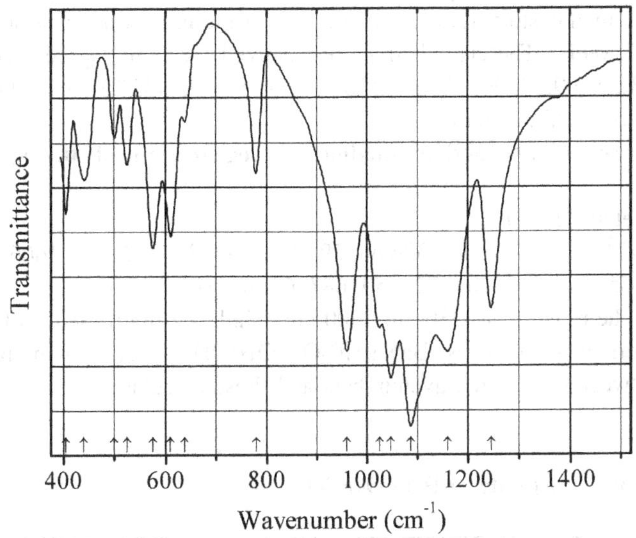

Origin: Synthetic.

Description: Green solid synthesized by the conventional solid-state reaction technique, by heating stoichiometric mixture of $BaCO_3$, Cr_2O_3, and $(NH_4)(H_2PO_4)$ up to 1200 °C. Characterized by powder X-ray diffraction data. Triclinic, space group *P*-1, a = 6.1408(5), b = 6.1898(4), c = 7.8027(6) Å, α = 96.692(5)°, β = 101.686(5)°, γ = 105.542(4)°, V = 275.19(4) Å^3, Z = 1. D_{meas} = 2.39(3) g/cm^3, D_{calc} = 2.391 g/cm^3.

Kind of sample preparation and/or method of registration of the spectrum: KBr disc. Transmission.

Source: Tao et al. (2014).

Wavenumbers (cm^{-1}): 1245, 1160s, 1087s, 1047, 1024, 960s, 780, 638w, 610, 575, 525, 500, 440, 405.

Note: The wavenumbers were partly determined by us based on spectral curve analysis of the published spectrum.

P670 Barium sodium cyclotriphosphate hydrate $BaNa(P_3O_9)\cdot 3H_2O$

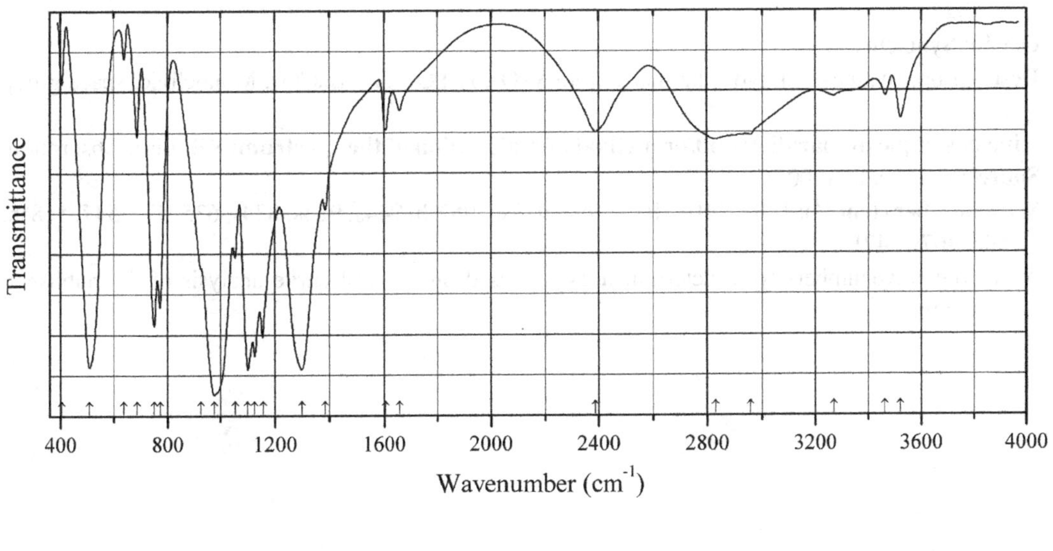

Origin: Synthetic.

Description: Pink crystals obtained by adding barium nitrate to a saturated aqueous solution of sodium cyclotriphosphate in the stoichiometric ratio. The resulting mixture was left to stand at room temperature for 2 weeks. The crystal structure is solved. Triclinic, space group *P*-1, $a = 7.0350(3)$, $b = 9.0470(3)$, $c = 9.8800(2)$ Å, $\alpha = 116.551(3)°$, $\beta = 95.932(2)°$, $\gamma = 74.088(3)°$, $V = 540.81(3)$ Å^3, $Z = 2$. $D_{calc} = 2.771$ g/cm^3.

Kind of sample preparation and/or method of registration of the spectrum: KBr disc. Transmission.

Source: Ezzaafrani et al. (2014).

Wavenumbers (cm^{-1}): 3522, 3465w, 3269w, 2959, 2829, 2387, 1659w, 1606, 1385, 1298s, 1157, 1126s, 1101s, 1055, 976s, 926sh, 772, 750, 687, 638w, 509s, 405w.

Note: The bands in the range from 2300 to 2900 cm^{-1} indicate the presence of acid P–OH groups. Possibly, the correct formula is $BaNa(HP_3O_9)(OH)·2H_2O$. This assumption could explain discrepancies between observed wavenumbers and those calculated according to the Libowitzky formula.

P671 Barium vanadyl phosphate α-$Ba(VO_2)(PO_4)$

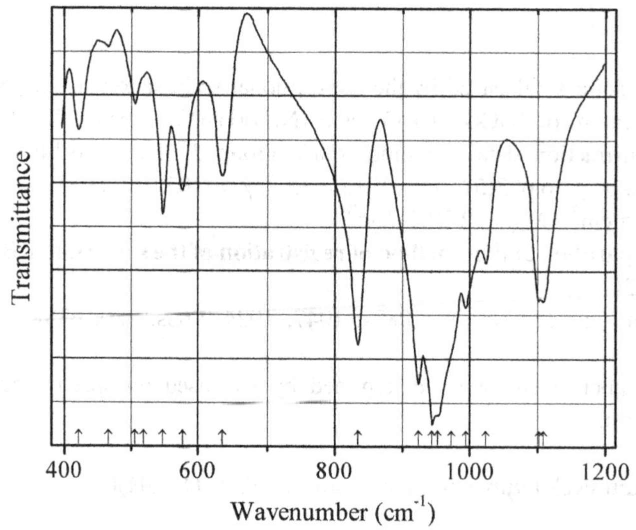

Origin: Synthetic.

Description: Synthesized hydrothermally from V_2O_5, H_3PO_4, and $BaCO_3$. Monoclinic, space group *P*2$_1$/*c*.

Kind of sample preparation and/or method of registration of the spectrum: KBr disc. Absorption.

Source: Borel et al. (2000).

Wavenumbers (cm^{-1}): 1109, 1102, 1024, 994, 973sh, 952sh, 944s, 924s, 834s, 635, 577, 547, 518sh, 506, 467w, 421.

Note: The wavenumbers were determined by us based on spectral curve analysis of the published spectrum.

P672 β-Vanadyl pyrophosphate β-$(VO)_2(P_2O_7)$

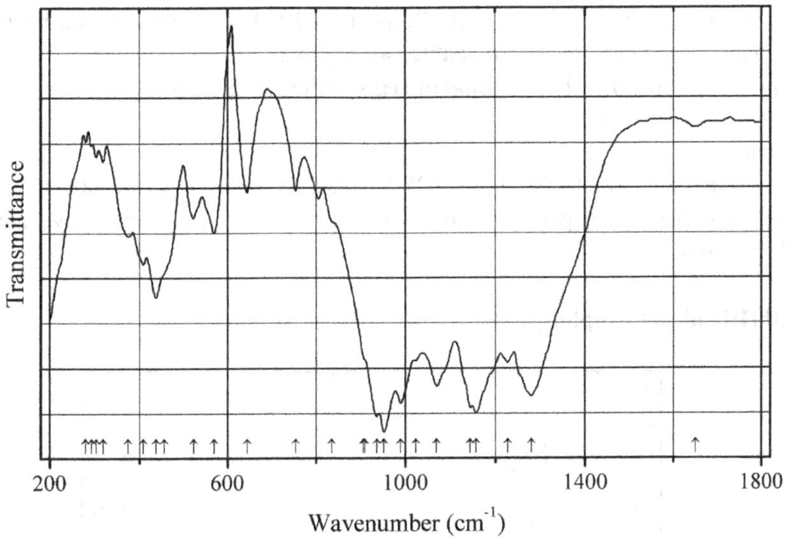

Origin: Synthetic.

Description: Orthorhombic, space group *Pca*2.

Kind of sample preparation and/or method of registration of the spectrum: KBr disc. Transmission.

Source: Bordes et al. (1984).

Wavenumbers (cm^{-1}): 1650w, 1280s, 1227, 1158s, 1145, 1070, 1024, 990, 953s, 937s, 910sh, 836sh, 906, 754, 643, 568, 523, 456sh, 438, 409, 376, 319w, 304w, 293w, 281w.

Note: The wavenumbers were determined by us based on spectral curve analysis of the published spectrum.

P673 Bismuth(III) calcium oxophosphate $BiCa_4(PO_4)_3O$

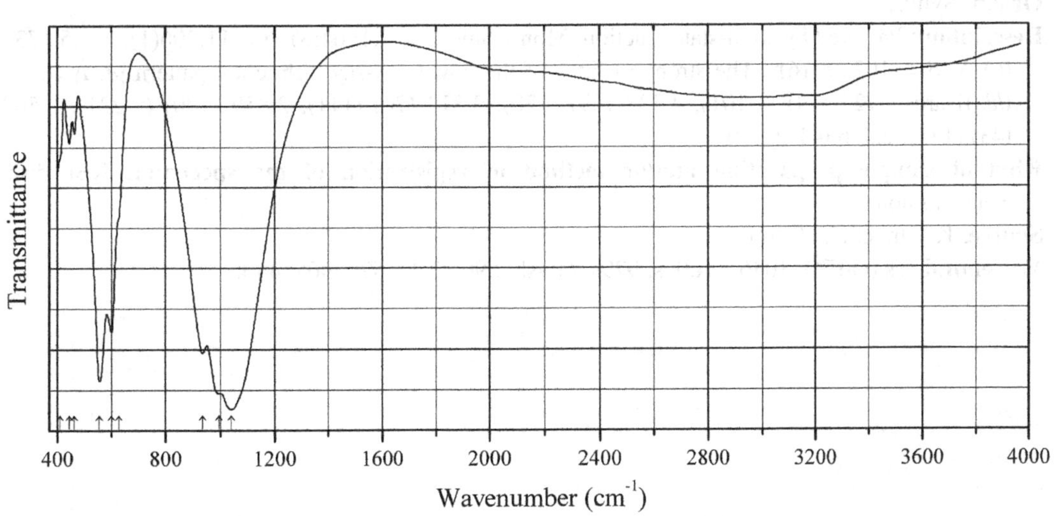

Origin: Synthetic.

Description: Prepared by high temperature solid-state method, by stepwise heating stoichiometric mixture of Bi_2O_3, $Ca(CO_3)$, and $(NH_4)(H_2PO_4)$ up to 1050 °C. Characterized by powder X-ray diffraction data. Structurally related to apatite, space group $P6_3/m$.

Kind of sample preparation and/or method of registration of the spectrum: Transmission. Kind of sample preparation is not indicated.

Source: Sumathi and Gopal (2015).

Wavenumbers (cm^{-1}): 1041s, 995sh, 935s, 628sh, 602s, 557s, 464w, 445w, 409sh.

Note: The wavenumbers were partly determined by us based on spectral curve analysis of the published spectrum.

P674 Bismuth(III) nickel oxophosphate BiNi(PO$_4$)O BiNi(PO$_4$)O

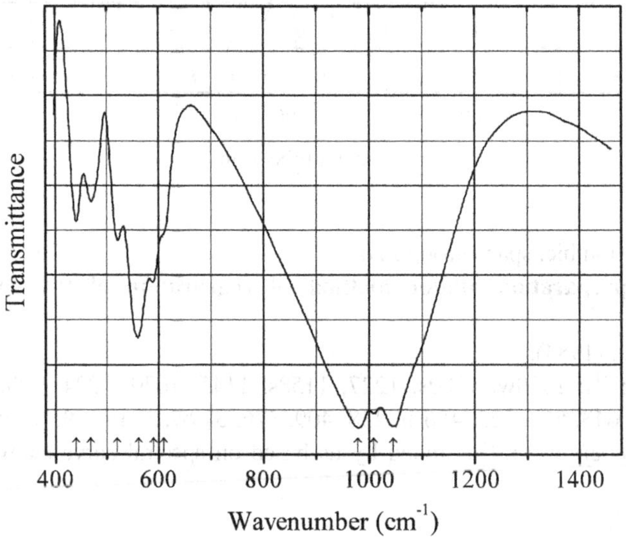

Origin: Synthetic.

Description: Prepared by solid-state reaction. Monoclinic, $a = 7.1664(8)$, $b = 11.206(1)$, $c = 5.1732(6)$ Å, $\beta = 107.281(6)°$. The strongest lines of the powder X-ray diffraction pattern [d, Å (I, %) (hkl)] are: 4.727 (44) (−101), 4.338 (69) (120), 3.372 (70) (111), 2.850 (100) (−221), 2.568 (43) (131), 2.516 (41) (230).

Kind of sample preparation and/or method of registration of the spectrum: KBr disc. Transmission.

Source: Ketatni et al. (1999).

Wavenumbers (cm^{-1}): 1046s, 1009s, 979s, 609sh, 590, 561s, 520, 468, 440.

P675 Boron phosphate BPO_4

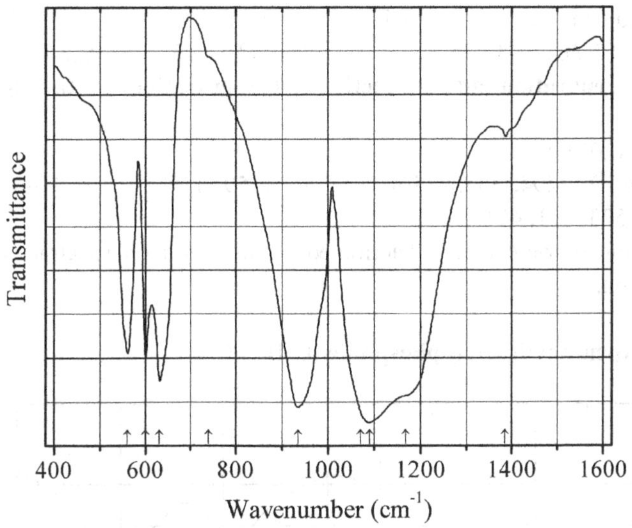

Origin: Synthetic.

Description: Structurally related to cristobalite.

Kind of sample preparation and/or method of registration of the spectrum: KBr disc. Transmission.

Source: Osaka et al. (1984).

Wavenumbers (cm^{-1}): 1386w, 1168sh, 1090s, 1070sh, 935s, 739sh, 630, 600, 560.

Note: The wavenumbers were partly determined by us based on spectral curve analysis of the published spectrum.

P676 Magnesium borophosphate $(H_3O)Mg(BP_2O_8)\cdot 3H_2O$

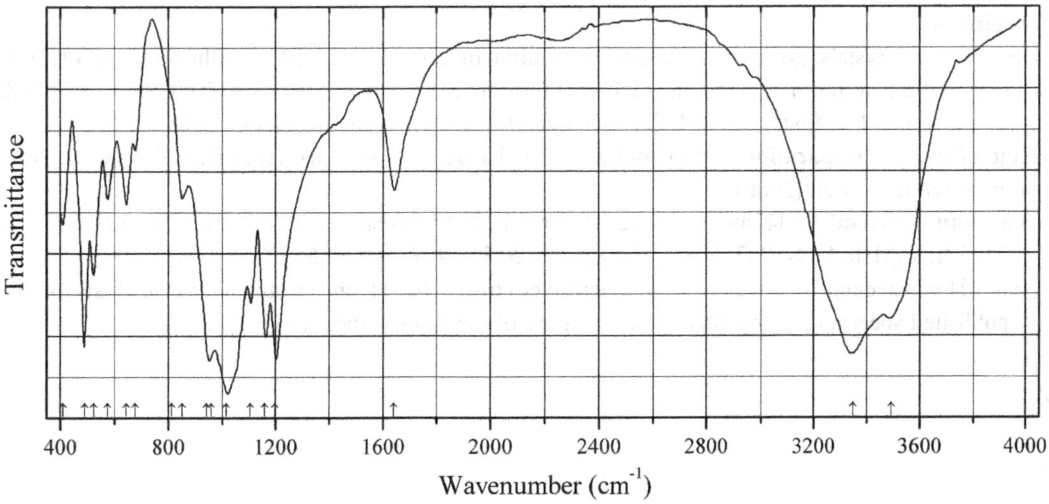

Origin: Synthetic.

Description: Synthesized from $MgCl_2$, B_2O_3, and H_3PO_4 in the presence of pyridine and HCl, under mild hydrothermal conditions (at 170 °C for 3 days). The crystal structure is solved. Hexagonal, space group *P*6(1)22, *a* = 9.4462(7), *c* = 15.759(2) Å, *V* =1217.8(2) Å^3, *Z* = 6. D_{calc} = 2.439(3) g/cm^3.

Kind of sample preparation and/or method of registration of the spectrum: KBr disc. Transmission.

Source: Yang et al. (2011f).

Wavenumbers (cm^{-1}): 3494, 3348s, 1643, 1198s, 1159, 1105, 1016s, 959s, 940sh, 851, 812sh, 678w, 645, 576, 523, 491, 411.

Note: The wavenumbers were partly determined by us based on spectral curve analysis of the published spectrum.

P677 Cesium acid (pentahydrogen) phosphate $CsH_5(PO_4)_2$

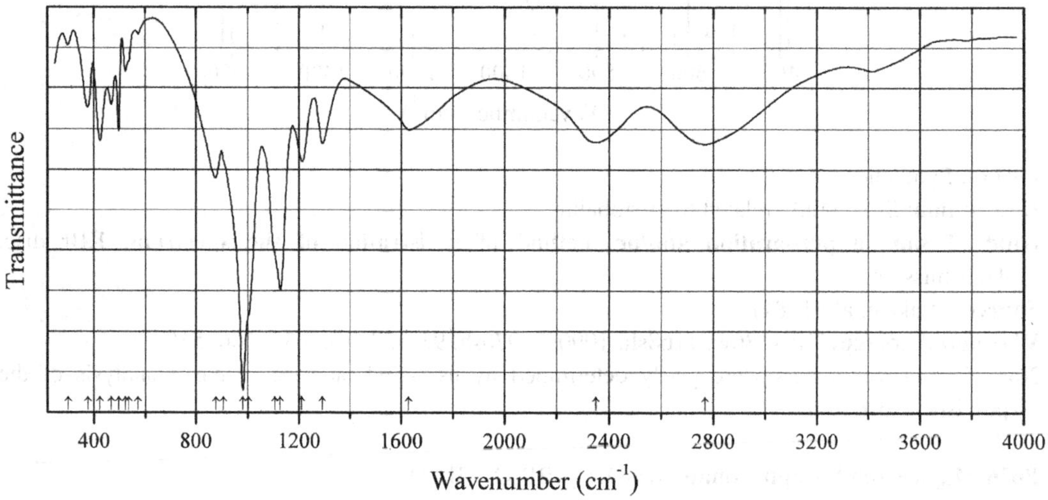

Origin: Synthetic.

Description: Crystals grown from an aqueous solution of Cs_2CO_3 and H_3PO_4 (at the mole ratio 1:2) by evaporation at room temperature. Monoclinic, space group *P*2$_1$/*c*, *a* = 10.879, *b* = 7.768, *c* = 9.526 Å, *β* = 96.60°, *Z* = 4. Characterized by powder X-ray diffraction data.

Kind of sample preparation and/or method of registration of the spectrum: KBr disc. Absorption.

Source: Lavrova et al. (2006).

Wavenumbers (cm^{-1}): 3414w (broad), 2770 (broad), 2350 (broad), 1630 (broad), 1293, 1212, 1127s, 1105sh, 1001sh, 980s, 907, 878, 571w, 536sh, 523w, 497, 468, 425, 377, 300w.

Note: The wavenumbers were partly determined by us based on spectral curve analysis of the published spectrum. In the cited paper, a figure of the Raman spectrum is given.

P678 Cesium manganese(II) pyrophosphate $Cs_2MnP_2O_7$

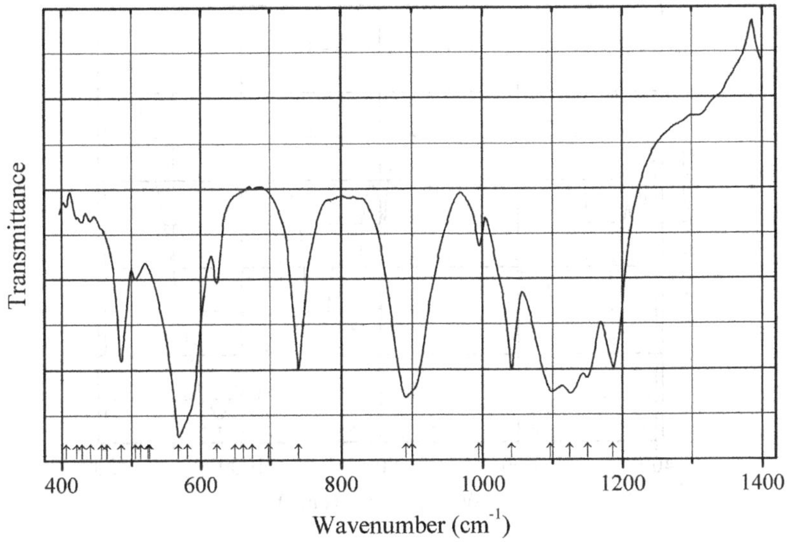

Origin: Synthetic.

Description: Pink solid prepared from aqueous solutions of $Mn(NO_3)_2{\cdot}4H_2O$, CsCl, and $NH_4(H_2PO_4)$ as starting materials by heating resulting powder progressively from 200 to 700 °C with intermediate regrindings. The crystal structure is solved. Orthorhombic, space group *Pnma*, $a = 16.3398(3)$, $b = 5.3872(1)$, $c = 9.8872(2)$ Å, $V = 870.33(3)$ Å^3, $Z = 4$. $D_{calc} = 3.775$ g/cm^3.

Kind of sample preparation and/or method of registration of the spectrum: Diffuse reflection of a mixture with KBr.

Source: Kaoua et al. (2013).

Wavenumbers (IR, cm^{-1}): 1186s, 1151s, 1125s, 1097s, 1041s, 995, 900sh, 891s, 740, 697sh, 674w, 661sh, 649sh, 623, 580sh, 568s, 525sh, 514sh, 506w, 485, 465sh, 457sh, 441w, 429w, 421w, 406w.

Note: The wavenumbers were partly determined by us based on spectral curve analysis of the published spectrum. The strong band at 891 cm^{-1} and shoulder at 900 cm^{-1} are indicated by Kaoua et al. (2013) as strong band at 896 cm^{-1}. In the cited paper, Raman spectrum is given.

Wavenumbers (Raman, cm^{-1}): 1197w, 1165w, 1154w, 1142w, 1129, 1128w, 1107w, 1098w, 1081w, 1060w, 1041w, 1021s, 954w, 934w, 898w, 700s, 635w, 598w, 576w, 562, 536w, 521w, 498w, 466w, 428, 364, 327, 269w, 245w, 205, 202, 166w, 140w, 129w, 124w.

P679 Cesium uranyl oxophosphate $Cs_3(UO_2)_2(PO_4)O_2$

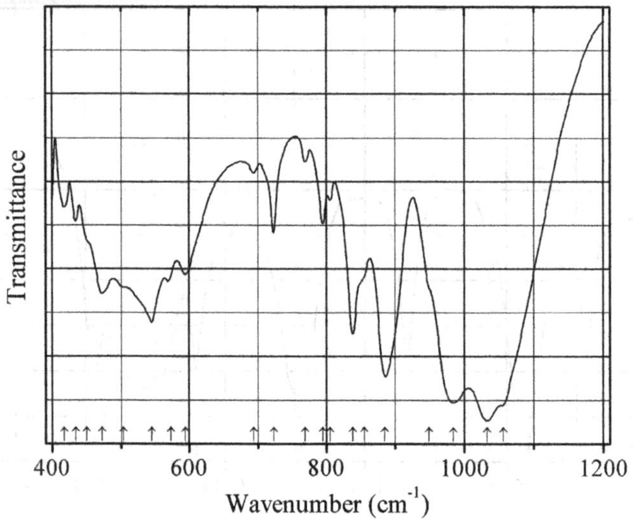

Origin: Synthetic.

Description: Yellow crystals obtained in the reaction of triuranyl diphosphate tetrahydrate with a CsI flux at 750 °C. The crystal structure is solved. Monoclinic, space group *C2/c*, $a = 13.6261(13)$, $b = 8.1081(8)$, $c = 12.3983(12)$ Å, $\beta = 114.61(12)°$, $V = 1245.41(20)$ Å^3, $Z = 4$. $D_{calc} = 2.684$ g/cm^3.

Kind of sample preparation and/or method of registration of the spectrum: KBr disc. Transmission.

Source: Yagoubi et al. (2013).

Wavenumbers (cm^{-1}): 1056sh, 1033s, 984s, 949sh, 885s, 855sh, 838, 805, 795, 769w, 723, 694w, 595, 574, 545, 504sh, 473, 451sh, 435w, 418w.

Note: The wavenumbers were partly determined by us based on spectral curve analysis of the published spectrum.

P680 Calcium chlorophosphate ("chlor-spodiosite") $Ca_2(PO_4)Cl$

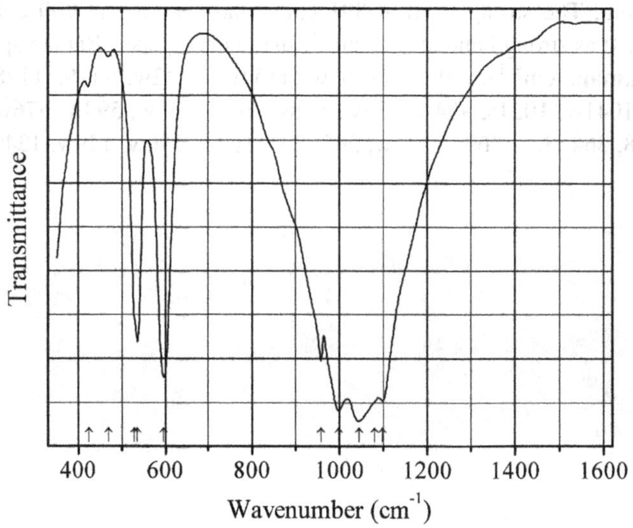

Origin: Synthetic.

Description: Crystals grown from melt by means of a reaction flux technique using $Ca_3(PO_4)_2$ and $CaCl_2$ as starting materials. Orthorhombic, space group *Pbcm*.

Kind of sample preparation and/or method of registration of the spectrum: KBr disc. Transmission.

Source: Kowalczyk and Condrate Sr (1974).

Wavenumbers (IR, cm^{-1}): 1099s, 1080sh, 1046s, 999s, 959, 596, 536, 529sh, 471w, 425w.

Note: In the cited paper, Raman spectrum is given.

Wavenumbers (Raman, cm^{-1}): 1077, 1063, 1048w, 1026w, 995, 958s, 627, 611, 551, 463, 397.

P681 Calcium dihydrophosphate monohydrate $Ca(H_2PO_4)_2{\cdot}H_2O$

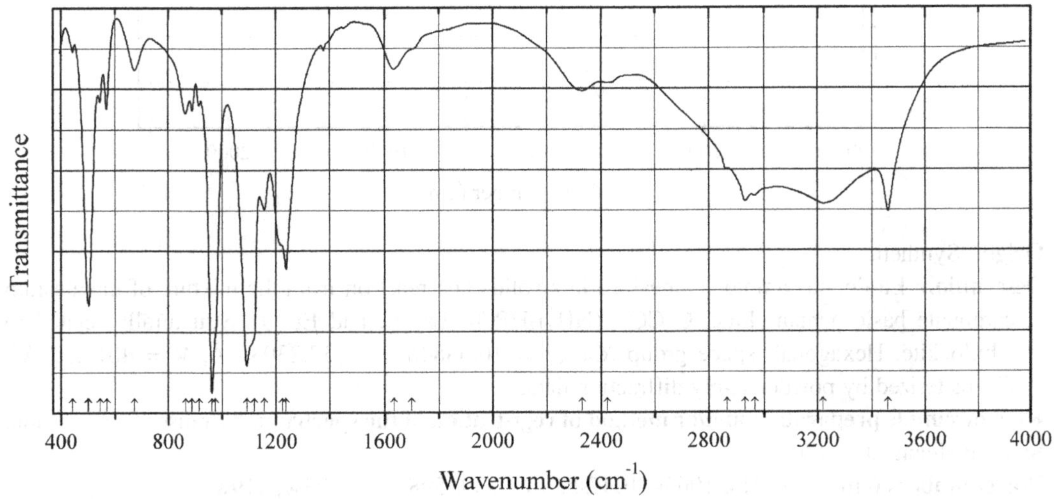

Origin: Synthetic.

Description: Commercial reactant. Triclinic, space group *P*-1, *Z* = 2.

Kind of sample preparation and/or method of registration of the spectrum: Absorption. Kind of sample preparation is not indicated.

Source: Xu et al. (1998).

Wavenumbers (IR, cm^{-1}): 3461, 3220, 2967, 2934, 2868sh, 2423w, 2330, 1700sh, 1635w, 1239s, 1225sh, 1158, 1120sh, 1092s, 975sh, 962s, 914w, 888w, 864w, 675w, 570w, 545w, 504s, 500sh, 444w, 355, 250sh, 230.

Note: The wavenumbers were partly determined by us based on spectral curve analysis of the published spectrum. In the cited paper, Raman spectrum is given.

Wavenumbers (Raman, cm^{-1}): 2810, 2386, 1217, 1159, 1112s, 1016, 988s, 956, 916s, 906, 580, 523, 499, 421, 366, 337, 251, 210, 170, 141, 124, 103, 85.

P682 Calcium magnesium lanthanum phosphate $Ca_8MgLa(PO_4)_7$

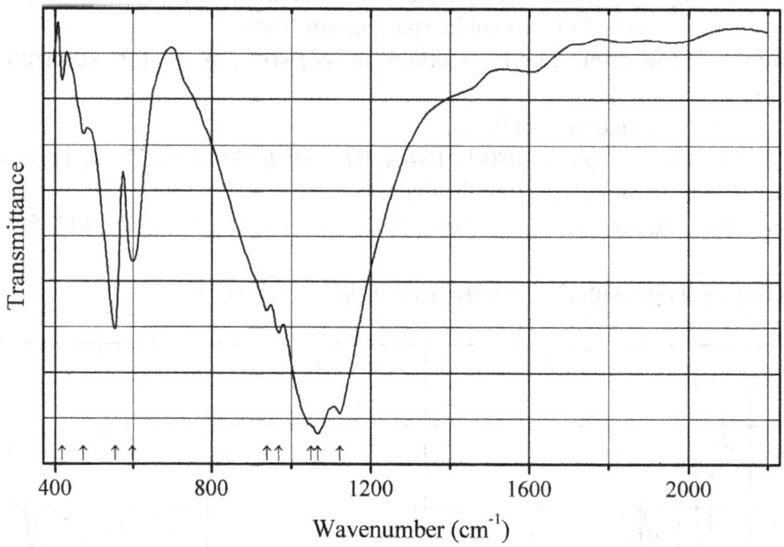

Origin: Synthetic.

Description: Eu-doped sample synthesized in a solid-state reaction from the mixture of magnesium carbonate basic pentahydrate, $CaCO_3$, $(NH_4)_2HPO_4$, La_2O_3, and Eu_2O_3. Structurally related to whitlockite. Hexagonal, space group *R*3*c* ,a = 10.38848, c = 37.23035 Å, V = 4017.89 Å^3. Characterized by powder X-ray diffraction data.

Kind of sample preparation and/or method of registration of the spectrum: KBr disc. Transmission.

Source: Huang et al. (2009).

Wavenumbers (cm^{-1}): 1121s, 1065s, 1047sh, 969, 939, 598, 552, 473w, 419w.

Note: The wavenumbers were determined by us based on spectral curve analysis of the published spectrum.

P683 Calcium magnesium yttrium phosphate $Ca_8MgY(PO_4)_7$

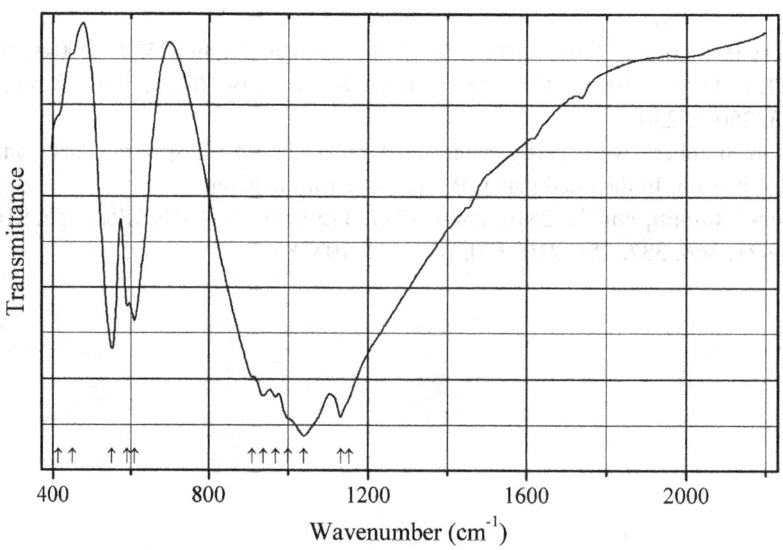

Origin: Synthetic.

Description: Eu-doped sample synthesized in a solid-state reaction from a mixture of magnesium carbonate basic pentahydrate, $CaCO_3$, $(NH_4)_2HPO_4$, Y_2O_3, and Eu_2O_3. Structurally related to whitlockite. Hexagonal, space group *R*3*c*, $a = 10.32966$, $c = 36.94593$ Å, $V = 3942.20$ Å^3. Characterized by powder X-ray diffraction data.

Kind of sample preparation and/or method of registration of the spectrum: KBr disc. Transmission.

Source: Huang et al. (2009).

Wavenumbers (cm^{-1}): 1151sh, 1130s, 1039s, 1000sh, 969s, 938s, 910sh, 608, 590, 551, 450sh, 415sh.

Note: The wavenumbers were partly determined by us based on spectral curve analysis of the published spectrum.

P684 Cerium(IV) pyrophosphate CeP_2O_7

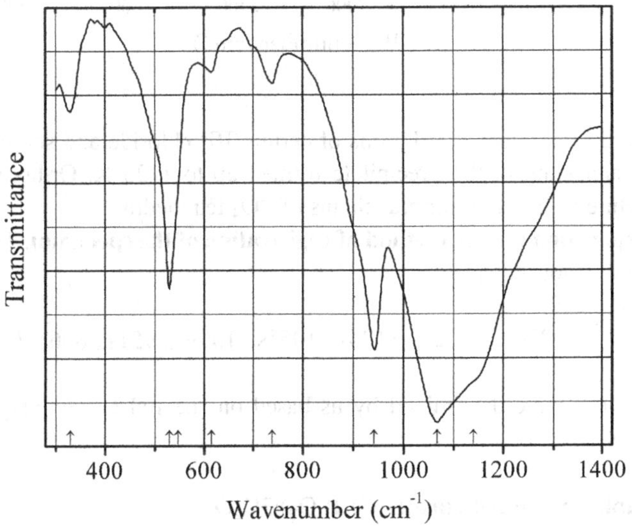

Origin: Synthetic.

Description: Obtained by adding aqueous solution of cerium(IV) sulfate to the solution of sodium pyrophosphate. Cubic, $a = 8.607(4)$ Å. The strongest lines of the powder X-ray diffraction pattern [*d*, Å (*I*, %) (*hkl*)] are: 4.97 (30) (111), 4.31 (100) (200), 3.85 (15) (210), 3.51 (15) (211), 3.057 (28) (220), 2.599 (41) (311), 1.925 (17) (420).

Kind of sample preparation and/or method of registration of the spectrum: KBr disc. Transmission.

Source: Botto and Baran (1977).

Wavenumbers (cm^{-1}): 1140sh, 1067s, 942s, 738w, 615w, 548sh, 530, 330.

P685 Cerium(III) polyphosphate $Ce(PO_3)_3$

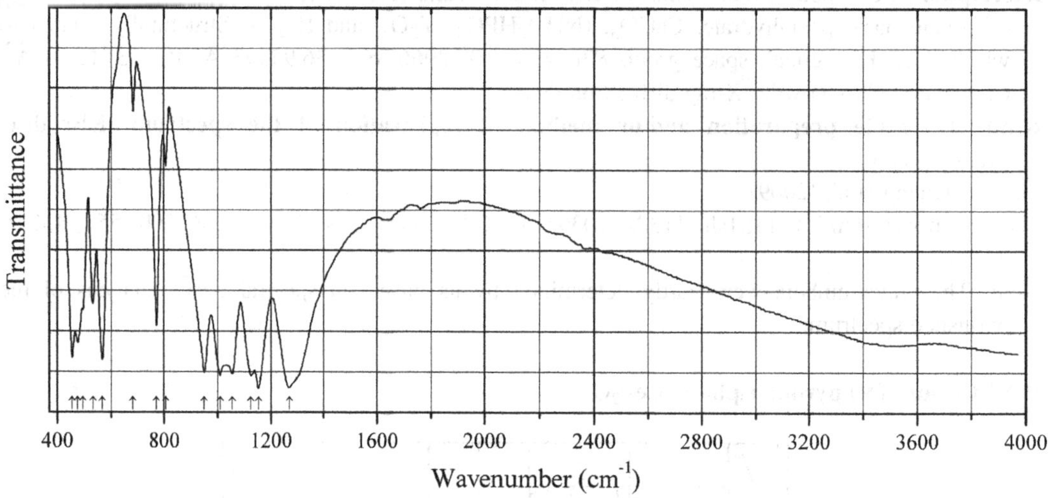

Origin: Synthetic.

Description: Prepared from aqueous solutions of cerium(III) chloride and sodium cyclotriphosphate with subsequent annealing of the precipitate formed up to 1123 K. Orthorhombic, space group $C222_1$. The structure is based on infinite chains of PO_4 tetrahedra.

Kind of sample preparation and/or method of registration of the spectrum: Transmission. Kind of sample preparation is not indicated.

Source: Ternane et al. (2008).

Wavenumbers (cm^{-1}): 1270s, 1152s, 1124s, 1055s, 1009s, 951s, 806, 770, 682w, 568, 532, 496, 476, 456.

Note: The wavenumbers were determined by us based on spectral curve analysis of the published spectrum.

P686 Cerium metaphosphate trihydrate $Ce(P_3O_9)\cdot 3H_2O$

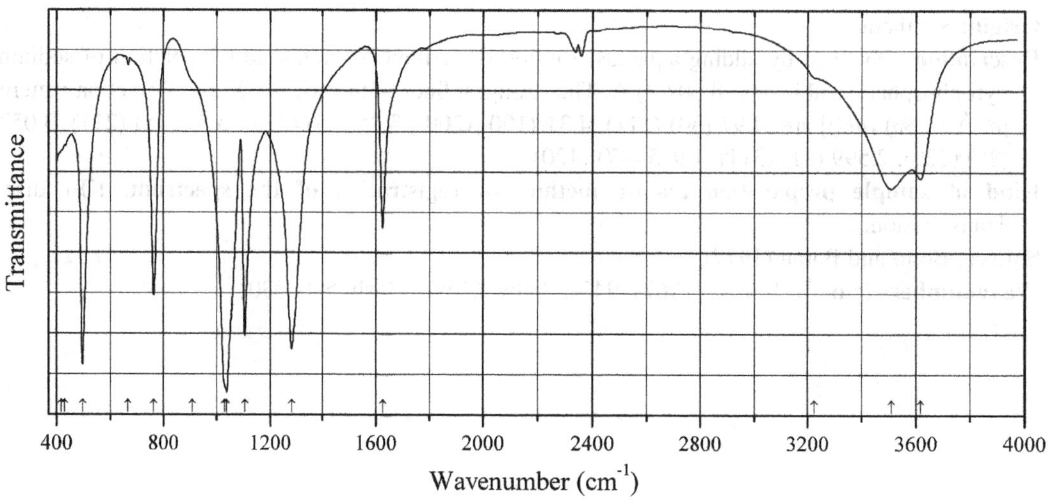

Origin: Synthetic.

Description: Prepared from aqueous solutions of cerium(III) chloride and sodium cyclotriphosphate. Hexagonal, space group *P*-6. The cyclotriphosphate anion $P_3O_9^{3-}$ has a benitoite-type planar configuration.

Kind of sample preparation and/or method of registration of the spectrum: Transmission. Kind of sample preparation is not indicated.

Source: Ternane et al. (2008).

Wavenumbers (cm^{-1}): 3618, 3510, 3225sh, 1627, 1284s, 1105s, 1039s, 1029sh, 909sh, 761, (667w), 497s, 430sh, 417sh.

Note: The wavenumbers were determined by us based on spectral curve analysis of the published spectrum. Weak bands in the range from 2300 to 2400 cm^{-1} and at 667 cm^{-1} correspond to atmospheric CO_2.

P687 Copper tinanium oxyphosphate α-$CuTi_2(PO_4)_2O_2$

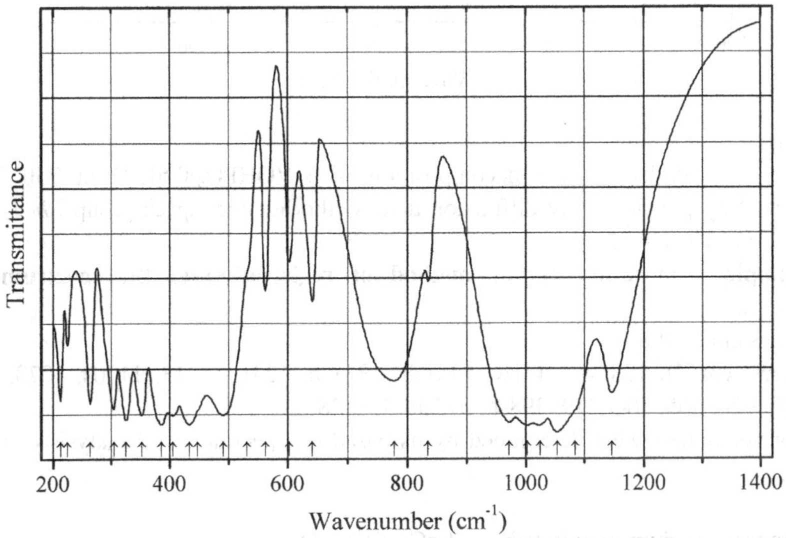

Origin: Synthetic.

Description: Prepared by coprecipitation from aqueous solutions of $Cu(NO_3)_2 \cdot 3H_2O$ and $NH_4(H_2PO_4)$ taken in of stoichiometric quantities and ethanol solution of $TiCl_4$. After evaporation of the solvent, the solid was stepwise heated up to 950 °C. The crystal structure is solved. Monoclinic, space group $P2_1/c$, $a = 7.5612(4)$, $b = 7.0919(4)$, $c = 7.4874(4)$ Å, $\beta = 122.25(1)°$, $V = 339.55(6)$ Å^3, $Z = 4$. $D_{meas} = 3.71(2)$ g/cm^3, $D_{calc} = 3.729$ g/cm^3. The strongest lines of the powder X-ray diffraction pattern [*d*, Å (*I*, %) (*hkl*)] are: 6.393 (30) (100), 4.746 (23) (110), 3.334 (27) (21−1), 3.307 (100) (11−2), 3.233 (49) (111), 3.198 (36) (200), 3.093 (40) (02−1), 2.585 (23) (22−1).

Kind of sample preparation and/or method of registration of the spectrum: KBr and polyethylene discs. Transmission.

Source: Benmokhtar et al. (2007a).

Wavenumbers (IR, cm^{-1}): 1146, 1085sh, 1054s, 1025s, 1003s, 972s, 835, 778, 641, 602, 562, 531sh, 490, 447sh, 433s, 405, 386s, 352, 325, 304, 264, 225, 213.

Note: The wavenumbers were partly determined by us based on spectral curve analysis of the published spectrum. In the cited paper, Raman spectrum is given.

Wavenumbers (Raman, cm^{-1}): 1119, 1069, 1057, 1036, 1026, 1007s, 986, 835, 732s, 571, 470, 445, 417, 392, 371, 353.

P688 γ-Vanadyl pyrophosphate γ-$(VO)_2(P_2O_7)$

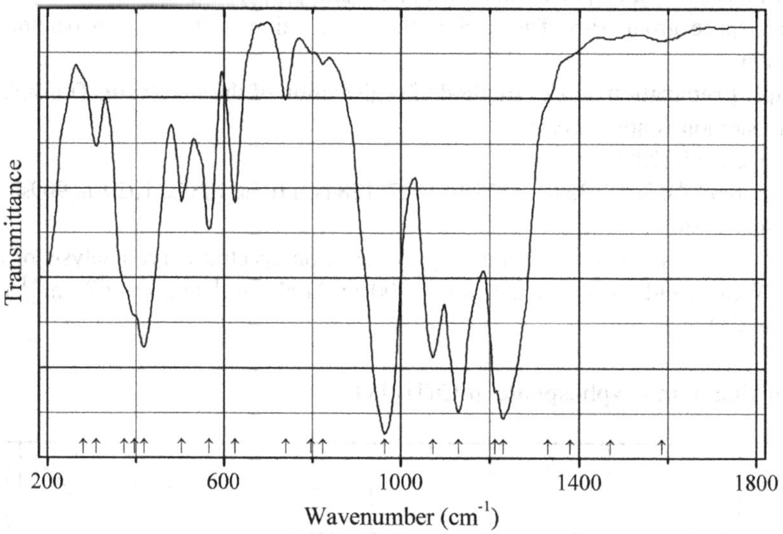

Origin: Synthetic.

Description: Prepared by thermal decomposition of γ-$VOHPO_4 \cdot 0.5H_2O$ at 750 °C for 3 h. Characterized by powder X-ray diffraction data. Orthorhombic, space group $Pbc2_1$, $a = 9.571$, $b = 7.728$, $c = 16.568$ Å.

Kind of sample preparation and/or method of registration of the spectrum: KBr disc. Transmission.

Source: Bordes et al. (1984).

Wavenumbers (cm^{-1}): 1585w, 1471w, 1381sh, 1331sh, 1231s, 1213, 1130s, 1073, 964s, 824w, 797w, 740, 624, 565, 504, 419, 398sh, 373sh, 311, 282sh.

Note: The wavenumbers were determined by us based on spectral curve analysis of the published spectrum.

P689 Lanthanum calcium oxophosphate $LaCa_4(PO_4)_3O$

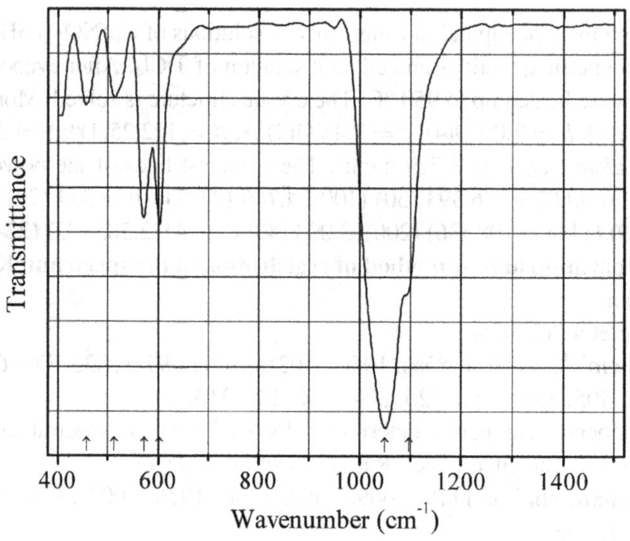

Origin: Synthetic.

Description: Apatite-type compound synthesized by high-temperature solid-state reaction from a stoichiometric mixture of preheated La_2O_3, $CaCO_3$, and $(NH_4)(H_2PO_4)$. Characterized by powder X-ray diffraction data. Hexagonal, space group $P6_3$, $a = 9.463(8)$, $c = 6.92(1)$ Å, $V = 536.64$ Å^3.

Kind of sample preparation and/or method of registration of the spectrum: KBr disc. Transmission.

Source: Buvaneswari and Varadaraju (2000).

Wavenumbers (cm^{-1}): 1097sh, 1051s, 602, 572, 513, 459.

P690 Lanthanum strontium oxophosphate $LaSr_4(PO_4)_3O$

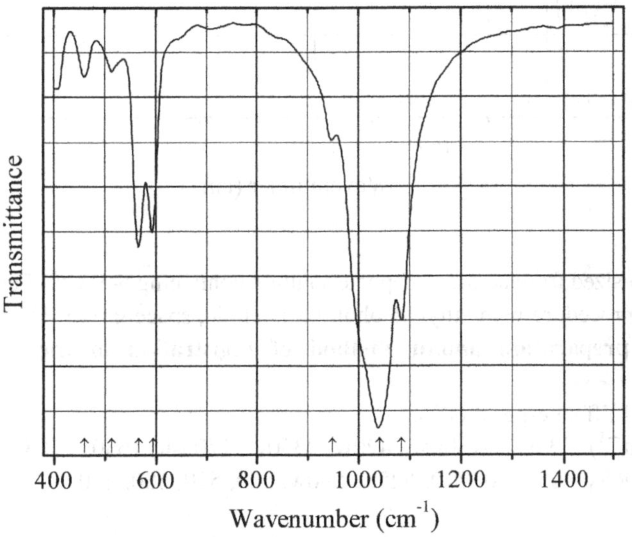

Origin: Synthetic.

Description: Apatite-type compound synthesized by high-temperature solid-state reaction from a stoichiometric mixture of preheated La_2O_3, $SrCO_3$, and $(NH_4)(H_2PO_4)$. Characterized by powder X-ray diffraction data. Hexagonal, space group $P6_3$, $a = 9.71(1)$, $c = 7.30(1)$ Å, $V = 596.05$ Å^3.

Kind of sample preparation and/or method of registration of the spectrum: KBr disc. Transmission.

Source: Buvaneswari and Varadaraju (2000).

Wavenumbers (cm^{-1}): 1085s, 1041s, 948sh, 591, 566, 513w, 460w.

P691 Iron(II) acid phosphate hydrate $Fe(H_2PO_4)_2{\cdot}2H_2O$

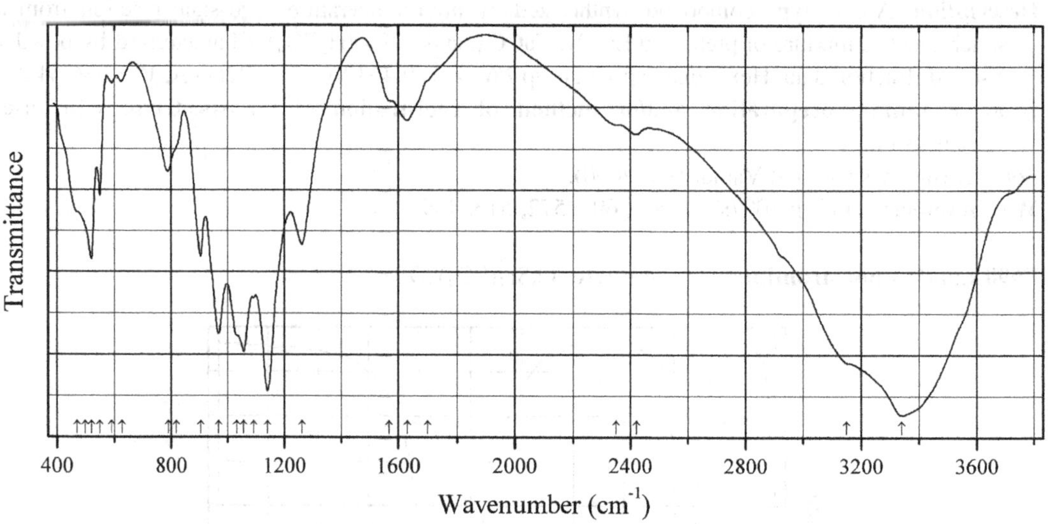

Origin: Synthetic.

Description: Synthesized from acidic phosphate solution containing 5–9 wt% FeO and 65 wt% H_3PO_4 by a salting out procedure with ethyl alcohol. Monoclinic, space group $P2_1/n$, $Z = 2$.

Kind of sample preparation and/or method of registration of the spectrum: KBr disc. Transmission.

Source: Koleva and Effenberger (2007).

Wavenumbers (cm^{-1}): 3340s, 3150sh, 2420w, 2350w, 1700sh, 1630, 1568w, 1262, 1140s, 1090, 1055s, 1030sh, 968s, 905, 820, 790, 625w, 590w, 550, 520, 500, 470.

P692 Lead beryllium phosphate hurlbutite-type $PbBe_2(PO_4)_2$

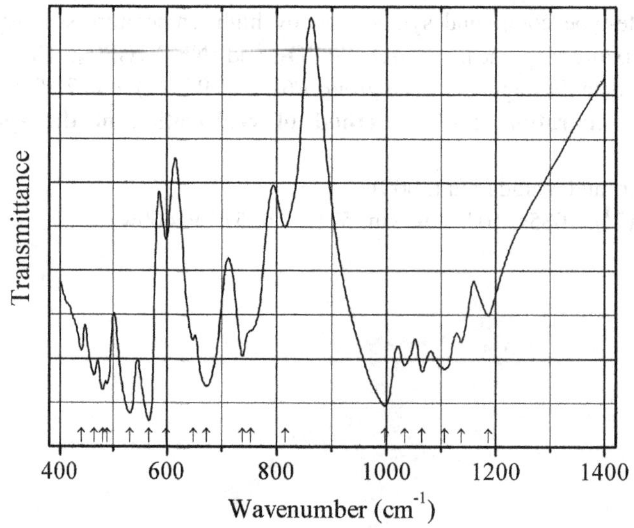

Origin: Synthetic.

Description: Synthesized hydrothermally from BeO, H_3PO_4 (85%), $(NH_4)(H_2PO_4)$, and $Pb(NO_3)_2$ at 200 °C for 7 days. The crystal structure is solved. Monoclinic, space group $P2_1/c$, $a = 8.088(1)$, $b = 9.019(1)$, $c = 8.391(1)$ Å, $\beta = 90.12(1)°$, $V = 612.22(1)$ Å^3, $Z = 4$. $D_{calc} = 4.504$ g/cm^3. Characterized by powder X-ray diffraction data.

Kind of sample preparation and/or method of registration of the spectrum: KBr disc. Transmission.

Source: Dal Bo et al. (2014).

Wavenumbers (cm^{-1}): 1187, 1137, 1106, 1065, 1033, 997s, 816, 753, 738, 674s, 649, 598, 566s, 531, 489, 481, 465, 441.

Note: The wavenumbers were partly determined by us based on spectral curve analysis of the published spectrum.

P693 Lead iron(III) phosphate $Pb_3Fe_2(PO_4)_4$

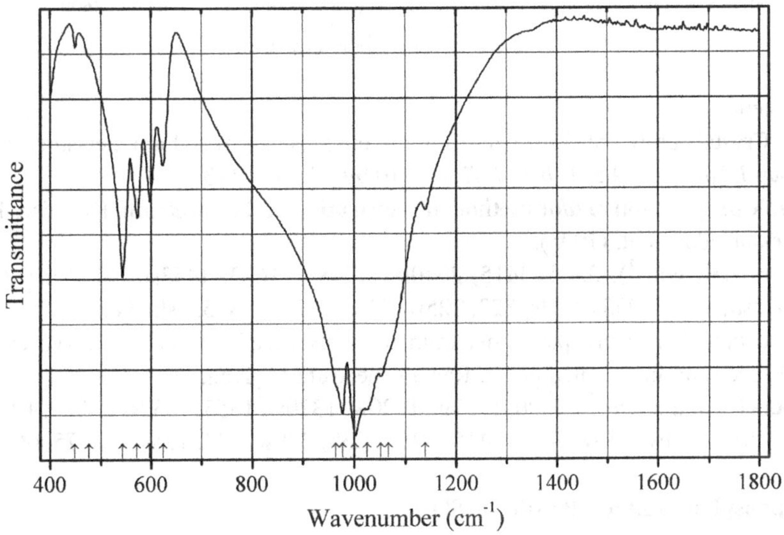

Origin: Synthetic.

Description: Prepared from a stoichiometric mixture of Fe_2O_3, $PbCO_3$, and $(NH_4)(H_2PO_4)$ by a ceramic technique. Characterized by elemental analysis, powder X-ray diffraction data, and Mössbauer spectrum. The crystal structure is solved. Monoclinic, space group $P2_1/c$, $a = 9.0065(6)$, $b = 9.0574(6)$, $c = 9.3057(6)$ Å, $\beta = 116.880(4)°$, $V = 677.10(8)$ Å^3, $Z = 2$. $D_{calc} = 5.412$ g/cm^3.

Kind of sample preparation and/or method of registration of the spectrum: KBr disc. Transmission.

Source: Malakho et al. (2005).

Wavenumbers (cm^{-1}): 1139, 1067sh, 1052s, 1025s, 1002s, 978s, 965sh, 624, 598, 573, 544, 477sh, 449w.

Note: The wavenumbers were determined by us based on spectral curve analysis of the published spectrum.

P694 Lead phosphate nitrate hydrate $Pb_2(NO_3)(PO_4)\cdot H_2O$

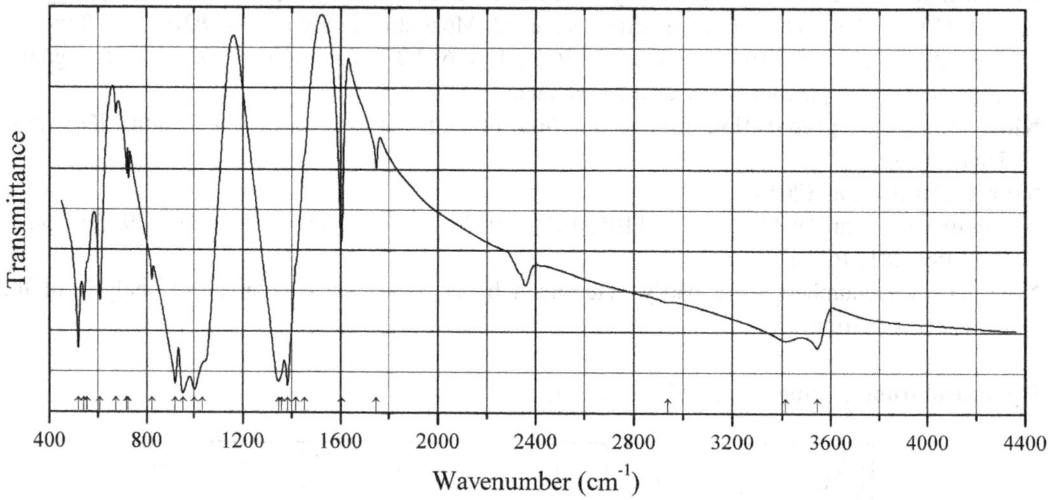

Origin: Synthetic.

Description: Crystals grown by a sol-gel method in the presence of sodium metasilicate. Monoclinic, space group $P2_1/c$, a =19.511, b = 7.37, c = 10.994 Å, β = 113°.

Kind of sample preparation and/or method of registration of the spectrum: KBr disc. Transmission.

Source: Vivekanandan et al. (1995).

Wavenumbers (IR, cm^{-1}): 3544, 3415, 2940sh, 1748w, 1603, 1453sh, 1419sh, 1384s, 1359sh, 1348s, 1037sh, 1004s, 955s, 923s, 823, 725w, 718w, 673, 608, 556sh, 542, 518.

Note: The wavenumbers were partly determined by us based on spectral curve analysis of the published spectrum. In the cited paper, Raman spectrum is given.

Wavenumbers (Raman, cm^{-1}): 3500, 1653w, 1620w, 1370w, 1353, 1062s, 1044w, 1016s, 967, 947, 733, 722w, 617, 574w, 535w, 444w, 415, 392w, 239, 205w, 144, 125s, 96s, 75, 66.

P695 Lead phosphate sulfate $Pb_4(PO_4)_2(SO_4)$

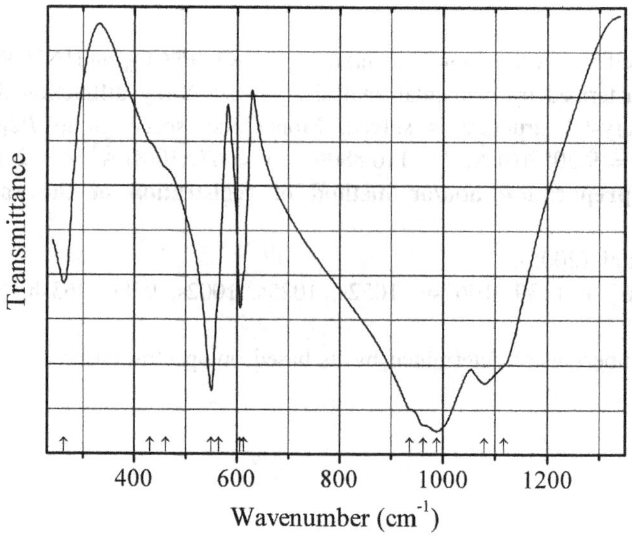

Origin: Synthetic.

Description: Prepared by a solid-state reaction from a stoichiometric mixture of $Pb_3(CO_3)_2(OH)_2$, $(NH_4)_2(SO_4)$, and $(NH_4)_2(HPO_4)$ in air at 700 °C for several days. Cubic, space group $I\bar{4}3d$, $Z = 4$.

Kind of sample preparation and/or method of registration of the spectrum: KBr disc. Transmission.

Source: Massaferro et al. (1999).

Wavenumbers (IR, cm^{-1}): 1118sh, 1080s, 989s, 936sh, 613sh, 606, 565sh, 551, 460sh, 430sh.

Note: The wavenumbers were partly determined by us based on spectral curve analysis of the published spectrum. In the cited paper, Raman spectrum is given.

Wavenumbers (Raman, cm^{-1}): 1120, 1085, 987s, 963sh, 944s, 609, 551, 459, 440, 430, 262.

P696 Lead silver phosphate apatite-type $Pb_4Ag(PO_4)_3$

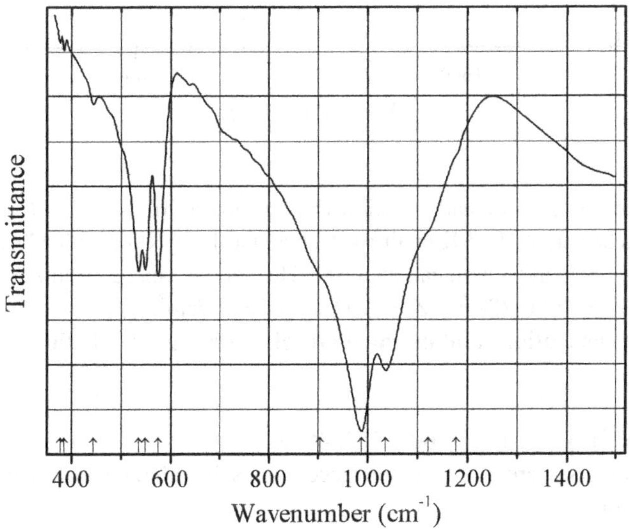

Origin: Synthetic.

Description: Prepared hydrothermally from a stoichiometric mixture of $Pb_3(PO_4)_2$ and $Ag_3(PO_4)$ at 215 °C for 1 day. Hexagonal, space group $P6_3/m$, $a = 9.772(4)$, $c = 7.210(3)$ Å. In the apatite-type structure, Ag^+ ions concentrate in the column positions. The strongest lines of the powder X-ray diffraction pattern [*d*, Å (*I*, %) (*hkl*)] are: 4.238 (53) (200), 4.051 (53) (111), 3.202 (52) (210), 2.926 (100) (211), 2.904 (71) (112), 2.823 (72) (300).

Kind of sample preparation and/or method of registration of the spectrum: Transmission. Kind of sample preparation is not indicated.

Source: Ternane et al. (2000).

Wavenumbers (IR, cm^{-1}): 1178sh, 1121sh, 1037s, 988s, 907sh, 574, 548, 535, 444w, 385w, 377w.

Note: The wavenumbers were partly determined by us based on spectral curve analysis of the published spectrum. In the cited paper, Raman spectrum is given.

Wavenumbers (Raman, cm^{-1}): 1017, 963s, 933s, 577s, 558, 421, 388.

P697 Lead sodium calcium phosphate apatite-type $Pb_3CaNa(PO_4)_3$

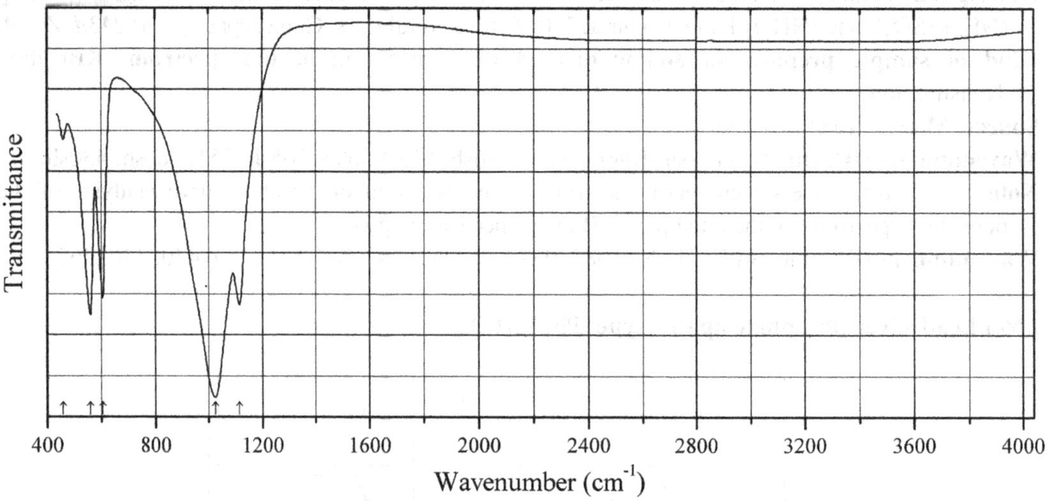

Origin: Synthetic.

Description: Prepared by a solid-state reaction from a mixture of Na_2CO_3, $(NH_4)_2(HPO_4)$, $CaCO_3$, and PbO powders heated at 1073 K in air for 12 h and at 1173 K for 12 h. Characterized by powder X-ray diffraction data and chemical analysis. Hexagonal, space group $P6_3/m$, $a = 9.658(8)$, $c = 7.081(6)$ Å, $V = 572.01(8)$ Å^3, $Z = 2$. $D_{calc} = 5.63$ g/cm^3.

Kind of sample preparation and/or method of registration of the spectrum: KBr disc. Transmission.

Source: Naddari et al. (2003).

Wavenumbers (cm^{-1}): 1113, 1024s, 604, 559, 459w.

Note: The wavenumbers were partly determined by us based on spectral curve analysis of the published spectrum.

P698 Lithium chromium pyrophosphate $LiCrP_2O_7$

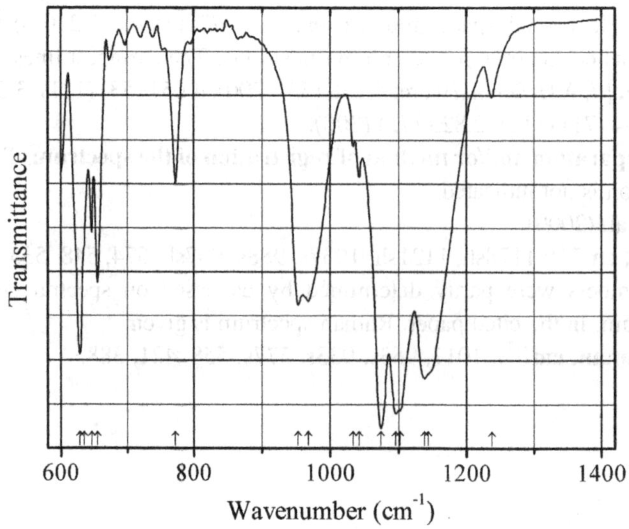

Origin: Synthetic.

Description: Prepared by solid-state reaction from a stoichiometric mixture of $Li(H_2PO_4)$, $NH_4(H_2PO_4)$, and $CrCl_3 \cdot 6H_2O$, first at 200 °C for 5 h, then at 700 °C for 20 h, and finally at 950 °C for 20 h with intermediate grindings and pelletizing. Characterized by powder X-ray diffraction data. The crystal structure is solved. Monoclinic, space group $P2_1$, $a = 4.7867(7)$, $b = 8.0049(11)$, $c = 6.9093(10)$ Å, $\beta = 109.003(2)°$, $V = 250.32(6)$ Å^3, $Z = 2$. $D_{calc} = 3.090$ g/cm^3.

Kind of sample preparation and/or method of registration of the spectrum: The spectrum was recorded using a Pike MIRacle micrometer pressure clamp.

Source: Pachoud et al. (2013).

Wavenumbers (cm^{-1}): 1238w, 1144sh, 1138s, 1102sh, 1096s, 1074s, 1042w, 1034w, 968, 953, 772, 655, 646, 634sh, 628s.

Note: The wavenumbers were partly determined by us based on spectral curve analysis of the published spectrum.

P699 Lithiumiron(III) pyrophosphate $LiFe^{3+}(P_2O_7)$

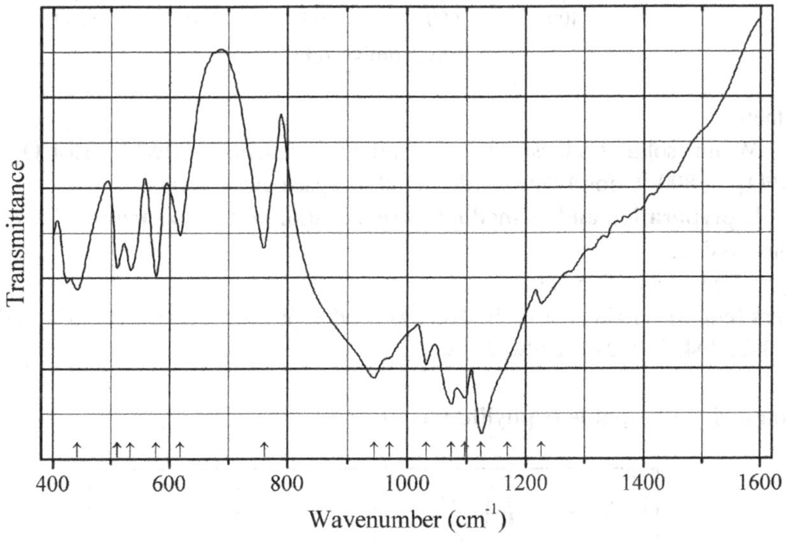

Origin: Synthetic.

Description: Obtained from Li and Fe nitrates, and $(NH_4)_2(HPO_4)$ by the co-precipitation method with subsequent heating of the obtained precipitate up to 750 °C with several intermediate grindings. Monoclinic space group $P2_1$, $Z = 4$.

Kind of sample preparation and/or method of registration of the spectrum: KBr disc. Transmission.

Source: Parajón-Costa et al. (2013).

Wavenumbers (IR, cm^{-1}): 1227, 1170sh, 1125s, 1098s, 1075s, 1032, 970sh, 945s, 761, 618, 577, 533, 510, 442.

Note: In the cited paper, Raman spectrum is given.

Wavenumbers (Raman, cm^{-1}): 1122s, 1105s, 1073s, 1037, 975, 941, 767, 761, 618, 559, 537, 520, 512, 448, 427, 415.

P700 Lithium magnesium phosphate olivine-type $LiMg(PO_4)$

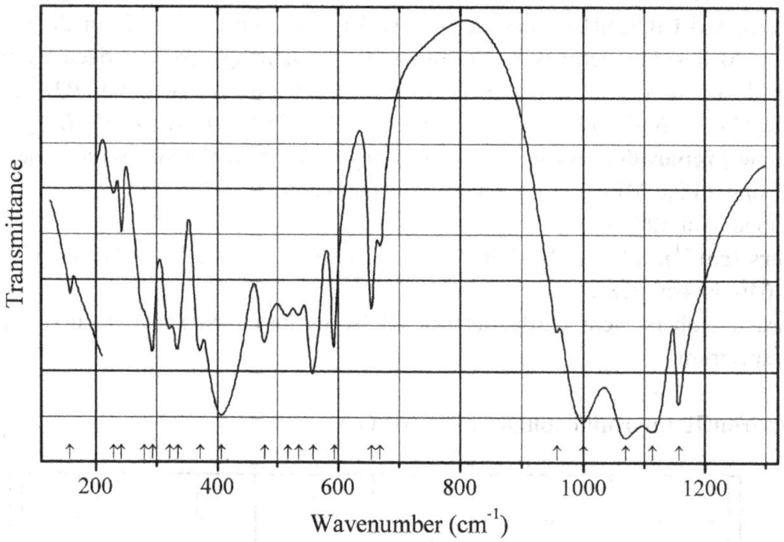

Origin: Synthetic.

Description: White solid synthesized by solid-state reaction between Li_2CO_3, MgO, and $(NH_4)_2(HPO_4)$ at 800 °C for 4 days. Orthorhombic, space group *Pnma*.

Kind of sample preparation and/or method of registration of the spectrum: KI and polyethylene discs. Transmission.

Source: Paques-Ledent and Tarte (1974).

Wavenumbers (cm^{-1}): 1157s, 1113s, 1070s, 1000s, 958, 669, 655, 593, 558s, 535, 517, 479, 407s, 372, 335, 322, 294, 280, 242, 229w, 158w.

P701 Lithium nickel phosphate triphylite-type $LiNi(PO_4)$

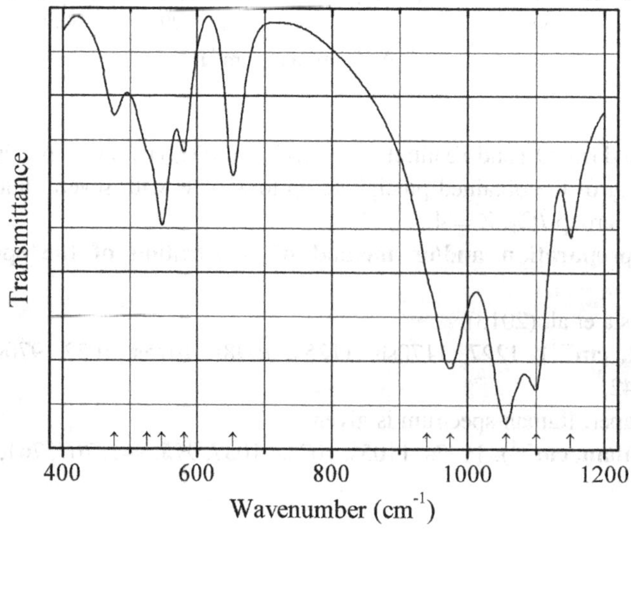

Origin: Synthetic.

Description: Synthesized by a solid-state reaction technique from the stoichiometric mixture of Li_2CO_3, $(NH_4)_2(HPO_4)$, and NiO at 773 K for 48 h. Characterized by powder X-ray diffraction data. Orthorhombic, space group *Pnma*, $a = 10.0252(7)$, $b = 5.8569(5)$, $c = 4.6758(4)$ Å, $V = 274.546$ Å^3.

Kind of sample preparation and/or method of registration of the spectrum: Absorption. Kind of sample preparation is not indicated.

Source: Bechir et al. (2014).

Wavenumbers (IR, cm^{-1}): 1150, 1100s, 1057s, 975, 940sh, 653, 580, 547, 525sh, 477.

Note: In the cited paper, Raman spectrum is given.

Wavenumbers (Raman, cm^{-1}): 1085, 1070, 1009s, 946s, 637, 597w, 590, 575w, 460w, 320w, 300, 280, 252w, 233w, 170, 165, 116w.

P702 Lithium cyclo-hexaphosphate trihydrate $Li_6P_6O_{18}{\cdot}3H_2O$

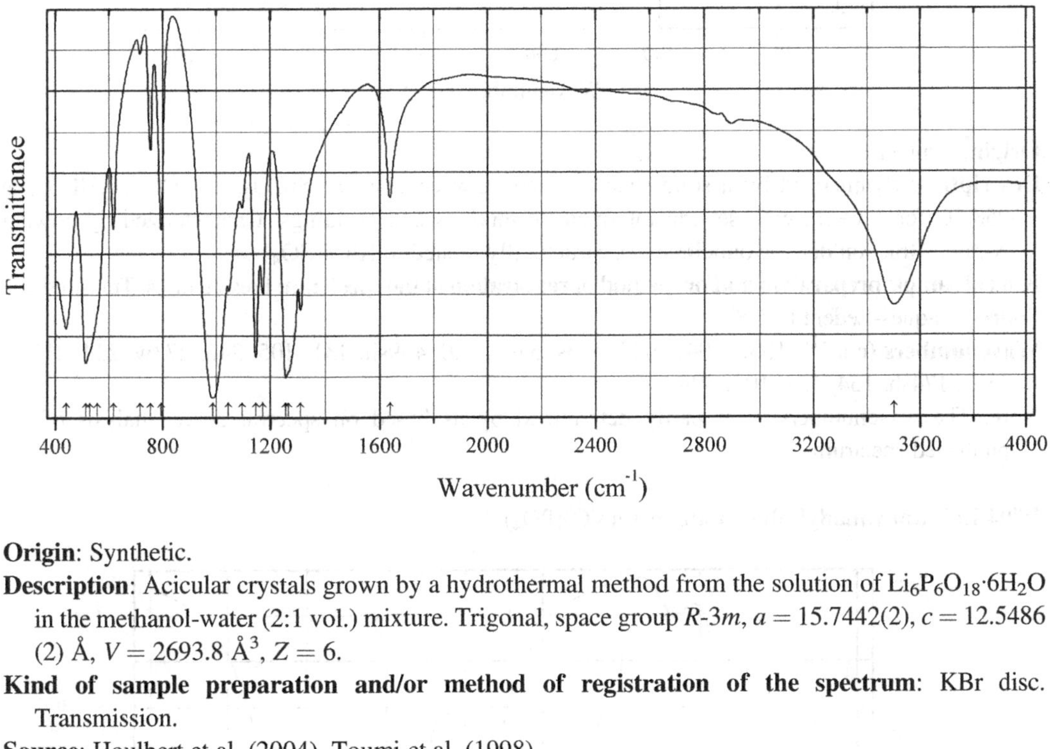

Origin: Synthetic.

Description: Acicular crystals grown by a hydrothermal method from the solution of $Li_6P_6O_{18}{\cdot}6H_2O$ in the methanol-water (2:1 vol.) mixture. Trigonal, space group *R*-3*m*, $a = 15.7442(2)$, $c = 12.5486(2)$ Å, $V = 2693.8$ Å^3, $Z = 6$.

Kind of sample preparation and/or method of registration of the spectrum: KBr disc. Transmission.

Source: Houlbert et al. (2004), Toumi et al. (1998).

Wavenumbers (cm^{-1}): 3501, 1637, 1309, 1267sh, 1257s, 1173, 1146, 1096, 1047, 987s, 794, 755, 716w, 617, 558sh, 530sh, 514, 440.

Note: The wavenumbers were determined by us based on spectral curve analysis of the published spectrum.

P703 Lithium strontium orthophosphate LiSr(PO_4)

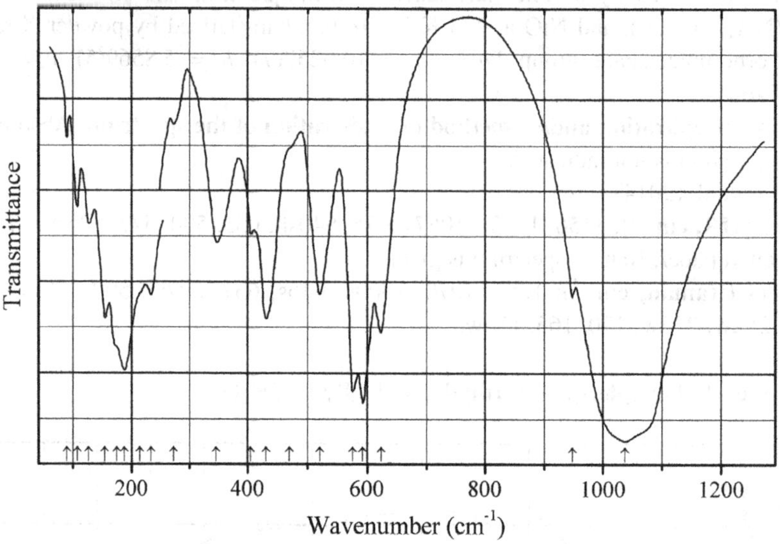

Origin: Synthetic.

Description: Synthetized by a solid-state reaction between Li_2CO_3, $SrCO_3$, and $(NH_4)_2(HPO_4)$ at 950 °C for 1 week, with several intermediate mixings and grindings. Characterized by powder X-ray diffraction data. Monoclinic (?). Structurally related to RbLi(SO_4).

Kind of sample preparation and/or method of registration of the spectrum: Pressed discs. Transmission.

Source: Paques-Ledent (1978).

Wavenumbers (cm^{-1}): 1038s, 948, 623, 593s, 576s, 520, 469sh, 431, 405, 346, 273w, 234, 215sh, 188s, 174sh, 154, 127, 107, 89w.

Note: The wavenumbers were partly determined by us based on spectral curve analysis of the published spectrum.

P704 Lithium vanadyl phosphate α-Li(VO)(PO_4)

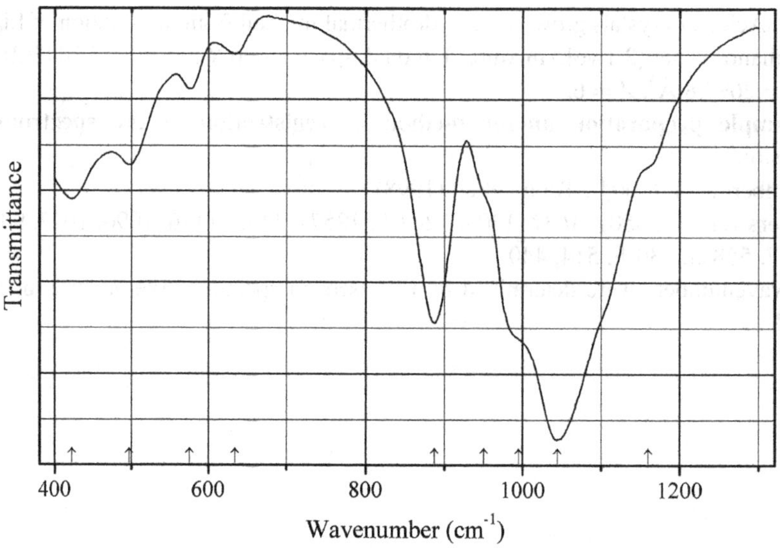

Origin: Synthetic.

Description: Synthesized from V_2O_5, LiOH, and H_3PO_4 in the ratio Li:V:P = 5:1:5 *via* intermediate vanadyl oxalate by a microwave-assisted solvothermal method. Characterized by powder X-ray diffraction data. Triclinic, a = 6.7872(3), b = 7.2152(2), c = 7.8861(3) Å, α = 89.904(2)°, β = 88.578(2)°, γ = 62.835(3)°, V = 343.46(2) $Å^3$.

Kind of sample preparation and/or method of registration of the spectrum: KBr disc. Transmission.

Source: Harrison and Manthiram (2013).

Wavenumbers (cm^{-1}): 1160sh, 1045s, 995sh, 950sh, 887s, 634w, 576w, 497, 423.

Note: The wavenumbers were determined by us based on spectral curve analysis of the published spectrum.

P705 Lithium vanadyl phosphate Li(VO)(PO_4)

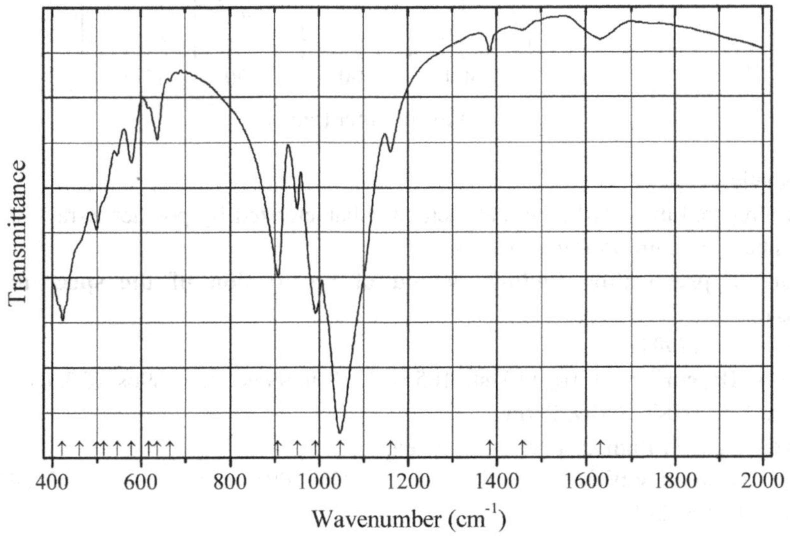

Origin: Synthetic.

Description: Synthesized using LiAc·$2H_2O$, V_2O_5, (NH_4)(H_2PO_4), and citric acid as the starting reagents. After evaporation the solvent, the product was sintered at 600 °C. Characterized by powder X-ray diffraction data. Triclinic, a = 6.731(5), b = 7.202(4), c = 7.923(2) Å, α = 89.859(8)°, β = 91.261(5)°, γ = 116.891(10)°, V = 342.507 $Å^3$, Z = 4.

Kind of sample preparation and/or method of registration of the spectrum: KBr disc. Transmission.

Source: Yang et al. (2008b).

Wavenumbers (cm^{-1}): 1633w, 1458w, 1385w, 1161, 1047s, 991s, 950, 907s, 666, 637, 617, 579, 546, 516sh, 500, 461sh, 423s.

Note: The wavenumbers were partly determined by us based on spectral curve analysis of the published spectrum. The band at 1385 cm^{-1} may correspond to NO_3^- admixture in KBr. The band at 1633 cm^{-1} corresponds to adsorbed (?) water.

P706 Lithium vanadyl phosphate β-$LiVOPO_4$

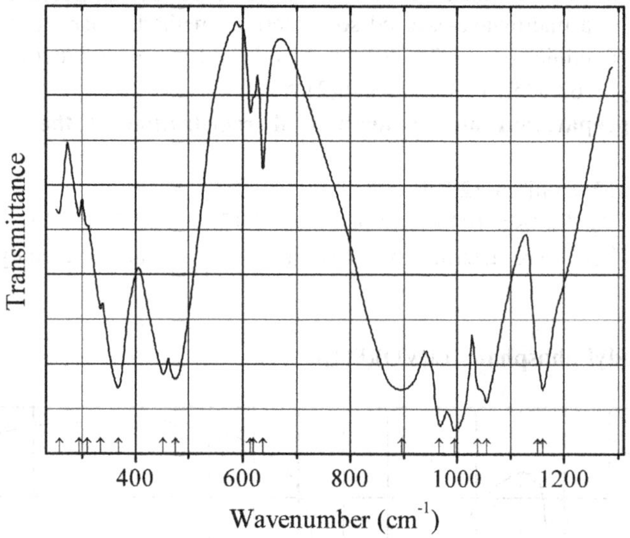

Origin: Synthetic.

Description: Prepared by a hydrothermal method. Characterized by powder X-ray diffraction data. Orthorhombic, space group *Pnma*, $Z = 4$.

Kind of sample preparation and/or method of registration of the spectrum: KBr disc. Transmission.

Source: Baran et al. (1994).

Wavenumbers (IR, cm^{-1}): 1160s, 1151sh, 1055s, 1038sh, 995s, 966s, 896s, 637, 619sh, 614, 476s, 453s, 368s, 335, 310sh, 294w, 256w.

Note: In the cited paper, Raman spectrum is given.

Wavenumbers (Raman, cm^{-1}): 1104, 1077, 1062, 1025, 1005, 979, 936, 884s, 629, 468, 429, 363, 332, 322, 310, 266, 250.

P707 Lithium zinc phosphate monohydrate α-$LiZn(PO_4)\cdot H_2O$

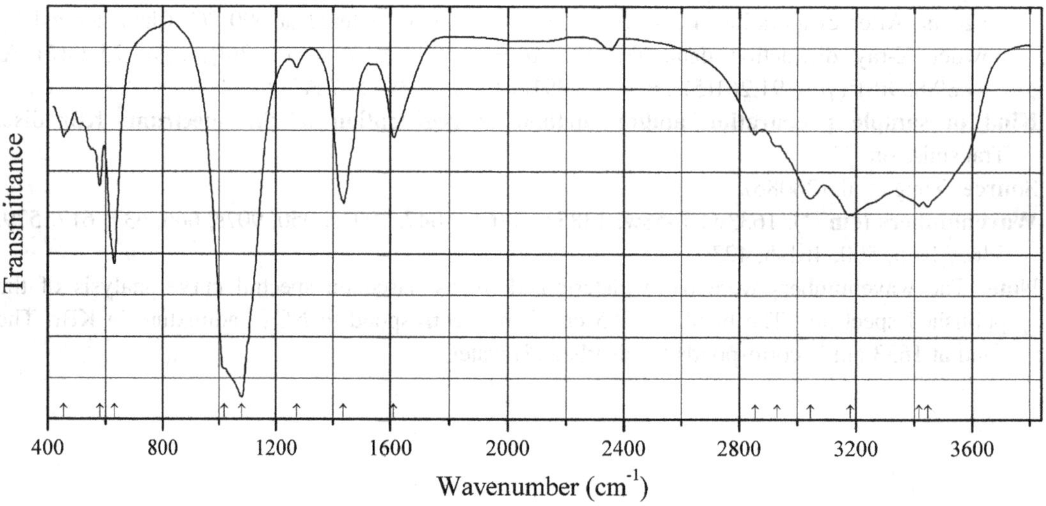

Origin: Synthetic.

Description: Prepared from $LiH_2(PO_4){\cdot}H_2O$ and $ZnSO_4{\cdot}7H_2O$ in the presence of $(NH_4)(HCO_3)$ and polyethylene glycol. Characterized by powder X-ray diffraction data. Orthorhombic, $a = 10.51848$ (8), $b = 8.12715(6)$, $c = 5.02215(5)$ Å.

Kind of sample preparation and/or method of registration of the spectrum: KBr disc. Transmission.

Source: Liao et al. (2009).

Wavenumbers (cm^{-1}): 3447, 3417, 3182, 3045, 2932sh, 2854w, 1612, 1436, 1272w, 1079s, 1018, 630s, 580, 455.

Note: The wavenumbers were partly determined by us based on spectral curve analysis of the published spectrum. In the cited paper, the wavenumber 455 cm^{-1} is erroneously indicated as 415 cm^{-1}. The bands in the range from 2800 to 3000 cm^{-1} correspond to the admixture of an organic substance. The band at 1436 cm^{-1} may correspond to a carbonate.

P708 Magnesium acid phosphate hydrate $Mg(H_2PO_4)_2{\cdot}2H_2O$

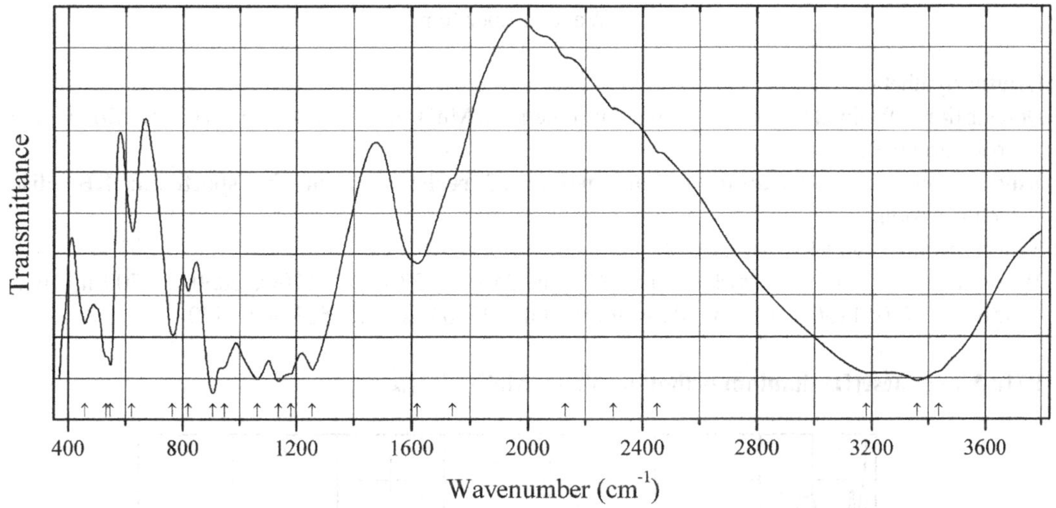

Origin: Synthetic.

Description: Obtained by salting out with acetone from MgO solution in 65 wt% H_3PO_4. Monoclinic, space group $P2_1/n$, $Z = 2$.

Kind of sample preparation and/or method of registration of the spectrum: KBr disc. Transmission.

Source: Koleva and Effenberger (2007).

Wavenumbers (cm^{-1}): 3435sh, 3360, (3180sh), 2450w, 2300w, 2130w, 1740sh, 1618, 1256s, 1180s, 1138s, 1063s, 946sh, 906s, 820, 765, 623, 548, 533sh, 458.

Note: Unlike structurally investigated Ni, Zn, and Cd analogues (Koleva and Effenberger 2007), bands of acid OH groups in the range 2200–2500 cm^{-1} are anomalously weak.

P709 Manganese acid phosphate hydrate $Mn(H_2PO_4)_2{\cdot}2H_2O$

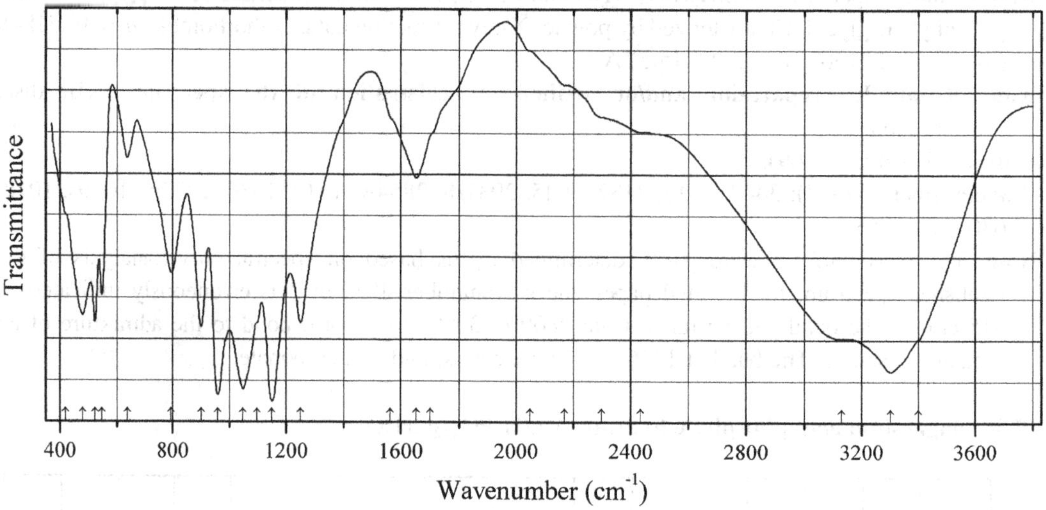

Origin: Synthetic.

Description: Obtained by salting out with acetone from MnO solution in 65 wt% H_3PO_4. Monoclinic, space group $P2_1/n$, $Z = 2$.

Kind of sample preparation and/or method of registration of the spectrum: KBr disc. Transmission.

Source: Koleva and Effenberger (2007).

Wavenumbers (cm^{-1}): 3400sh, 3300s, 3130sh, 2435sh, 2300sh, 2170w, 2050w, 1700sh, 1652, 1562sh, 1250, 1150s, 1095sh, 1046s, 960s, 900, 795, 637w, 547, 523, 480, 420.

P710 Manganese(II) titanium orthophosphate $MnTi_4(PO_4)_6$

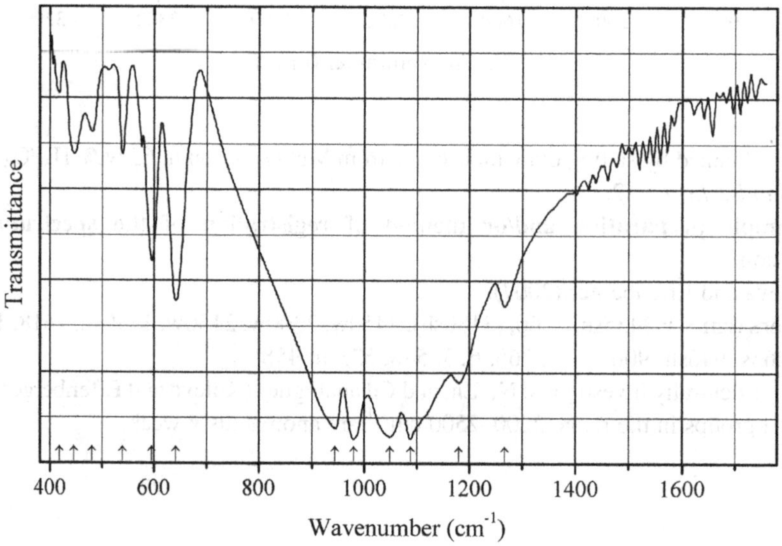

Origin: Synthetic.

Description: Orange crystals obtained hydrothermally from $TiCl_4$, Mn, and H_3PO_4 in the presence of H_2O_2, at 250 °C for 20 days. The crystal structure is solved. Hexagonal, space group *R*-3, $a = 8.51300(10)$, $c = 21.0083(3)$ Å, $V = 1318.52(3)$ Å^3, $Z = 6$. $D_{meas} = 2.39(3)$ g/cm^3, $D_{calc} = 3.083$ g/cm^3.

Kind of sample preparation and/or method of registration of the spectrum: KBr disc. Transmission.

Source: Essehli et al. (2009).

Wavenumbers (cm^{-1}): 1267, 1180, 1089s, 1049s, 980s, 944s, 642, 596, 539w, 482, 447w, 419.

P711 Manganese(II) titanium phosphate $MnTi_4(PO_4)_6$ $MnTi_4(PO_4)_6$

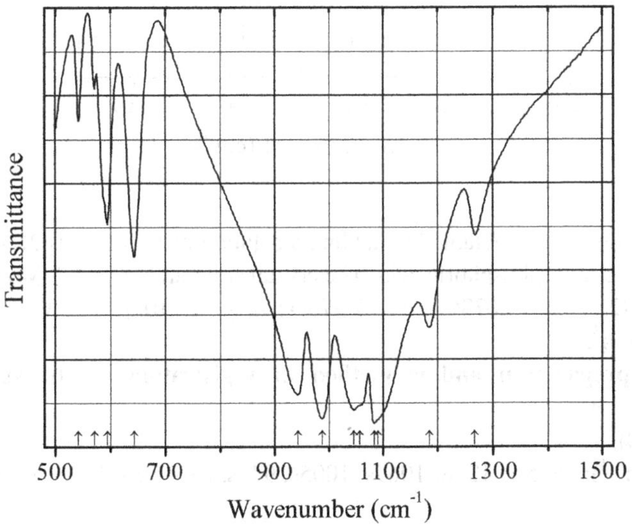

Origin: Synthetic.

Description: Prepared from the stoichiometric mixture of $MnCO_3$, TiO_2, and $(NH_4)(H_2PO_4)$ using a solid-state reaction technique. Characterized by powder X-ray diffraction data. Hexagonal, space group *R*-3.

Kind of sample preparation and/or method of registration of the spectrum: KBr and polyethylene discs. Transmission.

Source: Pikl et al. (1998).

Wavenumbers (IR, cm^{-1}): 1267, 1185, 1090sh, 1083s, 1057sh, 1046s, 988s, 944s, 643, 594, 571w, 542, 441, 382, 360, 319s, 283w, 264w, 255w, 186w.

Note: The wavenumbers were partly determined by us based on spectral curve analysis of the published spectrum. In the cited paper, Raman spectrum is given.

Wavenumbers (Raman, cm^{-1}): 1221w, 1091, 1083, 1050, 1034s, 1005, 976s, 939s, 696, 654, 604, 537, 454, 445, 438, 358, 350, 313, 285, 271, 259, 241, 199w.

P712 Mercury(I) acid phosphate $(Hg_2)_2(H_2PO_4)(PO_4)$

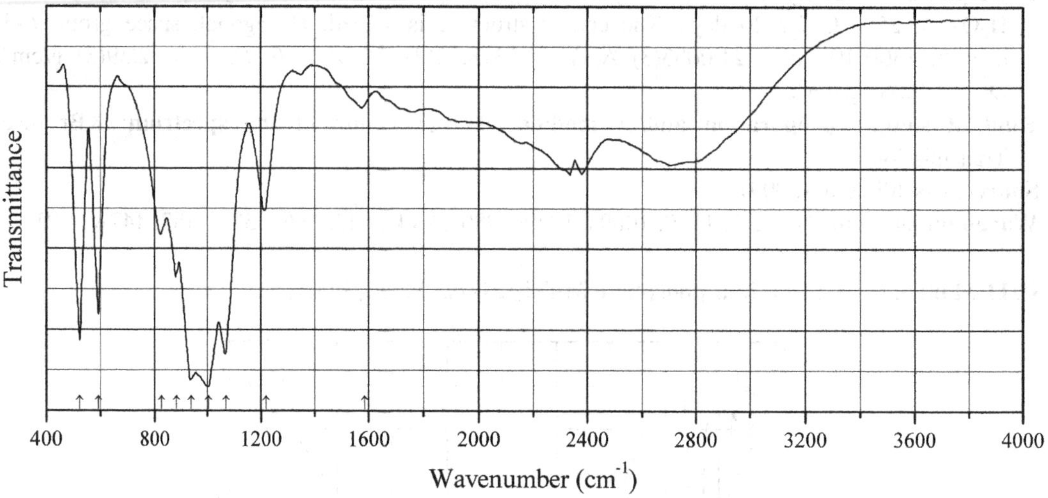

Origin: Synthetic.

Description: White precipitate obtained by adding a solution of $Hg_2(NO_3)_2 \cdot 2H_2O$ with minor HNO_3 to excess of diluted orthophosphoric acid. The crystal structure is solved. Monoclinic, space group *C*2/*c*, $a = 9.597(2)$, $b = 12.673(2)$, $c = 7.976(1)$ Å, $\beta = 110.91(1)°$, $V = 906.2(2)$ Å^3, $Z = 4$. $D_{calc} = 7.296$ g/cm^3.

Kind of sample preparation and/or method of registration of the spectrum: KBr disc. Transmission.

Source: Weil (2000).

Wavenumbers (cm^{-1}): 1585w, 1218, 1069s, 1005s, 940s, 886, 829, 592s, 523s.

P713 Molybdyl phosphate α-(MoO)(PO$_4$)

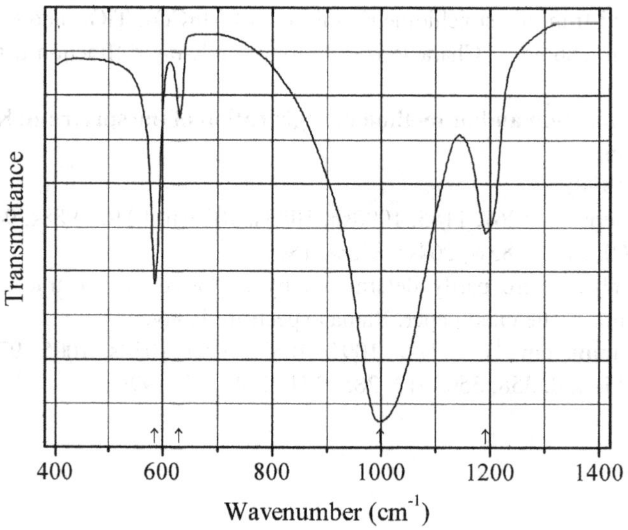

Origin: Synthetic.

Description: Obtained from the melt prepared from ammonium paramolybdate and H_3PO_4 at 950 °C. Tetragonal, space group *P*4/*n*, *Z* = 2.

Kind of sample preparation and/or method of registration of the spectrum: KBr disc. Transmission.

Source: Stranford and Condrate Sr (1984b).

Wavenumbers (IR, cm^{-1}): 1193, 998s, 631w, 585.

Note: In the cited paper, Raman spectrum is given.

Wavenumbers (Raman, cm^{-1}): 1079, 1013s, 947s, 621, 607, 447, 361, 292.

P714 Nickel vanadyl phosphate hydrate $Ni(VO)(PO_4)_2{\cdot}4H_2O$

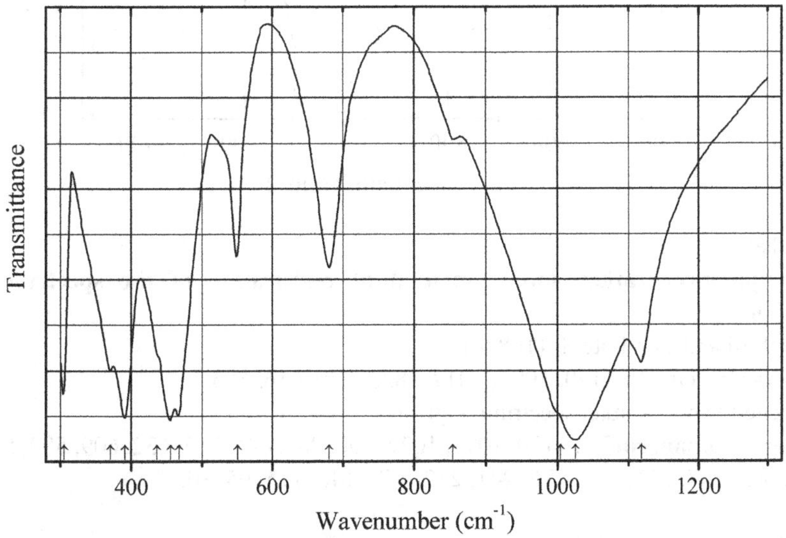

Origin: Synthetic.

Description: Tetragonal. The structure is based on VPO_5 layers linked by NiO_6 octahedra. V has fivefold coordination.

Kind of sample preparation and/or method of registration of the spectrum: KBr disc. Transmission.

Source: Baran et al. (1995).

Wavenumbers (cm^{-1}): 3505sh, 3347, 3302s, 3070sh, 1687, 1558w, 1119, 1026s, 1005sh, 854w, 681, 550, 467s, 455s, 436sh, 391s, 370, 306, 289, 255, 227, 190.

P715 Niobylphosphate *β*-(NbO)(PO_4)

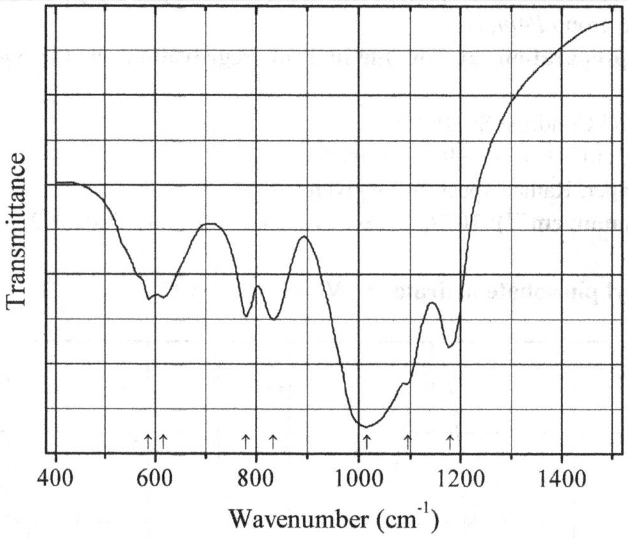

Origin: Synthetic.

Kind of sample preparation and/or method of registration of the spectrum: KBr disc. Transmission.

Source: Stranford and Condrate Sr (1984b).

Wavenumbers (IR, cm^{-1}): 1179, 1096, 1017s, 833, 779, 616, 585.

Note: In the cited paper, Raman spectrum is given.

Wavenumbers (Raman, cm^{-1}): 1120, 1097, 1022, 990, 969, 834, 785, 632, 609, 599, 583, 530, 442, 416, 383, 363, 347, 311, 288, 276, 241, 212, 177, 140, 115, 95, 80.

P716 Niobyl phosphate *α*-$NbPO_5$

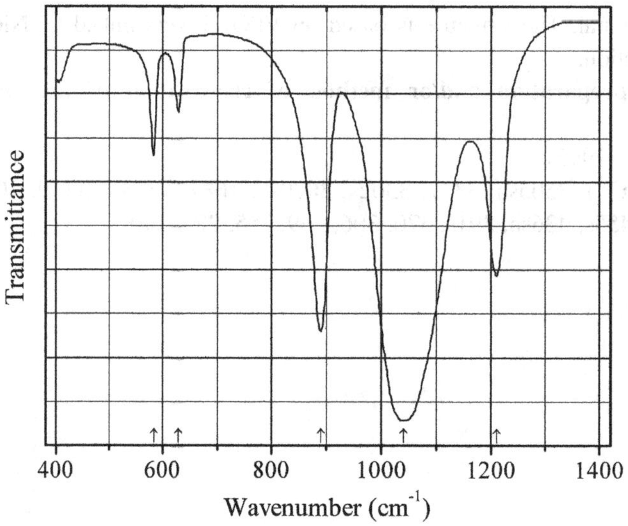

Origin: Synthetic.

Description: Tetragonal, space group *P4/n*, $Z = 2$.

Kind of sample preparation and/or method of registration of the spectrum: KBr disc. Transmission.

Source: Stranford and Condrate Sr (1984b).

Wavenumbers (IR, cm^{-1}): 1211, 1040s, 891s, 629, 583.

Note: In the cited paper, Raman spectrum is given.

Wavenumbers (Raman, cm^{-1}): 1113, 1014s, 984s, 800, 612, 467, 458, 376, 288, 200, 177, 160, 111.

P717 Ammonium manganese(II) borophosphate $[NH_4]_4[Mn_9B_2(OH)_2(HPO_4)_4(PO_4)_6]$

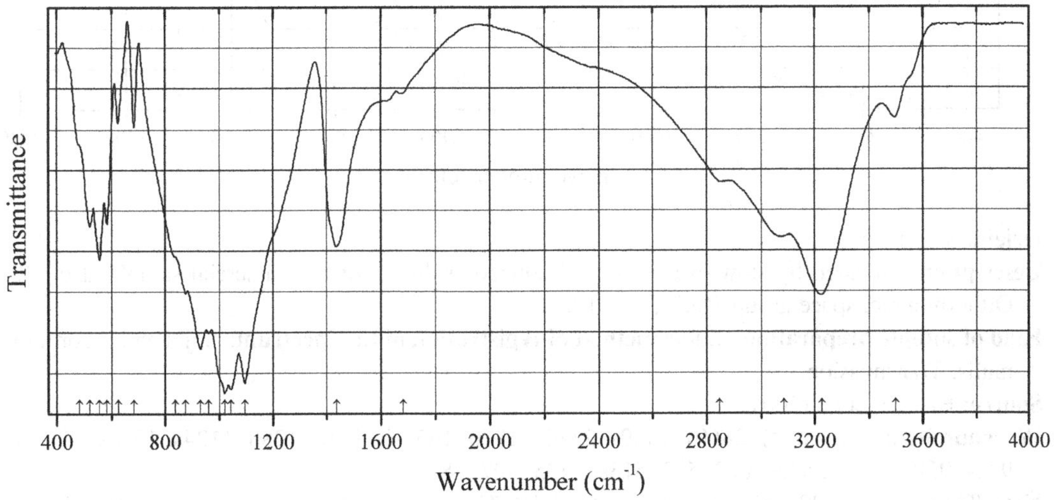

Origin: Synthetic.

Description: Pink stick-like crystals prepared hydrothermally from $Mn(OAc)_2{\cdot}4H_2O$, H_3BO_3, and $(NH_4)_2(HPO_4)$ at 200 °C for 5 days. Monoclinic, space group *C2/c*, $a = 32.603(7)$, $b = 10.617(2)$, $c = 10.718(2)$ Å, $\beta = 108.26(3)°$, $V = 3523.2(12)$ Å^3, $Z = 4$. $D_{calc} = 2.971$ g/cm^3. In the crystal structure, layers $[Mn_9(OH)_2(HPO_4)_4(PO_4)_6]$ are connected by B atoms having tetrahedral coordination to form 3D framework. Mn^{2+} has five- and sixfold coordination.

Kind of sample preparation and/or method of registration of the spectrum: KBr disc. Transmission.

Source: Yang et al. (2006).

Wavenumbers (cm^{-1}): 3498, 3226s, 3084, 2848, 1684w, 1437, 1096s, 1044s, 1021s, 960, 930s, 875, 837sh, 685, 626, 583, 557, 521, 482sh.

Note: The wavenumbers were partly determined by us based on spectral curve analysis of the published spectrum.

P718 Potassium acid phosphate $K_2(HPO_4)$

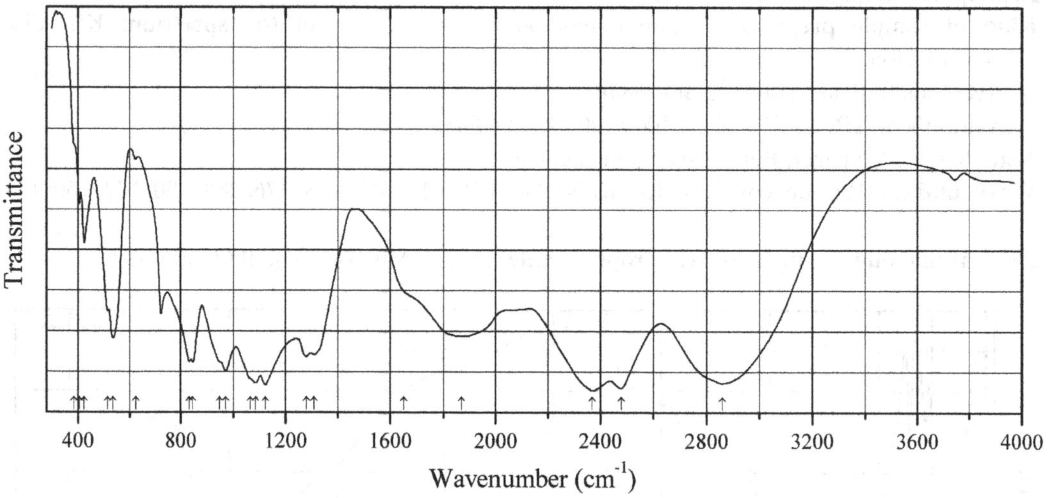

Origin: Synthetic.

Description: Prepared by slow evaporation of aqueous solution of a commercial sample at 60 °C. Orthorhombic, space group $Pna2_1$, $Z = 12$.

Kind of sample preparation and/or method of registration of the spectrum: Nujol and Fluorolube mulls. Transmission.

Source: Hadrich et al. (2001).

Wavenumbers (IR, cm^{-1}): 2860s, 2480s, 2370s, 1870, 1650sh, 1310, 1280, 1124s, 1085s, 1065sh, 972s, 950sh, 847s, 832s, 622, 535, 515sh, 425, 407, 385sh.

Note: The band at ~720 cm^{-1} corresponds to Nujol. The wavenumbers were partly determined by us based on spectral curve analysis of the published spectrum. In the cited paper, Raman spectrum is given.

Wavenumbers (Raman, cm^{-1}): 1129w, 1119w, 1111w, 1100w, 1081w, 1066w, 1000, 969s, 946s, 856, 839sh, 828w, 587w, 571w, 559, 547, 534, 511w, 427w, 405w, 389, 382sh.

P719 Potassium acid pyrophosphate hydrate $K_3(HP_2O_7)\cdot 3H_2O$

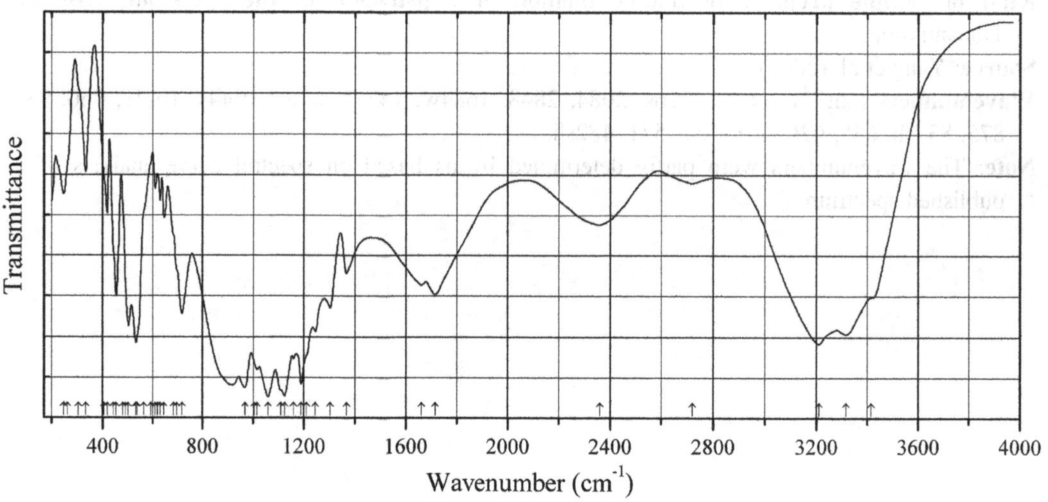

Origin: Synthetic.

Description: Prepared in the reaction between a concentrated aqueous solution of potassium pyrophosphate and equimolar quality of acetic acid at 0–5 °C. Monoclinic, space group $P2_1/c$, $Z = 4$.

Kind of sample preparation and/or method of registration of the spectrum: Nujol and hexachlorobutadiene mulls. Transmission.

Source: Sarr and Diop (1984).

Wavenumbers (IR, cm^{-1}): 3414, 3316s, 3210s, 2720w, 2360 (broad), 1714, 1660, 1366, 1304, 1246, 1210sh, 1190s, 1160, 1125s, 1110sh, 1060s, 1015, 1005, 966s, 932–870s (broad), 720, 700sh, 686sh, 648w, 630w, 620w, 610w, 590sh, 562sh, 536sh, 531, 502, 480sh, 454, 444sh, 418, 407sh, 334, 306sh, 264sh, 247.

Note: The wavenumbers were partly determined by us based on spectral curve analysis of the published spectrum.

P720 Potassium antimony(V) oxophosphate $K_5Sb_5P_2O_{20}$

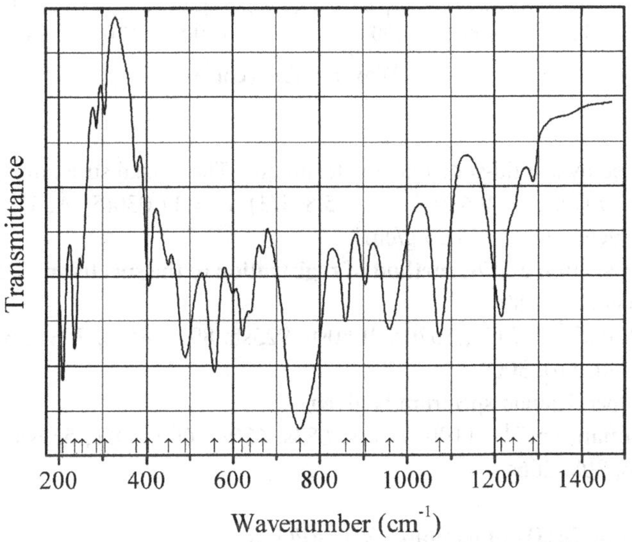

Origin: Synthetic.

Source: Husson et al. (1984).

Wavenumbers (IR, cm^{-1}): 1288w, 1244sh, 1216s, 1074s, 960s, 905, 860, 755s, 670, 640, 622s, 601w, 558s, 490s, 450, 405, 376, 304, 285, 252, 235s, 208s.

Note: The wavenumbers were partly determined by us based on spectral curve analysis of the published spectrum. In the cited paper, Raman spectrum is given.

Wavenumbers (Raman, cm^{-1}): 1240, 1180, 1070, 975, 821w, 775w, 647s, 599s, 566w, 543, 507, 489, 472, 440w, 380w, 339s, 322, 267, 215w, 161, 108sh, 74.

P721 Potassium antimony oxophosphate $K_2Sb(PO_4)O_2$

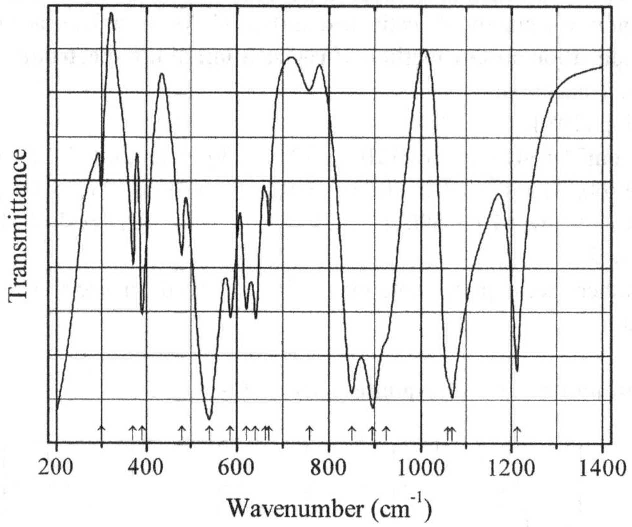

Origin: Synthetic.

Description: Obtained by a solid-state reaction technique. The crystal structure is solved. Orthorhombic, space group *Pnma*, $a = 9.429(4)$, $b = 5.891(3)$, $c = 11.030(5)$ Å, $V = 612.72$ Å^3, $Z = 4$. $D_{meas} = 3.50(5)$ g/cm^3, $D_{calc} = 3.53$ g/cm^3.

Kind of sample preparation and/or method of registration of the spectrum: KBr disc. Transmission.

Source: Botto and Garcia (1989).

Wavenumbers (IR, cm^{-1}): 1212s, 1070s, 1060sh, 925sh, 895s, 850s, 758w, 670, 662sh, 640, 620, 585, 540s, 480, 390, 370, 300w.

Note: In the cited paper, Raman spectrum is given.

Wavenumbers (Raman, cm^{-1}): 1190, 1071s, 653s, 620w, 600, 545s, 537sh, 508w, 496w, 487w, 473w, 434w, 326, 308, 266s.

P722 Potassium bismuth(III) phosphate $K_3Bi_2(PO_4)_3$

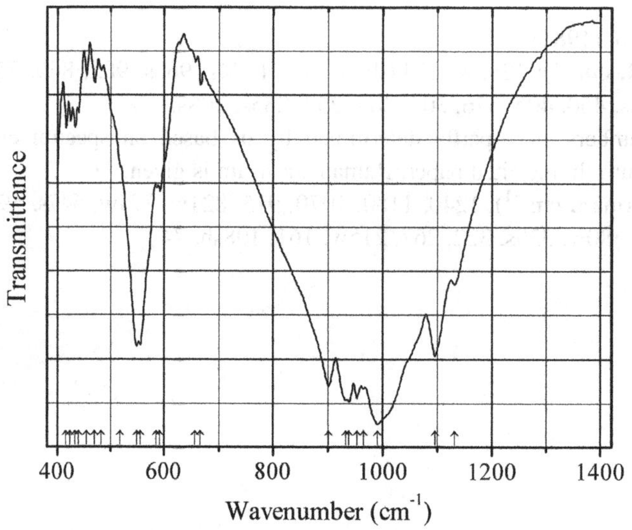

Origin: Synthetic.

Description: Prepared by crystallization from the melt obtained by heating stoichiometric mixture of K_2CO_3, Bi_2O_3, and $(NH_4)(H_2PO_4)$ to 1223 K. Characterized by powder X-ray diffraction data. Orthorhombic, possibly isostructural with $Na_3Bi_2(PO_4)_3$.

Kind of sample preparation and/or method of registration of the spectrum: KBr disc. Transmission.

Source: Mariappan et al. (2005).

Wavenumbers (cm^{-1}): 1132, 1095s, 990s, 964s, 952s, 938s, 933sh, 901s, (667w), (658w), 592, 585, 555s, 548s, (517), (482w), (470w), (455w), (440w), (434w), (425w), (417w).

Note: The wavenumbers were determined by us based on spectral curve analysis of the published spectrum.

P723 Potassium difluorphosphate $K(PO_2F_2)$

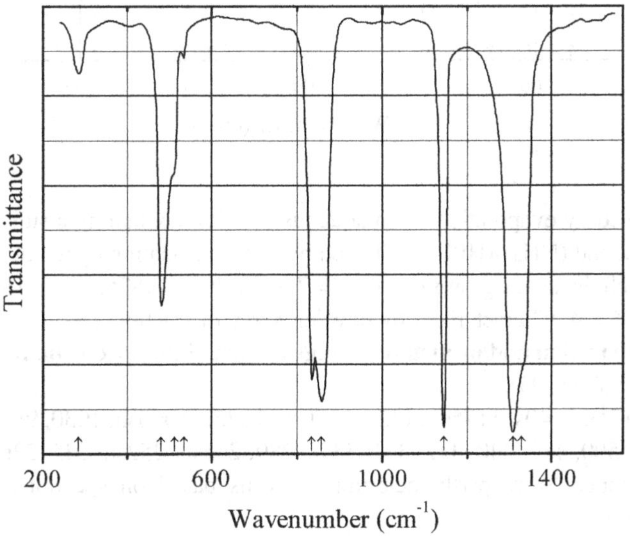

Origin: Synthetic.

Description: Obtained from a melt prepared from KPF_6 and KPO_3. Orthorhombic (?).

Kind of sample preparation and/or method of registration of the spectrum: KBr disc. Absorption.

Source: Bühler and Bues (1961).

Wavenumbers (cm^{-1}): 1330sh, 1311s, 1145s, 857s, 834, 535w, 512sh, 481, 286w.

P724 Potassium iron pyrophosphate $KFe(P_2O_7)$

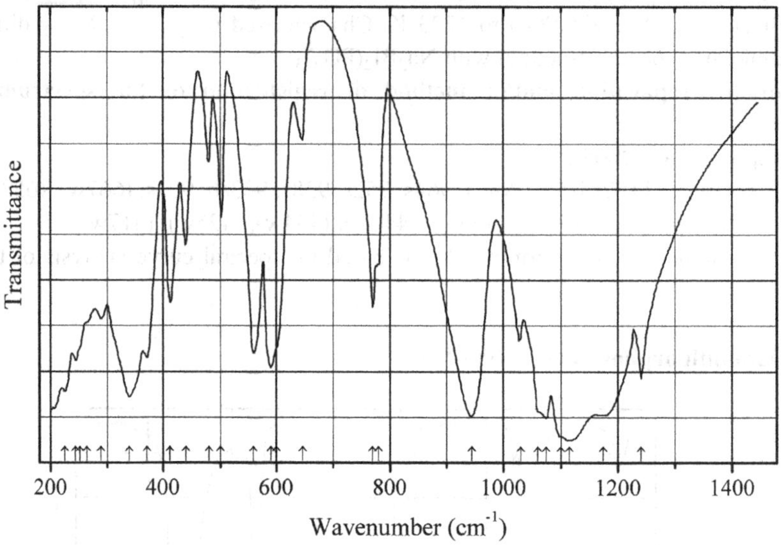

Origin: Synthetic.

Description: Obtained by evaporation of an aqueous solution containing a stoichiometric mixture of KNO_3, $Fe(NO_3)_3$, and $(NH_4)_2(HPO_4)$ followed by stepwise heating up to 750 °C with intermediate grindings. Monoclinic, space group *P*2$_1$/*c*, $a = 7.3523$, $b = 9.9875$, $c = 8.1872$ Å, $\beta = 106.499°$, $V = 576.45$ Å^3, $Z = 4$. Characterized by powder X-ray diffraction data.

Kind of sample preparation and/or method of registration of the spectrum: KBr disc. Transmission.

Source: Belkouch et al. (1995).

Wavenumbers (cm^{-1}): 1240s, 1175s, 1115s, 1100sh, 1075s, 1060sh, 1030, 945s, 780sh, 770, 645w, 600sh, 590, 560, 500, 480, 440, 412, 371, 339s, 289, 265sh, 252sh, 245, 226.

Note: The wavenumbers were partly determined by us based on spectral curve analysis of the published spectrum.

P725 Potassium lead borophosphate $KPb(BP_2O_8)$

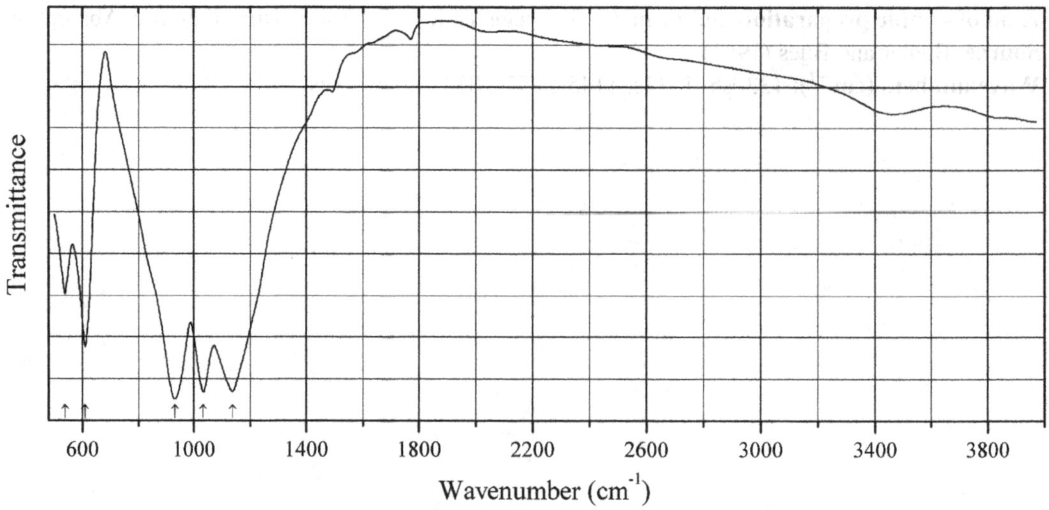

Origin: Synthetic.

Description: Colorless crystals grown by the top seed growth method from a K_2O–PbO–B_2O_3–P_2O_5 system. The melt was kept at 880 °C for 24 h and then cooled first to 865 °C at a rate of 0.5 °C/min and thereafter to 861 °C within 20 h. Characterized by powder X-ray diffraction data. The crystal structure is solved. Tetragonal, space group *I*-42*d*, $a = 7.1464(7)$, $c = 13.8917(16)$ Å, $V = 709.46(13)$ Å^3, $Z = 4$. $D_{calc} = 4.185$g/cm^3. The structure contains 12-membered rings, in which 6 PO_4 tetrahedra and 6 BO_4 tetrahedra are linked by O atoms.

Kind of sample preparation and/or method of registration of the spectrum: KBr disc. Transmission.

Source: Li et al. (2013).

Wavenumbers (cm^{-1}): 1139s, 1036s, 932s, 610, 538.

P726 Potassium lead phosphate $KPb_4(PO_4)_3$

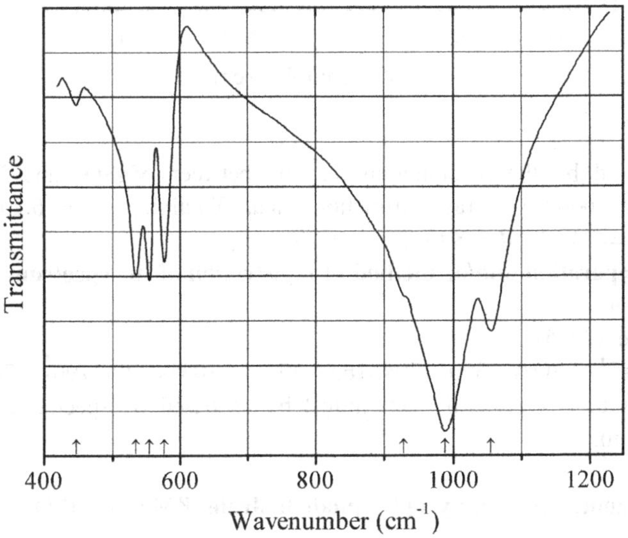

Origin: Synthetic.

Description: Prepared in a solid-state reaction, by heating a mixture of PbO, K_2CO_3, and $(NH_4)_2(HPO_4)$ first to 500 °C, and thereafter (after intermediate grinding) at 700 °C for 48 h. Characterized by powder X-ray diffraction data. Hexagonal, space group *P*6$_3$/*m*, $a = 9.8276(3)$, $c = 7.3010(4)$ Å, $V = 610.67(2)$ Å^3.

Kind of sample preparation and/or method of registration of the spectrum: KBr disc. Absorption.

Source: Azrour et al. (2011).

Wavenumbers (IR, cm^{-1}): 1055s, 988s, 928sh, 576, 554, 534, 447w.

Note: In the cited paper, Raman spectrum is given.

Wavenumbers (Raman, cm^{-1}): 1025, 976, 959s, 933s, 583s, 558, 418, 389, 235sh, 204, 159w, 130s.

P727 Potassium magnesium acid phosphate hydrate $KMg_2H(PO_4)_2 \cdot 15H_2O$

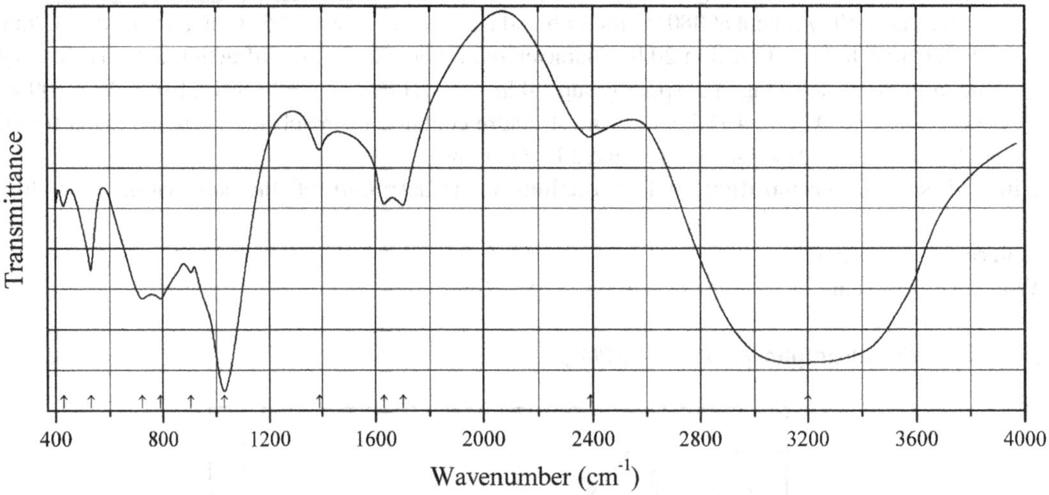

Origin: Synthetic.

Description: Prepared by the precipitation reaction between $MgSO_4$ and $K_2(HPO_4)$ solutions. Characterized by powder X-ray diffraction data. Triclinic, a = 6.2908, b = 12.2451, c = 6.5551 Å, α = 93.64°, β = 89.14°, γ = 94.73°.

Kind of sample preparation and/or method of registration of the spectrum: KBr disc and Nujol mull. Transmission.

Source: Koleva et al. (2015).

Wavenumbers (cm^{-1}): 3200s, 2390, 1704, 1633, 1390w, 1031s, 905, 790s, 720s, 530, 429w.

Note: The wavenumbers were partly determined by us based on spectral curve analysis of the published spectrum.

P728 Potassium magnesium acid pyrophosphate hydrate $KMg_{0.5}(H_2P_2O_7) \cdot H_2O$

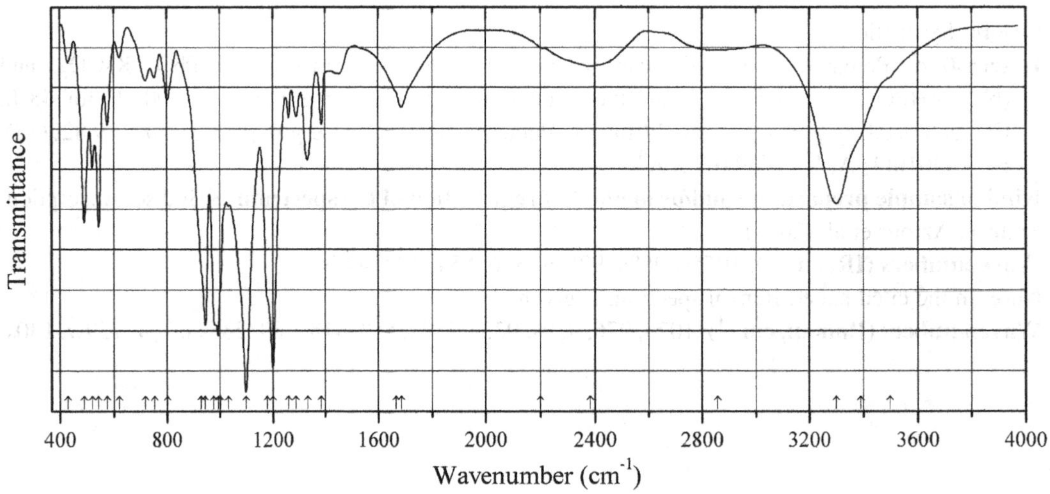

Origin: Synthetic.

Description: Prepared in the reaction of aqueous solutions of $MgCl_2$ and $K_4P_2O_7$ in the presence of HCl. The crystal structure is solved. Triclinic, space group *P*-1, $a = 6.8565(2)$, $b = 7.3621(3)$, $c = 7.6202(3)$ Å, $\alpha = 81.044(2)°$, $\beta = 72.248(2)°$, $\gamma = 83.314(3)°$, $V = 360.90(2)$ Å^3, $Z = 2$. $D_{calc} = 2.257$ g/cm^3.

Kind of sample preparation and/or method of registration of the spectrum: KBr disc. Transmission.

Source: Harcharras et al. (2003).

Wavenumbers (IR, cm^{-1}): 3498sh, 3389sh, 3299, 2858, 2383 (broad), 2201, 1687, 1669sh, 1385w, 1333, 1290, 1262, 1200s, 1179sh, 1097s, 1033s, 1002sh, 994s, 978sh, 946s, 931sh, 802, 751, 717, 622w, 577, 544, 520, 491, 431w.

Note: In the cited paper, Raman spectrum is given.

Wavenumbers (Raman, cm^{-1}): 3484w, 3387, 3318, 3281s, 2787w, 2405w, 2176w, 1705w, 1605w, 1332w, 1278w, 1217s, 1165, 1121, 1115w, 1081, 1048s, 1023, 992, 955w, 900w, 763s, 729, 589s, 555, 534, 512, 472, 452, 409s, 381, 373s, 365, 339, 311, 300s, 245.

P729 Potassium magnesium orthophosphate $KMg_4(PO_4)_3$ $KMg_4(PO_4)_3$

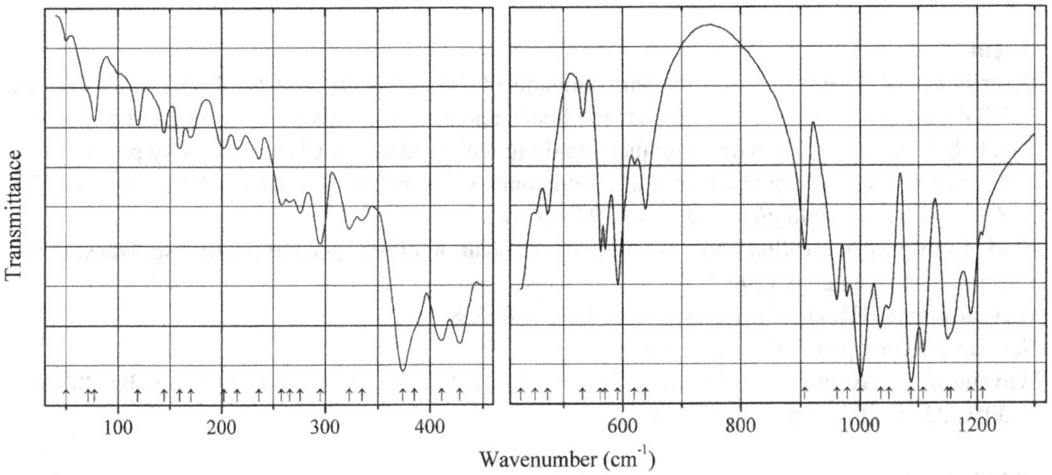

Origin: Synthetic.

Description: Obtained from K_2WO_4–WO_3 flux containing $K(H_2PO_4)$ and MgO in the molar ratio 2:1. The crystal structure is solved. Orthorhombic, space group *Pnnm*, $a = 16.361(3)$, $b = 9.562(19)$, $c = 6.171(12)$ Å, $V = 965.4(3)$ Å^3, $Z = 4$. $D_{calc} = 2.898$ g/cm^3.

Kind of sample preparation and/or method of registration of the spectrum: KBr disc (400–1500 cm^{-1}) and Nujol mull (30–500 cm^{-1}). Absorption.

Source: Tomaszewski et al. (2005).

Wavenumbers (IR, cm^{-1}): 1209w, 1189, 1155sh, 1149s, 1108s, 1087s, 1049s, 1035s, 1000s, 978, 961, 907, 638, 619, 591, 570, 562, 532, 473, 452, 428, 411, 385sh, 374, 335, 323, 295, 276, 266, 258, 236w, 215w, 202w, 170w, 159w, 144w, 119w, 77w, 71sh, 50w.

Note: The wavenumbers were partly determined by us based on spectral curve analysis of the published spectrum. In the cited paper, polarized Raman spectra are given.

Wavenumbers (Raman, [*z*(*xx*)–*z*] polarization, cm^{-1}): 1152w, 1141w, 1103w, 1086, 1079, 1016s, 982s, 956, 636, 624w, 604w, 589w, 565, 550w, 509w, 471w, 415, 298w, 283, 275w, 253w, 156, 132.

P730 Potassium magnesium yttrium phosphate (xenotime-type) $KMgY(PO_4)_2$

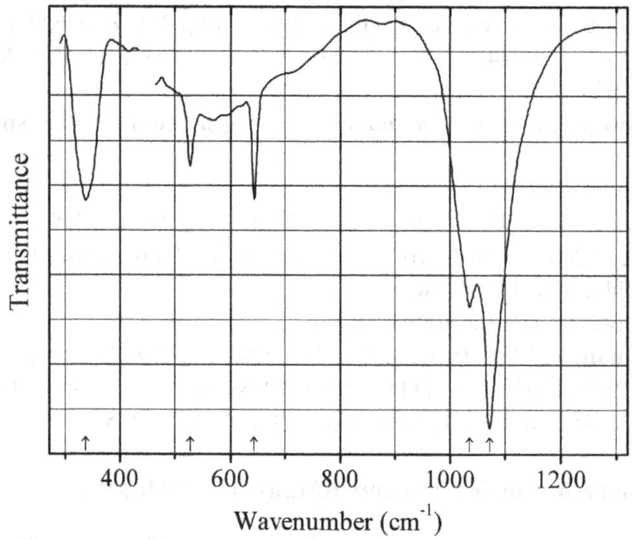

Origin: Synthetic.

Description: Obtained from stoichiometric amounts of $(NH_4)_2(HPO_4)$, K_2CO_3, MgO, and Y_2O_3, with NH_4Cl as a flux, first at 850 °C for several hours, and thereafter (after adding KCl and grinding) at 650–850 °C for 2 days, with subsequent washing with cold water. Characterized by powder X-ray diffraction data and elemental analysis. Tetragonal, space group $I4_1/amd$, a =6.886, c = 6.025 Å, Z = 2. D_{meas} = 3.940 g/cm^3, D_{calc} = 3.983 g/cm^3.

Kind of sample preparation and/or method of registration of the spectrum: KBr disc. Transmission.

Source: Shao-Long et al. (1996).

Wavenumbers (IR, cm^{-1}): 1072s, 1036s, 643, 528, 338.

Note: In the cited paper, Raman spectrum is given.

Wavenumbers (Raman, cm^{-1}): 1061s, 1030, 1004s, 669, 659w, 590, 569w, 553w, 533w, 488, 338, 305s, 225w, 205, 195, 165, 92w, 85w.

P731 Potassium monofluorphosphate $K_2(PO_3F)$

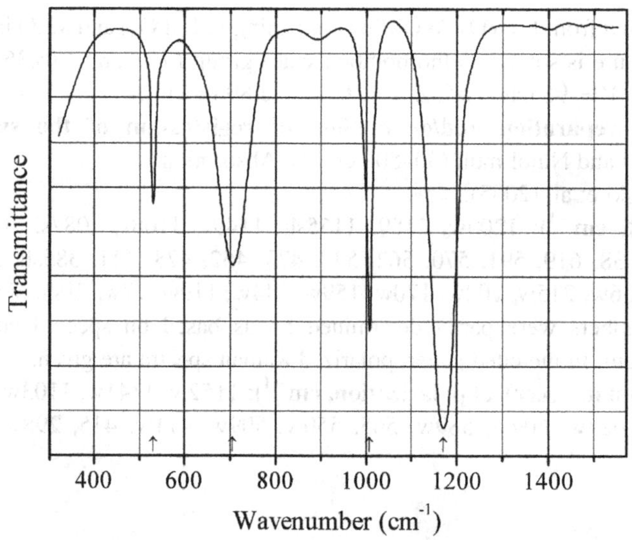

Origin: Synthetic.

Description: Obtained from the melt containing KF and KPO_3. Orthorhombic, $a = 7.554$, $b = 5.954$, $c = 10.171$ Å, $V = 457$ Å^3, $Z = 4$ (Payen et al. 1979).

Kind of sample preparation and/or method of registration of the spectrum: KBr disc. Absorption.

Source: Bühler and Bues (1961).

Wavenumbers (cm^{-1}): 1170s, 1008s, 705s, 530.

P732 Potassium niobium oxophosphate $K_3Nb_5O_{11}(PO_4)_2$

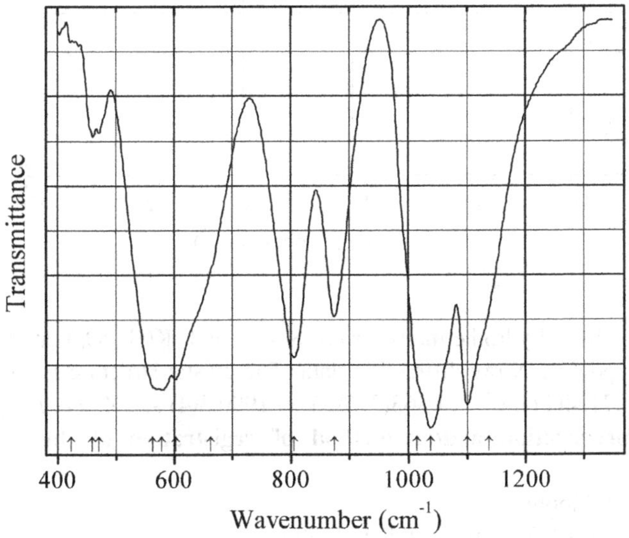

Origin: Synthetic.

Description: Prepared in the cation exchange reaction between $Tl_3Nb_5O_{11}(PO_4)_2$ and excess of KCl at 460 °C for 24 h. Characterized by powder X-ray diffraction data and electron microprobe analysis. Trigonal, space group *R*-3*c*, $a = 13.002(7)$, $c = 53.742(3)$ Å, $V = 7868(13)$ Å^3, $Z = 18$.

Kind of sample preparation and/or method of registration of the spectrum: KBr disc. Transmission.

Source: Fakhfakh et al. (2003).

Wavenumbers (cm^{-1}): 1136sh, 1099s, 1036s, 1013sh, 874, 806, 663sh, 603, 579s, 563s, 471, 460, 424w.

P733 Potassium tin orthophosphate $KSn_4(PO_4)_3$

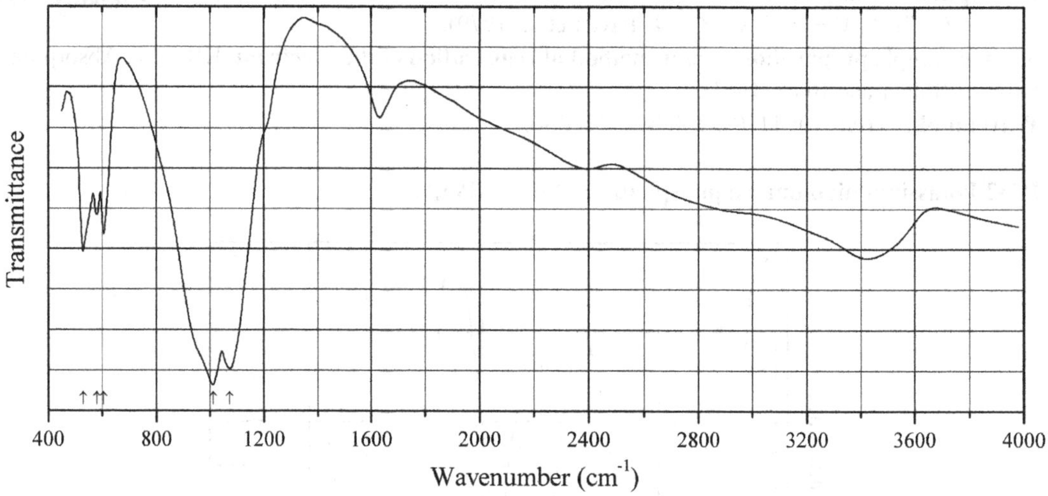

Origin: Synthetic.

Description: Synthesized hydrothermally from $SnCl_2$ and $K(H_2PO_4)$ at 170 °C for 3 days. Characterized by powder X-ray diffraction data. The crystal structure is solved. Trigonal, space group *R*3*c*, *a* = 9.7124(11), *c* = 24.363(3) Å, *V* = 1990.3(4) Å^3, *Z* = 6. D_{calc} = 3.999 g/cm^3.

Kind of sample preparation and/or method of registration of the spectrum: KBr disc. Transmission.

Source: Bontchev and Moore (2004).

Wavenumbers (cm^{-1}): 1073s, 1012s, 604, 579, 529.

Note: Bands above 1400 cm^{-1} indicate that the sample was contaminated with a compound containing acid OH groups and H_2O molecules.

P734 Potassium titanium oxophosphate $KTi(PO_4)O$

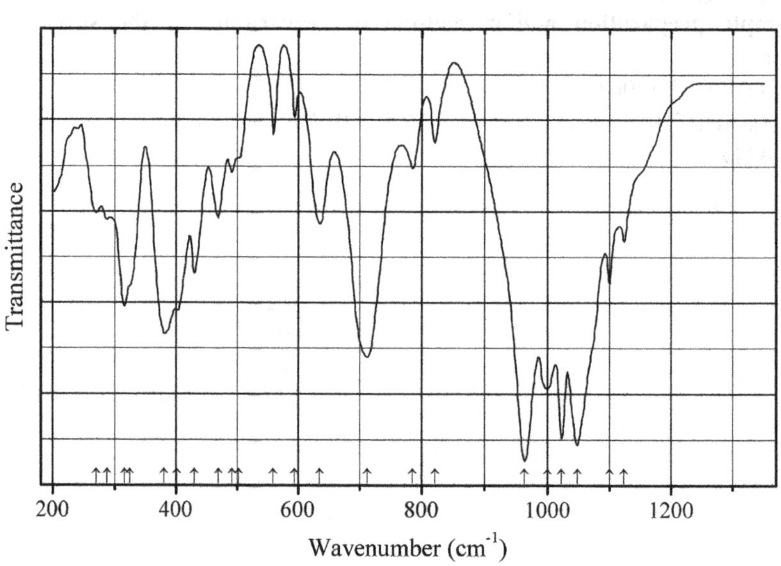

Origin: Synthetic.

Description: Orthorhombic, space group $Pn2_1a$. The structure contains distorted TiO_6 octahedra whereas PO_4 tetrahedra are stated as being undistorted. However the strong band at 964 cm^{-1} indicates that PO_4 tetrahedra are actually distorted.

Kind of sample preparation and/or method of registration of the spectrum: Nujol mull. Transmission.

Source: Jacco (1986).

Wavenumbers (cm^{-1}): 1124, 1100, 1048s, 1023s, 1000s, 964s, 820, 785, 712s, 635, 594w, 560, 503sh, 492, 469, 430, 402, 381s, 324sh, 315, 287, 270.

Note: The wavenumbers were partly determined by us based on spectral curve analysis of the published spectrum.

P735 Potassiumvanadyl phosphate $K_2(VO_2)(PO_4)$

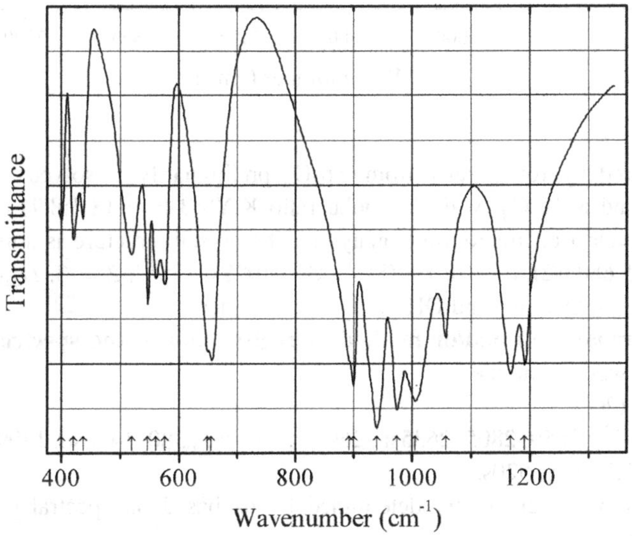

Origin: Synthetic.

Description: Yellow crystals prepared from the powders of NH_4VO_3, $(NH_4)_2(HPO_4)$, and KNO_3 mixed in the molar ratio 1:1:2 and heated first at 350 °C for 2 h and thereafter for 440 °C for 1 h. The crystal structure is solved. Monoclinic, space group $P2_1/n$, $a = 6.863(2)$, $b = 13.479(5)$, $c = 7.505$ (1) Å, $\beta = 111.02$ (10)°, $V = 648.0(3)$ Å^3, $Z = 4$. Vanadium has fivefold coordination with two short V–O distances.

Kind of sample preparation and/or method of registration of the spectrum: KBr disc. Transmission.

Source: Korthuis et al. (1993).

Wavenumbers (cm^{-1}): 1191s, 1167s, 1057, 1005s, 974s, 939s, 899s, 892sh, 655s, 648, 575, 560, 546, 518, 436, 420, 400.

P736 Potassium ytterbium acid orthoborate acid orthophosphate $K_3Yb[BO(OH)_2]_2(HPO_4)_2$

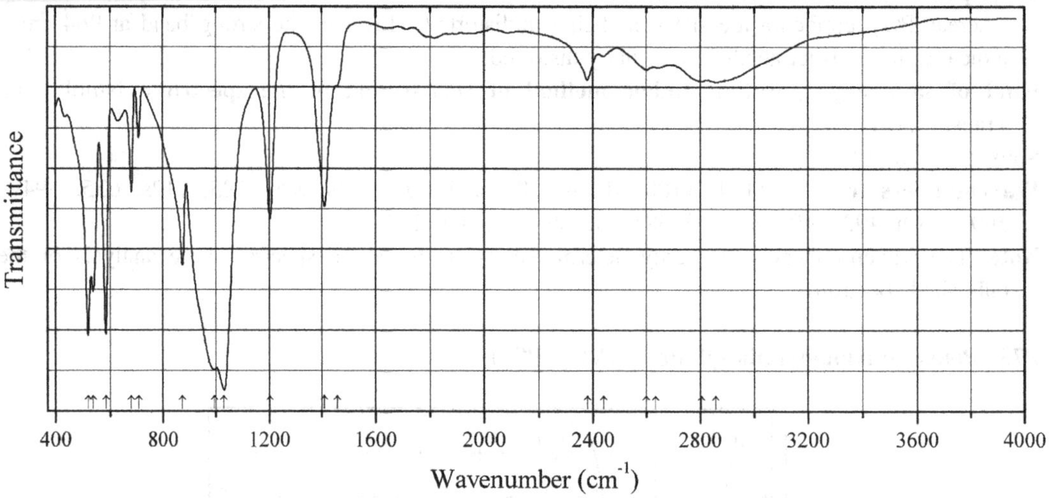

Origin: Synthetic.

Description: Prepared hydrothermally from Yb_2O_3 preliminarily dissolved in concentrated HCl, $K_2B_4O_7 \cdot 4H_2O$, and K_2HPO_4 (with the molar ratio K:Yb:B:P = 18:1:8:7) at 453 K for 5 days. Characterized by electron microprobe analyses. The crystal structure is solved. Trigonal, space group *R*-3, $a = 5.6809(2)$, $c = 36.594(5)$ Å, $V = 1022.8(2)$ Å^3, $Z = 3$. $D_{calc} = 2.942$ g/cm^3. In B-centered triangles, O atoms and OH groups are disordered.

Kind of sample preparation and/or method of registration of the spectrum: Attenuated total reflection of a powdered sample.

Source: Zhou et al. (2011).

Wavenumbers (cm^{-1}): 2859, 2805, 2635sh, 2600w, 2441w, 2380, 1456sh, 1409, 1204, 1031s, 994s, 875, 706, 680, 585s, 538, 520s.

Note: The wavenumbers were partly determined by us based on spectral curve analysis of the published spectrum.

P737 Potassium zinc acid pyrophosphate hydrate $K_2Zn(H_2P_2O_7)_2 \cdot 2H_2O$

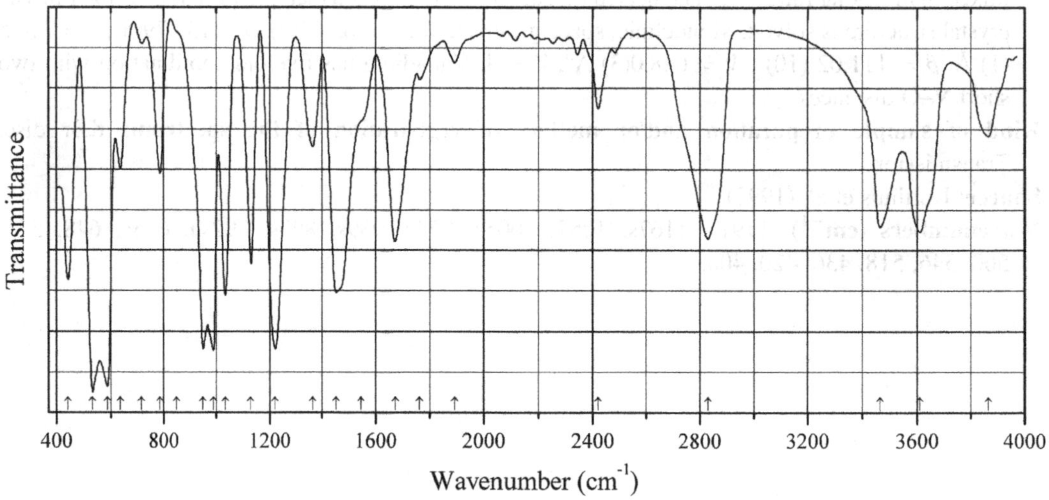

Origin: Synthetic.

Description: The crystal structure is solved. Orthorhombic, space group *Pnma*, $a = 9.901(17)$, $b = 11.071(14)$, $c = 13.65(4)$ Å, $V = 1496$ Å^3, $Z = 4$.

Kind of sample preparation and/or method of registration of the spectrum: KBr disc. Transmission.

Source: Khaoulaf et al. (2012).

Wavenumbers (IR, cm^{-1}): 3864, 3612, 3466, 2829, 2421, 1891w, 1764w, 1676, 1545sh, 1453, 1363, 1222s, 1129, 1036, 990s, 951s, 850sh, 785, 715w, 637, 589s, 534s, 444.

Note: The wavenumbers were partly determined by us based on spectral curve analysis of the published spectrum. In the cited paper, Raman spectrum is given.

Wavenumbers (Raman, cm^{-1}): 3544, 3401s, 3300, 2750, 2316w, 1787w, 1665w, 1312, 1190s, 1136, 1070s, 1006, 918, 755s, 823w, 609s, 609s, 572, 524w, 464s, 416s, 389, 341s, 314w, 267s.

P738 Potassium zinc cyclotriphosphate benitoite-type $KZn(P_3O_9)$

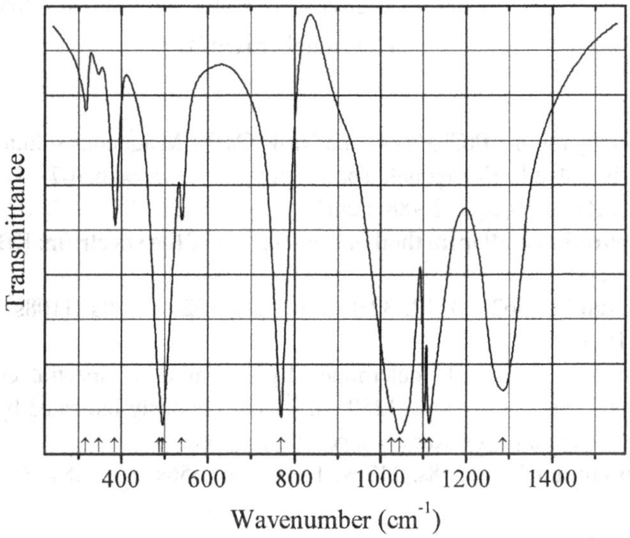

Origin: Synthetic.

Description: Synthesized by solid-state reaction techniques from a stoichiometric mixture of $(NH_4)(H_2PO_4)$, $KHCO_3$, and ZnO (or $ZnCO_3$) at 600 °C with several intermediate grindings. Characterized by powder X-ray diffraction data. Hexagonal, isostructural with benitoite.

Kind of sample preparation and/or method of registration of the spectrum: KBr disc. Transmission.

Source: Tarte et al. (1987).

Wavenumbers (cm^{-1}): 1287s, 1113s, 1102s, 1046s, 1027s, 769s, 540, 495s, 487sh, 385, 347w, 317.

P739 Praseodymium cyclotriphosphatetrihydrate $Pr(P_3O_9){\cdot}3H_2O$

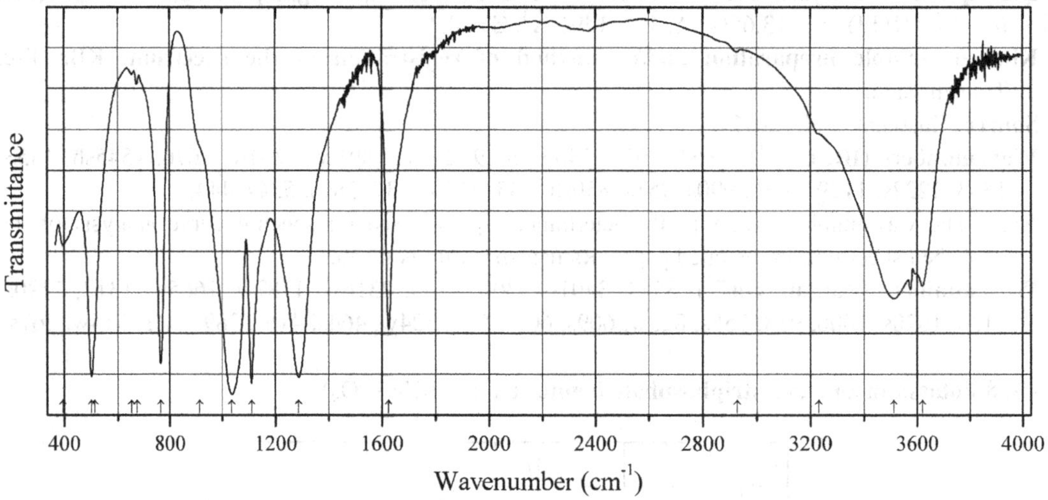

Origin: Synthetic.

Description: Prepared by mixing $PrCl_3{\cdot}6H_2O$ and $Na_3P_3O_9$ 0.1 M aqueous solutions in a 1:1 ratio. The crystal structure is solved. Hexagonal, space group *P*-6, a = 6.7677(4), c = 6.0501(4) Å, V = 239.98(3) Å^3, Z = 1. D_{calc} = 2.988 g/cm^3.

Kind of sample preparation and/or method of registration of the spectrum: KBr disc. Transmission.

Source: Jouini et al. (2006).

Wavenumbers (IR, cm^{-1}): 3620, 3512, 3230sh, (2928), 1626, 1289s, 1108s, 1036s, 915sh, 764s, (672), (653), (514), 501s, 394.

Note: The wavenumbers were partly determined by us based on spectral curve analysis of the published spectrum. The wavenumber 1289 cm^{-1} is erroneously indicated by Jouini et al. (2006) as 1298 cm^{-1}. In the cited paper, Raman spectrum is given.

Wavenumbers (Raman, cm^{-1}): 1248s, 1176s, 1103, 900, 656s, 662, 481, 357s, 306s, 269, 202w, 169, 131, 78.

P740 Rubidium iron(III) pyrophosphate $RbFe^{3+}(P_2O_7)$

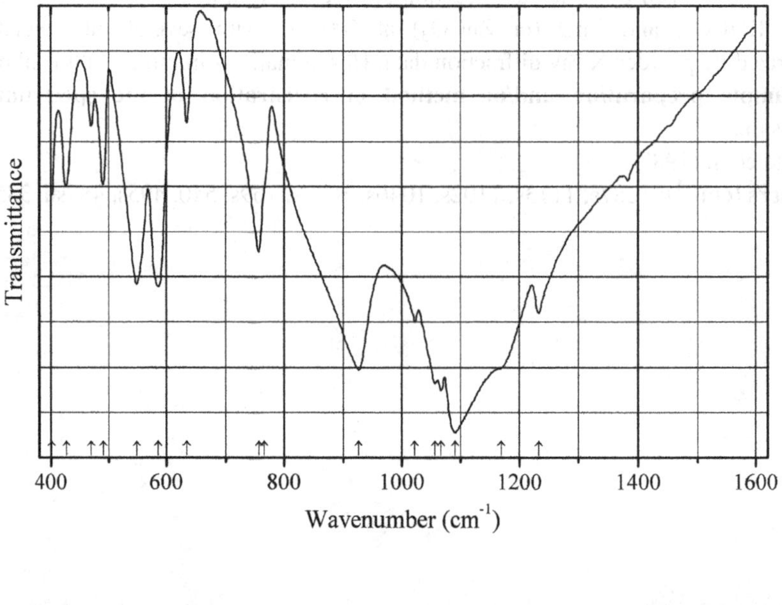

Origin: Synthetic.

Description: Prepared from aqueous solutions of corresponding nitrates and $(NH_4)_2(HPO_4)$ using a co-precipitation method, with subsequently heating precipitate at 750 °C at 15–20 h. Characterized by powder X-ray diffraction data. Triclinic, space group $P2_1/c$, $Z = 4$.

Kind of sample preparation and/or method of registration of the spectrum: KBr disc. Transmission.

Source: Parajón-Costa et al. (2013).

Wavenumbers (IR, cm^{-1}): 1233, 1170sh, 1091s, 1067s, 1057s, 1022, 927s, 766sh, 757, 634, 586, 548, 490, 470, 427, 402.

Note: In the cited paper, Raman spectrum is given.

Wavenumbers (Raman, cm^{-1}): 1156s, 1119s, 1094s, 1066s, 908, 765, 640, 598, 574, 548, 473, 431, 406.

P741 Rubidium vanadyl phosphate $Rb(VO)(PO_4)$

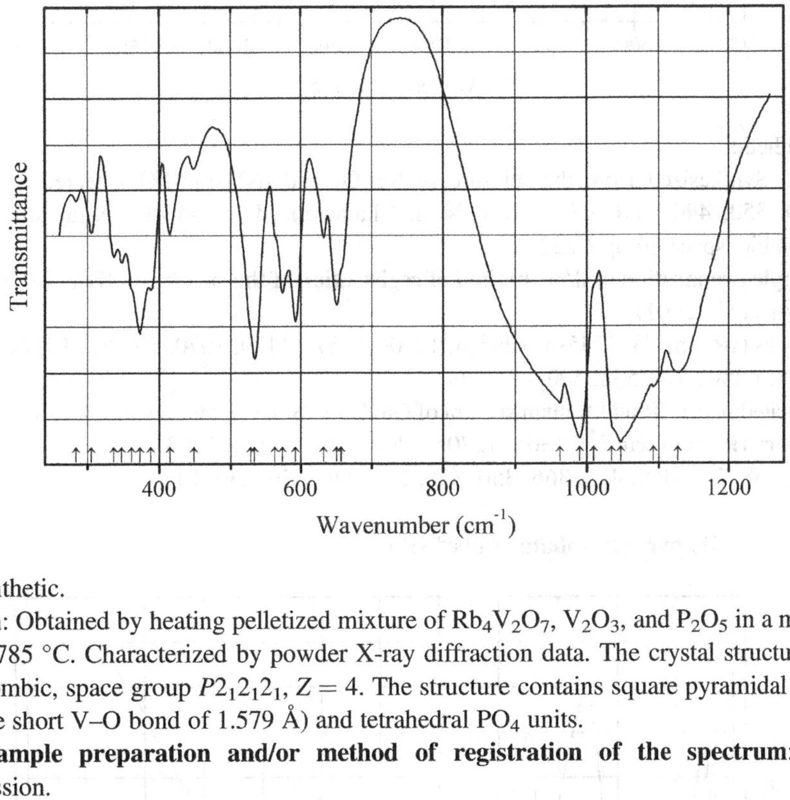

Origin: Synthetic.

Description: Obtained by heating pelletized mixture of $Rb_4V_2O_7$, V_2O_3, and P_2O_5 in a molar ratio of 1:1:2 at 785 °C. Characterized by powder X-ray diffraction data. The crystal structure is solved. Orthorhombic, space group $P2_12_12_1$, $Z = 4$. The structure contains square pyramidal VO_5 groups (with one short V–O bond of 1.579 Å) and tetrahedral PO_4 units.

Kind of sample preparation and/or method of registration of the spectrum: KBr disc. Transmission.

Source: Baran et al. (1996).

Wavenumbers (IR, cm^{-1}): 1128s, 1093s, 1047s, 1035sh, 1010sh, 990s, 962s, 657sh, 650, 632, 592, 574, 563, 534s, 529sh, 449w, 415, 389, 373, 362sh, 348, 337, 306, 284w.

Note: The wavenumbers were partly determined by us based on spectral curve analysis of the published spectrum. In the cited paper, Raman spectrum is given.

Wavenumbers (Raman, cm^{-1}): 1149, 1125, 1094, 1077, 1045, 1005s, 964, 664, 630, 593, 560, 523, 450, 405, 351, 320, 285s.

P742 Samarium metaphosphate $Sm(PO_3)_3$

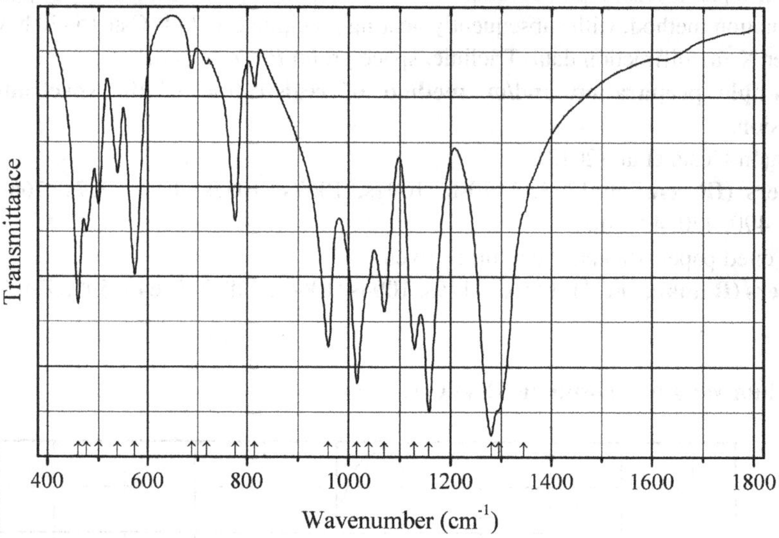

Origin: Synthetic.

Description: Synthesized from the mixture of Sm_2O_3 and $(NH_4)(H_2PO_4)$ heated successively at 170, 240, 350, 440, and 550 °C for 24 h. Characterized by powder X-ray diffraction data. Orthorhombic, space group $C222_1$.

Kind of sample preparation and/or method of registration of the spectrum: KBr disc. Transmission.

Source: Ilieva et al. (2001).

Wavenumbers (IR, cm^{-1}): 1345sh, 1295sh, 1280s, 1157s, 1129, 1070, 1040sh, 1017s, 960s, 816w, 776, 718w, 688w, 574, 538, 500, 477, 460.

Note: In the cited paper, IR and Raman spectra of Ga, In, Y, Sm, Gd, and Dy metaphosphates are given.

Wavenumbers (Raman, cm^{-1}): 1305, 1270s, 1200s, 1170, 1131, 1093, 1064, 983, 753, 720, 692s, 580, 565, 538, 501, 469, 405, 366, 350, 326, 294, 276, 250, 230, 145.

P743 Silver iron(III) pyrophosphate $AgFe^{3+}(P_2O_7)$

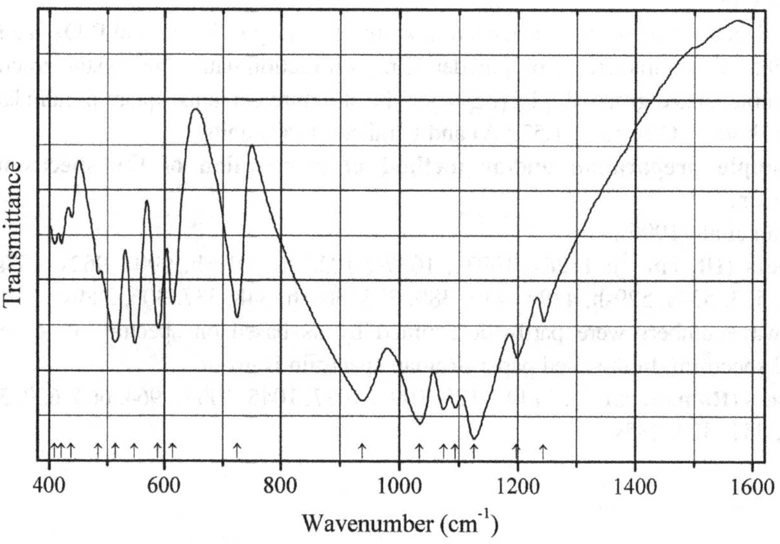

Origin: Synthetic.

Description: Obtained by heating precipitate obtained in the reaction between aqueous solutions of Ag and Fe nitrates and $(NH_4)_2(HPO_4)$ at 750 °C for 15–20 h with several grindings. Monoclinic, $a = 9.566(4)$, $b = 8.001(2)$, $c = 7.325(2)$ Å, $\beta = 111.86(1)°$.

Kind of sample preparation and/or method of registration of the spectrum: KBr disc. Transmission.

Source: Parajón-Costa et al. (2013).

Wavenumbers (IR, cm^{-1}): 1244, 1198, 1126s, 1094s, 1074s, 1034s, 939s, 725, 613, 588, 547, 513, 484, 438w, 421w, 409w.

Note: In the cited paper, Raman spectrum is given.

Wavenumbers (Raman, cm^{-1}): 1225, 1136, 1105s, 1084s, 1050, 1038, 1020, 933, 741, 625, 569, 535w, 511, 490, 443, 415w, 400w.

P744 Sodium acid pyrophosphate hydrate $Na_3(HP_2O_7)\cdot 9H_2O$

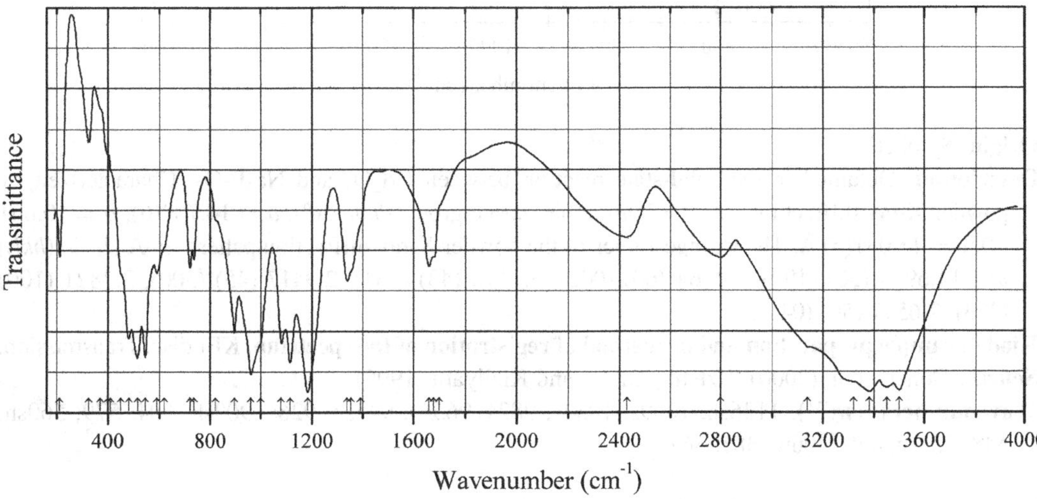

Origin: Synthetic.

Description: Needle-like crystals obtained at 0 °C by crystallization from an aqueous solution containing stoichiometric amounts of $H_4P_2O_7$ and NaOH. Monoclinic, space group $P2_1/c$, $Z = 4$.

Kind of sample preparation and/or method of registration of the spectrum: Nujol or hexachlorobutadiene mull. Transmission.

Source: Sarr and Diop (1987).

Wavenumbers (IR, cm^{-1}): 3500s, 3450s, 3380s, 3320sh, 3140sh, 2800, 2430, 1700, 1680, 1660, 1392sh, 1356sh, 1340, 1185s, 1115s, 1080, 1000sh, 960s, 895, 820sh, 735, 720, 620, 590, 545s, 515s, 480s, 450sh, 410sh, 400sh, 370sh, 325, (211).

Note: The wavenumbers were partly determined by us based on spectral curve analysis of the published spectrum. In the cited paper, Raman spectrum is given.

Wavenumbers (Raman, cm^{-1}): 3507, 3436, 3376, 3325, 2807w, 2360w, 1713w, 1670w, 1426w, 1351w, 1328w, 1189, 1102s, 1088, 1070, 976, 963, 945w, 940w, 732sh, 726, 552w, 529w, 488, 451w, 426w.

P745 Sodium gadolinium oxophosphate $Na_2GdO(PO_4)$

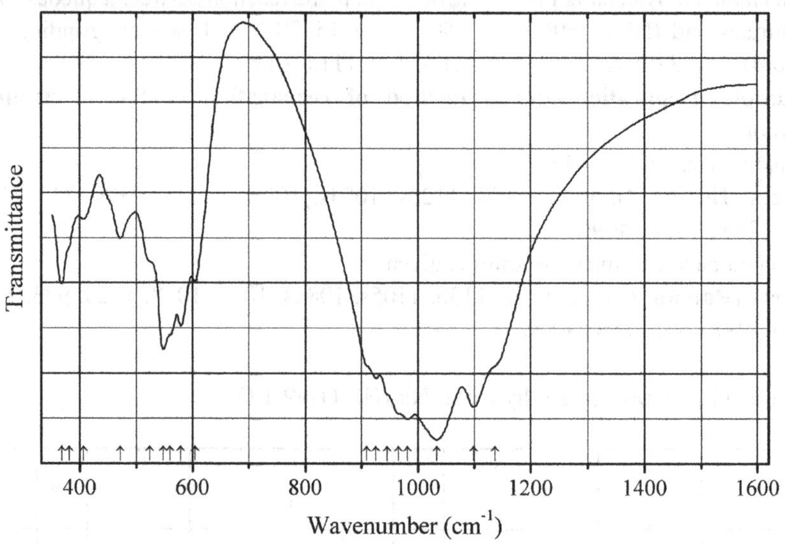

Origin: Synthetic.

Description: Obtained in the solid-state reaction between Gd_2O_3 and $Na_4P_2O_7$. Characterized by powder X-ray diffraction data. Orthorhombic, space group *Pmm*2 (?), $a = 14.709(6)$, $b = 10.661(4)$, $c = 13.081(6)$ Å. The strongest lines of the powder X-ray diffraction pattern [*d*, Å (*I*, %) (*hkl*)] are: 14.6929 (30) (100), 6.4964 (60) (002), 3.1178 (43) (231), 2.9417 (45) (500), 2.7821 (100) (323), 2.6547 (50) (040).

Kind of sample preparation and/or method of registration of the spectrum: KBr disc. Transmission.

Source: Gönen et al. (2000), Uztetik-Amour and Kizilyalli (1995).

Wavenumbers (cm^{-1}): 1136sh, 1099s, 1034s, 982s, 965sh, 945sh, 926s, 909sh, 604, 579, 560sh, 548, 523sh, 470, 406w, 381, 367.

P746 Sodium iron(II) iron(III) phosphate alluaudite-type $Na_2Fe^{2+}{}_2Fe^{3+}(PO_4)_3$

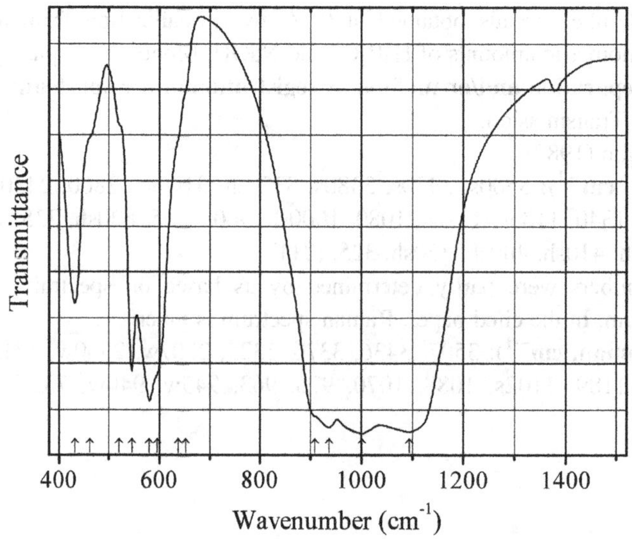

Origin: Synthetic.

Description: Synthesized hydrothermally from stoichiometric quantities of $NaH_2PO_4{\cdot}H_2O$, $FePO_4$, and FeO at 400 °C for 7 days. Characterized by powder X-ray diffraction data. Monoclinic, space group *C*2/*c*, $a = 11.849(2)$, $b = 12.539(1)$, $c = 6.486(1)$ Å, $\beta = 114.51(1)°$, $V = 876.8(1)$ Å^3, $Z = 4$.

Kind of sample preparation and/or method of registration of the spectrum: KBr disc. Transmission.

Source: Hatert et al. (2005).

Wavenumbers (cm^{-1}): 1094s, 1001s, 936s, 909sh, 653sh, 638sh, 595sh, 580s, 545, 520sh, 463sh, 433.

Note: A weak band between 1370 and 1380 cm^{-1} may correspond to the NO_3^- impurity.

P747 Sodium iron(II) pyrophosphate $Na_2Fe(P_2O_7)$

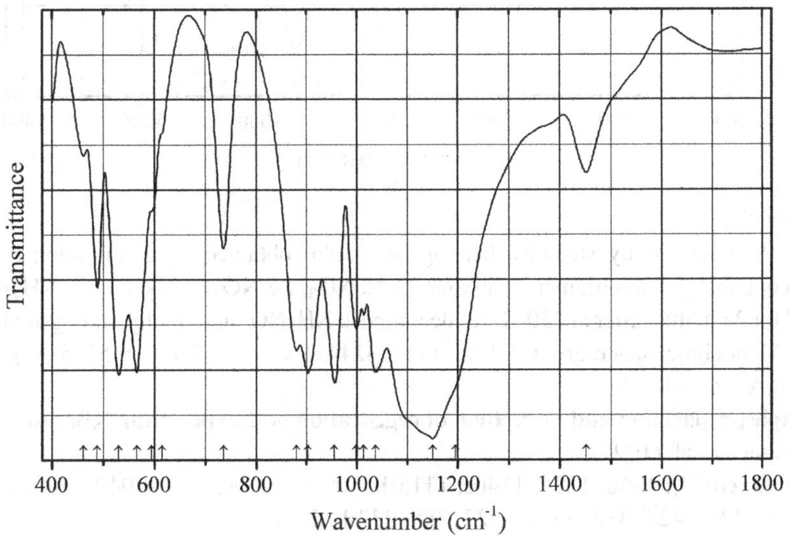

Origin: Synthetic.

Description: Synthesized by a solid-state route from a mixture containing stoichiometric molar amounts of $Na(HCO_3)$, $Fe(C_2O_4){\cdot}2H_2O$, and $(NH_4)_2(HPO_4)$ at 600 °C for 12 h in a reducing atmosphere. Characterized by powder X-ray diffraction data and Mössbauer spectrum. Triclinic, space group *P*-1, $a = 6.4415(3)$, $b = 9.4576(4)$, $c = 11.0076(5)$ Å, $\alpha = 64.685(2)°$, $\beta = 85.989(3)°$, $\gamma = 73.033(3)°$, $V = 578.64(4)$ Å^3. Fe^{2+} occupies two independent sites.

Kind of sample preparation and/or method of registration of the spectrum: KBr disc. Transmission.

Source: Barpanda et al. (2014).

Wavenumbers (cm^{-1}): 1452w, 1196sh, 1150s, 1038s, 1014, 999, 956s, 903s, 881, 736, 616sh, 595sh, 566s, 531s, 489, 462.

Note: The wavenumbers were determined by us based on spectral curve analysis of the published spectrum. The band at 1452 cm^{-1} may correspond to the admixture of a carbonate.

P748 Sodium iron(III) pyrophosphate $NaFe(P_2O_7)$

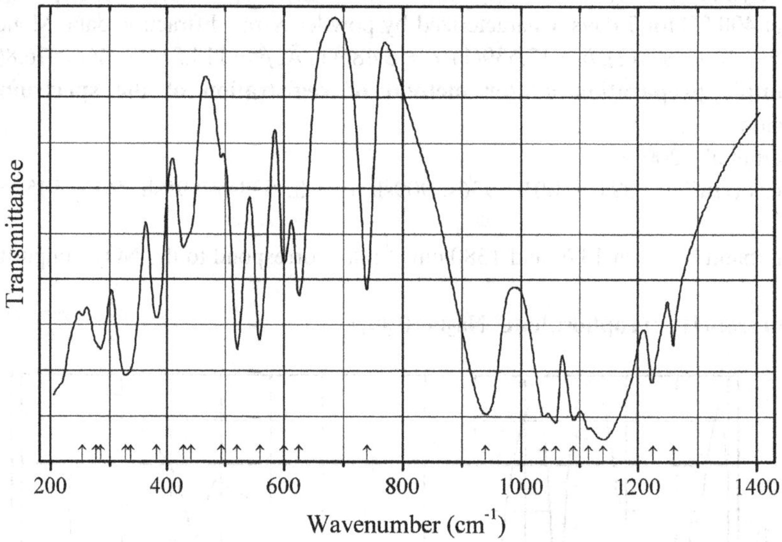

Origin: Synthetic.

Description: Synthesized by stepwise heating of a solid obtained by evaporation of an aqueous solution containing a stoichiometric mixture of $NaNO_3$, $Fe(NO_3)_3 \cdot nH_2O$, and $(NH_4)_2(HPO_4)$ first at 120 °C for 24 h, thereafter at 320 °C (to decompose NH_4NO_3 and finally, after grinding, at 750 °C for 16 h. Monoclinic, space group $P2_1/c$, $a = 7.3244$, $b = 7.9045$, $c = 9.5745$ Å, $\beta = 111.858°$, $V = 514.5$ Å^3, $Z = 4$.

Kind of sample preparation and/or method of registration of the spectrum: KBr disc. Transmission.

Source: Belkouch et al. (1995).

Wavenumbers (cm^{-1}): 1260, 1225, 1140s, 1115sh, 1090s, 1060s, 1040s, 940s, 740, 625, 600, 560, 520, 493w, 442sh, 428, 382, 336sh, 327, 285, 277sh, 254.

Note: The wavenumbers were partly determined by us based on spectral curve analysis of the published spectrum.

P749 Sodium lanthanum pyrophosphate $NaLaP_2O_7$

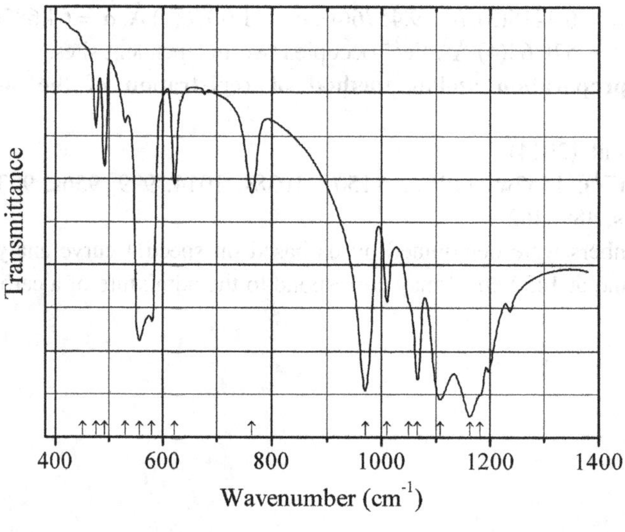

Origin: Synthetic.

Description: Crystals obtained by heating a mixture containing 3 g of $Na_3P_3O_9$ and 0.5 g of La_2O_3, first at 1000 °C for 20 days and thereafter at 600 °C for 10 h. The crystal structure is solved. Orthorhombic, space group *Pnma*, $a = 8.645(2)$, $b = 5.317(1)$, $c = 12.737(2)$ Å, $V = 585.5(2)$ Å^3, $Z = 4$. $D_{calc} = 3.810$ g/cm^3.

Kind of sample preparation and/or method of registration of the spectrum: KBr disc. Transmission.

Source: Férid and Horchani-Naifer (2004).

Wavenumbers (IR, cm^{-1}): 1182sh, 1163s, 1108s, 1066s, 1050sh, 1010, 970s, 763, 623, 580, 557s, 530, 492, 476, 451sh.

Note: The wavenumbers were partly determined by us based on spectral curve analysis of the published spectrum. In the cited paper, Raman spectrum is given.

Wavenumbers (Raman, cm^{-1}): 1152, 1132, 1086, 1072s, 1005, 939, 778s, 618, 580, 535, 523w, 486, 463, 367w, 324, 305w, 251, 221.

P750 Sodium magnesium orthophosphate pyrophosphate $Na_4Mg_3(PO_4)_2(P_2O_7)$ $Na_4Mg_3(PO_4)_2(P_2O_7)$

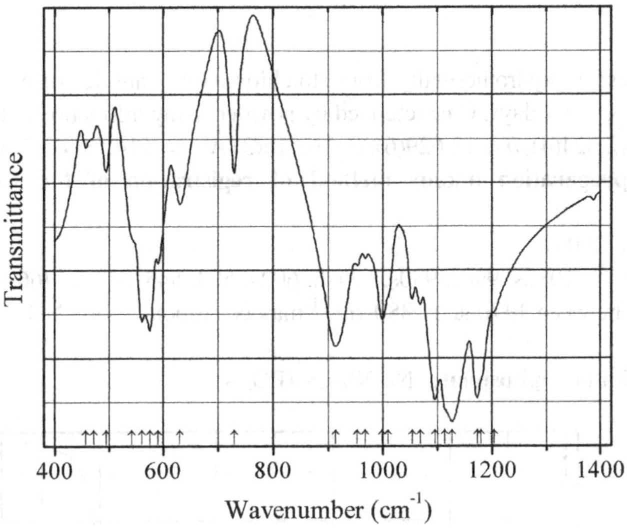

Origin: Synthetic.

Description: Crystals grown from the melt of a mixture of Na_2CO_3, MgO, and $(NH_4)(H_2PO_4)$ in the molar ratio Na:Mg:P = 4:3:4, by stepwise heating the mixture at 200, 500, and 900 °C followed by cooling down to 400 °C at the rate of 10 °C/h. Characterized by powder X-ray diffraction data. The crystal structure is solved. Orthorhombic, space group *Pn*2$_1$*a*, $a = 17.985(2)$, $b = 6.525(9)$, $c = 10.511(1)$ Å, $V = 1233.58(18)$ Å^3, $Z = 4$. $D_{calc} = 2.847$ g/cm^3. The structure is based on a 3D framework $[Mg_3P_2O_{13}]$ formed by PO_4^{3-} and $P_2O_7^{4-}$ groups and MgO_6 octahedra.

Kind of sample preparation and/or method of registration of the spectrum: KBr disc. Transmission.

Source: Essehli et al. (2010).

Wavenumbers (cm^{-1}): 1205sh, 1180sh, 1173s, 1126s, 1113sh, 1095s, 1068, 1053, 1009sh, 999s, 968, 953, 914s, 730, 631, 590, 575s, 561s, 542sh, 494, 471sh, 457w.

P751 Sodium manganese(II) iron(III) phosphate alluaudite-type $Na_2Mn^{2+}{}_2Fe^{3+}(PO_4)_3$

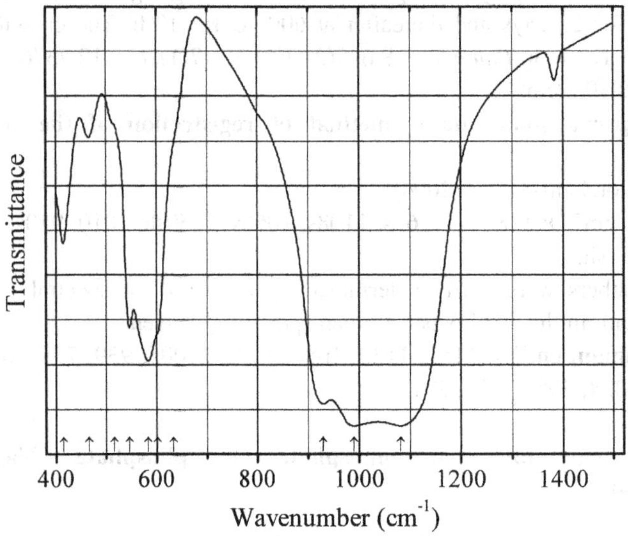

Origin: Synthetic.

Description: Synthesized hydrothermally from stoichiometric quantities of $NaH_2PO_4{\cdot}H_2O$, $FePO_4$, and MnO at 400 °C for 7 days. Characterized by powder X-ray diffraction data. Monoclinic, space group *C2/c*, $a = 12.024(4)$, $b = 12.629(6)$, $c = 6.515(3)$ Å, $\beta = 114.58(4)°$, $V = 899.6(5)$ Å^3, $Z = 4$.

Kind of sample preparation and/or method of registration of the spectrum: KBr disc. Transmission.

Source: Hatert et al. (2005).

Wavenumbers (cm^{-1}): 1082s, 990s, 930s, 634sh, 602s, 583, 545, 516sh, 466w, 415.

Note: A weak band between 1370 and 1380 cm^{-1} may correspond to the NO_3^- impurity.

P752 Sodium niobium oxophosphate $Na_3Nb_5O_{11}(PO_4)_2$

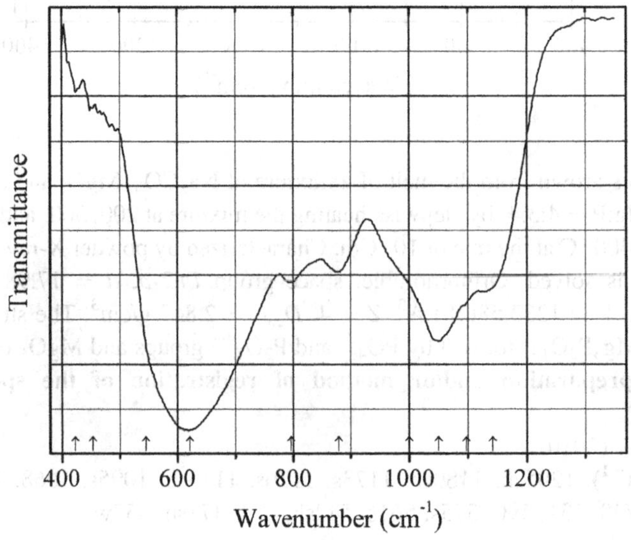

Origin: Synthetic.

Description: Prepared in the cation exchange reaction between $Tl_3Nb_5O_{11}(PO_4)_2$ and excess of KCl at 460 °C for 24 h. Characterized by powder X-ray diffraction data and electron microprobe analysis. Trigonal, space group *R*-3*c*, $a = 12.979(3)$, $c = 53.613(2)$ Å, $V = 7822(7)$ Å^3, $Z = 18$.

Kind of sample preparation and/or method of registration of the spectrum: KBr disc. Transmission.

Source: Fakhfakh et al. (2003).

Wavenumbers (cm^{-1}): 1142, 1097sh, 1050s, 999sh, 879, 797, 622s, 545, (454w), (424w).

P753 Sodium tin orthophosphate $NaSn_4(PO_4)_3$

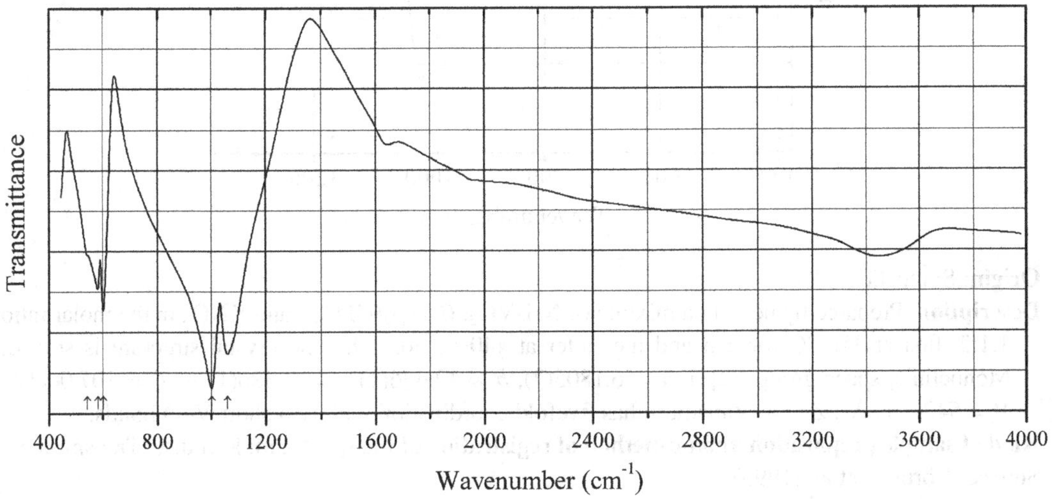

Origin: Synthetic.

Description: Prepared hydrothermally from $SnCl_2$, NaOH, and H_3PO_4 at 170 °C for 3 days. Characterized by powder X-ray diffraction data. The crystal structure is solved. Trigonal, space group *R*3*c*, $a = 9.5508(13)$, $c = 24.083(3)$ Å, $V = 1902.4(4)$ Å^3, $Z = 6$. $D_{calc} = 4.099$ g/cm^3.

Kind of sample preparation and/or method of registration of the spectrum: KBr disc. Transmission.

Source: Bontchev and Moore (2004).

Wavenumbers (cm^{-1}): 1059s, 1002s, 601, 581, 543sh.

P754 Sodium vanadyl phosphate $Na_2(VO_2)(PO_4)$ $Na_2(VO_2)(PO_4)$

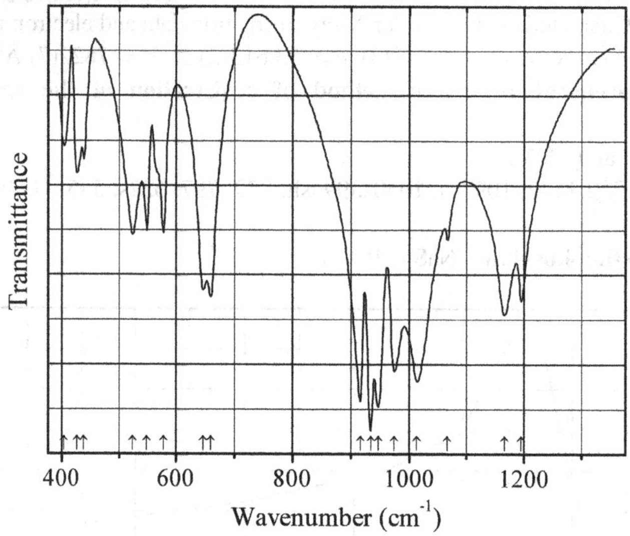

Origin: Synthetic.

Description: Prepared by heating a mixture of NH_4VO_3, $(NH_4)_2(HPO_4)$, and KNO_3 in the molar ratio 1:1:2, first at 350 °C for 2 h and thereafter at 440 °C for 1 h. The crystal structure is solved. Monoclinic, space group $P2_1/n$, $a = 6.1805(7)$, $b = 12.436(1)$, $c = 7.386(1)$ Å, $\beta = 107.00(1)°$, $V = 542.9(1)$ Å^3, $Z = 4$. Vanadium has fivefold coordination with two short V–O bonds.

Kind of sample preparation and/or method of registration of the spectrum: KBr disc. Transmission.

Source: Korthuis et al. (1993).

Wavenumbers (cm^{-1}): 1195, 1166, 1067, 1014s, 975s, 947s, 934s, 916s, 660, 647, 578, 565sh, 549, 524, 439, 427, 404.

Note: The wavenumbers were partly determined by us based on spectral curve analysis of the published spectrum.

P755 Sodium vanadyl phosphate $Na(VO)PO_4$ $Na(VO)PO_4$

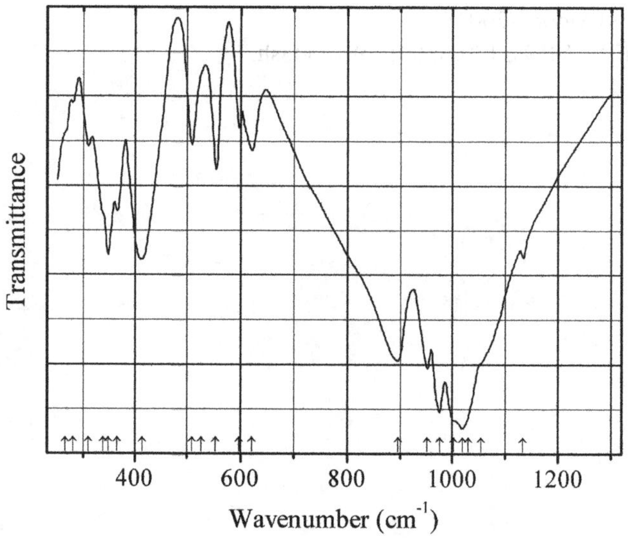

Origin: Synthetic.

Description: Prepared hydrothermally. Characterized by powder X-ray diffraction data. The crystal structure is solved. Monoclinic, space group $P2_1/c$, $Z = 4$. Vanadium has octahedral coordination.

Kind of sample preparation and/or method of registration of the spectrum: KBr disc. Transmission.

Source: Baran et al. (1994).

Wavenumbers (IR, cm^{-1}): 1134, 1055sh, 1030sh, 1019s, 1002sh, 975s, 952s, 896s, 621, 597w, 554, 527sh, 509, 413, 366, 348, 339sh, 310w, 281w, 265sh.

Note: In the cited paper, Raman spectrum is given.

Wavenumbers (Raman, cm^{-1}): 1025, 1012sh, 1000, 970, 930, 878s, 620, 600, 388, 340, 319, 268.

P756 Sodium zinc orthophosphate $NaZn(PO_4)$

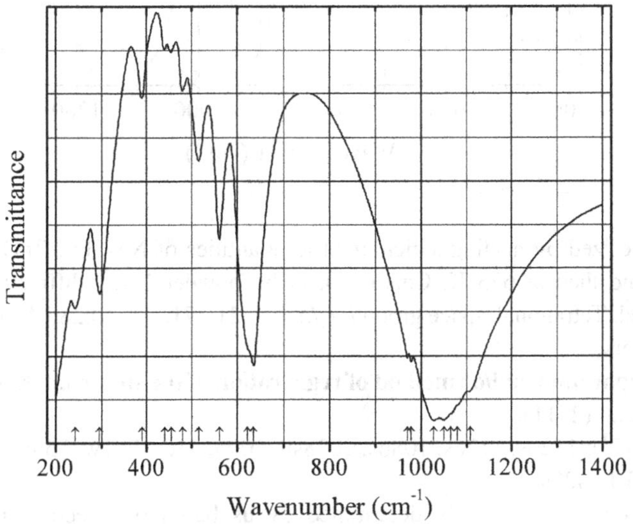

Origin: Synthetic.

Description: Single crystals obtained from a Na_2MoO_4 flux, decreasing the temperature. Monoclinic, space group $P2_1/n$, $Z = 12$. Structurally related to beryllonite.

Kind of sample preparation and/or method of registration of the spectrum: KBr disc. Transmission.

Source: Botto and Vassallo (1989).

Wavenumbers (IR, cm^{-1}): 1108sh, 1080sh, 1065sh, 1050s, 1028s, 978, 970, 635s, 622sh, 560, 515, 480, 454w, 440w, 390, 296, 242.

Note: In the cited paper, Raman spectrum is given.

Wavenumbers (Raman, cm^{-1}): 1096, 1066w, 1036, 1029, 1023, 975s, 969s, 634w, 564w, 505w, 384, 378, 222w.

P757 Sodium zinc pyrophosphate $Na_2ZnP_2O_7$

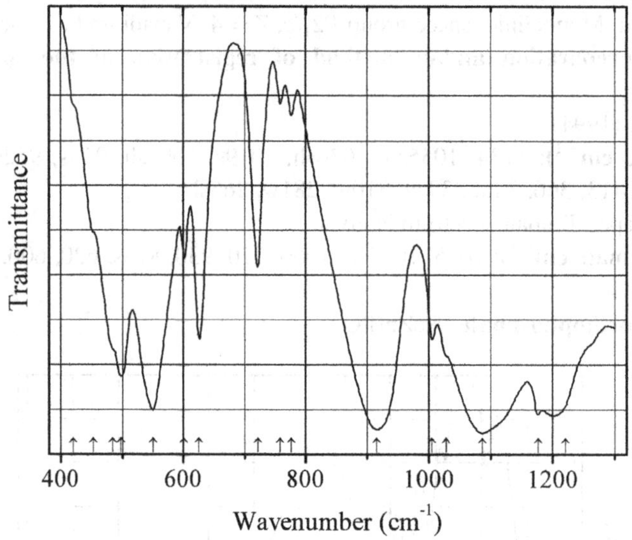

Origin: Synthetic.

Description: Synthesized by heating stoichiometric quantities of Na_2CO_3, ZnO, and $(NH_4)_2(HPO_4)$ first at 523 K and then at 623 K. Characterized by powder X-ray diffraction data. The crystal structure is solved. Tetragonal, space group $P4_2/n$, $a = 21.771$, $c = 10.285$ Å. The structure is based on $[ZnP_2O_7]$ layers.

Kind of sample preparation and/or method of registration of the spectrum: KBr disc. Transmission.

Source: Chouaib et al. (2011).

Wavenumbers (cm^{-1}): 1220s, 1176s, 1086s, 1028sh, 1005, 916s, 777w, 759w, 722, 626, 602, 550s, 498s, 485sh, 453sh, 420sh.

Note: The wavenumbers were partly determined by us based on spectral curve analysis of the published spectrum.

P758 Strontium iron phosphate whitlockite-related $Sr_9Fe(PO_4)_7$

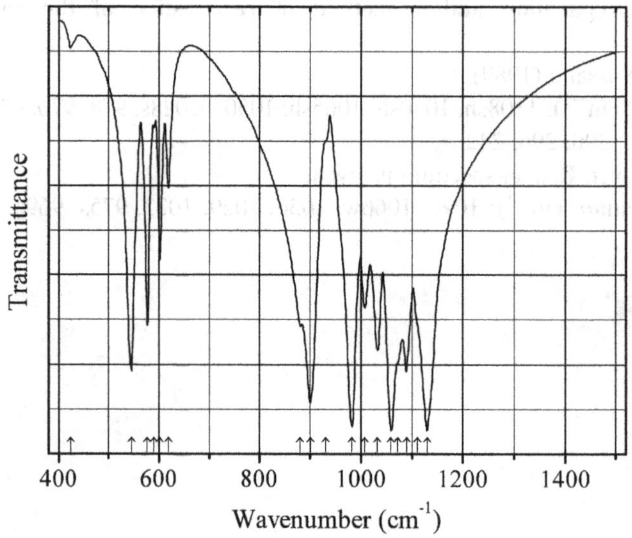

Origin: Synthetic.

Description: Prepared by heating a mixture of $SrCO_3$, Fe_2O_3, and $(NH_4)(H_2PO_4)$ with a weight ratio of 9:0.5:7, first at 900 K, and thereafter at 1370 K for 120 h with several intermediate grindings. Characterized by Mössbauer spectroscopy and powder X-ray diffraction data. The crystal structure is solved from powder neutron diffraction data. Monoclinic, space group *C2/c*, a = 14.4971(2), b =10.6005(13), c = 17.9632(3) Å, β = 112.5053(9)°, V = 2550.28(7) Å^3, Z = 4.

Kind of sample preparation and/or method of registration of the spectrum: KBr disc. Absorption.

Source: Belik et al. (2005).

Wavenumbers (IR, cm^{-1}): 1130s, 1111sh, 1090s, 1072sh, 1060s, 1033, 1008, 983s, 931w, 901s, 881, 619, 601, 590, 577, 545s, 424w.

Note: The wavenumbers were partly determined by us based on spectral curve analysis of the published spectrum. In the cited paper, a figure of the Raman spectrum is given.

P759 Strontium magnesium pyrophosphate $SrMgP_2O_7$

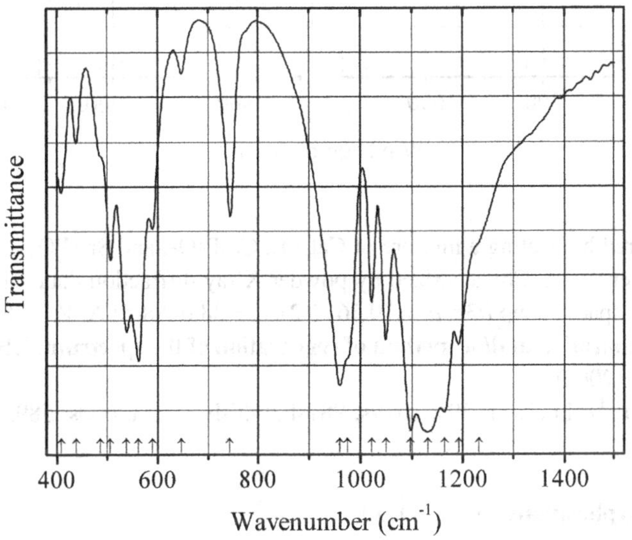

Origin: Synthetic.

Description: Prepared by stepwise heating a mixture of $SrCO_3$, $MgCO_3$, and $(NH_4)(H_2PO_4)$, taken in stoichiometric amounts, at 500, 700, and 900 °C, for 5 h at each temperature, with intermediate grindings. Monoclinic, a = 5.309, b = 8.299, c = 12.68 Å, β = 90.6°, V = 558.64 Å^3 (see JCPDS card No. 49-1027).

Kind of sample preparation and/or method of registration of the spectrum: KBr disc. Transmission.

Source: Velchuri et al. (2011b).

Wavenumbers (cm^{-1}): 1232sh, 1192, 1164s, 1132s, 1098s, 1050, 1022, 975sh, 960s, 744, 647w, 589, 562, 538, 506, 486sh, 439, 409.

Note: The wavenumbers were partly determined by us based on spectral curve analysis of the published spectrum.

P760 Calcium strontium orthophosphate whitlockite-type $Ca_2Sr(PO_4)_2$

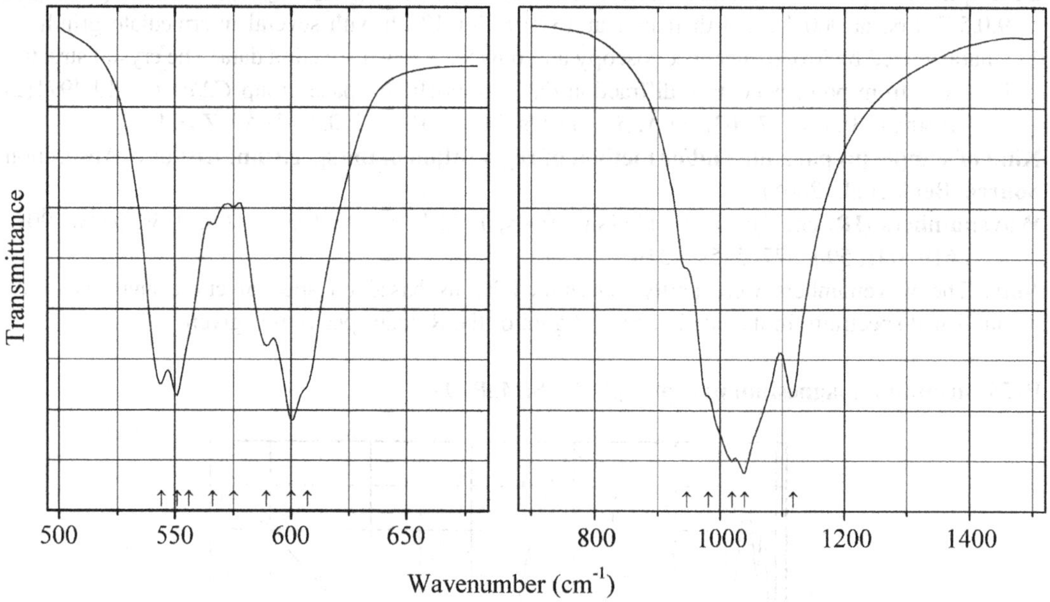

Origin: Synthetic.

Description: Prepared by heating a mixture of $CaCO_3$, $Ce_2P_2O_7$, and $Sr_3(PO_4)_2$ at 1000 °C for 120 h with grinding every 30 h. Characterized by powder X-ray diffraction data. The crystal structure is solved. Trigonal, space group *R*3*c*, $a = 10.5612(2)$, $c = 38.0588(5)$ Å, $V = 3676.32(9)$ Å^3, $Z = 21$.

Kind of sample preparation and/or method of registration of the spectrum: KBr disc. Transmission.

Source: Belik et al. (2002).

Wavenumbers (cm^{-1}): 1117s, 1039s, 1019s, 981sh, 946sh, 607sh, 600s, 589, 575w, 566w, 556sh, 551s, 544s.

P761 Tantalum oxyphosphate β-$Ta(PO_4)O$

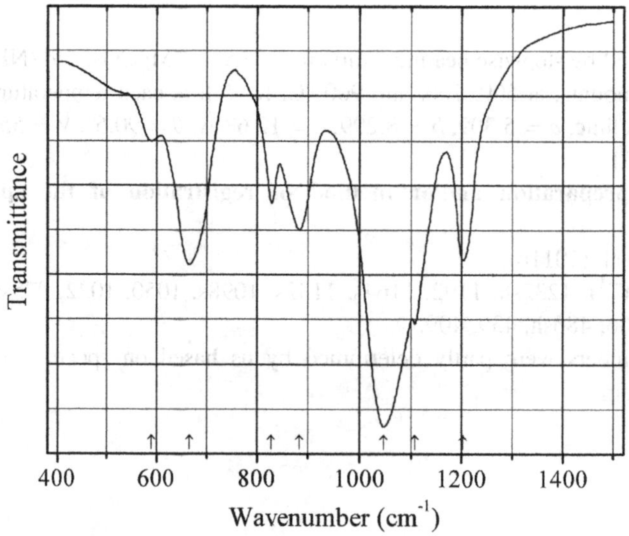

Origin: Synthetic.

Description: Prepared by dehydration of a Ta(PO_4)O·nH_2O precursor at 900 °C. For details of synthesis techniques see Hahn (1951). Monoclinic, space group $P2_1$ (?).

Kind of sample preparation and/or method of registration of the spectrum: KBr disc. Transmission.

Source: Stranford and Condrate Sr (1984c, 1990).

Wavenumbers (IR, cm^{-1}): 1202, 1108, 1048s, 883, 828, 666, 589w.

Note: In the cited papers, Raman spectra are given.

Wavenumbers (Raman, cm^{-1}): 1135, 1107, 1041, 1019, 997, 882w, 810w, 671w, 636w, 616w, 606w, 593w, 558w, 471w, 425w, 412w, 384w, 366, 334, 287, 269, 236, 214, 176, 118.

P763 Tellurium(IV) oxyphosphate $Te_2(PO_4)_2O$

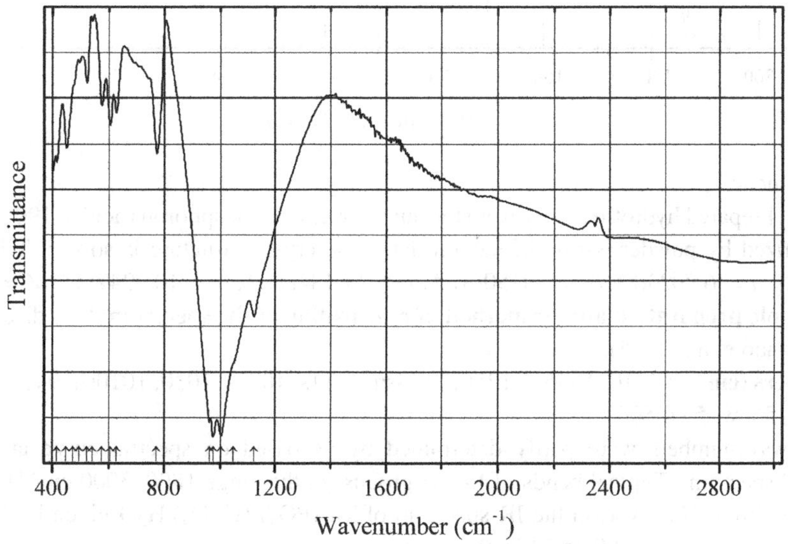

Origin: Synthetic.

Description: Colorless crystals prepared by heating a mixture of TeO_2 and P_4O_{10} at 550 °C for 24 h. Characterized by powder X-ray diffraction data. The crystal structure is solved. Monoclinic, space group *Cc*, $a = 5.3819(7)$, $b = 13.6990(19)$, $c = 9.5866(12)$ Å, $\beta = 103.682(2)°$, $V = 686.73(16)$ Å^3, $Z = 4$. Te has fivefold coordination.

Kind of sample preparation and/or method of registration of the spectrum: A sample pressed between two KBr pellets (Authors' wording, maybe erroneous). Transmission.

Source: Kim et al. (2010a).

Wavenumbers (cm^{-1}): 1121, 1047sh, 1004s, 975s, 961, 773, 628, 606, 577, 543w, 524, 498, 476sh, 451, (416).

Note: The wavenumbers were partly determined by us based on spectral curve analysis of the published spectrum.

P764 Thorium hydrogenphosphate $Th_2(PO_4)_2(HPO_4){\cdot}H_2O$

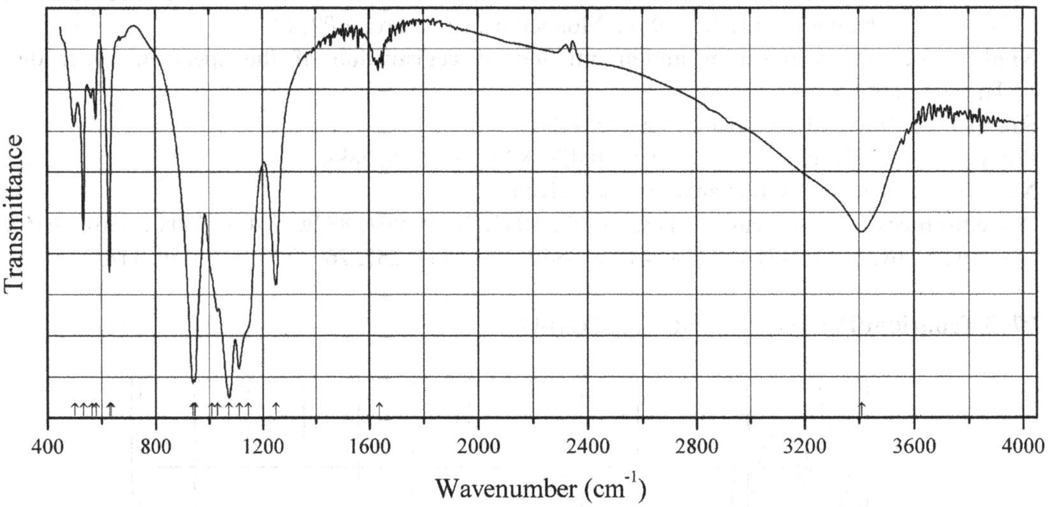

Origin: Synthetic.

Description: Prepared hydrothermally from thorium nitrate and phosphorous acid at 190 °C for 7 days. Characterized by powder X-ray diffraction data. The crystal structure is solved. Triclinic, space group $P2_1$, $a = 6.7023(8)$, $b = 7.0150(8)$ Å, $c = 11.184(1)$ Å, $\beta = 107.242(4)°$, $Z = 2$.

Kind of sample preparation and/or method of registration of the spectrum: KBr disc. Absorption.

Source: Salvadó et al. (2005).

Wavenumbers (cm^{-1}): 3410, 1636w, 1250, 1144sh, 1111s, 1075s, 1030, 1010sh, 947s, 940s, 635sh, 629, 580, 563w, 534, 500.

Note: The wavenumbers were partly determined by us based on spectral curve analysis of the published spectrum. Typical bands of P–OH groups (in the range 1800–3000 cm^{-1}) are absent in the IR spectrum. However, in the IR spectrum of $Th_2(PO_4)_2(HPO_4){\cdot}H_2O$ given by Brandel et al. (2001) a weak band at 2400 cm^{-1} is observed.

P765 Titanium(III) orthophosphate $Ti(PO_4)$

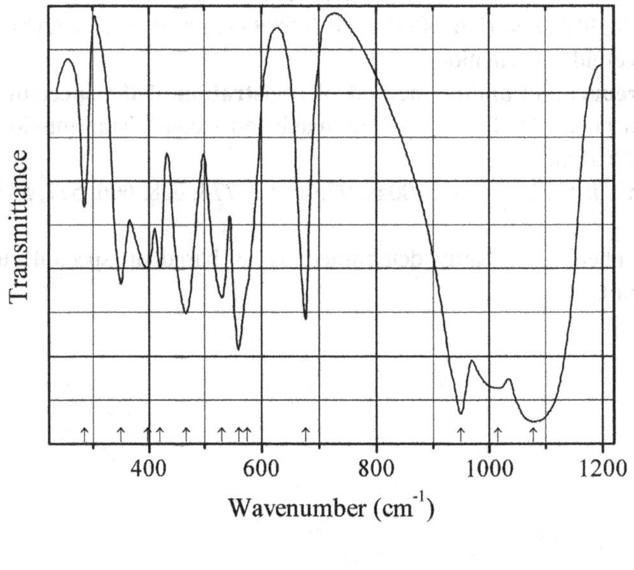

Origin: Synthetic.

Description: Prepared by heating a mixture of TiO_2 and $(NH_4)_2(HPO_4)$ (in a molar ratio of 1.0:1.1) at 950 °C under reducing conditions (in argon gas, in the presence of iron wires and porous titanium metal). Orthorhombic, isostructural with β-$Cr(PO_4)$.

Kind of sample preparation and/or method of registration of the spectrum: KBr disc. Transmission.

Source: Baran et al. (1989).

Wavenumbers (cm^{-1}): 1078s, 1015s, 949s, 677, 574sh, 559sh, 530, 467, 420, 397, 350, 285.

P766 Titanium acid phosphate monohydrate α-$Ti(HPO_4)_2{\cdot}H_2O$

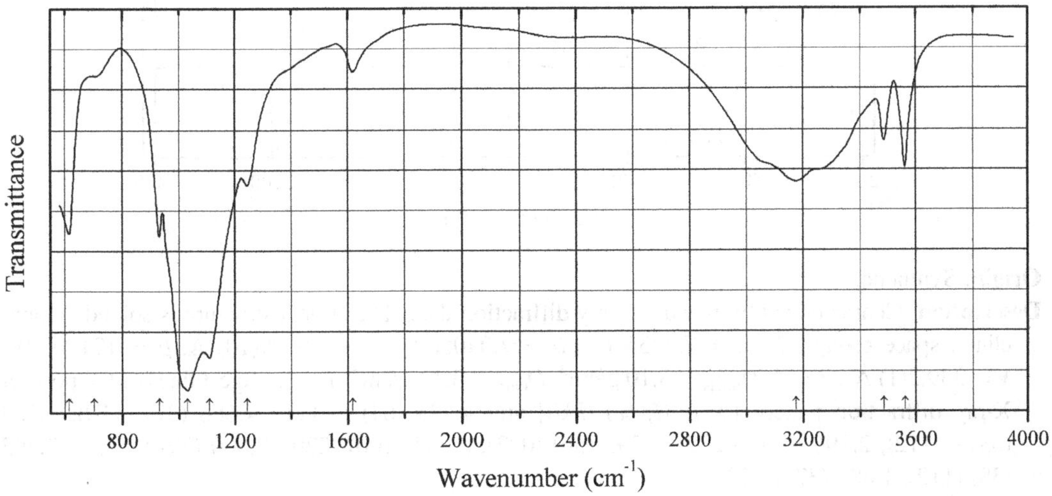

Origin: Synthetic.

Description: Prepared in the reaction of a solution of $TiCl_4$ in HCl(aq) with an aqueous solution of phosphoric acid. Characterized by powder X-ray diffraction data. Monoclinic, $a = 8.85(6)$, $b = 5.21(3)$, $c = 15.2(1)$ Å, $\beta = 115.8°$.

Kind of sample preparation and/or method of registration of the spectrum: KBr disc. Transmission.

Source: Slade et al. (1997).

Wavenumbers (IR, cm^{-1}): 3561, 3484, 3175 (broad), 1617w, 1242, 1108s, 1029s, 932, 702w, 615.

Note: The wavenumbers were partly determined by us based on spectral curve analysis of the published spectrum. In the cited paper, Raman spectrum is given.

Wavenumbers (Raman, cm^{-1}): 3559w, 3552, 3528, 3484, 3209, 3012, 1204, 1048sh, 1034sh, 1024s, 1016s, 975w, 588, 492w, 428, 329s, 233, 198, 181, 154w, 108w, 79w.

P767 Titanium oxophosphate hydrate $Ti_2(PO_4)_2O{\cdot}H_2O$

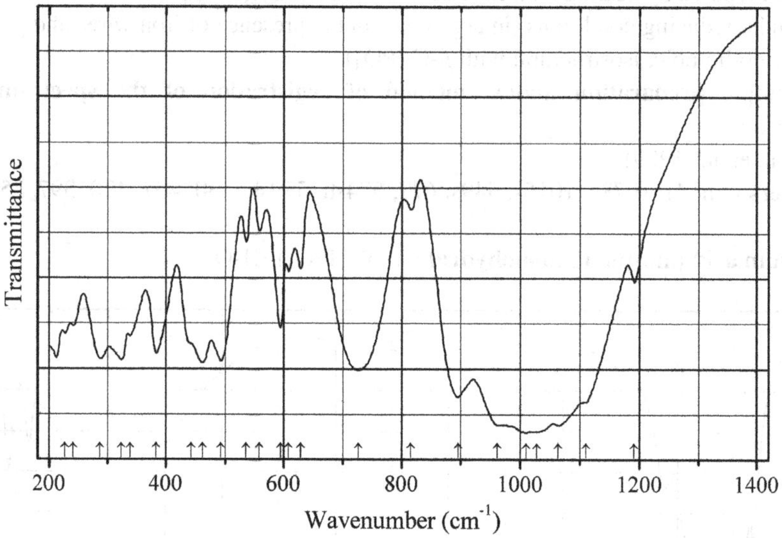

Origin: Synthetic.

Description: Characterized by powder X-ray diffraction data. The crystal structure is solved. Monoclinic, space group $P2_1$, a = 7.3735(12), b = 7.0405(10), c = 7.6609(10) Å, β = 121.48(2)°, V = 339.2(1) Å^3, Z = 4. D_{meas} = 3.10 g/cm^3, D_{calc} = 3.13 g/cm^3. The strongest lines of the powder X-ray diffraction pattern [d, Å (I, %) (hkl)] are: 4.795 (64) (−111), 4.688 (37) (110), 3.364 (33) (−112), 2.396 (56) (−222), 2.392 (56) (022), 2.344 (100) (220), 2.300 (39) (−312), 2.293 (38) (112), 1.682 (37) (−224).

Kind of sample preparation and/or method of registration of the spectrum: KBr and polyethylene discs. Transmission.

Source: Benmokhtar et al. (2007b).

Wavenumbers (IR, cm^{-1}): 3256, 1651w, 1514w, 1192, 1110s, 1064s, 1028sh, 1010s, 962sh, 896s, 816w, 727s, 628, 608, 594, 559w, 536w, 492, 461, 442sh, 383, 338, 323, 287, 242, 228, 196, 130.

Note: In the cited paper, Raman spectrum is given.

Wavenumbers (Raman, cm^{-1}): 1087, 1050w, 1015sh, 1006s, 972sh, 904w, 681s, 642, 614, 601, 585, 536, 498, 451, 410, 376, 348, 323, 287, 268, 245s, 213.

P768 Tungsten(VI) oxyphosphate $W_2O_3(PO_4)_2$

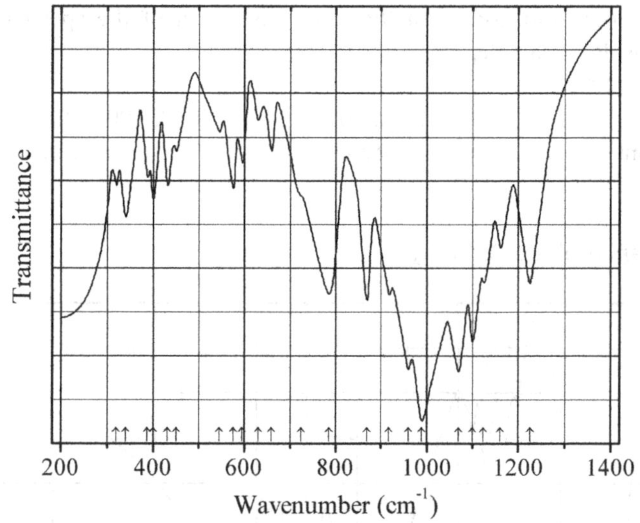

Wavenumbers (IR, cm^{-1}): 1225, 1162, 1125, 1100, 1070, 990, 960, 918, 869, 785, 725sh, 659, 629, 595, 575, 545, 450, 431, 400, 388, 341, 320.

Note: In the cited paper, Raman spectrum is given.

Wavenumbers (Raman, cm^{-1}): 1210, 1170, 1110, 1090, 1020, 999, 995, 980, 950, 912, 857, 810, 715, 660, 640, 613, 580, 448, 432, 421, 407, 393, 380, 350, 320, 300, 278, 257, 242, 222, 205, 172.

P769 Uranyl oxy-hydroxyphosphate $(UO_2)_3(PO_4)O(OH)\cdot 3H_2O$

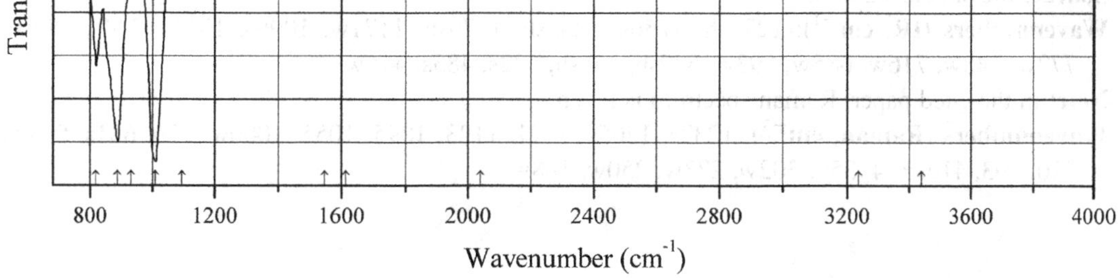

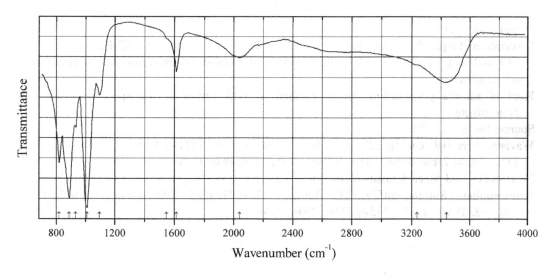

Origin: Synthetic.

Description: Crystals prepared by hydrothermal treatment of crystals of natural albite with inclusions of natural phosphates with 0.1 M solution of uranyl nitrate. Characterized by electron microprobe analysis. The crystal structure is solved. Tetragonal, space group $P4_2/mbc$, a = 14.015(1), c = 13.083(2) Å, V = 2575.6(4) Å^3, Z = 8. D_{calc} = 5.092 g/cm^3. The structure contains chains composed of uranyl pentagonal and hexagonal bipyramids and phosphate tetrahedra linked via common edges.

Kind of sample preparation and/or method of registration of the spectrum: Attenuated total reflection of a powdered sample.

Source: Burns et al. (2004).

Wavenumbers (cm^{-1}): 3439, 3237sh, 2040, 1614, 1547sh, 1093, 1008s, 931, 887s, 818.

Note: The wavenumbers were determined by us based on spectral curve analysis of the published spectrum.

P770 Yttrium metaphosphate $Y(PO_3)_3$

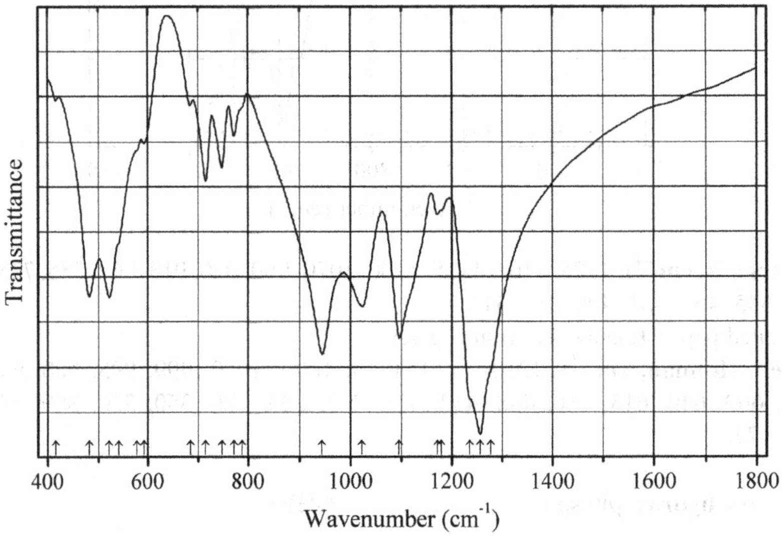

Origin: Synthetic.

Description: Prepared by stepwise heating a mixture of Y_2O_3 and $(NH_4)(H_2PO_4)$ at 170, 240, 350, 440, and 550 °C. Monoclinic. Powder X-ray diffraction pattern corresponds to JCPDS card no. 42-0501.

Kind of sample preparation and/or method of registration of the spectrum: KBr disc. Transmission.

Source: Ilieva et al. (2001).

Wavenumbers (IR, cm^{-1}): 1277sh, 1256s, 1235sh, 1179sh, 1171w, 1096s, 1023s, 944s, 789sh, 772w, 748w, 716w, 685w, 593w, 579sh, 541sh, 522s, 483s, 416w.

Note: In the cited paper, Raman spectrum is given.

Wavenumbers (Raman, cm^{-1}): 1237s, 1205s, 1171, 1123, 1095, 1055, 1000w, 726, 681s, 568w, 510, 493, 416, 374, 350, 302w, 273w, 250w, 188w.

P771 Zinc vanadyl phosphate $Zn_2(VO)(PO_4)_2$

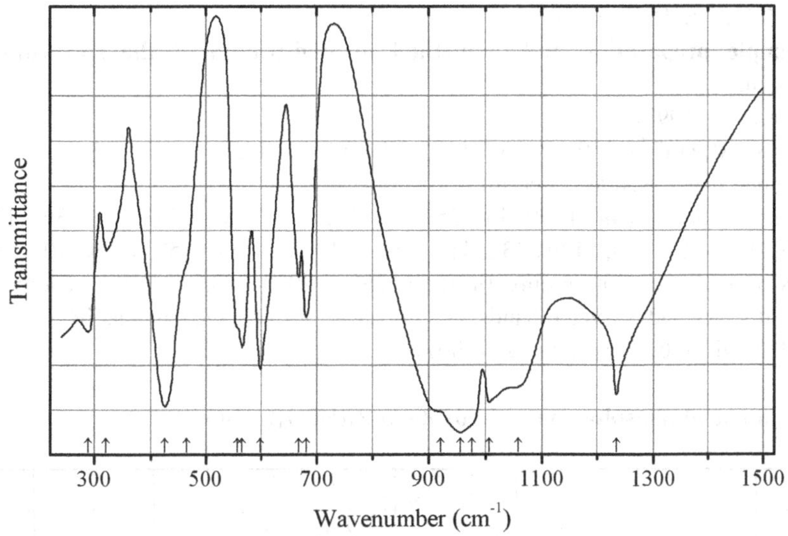

Origin: Synthetic.

Description: Prepared by heating a mixture of ZnO, VO_2, and P_2O_5, taken in stoichiometric amounts, at 850 °C for 2 days. Characterized by powder X-ray diffraction data. Tetragonal, space group *I4cm*, $a = 8.9227(13)$, $c = 9.039(3)$ Å, Z = 4. The Zn^{2+} ions exhibit a square pyramidal coordination.

Kind of sample preparation and/or method of registration of the spectrum: KBr disc. Transmission.

Source: Baran and Lii (1992).

Wavenumbers (IR, cm^{-1}): 1235s, 1058s, 1007s, 976sh, 956s, 921sh, 681, 668, 598, 565, 556sh, 465sh, 426s, 320, 288.

Note: The wavenumbers were partly determined by us based on spectral curve analysis of the published spectrum. In the cited paper, Raman spectrum is given.

Wavenumbers (Raman, cm^{-1}): 1125, 1067, 1010, 976s, 915, 662, 595, 570, 470sh, 430, 413, 325, 285, 248.

P772 Zirconium acid phosphate monohydrate α-$Zr(HPO_4)_2{\cdot}H_2O$

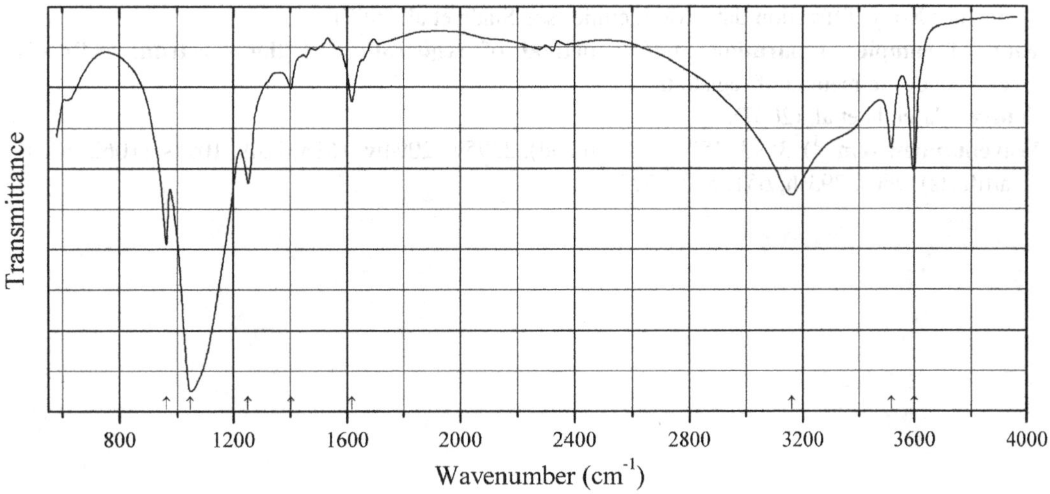

Origin: Synthetic.

Description: Characterized by DSC and powder X-ray diffraction data. Monoclinic, $a \approx 9.06–9.07$, $b \approx 5.26–5.31$, $c \approx 16.0–16.3$ Å, $\beta \approx 111°$.

Kind of sample preparation and/or method of registration of the spectrum: KBr disc. Transmission.

Source: Slade et al. (1997).

Wavenumbers (IR, cm^{-1}): 3598, 3515, 3159, 1617, 1401w, 1250, 1046s, 963.

Note: In the cited paper, Raman spectrum is given.

Wavenumbers (Raman, cm^{-1}): 3914w, 3592w, 3515w, 3142w, 2751w, 1613w, 1145w, 1081, 1055s, 990, 964, 591, 540, 517w, 432, 418, 399, 294, 214, 186w, 157, 118w, 107, 79, 63, 55w.

Note: Bands of acid phosphate groups (in the range of 2000–2400 cm^{-1}) are anomalously weak. However, distinct and stronger bands in this range are observed in the IR spectrum of α-Zr$(HPO_4)_2 \cdot H_2O$ given by Casciola et al. (2007).

P773 Zirconium acid phosphate monohydrate α-Zr$(HPO_4)_2 \cdot H_2O$

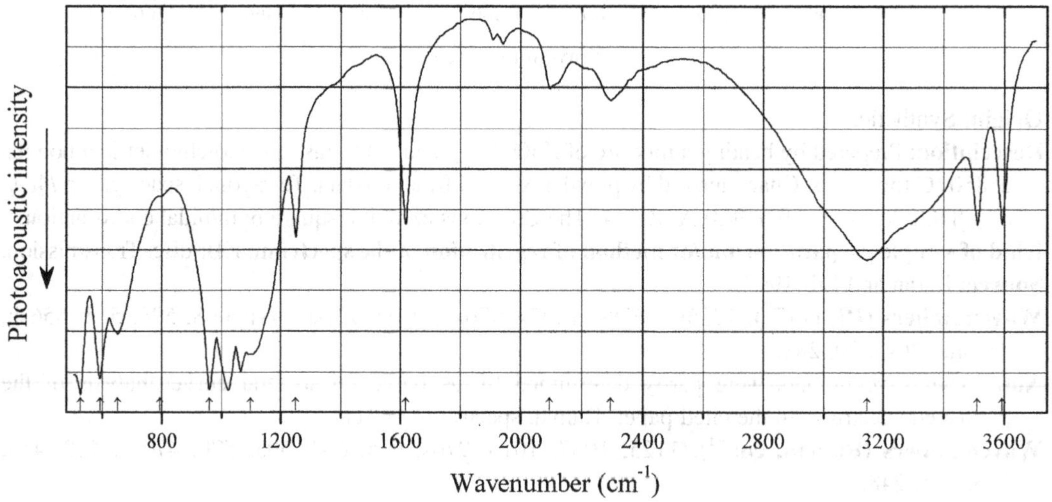

Origin: Synthetic.

Description: Prepared by the direct precipitation method in the presence of HF. Characterized by powder X-ray diffraction data. Monoclinic (see Slade et al. 1997).

Kind of sample preparation and/or method of registration of the spectrum: KBr disc. Photoacoustic method of registration.

Source: Casciola et al. (2007).

Wavenumbers (cm^{-1}): 3593, 3511, 3144 (broad), 2295w, 2094w, 1618, 1249, 1095s, (1062, 1024—artifacts), 960s, 793sh, 651, 592s, 527s.

P774 Alforsite $Ba_5(PO_4)_3Cl$

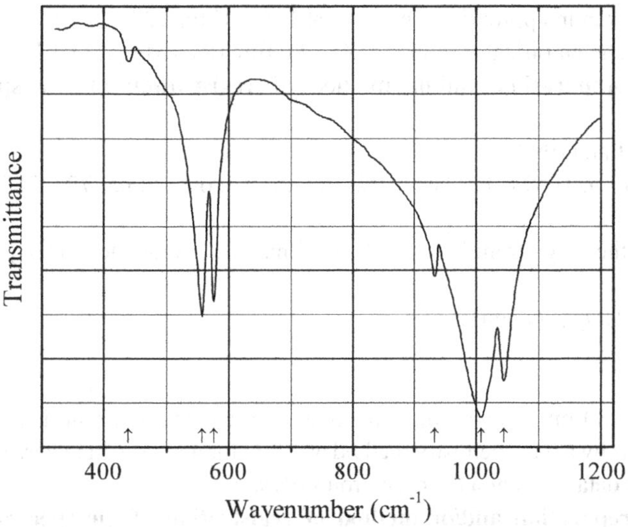

Origin: Synthetic.
Description: Hexagonal, space group *P*6$_3$/*m*, *Z* = 2.
Kind of sample preparation and/or method of registration of the spectrum: KBr disc. Transmission.
Source: Baran and Aymonino (1972).
Wavenumbers (cm^{-1}): 1045s, 1008s, 933, 576, 557, 440w.

P775 Alforsite F-analogue $Ba_5(PO_4)_3F$

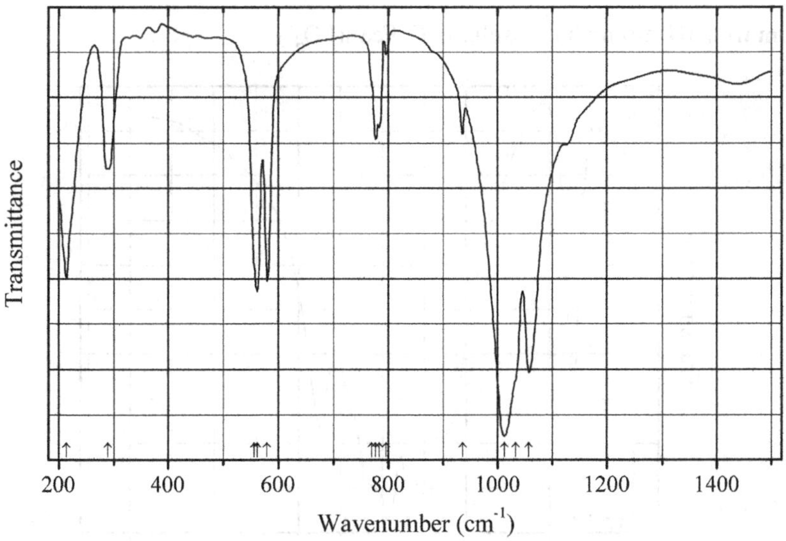

Origin: Synthetic.

Description: Mn-doped sample obtained by repeated heating a mixture of $BaCO_3$, Mn_2O_3, NH_4F, and $(NH_4)(H_2PO_4)$, taken in appropriate amounts, at 1250 °C for 12 h. Characterized by powder X-ray diffraction data. The empirical formula is (Z = 1): $Ba_{10}[P_{0.95}Mn_{0.05})O_4]_6F_2$.

Kind of sample preparation and/or method of registration of the spectrum: KBr disc. Transmission.

Source: Dardenne et al. (1998).

Wavenumbers (cm^{-1}): 1057s, 1033sh, 1012s, 935.5w, 796w, 783.5, 777, 769sh, 580, 561, 556sh, 290, (213).

Note: The bands in the range from 760 to 800 cm^{-1} may correspond to $[MnO_4]^{3-}$ vibrational modes.

P776 Ankoleite $K(UO_2)(PO_4){\cdot}nH_2O$

Origin: Synthetic.

Description: Obtained from uranyl nitrate, potassium nitrate, and phosphoric acid mixed in stoichiometric proportions by a wet chemistry method at 60 °C for 4 days. Characterized by TG and powder X-ray diffraction data. Tetragonal, space group *P4/ncc*.

Kind of sample preparation and/or method of registration of the spectrum: Attenuated total reflection of a powdered sample.

Source: Clavier et al. (2016).

Wavenumbers (IR, cm^{-1}): 3589, 3468, 3350, 3201, 2990, 1658, 1622, 1109, 1059, 985, 904, 866, 813, 666, 541, 529.

Note: The wavenumbers are taken from the table given in the cited paper. There are strong discrepancies between these values and the figure of the IR spectrum of ankoleite from this paper. In the cited paper, Raman spectrum is given.

Wavenumbers (Raman, cm^{-1}): 3805w, 3498w, 3375w, 3237w, 3110w, 2786w, 1004s, 994s, 831s, 826s, 400, 291, 195, 173, 113, 108.

P777 Calcium iron(III) tin orthophosphate $CaFeSn(PO_4)_3$

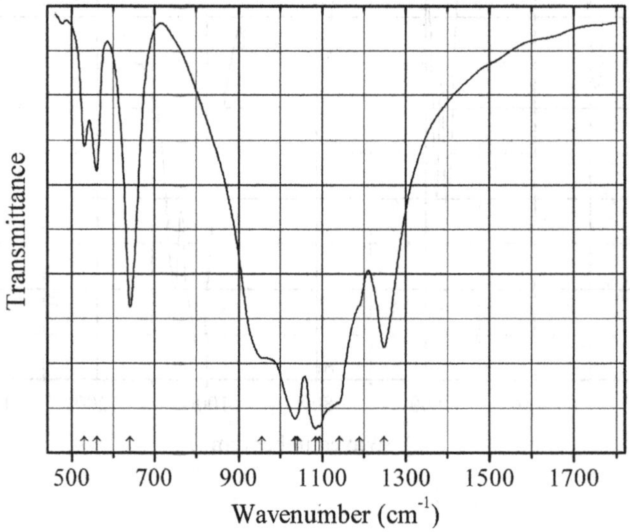

Origin: Synthetic.

Description: Synthesized by a solid-state reaction technique from a mixture of $CaCO_3$, SnO_2, Fe_2O_3, and $(NH_4)(H_2PO_4)$ taken in stoichiometric proportion. Trigonal, space group *R*-3*c*, $Z = 6$.

Kind of sample preparation and/or method of registration of the spectrum: KBr disc. Transmission.

Source: Antony et al. (2011).

Wavenumbers (IR, cm^{-1}): 1248s, 1189sh, 1140sh, 1092sh, 1083s, 1041sh, 1035s, 954s, 641, 561, 532.

Note: In the cited paper, Raman spectrum is given.

Wavenumbers (Raman, cm^{-1}): 1213, 1106s, 1046sh, 999sh, 977sh, 784, 657, 600, 576, 486sh, 436s, 380w, 348, 309, 254sh, 226, 201, 145.

P778 Jörgkellerite $(Na,\square)_3Mn^{3+}{}_3(PO_4)_2(CO_3)(O,OH)_2\cdot 5H_2O$

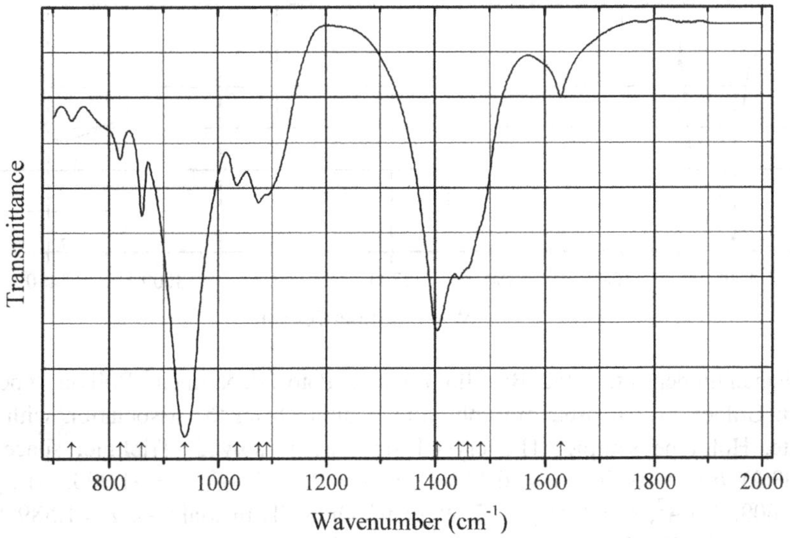

Origin: Oldoinyo Lengai volcano, Gregory Rift, northern Tanzania (type locality).

Description: Brown spherulites from the association with shortite, calcite, fluorite, magnetite, and khanneshite. Holotype sample. The crystal structure is solved. Trigonal, space group *P*-3, $a = 11.201(2)$, $c = 10.969(2)$ Å, $V = 1191.9(7)$ Å^3, $Z = 3$. $D_{calc} = 2.56$ g/cm^3. Optically uniaxial (−), $\omega = 1.700(2)$, $\varepsilon = 1.625(2)$. The empirical formula is (electron microprobe, H_2O and CO_2 calculated): $(Na_{2.46}K_{0.28}Ca_{0.08}Sr_{0.04}Ba_{0.02})(Mn^{3+}{}_{2.39}Fe^{3+}{}_{0.56})(PO_4)_{1.95}(SO_4)_{0.05}(CO_3)[O_{1.84}(OH)_{0.16}]\cdot 5H_2O$. The strongest lines of the powder X-ray diffraction pattern [*d*, Å (*I*, %) (*hkl*)] are: 10.970 (100) (001), 5.597 (15) (002), 4.993 (8) (111), 2.796 (14) (220), 2.724 (20) (004).

Kind of sample preparation and/or method of registration of the spectrum: Attenuated total reflection using IR microscope.

Source: Zaitsev et al. (2017).

Wavenumbers (cm^{-1}): 1629, 1483sh, 1459sh, 1443s, 1404s, 1091sh, 1075, 1035, 939s, 861, 821, 733w.

Note: The wavenumbers were partly determined by us based on spectral curve analysis of the published spectrum. The tentative assignment of the strongest band at 939 cm^{-1} to phosphate groups made in the cited paper is questionable.

P779 Ferrivauxite $Fe^{3+}Al_2(PO_4)_2(OH)_3{\cdot}5H_2O$

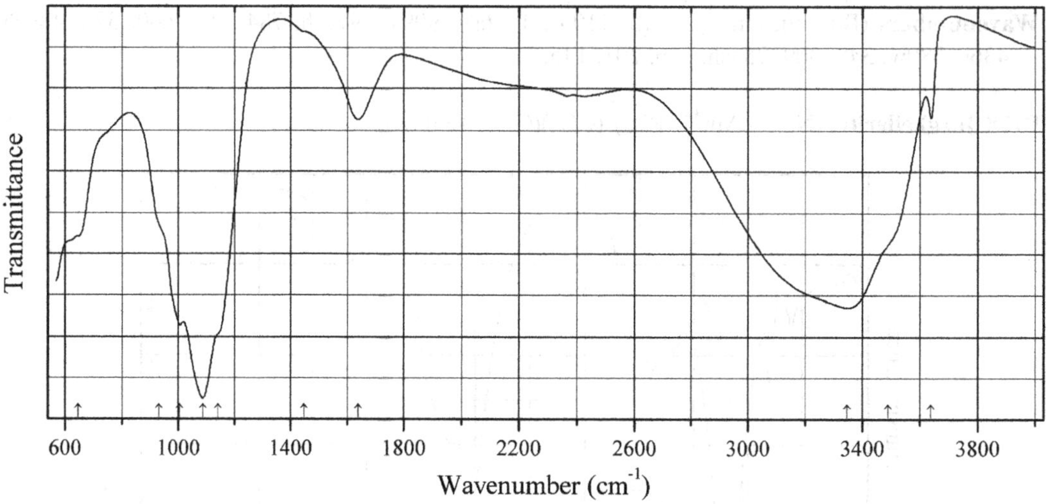

Origin: Llallagua tin deposit, Rafael Bustillo province, Potosí department, Bolivia (type locality).

Description: Golden brown pseudomorphs after vauxite from the association with sigloite and crandallite. Holotype sample. The crystal structure is solved. Triclinic, space group *P*-1, a = 9.198(2), b = 11.607(3), c = 6.112(2) Å, α = 98.237(9)°, β = 91.900(13)°, γ = 108.658 (9)°, V = 609.7(5) Å^3, Z = 2. D_{calc} = 2.39 g/cm^3. Optically biaxial (–), α = 1.589(1), β = 1.593 (1), γ = 1.596(1), $2V$ = 60(4)°. The empirical formula is (electron microprobe): $Fe^{3+}{}_{0.94}Mn_{0.01}Al_{1.98}P_{2.05}O_8(OH)_3{\cdot}5H_2O$. The strongest lines of the powder X-ray diffraction pattern [*d*, Å (*I*, %) (*hkl*)] are: 10.834 (100) (010), 8.682 (24) (100), 8.242 (65) (–110), 6.018 (28) (001), 5.918 (23) (110), 5.491 (30) (–120), 4.338 (26) (200), 2.898 (32) (300).

Kind of sample preparation and/or method of registration of the spectrum: Transmission. A diamond anvil microsample cell was used.

Source: Raade et al. (2016).

Wavenumbers (cm^{-1}): 3640, 3490sh, 3348s, 1638, 1447sh, 1142sh, 1087s, 1007s, 932sh, 645sh.

Note: The wavenumbers were partly determined by us based on spectral curve analysis of the published spectrum.

P780 Kosnarite NH4-analogue $(NH_4)Zr_2(PO_4)_3$

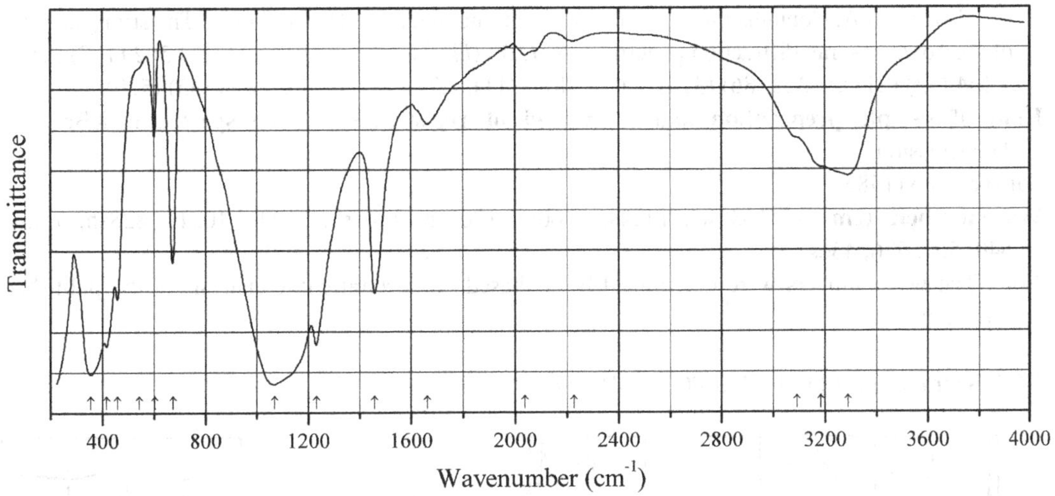

Origin: Synthetic.

Description: Prepared hydrothermally. Rhombohedral, $a = 8.676(1)$, $c = 24.288(5)$ Å. The strongest lines of the powder X-ray diffraction pattern [*d*, Å (*I*, %) (*hkl*)] are: 6.394 (67) (102), 4.721 (72) (104), 4.340 (80) (110), 3.825 (92) (113), 3.194 (57) (204), 2.960 (100) (116).

Kind of sample preparation and/or method of registration of the spectrum: Transmission. Kind of sample preparation is not indicated.

Source: Clearfield et al. (1984).

Wavenumbers (cm^{-1}): 3290, 3185sh, 3090sh, 2226w, 2037w, 1665, 1458, 1232s, 1069s, 672, 601, 541sh, 457, 416s, 356s.

Note: The wavenumbers were partly determined by us based on spectral curve analysis of the published spectrum. The band at 1665 cm^{-1} indicates the presence of H_2O molecules.

P781 Kosnarite NH4-analogue cubic polymorph $(NH_4)Zr_2(PO_4)_3$

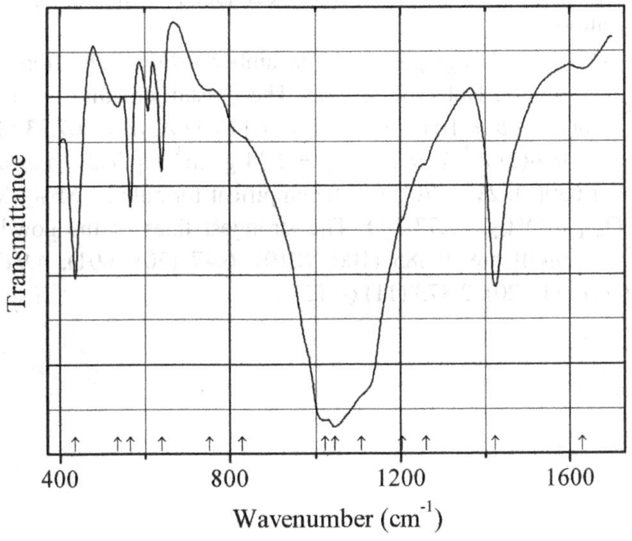

Origin: Synthetic.

Description: Prepared by heating a mixture of ZrO_2 and $(NH_4)(H_2PO_4)$ at 608 K for 72 h. Cubic, $a = 1.0186(3)$ Å. Contains minor admixture of rhombohedral $(NH_4)Zr_2(PO_4)_3$. The strongest lines of the powder X-ray diffraction pattern [*d*, Å (*I*, %) (*hkl*)] are: 5.893 (75) (111), 4.562 (100) (210), 4.164 (33) (211), 3.222 (86) (310), 3.072 (40) (311), 2.723 (75) (321), 1.895 (35) (520).

Kind of sample preparation and/or method of registration of the spectrum: KBr disc. Transmission.

Source: Ono (1985).

Wavenumbers (cm^{-1}): (1630w), 1424s, 1260w, 1203sh, 1107sh, 1044s, 1021s, 828sh, 752w, 640, 566, 536, 436s.

Note: The wavenumbers were determined by us based on spectral curve analysis of the published spectrum.

P782 Kummerite $Mn^{2+}Fe^{3+}Al(PO_4)_2(OH)_2{\cdot}8H_2O$

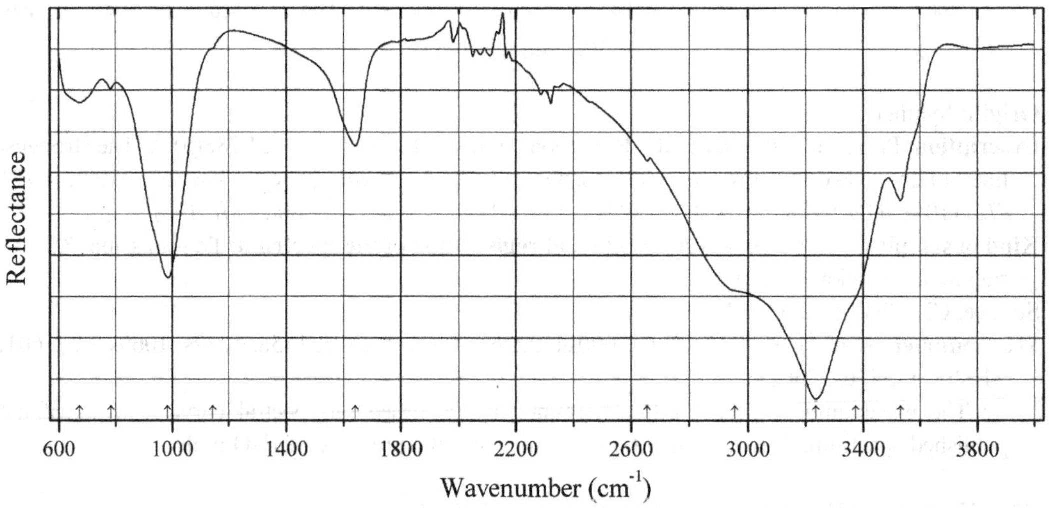

Origin: Hagendorf South pegmatite, Cornelia mine, Hagendorf, Waidhaus, Upper Palatinate, Bavaria, Germany (type locality).

Description: Sprays or rounded aggregates of thin amber yellow laths from the association with Zn-and Al-bearing beraunite. Holotype sample. The crystal structure is solved. Triclinic, space group *P*-1, $a = 5.316(1)$ Å, $b = 10.620(3)$ Å, $c = 7.118(1)$ Å, $\alpha = 107.33\ (3)°$, $\beta = 111.22\ (3)°$, $\gamma = 72.22\ (2)°$, $V = 348.4(2)$ Å^3, $Z = 1$. $D_{calc} = 2.34$ g/cm^3. Optically biaxial (−), $\alpha = 1.565(5)$, $\beta = 1.600(5)$, $\gamma = 1.630(5)$, $2V = 70(5)°$. The empirical formula is $(Mn_{0.37}Mg_{0.27}Zn_{0.03}Fe^{2+}_{0.33})(Fe^{3+}_{1.06}Al_{0.94})(PO_4)_{1.91}(OH)_{2.27}{\cdot}7.73H_2O$. The strongest lines of the powder X-ray diffraction pattern [*d*, Å (*I*, %) (*hkl*)] are: 9.885 (100) (010), 6.47 (20) (001), 4.942 (30) (020), 3.988 (9) (−110), 3.116 (18) (1−20), 2.873 (11) (−121).

Kind of sample preparation and/or method of registration of the spectrum: Attenuated total reflection of an individual crystal.

Source: Grey et al. (2016a).

Wavenumbers (cm^{-1}): 3530, 3375sh, 3235s, 2955sh, 2655, 1640, 1145sh, 985s, 783w, 674.

Note: The wavenumbers were partly determined by us based on spectral curve analysis of the published spectrum.

P783 Krásnoite $Ca_3Al_{7.7}Si_3P_4O_{22.9}(OH)_{13.3}F_2{\cdot}8H_2O$

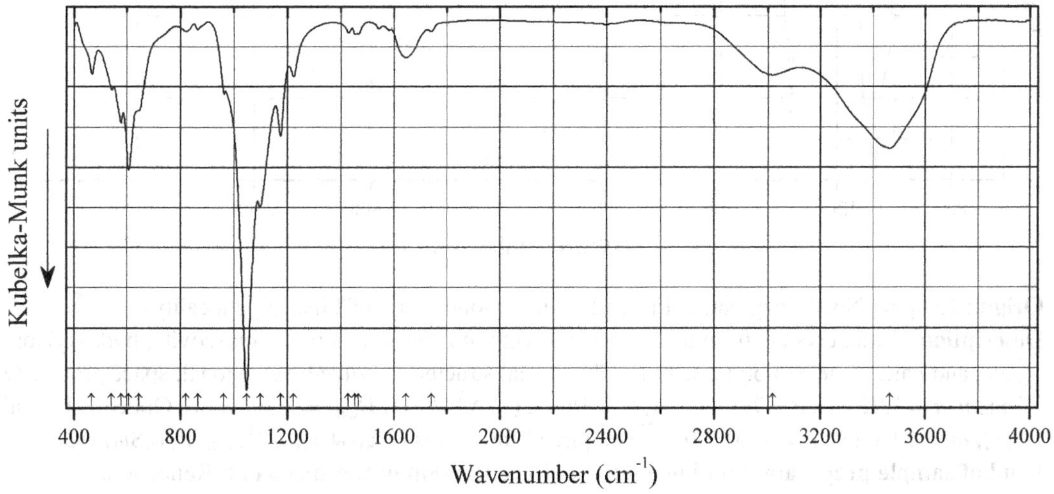

Origin: Huber open pit, Krásno ore district, Czech Republic (type locality).

Description: Aggregates of colorless platy crystals. Holotype sample. Trigonal, space group *P*-3*m*1, $a = 6.9956(4)$, $c = 20.200(2)$ Å, $V = 856.09(9)$ Å^3, $Z = 3$. $D_{meas} = 2.48(4)$ g/cm^3, $D_{calc} = 2.476$ g/cm^3. Optically uniaxial (+), $\omega = 1.548(2)$, $\varepsilon = 1.549(2)$. The empirical formula is $Ca_3Al_{7.7}Si_3P_4O_{22.9}(OH)_{12.1}F_2{\cdot}8H_2O$. The strongest lines of the powder X-ray diffraction pattern [*d*, Å (*I*, %) (*hkl*)] are: 20.186 (97) (001), 6.736 (100) (003), 5.800 (67) (101, 011), 3.496 (60) (110), 2.8730 (87) (114, 11–4), 2.7633 (73) (203), 2.1042 (75) (109).

Kind of sample preparation and/or method of registration of the spectrum: Micro-diffuse reflectance of a mixture with KBr recalculated in Kubelka-Munk units.

Source: Mills et al. (2012b).

Wavenumbers (IR, cm^{-1}): 3463s, 3017, 174w1, 1645, 1465w, 1453w, 1429w, 1223, 1175, 1098, 1048s, 963, 865w, 820w, 647sh, 608s, 580, 542, 467.

Note: In the cited paper, Raman spectrum is given.

Wavenumbers [Raman, cm^{-1}, for the wavelengths 532 nm (785 nm)]: 1425 (1422), (1289), 1190 (1196), 1091 (1091), 1032 (1032), 1009 (1007), 960 (962), 920 (920), 705 (706), 634 (638), 620 (621), 512 (508), 477 (460), 424 (430), 364 (363), 271, 190, 143.

P784 Minjiangite $BaBe_2(PO_4)_2$

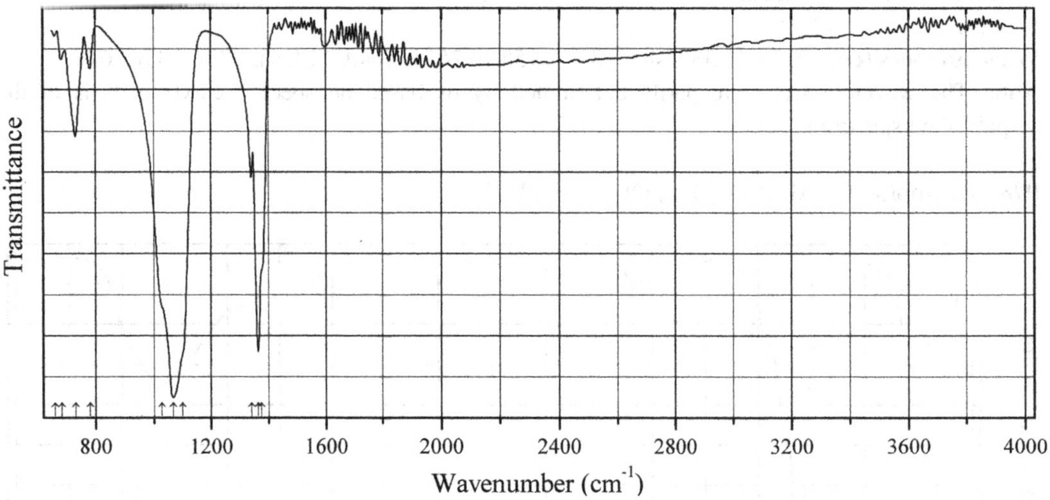

Origin: Nanping No. 31 pegmatite, Fujian Province, southeastern China (type locality).

Description: White crystals from the association with montebrasite, quartz, muscovite, hydroxylapatite, and palermoite. Holotype sample. The crystal structure is solved. Hexagonal, space group *P6/mmm*, $a = 5.029(1)$, $c = 7.466(1)$ Å, $V = 163.52(1)$ Å^3, $Z = 1$. $D_{calc} = 3.49$ g/cm^3. Optically biaxial (+), $\omega = 1.587(3)$, $\varepsilon = 1.602(2)$. The empirical formula is $(Ba_{0.99}Ca_{0.01})Be_{1.98}(P_{1.99}Si_{0.01})O_8$.

Kind of sample preparation and/or method of registration of the spectrum: Reflection.

Source: Rao et al. (2015).

Wavenumbers (IR, cm^{-1}): 1375s, 1363s, 1339s, 1101sh, 1068s, 1027sh, 781, 730, 683w, 660w.

Note: Possibly, an erroneous spectrum. In particular, assignment of the strong bands at 1375, 1363, and 1339 cm^{-1} to Be–O-stretching vibrations (Rao et al. 2015) is questionable. Dal Bo et al. (2014) give another IR spectrum for the synthetic analogue of minjiangite. The wavenumbers were partly determined by us based on spectral curve analysis of the published spectrum. In the cited paper, Raman spectrum is given.

Wavenumbers (Raman, cm^{-1}): 1233s, 1050s, 491, 478, 328w, 189w.

P785 Xanthoxenite $Ca_4Fe^{3+}{}_2(PO_4)_4(OH)_2\cdot 3H_2O$

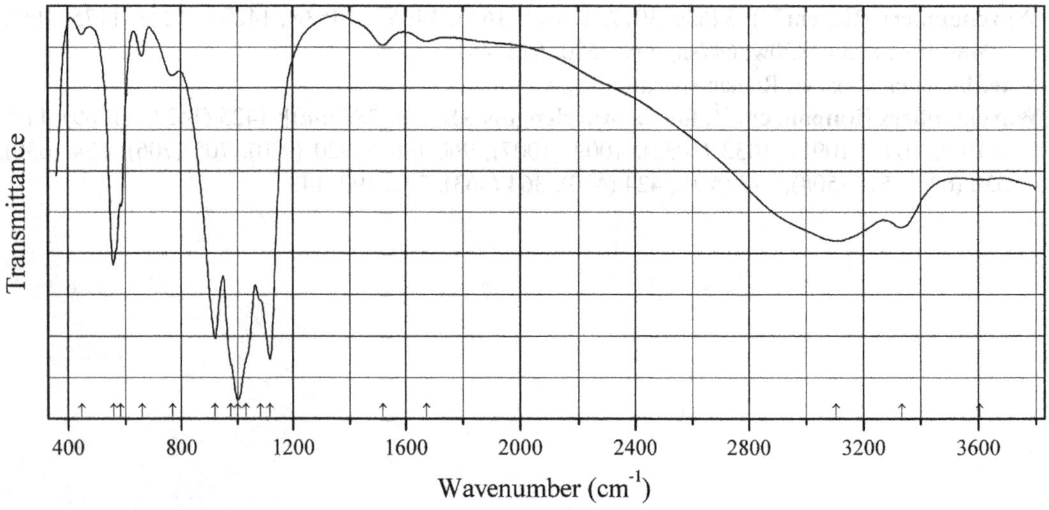

Origin: Palermo No. 1 mine, Groton, Grafton Co., New Hampshire, USA.
Description: Beige crust. Investigated by A.V. Kasatkin. The empirical formula is (electron microprobe): $(Ca_{3.49}Mn_{0.49})Fe_{2.10}(PO_4)_{3.92}(OH)_x{\cdot}nH_2O$.
Kind of sample preparation and/or method of registration of the spectrum: KBr disc. Absorption.
Wavenumbers (cm^{-1}): 3605w, 3334, 3103s, 1674w, 1518w, 1115s, 1080sh, 1030sh, 1001s, 977sh, 921s, 767, 658w, 585, 559s, 448w.
Note: The spectrum was obtained by N.V. Chukanov.

P786 Oxypyromorphite $Pb_{10}(PO_4)_6O$

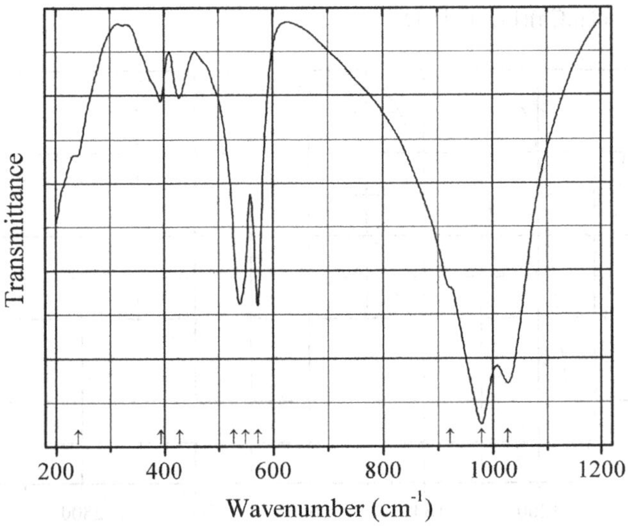

Origin: Synthetic.
Description: Hexagonal, space group *P*6$_3$/*m* or *P*-6, *a* = 9.826, *c* = 7.431 Å.
Kind of sample preparation and/or method of registration of the spectrum: RbI disc. Transmission.
Source: Engel (1973).
Wavenumbers (cm^{-1}): 3560w, 1028s, 979s, 921sh, 572s, 549sh, 538s, 428, 393, 241.

P787 Fluorwavellite $Al_3(PO_4)_2(OH)_2F{\cdot}5H_2O$

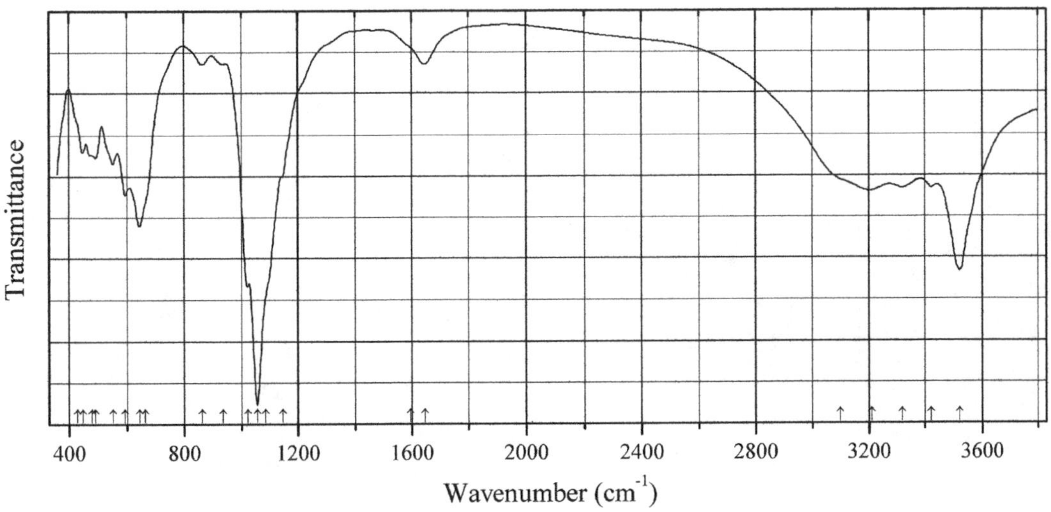

Origin: Baturovskiy stone quarry, Chelyabinsk region, South Urals, Russia.

Description: Pale green radiated aggregate from the association with quartz and crandallite. Investigated by A.V. Kasatkin. The empirical formula is (electron microprobe): $Al_{2.96}(PO_4)_{2.04}F_{0.95}(OH)_x{\cdot}nH_2O$.

Kind of sample preparation and/or method of registration of the spectrum: KBr disc. Absorption.

Wavenumbers (cm^{-1}): 3519s, 3422, 3320, 3210, 3100sh, 1647, 1595sh, 1145, 1085sh, 1055s, 1022s, 934w, 863w, 665sh, 644s, 595, 552, 493, 479, 448, 430sh.

Note: Many samples regarded earlier as wavellite are actually fluorwavellite. The spectrum was obtained by N.V. Chukanov.

P788 Smirnovskite $(Th,Ca)(PO_4){\cdot}nH_2O$

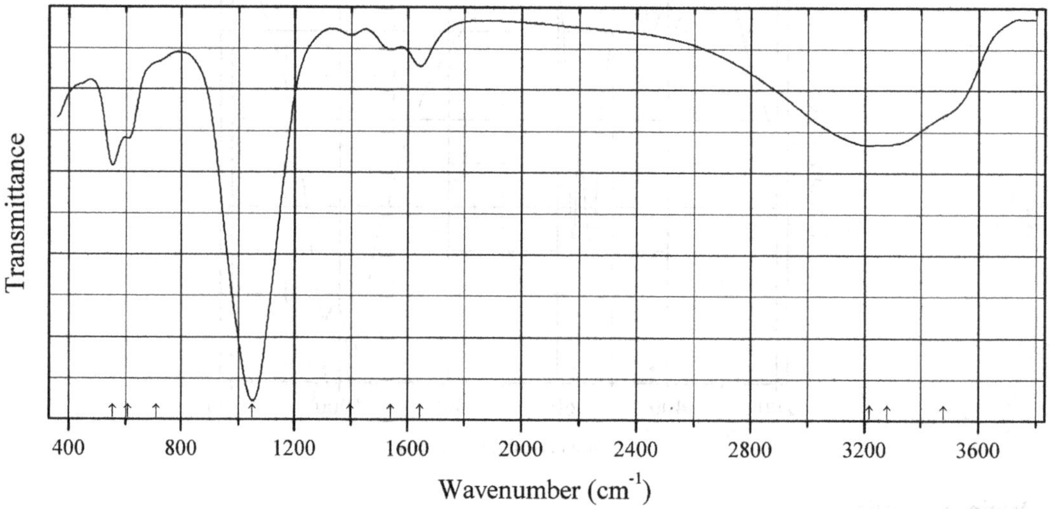

Origin: Etyka (Etykinskoe) Ta deposit, Baley district, Transbaikal area, Siberia, Russia.

Description: Dark red-brown grain. Investigated by A.V. Kasatkin. X-ray amorphous, metamict. The empirical formula is (electron microprobe): $(Th_{0.84}Ca_{0.22}Pb_{0.02})(PO_4)_{0.93}(H_2O,OH)_x$.

Kind of sample preparation and/or method of registration of the spectrum: KBr disc. Absorption.

Wavenumbers (cm^{-1}): 3480sh, 3280s, 3217s, 1645, 1543, 1398w, 1049s, 710sh, 608, 554s.

Note: The spectrum was obtained by N.V. Chukanov.

P789 Goryainovite $Ca_2(PO_4)Cl$

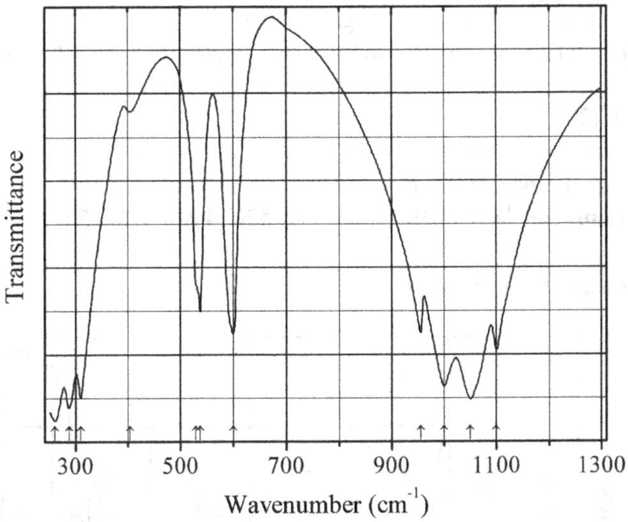

Origin: Synthetic.

Description: Crystals grown from the melt using excess $CaCl_2$ as flux ("chlorospodiosite").

Kind of sample preparation and/or method of registration of the spectrum: KBr disc. Transmission.

Source: Banks et al. (1967).

Wavenumbers (cm^{-1}): 1100, 1050s, 1000s, 955, 600, 538, 530sh, 404w, 309s, 287s, 260s.

Note: The wavenumbers were partly determined by us based on spectral curve analysis of the published spectrum.

P790 Sodium iron(III) tin orthophosphate $Na_2FeSn(PO_4)_3$

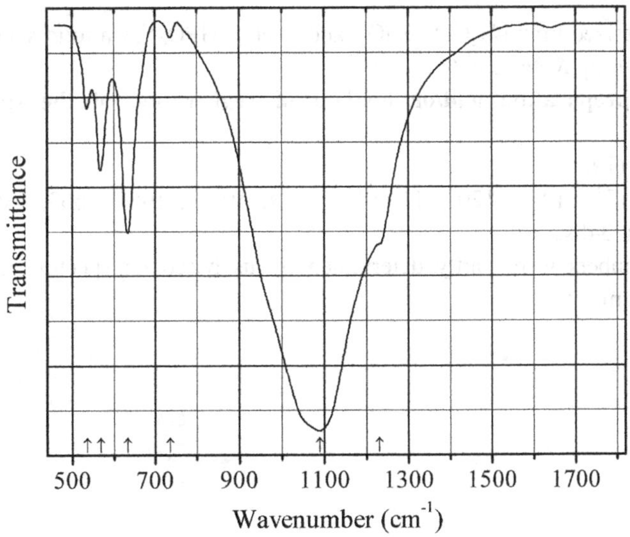

Origin: Synthetic.

Description: Powdery sample synthesized from Na_2CO_3, SnO_2, Fe_2O_3, and $(NH_4)(H_2PO_4)$ by a solid-state reaction technique. Hexagonal, space group *R*-3*C*, *Z* = 6.

Kind of sample preparation and/or method of registration of the spectrum: KBr disc. Transmission.

Source: Antony et al. (2011).

Wavenumbers (IR, cm^{-1}): 1230sh, 1088s, 735w, 633, 568, 536w.

Note: In the cited paper, Raman spectrum is given.

Wavenumbers (Raman, cm^{-1}): (1250), 1042s, 571, 530, 449s, 416, 250, 220, 168.

P791 Sodium tin phosphate $NaSn_2(PO_4)_3$

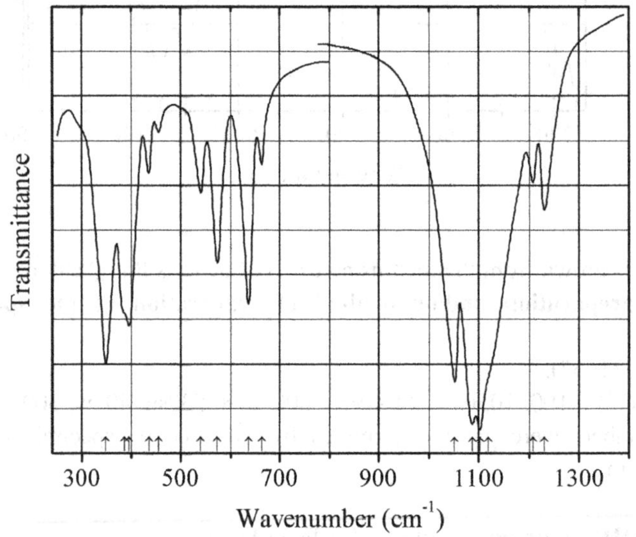

Origin: Synthetic.

Description: Synthesized from Na_2CO_3, SnO_2, and $(NH_4)_2(HPO_4)$ by a solid-state reaction technique. Trigonal, space group *R*-3*c*, *Z* = 6.

Kind of sample preparation and/or method of registration of the spectrum: KBr disc. Transmission.

Source: Tarte et al. (1986).

Wavenumbers (cm^{-1}): 1230, 1207, 1119sh, 1102s, 1087s, 1052s, 663, 636, 574, 541, 455w, 435, 395s, 385sh, 348s.

Note: The wavenumbers were partly determined by us based on spectral curve analysis of the published spectrum.

P792 Väyrynenite $BeMn^{2+}(PO_4)(OH)$

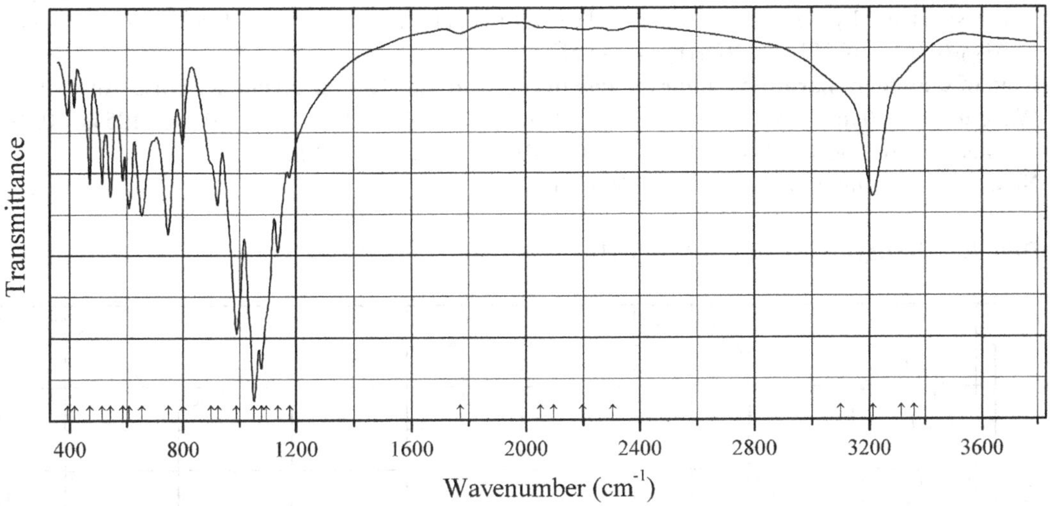

Origin: Chalot (Chalotuy) Be-Ta pegmatite deposit, Onon district, Transbaikal area, Siberia, Russia.

Description: Pink grains from the association with moraesite, eosphorite, and fluorapatite. Investigated by I.S. Lykova. Characterized by single-crystal X-ray diffraction data. Monoclinic, $a = 4.726(6)$, $b = 14.525(16)$, $c = 5.416(3)$, $\beta = 102.81(8)°$, $V = 362.6(7)$ Å^3. The empirical formula is (electron microprobe): $Be_{1.00}(Mn_{0.69}Fe_{0.22}Mg_{0.03}Ca_{0.03})_{\Sigma 0.97}P_{1.01}O_4(OH)$.

Kind of sample preparation and/or method of registration of the spectrum: KBr disc. Absorption.

Wavenumbers (cm^{-1}): 3360sh, 3315sh, 3213s, 3100sh, 2306w, 2199w, 2096w, 2050w, 1769w, 1177, 1136s, 1095sh, 1078s, 1051s, 989s, 922, 900sh, 799, 748s, 654, 609, 587, 544, 515, 471, 417, 393.

Note: The spectrum was obtained by N.V. Chukanov.

P793 Wavellite-(OH) $Al_3(PO_4)_2(OH)_2(OH,F)·5H_2O$

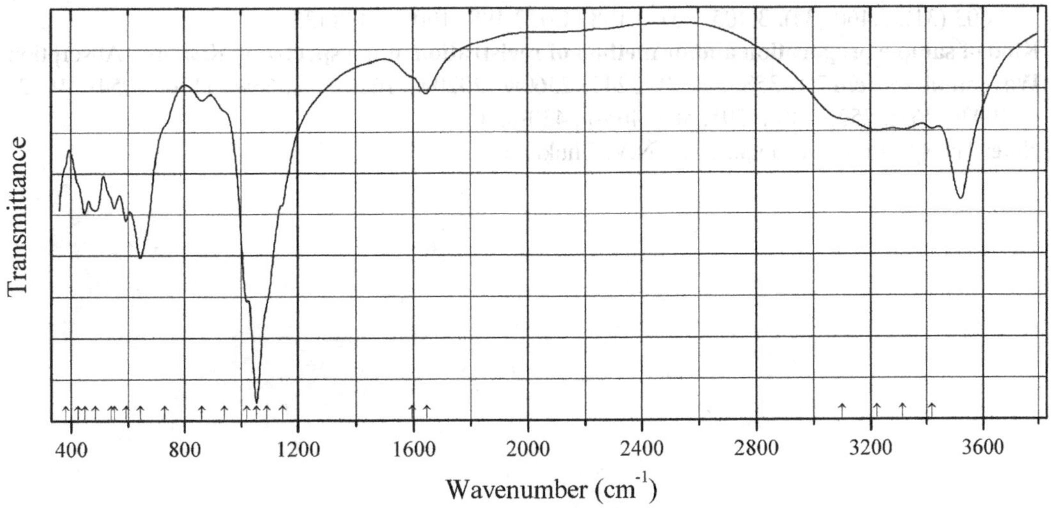

Origin: Mauldin Mt. quarries, Arkansas, USA.

Description: White radiated aggregate from the association with quartz. OH dominant sample. Investigated by A.V. Kasatkin. The empirical formula is (electron microprobe): $Al_{2.95}(PO_4)_{2.08}(OH,H_2O)_2[(OH)_{0.53}F_{0.47}]·5H_2O$.

Kind of sample preparation and/or method of registration of the spectrum: KBr disc. Absorption.

Wavenumbers (cm^{-1}): 3223s, 3419, 3315, 3223, 3100, 1647, 1595sh, 1145, 1090sh, 1054s, 1021s, 940sh, 861w, 730sh, 645s, 593s, 552, 540sh, 486, 448, 425sh, 380sh.

Note: The spectrum was obtained by N.V. Chukanov.

P794 Rockbridgeite $Fe^{2+}Fe^{3+}_4(PO_4)_3(OH)_5$

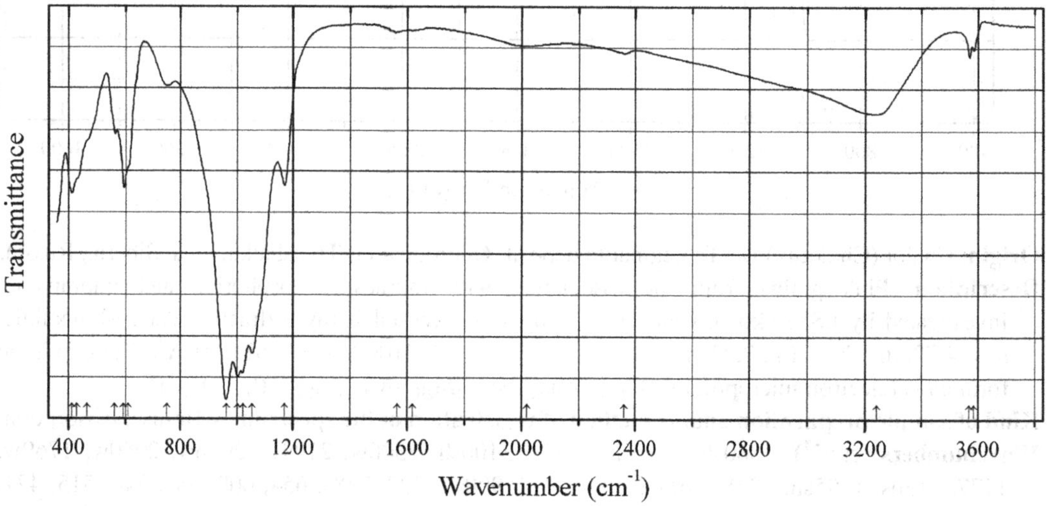

Origin: Kyz-Aul deposit, Naberezhnoe, Kerch Peninsula, Kerch iron-ore basin, Russia.

Description: Black crystalline crust from the association with leucophosphite. The empirical formula is (electron microprobe): $(Fe_{0.93}Mn_{0.03}Mg_{0.02}Ca_{0.02})Fe_4(PO_4)_{2.98}(SiO_4)_{0.02}(OH)_5$. The strongest lines of the powder X-ray diffraction pattern [*d*, Å (*I*, %)] are: 6.97 (29), 4.847 (28), 4.659 (21), 3.603 (31), 3.460 (34), 3.405 (43), 3.198 (43), 3.198 (100), 2.428 (33).

Kind of sample preparation and/or method of registration of the spectrum: KBr disc. Absorption.

Wavenumbers (cm^{-1}): 3585, 3569, 3241, 2360w, 2020w, 1620sh, 1566w, 1170, 1054s, 1022s, 1003s, 963s, 752, 610sh, 595, 563, 465sh, 430sh, 412.

Note: The spectrum was obtained by N.V. Chukanov.

P795 Sodium titanium phosphate $NaTi_2(PO_4)_3$

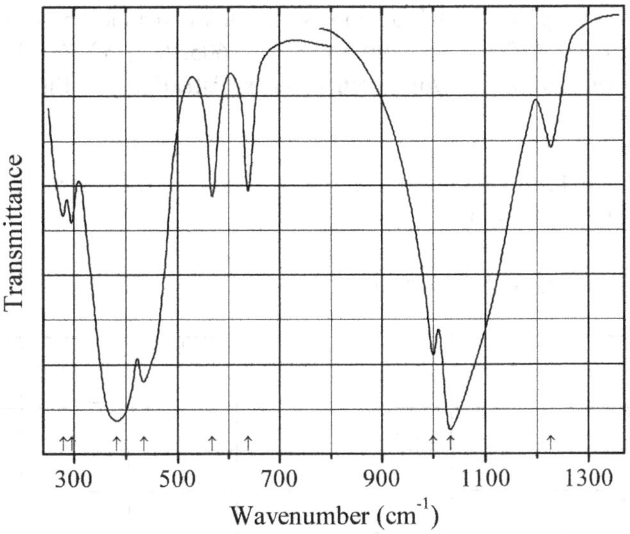

Origin: Synthetic.

Description: Synthesized from stoichiometric quantities of Na_2CO_3, TiO_2, and $(NH_4)_2(HPO_4)$ by conventional solid-state reaction techniques. Characterized by powder X-ray diffraction data. Hexagonal, space group *R*-3*c*, *Z* = 6

Kind of sample preparation and/or method of registration of the spectrum: KBr disc. Transmission.

Source: Tarte et al. (1986).

Wavenumbers (cm^{-1}): 1227, 1033s, 1000, 638, 568, 434, 382s, 295, 279.

P796 Strontiohurlbutite $SrBe_2(PO_4)_2$

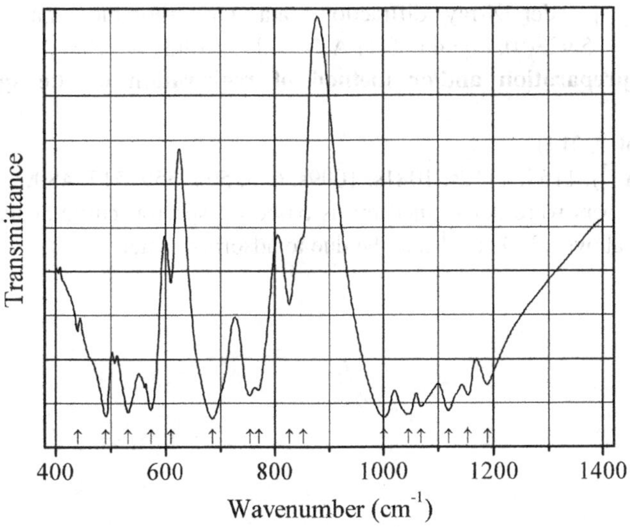

Origin: Synthetic.

Description: Synthesized hydrothermally from BeO, H_3PO_4, and $Sr(NO_3)_2$ at 200 °C for 7 days with subsequent rapid cooling. The crystal structure is solved. Monoclinic, space group $P2_1/c$, $a = 8.000(1)$, $b = 8.986(1)$, $c = 8.418(1)$ Å, $\beta = 90.22(1)°$, $V = 605.10(6)$ Å^3, $Z = 4$. $D_{calc} = 3.244$ g/cm^3.

Kind of sample preparation and/or method of registration of the spectrum: KBr disc. Transmission.

Source: Dal Bo et al. (2014).

Wavenumbers (cm^{-1}): 1189, 1153s, 1118s, 1068s, 1046s, 1001s, 853sh, 827, 770s, 754s, 686s, 609, 572s, 531s, 507, 491s, 440.

P798 Triphylite Mg-analogue $LiMg(PO_4)$

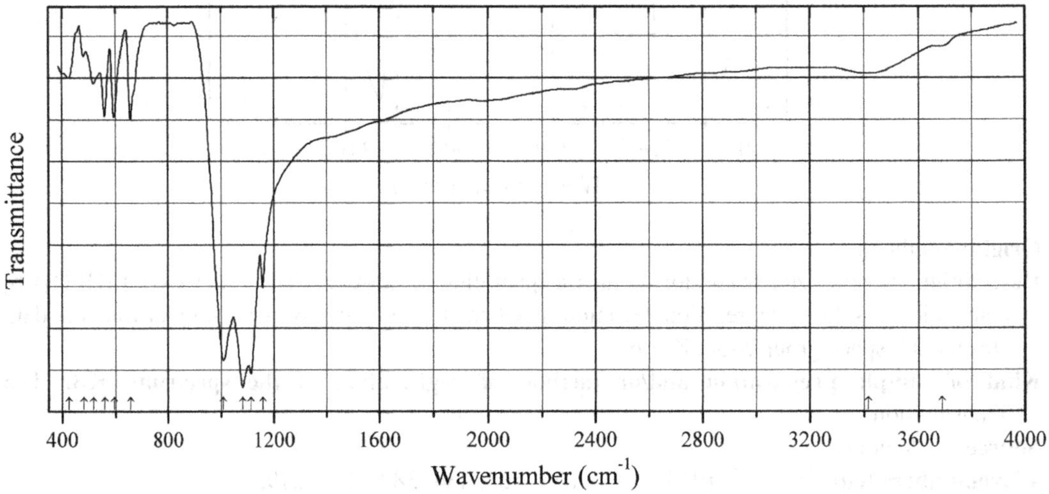

Origin: Synthetic.

Description: Obtained in a solid-state reaction between $(NH_4)Mg(PO_4)\cdot H_2O$ and $Li_2(CO_3)$. Characterized by powder X-ray diffraction data. Orthorhombic, space group *Pmn*21 (?), $a = 10.114(4)$, $b = 5.928(9)$, $c = 4.666(1)$ Å, $V = 279.813(3)$ Å^3, $Z = 2$.

Kind of sample preparation and/or method of registration of the spectrum: KBr disc. Transmission.

Source: Sronsri et al. (2014).

Wavenumbers (cm^{-1}): 1157, 1112s, 1081s, 1009s, 655, 594, 559, 517, 480w, 425.

Note: The wavenumbers were determined by us based on spectral curve analysis of the published spectrum. Bands above 3200 cm^{-1} may be due to adsorbed water.

P799 Tvrdýite $Fe^{2+}Fe^{3+}{}_2Al_3(PO_4)_4(OH)_5(H_2O)_4{\cdot}2H_2O$

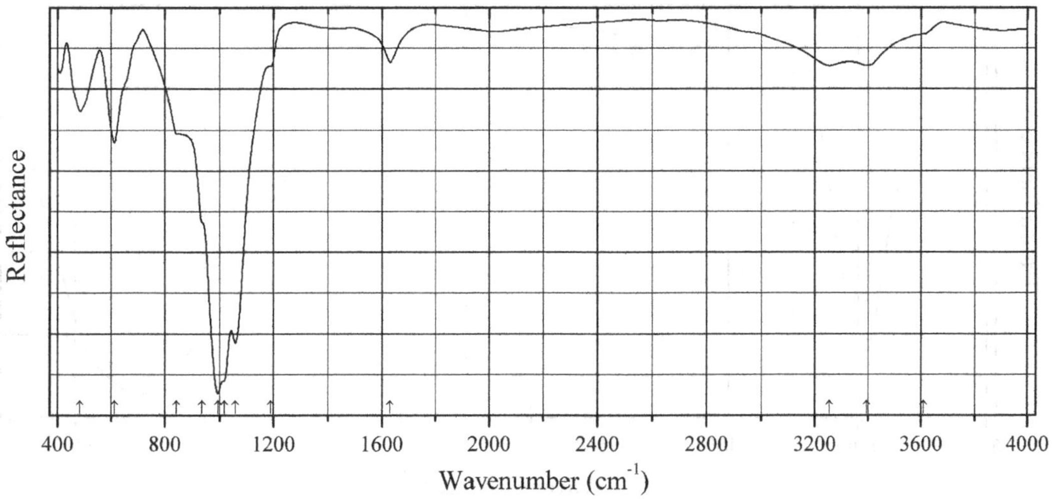

Origin: Krásno, near Horní Slavkov, Czech Republic (type locality).

Description: Aggregates of olive-greyish-green acicular crystals from the association with quartz, Al-rich beraunite, fluorapatite, and pharmacosiderite. Holotype sample. The crystal structure is solved. Triclinic, space group *C2/c*, $a = 20.564$, $b = 5.101(1)$, $c = 18.883(4)$ Å, $\beta = 93.68(3)°$, $V = 1976.7(7)$ Å^3, $Z = 4$. $D_{calc} = 2.834$ g/cm^3. Optically biaxial (–), $\alpha = 1.650(2)$, $\beta = 1.671(1)$, $\gamma = 1.677(1)$, $2V = 56(1)°$. The empirical formula based on electron microprobe analyses is $Zn_{0.52}Fe^{2+}{}_{0.50}Fe^{3+}{}_{2.21}Al_{2.75}(PO_4)_{3.86}(AsO_4)_{0.19}(OH)_{4.60}F_{0.23}{\cdot}nH_2O$. The strongest lines of the powder X-ray diffraction pattern [*d*, Å (*I*, %) (*hkl*)] are: 10.227 (100) (200), 9.400 (6) (002), 7.156 (14) (20–2), 5.120 (7) (400), 3.416 (11) (600), 3.278 (6) (60–2), 2.562 (5) (800), 2.0511 (3) (10.0.0).

Kind of sample preparation and/or method of registration of the spectrum: Attenuated total reflection of powdered mineral.

Source: Sejkora et al. (2016).

Wavenumbers (IR, cm^{-1}): 3610sh, 3394, 3255, 1631, 1191sh, 1058s, 1017sh, 994s, 936sh, 843sh, 613, 485.

Note: In the cited paper, Raman spectrum is given.

Wavenumbers (Raman, cm^{-1}): 1623, 1194, 1102, 1023s, 860s, 698,637, 586, 496, 415, 303, 281, 233, 143.

P800 Wilhelmgümbelite $ZnFe^{2+}Fe^{3+}{}_3(PO_4)_3(OH)_4{\cdot}7H_2O$

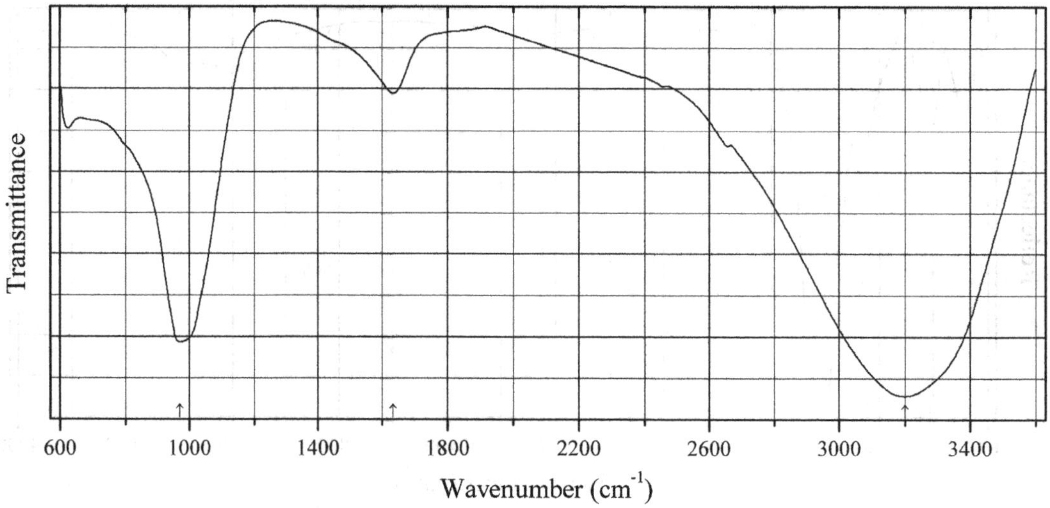

Origin: Hagendorf South pegmatite, Cornelia mine, Hagendorf, Waidhaus, Upper Palatinate, Bavaria, Germany (type locality).

Description: Radiating sprays of needle-like rectangular laths from the association with steinmetzite, chalcophanite, jahnsite, mitridatite, albite, apatite, muscovite, and quartz. Holotype sample. The crystal structure is solved. Orthorhombic, space group *Pmab*, $a = 10.987(7)$, $b = 25.378(13)$, $c = 6.387(6)$ Å, $V = 1781(2)$ Å^3, $Z = 4$. $D_{calc} = 2.82$ g/cm^3. Optically biaxial (+), $\alpha = 1.560(2)$, $\beta = 1.669(2)$, $\gamma = 1.718(2)$, $2V = 63(1)°$. The empirical formula is $Zn_{1.50}Mn^{2+}{}_{0.27}Fe^{2+}{}_{0.60}Fe^{3+}{}_{2.33}(PO_4)_3(OH)_{2.73}{\cdot}8.27H_2O$. The strongest lines of the powder X-ray diffraction pattern [*d*, Å (*I*, %) (*hkl*)] are: 12.65 (100) (020), 8.339 (5) (120), 6.421 (14) (001), 6.228 (8) (011), 4.223 (30) (120) and 2.111 (7) (0.12.0).

Kind of sample preparation and/or method of registration of the spectrum: Attenuated total reflection of powdered mineral.

Source: Grey et al. (2016c).

Wavenumbers (cm^{-1}): 3200s, 1635, 970s.

P801 Ximengite polymorph $Bi(PO_4)$

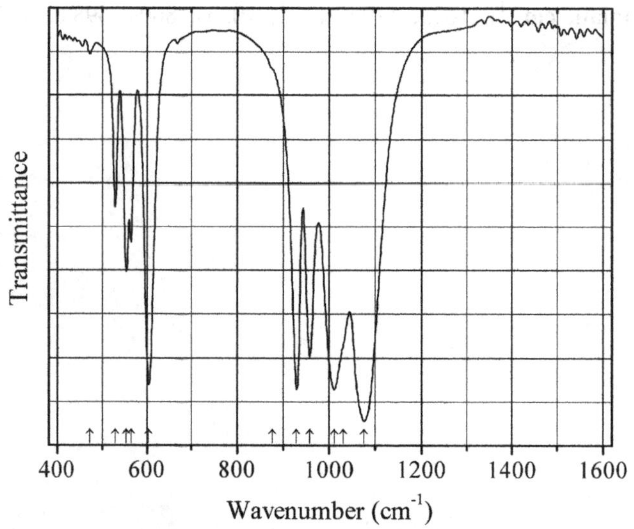

Origin: Synthetic.

Description: Obtained by heating trigonal $Bi(PO_4)$ (ximengite) at 673 K for 5 h. Characterized by powder X-ray and neutron diffraction. Monoclinic, space group $P2_1/n$, $a = 6.7552(1)$, $b = 6.9417(2)$, $c = 6.4772(2)$ Å, $\beta = 103.691(2)°$, $V = 295.10(1)$ Å^3, $Z = 4$.

Kind of sample preparation and/or method of registration of the spectrum: KBr disc. Transmission.

Source: Achary et al. (2013).

Wavenumbers (IR, cm^{-1}): 1076s, 1031sh, 1011s, 958s, 930s, 876sh, 604s, 564, 554, 529, 473w.

Note: In the cited paper, Raman spectrum is given.

Wavenumbers (Raman, cm^{-1}): 1050, 1039, 1021, 981, 970, 948, 926, 604, 598, 573, 557, 523, 496, 464, 457, 407, 388, 284, 273, 237, 230, 207, 183, 177, 170, 136, 131, 109, 97, 90, 70, 60, 51.

P802 Ximengite polymorph $Bi(PO_4)$

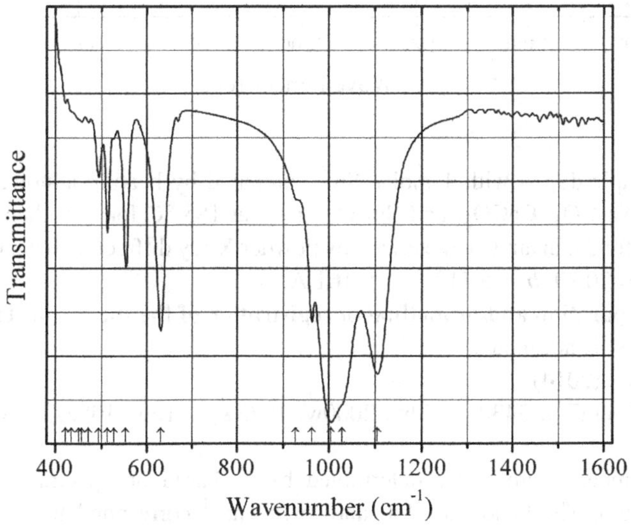

Origin: Synthetic.

Description: Obtained by heating trigonal $Bi(PO_4)$ (ximengite) at 973 K for 5 h. Characterized by powder X-ray and neutron diffraction data. Monoclinic, space group $P2_1/m$, $a = 4.8804(1)$, $b = 7.0684(2)$, $c = 4.7033(1)$ Å, $\beta = 96.285(3)°$, $V = 161.27(1)$ Å^3, $Z = 2$.

Kind of sample preparation and/or method of registration of the spectrum: KBr disc. Transmission.

Source: Achary et al. (2013).

Wavenumbers (IR, cm^{-1}): 1105s, 1029sh, 1005s, 963, 928sh, 630s, 554, 527sh, 513, 494, 472w, 457w, 451sh, 434sh, 421w.

Note: In the cited paper, Raman spectrum is given.

Wavenumbers (Raman, cm^{-1}): 1046, 1038, 983, 966, 610, 557, 548, 486, 354, 244, 214, 171, 144, 136, 92, 69, 56.

P803 Buchwaldite dimorph $NaCa(PO_4)$

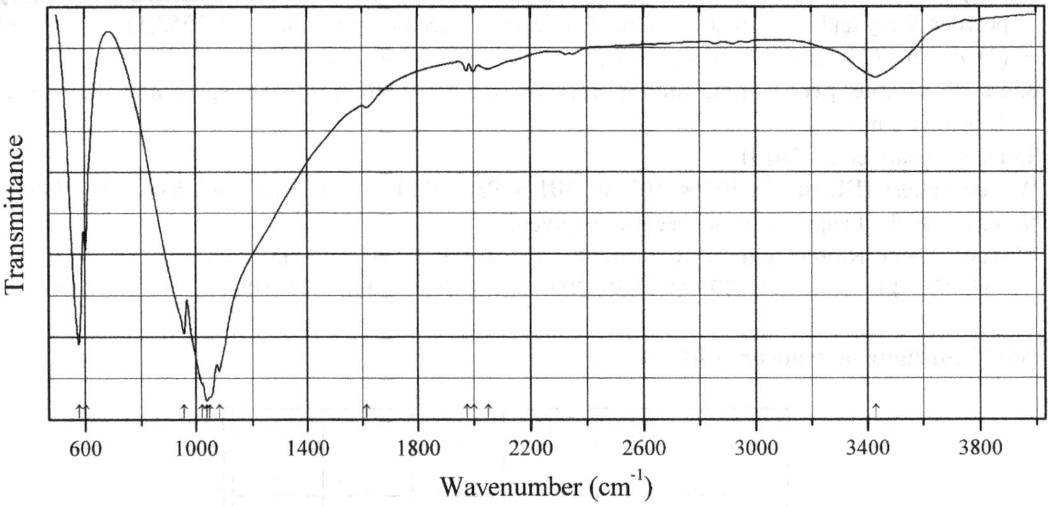

Origin: Synthetic.

Description: A sample doped with 1 mol% Sm^{3+} prepared by heating a stoichiometric mixture of $(NH_4)(H_2PO_4)$, Na_2CO_3, $CaCO_3$, and Sm_2O_3 firstly at 185 °C for 2 h, then 714 °C for 1 h and finally at 950 °C for 3 h in air. Characterized by powder X-ray diffraction data. Orthorhombic, space group $Pn2_1a$, $a = 20.39$, $b = 5.412$, $c = 9.161$ Å.

Kind of sample preparation and/or method of registration of the spectrum: Transmission. Kind of sample preparation is not indicated.

Source: Ratnam et al. (2014).

Wavenumbers (IR, cm^{-1}): 3430, 2050w, 2000w, 1975w, 1616w, 1082s, 1049sh, 1038s, 1022sh, 957s, 602, 578s.

Note: The wavenumbers were partly determined by us based on spectral curve analysis of the published spectrum. The bands at 3430 and 1616 cm^{-1} correspond to the admixture of water molecules. For the IR spectrum of Eu-doped buchwaldite dimorph see also Grandhe et al. (2012). In the cited paper, Raman spectrum is given.

Wavenumbers (Raman, cm^{-1}): 1158, 965s, 898.

P804 Nalipoite $NaLi_2(PO_4)$

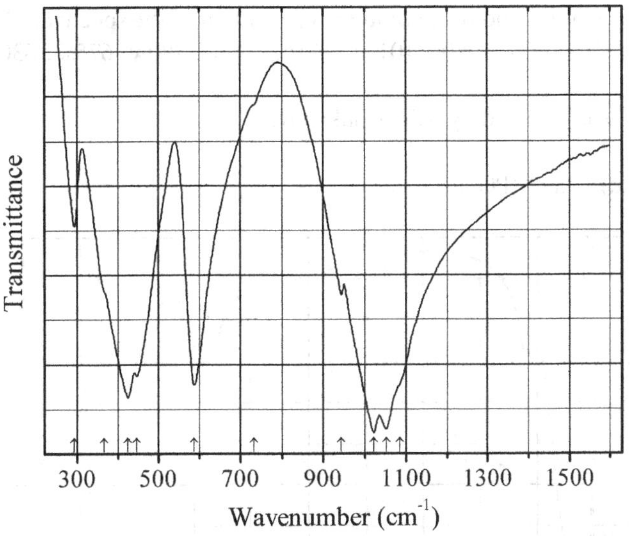

Origin: Synthetic.

Description: Obtained by mixing aqueous solutions containing stoichiometric amounts of NaOH, H_3PO_4, and LiOH, followed by drying at 100 °C in air. Characterized by powder X-ray diffraction data. Orthorhombic, space group *Pmnb*, a = 6.8751(1), b = 9.9888(3), c = 4.9315(6) Å, V = 338.66(8) Å^3.

Kind of sample preparation and/or method of registration of the spectrum: Transmission. Kind of sample preparation is not indicated.

Source: López et al. (2014a).

Wavenumbers (cm^{-1}): 1085sh, 1052s, 1022s, 944, 733sh, 588s, 447, 422s, 367sh, 293.

Note: The wavenumbers were determined by us based on spectral curve analysis of the published spectrum.

P806 Fluorcarmoite-(BaNa) Ba□Na_2Na_2□$CaMg_{13}Al(PO_4)_{11}(PO_3OH)F_2$

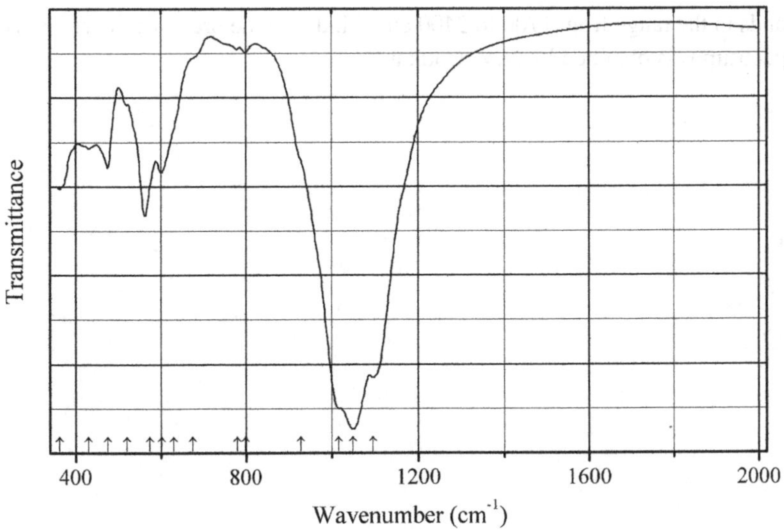

Origin: Costa Balzi Rossi, Magliolo, Savona, Liguria, Italy (type locality).
Description: Yellow grain.
Kind of sample preparation and/or method of registration of the spectrum: KBr disc. Absorption.
Wavenumbers (cm^{-1}): 1097s, 1050s, 1017s, 930sh, 799w, 780w, 675sh, 630sh, 602, 574, 520w, 475, 430, 363.
Note: The spectrum was obtained by N.V. Chukanov.

P807 Florencite-(Nd) $NdAl_3(PO_4)_2(OH)_6$

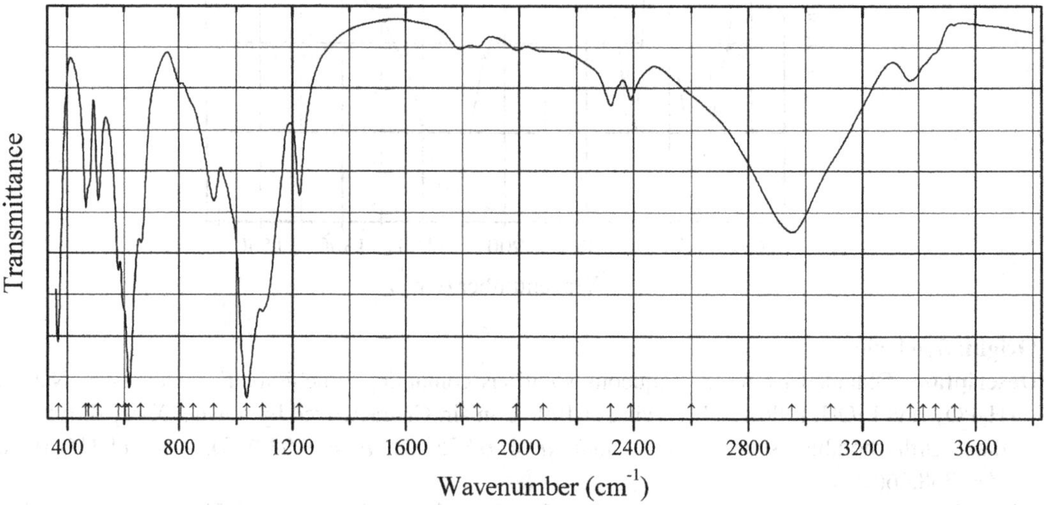

Origin: Svodovy area, Maldynyrd Ridge, Subpolar Urals, Russia.
Description: Pink crystals from the association with xenotime-(Y) and quartz. The empirical formula is (electron microprobe): $(Nd_{0.36}Sm_{0.23}Ce_{0.23}La_{0.05}Pr_{0.05}Sr_{0.05}Ca_{0.01})Al_{1.99}Fe_{0.02}(PO_4)_{2.00}(OH,H_2O)_6$.
Kind of sample preparation and/or method of registration of the spectrum: KBr disc. Absorption.
Wavenumbers (cm^{-1}): 3527w, 3460sh, 3415sh, 3370, 3090sh, 2951s, 2600sh, 2389, 2320, 2085w, 1991w, 1853w, 1792w, 1224, 1092s, 1036s, 923, 850sh, 806w, 660, 619s, 605sh, 581, 510, 475sh, 466, 368s.
Note: The bands in the range from 1700 to 2400 cm^{-1} indicate the presence of acid phosphate groups.
Note: The spectrum was obtained by N.V. Chukanov.

P808 Lulzacite $Sr_2Fe^{2+}{}_3Al_4(PO_4)_4(OH)_{10}$

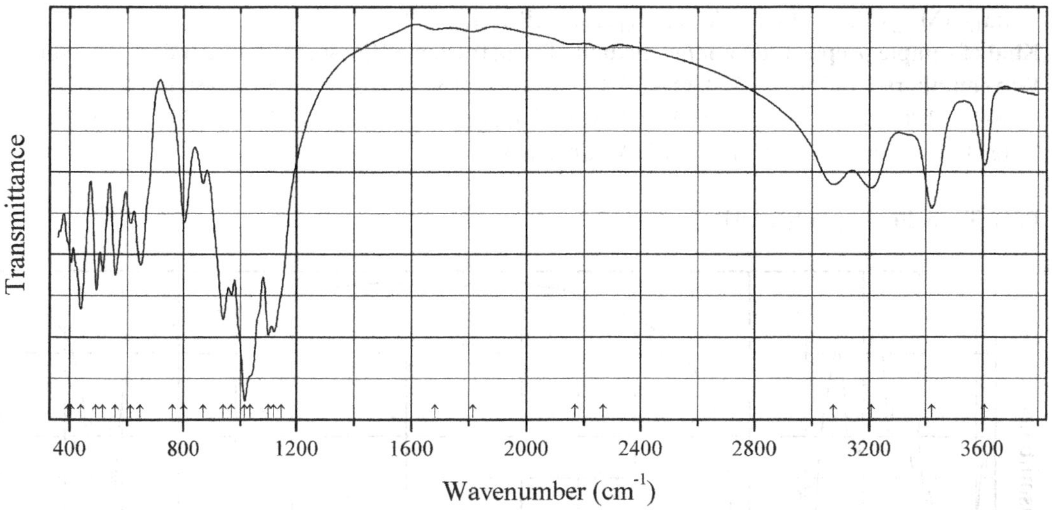

Origin: Bois-de-la-Roche quarry, Saint-Aubin-des-Châteaux, Loire-Atlantique, Pays de la Loire, France (type locality).

Description: Light greenish-gray columnar aggregate. The empirical formula is (electron microprobe): $(Sr_{1.9}Ca_{0.1})(Fe_{1.9}Mg_{0.9}Zn_{0.1}Mn_{0.1})(Al_{3.7}Fe_{0.3})(PO_4)_{4.0}(OH)_{10}$.

Kind of sample preparation and/or method of registration of the spectrum: KBr disc. Absorption.

Wavenumbers (cm^{-1}): 3607, 3422s, 3209, 3074, 2270w, 2170w, 1814w, 1683w, 1145sh, 1120s, 1099s, 1035sh, 1015s, 968, 939s, 869, 803, 760sh, 648, 613, 560, 516, 493, 438s, 404, 395sh.

Note: The spectrum was obtained by N.V. Chukanov.

P809 Penikisite $BaMg_2Al_2(PO_4)_3(OH)_3$

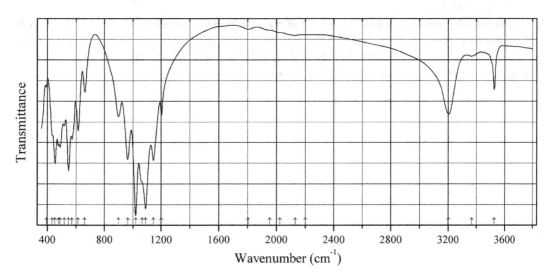

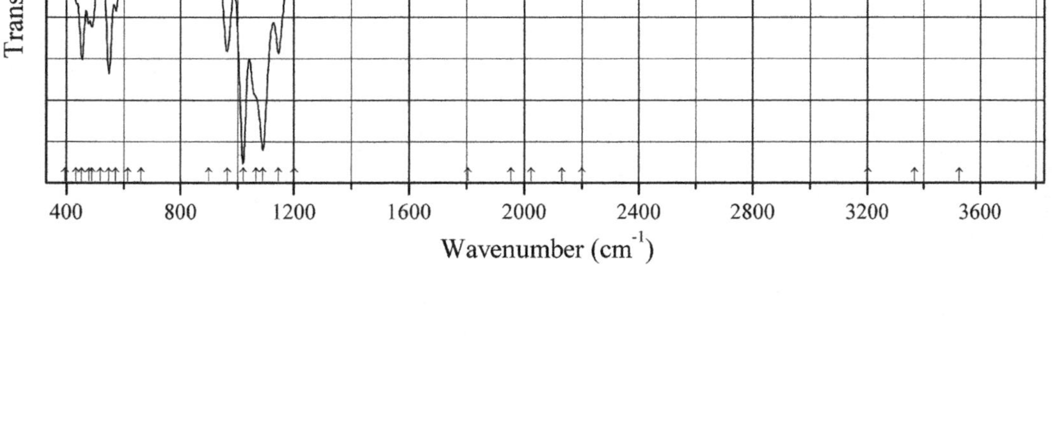

Origin: Blow River, Yukon, Canada.

Description: Blue crystals. The empirical formula is (electron microprobe): $Ba_{1.01}(Mg_{0.99}Fe_{0.92}Mn_{0.08})(Al_{1.84}Fe_{0.13}Ti_{0.03})(PO_4)_{3.00}(OH)_3$.

Kind of sample preparation and/or method of registration of the spectrum: KBr disc. Absorption.

Wavenumbers (cm^{-1}): 3525, 3368w, 3203, 2200sh, 2130w, 2025w, 1953w, 1805w, 1202, 1146s, 1090s, 1065sh, 1019s, 963s, 899, 662, 616, 573, 548s, 519, 489, 480, 454s, 435sh, 395w.

Note: The spectrum was obtained by N.V. Chukanov.

P810 Trolleite $Al_4(PO_4)_3(OH)_3$

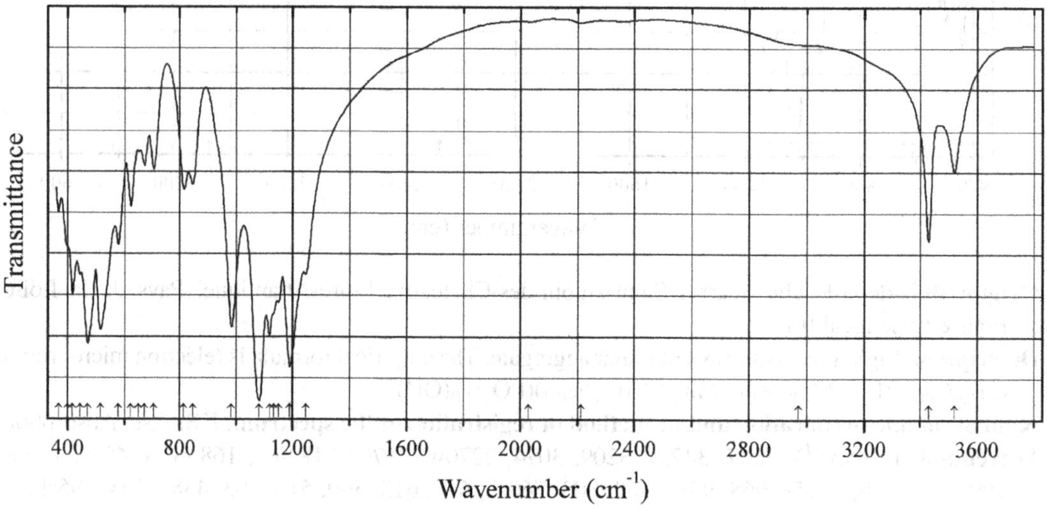

Origin: Hålsjöberg (Horrsjöberg), Torsby, Värmland, Sweden.

Description: Pale bluish-green grains from the association with scorzalite, kyanite, and rutil. Confirmed by the IR spectrum.

Kind of sample preparation and/or method of registration of the spectrum: KBr disc. Absorption.

Wavenumbers (cm^{-1}): 3519, 3430, 2970w, 2213w, 2030w, 1247, 1190s, 1150, 1133, 1118, 1081s, 984, 847, 815, 706, 671, 650w, 624, 580, 516s, 471s, 444, 417, 400sh, 369.

Note: The spectrum was obtained by N.V. Chukanov.

P811 Petitjeanite $Bi_3O(PO_4)_2(OH)$

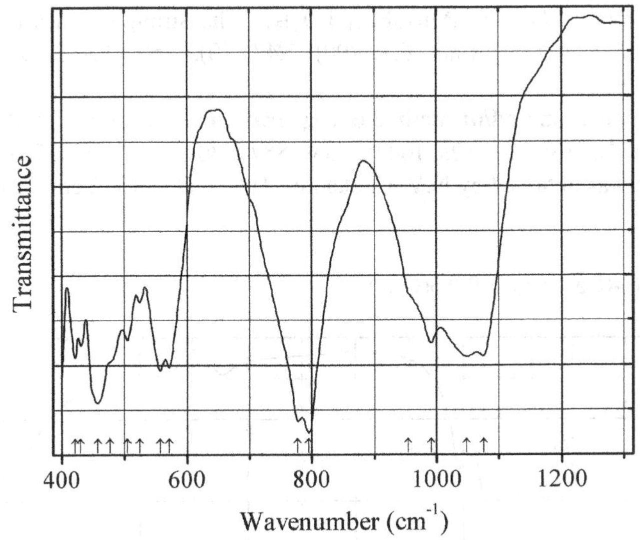

Origin: Schneeberg District, Erzgebirge (Ore Mts.), Saxony, Germany.

Description: Spherulitic crust. The empirical formula is (electron microprobe): $(Bi_{2.85}Pb_{0.1}Ca_{0.1})[(PO_4)_{0.9}(AsO_4)_{0.4}(VO_4)_{0.3}](OH)$.

Kind of sample preparation and/or method of registration of the spectrum: KBr disc. Absorption.

Wavenumbers (cm^{-1}): 1076s, 1049s, 992, 955sh, 795s, 777s, 570, 556, 523, 504, 477sh, 457s, 430, 421.

Note: The spectrum was obtained by N.V. Chukanov.

P812 Varulite $NaCaMn^{2+}_3(PO_4)_3$

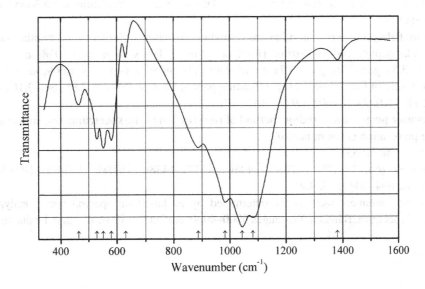

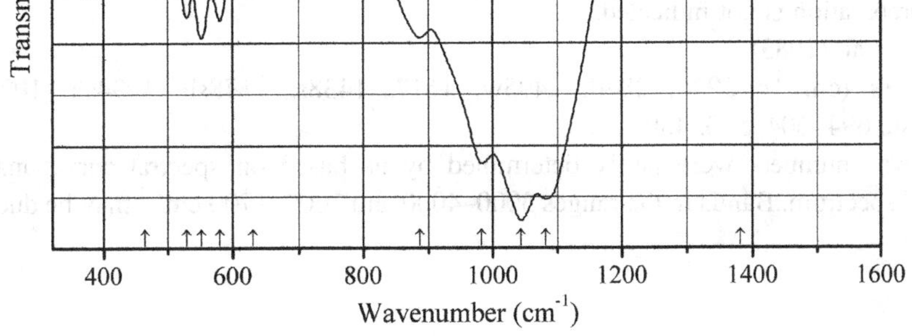

Origin: Solleftea, Ångermanland, Sweden.

Description: Anhedral grains. Ca-deficient variety or analogue of varulite. The empirical formula is (electron microprobe): $Na_{1.5}Ca_{0.3}Mn_{2.4}Fe_{0.8}(PO_4)_{3.0}$. The strongest lines of the powder X-ray diffraction pattern [*d*, Å (*I*, %)] are: 6.12 (90), 5.47 (40), 3.50 (70), 3.146 (100), 2.736 (100), 2.560 (30).

Kind of sample preparation and/or method of registration of the spectrum: KBr disc. Absorption.

Wavenumbers (cm^{-1}): (1382), 1082s, 1044s, 983s, 887, 630w, 580, 552s, 530, 465.

Note: The spectrum was obtained by N.V. Chukanov. The band at 1382 cm^{-1} may correspond to an impurity.

P813 Daqingshanite-(Ce) $Sr_3Ce(PO_4)(CO_3)_3$

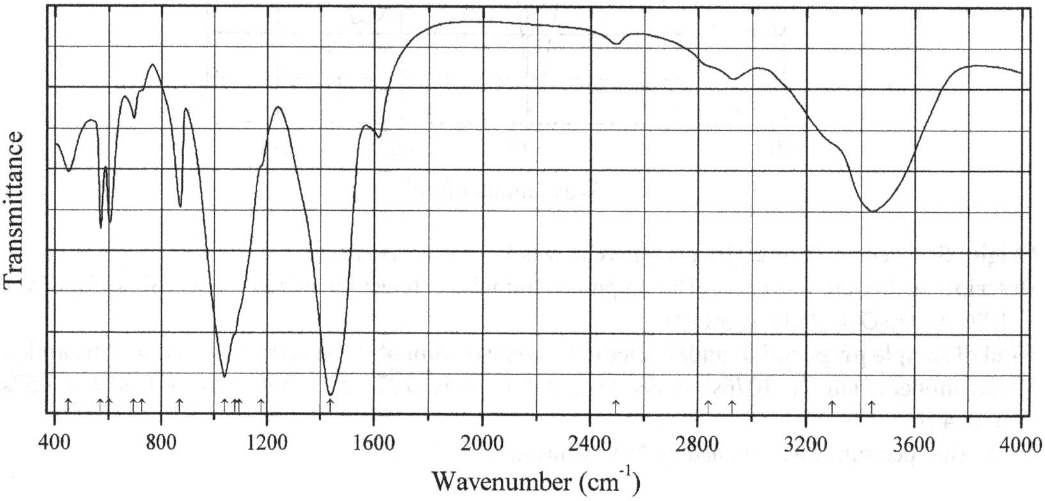

Origin: Bayan Obo deposit, Bayan Obo Mining District, Baotou Prefecture, Inner Mongolia, China (type locality).

Description: Pale yellow crystalsand grains from the association with benstonite, huntite, strontianite, pyrite, phlogopite, and monazite. Holotype sample. Trigonal, a = 10.058, c = 9.225 Å. D_{meas} = 3.81 g/cm^3, D_{calc} = 3.71 g/cm^3. Optically uniaxial (−), ε = 1.609, ω = 1.708. The strongest lines of the powder X-ray diffraction pattern [*d*, Å (*I*, %)] are: 3.95 (60), 3.16 (100), 2.52 (70), 2.110 (50), 2.040 (60), 1.941 (60).

Kind of sample preparation and/or method of registration of the spectrum: Transmission. Kind of sample preparation is not indicated.

Source: Ren et al. (1983).

Wavenumbers (cm^{-1}): 2930, 2840, 2495w, 1617, 1438s, 1178sh, 1094sh, 1078sh, 1040s, 872, 724sh, 694, 604, 570, 450.

Note: The wavenumbers were partly determined by us based on spectral curve analysis of the published spectrum. Bands in the ranges 3000–4000 and 1600–1700 cm^{-1} may be due to absorbed water.

P814 Sodium calcium silicophosphate $Na_2Ca_4(PO_4)_2SiO_4$ (apatite-type) $Na_2Ca_4(PO_4)_2SiO_4$

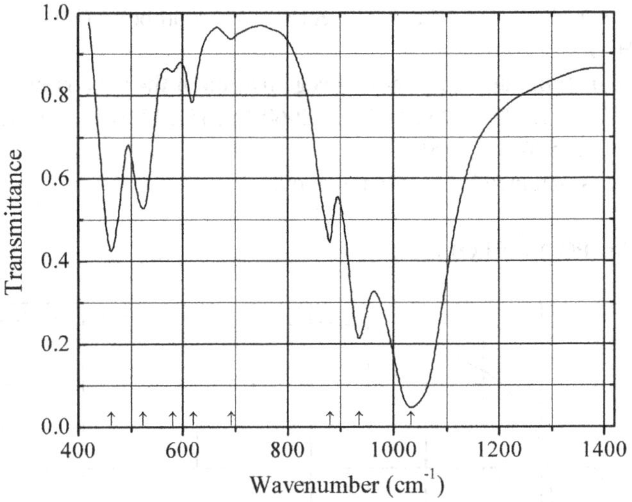

Origin: Synthetic.

Description: Fine powder. Characterized by thermal and powder X-ray diffraction data.

Kind of sample preparation and/or method of registration of the spectrum: Absorption. Kind of sample preparation is not indicated.

Source: Pirayesh and Nychka (2013).

Wavenumbers (cm^{-1}): 1033s, 935s, 879, 692w, 619, 580w, 523, 462.

Note: The wavenumbers were determined by us based on spectral curve analysis of the published spectrum.

P815 Mangangordonite $Mn^{2+}Al_2(PO_4)_2(OH)_2{\cdot}8H_2O$

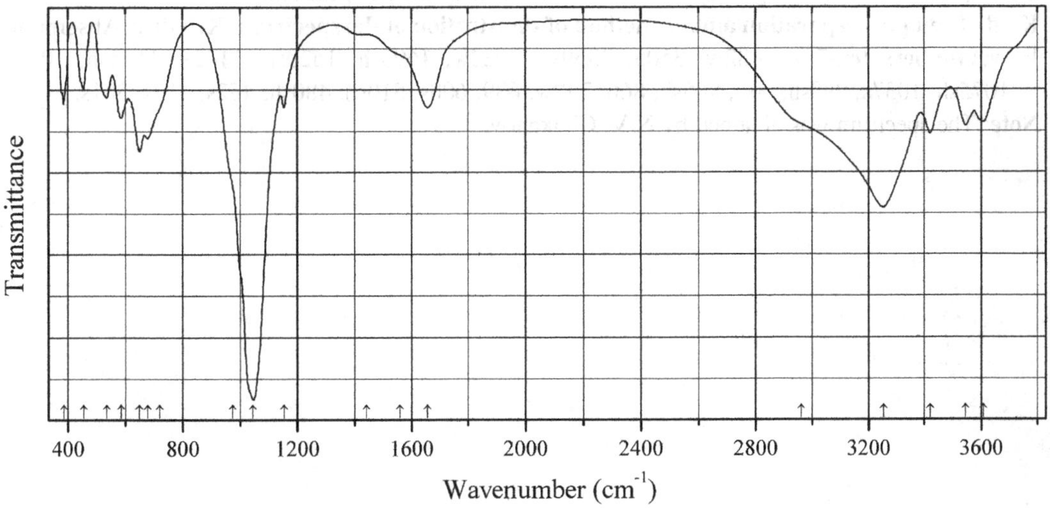

Origin: Foote Mine, Kings Mountain, Cleveland Co., North Carolina, USA (type locality).
Description: Pale yellow prismatic crystals from the association with whiteite-(MnFeMg) and birnessite. The empirical formula is (electron microprobe): $(Mn_{0.7}Fe_{0.2}Mg_{0.1})(Al_{0.8}Fe_{0.2})(PO_4)_{2.0}(OH)_2{\cdot}8H_2O$.
Kind of sample preparation and/or method of registration of the spectrum: KBr disc. Absorption.
Wavenumbers (cm^{-1}): 3607, 3545, 3420, 3253s, 2960sh, 1657, 1560sh, 1442w, 1157, 1045s, 975sh, 720sh, 679, 650s, 587, 536, 456, 386.
Note: The spectrum was obtained by N.V. Chukanov.

P817 Althausite $Mg_4(PO_4)_2(OH,O)(F,\square)$

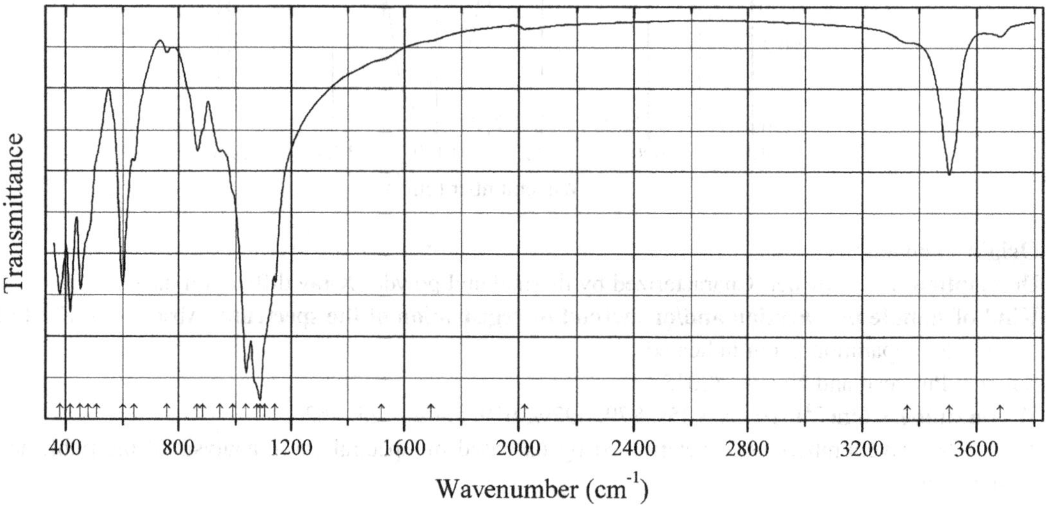

Origin: Tingelstadtjern quarry, Modum, Buskerud, Norway (type locality).
Description: Brownish single-crystal grain. Confirmed by the IR spectrum.
Kind of sample preparation and/or method of registration of the spectrum: KBr disc. Absorption.
Wavenumbers (cm^{-1}): 3682w, 3505, 3359w, 2022w, 1695sh, 1520sh, 1142s, 1105sh, 1087s, 1075sh, 1037s, 990sh, 946, 885sh, 866, 759w, 639, 600s, 510sh, 480sh, 452s, 417s, 381s.
Note: The spectrum was obtained by N.V. Chukanov.

P818 Cu,Al-hydroxyphosphate $CuAl_5(PO_4)(OH)_{13}F \cdot nH_2O$

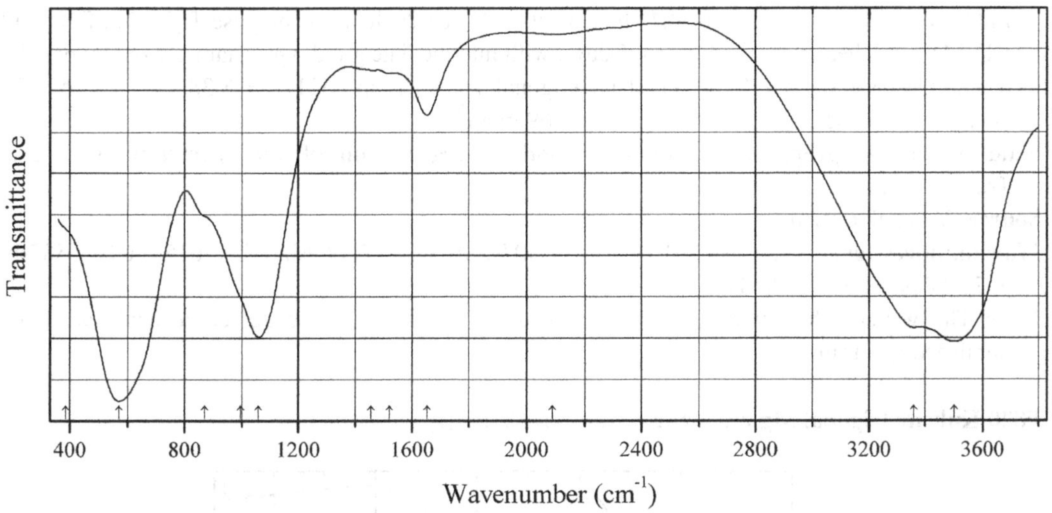

Origin: West Caradon Adit, Cornwall, GB.

Description: Blue collophorm crust from the association with fluorite. X-ray amorphous. The empirical formula is $(Cu_{0.89}Mg_{0.03}Zn_{0.02}Cu_{0.02})Al_{5.03}[(PO_4)_{0.39}(SiO_{34})_{0.33}(AsO_4)_{0.18}(SO_4)_{0.12}](OH)_{12.74}F_{1.00} \cdot nH_2O$.

Kind of sample preparation and/or method of registration of the spectrum: KBr disc. Absorption.

Wavenumbers (cm^{-1}): 3500s, 3360s, 2090w, 1653, 1520w, 1455w, 1059s, 995sh, 870sh, 571s, (385sh).

Note: The spectrum was obtained by N.V. Chukanov.

P819 Kuksite trigonal dimorph $Pb_3Zn_3TeO_6(PO_4)_2$

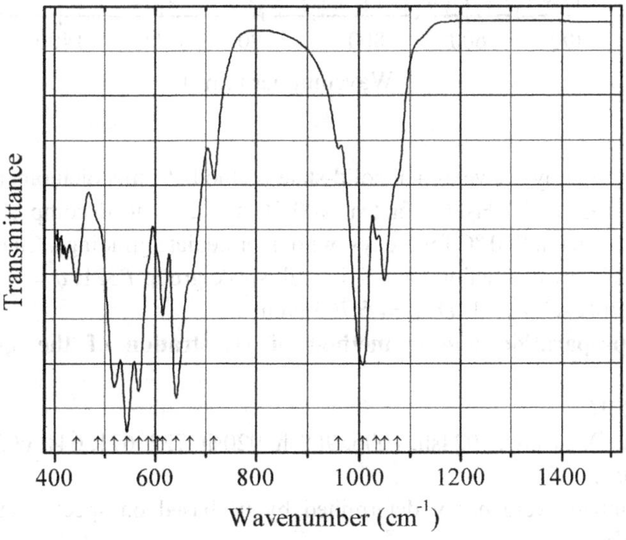

Origin: Synthetic.

Description: Synthesized by conventional solid-state methods from stoichiometric amounts of PbO, ZnO, $H_2TeO_4{\cdot}2H_2O$, and KH_2PO_4 first at 400 °C for 20 h to decompose $H_2TeO_4{\cdot}2H_2O$ and KH_2PO_4, and thereafter at 700 °C for 5 days, with intermediate grindings. Characterized by powder and single-crystal X-ray diffraction data. Trigonal, space group *P*321, $a = 8.3831(3)$, $c = 5.1930(4)$ Å, $V = 316.05(3)$ Å^3, $Z = 1$. $D_{calc} = 6.469$ g/cm^3.

Kind of sample preparation and/or method of registration of the spectrum: KBr disc. Transmission.

Source: Yu et al. (2016).

Wavenumbers (cm^{-1}): 1075sh, 1050, 1034, 1007s, 961w, 717, 669sh, 642s, 615, 604sh, 587sh, 567s, 544s, 519s, 486sh, 443.

Note: The wavenumbers were partly determined by us based on spectral curve analysis of the published spectrum.

P820 Kuksite trigonal Mg analogue $Pb_3Mg_3TeO_6(PO_4)_2$

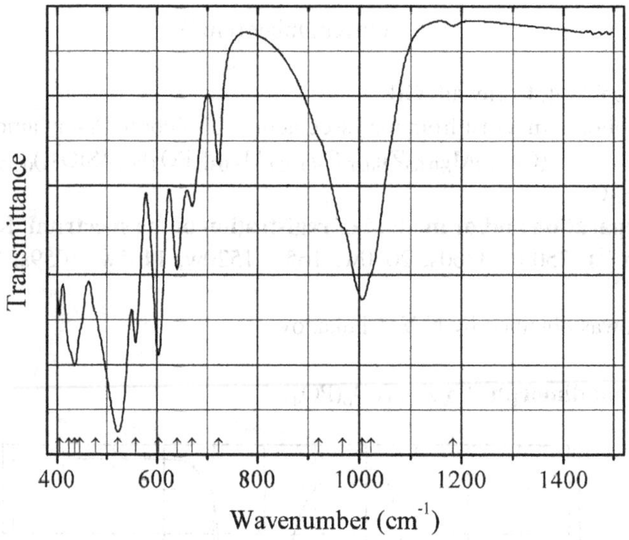

Origin: Synthetic.

Description: Synthesized by conventional solid-state methods from stoichiometric amounts of PbO, MgO, $H_2TeO_4{\cdot}2H_2O$, and KH_2PO_4 first at 400 °C for 20 h to decompose $H_2TeO_4{\cdot}2H_2O$ and KH_2PO_4, and thereafter at 850 °C for 5 days, with intermediate grindings. Characterized by powder and single-crystal X-ray diffraction data. Trigonal, space group *P*321, $a = 8.4072(4)$, $c = 5.2158(5)$ Å, $V = 319.27(4)$ Å^3, $Z = 1$. $D_{calc} = 5.763$ g/cm^3.

Kind of sample preparation and/or method of registration of the spectrum: KBr disc. Transmission.

Source: Yu et al. (2016).

Wavenumbers (cm^{-1}): 1183w, 1024sh, 1006, 968sh, 920sh, 722, 671, 641, 603s, 557, 522s, 477sh, 447sh, 436s, 425sh, 405.

Note: The wavenumbers were partly determined by us based on spectral curve analysis of the published spectrum.

P821 Natrophilite $NaMn^{2+}(PO_4)$

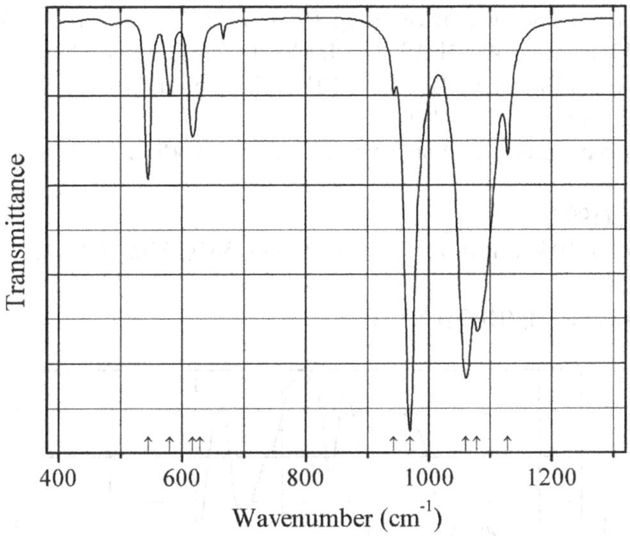

Origin: Synthetic.

Description: Prepared hydrothermally from $KMnPO_4 \cdot H_2O$ and $NaCH_3COO \cdot 3H_2O$ at a ratio of 1:10 at 200 °C for 15 h. Characterized by powder X-ray diffraction data. Orthorhombic, $a = 10.5177(3)$, $b = 6.3144(2)$, $c = 4.9873(2)$ Å, $V = 331.227(22)$ Å^3.

Kind of sample preparation and/or method of registration of the spectrum: KBr disc. Absorption.

Source: Boyadzhieva et al. (2015).

Wavenumbers (IR, cm^{-1}): 1129, 1078s, 1060s, 969s, 943w, 630sh, 618, 580, 545.

Note: In the cited paper, Raman spectrum is given.

Wavenumbers (Raman, cm^{-1}): 1048w, 1006w, 946s, 650s, 577.

P822 Paganoite phosphate analogue $NiBi^{3+}O(PO_4)$

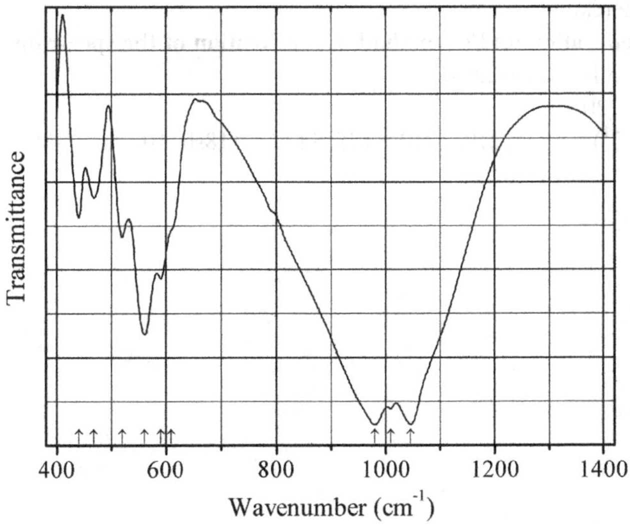

Origin: Synthetic.

Description: Prepared by solid-state reaction from Bi_2O_3, CoO, and $(NH_4)_2(HPO_4)$. The crystal structure is solved. Monoclinic, space group $P2_1/n$, a = 7.2470(1), b =11.2851(2), c = 5.2260 (1) Å, β = 107.843(1)°, V = 406.91 Å^3, Z = 4. The strongest lines of the powder X-ray diffraction pattern [d, Å (I, %) (hkl)] are: 4.727 (44) (−101), 4.338 (69) (120), 3.372 (70) (111), 2.850 (100) (−221), 2.568 (43) (131), 2.516 (41) (230).

Kind of sample preparation and/or method of registration of the spectrum: KBr disc. Transmission.

Source: Ketani et al. (1999).

Wavenumbers (cm^{-1}): 1046s, 1009s, 979s, 609sh, 590, 561s, 520, 468, 440.

P823 Phosphorrösslerite $Mg(HPO_4)\cdot 7H_2O$

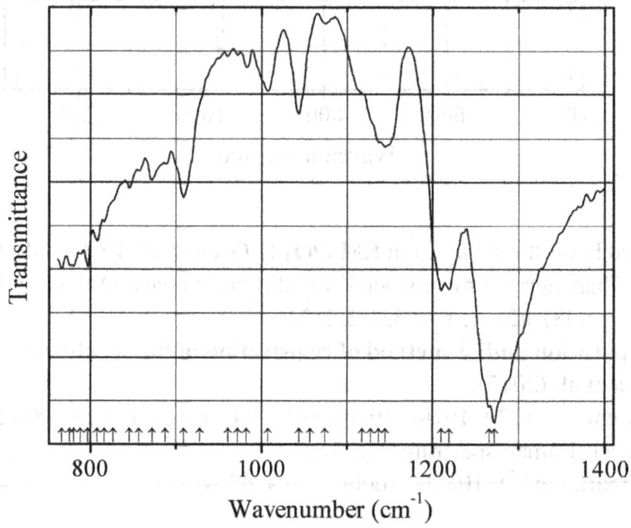

Origin: Synthetic.

Description: Commercial reactant (?).

Kind of sample preparation and/or method of registration of the spectrum: Absorption. Kind of sample preparation is not indicated.

Source: Pucka et al. (2000).

Wavenumbers (cm^{-1}): 1272, 1219, 1210, 1145, 1137, 1118sh, 1074w, 1044, 1007, 982w, 909, 887, 871, (796).

P824 Potassium zinc hydrogen phosphate $KZn_2(PO_4)(PO_3OH)$

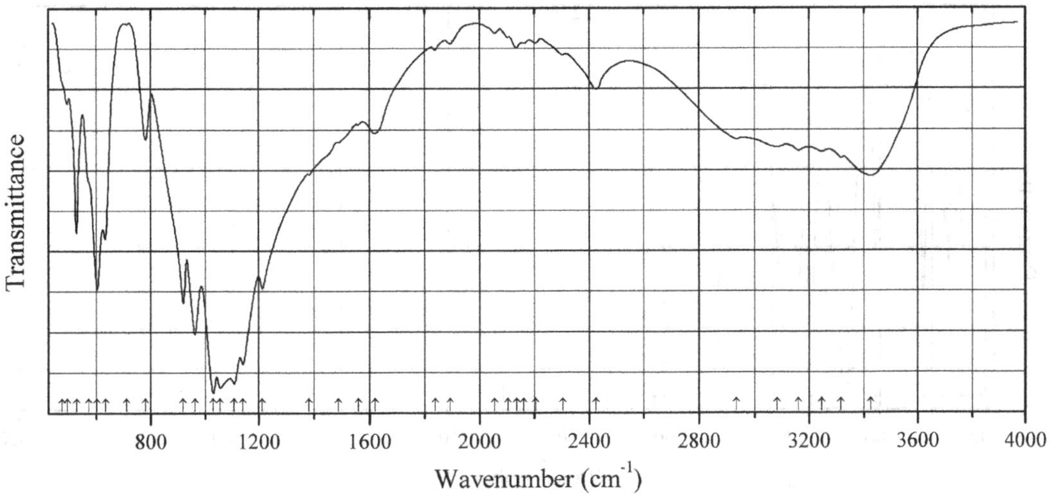

Origin: Synthetic.

Description: The sample may contain $KZn(PO_4)$ impurity.

Kind of sample preparation and/or method of registration of the spectrum: KBr disc. Transmission.

Source: Alibakhshi et al. (2012).

Wavenumbers (IR, cm^{-1}): 3424, 3316, 3247, 3160, 3081, 2938, 2426, 2304w, 2204w, 2161w, 2136w, 2103w, 2055w, 1894w, 1838w, 1618, 1559, 1488, 1383, 1212, 1140s, 1108s, 1056s, 1031s, 963s, 920s, 782, 712, 634, 604s, 574sh, 528, 493w, 474sh.

Note: The wavenumbers were partly determined by us based on spectral curve analysis of the published spectrum. In the cited paper, Raman spectrum is given.

Wavenumbers (Raman, cm^{-1}): 3429s (broad), 3315, 3093, 2451, 2399, 2162w, 1979w, 1874w, 1813w, 1322w, 1240w, 1137, 1074, 1013s, 910s, 766w, 590, 490, 303s.

P825 Raadeite $Mg_7(PO_4)_2(OH)_8$

Origin: Tingelstadtjern quarry, Modum, Buskerud, Norway (type locality).

Description: Anhedral inclusion in holtedahlite. Holotype sample. The crystal structure is solved. Monoclinic, space group $P2_1/n$, $a = 5.250(1)$, $b = 11.647(2)$, $c = 9.655(2)$ Å, $\beta = 95.94(1)°$, $Z = 2$. Optically biaxial (−), $\alpha = 1.5945(5)$, $\beta = 1.6069(5)$, $\gamma = 1.6088(5)$, $2V = 45.6(1)°$.

Kind of sample preparation and/or method of registration of the spectrum: Reflection of a single-crystal grain.

Source: Chopin et al. (2001).

Wavenumbers (cm^{-1}): 3580, 3540, 3475, 3375.

P826 Vyacheslavite $U^{4+}(PO_4)(OH)\cdot 2.5H_2O$

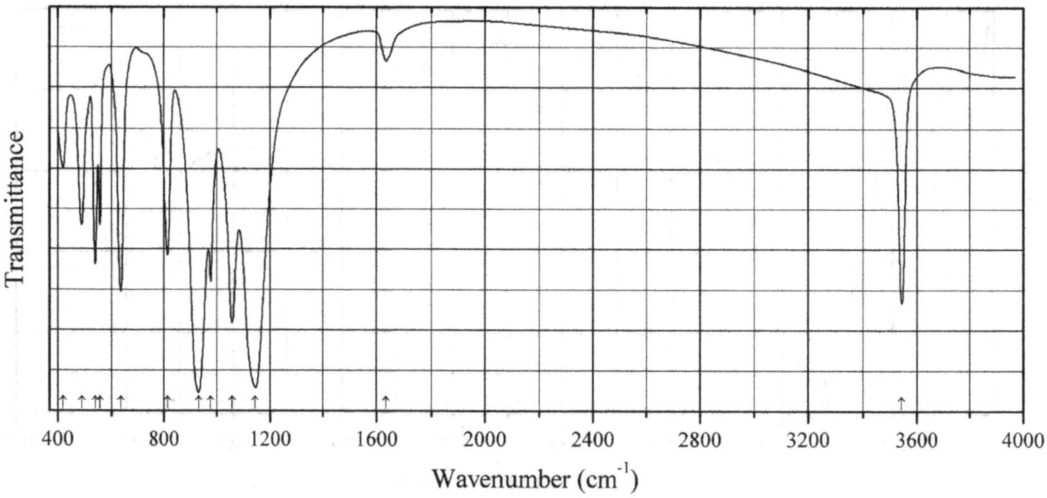

Origin: Synthetic.

Description: Prepared hydrothermally using hydrolyzed uranium bromide phosphate. Characterized by powder X-ray diffraction data.

Kind of sample preparation and/or method of registration of the spectrum: KBr disc. Transmission.

Source: Brandel et al. (2001).

Wavenumbers (cm^{-1}): 3544, 1636w, 1144s, 1058s, 978, 932s, 814, 636s, 558, 540, 490, 420.

P827 Vyacheslavite anhydrous Th analogue $Th^{4+}(PO_4)(OH)$

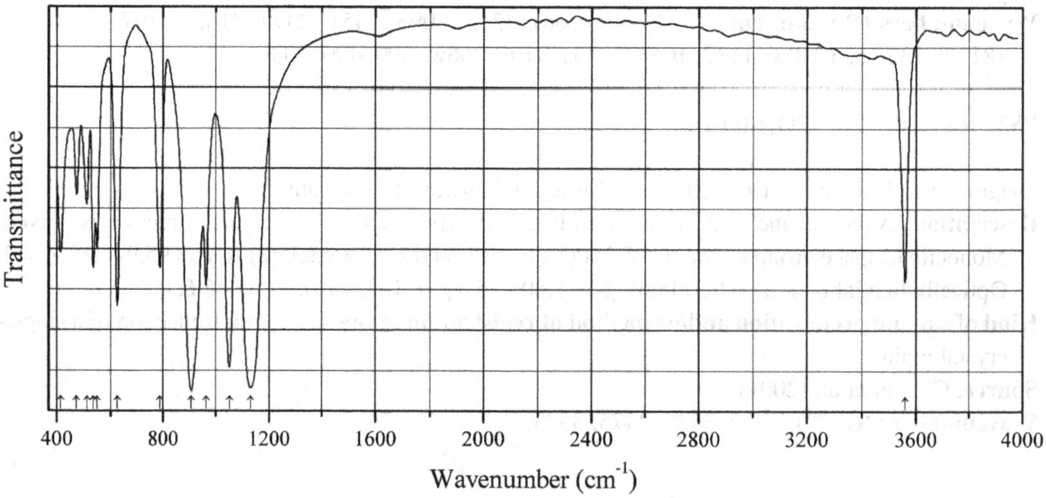

Origin: Synthetic.

Description: Characterized by powder X-ray diffraction data. The crystal structure is solved. Orthorhombic, space group *Cmca*, $a = 7.1393(2)$, $b = 9.2641(2)$, $c = 12.5262(4)$ Å, $V = 828.46(4)$ Å^3, $Z = 8$.

Kind of sample preparation and/or method of registration of the spectrum: KBr disc. Transmission.

Source: Brandel et al. (2001), Dacheux et al. (2007).

Wavenumbers (IR, cm^{-1}): 3560, 1130s, 1052s, 964, 908s, 787, 626, 550, 536, 512, 474, 414.

Note: Raman spectrum is given by Dacheux et al. (2007).

Wavenumbers (Raman, cm^{-1}): 3568s, 1195w, 1078s, 1060s, 989s, 799w, 789w, 618, 568, 556, 449, 416, 368, 282, 236.

P828 Fupingqiuite $(Na,Mn^{2+},\square)_2Mn^{2+}{}_2Fe^{3+}(PO_4)_3$

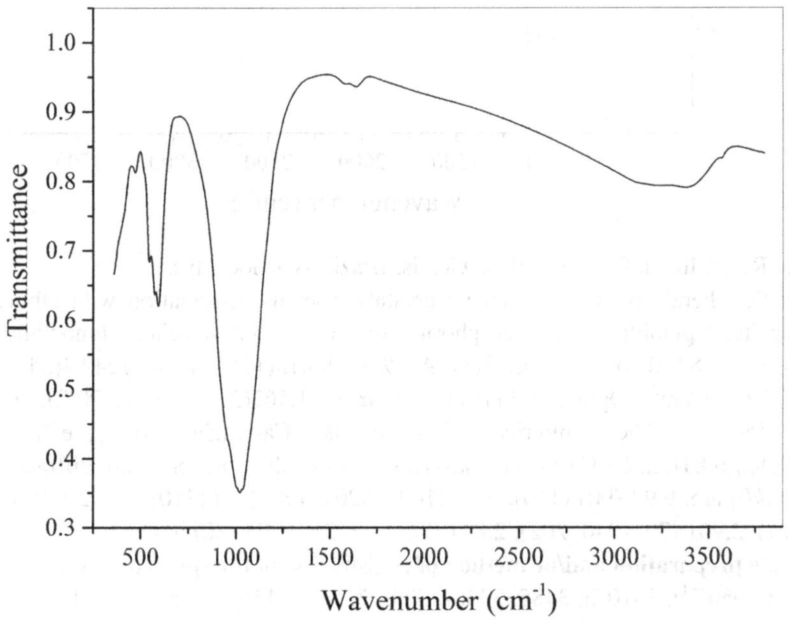

Origin: Nancy pegmatite, Chacabuco department, San Luis, Argentina (type locality).

Description: Dark brownish-gray grains with perfect cleavage. Partly altered and contaminated by a hydrous phosphate.

Kind of sample preparation and/or method of registration of the spectrum: KBr disc. Absorption.

Wavenumbers (cm^{-1}): (3563w), (3384), (3250), 1636w, 1587w, 1040sh, 1023s, 1015sh, 965sh, 594s, 578, 549, 474w, 390sh.

Note: The spectrum was obtained by N.V. Chukanov.

P829 Guimarãesite $Ca_2Be_4Zn_5(PO_4)_6(OH)_4{\cdot}6H_2O$

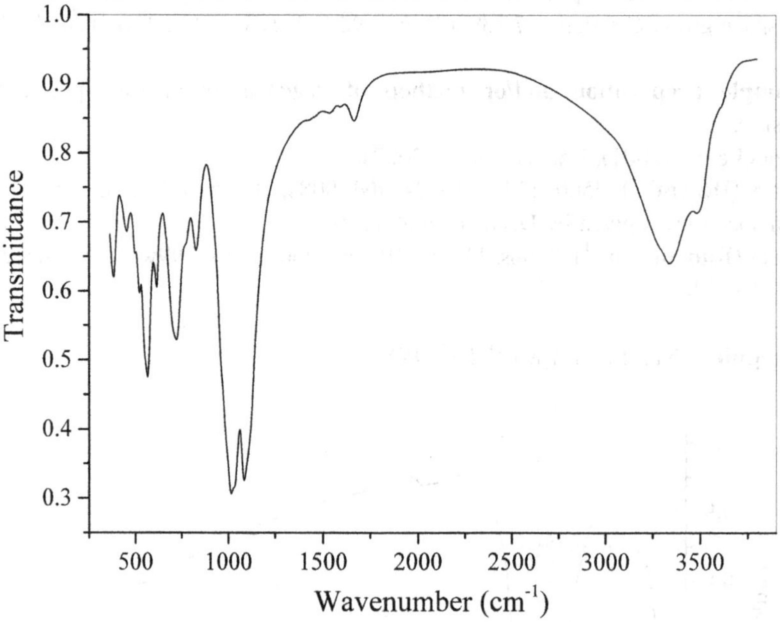

Origin: Piauí River, Itinga County, Minas Gerais, Brazil (type locality).

Description: Peripheral zones of zanazziite crystals from the association with albite, microcline, quartz, elbaite, lepidolite, schorl, eosphorite, moraesite, and saleeite. Monoclinic, a = 15.98 (1) Å, b = 11.84(2) Å, c = 6.63(1) Å, β = 95.15(15)°, V = 1249.4(34) Å^3, Z = 2. D_{calc} = 2.963 g/cm^3. Optically biaxial (–), α = 1.562(2), β = 1.600(2), γ = 1.602(2), 2V = 55–75°. The empirical formula is $Ca_{1.93}(Zn_{2.61}Mg_{1.11}Fe^{2+}_{0.41}Al_{0.37}Mn_{0.34})Be_{4.00}(PO_4)_{6.00}(OH)_{3.90}{\cdot}6.41H_2O$. The strongest lines of the powder X-ray diffraction pattern [d, Å (I, %) (hkl)] are: 9.98 (90) (110), 5.98 (100) (020), 4.82 (80) (310), 3.152 (90) (–202), 3.052 (70) (–421), 2.961 (70) (040, 202), 2.841 (70) (–312), 2.708 (80) (041).

Kind of sample preparation and/or method of registration of the spectrum: KBr disc. Absorption.

Wavenumbers (cm^{-1}): 3610sh, 3485, 3336s, 3295sh, 1667, 1594w, 1537w, 1416w, 1110sh, 1087s, 1030sh, 1016s, 827, 770sh, 723s, 616, 567s, 523, 502, 452, 380.

Note: The spectrum was obtained by N.V. Chukanov.

P830 Drugmanite $Pb_2Fe^{3+}(PO_4)(PO_3OH)(OH)_2$

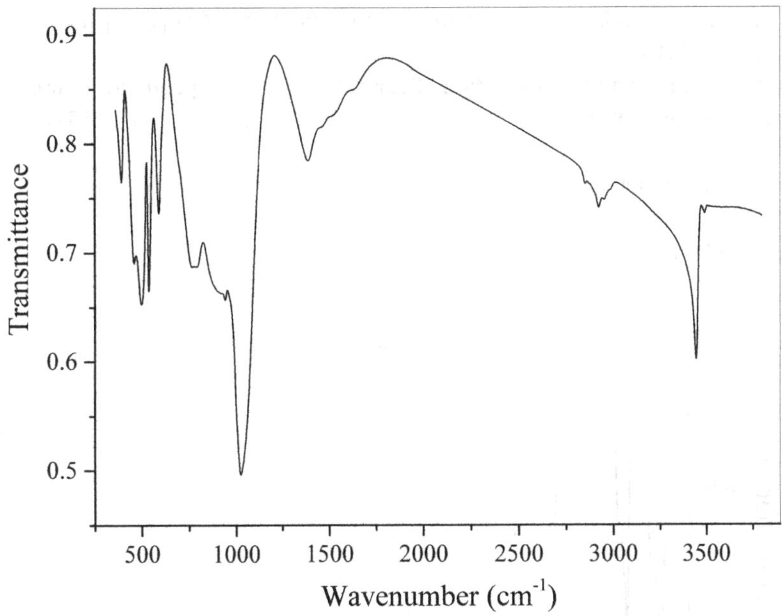

Origin: Bleialf, Prüm, Eifel, Germany.

Description: Spherulitic crust on galena. The empirical formula is (electron microprobe): $H_xPb_{2.18}(Fe_{0.99}Al_{0.01})(PO_4)_{2.00}(OH)_2$.

Kind of sample preparation and/or method of registration of the spectrum: KBr disc. Absorption.

Wavenumbers (cm^{-1}): 3491w, 3445, 1387, 1025s, 942, 915sh, 790, 766, 590, 537, 498, 459, 392.

Note: The spectrum was obtained by N.V. Chukanov. The band at 1387 cm^{-1} corresponds to isolated H+ cation. Weak bands between 1400 and 3000 cm^{-1} are due to an organic impurity.

P831 Roscherite $Ca_2(Mn,Fe^{2+},Fe^{3+},Mg,Al,Zn)_5Be_4(PO_4)_6(OH)_4 \cdot 6H_2O$

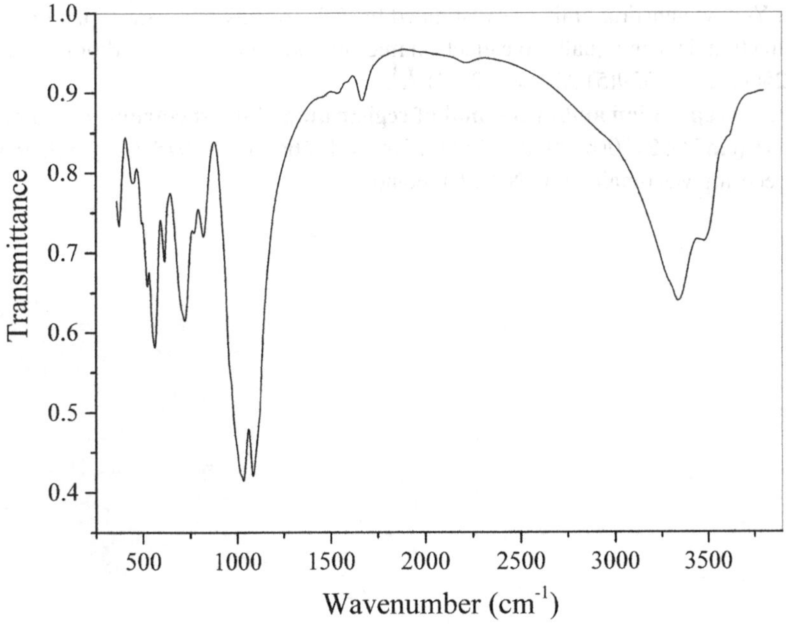

Origin: Taquaral, Itinga, Minas Gerais, Brazil.

Description: Olive-green spherulite from the association with eosphorite, feldspar, and metaautunite. The empirical formula is (electron microprobe): $Ca_{2.0}(Mn_{1.5}Fe_{1.3}Zn_{1.1}Mg_{0.5}Al_{0.1})Be_4(PO_4)_{6.0}(OH)_4 \cdot 6H_2O$.

Kind of sample preparation and/or method of registration of the spectrum: KBr disc. Absorption.

Wavenumbers (cm^{-1}): 3605sh, 3439, 3336s, 3290sh, 2950sh, 2217w, 1667, 1538w, 1450sh, 1084s, 1033s, 1020sh, 820, 771, 722s, 615, 563s, 523, 497, 451, 374.

Note: The spectrum was obtained by N.V. Chukanov.

P832 Thadeuite $Ca(Mg,Fe^{2+})_3(PO_4)_2(OH,F)_2$

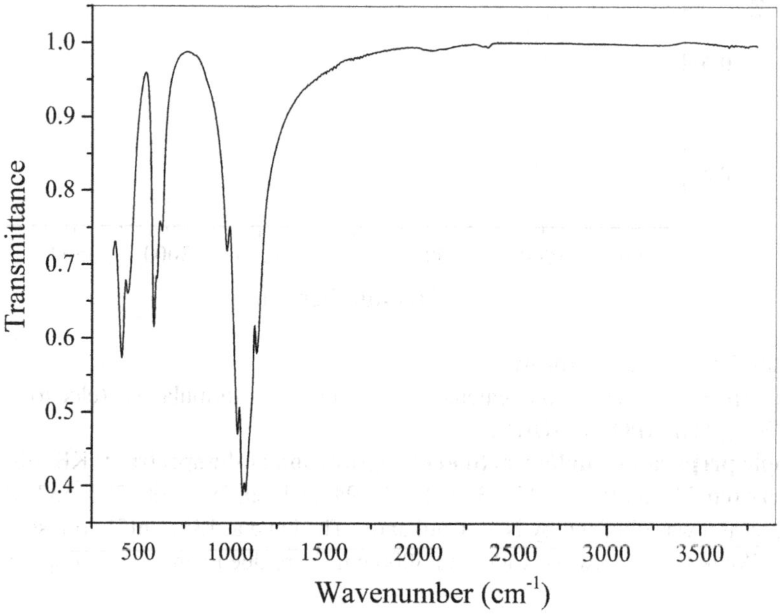

Origin: Panasqueira Mines, Covilhã, Castelo Branco district, Portugal (type locality).

Description: Yellow anhedral grains. Investigated by A.V. Kasatkin. Characterized by single-crystal X-ray diffraction data and qualitative electron microprobe analyses. Orthorhombic, $a = 6.465(14)$, $b = 13.525(7)$, $c = 8.539(5)$ Å, $V = 727(2)$ Å^3.

Kind of sample preparation and/or method of registration of the spectrum: KBr disc. Absorption.

Wavenumbers (cm^{-1}): 2120sh, 2060w, 1141s, 1083s, 1066s, 1038s, 978, 629, 599, 584s, 440, 407s.

Note: The spectrum was obtained by N.V. Chukanov.

P833 Minjiangite $BaBe_2(PO_4)_2$

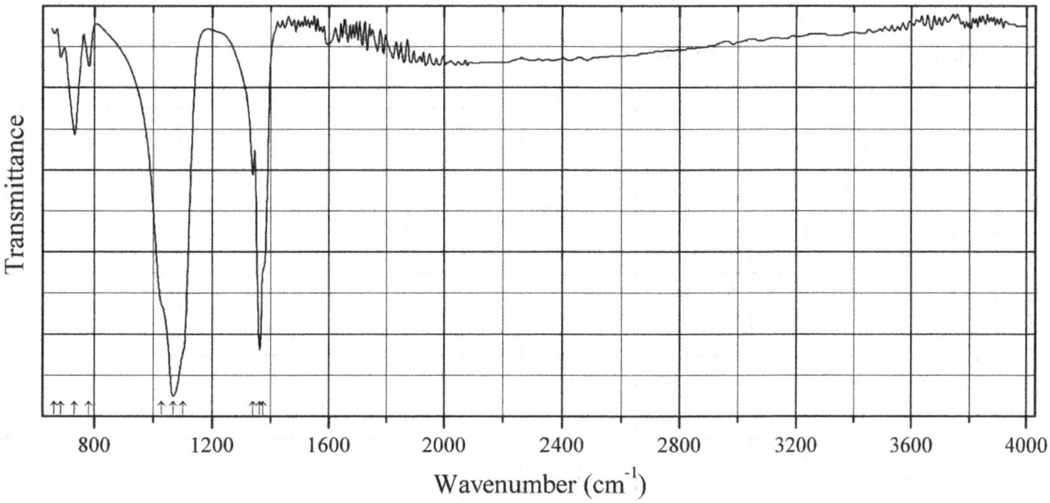

Origin: Nanping No. 31 pegmatite, Fujian Province, southeastern China (type locality).

Description: White crystals from the association with montebrasite, quartz, muscovite, hydroxylapatite, and palermoite. Holotype sample. The crystal structure is solved. Hexagonal, space group *P6/mmm*, $a = 5.029(1)$, $c = 7.466(1)$ Å, $V = 163.52(1)$ Å^3, $Z = 1$. $D_{calc} = 3.49$ g/cm^3. Optically biaxial (+), $\omega = 1.587(3)$, $\varepsilon = 1.602(2)$. The empirical formula is $(Ba_{0.99}Ca_{0.01})Be_{1.98}(P_{1.99}Si_{0.01})O_8$.

Kind of sample preparation and/or method of registration of the spectrum: Reflection.

Source: Rao et al. (2015).

Wavenumbers (IR, cm^{-1}): 1375s, 1363s, 1339s, 1101sh, 1068s, 1027sh, 781, 730, 683w, 660w.

Note: Possibly, an erroneous spectrum. In particular, assignment of the strong bands at 1375, 1363, and 1339 cm^{-1} to Be–O-stretching vibrations (Rao et al. 2015) is questionable. Dal Bo et al. (2014) give another IR spectrum for the synthetic analogue of minjiangite. The wavenumbers were partly determined by us based on spectral curve analysis of the published spectrum. In the cited paper, Raman spectrum is given.

Wavenumbers (Raman, cm^{-1}): 1233s, 1050s, 491, 478, 328w, 189w.

2.9 Sulfides, Sulfites, Sulfates, Carbonato-Sulfates, Phosphato-Sulfates, and Tellurato-Sulfates

S554 Eleomelanite $(K_2Pb)Cu_4O_2(SO_4)_4$

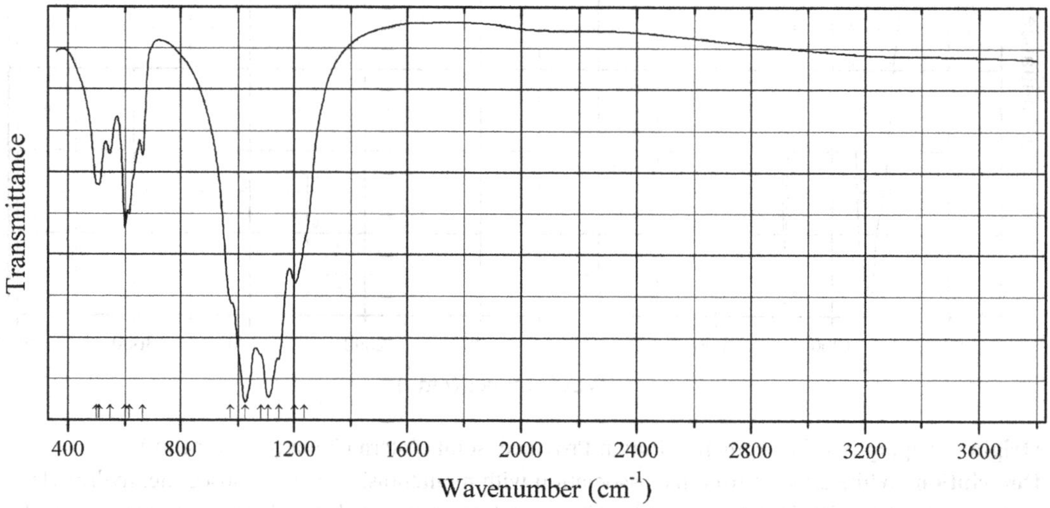

Origin: Arsenatnaya fumarole, Second scoria cone of the Northern Breakthrough of the Great Tolbachik Fissure Eruption, Tolbachik volcano, Kamchatka Peninsula, Far-Eastern Region, Russia (type locality).

Description: Dark green crystalline crust from the association with euchlorine, wulffite, klyuchevskite, alumoklyuchevskite, fedotovite, anglesite, cryptochalcite, langbeinite, aphthitalite, chalcocyanite, dolerophanite, piypite, anhydrite, steklite, etc. Holotype sample. The crystal structure is solved. Monoclinic, space group $P2_1/n$, $a = 9.3986(3)$, $b = 4.9811(1)$, $c = 18.2293(5)$ Å, $\beta = 104.409(3)°$, $V = 811.63(4)$ Å^3, $Z = 2$. $D_{calc} = 3.790$ g/cm^3. Optically biaxial (–), $\alpha = 1.646(3)$, $\beta = 1.715(6)$, $\gamma = 1.734(6)$, $2V = 60(15)°$. The empirical formula is (electron microprobe): $(K_{1.88}Pb_{0.79}Ca_{0.20}Rb_{0.05}Cs_{0.02})_{\Sigma 2.94}Cu_{4.07}S_{3.99}O_{18}$. The strongest lines of the powder X-ray diffraction pattern [*d*, Å (*I*, %) (*hkl*)] are: 9.07 (63) (–101), 7.38 (44) (101), 3.699 (78) (112, 202), 3.658 (100) (–204), 3.173 (40) (211, –213), 2.576 (51) (310, –116).

Kind of sample preparation and/or method of registration of the spectrum: KBr disc. Absorption.

Wavenumbers (cm^{-1}): 1235sh, 1202s, 1145s, 1108s, 1080sh, 1025s, 975sh, 662, 615, 601s, 547, 508, 500sh.

Note: The spectrum was obtained by N.V. Chukanov.

S555 Kottenheimite dimorph $Ca_3Si(SO_4)_2(OH)_6{\cdot}12H_2O$

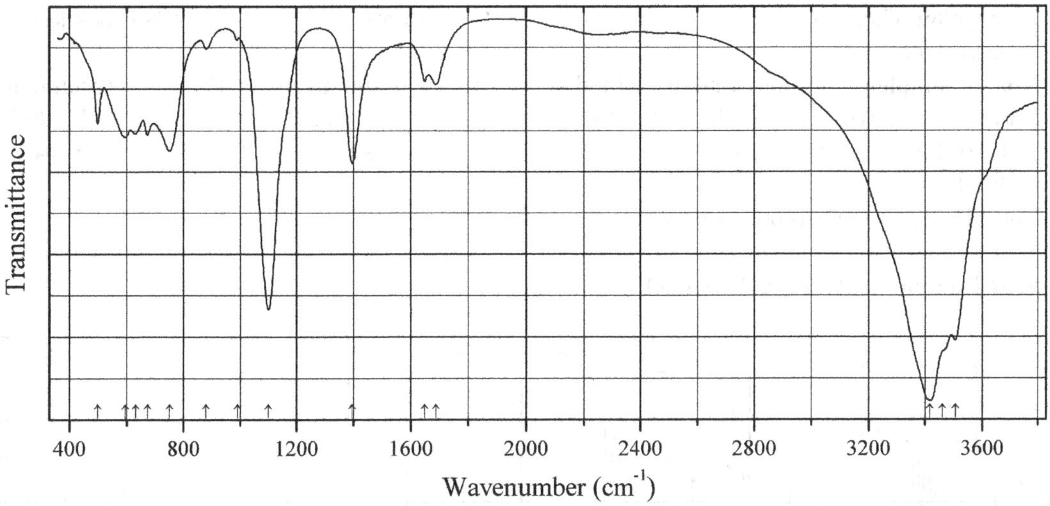

Origin: Bellerberg, near Mayen, Eifel, Rheinland-Pfalz (Rhineland-Palatinate), Germany.

Description: White random aggregate of acicular crystals. Isostructural with thaumasite. The empirical formula is (electron microprobe): $Ca_{3.05}(Si_{0.9}Al_{0.1})(SO_4)[(SO_4)_{0.6}(CO_3)_{0.4}](OH)_6{\cdot}nH_2O$.

Kind of sample preparation and/or method of registration of the spectrum: KBr disc. Absorption.

Wavenumbers (cm^{-1}): 3505s, 3460sh, 3417s, 1688, 1650, 1396s, 1100s, 990w, 880, 750s, 673, 632, 596, 499.

Note: The spectrum was obtained by N.V. Chukanov.

S556 Bobcookite $NaAl(UO_2)_2(SO_4)_4{\cdot}18H_2O$

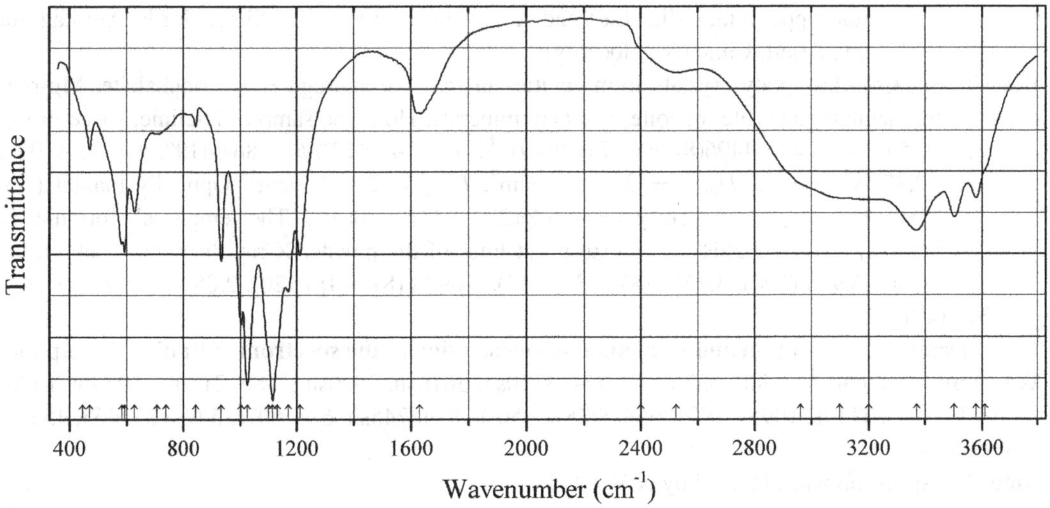

Origin: Blue Lizard Mine, Red Canyon, White Canyon District, San Juan Co., Utah, USA (type locality).

Description: Green-yellow crystals. Investigated by A.V. Kasatkin, the coauthor of bobcookite first description.

Kind of sample preparation and/or method of registration of the spectrum: KBr disc. Absorption.

Wavenumbers (cm^{-1}): 3610sh, 3578, 3502, 3370, 3200sh, 3100sh, 2960sh, 2523w, 2400sh, 1630, 1210, 1169s, 1130sh, 1114s, 1100sh, 1024s, 1004s, 933, 843w, 740sh, 708w, 628, 594, 583, 471, 445sh.

Note: The spectrum was obtained by N.V. Chukanov.

S557 Riotintoite $Al(SO_4)(OH){\cdot}3H_2O$

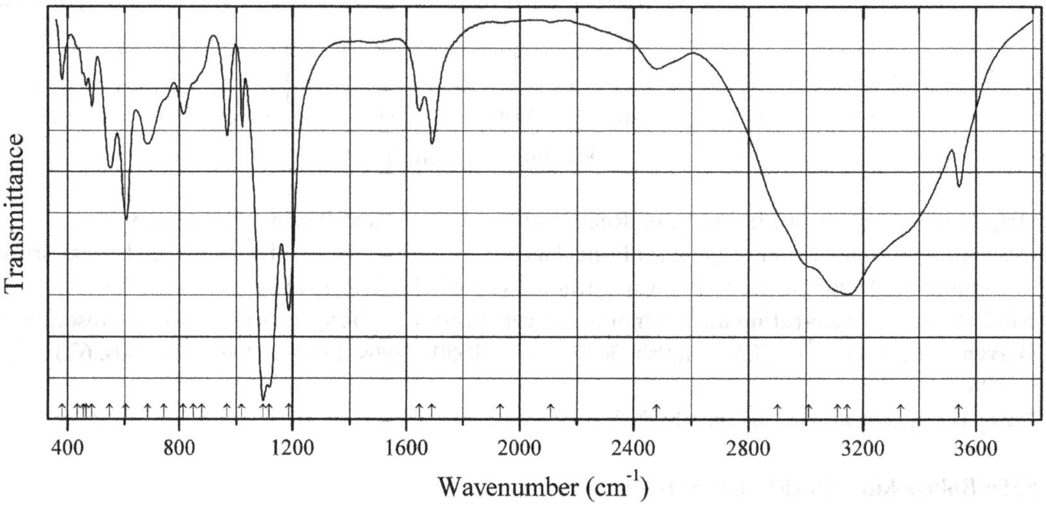

Origin: La Vendida copper mine (Mina La Vendida), about 5 km WNW of Sierra Gorda, Antofagasta Region, Atacama desert, Chile (type locality).

Description: Colorless platy crystals from cavities in massive aggregates of eriochalcite, Mg-rich aubertite, magnesioaubertite, belloite, and clay minerals. Holotype sample. Triclinic, space group *P*-1, $a = 5.6000$, $b = 7.4496(8)$, $c = 7.6709(9)$ Å, $\alpha = 74.7847°$, $\beta = 86.0419°$, $\gamma = 75.8103°$, $V = 299.37$ Å^3, $Z = 2$. $D_{meas} = 2.11(2)$ g/cm^3, $D_{calc} = 2.129$ g/cm^3. Optically biaxial (–), $\alpha = 1.513(2)$, $\beta = 1.522(2)$, $\gamma = 1.526(2)$, $2V = 70(5)°$. The empirical formula is $Al_{0.93}(SO_4)_{0.99}(OH)_{0.81}{\cdot}3.25H_2O$. The strongest lines of the powder X-ray diffraction pattern [*d*, Å (*I*, %)] are: 6.975 (100), 4.466 (18), 4.379 (19), 3.698 (18), 3.487 (20), 2.882 (17), 2.669 (54), 2.397 (40).

Kind of sample preparation and/or method of registration of the spectrum: KBr disc. Absorption.

Wavenumbers (cm^{-1}): 3540, 3335sh, 3145s, 3110sh, 3010sh, 2900sh, 2480, 2110w, 1933w, 1690, 1647, 1186s, 1117s, 1095s, 1021, 968, 880sh, 850sh, 814, 745sh, 683, 607s, 549, 485, 465, 455sh, 435sh, 382.

Note: The spectrum was obtained by N.V. Chukanov.

S558 Rhomboclase $(H_5O_2)Fe^{3+}(SO_4)_2 \cdot 2H_2O$

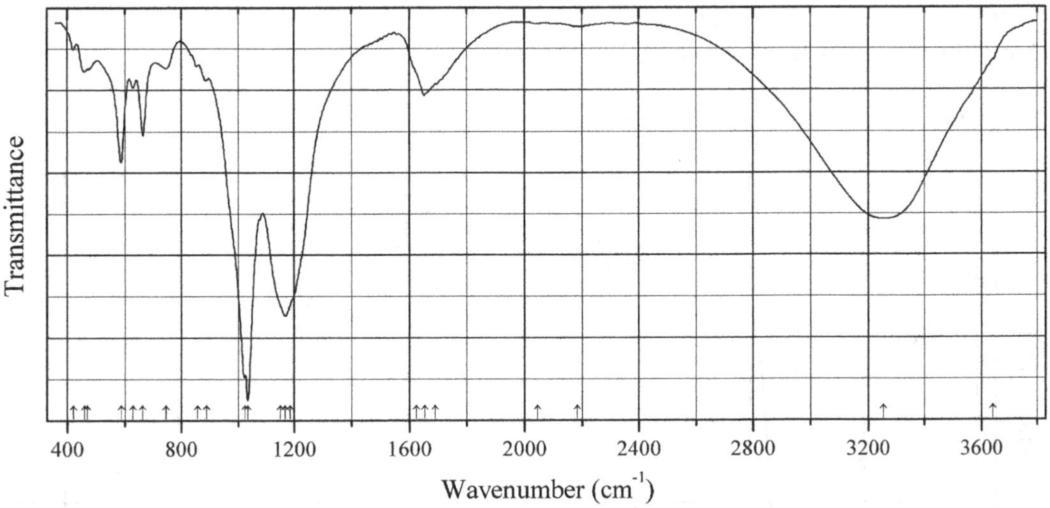

Origin: Alcaparrosa mine, Cerritos Bayos, Calama, El Loa Province, Antofagasta, Chile.

Description: Light gray grains. Investigated by I.V. Pekov. Characterized by single-crystal X-ray diffraction data. Orthorhombic, $a = 5.426(3)$, $b = 9.470(7)$, $c = 18.333(17)$ Å.

Kind of sample preparation and/or method of registration of the spectrum: KBr disc. Absorption.

Wavenumbers (cm^{-1}): 3640sh, 3255s (broad), 2185w, 2045w, 1690sh, 1653, 1625sh, 1185sh, 1168s, 1150sh, 1034s, 1025s, 888w, 857w, 747w, 665, 630, 588, 471, 460, 421w.

Note: The sample has altered as a result of a reaction with KBr. The spectrum was obtained by N.V. Chukanov.

S559 Ferrinatrite $Na_3Fe^{3+}(SO_4)_3 \cdot 3H_2O$

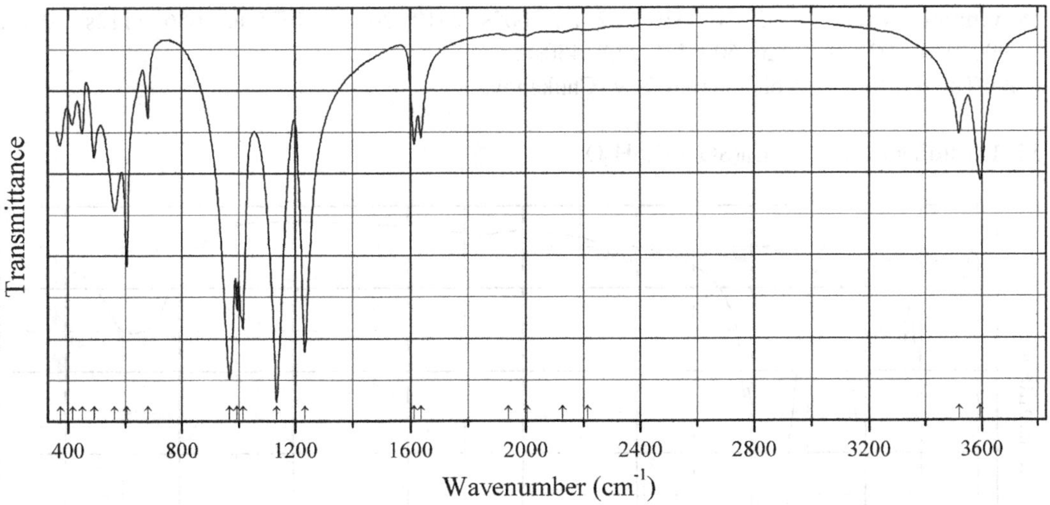

Origin: Coronel Manuel Rodríguez mine, Mejillones peninsula, Mejillones, Antofagasta Province, Antofagasta Region, Chile.

Description: White aggregate of acicular crystals. Investigated by I.V. Pekov. Characterized by powder X-ray diffraction data and electron microprobe analyses.

Kind of sample preparation and/or method of registration of the spectrum: KBr disc. Absorption.
Wavenumbers (cm^{-1}): 3594, 3517, 2215w, 2128w, 2005w, 1939w, 1637, 1614, 1232s, 1133s, 1014s, 994s, 966s, 680, 606, 563, 491, 450, 416, 373.
Note: The spectrum was obtained by N.V. Chukanov.

S560 Magnesioaubertite $MgAl(SO_4)_2Cl{\cdot}14H_2O$

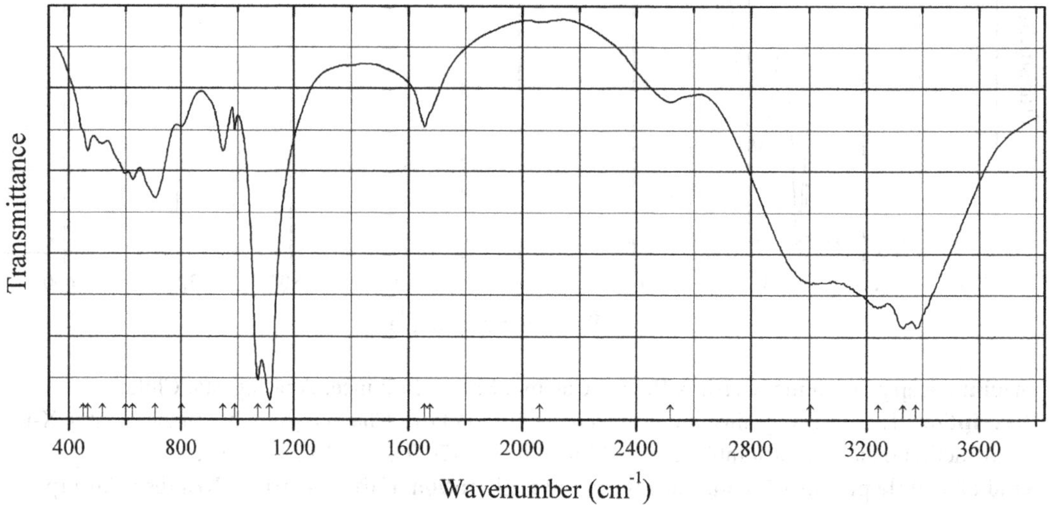

Origin: La Vendida copper mine, about 5 km WNW of Sierra Gorda, Antofagasta Region, Atacama desert, Chile.
Description: Turquoise-blue granular aggregate from the association with vendidaite and eriochalcite. Investigated by I.V. Pekov. Characterized by powder X-ray diffraction data. The empirical formula is (electron microprobe): $Mg_{0.56}Cu_{0.39}Al_{1.09}(SO_4)_{2.00}Cl_{0.65}(OH)_x{\cdot}nH_2O$.
Kind of sample preparation and/or method of registration of the spectrum: KBr disc. Absorption.
Wavenumbers (cm^{-1}): 3376s, 3330s, 3241s, 3005s, 2519, 2060w, 1675sh, 1656, 1112s, 1070s, 990, 950, 802, 706s, 626, 600, 518, 468, 450sh.
Note: The spectrum was obtained by N.V. Chukanov.

S561 Antofagastaite $Na_2Ca(SO_4)_2{\cdot}1.5H_2O$

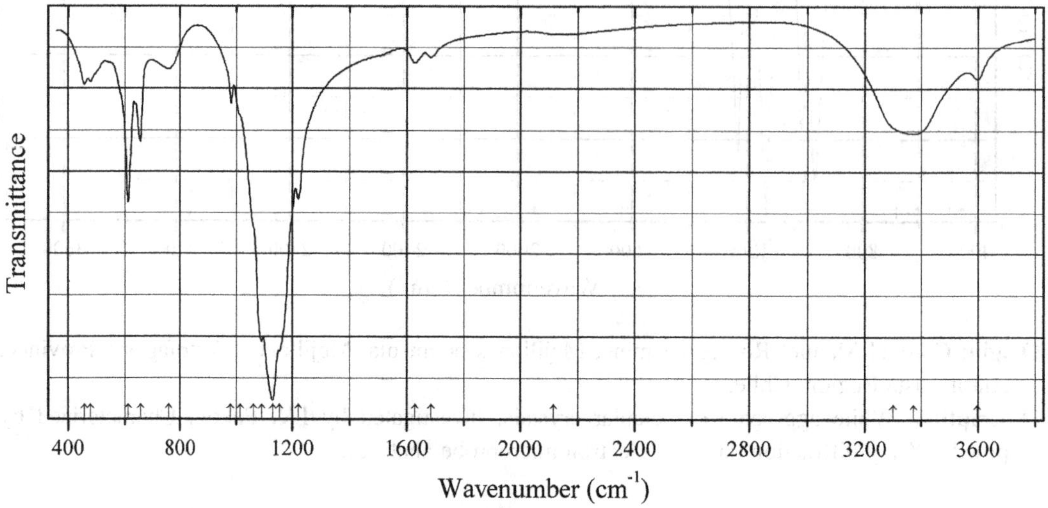

Origin: Coronel Manuel Rodríguez mine, Mejillones peninsula, Mejillones, Antofagasta region, Chile (type locality).

Description: Colorless prismatic crystals from the association with sideronatrite, metasideronatrite, aubertite, gypsum, ferrinatrite, glauberite, and amarillite. Holotype sample. The crystal structure is solved. Monoclinic, space group $P2_1/m$, a = 6.4596(4), b = 6.8703(5), c = 9.4685(7) Å, β = 104.580(4)°, V = 406.67(5) Å^3, Z = 2. D_{meas} = 2.42(1) g/cm^3, D_{calc} = 2.465 g/cm^3. Optically biaxial (−), α = 1.489 (2), β = 1.508 (2), γ = 1.510 (2), 2V = 40(10)°. The empirical formula is $Na_{2.06}Ca_{0.95}S_{2.01}O_8 \cdot 1.35H_2O$. The strongest lines of the powder X-ray diffraction pattern [d, Å (I, %) (hkl)] are: 9.17 (100) (001), 5.501 (57) (011), 4.595 (32) (002), 3.437 (59) (020), 3.058 (43) (−103, 003), 2.918 (50) (−211), 2.795 (35) (−113, 013).

Kind of sample preparation and/or method of registration of the spectrum: KBr disc. Absorption.

Wavenumbers (cm^{-1}): 3598, 3373, 3300sh, 1685w, 1629w, 1220, 1153s, 1128s, 1090s, 1060sh, 1015sh, 982, 758, 656, 613, 480, 458.

Note: The spectrum was obtained by N.V. Chukanov. Very weak absorptions in the range 2100–2250 cm^{-1} correspond to overtones and combination modes.

S562 Römerite $Fe^{2+}Fe^{3+}_2(SO_4)_4 \cdot 14H_2O$

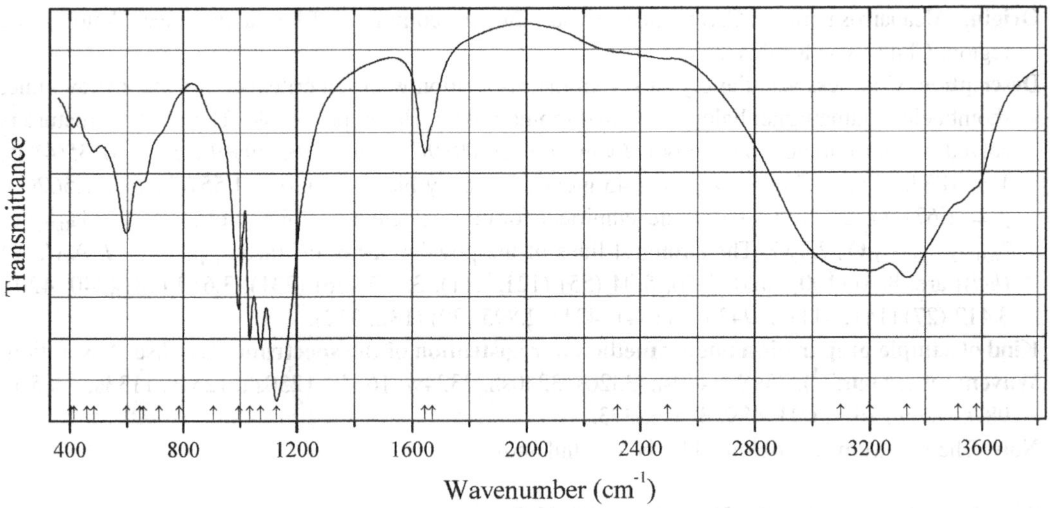

Origin: Alcaparrosa mine, Cerritos Bayos, Calama, El Loa Province, Antofagasta, Chile.

Description: Reddish-brown crystals from the association with coquimbite, metavoltine, and voltaite. Investigated by I.V. Pekov. Characterized by single-crystal X-ray diffraction data. Triclinic, a = 6.317(4), b = 6.453(4), c = 15.318(10) Å, α = 85.61(5)°, β = 89.78(5)°, γ = 79.06(5)°, V = 611.2(7) Å^3. Only Fe and S have been found by means of electron microprobe analyses.

Kind of sample preparation and/or method of registration of the spectrum: KBr disc. Absorption.

Wavenumbers (cm^{-1}): 3580sh, 3515sh, 3334s, 3200s, 3100sh, 2495w, 2320sh, 1670sh, 1645, 1129s, 1072s, 1034s, 995s, 905sh, 785sh, 715sh, 660sh, 646, 599s, 484, 460sh, 414, 405sh.

Note: The spectrum was obtained by N.V. Chukanov.

S563 Calamaite $Na_2TiO(SO_4)_2{\cdot}2H_2O$

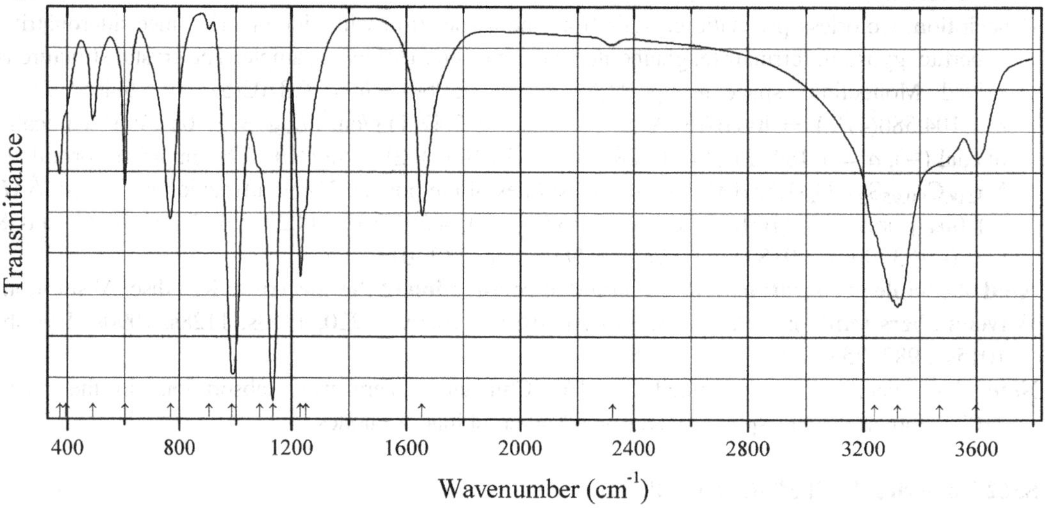

Origin: Alcaparrosa mine, Cerro Alcaparrosa, Calama commune, El Loa province, Antofagasta region, Chile (type locality).

Description: Colorless acicular crystals from the association with römerite, coquimbite, metavoltine, rhomboclase, tamarugite, halotrichite, and szomolnokite. Holotype sample. The crystal structure is solved. Orthorhombic, space group *Ibam*, a = 16.0989(11), b = 16.2399(9), c = 7.0135(4) Å, V = 1833.6(2) Å^3, Z = 8. D_{calc} = 2.45 g/cm^3. Optically biaxial (+), α = 1.557(2), β = 1.562(2), γ = 1.671(3), 2V = 30(10)°. The empirical formula is (electron microprobe): $Na_{1.97}(Ti_{0.92}Fe^{3+}_{0.07})_{\Sigma0.99}S_{2.02}O_9{\cdot}2H_2O$. The strongest lines of the powder X-ray diffraction pattern [d, Å (I, %) (hkl)] are: 8.10 (100) (020, 200), 5.04 (55) (121, 211), 3.787 (26) (231), 3.619 (18) (240, 420), 3.417 (27) (141, 411), 2.943 (20) (341, 431), 2.895 (20) (132, 312).

Kind of sample preparation and/or method of registration of the spectrum: KBr disc. Absorption.

Wavenumbers (cm^{-1}): 3598, 3470sh, 3320s, 3240sh, 2324w, 1655s, 1250sh, 1231s, 1133s, 1085sh, 987s, 906w, 767s, 603, 489, 395sh, 373.

Note: The spectrum was obtained by N.V. Chukanov.

S564 Metasideronatrite $Na_2Fe^{3+}(SO_4)_2(OH){\cdot}H_2O$

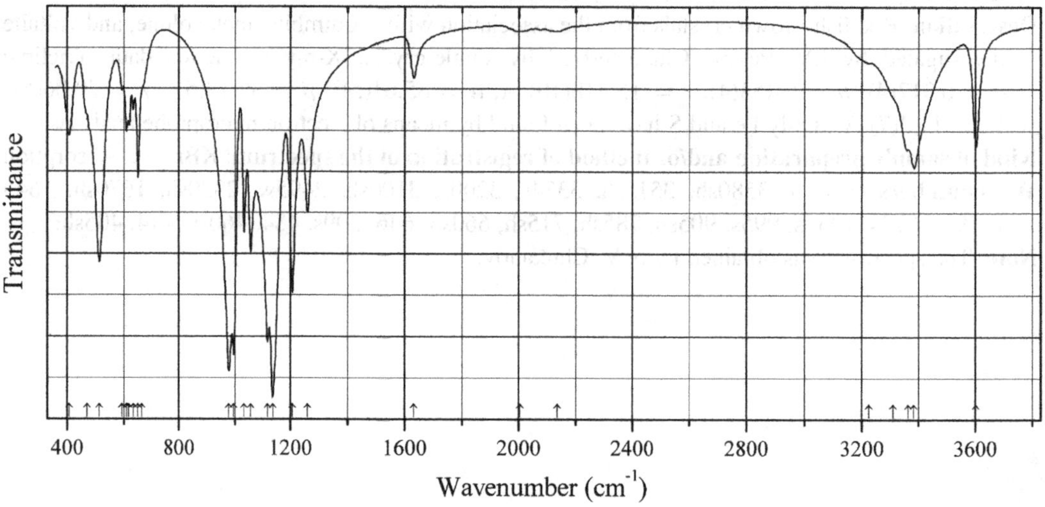

Origin: Coronel Manuel Rodríguez mine, Mejillones peninsula, Mejillones, Antofagasta Province, Antofagasta Region, Chile.

Description: Orange-beige pseudomorphs after prismatic sideronatrite crystals. Investigated by I.V. Pekov. Characterized by powder X-ray diffraction data and electron microprobe analyses.

Kind of sample preparation and/or method of registration of the spectrum: KBr disc. Absorption.

Wavenumbers (cm^{-1}): 3603, 3383, 3362, 3310sh, 3225sh, 2135w, 2004w, 1632, 1260, 1206s, 1135s, 1117s, 1056, 1032, 996s, 979s, 665sh, 650, 634, 618, 610, 595w, 514s, 470sh, 407.

Note: The spectrum was obtained by N.V. Chukanov.

S565 Parabutlerite $Fe^{3+}(SO_4)(OH)\cdot 2H_2O$

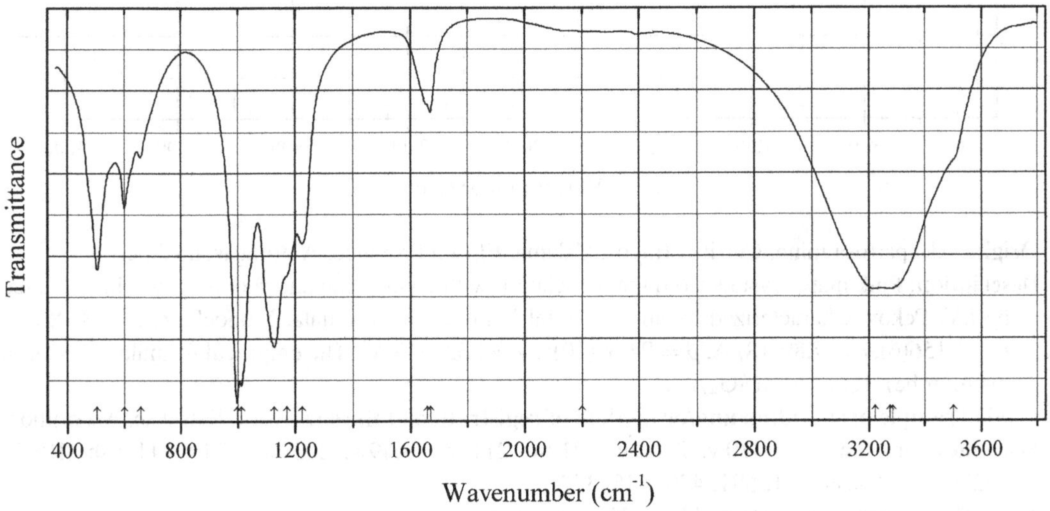

Origin: Coronel Manuel Rodríguez mine, Mejillones peninsula, Mejillones, Antofagasta Province, Antofagasta Region, Chile.

Description: Orange-brown crystals from the association with gypsum. Investigated by I.V. Pekov. Characterized by single-crystal X-ray diffraction data. Orthorhombic, $a = 7.386(3)$, $b = 7.405(4)$, $c = 20.091(10)$ Å, $V = 1072(1)$ Å^3.

Kind of sample preparation and/or method of registration of the spectrum: KBr disc. Absorption.

Wavenumbers (cm^{-1}): 3500sh, 3276s, 3223s, 3285w, (2200sh), 1670, 1660sh, 1222s, 1170sh, 1126s, 1012s, 997s, 658, 601, 505s.

Note: The spectrum was obtained by N.V. Chukanov.

S566 Yavapaiite $KFe^{3+}(SO_4)_2$

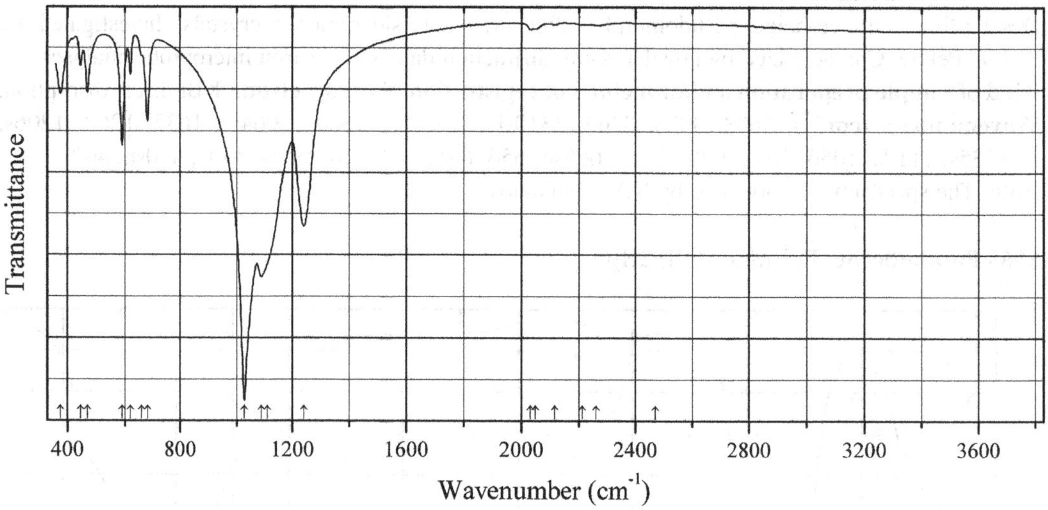

Origin: Alcaparrosa mine, Cerritos Bayos, Calama, El Loa Province, Antofagasta, Chile.

Description: Pink platy crystals from the association with coquimbite and rhomboclase. Investigated by I.V. Pekov. Characterized by single-crystal X-ray diffraction data. Monoclinic, $a = 8.186(8)$, $b = 5.156(6)$, $c = 7.893(8)$ Å, $\beta = 94.69(10)°$, $V = 332.0(6)$ Å^3. The empirical formula is (electron microprobe): $K_{0.98}Fe_{1.01}(SO_4)_{2.00}$.

Kind of sample preparation and/or method of registration of the spectrum: KBr disc. Absorption.

Wavenumbers (cm^{-1}): 2470w, 2262w, 2212w, 2117w, 2049w, 2032w, 1240s, 1110sh, 1087s, 1027s, 680, 660sh, 621, 591, 470, 446, 377.

Note: The spectrum was obtained by N.V. Chukanov.

S567 Szomolnokite $Fe(SO_4)·H_2O$

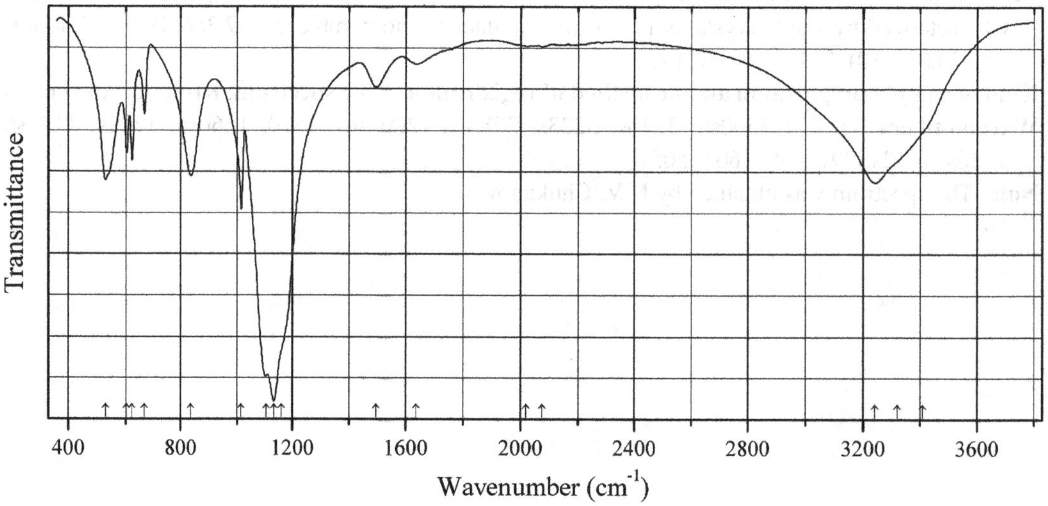

Origin: Alcaparrosa mine, Cerritos Bayos, Calama, El Loa Province, Antofagasta, Chile.
Description: Pale greenish-yellow crystals from the association with metavoltine, coquimbite, and römerite. Investigated by I.V. Pekov. Characterized by single-crystal X-ray diffraction data. Monoclinic, $a = 7.66$, $b = 7.53$, $c = 7.09$ Å, $\beta = 116.66°$, $V = 365.7$ Å^3. The empirical formula is (electron microprobe): $(Fe_{0.89}Zn_{0.05}Mg_{0.03})S_{1.01}O_4 \cdot H_2O$.
Kind of sample preparation and/or method of registration of the spectrum: KBr disc. Absorption.
Wavenumbers (cm^{-1}): 3410sh, 3320sh, 3243s, 2076w, 2020w, 1637, 1497, 1160sh, 1133s, 1105s, 1016s, 837s, 667, 624, 604, 530s.
Note: The spectrum was obtained by N.V. Chukanov.

S568 Metathénardite $Na_2(SO_4)$

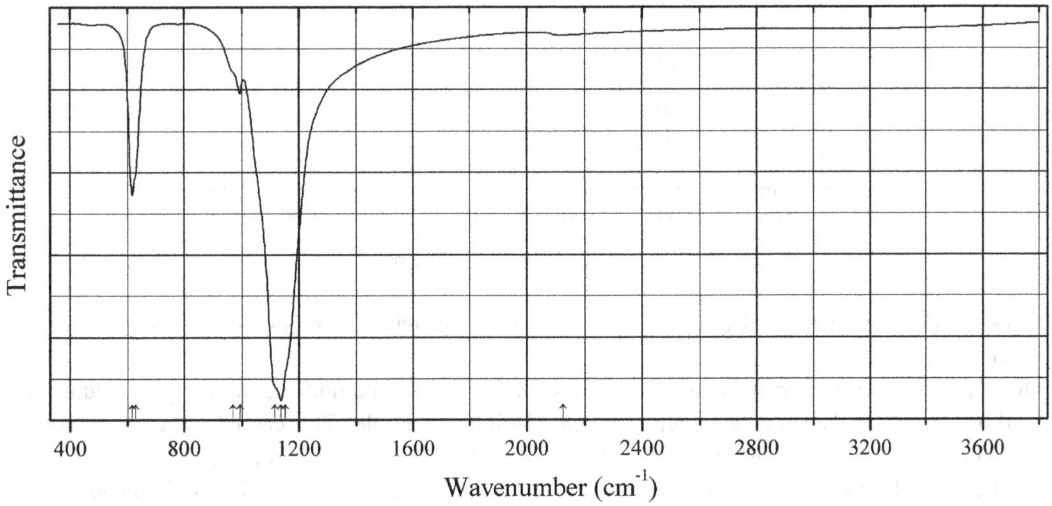

Origin: Yadovitaya (Poisonous) fumarole, Second scoria cone, Tolbachik volcano, Kamchatka peninsula, Far-Eastern Region, Russia.
Description: Pale blue crystals. Investigated by I.V. Pekov, the author of the first description of metathénardite.
Kind of sample preparation and/or method of registration of the spectrum: KBr disc. Absorption.
Wavenumbers (cm^{-1}): 2125w, 1150sh, 1136s, 1115sh, 993, 970sh, 630sh, 618s.
Note: The spectrum was obtained by N.V. Chukanov.

S569 Magnesiovoltaite $K_2Mg_5Fe^{3+}{}_3Al(SO_4)_{12}\cdot 18H_2O$

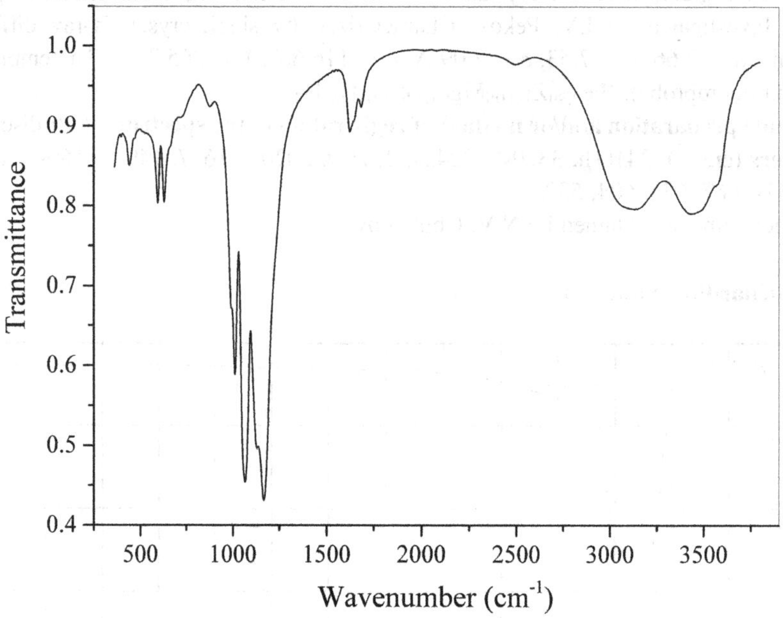

Origin: Alcaparrosa mine, Cerro Alcaparrosa, El Loa province, Antofagasta region, Chile (type locality).

Description: Yellow crystals from the association with coquimbite, tamarugite, alum-(Na), rhomboclase, yavapaiite, voltaite, and opal. Holotype sample. The crystal structure is solved. Cubic, space group *Fd*-3*c*, $a = 27.161(1)$ Å, $V = 20038(2)$ Å^3, $Z = 16$. $D_{meas} = 2.51(2)$ g/cm^3, $D_{calc} = 2.506$ g/cm^3. Optically anomalously anisotropic, uniaxial with $\varepsilon = 1.584$ (2) and $\omega = 1.588$ (2), or biaxial (−) with $\alpha = 1.584$ (2), $\beta = 1.587$ (2), and $\gamma = 1.588$ (2). The empirical formula is $(K_{1.85}Na_{0.08})(Mg_{4.25}Mn_{0.46}Zn_{0.14})Fe^{3+}{}_{3.14}Al_{0.91}(SO_4)_{11.91}(H_2O)_{18.325}O_{0.035}$. The strongest lines of the powder X-ray diffraction pattern [*d*, Å (*I*, %) (*hkl*)] are: 9.56 (29) (022), 6.77 (37) (004), 5.53 (61) (224), 3.532 (68) (137), 3.392 (100) (008), 3.034 (45) (048), 2.845 (30) (139).

Kind of sample preparation and/or method of registration of the spectrum: KBr disc. Absorption.

Wavenumbers (cm^{-1}): 3565sh, 3480sh, 3441, 3134, (3070sh), 2496w, 1684w, 1640sh, 1624, 1168s, 1133s, 1067s, 1011s, 995sh, 876w, 718sh, 660sh, 629, 596, 440.

Note: The spectrum was obtained by N.V. Chukanov.

S570 Barium titanium sulfide Ba_2TiS_4

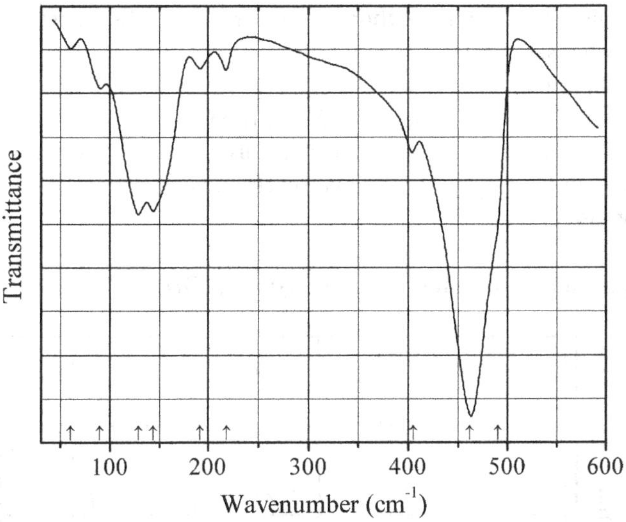

Origin: Synthetic.

Description: Orthorhombic, space group *Pnma*. The crystal structure contains TiS_4 tetrahedra.

Kind of sample preparation and/or method of registration of the spectrum: Transmission of a polycrystalline powder sample.

Source: Ishii and Saeki (1992).

Wavenumbers (IR, cm^{-1}): 491sh, 462s, 405, 218w, 192w, 144, 129, 90, 60w.

Note: The wavenumbers were partly determined by us based on spectral curve analysis of the published spectrum. In the cited paper, Raman spectrum is given.

Wavenumbers (Raman, cm^{-1}): 477, 457w, 441w, 445sh, 403s.

S571 Barium titanium sulfide Ba_3TiS_5

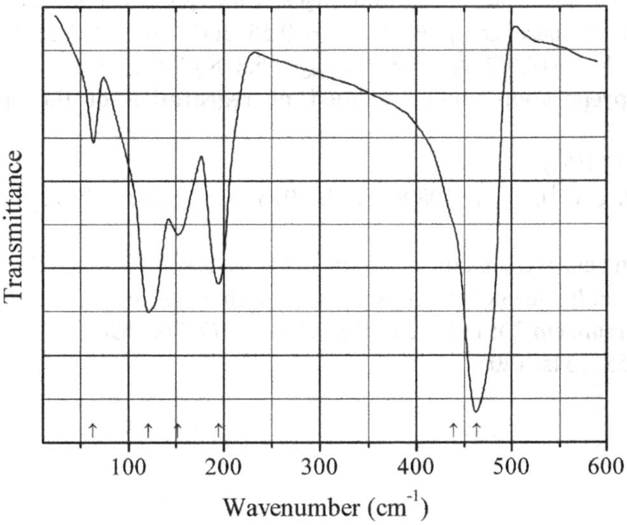

Origin: Synthetic.

Description: Tetrahedral, space group *I4/mcm*. The crystal structure contains TiS_4 tetrahedra.

Kind of sample preparation and/or method of registration of the spectrum: Transmission of a polycrystalline powder sample.

Source: Ishii and Saeki (1992).

Wavenumbers (IR, cm^{-1}): 463s, 439sh, 195, 152, 121, 63w.

Note: The wavenumbers were partly determined by us based on spectral curve analysis of the published spectrum. In the cited paper, Raman spectrum is given.

Wavenumbers (Raman, cm^{-1}): 478, 462, 416s.

S572 Bismuth copper sulfate tellurite $BiCu_2(TeO_3)(SO_4)(OH)_3$

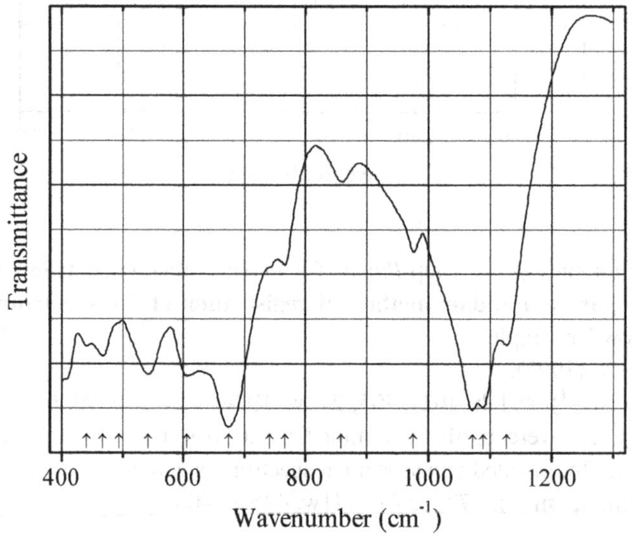

Origin: Synthetic.

Description: Synthesized hydrothermally from Bi_2O_3, $CuSO_4 \cdot 5H_2O$, TeO_2, and H_2SO_4 at 230 °C for 3 days. Monoclinic, space group $P2_1/n$, a = 9.5513(15), b = 6.3022(10), c =13.955(2) Å, β = 102.845(3)°, V = 819.0(2) Å^3, Z = 4. D_{calc} = 5.318 g/cm^3.

Kind of sample preparation and/or method of registration of the spectrum: KBr disc. Transmission.

Source: Chen et al. (2015a).

Wavenumbers (IR, cm^{-1}): 1127, 1089s, 1071s, 976, 858w, 767, 742sh, 675s, 607, 542, 494sh, 468, 441.

Note: The wavenumbers were partly determined by us based on spectral curve analysis of the published spectrum. In the cited paper, Raman spectrum is given.

Wavenumbers (Raman, cm^{-1}): 1110w, 1092w, 1071w, 977, 760, 690w, 630, 563w, 438, 412, 354s, 273, 228, 211, 155, 134s, 120s.

S573 Bismuth sulfate $Bi_2(SO_4)_3$

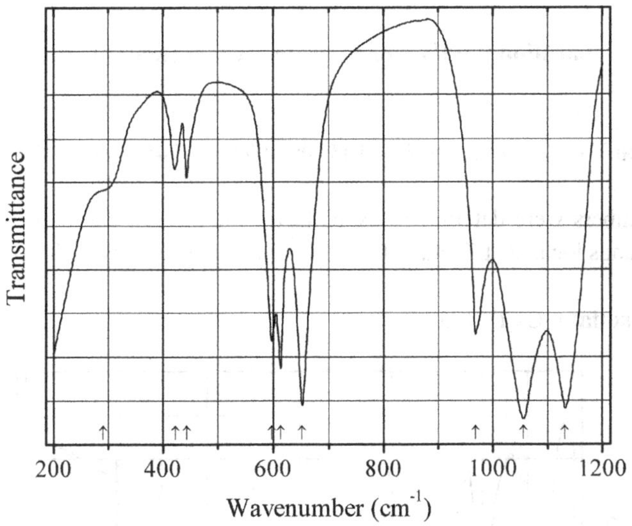

Origin: Synthetic.
Kind of sample preparation and/or method of registration of the spectrum: KBr disc. Transmission.
Source: Botto et al. (1995).
Wavenumbers (cm^{-1}): 1131s, 1056s, 968, 652s, 613, 597, 444w, 422w, 290sh.
Note: The wavenumbers were determined by us based on spectral curve analysis of the published spectrum.

S574 Bismuthyl sulfate $(BiO)_2(SO_4)$

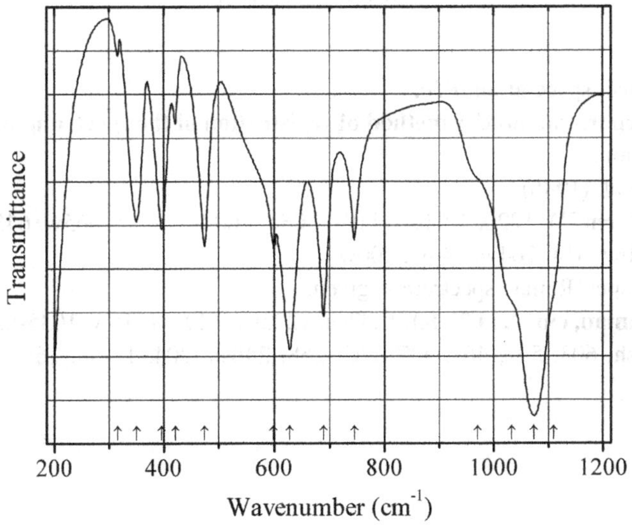

Origin: Synthetic.

Description: Product of heating of tetradymite at 500 °C in air. The sample contains admixture of tellurium oxide.

Kind of sample preparation and/or method of registration of the spectrum: KBr disc. Transmission.

Source: Botto et al. (1995).

Wavenumbers (cm^{-1}): 1110sh, 1074s, 1033sh, 971sh, 745, 689, 628s, 599, 475, 421w, 396, 350, 316w.

Note: The wavenumbers were determined by us based on spectral curve analysis of the published spectrum. The bands located at 745 and 689 cm^{-1} can be tentatively assigned to a tellurium oxide.

S575 Cesium iron sulfate $Cs_3Fe(SO_4)_3$

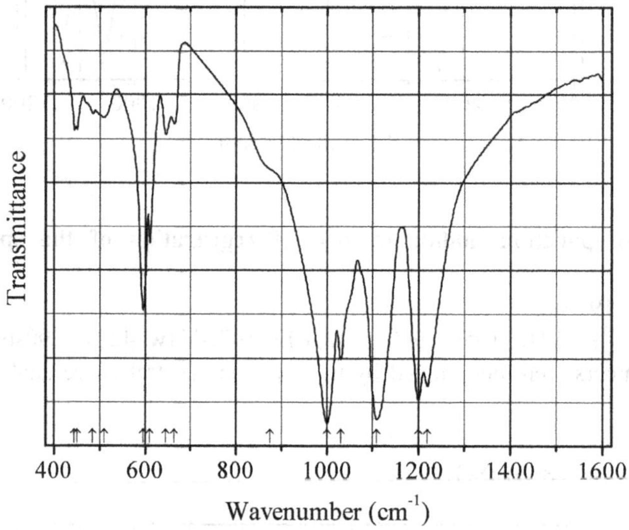

Origin: Synthetic.

Description: Trigonal, space group *R*3*c*.

Kind of sample preparation and/or method of registration of the spectrum: KBr and polyethylene discs. Transmission.

Source: Bremard et al. (1986).

Wavenumbers (IR, cm^{-1}): 1220, 1200s, 1110s, 1030, 1000s, 875sh, 665w, 645w, 610, 595, 510w, 485w, 445w, 450w, 314, 264w, 244w, 200w.

Note: In the cited paper, Raman spectrum is given.

Wavenumbers (Raman, cm^{-1}, at 77 K): 1237sh, 1225sh, 1205, 1115w, 1035sh, 1030w, 1010, 990s, 650w, 620, 615sh, 603, 597, 463, 447, 263, 258, 246sh, 204, 178sh, 168w, 155w, 138w, 57w, 37.5w.

S576 Dysprosium copper hydroxysulfate $Dy_2Cu(SO_4)_2(OH)_4$ $Dy_2Cu(SO_4)_2(OH)_4$

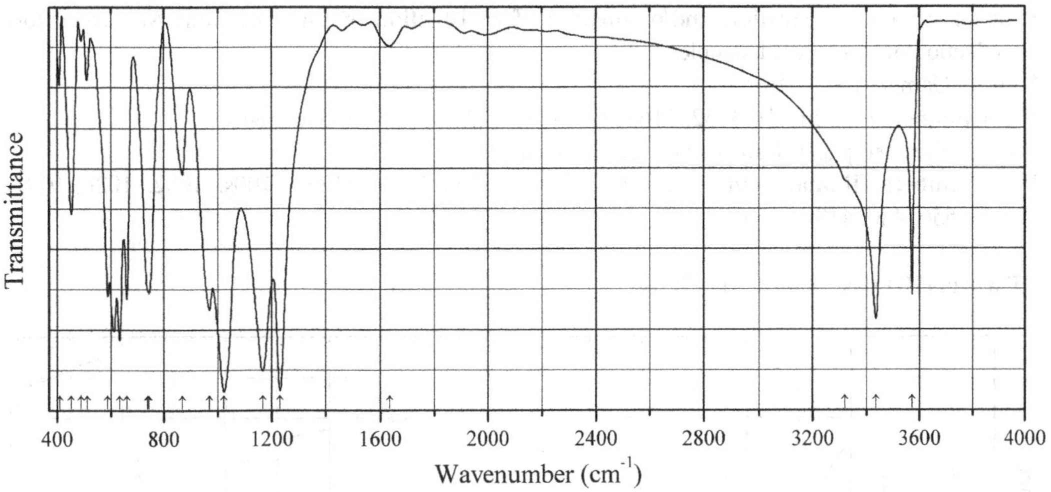

Origin: Synthetic.

Description: Synthesized by a hydrothermal method. Monoclinic, space group $P2_1/c$, $a = 6.304(4)$, $b = 6.663(4)$, $c = 10.724(6)$ Å, $\beta = 98.527(1)°$, $V = 445.5(5)$ Å^3, $Z = 2$. $D_{calc} = 4.806$ g/cm^3.

Kind of sample preparation and/or method of registration of the spectrum: Transmission. Kind of sample preparation is not indicated.

Source: Tang et al. (2015).

Wavenumbers (cm^{-1}): 3573, 3435, 3320sh, 1635w, 1232s, 1168s, 1023s, 970, 870, 744, 739sh, 661, 634s, 612s, 589, 512w, 491w, 454, 409w.

Note: The wavenumbers were partly determined by us based on spectral curve analysis of the published spectrum.

S577 Iron(III) basic sulfate $Fe(SO_4)(OH)$

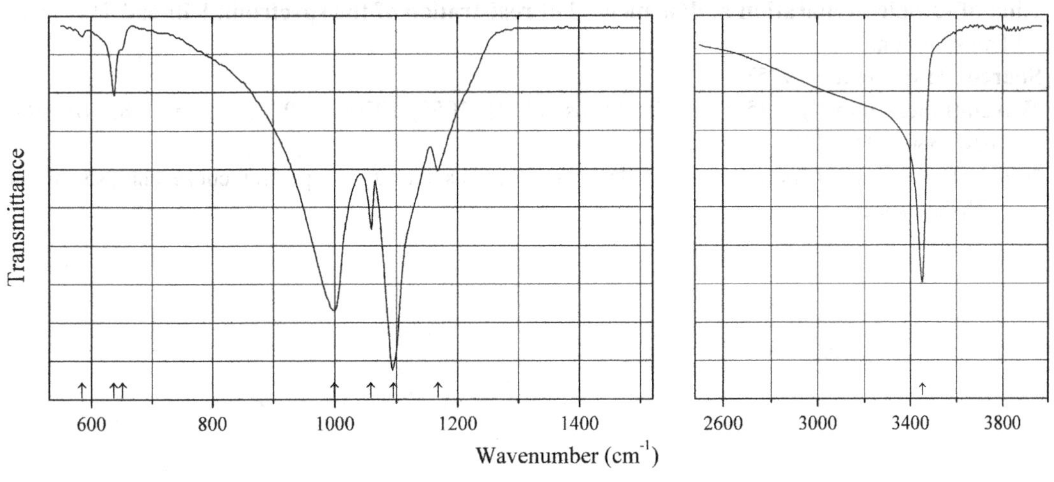

Origin: Synthetic.
Description: Prepared hydrothermally from $Fe_2(SO_4)_3{\cdot}nH_2O$. Monoclinic, space group $P2_1/c$.
Kind of sample preparation and/or method of registration of the spectrum: Attenuated total reflection of a powdered sample.
Source: Gomez et al. (2013).
Wavenumbers (IR, cm^{-1}): 3452, 1168, 1095s, 1058, 999s, 651sh, 636, 584w.
Note: In the cited paper, Raman spectrum is given.
Wavenumbers (Raman, cm^{-1}): 3077, 3453s, 3587w, 1183, 1122, 1100s, 1062, 1026, 914w, 645, 556, 480, 418w, 370, 231.

S578 Iron(III) basic sulfate $Fe(SO_4)(OH)$

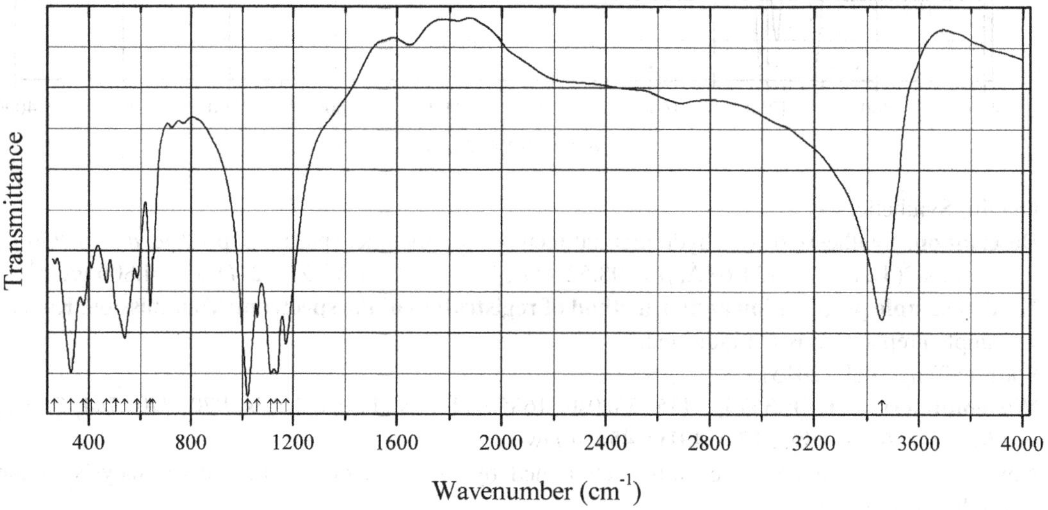

Origin: Synthetic.
Description: Orthorhombic, space group *Pnma*, $a = 7.33$, $b = 6.42$, $c = 7.14$ Å.
Kind of sample preparation and/or method of registration of the spectrum: KBr and TlBr discs. Transmission.
Source: Powers et al. (1975).
Wavenumbers (cm^{-1}): 3458s, 1172s, 1138s, 1112s, 1058, 1020s, 650, 638, 585, 538, 505, 468, 410w, 380, 331s, 270w.
Note: The wavenumbers were partly determined by us based on spectral curve analysis of the published spectrum.

S579 Lanthanum oxosulfate $La(SO_4)O_2$

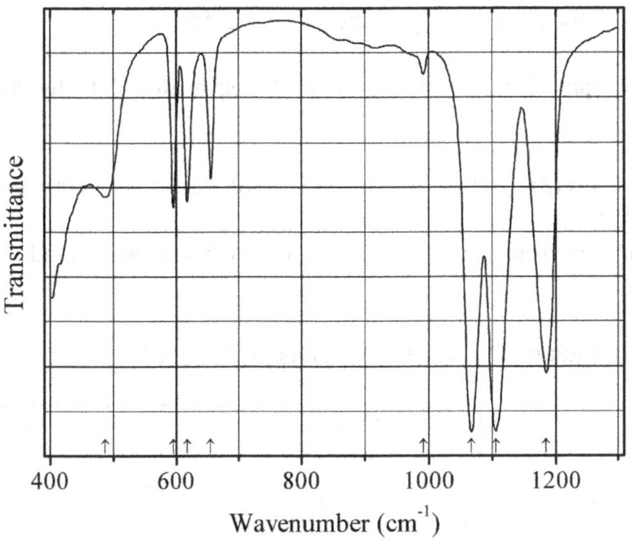

Origin: Synthetic.

Description: Synthesized by a template-assisted route described elsewhere. Characterized by powder X-ray diffraction data. Monoclinic, $a = 14.354(3)$, $b = 4.2862(6)$, $c = 8.388(2)$ Å, $\beta = 107.16(2)$.

Kind of sample preparation and/or method of registration of the spectrum: KBr disc. Absorption.

Source: Zhang et al. (2008).

Wavenumbers (cm^{-1}): 1185s, 1106s, 1067s, 992w, 655, 618, 595, 487.

Note: The wavenumbers were determined by us based on spectral curve analysis of the published spectrum.

S580 Lead(II) oxysulfate $Pb_5(SO_4)O_4$

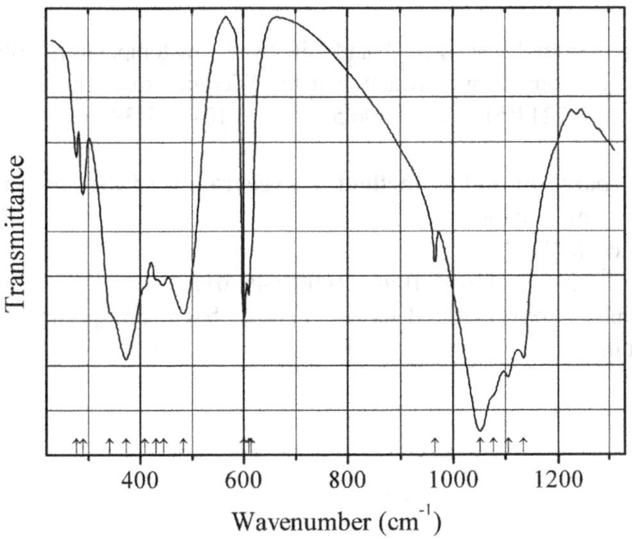

Origin: Synthetic.

Description: Microcrystalline powder obtained by the reaction of PbO with diluted H_2SO_4 at 80 °C during 4–6 h and subsequent heating of the product at 550 °C. Monoclinic, space group $P2_1/c$, $Z = 4$.

Kind of sample preparation and/or method of registration of the spectrum: KBr disc. Transmission.

Source: Grasselli and Baran (1984).

Wavenumbers (cm^{-1}): 1135s, 1105s, 1077sh, 1051s, 965, 614sh, 609, 600, 483, 445, 430sh, 408sh, 372s, 341sh, 290, 278w.

Note: The wavenumbers were partly determined by us based on spectral curve analysis of the published spectrum.

S581 Lithium iron(II) sulfate fluoride tavorite-type $LiFe(SO_4)F$

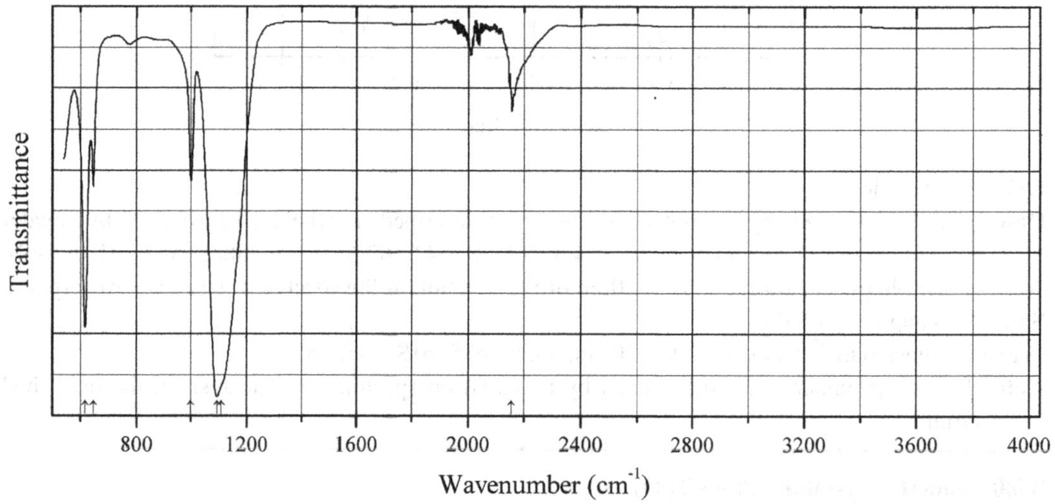

Origin: Synthetic.

Description: Prepared from $Fe(SO_4)\cdot H_2O$ and LiF by a low-temperature solvothermal approach. Characterized by powder X-ray diffraction data. Triclinic, space group P-1, a = 5.1760(4), b = 5.4909(4), c = 7.2214(5) Å, α = 106.511°, β = 107.187(3)°, γ = 97.847(2)°, V = 182.46 (2) Å^3.

Kind of sample preparation and/or method of registration of the spectrum: Attenuated total reflection of a powdered sample.

Source: Sobkowiak et al. (2013).

Wavenumbers (cm^{-1}): 2156, 1110sh, 1094s, 1000, 646, 617s.

Note: The wavenumbers were partly determined by us based on spectral curve analysis of the published spectrum.

S582 Magnesium hydroxysulfate hydrate $Mg_3(SO_4)_2(OH)_2{\cdot}2H_2O$

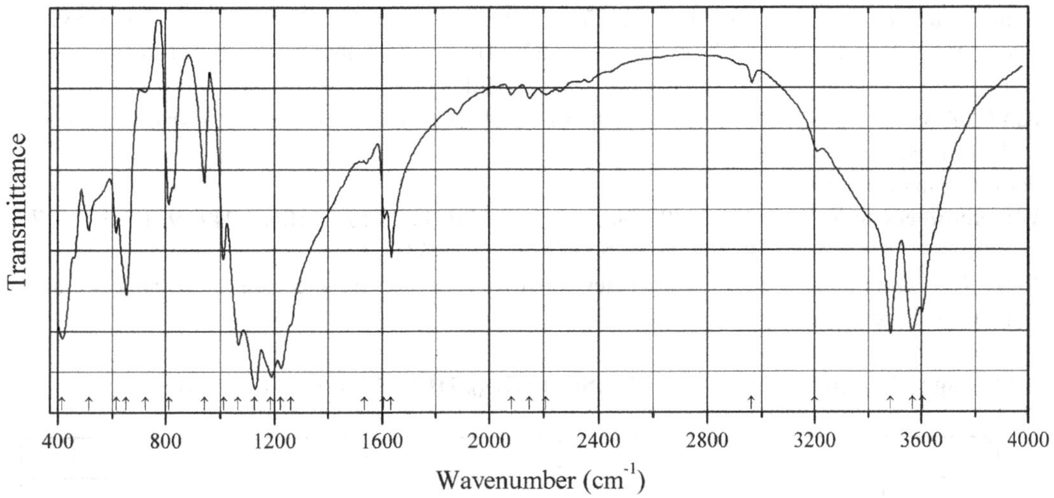

Origin: Synthetic.

Description: Crystals grown hydrothermally from NaOH and $MgSO_4$ at 160 °C for 21 days. The crystal structure is solved. Orthorhombic, space group *Pbcm*, a = 7.177(1), b = 9.804(2), c = 12.775(2) Å, V = 898.9(2) Å^3, Z = 4. D_{calc} = 2.476 g/cm^3.

Kind of sample preparation and/or method of registration of the spectrum: KBr disc. Transmission.

Source: Tao et al. (2002).

Wavenumbers (IR, cm^{-1}): 3607, 3567, 3483, 3200w, 2964w, 2207w, 2148w, 2081w, 1633, 1606, 1534w, 1261sh, 1224s, 1187s, 1126s, 1066s, 1012, 943, 813, 725w, 652, 616, 516, 416s.

Note: In the cited paper, Raman spectrum is given.

Wavenumbers (Raman, cm^{-1}): 3604s, 3565s, 3484s, 1215, 1103, 1027s, 653, 634, 494, 459, 270.

S583 Manganese hydroxysulfate $Mn_5(SO_4)(OH)_8$ $Mn_5(SO_4)(OH)_8$

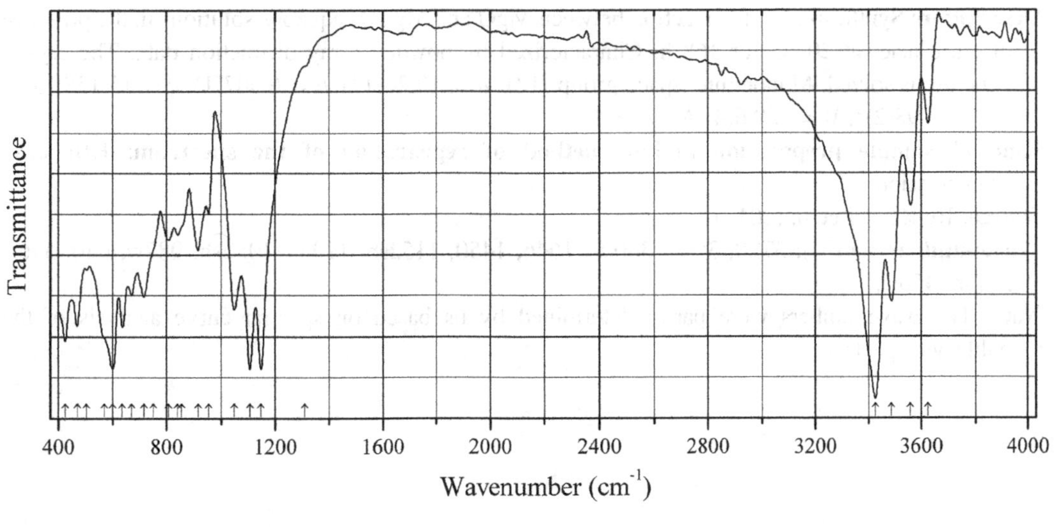

Origin: Synthetic.

Description: Prepared under mild hydrothermal conditions. Characterized by powder X ray diffraction data. The crystal structure is solved. Triclinic, space group *P*-1, $a = 7.5501(5)$, $b = 8.5558(6)$, $c = 8.6059(5)$ Å, $\alpha = 98.122(4)°$, $\beta = 102.370(4)°$, $\gamma = 99.646(4)°$, $V = 526.19(6)$ Å^3, $Z = 2$. $D_{calc} = 3.199$ g/cm^3. Mn atoms have five- and sixfold coordination.

Kind of sample preparation and/or method of registration of the spectrum: KBr disc. Transmission.

Source: Fan et al. (2005).

Wavenumbers (cm^{-1}): 3626, 3560, 3487, 3424s, 1310sh, 1148s, 1106, 1049, 954, 916, 857sh, 841, 809, 753sh, 716, 670, 637, 601s, 569sh, 505, 470, 425.

Note: The wavenumbers were partly determined by us based on spectsral curve analysis of the published spectrum.

S584 Magnesium sulfate hydroxide $Mg_6(SO_4)(OH)_{10}·7H_2O$ $Mg_6(SO_4)(OH)_{10}·7H_2O$

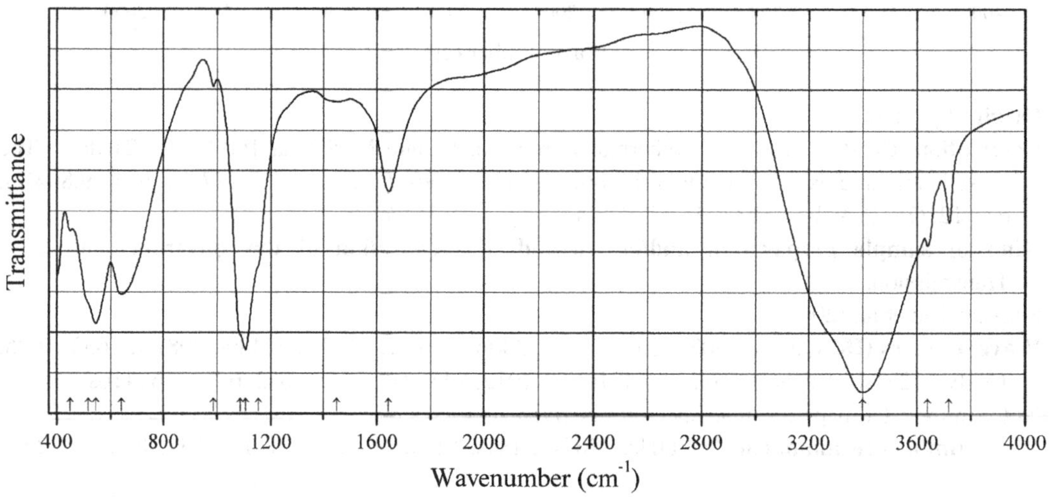

Origin: Synthetic.

Description: Synthesized in the reaction between MgO and $MgSO_4$ aqueous solutions in the presence of citric acid, at 20 °C for 168 h. Characterized by powder X-ray diffraction data. The crystal structure is solved. Monoclinic, space group *I*121, $a = 10.260(3)$, $b = 6.307(1)$, $c = 15.138(3)$ Å, $\beta = 103.98(2)°$, $V = 950.6(4)$ Å^3, $Z = 4$.

Kind of sample preparation and/or method of registration of the spectrum: KBr disc. Transmission.

Source: Runčevski et al. (2013).

Wavenumbers (cm^{-1}): 3720, 3640, 3400s, 1646, 1450, 1152sh, 1105s, 1086sh, 987w, 640, 546s, 517sh, 450w.

Note: The wavenumbers were partly determined by us based on spectral curve analysis of the published spectrum.

S585 Nickel hydroxysulfate hydrate $Ni_3(SO_4)_2(OH)_2{\cdot}2H_2O$ $Ni_3(SO_4)_2(OH)_2{\cdot}2H_2O$

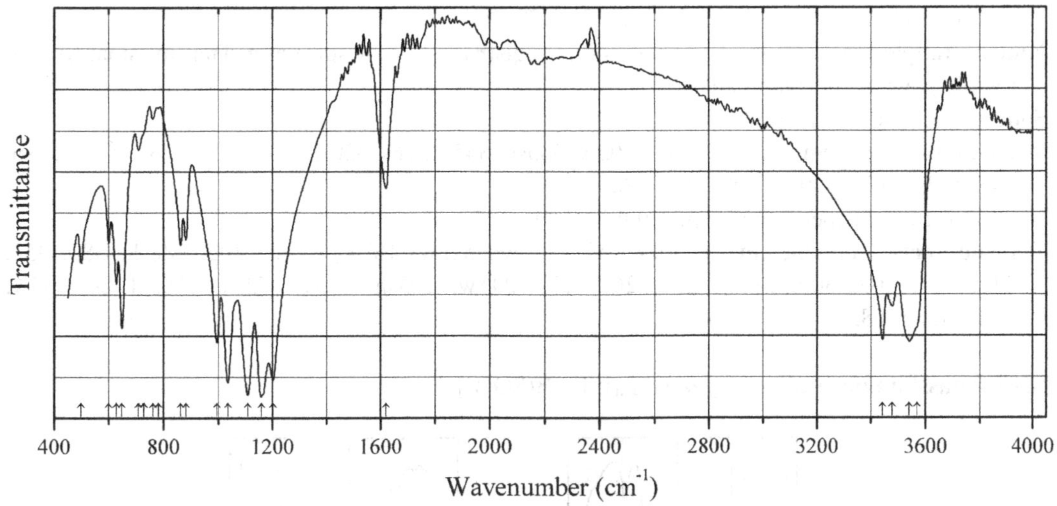

Origin: Synthetic.

Description: Green crystals prepared hydrothermally from $Ni(SO_4){\cdot}7H_2O$ and NaOH with the ratio Ni:Na:H_2O = 1:0.25:250 at 215–240 °C for 1–4 days. Orthorhombic, space group *Pbcm*, $a = 7.1485(3)$, $b = 9.6844(4)$, $c = 12.6643(3)$ Å, $V = 876.74(6)$ Å^3, $Z = 4$.

Kind of sample preparation and/or method of registration of the spectrum: KBr disc. Transmission.

Source: Vilminot et al. (2003).

Wavenumbers (cm^{-1}): 3570sh, 3541, 3477, 3442, 1619, 1202s, 1159s, 1109s, 1035s, 995, 881, 862, 780w, 760w, 727sh, 709w, 648, 627, 600, 499.

Note: The wavenumbers were partly determined by us based on spectral curve analysis of the published spectrum.

S586 Niobium sulfide NbS_3 NbS_3

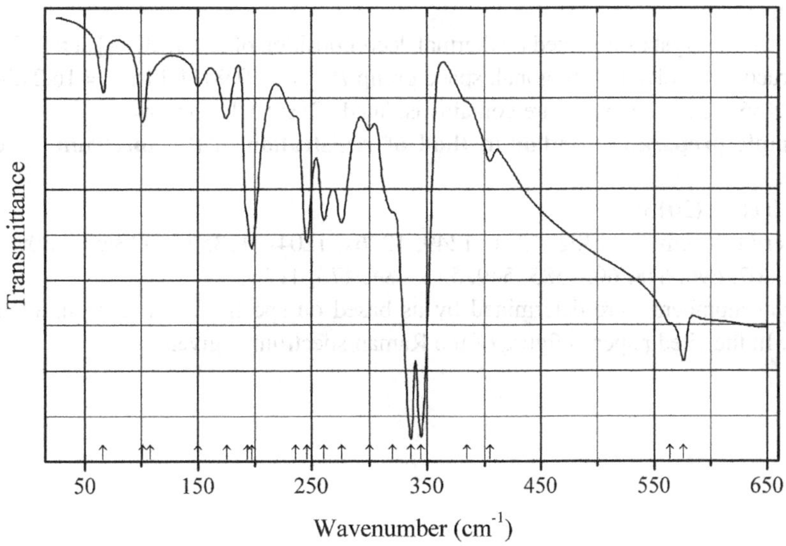

Origin: Synthetic.

Description: Obtained by a chemical vapor transport method. Triclinic, space group $P1$, $a = 4.963(2)$, $b = 6.730(2)$, $c = 9.144(4)$ Å, $\alpha = 90°$, $\beta = 97.17(1)°$, $\gamma = 90°$.

Kind of sample preparation and/or method of registration of the spectrum: Transmission. Kind of sample preparation is not indicated.

Source: Sourisseau et al. (1990).

Wavenumbers (IR, cm^{-1}): 576, 564w, 405w, 385sh, 345s, 336s, 320sh, 300w, 276, 261, 246, 236sh, 197, 193sh, 175, 150w, 108w, 101, 66.

Note: In the cited paper, Raman spectrum is given.

Wavenumbers (Raman, cm^{-1}): 602w, 573, 559w, 522w, 462w, 400sh, 392, 388sh, 380, 352, 341, 323, 303, 300, 288w, 281, 263, 257, 241w, 203sh, 195s, 172sh, 160, 152s, 133w, 108, 94, 85, 68.

S587 Potassium borosulfate $K_5[B(SO_4)_4]$ $K_5[B(SO_4)_4]$

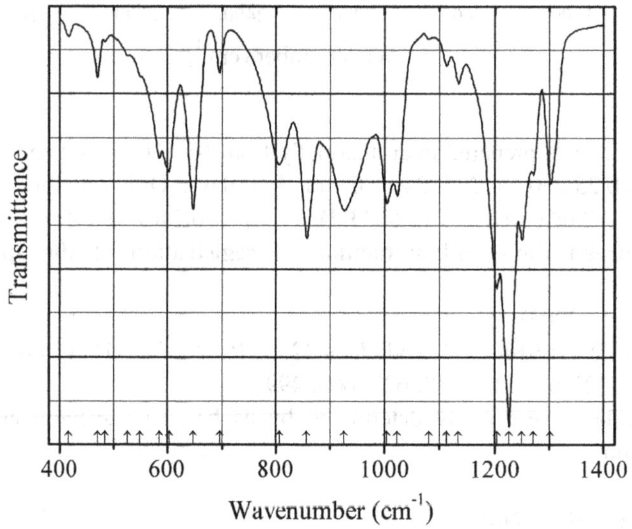

Origin: Synthetic.

Description: Single crystals obtained by thermal decomposition of $K_3[B(SO_4)_3]$ at 673 K for 12 h. The crystal structure is solved. Tetragonal, space group $P4_1$, $a = 9.9044(14)$, $c = 16.215(3)$ Å, $Z = 4$. $D_{calc} = 2.466$ g/cm^3. The structure contains isolated $[B(SO_4)_4]^{5-}$ anions.

Kind of sample preparation and/or method of registration of the spectrum: Attenuated total reflection.

Source: Daub et al. (2013).

Wavenumbers (IR, cm^{-1}): 1302, 1271, 1249, 1226s, 1204s, 1135w, 1113w, 1080w, 1023, 1004, 926, 857, 807, 696, 648, 603, 585, 549, 526, 484, 470, 417w.

Note: The wavenumbers were determined by us based on spectral curve analysis of the published spectrum. In the cited paper, a figure of the Raman spectrum is given.

S588 Potassium sodium vanadyl sulfate $K_6Na_2(VO)_2(SO_4)_7$

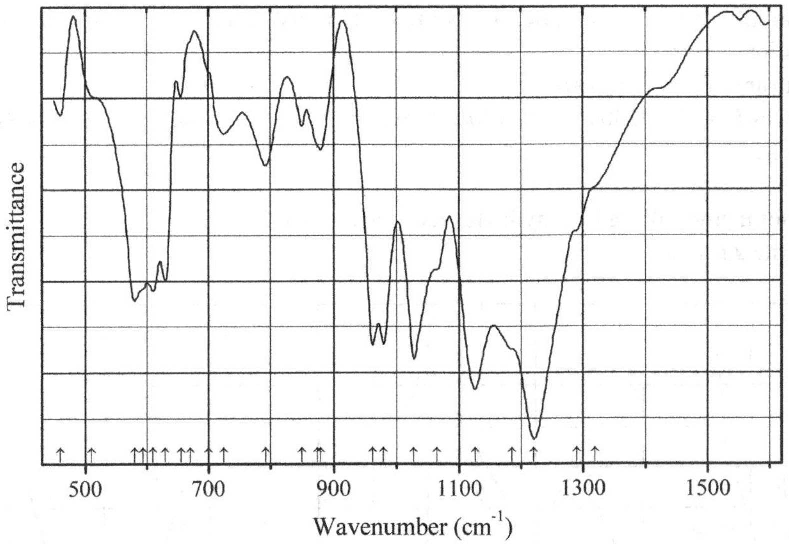

Origin: Synthetic.

Description: Prepared from the mixture of $Na_2S_2O_7$, $K_2S_2O_7$, and V_2O_5 at 325 °C. The crystal structure is solved. Tetragonal, space group $P4_32_12$, $a = 9.540(3)$, $c = 29.551(5)$ Å, $V = 2689.5$ (13) Å^3, $Z = 4$. $D_{calc} = 2.684$ g/cm^3. The shortest V–O bond length is equal to 1.552(6) Å.

Kind of sample preparation and/or method of registration of the spectrum: KBr disc. Transmission.

Source: Karydis et al. (2002).

Wavenumbers (IR, cm^{-1}): 1320sh, 1290sh, 1220s, 1185sh, 1126s, 1065sh, 1028s, 980s, 962s, 880, 875sh, 850, 792, 725, 700sh, 670w, 655, 630, 610, 594sh, 580, 510sh, 460w.

Note: In the cited paper, Raman spectrum is given.

Wavenumbers (Raman, cm^{-1}): 1310w, 1230w, 1165w, 1028, 1005, 968, 895w, 860w, 790, 685w, 630s, 605s, 570w, 490w, 445w, 396s, 302, 260w, 174s, 140s.

S589 Potassium zinc sulfate chloride trihydrate $KZn(SO_4)Cl{\cdot}3H_2O$

Kainite Zn-analogue

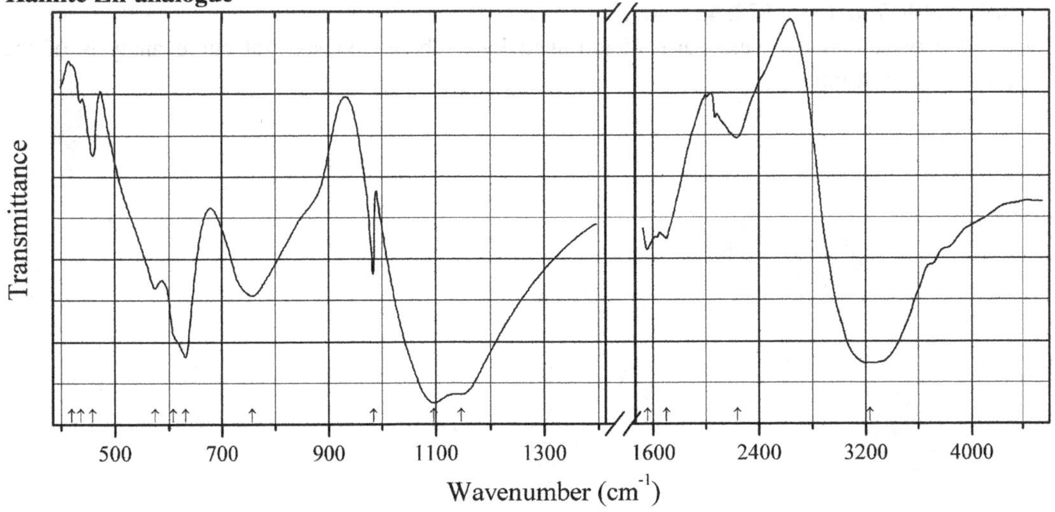

Origin: Synthetic.

Description: Single crystals doped by Cu^{2+} ions. Monoclinic, space group *C*2/*m*.

Kind of sample preparation and/or method of registration of the spectrum: KBr disc. Transmission.

Source: Narasimhulu et al. (2000).

Wavenumbers (cm^{-1}): 3230s, 2236, 1700, 1559, 1146s, 1096s, 984, 756, 631s, 608sh, 575, 459, 436w, 420sh.

S590 Potassium zinc sulfate hexahydrate $K_2Zn(SO_4)_2·6H_2O$
Picromerite Zn-analogue

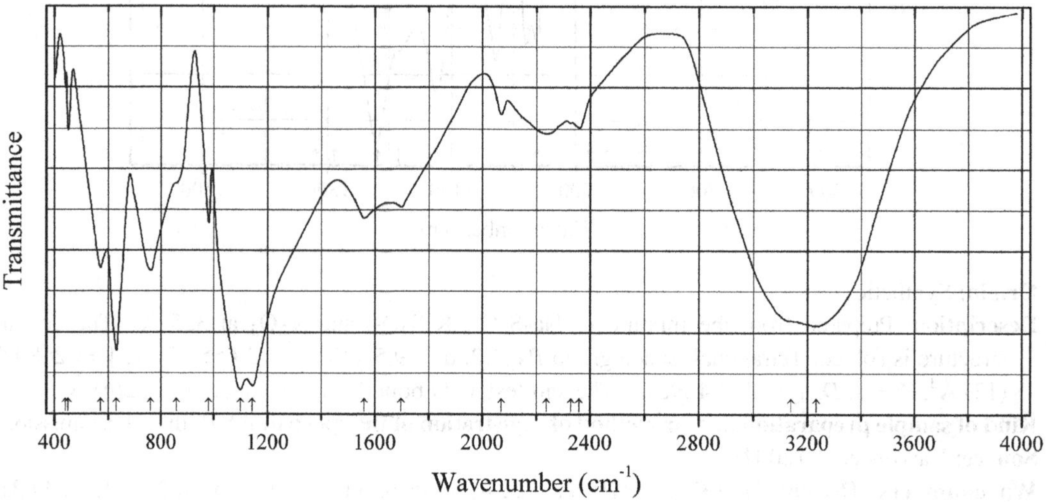

Origin: Synthetic.

Description: Crystals grown from aqueous solution by slow evaporation. Characterized by powder X-ray diffraction data. Monoclinic, space group *P*2$_1$/*c*. Isostructural with picromerite.

Kind of sample preparation and/or method of registration of the spectrum: KBr disc. Transmission.

Source: Manonmoni et al. (2014).

Wavenumbers (cm^{-1}): 3233s, 3140sh, 2360, 2328w, 2238, 2070w, 1699, 1559, 1142s, 1098s, 983, 862sh, 760, 631s, 572, 451, 441.

Note: The wavenumbers were partly determined by us based on spectral curve analysis of the published spectrum.

S591 Rubidium beryllium sulfate hydrate $Rb_2Be(SO_4)_2{\cdot}2H_2O$

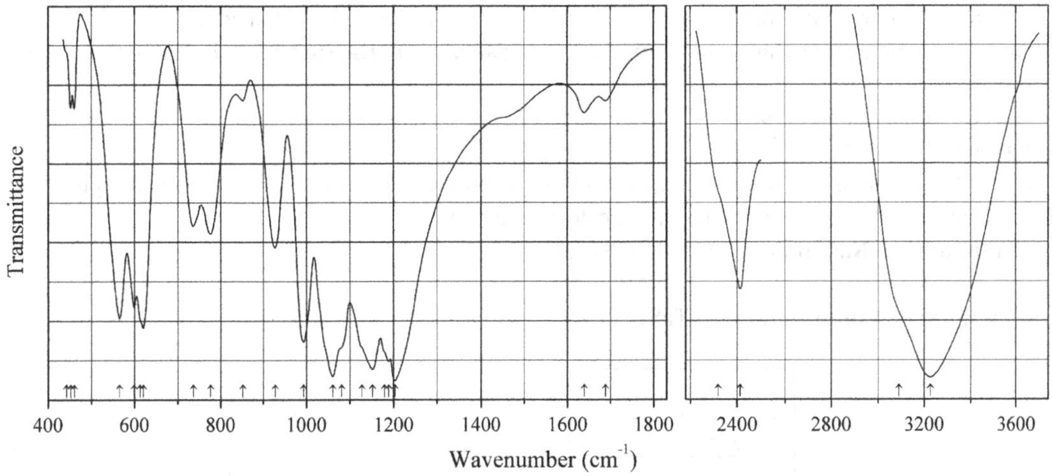

Origin: Synthetic.

Description: Obtained from the three-component system Rb_2SO_4–$BeSO_4$–H_2O by the method of isothermal decrease of supersaturation. The crystal structure is solved. Monoclinic, space group $P2_1/c$, $a = 11.371(2)$, $b = 11.858(2)$, $c = 7.431(1)$ Å, $\beta = 96.33(1)°$, $V = 996.0$ Å^3, $Z = 4$. $D_{calc} = 2.722$ g/cm^3. In the structure, the $[Be(SO_4)_2(H_2O)_2]^{2-}$ units are arranged to form double layers.

Kind of sample preparation and/or method of registration of the spectrum: KBr disc. Transmission.

Source: Georgiev et al. (2007).

Wavenumbers (IR, cm^{-1}): 3226s, 3090sh, 2413, 2320sh, 1688w, 1639, 1203s, 1188s, 1178sh, 1151s, 1127sh, 1081sh, 1060s, 993s, 928, 852w, 776, 736, 620s, 613sh, 599, 565s, 460, 452, 442sh.

Note: The bands indicated by Georgiev et al. (2007) at 3054 and 2294 cm^{-1} are observed as shoulders at 3090 and 2320 cm^{-1}. In the cited paper, Raman spectrum is given.

Wavenumbers (Raman, cm^{-1}): 1212, 1185, 1156, 1120w, 1078s, 1007s, 993sh, 924w, 757, 633, 606, 589, 557w, 497, 461, 436, 387, 311w, 266.

S592 Silver indium sulfide $AgIn_5S_8$ $AgIn_5S_8$

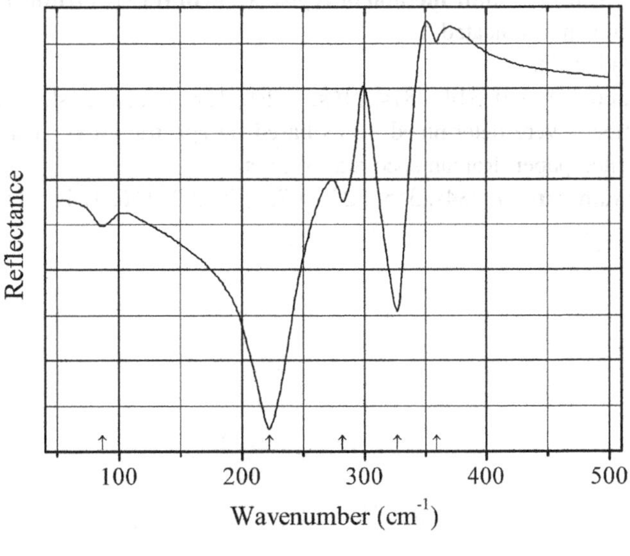

Origin: Synthetic.

Description: A compound with cubic spinel-type structure. Space group $Fd\bar{3}m$, $Z = 2$. One indium atom has tetrahedral coordination, and four indium atoms have octahedral coordination.

Kind of sample preparation and/or method of registration of the spectrum: Reflection of a single crystal.

Source: Gasanly et al. (1993).

Wavenumbers (IR, cm^{-1}): 359w, 327s, 282, 222s, 87.

Note: The wavenumbers were determined by us based on spectral curve analysis of the published spectrum. In the cited paper, Raman spectrum is given.

Wavenumbers (Raman, cm^{-1}): 354, 328s, 292, 181, 69.

S593 Silver tantalum sulfide $AgTaS_3$

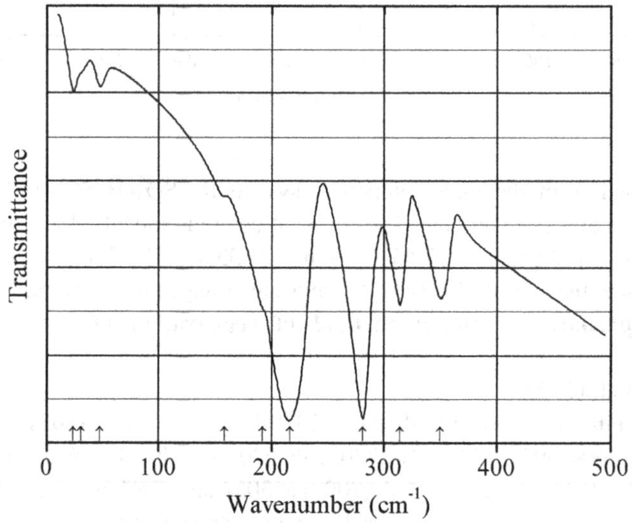

Origin: Synthetic.

Description: Prepared by heating a mixture of Ta, S, and Ag_2S powders at 500 °C for 4 days. Orthorhombic, with a layered structure; $a = 3.3755(2)$, $b = 14.0608(11)$, $c = 7.7486(7)$ Å, $Z = 4$. $D_{meas} = 6.82(3)$ g/cm^3.

Kind of sample preparation and/or method of registration of the spectrum: Transmission. Kind of sample preparation is not indicated.

Source: Ishii and Wada (2000).

Wavenumbers (IR, cm^{-1}): 350, 314, 281s, 216s, 192sh, 158sh, 48w, 31sh, 24w.

Note: The wavenumbers were determined by us based on spectral curve analysis of the published spectrum. In the cited paper, Raman spectrum is given.

Wavenumbers (Raman, cm^{-1}): 384s, 355, 322, 307, 260, 209, 166, 124s, 111s, 30.

S594 Sodium cadmium sulfate hydrate $NaCd(SO_4)_2 \cdot 2H_2O$

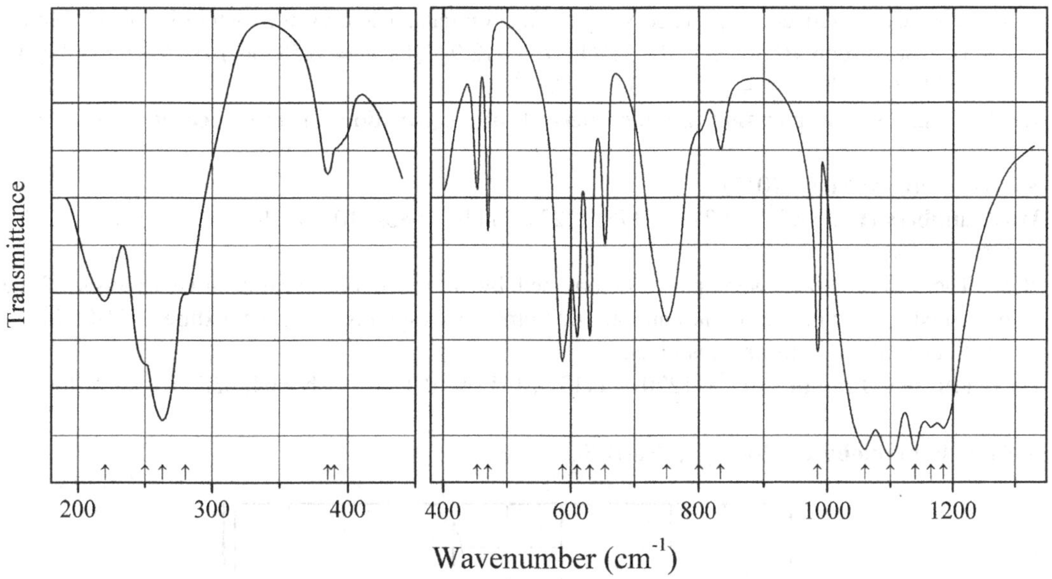

Origin: Synthetic.

Description: Monoclinic, space group *P*2$_1$/*c*.

Kind of sample preparation and/or method of registration of the spectrum: CsI and/or KBr disc. Transmission.

Source: Peytavin et al. (1972a).

Wavenumbers (IR, cm^{-1}): 1185s, 1165s, 1140s, 1100s, 1060s, 985, 834w, 800sh, 750, 654, 630, 610, 587, 470, 453, 390sh, 385w, 280sh, 263s, 250sh, 220.

Note: In the cited paper, Raman spectrum is given.

Wavenumbers (Raman, cm^{-1}): 1174, 1165sh, 1132s, 1120sh, 1047s, 991s, 800w, 750w, 652, 633, 619, 590, 467s, 446s, 358w, 310sh, 270, 260, 220sh, ~150sh.

S595 Sodium manganese(II) sulfate alluadite-type $Na_{2+x}Mn_{2-x/2}(SO_4)_3$

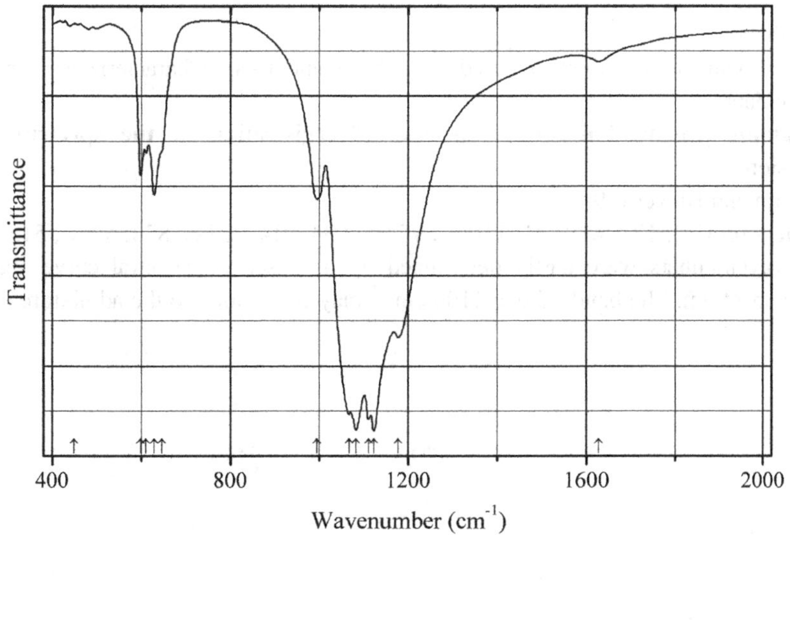

Origin: Synthetic.

Description: Prepared by dehydratation of the compound $Na_2Mn(SO_4)_3 \cdot 2H_2O$ with a kröhnkite-type structure. Characterized by powder X-ray diffraction data. The crystal structure is solved. Monoclinic, space group $P2_1/c$, $a = 11.541(1)$, $b = 12.944(1)$, $c = 6.5875(6)$ Å, $\beta = 95.149(3)°$, $V = 980.13(26)$ Å^3, $Z = 4$. $D_{calc} = 3.078$ g/cm^3.

Kind of sample preparation and/or method of registration of the spectrum: KBr disc. Transmission.

Source: Marinova et al. (2015).

Wavenumbers (IR, cm^{-1}): 1627w, 1178, 1123s, 1111s, 1083s, 1067s, 999, 991, 647sh, 629, 611, 599, ~450w.

Note: The wavenumbers were partly determined by us based on spectral curve analysis of the published spectrum. The weak band at 1627 cm^{-1} corresponds to the admixture of H_2O. In the cited paper, Raman spectrum is given.

Wavenumbers (Raman, cm^{-1}): 1220w, 1117w, 1051w, 1012s, 993sh, 664, 635, 616, 602, 466.

S596 Sodium thioborate $Na_3B_3S_6$ $Na_3B_3S_6$

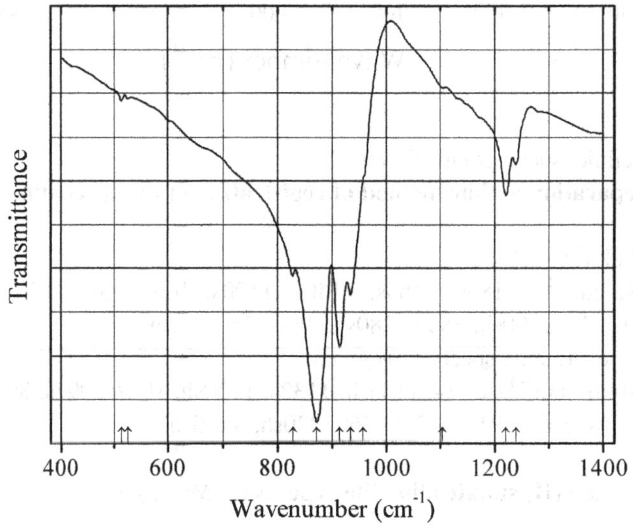

Origin: Synthetic.

Description: Obtained from melt prepared from Na_2S and B_2S_3. Characterized by powder X-ray diffraction data.

Kind of sample preparation and/or method of registration of the spectrum: KBr disc. Transmission.

Source: Martin and Bloyer (1991).

Wavenumbers (cm^{-1}): 1239w, 1221, 1104w, 957sh, 934, 914s, 872s, 828, 525w, 513w.

Note: The wavenumbers were partly determined by us based on spectral curve analysis of the published spectrum. The bands above 1100 cm^{-1} may correspond to the admixture of a borate.

S597 Sodium thioborate Na_3BS_3 Na_3BS_3

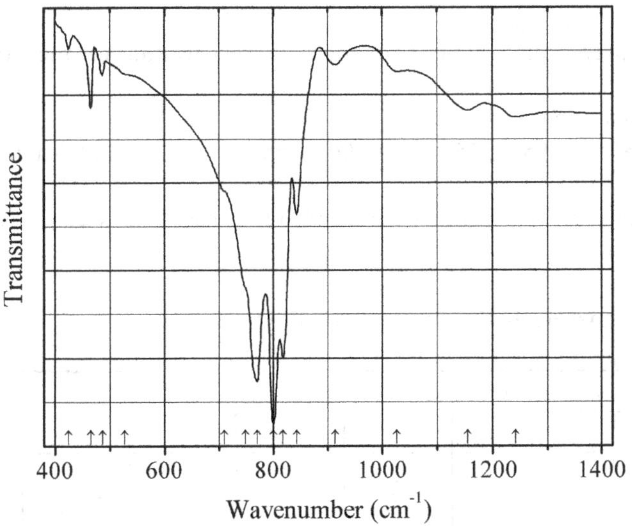

Origin: Synthetic.
Description: Obtained from melt prepared from Na_2S and B_2S_3.
Kind of sample preparation and/or method of registration of the spectrum: KBr disc. Transmission.
Source: Martin and Bloyer (1991).
Wavenumbers (cm^{-1}): 1242, 1155, 1027w, 914w, 843, 818s, 800s, 770s, 749sh, 710sh, 527sh, 486w, 465, 425w.
Note: The wavenumbers were partly determined by us based on spectral curve analysis of the published spectrum. The bands above 1100 cm^{-1} may correspond to the admixture of a borate.

S598 Tellurium(IV) oxosulfate $Te_2(SO_4)O_3$

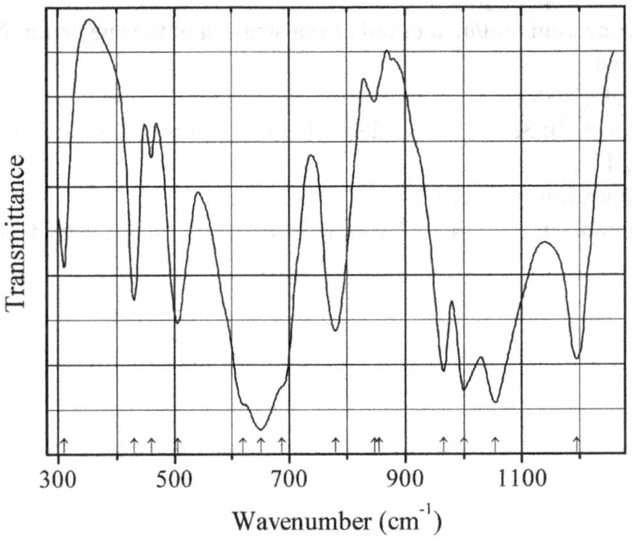

Origin: Synthetic.

Description: Orthorhombic, space group $P2_1mn$, $a = 4.676(2)$, $b = 8.911(3)$, $c = 6.879(4)$ Å, $V = 286.63$ Å^3, $Z = 2$. $D_{calc} = 4.61$ g/cm^3.

Kind of sample preparation and/or method of registration of the spectrum: KBr disc. Transmission.

Source: Gaitán et al. (1985).

Wavenumbers (cm^{-1}): 1195, 1055s, 1000s, 965, 855sh, 848w, 780, 687sh, 650s, 620, 505, 460w, 430, 310.

Note: The wavenumbers were partly determined by us based on spectral curve analysis of the published spectrum.

S599 Triammoniun hydrogen disulfate $(NH_4)_3H(SO_4)_2$

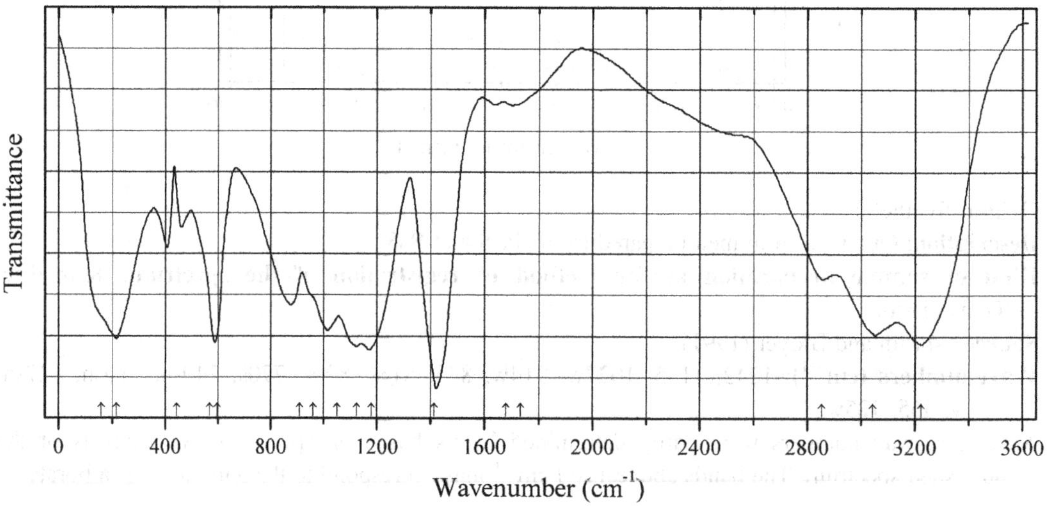

Origin: Synthetic.

Description: Monoclinic, space group $C2/c$, $Z = 4$.

Kind of sample preparation and/or method of registration of the spectrum: Nujol and Fluorolube mulls. Transmission.

Source: Kamoun et al. (1988).

Wavenumbers (IR, cm^{-1}): 3223s, 3043s, 2855, 1733w, 1680w, 1414s, 1180s, 1125s, 1080, 960sh, 910, 597s, (570), 444.

Note: In the cited paper, Raman spectrum is given.

Wavenumbers (Raman, cm^{-1}): 3145, 2870sh, 1674w, 1415w, 1078s, 966s, 619, 606, 590, 467, 442.

S600 Alpersite (Mg,Cu)$(SO_4)\cdot 7H_2O$

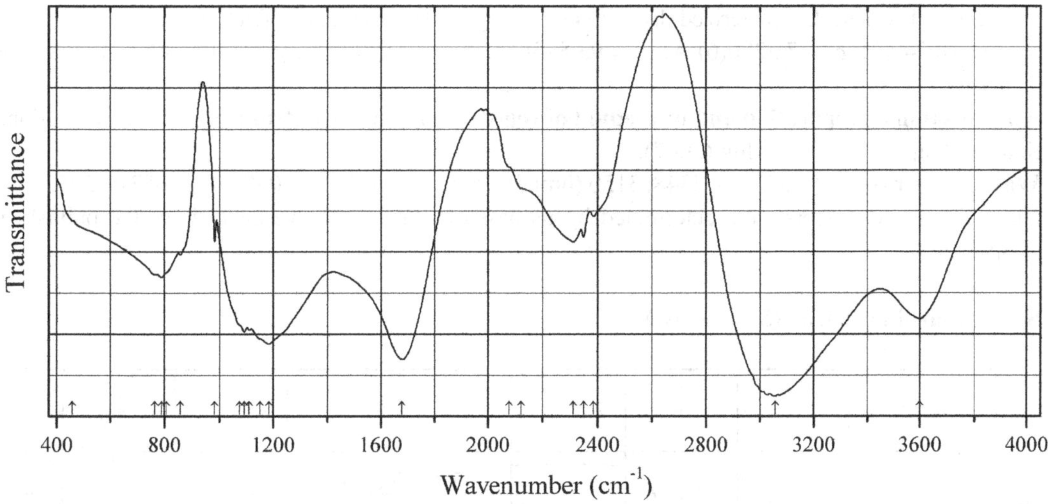

Origin: Malanjkhand porphyry copper mine, near Balaghet, Madhya Pradesh, India.

Description: Light blue crusts associated with epsomite, hexahydrite, and gypsum. Characterized by powder X-ray diffraction data and semiquantitative electron microprobe analysis.

Kind of sample preparation and/or method of registration of the spectrum: KBr disc. Transmission.

Source: Equeenuddin (2015).

Wavenumbers (cm^{-1}): 3600, 3054s, 2312, 2121sh, 2077sh, 1679s, 1184s, (1152sh), (1111), (1092), (1074sh), 984w, 859, (809sh), 790, (765sh), 460sh.

Note: The wavenumbers were partly determined by us based on spectral curve analysis of the published spectrum. The bands of H_2O (at 3054 and 1679 cm^{-1}) are anomalously strong and may be due to adsorbed water.

S601 Aluminocopiapite (Al,Mg)$Fe^{3+}{}_4(SO_4)_6(OH,O)_2\cdot 20H_2O$

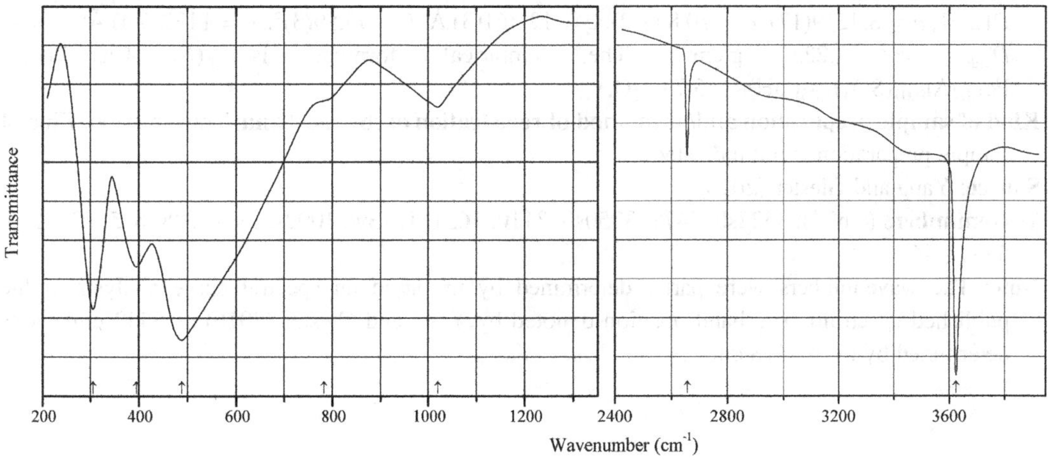

Origin: Synthetic.

Description: Prepared from $Fe_2(SO_4)_3{\cdot}6.25H_2O$ and $Al_2(SO_4)_3{\cdot}17H_2O$ at 25°, in the presence of excess of water. Characterized by powder X-ray diffraction data. Triclinic, a = 7.3853(7), b = 18.249(2), c = 7.3280(6) Å, α = 93.873(7)°, β = 102.221(6)°, γ = 99.163(6)°, V = 947.7 (2) Å^3.

Kind of sample preparation and/or method of registration of the spectrum: KCl disc. Absorption.

Source: Majzlan and Michallik (2007).

Wavenumbers (cm^{-1}): 3528sh, 3348, 3150 (broad), 1630, 1127s, 1075, 1030, 988s, 657sh, 595, 551.

Note: The wavenumbers were determined by us based on spectral curve analysis of the published spectrum.

S602 Amarillite $NaFe^{3+}(SO_4)_2{\cdot}6H_2O$

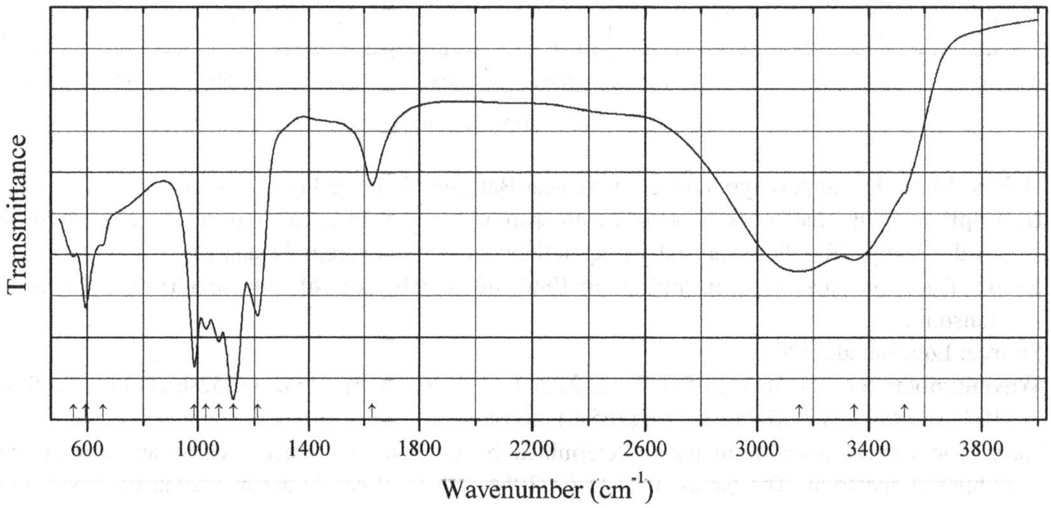

Origin: Xitieshan Pb-Zn deposit, Qaidam basin, Qinghai province, China.

Description: Pale yellow or white, fibrous and tabular aggregates, from the association with copiapite, römerite, coquimbite, and melanterite. The crystal structure is solved. Monoclinic, space group $P12/c1$, a = 8.4219(17), b = 10.844(2), c = 12.461(3) Å, β = 95.59(3)°, V = 1132.6(4) Å^3, Z = 4. D_{calc} = 2.223 g/cm^3. The empirical formula is $(Na_{0.97}Ca_{0.01}Pb_{0.01})Fe_{1.04}Al_{0.07}(SO_4)_{1.05}(OH)_{0.42}{\cdot}5.78H_2O$.

Kind of sample preparation and/or method of registration of the spectrum: Transmission. Kind of sample preparation is not indicated.

Source: Yang and Giester (2016).

Wavenumbers (cm^{-1}): 3522sh, 3478, 3350sh, 3110, 1631, 1213w, 1092s, 989s, 708, 675, 634, 595, 549s, 446.

Note: The wavenumbers were partly determined by us based on spectral curve analysis of the published spectrum. The band position denoted by Yang and Giester (2016) as 3442 cm^{-1} was determined by us at 3478 cm^{-1}.

S603 Campostriniite $(Bi_{2.5}Na_{0.5})(NH_4)_2Na_2(SO_4)_6{\cdot}H_2O$

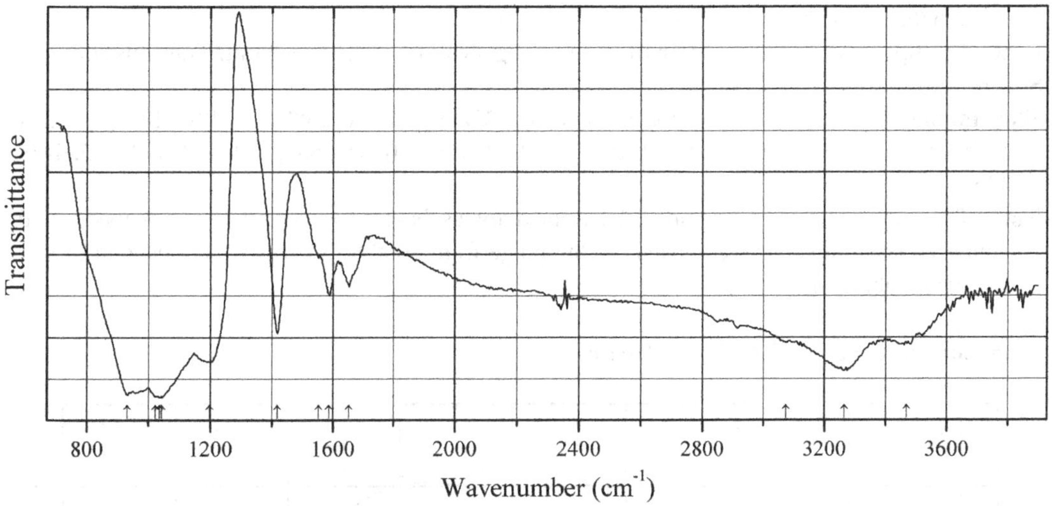

Origin: La Fossa crater, Vulcanoisland, Lipari, Eolie (Aeolian) islands, Messina province, Sicily, Italy (type locality).

Description: White prismatic crystals from the association with adranosite, demicheleite-(Br), demicheleite-(I), argesite, and sassolite. Holotype sample. The crystal structure is solved. Monoclinic, space group *C*2/*c*, $a = 17.748(3)$, $b = 6.982(1)$, $c = 18.221(3)$ Å, $\beta = 113.97(1)°$, $V = 2063(1)$ Å^3, $Z = 4$. $D_{calc} = 3.87$ g/cm^3. The empirical formula is $Bi_{2.43}N_{1.52}Na_{2.41}K_{0.48}S_{6.07}H_{8.08}O_{25}$. The strongest lines of the powder X-ray diffraction pattern [*d*, Å (*I*, %) (*hkl*)] are: 6.396 (100) (110), 7.507 (75) (−202), 2.766 (60) (−316), 3.380 (57) (312), 5.677 (55) (111), 3.166 (50) (402).

Kind of sample preparation and/or method of registration of the spectrum: No data.

Source: Demartin et al. (2015).

Wavenumbers (cm^{-1}): 3470, 3265s, 3071, 1654, 1588, 1553sh, 1418s, 1198s, 1043s, 1036s, 1021sh, 932s.

Note: The wavenumbers were partly determined by us based on spectral curve analysis of the published spectrum. Weak bands in the range from 2800 to 3000 cm^{-1} correspond to the admixture of an organic substance. Weak bands in the range from 2300 to 2400 cm^{-1} correspond to atmospheric CO_2.

S604 Changoite (slightly deuterated) $Na_2Zn(SO_4)_2{\cdot}4H_2O$

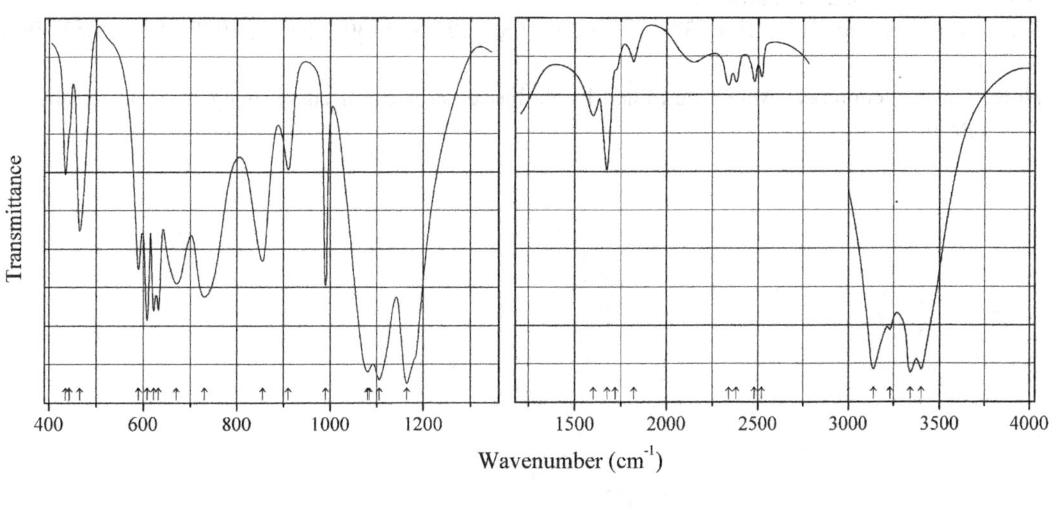

Origin: Synthetic.

Description: Prepared by crystallization from aqueous solurion. Monoclinic, space group $P2_1/a$, $a = 5.536$, $b = 8.249$, $c = 11.078$ Å, $\beta = 100.25°$.

Kind of sample preparation and/or method of registration of the spectrum: Transmission.

Source: Peytavin et al. (1972b).

Wavenumbers (cm^{-1}): 3400s, 3340s, 3230, 3140s, 2520w, 2480w, 2380w, 2340w, 1820w, 1720sh, 1675, 1600, 1165s, 1105s, 1084sh, 1080s, 990, 910, 855, 730, 670, 632, 622, 608, 590, 465, 443sh, 435.

Note: The wavenumbers were partly determined by us based on spectral curve analysis of the published spectrum. The weak bands in the range from 2000 to 2600 cm^{-1} correspond to D–O-stretching vibrations.

S605 Copiapite $Fe^{2+}Fe^{3+}{}_4(SO_4)_6(OH)_2{\cdot}20H_2O$

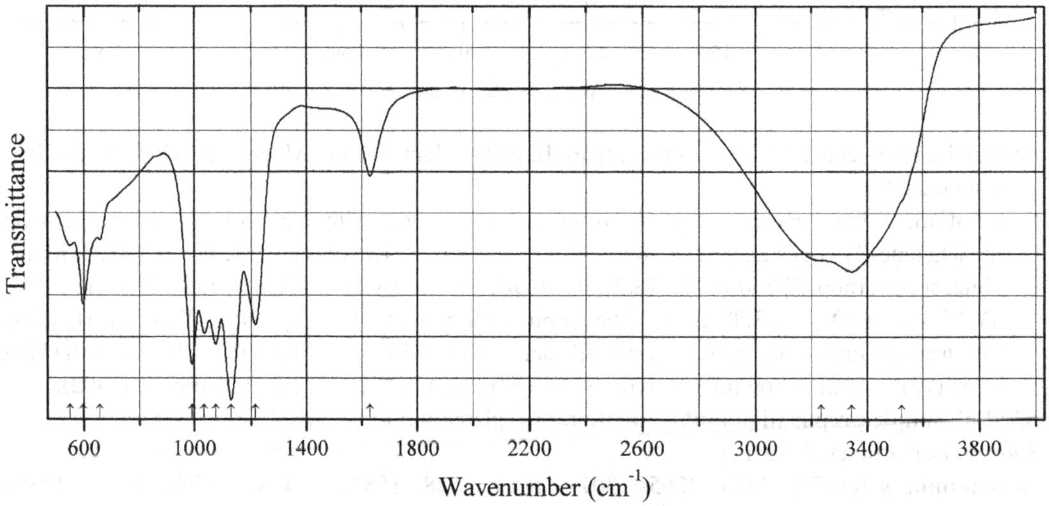

Origin: Synthetic.

Description: Prepared from $Fe_2(SO_4)_3{\cdot}6.25H_2O$ and $Fe(SO_4){\cdot}7H_2O$ at 25°, in the presence of excess of water. Characterized by powder X-ray diffraction data. Triclinic, $a = 7.3858(9)$, $b = 18.592(3)$, $c = 7.3543(8)$ Å, $\alpha = 92.273(9)°$, $\beta = 102.274(8)°$, $\gamma = 98.290(9)°$, $V = 973.9(2)$ Å^3.

Kind of sample preparation and/or method of registration of the spectrum: KCl disc. Absorption.

Source: Majzlan and Michallik (2007).

Wavenumbers (cm^{-1}): 3523sh, 3344, 3234sh, 1630, 1216s, 1130s, 1075s, 1035s, 991s, 657, 597, 551.

Note: The wavenumbers were determined by us based on spectral curve analysis of the published spectrum.

S606 Ferricopiapite $Fe^{3+}_{0.67}Fe^{3+}_4(SO_4)_6(OH)_2 \cdot 20H_2O$

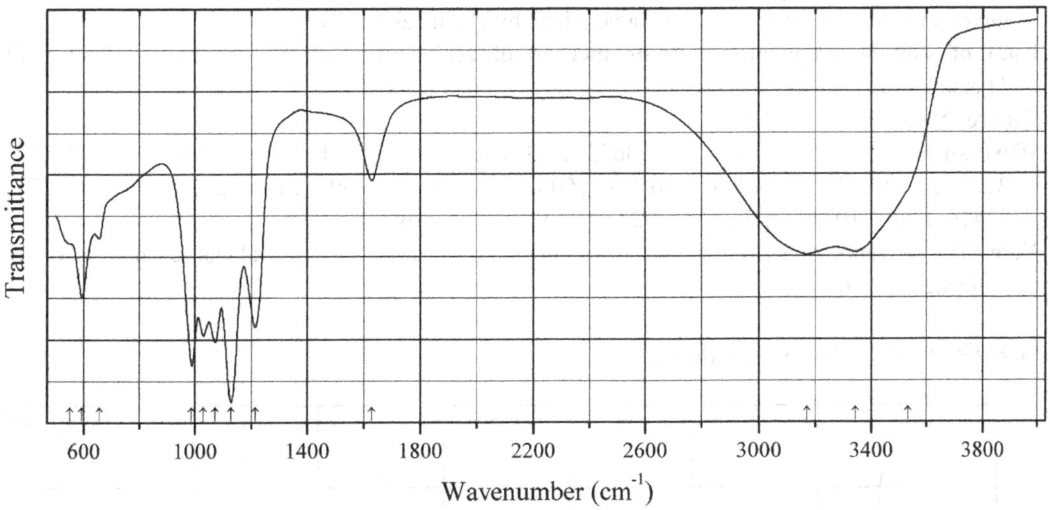

Origin: Synthetic.

Description: Prepared from $Fe_2(SO_4)_3 \cdot 6.25H_2O$ in the presence of excess of water at 25 °C. Characterized by powder X-ray diffraction data. Triclinic, a = 7.3871(5), b = 18.362(1), c = 7.3286(4) Å, α = 93.938(5)°, β = 102.208(4)°, γ = 98.920(4)°, V = 954.5(1) Å^3.

Kind of sample preparation and/or method of registration of the spectrum: KCl disc. Absorption.

Source: Majzlan and Michallik (2007).

Wavenumbers (cm^{-1}): 3532sh, 3343, 3170, 1628, 1215s, 1127s, 1071s, 1030s, 988s, 656w, 551sh.

Note: The wavenumbers were determined by us based on spectral curve analysis of the published spectrum.

S607 Geschieberite (?) $K_2(UO_2)(SO_4)_2 \cdot 2H_2O$

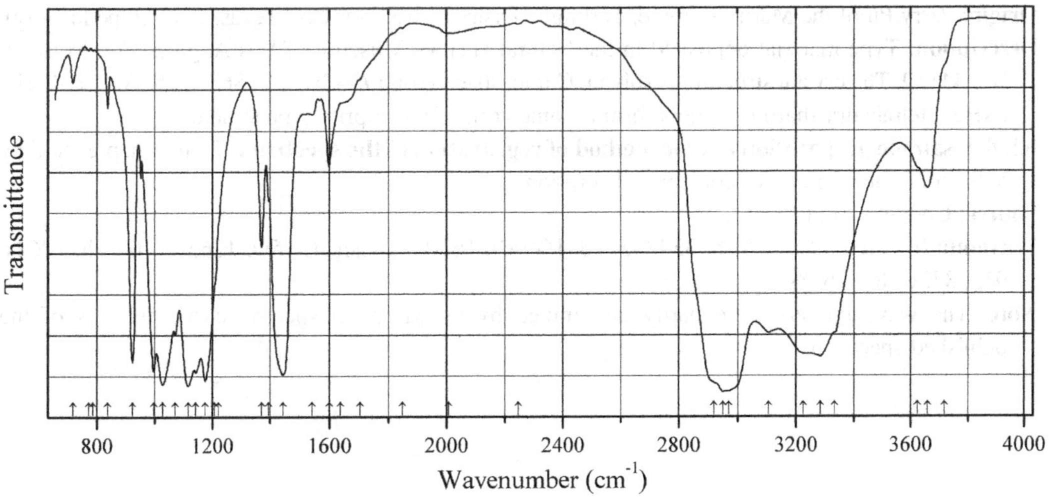

Origin: Synthetic.

Description: Prepared by cooling a hot saturated solution of potassium sulfate and uranyl sulfate mixed in equimolar proportions. Characterized by chemical analyses.

Kind of sample preparation and/or method of registration of the spectrum: Nujol mull. Transmission.

Source: Narasimham and Girija (1967).

Wavenumbers (cm^{-1}): 3717sh, 3658, 3623sh, 3334sh, 3284s, 3226sh, 3106, 2970sh, 2950s, 2920sh, 2247w, 2008w, 1850w, 1704sh, 1638sh, 1600, 1543w, 1444s, 1395, 1370, 1220sh, 1207sh, 1173s, 1141s, 1114s, 1070, 1027s, 996s, 924s, 837w, 785w, 771w, 717w.

Note: The wavenumbers were partly determined by us based on spectral curve analysis of the published spectrum.

S608 Gianellaite $(Hg_2N)_2(SO_4){\cdot}nH_2O$

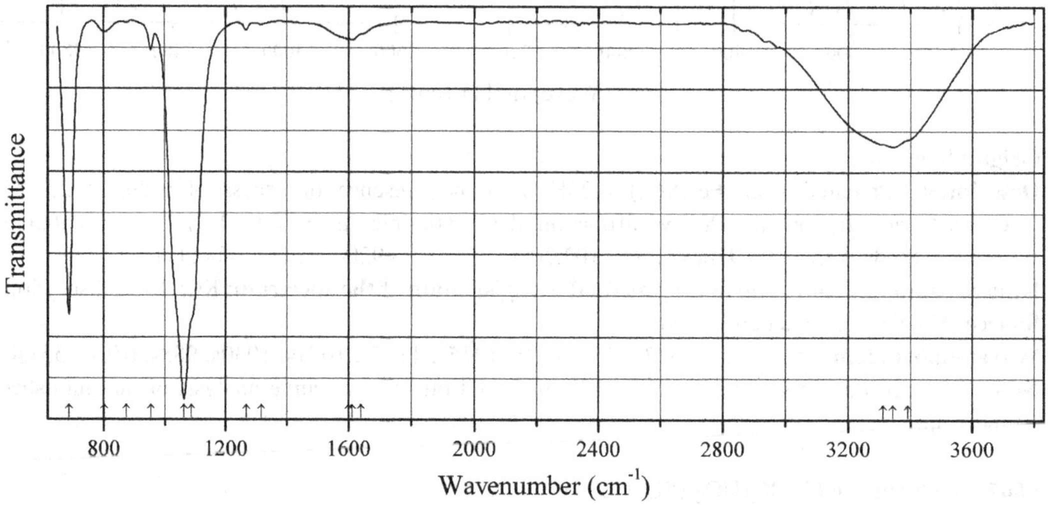

Origin: Perry Pit of the Mariposa mine, Terlingua District, Brewster Co., Texas, USA (type locality).

Description: Type material deposited in the Natural History Museum of Los Angeles Co., Museum No. 44159. The crystal structure is solved. Cubic, space group F-43m, a = 863.1(16) Å, Z = 4. The (NHg_4) tetrahedra sharing corners form a framework of the cuprite-type structure.

Kind of sample preparation and/or method of registration of the spectrum: Thin film, prepared by a diamond micro-compression cell. Absorption.

Source: Cooper et al. (2016a).

Wavenumbers (cm^{-1}): 3390sh, 3342, 3310, 1640sh, 1610, 1600sh, 1315w, 1265w, 1088sh, 1064s, 953, 872w, 803, 688s.

Note: The wavenumbers were partly determined by us based on spectral curve analysis of the published spectrum.

S609 Gordaite $NaZn_4(SO_4)(OH)_6Cl{\cdot}6H_2O$

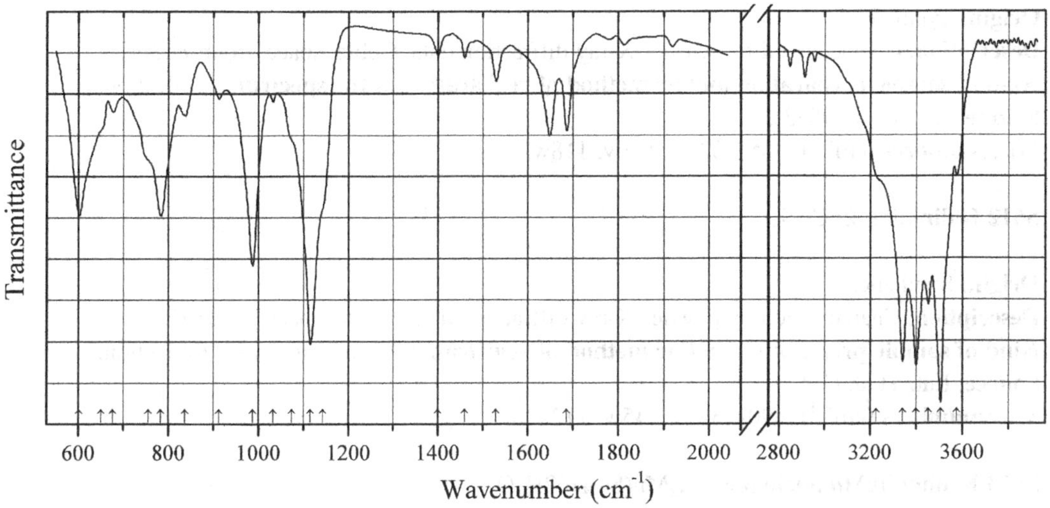

Origin: Juan de Fuca Ridge, northeastern Pacific Ocean (130° 22′ 34″ W, 44° 38′ 53″ N).

Description: Tabular crystals from the association with sphalerite, baryte, with minor pyrite, pyrrhotite, sulfur, and Fe-hydroxides. Characterized by electron microprobe analysis. Trigonal, $a = 8.353$ (2), $c = 13.087(8)$ Å. The strongest lines of the powder X-ray diffraction pattern [*d*, Å (*I*, %) (*hkl*)] are: 13.19 (100) (001), 3.737 (24) (103), 2.967 (30) (104), 2.737 (24) (120), 2.675 (34) (121), 2.523 (30) (122), 2.098 (24) (124).

Kind of sample preparation and/or method of registration of the spectrum: Thin-tabular chip. Transmission.

Source: Nasdala et al. (1998).

Wavenumbers (IR, cm^{-1}): 3577w, 3508s, 3454, 3401s, 3342s, 3235sh, 1687, 1649, 1530, 1460w, 1400w, 1144sh, 1116s, 1074sh, 1033w, 988s, 913w, 837, 784s, 756sh, 678w, 651sh, 603s.

Note: The wavenumbers were partly determined by us based on spectral curve analysis of the published spectrum. In the cited paper, Raman spectrum is given.

Wavenumbers (Raman, cm^{-1}): 3508, 3422, 1099, 973s, 598, 394.

S610 Ivsite $Na_3H(SO_4)_2$

Origin: Synthetic.

Description: Monoclinic, space group $P2_1/c$, $Z = 4$.

Kind of sample preparation and/or method of registration of the spectrum: Nujol and Fluorolube mulls at 20 K. Transmission.

Source: Fillaux et al. (1991).

Wavenumbers (IR, cm^{-1}): 1635, 1535, 1400, 1236, 1197, 1170, 1095, 972, 940, 850sh, 772, 658, 610, 580, 530, 500, 458, 438, 310.

Note: These wavenumbers given by Fillaux et al. (1991) in a table don't conform to the figure from this paper. In the cited paper, Raman spectrum is given.

Wavenumbers (Raman, cm^{-1}): 1242, 1198, 1162, 1154, 1115, 973, 639, 613, 604, 522, 479, 445, 437, 308, 182, 155, 126, 95, 76, 51.

S611 Joegoldsteinite $MnCr_2S_4$

Origin: Synthetic.
Description: Characterized by powder X-ray diffraction data. Cubic, space group *Fd*3*m*.
Kind of sample preparation and/or method of registration of the spectrum: Reflection.
Source: Lutz et al. (1983).
Wavenumbers (cm^{-1}): 385s, 321s, 257w, 118w.

S612 Kalininite $ZnCr_2S_4$

Origin: Synthetic.
Description: Characterized by powder X-ray diffraction data. Cubic, space group *Fd*3*m*.
Kind of sample preparation and/or method of registration of the spectrum: Reflection.
Source: Lutz et al. (1983).
Wavenumbers (cm^{-1}): 390s, 342s, 245w, 112w.

S613 KröhnkiteMn analogue $Na_2Mn(SO_4)_2{\cdot}2H_2O$

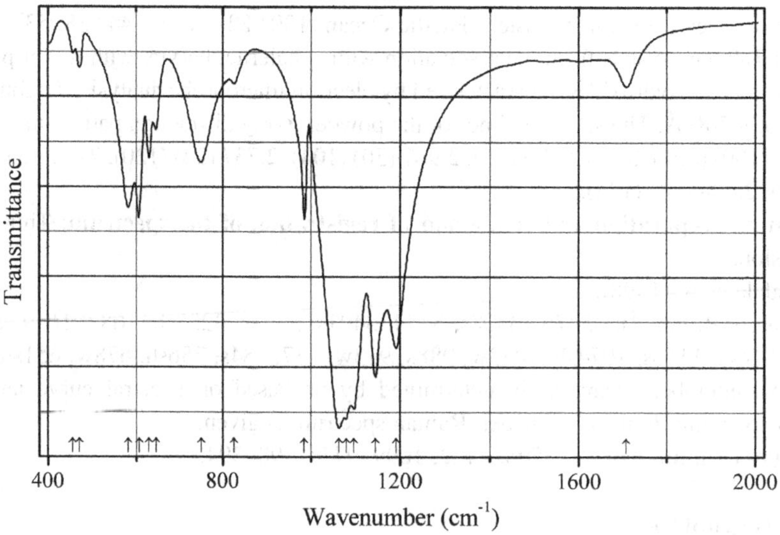

Origin: Synthetic.
Description: Prepared by crystallization from the Na_2SO_4–$MnSO_4$–H_2O system at 25 °C using the method of isothermal decrease in super-saturation. Characterized by DTA, TG, and powder X-ray diffraction data. Monoclinic, space group $P2_1/c$, a = 5.8206(2), b = 12.9958(21), c = 5.4920 (18) Å, β = 106.10(4)°, V = 399.1 Å^3, Z = 4.
Kind of sample preparation and/or method of registration of the spectrum: KBr disc. Absorption.
Source: Marinova et al. (2015).
Wavenumbers (IR, cm^{-1}): 1708w, 1192s, 1146s, 1097s, 1079s, 1063s, 985, 825w, 750, 646, 630, 607, 582, 472w, 456w.
Note: The wavenumbers were partly determined by us based on spectral curve analysis of the published spectrum. In the cited paper, Raman spectrum is given.
Wavenumbers (Raman, cm^{-1}): 1170w, 1128, 1045, 1020, 988s, 646w, 632w, 619, 463, 446w.

S614 Lishizhenite $ZnFe^{3+}_2(SO_4)_4 \cdot 14H_2O$

Origin: Xitieshan, Qinghai Province, China (type locality).

Description: Pale violet tabular crystals from the association with römerite, copiapite, sulfur, gypsum, pyrite, and quartz. Holotype sample. Triclinic, space group *P*, $a = 6.477(1)$, $b = 15.298(3)$, $c = 6.309(1)$, $\alpha = 90.20(1)°$, $\beta = 101.11(1)°$, $\gamma = 93.97(1)°$, $V = 611.9(1)$ Å^3, $Z = 1$. $D_{meas} = 2.206$ (4) g/cm^3, $D_{calc} = 2.201$ g/cm^3. Optically biaxial (–), $\alpha = 1.522(2)$, $\beta = 1.568(1)$, $\gamma = 1.578(4)$, $2V = 70(5)°$.

Source: Li and Chen (1990).

Wavenumbers (cm^{-1}, for absorption intervals): 3351–3035s, 1658–1651, 1131–997s, 667–537, 481.

S615 Magnesiocopiapite $MgFe^{3+}_4(SO_4)_6(OH)_2 \cdot 20H_2O$

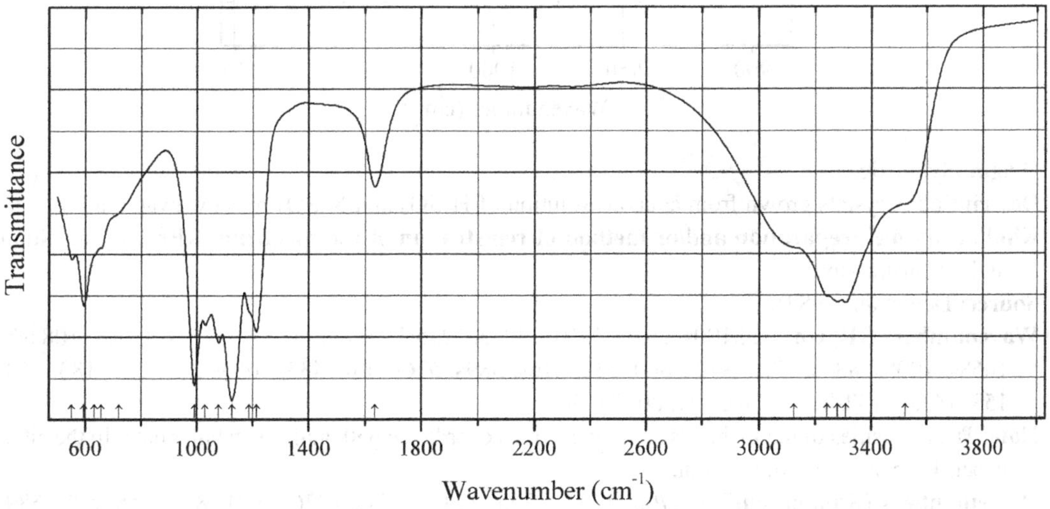

Origin: Synthetic.

Description: Prepared from $Fe_2(SO_4)_3 \cdot 6.25H_2O$ and $Mg(SO_4) \cdot 7H_2O$ at 25°, in the presence of excess of water. Characterized by powder X-ray diffraction data. Triclinic, $a = 7.3451(4)$, $b = 18.794(1)$, $c = 7.3891(4)$ Å, $\alpha = 91.369(5)°$, $\beta = 102.169(4)°$, $\gamma = 98.831(4)°$, $V = 983.6(1)$ Å^3.

Kind of sample preparation and/or method of registration of the spectrum: KCl disc. Absorption.

Source: Majzlan and Michallik (2007).

Wavenumbers (cm^{-1}): 3523sh, 3308s, 3278s, 3242sh, 3125sh, 1634, 1215s, 1188sh, 1127s, 1078, 1030, 991s, 721sh, 657sh, 633sh, 595s, 552.

Note: The wavenumbers were determined by us based on spectral curve analysis of the published spectrum.

S616 Mercallite $KHSO_4$

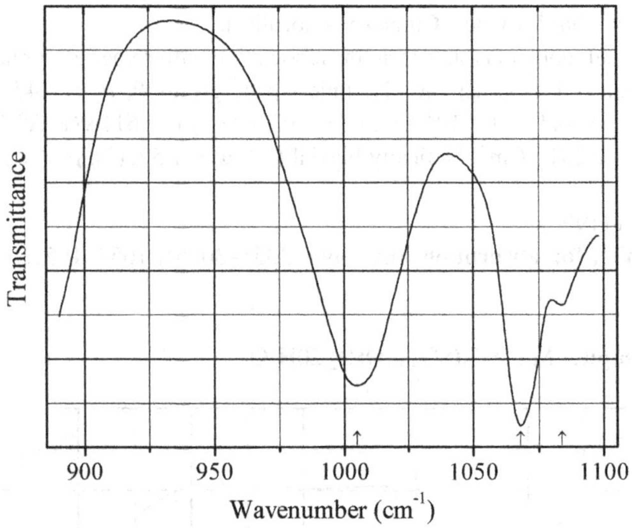

Origin: Synthetic.

Description: Crystals grown from aqueous solution of H_2SO_4 and K_2SO_4 by slow evaporation.

Kind of sample preparation and/or method of registration of the spectrum: KBr disc and Nujol mull. Transmission.

Source: Dey et al. (1982).

Wavenumbers (IR, cm^{-1}): 3100s, 2930, 2510, 1328s, 1295sh, 1284s, 1255s, 1228s, 1170s, 1084sh, 1068s, 1005s, 884s, 872s, 849s, 660, 632, 616s, 589s, 576s, 454, 435, 405w, 265, 220, 183, 160, 153, 141sh, 132sh, 112, 103, 94, 80, 54, 36.

Note: The intensities of the IR bands are indicated in accordance with authors' tabular data. In the cited paper, Raman spectrum is given.

Wavenumbers (Raman, cm^{-1}): 2700, 1337, 1265, 1242, 1170, 1026, 1001, 872, 855, 598, 589, 581, 572, 452, 445, 182, 139, 126, 102, 82, 50, 46.

S617 Mercallite $KHSO_4$

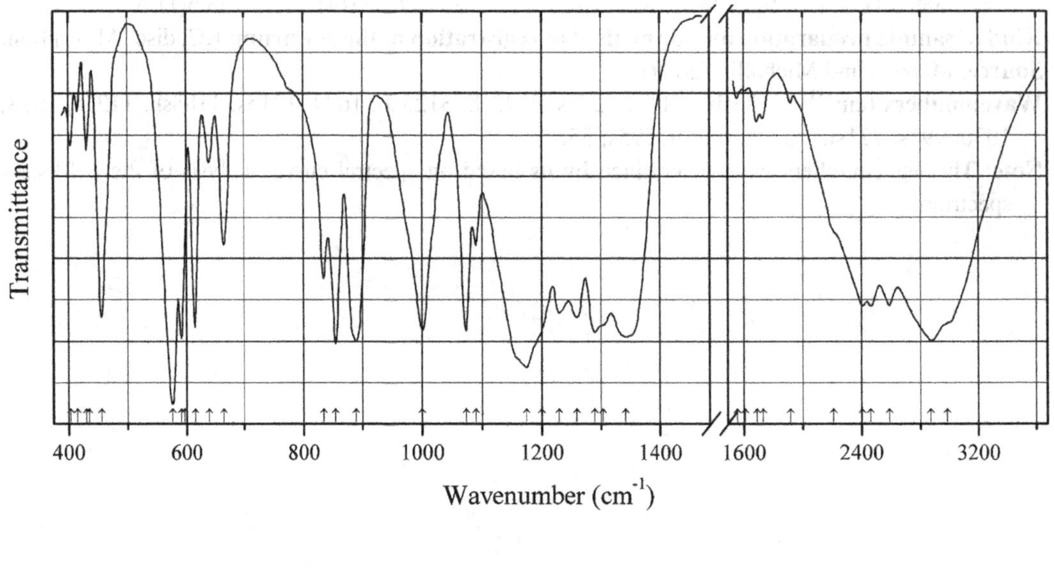

Origin: Synthetic.

Description: Pyramidal crystals grown from aqueous solution of H_2SO_4 and K_2SO_4. Orthorhombic, space group *Pbca*.

Kind of sample preparation and/or method of registration of the spectrum: Absorption of a polycrystalline sample at 90 K. Kind of sample preparation is not indicated.

Source: Goypiron et al. (1980).

Wavenumbers (IR, cm^{-1}): 2984sh, 2873, 2587, 2460, 2405, 2210sh, 1916w, 1730w, 1687w, 1608w, 1555w, 1342s, 1303sh, 1290s, 1259, 1230, 1200sh, 1175s, 1089, 1073s, 1000s, 889s, 855s, 835, 665, 640, 615s, 597s, 591sh, 576s, 456s, 435, 431sh, 416w, 404

Note: The wavenumbers were determined by us based on spectral curve analysis of the published spectrum. In the cited paper, Raman spectrum is given.

Wavenumbers (Raman, for a polycrystalline sample at 300 K, cm^{-1}): 2860w, 1257sh, 1244w, 1219w, 1171w, 1026s, 1001s, 883w, 870w, 855, 837w, 625w, 600sh, 596, 589, 581, 573sh, 455sh, 444, 419sh, 411, 192, 182, 139, 128, 126, 124, 116, 108, 102, 96, 84, 83, 82, 76, 75, 60, 50, 47, 45, 37.

S618 Plášilite $Na(UO_2)(SO_4)(OH){\cdot}2H_2O$

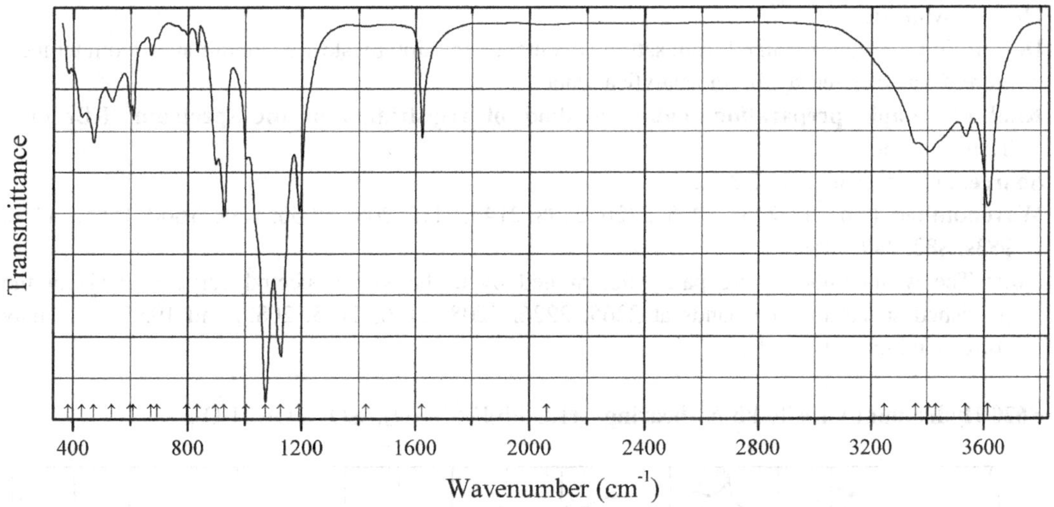

Origin: Blue Lizard mine, White Canyon District, San Juan County, Utah, USA (type locality).

Description: Yellow platelets. Investigated by A.V. Kasatkin. Characterized by single-crystal X-ray diffraction data. Monoclinic, $a = 8.702$, $b = 13.822$, $c = 7.042$ Å, $\beta = 112.08°$, $V = 384.9$ Å^3. The empirical formula is (electron microprobe): $Na_{0.88}(UO_2)_{1.06}(SO_4)_{1.06}(OH)_x{\cdot}nH_2O$.

Kind of sample preparation and/or method of registration of the spectrum: KBr disc. Absorption.

Wavenumbers (cm^{-1}): 3614s, 3535, 3430sh, 3405, 3360, 3250sh, 2060w, 1624, 1426w, 1191s, 1127s, 1073s, 1006, 927s, 898, 833, 797w, 693w, 671, 611, 601, 536, 472, 431, 384.

Note: The spectrum was obtained by N.V. Chukanov.

S619 Picromerite dimorph (?) $K_2Mg(SO_4)_2 \cdot 6H_2O$

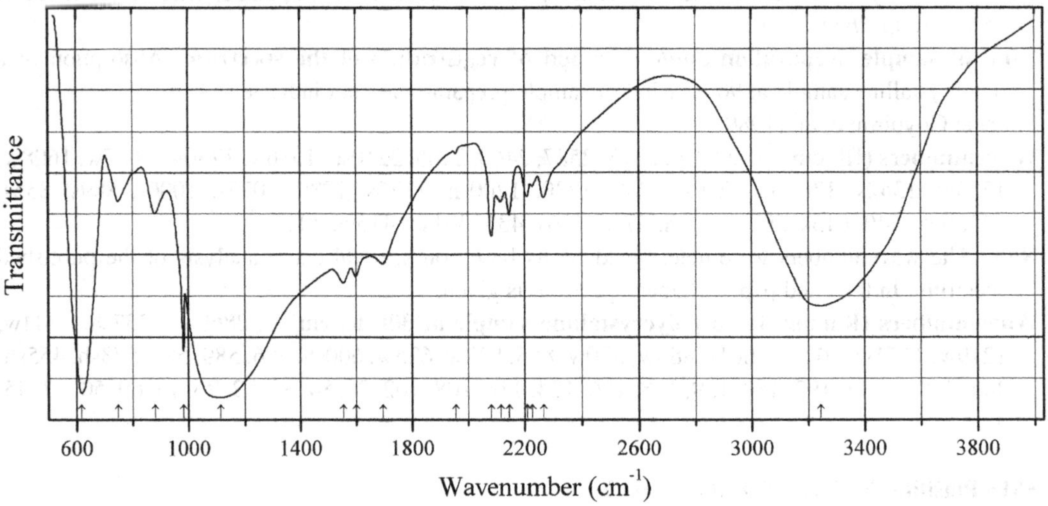

Origin: Synthetic.

Description: Crystals obtained from saturated aqueous solution by slow evaporation at room temperature. Characterized by thermoanalytical data.

Kind of sample preparation and/or method of registration of the spectrum: KBr disc. Transmission.

Source: Dhandapani et al. (2006).

Wavenumbers (cm^{-1}): 3243s, 2265, 2226, 2208, 2146, 2115, 2083, 1955, 1698, 1600, 1555, 1115s, 983s, 882, 749, 618s.

Note: The wavenumbers were partly determined by us based on spectral curve analysis of the published spectrum. The bands at 2265, 2226, 2208, 2146, 2115, 2083, and 1955 cm^{-1} may correspond to acid groups.

S620 Hydronium jarosite Pb,As-bearing $(H_3O,Pb)Fe^{3+}{}_3(SO_4,AsO_4)_2(OH,H_2O)_6$

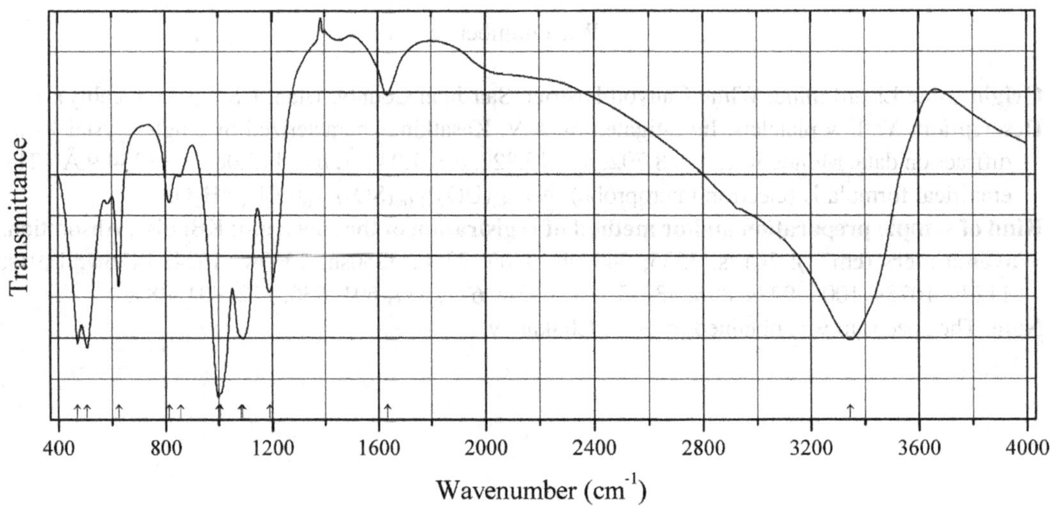

Origin: Synthetic.

Description: Prepared from an aqueous solution containing $Pb(NO_3)_2$, $Fe_2(SO_4)_3·5H_2O$, and H_3AsO_4 at 95 °C. Characterized by elemental analysis and powder X-ray diffraction data. Trigonal, $a = 7.3417(8)$, $c = 16.9213(6)$ Å. The empirical formula is $(H_3O)_{0.68}Pb_{0.32}Fe_{2.86}(SO_4)_{1.69}(AsO_4)_{0.31}(OH)_{5.59}(H_2O)_{0.41}$.

Kind of sample preparation and/or method of registration of the spectrum: KBr disc. Transmission.

Source: Forray et al. (2014).

Wavenumbers (cm^{-1}): 3343s, 2930sh, 1634, 1189, 1090s, 1083sh, 1005sh, 999s, 855, 814, 625, 507s, 472s.

Note: The wavenumbers were partly determined by us based on spectral curve analysis of the published spectrum.

S621 Hydronium jarosite Pb,Cu-bearing $(H_3O,Pb)(Fe^{3+},Cu^{2+})_3(SO_4)_2(OH)_6$

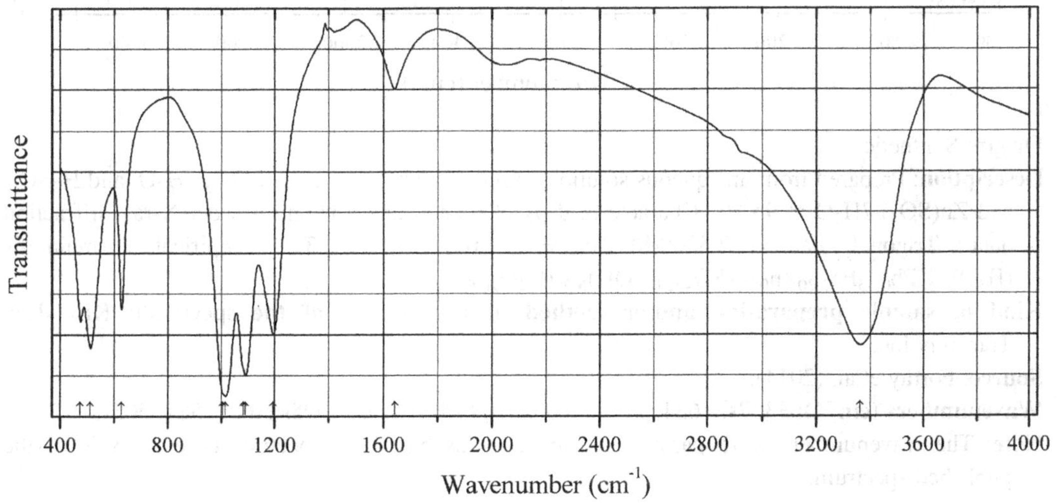

Origin: Synthetic.

Description: Prepared from an aqueous solution containing $Pb(NO_3)_2$, $Fe_2(SO_4)_3·5H_2O$, and H_2SO_4, and $Cu(SO_4)·5H_2O$ at 95 °C. Characterized by elemental analysis and powder X-ray diffraction data. Trigonal, $a = 7.3208(8)$, $c = 17.0336(7)$ Å. The empirical formula is $(H_3O)_{0.67}Pb_{0.33}Fe_{2.71}Cu_{0.25}(SO_4)_{2.00}(OH)_{5.96}(H_2O)_{0.04}$.

Kind of sample preparation and/or method of registration of the spectrum: KBr disc. Transmission.

Source: Forray et al. (2014).

Wavenumbers (cm^{-1}): 3362s, 1642, 1195, 1092s, 1083sh, 1016s, 1005sh, 628, 513s, 475.

Note: The wavenumbers were partly determined by us based on spectral curve analysis of the published spectrum.

S622 Hydronium jarosite Pb,Zn-bearing $(H_3O,Pb)(Fe^{3+},Zn)_3(SO_4)_2(OH)_6$

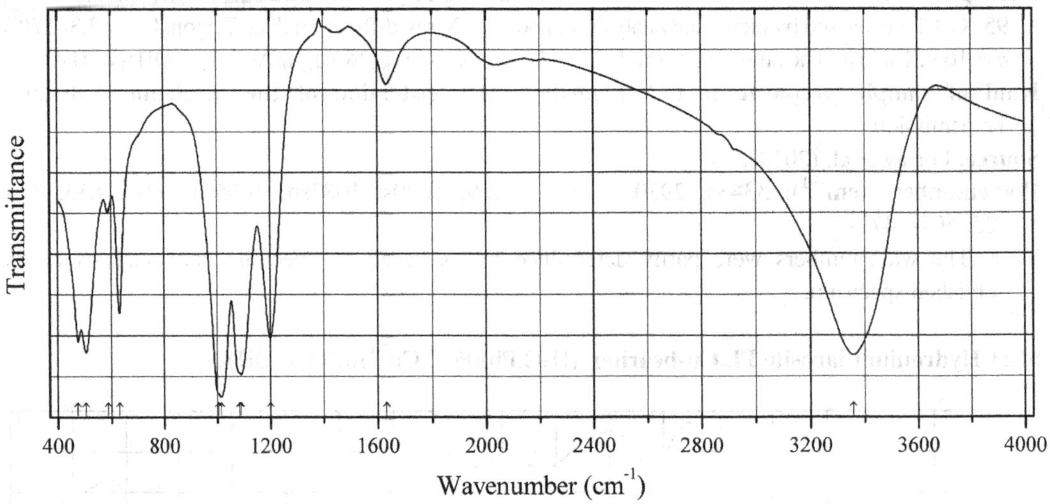

Origin: Synthetic.

Description: Prepared from an aqueous solution containing $Pb(NO_3)_2$, $Fe_2(SO_4)_3{\cdot}5H_2O$, and H_2SO_4, and $Zn(SO_4){\cdot}7H_2O$ at 95 °C. Characterized by elemental analysis and powder X-ray diffraction data. Trigonal, a = 7.3208(8), c = 17.0336(7) Å. The empirical formula is $(H_3O)_{0.57}Pb_{0.43}Fe_{2.70}Zn_{0.21}(SO_4)_{2.00}(OH)_{5.95}(H_2O)_{0.05}$.

Kind of sample preparation and/or method of registration of the spectrum: KBr disc. Transmission.

Source: Forray et al. (2014).

Wavenumbers (cm^{-1}): 3357s, 1634, 1199, 1089s, 1083sh, 1015s, 1005sh, 631, 587, 505s, 475.

Note: The wavenumbers were partly determined by us based on spectral curve analysis of the published spectrum.

S623 Shumwayite $(UO_2)_2(SO_4)_2{\cdot}5H_2O$

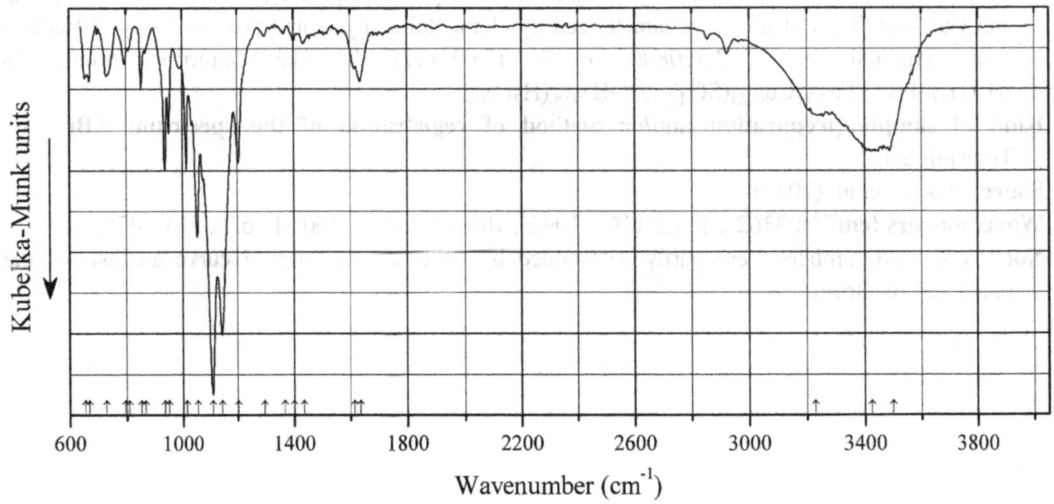

Origin: White Canyon district, San Juan Co., Utah, USA (type locality).

Description: Greenish-yellow prisms from the association with other secondary sulfates. Holotype sample. The crystal structure is solved. Monoclinic, space group $P2_1/c$, a = 6.74747(15), b = 12.5026(3), c = 16.9032(12) Å, β = 90.919(6)°, V = 1425.79(11) Å^3, Z = 4. D_{calc} = 3.831 g/cm^3. Optically biaxial (+/–), α = 1.581(1), β = 1.588(1), γ = 1.595(1), $2V$ = 89.8(8)°. The empirical formula is (electron microprobe): $U_{2.01}S_{1.99}O_{12.00}·5H_2O$. The strongest lines of the powder X-ray diffraction pattern [d, Å (I, %) (hkl)] are: 5.58 (48) (−111, 111), 5.11 (100) (013), 4.86 (44) (−112, 112), 4.04 (47) (031), 3.459 (42) (−131, –114, 114), 3.373 (50) (200, 033, –132).

Kind of sample preparation and/or method of registration of the spectrum: Reflection of powdered mineral mixed with KBr.

Source: Kampf et al. (2016f).

Wavenumbers (IR, cm^{-1}): 3500, 3425, 3230sh, 1635, 1615w, 1435w, 1400w, 1365w, 1295w, 1202, 1143s, 1110s, 1055s, 1015, 951, 937, 868, 854, 810, 795, 730, 670, 655.

Note: The band position denoted by Kampf et al. (2016f) as 927 cm^{-1} was determined by us at 937 cm^{-1}. In the cited paper, Raman spectrum is given.

Wavenumbers (Raman, cm^{-1}): 1185, 1155, 1100, 1073s, 1050, 1035, 1015, 930, 865s, 850s, 645, 615, 470, 430, 273sh, 255, 210, 200s, 160, 150.

S624 Carlsonite $(NH_4)_5Fe^{3+}{}_3(SO_4)_6O·7H_2O$

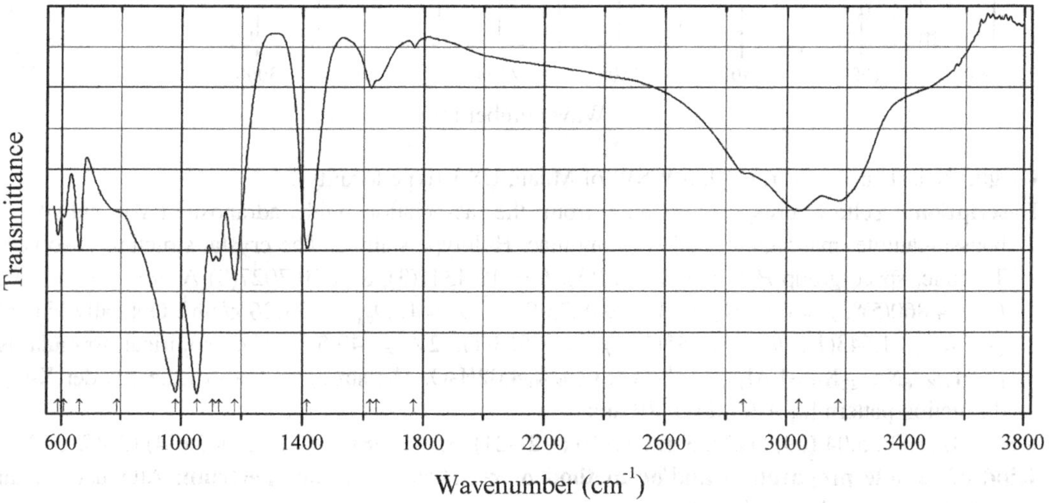

Origin: Near Huron River, 6.1 km WSW of Milan, USA (type locality).

Description: Yellow to orange-brown crystals from the association with anhydrite, boussingaultite, gypsum, and lonecreekite. Holotype sample. The crystal structure is solved. Triclinic, space group P-1, a = 9.5927(2), b = 9.7679(3), c = 18.3995(13) Å, α = 93.250(7)°, β = 95.258(7)°, γ = 117.993(8)°, V = 1506.15(16) Å^3, Z = 2. D_{calc} = 2.167 g/cm^3. Optically biaxial (–), α = 1.576(1), β = 1.585(1), γ = 1.591(1), $2V$ = 80(1)°. The empirical formula is $[(NH_4)_{4.64}Na_{0.24}K_{0.12}]Fe^{3+}{}_{3.05}O(SO_4)_6·6.93H_2O$. The strongest lines of the powder X-ray diffraction pattern [d, Å (I, %) (hkl)] are: 9.23 (100) (002), 8.26 (40) (100, 011), 7.57 (43) (−111, 1–11, 011), 4.93 (23) (−1–11, –120), and 3.144 (41) (multiple).

Kind of sample preparation and/or method of registration of the spectrum: Attenuated total reflection of powdered mineral.

Source: Kampf et al. (2016h).

Wavenumbers (IR, cm^{-1}): 3176, 3044, 2858, 1768w, 1644sh, 1624, 1416, 1179s, 1127, 1107, 1054s, 983s, 788sh, 660, 610, 588.

Note: The wavenumbers were partly determined by us based on spectral curve analysis of the published spectrum. In the cited paper, Raman spectrum is given.

Wavenumbers (Raman, cm^{-1}): 1219, 1188, 1160, 1140, 1104, 1066, 1015s, 670w, 629w, 617w, 576, 552, 514, 487, 436, 275s, 245s.

S625 Huizingite-(Al) $(NH_4)_9(Al,Fe^{3+})_3(SO_4)_8(OH)_2 \cdot 4H_2O$

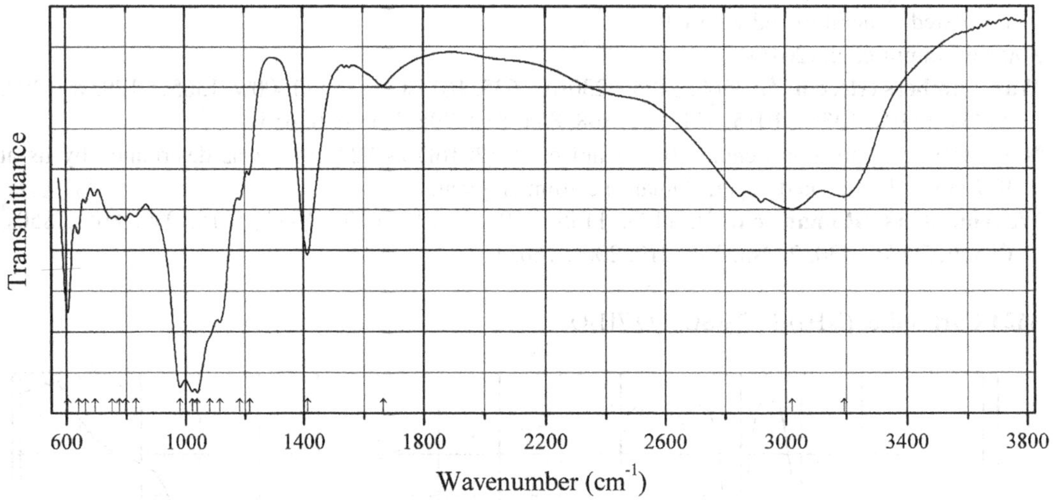

Origin: Near Huron River, 6.1 km WSW of Milan, USA (type locality).

Description: Yellow drusy aggregates from the association with adranosite-(Al), anhydrite, boussingaultite, mascagnite, and salammoniac. Holotype sample. The crystal structure is solved. Triclinic, space group *P*-1, a = 9.7093(3), b = 10.4341(3), c = 10.7027(8) Å, α = 77.231(5)°, β = 74.860(5)°, γ = 66.104(5)°, V = 948.73(9) Å^3, Z = 1. D_{calc} = 2.026 g/cm^3. Optically biaxial (+), α = 1.543(1), β = 1.545(1), γ = 1.563(1), 2V = 40(3)°. The empirical formula is $[(NH_4)_{8.76}Na_{0.22}K_{0.02}](Al_{1.65}Fe^{3+}_{3.05})(SO_4)_{8.00} \cdot 4.02H_2O$. The strongest lines of the powder X-ray diffraction pattern [*d*, Å (*I*, %) (*hkl*)] are:

8.82 (60) (100), 5.04 (69) (121), 3.427 (100) (−2−21), 3.204 (68) (−211), 3.043 (94) (2–12, 312).

Kind of sample preparation and/or method of registration of the spectrum: Attenuated total reflection of powdered mineral.

Source: Kampf et al. (2016h).

Wavenumbers (IR, cm^{-1}): 3192, 3022, 1667w, 1413, 1215w, 1183w, 1117, 1083sh, 1040s, 1023s, 980s, 831w, 798w, 776w, 751w, 694w, 664w, 640, 605.

Note: Bands in the range from 2800 to 3000 cm^{-1} correspond to the admixture of an organic substance. The IR bands at 1083, 798, 776, and 694 cm^{-1} are close to those of quartz. The wavenumbers were partly determined by us based on spectral curve analysis of the published spectrum. In the cited paper, Raman spectrum is given.

Wavenumbers (Raman, cm^{-1}): 1205, 1151, 1123, 1064, 1027s, 1010s, 1003s, 980s, 673, 641, 618, 478, 468, 448, 263, 223.

S626 Alunogen $Al_2(SO_4)_3(H_2O)_{12}·5H_2O$

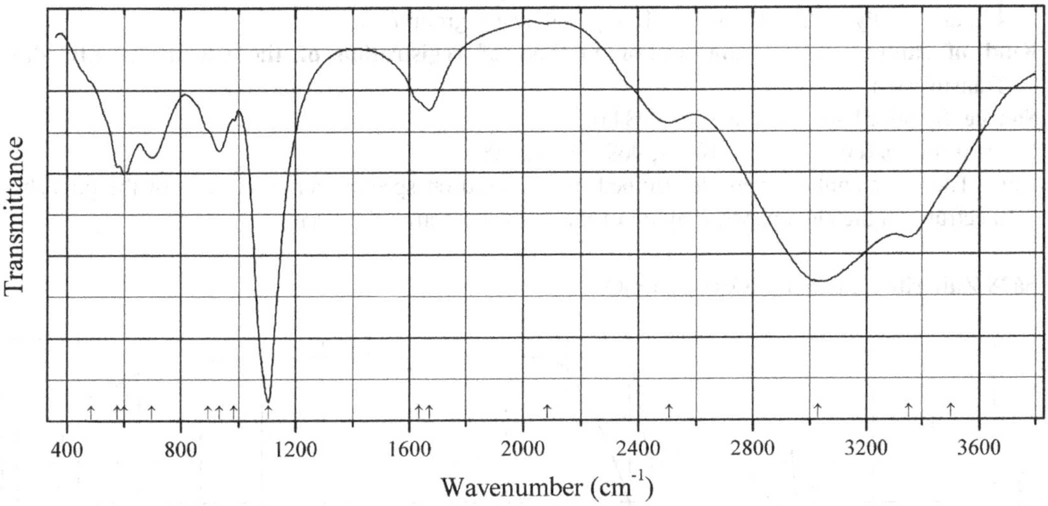

Origin: Kalamos fumarole field, Milos Island, Greece.

Description: White sugar-like aggregate. Characterized by powder X-ray diffraction data. The empirical formula is (electron microprobe): $(Al_{1.97}Fe_{0.03})(SO_4)_3·nH_2O$.

Kind of sample preparation and/or method of registration of the spectrum: KBr disc. Absorption.

Wavenumbers (cm^{-1}): 3500sh, 3353s, 3030s, 2505, 2085w, 1670, 1635sh, 1104s, 983, 933, 895sh, 699, 600, 578, 485sh.

Note: The spectrum was obtained by N.V. Chukanov.

S627 Vanadyl sulfate (VO)(SO_4)
Paufierite tetragonal dimorph

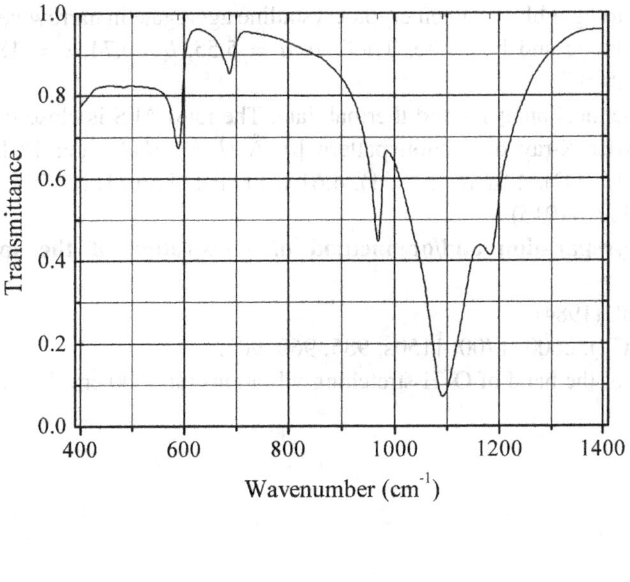

Origin: Synthetic.

Description: Obtained by heating commercial $VSO_5 \cdot xH_2O$ first at 165 °C for 12 h, then at 260 °C for 4 h, and finally at 330 °C for 1 h. Tetragonal, space group *P*4/*n*.

Kind of sample preparation and/or method of registration of the spectrum: KBr disc. Transmission.

Source: Stranford and Condrate Sr (1984a).

Wavenumbers (cm^{-1}): 1183, 1091s, 969, 687w, 588.

Note: The wavenumbers were determined by us based on spectral curve analysis of the published spectrum. In the cited paper, a figure of the Raman spectrum is given.

S628 Zaherite $Al_{12}(SO_4)_5(OH)_{26} \cdot 20H_2O$

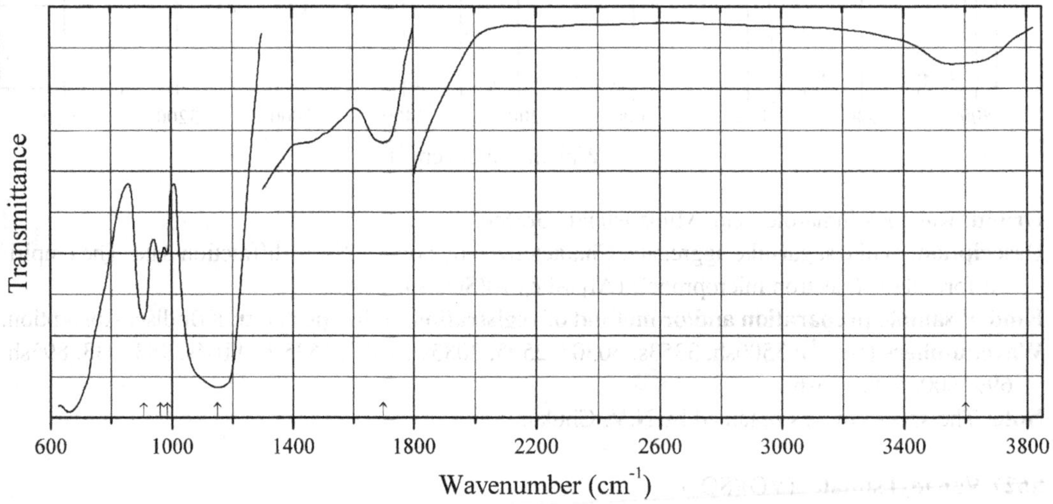

Origin: Pofadder, Bushmanland, South Africa.

Description: White to light bluish-green cryptocrystalline aggregate in narrow veins, in close association with natro-alunite and hotsonite. Triclinic, a = 5.55, b = 9.74, c = 18.43 Å, α = 99.71°, β = 89.13°, γ = 94.97°.

Characterized by chemical analyses and thermal data. The ratio Al:S is close to 12:5. The strongest lines of the powder X-ray diffraction pattern [*d*, Å (*I*, %) (*hkl*)] are: 18.12 (100) (001), 9.56 (5) (010), 9.08 (4) (002), 4.82 (6) (0−21), 4.61 (8) (110), 4.56 (4) (0−22), 4.44 (4) (021), 3.61 (4) (1−2−1), 3.33 (8) (015).

Kind of sample preparation and/or method of registration of the spectrum: KBr disc. Transmission.

Source: Beukes et al. (1984).

Wavenumbers (cm^{-1}): 3600, 1700, 1150s, 985, 960, 905.

Note: The intensity of the band of O–H-stretching vibrations (at 3600 cm^{-1}) is anomalously low.

S629 Fibroferrite $Fe^{3+}(SO_4)(OH)\cdot 5H_2O$

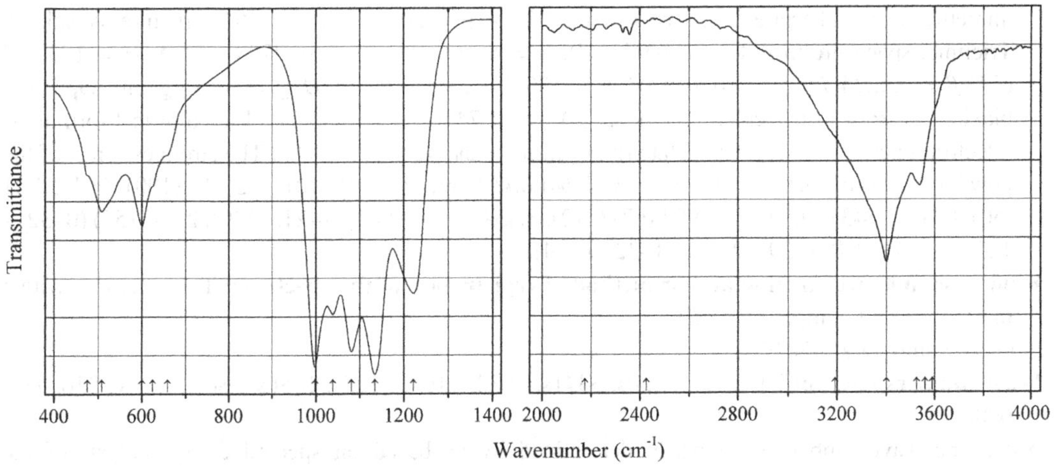

Origin: Ancient Pb-Zn mine of Saint Felix de Paillères, Anduze, Gard, Languedoc-Roussillon, France.

Description: Hand-picked crystals. The crystal structure is solved. Trigonal, space group *R*-3, $a = 24.199(3)$, $c = 7.6476(9)$ Å, $V = 3878.4(8)$ Å^3.

Kind of sample preparation and/or method of registration of the spectrum: KBr disc. Transmission.

Source: Ventruti et al. (2016).

Wavenumbers (IR, cm^{-1}): 5180w, 4487w, 4234w, 3590sh, 3561sh, 3526, 3400s, 3192sh, 2427w, 2175w, 1662, 1612, 1426w, 1221s, 1134s, 1081s, 1038s, 998s, 659sh, 625sh, 600, 508, 475sh.

Note: In the cited paper, Raman spectrum is given.

Wavenumbers (Raman, cm^{-1}): 3590, 3522, 3140, 3411, 1175, 1135, 1097, 1073, 1031, 998s, 613, 590, 523, 488, 427, 390, 297, 287, 272, 256, 219, 187, 173, 133, 114.

S630 Pauladamsite $Cu_4(SeO_3)(SO_4)(OH)_4\cdot 2H_2O$

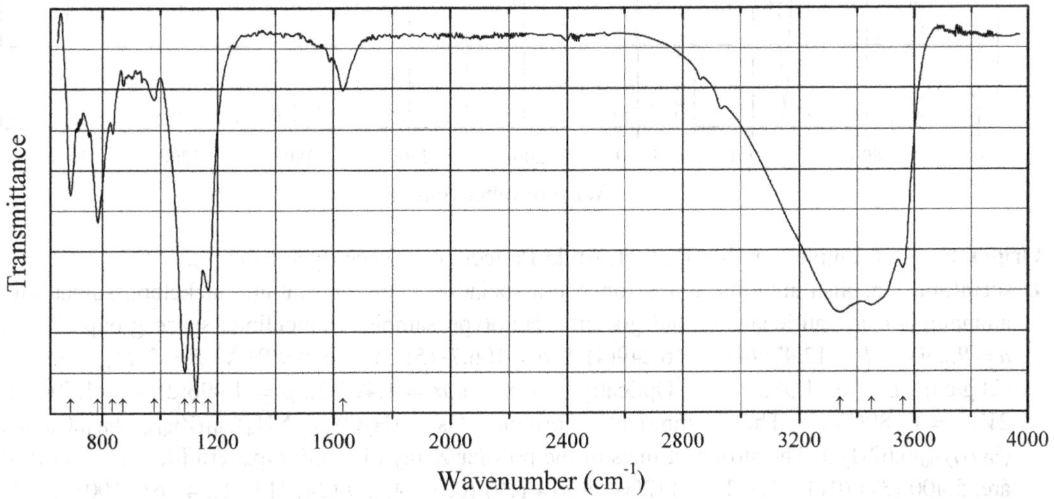

Origin: Santa Rosa mine, Darwin district, Inyo Co., California, USA (type locality).

Description: Green crystals from the association with brochantite, chalcanthite, gypsum, ktenasite, mimetite, schulenbergite, and smithsonite. Holotype sample. The crystal structure is solved. Triclinic, space group *P*-1, $a = 6.0742(7)$, $b = 8.4147(11)$, $c = 10.7798$ (15) Å, $\alpha = 103.665$ (7)°, $\beta = 95.224(7)°$, $\gamma = 90.004(6)°$, $V = 533.03(12)$ Å^3, $Z = 2$. $D_{calc} = 3.535$ g/cm^3. Optically biaxial (–), $\alpha = 1.667$ (calc.), $\beta = 1.723(2)$, $\gamma = 1.743(2)$, $2V = 60(2)°$. The empirical formula is (electron microprobe, H_2O calculated): $H_{8.50}Cu_{3.55}Zn_{0.25}Se_{0.98}S_{1.00}O_{13}$. The strongest lines of the powder X-ray diffraction pattern [*d*, Å (*I*, %) (*hkl*)] are: 10.5 (46) (011), 3.245 (100) (001), 5.81 (50) (011), 2.743 (49) (112), 3.994 (67) (012), 3.431 (23) (–112, –1–21, –120), 2.692 (57) (0–32, –122, –2–12), 2.485 (39) (2–12, –1–32, 0–24).

Kind of sample preparation and/or method of registration of the spectrum: Transmission, with a micro diamond compression cell.

Source: Kampf et al. (2016d).

Wavenumbers (IR, cm^{-1}): 3560, 3450s, 3341s, 1633, 1167, 1125s, 1086s, 980w, 873w, 836, 784, 690.

Note: The wavenumbers were partly determined by us based on spectral curve analysis of the published spectrum. In the cited paper, Raman spectrum is given.

Wavenumbers (Raman, cm^{-1}): 1166, 1076, 989s, 839s, 745, 679, 638w, 610w, 487, 412, 396, 299, 270, 222, 166, 153.

S631 Katerinopoulosite $(NH_4)_2Zn(SO_4)_2 \cdot 6H_2O$

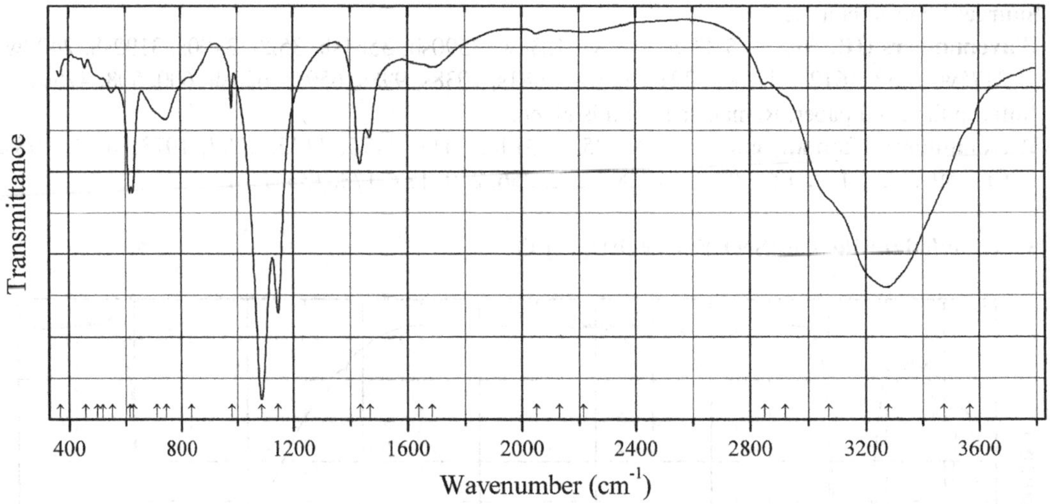

Origin: Esperanza mine, Lavrion District, Attikí Prefecture, Greece (type locality).

Description: Greenish antholite crust from the association with chalcanthite, nickelboussingaultite, ammoniojarosite, aurichalcite, and goethite. Holotype sample. Monoclinic, space group: $P2_1/a$, $a = 9.230(6)$, $b = 12.476(4)$, $c = 6.249(4)$ Å, $\beta = 106.79(5)°$, $V = 688.9(9)$ Å^3, $Z = 2$. $D_{meas} = 1.97$ (2) g/cm^3, $D_{calc} = 1.986$ g/cm^3. Optically biaxial (+), $\alpha = 1.492(2)$, $\beta = 1.496(2)$, $\gamma = 1.502(2)$, $2V = 80(5)°$. The empirical formula is $(H_3O)_{0.13}(NH_4)_{1.91}(Zn_{0.86}Ni_{0.10}Cu_{0.02})(SO_4)_{2.00} \cdot 6.62H_2O$. The strongest lines of the powder X-ray diffraction pattern [*d*, Å (*I*, %) (*hkl*)] are: 5.400 (37) (011), 4.411 (19) (200), 4.314 (19) (021), 4.229 (24) (12–1), 4.161 (100) (20–1, 210, 111), 3.749 (53) (130), 3.034 (29) (211, 11–2).

Kind of sample preparation and/or method of registration of the spectrum: KBr disc. Absorption.

Wavenumbers (cm^{-1}): 3565sh, 3475sh, 3278s, 3070sh, 2920sh, 2850, 2215w, 2130w, 2050w, 1685, 1640sh, 1468, 1433, 1144s, 1086s, 981, 835sh, 743, 710sh, 627, 615, 552w, 519w, 500sh, 459w, 368w.
Note: The spectrum was obtained by N.V. Chukanov.

S632 Charlesite $Ca_6Al_2(SO_4)_2B(OH)_4(OH,O)_{12}·26H_2O$

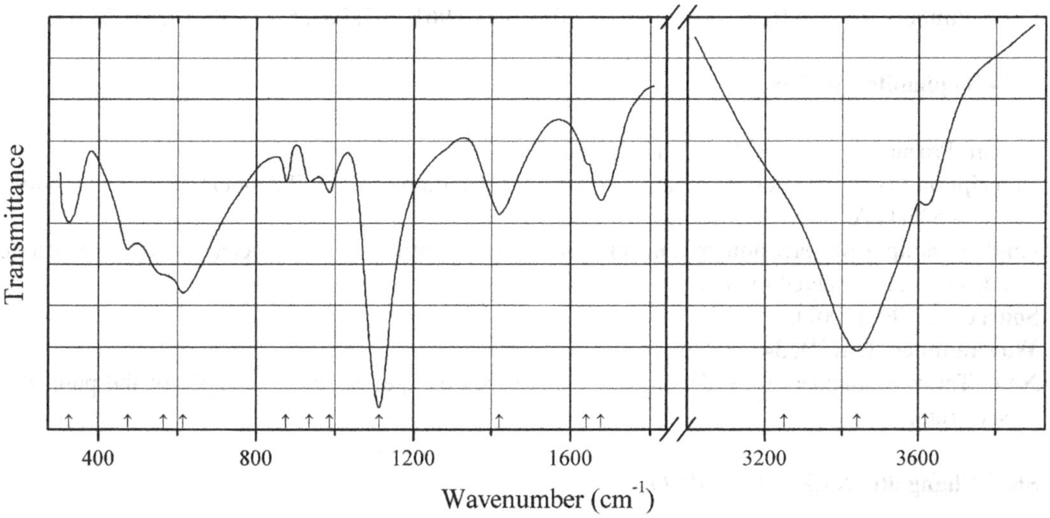

Origin: Wessels mine, Hotazel, Kalahari manganese fields, Northern Cape province, South Africa.
Description: Yellow. Intermediate zone of a mixed ettringite-charlesite-sturmanite crystal. The empirical formula is (electron microprobe): $Ca_{6.00}(Al_{0.7}Fe_{0.6}Si_{0.5}Mn_{0.2})(SO_4)_{2.27}(CO_3)_x[B(OH)_4]_{\sim 1}(OH,O)_{12}·nH_2O$ ($x \ll 1$).
Kind of sample preparation and/or method of registration of the spectrum: KBr disc. Absorption.
Wavenumbers (cm^{-1}): 3620, 3440s, 3250sh, 1678, 1641sh, 1420, 1112s, 986, 935, 876, 615s, 565sh, 475, 324.
Note: The spectrum was obtained by N.V. Chukanov.

S633 Osakaite $Zn_4(SO_4)(OH)_6·5H_2O$

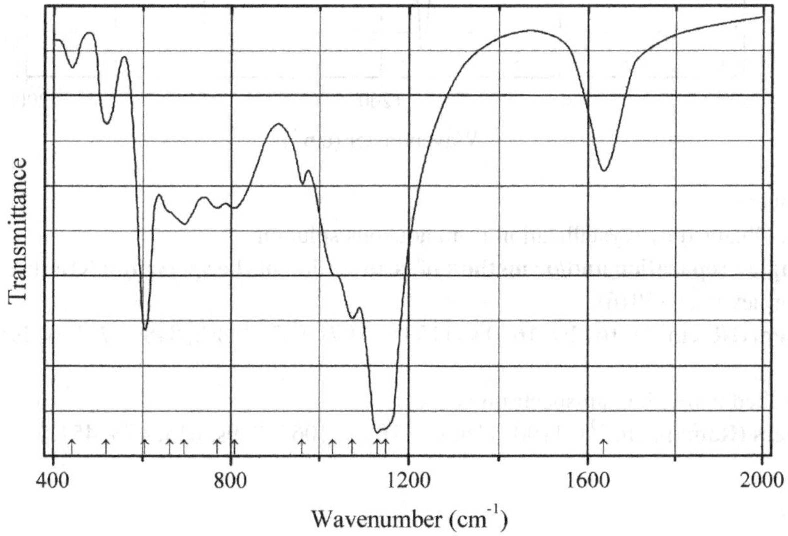

Origin: Synthetic.
Description: Obtained by mixing of 1 g ZnO powder with 30 ml 0.5 M solution of $ZnSO_4$ for 72 h. Characterized by thermal and powder X-ray diffraction data.
Kind of sample preparation and/or method of registration of the spectrum: KBr disc. Transmission.
Source: Stanimirova et al. (2016).
Wavenumbers (cm^{-1}): 1636, 1149s, 1130s, 1074, 1031, 961, 807, 768, 694, 661sh, 606s, 519, 442w.

S634 Stephanite Ag_5SbS_4

Origin: Proano mine, Fresnillo, Zacatecas, Mexico.
Description: Black pseudohexagonal crystals. Orthorhombic, a = 7.8396(7), b = 12.4684(9), c = 8.536(1) Å.
Kind of sample preparation and/or method of registration of the spectrum: Attenuated total reflection of powdered mineral.
Source: RRUFF (2007).
Wavenumbers (cm^{-1}): 343s, 322s, 262, 235.
Note: The wavenumbers were determined by us based on spectral curve analysis of the published spectrum.

S635 Changoite $Na_2Zn(SO_4)_2{\cdot}4H_2O$

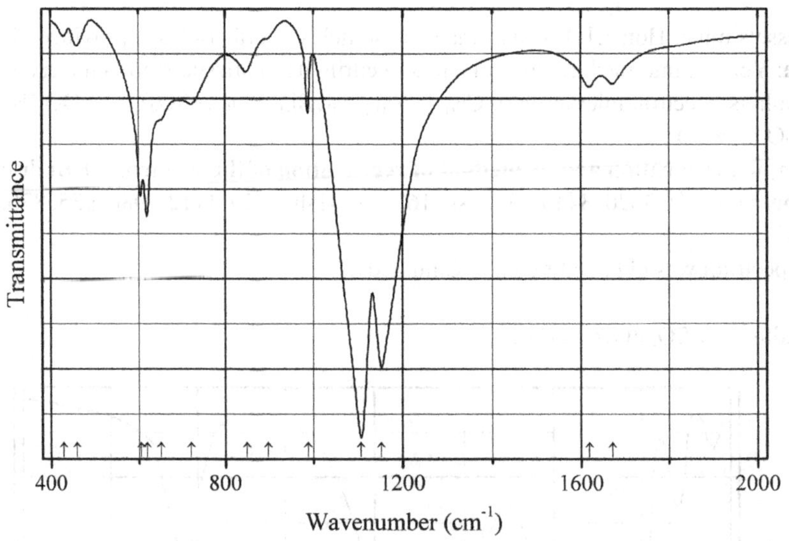

Origin: Synthetic.
Description: Obtained by crystallization from aqueous solution.
Kind of sample preparation and/or method of registration of the spectrum: KBr disc. Absorption.
Source: Georgiev et al. (2016).
Wavenumbers (IR, cm^{-1}): 1672w, 1620w, 1153s, 1107s, 987, 898sh, 848w, 721, 652sh, 620s, 605s, 460, 430.
Note: In the cited paper, Raman spectrum is given.
Wavenumbers (Raman, cm^{-1}): 1190, 1160w, 1101w, 1067, 989s, 615, 473, 451.

S636 Galeite $Na_{15}(SO_4)_5ClF_4$

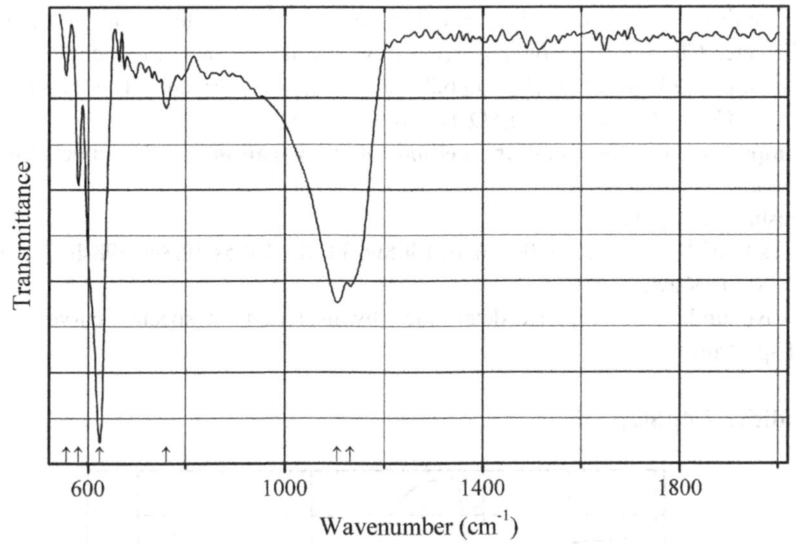

Origin: Synthetic.

Description: Ce^{3+}-doped sample prepared using a wet chemical method. Characterized by powder X-ray diffraction data. The crystal structure is solved. Trigonal, space group *P*31*m*, *a* = 12.19, *c* = 13.95 Å.

Kind of sample preparation and/or method of registration of the spectrum: Transmission. Kind of sample preparation is not indicated.

Source: Bhake et al. (2016).

Wavenumbers (cm^{-1}): 1132, 1106s, 760w, 624s, 580, 555.

S637 Ktenasite $(Cu,Zn)_5(SO_4)_2(OH)_6 \cdot 6H_2O$

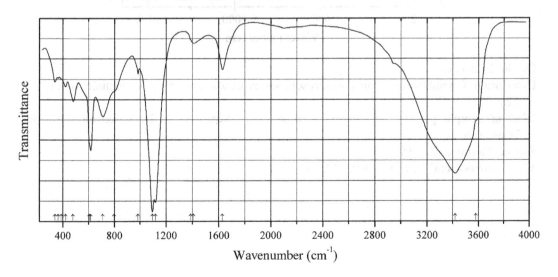

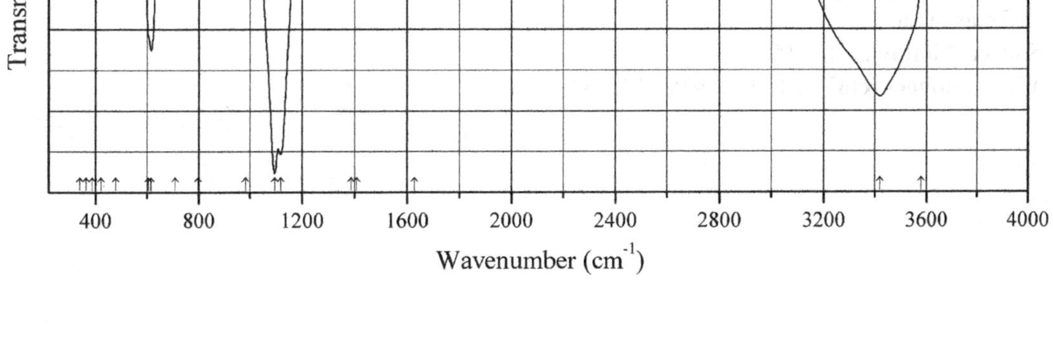

Origin: Hirao mine, Minoo, Osaka, Japan.

Description: Aggregates of flattened prismatic crystals from the association with primary sulfides, gypsum, smithsonite, hydrozincite, aurichalcite, schulenbergite, brianyoungite, serpierite, brochantite, etc. Characterized by powder X-ray diffraction data. Monoclinic, a = 5.590(1), b = 6.161(1), c = 23.741(3) Å, β = 95.628(3)°. D_{meas} = 2.93 g/cm^3. The empirical formula is $(Cu_{3.446}Zn_{1.451}Co_{0.080}Pb_{0.018}Ni_{0.007})(SO_4)_{2.003}(OH)_{5.998}\cdot5.99H_2O$.

Kind of sample preparation and/or method of registration of the spectrum: KBr disc. Transmission.

Source: Ohnishi et al. (2002).

Wavenumbers (cm^{-1}): 3580sh, 3420s, 1630, 1408w, 1118s, 1095s, 983w, 798sh, 710, 615s, 605sh, 480, 420, 387sh, 364w, 340w.

Note: The wavenumbers were partly determined by us based on spectral curve analysis of the published spectrum.

S638 Lazaridisite $Cd_3(SO_4)_3\cdot8H_2O$

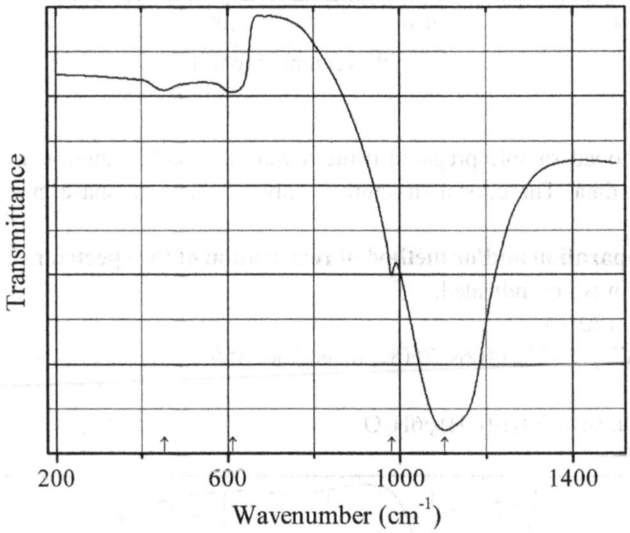

Origin: Synthetic.

Description: Crystals grown from aqueous solution by slow evaporation. Monoclinic, space group *C*2/*c*.

Kind of sample preparation and/or method of registration of the spectrum: KBr disc. Transmission.

Source: Murthy et al. (1992).

Wavenumbers (cm^{-1}): 1103s, 980, 610, 453.

S639 Pyracmonite $(NH_4)_3Fe(SO_4)_3$

Origin: La Fossa crater, Vulcano Island, Lipari, Eolie (Aeolian) islands, Messina province, Sicily, Italy (type locality).

Description: Holotype sample. Trigonal, space group *R*3*c*, *a* = 15.2171(14), *c* = 8.9323(8) Å, *V* = 1791.3(3) Å^3, *Z* = 6. The empirical formula is $[(NH_4)_{2.74}K_{0.23}](Fe_{0.94}Al_{0.04})S_{3.02}O_{12}$. The strongest lines of the powder X-ray diffraction pattern [*d*, Å (*I*, %) (*hkl*)] are: 7.596 (100) (110), 3.320 (30) (122), 3.371 (26) (131), 4.358 (23) (12–1), 2.829 (14) (312), 2.863 (8) (321).

Kind of sample preparation and/or method of registration of the spectrum: No data.

Source: Demartin et al. (2010).

Wavenumbers (cm^{-1}): 3203s, 3064s, 1430s.

S640 Rhodium sulfate $Rh_2(SO_4)_3$

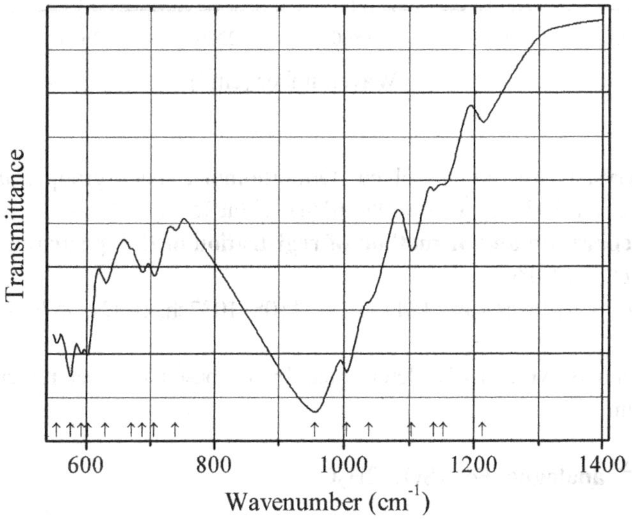

Origin: Synthetic.

Description: The crystal structure is solved. Trigonal, space group *R*-3, *a* = 8.068(1), *c* = 22.048(4) Å, *V* = 1242.8(4) Å^3 (at 153 K), *Z* = 6.

Kind of sample preparation and/or method of registration of the spectrum: No data.

Source: Wickleder et al. (2016).

Wavenumbers (cm^{-1}): 1214w, 1138w, 1152sh, 1104, 1038sh, 1004s, 956s, 738w, 705, 687, 670sh, 630, 602, 592, 575s, 554.

Note: The wavenumbers were partly determined by us based on spectral curve analysis of the published spectrum.

S641 Rhodium sulfate hydrate $Rh_2(SO_4)_3 \cdot 2H_2O$

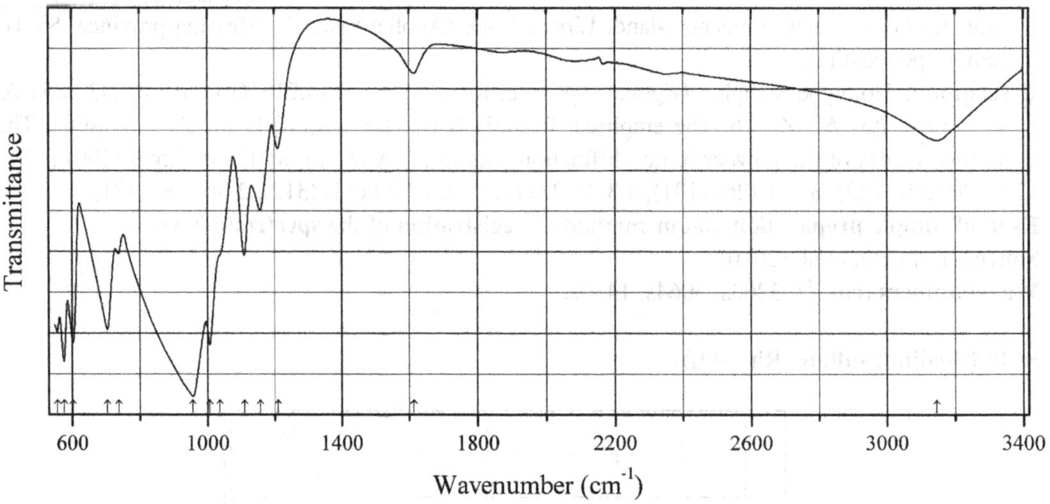

Origin: Synthetic.

Description: The crystal structure is solved. Orthorhombic, space group *Pnma*, $a = 9.2046(2)$, $b = 12.4447(3)$, $c = 8.3337(2)$ Å, $V = 954.61(4)$ Å^3 (at 153 K), $Z = 4$.

Kind of sample preparation and/or method of registration of the spectrum: No data.

Source: Wickleder et al. (2016).

Wavenumbers (cm^{-1}): 3146, 1614w, 1211, 1157, 1108, 1037sh, 1007s, 957s, 737, 704s, 602s, 575s, 555.

Note: The wavenumbers were partly determined by us based on spectral curve analysis of the published spectrum.

S642 Sanderite Fe^{2+} analogue $Fe^{2+}(SO_4) \cdot 2H_2O$

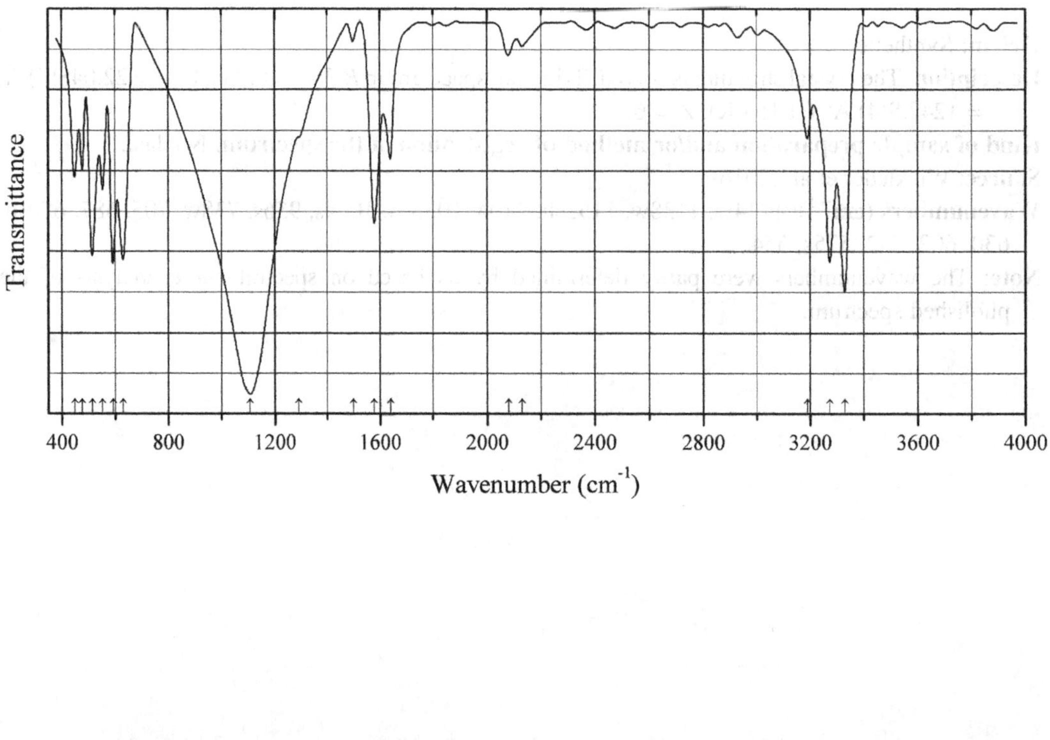

Origin: Synthetic.

Description: Synthesized by the hydro/solvothermal method. The crystal structure is solved. Orthorhombic, space group *Pccn*, $a = 6.3160$, $b = 7.7550$, $c = 8.9880$ Å, $V = 440.2$ Å^3, $Z = 4$.

Kind of sample preparation and/or method of registration of the spectrum: Transmission. Kind of sample preparation is not indicated.

Source: Zhao et al. (2015).

Wavenumbers (cm^{-1}): 3328s, 3273s, 3190, 2132w, 2081w, 1640, 1578s, 1498w, 1292sh, 1108s, 630s, 592s, 553, 514s, 477, 448.

Note: The wavenumbers were partly determined by us based on spectral curve analysis of the published spectrum.

S643 Schairerite $Na_{21}(SO_4)_7ClF_6$

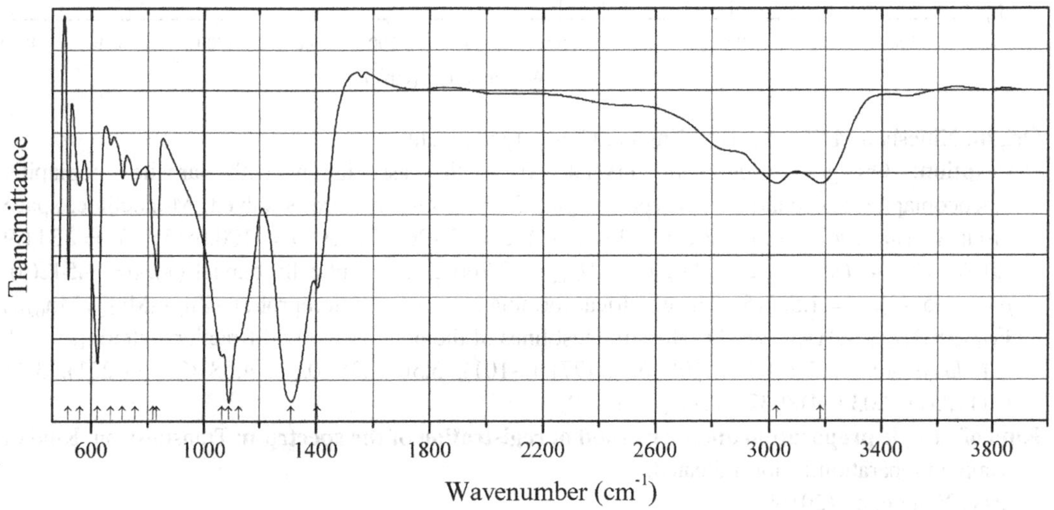

Origin: Synthetic.

Description: Ce-doped sample synthesized by a wet method. Characterized by powder X-ray diffraction data.

Kind of sample preparation and/or method of registration of the spectrum: Transmission. Kind of sample preparation is not indicated.

Source: Shinde and Dhoble (2015), Shinde et al. (2015).

Wavenumbers (cm^{-1}): 3183, 3026, 1405, 1313s, 1125sh, 1090s, 1064, 833, 820sh, 755, 710, 667w, 622s, 557, 517.

Note: The wavenumbers were partly determined by us based on spectral curve analysis of the published spectrum. The bands in the range from 2800 to 3200 cm^{-1} indicate the presence of covalent O–H-bonds. The band at 833 cm^{-1} may correspond to Na···O–H bending vibrations. The band at 1313 cm^{-1} may be due to an impurity.

S644 Zincobotryogen $ZnFe^{3+}(SO_4)_2(OH)\cdot 7H_2O$

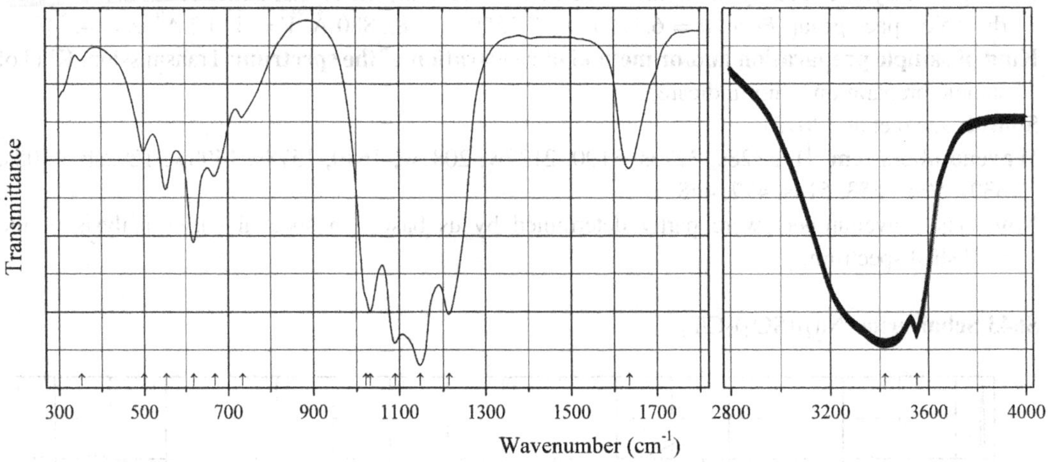

Origin: Xitieshan Pb-Zn deposit, Qinghai, China (type locality).

Description: Orange-red prismatic crystals from the association with jarosite, copiapite, zincocopiapite, and quartz. Holotype sample. The crystal structure is solved. Monoclinic, space group $P12_1/n1$, a = 10.504(2), b = 17.801(4), c = 7.1263(14) Å, β = 100.08(3)°, V = 1311.9 (5) Å^3, Z = 4. D_{meas} = 2.20(1) g/cm^3, D_{calc} = 2.266 g/cm^3. Optically biaxial (+), α= 1.542(5), β = 1.551(5), γ = 1.587(5). The empirical formula is (electron microprobe): $(Zn_{0.73}Mg_{0.16}Mn_{0.08})Fe_{0.99}(SO_4)_{2.04}(OH)_{0.82}\cdot 7H_2O$. The strongest lines of the powder X-ray diffraction pattern [d, Å (I, %) (hkl)] are: 8.92 (100) (110), 6.32 (77) (−101), 5.56 (23) (021), 4.08 (22) (−221), 3.21 (31) (231), 3.03 (34) (032), 2.77 (22) (042).

Kind of sample preparation and/or method of registration of the spectrum: Transmission. Kind of sample preparation is not indicated.

Source: Yang et al. (2016b).

Wavenumbers (cm^{-1}): 3550s, 3420s, 1635, 1213s, 1147s, 1090s, 1032s, 1022sh, 732, 668, 618s, 553, 499, 352w.

Note: The wavenumbers were partly determined by us based on spectral curve analysis of the published spectrum.

S645 Jurbanite $Al(SO_4)(OH){\cdot}5H_2O$

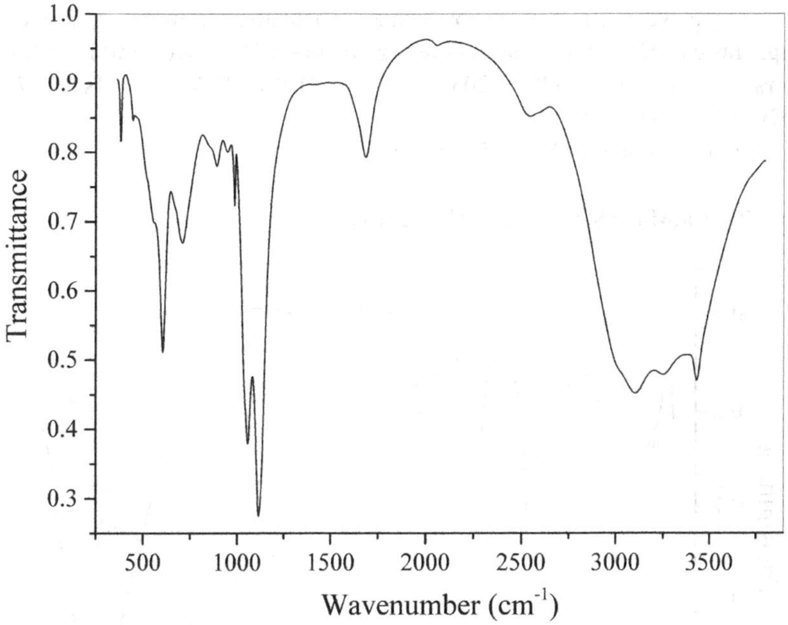

Origin: Le Cetine di Cotorniano mine, Chiusdino, Siena province, Tuscany, Italy.

Description: White aggregate. Investigated by A.V. Kasatkin. Characterized by powder X-ray diffraction data and electron microprobe analyses.

Kind of sample preparation and/or method of registration of the spectrum: KBr disc. Absorption.

Wavenumbers (cm^{-1}): 3432s, 3252s, 3104s, 3025sh, 2519, 2061w, 1685, 1114s, 1057s, 987, 949w, 891, 885sh, 709, 605s, 560sh, 447w, 381.

Note: The spectrum was obtained by N.V. Chukanov.

S646 Beaverite-(Zn) $Pb(Fe^{3+}{}_2Zn)(SO_4)_2(OH)_6$

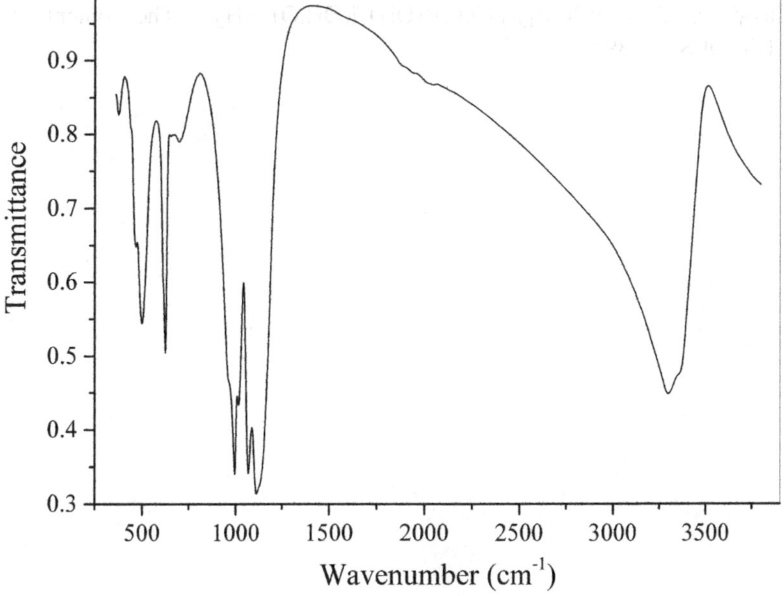

Origin: San Francisco mine, Sierra Gorda, Chile.

Description: Brown crystalline crusts from the association with atacamite and paratacamite. Characterized by powder X-ray diffraction data and qualitative electron microprobe analyses.

Kind of sample preparation and/or method of registration of the spectrum: KBr disc. Absorption.

Wavenumbers (cm^{-1}): 3350sh, 3296s, 2032w, 1947w, 1888w, 1130sh, 1111s, 1067s, 1015, 995s, 965sh, 697w, 655w, 625, 500, 468, 440sh, 346w.

Note: The spectrum was obtained by N.V. Chukanov.

S647 Jouravskite $Ca_3Mn^{4+}(SO_4)(CO_3)(OH)_6 \cdot 12H_2O$

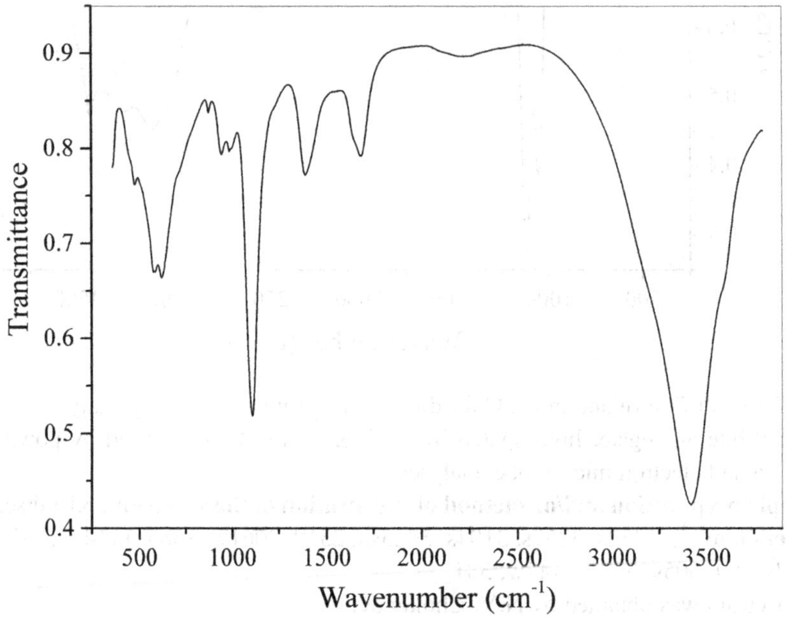

Origin: N'Chwaning 3 mine, Kuruman, Kalahari manganese field, Northern Cape province, South Africa.

Description: Yellow crystals. A boron-bearing variety. The empirical formula is (electron microprobe): $Ca_{3.0}(Mn_{0.95}Fe_{0.04})(SO_4)_{1.00}[CO_3,B(OH)_4](OH,O)_6 \cdot nH_2O$. The content of B_2O_3 determined by ICP-OES is 0.39 wt%.

Kind of sample preparation and/or method of registration of the spectrum: KBr disc. Absorption.
Wavenumbers (cm^{-1}): 3485sh, 3421s, 3385sh, 3245sh, 3085sh, 2880sh, 2455w, 2257w, 1696, 1645, 1392s, 1104s, 1004w, 962, 887, 719, 639, 600sh, 578s, 550sh, 485sh, 460sh, 374.
Note: The spectrum was obtained by N.V. Chukanov.

S648 Zincovoltaite $K_2Zn_5Fe^{3+}{}_3Al(SO_4)_{12}\cdot 18H_2O$

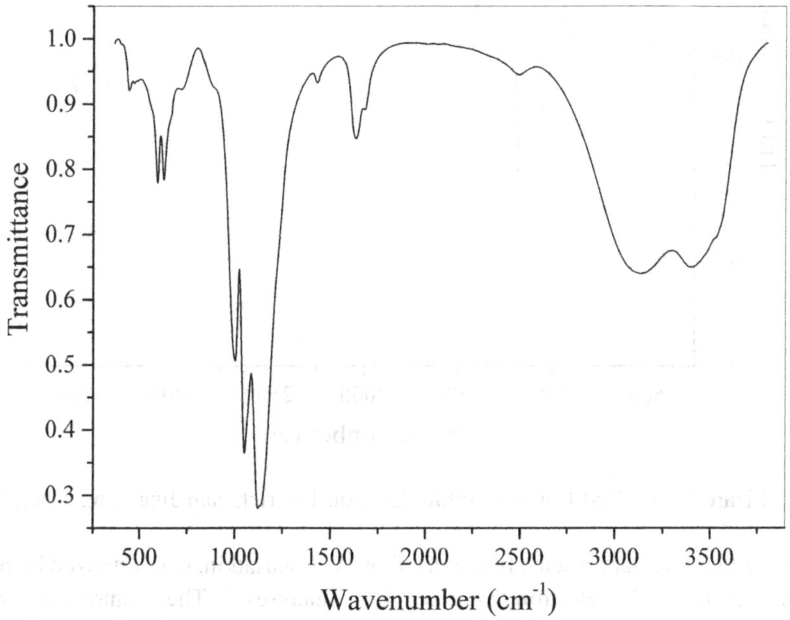

Origin: Muzhievskoe deposit, Transcarpathian Region, Ukraine.
Description: Black crystals from the association with zincocopiapite, bianchite, and boyleite. A NH_4-bearing variety. Investigated by A.V. Kasatkin. Characterized by single-crystal X-ray diffraction data and electron microprobe analyses. Cubic, a = 27.2450(14) Å. The empirical formula is $K_{1.57}(NH_4)_x(Zn_{3.54}Mg_{0.49}Cu_{0.44}Fe^{2+}{}_{0.36}Mn_{0.08})Fe^{3+}{}_{2.82}Al_{1.30}S_{11.97}O_{48}\cdot 18H_2O$.
Kind of sample preparation and/or method of registration of the spectrum: KBr disc. Absorption.
Wavenumbers (cm^{-1}): 3520sh, 3406 (broad), 3134 (broad), 2493w, 1677, 1635, 1429w, 1134s, 1121s, 1051s, 1003, 890sh, 718w, 655sh, 626, 593, 468w, 441w.
Note: The spectrum was obtained by N.V. Chukanov. The band at 1429 cm^{-1} corresponds to NH_4^+ cations.

S649 Ammoniozippeite $(NH_4)_2[(UO_2)_2(SO_4)O_2]\cdot H_2O$

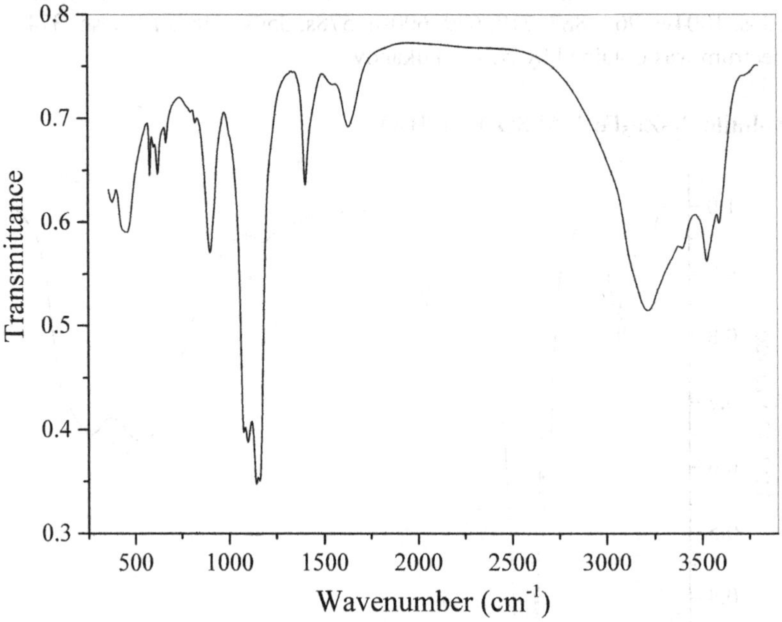

Origin: Blue Lizard Mine, Red Canyon, White Canyon District, San Juan Co., Utah, USA (cotype locality).

Description: Yellow acicular crystals. Investigated by A.V. Kasatkin. Characterized by powder X-ray diffraction data and electron microprobe analyses. The empirical formula is: $(NH_4)_x[(UO_2)_{2.06}(SO_4)_{0.94}O_2]\cdot nH_2O$. The observed lines of the powder X-ray diffraction pattern (*d*, Å) are: 8.42, 8.10s, 5.46w, 4.24, 3.65, 3.48s, 3.12s, 2.85, 2.65w, 2.37w, 2.20w, 2.11w, 2.04w, 1.96w, 1.87w, 1.74w, 1.70w.

Kind of sample preparation and/or method of registration of the spectrum: KBr disc. Absorption.

Wavenumbers (cm^{-1}): 3592, 3526, 3397, 3217s, 1635, 1551w, 1402, 1162s, 1145s, 1100s, 1081s, 901, 825w, 801w, 668, 625, 605w, 583, 462, 440sh, 380.

Note: The spectrum was obtained by N.V. Chukanov. The bands at 1402 cm^{-1} correspond to NH_4^+.

S650 Motukoreaite-related mineral $[(Mg,Al)_9(OH)_{18}][Na_x(SO_4,CO_3)_2(H_2O)_{12}]$ (?)

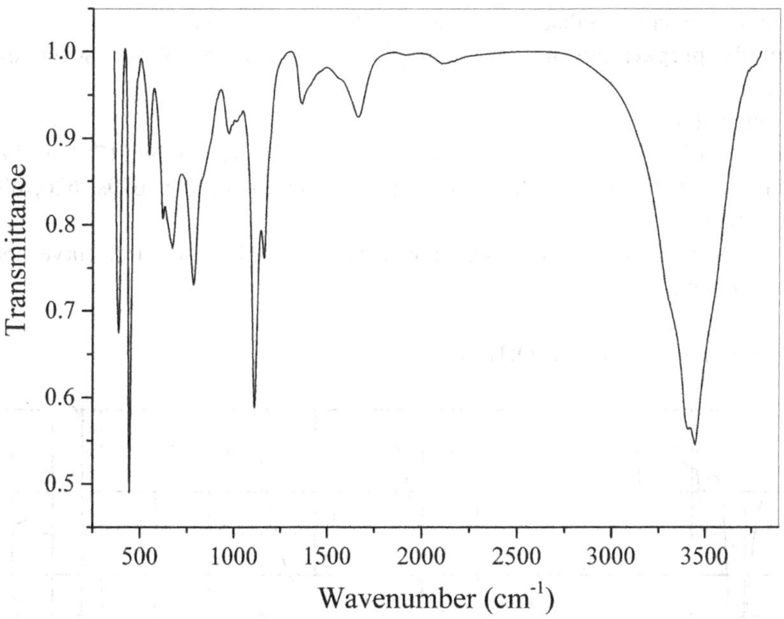

Origin: Verkhnekamskoe salt deposit, western Urals, Russia.

Description: Colorless grains from the association with halite. Investigated by I.V. Pekov. Characterized by X-ray diffraction.

Kind of sample preparation and/or method of registration of the spectrum: KBr disc. Absorption.

Wavenumbers (cm^{-1}): 3520sh, 3449s, 3412s, 3320sh, 2104w, 1918w, 1662, 1565sh, 1363, 1162, 1110s, 1010w, 972, 830sh, 784, 670, 620, 548, 444s, 387s.

Note: The spectrum was obtained by N.V. Chukanov.

SC18 Leadhillite $Pb_4(SO_4)(CO_3)_2(OH)_2$

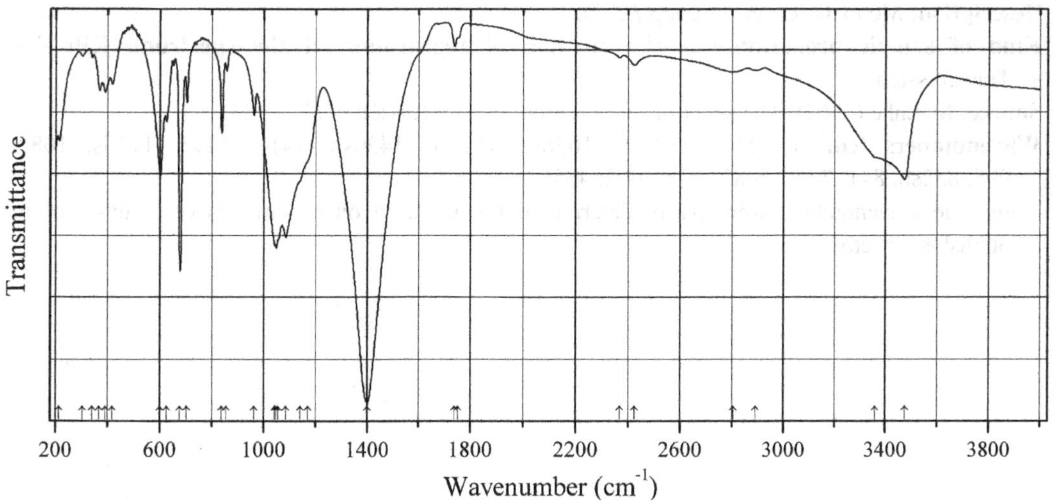

Origin: Leadhills, South Lanarkshire, Strathclyde, Scotland, UK (type locality).
Description: Characterized by chemical analyses and powder X-ray diffraction data. The chemical composition is very close to that calculated from the ideal formula.
Kind of sample preparation and/or method of registration of the spectrum: CsI disc. Transmission.
Source: Russell et al. (1983).
Wavenumbers (cm^{-1}): 3475, 3360sh, 2890w, 2805w, 2425w, 2368w, 1751sh, 1738w, 1399s, 1170sh, 1140sh, 1085s, 1055s, 1049s, 1043sh, 962, 857, 838, 705, 679s, 626, 602, 419, 392, 370, 342w, 304w, 215.
Note: The wavenumbers were partly determined by us based on spectral curve analysis of the published spectrum.

SC19 Leadhillite $Pb_4(SO_4)(CO_3)_2(OH)_2$

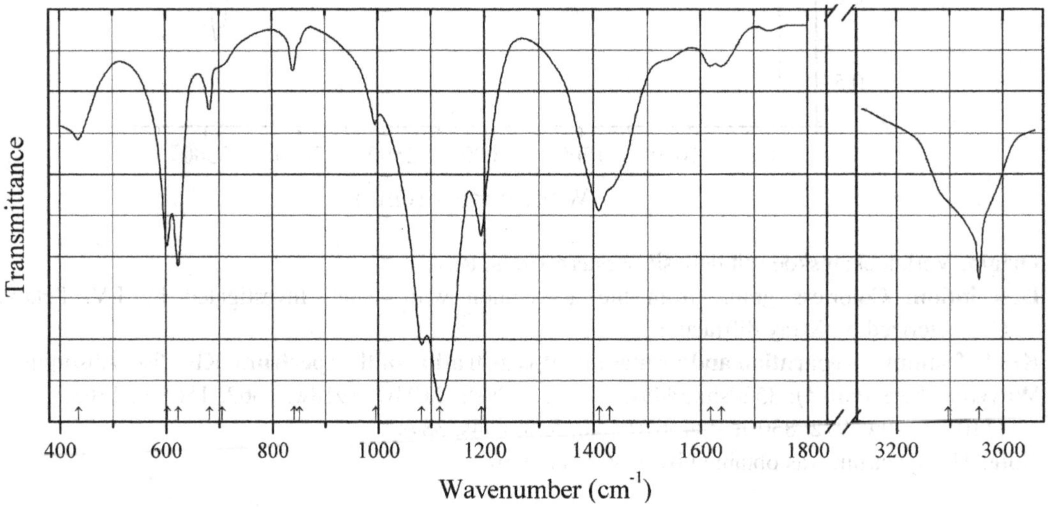

Origin: Leadhills, South Lanarkshire, Strathclyde, Scotland, UK (type locality).
Description: Monoclinic, space group $P2_1/a$.
Kind of sample preparation and/or method of registration of the spectrum: KBr disc. Transmission.
Source: Moenke (1962). The spectrum was reproduced by Russell et al. (1983).
Wavenumbers (cm^{-1}): 3508, 3392sh, 1638w, 1618w, 1430sh, 1410, 1193, 1115s, 1082s, 995, 852sh, 841, 705sh, 682, 623s, 602, 435.
Note: The wavenumbers were partly determined by us based on spectral curve analysis of the published spectrum.

SP25 Arangasite $Al_2(SO_4)(PO_4)F{\cdot}9H_2O$

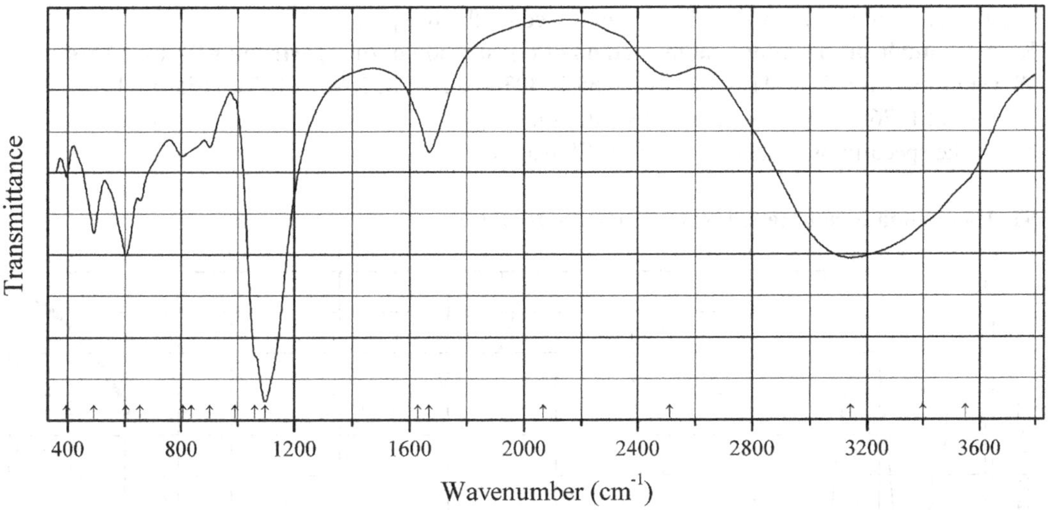

Origin: Alyaskitovoye Sn-W deposit, Ust'-Nera, Indigirka River Basin, Sakha Republic (Yakutia), Russia (type locality).

Description: White granular aggregate. Investigated by I.V. Pekov. The empirical formula is (electron microprobe): $Al_{2.09}(SO_4)_{1.00}(PO_4)_{0.89}(AsO_4)_{0.105}(SiO_4)_{0.005}F_{1.41}{\cdot}nH_2O$. The strongest lines of the powder X-ray diffraction pattern [*d*, Å (*I*, %)] are: 10.68 (55), 9.66 (100), 5.33 (26), 4.21 (41), 3.491 (22), 3.145 (22).

Kind of sample preparation and/or method of registration of the spectrum: KBr disc. Absorption.

Wavenumbers (cm^{-1}): 3550sh, 3400sh, 3143s (broad), 2511, 2068w, 1669, 1630sh, 1098s, 1060sh, 990sh, 901, 835sh, 806, 655, 605s, 491, 396.

Note: The spectrum was obtained by N.V. Chukanov.

SP26 Ardealite $Ca_2(PO_3OH)(SO_4){\cdot}4H_2O$

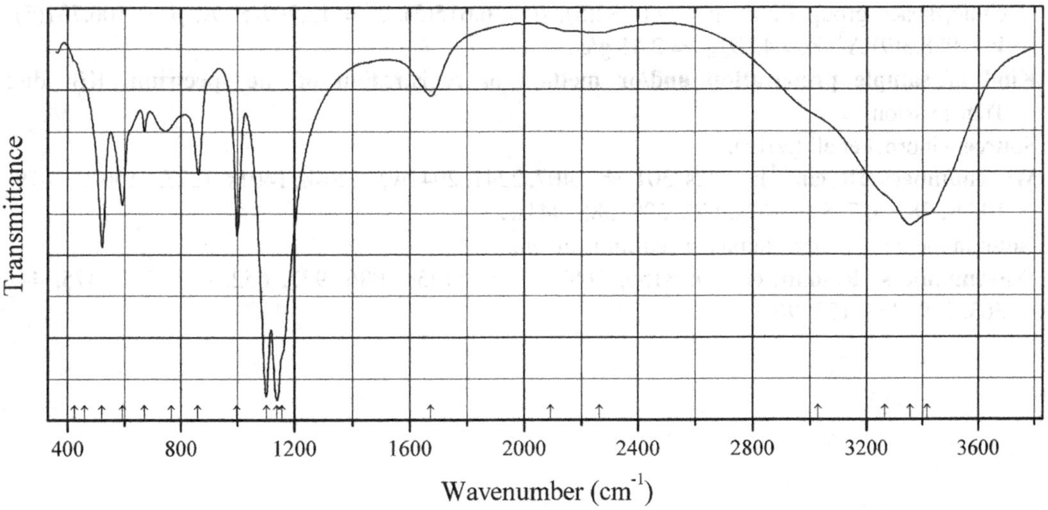

Origin: Cerro Mejillones, Mejillones Peninsula, Mejillones, Antofagasta, II Region, Chile.
Description: White soft fine-granular aggregate. Investigated by I.V. Pekov. Characterized by qualitative electron microprobe analyses. Confirmed by the IR spectrum.
Kind of sample preparation and/or method of registration of the spectrum: KBr disc. Absorption.
Wavenumbers (cm^{-1}): 3415sh, 3355s, 3265sh, 3030sh, 2265w, 2092w, 1673, 1155sh, 1139s, 1101s, 999s, 861, 765, 671, 593, 522s, 460sh, 425sh.
Note: The spectrum was obtained by N.V. Chukanov.

STe1 Ammonium sulfate tellurate $(NH_4)_2(SO_4)\cdot Te(OH)_6$

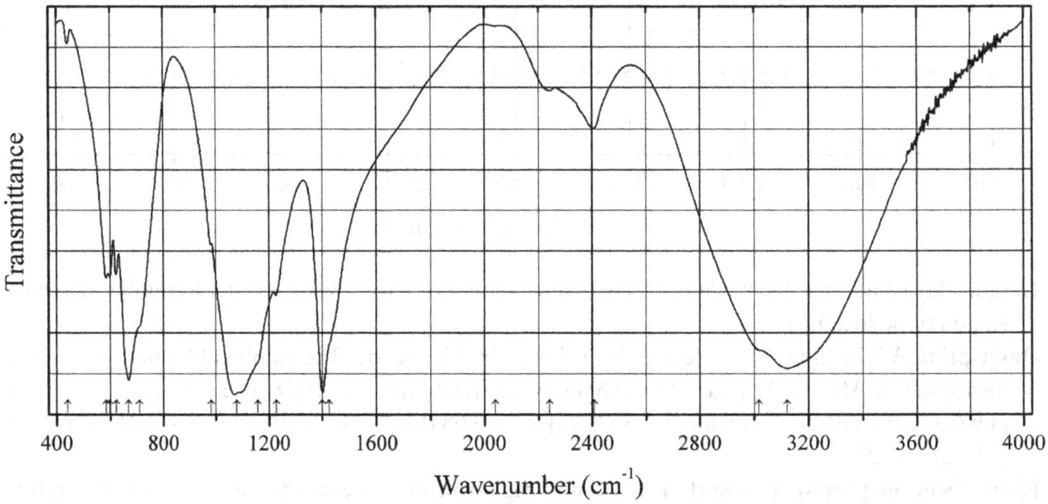

Origin: Synthetic.
Description: As-doped sample produced from an aqueous stoichiometric solution of telluric acid, ammonium sulfate, ammonium carbonate, and arsenic acid. The empirical formula $(NH_4)_2(SO_4)_{0.92}H(AsO_4)_{0.08}Te(OH)_6$ given in the original paper isn't charge-balanced. The correct formula should be $(NH_4)_2[(SO_4)_{0.92}(HAsO_4)_{0.08}]\cdot Te(OH)_6$. The crystal structure is solved. Monoclinic, space group $P2_1/c$, $a = 11.382(5)$, $b = 6.615(5)$, $c = 13.707(5)$ Å, $\beta = 106.731(5)°$, $V = 988.3(9)$ Å^3, $Z = 4$. $D_{calc} = 2.41$ g/cm^3.
Kind of sample preparation and/or method of registration of the spectrum: KBr disc. Transmission.
Source: Ghorbel et al. (2015).
Wavenumbers (IR, cm^{-1}): 3122s, 3017sh, 2407, 2241, 2041w, 1426sh, 1401s, 1227, 1156sh, 1079s, 1070s, 983sh, 708sh, 669s, 622, 599, 585, 441w.
Note: In the cited paper, Raman spectrum is given.
Wavenumbers (Raman, cm^{-1}): 3156, 1687, 1428, 1175, 1086, 977, 652, 622, 600, 475, 443, 363, 339, 323, 135, 96.

2.10 Chlorides and Hydroxychlorides

Cl71 Schwartzembergite $Pb_5H_2(IO_2)O_4Cl_3$

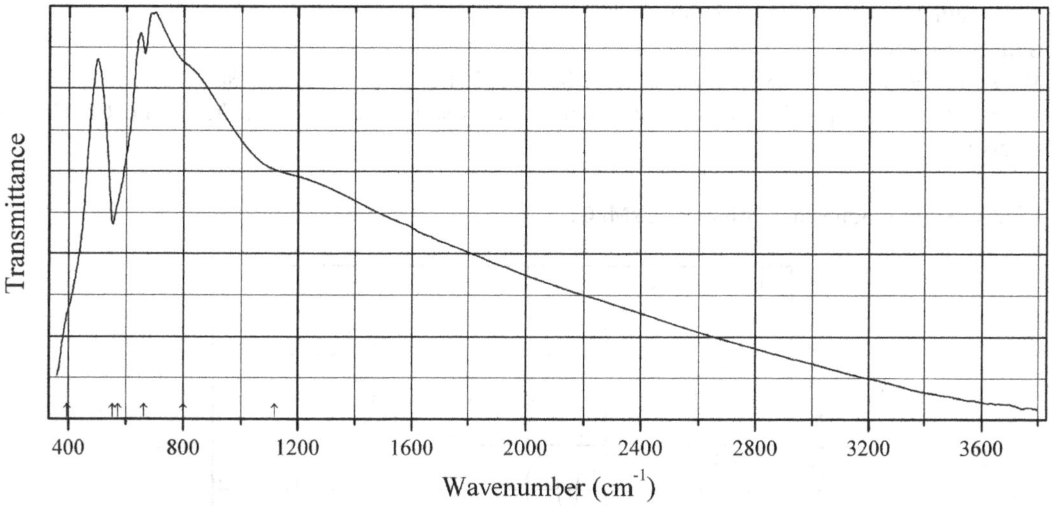

Origin: San Francisco (Beatrix) mine, Caracoles, Sierra Gorda district, Antofagasta Region, Chile.

Description: Orange lenticular crystals from the association with paralaurionite. Investigated by I.V. Pekov. Characterized by powder X-ray diffraction data and electron microprobe analyses.

Kind of sample preparation and/or method of registration of the spectrum: KBr disc. Absorption.

Wavenumbers (cm^{-1}): 1120sh, 800sh, 665, 575sh, 555, 395sh.

Note: The spectrum was obtained by N.V. Chukanov. No bands corresponding to covalent O–H bonds are observed. The shoulder at 1120 cm^{-1} may correspond to the essentially ionic bond $Cl^{-}\cdots H^{+}$.

Cl72 Cesium copper chloride $CsCuCl_3$

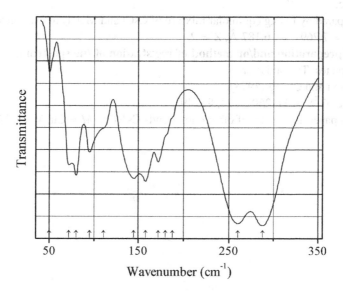

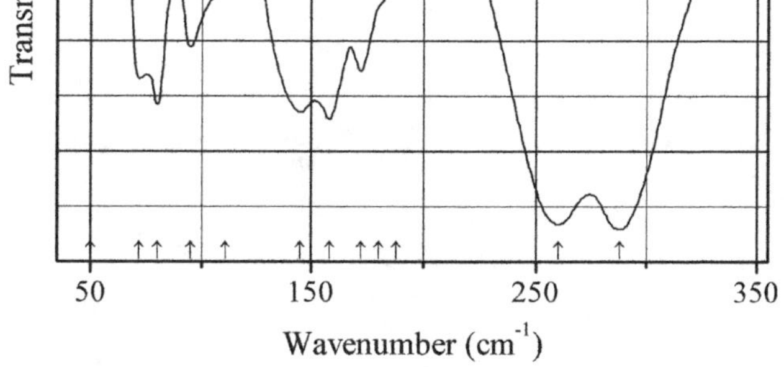

Origin: Synthetic.

Description: Crystallized by slow evaporation from hot concentrated aqueous solutions of CsCl and $CuCl_2$ in a 1:1 mole ratio. Hexagonal, space group $P6_122$, $a = 7.2157$, $c = 18.1777$ Å, $Z = 6$.

Kind of sample preparation and/or method of registration of the spectrum: Nujol mull between polyethylene plates. Transmission.

Source: McPherson and Chang (1973).

Wavenumbers (cm^{-1}): 288s, 260s, 188sh, 180sh, 172, 158, 145, 111sh, 95, 80, 72, 50w.

Note: In the cited paper, IR spectra of other compounds $CsMCl_3$ (M = Mg, V, Cr, Mn, Fe, Co, Ni) are given.

Cl73 Cesium magnesium chloride $CsMgCl_3$

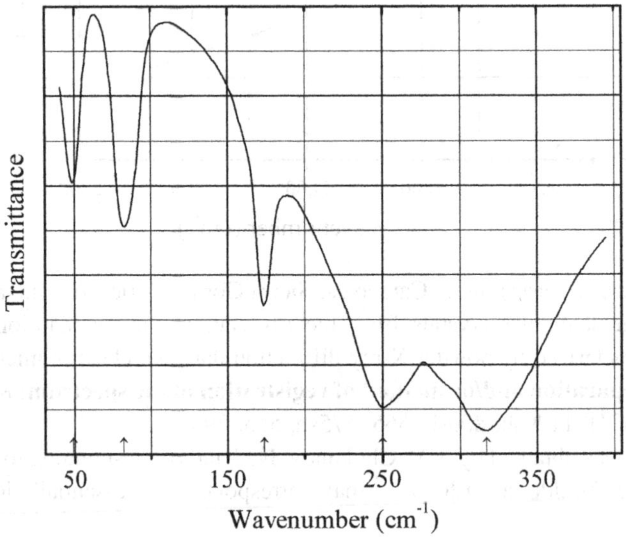

Origin: Synthetic.

Description: Prepared by fusing equimolar mixture of CsCl and $MgCl_2$ in evacuated quartz ampoule. Hexagonal, $a = 7.269$, $c = 6.187$ Å, $Z = 2$.

Kind of sample preparation and/or method of registration of the spectrum: Nujol mull between polyethylene plates. Transmission.

Source: McPherson and Chang (1973).

Wavenumbers (cm^{-1}): 317s, 250s, 174, 82, 49.

Note: In the cited paper, IR spectra of other compounds $CsMCl_3$ (M = Cu, V, Cr, Mn, Fe, Co, Ni) are given.

Cl74 Cesium sodium stibiochloride $Cs_2NaSbCl_6$

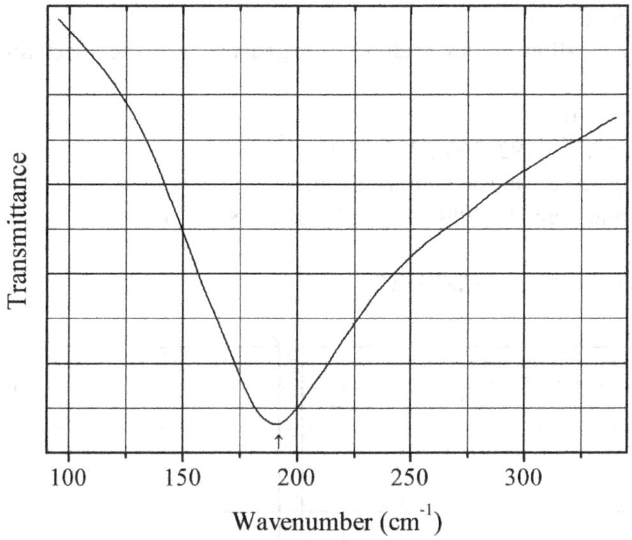

Origin: Synthetic.

Description: Prepared by heating a mixture of stoichiometric quantities of $SbCl_3$, CsCl, and NaCl at 800 °C. Characterized by powder X-ray diffraction data. Cubic, $a = 10.770$.

Kind of sample preparation and/or method of registration of the spectrum: Polyethylene disc. Absorption.

Source: Smit et al. (1990).

Wavenumbers (IR, cm^{-1}): 192.

Note: In the cited paper, Raman spectrum is given.

Wavenumbers (Raman, cm^{-1}): 314, 284, 228, 117, 68, 47, 34.

Cl75 Cesium antimony chloride $Cs_3Sb_2Cl_9$

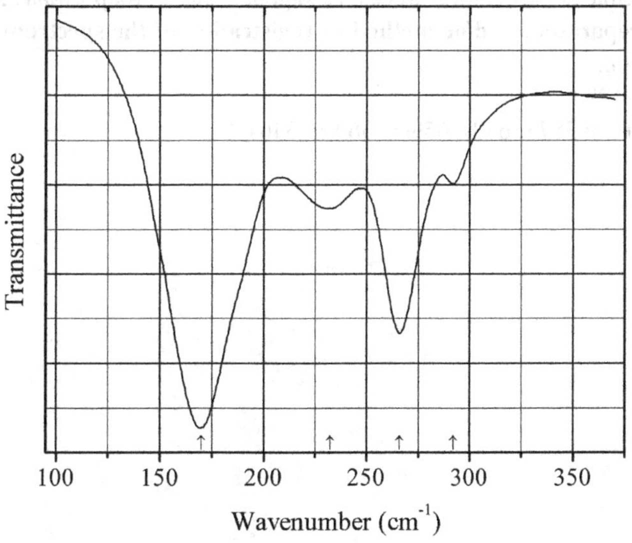

Origin: Synthetic.

Description: Trigonal α-modification prepared by evaporating to dryness a hot aqueous HCI solution containing appropriate cations.

Kind of sample preparation and/or method of registration of the spectrum: Polyethylene disc. Transmission.

Source: Smit et al. (1990).

Wavenumbers (IR, cm^{-1}): 292, 266s, 232, 170s.

Note: In the cited paper, Raman spectrum is given.

Wavenumbers (Raman, cm^{-1}): 305s, 257s, 127, 102, 88, 55, 48, 42.

Cl76 Calcium hydroxychloride Ca(OH)Cl

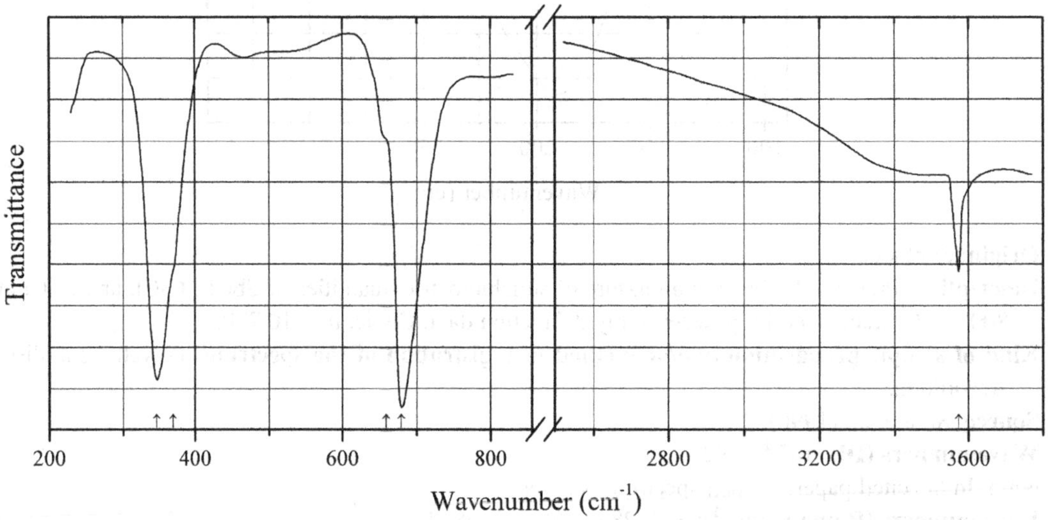

Origin: Synthetic.

Description: Prepared by heating stoichiometric amounts of anhydrous $CaCl_2$ and $Ca(OH)_2$ in a carbon glass crucible at 610 K for 3 weeks. Hexagonal, space group $P6_3mc$, $Z = 4$.

Kind of sample preparation and/or method of registration of the spectrum: CsI disc and Nujol mull. Transmission.

Source: Lutz et al. (1993).

Wavenumbers (cm^{-1}): 3573, 680s, 659sh, 369sh, 346s.

Cl77 Copper oxychloride hydrate $Cu_2OCl_2{\cdot}2H_2O$.

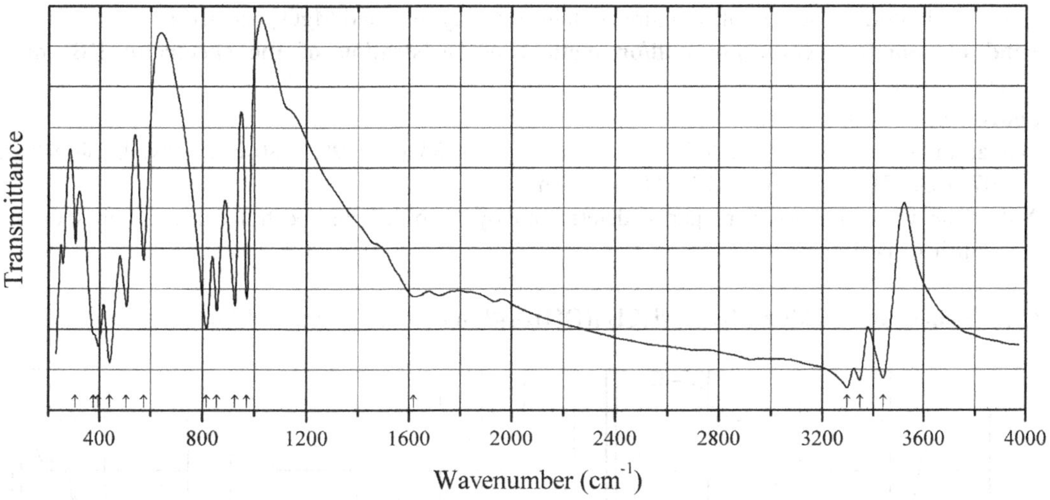

Origin: Synthetic.

Description: Precipitate prepared by mixing equal volumes of aqueous solutions of 0.1 M $CuCl_2{\cdot}2H_2O$ and 0.5 M of urea. The mixture was heated to ~75 °C for 4–6 h. Characterized by the elemental analysis.

Kind of sample preparation and/or method of registration of the spectrum: KBr disc. Transmission.

Source: El-Metwally and Al Thani (1989).

Wavenumbers (cm^{-1}): 3440s, 3347s, 3298s, 1620, 970, 925, 855s, 815s, 572, 505, 440s, 395sh, 375, 305.

Cl78 Magnesium oxychloride hydrate $Mg_3Cl_2(OH)_4{\cdot}4H_2O$ $Mg_3Cl_2(OH)_4{\cdot}4H_2O$

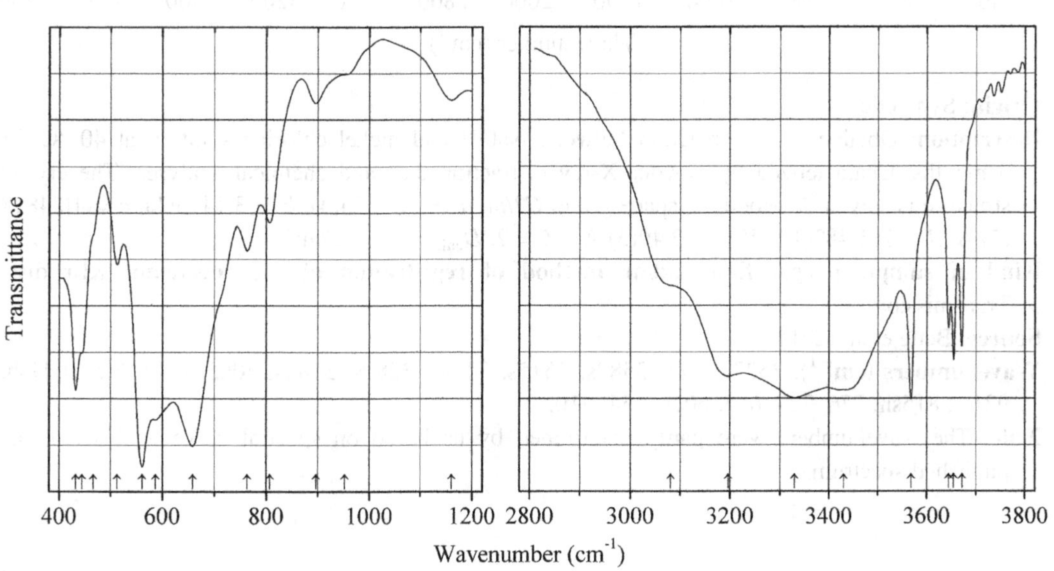

Origin: Synthetic.

Description: Monoclinic, space group *C2/m*, $a = 15.1263(3)$, $b = 3.1707(1)$, $c = 10.5236(2)$ Å, $\beta = 101.546(2)$. The crystal structure contains strongly distorted MgO_6 octahedra.

Kind of sample preparation and/or method of registration of the spectrum: KBr disc. Transmission.

Source: Bette et al. (2014).

Wavenumbers (cm^{-1}): 3671, 3654, 3644, 3568s, 3430s, 3330s, 3196, 3080sh, 1160w, 952sh, 896w, 807, 763, 658s, 587, 561s, 513, 466sh, 444sh, 432.

Note: The wavenumbers were partly determined by us based on spectral curve analysis of the published spectrum.

Cl79 Nickel oxychloride hydrate $Ni_3Cl_{2.1}(OH)_{3.9}{\cdot}4H_2O$ $Ni_3Cl_{2.1}(OH)_{3.9}{\cdot}4H_2O$

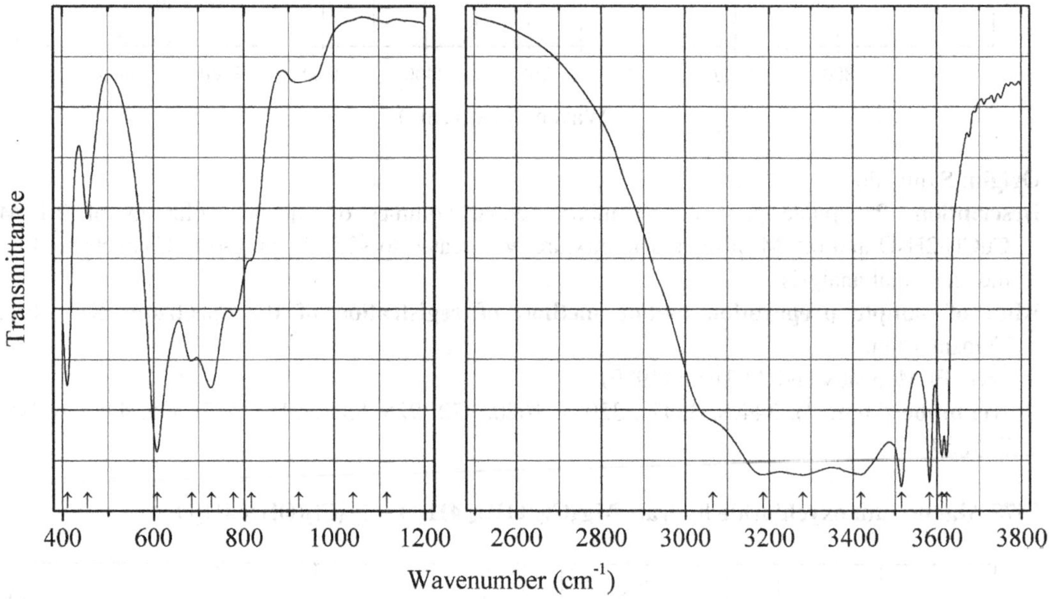

Origin: Synthetic.

Description: Obtained in the reaction between NaOH and nickel chloride solution at 40 °C for 3 months. Characterized by powder X-ray diffraction data and chemical analyses. The crystal structure is solved. Monoclinic, space group *C2/m*, $a = 14.9575(4)$, $b = 3.1413(1)$, $c = 10.4818(5)$ Å, $\beta = 101.482(1)°$, $V = 482.49(3)$ Å^3, $Z = 2$. $D_{calc} = 2.67$ g/cm^3.

Kind of sample preparation and/or method of registration of the spectrum: KBr disc. Transmission.

Source: Bette et al. (2014).

Wavenumbers (cm^{-1}): 3622, 3611, 3582s, 3515s, 3419s, 3280s, 3185s, 3065sh, 1116w, 1041sh, 921w, 815sh, 776, 727, 684, 607s, 454, 410.

Note: The wavenumbers were partly determined by us based on spectral curve analysis of the published spectrum.

Cl80 Potassium mercury chloride hydrate $K_2HgCl_4{\cdot}H_2O$

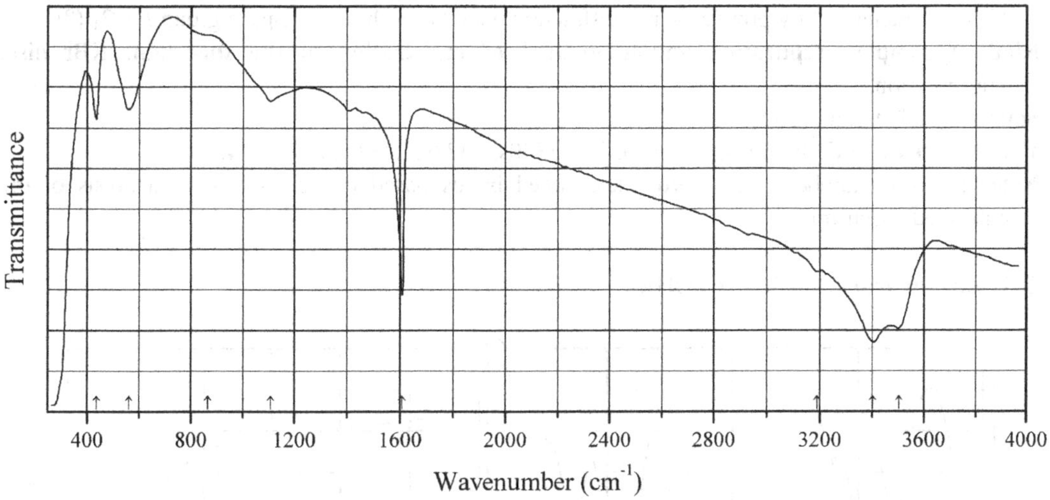

Origin: Synthetic.

Description: Characterized by powder X-ray diffraction data. Orthorhombic, space group *Pbam*, $a = 5.258$, $b = 11.662$, $c = 8.925$ Å, $Z = 4$.

Kind of sample preparation and/or method of registration of the spectrum: KCl disc. Transmission.

Source: Falk and Knop (1977).

Wavenumbers (cm^{-1}): 3505, 3404s, 3191w, 1610s, 1108, 865sh, 560, 436.

Note: The wavenumbers were partly determined by us based on spectral curve analysis of the published spectrum.

Cl81 Yttrium hydroxychloride hydrate $Y_2(OH)_5Cl{\cdot}1.5H_2O$

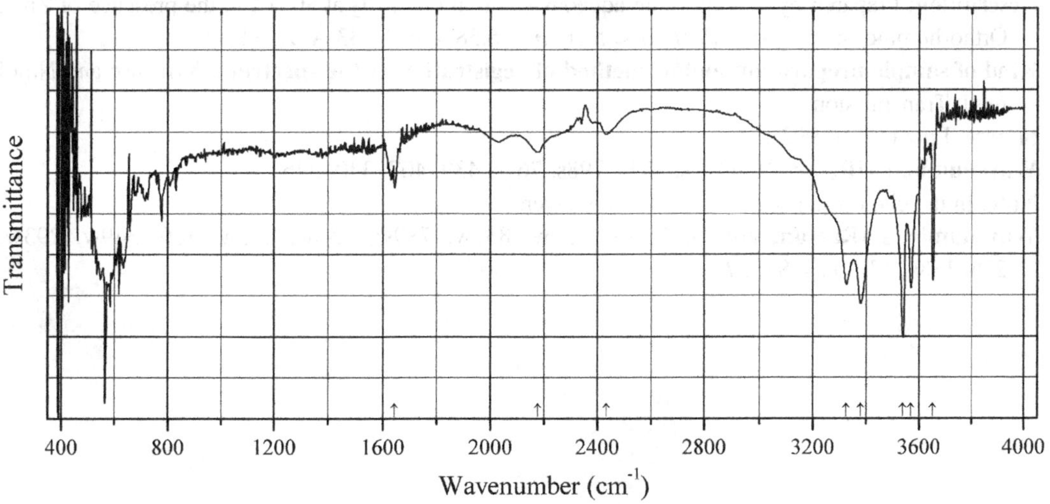

Origin: Synthetic.

Description: Prepared hydrothermally from YCl_3 in the presence of NaOH and NaCl at 150 °C for 12 h. Characterized by powder X-ray diffraction data. Orthorhombic. Space group *Pca*2_1 (?).

Kind of sample preparation and/or method of registration of the spectrum: KBr disc. Transmission.

Source: Poudret et al. (2008).

Wavenumbers (cm^{-1}): 3652s, 3570s, 3539s, 3380s, 3327s, 2433, 2177, 1644.

Note: The wavenumbers were partly determined by us based on spectral curve analysis of the published spectrum.

Cl82 Zinc hydroxychloride Zn(OH)Cl

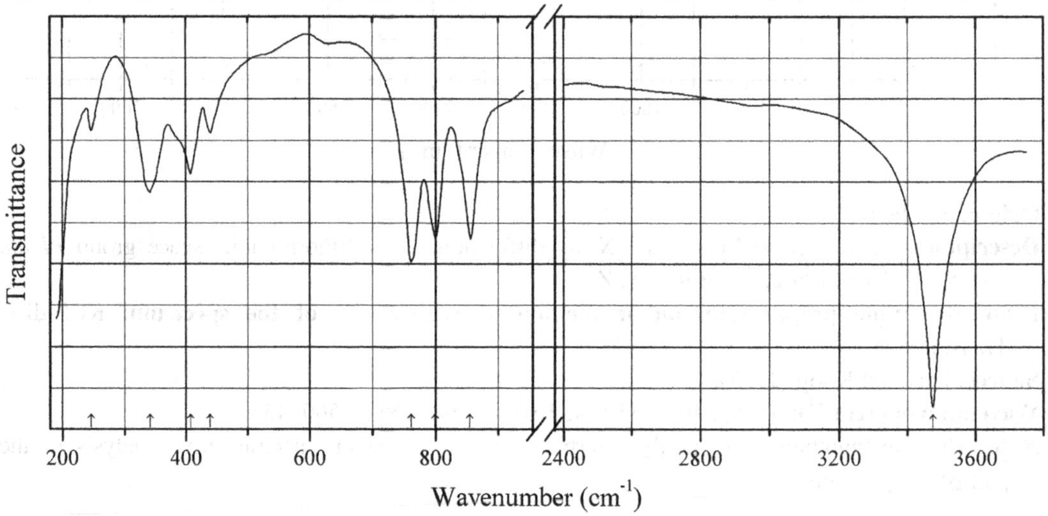

Origin: Synthetic.

Description: Prepared by heating of an aqueous solution of $ZnCl_2$ at 80 °C, in the presence of ZnO. Orthorhombic, space group *Pcab*, $a \approx 5.86$, $b \approx 6.58$, $c \approx 11.33$ Å, Z = 8.

Kind of sample preparation and/or method of registration of the spectrum: KBr disc and Nujol mull. Transmission.

Source: Lutz et al. (1993).

Wavenumbers (IR, cm^{-1}): 3476s, 853s, 798s, 761s, 439, 407, 340, 246.

Note: In the cited paper, Raman spectrum is given.

Wavenumbers (Raman, cm^{-1}): 3456s, 850w, 837w, 789w, 756w, 455w, 386, 349w, 293w, 216, 188s, 171, 114, 87, 57.

Cl83 Centennialite $CaCu_3Cl_2(OH)_6 \cdot nH_2O$ ($n \sim 0.7$)

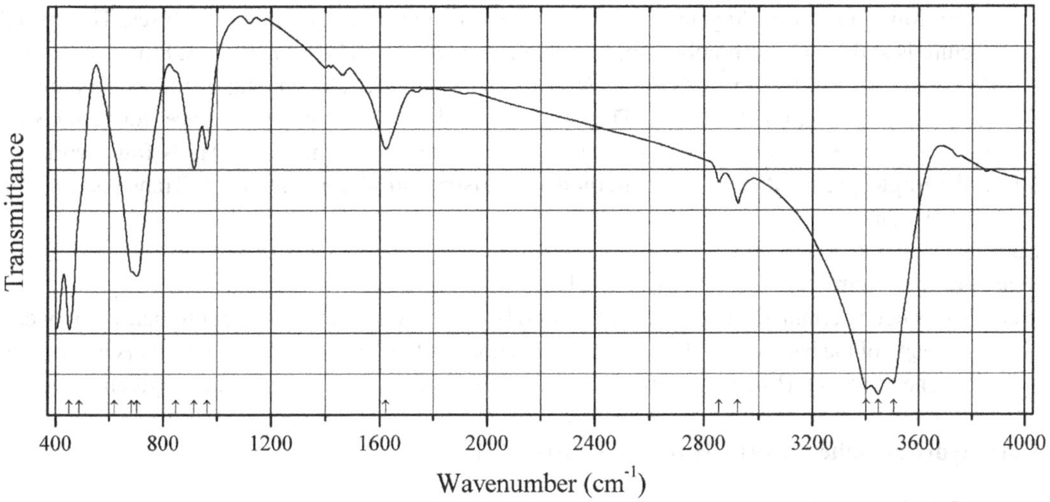

Origin: Synthetic.

Description: Synthesized from $CuCl_2 \cdot 2H_2O$, $CaCl_2$, $LiCl \cdot H_2O$, and $LiOH \cdot H_2O$ through a solid-state reaction method. Characterized by electron microprobe analyses. The crystal structure is solved. Trigonal, space group P-3m1, a = 6.6475(9), c = 5.7600(12) Å, V = 220.43(8) Å^3, Z = 1. D_{calc} = 3.108 g/cm^3.

Kind of sample preparation and/or method of registration of the spectrum: KBr disc. Transmission.

Source: Sun et al. (2015).

Wavenumbers (cm^{-1}): 3505s, 3447s, 3402s, (2925), (2855), 1623, 964, 916, 848sh, 703s, 685sh, 621sh, 490sh, 452s.

Note: The wavenumbers were partly determined by us based on spectral curve analysis of the published spectrum. Weak bands in the range from 2800 to 3000 cm^{-1} correspond to the admixture of an organic substance.

Cl84 Comancheite $Hg^{2+}{}_{55}N^{3-}{}_{24}(Cl,Br)_{34}(OH,NH_2)_4$

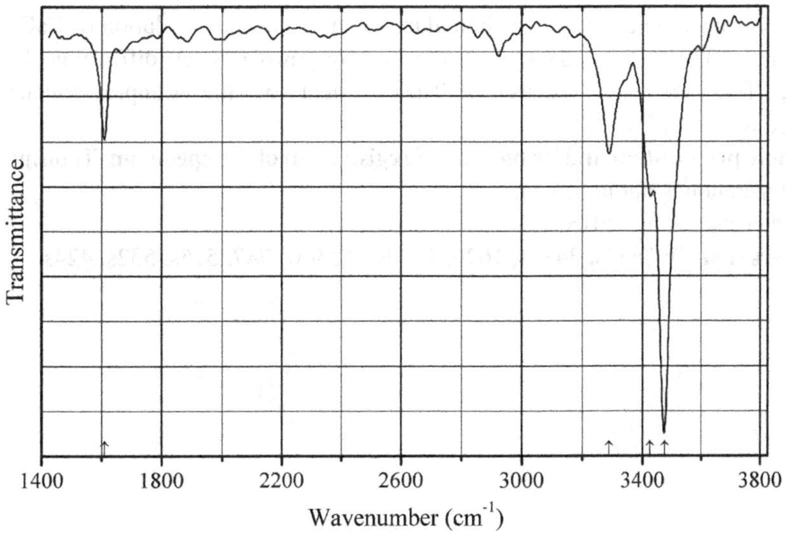

Origin: Mariposa mine, Terlingua district, Brewster Co., Texas, USA (type locality).

Description: Orange aggregate on calcite. Specimen No. 26686 from the collections of the Natural History Museum of Los Angeles Co. Characterized by electron microprobe analyses. The crystal structure is solved. Orthorhombic, space group *Pnnm*, a = 18.414(5), b = 21.328(6), c = 6.6976 (19) Å, V = 2630(2) Å^3, Z = 1. D_{calc} = 8.25 g/cm^3. The crystal-chemical formula is $Hg_{55}(Cl_{24.5}Br_{9.5})N_{24}(OH,O,NH_2)_4$. The N^{3-} anion shows a strong preference for tetrahedral coordination by Hg^{2+}, which results in a strongly bonded three-dimensional Hg-N framework.

Kind of sample preparation and/or method of registration of the spectrum: Transmission of a single crystal.

Source: Cooper et al. (2013a).

Wavenumbers (cm^{-1}): 3475s, 3426, 3288, 1610.

Note: N was not determined chemically. Despite NH_2 group is considered as a subordinate component, the intensity of the band of H–N–H bending vibrations at 1610 cm^{-1} is rather high. This band could be assigned to H–O–H bending vibrations. In the cited paper, Raman spectrum is given.

Cl85 Hydrocalumite $Ca_2Al(OH)_6(Cl,CO_3,OH)_{1-x}·2H_2O$

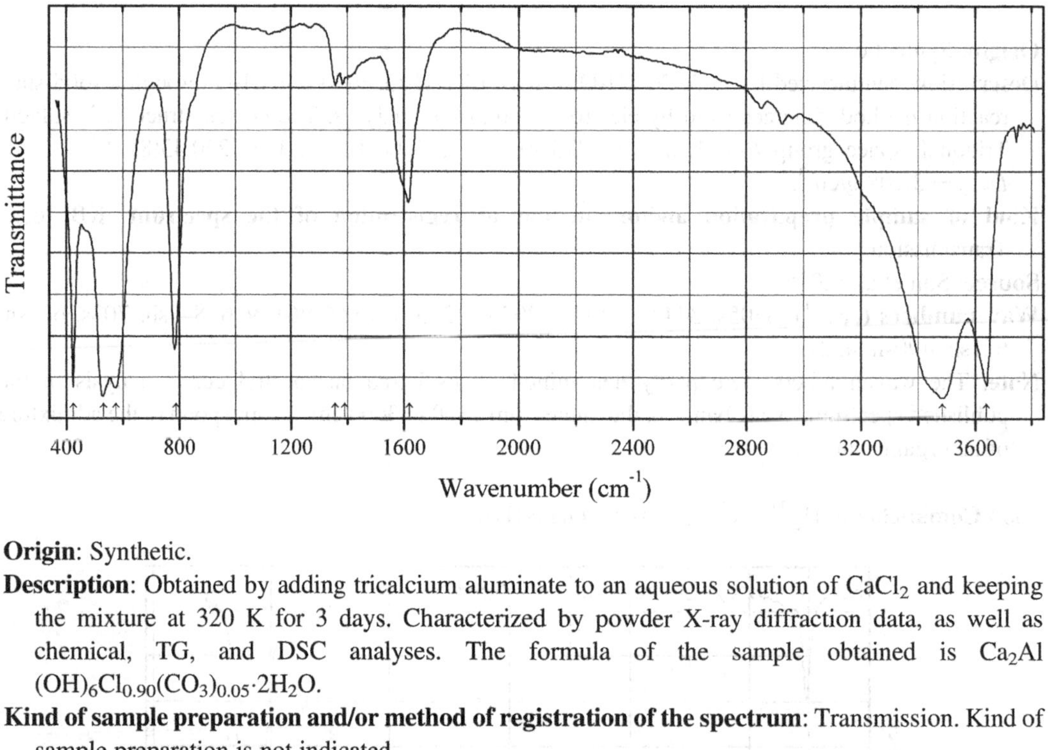

Origin: Synthetic.

Description: Obtained by adding tricalcium aluminate to an aqueous solution of $CaCl_2$ and keeping the mixture at 320 K for 3 days. Characterized by powder X-ray diffraction data, as well as chemical, TG, and DSC analyses. The formula of the sample obtained is $Ca_2Al(OH)_6Cl_{0.90}(CO_3)_{0.05}·2H_2O$.

Kind of sample preparation and/or method of registration of the spectrum: Transmission. Kind of sample preparation is not indicated.

Source: Grishchenko et al. (2013).

Wavenumbers (cm^{-1}): 3639s, 3484s, 1620, 1390w, 1356w, 787, 575s, 532s, 424s.

Cl86 Laurionite Ba-analogue Ba(OH)Cl

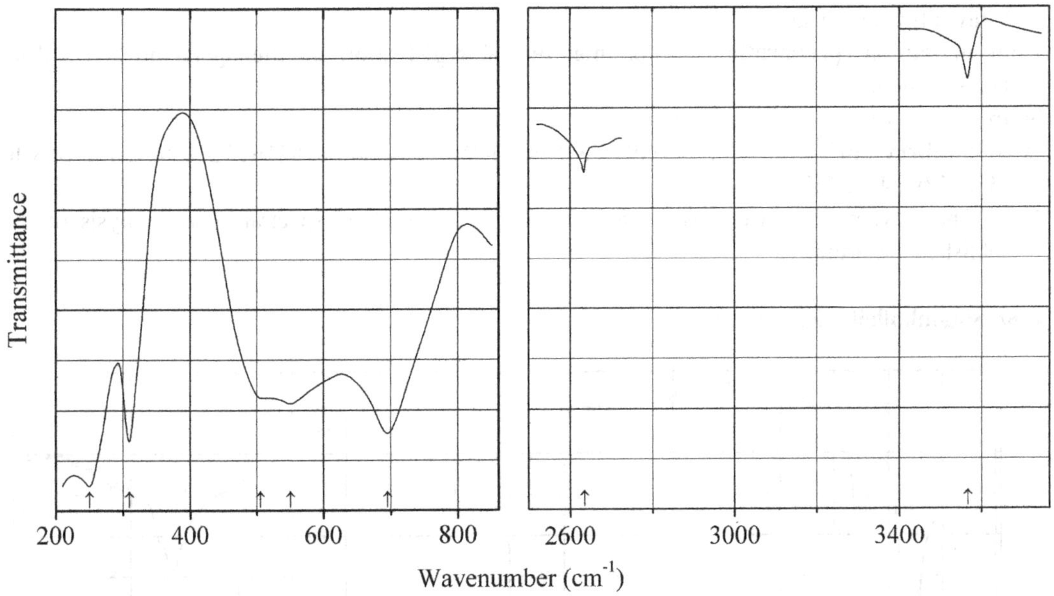

Origin: Synthetic.

Description: Characterized by powder X-ray diffraction data and thermoanalytical methods. Isostructural with laurionite.

Kind of sample preparation and/or method of registration of the spectrum: KBr or CsI disc, and Nujol or poly(chlortrifluorethen) mull. Transmission.

Source: Lutz et al. (1995).

Wavenumbers (IR, cm^{-1}): 3566, 695s, 550, 505sh, 309s, 250s.

Note: In the cited paper, Raman spectrum is given.

Wavenumbers (Raman, cm^{-1}): 3594s, 480s, 265s, 249, 195, 142, 130sh, 93w, 78, 64.

Cl87 Simonkolleite $Zn_5(OH)_8Cl_2{\cdot}H_2O$

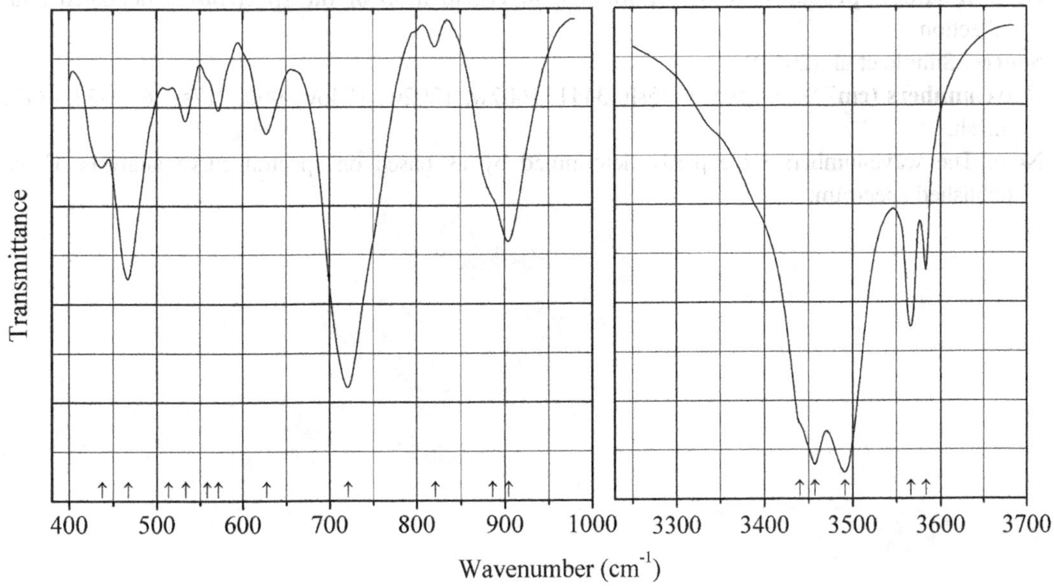

Origin: Synthetic.

Description: Prepared by precipitation method at about 80–90 °C. Characterized by DTA and powder X-ray diffraction data.

Kind of sample preparation and/or method of registration of the spectrum: KBr disc. Transmission.

Source: Stoilova and Vassileva (2002).

Wavenumbers (cm^{-1}): 3583, 3566, 3491s, 3457s, 3440sh, 904, 886sh, 821w, 721s, 627, 571, 558sh, 533, 513w, 467, 437.

Note: The wavenumbers were partly determined by us based on spectral curve analysis of the published spectrum.

Cl88 Simonkolleite $Zn_5(OH)_8Cl_2 \cdot H_2O$

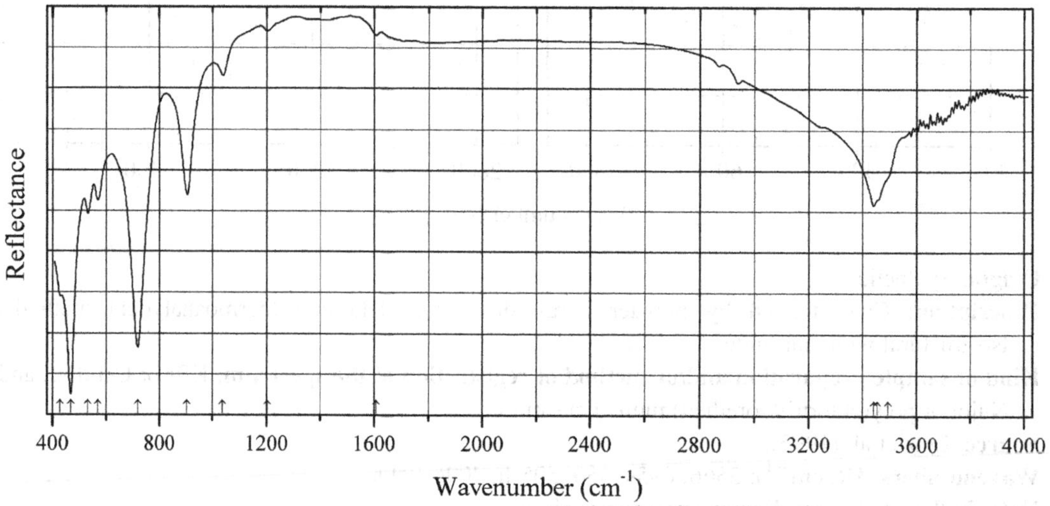

Origin: Synthetic.

Description: Micro-platelets prepared hydrothermally from $Zn(NO_3)_2 \cdot 6H_2O$ and NaCl in the presence of hexamethylenetetramine, at 85 °C. Characterized by powder X-ray diffraction data.

Kind of sample preparation and/or method of registration of the spectrum: Attenuated total reflection.

Source: Sithole et al. (2012).

Wavenumbers (cm^{-1}): 3495sh, 3455sh, 3441, 1607w, 1202w, 1040w, 906, 717s, 569, 532, 468s, 427sh.

Note: The wavenumbers were partly determined by us based on spectral curve analysis of the published spectrum.

Cl89 Terlinguacreekite $Hg_3O_2Cl_2$

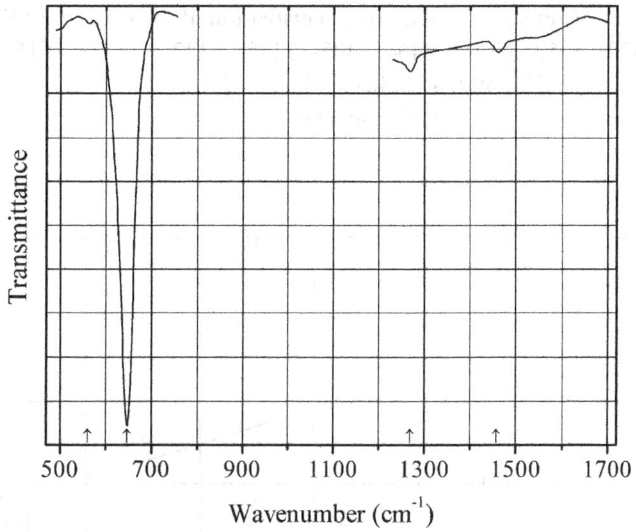

Origin: McDermitt mine, Opalite district, Humboldt Co., Nevada, USA.

Description: Orange powdery from the association with quartz and kleinite. Investigated by A.V. Kasatkin. The empirical formula is (electron microprobe): $Hg_{3.00}O_{2.2}Cl_{1.6}$. The observed lines of the powder X-ray diffraction pattern [*d*, Å (*I*, %)] are: 12.8 (2), 6.42 (8), 5.41 (5), 5.27 (6), 4.64 (3), 4.26 (16), 4.06 (6), 3.75 (5), 3.45 (10), 3.34 (63), 3.25 (14), 3.21 (100), 3.02 (12), 2.97 (10).

Kind of sample preparation and/or method of registration of the spectrum: KBr disc. Absorption.

Wavenumbers (cm^{-1}): 647, 561w.

Note: The spectrum was obtained by N.V. Chukanov.

Cl90 Cumengeite $Pb_{21}Cu_{20}Cl_{42}(OH)_{40}·6H_2O$

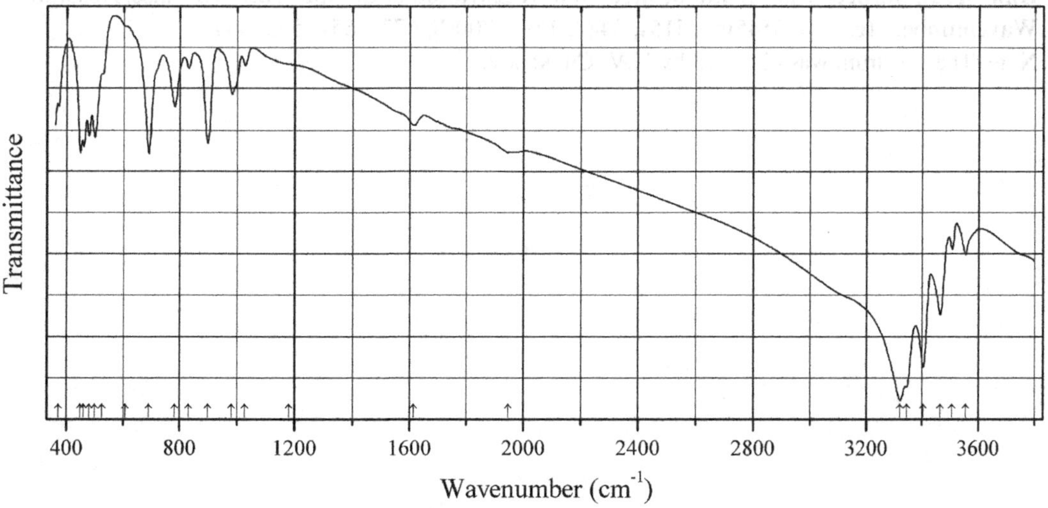

Origin: Boleo district, Santa Rosalía, Baja California, Mexico (type locality).
Description: Bluc crystals.
Kind of sample preparation and/or method of registration of the spectrum: KBr disc. Absorption.
Wavenumbers (cm^{-1}): 3553, 3506w, 3464, 3405s, 3445s, 3322s, 1948w, 1616w, 1028w, 982, 898s, 831w, 782, 690s, 609w, 527, 499, 479, 461s, 448s, (372).
Note: The spectrum was obtained by N.V. Chukanov.

Cl91 Fiedlerite-1*A* $Pb_3Cl_4F(OH)·H_2O$

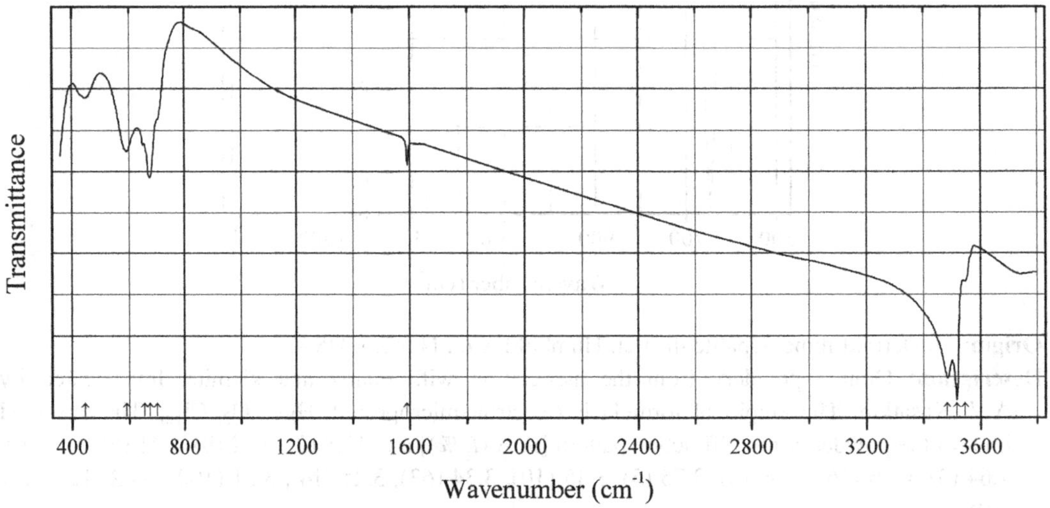

Origin: Pacha Limani (Passa Limani), Lavrion mining district, Attiki Prefecture, Greece.
Description: Colorless platy crystals from the association with phosgenite in ancient metallurgical slag. The crystal structure is solved. Triclinic, space group *P*-1, $a = 8.5741(7)$, $b = 8.0480(5)$, $c = 7.2695(4)$ Å, $\alpha = 90.087(5)$, $\beta = 102.126(6)$, $\gamma = 103.424(6)°$, $V = 476.37(6)$ Å^3, $Z = 2$. The empirical formula is (electron microprobe): $Pb_{3.00}Cl_{3.98}F_{0.96}(OH)_{1.06}·H_2O$.
Kind of sample preparation and/or method of registration of the spectrum: KBr disc. Absorption.
Wavenumbers (cm^{-1}): 3545w, 3515s, 3482, 1592, 700sh, 673s, 654, 592, 447.
Note: The spectrum was obtained by N.V. Chukanov.

Cl92 Kuliginite $Fe_3Mg(OH)_6Cl_2$

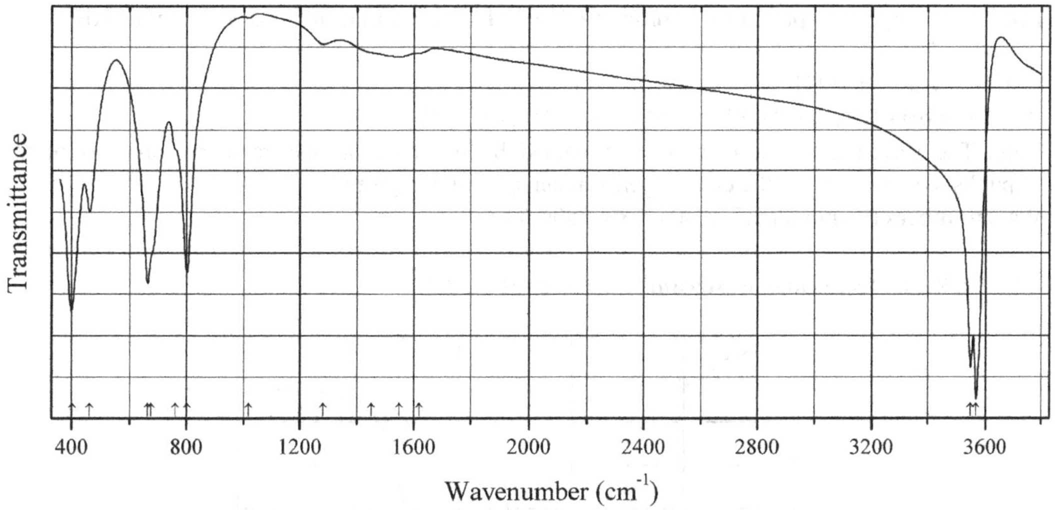

Origin: Udachnaya kimberlite pipe, Yakutia, Russia (type locality).

Description: Green crystals from the association with iowaite, gypsum, calcite, halite, baryte, celestine, etc. Holotype sample. The crystal structure is solved. Trigonal, space group *R*-3, $a = 6.9512(1)$, $c = 14.5713(3)$ Å, $V = 609.74(2)$ Å^3, $Z = 3$. $D_{meas} = 3.1(1)$ g/cm^3, $D_{calc} = 3.01$ g/cm^3. Optically biaxial (+), $\alpha = 1.709(3)$, $\beta = 1.709(3)$, $\gamma = 1.718$, $2V = 10(5)°$. The empirical formula is $(Fe_{2.99}Mn_{0.01})(Mg_{0.90}Mn_{0.10})(OH_{5.94}F_{0.03}Cl_{0.03})Cl_2$. The strongest lines of the powder X-ray diffraction pattern [*d*, Å (*I*, %) (*hkl*)] are: 5.569 (54) (01−1), 2.949 (16) (021), 2.831 (35) (113), 2.324 (100) (024), 2.098 (18) (02−5), 1.856 (13) (033), 1.739 (36) (220).

Kind of sample preparation and/or method of registration of the spectrum: KBr disc. Absorption.

Wavenumbers (cm^{-1}): 3567s, 3548s, 1620w, 1550 (broad), 1450sh, 1278w, 1018w, 801s, 760sh, 675sh, 663s, 464, 402s.

Note: The spectrum was obtained by N.V. Chukanov.

Cl93 Magnesium hydroxychlorite atacamite-type $Mg_2(OH)_3Cl$

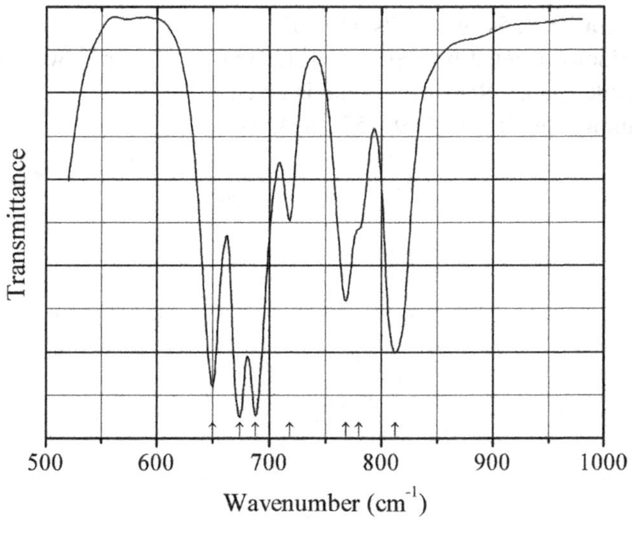

Origin: Synthetic.
Description: Prepared hydrothermally. Characterized by thermal and powder X-ray diffraction data.
Kind of sample preparation and/or method of registration of the spectrum: KBr disc. Transmission.
Source: Bette et al. (2015).
Wavenumbers (IR, cm^{-1}): 812s, 780sh, 768, 718, 688s, 674s, 650s.
Note: The wavenumbers were partly determined by us based on spectral curve analysis of the published spectrum. In the cited paper, Raman spectrum is given.
Wavenumbers (Raman, cm^{-1}): 804, 786, 750s, 712, 635.

Cl94 Nickel hydroxychlorite atacamite-type $Ni_2(OH)_3Cl$

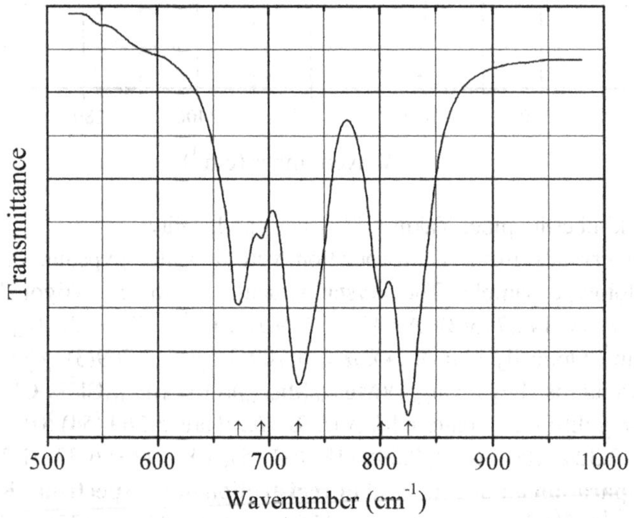

Origin: Synthetic.
Description: Prepared hydrothermally. Characterized by thermal and powder X-ray diffraction data.
Kind of sample preparation and/or method of registration of the spectrum: KBr disc. Transmission.
Source: Bette et al. (2015).
Wavenumbers (IR, cm^{-1}): 825s, 801, 727s, 693, 672.
Note: The band positions denoted by Bette et al. (2015) as 627 cm^{-1} were determined by us at 672 cm^{-1}. In the cited paper, Raman spectrum is given.
Wavenumbers (Raman, cm^{-1}): 823, 799, 757s, 694, 675, 624.

Cl95 Hydrohalite $NaCl{\cdot}2H_2O$

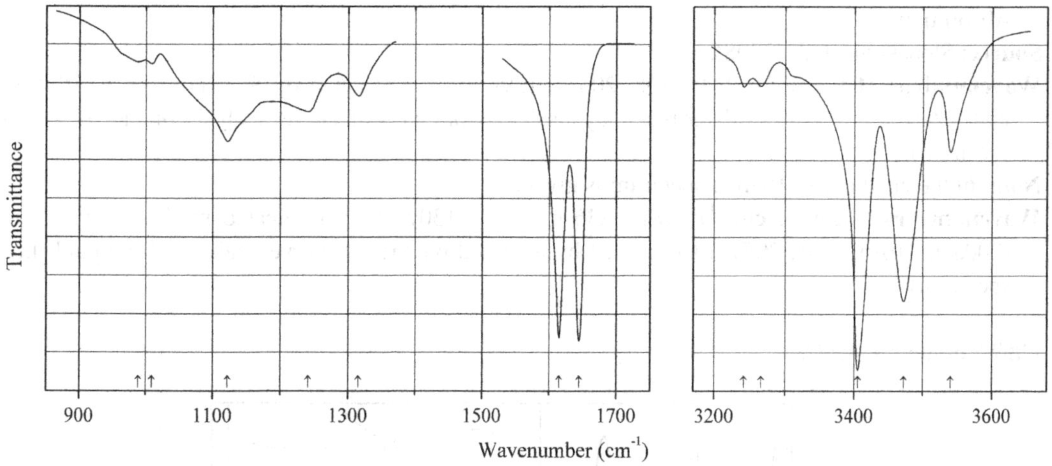

Origin: Synthetic.

Description: A film obtained by slow condensation of H_2O vapor is on a cold NaCl plate with subsequent heating up to –20 °C.

Kind of sample preparation and/or method of registration of the spectrum: Transmission of a thin film.

Source: Schiffer and Hornig (1961).

Wavenumbers (cm^{-1}): 3540, 3472s, 3405s, 3266w, 3242w, 1615s, 1645s, 1315, 1241, 1122, 1009w, 989w.

Note: The wavenumbers were determined by us based on spectral curve analysis of the published spectrum.

Cl96 Sanguite $KCuCl_3$

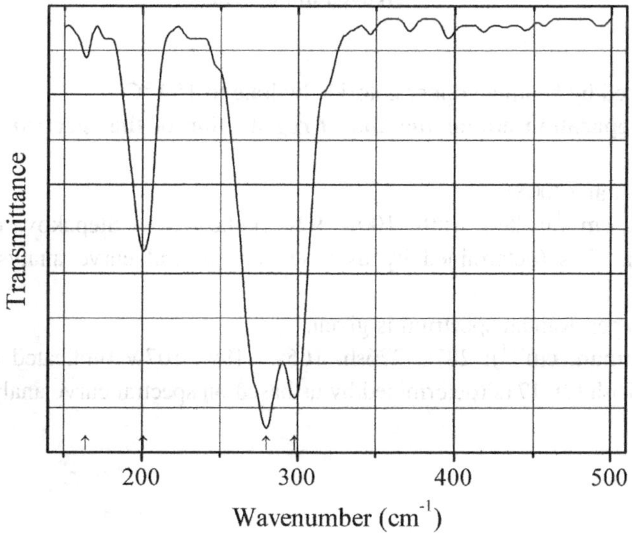

Origin: Synthetic.
Kind of sample preparation and/or method of registration of the spectrum: Polyethylene disc. Absorption.
Source: Stepakova et al. (2008).
Wavenumbers (IR, cm^{-1}): 284s, 208, 205sh, 129w, 94w [indicated by Stepakova et al. (2008)]; 298s, 280s, 201s, 164w (determined by us based on spectral curve analysis of the published spectrum).
Note: In the cited paper, Raman spectrum is given.
Wavenumbers (Raman, cm^{-1}): 274s, 236sh, 146w, 130w, 97w [indicated by Stepakova et al. (2008)]; 310sh, 274s, 205w (determined by us based on spectral curve analysis of the published spectrum).

Cl97 Tolbachite $CuCl_2$

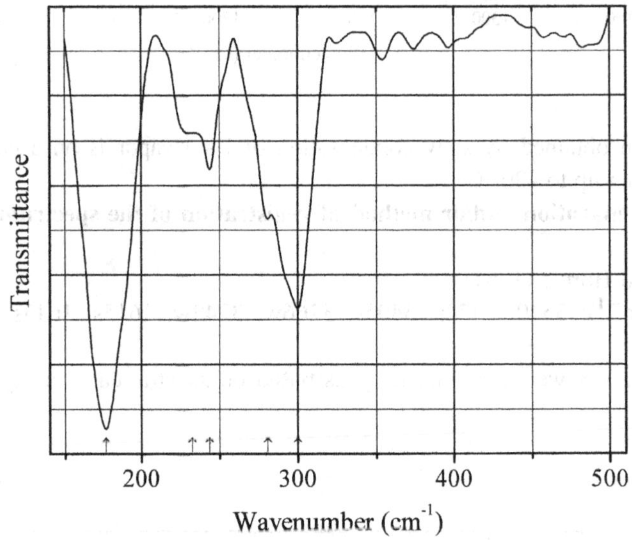

Origin: Synthetic.
Description: Obtained by heating copper chloride hydrate to 150 °C.
Kind of sample preparation and/or method of registration of the spectrum: Polyethylene disc. Absorption.
Source: Stepakova et al. (2008).
Wavenumbers (IR, cm^{-1}): 284s, 189s, 100w, 92w [indicated by Stepakova et al. (2008)]; 300s, 281, 244, 233sh, 177s (determined by us based on spectral curve analysis of the published spectrum).
Note: In the cited paper, Raman spectrum is given.
Wavenumbers (Raman, cm^{-1}): 287s, 276sh, 166s, 119w, 107w [indicated by Stepakova et al. (2008)]; 287s, 276sh (?), 171s (determined by us based on spectral curve analysis of the published spectrum).

Cl98 Telluroperite $Pb(Te_{0.5}Pb_{0.5})O_2Cl$

Origin: Synthetic.

Description: Crystals grown by heating a mixture of $Pb_3O_2Cl_2$ and TeO_2 at 550 °C for 1 day. The crystal structure is solved. Orthorhombic, space group *Bmmb*, $a = 5.576(1)$, $b = 5.559(1)$, $c = 12.4929(6)$ Å, $Z = 4$.

Kind of sample preparation and/or method of registration of the spectrum: KBr disc. Transmission.

Source: Porter and Halasyamani (2003).

Wavenumbers (cm^{-1}): 661, 628, 509, 439.

2.11 Vanadates and Vanadium Oxides

V116 Schäferite $(NaCa_2)Mg_2(VO_4)_3$

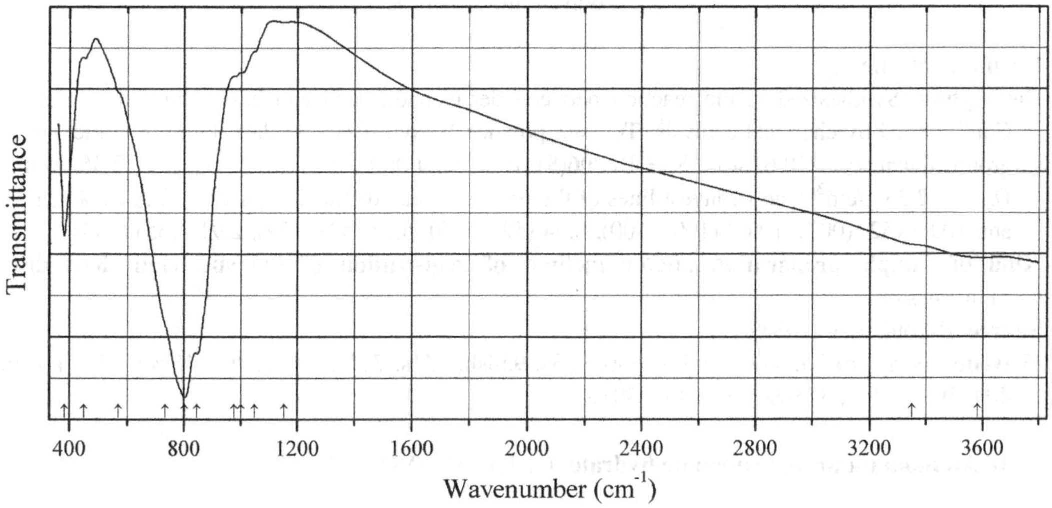

Origin: Slag dump near the Kamariza mine, Lavrion, mining district, Attikí (Attika, Attica) Prefecture, Greece.

Description: Brown crystals from the association with minerals of the forsterite–liebenbergite series, trevorite, albite, nosean, haüyne, bannermanite, a Ni-Mg-analogue of lyonsite, etc. The crystal structure is solved. Cubic, space group *Ia*3*d*, $a = 12.388(3)$ Å, $V = 1901.1(14)$ Å^3, $Z = 8$. The crystal-chemical formula is $(Na_{1.5}Ca_{1.5})(Mg_{1.1}Fe_{0.5}Ni_{0.4})(V_{2.8}P_{0.2})(O,OH)_{12}$.

Kind of sample preparation and/or method of registration of the spectrum: KBr disc. Absorption.

Wavenumbers (cm^{-1}): 3580w, 3350w, 1152w, 1049w, 1000w, 977w, 846s, 803s, 735sh, 570sh, 451w, 383s.

Note: The spectrum was obtained by N.V. Chukanov.

V117 Aluminium decavanadate hydrate $Al_2V_{10}O_{28}·22H_2O$

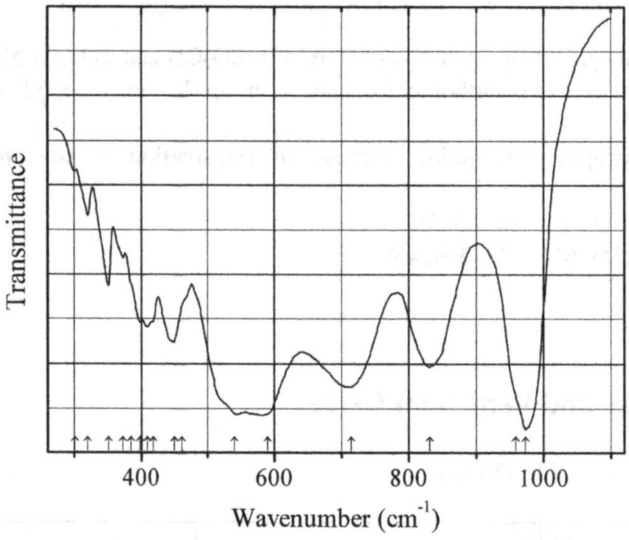

Origin: Synthetic.

Description: Synthesized in the reaction between decavanadic acid and basic aluminium acetate. Confirmeded by chemical analysis, TG, and powder X-ray diffraction data. Orthorhombic, space group *Acmm*, $a = 10.618(5)$, $b = 18.296(8)$, $c = 21.560(10)$ Å, $Z = 2$. $D_{meas} = 2.35$ g/cm^3, $D_{calc} = 2.23$ g/cm^3. The strongest lines of the powder X-ray diffraction pattern [*d*, Å (*I*, %) (*hkl*)] are: 10.83 (52) (002), 10.62 (100) (100), 6.94 (52) (120), 5.81 (42) (122), 2.914 (35) (244).

Kind of sample preparation and/or method of registration of the spectrum: KBr disc. Transmission.

Source: Rigotti et al. (1983).

Wavenumbers (cm^{-1}): 3440, 2930, 1630, 973s, 958sh, 831s, 715s, 590s, 540s, 462sh, 450, 418sh, 409, 398, 385sh, 373w, 351, 320w, 301w.

V118 Ammonium uranyl vanadate hydrate $(NH_4)(UO_2)(VO_4)·2.5H_2O$

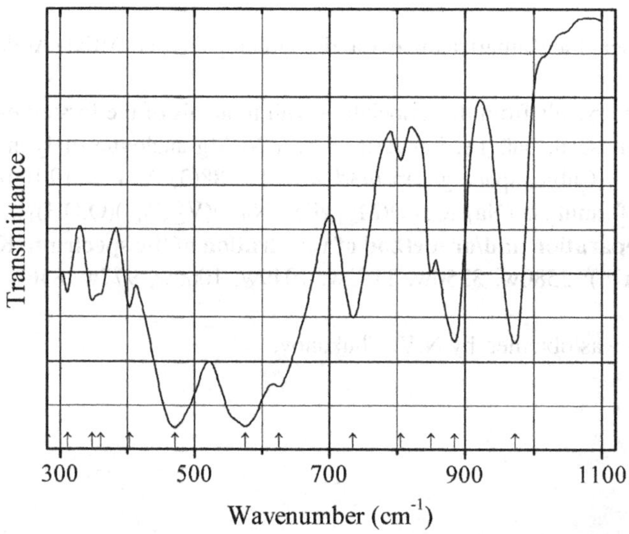

Origin: Synthetic.

Description: Synthesized from $(NH_4)(VO_3)$ and $(UO_2)(NO_3)_2$ with subsequent heating of the precipitate at 60 °C for 3 h. Orthorhombic, $a = 13.29(1)$, $b = 16.21(2)$, $c = 12.05(1)$ Å. The strongest lines of the powder X-ray diffraction pattern [d, Å (I, %)] are: 6.78 (90), 4.23 (40), 3.51 (40), 3.15 (100), 2.15 (20).

Kind of sample preparation and/or method of registration of the spectrum: KBr disc. Transmission.

Source: Botto and Baran (1976).

Wavenumbers (IR, cm^{-1}): 973, 885, 850, 805w, 735, 625s, 575s, 470s, 403, 360sh, 348, 310, 280.

Note: In the cited paper, Raman spectrum is given.

Wavenumbers (Raman, cm^{-1}): 975, 822, 738, 645, 580, 540, 482, 410, 375, 360sh, 255, 230.

V119 Ammonium vanadyl compound $(NH_4)_{0.5}V_2O_5{\cdot}nH_2O$ $(NH_4)_{0.5}V_2O_5{\cdot}nH_2O$

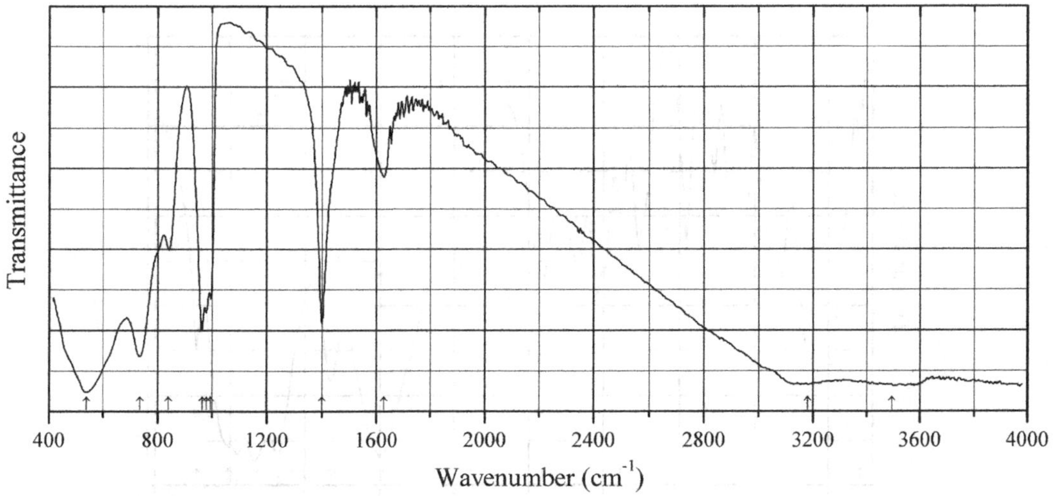

Origin: Synthetic.

Description: Synthesized using a surfactant-free hydrothermal method. Characterized by powder X-ray diffraction, TG, and EDX spectroscopy.

Kind of sample preparation and/or method of registration of the spectrum: KBr disc. Transmission.

Source: Chandrappa et al. (2011).

Wavenumbers (cm^{-1}): 3495, 3180, 1628, 1400, 996, 981, 965, 841, 736, 539s.

Note: The formula is questionable and is to be checked.

V120 Barium lanthanum thorium orthovanadate $BaLaTh(VO_4)_3$

Origin: Synthetic.

Description: Obtained by treating stoichiometric mixture of corresponding metal nitrates with ammonium metavanadate in aqueous medium for 1 h followed by evaporation and calcination. Isostructural with monazite. Monoclinic, $a = 7.070(5)$, $b = 7.323(8)$, $c = 6.810(6)$ Å, $\beta = 104.96(7)°$, $V = 340.8$ Å^3. $D_{meas} = 5.52$ g/cm^3, $D_{calc} = 5.54$ g/cm^3.

Kind of sample preparation and/or method of registration of the spectrum: Attenuated total reflection of a powdered sample. KBr disc. Transmission.

Source: Nabar and Mhatre (2001).

Wavenumbers (cm^{-1}): 839sh, 810s, 780sh, 770sh, 750sh, 735sh, 478sh, 421, 412sh, 382w, 372sh, 350sh.

V121 Barium vanadyl vanadate $Ba_2(VO)(V_2O_8)$

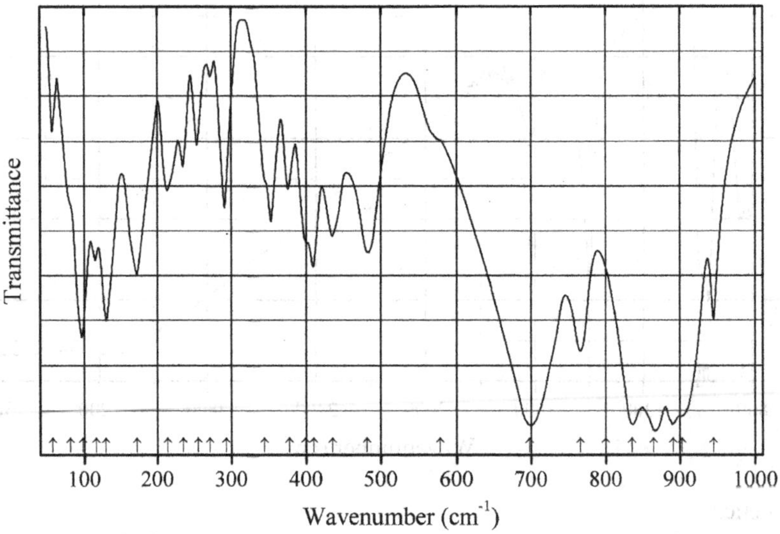

Origin: Synthetic.

Description: Prepared by the solid-state reaction of an intimate 4:1:1 mixture of $Ba_2V_2O_7$, V_2O_3, and V_2O_5 at 950 °C. Monoclinic, space group $P2_1$, $Z = 2$. The crystal structure is built up of infinite chains of strongly distorted edge-sharing $V^{IV}O_6$ octahedra, connected with $V^{V}O_4$ tetrahedra.

Kind of sample preparation and/or method of registration of the spectrum: KBr and polyethylene discs. Transmission.

Source: Baran (1997).

Wavenumbers (IR, cm^{-1}): 945, 903sh, 891s, 864s, 835s, (800sh), 766s, 698s, 578sh, 482, 436, 411, 399, 378, 345sh, 293, 271w, 255, 235, 214, 172s, 129s, 116, 98s, 81sh, 57w.

Note: The wavenumbers were partly determined by us based on spectral curve analysis of the published spectrum. In the cited paper, Raman spectrum is given.

Wavenumbers (Raman, cm^{-1}): 902s, 872, 860sh, 830w, 802s, 762w, 737w, 696w, 676sh, 563w, 499w, 453w, 435sh, 397sh, 371, 345, 294, 268, 243, 212w, 163, 121, 85s.

V122 Bismuth(III) magnesium oxovanadate BiMg(VO$_4$)O BiMg(VO$_4$)O

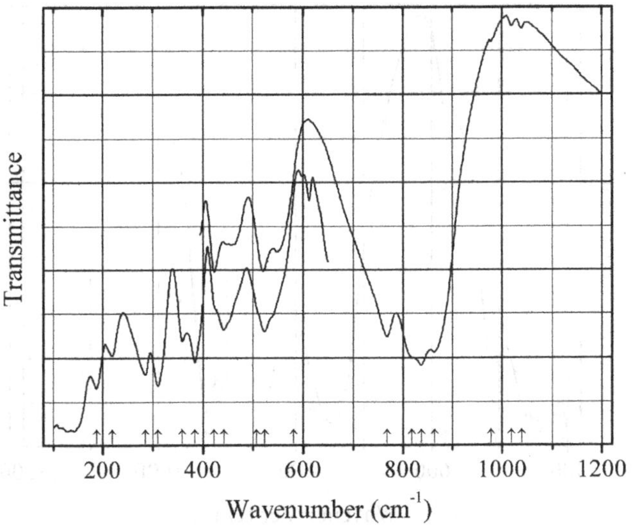

Origin: Synthetic.

Description: Prepared by solid-state reaction from a stoichiometric mixture of Bi_2O_3, MgO, and NH_4VO_3 gradually heated at 200, 500, and finally 850 °C for 18 h with intermediate grindings. Characterized by powder X-ray diffraction data. The crystal structure is solved. Monoclinic, space group $P2_1/n$, $a = 7.542(6)$, $b = 11.615(5)$, $c = 5.305(3)$ Å, $\beta = 107.38(5)°$, $V = 443.5(5)$ Å^3, $Z = 4$. $D_{calc} = 5.455$ g/cm^3.

Kind of sample preparation and/or method of registration of the spectrum: KBr and polyethylene discs. Transmission.

Source: Benmokhtar et al. (2004).

Wavenumbers (IR, cm^{-1}): 1040w, 1019w, 978w, 864s, 837s, 819s, 768s, 581, 523, 507, 442, 423, 384, 359, 311, 285, 219, 188.

Note: Weak bands in the range from 900 to 1100 cm^{-1} may correspond to the admixture of PO_4^{3-} groups. In the cited paper, Raman spectrum is given.

Wavenumbers (Raman, cm^{-1}): 852s, 805s, 748sh, 570, 389w, 340, 303, 250w, 179, 133, 108.

V123 Bismuth(III) magnesium oxovanadate $BiMg_2(VO_4)O_2$

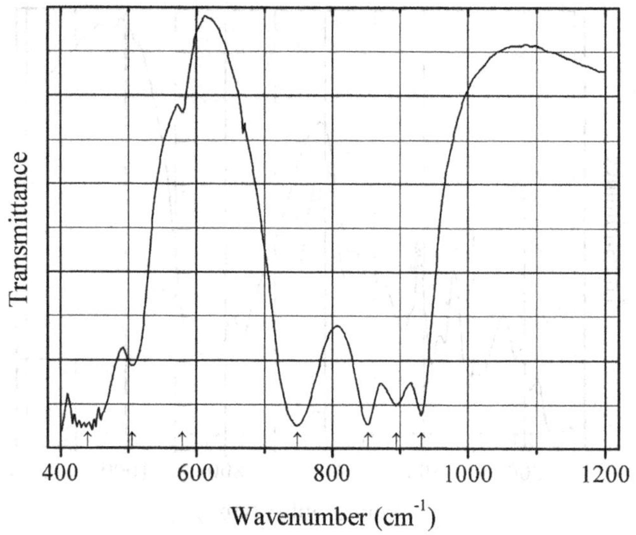

Origin: Synthetic.

Description: Synthesized by heating a mixture of Bi_2O_3, MgO, and NH_4VO_3 in the molar ratio Bi:Mg:V = 2:2:1 first at 700 °C for 12 h, then at 800 °C for 6 h, and finally at 1000 °C for 5 min. The product was structurally characterized from single crystal X-ray diffraction data. Orthorhombic, space group *Cmcm*, $a = 7.9136(6)$, $b = 12.246(2)$, $c = 5.444(2)$ Å, $V = 527.6(2)$ Å^3, Z = 4. $D_{calc} = 5.093$ g/cm^3.

Kind of sample preparation and/or method of registration of the spectrum: KBr disc. Transmission.

Source: Huang and Sleight (1992).

Wavenumbers (cm^{-1}): 932s, 895s, 853s, 749s, 579w, 505, 440s.

V124 Calcium orthovanadate trigonal polymorph $Ca_3(VO_4)_2$

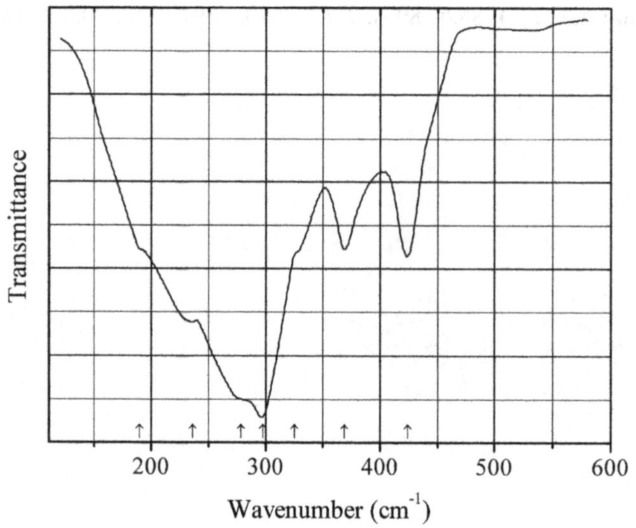

Origin: Synthetic.
Description: Prepared in a solid-state reaction, from the mixture of $CaCO_3$ and As_2O_5 at 700 °C for 4 h. Trigonal, space group *R*3*c*, *Z* = 7.
Kind of sample preparation and/or method of registration of the spectrum: CsI disc. Transmission.
Source: Baran (1976).
Wavenumbers (IR, cm^{-1}): 872, 910sh, 841sh, 810, 760sh, 424, 369, 325sh, 297s, 278sh, 236, 190sh.
Note: In the cited paper, Raman spectrum is given.
Wavenumbers (Raman, cm^{-1}): 930, 912w, 865s, 850s, 825, 790sh, 770, 410w, 360s, 337s, 285, 225, 195, 163sh, 150.

V125 Chromium iron(III) orthovanadate $CrFe(VO_4)_2$

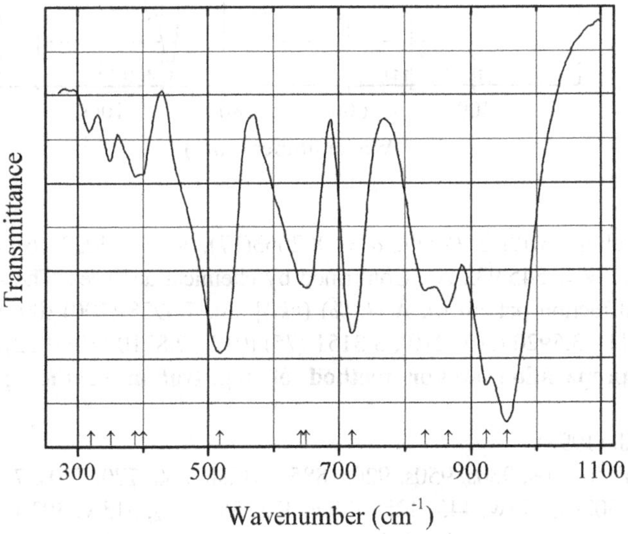

Origin: Synthetic.
Description: Prepared by a solid-state reaction. Characterized by powder X-ray diffraction data. Monoclinic, space group *C*2/*m*. Isostructural with α-$MnMoO_4$.
Kind of sample preparation and/or method of registration of the spectrum: KBr disc. Transmission.
Source: Lavat et al. (1989).
Wavenumbers (cm^{-1}): 956s, 924s, 865, 830, 720s, 650, 642sh, 518s, 400sh, 388, 350w, 320w.

V126 Chromium vanadate $Cr_2V_4O_{13}$ $Cr_2V_4O_{13}$

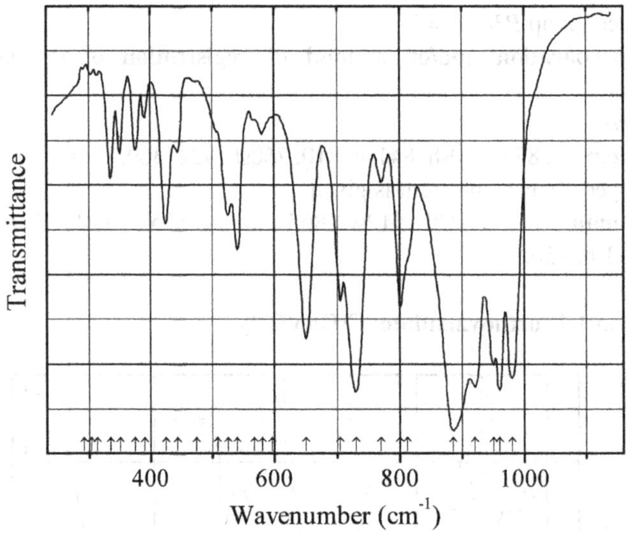

Origin: Synthetic.

Description: Monoclinic, with a *P*-cell, $a = 8.2663(17)$, $b = 9.3033(26)$, $c = 7.5373(16)$ Å, $\beta = 109.638(37)°$, $V = 545.932$ Å^3. Confirmed by chemical analyses. The strongest lines of the powder X-ray diffraction pattern [*d*, Å (*I*, %) (*hkl*)] are: 7.0928 (100) (001), 3.8938 (60) (200), 3.7026 (30) (−211), 3.5920 (50) (210), 3.3151 (75) (012), 2.8210 (30) (022).

Kind of sample preparation and/or method of registration of the spectrum: KBr disc. Transmission.

Source: Filipek et al. (1998).

Wavenumbers (cm^{-1}): 980s, 960s, 950s, 920s, 885s, 812sh, 800, 770, 730s, 705, 650, 596w, 580w, 567sh, 540, 525, 508sh, 475w, 445, 425, 390w, 375, 350, 335, 313w, 303w, 291w.

Note: The wavenumbers were partly determined by us based on spectral curve analysis of the published spectrum. In the cited paper, the wavenumber 650 cm^{-1} is erroneously indicated as 605 cm^{-1}.

V127 Copper divanadate hydroxide hydrate $Cu_3(V_2O_7)(OH)_2{\cdot}nH_2O$

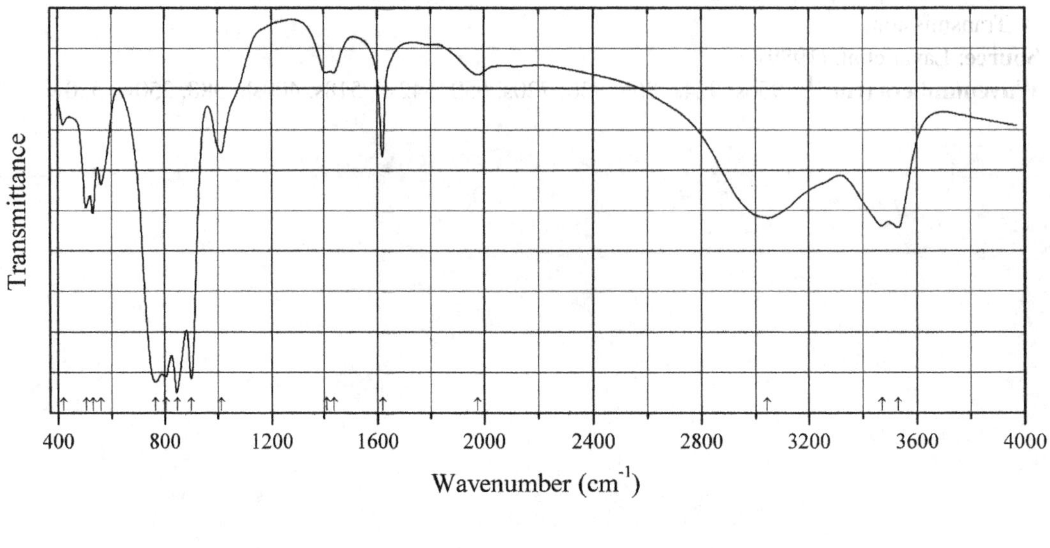

Origin: Synthetic.

Description: Nanoparticles obtained by heating (at 140 °C for 24 h) of a precipitate formed in the reaction between V_2O_5 and $CuSO_4·7H_2O$ in the presence of hexamethylenetetramine, Na_2SO_4, and H_2O. Characterized by EDS analysis and powder X-ray diffraction data. Monoclinic, $a = 10.61$, $b = 5.86$, $c = 7.205$ Å, $\beta = 94.86°$ (see JCPDS, No. 46-1443).

Kind of sample preparation and/or method of registration of the spectrum: Transmission. Kind of sample preparation is not indicated.

Source: Ni et al. (2010a).

Wavenumbers (IR, cm^{-1}): 3531, 3470, 3044, 1974w, 1620, 1437w, 1409w, 1012, 900s, 847s, 804s, 763s, 562, 531, 505, 419w.

Note: The wavenumber 1620 cm^{-1} is erroneously indicated by Ni et al. (2010a) as 1920 cm^{-1}. The weak bands at 1437 and 1409 cm^{-1} may correspond to the admixture of a carbonate. In the cited paper, Raman spectrum is given.

Wavenumbers (Raman, cm^{-1}): 894s, 820s, 758, 476, 438, 342, 236, 164w.

V128 Dysprosium decavanadate hydrate $Dy_2V_{10}O_{28}·24H_2O$

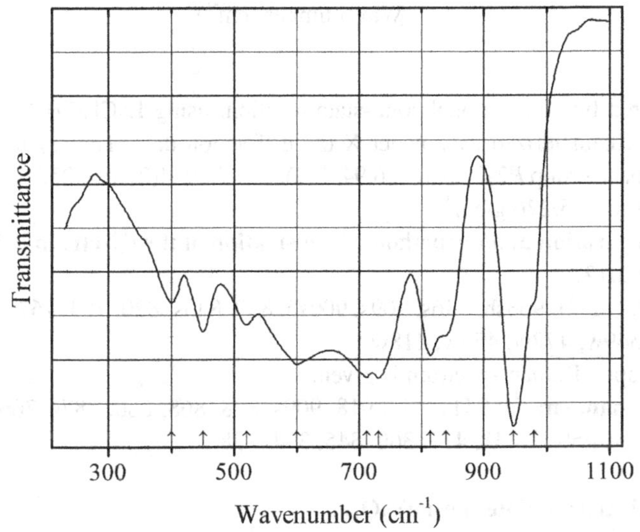

Origin: Synthetic.

Description: Obtained by slow evaporation of an aqueous solution containing decavanadic acid and dysprosium acetate. Triclinic, space group *P*-1, $a = 9.22(2)$, $b = 9.99(7)$, $c = 13.98(6)$ Å, $\alpha = 108.2(7)°$, $\beta = 62.3(5)°$, $\gamma = 89.1(3)°$, $V = 1063$ Å^3, $Z = 1$.

Kind of sample preparation and/or method of registration of the spectrum: KBr disc. Transmission.

Source: Rigotti et al. (1981).

Wavenumbers (IR, cm^{-1}): 980sh, 948s, 840sh, 815, 731s, 712s, 600, 520, 450, 400.

Note: The band at 725 cm^{-1} indicated by Rigotti et al. (1981) is a doublet (731+712 cm^{-1}). In the cited paper, Raman spectrum is given as a figure, without indication of positions of the bands.

V129 Lanthanum uranyl orthovanadate divanadate $La(UO_2)_2(VO_4)(V_2O_7)$

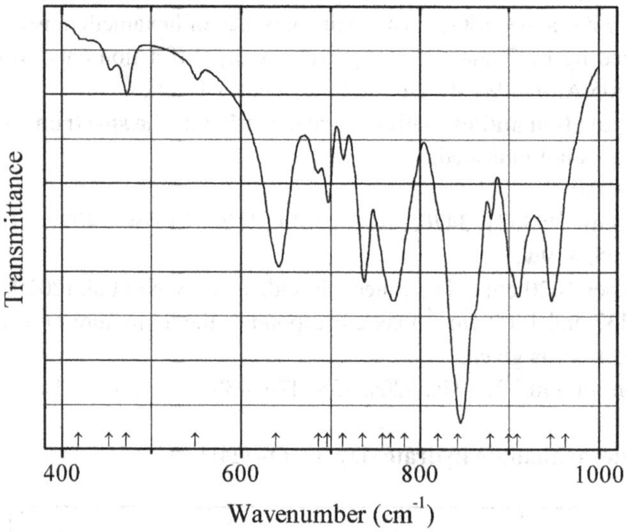

Origin: Synthetic.

Description: Prepared by conventional solid-state reaction, using $LaCl_3·7H_2O$, U_3O_8, and V_2O_5 as initial materials. Characterized by powder X-ray diffraction data. The crystal structure is solved. Orthorhombic, space group $P2_12_12_1$, $a = 6.9470(2)$, $b = 7.0934(2)$, $c = 25.7464(6)$ Å, $V = 1268.73(5)$ Å^3, $Z = 4$. $D_{calc} = 5.276$ g/cm^3.

Kind of sample preparation and/or method of registration of the spectrum: KBr disc. Absorption.

Source: Mer et al. (2012).

Wavenumbers (IR, cm^{-1}): 963sh, 946s, 909s, 900sh, 879, 843s, 820, 783, 767s, 758sh, 736s, 713w, 696, 686, 639s, 549w, 472w, 453w, 418w.

Note: In the cited paper, Raman spectrum is given.

Wavenumbers (Raman, cm^{-1}): 951s, 943, 918, 909s, 898, 868, 860, 787s, 766s, 753, 739s, 711w, 592w, 562w, 516w, 450sh, 431, 413, 360, 345, 334, 328.

V130 Lead iron(III) trivanadate $Pb_2FeV_3O_{11}$

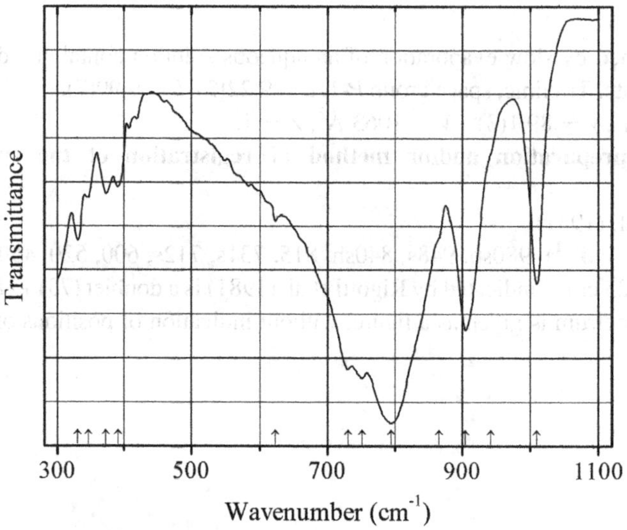

Origin: Synthetic.

Description: Yellow solid formed in the solid-state reaction between $FeVO_4$ and $Pb_2V_2O_7$. Monoclinic, $a = 11.385(13)$, $b = 5.6414(7)$, $c = 7.4970(9)$ Å, $\beta = 81.72(1)°$. $D_{meas} = 5.52(5)$ g/cm^3. The strongest lines of the powder X-ray diffraction pattern [*d*, Å (*I*, %) (*hkl*)] are: 11.277 (27) (100), 3.372 (27) (−211), 3.126 (100) (310), 3.086 (52) (112), 2.821 (38) (020), 2.767 (26) (401).

Kind of sample preparation and/or method of registration of the spectrum: KBr disc. Transmission.

Source: Blonska-Tabero (2009).

Wavenumbers (cm^{-1}): 1010, 943sh, 905s, 865sh, 795s, 751s, 730s, 623w, 391w, 373w, 346w, 330.

Note: The wavenumbers were partly determined by us based on spectral curve analysis of the published spectrum.

V131 Lead uranyl divanadate $Pb(UO_2)(V_2O_7)$

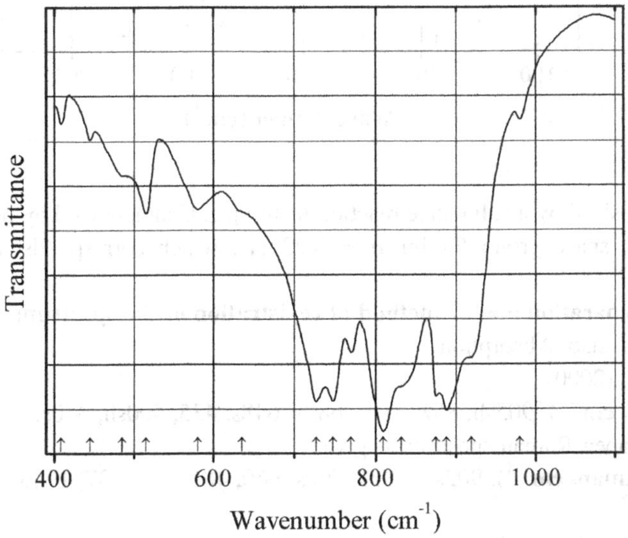

Origin: Synthetic.

Description: Synthesized by solid-state reaction of PbO, V_2O_5, and U_3O_8 in the metallic ratio Pb:V:U = 1:6:2 in air, at 680 °C, for 2 h. Characterized by powder X-ray diffraction data. The crystal structure is solved. Monoclinic, space group $P2_1/n$, $a = 6.9212(9)$, $b = 9.6523(13)$, $c = 11.7881(16)$ Å, $\beta = 91.74(1)°$, $V = 787.2(2)$ Å^3, $Z = 4$. $D_{meas} = 5.82(3)$ g/cm^3, $D_{calc} = 5.81(1)$ g/cm^3. The structure is based on a three-dimensional framework composed by edge- and corner-sharing U- and V-centered polyhedra forming elliptic tunnels occupied by Pb^{2+} ions.

Kind of sample preparation and/or method of registration of the spectrum: KBr disc. Transmission.

Source: Obbade et al. (2004).

Wavenumbers (cm^{-1}): 980w, 920sh, 888s, 875, 832sh, 810s, 770, 747s, 727s, 635sh, 580w, 515, 485sh, 445w, 409w.

Note: The wavenumbers were partly determined by us based on spectral curve analysis of the published spectrum.

V132 Lithium nickel vanadate $LiNi(VO_4)$

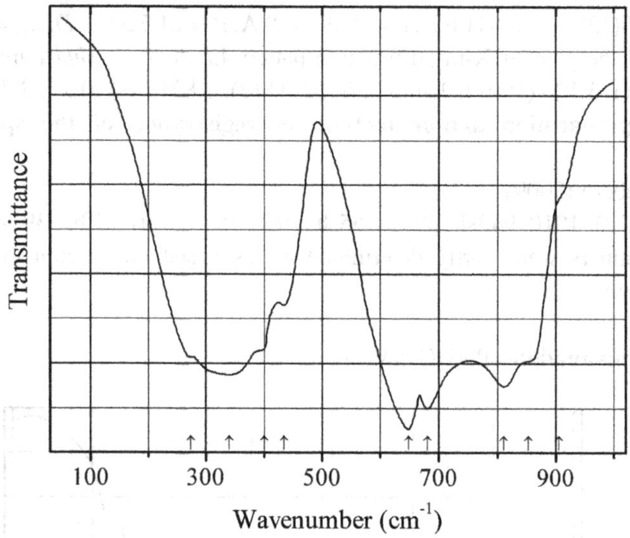

Origin: Synthetic.

Description: Synthesized by a solid-state reaction technique. Characterized by powder X-ray diffraction data. Cubic, space group *Fd*-3*m*, *a* = 8.221(1), which corresponds to the inverse spinel structure.

Kind of sample preparation and/or method of registration of the spectrum: Fine powder painted onto polyethylene slab. Absorption.

Source: Chitra et al. (2000).

Wavenumbers (IR, cm^{-1}): 905sh, 852, 810s, 680s, 648s, 435, 400sh, 340s, 275sh.

Note: In the cited paper, Raman spectrum is given.

Wavenumbers (Raman, cm^{-1}): 902sh, 823s, 790s, 660, 481, 420, 337, 190w.

V133 Lithium trivanadate LiV_3O_8

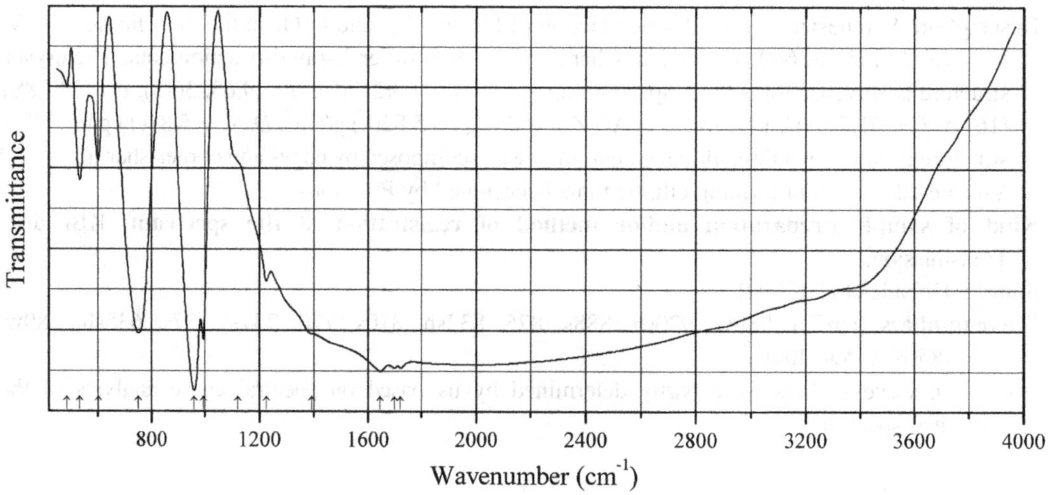

Origin: Synthetic.

Description: Prepared by stepwise heating a mixture of Li_2CO_3 and V_2O_5 powders up to 700 °C. Characterized by powder X-ray diffraction data. Monoclinic, space group $P2_1/m$, a = 6.68, b = 3.60, c = 12.03 Å, β = 107°, which corresponds to the JCPDS card No. 72-1193.

Kind of sample preparation and/or method of registration of the spectrum: KBr disc. Transmission.

Source: Ramaraghavulu et al. (2012).

Wavenumbers (IR, cm^{-1}): 1723w, 1700w, 1646w, 1392sh, 1224w, 1120sh, 994s, 960s, 750s, 602, 535, 488w.

Note: In the cited paper, Raman spectrum is given.

Wavenumbers (Raman, cm^{-1}): 999w, 782s, 555, 491, 395w, 295.

Note: The wavenumbers were partly determined by us based on spectral curve analysis of the published spectrum.

V134 Lithium tungstate vanadate brannerite-type $LiWVO_6$

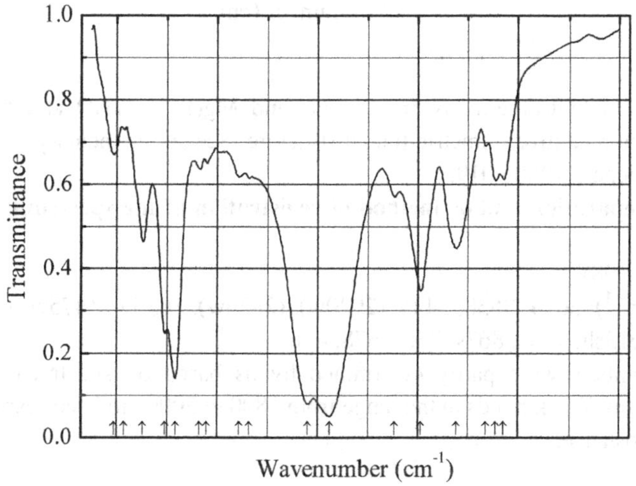

Origin: Synthetic.

Description: Synthesized by heating to 400 °C of a precursor formed in the reaction between $Li(NO_3)$, $(NH_4)(VO_3)$, and tungstic acid in aqueous solution, in the presence of glycine, with subsequent calcination of the product at 550 °C. Characterized by powder X-ray diffraction data. Monoclinic, space group $C2$, a = 9.347, b = 3.670, c = 6.593 Å, β = 111.83°.

Kind of sample preparation and/or method of registration of the spectrum: CsI disc. Absorption.

Source: Amdouni et al. (2003).

Wavenumbers (IR, cm^{-1}): 972, 956, 936w, 877, 807, 754, 624s, 581s, (463), 443w, 379w, 365w, 316s, 295, 252, (215), 195.

Note: The wavenumbers were partly determined by us based on spectral curve analysis of the published spectrum. In the cited paper, Raman spectrum is given.

Wavenumbers (Raman, cm^{-1}): 970s, 860sh, 826, 743, 524w, 449, 324, 268, 238, 207, 146s, 117.

V135 Magnesium vanadate $Mg_7V_4O_{16}(OH)_2{\cdot}H_2O$ $Mg_7V_4O_{16}(OH)_2{\cdot}H_2O$

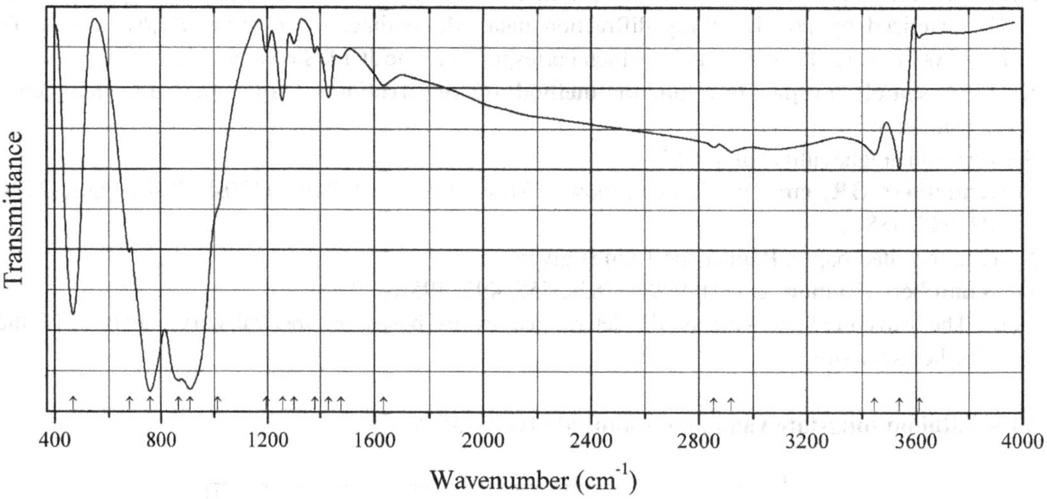

Origin: Synthetic.

Description: Prepared hydrothermally from V_2O_5 and $Mg(BO_2)_2{\cdot}H_2O$ at 200 °C for 5 days. A compound with non-centrosymmetric tunnel structure. See supplementary data at doi: https://doi.org/10.1016/j.inoche.2008.05.019.

Kind of sample preparation and/or method of registration of the spectrum: No data in the cited paper.

Source: Hu et al. (2008).

Wavenumbers (cm^{-1}): 3614, 3538, 3445, (2920w), (2856w), 1635w, 1475w, 1430, 1379w, 1302w, 1258, 1197w, 1013sh, 910s, 867s, 756s, 678, 468.

Note: The wavenumbers were partly determined by us based on spectral curve analysis of the published spectrum. Weak bands in the range from 2800 to 3000 cm^{-1} correspond to the admixture of an organic substance.

V136 Potassium chromium divanadate $KCrV_2O_7$

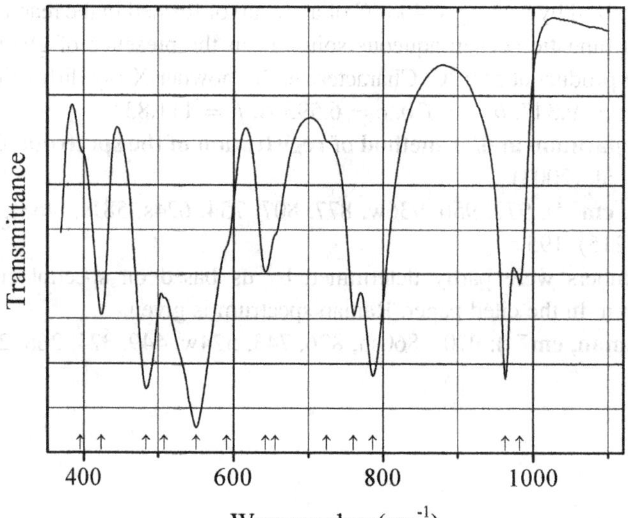

Origin: Synthetic.

Description: Prepared by solid-state reaction in air, by stepwise heating a stoichiometric mixture of $K_2Cr_2O_7$ and V_2O_5 up to 600 °C with intermediate regrindings. Characterized by powder X-ray diffraction data. The crystal structure is solved. Monoclinic, space group *P2/c*, $a = 7.9526(1)$, $b = 4.87543(5)$, $c = 6.8910(1)$ Å, $\beta = 101.162(1)°$, $V = 262.1(1)$ Å^3, $Z = 2$. Vanadium has sixfold coordination with one long and two short V–O bonds.

Kind of sample preparation and/or method of registration of the spectrum: Nujol mull. Transmission.

Source: Tyutyunnik et al. (2006).

Wavenumbers (IR, cm^{-1}): 982, 963s, 786s, 760, 724sh, 655sh, 642, 591sh, 551s, 508sh, 484s, 424, 395sh.

Note: For another treatment of the crystal structure see Wang et al. (2012). In the cited paper, Raman spectrum is given.

Wavenumbers (Raman, cm^{-1}): 985s, 960w, 890, 868, 836, 768s, 712, 685, 663, 567w, 562, 542s, 483, 397, 342, 316, 283, 257, 218, 201, 173, 142.

V137 Potassium chromium divanadate $KCrV_2O_7$

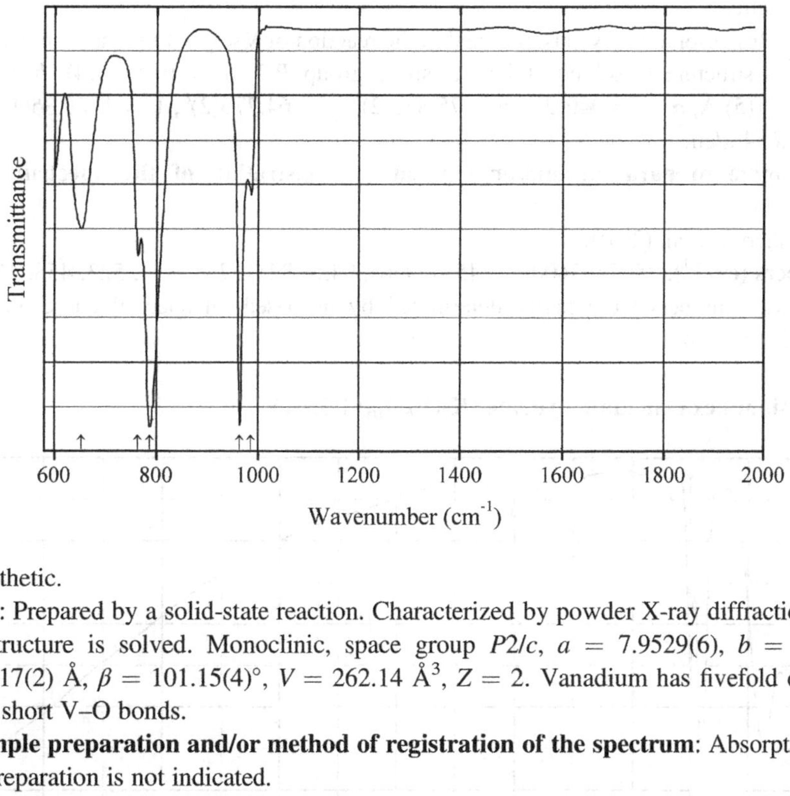

Origin: Synthetic.

Description: Prepared by a solid-state reaction. Characterized by powder X-ray diffraction data. The crystal structure is solved. Monoclinic, space group *P2/c*, $a = 7.9529(6)$, $b = 4.87548(5)$, $c = 6.8917(2)$ Å, $\beta = 101.15(4)°$, $V = 262.14$ Å^3, $Z = 2$. Vanadium has fivefold coordination with two short V–O bonds.

Kind of sample preparation and/or method of registration of the spectrum: Absorption. Kind of sample preparation is not indicated.

Source: Wang et al. (2012).

Wavenumbers (cm^{-1}): 986, 964s, 787s, 764, 654.

Note: For another treatment of the crystal structure see Tyutyunnik et al. (2006). The wavenumbers were partly determined by us based on spectral curve analysis of the published spectrum.

V138 Potassium decavanadate decahydrate $K_6(V_{10}O_{28})\cdot 10H_2O$

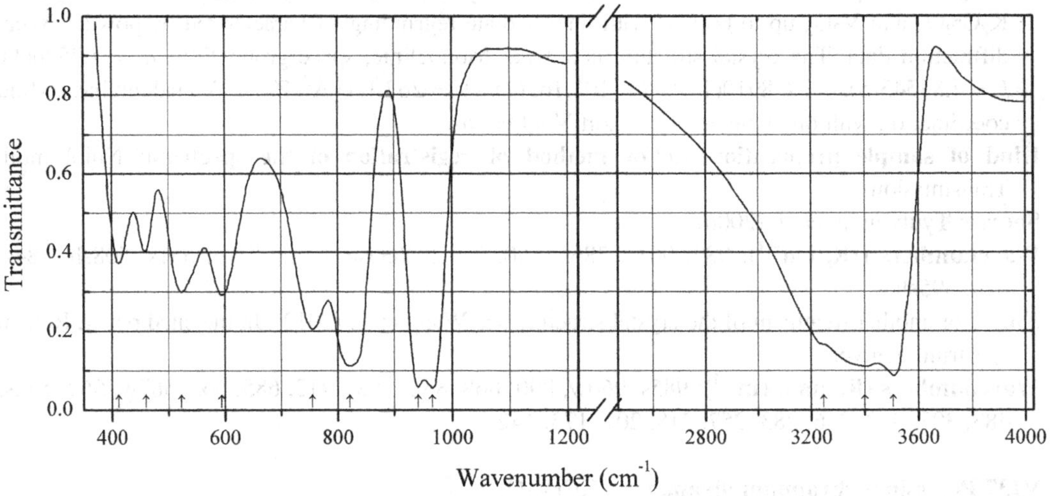

Origin: Synthetic.

Description: Brown-orange crystals obtained in the reaction of V_2O_5 with potassium malate solution. The crystal structure is solved. Triclinic, space group *P*-1, $a = 10.5334(4)$, $b = 10.6600(4)$, $c = 17.7351(5)$ Å, $\alpha = 76.940(2)°$, $\beta = 75.836(2)°$, $\gamma = 64.776(2)°$, $V = 1729.86(10)$ Å^3, $Z = 2$. $D_{calc} = 2.634$ g/cm^3.

Kind of sample preparation and/or method of registration of the spectrum: KBr disc. Transmission.

Source: Guilherme et al. (2010).

Wavenumbers (cm^{-1}): 3507s, 3400s, 3245sh, 966s, 942s, 821s, 755, 594, 523, 458, 411.

Note: The wavenumbers were partly determined by us based on spectral curve analysis of the published spectrum.

V139 Potassium hexavanadate hydrate $K_2(V_6O_{16})\cdot 1.5H_2O$

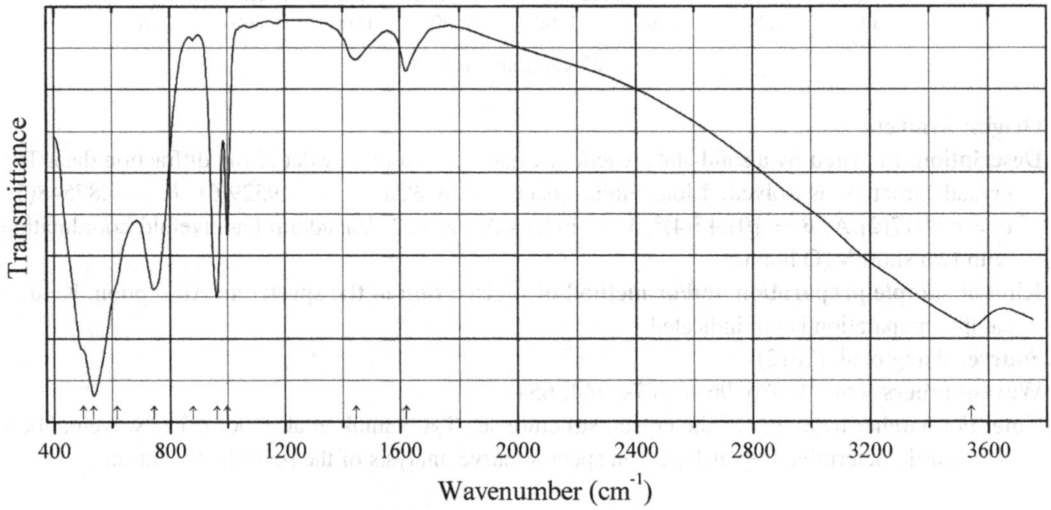

Origin: Synthetic.

Description: Nanobelts prepared by a low-temperature hydrothermal method. Monoclinic, space group $P2_1/c$, a = 12.29 Å, b = 3.60 Å, c =16.01 Å, β = 93.89°. Characterized by powder X-ray diffraction data.

Kind of sample preparation and/or method of registration of the spectrum: KBr disc. Transmission.

Source: Bai et al. (2013).

Wavenumbers (IR, cm^{-1}): 3542, 1620, 1445, 999, 962s, 880w, 741s, 617sh, 537s, 500sh.

Note: The wavenumbers were partly determined by us based on spectral curve analysis of the published spectrum. In the cited paper, Raman spectrum is given.

Wavenumbers (Raman, cm^{-1}): 1011w, 774, 686, 507, 270, 154s.

V140 Strontium vanadyl vanadate $Sr_2(VO)(VO_4)_2$

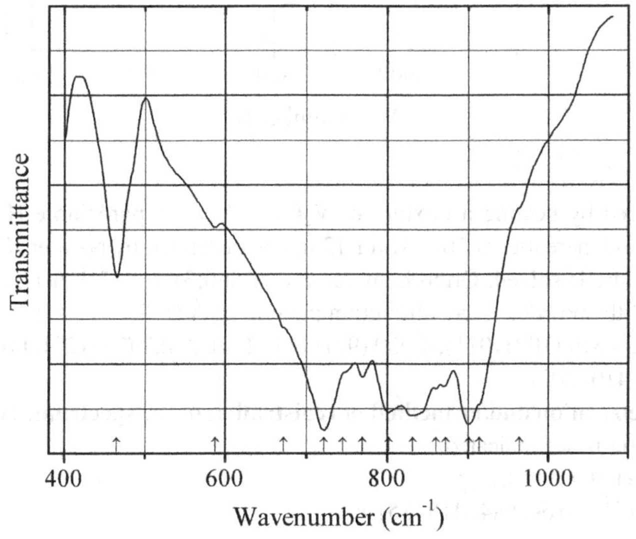

Origin: Synthetic.

Description: Prepared from a mixture of $SrCO_3$, V_2O_5, and VO_2 pressed into pellet and heated with a CO_2 laser in a nitrogen atmosphere. Monoclinic, space group $I2/a$, a = 6.929, b = 16.246, c = 7.260 Å, β = 115.82°, Z = 4.

Kind of sample preparation and/or method of registration of the spectrum: KBr disc. Transmission.

Source: Baran (1996).

Wavenumbers (IR, cm^{-1}): 965sh, 914sh, 901s, 872, 860, 831s, 802s, 769, 745sh, 721s, 672sh, 588w, 465.

Note: The wavenumbers were partly determined by us based on spectral curve analysis of the published spectrum. In the cited paper, Raman spectrum is given.

Wavenumbers (Raman, cm^{-1}): 912, 893, 870sh, 860s, 831s, 791w, 772, 720w, 468w, 430s, 400w, 370sh, 358w.

V141 Tantalum oxyvanadate $Ta(VO_4)O$

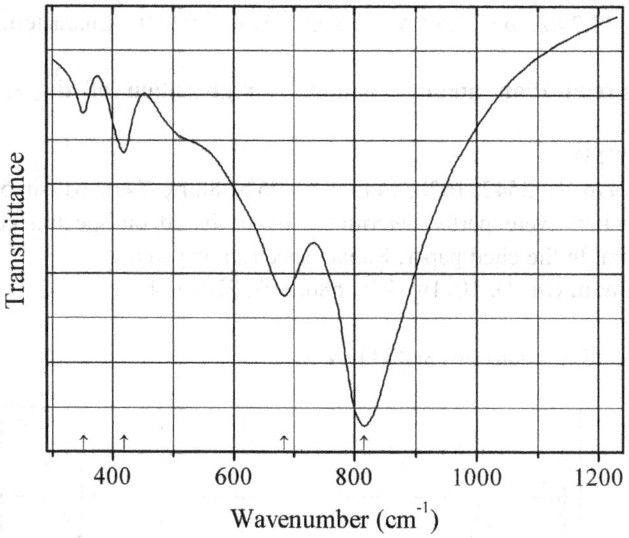

Origin: Synthetic.

Description: Prepared by heating a mixture of V_2O_5 and defect pyrochlore $H_2Ta_6O_6 \cdot H_2O$, first at 873 K for 24 h, and thereafter at 1073 K for 12 h. Characterized by powder X-ray diffraction data. The crystal structure is solved. Orthorhombic, $a = 11.860(3)$, $b = 5.516(1)$, $c = 6.928(1)$ Å. The strongest lines of the powder X-ray diffraction pattern [*d*, Å (*I*, %) (*hkl*)] are: 5.96 (94) (101, 200), 4.491 (84) (201), 4.310 (100) (011), 4.050 (92) (111, 210), 3.487 (71) (211, 002), 2.845 (71) (112), 2.609 (68) (212, 410, 302).

Kind of sample preparation and/or method of registration of the spectrum: Transmission. Kind of sample preparation is not indicated.

Source: Chahboun et al. (1988).

Wavenumbers (cm^{-1}): 816s, 684, 418, 351w.

V142 Tellurium(IV) oxovanadate $Te_2V_2O_9$

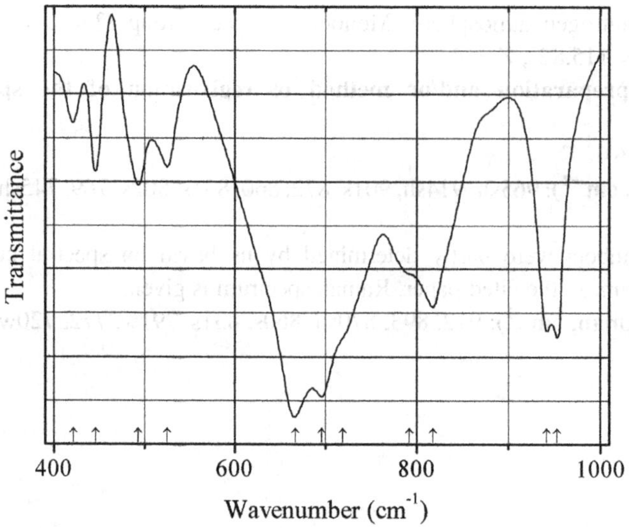

Origin: Synthetic.

Description: Polycrystalline sample prepared by heating a mixture of TeO_2 and V_2O_5 at 450 °C for 24 h with several intermediate grindings and mixings. Characterized by powder X-ray diffraction data. Orthorhombic. V is in tetrahedral coordination environment with V–O bond lengths ranging from 1.633 to 1.946 Å. Te has threefold coordination.

Kind of sample preparation and/or method of registration of the spectrum: KBr disc. Transmission.

Source: Zhang et al. (2012b).

Wavenumbers (cm^{-1}): 953, 942, 818, 792sh, 719sh, 696s, 667s, 525, 493, 446, 421w.

Note: The wavenumbers were partly determined by us based on spectral curve analysis of the published spectrum.

V143 Thallium(I) selenite vanadate $TlSeVO_5$ $TlSeVO_5$

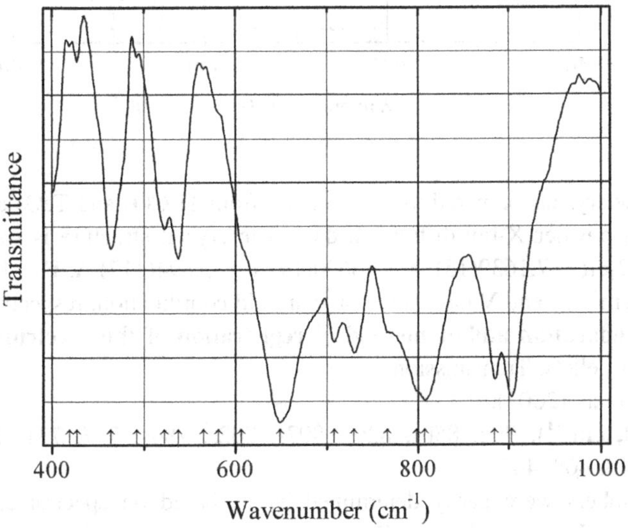

Origin: Synthetic.

Description: Yellow crystals prepared hydrothermally from Tl_2CO_3 and SeO_2 at 230 °C for 3 days. Characterized by powder X-ray diffraction data. The crystal structure is solved. Orthorhombic, space group $Pna2_1$, $a = 7.1639(15)$, $b = 8.6630(19)$, $c = 7.8946(17)$ Å, $V = 489.95(18)$ Å^3, $Z = 4$. $D_{calc} = 5.616$ g/cm^3. Se and V have three- and sixfold coordination, respectively.

Kind of sample preparation and/or method of registration of the spectrum: A sample pressed between two KBr pellets. Transmission.

Source: Sivakumar et al. (2007).

Wavenumbers (IR, cm^{-1}): 903s, 885s, 808s, 799sh, 767sh, 730, 708, 649s, 606, 581w, 565w, 537, 522, 492, 464, 427w, 419w.

Note: The wavenumber 649 cm^{-1} is erroneously indicated by Sivakumar et al. (2007) as 657 cm^{-1}. The wavenumbers were partly determined by us based on spectral curve analysis of the published spectrum. In the cited paper, Raman spectrum is given.

V144 Thallium(I) tellurite vanadate TlTeVO₅ $TlTeVO_5$

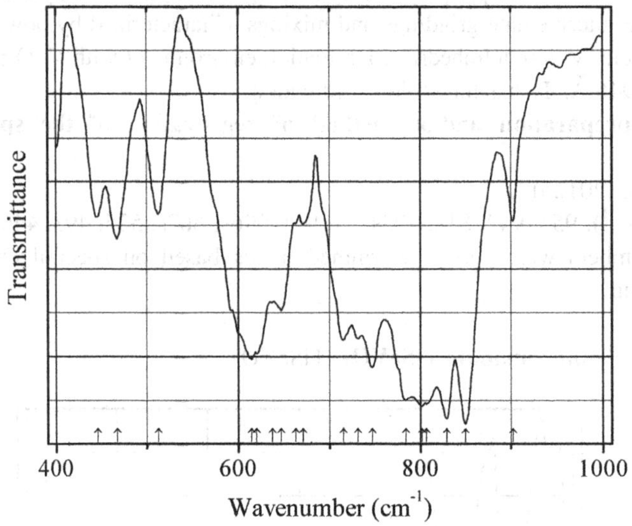

Origin: Synthetic.

Description: Yellow crystals prepared hydrothermally from Tl_2CO_3 and TeO_2 at 230 °C for 3 days. Characterized by powder X-ray diffraction data. The crystal structure is solved. Orthorhombic, space group $Pna2_1$, a = 7.1639(15), b =8.6630(19), c = 7.8946(17) Å, V = 489.95(18) Å^3, Z = 4. D_{calc} = 6.103 g/cm^3. Te and V have three- and sixfold coordination, respectively.

Kind of sample preparation and/or method of registration of the spectrum: A sample pressed between two KBr pellets. Transmission.

Source: Sivakumar et al. (2007).

Wavenumbers (IR, cm^{-1}): 902, 850s, 829s, 807s, 802s, 784s, 747, 731, 715, 671, 663, 647, 637, 620, 614, 513, 468, 446.

Note: The wavenumbers were partly determined by us based on spectral curve analysis of the published spectrum. In the cited paper, Raman spectrum is given.

V145 Thorium divanadate cubic polymorph α-$Th(V_2O_7)$

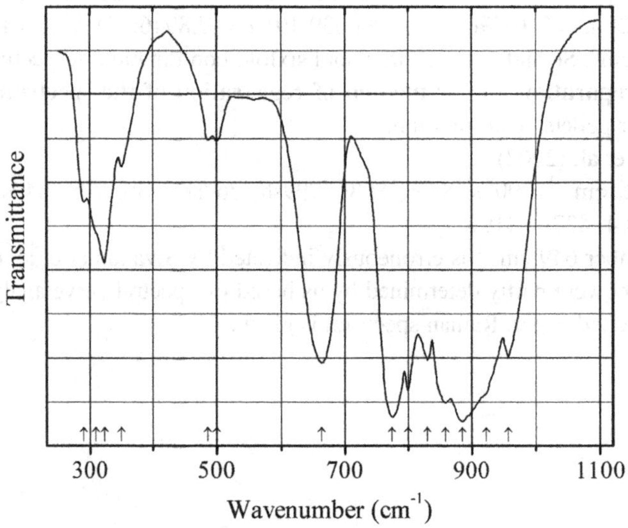

Origin: Synthetic.

Description: Prepared by short-time heating a mixture of ThO_2 and P_2O_5 between 600 and 1000 °C. Characterized by powder X-ray diffraction data. Cubic, $a = 8.72$, $Z = 4$.

Kind of sample preparation and/or method of registration of the spectrum: KBr disc. Transmission.

Source: Baran et al. (1974).

Wavenumbers (cm^{-1}): 958, 923sh, 885s, 858, 830, 799, 774s, 663, 500w, 486w, 350, 323, 310, 290.

V146 Thorium divanadate orthorhombic polymorph β-$Th(V_2O_7)$

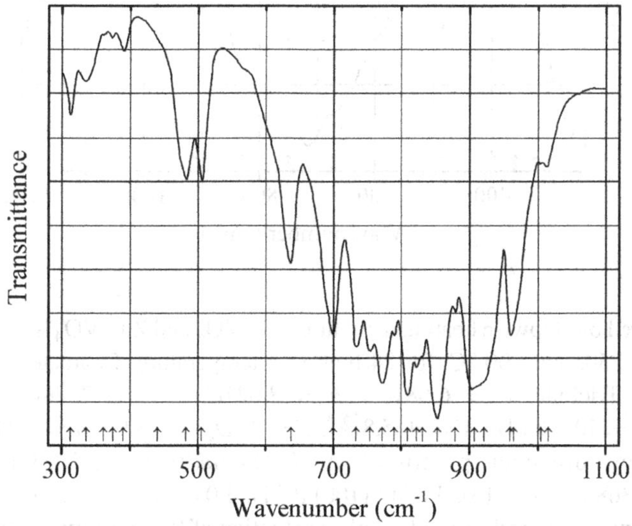

Origin: Synthetic.

Description: Prepared by long-time heating a mixture of ThO_2 and V_2O_5 at 600 °C. Characterized by powder X-ray diffraction data. Orthorhombic, $a = 7.216$, $b = 6.964$, $c = 22.800$ Å.

Kind of sample preparation and/or method of registration of the spectrum: KBr disc. Transmission.

Source: Baran et al. (1974).

Wavenumbers (cm^{-1}): 1015w, 1005w, 965sh, 960, 922sh, 908s, 880, 853s, 832sh, 823, 810s, 790, 773s, 754, 734, 700, 638, 505, 482, 440, 390w, 375w, 360w, 335w, 313w.

V147 Zinc iron(III) orthovanadate $Zn_3Fe_4(VO_4)_6$

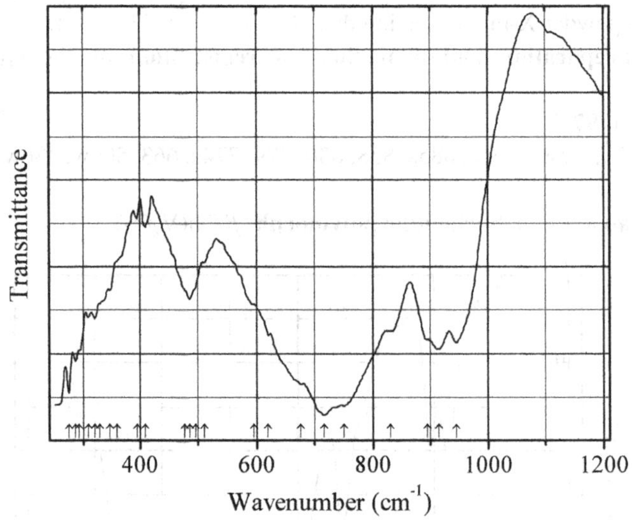

Origin: Synthetic.

Description: Prepared by stepwise heating a mixture of $FeVO_4$ and $Zn_3(VO_4)_2$, taken in stoichiometric amounts, at 700, 750, and 800 °C, for 24 h at each temperature. Characterized by powder X-ray diffraction data. Triclinic, a = 6.681(1), b = 8.021(2), c = 9.778(4) Å, α = 105.25(4)°, β = 105.00°4, γ = 102.20(4)°, V = 465.8 Å^3, Z = 1. D_{calc} = 3.95 g/cm^3. The strongest lines of the powder X-ray diffraction pattern [d, Å (I, %) (hkl)] are: 3.2713 (34) (−201), 3.1997 (45) (2−10), 3.1368 (56) (0−13), 3.0747 (100) (021), 3.0442 (36) (−202).

Kind of sample preparation and/or method of registration of the spectrum: KBr disc. Transmission.

Source: Kurzawa and Blonska-Tabero (2002).

Wavenumbers (cm^{-1}): 945s, 915s, 895sh, 830sh, 750s, 716s, 676, 511, 486, 408w, (394w).

Note: The wavenumbers were determined by us based on spectral curve analysis of the published spectrum.

V148 Zinc orthovanadate $Zn_3(VO_4)_2$

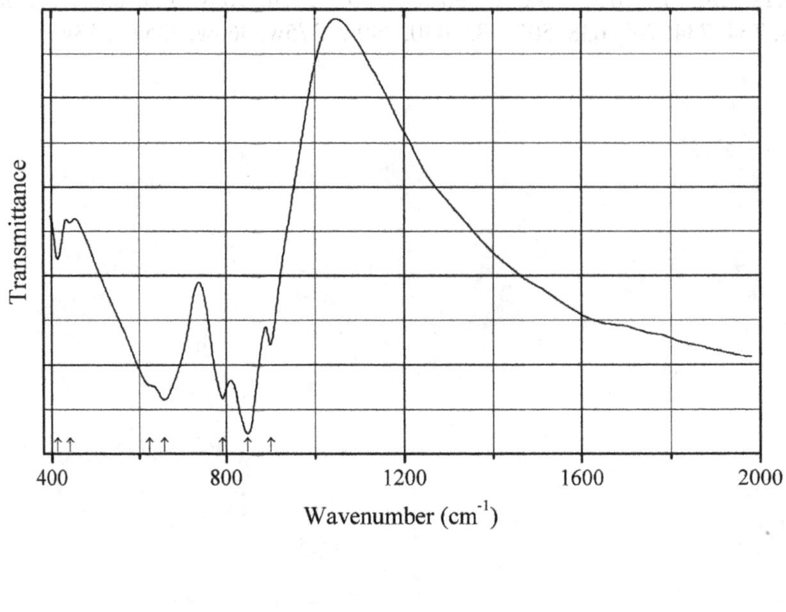

Origin: Synthetic.

Description: Prepared hydrothermally from $Zn(NO_3)_2$ and V_2O_5 in the presence of hexamethylenetetramine at 120 °C for 24 h with subsequent annealing of the $Zn_3(OH)_2V_2O_7{\cdot}nH_2O$ precursor at 600 °C for 10 h. Characterized by powder X-ray diffraction data. Orthorhombic, $a = 8.299$, $b = 11.52$, $c = 6.111$ Å (see JCPDS card no. 34-0378).

Kind of sample preparation and/or method of registration of the spectrum: Transmission. Kind of sample preparation is not indicated.

Source: Ni et al. (2010b).

Wavenumbers (IR, cm^{-1}): 901, 848s, 791s, 658s, 624, 442w, 414.

Note: In the cited paper, Raman spectrum is given.

Wavenumbers (Raman, cm^{-1}): 961s, 816, 797, 692w, 633w, 457w, 394, 374, 318s, 261, 224, 200w, 179w, 156.

V149 Zinc basic pyrovanadate hydrate $Zn_3(V_2O_7)(OH)_2{\cdot}2.5H_2O$

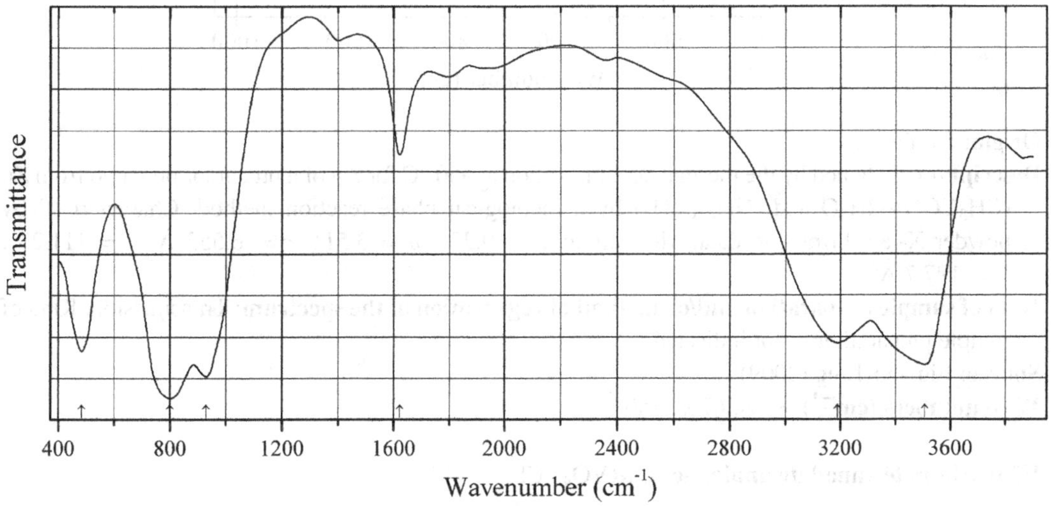

Origin: Synthetic.

Description: Precipitate obtained by adding 10% aqueous $(NH_4)(OH)$ to the solution prepared from 30% aqueous H_2O_2, V_2O_5, and of $Zn(NO_3)_2{\cdot}6H_2O$. Characterized by powder X-ray diffraction data and TG analysis. Hexagonal.

Kind of sample preparation and/or method of registration of the spectrum: KBr disc. Transmission.

Source: Melghit et al. (2007).

Wavenumbers (cm^{-1}): 3510s, 3190, 1622, 930s, 800s, 484.

V150 Zinc vanadyl oxide $Zn(VO_2)_2O_2$ $Zn(VO_2)_2O_2$

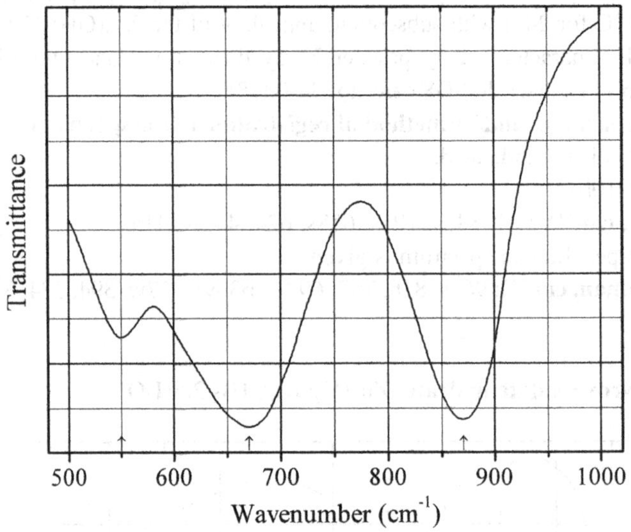

Origin: Synthetic.

Description: Obtained by the thermal decomposition at 550 °C for 8 h of a precursor prepared from Zn $(CH_3COO)_2{\cdot}2H_2O$ and $(NH_4)(VO_3)$ by a rheological phase reaction method. Characterized by powder X-ray diffraction data. Monoclinic, $a = 9.223$, $b = 3.511$, $c = 6.552$ Å, $\beta = 111.23°$, $V = 197.7$ Å^3.

Kind of sample preparation and/or method of registration of the spectrum: Transmission. Kind of sample preparation is not indicated.

Source: Liu and Tang (2009).

Wavenumbers (cm^{-1}): 870s, 670s, 550.

V151 Alforsite vanadate analogue $Ba_5(VO_4)_3Cl$

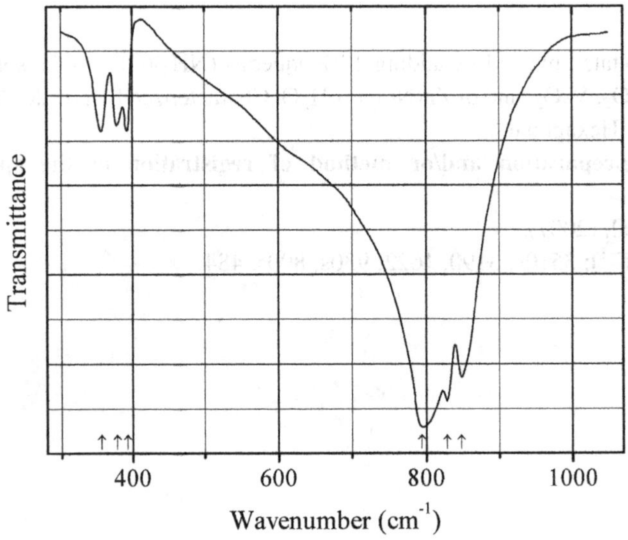

Origin: Synthetic.

Description: Obtained by double-ply heating a mixture of $Ba_3(VO_4)_2$ and $BaCl_2$, taken in stoichiometric molar ratio, at 950–1000 °C for 1–2 h with intermediate grinding. Characterized by powder X-ray diffraction data. Hexagonal, space group $P6_3/m$, $Z = 2$.

Kind of sample preparation and/or method of registration of the spectrum: KBr disc. Transmission.

Source: Baran and Aymonino (1972).

Wavenumbers (cm^{-1}): 848s, 828w, 794s, 393, 378, 357.

V152 Clinobisvanite $BiVO_4$

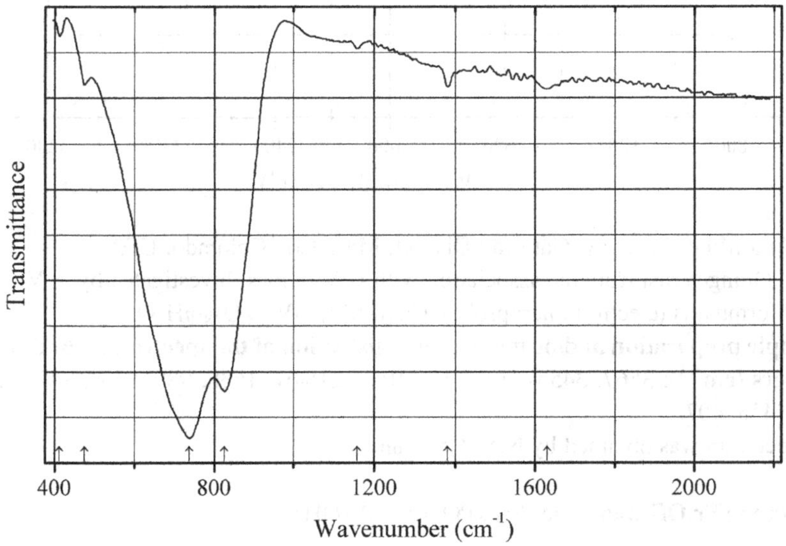

Origin: Synthetic.

Description: Prepared from ammonium vanadate and bismuth nitrate using a complex sol-gel procedure followed by calcination at 600 °C for 2 h. Monoclinic. Characterized by powder X-ray diffraction data and EDX spectroscopy.

Kind of sample preparation and/or method of registration of the spectrum: No data.

Source: Pookmanee et al. (2013).

Wavenumbers (cm^{-1}): 1629w, 1382w, 1158w, 826s, 737s, 474w, 412w.

Note: The wavenumbers were partly determined by us based on spectral curve analysis of the published spectrum. Weak bands with wavenumbers above 1000 cm^{-1} may correspond to impurities.

V153 Magnesiopascoite $Ca_2MgV^{5+}{}_{10}O_{28}\cdot16H_2O$

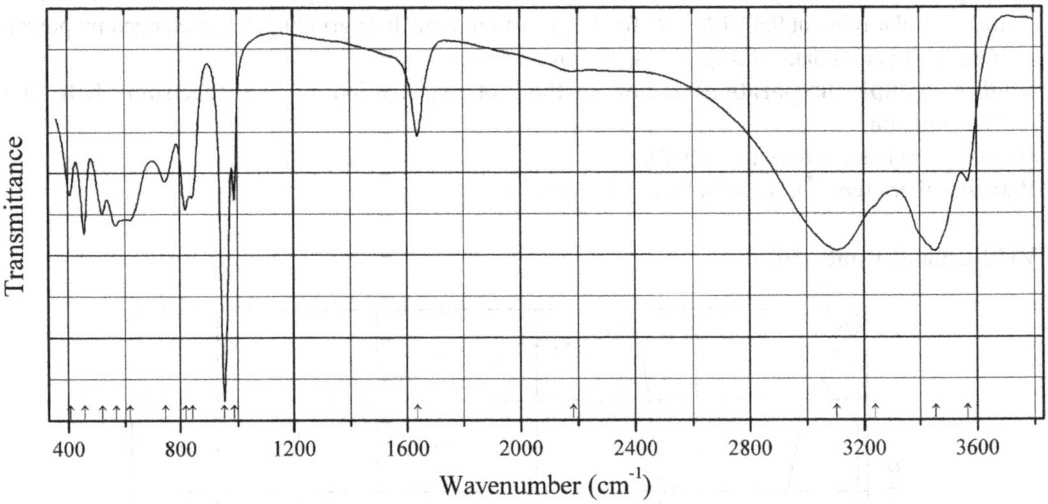

Origin: Packrat Mine, Gateway, Gateway District, Mesa Co., Colorado, USA.
Description: Orange crust from the association with U,V-oxides. Investigated by A.V. Kasatkin. The empirical formula is (electron microprobe): $Ca_{2.07}Mg_{1.03}V_{9.90}O_{28}\cdot nH_2O$.
Kind of sample preparation and/or method of registration of the spectrum: KBr disc. Absorption.
Wavenumbers (cm^{-1}): 3567, 3455s, 3240sh, 3105s, 2184w, 1638, 989, 955s, 842, 819, 746, 619, 569, 522, 458, 407.
Note: The spectrum was obtained by N.V. Chukanov.

V154 Vanadinite Sr,OH-analogue $Sr_{10}(PO_4)(VO_4)_5(OH)_2$

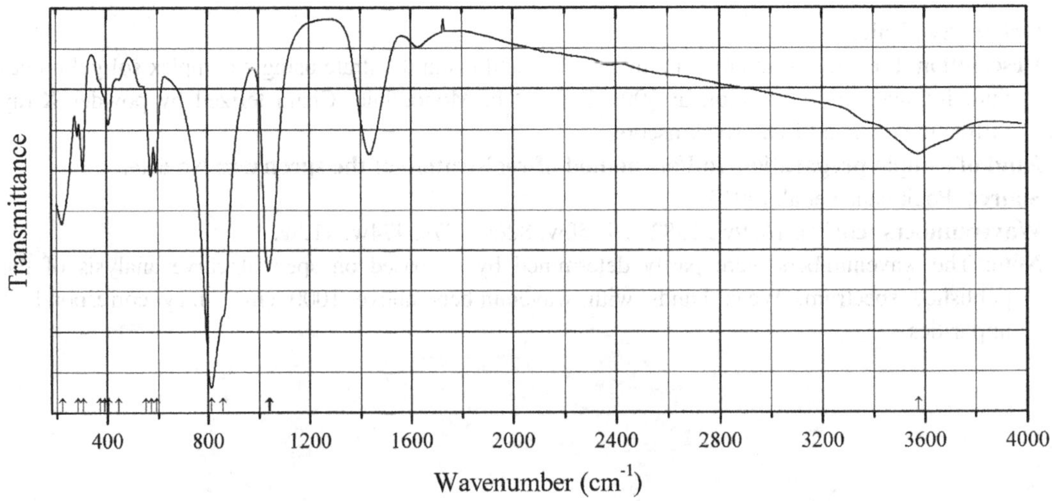

Origin: Synthetic.

Description: Obtained in the reaction between aqueous solution of $Sr(NO_3)_2$, containing NH_4OH, and a solution containing $(NH_4)_2(HPO_4)$ and $(NH_4)(VO_3)$ with subsequently heating a precipitate formed first at 100 °C for 2 h and thereafter at 850 °C for 2 h. Characterized by powder X-ray diffraction data.

Kind of sample preparation and/or method of registration of the spectrum: KBr disc. Transmission.

Source: Galera-Gomez et al. (1982).

Wavenumbers (cm^{-1}): 3575, 1042s, 858sh, 815s, 592, 572, 550, 443, 405, 390sh, 370sh, 305, 283, 222s.

Note: The wavenumbers were partly determined by us based on spectral curve analysis of the published spectrum. Bands in the range from 1400 to 1700 cm^{-1} correspond to impurities.

V155 Wakefieldite-(Pr) $PrVO_4$

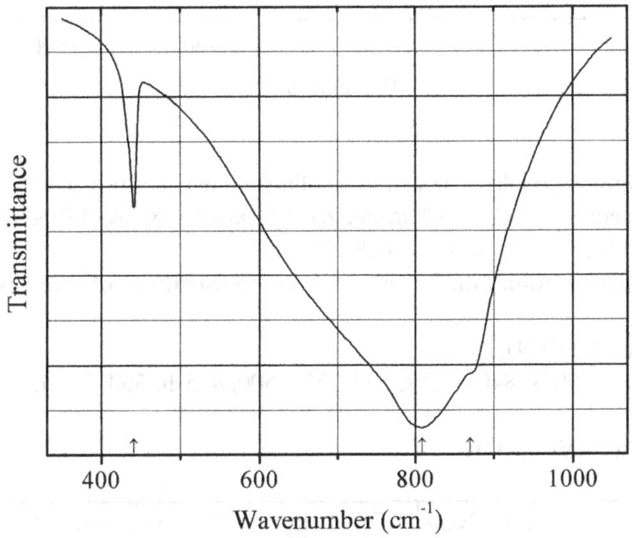

Origin: Synthetic.

Description: Obtained by heating a stoichiometric mixture of Pr_2O_3 and V_2O_5 powders in air for several hours, first at 750 °C and thereafter at 1000 °C with intermediate grinding. Tetragonal, $Z = 4$.

Kind of sample preparation and/or method of registration of the spectrum: KBr disc. Transmission.

Source: Baran and Aymonino (1971).

Wavenumbers (cm^{-1}): 870sh, 808s, 441.

V156 Ziesite and blossite polymorph $Cu_2V_2O_7$

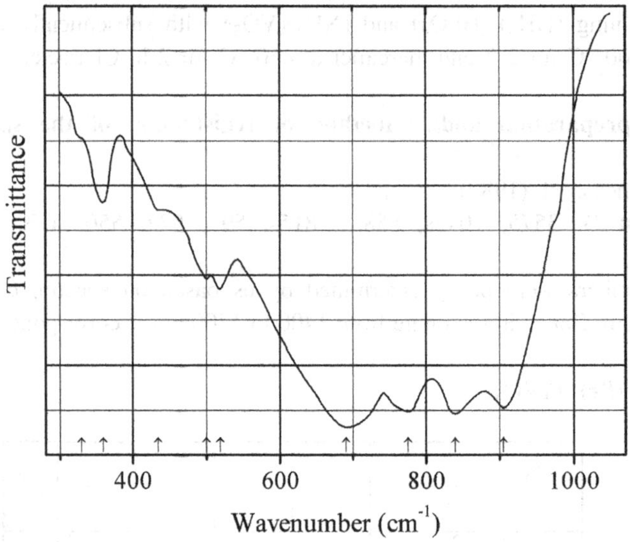

Origin: Synthetic.

Description: Dark brown powder obtained in a solid-state reaction of CuO and V_2O_5 at 600 °C for 16 h with intermediate grinding. Characterized by powder X-ray diffraction data. Monoclinic, $a = 6.87$, $b = 8.11$, $c = 9.16$ Å, $\beta = 109.5°$, $Z = 4$.

Kind of sample preparation and/or method of registration of the spectrum: KBr disc. Transmission.

Source: Pedregosa et al. (1974).

Wavenumbers (cm^{-1}): 905s, 840s, 775s, 690s, 518, 500, 435sh, 360, 330sh.

V157 Reppiaite $Mn^{2+}{}_5(VO_4)_2(OH)_4$

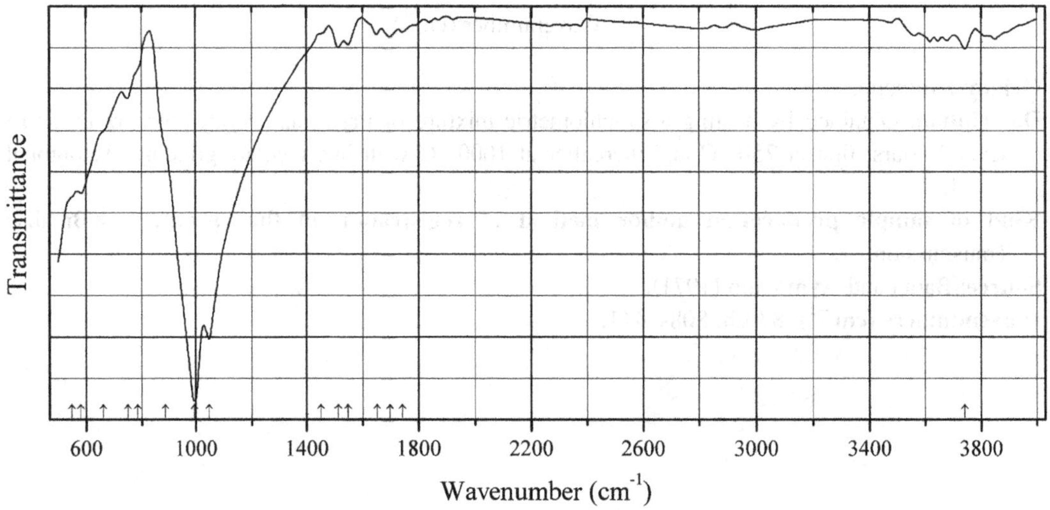

Origin: Synthetic.

Description: Dark red columnar crystals hydrothermally grown from Mn_2O_3 and V_2O_5 in 3 M CsOH at 580 °C and 1.5 kbar. The crystal structure is solved. Monoclinic, space group *C2/m*, $a = 9.6568(9)$ Å, $b = 9.5627(9)$, $c = 5.4139(6)$ Å, $\beta = 98.529(8)°$, $V = 494.42(9)$ Å^3, $Z = 2$. $D_{calc} = 3.846$ g/cm^3.

Kind of sample preparation and/or method of registration of the spectrum: KBr disc. Transmission.

Source: Sanjeewa et al. (2016).

Wavenumbers (IR, cm^{-1}): 3743w, 1743w, 1698w, 1651w, 1549w, 1513w, 1452sh, 1048s, 995s, 886sh, 785sh, 750w, 663sh, 582, 550sh.

Note: The wavenumbers were partly determined by us based on spectral curve analysis of the published spectrum. The intensity of the band of O–H-stretching vibrations at 3743 cm^{-1} is anomalously low, and the wavenumber is anomalously high. In the cited paper, Raman spectrum is given.

Wavenumbers (Raman, cm^{-1}): 789s, 749w, 402w, 315w.

V158 Ziminaite monoclinic polymorph $Fe^{3+}(VO_4)$

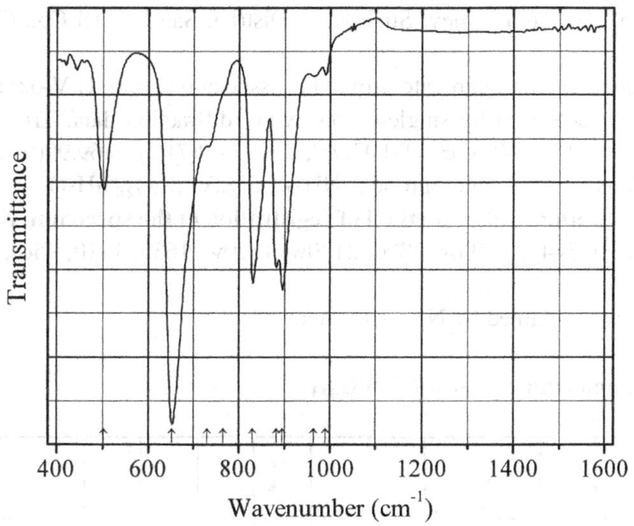

Origin: Synthetic.

Description: Obtained using a sol-gel technique with subsequently heating the sol at 500 °C. Characterized by electron diffraction. Monoclinic.

Kind of sample preparation and/or method of registration of the spectrum: Film on a Si wafer. Absorption.

Source: Vuk et al. (2001).

Wavenumbers (cm^{-1}): 990w, 965w, 896, 882s, 830s, 766sh, 730sh, 652s, 502.

V159 Wernerbaurite $\{(NH_4)_2[Ca_2(H_2O)_{14}](H_2O)_2\}\{V_{10}O_{28}\}$

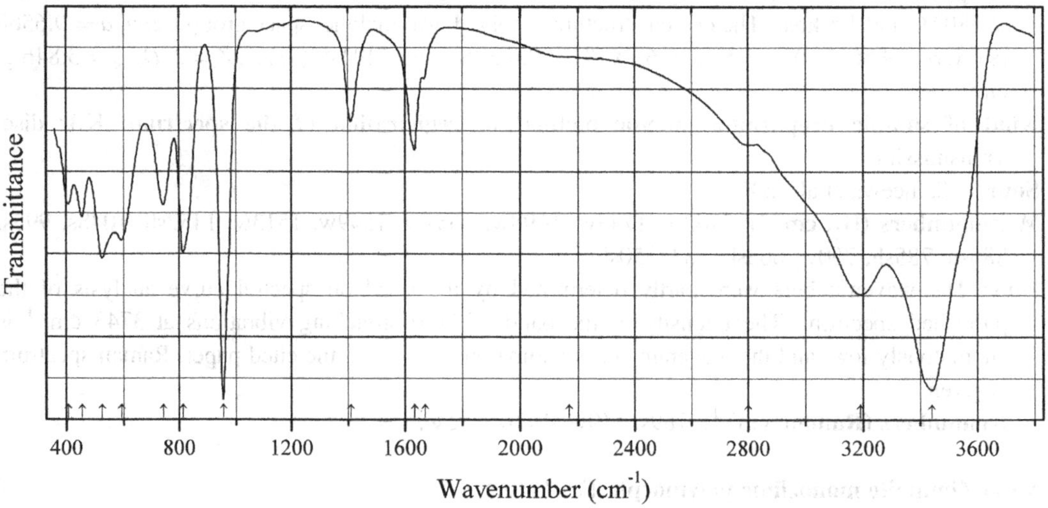

Origin: St Jude Mine, Gypsum Valley, Slick Rock District, San Miguel Co., Colorado, USA (type locality).

Description: Orange granular aggregate from the association with U,V-oxides. Investigated by A.V. Kasatkin. Characterized by single-crystal X-ray diffraction data. Triclinic, $a = 9.709(10)$, $b = 10.272(11)$, $c = 10.599(7)$ Å, $\alpha = 90.05(7)°$, $\beta = 77.09(7)°$, $\gamma = 69.90(9)°$, $V = 964(1)$ Å^3. The empirical formula is (electron microprobe): $(NH_4)_xCa_{1.96}V_{10.00}O_{28}\cdot nH_2O$.

Kind of sample preparation and/or method of registration of the spectrum: KBr disc. Absorption.

Wavenumbers (cm^{-1}): 3441s, 3190s, 2800, 2170w, 1670w, 1635, 1410, 956s, 814, 744, 593, 526, 456, 408.

Note: The spectrum was obtained by N.V. Chukanov.

V160 Schäferite Ni analogue $(Ca_2Na)Ni_2(VO_4)_3$

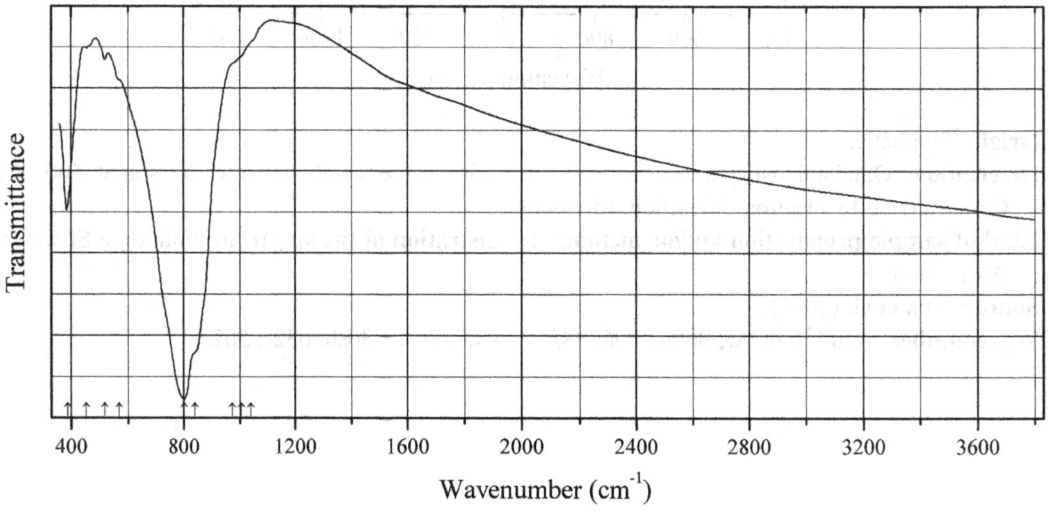

Origin: Slag dump near the Kamariza mine, Lavrion, mining district, Attikí (Attika, Attica) Prefecture, Greece.
Description: Dark olive-green crust from the association with trevorite and liebenbergite. The empirical formula is (electron microprobe): $(Ca_{1.93}Na_{1.04})(Ni_{1.32}Mg_{0.44}Fe_{0.25})(V_{2.93}P_{0.06}Cr_{0.03}Si_{0.01})O_{12}$.
Kind of sample preparation and/or method of registration of the spectrum: KBr disc. Absorption.
Wavenumbers (cm^{-1}): 1040sh, 1005sh, 975sh, 840sh, 801s, 570sh, 519w, 454w, 387.
Note: The spectrum was obtained by N.V. Chukanov.

V161 Pucherite $Bi(VO_4)$

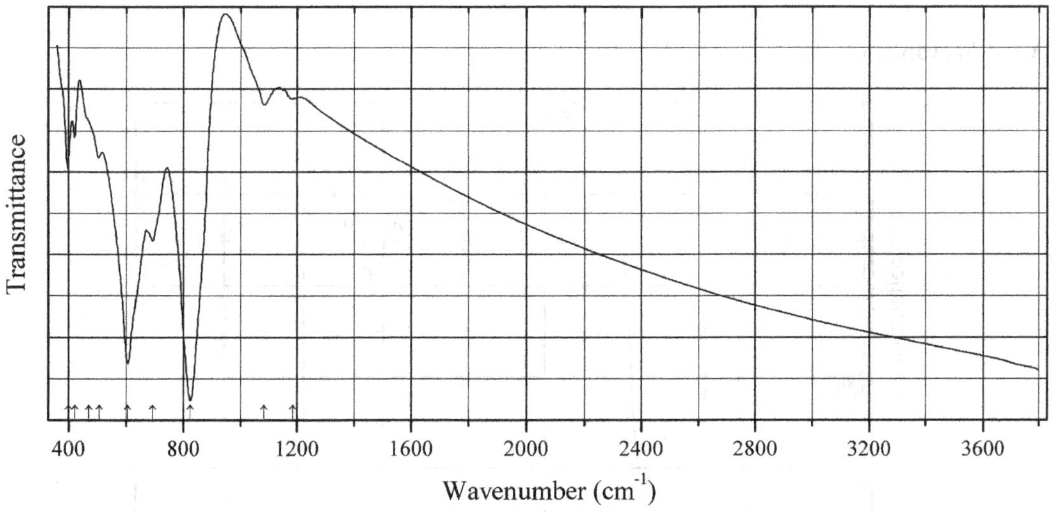

Origin: Neustädtel, Erzgebirge (Ore Mts.), Saxony, Germany.
Description: Yellowish-brown crystals.
Kind of sample preparation and/or method of registration of the spectrum: KBr disc. Absorption.
Wavenumbers (cm^{-1}): 1184w, 1083w, 824s, 693, 606s, 507, 470sh, 421, 400.
Note: The spectrum was obtained by N.V. Chukanov.

V162 Cheremnykhite trigonal dimorph $Pb_3Zn_3(TeO_6)(VO_4)_2$

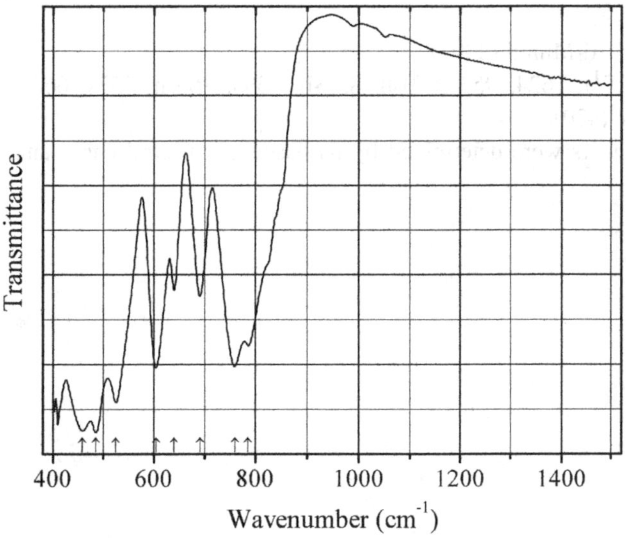

Origin: Synthetic.

Description: Synthesized by conventional solid-state methods from stoichiometric amounts of PbO, ZnO, $H_2TeO_4·2H_2O$, and V_2O_5 first at 400 °C for 20 h to decompose $H_2TeO_4·2H_2O$ and thereafter at 700 °C for 5 days, with intermediate grindings. Characterized by powder and single-crystal X-ray diffraction data. Trigonal, space group *P*321, $a = 8.608(2)$, $c = 5.186(3)$ Å, $V = 332.8(2)$ Å^3, Z = 1. $D_{calc} = 6.343$ g/cm^3.

Kind of sample preparation and/or method of registration of the spectrum: KBr disc. Transmission.

Source: Yu et al. (2016).

Wavenumbers (cm^{-1}): 785, 759s, 691, 640, 604s, 525, 485s, 459s.

V163 Fervanite (?) $Fe^{3+}{}_4V^{5+}{}_4O_{16}·5H_2O$

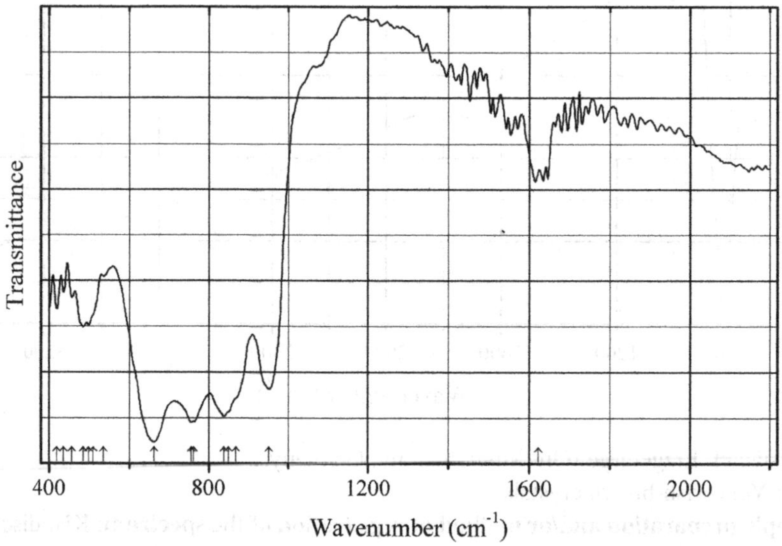

Origin: Synthetic.

Description: Poor-crystallized yellow powder prepared in the reaction between boiling aqueous solutions of ferric iron nitrate and V_2O_5. The empirical formula is $FeVO_4·1.1H_2O$.

Kind of sample preparation and/or method of registration of the spectrum: KBr disc. Transmission.

Source: Melghit and Al-Mungi (2007).

Wavenumbers (cm^{-1}): 1624, 953, 870sh, 850sh, 839s, 763sh, 757s, 663s, 537w, 511sh, (500), 486, (458), (437), (420).

Note: The wavenumbers were determined by us based on spectral curve analysis of the published spectrum.

V164 Ronneburgite $K_2MnV_4O_{12}$

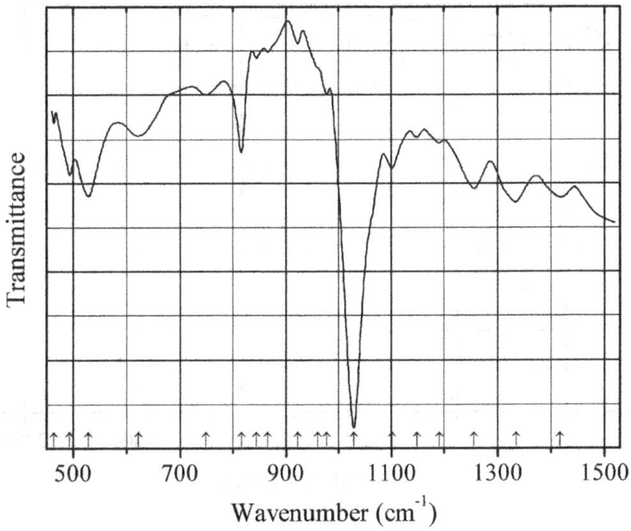

Origin: Ronneburg, Thuringia, Germany (type locality).

Description: Reddish-brown crystals from the association with hummerite, gypsum, epsomite, picromerite and hematite. Holotype sample. The crystal structure contains infinite metavanadate chains of corner-sharing VO_4 tetrahedra. Monoclinic, space group $P2_1/n$, $a = 8.183(3)$, $b = 9.247(3)$, $c = 8.651(2)$ Å, $\beta = 109.74(2)°$, $Z = 2$. $D_{meas} = 2.84$ g/cm^3, $D_{calc} = 2.85$ g/cm^3. Optically biaxial (–), $\alpha = 1.925(5)$, $\beta = 1.960(10)$, $\gamma = 1.988(4)$, $2V = 82°$. The empirical formula is $K_{1.91}Mn_{0.93}Mg_{0.08}V_{4.00}O_{11.96}$. The strongest lines of the powder X-ray diffraction pattern [*d*, Å (*I*, %) (*hkl*)] are: 3.701 (55) (–211), 3.336 (100) (121), 3.118 (50) (–122), 3.000 (36) (112), 2.878 (64) (–103, 031), 2.752 (68) (–222).

Kind of sample preparation and/or method of registration of the spectrum: Transmission of a small plate-like chip using an IR microscope.

Source: Witzke et al. (2001).

Wavenumbers (IR, cm^{-1}): (1417), (1335), (1256), (1191), (1149), 1102, 1029s, 978w, 961sh, 923w, 866w, 845w, 816, 749, 622, 529, 494, (464).

Note: The wavenumbers were partly determined by us based on spectral curve analysis of the published spectrum. Peaks above 1102 cm^{-1} may be due to interference. In the cited paper, Raman spectrum is given.

Wavenumbers (Raman, cm^{-1}): 952s, 942sh, 911, 878s, 830w, 658w, 461, 350, 336, 261.

V165 Wakefieldite-(Y) YVO_4

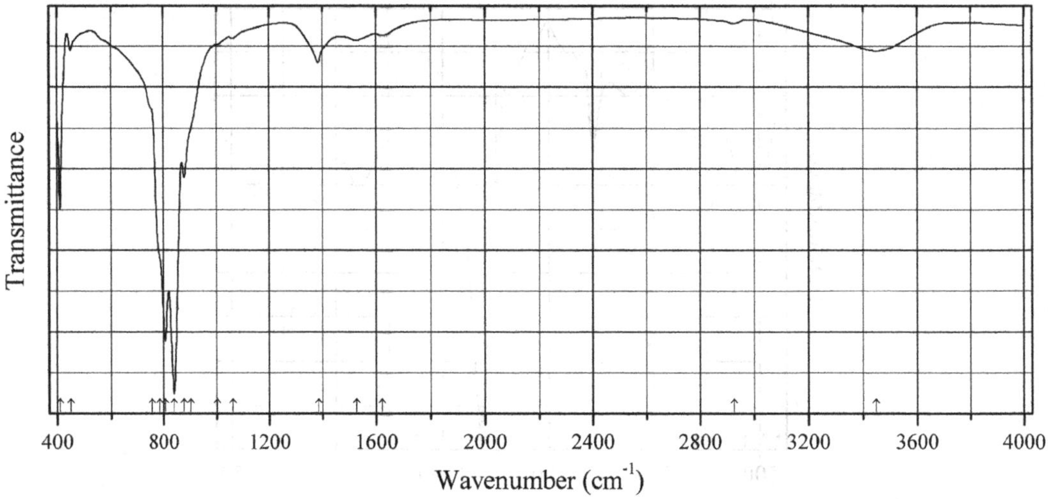

Origin: Synthetic.

Description: Eu-doped sample prepared hydrothermally from sodium orthovanadate and corresponding nitrates at Y: Eu = 9:1. Characterized by powder X-ray diffraction data.

Kind of sample preparation and/or method of registration of the spectrum: Absorption. Kind of sample preparation is not indicated.

Source: Tran et al. (2012).

Wavenumbers (cm^{-1}): 3449, 2926w, 1624w, 1526w, 1384, 1061w, 1003sh, 905sh, 880, 841s, 808s, 785sh, 755sh, 451w, 426, 411.

Note: The wavenumbers were partly determined by us based on spectral curve analysis of the published spectrum. The band at 2926 cm^{-1} corresponds to the admixture of an organic substance. The band at 1384 cm^{-1} indicates possible admixture of nitrate anions. The bands at 3449 and 1624 cm^{-1} correspond to adsorbed water molecules.

V166 Ziminaite $Fe^{3+}{}_6(VO_4)_6$

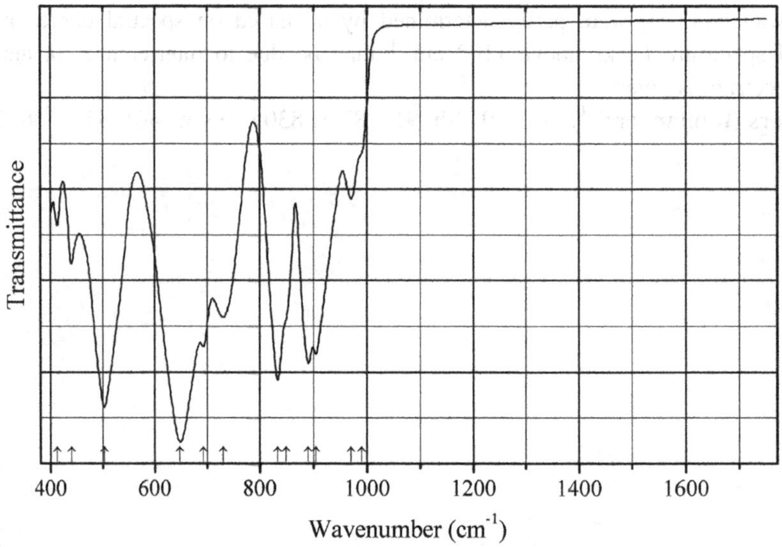

Origin: Synthetic.

Description: Nanorods obtained by dehydration of synthetic fervanite at 500 °C. Characterized by powder X-ray diffraction data and Mössbauer spectroscopy. Triclinic.

Kind of sample preparation and/or method of registration of the spectrum: Transmission. Kind of sample preparation is not indicated.

Source: Lehnen et al. (2014).

Wavenumbers (IR, cm^{-1}): 990sh, 970, 904s, 890s, 849sh, 833s, 730, 692, 648s, 502s, 440, 413w.

Note: The wavenumbers were partly determined by us based on spectral curve analysis of the published spectrum. In the cited paper, Raman spectrum is given.

Wavenumbers (Raman, cm^{-1}): 965, 931s, 907, 895s, 845s, 832s, 770, 736s, 660w, 633w, 502w, 450, 408, 391, 371, 329, 317.

V167 Janchevite $Pb_7V^{5+}(O_{8.5}\square_{0.5})Cl_2$

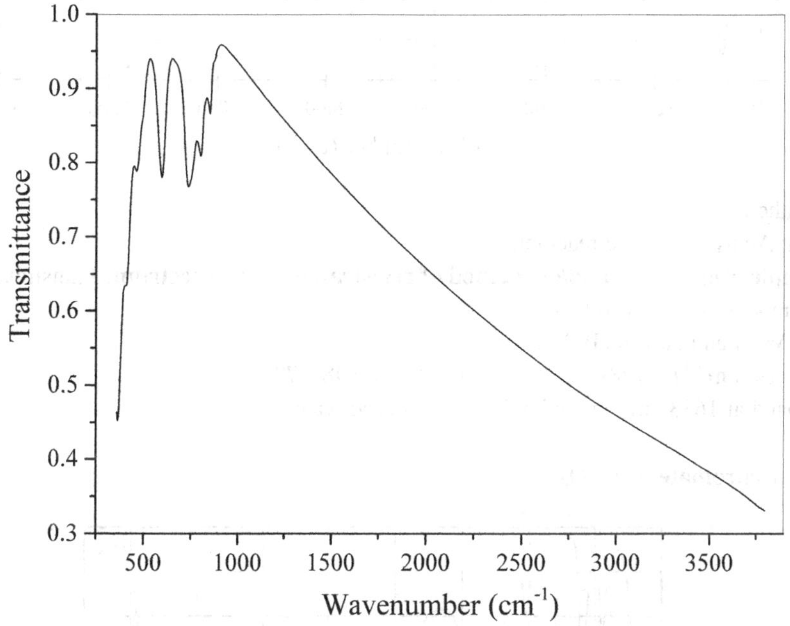

Origin: Kombat mine, Grootfontein district, Otjozondjupa region, Namibia (type locality).

Description: Orange-red, thick tabular anhedral grains from the association with baryte, hausmannite, calcite, magnesite, and kombatite. Holotype sample. Tetragonal, space group *I*4/*mmm*, $a = 3.9591$ (5) Å, $c = 22.6897(3)$ Å, $V = 355.65(1)$ Å^3; $Z = 1$. $D_{calc} = 8.18$ g/cm^3. The empirical formula is (electron microprobe): $Pb_{7.20}V^{5+}{}_{0.38}Mo^{6+}{}_{0.29}Si_{0.13}Cl_{2.06}O_{8.25}$. The strongest lines of the powder X-ray diffraction pattern [*d*, Å (*I*, %) (*hkl*)] are: 3.889 (24) (011), 3.501 (31) (013), 2.979 (86) (015), 2.833 (25) (008), 2.794 (100) (110), 1.992 (26) (118), 1.988 (49) (020), 1.649 (46) (215).

Kind of sample preparation and/or method of registration of the spectrum: KBr disc. Absorption.

Wavenumbers (cm^{-1}): 870w, 850, 802, 736, 595, 462, 405sh, 366s.

Note: The spectrum was obtained by N.V. Chukanov.

2.12 Chromates

Cr21 Ammonium dichromate $(NH_4)_2Cr_2O_7$

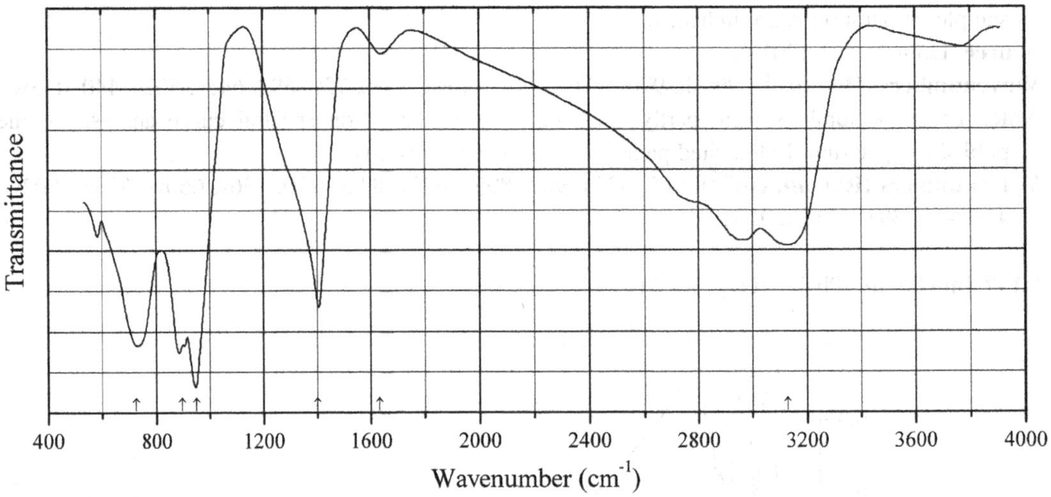

Origin: Synthetic.
Description: Analytical grade reactant.
Kind of sample preparation and/or method of registration of the spectrum: Transmission. Kind of sample preparation is not indicated.
Source: De Waal and Heyns (1992).
Wavenumbers (cm^{-1}): 3128s, 1633w, 1402s, 949s, 898s, 724s.
Note: The band at 1633 cm^{-1} may be due to adsorbed water.

Cr22 Copper chromate $CuCrO_4$

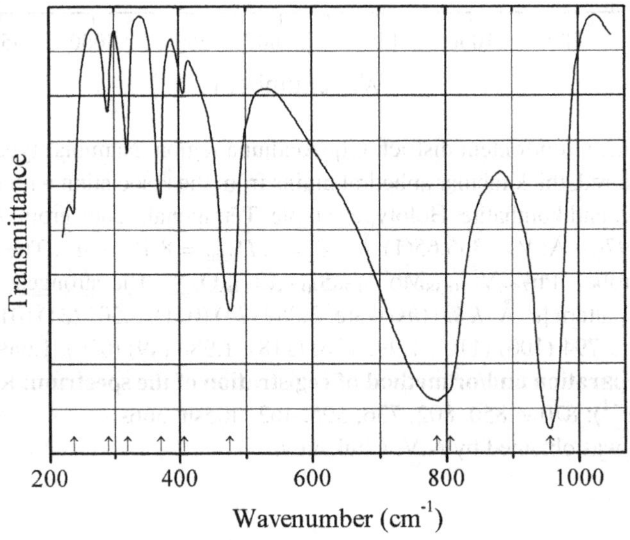

Origin: Synthetic.

Description: Prepared hydrothermally from $CuCO_3$, $Cu(OH)_2$, and CrO_3 at 220 °C for 24 h. Orthorhombic, space group *Cmcm*, $Z = 4$. In the crystal structure, strongly distorted CuO_6 octahedra are present.

Kind of sample preparation and/or method of registration of the spectrum: KBr disc. Transmission.

Source: Baran (1994).

Wavenumbers (IR, cm^{-1}): 956s, 805sh, 785s, 475s, 406w, 370, 320, 290, 237.

Note: In the cited paper, Raman spectrum is given.

Wavenumbers (Raman, cm^{-1}): 966, 944s, 928, 806s, 412, 386, 342, 254.

Cr23 Lead orthoborate chromate $Pb_6(BO_3)_2(CrO_4)O_2$

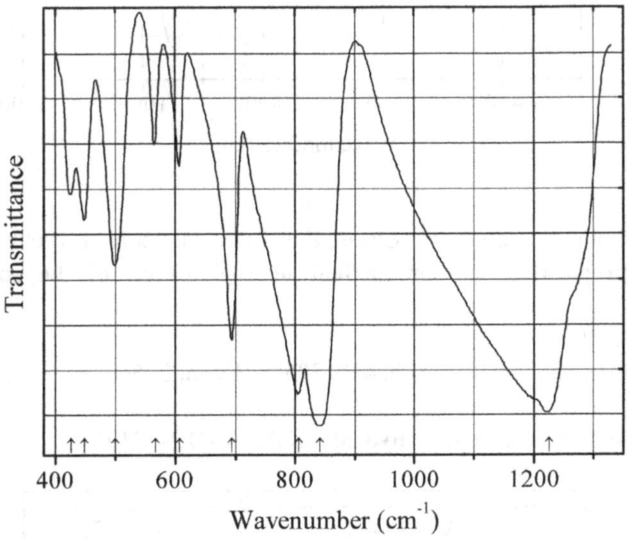

Origin: Synthetic.

Description: Prepared in a solid-state reaction from the powder mixture of PbO, CrO_3, and B_2O_3 with the molar ratio 15:2:3. Characterized by powder X-ray diffraction data. The crystal structure is solved. Orthorhombic, space group *Pnma*, $a = 6.4160(13)$, $b = 11.635(2)$, $c = 18.164(4)$ Å, $V = 1356.0(5)$ Å^3, $Z = 4$. $D_{calc} = 7.391$ g/cm^3.

Kind of sample preparation and/or method of registration of the spectrum: KBr disc. Transmission.

Source: Chen et al. (2009).

Wavenumbers (cm^{-1}): 1225s, 841s, 806s, 694s, 608, 567, 500, 449, 426.

Cr24 Magnesium chromate α-Mg(CrO_4)

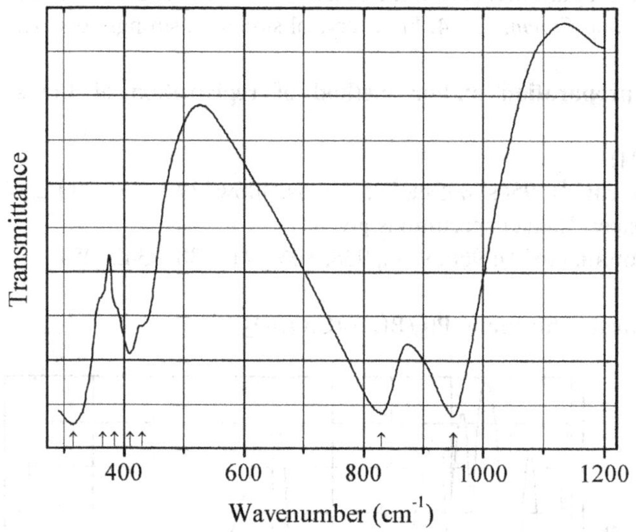

Origin: Synthetic.
Description: Orthorhombic, space group *Cmcm*, $Z = 4$ (see Muller et al. 1969b).
Kind of sample preparation and/or method of registration of the spectrum: KBr disc. Transmission.
Source: Muller et al. (1969a).
Wavenumbers (**cm**$^{-1}$): 950s, 830s, 430sh, 410, 385sh, 365sh, 315s.

Cr25 Potassium magnesium chromate hydrate $K_2Mg(CrO_4)_2{\cdot}2H_2O$

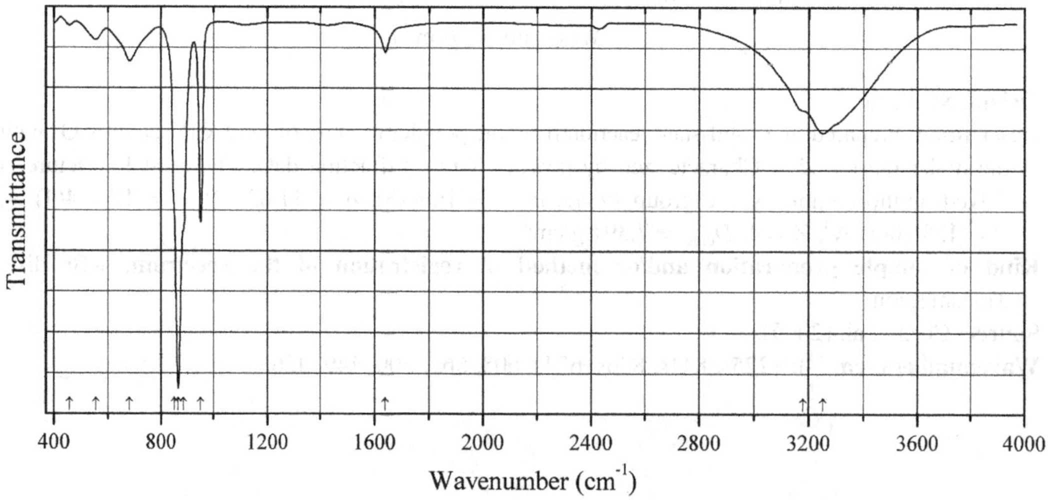

Origin: Synthetic.
Description: Prepared by precipitation from aqueous solution of magnesium acetate and potassium chromate. Triclinic, space group *P*-1, $Z = 1$. Structurally related to kröhnkite.

Kind of sample preparation and/or method of registration of the spectrum: KBr disc. Absorption.
Source: Stoilova et al. (2009).
Wavenumbers (cm^{-1}): 3251s, 3180sh, 1640, 953s, 887sh, 867s, 852sh, 682, 556, 458w.

Cr26 Potassium nickel chromate hydrate $K_2Ni(CrO_4)_2{\cdot}2H_2O$

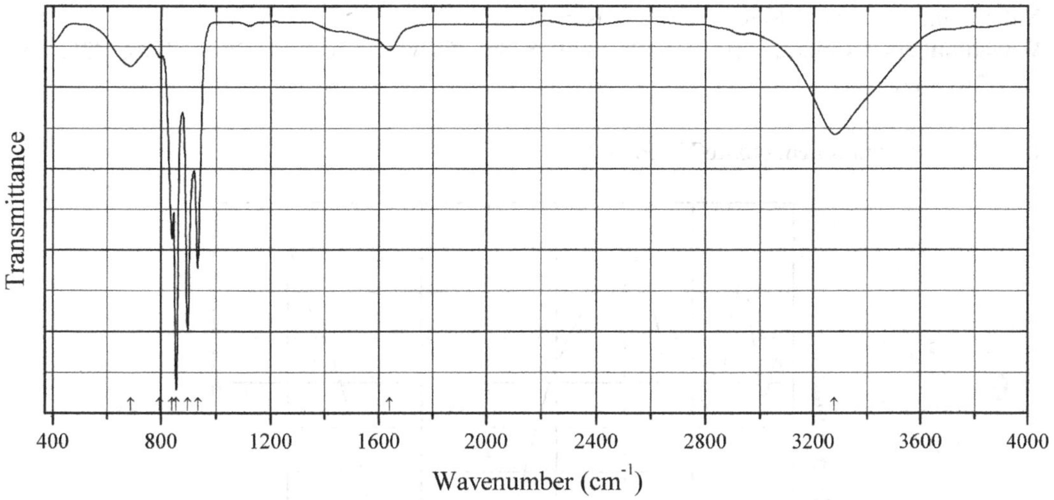

Origin: Synthetic.
Description: Prepared by precipitation from aqueous solution of nickel acetate and potassium chromate. Triclinic, space group *P*-1, *Z* = 1. Structurally related to kröhnkite.
Kind of sample preparation and/or method of registration of the spectrum: KBr disc. Absorption.
Source: Stoilova et al. (2009).
Wavenumbers (cm^{-1}): 3278 (broad), 1640w, 934s, 897s, 855s, 840, 793, 688 (broad).

Cr27 Potassium peroxochromate $K_3[Cr(O_2)_4]$

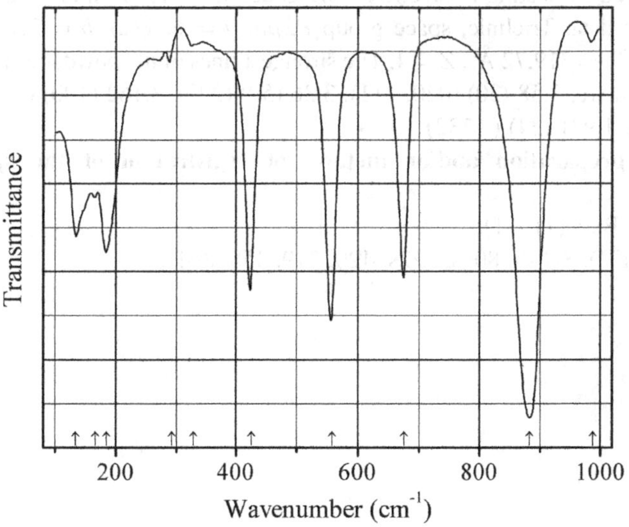

Origin: Synthetic.
Description: Tetragonal, space group *I*-42*m*.
Kind of sample preparation and/or method of registration of the spectrum: KBr disc. Transmission.
Source: Haeuseler and Haxhillazi (2003).
Wavenumbers (IR, cm^{-1}): 988w, 883s, 676s, 558s, 426s, 330w, 293w, 185, 166w, 134.
Note: In the cited paper, Raman spectrum is given.
Wavenumbers (Raman, cm^{-1}): 919s, 879, 838w, 682w, 564s, 526, 464s, 430s, 336, 286, 217w, 183w.

Cr28 Praseodymium chromate(V) $PrCrO_4$

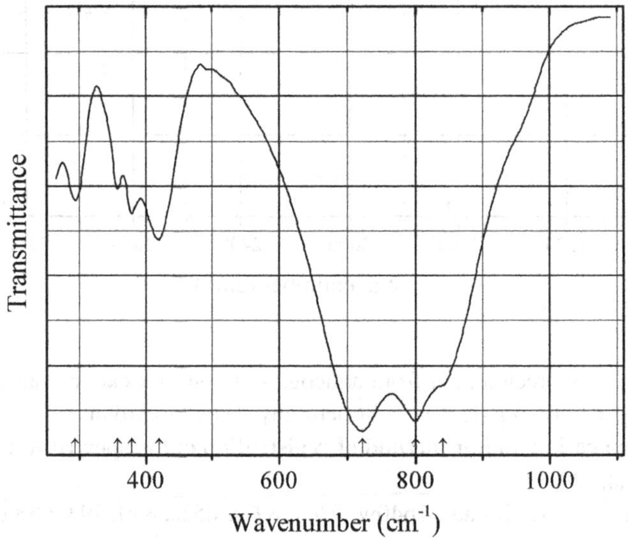

Origin: Synthetic.
Description: Prepared by heating $PrCr(C_2O_4)_3 \cdot 8H_2O$ at 500 °C for 10 min. Characterized by powder X-ray diffraction data. Triclinic, space group *P*2$_1$/*n*, $a = 6.98(2)$, $b = 7.16(1)$, $c = 6.63(1)$ Å, $\beta = 105.22(10)°$, $V = 319.72$ Å^3, $Z = 4$. The strongest lines of the powder X-ray diffraction pattern [*d*, Å (*I*, %) (*hkl*)] are: 3.58 (30) (020, 111), 3.36 (59) (200), 3.162 (100) (120), 2.925 (74) (012), 1.989 (30) (212), 1.921 (34) (−132).
Kind of sample preparation and/or method of registration of the spectrum: CsI disc. Transmission.
Source: Manca and Baran (1981).
Wavenumbers (cm^{-1}): 840sh, 800s, 722s, 420, 379, 358, 294.

Cr29 Embreyite $(Pb,Cu,\square)_2Pb[(Cr,P)O_4]_2{\cdot}nH_2O$

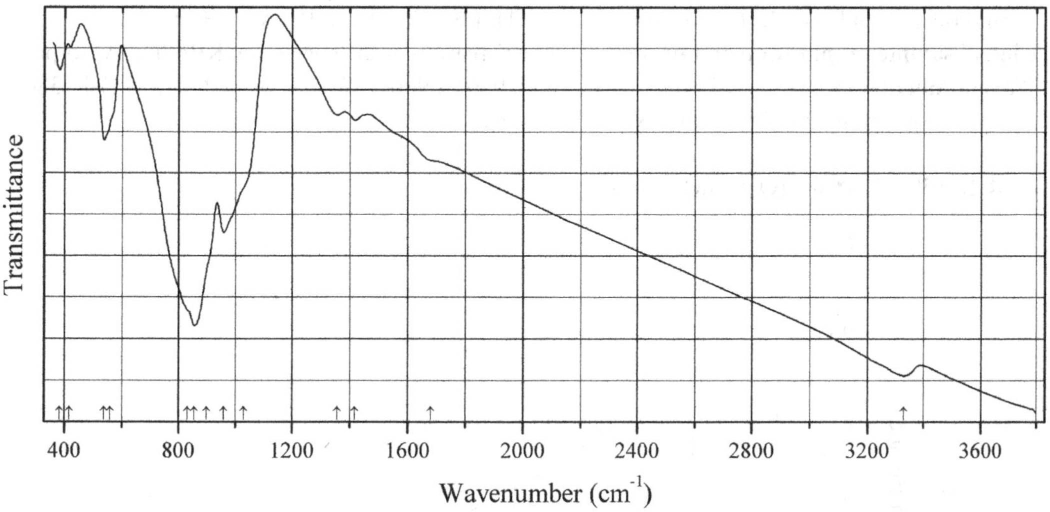

Origin: Krokoitovyi Shurf (Crocoite Pit), Uspenskaya Mt., Berezovskoe ore field, Middle Urals.

Description: Brownish-orange flattened crystals from the association with crocoite, vauquelinite pyromorphite, and goethite. The crystal structure is solved. Monoclinic, space group *C2/m*, $a = 9.802(16)$, $b = 5.603(9)$, $c = 7.649(12)$ Å, $\beta = 114.85(3)°$, $V = 381.2(11)$ Å^3. The empirical formula is (electron microprobe, Z = 2): $Pb_{1.29}Cu_{0.07}Cr_{0.52}P_{0.43}O_4{\cdot}nH_2O$.

Kind of sample preparation and/or method of registration of the spectrum: KBr disc. Absorption.

Wavenumbers (cm^{-1}): 3329, 1680w, 1418w, 1356w, 1030sh, 959, 900sh, 855s, 830sh, 560sh, 538, 418w, 383.

Note: The spectrum was obtained by N.V. Chukanov.

Cr30 Embreyite $(Pb,Cu,\square)_2Pb[(Cr,P)O_4]_2{\cdot}nH_2O$

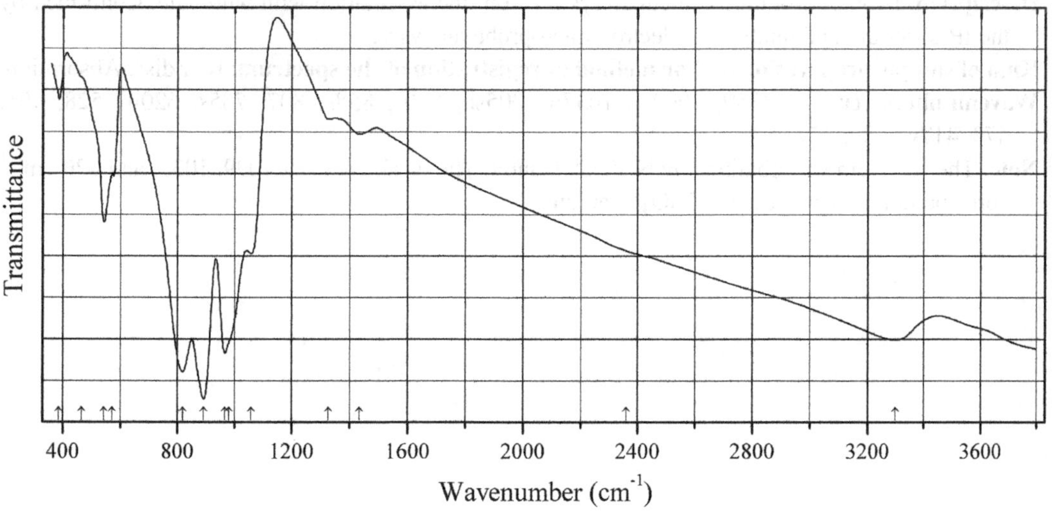

Origin: Krokoitovyi Shurf (Crocoite Pit), Uspenskaya Mt., Berezovskoe ore field, Middle Urals.
Description: Reddish-brown crystals from the association with vauquelinite. A Cu-rich variety. The empirical formula is (electron microprobe, $Z = 1$): $Pb_{2.5}Cu_{0.3}Cr_{1.05}P_{0.95}O_8{\cdot}nH_2O$.
Kind of sample preparation and/or method of registration of the spectrum: KBr disc. Absorption.
Wavenumbers (cm^{-1}): 3301, 1433, 1328w, 1059, 980sh, 966s, 891s, 819s, 572, 543, 465sh, 386.
Note: The spectrum was obtained by N.V. Chukanov.

Cr31 Iranite $Pb_{10}Cu(CrO_4)_6(SiO_4)_2(OH)_2$

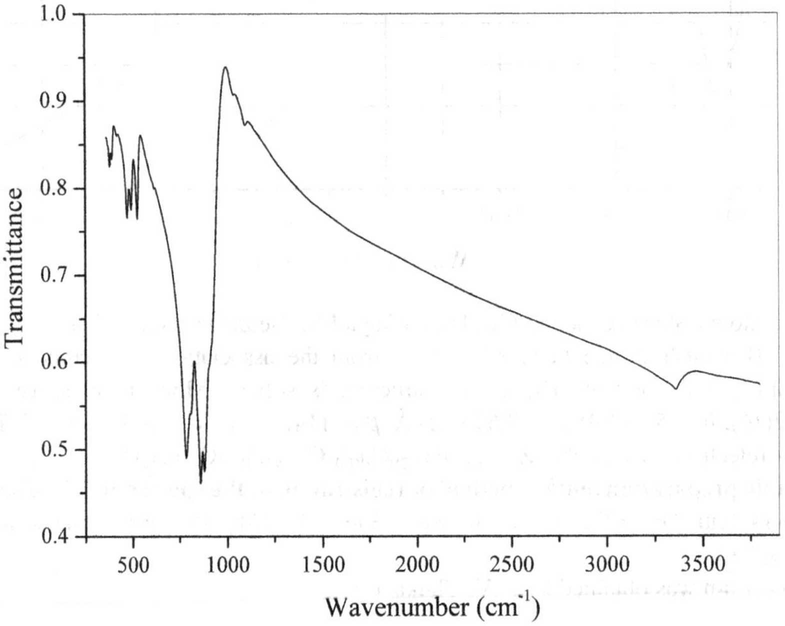

Origin: Santa Ana mine, Caracoles, Sierra Gorda district, Antofagasta Region, Chile.
Description: Brownish-orange lenticular crystals from the association with wulfenite. Confirmed by the IR spectrum and qualitative electron microprobe analyses.
Kind of sample preparation and/or method of registration of the spectrum: KBr disc. Absorption.
Wavenumbers (cm^{-1}): 3359, 1099w, 1037w, 905sh, 879s, 858s, 807, 785s, 620w, 528, 495, 473, 417w, 389, 379.
Note: The spectrum was obtained by N.V. Chukanov. The weak bands at 1099, 1037, and 620 cm^{-1} correspond to trace amounts of SO_4^{2-} anions.

2.13 Germanates

Ge4 Brunogeierite $Fe^{2+}{}_2Ge^{4+}O_4$

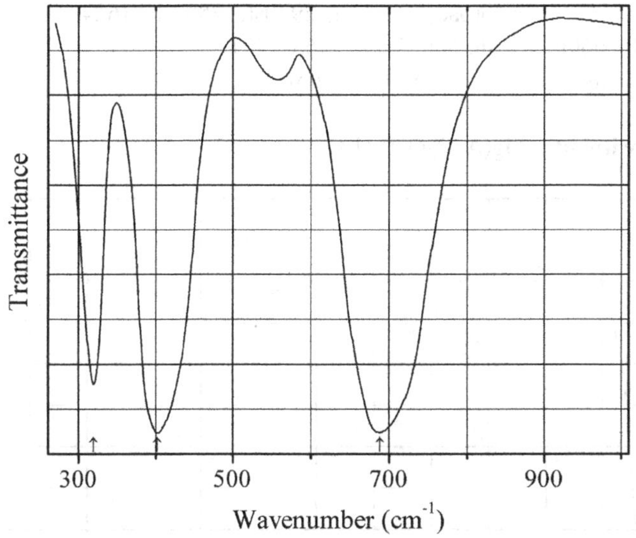

Origin: Synthetic.
Description: Prepared in the solid-state reaction between GeO_2 and FeO at 1000 °C.
Kind of sample preparation and/or method of registration of the spectrum: No data.
Source: Tarte (1963).
Wavenumbers (cm^{-1}): 688s, 402s, 319.

2.14 Arsenides, Arsenites, Arsenates, and Sulfato-Arsenates

As316 Castellaroite $Mn^{2+}{}_3(AsO_4)_2 \cdot 4H_2O$

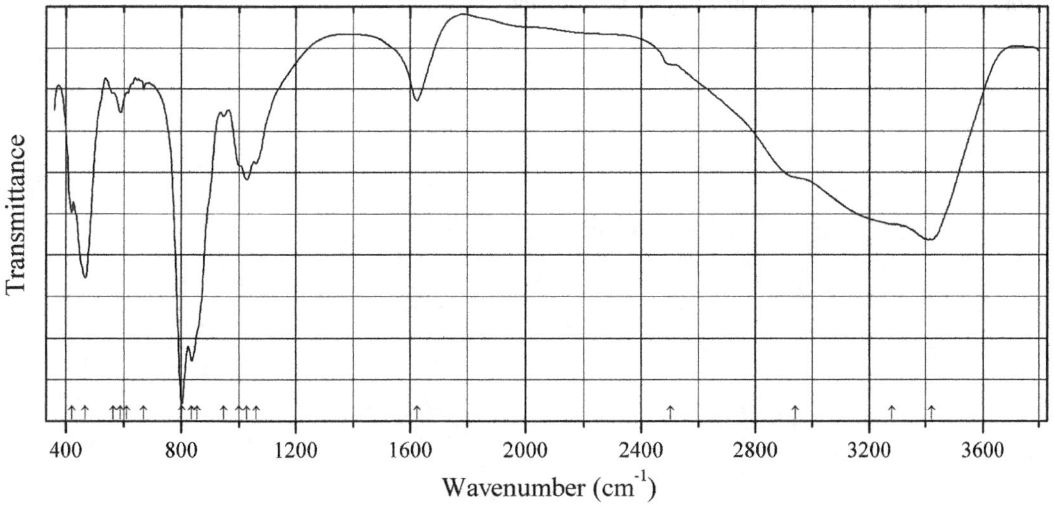

Origin: Monte Nero Mine, Rocchetta Vara, La Spezia Province, Liguria, Italy (type locality).
Description: White radiated aggregates from the association with rhodochrosite. The sample was received from L. Chiappino, a coauthor of the first description of castellaroite.
Kind of sample preparation and/or method of registration of the spectrum: KBr disc. Absorption.
Wavenumbers (cm^{-1}): 3421s (broad), 3280sh, 2940sh, 2500sh, 1624, 1061, 1027, 1001, 948w, 855sh, 836s, 801s, 669w, 610sh, 589, 565sh, 468s, 421.
Note: The spectrum was obtained by N.V. Chukanov.

As317 Magnesiokoritnigite $Mg(AsO_3OH) \cdot H_2O$

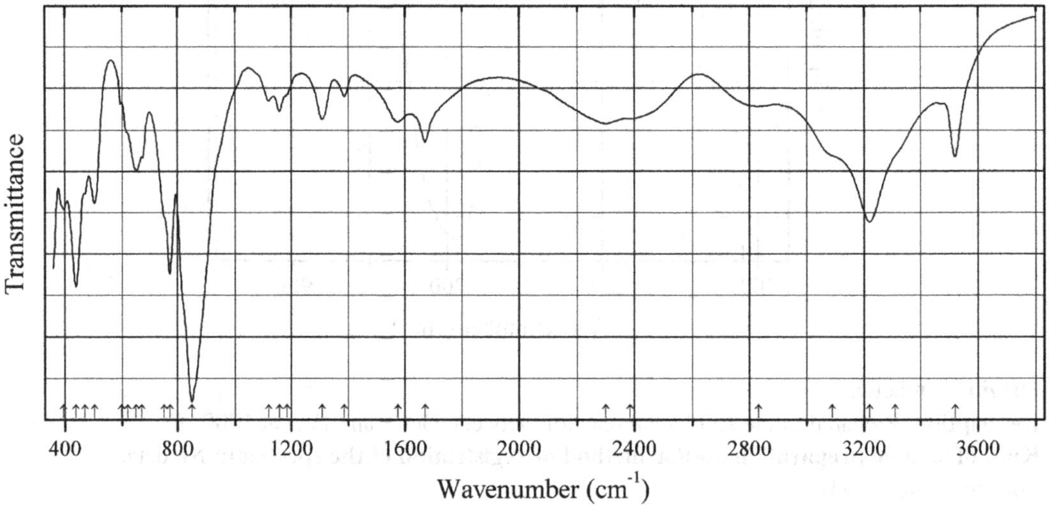

Origin: Torrecillas mine, Salar Grande, El Tamarugal Province, Tarapacá Region, Chile (type locality).
Description: Pink crystals from the association with magnesiocanutite. Investigated by I.V. Pekov. The empirical formula is (electron microprobe): $(Mg_{0.99}Mn_{0.01})(HAsO_4) \cdot H_2O$.
Kind of sample preparation and/or method of registration of the spectrum: KBr disc. Absorption.
Wavenumbers (cm^{-1}): 3519, 3310sh, 3220s, 3090sh, 2830, 2385, 2301, 1670, 1576, 1386w, 1307, 1185sh, 1158, 1120w, 850s, 772s, 750sh, 670, 650, 620sh, 599w, 504, 470sh, 439s, 395.
Note: The spectrum was obtained by N.V. Chukanov.

As318 Chudobaite $(Mg,Zn)_5(AsO_4)_2(HAsO_4)_2{\cdot}10H_2O$

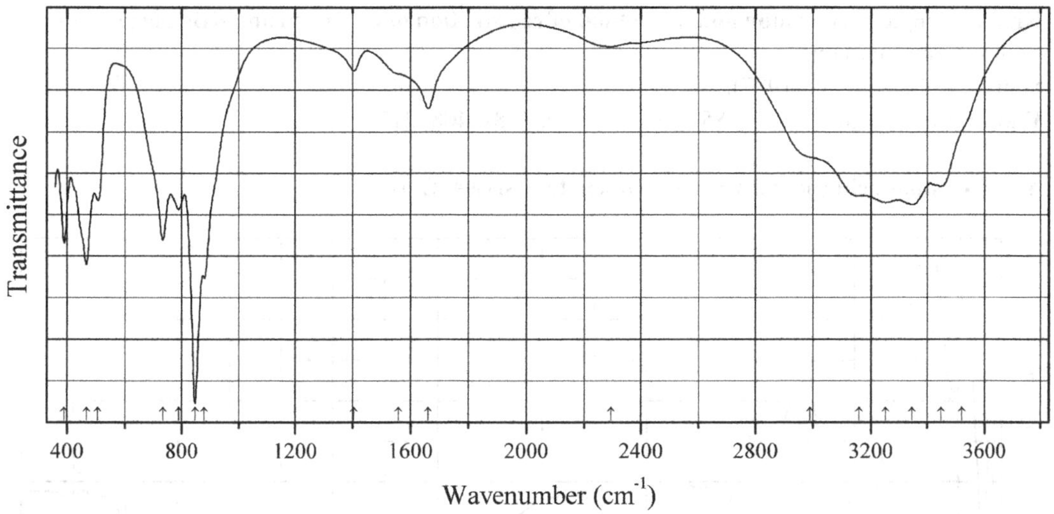

Origin: Torrecillas mine, Salar Grande, El Tamarugal Province, Tarapacá Region, Chile.

Description: White granular aggregate from the association with hörnesite, gypsum, arsenic, pyrite, dolomite, and quartz. Investigated by I.V. Pekov. Identified by qualitative electron microprobe analyses and powder X-ray diffraction data. The strongest lines of the powder X-ray diffraction pattern [*d*, Å (*I*, %)] are: 10.26 (100), 7.70 (11), 4.79 (15), 3.423 (15), 2.973 (22), 2.735 (11).

Kind of sample preparation and/or method of registration of the spectrum: KBr disc. Absorption.

Wavenumbers (cm^{-1}): 3520sh, 3447, 3346s, 3252, 3160, 2990sh, 2295w, 1662, 1560sh, 1404w, 879s, 847s, 789, 734s, 507, 467s, 301s.

Note: The spectrum was obtained by N.V. Chukanov.

As319 Boron arsenate $B(AsO_4)$

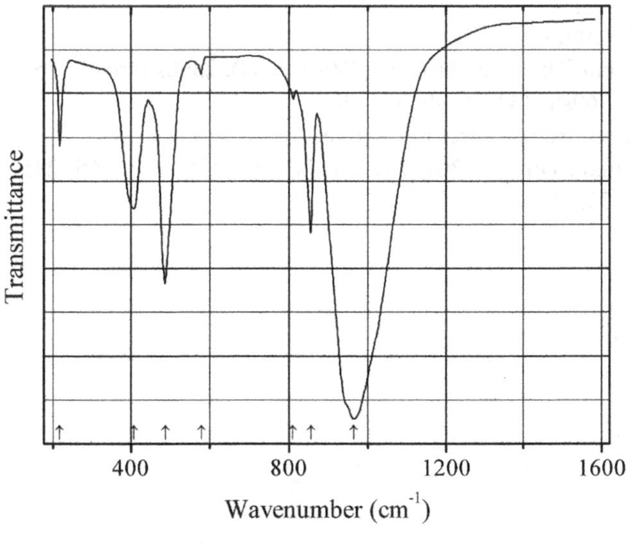

Origin: Synthetic.

Description: Structurally related to cristobalite.

Kind of sample preparation and/or method of registration of the spectrum: KBr and polyethylene discs. Transmission.

Source: Rulmont et al. (1987).

Wavenumbers (cm^{-1}): 965s, 857, 809w, 580w, 488s, 408, 217.

As320 Cesium acid (pentahydrogen) arsenate $CsH_5(AsO_4)_2$

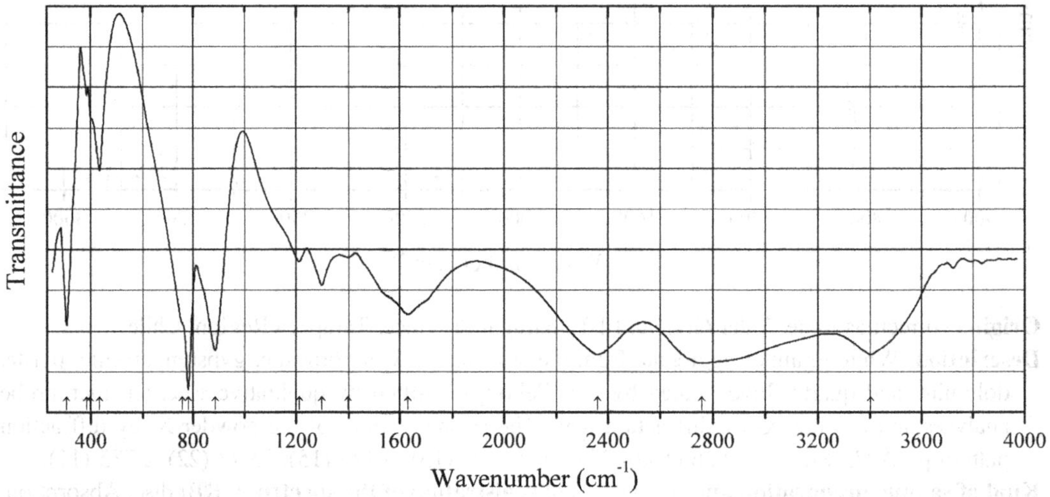

Origin: Synthetic.

Description: Produced from an aqueous stoichiometric solution of cesium carbonate and orthoarsenic acid. The crystal structure is solved. Monoclinic, space group $P2_1/c$, $a = 10.983(1)$, $b = 7.943(1)$, $c = 9.844(1)$ Å, $\beta = 96.15(1)°$, $V = 853.82(6)$ Å^3, $Z = 4$. $D_{calc} = 3.235$ g/cm^3.

Kind of sample preparation and/or method of registration of the spectrum: KBr disc. Transmission.

Source: Naïli et al. (2001).

Wavenumbers (IR, cm^{-1}): 3402s (broad), 2760s (broad), 2360s (broad), 1631 (broad), 1399, 1299, 1212, 887s, 782s, 760sh, 434, 398sh, 384, 306.

Note: In the cited paper, Raman spectrum is given.

Wavenumbers (Raman, cm^{-1}): 830s, 767s, 415s, 370, 335, 290, 265, 235, 200sh, 170w, 125w, 100w, 75, 65, 42, 39, 25.

As321 Cesium iron arsenate $Cs_7Fe_7O_2(AsO_4)_8$ $Cs_7Fe_7O_2(AsO_4)_8$

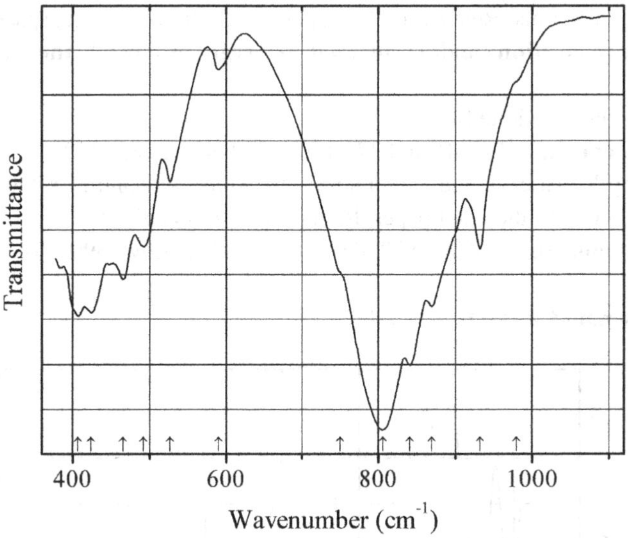

Origin: Synthetic.

Description: Prepared from the mixture of Cs_2CO_3, Fe_2O_3, and $(NH_4)(H_2AsO_4)$ in the molar ratio 3:3:4 heated first at 450 °C for 12 h and then at 800 °C for 10 days. The crystal structure is solved. Monoclinic, space group $P2_1/c$, a = 8.464(2), b = 23.146(5), c = 10.214(3) Å, β = 107.87(2)°, V = 1904.5 Å^3, Z = 2. D_{calc} = 4.298 g/cm^3.

Kind of sample preparation and/or method of registration of the spectrum: KBr disc. Transmission.

Source: Fitouri et al. (2015).

Wavenumbers (cm^{-1}): 980sh, 933, 870, 841s, 806s, 750sh, 590w, 526, 492, 465, 424, 407.

Note: The wavenumbers were partly determined by us based on spectral curve analysis of the published spectrum.

As322 Calcium chlorarsenate $Ca_2(AsO_4)Cl$

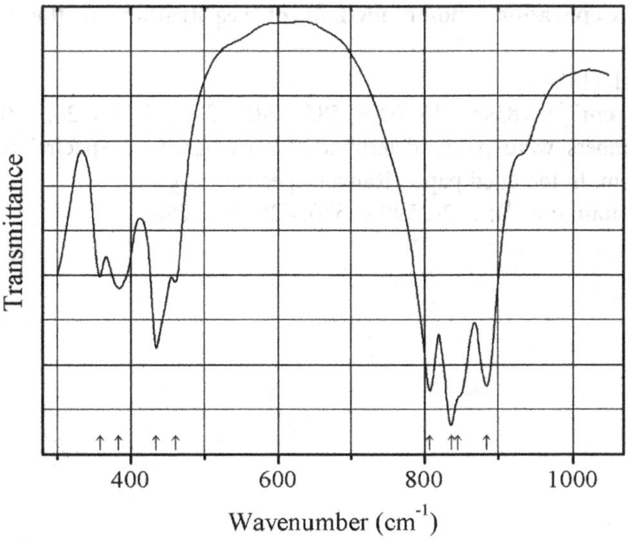

Origin: Synthetic.
Description: Crystals grown from melt by means of a reaction flux technique using As_2O_3, $CaCl_2$, and $CaCO_3$ as starting materials. Related to chlor-spodiosite. Orthorhombic, space group *Pbcm.*
Kind of sample preparation and/or method of registration of the spectrum: KBr disc. Transmission.
Source: Kowalczyk and Condrate Sr (1974).
Wavenumbers (IR, cm^{-1}): 885s, 845sh, 837s, 807s, 460, 434s, 383, 359.
Note: In the paper by Kowalczyk and Condrate Sr (1974) the wavenumber 460 cm^{-1} is erroneously indicated as 470 cm^{-1}. In the cited paper, Raman spectrum is given.
Wavenumbers (Raman, cm^{-1}): 878s, 870, 838s, 813, 497w, 466w, 390, 314.

As323 Calcium arsenate $CaAs_2O_6$ $CaAs_2O_6$

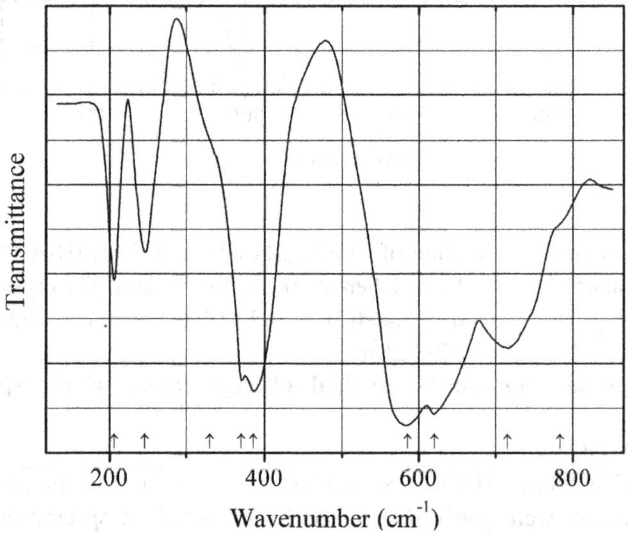

Origin: Synthetic.
Description: Obtained in a solid-state reaction between $CaCO_3$ and As_2O_3. In the crystal structure, AsO_6 octahedra are present. Trigonal, space group *P*-31/*m*, $a = 4.82$, $c = 5.07$ Å.
Kind of sample preparation and/or method of registration of the spectrum: CsI disc. Transmission.
Source: Husson et al. (1984).
Wavenumbers (IR, cm^{-1}): 783sh, 715, 620s, 585s, 385s, 370s, 330sh, 245, 205.
Note: The wavenumbers were partly determined by us based on spectral curve analysis of the published spectrum. In the cited paper, Raman, spectrum is given.
Wavenumbers (Raman, cm^{-1}): 762s, 590w, 570, 426, 397, 286s.

As324 Calcium orthoarsenate trigonal polymorph $Ca_3(AsO_4)_2$

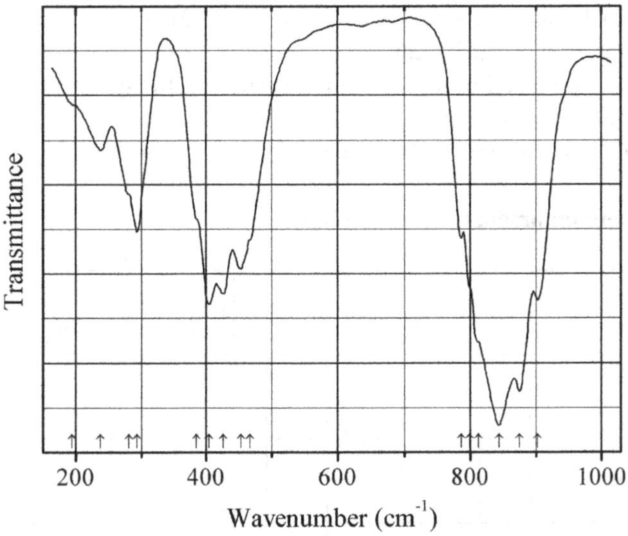

Origin: Synthetic.

Description: Prepared in a solid-state reaction, from a mixture of $CaCO_3$ and As_2O_5, first at 700 °C for 7 h and thereafter (after trituration of the product) at 800 °C for 3 h. Trigonal, space group *R*3*c*, $Z = 7$.

Kind of sample preparation and/or method of registration of the spectrum: CsI disc. Transmission.

Source: Baran (1976).

Wavenumbers (cm^{-1}): 902, 875s, 844s, 812sh, 799sh, 786, 467sh, 453, 426, 404, 385, 294, 281sh, 238, 195w.

As325 Calcium samarium thorium arsenate $CaSmTh(AsO_4)_3$

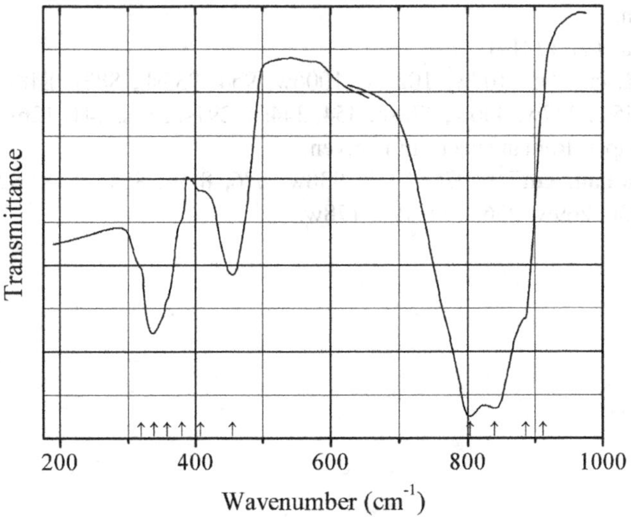

Origin: Synthetic.

Description: Prepared by a standard solid-state method. Structurally related to xenotime-group minerals. Tetragonal, space group $I4_1/amd$, a = 7.175(2), c = 6.409(3) Å, V = 330.0 Å^3. D_{meas} = 5.61 g/cm^3, D_{calc} = 5.63 g/cm^3.

Kind of sample preparation and/or method of registration of the spectrum: CsBr disc. Absorption.

Source: Nabar and Sakhardande (1985).

Wavenumbers (cm^{-1}): 911sh, 886sh, 840s, 803s, 455, 408sh, 380sh, 358sh, 338, 319sh.

As326 Lithium zirconium arsenate $LiZr_2(AsO_4)_3$

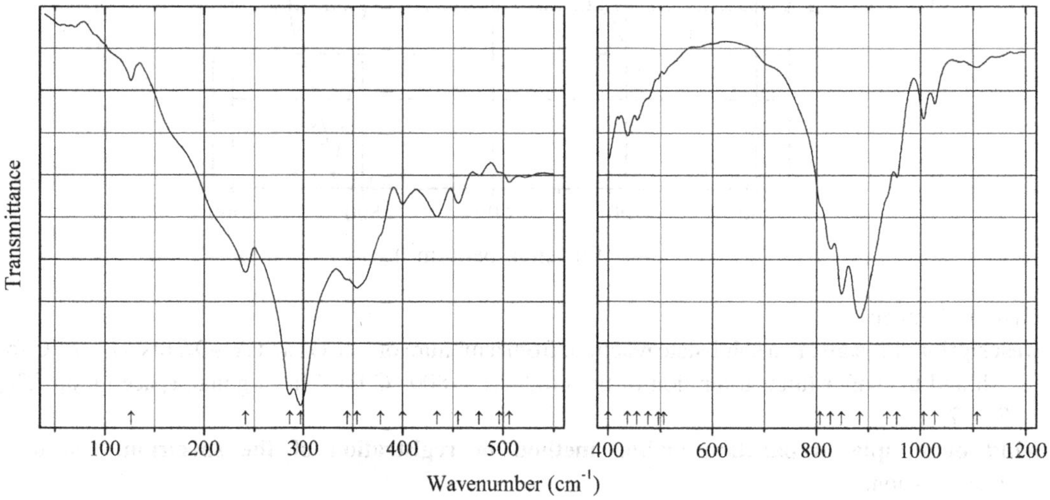

Origin: Synthetic.

Description: Synthesized from stoichiometric amounts of $LiNO_3$, $ZrOCl_2$, and arsenic acid using a precipitation method. In the structure which has a $P112_1/n$ space group, the arsenic atoms occupy three independent positions.

Kind of sample preparation and/or method of registration of the spectrum: KBr and polyethylene discs. Absorption.

Source: Borovikova et al. (2014).

Wavenumbers (IR, cm^{-1}): 1107w, 1027w, 1006w, 954, 935sh, 883s, 848s, 827, 807sh, 506w, 496sh, 476sh, 455w, 437w, 400w, 378sh, 354, 344sh, 297s, 286s, 241, 126w.

Note: In the cited paper, Raman spectrum is given.

Wavenumbers (Raman, cm^{-1}): 976s, 953, 938w, 876, 869s, 854s, 848, 820, 805w, 474, 430, 388, 364, 354, 336, 269w, 256, 230, 194, 178w.

As327 Lithium zirconium arsenate $LiZr_2(AsO_4)_3$

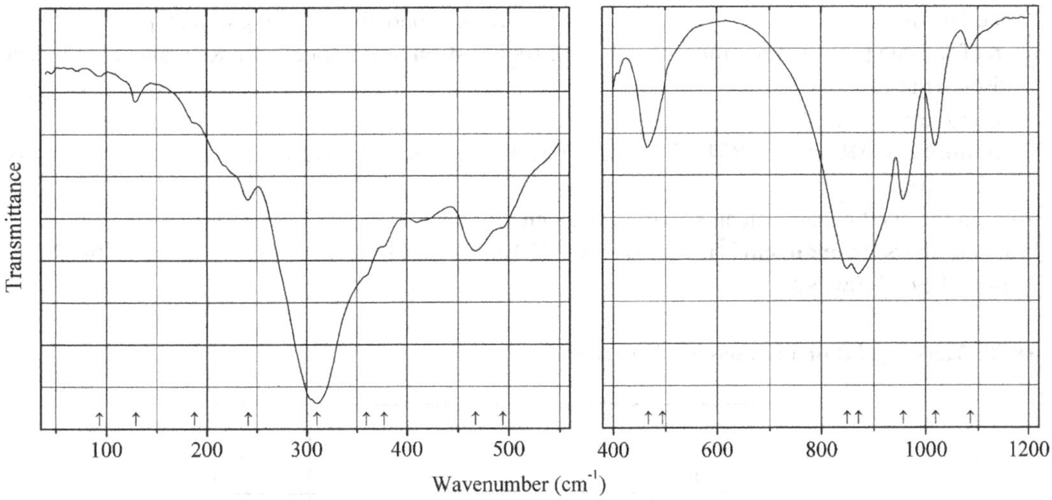

Origin: Synthetic.
Description: Synthesized from stoichiometric amounts of $LiNO_3$, $ZrOCl_2$, and arsenic acid using a precipitation method. Hexagonal, space group *R3c*.
Kind of sample preparation and/or method of registration of the spectrum: KBr and polyethylene discs. Absorption.
Source: Borovikova et al. (2014).
Wavenumbers (IR, cm^{-1}): 1084w, 1018, 956, 870s, 849s, 494sh, 467, 377sh, 359sh, 310s, 241, 187sh, 129w, 93w.
Note: In the cited paper, Raman spectrum is given.
Wavenumbers (Raman, cm^{-1}): 979s, 951, 864s, 857, 473, 445w, 380, 359, 346, 333, 253, 189, 176.

As328 Mercury(I) orthoarsenate $(Hg_2)_3(AsO_4)_2$

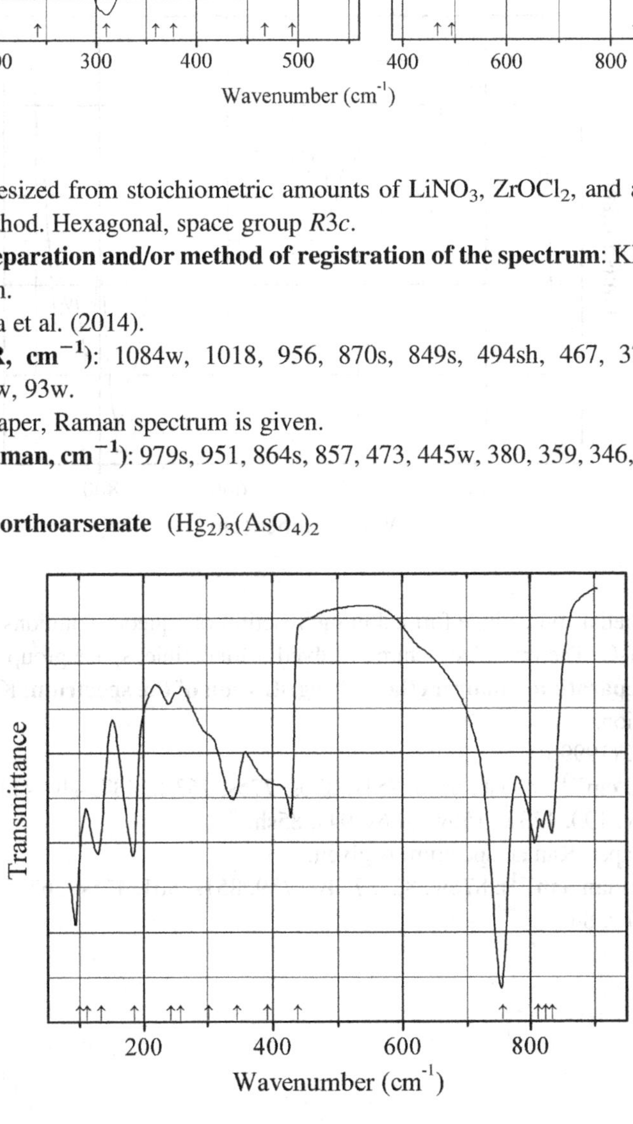

Origin: Synthetic.

Description: Obtained as precipitate formed in the reaction of aqueous solutions of orthoarsenic acid and $Hg_2(NO_3)_2 \cdot 2H_2O$. The crystal structure is solved. Monoclinic, space group $P2_1/c$, $Z = 2$.

Kind of sample preparation and/or method of registration of the spectrum: KBr and polyethylene discs. Transmission.

Source: Baran et al. (1999b).

Wavenumbers (IR, cm^{-1}): 834, 823, 811, 757s, 438, 392sh, 345, 302sh, 258sh, 242w, 185s, 134s, 112sh, 100s.

Note: In the cited paper, Raman spectrum is given.

Wavenumbers (Raman, cm^{-1}): ~845sh, 789, 770sh, 814s, 432s, 390sh, 368w, 312w, 253w, 225w, 148s, 129s, 110w, 98s.

As329 Mercury(II) orthoarsenate $Hg_3(AsO_4)_2$

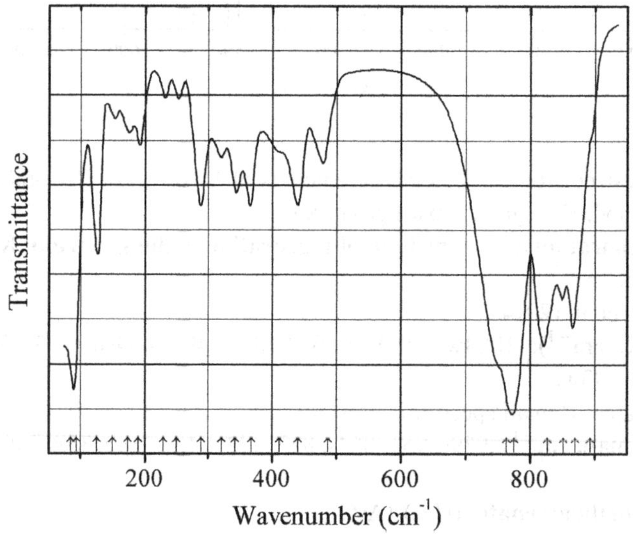

Origin: Synthetic.

Description: Obtained as precipitate formed in the reaction of aqueous solutions of orthoarsenic acid and $Hg(NO_3)_2 \cdot H_2O$. The crystal structure is solved. Monoclinic, space group $P2_1/c$, $Z = 4$.

Kind of sample preparation and/or method of registration of the spectrum: KBr and polyethylene discs. Transmission.

Source: Baran et al. (1999b).

Wavenumbers (IR, cm^{-1}): 893sh, 869s, 851s, 826s, 775s, 763sh, 487, 440, 411sh, 367, 342, 320w, 289, 250w, 230w, 190, 173w, 150w, 126s, 94s, 85sh.

Note: In the cited paper, Raman spectrum is given.

Wavenumbers (Raman, cm^{-1}): 876w, 825, 784w, 759, 851s, 501, 415s, 362, 327, 278, 227, 195sh, 170sh, 140s, 110w, 86.

As330 Potassium antimony oxoarsenate $K_2Sb(AsO_4)O_2$

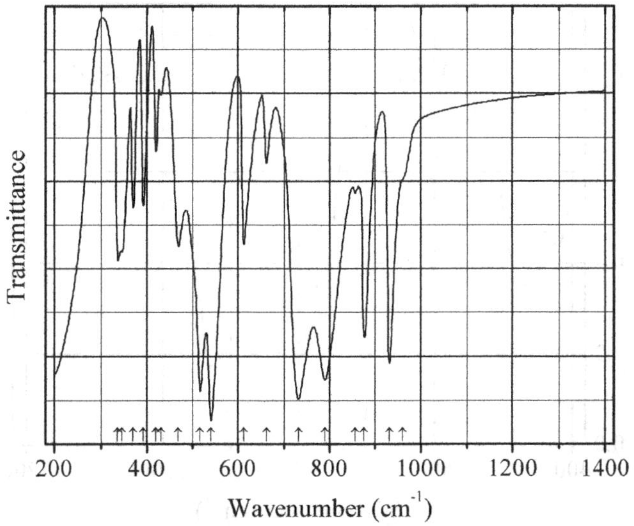

Origin: Synthetic.

Description: Obtained in the solid-state reaction between Sb_2O_3, As_2O_5, and K_2CO_3 at 900 °C for 1 day. Characterized by chemical analyses. The crystal structure is solved. Orthorhombic, space group *Pnma*, $a = 9.603(6)$, $b = 5.972(5)$, $c = 11.304(8)$ Å, $V = 648.27$ Å^3, $Z = 4$. $D_{meas} = 3.76$ (5) g/cm^3, $D_{calc} = 3.79$ g/cm^3.

Kind of sample preparation and/or method of registration of the spectrum: KBr disc. Transmission.

Source: Botto and Garcia (1989).

Wavenumbers (IR, cm^{-1}): 960sh, 932s, 876s, 856w, 790s, 732s, 662, 612, 540s, 526s, 468, 430w, 420, 392, 370, 345, 336.

Note: The wavenumber 516 cm^{-1} is erroneously indicated by Botto and Garcia (1989) as 526 cm^{-1}. In the cited paper, Raman spectrum is given.

Wavenumbers (Raman, cm^{-1}): 984w, 961, 915, 892w, 872s, 644s, 603w, 533s, 508sh, 475, 419w, 404w, 373, 340, 297, 279, 251.

As331 Potassium iron diarsenate (pyroarsenate) $KFe(As_2O_7)$

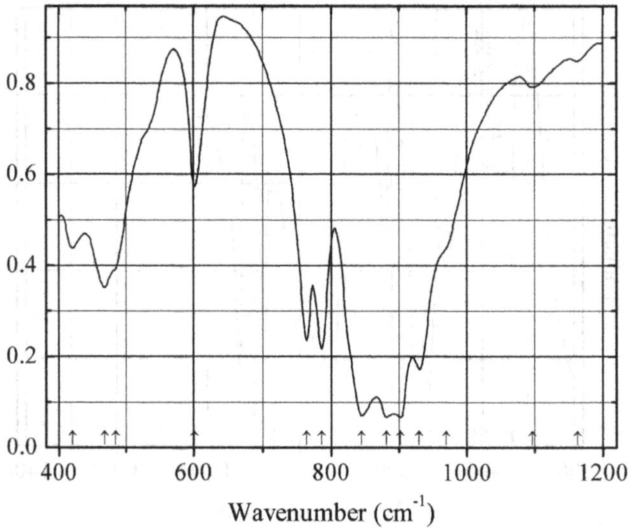

Origin: Synthetic.

Description: Beige single crystals grown from aqueous solution of KNO_3, $Fe(NO_3)_3{\cdot}9H_2O$, and H_3AsO_4 with the molar ratio of 10:1:20, with subsequent heating up to 700 °C in order to avoid volatile products. Characterized by qualitative EDX analysis. The crystal structure is solved. Triclinic, space group *P*-1, $a = 7.662(1)$, $b = 8.402(2)$, $c = 10.100(3)$ Å, $\alpha = 90.42(3)°$, $\beta = 89.74(2)°$, $\gamma = 106.39(2)°$, $V = 623.8(3)$ Å^3, $Z = 4$. $D_{calc} = 3.799$ g/cm^3.

Kind of sample preparation and/or method of registration of the spectrum: KBr disc. Transmission.

Source: Ouerfelli et al. (2007).

Wavenumbers (cm^{-1}): 1163w, 1097w, 970sh, 930s, 902s, 882s, 845sh, 786s, 764s, 601, 530sh, 484sh, 468, 420.

Note: The wavenumbers were partly determined by us based on spectral curve analysis of the published spectrum.

As332 Potassium magnesium arsenate hexahydrate $KMg(AsO_4){\cdot}6H_2O$

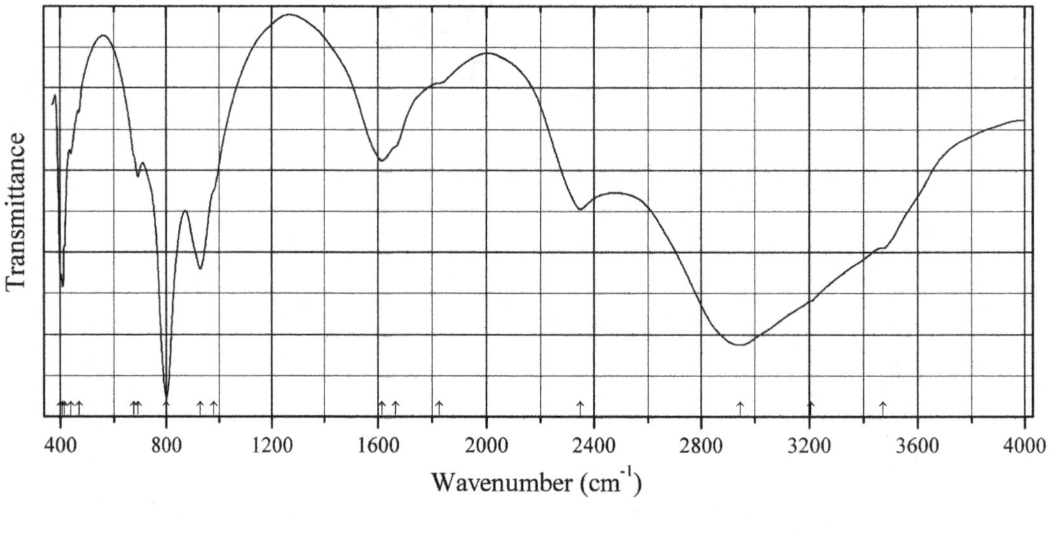

Origin: Synthetic.

Description: Prepared by a simple precipitation procedure of mixing $MgSO_4 \cdot 7H_2O$ and $K(H_2AsO_4)$ solutions at room temperature. Characterized by powder X-ray diffraction data. The crystal structure is solved. Orthorhombic, space group *Pmn*2$_1$, a = 6.99(3), b = 6.22(2), c = 11.26 (4) Å, V = 490.63(3) Å^3, Z = 2. Isostructural with struvite.

Kind of sample preparation and/or method of registration of the spectrum: KBr disc (?). Absorption.

Source: Abdija et al. (2014).

Wavenumbers (IR, cm^{-1}): 3470sh, 3205sh, 2945s, 2348, 1825sh, 1667sh, 1616, 980sh, 930s, 800s, 690, 675sh, 470, 440, 417, 409s, 402s.

Note: In the cited paper, Raman spectrum is given.

Wavenumbers (Raman, cm^{-1}): 3500–2200s, 1760–1500w, 819s, 458w, 414w, 382, 350w.

As333 Potassium manganese arsenate $K_2Mn^{2+}{}_2Mn^{3+}(AsO_4)_3$

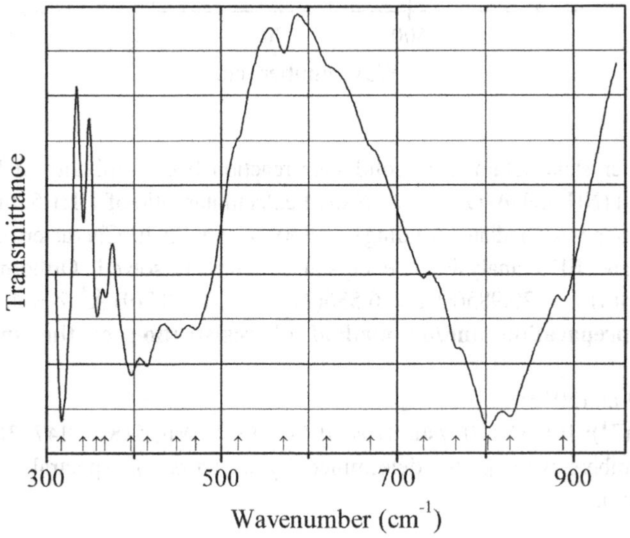

Origin: Synthetic.

Description: Obtained by heating a stoichiometric mixture of manganese oxide, ammonium dihydrogen arsenate, and potassium carbonate first at 400 °C for 4 h and thereafter at 800 °C for 48 h. The crystal structure is solved. Monoclinic, space group *C*2/*c*, a = 12.490(1), b = 13.013(1), c = 6.888(1) Å, β = 114.46(2)°, V = 1019.2(8) Å^3, Z = 4. D_{meas} = 4.28(4) g/cm^3, D_{calc} = 4.30 g/cm^3.

Kind of sample preparation and/or method of registration of the spectrum: KBr disc. Transmission.

Source: Chaalia et al. (2012).

Wavenumbers (cm^{-1}): 888, 828s, 803s, 766sh, 730, 670sh, 621sh, 572w, 520sh, 471, 449, 415, 397, 367, 357, 342, 317s.

Note: The wavenumbers were partly determined by us based on spectral curve analysis of the published spectrum.

As334 Potassium sodium iron arsenate $Na_{2.77}K_{1.52}Fe_{2.57}(AsO_4)_4$ $Na_{2.77}K_{1.52}Fe_{2.57}(AsO_4)_4$

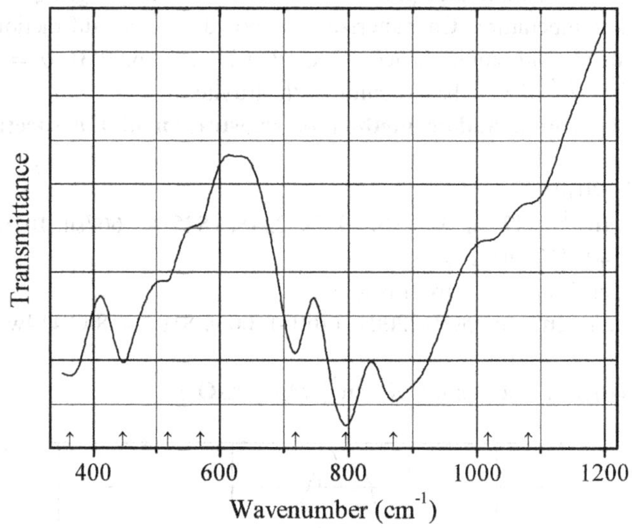

Origin: Synthetic.

Description: Green crystals obtained by solid-state reaction from a mixture of Na_2CO_3, K_2CO_3, Fe $(NO_3)_3{\cdot}9H_2O$, and $(NH_4)(H_2AsO_4)$ with a Na:K:Fe:As molar ratio of 1:1:1:5, first at 400 °C for 24 h and thereafter (after intermediate grinding) at 850 °C for 72 h. Characterized by powder X-ray diffraction data and EDS analysis. The crystal structure is solved. Orthorhombic, space group *Cmce*, $a = 10.854(4)$, $b = 20.985(8)$, $c = 6.536(2)$ Å, $V = 1488.7(9)$ Å^3, $Z = 4$. $D_{calc} = 3.669$ g/cm^3.

Kind of sample preparation and/or method of registration of the spectrum: KBr disc. Transmission.

Source: Ouerfelli et al. (2015).

Wavenumbers (cm^{-1}): 1081sh, 1017sh, 870s, 796s, 718, 570sh, 518sh, 447, 362s.

Note: The wavenumbers were partly determined by us based on spectral curve analysis of the published spectrum.

As335 Potassium zirconium arsenate $KZr_2(AsO_4)_3$

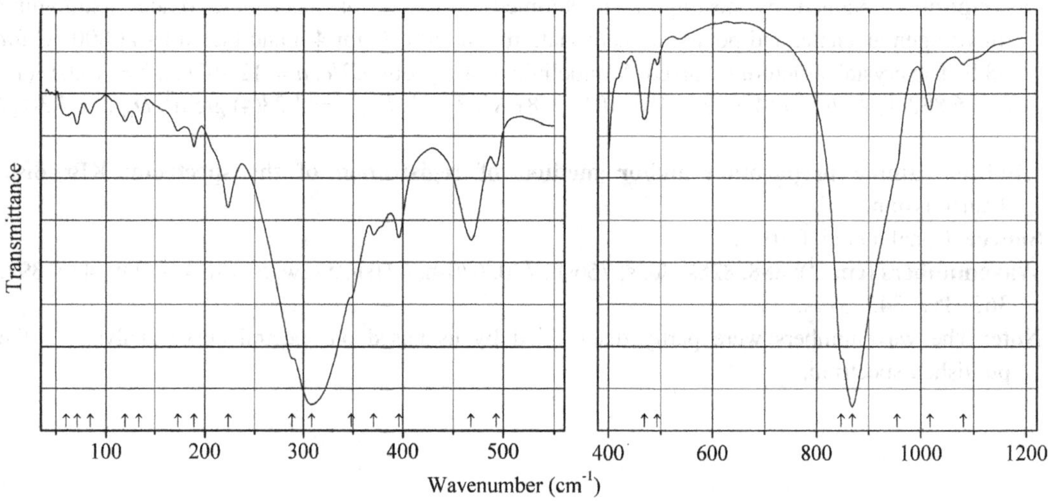

Origin: Synthetic.

Description: Colorless polycrystalline powder obtained by evaporation of aqueous solution containing stoichiometric amounts of KNO_3, $ZrOCl_2 \cdot 8H_2O$, and H_3AsO_4 at 90 °C, drying at 270 °C and sintering at 600 and 850–950 °C with intermediate grindings. Characterized by powder X-ray diffraction data. Trigonal, space group *R*-3*c*.

Kind of sample preparation and/or method of registration of the spectrum: KBr and polyethylene discs. Absorption.

Source: Borovikova et al. (2014).

Wavenumbers (IR, cm^{-1}): 1080w, 1017, 954sh, 868s, 847sh, 493w, 468, 396, 371, 348sh, 308s, 288sh, 223, 189w, 173w, 133w, 119w, 84w, 71w, 60w.

Note: In the cited paper, Raman spectrum is given.

Wavenumbers (Raman, cm^{-1}): 982s, 949, 862s, 857s, 842, 468w, 437, 381w, 358, 255, 237.

As336 Scandium arsenate monohydrate $Sc(AsO_4) \cdot H_2O$

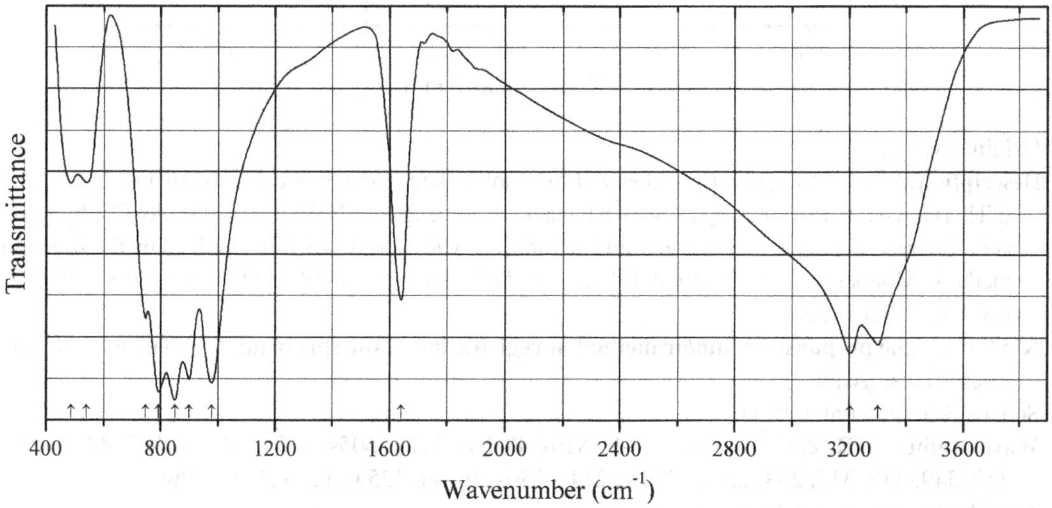

Origin: Synthetic.

Description: Colorless platy crystals prepared from Sc_2O_3, hydrated arsenic acid, and Li_2CO_3 by hydrothermal synthesis at 493 K for 7 days. Triclinic, space group *P*-1, *Z* = 2.

Kind of sample preparation and/or method of registration of the spectrum: KBr disc. Transmission.

Source: Baran et al. (2006).

Wavenumbers (IR, cm^{-1}): 3300s, 3203s, 1641, 978s, 899s, 849s, 793s, 746, 540, 486.

Note: In the cited paper, Raman spectrum is given.

Wavenumbers (Raman, cm^{-1}): 1638, 935s, 866s, 832s, 805, 744w, 484, 385, 347, 323w, 287, 244w, 188w, 167, 138.

As337 Sodium indium arsenate (alluaudite-type) $Na_3In_2(AsO_4)_3$

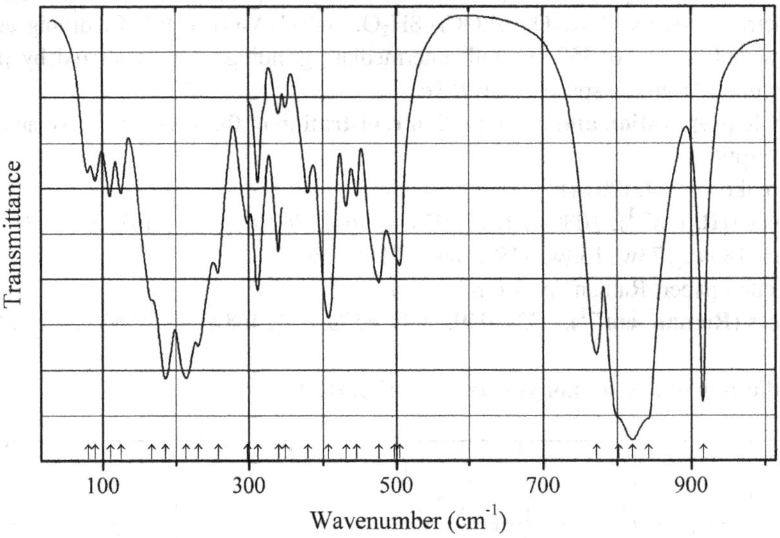

Origin: Synthetic.

Description: $Na_3In_2(AsO_4)_3$ was synthesized by a solid-state reaction between $NaHCO_3$, In_2O_3, and $(NH_4)(H_2AsO_4)$, as well as by a chemical attack of the reagents ($NaHCO_3$, In_2O_3, As_2O_3) by nitric acid. Characterized by powder X-ray diffraction data. The crystal structure is solved by the Rietveld method. Monoclinic, space group *C*2/*c*, $a = 12.6025(1)$, $b = 13.1699(1)$, $c = 6.8335(1)$ Å, $\beta = 113.7422(5)°$, $Z = 4$.

Kind of sample preparation and/or method of registration of the spectrum: KBr and polyethylene discs. Transmission.

Source: Khorari et al. (1997).

Wavenumbers (IR, cm^{-1}): 917s, 842sh, 820s, 802sh, 772s, 505s, 497sh, 477s, 447, 432, 407s, 379, 349, 340, 312, 298, 257w, 230w, 214s, 186s, 167sh, 125w, 110w, 90w, 80w.

Note: In the cited paper, Raman spectrum is given.

Wavenumbers (Raman, cm^{-1}): 934s, 907w, 990, 866s, 855, 837, 832sh, 806sh, 796, 773, 480s, 472sh, 419, 397, 378w, 369w, 354w, 328, 284, 271, 231w, 168, 149, 133, 122, 96w, 90w, 86w, 78.

As338 Sodium lead neodymium arsenate chloride (apatite-type) $Na_2Pb_6Nd_2(AsO_4)_6Cl_2$

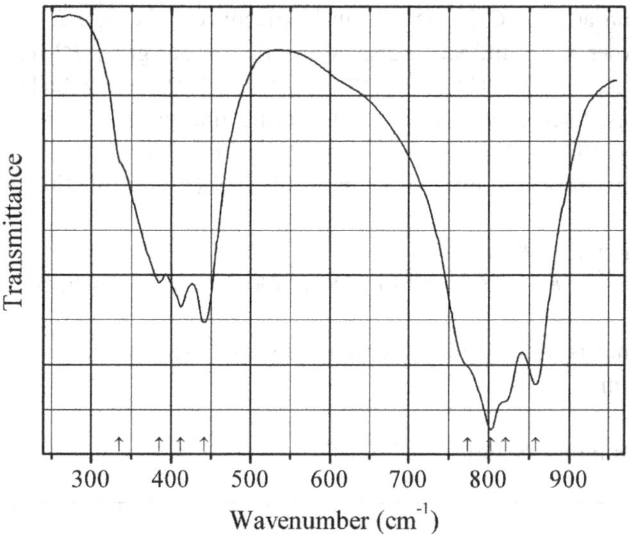

Origin: Synthetic.

Description: Prepared by stepwise heating of the mixture of NaCl, PbO, Nd_2O_3, and As_2O_5, first at 350 °C and thereafter (after intermediate grindings) at 650, 800, and 850 °C for 2 h at each temperature. Characterized by powder X-ray diffraction data. Hexagonal, space group $P6_3/m$, $a = 10.08(1)$, $c = 7.21(1)$ Å, $V = 634.4$ Å^3, $Z = 1$. $D_{meas} = 6.3$ g/cm^3, $D_{calc} = 6.49$ g/cm^3. The strongest lines of the powder X-ray diffraction pattern [*d*, Å (*I*, %) (*hkl*)] are: 4.12 (18) (111), 3.60 (19) (002), 3.294 (27) (210), 2.999 (100) (211), 2.929 (44) (112), 2.906 (53) (300), 1.943 (19) (213), 1.928 (18) (321).

Kind of sample preparation and/or method of registration of the spectrum: KBr disc. Transmission.

Source: Escobar and Baran (1982).

Wavenumbers (IR, cm^{-1}): 858s, 822sh, 803s, 773sh, 441, 412, 385, 335sh.

Note: In the cited paper, Raman spectrum is given.

Wavenumbers (Raman, cm^{-1}): 850, 822s, 800, 772, 410, 330.

As339 Sodium nickel iron(III) arsenate $NaNiFe_2(AsO_4)_3$

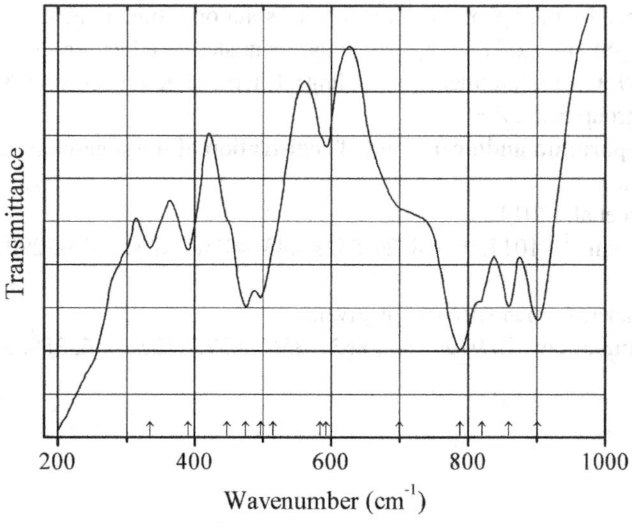

Origin: Synthetic.

Description: Crystals obtained from a stoichiometric mixture of Na_2CO_3, NiO, Fe_2O_3, and As_2O_5 by solid-state reaction at 800 °C for 15 h with intermediate grindings. Characterized by chemical analyses and powder X-ray diffraction data. Monoclinic, space group $P2_1/c$, $a = 7.06(1)$, $b = 9.38(1)$, $c = 19.63(1)$ Å, $\beta = 114.2(1)°$, $V = 1186(2)$ Å^3, $Z = 4$. $D_{meas} = 3.40(1)$ g/cm^3, $D_{calc} = 3.42$ g/cm^3. The strongest lines of the powder X-ray diffraction pattern [*d*, Å (*I*, %) (*hkl*)] are: 5.63 (56) (−111), 2.847 (98) (−204), 2.815 (100) (−222), 2.654 (56) (220), 2.263 (57) (−227).

Kind of sample preparation and/or method of registration of the spectrum: KBr disc. Transmission.

Source: Augsburger et al. (1992).

Wavenumbers (cm^{-1}): 901s, 860s, 820sh, 789s, 700sh, 592w, 584sh, 515sh, 496, 474, 447sh, 390, 334.

Note: The wavenumbers were partly determined by us based on spectral curve analysis of the published spectrum.

As340 Sodium zirconium arsenate $NaZr_2(AsO_4)_3$

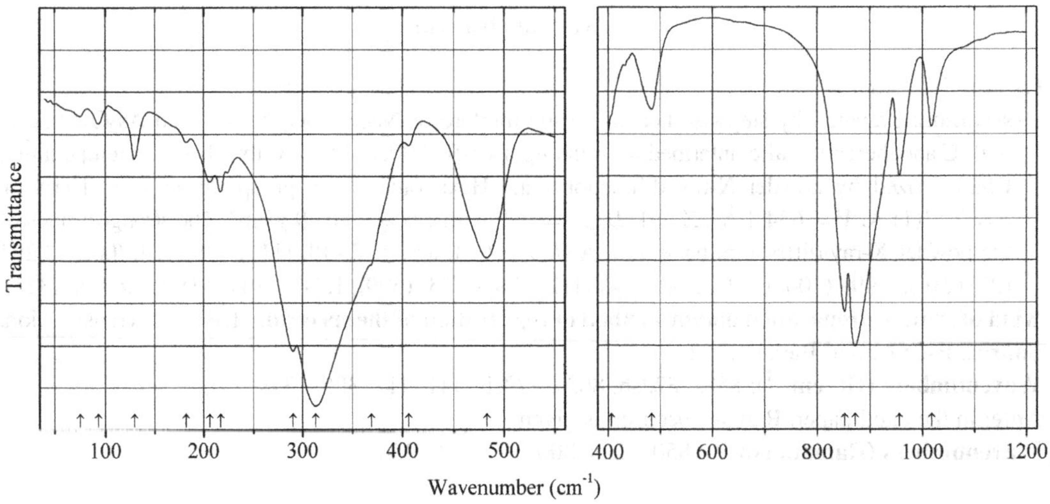

Origin: Synthetic.

Description: Obtained by precipitation from aqueous solutions containing stoichiometric amounts of $NaNO_3$ + $ZrOCl_2{\cdot}8H_2O$ and H_3AsO_4 with subsequent stepwise heating the precipitate at 90, 270, 600, and 850–950 °C with intermediate grindings. Characterized by powder X-ray diffraction data. Trigonal, space group *R*-3*c*, Z = 6.

Kind of sample preparation and/or method of registration of the spectrum: KBr and polyethylene discs. Absorption.

Source: Borovikova et al. (2014).

Wavenumbers (IR, cm^{-1}): 1017, 955, 872s, 852s, 483, 406w, 369sh, 313s, 290, 217w, 206w, 182w, 129w, 93w, 75w.

Note: In the cited paper, Raman spectrum is given.

Wavenumbers (Raman, cm^{-1}): 979s, 948, 863s, 856s, 837, 472w, 446, 389, 363, 340, 256.

As341 Sodium zirconium arsenate $NaZr_2(AsO_4)_3$

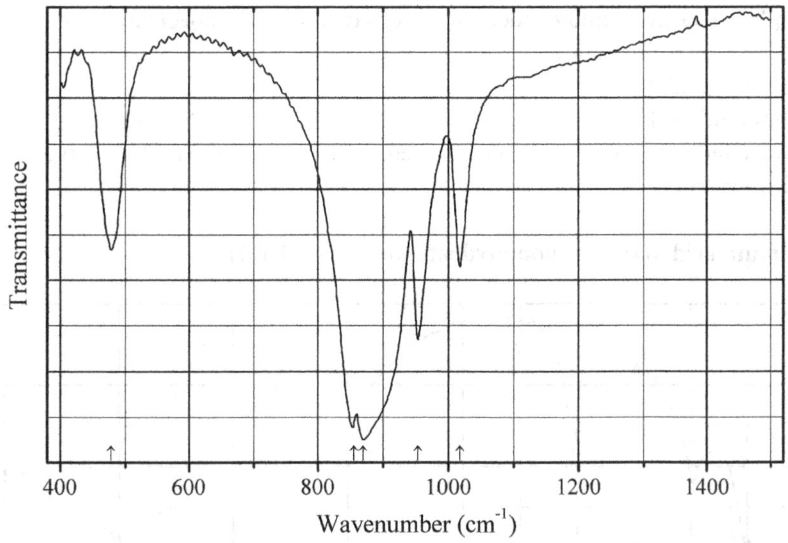

Origin: Synthetic.

Description: Obtained by evaporation of an aqueous solutions containing stoichiometric amounts of $NaNO_3$, $ZrOCl_2{\cdot}8H_2O$, and $(NH_4)(H_2AsO_4)$ and subsequent stepwise heating the resulting powder up to 800 °C with intermediate grindings. Characterized by powder X-ray diffraction data. The crystal structure is solved. Trigonal, space group *R*-3*c*, a = 9.1518(2), c = 23.1097(4) Å, V = 1676.26(1) Å^3, Z = 6.

Kind of sample preparation and/or method of registration of the spectrum: KBr disc. Transmission.

Source: Chakir et al. (2003).

Wavenumbers (IR, cm^{-1}): 1017, 953, 870s, 855s, 478.

Note: In the cited paper, Raman spectrum is given.

Wavenumbers (Raman, cm^{-1}): 978.5s, 949, 863s, 837, 470w, 446, 390w, 362w, 339, 300w, 255, 236w, 179w, 153w, 115, 63.

As342 Tantalum oxyarsenate $Ta(AsO_4)O$

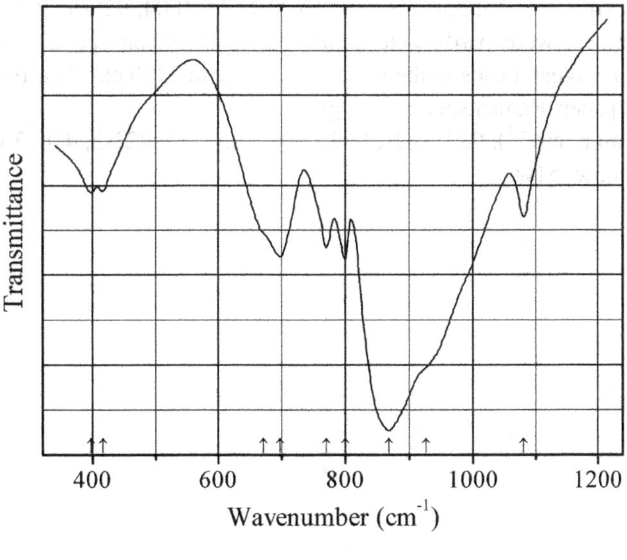

Origin: Synthetic.

Description: Orthorhombic, $a = 11.57$, $b = 5.31$, $c = 6.66$ Å.

Kind of sample preparation and/or method of registration of the spectrum: Transmission. Kind of sample preparation is not indicated.

Source: Chahboun et al. (1988).

Wavenumbers (cm^{-1}): 1082, 927sh, 868s, 800, 770, 698, 672sh, 416, 398.

Note: The wavenumbers were partly determined by us based on spectral curve analysis of the published spectrum.

As343 Zirconium acid arsenate monohydrate α-$Zr(HAsO_4)_2{\cdot}H_2O$

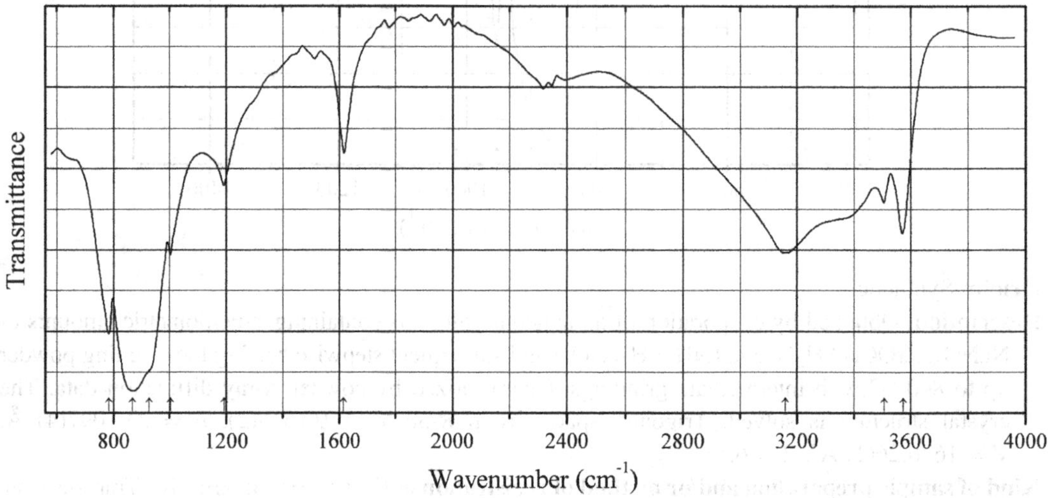

Origin: Synthetic.

Description: Prepared in the reaction of zirconyl chloride with a mixture of arsenic and hydrochloric acids with subsequent refluxing and drying at 110 °C. Characterized by DSC and powder X-ray diffraction data. Monoclinic, $a = 9.146(1)$, $b = 5.381(5)$, $c = 16.61(2)$ Å, $\beta = 111.5°$.

Kind of sample preparation and/or method of registration of the spectrum: KBr disc. Transmission.

Source: Slade et al. (1997).

Wavenumbers (IR, cm^{-1}): 3573, 3508, 3160s, 1615, 1192, 1004, 926sh, 868s, 784.

Note: The wavenumbers were partly determined by us based on spectral curve analysis of the published spectrum. Weak bands in the range from 2320 to 2380 cm^{-1} correspond to atmospheric CO_2. In the cited paper, Raman spectrum is given.

Wavenumbers (Raman, cm^{-1}): 983, 882s, 869s, 808, 785, 775, 423sh, 410, 363w, 326, 287, 165w, 136, 107w, 78w, 65w, 51w.

As344 Allactite $Mn^{2+}{}_7(AsO_4)_2(OH)_8$

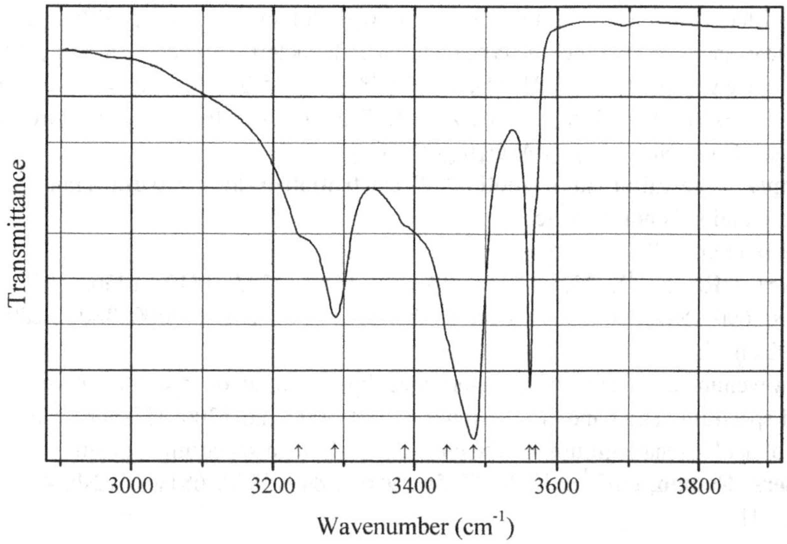

Origin: Långban deposit, Bergslagen ore region, Filipstad district, Värmland, Sweden.
Description: Red crystals from the association with native lead, calcite, dolomite, domeykite, and pyrochroite. The crystal structure is solved for two prismatic crystals. Monoclinic, space group $P2_1/n$, a = 5.482–5.5225, b = 12.153–12.276, c = 10.014–10.123 Å, β = 95.55–95.63°, Z = 2. D_{calc} = 3.856 g/cm^3. Optically biaxial (−), α = 1.554(2), β = 1.558(2), γ = 1.566(2), $2V$ = 70(5)°. The empirical formula is (electron microprobe): $Mn^{2+}{}_{6.73}Ca_{0.13}Mg_{0.12}Zn_{0.02})As^{5+}{}_{2.00}O_{16}H_8$.
Kind of sample preparation and/or method of registration of the spectrum: KBr disc. Absorption.
Source: Gatta et al. (2016).
Wavenumbers (cm^{-1}): 3570sh, 3562, 3484s, 3446sh, 3387sh, 3288, 3236sh.

As345 Arhbarite $Cu_2Mg(AsO_4)(OH)_3$

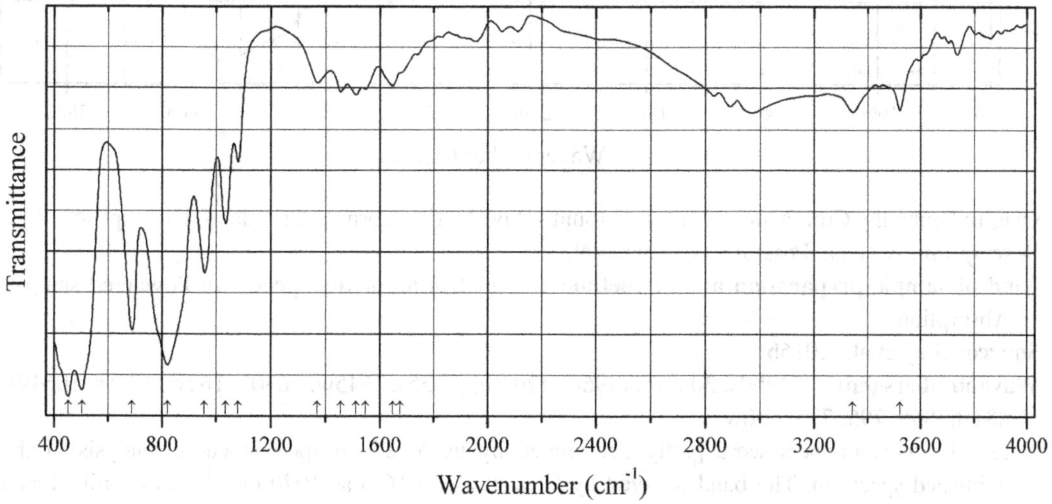

Origin: Aghbar Mine (Arhbar Mine), Aghbar, Bou Azzer District, Tazenakht, Ouarzazate Province, Morocco (type locality).

Description: Blue aggregates from the association with dolomite, hematite, löllingite, pharmacolite, erythrite, talc, and mcguinessite (the type material). Characterized by powder X-ray diffraction data. Triclinic, space group *P*1, $a = 5.315(4)$, $b = 5.978(6)$, $c = 5.030(6)$ Å, $\alpha = 113.58(6)°$, $\beta = 97.14(7)°$, $\gamma = 89.30(8)°$, $V = 145.2(1)$ Å^3, $Z = 1$. The empirical formula is (electron microprobe): $Cu_{1.98}(Mg_{0.88}Cu_{0.09}Ni_{0.01}Co_{0.01})(AsO_4)_{1.02}(OH)_{2.92}$.

Kind of sample preparation and/or method of registration of the spectrum: Transmission. Kind of sample preparation is not indicated.

Source: Krause et al. (2003).

Wavenumbers (IR, cm^{-1}): 3520, 3345, 1676sh, 1647, 1547, 1512, 1456, 1371, 1081, 1036, 958, 820s, 690, 504s, 453s, as well as bands in the ranges 3600–3800, 2800–3200, and 1800–2200 cm^{-1}.

Note: The wavenumbers were partly determined by us based on spectral curve analysis of the published spectrum. The bands in the ranges 1800–3000 and 1300–1500 cm^{-1} indicate possible presence of acid arsenate groups. In the cited paper, Raman spectrum is given.

Wavenumbers (Raman, cm^{-1}): 3525, 3355w, 956w, 840s, 820, 681w, 512sh, 489, 464s, 409w, 331, 318, 141.

As346 Bettertonite $Al_6(AsO_4)_3(OH)_9 \cdot 16H_2O$

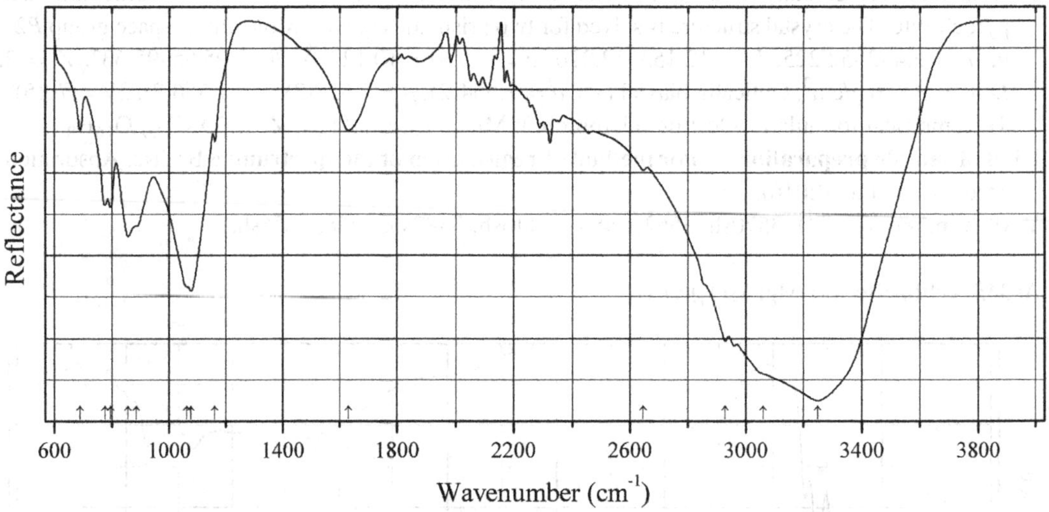

Origin: Penberthy Croft Mine, St Hilary, Mount's Bay District, Cornwall, England, UK (type locality).

Description: No data. Possibly, type material.

Kind of sample preparation and/or method of registration of the spectrum: Powdered sample. Absorption.

Source: Grey et al. (2016b).

Wavenumbers (cm^{-1}): 3250s, 3060sh, (2960), (2930), (2855), 2645w, 1630, 1162w, 1078s, 1064sh, 884sh, 855, 795, 775, 690w.

Note: The wavenumbers were partly determined by us based on spectral curve analysis of the published spectrum. The band denoted by Grey et al. (2016b) as 1070 cm^{-1} was identified as a doublet (1078+1064 cm^{-1}).

As347 Canosioite $Ba_2Fe^{3+}(AsO_4)_2(OH)$

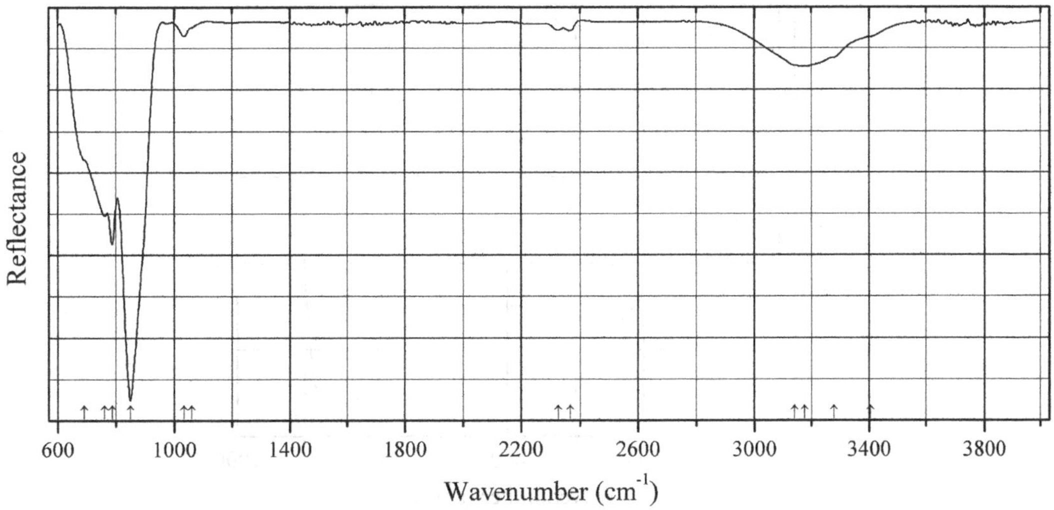

Origin: Valletta mine, Maira Valley, Cuneo Province, Piedmont, Italy (type locality).

Description: Reddish-brown granules from the association with aegirine, baryte, calcite, hematite, Mn-bearing muscovite, as well as unidentified Mn oxides and arsenates. Holotype sample. The crystal structure is solved. Monoclinic, space group $P2_1/m$, a = 7.8642(4), b = 6.1083(3), c = 9.1670(5) Å, β = 112.874(6)°, V = 405.73(4) Å^3, Z = 2. D_{calc} = 4.943 g/cm^3. Optically biaxial (+), $2V$ = 84(2)°. The empirical formula is (electron microprobe): $(Ba_{1.92}Pb_{0.05}Sr_{0.02}Na_{0.01})(Fe^{3+}{}_{0.52}Mn^{3+}{}_{0.29}Al_{0.16}Mg_{0.06})[(As_{0.64}V_{0.36})O_4]_2(OH)_{0.92}F_{0.01}\cdot0.07H_2O$. The strongest lines of the powder X-ray diffraction pattern [d, Å (I, %) (hkl)] are: 3.713 (18) (111), 3.304 (100) (21–1), 3.058 (31) (020), 3.047 (59) (10–3), 2.801 (73) (112), 2.337 (24) (220), 2.158 (24) (12–3).

Kind of sample preparation and/or method of registration of the spectrum: Reflection using IR microscope. Kind of sample preparation is not indicated.

Source: Cámara et al. (2016a).

Wavenumbers (IR, cm^{-1}): 3405sh, 3278sh, 3175, 3139sh, (2366), (2326), 1061sh, 1035, 849s, 787s, 762, 692sh.

Note: The wavenumbers were partly determined by us based on spectral curve analysis of the published spectrum. Weak bands in the range from 2300 to 2400 cm^{-1} correspond to atmospheric CO_2. In the cited paper, Raman spectrum is given.

Wavenumbers (Raman, cm^{-1}): 896s, 862s, 838s, 820sh, 779w, 719sh, 686, 595w, 507, 478, 457, 368s, 326, 282, 234, 187, 163, 147, 133.

As348 Cheralite La-bearing $CaLaTh(AsO_4)_3$

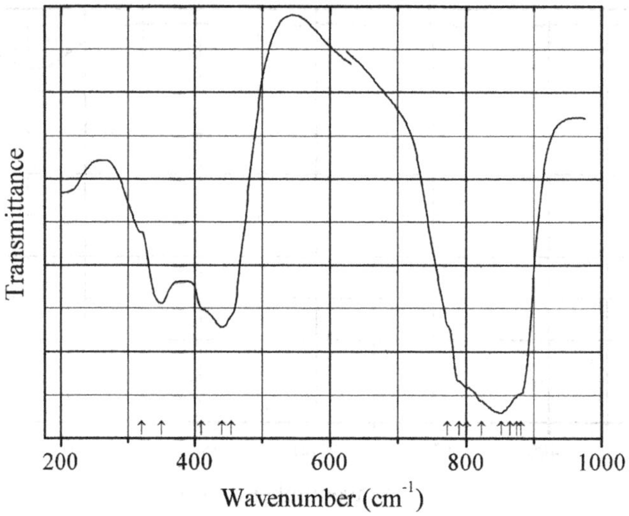

Origin: Synthetic.

Description: Prepared using a solid-state reaction technique. Characterized by powder X-ray diffraction data. Monoclinic, isostructural with monazite, space group $P2_1/m$, $a = 6.883(5)$, $b = 7.070(6)$, $c = 6.674(7)$ Å, $\beta = 104.74(8)°$, $V = 314.1$ Å^3, $Z = 4$. $D_{meas} = 5.80$ g/cm^3, $D_{calc} = 5.83$ g/cm^3.

Kind of sample preparation and/or method of registration of the spectrum: CsBr disc. Transmission.

Source: Nabar and Sakhardande (1985).

Wavenumbers (cm^{-1}): 880sh, 874sh, 864sh, 851s, 823sh, 801sh, 790sh, 772sh, 454sh, 440s, 410sh, 350, 320sh.

As349 Fetiasite $(Fe^{2+},Fe^{3+},Ti^{4+})_3(As^{3+}{}_2O_5)O_2$

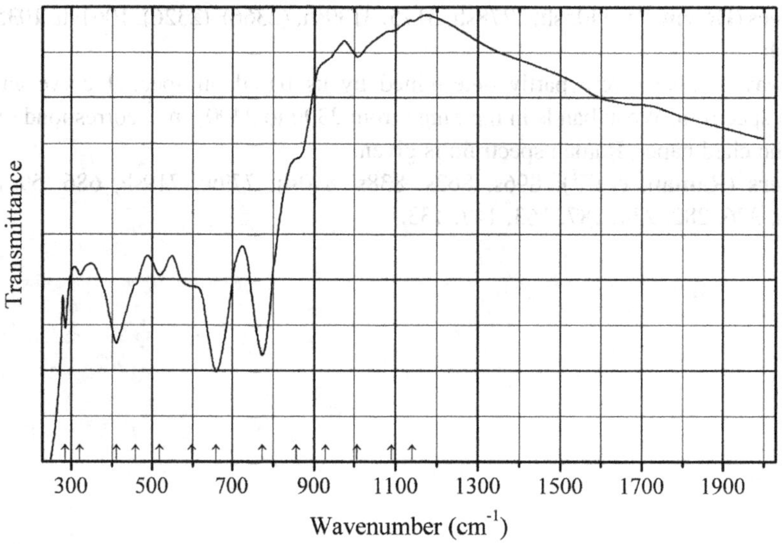

Origin: Cervandone Mt., Val Devero, Baceno, Verbano-Cusio-Ossola province, Piedmont, Italy (type locality).

Description: Brown to black aggregates from the association with asbecasite, cafarsite, cervandonite, etc. Holotype sample. The crystal structure is solved. Monoclinic, space group $P2_1/m$, $a = 10.614(2)$, $b = 3.252(1)$, $c = 8.945(1)$ Å, $\beta = 108.95(2)°$, $V = 291.9(2)$ Å^3, $Z = 2$. $D_{meas} = 4.6$ g/cm^3, $D_{calc} = 4.76–4.80$ g/cm^3. The empirical formula is (electron microprobe): $(Fe^{2+}_{1.38}Fe^{3+}_{0.92}Ti_{0.54}Mn_{0.08})(As^{3+}_2O_5)O_2$. The strongest lines of the powder X-ray diffraction pattern [*d*, Å (*I*, %) (*hkl*)] are: 2.985 (67) (−103), 2.811 (94) (202, 301), 2.749 (100) (−211, 210), 2.391 (85) (112), 1.779 (48) (−504, –511), 1.709 (35) (510, –603).

Kind of sample preparation and/or method of registration of the spectrum: Transmission. Kind of sample preparation is not indicated.

Source: Graeser et al. (1994).

Wavenumbers (cm^{-1}): 1142w, 1090sh, 1007w, 929sh, 856sh, 774s, 659s, 599sh, 519w, 459sh, 410, 322w, (286).

Note: The wavenumbers were determined by us based on spectral curve analysis of the published spectrum. No data on absorptions above 2000 cm^{-1} are given. Consequently, the presence of OH groups in fetiasite cannot be excluded.

As350 Johnbaumite Sr-analogue $Sr_5[(AsO_4)_2(PO_4)](OH)$

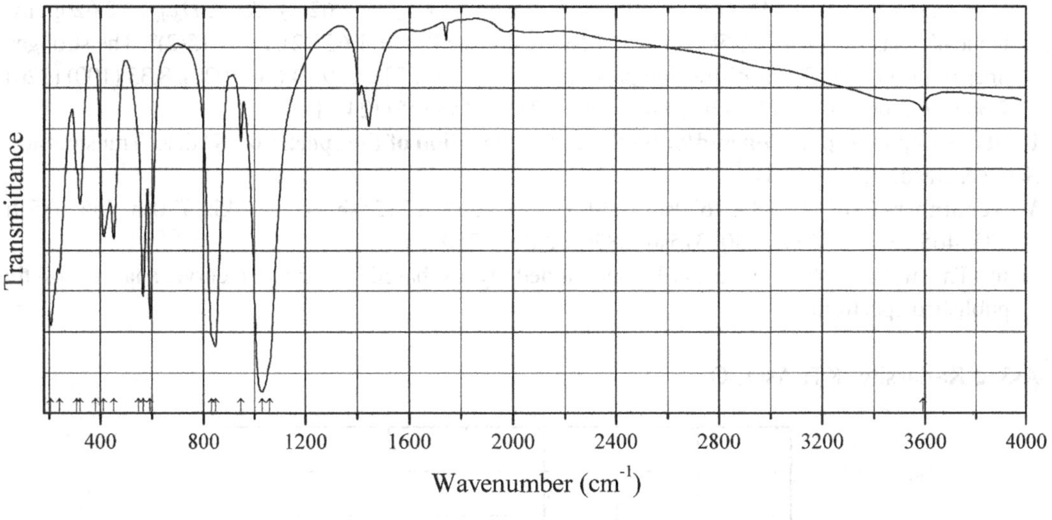

Origin: Synthetic.

Description: Apatite-type compound prepared from aqueous solutions of $Sr(NO_3)_2$, $(NH_4)_2(HPO_4)$, and $Na_3(AsO_4)·H_2O$ and $(NH_4)(OH)$ with subsequently heating a precipitate first at 100 °C for 2 h and thereafter at 850 °C for 2 h. Characterized by powder X-ray diffraction data.

Kind of sample preparation and/or method of registration of the spectrum: KBr disc. Transmission.

Source: Galera-Gomez et al. (1982).

Wavenumbers (cm^{-1}): 3593w, 1061sh, 1030s, 946, 848s, 832sh, 593s, 565s, 548sh, 452, 413, 380, 320, 307, (240), (207s), and bands in the range of 1400–1500 cm^{-1}.

Note: The wavenumbers were partly determined by us based on spectral curve analysis of the published spectrum. Bands in the range of 1400–1500 cm^{-1} correspond to carbonate groups.

As351 Kaatialaite $Fe^{3+}(H_2AsO_4)_3 \cdot 5H_2O$

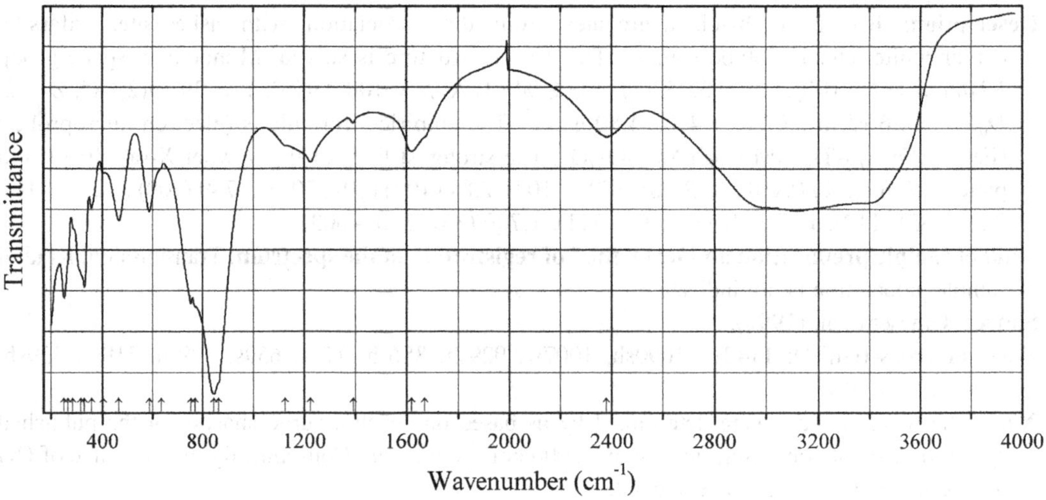

Origin: Synthetic.

Description: Greenish blue aggregates. Monoclinic, space group $P2_1$ or $P2_1/m$, a = 15.363(5), b = 19.844(5), c = 4.736(2) Å, β = 91.77(3)°, Z = 4. D_{meas} = 2.62(3) g/cm³, D_{calc} = 2.62 g/cm³. Optically biaxial (+), α = 1.581(2), β = 1.582 (calculated), γ = 1.625(2), $2V$ = 15(2)°. The strongest lines of the powder X-ray diffraction pattern [d, Å (I, %) (hkl)] are: 9.94 (50) (020), 8.33 (100) (120), 7.68 (70) (200), 6.08 (40) (130), 3.410 (40) (231), 3.153 (45) (24–1).

Kind of sample preparation and/or method of registration of the spectrum: KI disc. Transmission.

Source: Raade et al. (1984).

Wavenumbers (cm^{-1}): 2380, 1670sh, 1620, 1390w, 1225w, 1125sh, 865sh, 845s, 770sh, 755, 635sh, 585, 465, 403sh, 358w, 330, 315sh, 285sh, 265sh, 250.

Note: The wavenumbers were partly determined by us based on spectral curve analysis of the published spectrum.

As352 Katiarsite $KTi(AsO_4)O$

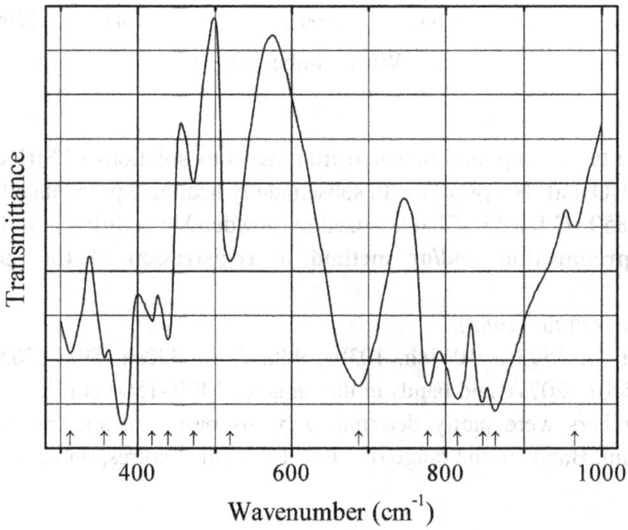

Origin: Synthetic.

Description: Prepared by heating a mixture of K_2CO_3/KNO_3, TiO_2, and $(NH_4)(H_2AsO_4)$ powders taken in stoichiometric amounts first at 500 °C for 12 h and thereafter at 900 °C for 24 h with intermediate grinding. Characterized by powder X-ray diffraction data. Orthorhombic, space group $Pna2_1$, $a = 12.815(8)$, $b = 6.402(4)$, $c = 10.589(6)$ Å, $V = 868.7$ Å^3.

Kind of sample preparation and/or method of registration of the spectrum: KBr disc. Transmission.

Source: Rangan et al. (1993).

Wavenumbers (cm^{-1}): 965w, 864s, 847s, 815s, 777s, 687s, 520, 473w, 440, 419, 381s, 357, 313.

Note: The wavenumbers were determined by us based on spectral curve analysis of the published spectrum.

As353 Lemanskiite $NaCaCu_5(AsO_4)_4Cl{\cdot}5H_2O$

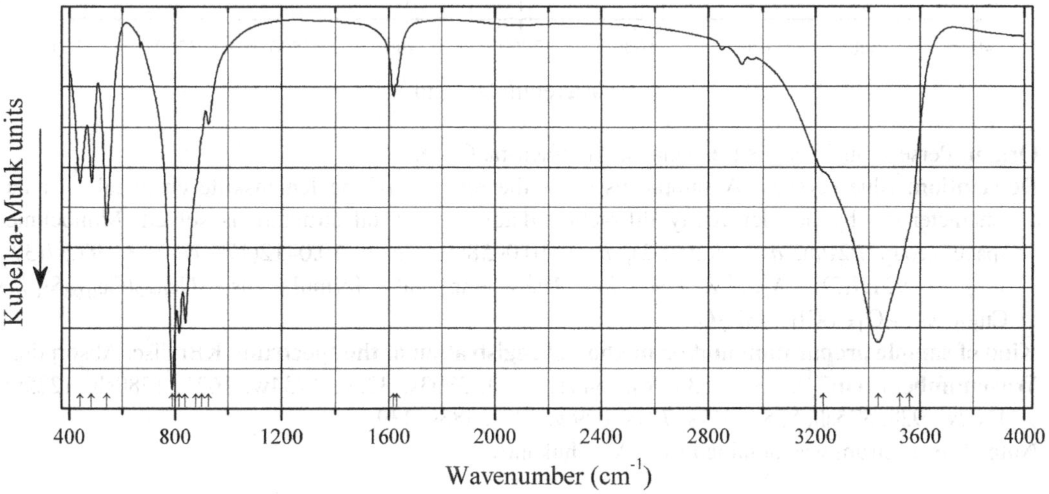

Origin: Abundancia mine, El Guanaco miningdistrict, Region II, Antofagasta province, Chile (type locality).

Description: Dark sky blue massive nodule from the association with lammerite, olivenite, mansfieldite, sénarmontite, a mineral of the crandallite group, rutile, anatase, and talc. Holotype sample. Described as tetragonal ($P4_122$ or $P4_322$) mineral with $a = 9.9758(4)$, $c = 36.714(1)$ Å, $V = 653.6(2)$ Å^3, $Z = 8$. $D_{meas} = 3.78(1)$ g/cm^3, $D_{calc} = 3.863(5)$ g/cm^3. Optically uniaxial (–), $\varepsilon = 1.647(2)$, $\omega = 1.749(2)$. The empirical formula is $Na_{1.04}Ca_{1.00}Cu_{5.01}(AsO_4)_{4.00}Cl_{0.96}(OH)_{0.11}{\cdot}4.93H_2O$. The strongest lines of the powder X-ray diffraction pattern [*d*, Å (*I*, %) (*hkl*)] are: 9.60 (9) (101), 9.177 (100) (004), 4.588 (32) (008), 4.167 (10) (108), 3.059 (15) (0.0.12).

Kind of sample preparation and/or method of registration of the spectrum: KBr disc. Transmission.

Source: Ondruš et al. (2006).

Wavenumbers (cm^{-1}): 3559sh, 3521sh, 3437s, 3229sh, 1630sh, 1618, 927, 902sh, 880sh, 839s, 815s, 789s, 541, 485, 440.

Note: Weak bands in the range from 2800 to 3000 cm^{-1} correspond to the admixture of an organic substance.

As354 Lemanskiite $NaCaCu_5(AsO_4)_4Cl \cdot 3H2O$

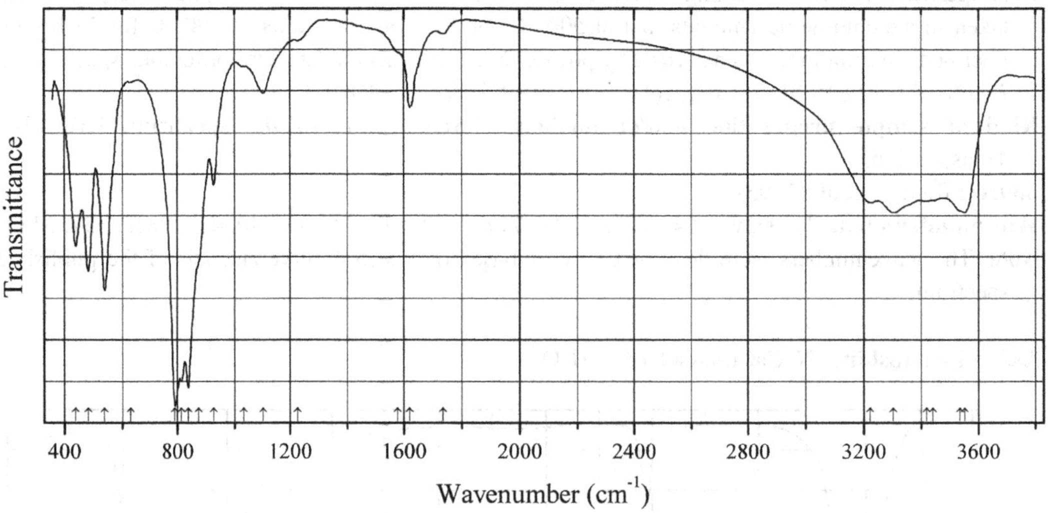

Origin: Perseverancia deposit, Guanaco, Antofagasta, Chile.

Description: Blue crystals. A sample used for the revision of the lemanskiite chemical formula. Characterized by powder X-ray diffraction data. The crystal structure is solved. Monoclinic, space group $P2_1/m$, a = 9.250(2), b = 10.0058(10), c = 10.0412(17) Å, β = 97.37(3)°, V = 921.7(3) Å^3, Z = 2. The empirical formula is $Na_{0.98}(Ca_{0.98}Sr_{0.03})Cu_{5.07}As_{3.97}O_{15.97}Cl_{1.03} \cdot 3H_2O$.

Kind of sample preparation and/or method of registration of the spectrum: KBr disc. Absorption.

Wavenumbers (cm^{-1}): 3551s, 3535sh, 3441, 3419, 3303s, 3224, 1734w, 1621, 1580sh, 1225w, 1102w, 928, 875sh, 838s, 813s, 791s, 629w, 539s, 483s, 439.

Note: The spectrum was obtained by N.V. Chukanov.

As355 Penberthycroftite $Al_6(AsO_4)_3(OH)_9 \cdot 13H_2O$

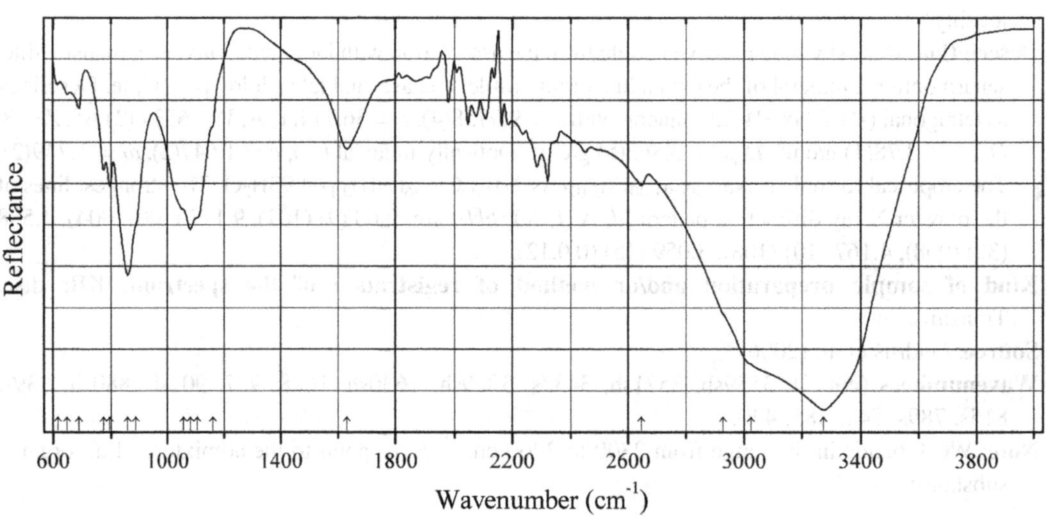

Origin: Penberthy Croft mine, St. Hilary, Cornwall, England, UK (type locality).

Description: White rectangular laths from the association with arsenopyrite, bettertonite, bulachite, cassiterite, chalcopyrite, chamosite, goethite, liskeardite, pharmacoalumite–pharmacosiderite, and quartz. Holotype sample. The crystal structure is solved. Monoclinic, space group $P2_1/c$, $a = 7.753$ (2) Å, $b = 24.679(5)$ Å, $c = 15.679(3)$ Å, $\beta = 94.19(3)°$, $V = 2991.9(12)$ Å^3, $Z = 4$. $D_{calc} = 2.18$ g/cm^3. The empirical formula is $Al_{5.96}Fe_{0.04}[(As_{0.97}Al_{0.03})O_4]_3(SO_4)_{0.26}(OH)_{8.30}·13.24H_2O$. The strongest lines of the powder X-ray diffraction pattern [*d*, Å (*I*, %) (*hkl*)] are: 13.264 (46) (011), 12.402 (16) (020), 9.732 (100) (021), 7.420 (28) (110), 5.670 (8) (130), 5.423 (6) (−131).

Kind of sample preparation and/or method of registration of the spectrum: Attenuated total reflection of powdered mineral.

Source: Grey et al. (2016a).

Wavenumbers (cm^{-1}): 3275s, 3025sh, 2930sh, 2645w, 1630, 1160sh, 1105sh, 1082, 1057sh, 885sh, 858, 796, 774, 690w, 650w, 617w.

Note: The wavenumbers were partly determined by us based on spectral curve analysis of the published spectrum.

As356 Petewilliamsite-related Cd diarsenate $Cd_2As_2O_7$

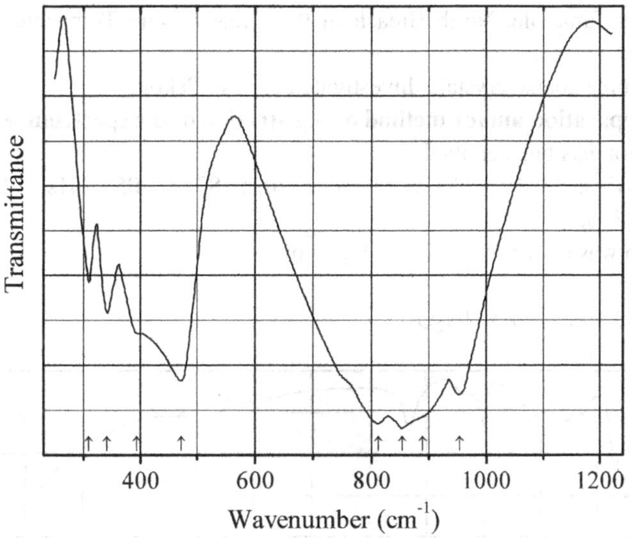

Origin: Synthetic.

Description: Prepared by solid-state reaction of CdO and As_2O_5 at 873 K for 10–12 days. Monoclinic, space group *C*2/*m*, $Z = 2$.

Kind of sample preparation and/or method of registration of the spectrum: KBr disc. Transmission.

Source: Baran and Weil (2004).

Wavenumbers (IR, cm^{-1}): 953, 890sh, 854s, 812s, 472, 393, 341, 310.

Note: In the cited paper, Raman spectrum is given.

Wavenumbers (Raman, cm^{-1}): 880s, 810w, 489w, 358, 423, 323, 294, 216w.

As357 Bradaczekite $NaCu_4(AsO_4)_3$

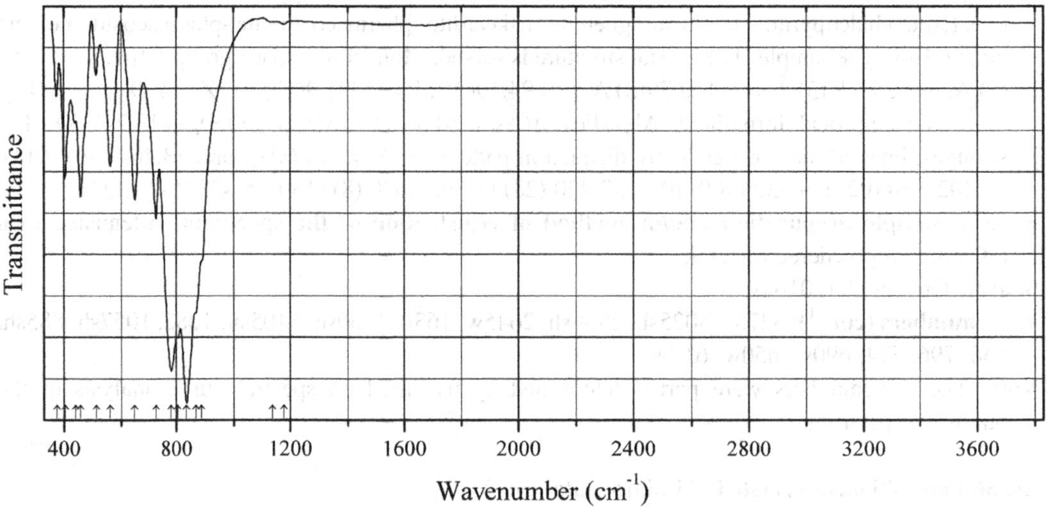

Origin: Arsenatnaya fumarole, North Breach of the Great Fissure Tolbachik volcano, Kamchatka peninsula, Russia.

Description: Deep blue coarse crystals. Investigated by I.V. Pekov.

Kind of sample preparation and/or method of registration of the spectrum: KBr disc. Absorption. Baseline correction has been applied.

Wavenumbers (cm^{-1}): 1176w, 1135w, 890sh, 870sh, 837s, 805, 781s, 724, 648, 563, 513w, 459, 442w, 404, 375.

Note: The spectrum was obtained by N.V. Chukanov.

As358 Vysokýite $U^{4+}(H_2AsO_4)_4 \cdot 4H_2O$

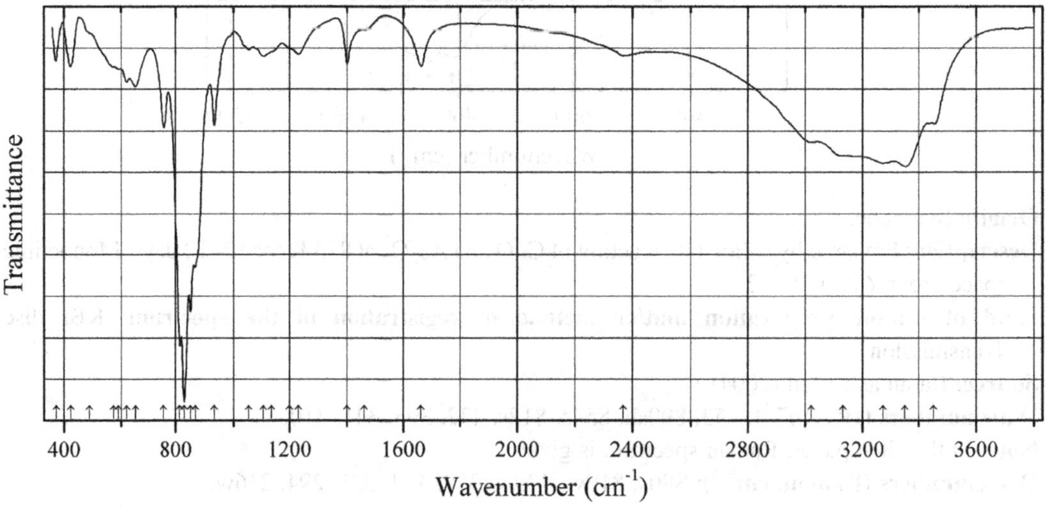

Origin: Geschieber vein, Svornost shaft, Jáchymov, Krušné Hory Mts. (Ore Mts.), Czech Republic (type locality).

Description: Aggregates of green acicular crystals from the association with štěpite. The sample was received from the authors of the first description of vysokýite.

Kind of sample preparation and/or method of registration of the spectrum: KBr disc. Absorption.

Wavenumbers (cm^{-1}): 3452, 3350, 3273, 3132, 3022, 2368w, 1662, 1463w, 1401, 1231, 1200sh, 1135sh, 1108, 1053w, 936, 869, 853s, 831s, 815s, 757, 651, 621, 591, 575sh, 490sh, 424, 373.

Note: The spectrum was obtained by N.V. Chukanov.

As359 Ludlockite $PbFe^{3+}{}_4As^{3+}{}_{10}O_{22}$

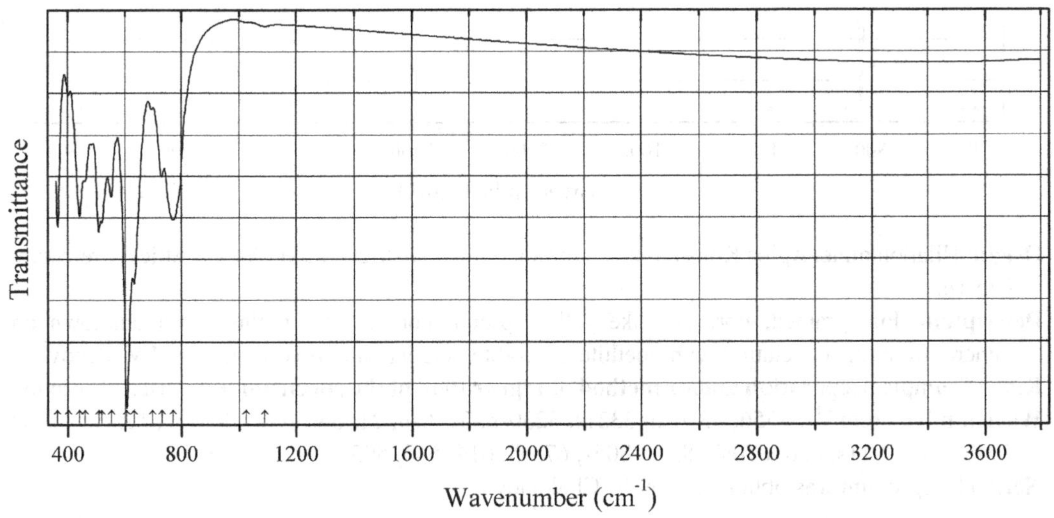

Origin: Tsumeb mine, Tsumeb, Namibia (type locality).

Description: Brownish-red acicular crystals from a sulfide aggregate. Investigated by A.V. Kasatkin. Holotype sample. The empirical formula is (electron microprobe): $Pb_{1.00}Fe_{3.65}As_{10.35}O_{22}$.

Kind of sample preparation and/or method of registration of the spectrum: KBr disc. Absorption.

Wavenumbers (cm^{-1}): 1090w, 1025w, 770, 731, 697w, 632s, 606s, 552, 519, 508, 460, 441, 403w, 365.

Note: The spectrum was obtained by N.V. Chukanov.

As360 Kamarizaite $Fe^{3+}{}_3(AsO_4)_2(OH)_3 \cdot 3H_2O$

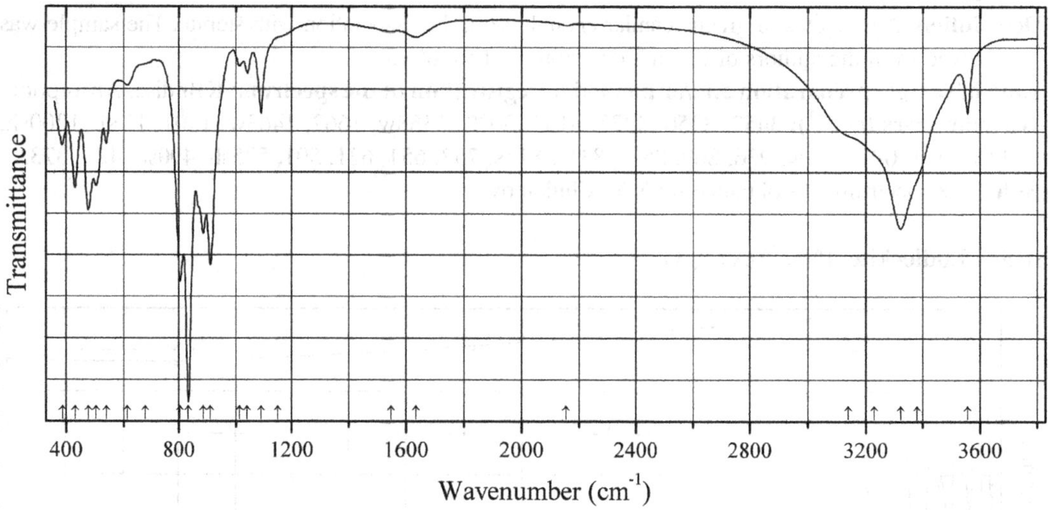

Origin: Hilarion mine, Agios Konstantinos, Lavrion mining District, Attikí (Attika, Attica) Prefecture, Greece.

Description: Fine-grained, porcelain-like yellow pseudomorphs after grains of an unknown ore mineral from the association with goethite, scorodite, and jarosite. Investigated by I.V. Pekov.

Kind of sample preparation and/or method of registration of the spectrum: KBr disc. Absorption.

Wavenumbers (cm^{-1}): 3556, 3380sh, 3321s, 3230sh, 3140sh, 2156w, 1635, 1550w, 1150sh, 1089, 1040, 1013, 911s, 886, 870sh, 834s, 805s, 675sh, 614, 540, 505, 477, 431, 387.

Note: The spectrum was obtained by N.V. Chukanov.

As361 Symplesite $Fe^{2+}{}_3(AsO_4)_2 \cdot 8H_2O$

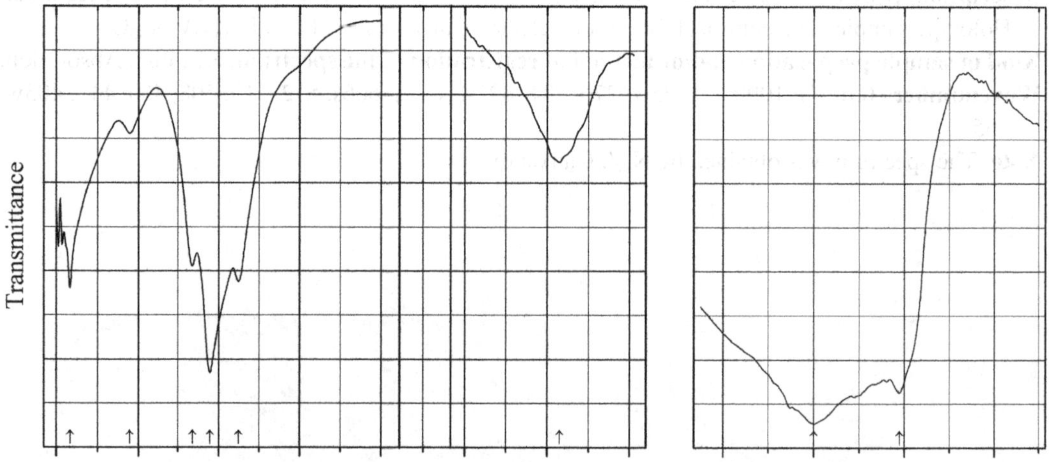

Origin: Saubach, near Mildenberg, Vogtland, Saxony, Germany.

Description: Black acicular crystals. Characterized by powder X-ray diffraction data. Triclinic, space group *P*-1, $a = 7.785$, $b = 9.259$, $c = 4.751$ Å, $\alpha = 93.053°$, $\beta = 98.139°$, $\gamma = 106.379°$.

Kind of sample preparation and/or method of registration of the spectrum: KBr disc. Absorption.

Source: Makreski et al. (2015b).

Wavenumbers (cm^{-1}): 3377, 3000s, 1630, 848, 779s, 736, 578w, 433.

As362 Agardite-(Ce) $CeCu^{2+}_6(AsO_4)_3(OH)_6 \cdot 3H_2O$

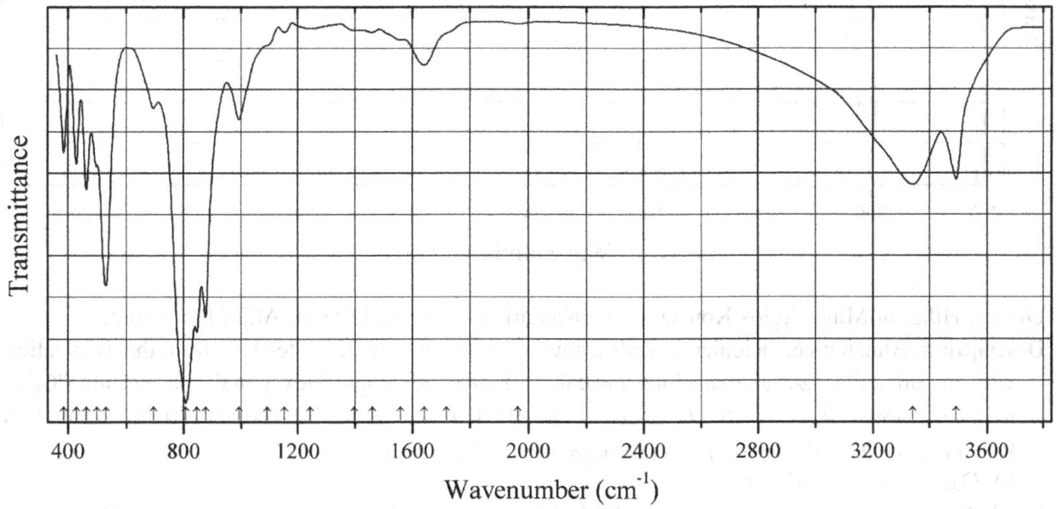

Origin: Clara Mine, Schwarzwald, Germany (type locality).

Description: Random aggregates of olive green acicular hexagonal crystals growing on fluorite. The crystal structure is solved. Hexagonal, space group *P*6$_3$/*m*, $a = 13.598(6)$ Å, $c = 5.954(3)$ Å, $V = 953.5(2)$ Å^3, $Z = 2$. The empirical formula is (electron microprobe): $[(Ce_{0.32}La_{0.19}Nd_{0.15}Pr_{0.06}Gd_{0.04}Y_{0.04}Sm_{0.03}Eu_{0.02})Ca_{0.20}Sr_{0.06}](Cu_{5.74}Fe^{3+}_{0.16}Mn^{2+}_{0.02})$ $[(AsO_4)_{2.89}(PO_4)_{0.04}(SiO_4)_{0.04}(SbO_4)_{0.03}](OH)_{5.97}O_{0.03} \cdot 3H_2O$.

Kind of sample preparation and/or method of registration of the spectrum: KBr disc. Absorption.

Wavenumbers (cm^{-1}): 3492, 3340, 1965w, 1720sh, 1641, 1560w, 1460w, 1397w, 1243w, 1154w, 1092w, 993, 876s, 846s, 807s, 695, 530s, 499, 464, 429, 385.

Note: The spectrum was obtained by N.V. Chukanov. The weak bands in the range from 1092 to 1560 cm^{-1} correspond to isolated H$^+$ cations that do not form strong covalent bonds with coordinating O atoms.

As363 Agardite-(Nd) $NdCu^{2+}{}_6(AsO_4)_3(OH)_6 \cdot 3H_2O$

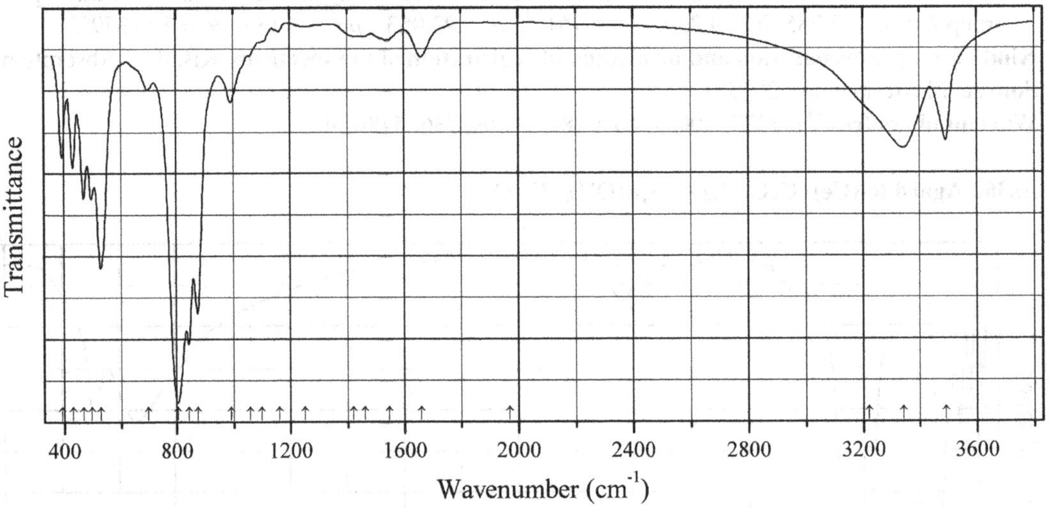

Origin: Hilarion Mine, Agios Konstantinos (Kamariza), Lavrion District, Attikí Prefecture.

Description: Bluish green acicular crystals growing [with zones of agardite-(La)] from the association with zincolivenite, azurite, malachite, and calcite. Holotype sample. Hexagonal, space group *P*6$_3$/*m*, a = 13.548(8) Å, c = 5.894(6) Å, V = 937(2) Å^3, Z = 2. The empirical formula is $[(Nd_{0.19}La_{0.14}Y_{0.12}Pr_{0.05}Gd_{0.02}Ce_{0.02}Sm_{0.02}Dy_{0.02})Ca_{0.39}](Cu_{5.49}Zn_{0.44})(AsO_4)_3(OH)_{5.38} \cdot 2.64H_2O$.

Kind of sample preparation and/or method of registration of the spectrum: KBr disc. Absorption.

Wavenumbers (cm^{-1}): 3491, 3340, 3220sh, 1968w, 1660, 1547w, 1460w, 1420w, 1250w, 1160w, 1098w, 1057w, 991, 874s, 843s, 804s, 695, 529s, 496, 469, 432, 393.

Note: The spectrum was obtained by N.V. Chukanov. The weak bands in the range from 1057 to 1547 cm^{-1} correspond to isolated H^+ cations that do not form strong covalent bonds with coordinating O atoms.

As364 Chlorophoenicite $Mn_3Zn_2(HAsO_4)(OH)_8$

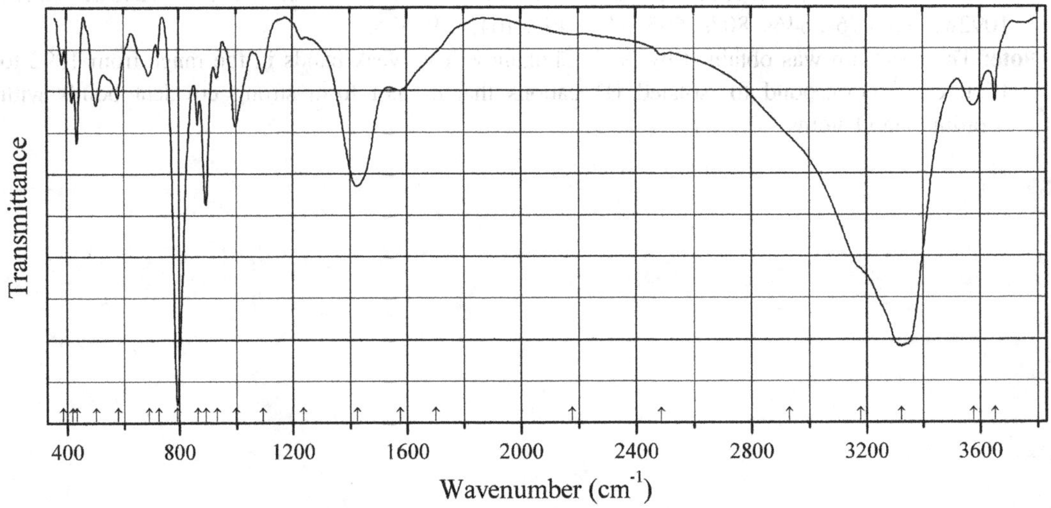

Origin: Sterling Hill, New Jersey, USA.

Description: White fibrous aggregate on rhodochrosite. The empirical formula is (electron microprobe): $(Mn_{2.68}Mg_{0.24}Fe_{0.02})Zn_{2.06}[H(As_{0.98}S_{0.02})O_4](OH)_8$.

Kind of sample preparation and/or method of registration of the spectrum: KBr disc. Absorption.

Wavenumbers (cm^{-1}): 3651, 3575, 3322s, 3180sh, 2930sh, 2487w, 2186w, 1700sh, 1577, 1235w, 1092, 998, 933w, 894, 793s, 688, 580, 503, 435, 419, 385w.

Note: The spectrum was obtained by N.V. Chukanov. Additional bands at 1424, 866, and 724 cm^{-1} correspond to admixed rhodochrosite.

As365 Gerdtremmelite $ZnAl_2(AsO_4)(OH)_5$

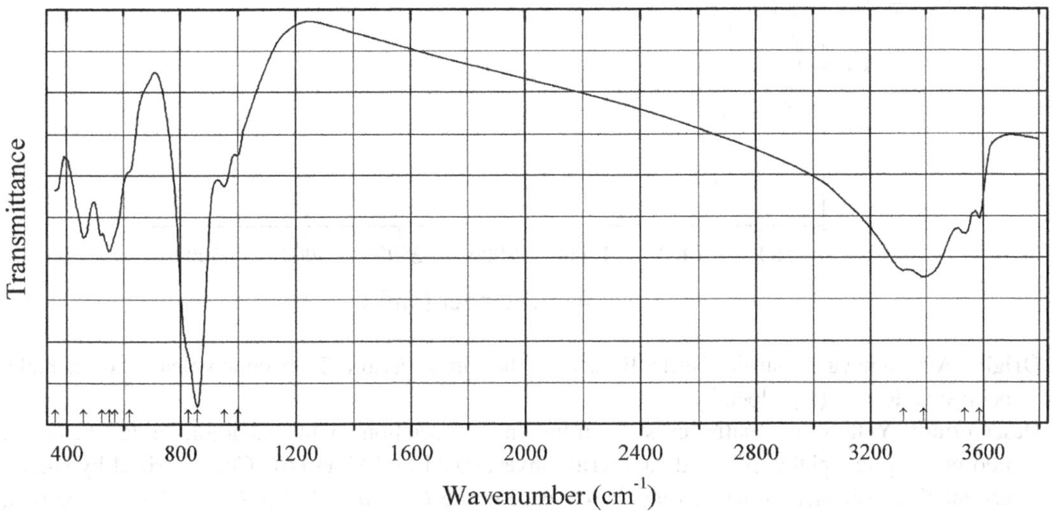

Origin: Tsumeb Mine, Tsumeb, Oshikoto Region, Namibia (type locality).

Description: Brown crust. Cotype sample received from Gerd Tremmel.

Kind of sample preparation and/or method of registration of the spectrum: KBr disc. Absorption.

Wavenumbers (cm^{-1}): 3586, 3535, 3389s, 3318, 999, 953, 857s, 825sh, 620sh, 570sh, 550s, 523, 459, (360).

Note: The spectrum was obtained by N.V. Chukanov.

As367 Badalovite $Na_2Mg_2Fe^{3+}(AsO_4)_3$

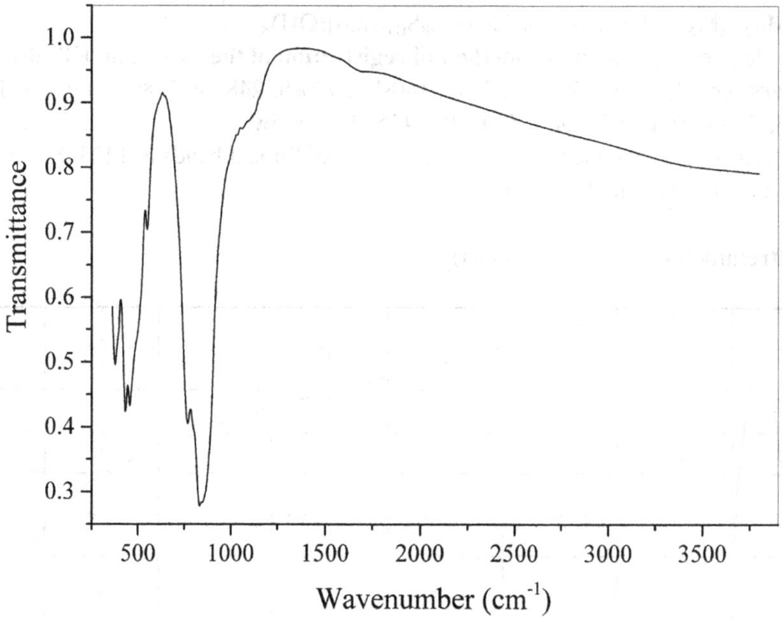

Origin: Arsenatnaya fumarole, North Breach of the Great Fissure Tolbachik volcano, Kamchatka peninsula, Russia (type locality).

Description: Yellow prismatic crystals from the association with calciojohillerite, hematite, fluorphlogopite, aphthitalite, and cassiterite. Investigated by I.V. Pekov. Characterized by single-crystal X-ray diffraction data. Monoclinic, space group *C*2/*c*, $a = 11.90$, $b = 12.78$, $c = 6.66$ Å, $\beta = 112.52°$, $V = 936.6$ Å^3, $Z = 4$. The empirical formula is (electron microprobe): $(Na_{1.61}Ca_{0.33}K_{0.03})(Mg_{1.78}Zn_{0.05}Mn_{0.03}Cu_{0.01})(Fe_{0.85}Al_{0.13})$ $[(AsO_4)_{2.94}(PO_4)_{0.04}(VO_4)_{0.01}(SO_4)_{0.01}]$.

Kind of sample preparation and/or method of registration of the spectrum: KBr disc. Absorption.

Wavenumbers (cm^{-1}): 1695w, 1160sh, 1125sh, 1063w, 1033w, 850s, 832s, 805sh, 769s, 551, 490sh, 455s, 431s, 390sh, 375.

Note: The spectrum was obtained by N.V. Chukanov.

As368 Svabite $Ca_5(AsO_4)_3F$

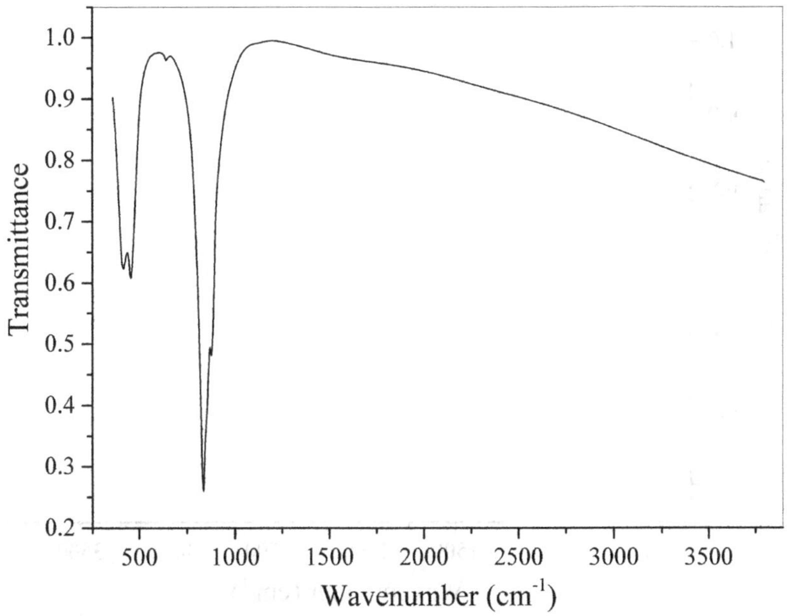

Origin: Arsenatnaya fumarole, North Breach of the Great Fissure Tolbachik volcano, Kamchatka peninsula, Russia.

Description: White radiated aggregates from the association with calciojohillerite, anhydrite, berzeliite, diopside, and hematite. Investigated by I.V. Pekov. A V-bearing variety (As:V ≈ 94:6). Characterized by single-crystal X-ray diffraction data. Hexagonal, space group $P6_3/m$, $a = 9.785$, $c = 6.946$ Å, $V = 576.1$ Å^3, $Z = 2$.

Kind of sample preparation and/or method of registration of the spectrum: KBr disc. Absorption.

Wavenumbers (cm^{-1}): 877, 835s, 643w, 456, 418.

Note: The spectrum was obtained by N.V. Chukanov.

As369 Calciojohillerite $NaCaMg_3(AsO_4)_3$

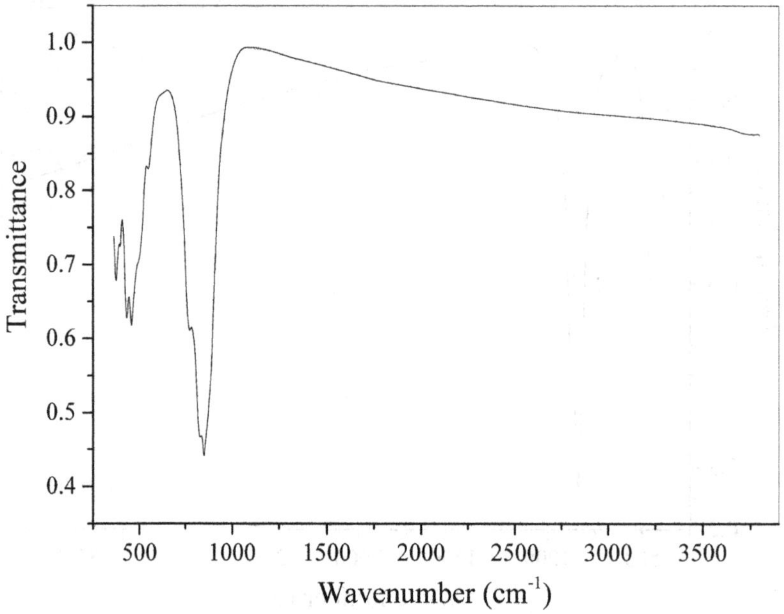

Origin: Arsenatnaya fumarole, North Breach of the Great Fissure Tolbachik volcano, Kamchatka peninsula, Russia (type locality).

Description: Greenish-brown prismatic crystals from the association with hematite, tenorite, cassiterite, johillerite, bradaczekite, hatertite, nickenichite, badalovite (IMA2016-053), aphthitalite, langbeinite, calciolangbeinite, etc. Investigated by I.V. Pekov.

Kind of sample preparation and/or method of registration of the spectrum: KBr disc. Absorption.

Wavenumbers (cm^{-1}): 848s, 830s, 549w, 495sh, 455, 429, 393, 373.

Note: The spectrum was obtained by N.V. Chukanov.

As370 Nickenichite Na(□,Ca)(□,Cu)(Mg,Fe^{3+})$_3$(AsO_4)$_3$

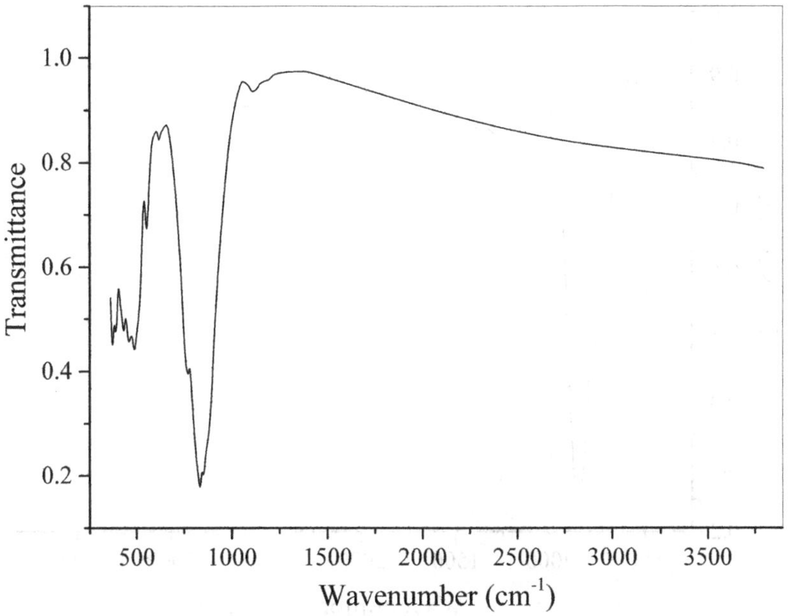

Origin: Arsenatnaya fumarole, North Breach of the Great Fissure Tolbachik volcano, Kamchatka peninsula, Russia (type locality).

Description: Violet prismatic crystals from the association with hematite, tenorite, johillerite, aphthitalite, etc. Investigated by I.V. Pekov. The empirical formula is (electron microprobe): $Na_{1.41}K_{0.03}Ca_{0.38}Cu_{0.17}Mg_{2.89}Fe_{0.31}Al_{0.09}[(As_{2.86}P_{0.03}V_{0.01}Si_{0.01})O_{12}]$.

Kind of sample preparation and/or method of registration of the spectrum: KBr disc. Absorption.

Wavenumbers (cm^{-1}): 1175sh, 1109w, 852s, 835s, 772, 619w, 554, 489, 463, 432, 390, 374.

Note: The spectrum was obtained by N.V. Chukanov. The bands at 1175, 1109, and 619 cm^{-1} correspond to trace amounts of SO_4^{2-} groups.

As371 Magnesiocanutite $NaMnMg_2[AsO_4]_2[AsO_2(OH)_2]$

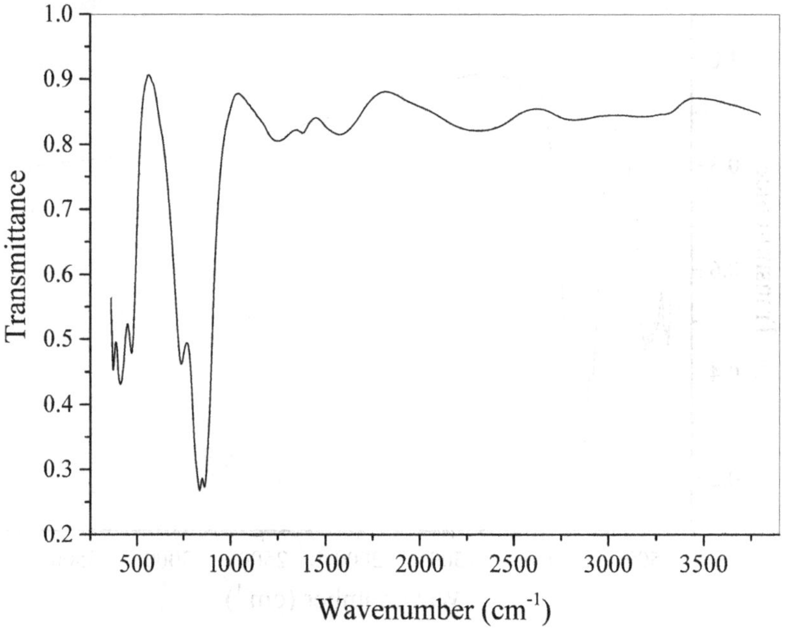

Origin: Torrecillas mine, Salar Grande, El Tamarugal Province, Tarapacá Region, Chile (type locality).
Description: Brown crystals from the association with magnesiokoritnigite and lavendulan. The empirical formula is (electron microprobe): $H_2Na_{1.0}Mn_{1.0}(Mg_{1.8}Mn_{0.15}Cu_{0.05})As_{3.0}O_{12}$.
Kind of sample preparation and/or method of registration of the spectrum: KBr disc. Absorption.
Wavenumbers (cm^{-1}): 3290sh, 3180, 2820, 2309, 1577, 1380, 1254, 863s, 837s, 738, 470, 411, 375.
Note: The spectrum was obtained by N.V. Chukanov.

As372 Magnesiocanutite $NaMnMg_2[AsO_4]_2[AsO_2(OH)_2]$

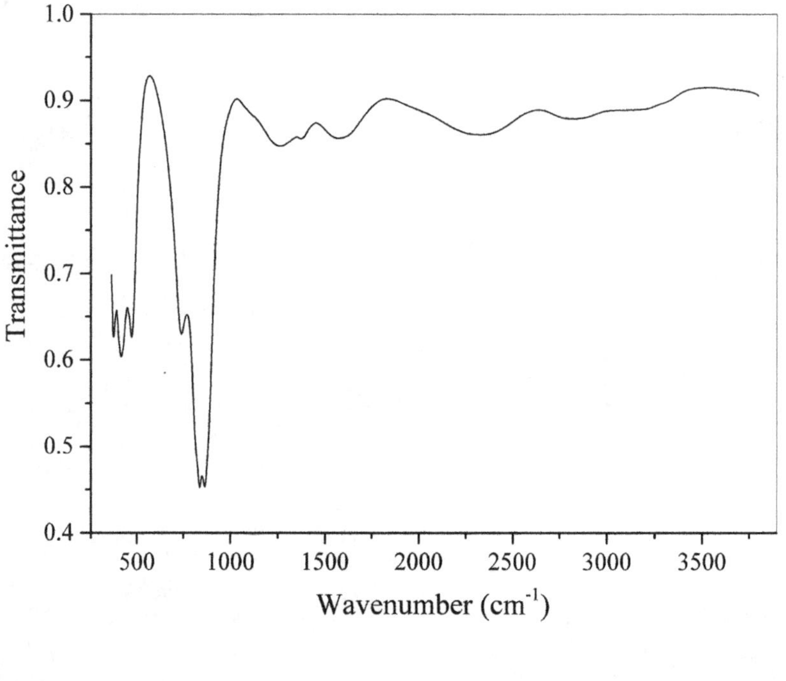

Origin: Torrecillas mine, Salar Grande, El Tamarugal Province, Tarapacá Region, Chile (type locality).

Description: Light brown crystals from the association with magnesiokoritnigite. A Mn-rich variety. The empirical formula is (electron microprobe): $H_2Na_{1.0}Mn_{1.0}(Mg_{1.2}Mn_{0.8})As_{3.0}O_{12}$.

Kind of sample preparation and/or method of registration of the spectrum: KBr disc. Absorption.

Wavenumbers (cm^{-1}): 3290sh, 3110sh, 2825, 2338, 1577, 1372, 1263, 1120sh, 864s, 837s, 740, 472, 414, 374.

Note: The spectrum was obtained by N.V. Chukanov.

As373 Canutite $NaMnMn_2[AsO_4]_2[AsO_2(OH)_2]$

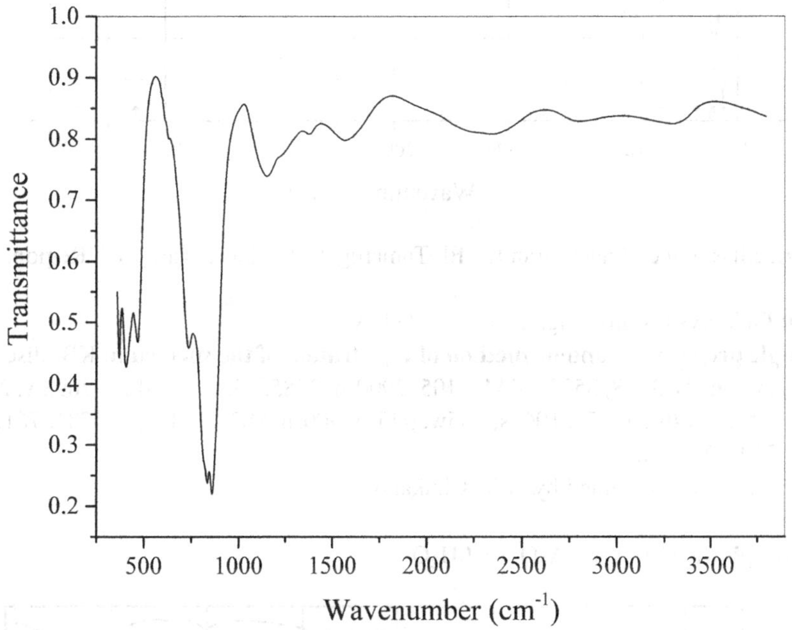

Origin: Torrecillas mine, Salar Grande, El Tamarugal Province, Tarapacá Region, Chile (type locality).

Description: Brown crystals. The empirical formula is (electron microprobe): $H_2Na_{1.0}Mn_{1.0}(Mn_{1.2}Mg_{0.7}Cu_{0.2})As_{3.0}O_{12}$.

Kind of sample preparation and/or method of registration of the spectrum: KBr disc. Absorption.

Wavenumbers (cm^{-1}): 3290sh, 3110sh, 2825, 2338, 1577, 1372, 1263, 1120sh, 864s, 837s, 740, 472, 414, 374.

Note: The spectrum was obtained by N.V. Chukanov.

AsS27 Juansilvaite $Na_5Al_3(HAsO_4)_2(H_2AsO_4)_2(SO_4)_2 \cdot 4H_2O$

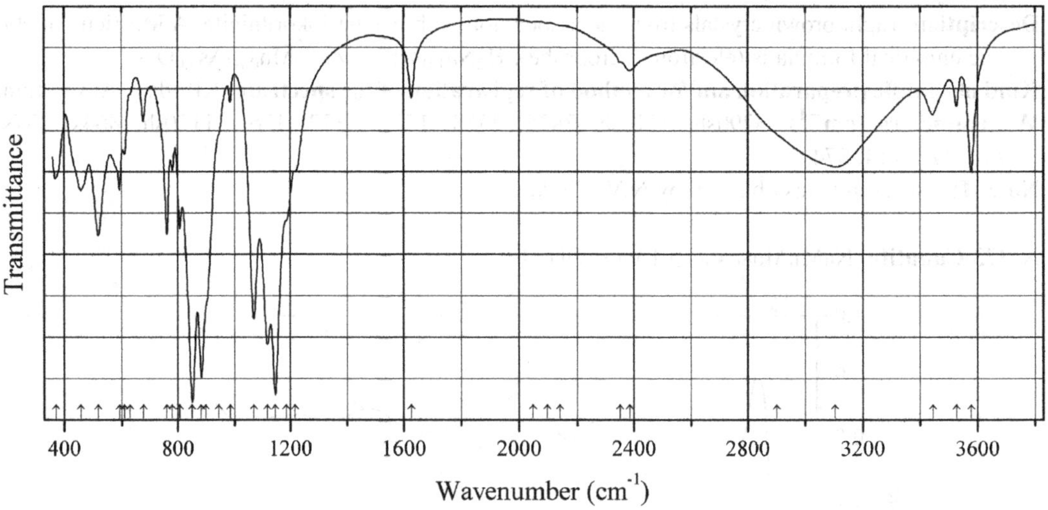

Origin: Torrecillas mine, Salar Grande, El Tamarugal Province, Tarapacá Region, Chile (type locality).

Description: Pink crystals. Investigated by I.V. Pekov.

Kind of sample preparation and/or method of registration of the spectrum: KBr disc. Absorption.

Wavenumbers (cm^{-1}): 3578, 3527, 3444, 3105, 2900sh, 2385, 2350w, 2140w, 2098w, 2049w, 1627, 1215sh, 1185sh, 1146s, 1117s, 1068s, 985w, 945sh, 900sh, 883s, 851s, 807, 781, 761, 675, 630sh, 611, 593, 519, 459, 372.

Note: The spectrum was obtained by N.V. Chukanov.

UAs23 Uranospinite $Ca(UO_2)_2(AsO_4)_2 \cdot 10H_2O$

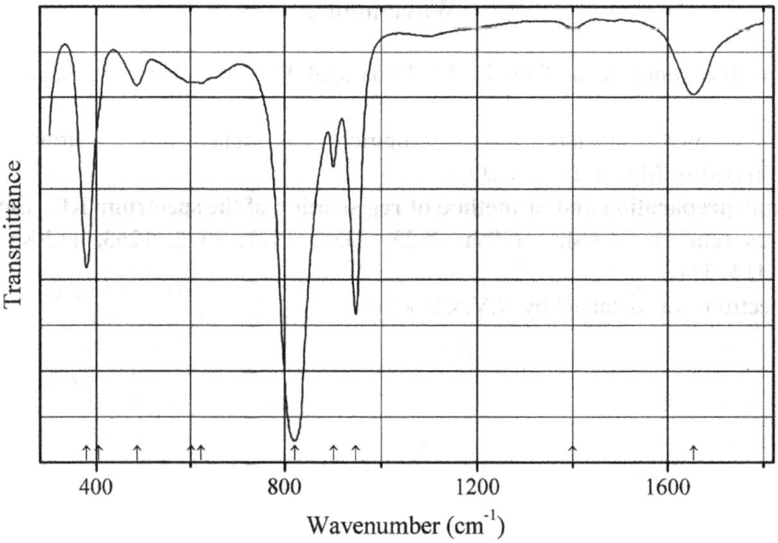

Origin: Synthetic.

Kind of sample preparation and/or method of registration of the spectrum: KBr disc. Transmission.

Source: Wilkins (1971).

Wavenumbers (cm^{-1}): 1654, 1400w, 948s, 901, 820s, 623, 602, 487, 405sh, 379s.

2.15 Selenides, Selenites, and Selenates

Se51 Aluminium acid selenite hydrate $AlH(SeO_3)_2 \cdot 2H_2O$

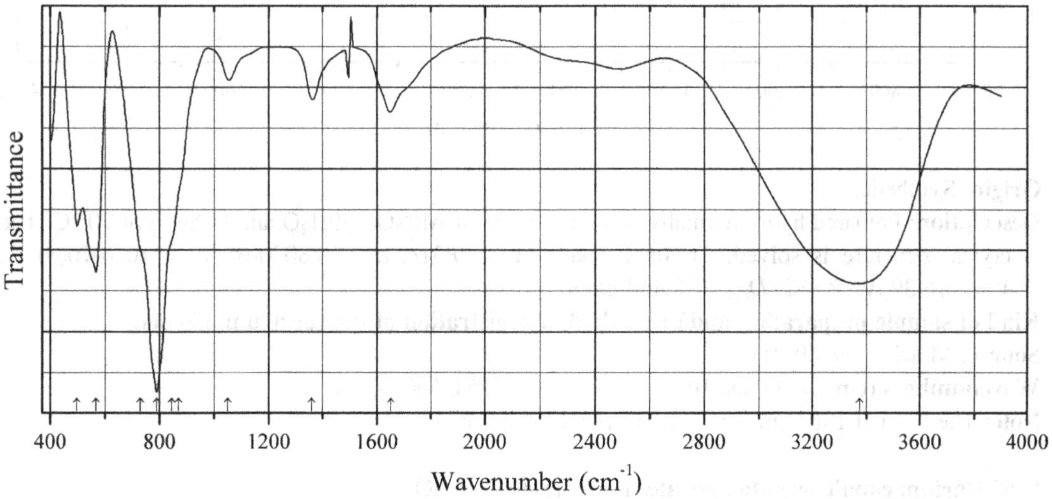

Origin: Synthetic.

Description: Prepared hydrothermally from a mixture of $Al(NO_3)_3 \cdot 9H_2O$ and H_2SeO_3 at 70 °C. The crystal structure is solved. Monoclinic, space group $P2_1/n$, a = 7.3853(5), b = 6.4895(6), c = 7.3958(7) Å, β =106.28(9)°, V = 340.24 Å^3, Z = 2. D_{calc} = 3.054 g/cm^3.

Kind of sample preparation and/or method of registration of the spectrum: No data.

Source: Morris et al. (1991).

Wavenumbers (cm^{-1}): 3376s, 1654, 1360, 1051, 870sh, 845sh, 790s, 730sh, 569s, 499, and a series of bands in the range from 2000 to 2800 cm^{-1}.

Se52 Aluminium selenite hydrate $Al_2(SeO_3)_3{\cdot}6H_2O$

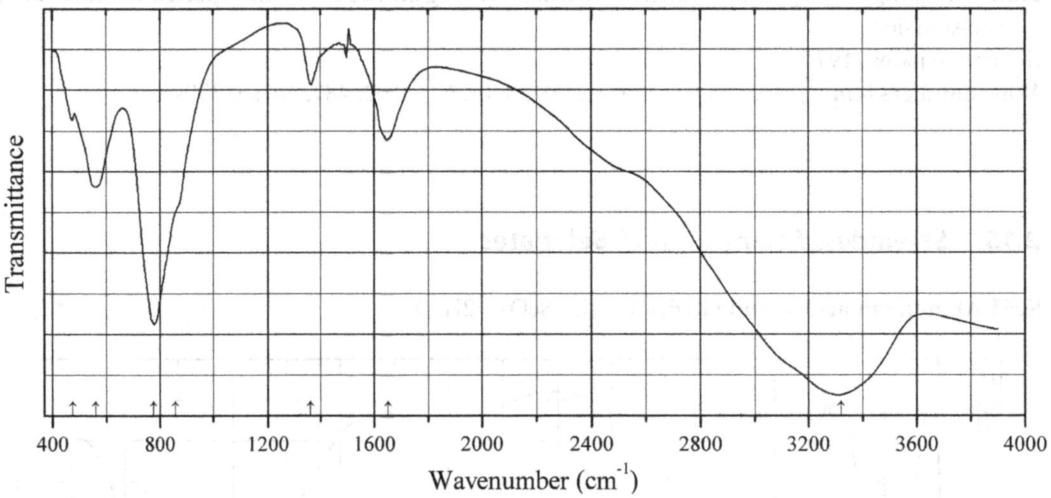

Origin: Synthetic.

Description: Prepared hydrothermally from a mixture of $Al(NO_3)_3{\cdot}9H_2O$ and H_2SeO_3 at 70 °C. The crystal structure is solved. Trigonal, space group *P*31*c*, $a = 8.8020(6)$, $c = 10.7070(8)$ Å, $V = 718.39$ Å^3, $Z = 2$. $D_{calc} = 2.468$ g/cm^3.

Kind of sample preparation and/or method of registration of the spectrum: No data.

Source: Morris et al. (1991).

Wavenumbers (cm^{-1}): 3320s, 1652, 1362, 860sh, 775s, 560, 475w.

Note: The band at 1362 cm^{-1} may correspond to an impurity.

Se53 Barium cobalt selenite hydrate $BaCo_2(SeO_3)_3{\cdot}3H_2O$

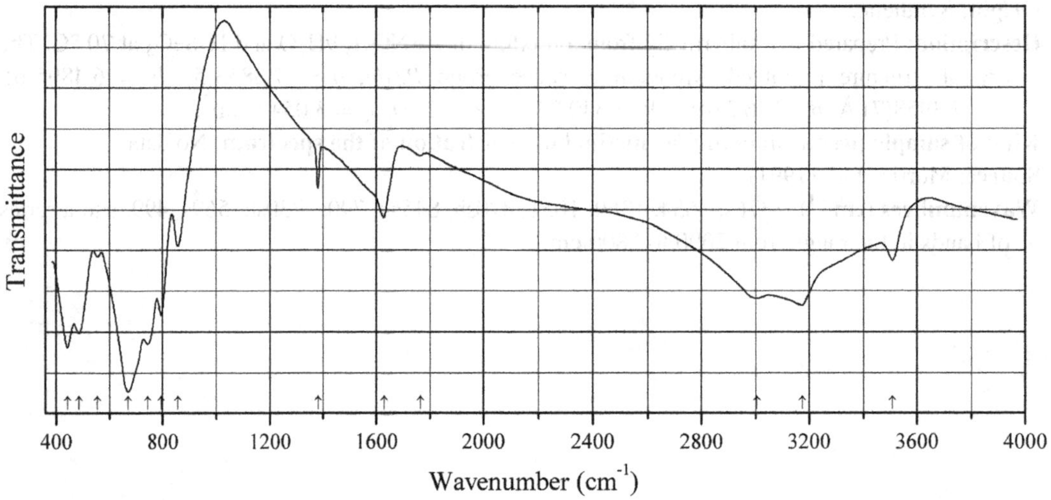

Origin: Synthetic.

Description: Purple hexagonal prismatic crystals. Structurally related to zemannite. Hexagonal, space group $P6_3$, $a = 18.0430(6)$, $c = 7.6120(2)$ Å, $V = 2146.08(12)$ Å^3, $Z = 8$. $D_{calc} = 4.272$ g/cm^3.

Kind of sample preparation and/or method of registration of the spectrum: KBr disc. Transmission.

Source: Johnston and Harrison (2011).

Wavenumbers (cm^{-1}): 3510, 3175, 3005, 1764w, 1630, (1381), 859, 795, 743s, 667s, 555w, 486s, 442s.

Note: The wavenumbers were partly determined by us based on spectral curve analysis of the published spectrum. The band at 1381 cm^{-1} may be due to the admixture of potassium nitrate in the KBr disc.

Se54 Baryte selenate analogue Ba(SeO$_4$)

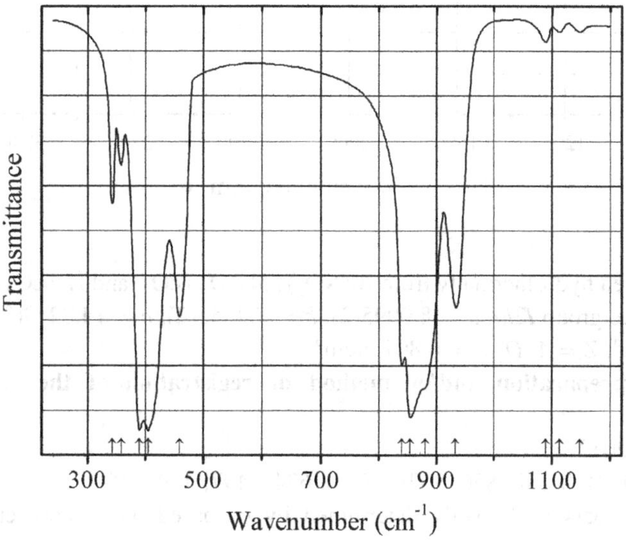

Origin: Synthetic.

Description: Prepared by precipitation from aqueous solutions of sodium selenate and strontium chloride. The precipitate was dried at 150 °C. Isostructural with baryte. Orthorhombic, space group *Pnma*, $a = 9.006$, $b = 5.690$, $c = 7.353$ Å, $Z = 2$.

Kind of sample preparation and/or method of registration of the spectrum: Thin film on a CsBr plate. Transmission.

Source: Scheuermann and Schutte (1973b).

Wavenumbers (IR, cm^{-1}): 1148w, 1112w, 1089w, 933, 880sh, 855s, 840s, 459, 405s, 390s, 358, 343.

Note: In the cited paper, Raman spectrum is given.

Wavenumbers (Raman, cm^{-1}): 915, 905sh, 901, 898, 874, 866, 846s, 465, 437w, 423, 421, 418sh, 352s, 338s, 333.

Note: The authors of the cited paper write: "There appears to be no combination of observed bands which would explain the three bands at 1089, 1112, and 1148 cm^{-1} satisfactorily." However, these bands may correspond to a sulfate impurity, which is typical for selenate reactants.

Se55 Bismuth(III) tellurite selenate $Bi_2(TeO_3)_2(SeO_4)$

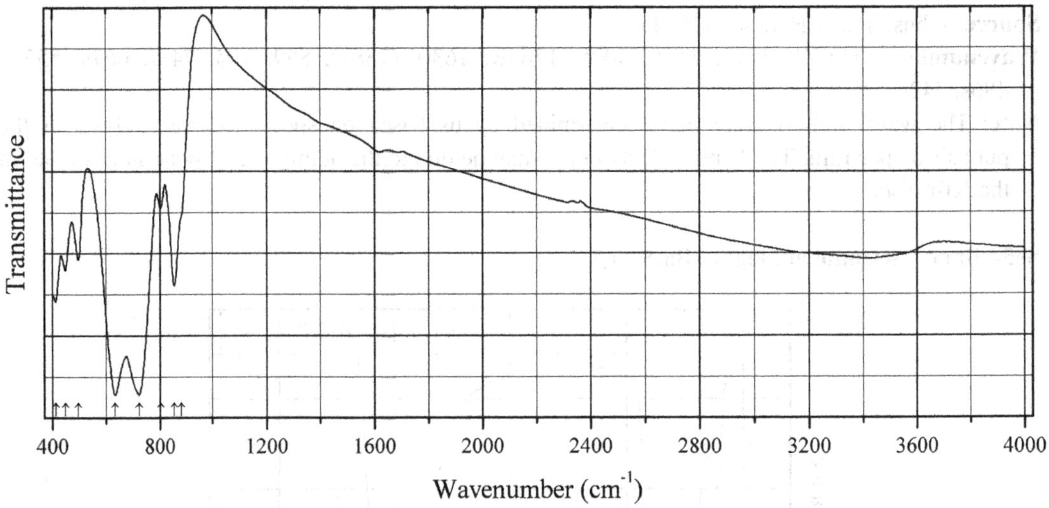

Origin: Synthetic.

Description: Prepared hydrothermally from $Bi(NO_3)_3{\cdot}5H_2O$, TeO_2, and H_2SeO_4 at 230 °C for 4 days. Monoclinic, space group *I*2/*a*, $a = 8.0995(2)$, $b = 7.4835(2)$, $c = 14.8219(5)$ Å, $\beta = 97.824(3)°$, $V = 890.03(4)$ Å^3, $Z = 4$. $D_{calc} = 6.807$ g/cm^3.

Kind of sample preparation and/or method of registration of the spectrum: KBr disc. Transmission.

Source: Lee et al. (2013).

Wavenumbers (cm^{-1}): 883sh, 856, 802w, 721s, 634s, 497, 449, 412.

Note: The wavenumbers were partly determined by us based on spectral curve analysis of the published spectrum.

Se56 Cesium acid arsenate selenate $Cs_4(SeO_4)(HSeO_4)_2(H_3AsO_4)$

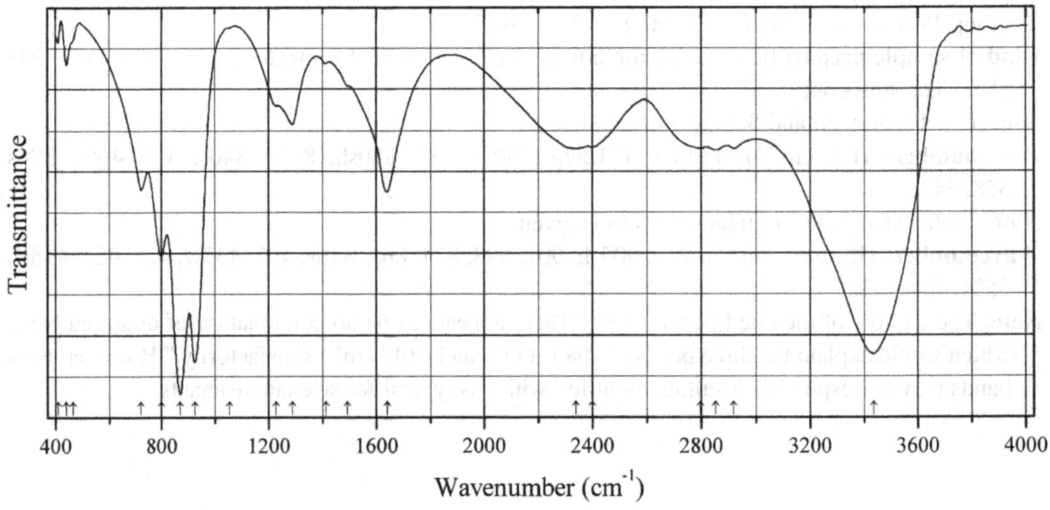

Origin: Synthetic.

Description: Synthesized by evaporation of aqueous solution of $CsHSeO_4$, $Cs_3H(SeO_4)_2$, and H_3AsO_4 at room temperature. Monoclinic, space group $P2_1$, $a = 5.973(1)$, $b = 13.691(3)$, $c = 11.910(3)$ Å, $\beta = 94.867(1)°$, $V = 970.39(4)$ Å^3, $Z = 2$. $D_{meas} = 3.782$ g/cm^3, $D_{calc} = 3.780$ g/cm^3. Characterized by DSC and TG data.

Kind of sample preparation and/or method of registration of the spectrum: KBr disc. Transmission.

Source: Amri et al. (2009).

Wavenumbers (cm^{-1}): 3434s (broad), 2920, 2854, 2798, 2400, 2338 (broad), 1640, 1492sh, 1414w, 1290, 1228sh, 1054w, 926s, 872s, 800s, 720, 465sh, 442w, 410w.

Note: The wavenumbers were partly determined by us based on spectral curve analysis of the published spectrum.

Se58 Lead selenate $PbSeO_4$

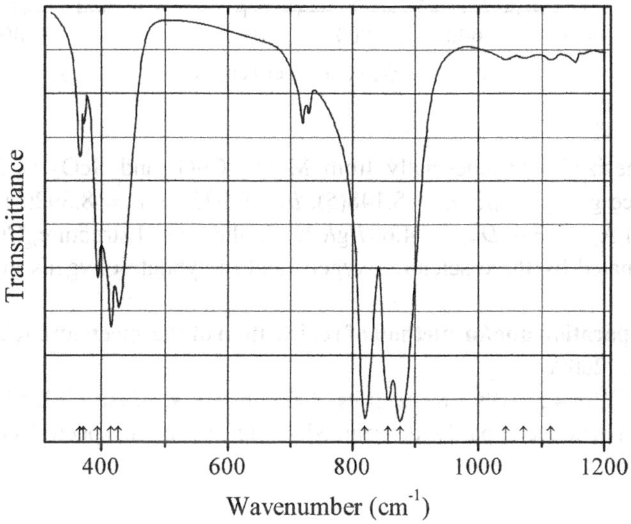

Origin: Synthetic.

Description: Prepared by precipitation from aqueous solutions of sodium selenate and lead acetate. Monoclinic, space group $P2_1/n$, $a = 7.153$, $b = 7.403$, $c = 6.957$ Å, $\beta = 103.27°$, $Z = 4$.

Kind of sample preparation and/or method of registration of the spectrum: Nujol mull. Transmission.

Source: Scheuermann and Schutte (1973b).

Wavenumbers (IR, cm^{-1}): 1114w, 1073w, 1044w, 876s, 857s, 820s, 427, 415, 394, 372, 366.

Note: The weak bands between 700 and 730 cm^{-1} may correspond to Nujol. The weak bands at 1114, 1073, and 1044 cm^{-1} assigned by Scheuermann and Schutte (1973b) to combination modes may actually correspond to the admixture of SO_4^{2-} anions. In the cited paper, Raman spectrum is given.

Wavenumbers (Raman, cm^{-1}): 861, 844, 829, 818, 435, 428, 422, 411, 402, 385, 357, 326, 314.

Se59 Copper molybdate selenite $Cu_2(MoO_4)(SeO_3)$

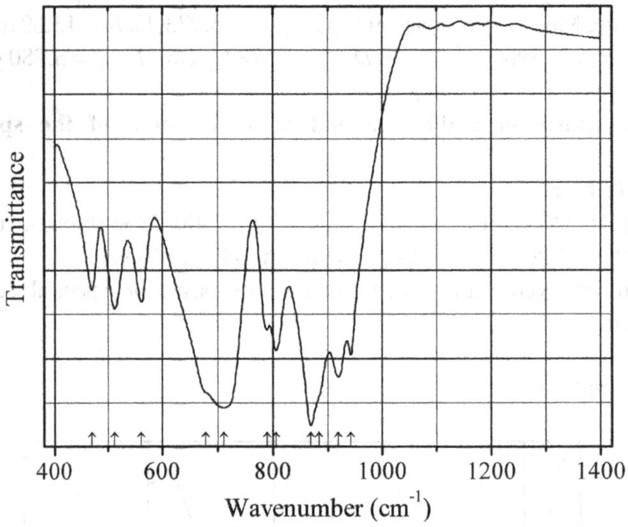

Origin: Synthetic.

Description: Synthesized hydrothermally from MoO_3, CuO, and SeO_2 at 210 °C for 4 days. Monoclinic, space group $P2_1/c$, $a = 8.148(5)$, $b = 9.023(5)$, $c = 8.392(5)$ Å, $\beta = 91.141(6)°$, $V = 104.675(12)$ Å^3, $Z = 4$. $D_{calc} = 4.607$ g/cm^3. In the crystal structure, Cu occupies two sites, each five-coordinated by three selenite oxygens and molybdate oxygens in a square pyramidal geometry.

Kind of sample preparation and/or method of registration of the spectrum: KBr disc. Transmission.

Source: Zhang et al. (2009c).

Wavenumbers (cm^{-1}): 942, 919s, 884sh, 869s, 806, 789, 711s, 679sh, 561, 512, 470.

Note: The wavenumbers were partly determined by us based on spectral curve analysis of the published spectrum.

Se60 Indium vanadate selenite In(VSe$_2$O$_8$) In(VSe$_2$O$_8$)

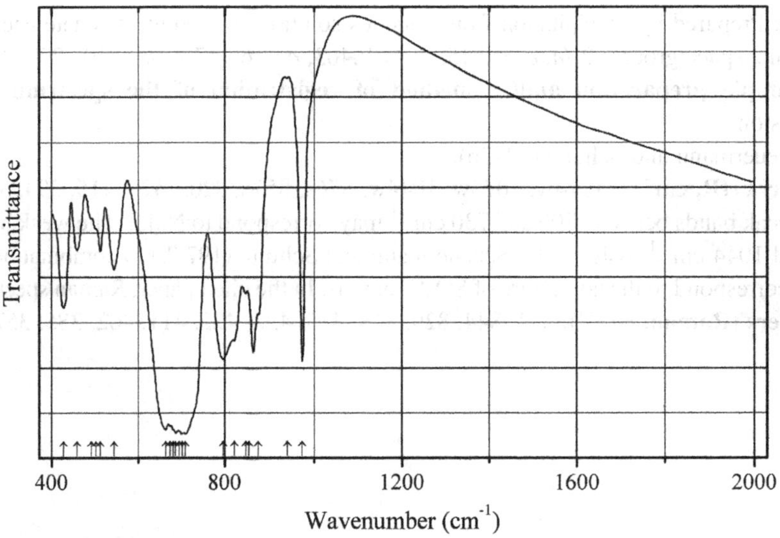

Origin: Synthetic.

Description: Prepared from In_2O_3, V_2O_5, and SeO_2 by a standard solid-state reaction technique. The crystal structure is silved. Monoclinic, space group *Pm*, $a = 4.6348(9)$, $b = 6.9111(14)$, $c = 10.507(2)$ Å, $\beta = 97.77(3)°$, $V = 333.48(11)$ Å^3, $Z = 2$. $D_{calc} = 4.498$ g/cm^3.

Kind of sample preparation and/or method of registration of the spectrum: KBr disc. Transmission.

Source: Lee et al. (2011).

Wavenumbers (cm^{-1}): 974, 940, 876sh, 854, 847, 822, 796, 709s, 701s, 695sh, 687s, 681sh, 674sh, 663, 543, 511, 501sh, 491sh, 458, 427.

Note: The wavenumbers were determined by us based on spectral curve analysis of the published spectrum.

Se61 Indium zinc selenite $In_2Zn(SeO_3)_4$ $In_2Zn(SeO_3)_4$

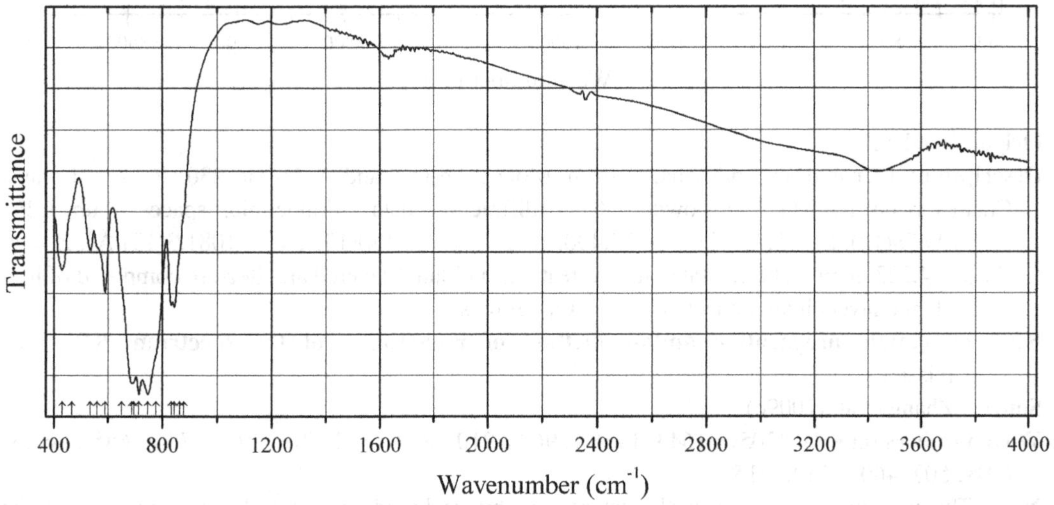

Origin: Synthetic.

Description: Obtained by a standard solid-state reaction from ZnO, In_2O_3, and SeO_2. Monoclinic, space group $P2_1/n$, $a = 8.4331(7)$, $b = 4.7819(4)$, $c = 14.6583(13)$ Å, $\beta = 101.684(6)°$, $V = 578.87$ (9) Å^3, $Z = 2$. $D_{calc} = 4.606$ g/cm^3.

Kind of sample preparation and/or method of registration of the spectrum: KBr disc. Transmission.

Source: Lee et al. (2012).

Wavenumbers (cm^{-1}): 878sh, 864sh, 844, 832, 777sh, 747s, 715s, 696sh, 690s, 650sh, 589, 561sh, 535, 465sh, 430.

Note: The wavenumbers were partly determined by us based on spectral curve analysis of the published spectrum.

Se62 Iron dimolybdate selenite hydrate $Fe_2(Mo_2O_7)(SeO_3)_2{\cdot}H_2O$ $Fe_2(Mo_2O_7)(SeO_3)_2{\cdot}H_2O$

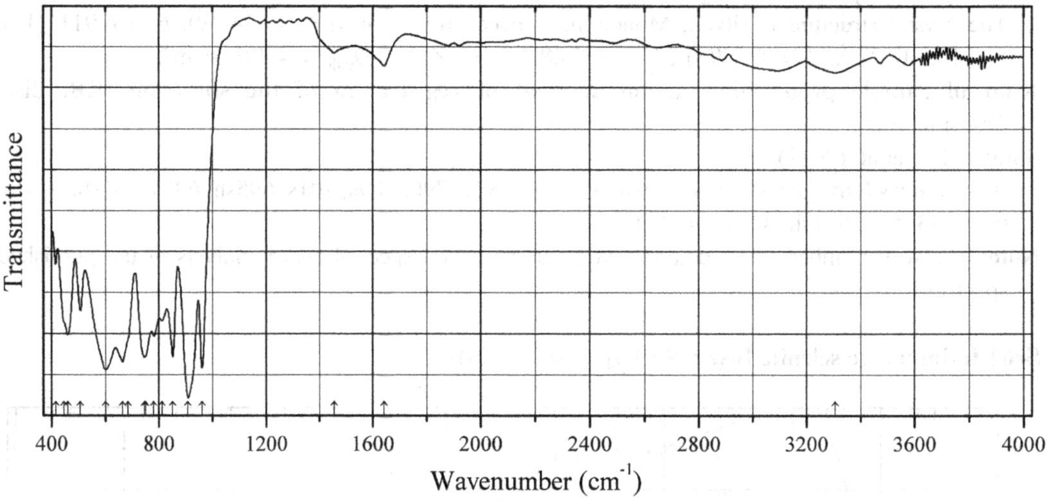

Origin: Synthetic.

Description: Prepared hydrothermally from MoO_3, Fe_2O_3, and SeO_2 at 230 °C for 4 days. Characterized by TG and powder X-ray diffraction data. Monoclinic, space group *C*2/*c*, $a = 19.898(12)$, $b = 5.469(3)$, $c = 13.400(9)$ Å, $\beta = 132.140(13)°$, $V = 1081.3(12)$ Å^3, $Z = 4$. $D_{calc} = 4.223$ g/cm^3. The crystal structure features a pillared-layered architecture composed of iron (III) selenite layers interconnected by Mo_2O_{10} dimers.

Kind of sample preparation and/or method of registration of the spectrum: KBr disc. Transmission.

Source: Zhang et al. (2009c).

Wavenumbers (cm^{-1}): 3305w, 1643, 1456w, 962s, 910s, 854s, 812, 781, 750sh, 744s, 685sh, 663s, 601s, 507, 460, 446sh, 415.

Note: The wavenumbers were partly determined by us based on spectral curve analysis of the published spectrum.

Se63 Lanthanum selenite $La_2(SeO_3)_3$

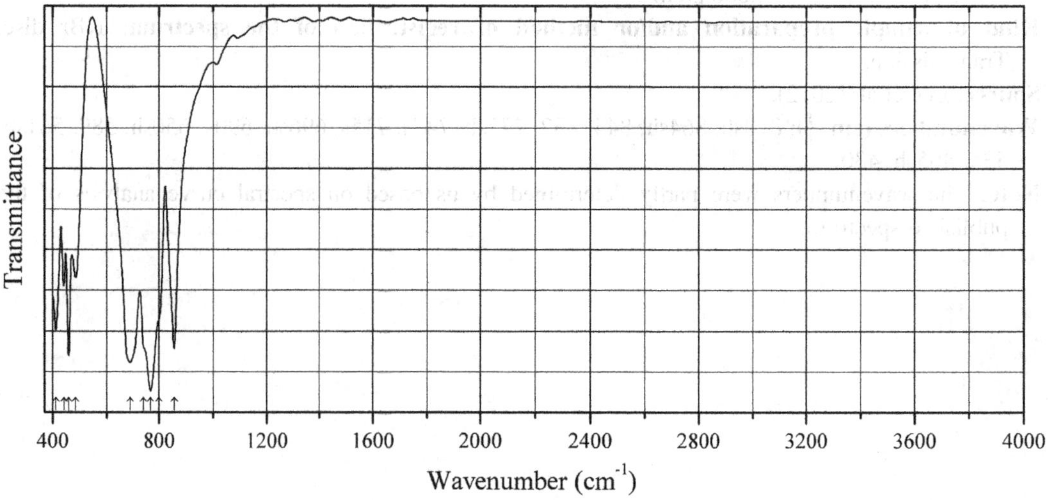

Origin: Synthetic.

Description: Obtained by thermal decomposition of $La_2(Se_2O_5)(SeO_3)_2$. Monoclinic, space group $P2_1/m$.

Kind of sample preparation and/or method of registration of the spectrum: KBr disc (above 400 cm^{-1}) and polyethylene disc (below 400 cm^{-1}). Transmission.

Source: Gopinath et al. (1998).

Wavenumbers (IR, cm^{-1}): 856s, 800sh, 767s, 740sh, 688s, 486, 460s, 442, 412, 382, 368, 241, 201, 157, 96w, 79w.

Note: In the cited paper, Raman spectrum is given.

Wavenumbers (Raman, cm^{-1}): 852s, 812s, 790w, 744s, 480, 450s, 430sh, 408, 389, 365w, 250, 219sh, 203, 176, 167, 137w, 89w.

Se64 Lanthanum selenite $La_2Se_3O_9$

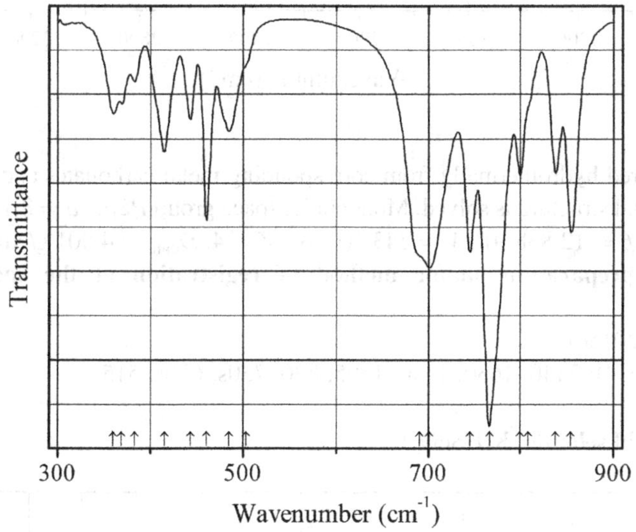

Origin: Synthetic.

Description: Monoclinic, space group $P2_1/m$, $a = 14.47(8)$, $b = 6.98(4)$, $c = 8.21(5)$ Å, $\beta = 91.0(7)°$, $Z = 4$. Characterized by powder X-ray diffraction data.

Kind of sample preparation and/or method of registration of the spectrum: Absorption. Kind of sample preparation is not indicated.

Source: Pedro et al. (1995).

Wavenumbers (cm^{-1}): 854s, 838, 809sh, 800, 767s, 745s, 701s, 691sh, 503sh, 484, 460, 443, 415, 383w, 369, 360.

Note: The wavenumbers were determined by us based on spectral curve analysis of the published spectrum.

Se65 Lithium zinc selenite $Li_2Zn_3(SeO_3)_4{\cdot}2H_2O$ $Li_2Zn_3(SeO_3)_4{\cdot}2H_2O$

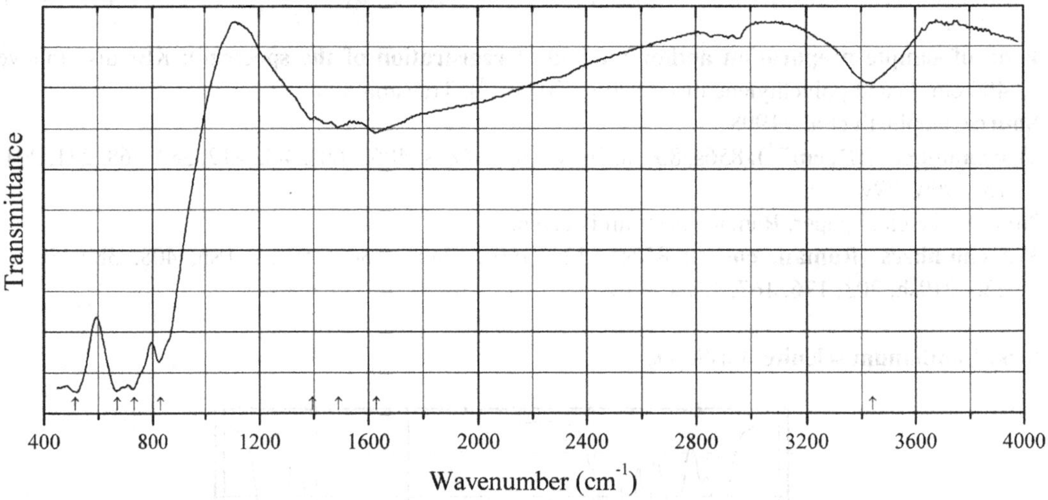

Origin: Synthetic.

Description: Prepared hydrothermally from corresponding metal carbonates and SeO_2 at 230 °C for 6 days. The crystal structure is solved. Monoclinic, space group $P2_1/c$, $a = 8.123(4)$, $b = 9.139(4)$, $c = 7.938(3)$ Å, $\beta = 12.838(9)°$, $V = 543.1(4)$ Å^3, $Z = 4$. $D_{calc} = 4.501$ g/cm^3.

Kind of sample preparation and/or method of registration of the spectrum: KBr disc. Transmission.

Source: Liu et al. (2015c).

Wavenumbers (cm^{-1}): 3440, 1630, 1490, 1395, 830, 730s, 670s, 515s.

Se66 Potassium acid selenite $K(HSeO_3)$

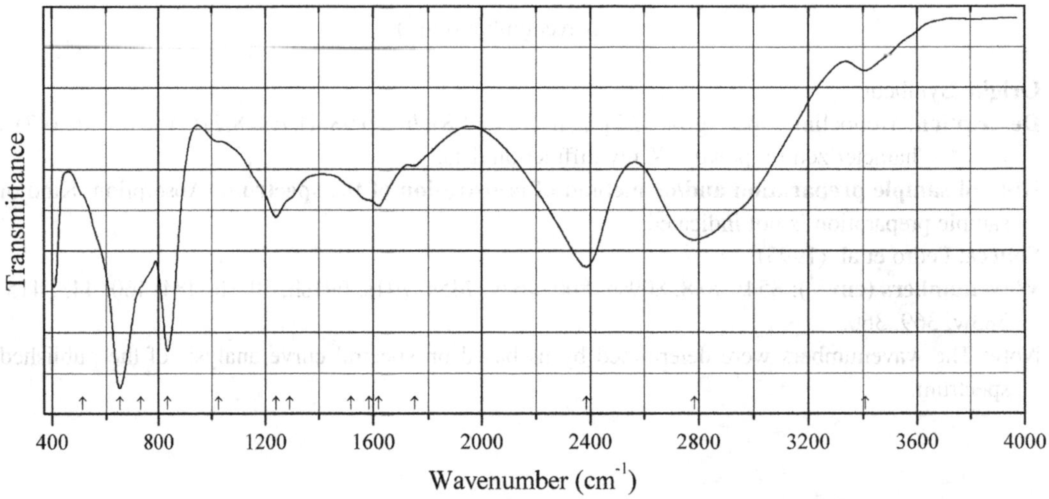

Origin: Synthetic.

Description: Crystals grown by slow evaporation of an aqueous solution formed by dissolving 2 moles of SeO and 1 mole of K_2CO_3.

Kind of sample preparation and/or method of registration of the spectrum: KBr disc. Transmission.

Source: Cody et al. (1978).

Wavenumbers (IR, cm^{-1}): 3408w, 2784s, 2385s, 1750w, 1621, 1584sh, 1517sh, 1292sh, 1240, 1024, 835s, 728sh, 650s, 512sh, 408.

Note: The wavenumbers were partly determined by us based on spectral curve analysis of the published spectrum. In the cited paper, Raman spectrum is given.

Wavenumbers (Raman, cm^{-1}): 826s, 665, 423, 403, 351, 341.

Se67 Potassium hydronium uranyl selenate hydrate $K_3(H_3O)(UO_2)_4(SeO_4)_6{\cdot}9H_2O$

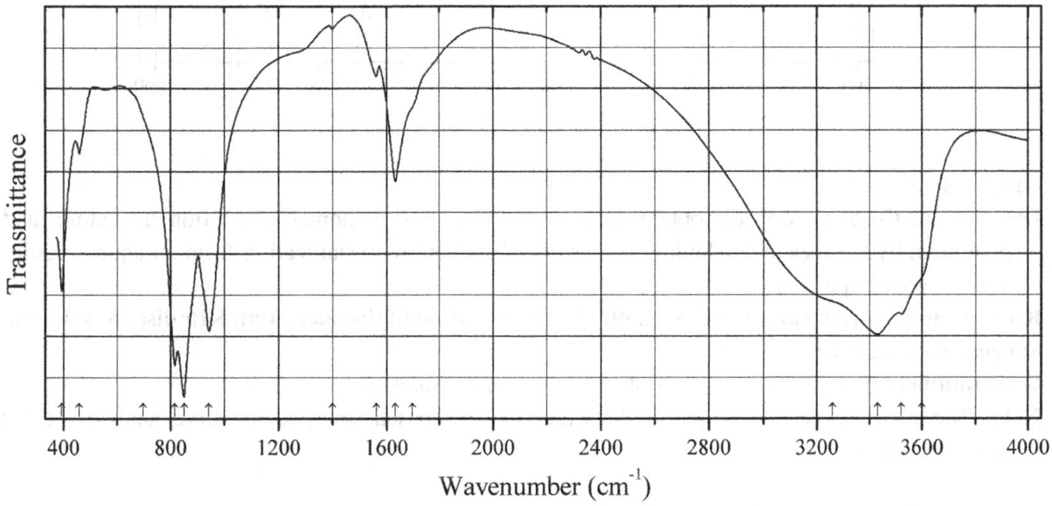

Origin: Synthetic.

Description: Yellow-green crystals prepared by evaporation of the mixture of an aqueous solution containing uranyl nitrate, selenic acid, potassium carbonate, and carbamide. The crystal structure is solved. Triclinic, space group $P2_1/m$, $a = 12.001(3)$, $b = 13.613(3)$, $c = 13.753(3)$ Å, $\beta = 109.187(4)°$, $V = 2122.0(8)$ Å^3, $Z = 2$. $D_{calc} = 3.467$ g/cm^3. The structure contains two kinds of uranyl cations which are coordinated by five and three O atoms.

Kind of sample preparation and/or method of registration of the spectrum: KBr disc. Absorption.

Source: Gurzhiy et al. (2014).

Wavenumbers (cm^{-1}): 3600sh, 3520, 3430s, 3260sh, 1698sh, 1635, 1561w, 1399w, 943s, 851s, 816s, ~680sh, 459, 392.

Note: The wavenumbers were partly determined by us based on spectral curve analysis of the published spectrum. The weak band 1399 cm^{-1} indicates that H_3O^+ groups are partly dissociated into H_2O and H^+.

Se68 Potassium sodium selenate $K_3Na(SeO_4)_2$

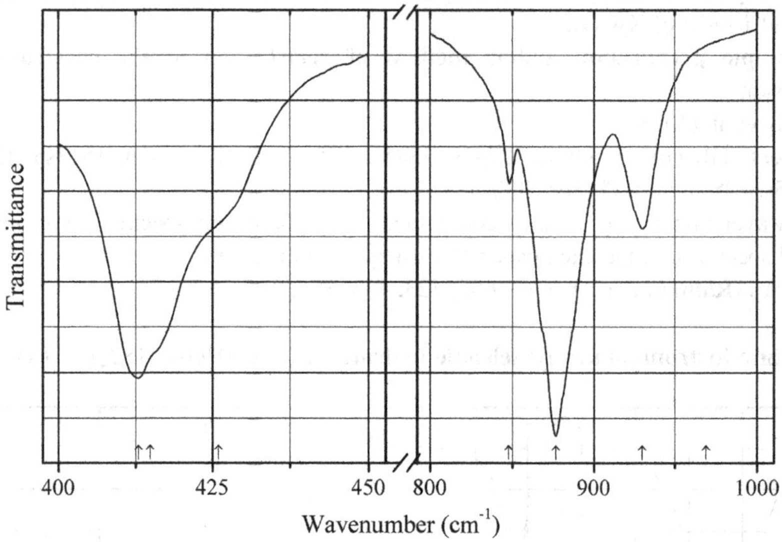

Origin: Synthetic.

Description: Colorless crystals obtained at 300 K from a stoichiometric solution of sodium and potassium hydroxides and selenium acid with subsequent recrystallization from an aqueous solution. Trigonal, space group *P*-3*m*1.

Kind of sample preparation and/or method of registration of the spectrum: KBr disc. Absorption.

Source: Kaczmarski et al. (2000).

Wavenumbers (cm^{-1}): 930, 877, 869sh, 848, 426sh, 415sh, 413.

Note: The wavenumbers were partly determined by us based on spectral curve analysis of the published spectrum.

Se69 Potassium yttrium selenite $KY(SeO_3)_2$

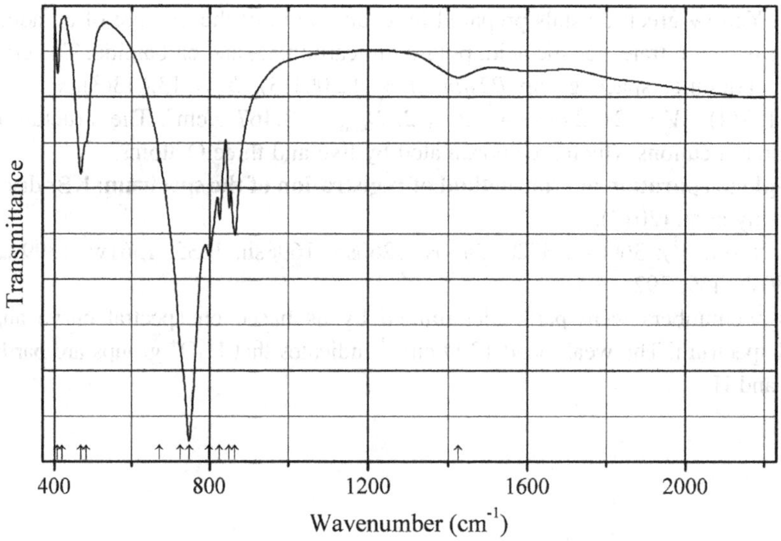

Origin: Synthetic.

Description: Synthesized hydrothermally from K_2CO_3, $Y(NO_3)_3·6H_2O$, and SeO_2 at 230 °C for 4 days. Characterized by powder X-ray diffraction data. The crystal structure is solved. Orthorhombic, space group *Pnma*, a = 13.3838(2), b = 5.70270(10), c = 8.6759(2) Å, V = 662.18(2) $Å^3$, Z = 4. D_{calc} = 3.831 g/cm^3.

Kind of sample preparation and/or method of registration of the spectrum: KBr disc. Transmission.

Source: Bang et al. (2014).

Wavenumbers (cm^{-1}): 1425, 865, 850, 825, 797, 747s, 725sh, 670, 482sh, 468, 420sh, 409w.

Note: The wavenumbers were partly determined by us based on spectral curve analysis of the published spectrum.

Se70 Potassium zinc selenite $K_2Zn_3(SeO_3)_4$ $K_2Zn_3(SeO_3)_4$

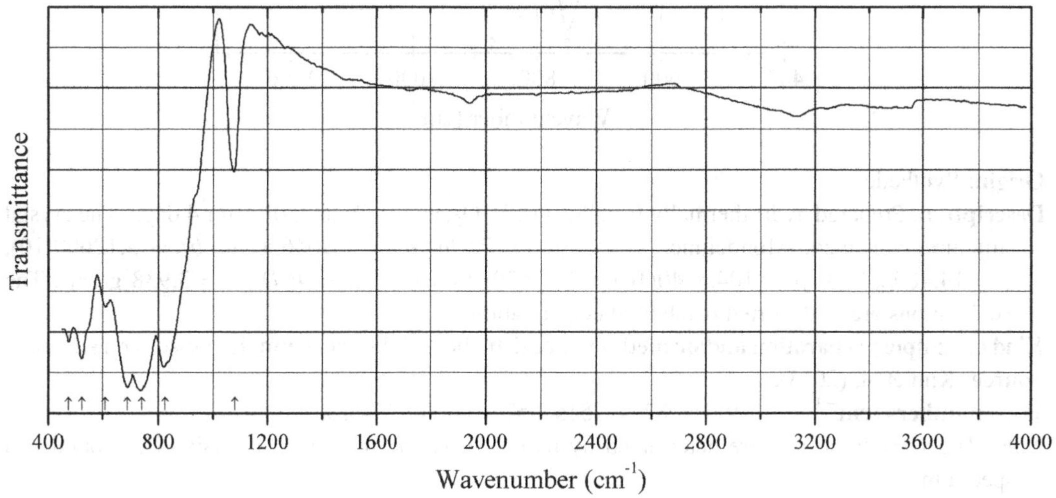

Origin: Synthetic.

Description: Prepared hydrothermally from K_2CO_3, $ZnCO_3$, and SeO_2 at 230 °C for 6 days. The crystal structure is solved. Monoclinic, space group *C*2/*c*, a = 11.3584(12), b = 8.6091(9), c = 13.6816(14) Å, β = 93.456(2)°, V = 1335.4(2) $Å^3$, Z = 4. D_{calc} = 3.890 g/cm^3.

Kind of sample preparation and/or method of registration of the spectrum: KBr disc. Transmission.

Source: Liu et al. (2015c).

Wavenumbers (cm^{-1}): 1080, 825s, 740s, 690s, 609, 525s, 474.

Note: The wavenumbers were partly determined by us based on spectral curve analysis of the published spectrum. The band at 1080 cm^{-1} was assigned by the authors to Se–O vibrations. However this band as well as weak bands above 1080 cm^{-1} should be assigned to impurities.

Se71 Scandium vanadyl selenite α-$ScVSe_2O_8$

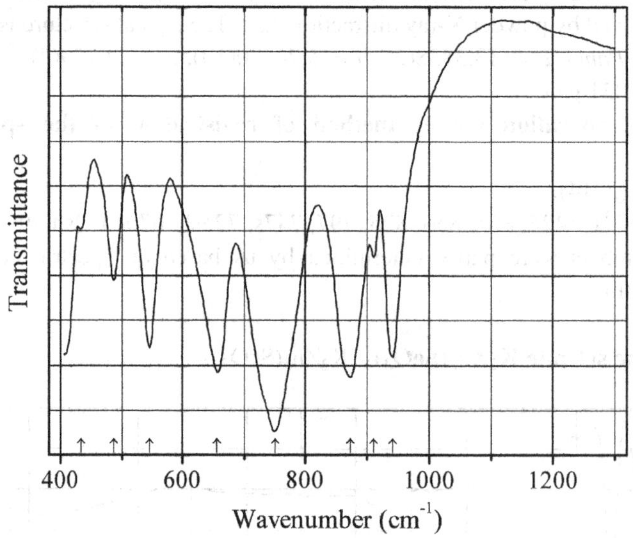

Origin: Synthetic.

Description: Prepared hydrothermally from Sc_2O_3, V_2O_5, and SeO_2 at 230 °C for 4 days. The crystal structure is solved. Monoclinic, space group $P2_1/n$, a = 8.96460(10), b = 5.12600(10), c = 14.4802(2) Å, β = 104.5740(10)°, V = 570.09(10) Å^3, Z = 4. D_{calc} = 3.938 g/cm^3. The Sc^{3+} cations are in distorted octahedral coordination.

Kind of sample preparation and/or method of registration of the spectrum: KBr disc. Transmission.

Source: Kim et al. (2013).

Wavenumbers (cm^{-1}): 941, 911w, 873s, 751s, 657s, 545, 487, 433w.

Note: The wavenumbers were determined by us based on spectral curve analysis of the published spectrum.

Se72 Scandium vanadyl selenite β-$ScVSe_2O_8$

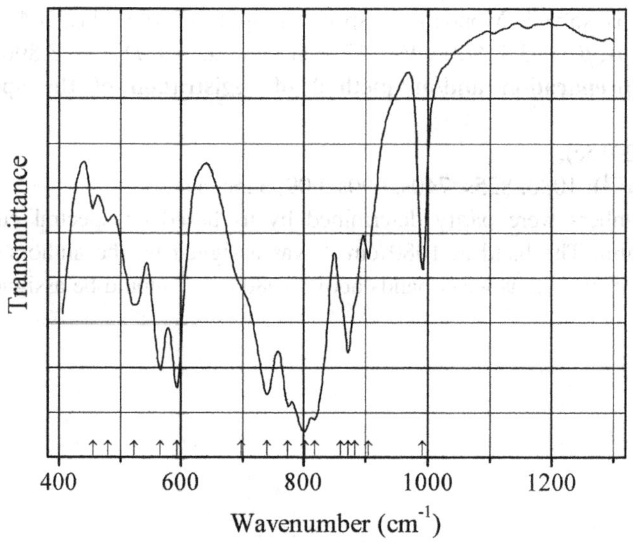

Origin: Synthetic.

Description: Prepared by a solid-state reaction from Sc_2O_3, V_2O_5, and SeO_2, first at 350 °C for 5 h, and thereafter at 450 °C for 48 h. The crystal structure is solved. Monoclinic, space group $P2_1/c$, $a = 6.59040(10)$, $b = 15.9098(3)$, $c = 6.63740(10)$ Å, $\beta = 92.2790(10)°$, $V = 695.39(2)$ Å^3, $Z = 4$. $D_{calc} = 3.647 g/cm^3$. The structure is composed of ScO_7, VO_5, and SeO_3 coordination polyhedra.

Kind of sample preparation and/or method of registration of the spectrum: KBr disc. Transmission.

Source: Kim et al. (2013).

Wavenumbers (cm^{-1}): 991, 905, 883sh, 872, 860sh, 818s, 802s, 775s, 740s, 698sh, 594s, 566, 523, 480w, 455w.

Note: The wavenumbers were determined by us based on spectral curve analysis of the published spectrum.

Se73 Sodium acid selenite $Na(HSeO_3)$

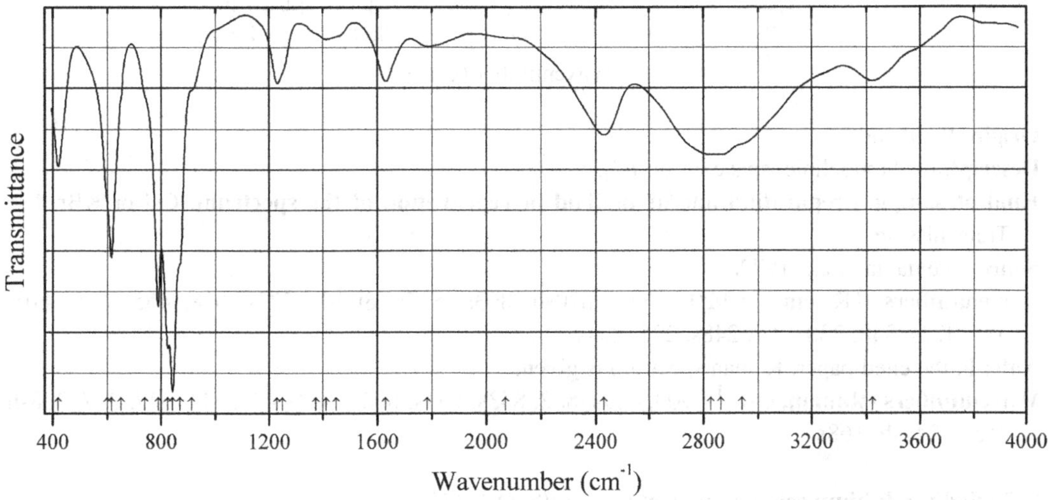

Origin: Synthetic.

Description: Crystals grown by slow evaporation of an aqueous solution containing 2 moles of SeO_2 and 1 mole of Na_2CO_3.

Kind of sample preparation and/or method of registration of the spectrum: KBr disc. Transmission.

Source: Cody et al. (1978).

Wavenumbers (IR, cm^{-1}): 3423, 2942sh, 2865, 2827sh, 2430, 2070sh, 1783w, 1630, 1450sh, 1412w, 1375sh, 1252sh, 1231, 912sh, 870sh, 842s, 825sh, 788s, 738sh, 650sh, 615s, 600sh, 420.

Note: The wavenumbers were partly determined by us based on spectral curve analysis of the published spectrum. In the cited paper, Raman spectrum is given.

Wavenumbers (Raman, cm^{-1}): 871s, 845s, 813s, 785, 601, 583, 449, 439, 378, 353, 337, 319.

Se74 Sodium cadmium selenate hydrate $NaCd(SeO_4)_2·2H_2O$

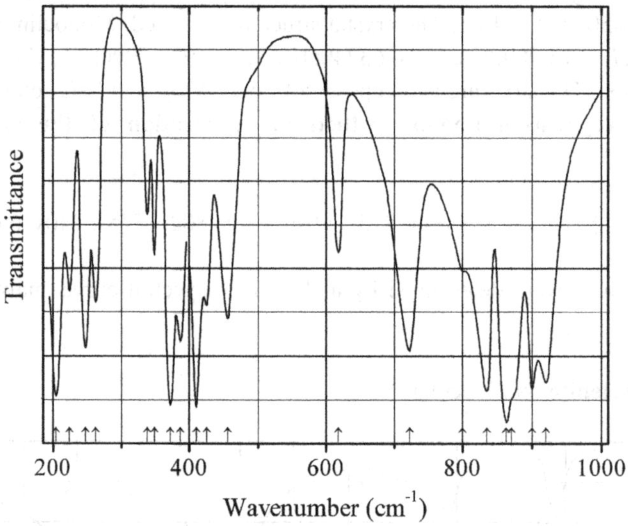

Origin: Synthetic.
Description: Monoclinic, space group *P*2₁/*c*.
Kind of sample preparation and/or method of registration of the spectrum: CsI or KBr disc. Transmission.
Source: Peytavin et al. (1972a).
Wavenumbers (IR, cm^{-1}): 920s, 900s, 870sh, 863s, 835s, 800sh, 722s, 618, 455, 425, 410s, 387, 372s, 348, 338, 263, 248s, 225, 204s.
Note: In the cited paper, Raman spectrum is given.
Wavenumbers (Raman, cm^{-1}): 909, 889s, 882, 843s, 831s, 600w, 460, 433, 411, 357, 337, 308sh, 255w, 241sh, 168s.

Se75 Sodium lithium selenate hydrate $Na_3Li(SeO_4)_2·6H_2O$

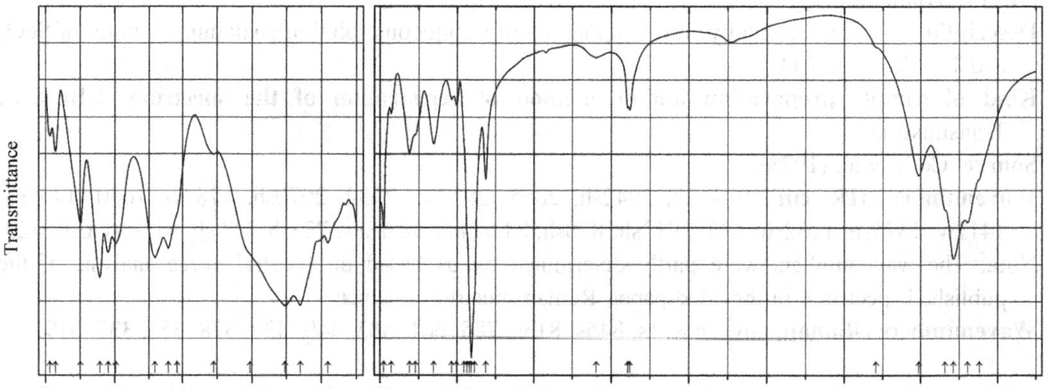

Origin: Synthetic.

Description: Colorless crystals grown by evaporation of an aqueous solution of Na_2SeO_4 and Li_2SeO_4 with a 1:1 ratio at 300 K. Trigonal, space group *R3c*, $Z = 2$. Confirmed by powder X-ray diffraction data.

Kind of sample preparation and/or method of registration of the spectrum: KBr disc and Nujol mull. Absorption.

Source: Hanuza et al. (2008a).

Wavenumbers (IR, cm^{-1}): 3505sh, 3443, 3370s, 3327, 3190, 2964sh, 1696, 1686, 1522w, 950, 892sh, 875s, 864sh, 852sh, 836sh, 796w, 770w, 677, 583sh, 553, 459w, 420, 398, 347sh, 295w, 242sh, 229, 209, 151, 140, 128, 100, 64w, 56w.

Note: In the cited paper, polarized Raman spectra are given.

Wavenumbers (Raman, with the *z*(*xx*)*z* polarization, cm^{-1}): 3440, 3372, 3215, 1690w, 903w, 866sh, 853s, 835s, 807, 454, 419sh, 398, 354, 346sh, 283, 205, 190, 167, 153, 144, 114.

Se76 Sodium yttrium selenite $NaY(SeO_3)_2$

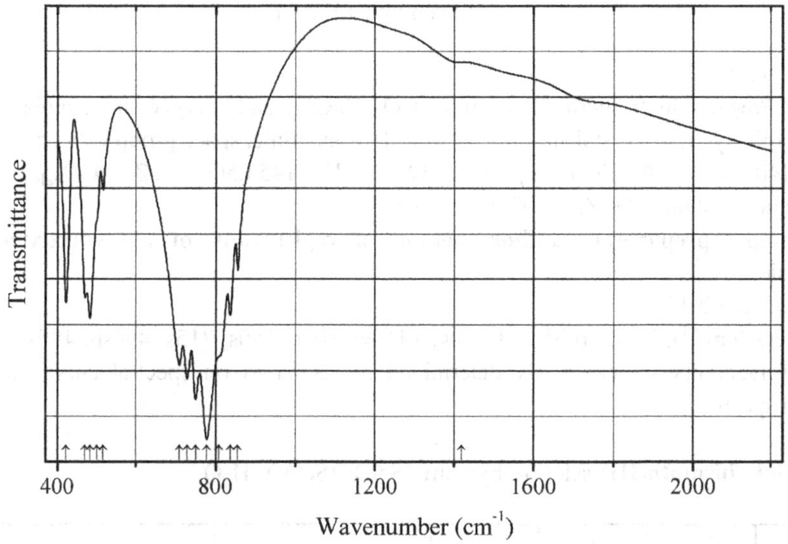

Origin: Synthetic.

Description: Colorless crystals prepared hydrothermally from Na_2CO_3, $Y(NO_3)_3 \cdot 6H_2O$, and SeO_2 at 230 °C for 4 days. Characterized by powder X-ray diffraction data. The crystal structure is solved. Orthorhombic, space group $P2_1cn$, $a = 5.397(2)$, $b = 8.525(2)$, $c = 12.765(2)$ Å, $V = 587.3(3)$ Å^3, $Z = 4$. $D_{calc} = 4.132$ g/cm^3. The structure is based on a 3D framework consisting of YO_7 monocapped trigonal prisms and SeO_3 trigonal pyramids.

Kind of sample preparation and/or method of registration of the spectrum: KBr disc. Transmission.

Source: Bang et al. (2014).

Wavenumbers (cm^{-1}): (1420sh), 854, 835, 808sh, 777s, 749s, 728s, 708s, 516w, 500sh, 483, 471, 422.

Note: The wavenumbers were partly determined by us based on spectral curve analysis of the published spectrum.

Se77 Sodium zinc selenite $Na_2Zn_3(SeO_3)_4 \cdot 2H_2O$ $Na_2Zn_3(SeO_3)_4 \cdot 2H_2O$

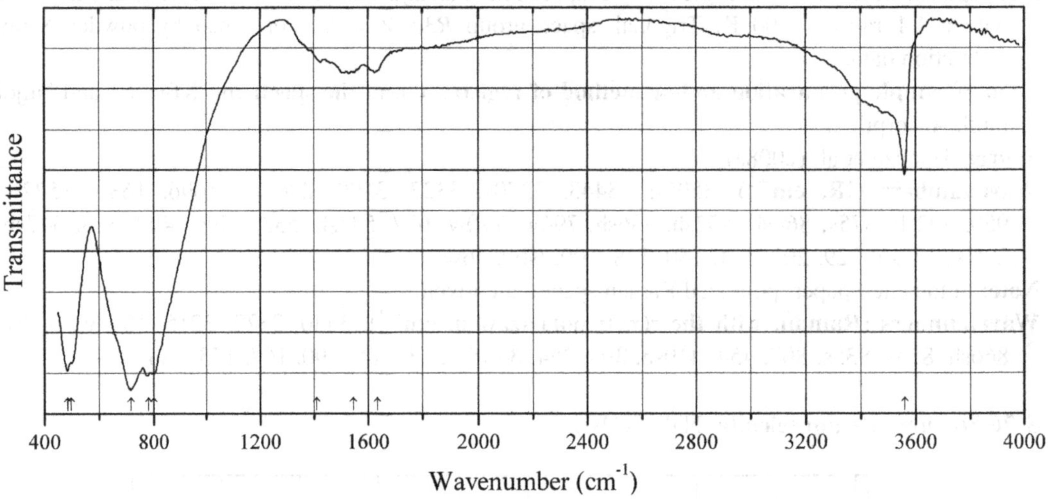

Origin: Synthetic.

Description: Prepared hydrothermally from Na_2CO_3, $ZnCO_3$, and SeO_2 (with the molar ratio 1:1:2) at 230 °C for 6 days. The crystal structure is solved. Monoclinic, space group *C*2/*c*, $a = 15.7940(18)$, $b = 6.5744(8)$, $c = 14.6787(17)$ Å, $\beta = 107.396(3)°$, $V = 1454.5(3)$ Å^3, $Z = 4$. $D_{calc} = 3.589$ g/cm^3. The structure contains 2D $[Zn_3(SeO_3)_4]^{2-}$ sheets.

Kind of sample preparation and/or method of registration of the spectrum: KBr disc. Transmission.

Source: Liu et al. (2015c).

Wavenumbers (cm^{-1}): 3560, 1635w, 1545w, 1410w, 800s, 780s, 715s, 496sh, 485s.

Note: The wavenumbers were partly determined by us based on spectral curve analysis of the published spectrum.

Se78 Strontium bismuth(III) selenite hydrate $Sr_3Bi_2(SeO_3)_6 \cdot H_2O$

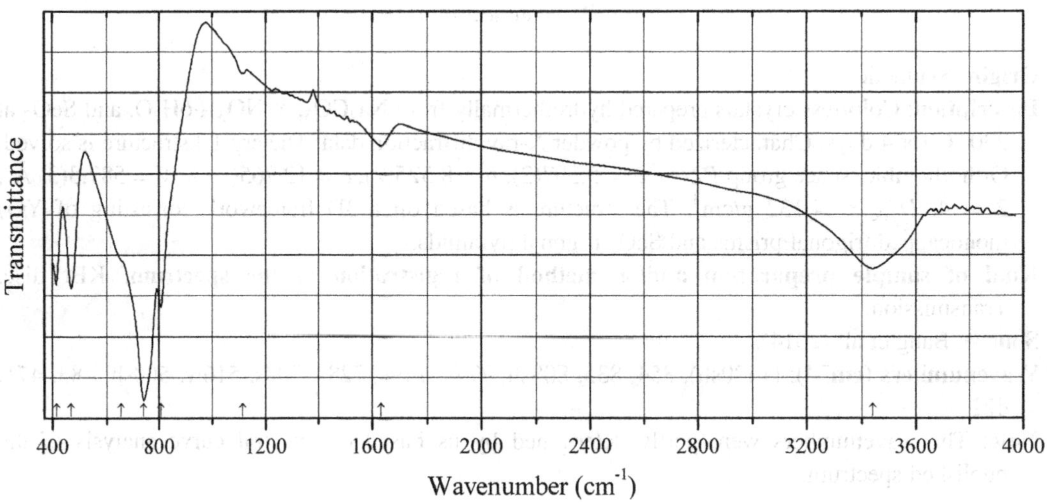

Origin: Synthetic.

Description: Crystals grown hydrothermally using $SrCO_3$, Bi_2O_3, and SeO_2 as starting reactants, at 230 °C for 4 days. Characterized by powder X-ray diffraction and TG data. The crystal structure is solved. Monoclinic, space group $P2_1/m$, a = 7.0054(10), b = 17.5092(3), c = 7.3053(10) Å, β = 92.299(10)°, V = 895.34(2) Å^3, Z = 2. D_{calc} = 5.418 g/cm^3. Bi^{3+} has sevenfold coordination.

Kind of sample preparation and/or method of registration of the spectrum: KBr disc. Transmission.

Source: Ahn et al. (2015).

Wavenumbers (cm^{-1}): 3440, 1633w, 1114w, 807, 740s, 654sh, 469, 414.

Note: The wavenumbers were partly determined by us based on spectral curve analysis of the published spectrum.

Se79 Strontium selenate $SrSeO_4$

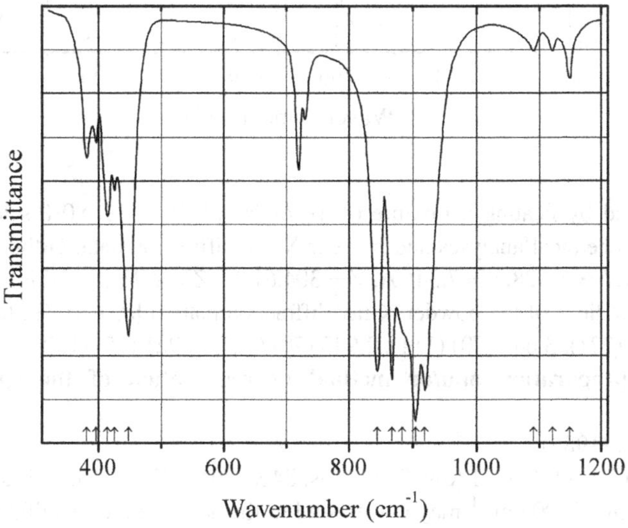

Origin: Synthetic.

Description: Prepared by precipitation from hot aqueous solutions of sodium selenate and lead acetate. Monoclinic, space group $P2_1/n$, a = 7.087, b = 7.317, c = 6.862 Å, β = 103.55°.

Kind of sample preparation and/or method of registration of the spectrum: Nujol mull. Transmission.

Source: Scheuermann and Schutte (1973b).

Wavenumbers (IR, cm^{-1}): 1148, 1121w, 1091w, 919s, 904s, 884sh, 867s, 844s, 447s, 425, 414, 396, 381.

Note: The bands in the range from 1000 to 1200 cm^{-1} may correspond to the admixture of a sulfate. Bands near 720 cm^{-1} may correspond to Nujol. In the cited paper, powder Raman spectrum is given.

Wavenumbers (Raman, cm^{-1}): 913, 905, 898, 887, 882, 852, 464, 465, 445, 424, 396, 376, 340, 326.

Se80 Tellurium(IV) oxyselenate $Te_2(SeO_4)O_3$

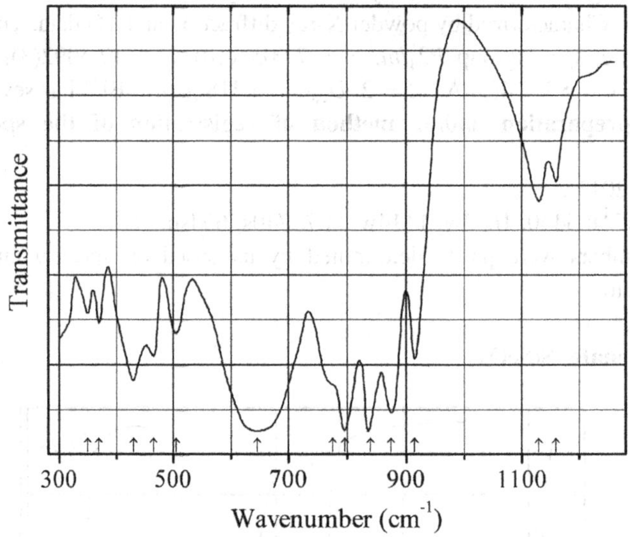

Origin: Synthetic.

Description: Prepared by heating a mixture of Te or $Na_2(TeO_3)$ with 80% selenic acid at 160 °C. Characterized by chemical analyses and powder X-ray diffraction data. Orthorhombic, space group $P2_1mn$, $a = 4.807$, $b = 8.628$, $c = 7.346$ Å, $V = 304.67$ Å^3, $Z = 2$. $D_{meas} = 4.82$ g/cm^3, $D_{calc} = 4.85$ g/cm^3. The strongest lines of the powder X-ray diffraction pattern [*d*, Å (*I*, %) (*hkl*)] are: 4.028 (100) (101), 3.724 (80) (021), 3.649 (70) (111), 2.943 (70) (121), 2.921 (75) (102), 2.450 (75) (003).

Kind of sample preparation and/or method of registration of the spectrum: KBr disc. Transmission.

Source: Gaitán et al. (1985).

Wavenumbers (cm^{-1}): 1160w, 1130w, 915, 875s, 840s, 795s, 775sh, 645s, 505, 465, 430, 370, 350.

Note: The bands above 1100 cm^{-1} may correspond to the admixture of a sulfate.

Se81 Tellurium oxyselenite $Te(SeO_3)O$

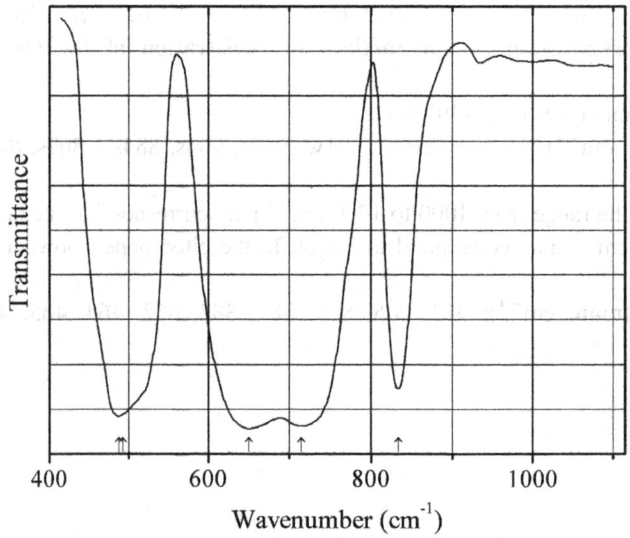

Origin: Synthetic.

Description: Prepared by heating a mixture of TeO_2 and SeO_2 (with the 1:1.15 molar ratio) at 370 °C for 3 days. Characterized by powder X-ray diffraction data. The crystal structure is solved. Monoclinic, space group *Ia*, $a = 4.3568(8)$, $b = 12.465(3)$, $c = 6.7176(15)$ Å, $\beta = 90.825(4)°$, $V = 364.77(14)$ Å^3, $Z = 4$. $D_{calc} = 4.927$ g/cm^3. Te has fivefold coordination.

Kind of sample preparation and/or method of registration of the spectrum: KBr disc. Transmission.

Source: Porter et al. (2001).

Wavenumbers (cm^{-1}): 833, 715s, 650s, ~510, 488s.

Se82 Vanadium(III) antimony(V) selenite $VSb(SeO_3)_4$

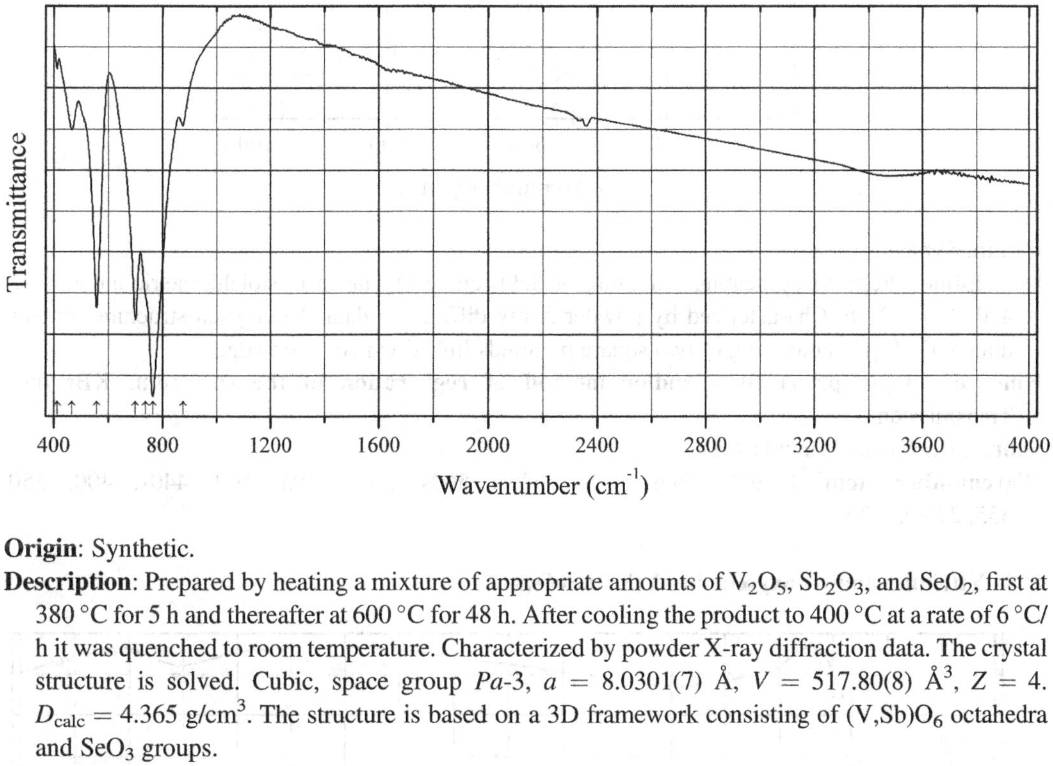

Origin: Synthetic.

Description: Prepared by heating a mixture of appropriate amounts of V_2O_5, Sb_2O_3, and SeO_2, first at 380 °C for 5 h and thereafter at 600 °C for 48 h. After cooling the product to 400 °C at a rate of 6 °C/h it was quenched to room temperature. Characterized by powder X-ray diffraction data. The crystal structure is solved. Cubic, space group *Pa*-3, $a = 8.0301(7)$ Å, $V = 517.80(8)$ Å^3, $Z = 4$. $D_{calc} = 4.365$ g/cm^3. The structure is based on a 3D framework consisting of $(V,Sb)O_6$ octahedra and SeO_3 groups.

Kind of sample preparation and/or method of registration of the spectrum: KBr disc. Transmission.

Source: Shin et al. (2013).

Wavenumbers (cm^{-1}): 875w, 764s, 738sh, 700s, 557s, 466, 412w.

Note: The wavenumbers were partly determined by us based on spectral curve analysis of the published spectrum.

Se83 Vanadyl selenite $(VO)(SeO_3)$

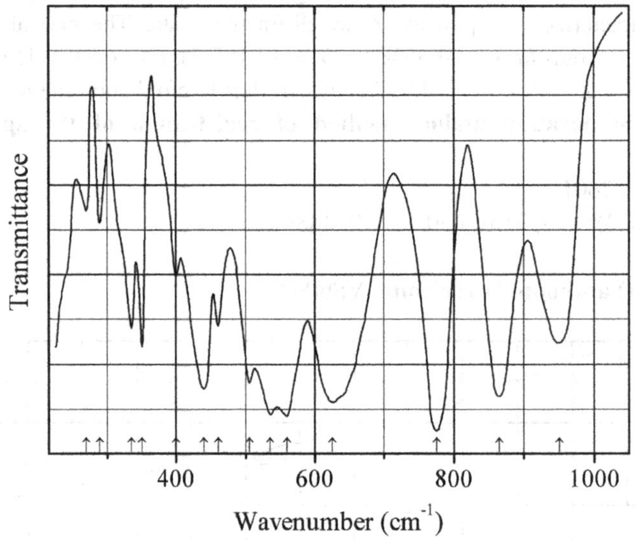

Origin: Synthetic.

Description: Prepared by heating a mixture of SeO_2 and VO_2, taken in stoichiometric amounts, at 400 °C for 24 h. Characterized by powder X-ray diffraction data. The crystal structure contains dimers $[V_2O_8]^{8-}$ consisting of two square pyramids linked via common edge.

Kind of sample preparation and/or method of registration of the spectrum: KBr disc. Transmission.

Source: Rocha and Baran (1988).

Wavenumbers (cm^{-1}): 950, 865s, 775s, 625s, 560s, 535s, 505, 460, 440s, 400, 350, 335, 290w, 270w.

Se84 Yttrium vanadyl oxyselenite $Y(VO)(SeO_3)_2O$

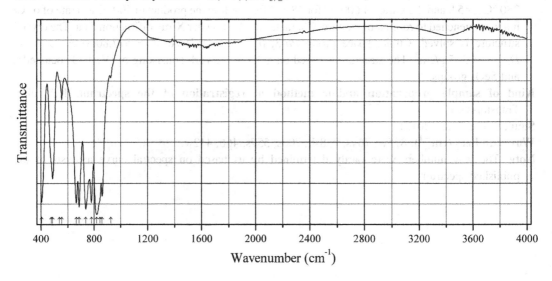

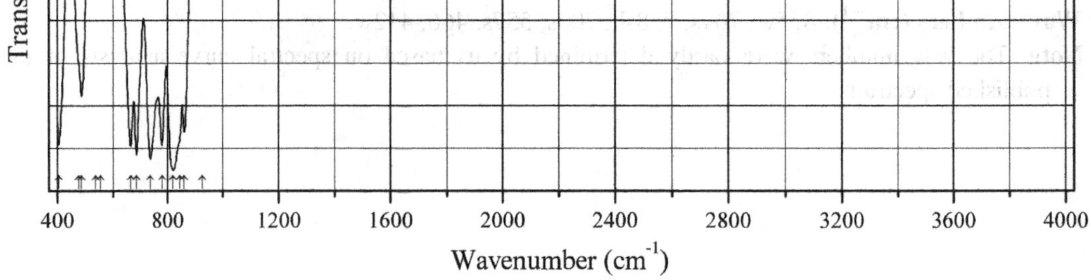

Origin: Synthetic.

Description: Crystals grown hydrothermally from Y_2O_3, V_2O_5, and SeO_2 at 450 °C for 48 h with three intermediate regrindings. Characterized by powder X-ray diffraction data. The crystal structure is solved. Orthorhombic, space group *Amb*2, $a = 10.4036(4)$, $b = 7.5904(3)$, $c = 7.8341(3)$ Å, $V = 618.64(4)$ Å^3, $Z = 4$. $D_{calc} = 4.571$ g/cm^3. V^{5+} has sixfold coordination with one short V–O bond.

Kind of sample preparation and/or method of registration of the spectrum: KBr disc. Transmission.

Source: Kim et al. (2014).

Wavenumbers (cm^{-1}): 926w, 862s, 845sh, 821s, 779s, 735s, 684s, 663s, 555, 538sh, 487, 477sh, 406s.

Note: The wavenumbers were partly determined by us based on spectral curve analysis of the published spectrum.

Se85 Alfredopetrovite $Al_2(Se^{4+}O_3)_3 \cdot 6H_2O$

Origin: Synthetic.

Description: Hexagonal, space group *P*62*c*, $Z = 2$.

Kind of sample preparation and/or method of registration of the spectrum: KBr disc. Transmission.

Source: Ratheesh et al. (1997).

Wavenumbers (IR, cm^{-1}): 3286–2900s (broad), 1638 (broad), 1372w, 857sh, 770s, 570, 546 (broad), 492w (broad), 460, 421w, 372sh, 360, 310, 295sh, 241w, 221w.

Note: The band at 1372 cm^{-1} may correspond to the admixture of NO_3^- in KBr. In the cited paper, Raman spectrum is given.

Wavenumbers (Raman, cm^{-1}): 3165s, 3024s, 1640w, 1329w, 874s, 831s, 750w, 722w, 704, 566w, 542, 532, 478w, 439w, 412, 347, 329w, 310w, 232, 191w, 174w, 123, 110, 92w, 74w.

Se86 Cobaltomenite $Co(SeO_3) \cdot 2H_2O$

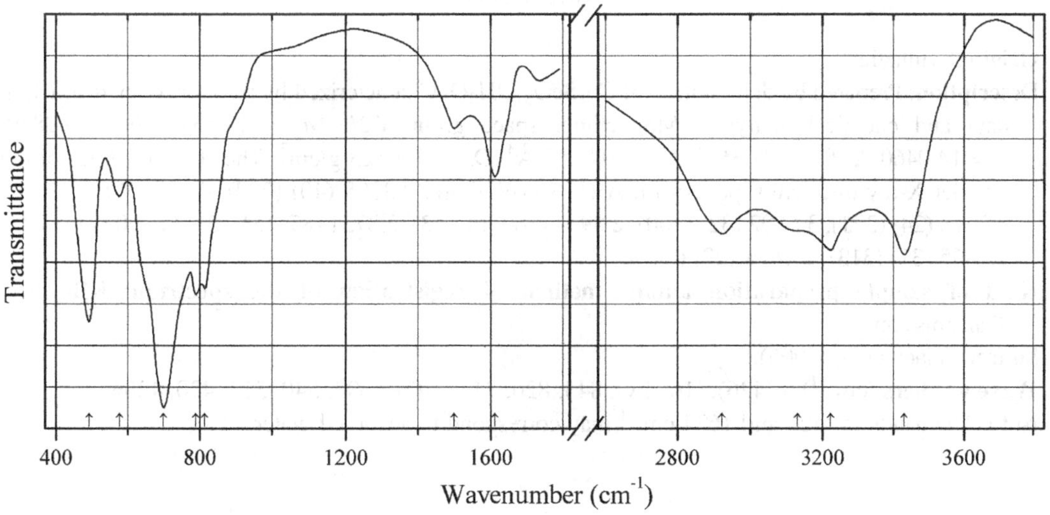

Origin: Synthetic.

Description: Prepared by precipitating an aqueous solution of cobalt(II) nitrate with aqueous solution of sodium selenite at 298 K. Characterized by powder X-ray diffraction data and chemical analyses. Monoclinic, space group *P*2$_1$/*n*, a = 6.5322, b = 8.8251, c = 7.6455 Å, β = 80.478°, V = 434.67 Å^3, Z = 4. D_{calc} = 3.392g/cm^3. The strongest lines of the powder X-ray diffraction pattern [*d*, Å (*I*, %) (*hkl*)] are: 5.7388 (100) (011), 4.4125 (29) (020), 3.7770 (50) (002), 3.2488 (39) (112), 2.7393 (55) (122), 2.6546 (26) (202), 2.5421 (30) (212), 2.4740 (27) (103), 2.3731 (41) (22−1).

Kind of sample preparation and/or method of registration of the spectrum: KBr disc. Transmission.

Source: Vlaev et al. (2005).

Wavenumbers (cm^{-1}): 3430s, 3224s, 3129sh, 2924s, 1613, 1500, 815s, 790s, 700s, 576, 492s.

Se87 Cobalt selenite Co(SeO$_3$)

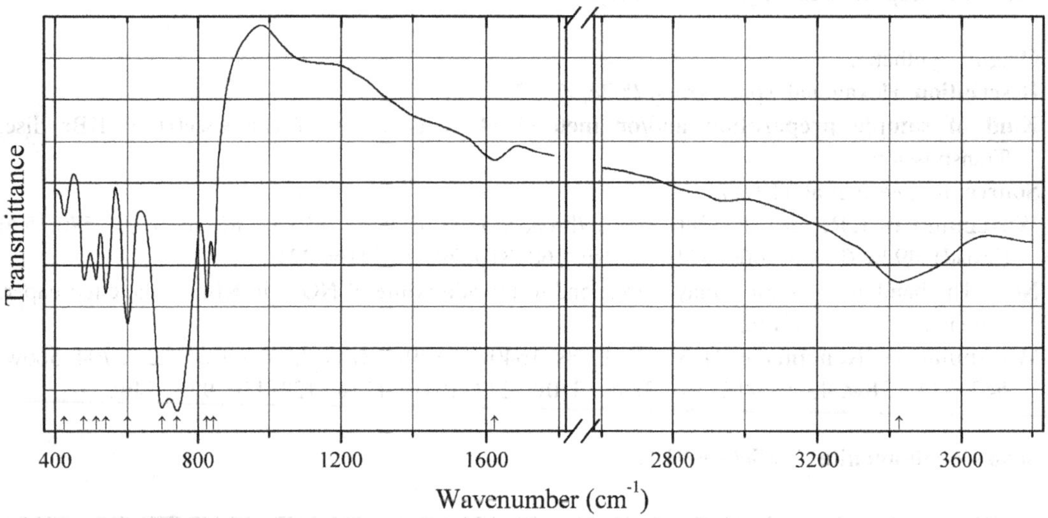

Origin: Synthetic.

Description: Prepared by dehydration of Co(SeO$_3$)·2H$_2$O. Characterized by powder X-ray diffraction data and chemical analyses. Monoclinic, space group *C*2/*c*, a = 14.5378, b = 9.9880, c = 14.0460 Å, β = 107.369°, V = 1946.53 Å^3. D_{calc} = 4.525 g/cm^3. The strongest lines of the powder X-ray diffraction pattern [*d*, Å (*I*, %) (*hkl*)] are: 7.3205 (40) (200), 4.1775 (26) (22−1), 3.7518 (24) (221), 3.6602 (32) (400), 2.8914 (100) (42−3), 2.8751 (87) (51−3), 2.8090 (77) (510), 2.7865 (31) (313), 2.4868 (29) (040).

Kind of sample preparation and/or method of registration of the spectrum: KBr disc. Transmission.

Source: Vlaev et al. (2005).

Wavenumbers (cm^{-1}): (3426), (1624w), 845, 826, 742s, 700s, 600, 540, 513, 480, 426w.

Note: The bands at 3426 and 1624 cm^{-1} may correspond to adsorbed water.

Se88 Cobalt selenite hydrate $Co(SeO_3)\cdot 1/3H_2O$

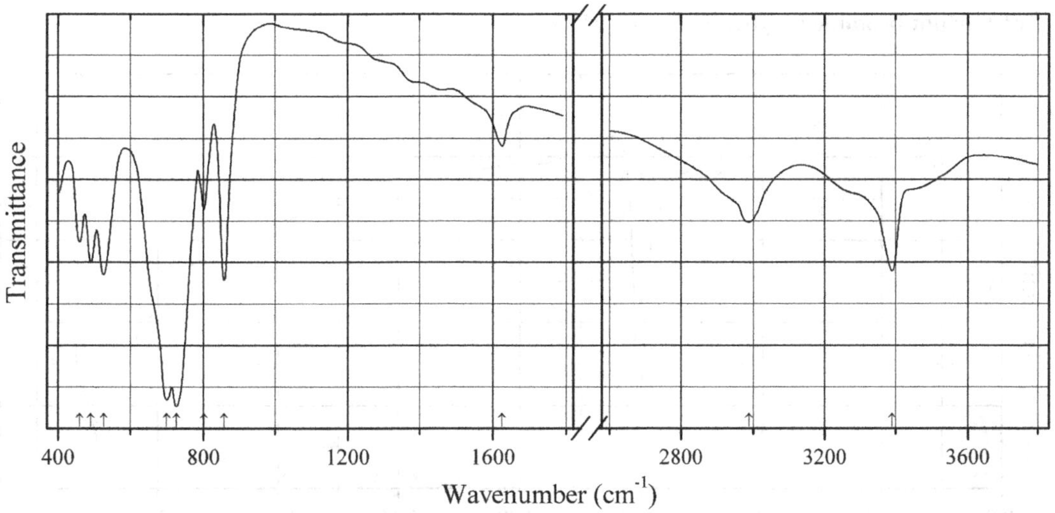

Origin: Synthetic.

Description: Prepared by partial dehydration of $Co(SeO_3)\cdot 2H_2O$. Characterized by powder X-ray diffraction data and chemical analyses. Triclinic, space group *P*-1, $a = 8.1197$, $b = 8.4383$, $c = 8.5345$ Å, $\alpha = 123.816°$, $\beta = 90.538°$, $\gamma = 111.591°$, $V = 434.02$ Å^3, $Z = 4$. $D_{calc} = 3.392$ g/cm^3.

Kind of sample preparation and/or method of registration of the spectrum: KBr disc. Transmission.

Source: Vlaev et al. (2005).

Wavenumbers (cm^{-1}): 3390, 2988, 1626w, 858, 802, 726s, 700s, 526, 491, 460.

Se90 Orlandiite $Pb_3Cl_4(Se^{4+}O_3)\cdot H_2O$

Origin: Baccu Locci mine, near Villaputzu, Sardinia, Italy (type locality).

Description: White aggregates from the association with chalcomenite, pseudoboléite, anglesite, etc. Holotype sample. Triclinic, $a = 8.290(8)$, $b = 10.588(13)$, $c = 13.587(15)$ Å, $\alpha = 124.47(8)°$, $\beta = 110.60(9)°$, $\gamma = 63.26(9)°$, $Z = 2$. $D_{calc} = 5.55$ g/cm^3. The empirical formula is $Pb_3[Cl_{3.68}(OH)_{0.32}](SeO_3)\cdot H_2O$. The strongest lines of the powder X-ray diffraction pattern [*d*, Å (*I*, %) (*hkl*)] are: 4.000 (100) (002), 3.258 (75) (−121), 3.188 (75) (−201), 3.818 (55) (201), 3.731 (44) (122), 2.103 (40) (142).

Kind of sample preparation and/or method of registration of the spectrum: Transmission. Kind of sample preparation is not indicated.

Source: Campostrini et al. (1999).

Wavenumbers (cm^{-1}): 3410–3160 (broad), 1586, 788, 724.

2.16 Bromides

Br3 Barium bromide dihydrate $BaBr_2{\cdot}2H_2O$

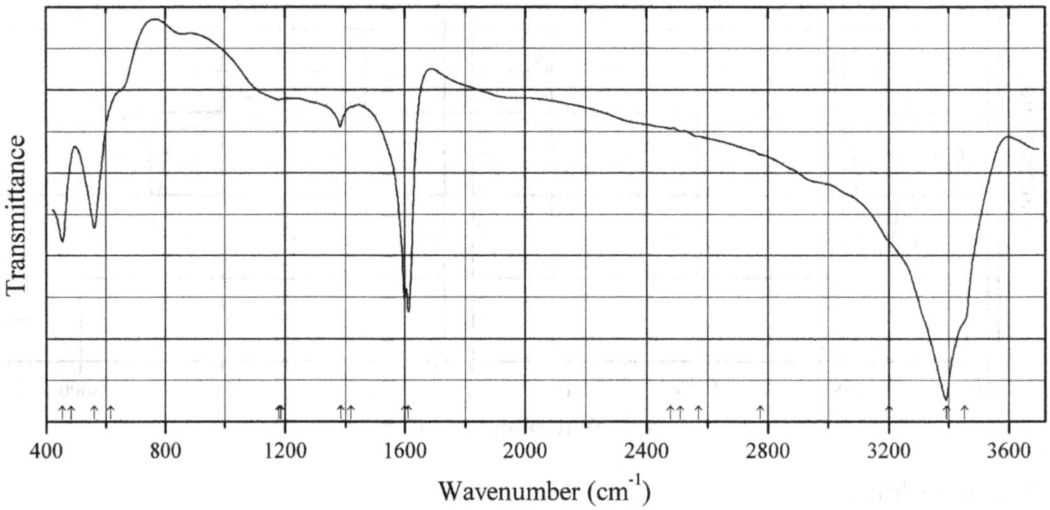

Origin: Synthetic.

Description: Obtained by crystallization from aqueous solution at room temperature. Monoclinic, space group *C*2/*c*, $a = 10.449$, $b = 7.204$, $c = 8.385$ Å, $\beta = 113.48°$, $Z = 4$.

Kind of sample preparation and/or method of registration of the spectrum: KBr disc. Transmission.

Source: Lutz et al. (1978).

Wavenumbers (IR, cm^{-1}): 3453sh, 3387s, 3200sh, 2775sh, 1613s, 1600s, 1417–1419, 1178–1186, 616sh, 561–566, (452), 420s.

Note: In the cited paper, Raman spectrum is given.

2.17 Molybdates

Mo40 Ammonium cuprooxopolymolybdate $(NH_4)_4[H_6CuMo_6O_{24}]\cdot 4H_2O$ $(NH_4)_4[H_6CuMo_6O_{24}]\cdot 4H_2O$

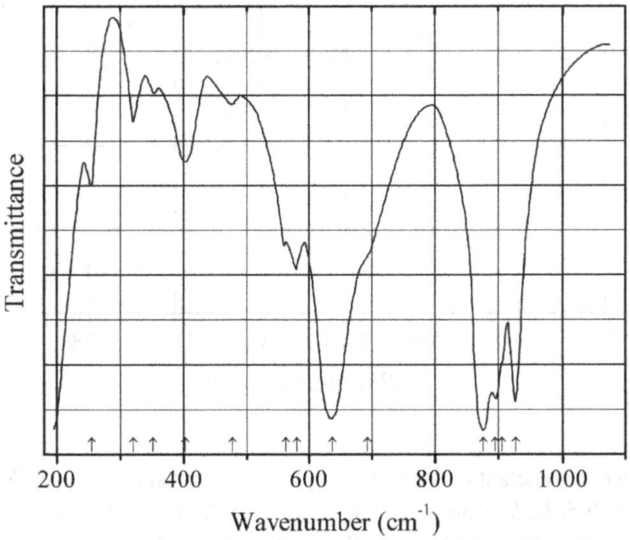

Origin: Synthetic.

Description: Obtained by precipitation in aqueous solution. Characterized by powder X-ray diffraction data. The crystal structure has been published elsewhere. In the cluster $[H_6CuMo_6O_{24}]$, both Cu and Mo have octahedral coordination. Six protons are bonded to the O atoms of the CuO_6 central polyhedron.

Kind of sample preparation and/or method of registration of the spectrum: KBr disc. Transmission.

Source: Botto et al. (1994).

Wavenumbers (IR, cm^{-1}): 3388s, 3144s, 2768sh, 2070w, 1631w, 1399s, 927s, 906, 895sh, 876s, 692sh, 637s, 581, 562, 478w, 403, 352w, 321, 255.

Note: In the cited paper, Raman spectrum is given.

Wavenumbers (Raman, cm^{-1}): 3120w (broad), 936s, 916, 873, 690w, 574w, 495w, 344, 252.

Note: The wavenumbers were partly determined by us based on spectral curve analysis of the published spectra.

Mo41 Ammonium heptamolybdate $(NH_4)_6Mo_7O_{24}\cdot 4H_2O$

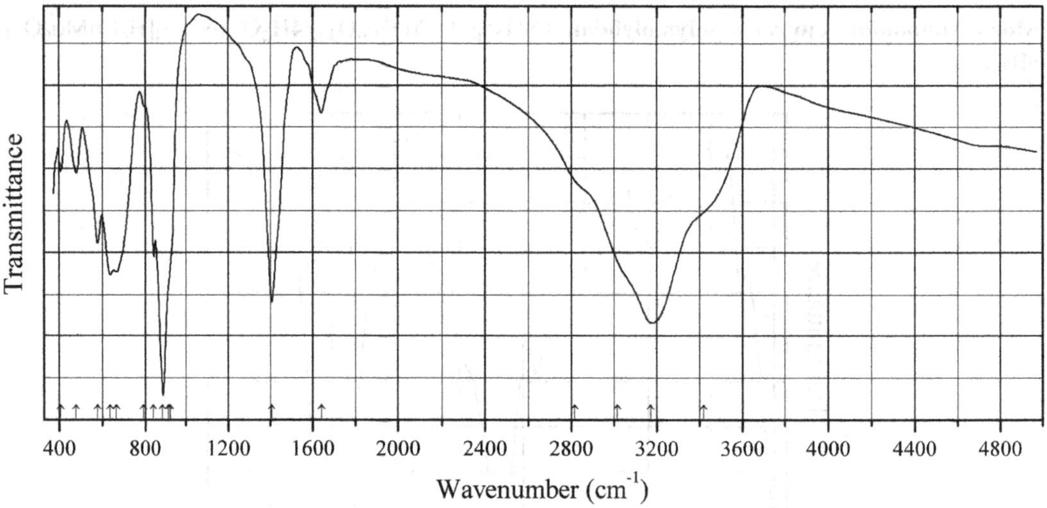

Origin: Synthetic.

Description: Commercial reactant characterized by powder and single-crystal X-ray diffraction data. Monoclinic, $a = 8.395(7)$, $b = 36.204(3)$, $c = 10.4765$ Å, $\beta = 115.884°$.

Kind of sample preparation and/or method of registration of the spectrum: KBr disc. Transmission.

Source: Wienold et al. (2003).

Wavenumbers (cm^{-1}): 3420sh, 3173s, 3017sh, 2817sh, 1642, 1406s, 925sh, 915sh, 887s, 840, 792w, 667, 636s, 578, 479, 406.

Mo42 Ammonium nickel molybdate $(NH_4)Ni_2(HMoO_4)(MoO_4)(OH)_2$ $(NH_4)Ni_2(HMoO_4)(MoO_4)(OH)_2$

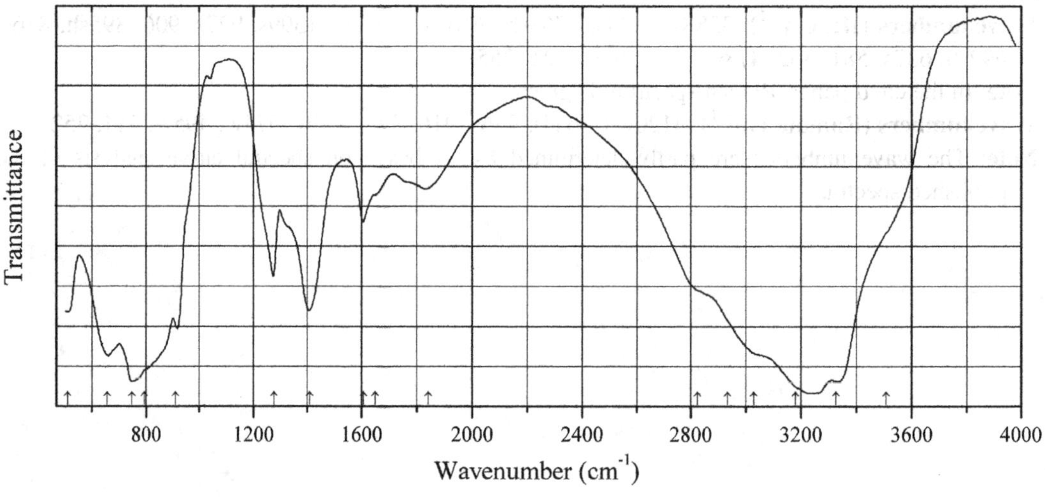

Origin: Synthetic.

Description: Prepared as a green precipitate by adding concentrated ammonium hydroxide to the solution of $(NH_4)_6Mo_7O_{24}·4H_2O$ and nickel nitrate containing 0.10 mol of Mo and 0.10 mol of Ni. The crystal structure is solved. Trigonal, space group *R*-3*m*, $a = 6.0147(4)$, $c = 21.8812(13)$ Å, $V = 685.53(7)$ Å^3, $Z = 3$. $D_{calc} = 3.446$ g/cm^3.

Kind of sample preparation and/or method of registration of the spectrum: Photoacoustic Fourier-transform IR spectroscopy.

Source: Levin et al. (1996).

Wavenumbers (IR, cm^{-1}): 3508sh, 3326, 3180, 3028, 2933, 2823, 1842, 1651, 1608, 1410, 1277, 913, 793, 746, 655, 512.

Note: In the cited paper, Raman spectrum is given.

Wavenumbers (Raman, cm^{-1}): 904, 321.

Mo43 Bismuth(III) ferrite dimolybdate $Bi_3(FeO_4)(MoO_4)_2$

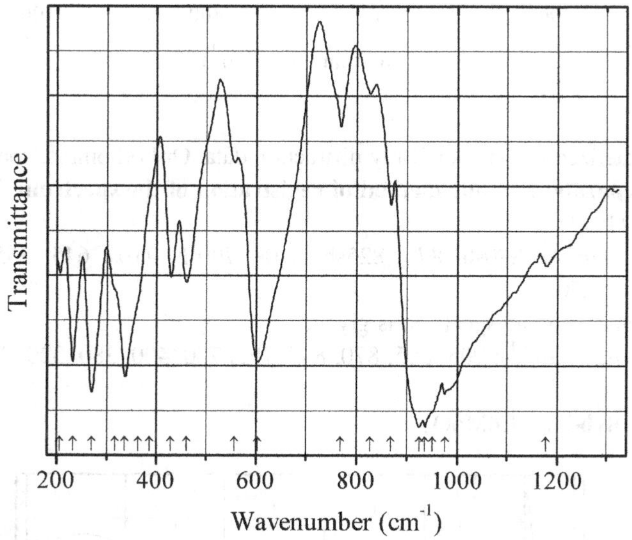

Origin: Synthetic.

Description: Synthesized from aqueous solution of ferric nitrate, bismuth nitrate, and ammonium molybdate heated first at 250 °C and then from 600 to 920 °C for 10 h. Characterized by powder X-ray diffraction data and Mössbauer spectrum. Monoclinic, $a = 16.904(1)$, $b = 11.653(1)$, $c = 5.2544(6)$ Å, $\beta = 107.15(1)°$, $V = 989.0(1)$ Å^3. $D_{calc} = 7.16$ g/cm^3. In the crystal structure, Fe and Mo are ordered.

Kind of sample preparation and/or method of registration of the spectrum: Nujol mull. Transmission.

Source: Jeitschko et al. (1976).

Wavenumbers (cm^{-1}): 1178w, 977s, 951sh, (936s), 925s, 868w, 827w, 769w, 603s, 557w, 461, 428, 386sh, 364sh, 337s, 316sh, 270s, 232, 206.

Note: The wavenumbers were determined by us based on spectral curve analysis of the published spectrum.

Mo44 Bismuth molybdate Bi_2MoO_6

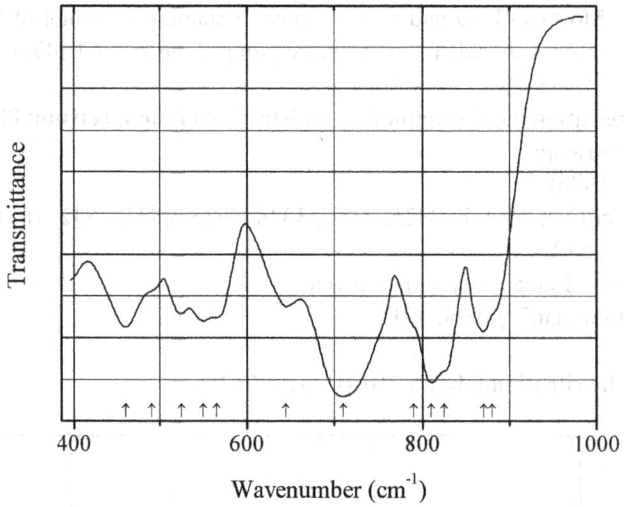

Origin: Synthetic.
Description: Characterized by powder X-ray diffraction data. Orthorhombic, space group *B2cb*.
Kind of sample preparation and/or method of registration of the spectrum: Transmission.
Source: Bode et al. (1973).
Wavenumbers (IR, cm^{-1}): 880sh, 870, 825sh, 810s, 790sh, 710s, 645, 565, 550, 525, 490sh, 460, 380, 335, 295, 250.
Note: In the cited paper, Raman spectrum is given.
Wavenumbers (Raman, cm^{-1}): 905, 885, 870, 830, 795, 730, 400, 380, 320, 265, 230, 210, 180.

Mo45 Cadmium molybdate $CdMoO_4$

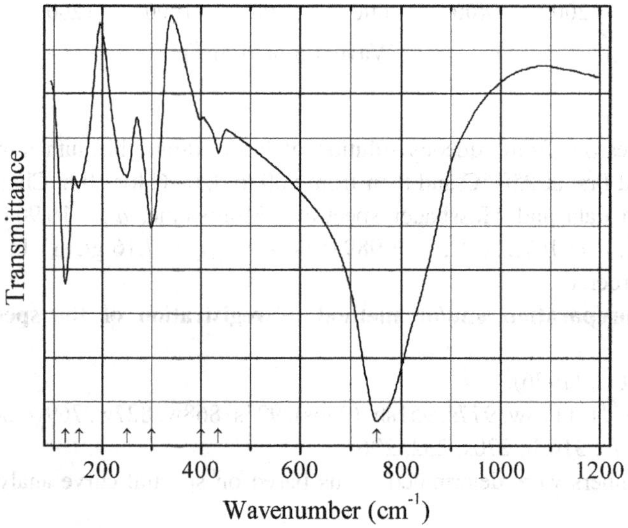

Origin: Synthetic.

Description: Prepared hydrothermally from corresponding oxides at 473 K for 48 h. Characterized by powder X-ray diffraction data and Ritveld crystal structure refinement. Isostructural with scheelite. Tetragonal, space group $I4_1/a$, $a = 5.156(1)$, $c = 11.196(1)$ Å.

Kind of sample preparation and/or method of registration of the spectrum: KBr and polyethylene discs. Transmission.

Source: Daturi et al. (1997).

Wavenumbers (IR, cm^{-1}): 752s, 435w, 400w, 300, 250, 152, 125.

Note: In the cited paper, Raman spectrum is given.

Wavenumbers (Raman, cm^{-1}): 863s, 822, 759, 403, 392, 323s, 268, 205, 191, 144, 112, 83.

Mo46 Cesium fluormolybdate $CsMoO_2F_3$ $CsMoO_2F_3$

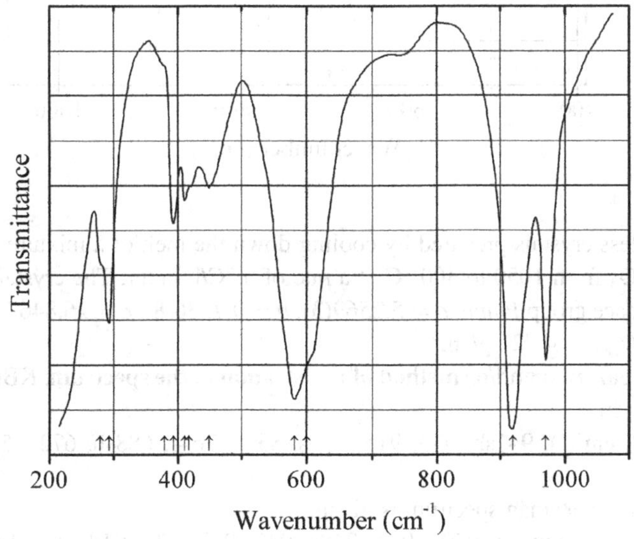

Origin: Synthetic.

Description: Produced from Cs_2CO_3, $MoO_3 \cdot H_2O$, and HF. Characterized by powder X-ray diffraction data. Orthorhombic, $a = 5.492(1)$, $b = 6.457(1)$, $c = 14.124(2)$ Å, $Z = 4$. $D_{meas} = 4.19$ g/cm^3, $D_{calc} = 4.21$ g/cm^3.

Kind of sample preparation and/or method of registration of the spectrum: KBr disc. Transmission.

Source: Mattes et al. (1972).

Wavenumbers (IR, cm^{-1}): 970, 919s, 600sh, 581s, 449, 418, 411, 393, 380sh, 293, 279sh.

Note: In the cited paper, Raman spectrum is given.

Wavenumbers (Raman, cm^{-1}): 974s, 912s, 580w, 403, 308, 282, 268, 242.

Mo47 Cesium thorium molybdate $Cs_2Th(MoO_4)_3$

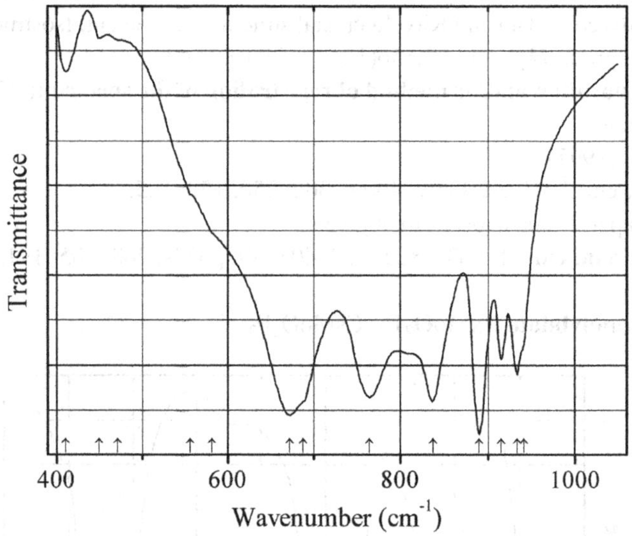

Origin: Synthetic.

Description: Colorless crystals prepared by cooling down the melt of a mixture of $Th(NO_3)_4 \cdot 5H_2O$, $CsNO_3$, and MoO_3 from 1050 to 400 °C at a rate of 5 °C/h in air. The crystal structure is solved. Orthorhombic, space group *Pnnm*, $a = 5.2569(3)$, $b = 9.7336(8)$, $c = 26.8467(16)$ Å, $V = 1373.71(16)$ Å^3, $Z = 4$. $D_{calc} = 4.727$ g/cm^3.

Kind of sample preparation and/or method of registration of the spectrum: KBr disc. Transmission.

Source: Xiao et al. (2014).

Wavenumbers (IR, cm^{-1}): 941sh, 933, 915, 890s, 837s, 764s, 688sh, 672s, 582sh, 557sh, 472w, 450w, 411w.

Note: In the cited paper, Raman spectrum is given.

Wavenumbers (Raman, cm^{-1}): 959s, 954, 948s, 942, 930, 924, 918, 911, 885, 866s, 828, 775, 751, 693, 465, 421, 405, 373, 360, 344, 334, 300, 292, 276, 187, 168, 144, 136, 124, 110, 98.

Mo48 Lanthanum molybdate La_2MoO_6

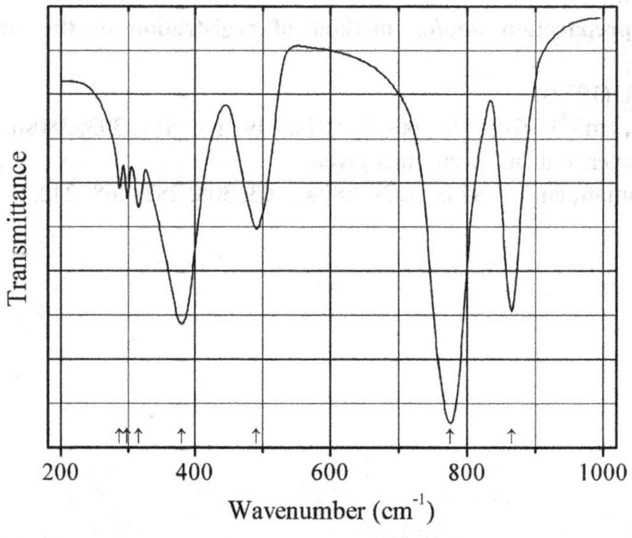

Origin: Synthetic.

Description: Tetragonal, space group *I*-42*m*. The crystal structure can be described as a succession of La_2O_2 and MoO_4 layers. The MoO_4 layer consists of MoO_4 tetrahedra.

Kind of sample preparation and/or method of registration of the spectrum: Transmission. Kind of sample preparation is not indicated.

Source: Bode et al. (1973).

Wavenumbers (IR, cm^{-1}): 865s, 775s, 490, 380s, 315w, 298w, 287w.

Note: In the cited paper, Raman spectrum is given.

Wavenumbers (Raman, cm^{-1}): 875s, 860sh, 800, 765, 695, 490, 455, 435, 375w, 325, 295, 275s, 240, 220.

Mo49 Lead orthoborate molybdate $Pb_6(BO_3)_2(MoO_4)O_2$

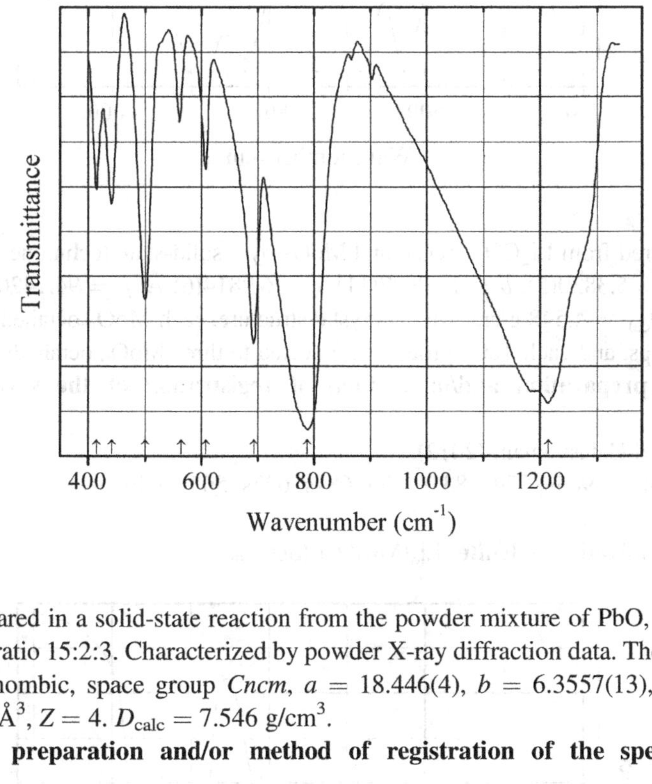

Origin: Synthetic.

Description: Prepared in a solid-state reaction from the powder mixture of PbO, MoO_3, and H_3BO_3 with the molar ratio 15:2:3. Characterized by powder X-ray diffraction data. The crystal structure is solved. Orthorhombic, space group *Cncm*, a = 18.446(4), b = 6.3557(13), c = 11.657(2) Å, V = 1366.6(5) Å^3, Z = 4. D_{calc} = 7.546 g/cm^3.

Kind of sample preparation and/or method of registration of the spectrum: KBr disc. Transmission.

Source: Chen et al. (2009).

Wavenumbers (cm^{-1}): 1215s, 788s, 693, 608, 564, 500, 441, 414.

Mo50 Lithium molybdate tellurite $Li_2(MoTeO_6)$

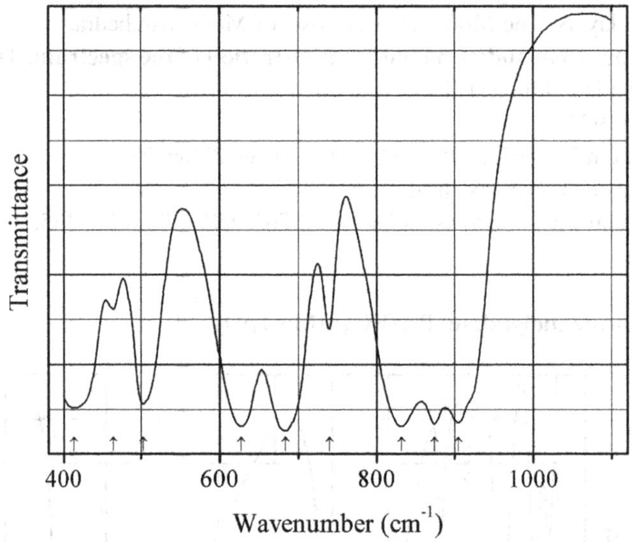

Origin: Synthetic.

Description: Prepared from Li_2CO_3, TeO_2, and MoO_3 by a solid-state technique. Monoclinic, space group $P2_1/n$, $a = 5.3830(5)$, $b = 13.0027(11)$, $c = 6.9814(6)$ Å, $\beta = 94.7420(10)°$, $V = 486.97(7)$ Å^3, $Z = 4$. $D_{calc} = 4.548$ g/cm^3. In the crystal structure, each MoO_6 octahedron is connected to three TeO_3 groups, and each TeO_3 group is connected to three MoO_6 octahedra to form a layer.

Kind of sample preparation and/or method of registration of the spectrum: KBr disc. Transmission.

Source: Nguyen and Halasyamani (2012).

Wavenumbers (cm^{-1}): 904s, 874s, 831s, 739, 684s, 627s, 503, 464w, 414.

Mo51 Lithium dimolybdate selenite $Li_6(Mo_2O_5)_3(SeO_3)_6$

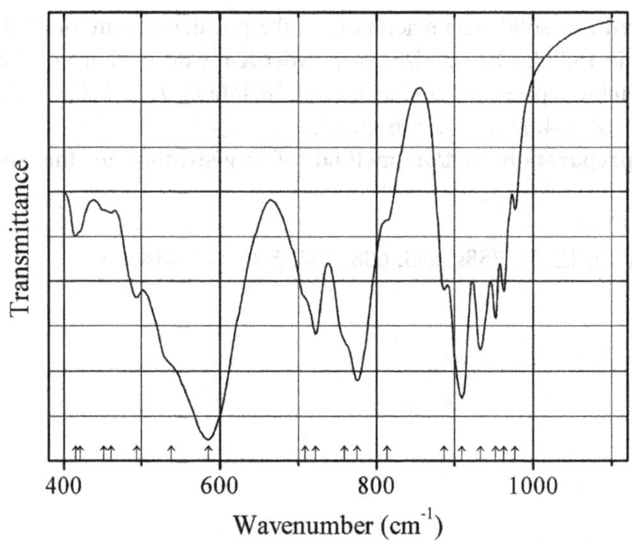

Origin: Synthetic.

Description: Crystals prepared from a mixture of Li_2MoO_4 and SeO_2 by a solid-state technique. Orthorhombic, space group $Pmn2_1$, a = 8.2687(4), b = 16.6546(7), c = 19.2321(8) Å, V = 2648.5(2) $Å^3$, Z = 4. D_{calc} = 4.060 g/cm^3. In the crystal structure, the Mo_2O_{10} dimers are connected by SeO_3 groups to form a layer.

Kind of sample preparation and/or method of registration of the spectrum: KBr disc. Transmission.

Source: Nguyen and Halasyamani (2012).

Wavenumbers (cm^{-1}): 977w, 963, 952, 933, 909s, 886, 813, 775s, 759sh, 722, 709sh, 586s, 539sh, 494, 461, 452sh, 421sh, 415w.

Note: The wavenumbers were partly determined by us based on spectral curve analysis of the published spectrum.

Mo52 Potassium aluminium molybdate $KAl(MoO_4)_2$

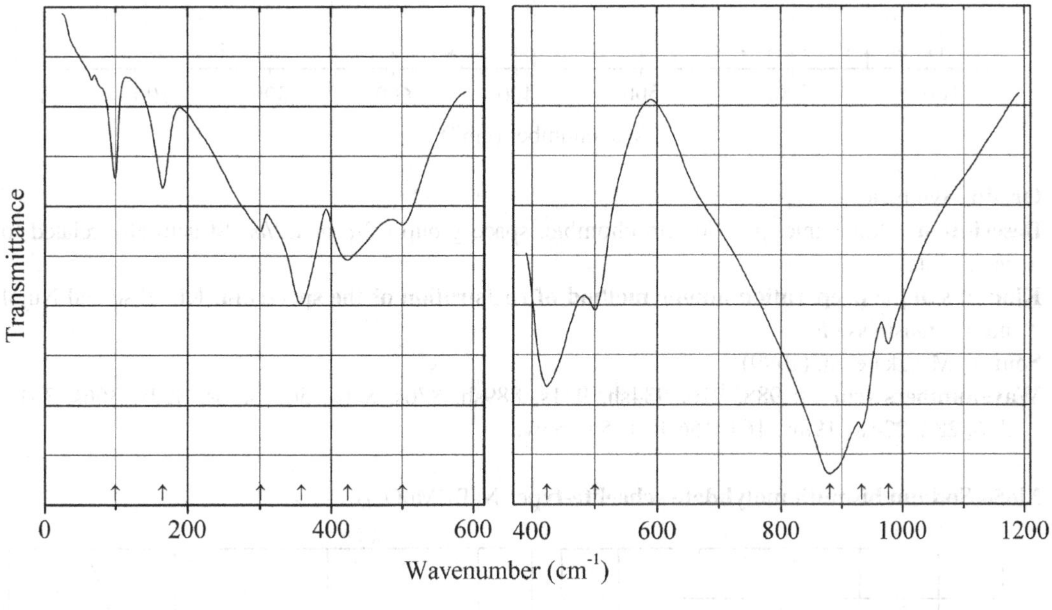

Origin: Synthetic.

Description: Trigonal, space group P-3m1, Z = 1.

Kind of sample preparation and/or method of registration of the spectrum: KBr disc. Transmission.

Source: Maczka et al. (1999).

Wavenumbers (cm^{-1}): 976, 932s, 880s, 500, 423s, 358s, 301, 165, 99.

Mo53 Sodium aluminium molybdate $NaAl(MoO_4)_2$

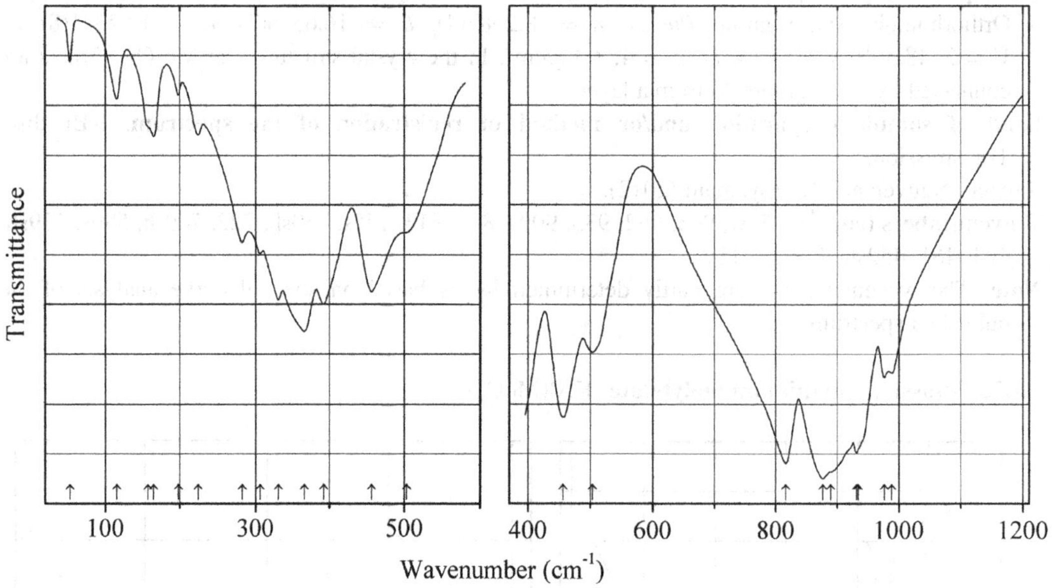

Origin: Synthetic.

Description: Monoclinic, pseudo-orthorhombic, space group *C2/c* or *C2/m*. Structurally related to yavapaiite.

Kind of sample preparation and/or method of registration of the spectrum: KBr disc and Nujol mull. Transmission.

Source: Maczka et al. (1999).

Wavenumbers (cm^{-1}): 988, 976, 934sh, 931s, 889sh, 876s, 816s, 503, 456s, 392s, 366s, 331s, 306, 282, 224w, 198w, 164, 156sh, 115w, 53w.

Mo54 Sodium bismuth molybdate scheelite-type $NaBi(MoO_4)_2$

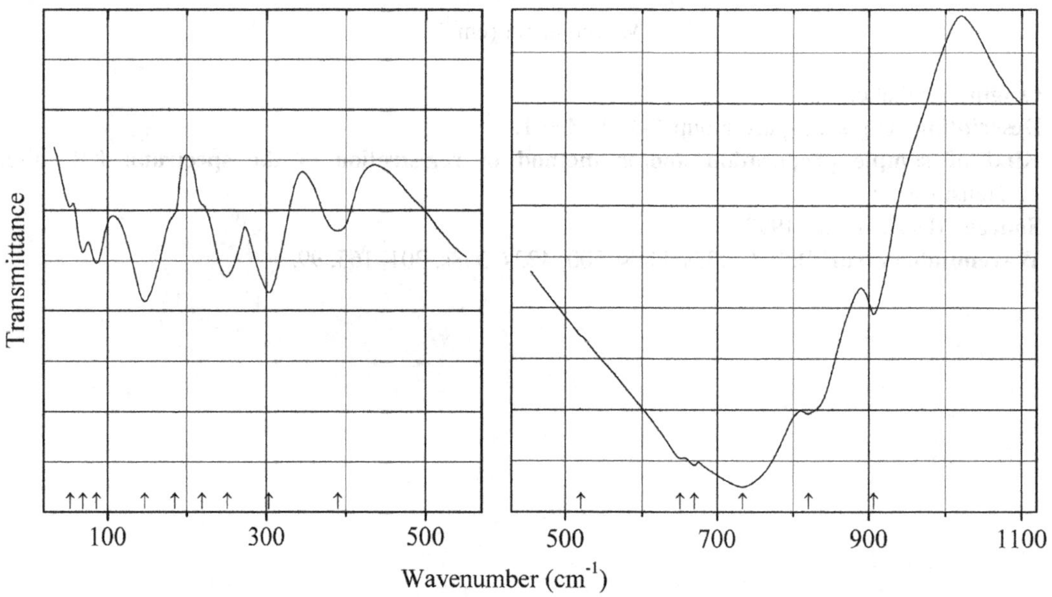

Origin: Synthetic.

Description: Colorless crystals grown at *ca.* 1100 K from the melt prepared from a stoichiometric mixture of Na_2CO_3, Bi_2O_3, and MoO_3. The crystal structure is solved. Tetragonal, space group *I*-4, $a = 5.267$, $c = 11.565$ Å, $V = 320.83$ (10) Å^3, $Z = 2$. $D_{calc} = 5.713$ g/cm^3. Structurally related to scheelite.

Source: Hanuza et al. (1997).

Wavenumbers (IR, cm^{-1}): 906, 820, 733s, 669, 650sh, 520sh, 390, 303, 250, 219sh, 185sh, 147, 86, 69, 53w.

Note: The wavenumbers were partly determined by us based on spectral curve analysis of the published spectrum. In the cited paper, polarized Raman spectra are given.

Wavenumbers (Raman, for the *x*(*zz*)–*x* polarization, cm^{-1}): 924sh, 909sh, 878s, 857sh, 760w, 409, 320, 192w, 130w, 88w, 54w.

Mo55 Tellurium oxumolybdate α-Te_2MoO_7

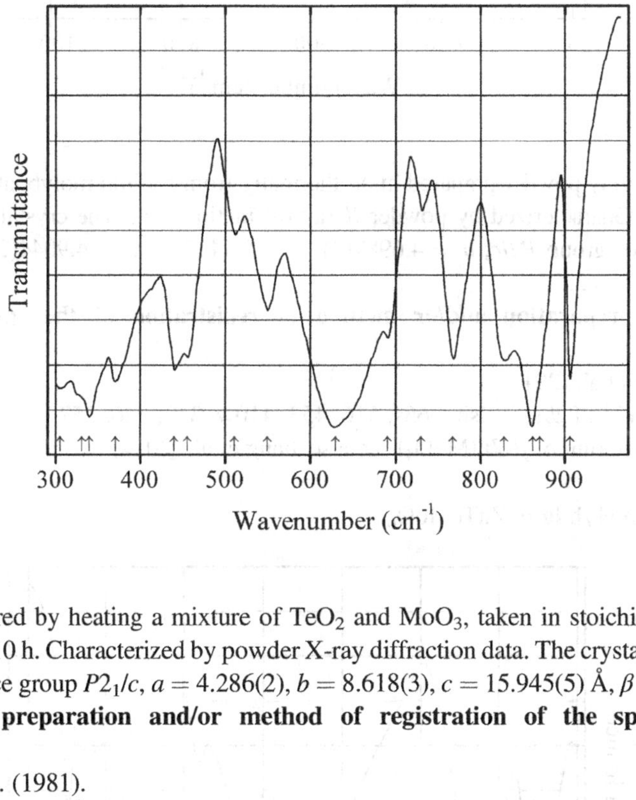

Origin: Synthetic.

Description: Prepared by heating a mixture of TeO_2 and MoO_3, taken in stoichiometric amounts, at 550–600 °C for 10 h. Characterized by powder X-ray diffraction data. The crystal structure is solved. Monoclinic, space group $P2_1/c$, $a = 4.286(2)$, $b = 8.618(3)$, $c = 15.945(5)$ Å, $\beta = 95.68(1)°$, $Z = 4$.

Kind of sample preparation and/or method of registration of the spectrum: CsI disc. Transmission.

Source: Baran et al. (1981).

Wavenumbers (IR, cm^{-1}): 906, 870sh, 862s, 829, 767, 730w, 690, 629s, 549, 511w, 456sh, 440, 371, 340, 331, 305, 296, 244, 230sh, 218.

Note: In the cited paper, Raman spectrum is given.

Wavenumbers (Raman, cm^{-1}): 911s, 867s, 818s, 737, 619, 548w, 513w, 460sh, 439, 365, 335w, 312w, 290sh, 283, 261w, 218s, 199w, 183w, 183sh, 173s.

Mo56 Zinc molybdate β-Zn(MoO$_4$)

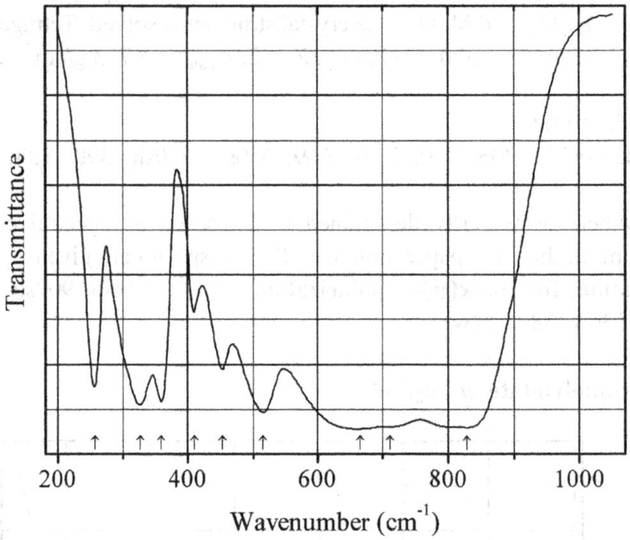

Origin: Synthetic.

Description: Light gray powder prepared hydrothermally from sodium molybdate and zinc nitrate at 140 °C for 8 h. Characterized by powder X-ray diffraction data. The crystal structure is solved. Monoclinic, space group *P2/c*, $a = 4.6987(3)$, $b = 5.7487(2)$, $c = 4.9044(2)$ Å, $\beta = 90.3312°$, $V = 132.47$ Å^3, $Z = 2$.

Kind of sample preparation and/or method of registration of the spectrum: KBr disc. Transmission.

Source: Cavalcante et al. (2013).

Wavenumbers (cm^{-1}): 829s, 712sh, 665s, 516, 453, 410w, 359, 326, 257.

Note: For the IR spectrum of β-Zn(MoO$_4$) see also Jiang et al. (2014).

Mo57 Zinc telluromolybdate ZnTeMoO$_6$

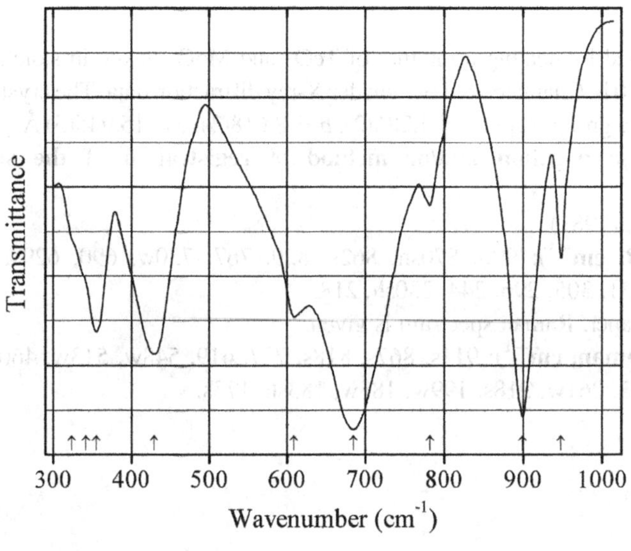

Origin: Synthetic.

Description: Orthorhombic, space group $P2_12_12_1$, $a = 5.255$, $b = 5.044$, $c = 8.909$ Å.

Kind of sample preparation and/or method of registration of the spectrum: CsI disc. Transmission.

Source: Baran et al. (1981).

Wavenumbers (cm^{-1}): 948, 899s, 782w, 685s, 609, 430, 356, 342sh, 324sh.

Mo58 Zirconium molybdenum oxide (monoclinic) $ZrMo_2O_8$

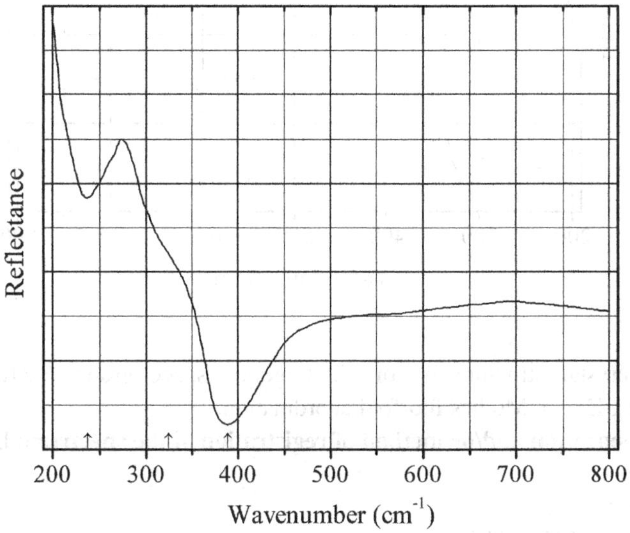

Origin: Synthetic.

Description: Prepared by heating a mixture of ZrO_2 and MoO_3 taken in the molar ratio of 1:2, at 600 °C for 64 h. Characterized by powder X-ray diffraction data. The crystal structure is solved. Monoclinic, space group *C2/c*, $a = 11.4243(19)$, $b = 7.9297(6)$, $c = 7.4610(14)$ Å, $\beta = 122.15(2)°$, $V = 572.3(2)$ Å^3, $Z = 4$. $D_{calc} = 4.771$g/cm^3. Mo has fivefold coordination. Two MoO_5 polyhedra share edges with each other forming Mo_2O_8 moieties.

Kind of sample preparation and/or method of registration of the spectrum: Diffuse reflectance of a powdered sample.

Source: Sahoo et al. (2009).

Wavenumbers (cm^{-1}): 388s, 324sh, 237.

Note: The wavenumbers were determined by us based on spectral curve analysis of the published spectrum.

Mo59 Zirconium molybdenum oxide (trigonal) $ZrMo_2O_8$

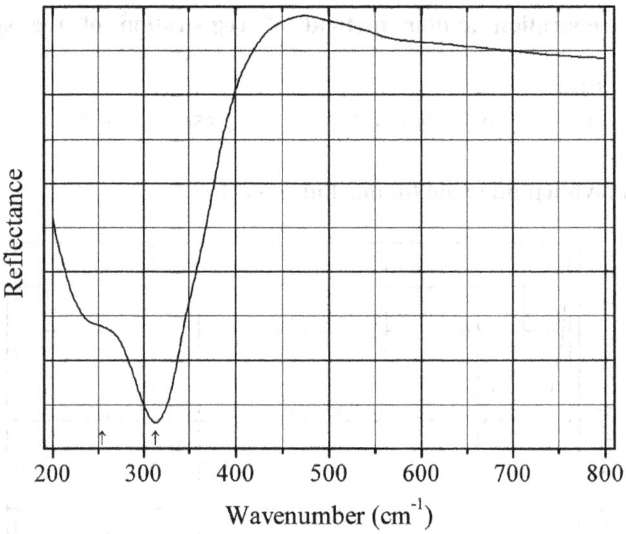

Origin: Synthetic.

Description: The crystal structure is solved. Trigonal, space group *P*-31*c*, a = 10.1391(6), c = 11.7084(8) Å, Z = 6. Mo has fourfold coordination.

Kind of sample preparation and/or method of registration of the spectrum: Diffuse reflectance of a powdered sample.

Source: Sahoo et al. (2009).

Wavenumbers (cm^{-1}): 313, 254sh.

Note: The wavenumbers were determined by us based on spectral curve analysis of the published spectrum. The wavenumber of the main band (313 cm^{-1}) is anomalously low for the MoO_4 tetrahedra.

Mo60 Vanadyl molybdate $(VO)(MoO_4)$

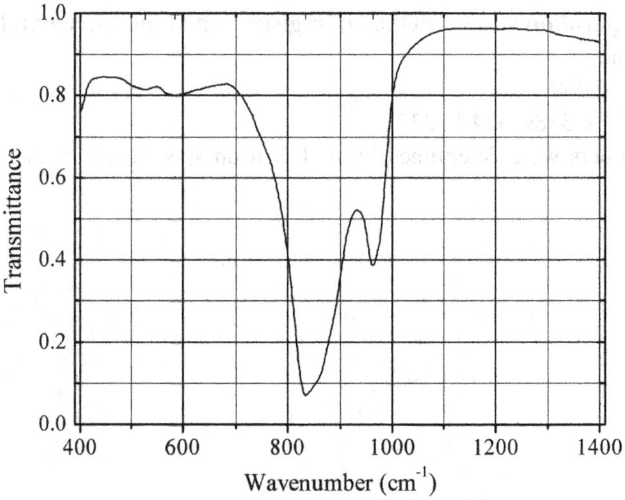

Origin: Synthetic.

Description: Obtained by heating a stoichiometric mixture of VO_2 and MoO_3 at 700 °C for 48 h. Tetragonal, space group *P4/n*.

Kind of sample preparation and/or method of registration of the spectrum: KBr disc. Transmission.

Source: Stranford and Condrate Sr (1984a).

Wavenumbers (IR, cm^{-1}): 963, 834s, 586w, 528w.

Note: The wavenumbers were determined by us based on spectral curve analysis of the published spectrum. In the cited paper, a figure of Raman spectrum is given.

Mo61 Kamiokite $Fe^{2+}{}_2Mo^{4+}{}_3O_8$

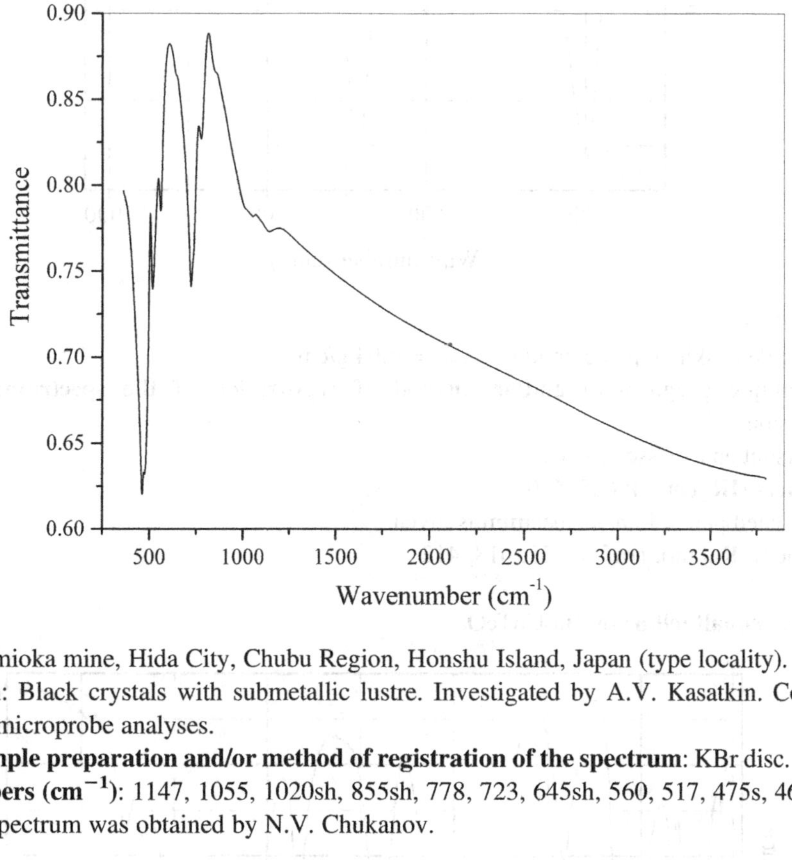

Origin: Kamioka mine, Hida City, Chubu Region, Honshu Island, Japan (type locality).

Description: Black crystals with submetallic lustre. Investigated by A.V. Kasatkin. Confirmed by electron microprobe analyses.

Kind of sample preparation and/or method of registration of the spectrum: KBr disc. Absorption.

Wavenumbers (cm^{-1}): 1147, 1055, 1020sh, 855sh, 778, 723, 645sh, 560, 517, 475s, 461s.

Note: The spectrum was obtained by N.V. Chukanov.

2.18 Tellurides, Tellurites, and Tellurates

Te52 Barium calcium tellurate Ba_2CaTeO_6

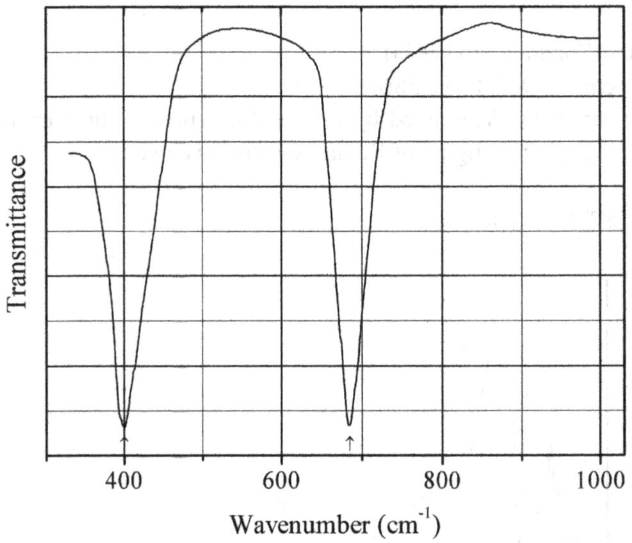

Origin: Synthetic.
Description: Perovskite-type compound. $D_{calc} = 6.04$ g/cm^3.
Kind of sample preparation and/or method of registration of the spectrum: KBr disc. Transmission.
Source: Corsmit and Blasse (1974).
Wavenumbers (IR, cm^{-1}): 685, 400.
Note: In the cited paper, Raman spectrum is given.
Wavenumbers (Raman, cm^{-1}): 752, 618, 412.

Te53 Barium cobalt tellurate Ba_2CoTeO_6

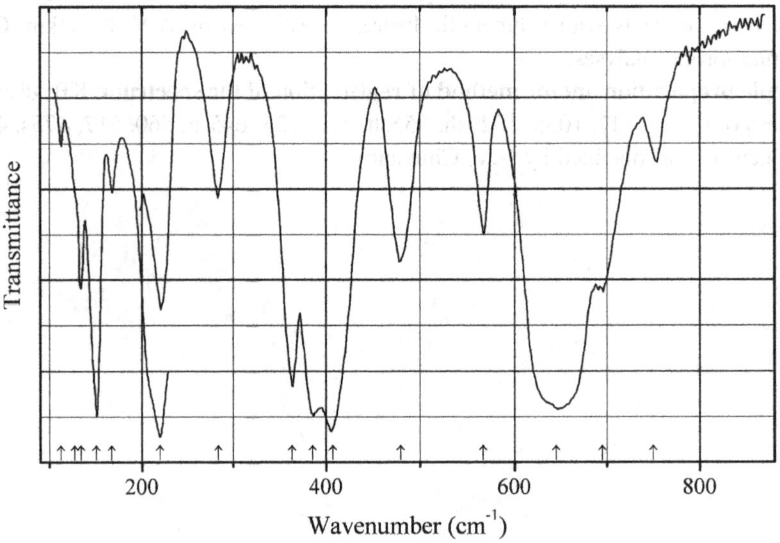

Origin: Synthetic.
Description: A compound with ordered hexagonal perovskite-type structure.
Kind of sample preparation and/or method of registration of the spectrum: KBr and polyethylene discs. Transmission.
Source: Liegeois-Duyckaerts (1985).
Wavenumbers (cm^{-1}): 750, 695, 645s, 567, 479, 407s, 385s, 363s, 284, 220, 167w, 150, (133), (127), 112w.

Te54 Barium copper tellurate tellurite $BaCuTe_2O_7$

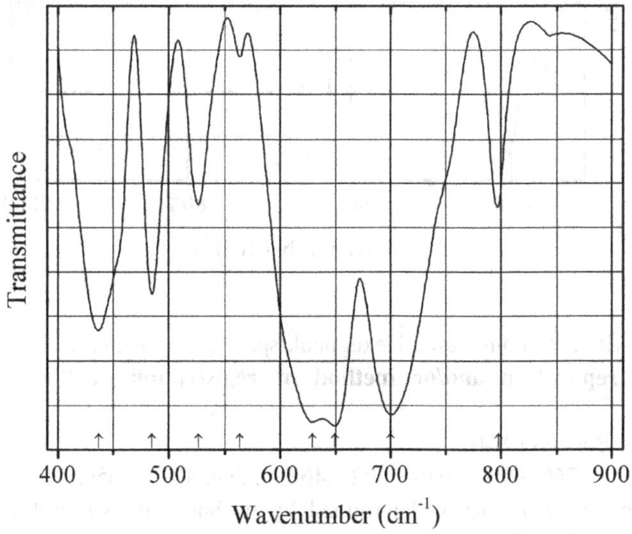

Origin: Synthetic.
Description: Prepared by solid-state method from the stoichiometric mixture of $BaCO_3$, CuO, TeO_2, and $H_2TeO_4{\cdot}2H_2O$ at 650 °C. Orthorhombic, space group *Ama*2, $a = 5.4869(8)$, $b = 15.4120(8)$, $c = 7.2066(4)$ Å, $V = 609.42(10)$ Å^3, $Z = 4$. $D_{meas} = 2.39(3)$ g/cm^3, $D_{calc} = 2.391$ g/cm^3. Characterized by powder X-ray diffraction, piezoelectric, and polarization measurements.
Kind of sample preparation and/or method of registration of the spectrum: Transmission. Kind of sample preparation is not indicated.
Source: Yeon et al. (2011).
Wavenumbers (cm^{-1}): 798, 700s, 649s, 629s, 564w, 527, 485, 437.

Te55 Barium nickel tellurate Ba_2NiTeO_6

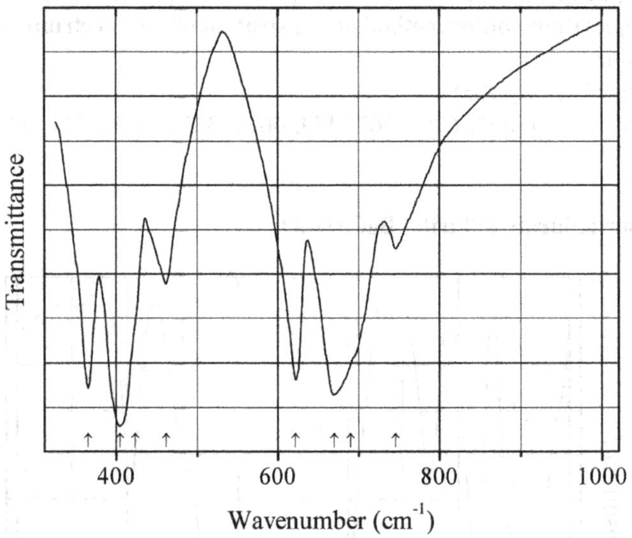

Origin: Synthetic.

Description: Perovskite-type compound. Hexagonal, space group *R*-3*m*, *Z* = 2.

Kind of sample preparation and/or method of registration of the spectrum: KBr disc. Transmission.

Source: Corsmit and Blasse (1974).

Wavenumbers (cm^{-1}): 746, 690sh, 670s, 622s, 462, 423sh, 405s, 365s.

Note: The wavenumbers were partly determined by us based on spectral curve analysis of the published spectrum.

Te56 Barium zinc tellurate Ba_2ZnTeO_6

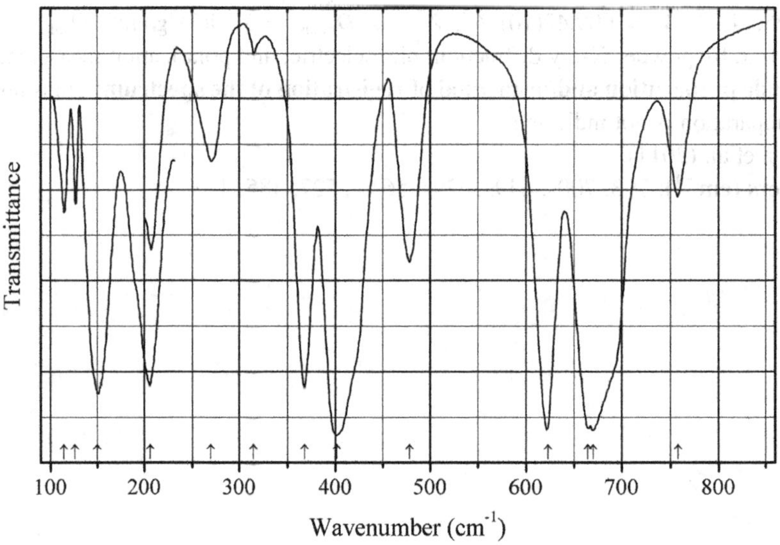

Origin: Synthetic.

Description: Synthesized by solid-state reaction. The stoichiometric mixture of the necessary oxides and carbonates was heated at 600 °C for about 1 night, then reground and heated up to 1100 °C for 1 day. A compound with perovskite-type cubic structure.

Kind of sample preparation and/or method of registration of the spectrum: Transmission. Kind of sample preparation is not indicated.

Source: Liegeois-Duyckaerts (1975).

Wavenumbers (IR, cm^{-1}): 758, 670s, 665s, 623s, 478, 402s, 367, 315w, 269, 206sh, 150, 127w, 115w.

Note: The wavenumbers were partly determined by us based on spectral curve analysis of the published spectrum. In the cited paper, Raman spectrum is given.

Wavenumbers (Raman, cm^{-1}): 769s, 691s, 620, 575, 473w, 406, 399, 385, 122, 105.

Te57 Calcium tellurite monohydrate $Ca(TeO_3) \cdot H_2O$

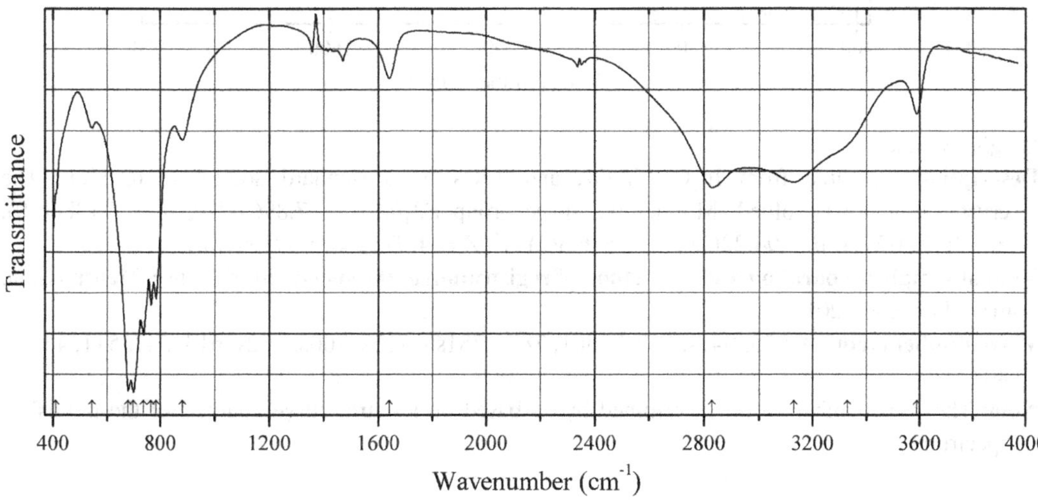

Origin: Synthetic.

Description: Obtained from $Ca(NO_3)_2$ and TeO_2, in the presence of NaOH, by microwave-assisted hydrothermal synthesis at 185 °C for 1 h. The crystal structure is solved. Orthorhombic, space group $P2_1cn$, $a = 14.78549(4)$, $b = 6.79194(3)$, $c = 8.06261(3)$ Å, $V = 809.665(6)$ Å^3, $Z = 8$.

Kind of sample preparation and/or method of registration of the spectrum: KBr disc. Transmission.

Source: Poupon et al. (2015).

Wavenumbers (cm^{-1}): 3590, 3330sh, 3130, 2830, 1641w, 882, 783, 765, 737s, 700s, 680s, 545w, 411sh.

Note: The strong band centered at 698 cm^{-1} and the shoulders at 740 and 770 cm^{-1} indicated by the authors of the cited paper were determined by us as bands at 680, 700, 737, 765, and 783 cm^{-1}.

Te58 Indium vanadate tellurite In(VTe$_2$O$_8$) In(VTe$_2$O$_8$)

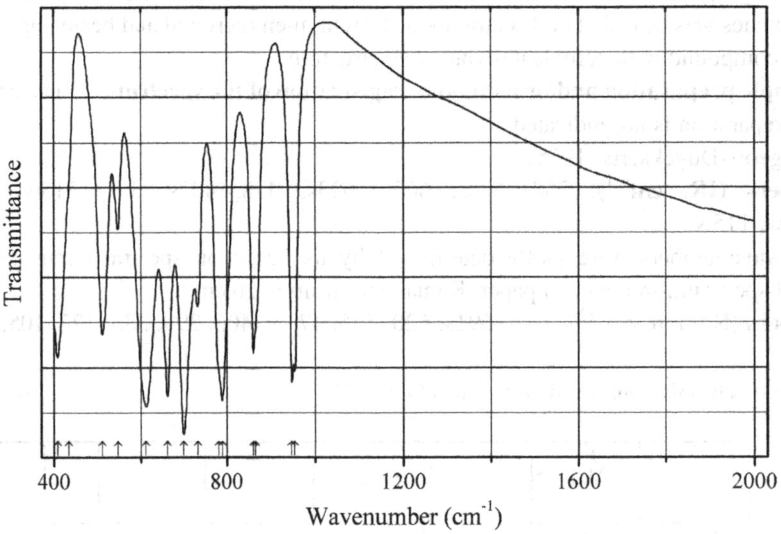

Origin: Synthetic.

Description: Prepared from In_2O_3, V_2O_5, and TeO_2 by a standard solid-state reaction. The crystal structure is solved. Monoclinic, space group $P2_1/n$, $a = 7.8967(16)$, $b = 5.1388(10)$, $c = 16.711(3)$ Å, $\beta = 94.22(3)°$, $V = 676.3(2)$ Å^3, $Z = 4$. $D_{calc} = 5.391$ g/cm^3.

Kind of sample preparation and/or method of registration of the spectrum: KBr disc. Transmission.

Source: Lee et al. (2011).

Wavenumbers (cm^{-1}): 955, 948s, 867sh, 861, 789s, 781sh, 732s, 700s, 662s, 613, 547, 511, 433sh, 407.

Note: The wavenumbers were determined by us based on spectral curve analysis of the published spectrum.

Te59 Lead copper tellurate tellurite PbCuTe$_2$O$_7$

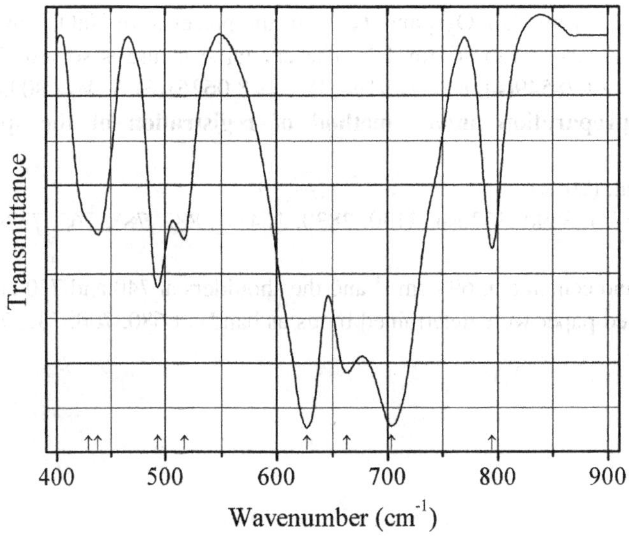

Origin: Synthetic.

Description: Prepared by a conventional solid-state method from stoichiometric amounts of PbO, CuO, TeO_2, and $H_2TeO_4{\cdot}2H_2O$. Characterized by powder X-ray diffraction data. The crystal structure is solved. Orthorhombic, space group *Pbcm*, a = 7.2033(5), b = 15.0468(10), c = 5.4691(4) Å, V = 592.78(7) Å^3, Z = 4.

Kind of sample preparation and/or method of registration of the spectrum: Transmission. Kind of sample preparation is not indicated.

Source: Yeon et al. (2011).

Wavenumbers (cm^{-1}): 795, 703s, 663, 627s, 517, 493, 438, 429sh.

Note: The wavenumbers were partly determined by us based on spectral curve analysis of the published spectrum.

Te60 Lithium tungstate tellurite $Li_2(WTeO_6)$

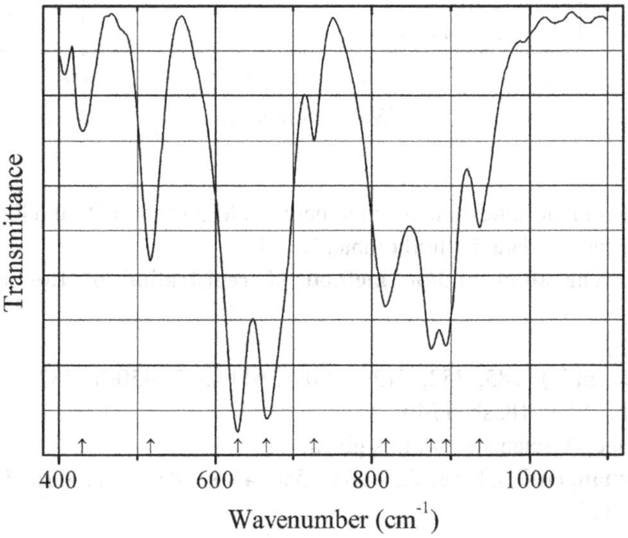

Origin: Synthetic.

Description: Prepared from Li_2CO_3, TeO_2, and WO_3 by a solid-state technique. Monoclinic, space group $P2_1/n$, a = 5.3950(5), b = 12.9440(12), c = 7.0149(7) Å, β = 94.2510(10)°, V = 488.52 (8) Å^3, Z = 4. D_{calc} = 5.729 g/cm^3. In the crystal structure, each WO_6 octahedron is connected to three TeO_3 groups, and each TeO_3 group is connected to three WO_6 octahedra to form a layer.

Kind of sample preparation and/or method of registration of the spectrum: KBr disc. Transmission.

Source: Nguyen and Halasyamani (2012).

Wavenumbers (cm^{-1}): 936, 894s, 875s, 818, 727s, 666s, 629, 517, 430w.

Te61 Magnesium tellurite $MgTe_2O_5$ $MgTe_2O_5$

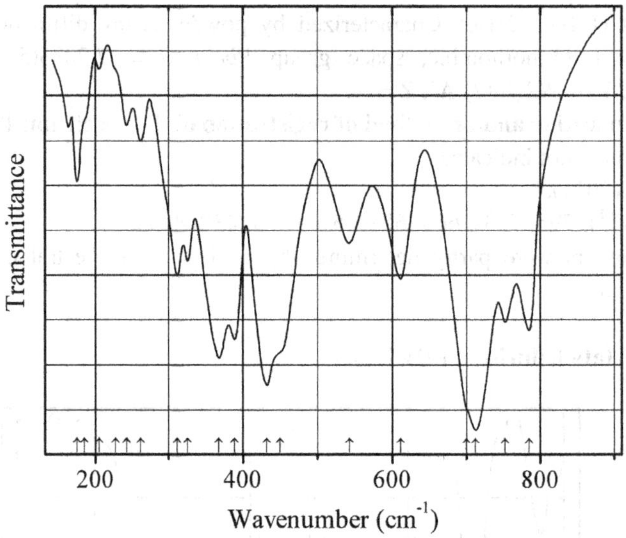

Origin: Synthetic.

Description: Obtained in the solid-state reaction between MgO and TeO_2 at 680 °C. Characterized by powder X-ray diffraction data. Orthorhombic, $Z = 4$.

Kind of sample preparation and/or method of registration of the spectrum: CsI disc. Transmission.

Source: Baran (1978).

Wavenumbers (IR, cm^{-1}): 785, 752, 712s, 700sh, 611, 542, 450sh, 432s, 389, 367s, 326, 312, 261, 242w, 227sh, 205w, 185sh, 174w.

Note: In the cited paper, Raman spectrum is given.

Wavenumbers (Raman, cm^{-1}): 810s, 725, 698s, 540, 437s, 405, 385, 345sh, 330, 267s, 253, 237w, 222s, 200, 185w, 115.

Te62 Potassium acid tellurate hydrate $K_2[TeO_2(OH)_4]\cdot 3H_2O$ $K_2[TeO_2(OH)_4]\cdot 3H_2O$

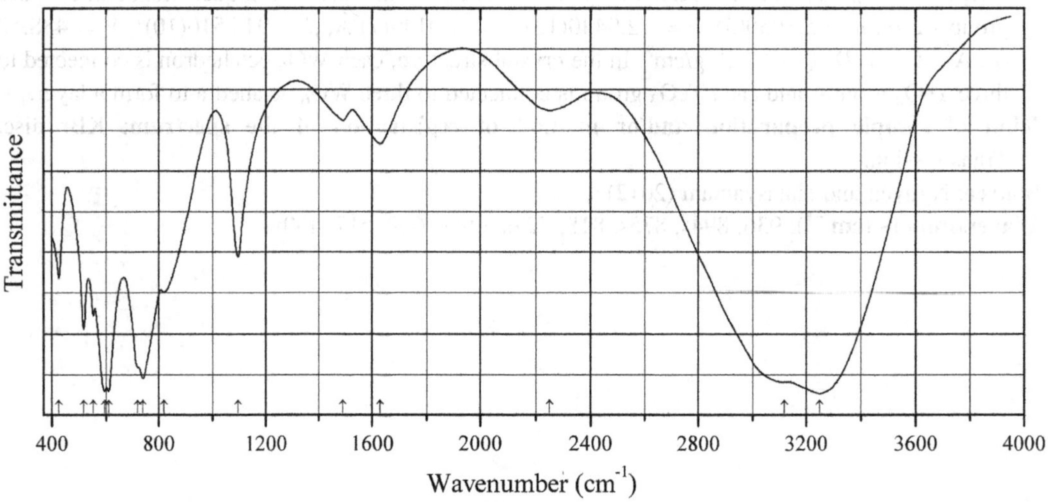

Origin: Synthetic.

Kind of sample preparation and/or method of registration of the spectrum: KBr disc. Transmission.

Source: Siebert (1959).

Wavenumbers (cm^{-1}): 3250s, 3120sh, 2250, 1630, 1490w, 1095, 820, 740s, 720sh, 610s, 595s, 554, 518, 426.

Te63 Scandium vanadate tellurite $ScVTe_2O_8$

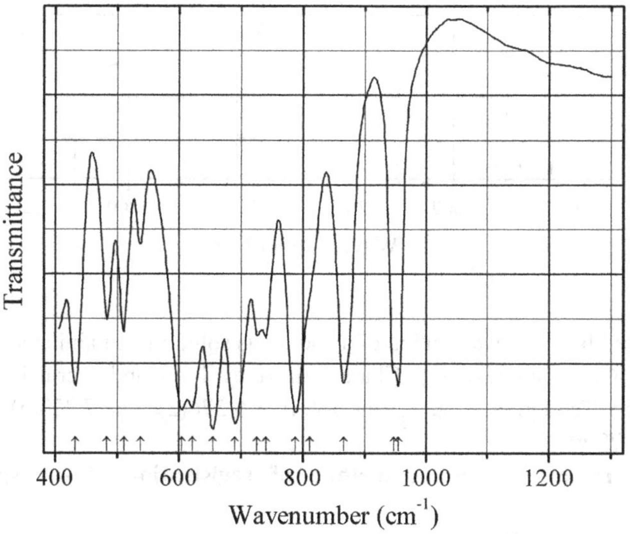

Origin: Synthetic.

Description: Prepared by a solid-state reaction from Sc_2O_3, V_2O_5, and TeO_2, first at 350 °C for 5 h, and thereafter at 450 °C for 48 h. The crystal structure is solved. Monoclinic, space group $P2_1/n$, $a = 7.9774(2)$, $b = 5.08710(10)$, $c = 16.5654(4)$ Å, $\beta = 93.400(2)°$, $V = 671.07(3)$ Å^3, Z = 4. $D_{calc} = 4.742$ g/cm^3. The structure is composed of ScO_6 octahedra, VO_4 tetrahedra, and asymmetric TeO_4 polyhedra that are connected *via* common O atoms.

Kind of sample preparation and/or method of registration of the spectrum: KBr disc. Transmission.

Source: Kim et al. (2013).

Wavenumbers (cm^{-1}): 955s, 947, 866s, 811sh, 788s, 740, 726, 691s, 655s, 622s, 605s, 538w, 511, 483, 432s.

Note: The wavenumbers were determined by us based on spectral curve analysis of the published spectrum.

Te64 Sodium acid diarsenite tellurite $Na_2(H_4As_2O_5)(H_2TeO_4)$

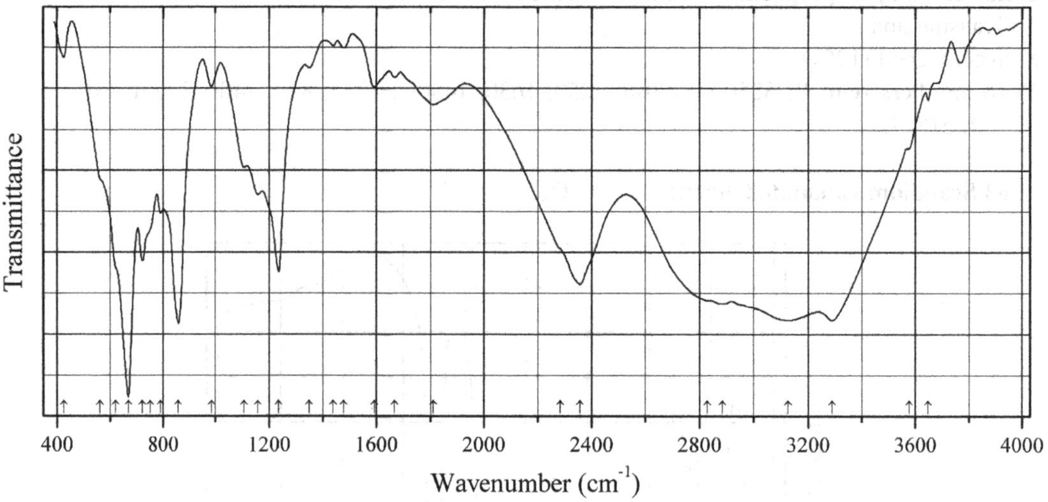

Origin: Synthetic.

Description: Prepared by slow evaporation of an aqueous solution containing stoichiometric amounts of $Te(OH)_6$, Na_2CO_3, and H_3AsO_4. Characterized by EDS and thermal analysis. The crystal structure is solved. Tetragonal, space group *I*-4, $a = 5.576(2)$, $c = 7.773(5)$ Å, $V = 241.8(2)$ Å^3, $Z = 2$. $D_{calc} = 5.72$ g/cm^3.

Kind of sample preparation and/or method of registration of the spectrum: KBr disc. Transmission.

Source: Bechibani et al. (2014).

Wavenumbers (IR, cm^{-1}): 3649w, 3578w, 3289s, 3127s, 2885s, 2830sh, 2358s, 2285sh, 1811, 1669w, 1592, 1478w, 1440w, 1350w, 1236s, 1158, 1106, 985, 860s, 790, 750, 720, 667s, 620sh, 560sh, 427w.

Note: The wavenumbers were partly determined by us based on spectral curve analysis of the published spectrum. In the cited paper, Raman spectrum is given.

Wavenumbers (Raman, cm^{-1}): 1327w, 1237w, 835, 718, 664sh, 650s, 623, 429, 423, 415, 372, 327, 321, 295w, 244w.

Te65 Sodium acid tellurate $Na_2[TeO_2(OH)_4]$ $Na_2[TeO_2(OH)_4]$

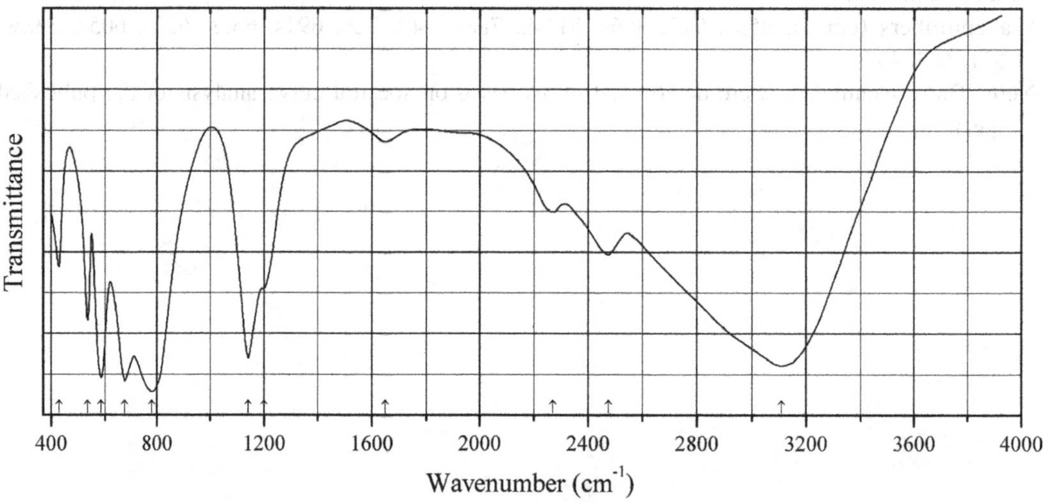

Origin: Synthetic.

Description: Obtained in the reaction of $Te(OH)_6$ with excess of NaOH in aqueous solution. Characterized by powder X-ray diffraction data.

Kind of sample preparation and/or method of registration of the spectrum: KBr disc. Absorption.

Source: Siebert (1959).

Wavenumbers (cm^{-1}): 3110s, 2475, 2270, 1650w, 1200sh, 1141s, 780s, 675s, 587s, 536, 429.

Te66 Sodium molybdenum(VI) tellurite $Na_2MoTe_4O_{12}$

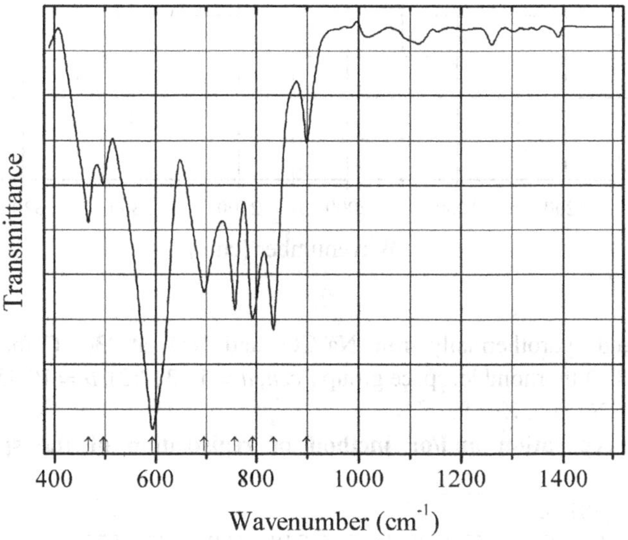

Origin: Synthetic.

Description: Synthesized hydrothermally from $Na_2MoO_4{\cdot}2H_2O$ and TeO_2 (with the ratio Na:Mo:Te = 6:3:1) at 225 °C for 4 days with subsequent cooling to room temperature over a period of 36 h. Characterized by powder X-ray diffraction data. The crystal structure is solved. Monoclinic, space group *C*2/*c*, $a = 17.341(4)$, $b = 5.8262(11)$, $c = 11.268(2)$ Å, $\beta = 104.38(2)°$, $V = 1102.7(4)$ Å^3, $Z = 4$. $D_{calc} = 5.086$ g/cm^3. The strongest lines of the powder X-ray diffraction pattern [*d*, Å (*I*, %) (*hkl*)] are: 5.500 (28) (110), 4.194 (29) (400), 4.028 (77) (31−1), 2.976 (49) (51−1), 2.911 (100) (020), 2.618 (29) (221), 2.085 (29) (422).

Kind of sample preparation and/or method of registration of the spectrum: KBr disc. Transmission.

Source: Balraj and Vidyasagar (1999).

Wavenumbers (cm^{-1}): 899, 833s, 792s, 757s, 696s, 593s, 496, 467.

Te67 Sodium tellurite β-$Na_2Te_4O_9$ β-$Na_2Te_4O_9$

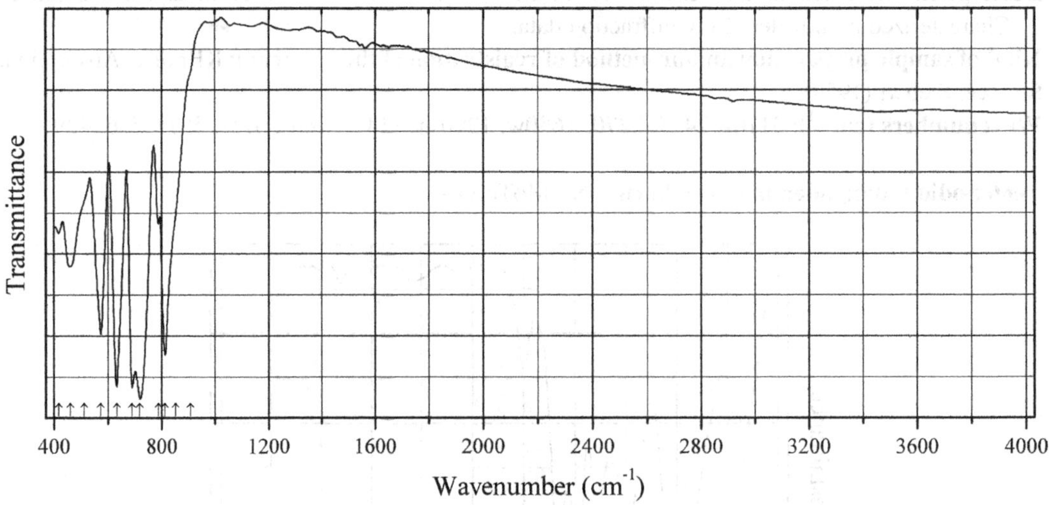

Origin: Synthetic.

Description: Prepared hydrothermally from Na_2CO_3 and TeO_2 at 230 °C for 4 days. The crystal structure is solved. Orthorhombic, space group *Pccn*, $a = 16.317(2)$, $b = 10.4544(10)$, $c = 10.8874(10)$ Å, $V = 1857.2(3)$ Å^3, $Z = 8$.

Kind of sample preparation and/or method of registration of the spectrum: KBr disc. Transmission.

Source: Lee and Ok (2014).

Wavenumbers (cm^{-1}): 908sh, 853sh, 814s, 787, 718s, 689s, 632s, 573, 512sh, 460, 419w.

Note: The wavenumbers were determined by us based on spectral curve analysis of the published spectrum.

Te68 Sodium tellurate tellurite hydrate $Na_2Te_2O_6{\cdot}1.5H_2O$ $Na_2Te_2O_6{\cdot}1.5H_2O$

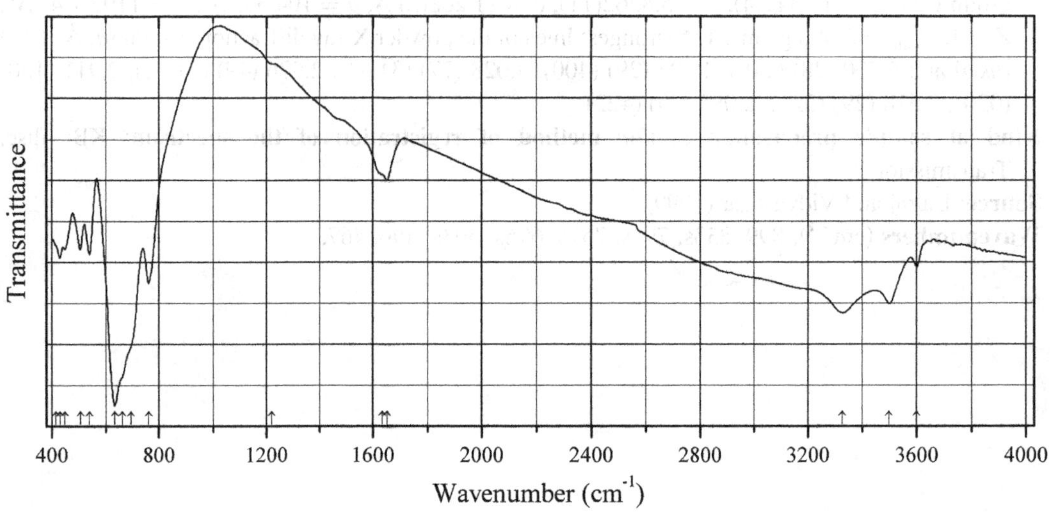

Origin: Synthetic.

Description: Prepared hydrothermally from Na_2CO_3 and TeO_2 at 230 °C for 4 days. The crystal structure is solved. Monoclinic, space group *C*2/*c*, $a = 8.9884(19)$, $b = 14.3739(19)$, $c = 10.387$ (3) Å, $\beta = 99.429(11)°$, $V = 1323.9(5)$ Å^3, $Z = 4$.

Kind of sample preparation and/or method of registration of the spectrum: KBr disc. Transmission.

Source: Lee and Ok (2014).

Wavenumbers (cm^{-1}): 3602w, 3498, 3327, 1651, 1635sh, 1221sh, 760s, 692sh, 659sh, 632s, 539, 505, 449, 430, 415sh.

Note: The wavenumbers were partly determined by us based on spectral curve analysis of the published spectrum.

Te69 Sodium tungsten tellurite $Na_2WTe_4O_{12}$

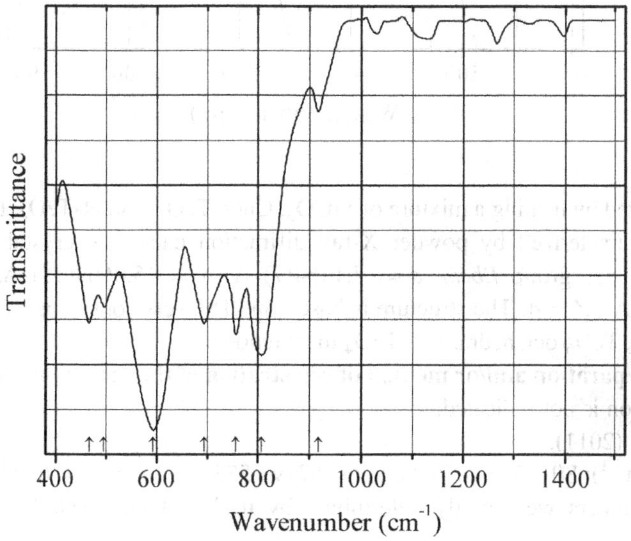

Origin: Synthetic.

Description: Tiny pale yellow crystals. Synthesized hydrothermally from $Na_2WO_4 \cdot 2H_2O$ and TeO_2 (with the ratio Na:W:Te = 6:3:1) at 225 °C for 4 days with subsequent cooling to room temperature over a period of 36 h. Characterized by powder X-ray diffraction data. The crystal structure is solved. Monoclinic, space group *C*2/*c*, $a = 17.348(3)$, $b = 5.7755(10)$, $c = 11.269(3)$ Å, $\beta = 104.33(2)°$, $V = 1094.0(4)$ Å^3, $Z = 4$. $D_{calc} = 5.660$ g/cm^3. The strongest lines of the powder X-ray diffraction pattern [*d*, Å (*I*, %) (*hkl*)] are: 4.007 (100) (31−1), 3.144 (49) (11−3), 2.966 (72) (51−1), 2.883 (48) (020), 2.597 (59) (221).

Kind of sample preparation and/or method of registration of the spectrum: KBr disc. Transmission.

Source: Balraj and Vidyasagar (1999).

Wavenumbers (cm^{-1}): 917w, 807s, 756, 694, 593s, 495, 467.

Te70 Strontium copper tellurate tellurite $SrCuTe_2O_7$

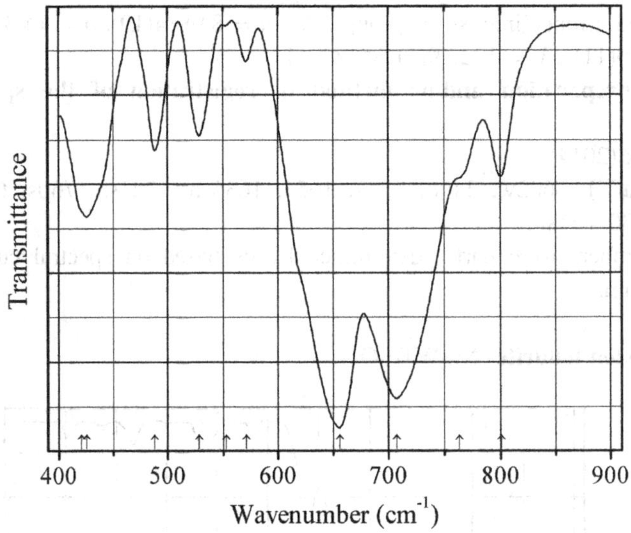

Origin: Synthetic.

Description: Prepared by heating a mixture of $SrCO_3$, CuO, TeO_2, and H_2TeO_4, taken in stoichiometric amounts. Characterized by powder X-ray diffraction data. The crystal structure is solved. Orthorhombic, space group *Pbcm*, $a = 7.1464(7)$ Å, $b = 15.0609(15)$ Å, $c = 5.4380(5)$ Å, $V = 585.30(10)$ Å^3, $Z = 4$. The structure is based on 2D layers consisting of corner-shared CuO_5 square pyramids, TeO_6 octahedra, and TeO_4 dispheniods.

Kind of sample preparation and/or method of registration of the spectrum: Transmission. Kind of sample preparation is not indicated.

Source: Yeon et al. (2011).

Wavenumbers (cm^{-1}): 801, 763sh, 707s, 656s, 571w, 553w, 529, 488, 426, 421sh.

Note: The wavenumbers were partly determined by us based on spectral curve analysis of the published spectrum.

Te71 Thallium tellurite β-Tl_2TeO_3

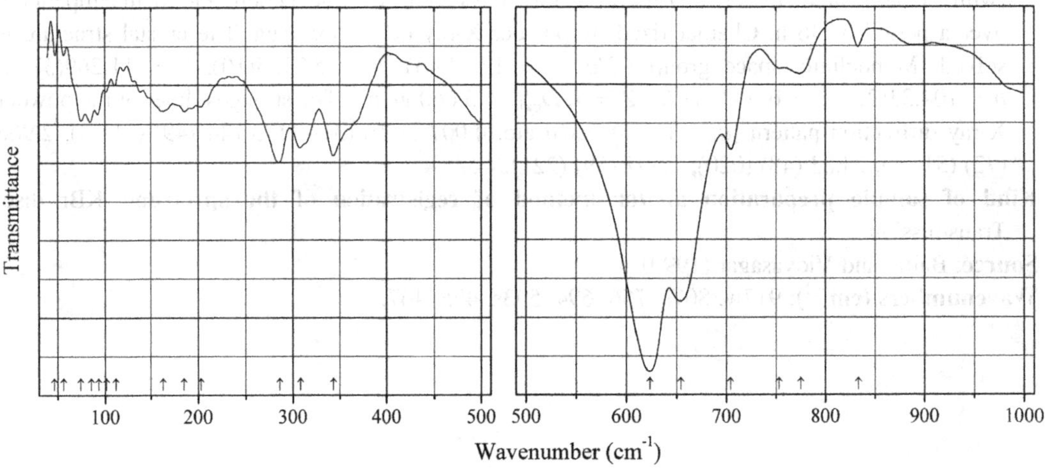

Origin: Synthetic.

Description: Light yellow to brown aggregates of crystals prepared by heating a mixture of Tl_2CO_3 and TeO_2, taken in stoichiometric amounts, at 160 °C for 3 days. Characterized by powder X-ray diffraction data. The crystal structure is solved. Monoclinic, space group $P2_1/c$, $a = 8.9752(18)$, $b = 4.8534(6)$, $c = 11.884(2)$ Å, $\beta = 109.67(2)°$, $V = 487.47(15)$ Å^3, $Z = 4$.

Kind of sample preparation and/or method of registration of the spectrum: KBr and polyethylene discs. Transmission.

Source: Rieger and Mudring (2007).

Wavenumbers (IR, cm^{-1}): 833w, 775w, 753w, 704, 654s, 623s, 343, 308, 286, 203, 185, 163, 111w, 101, 93, 85, 74, 56w, 46w.

Note: In the cited paper, Raman spectrum is given.

Wavenumbers (Raman, cm^{-1}): 708s, 646, 346w, 296, 172sh, 150, 117, 93s, 66w.

Te72 Thorium tellurite $ThTe_2O_6$

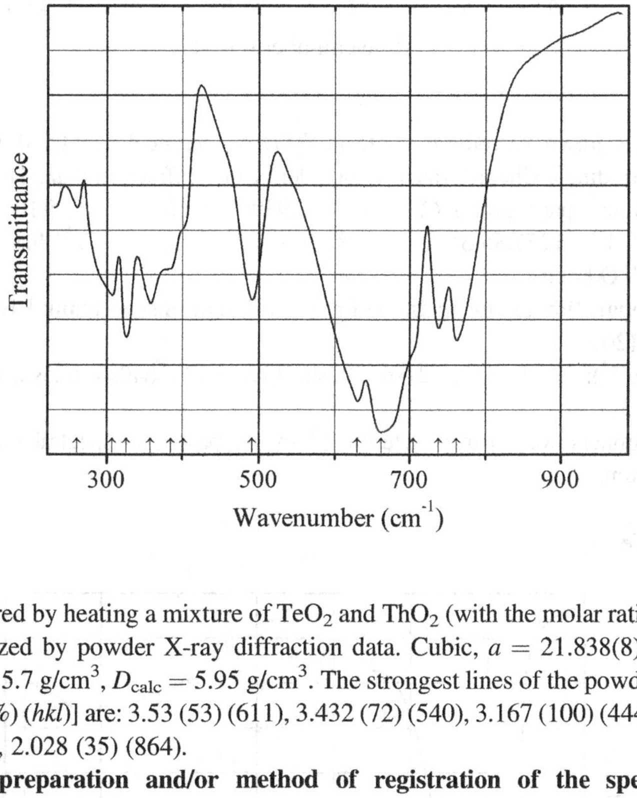

Origin: Synthetic.

Description: Prepared by heating a mixture of TeO_2 and ThO_2 (with the molar ratio 1:2) at 600 °C for 48 h. Characterized by powder X-ray diffraction data. Cubic, $a = 21.838(8)$, $V = 10414.5$ Å^3, $Z = 64$. $D_{meas} = 5.7$ g/cm^3, $D_{calc} = 5.95$ g/cm^3. The strongest lines of the powder X-ray diffraction pattern [*d*, Å (*I*, %) (*hkl*)] are: 3.53 (53) (611), 3.432 (72) (540), 3.167 (100) (444), 2.977 (41) (721), 2.849 (53) (731), 2.028 (35) (864).

Kind of sample preparation and/or method of registration of the spectrum: KBr disc. Transmission.

Source: Botto and Baran (1982).

Wavenumbers (cm^{-1}): 762, 738, 705sh, 662s, 630s, 491, 402sh, 385sh, 358, 326, 307, 260w.

Te73 Yttrium vanadyl oxytellurite Y(VO)$(TeO_3)_2$O

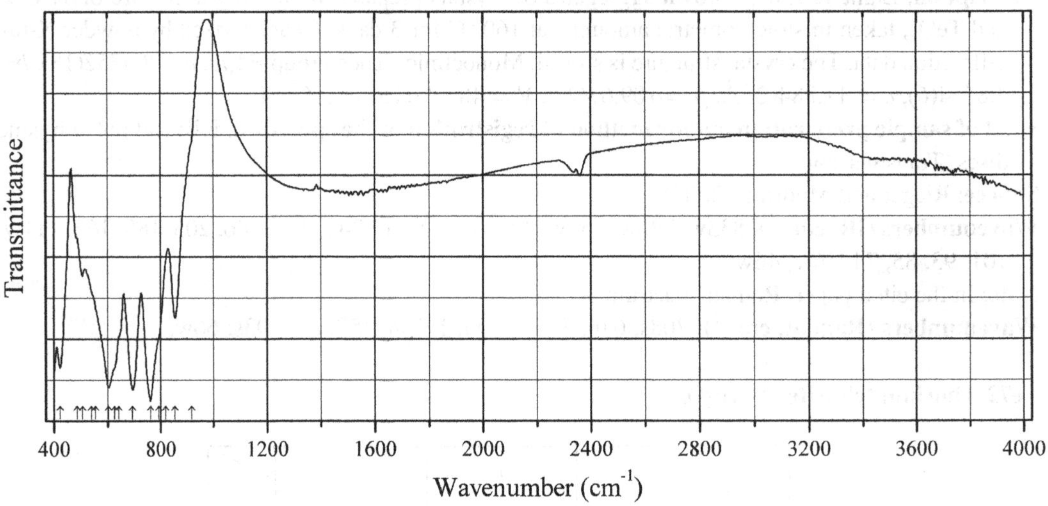

Origin: Synthetic.

Description: Crystals grown hydrothermally from Y_2O_3, V_2O_5, and TeO_2 at 550 °C for 48 h with three intermediate regrindings. Characterized by powder X-ray diffraction data. The crystal structure is solved. Monoclinic, space group *C*2/*m*, *a* = 7.9396(10), *b* = 7.5625(10), *c* = 21.282(2) Å, β = 90.010(10)°, *V* = 1277.85(3) Å^3, *Z* = 8. D_{calc} = 5.438 g/cm^3. V^{5+} has sixfold coordination with one short V–O bond.

Kind of sample preparation and/or method of registration of the spectrum: KBr disc. Transmission.

Source: Kim et al. (2014).

Wavenumbers (cm^{-1}): 920sh, 857, 821sh, 792sh, 759s, 692s, 640sh, 625sh, 603s, 555sh, 540sh, 507w, 487sh, 423.

Note: The wavenumbers were partly determined by us based on spectral curve analysis of the published spectrum.

Te74 Tellurium Te

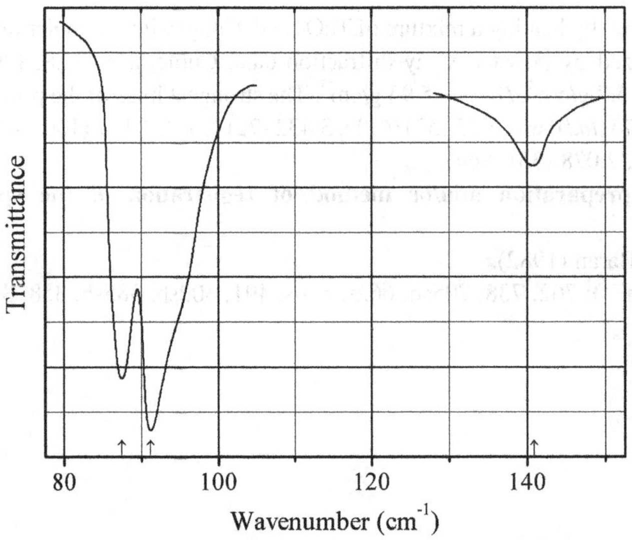

Origin: Synthetic.
Description: A sample prepared by deposition of evaporated Te on crystalline quartz.
Kind of sample preparation and/or method of registration of the spectrum: Polycrystalline film 1500 Å thick. Transmission.
Source: Grosse and Richter (1970).
Wavenumbers (cm^{-1}): 140.8, 91.2, 87.5.

Te75 Yafsoanite $Ca_3Zn_3Te^{6+}{}_2O_{12}$

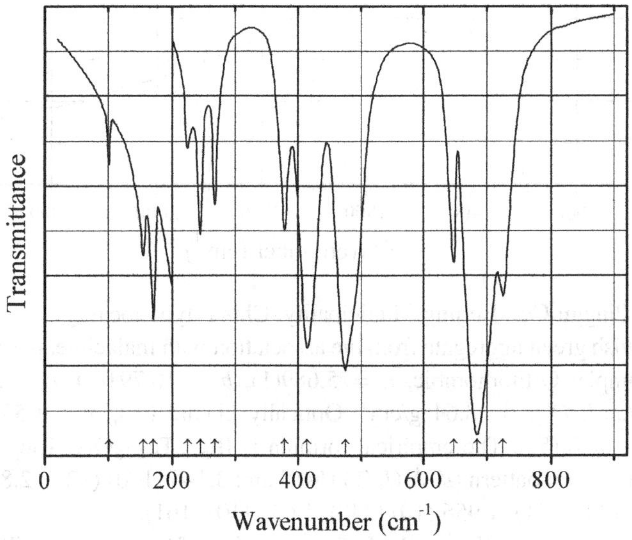

Origin: Synthetic.
Description: Synthesized by solid-state reaction, starting from $CaCO_3$, ZnO, and TeO_2. Confirmed by powder X-ray diffraction data.
Kind of sample preparation and/or method of registration of the spectrum: KBr and polyethylene discs. Transmission.
Source: Rulmont et al. (1992).
Wavenumbers (cm^{-1}): 725, 684s, 647, 475s, 415s, 379, 268, 245, 225w, 171, 155, 101w.

Te76 Pingguite $Bi_6Te^{4+}{}_2O_{13}$

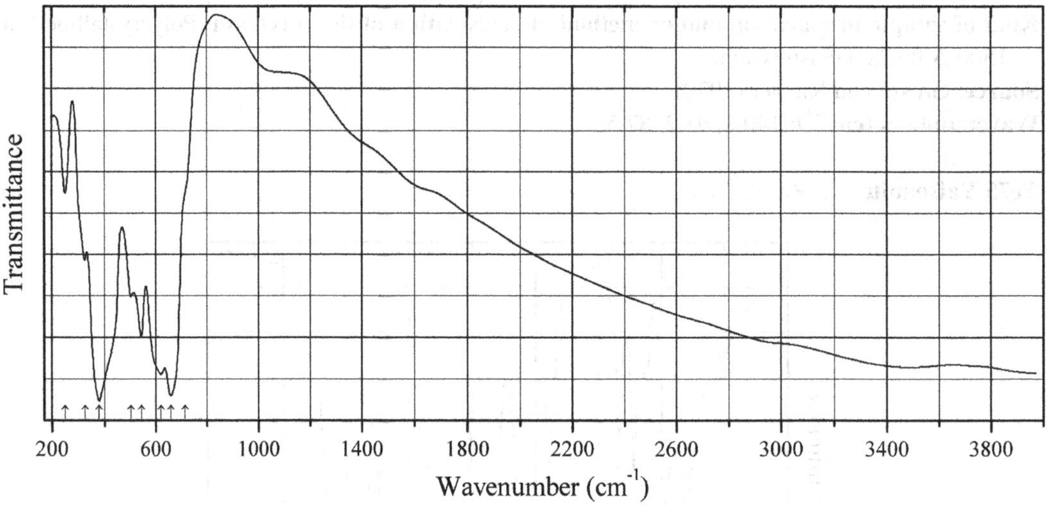

Origin: Yangjiava, Pinggu Co., Beijing Municipality, China (type locality).

Description: Yellowish green aggregate from the association with malachite, pyromorphite, bismutite, etc. Holotype sample. Orthorhombic, $a = 5.689(1)$, $b = 10.791(1)$, $c = 5.308(1)$ Å, $Z = 1$. $D_{meas} = 8.44$ g/cm^3, $D_{calc} = 8.64$ g/cm^3. Optically biaxial (−), $\alpha = 1.554(2)$, $\beta = 1.558(2)$, $\gamma = 1.566(2)$, $2V = 70(5)°$. The empirical formula is $Bi_{5.80}Te_{2.15}O_{13}$. The strongest lines of the powder X-ray diffraction pattern [*d*, Å (*I*, %) (*hkl*)] are: 3.146 (100) (121), 2.841 (80) (200), 2.694 (20) (040), 1.695 (20) (321), 1.956 (10) (240), 1.631 (10) (161).

Kind of sample preparation and/or method of registration of the spectrum: Transmission. Kind of sample preparation is not indicated.

Source: Sun et al. (1994).

Wavenumbers (cm^{-1}): 715sh, 660s, 620, 540, 500, 380s, 325, 250.

Note: The wavenumbers were partly determined by us based on spectral curve analysis of the published spectrum.

Te77 Denningite $CaMn^{2+}Te^{4+}{}_4O_{10}$

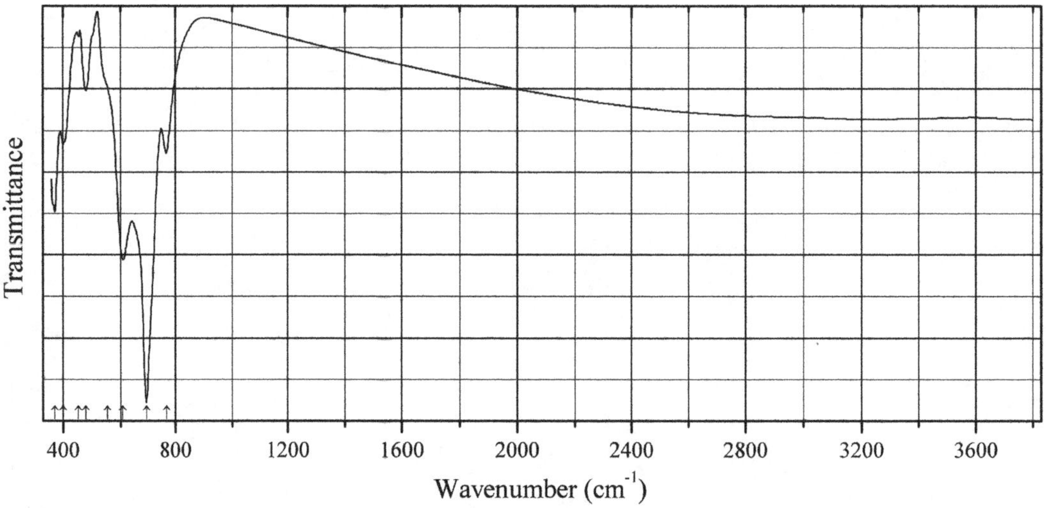

Origin: Moctezuma (La Bambolla) mine, Moctezuma, Sonora, Mexico (type locality).

Description: Pale yellowish-green granular aggregate from the association with muscovite. The empirical formula is (electron microprobe): $Ca_{0.79}Mn_{1.15}Mg_{0.05}Te_{4.00}O_5$.

Kind of sample preparation and/or method of registration of the spectrum: KBr disc. Absorption.

Wavenumbers (cm^{-1}): 767, 692s, 609s, 555sh, 481, 454w, 401, 372.

Note: The spectrum was obtained by N.V. Chukanov.

Te78 Poughite $Fe^{3+}_2(Te^{4+}O_3)_2(SO_4)\cdot 3H_2O$

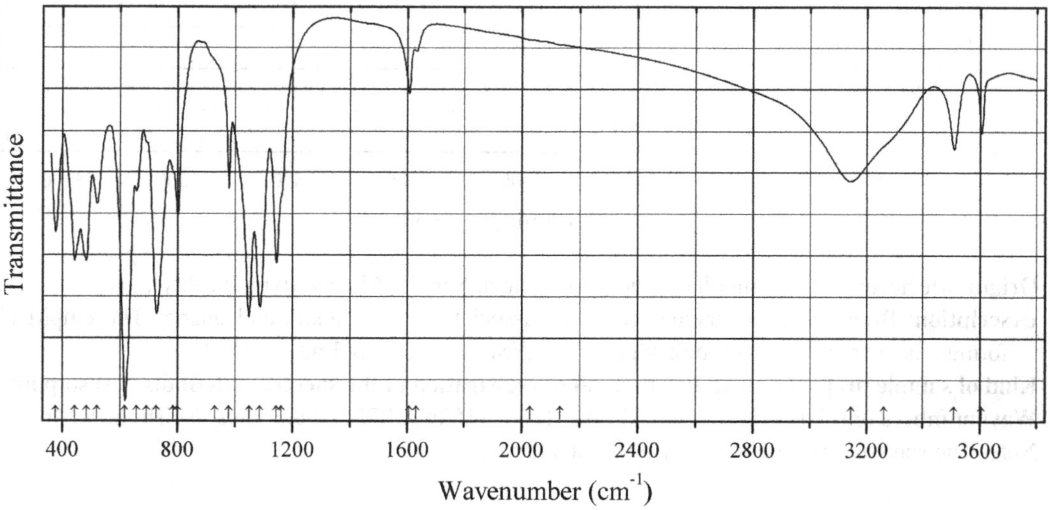

Origin: Moctezuma (La Bambolla) mine, Moctezuma, Sonora, Mexico (type locality).

Description: Yellow spherulites on quartz. The empirical formula is (electron microprobe): $(Fe_{0.92}Al_{0.08})(TeO_3)_{1.99}(SO_4)_{1.02}\cdot nH_2O$.

Kind of sample preparation and/or method of registration of the spectrum: KBr disc. Absorption.

Wavenumbers (cm^{-1}): 3606, 3508, 3260sh, 3144, 2134w, 2028w, 1633, 1608, 1160sh, 1144s, 1086s, 1048s, 979, 930sh, 800, 785sh, 728s, 693sh, 659, 619s, 522, 484s, 443s, 375.

Note: The spectrum was obtained by N.V. Chukanov.

Te79 Zemannite $Mg_{0.5}ZnFe^{3+}(Te^{4+}O_3)_3 \cdot 4.5H_2O$

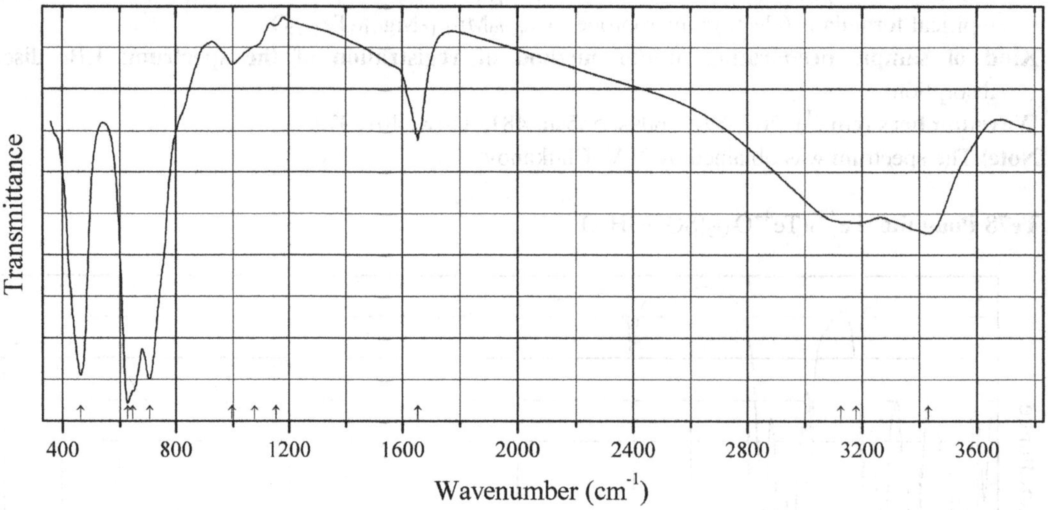

Origin: Moctezuma (La Bambolla) mine, Moctezuma, Sonora, Mexico (type locality).

Description: Brown acicular crystals from the association with dickite and quartz. The empirical formula is (electron microprobe): $Mg_{0.63}Zn_{0.84}Fe_{0.98}Al_{0.02}Ti_{0.02}(TeO_3)_3 \cdot nH_2O$.

Kind of sample preparation and/or method of registration of the spectrum: KBr disc. Absorption.

Wavenumbers (cm^{-1}): 3430s, 3180, 3125sh, 1654, 1152w, 1075sh, 996, 704s, 645sh, 627s, 464s.

Note: The spectrum was obtained by N.V. Chukanov.

Te80 Plumbotellurite $Pb(Te^{4+}O_3)$

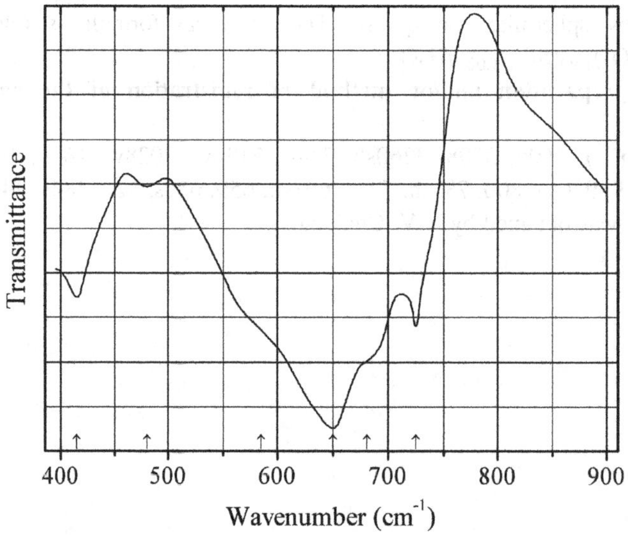

Origin: Synthetic.

Description: Synthesized by crystallization above the solidus temperature of the PbO–TeO_2 system. Characterized by powder X-ray diffraction data and electron microprobe analysis.

Kind of sample preparation and/or method of registration of the spectrum: KBr disc. Transmission.

Source: Stavrakieva et al. (1988).

Wavenumbers (cm^{-1}): 725, 681sh, 650s, 585sh, 480w, 415.

Note: The wavenumbers were partly determined by us based on spectral curve analysis of the published spectrum.

2.19 Iodides, Iodites, and Iodates

I15 Copper iodate α-$Cu(IO_3)_2$

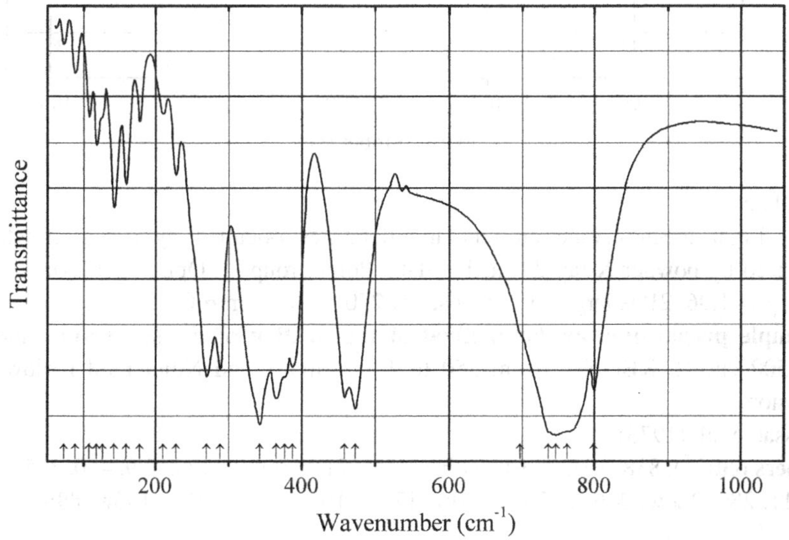

Origin: Synthetic.

Description: Light yellow green crystal clusters and spherulites obtained by heating bellingerite $Cu(IO_3)_2 \cdot 2H_2O$ to 110 °C. Characterized by powder X-ray diffraction data. Point group 2. Optically biaxial (+) or (−), $\alpha = 1.88$, $\beta = 1.94$, $\gamma = 2.00$, $2V$ is medium. Decomposes at 460 °C.

Kind of sample preparation and/or method of registration of the spectrum: Fluorolube mull (above 1500 cm^{-1}), KBr disc (from 550 to 1500 cm^{-1}), and Nujol mull (below 550 cm^{-1}). Transmission.

Source: Nassau et al. (1973).

Wavenumbers (cm^{-1}): 799, 762sh, 746s, 736sh, 697, 472s, 458s, 388, 377sh, 366s, 343s, 288, 270s, 228, 210w, 178, 159, 142, 126sh, 118, 108, 89w, 73w.

I16 Copper iodate β-$Cu(IO_3)_2$

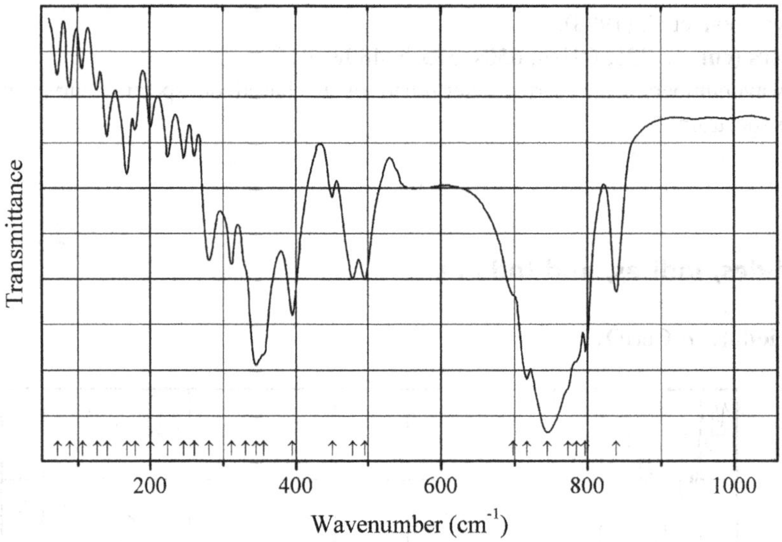

Origin: Synthetic.

Description: Light green crystal clusters and dendrites obtained by a gel growth technique. Characterized by powder X-ray diffraction data. Point group -1. Optically biaxial (–), $\alpha = 1.90$, $\beta = 1.94$, $\gamma = 1.96$, $2V$ is large. Decomposes at 250 °C to form α-$Cu(IO_3)_2$.

Kind of sample preparation and/or method of registration of the spectrum: Fluorolube mull (above 1500 cm^{-1}), KBr disc (from 550 to 1500 cm^{-1}), and Nujol mull (below 550 cm^{-1}). Transmission.

Source: Nassau et al. (1973).

Wavenumbers (cm^{-1}): 838, 797, 785sh, 774sh, 745s, 717s, 698sh, 495, 479, 450, 395s, 356sh, 346s, 331sh, 311, 280, 260w, 246w, 224w, 200w, 179w, 167, 140w, 126w, 106w, 89w, 72w.

I17 Copper iodate γ-$Cu(IO_3)_2$

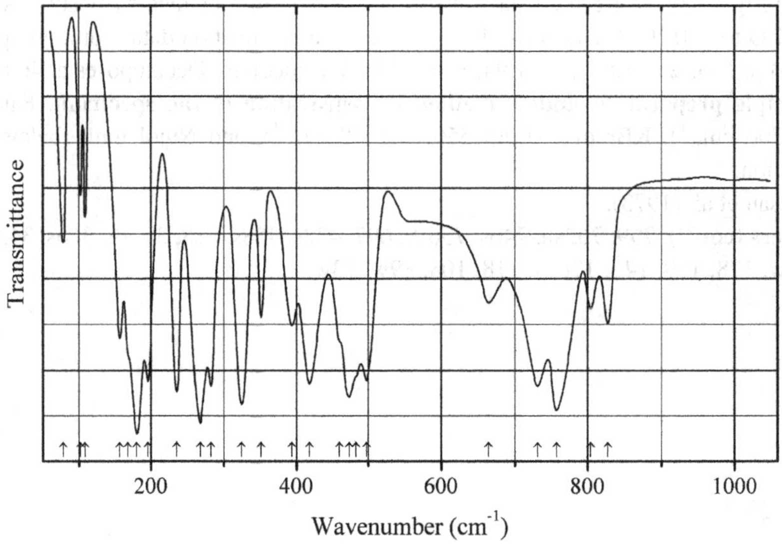

Origin: Synthetic.

Description: Dark yellow crystals and crystal clusters obtained by a gel growth technique. Characterized by powder X-ray diffraction data. Point group *2/m*. Optically biaxial (–), $\alpha = 1.89$, $\beta = 1.96$, $\gamma = 1.99$, $2V$ is medium. Decomposes at 460 °C to form CuO.

Kind of sample preparation and/or method of registration of the spectrum: Fluorolube mull (above 1500 cm^{-1}), KBr disc (from 550 to 1500 cm^{-1}), and Nujol mull (below 550 cm^{-1}). Transmission.

Source: Nassau et al. (1973).

Wavenumbers (cm^{-1}): 827, 804, 758s, 732, 665, 497, 483sh, 473s, 460sh, 418, 394, 352, 325s, 282, 267s, 235, 195, 180s, 167, 156, 109, 102, 78.

Note: In the cited paper, the wavenumber 180 cm^{-1} is erroneously indicated as 170 cm^{-1}.

I18 Potassium titanium iodate $K_2Ti(IO_3)_6$

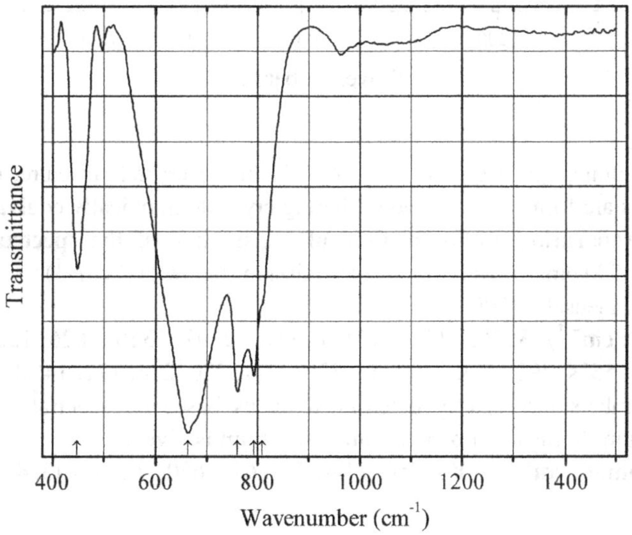

Origin: Synthetic.

Description: Synthesized hydrothermally from K_2CO_3, TiO_2, and HIO_3 at 230 °C for 4 days. Characterized by powder X-ray diffraction data. The crystal structure is solved. Trigonal, space group *R*-3, $a = 11.2703(6)$, $c = 11.3514(11)$ Å, $V = 1248.68(15)$ Å^3, $Z = 3$. $D_{calc} = 4.690$ g/cm^3.

Kind of sample preparation and/or method of registration of the spectrum: Transmission. Kind of sample preparation is not indicated.

Source: Chang et al. (2009).

Wavenumbers (cm^{-1}): 808sh, 792s, 760s, 664s, 448.

I19 Copper acid diperiodate hydrate $Cu(H_4I_2O_{10})\cdot 6H_2O$

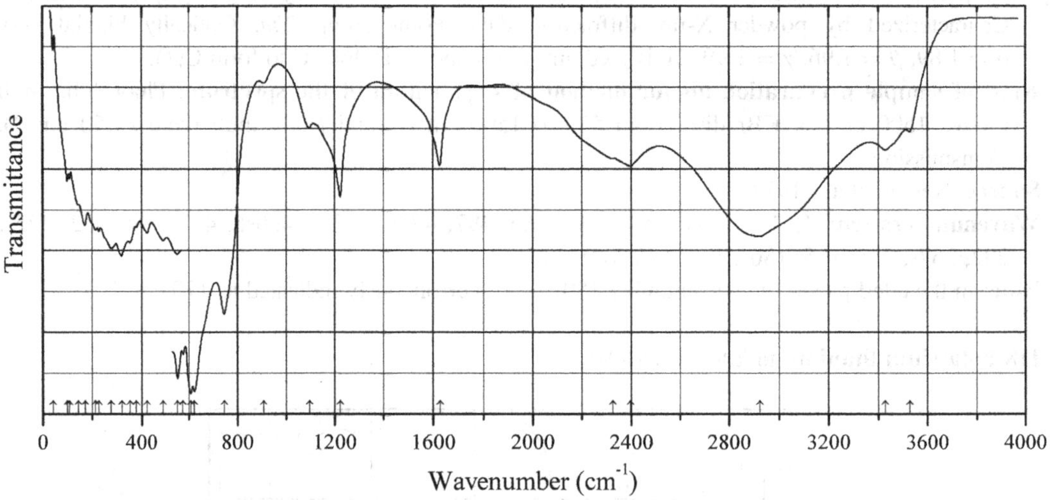

Origin: Synthetic.

Description: Monoclinic, space group *P*2₁/*c*, *Z* = 2. In the crystal structure, the centrosymmetric $H_4I_2O_{10}^{2-}$ anions are formed by two edge-sharing crystallographically equivalent IO_6 octahedra.

Kind of sample preparation and/or method of registration of the spectrum: Attenuated total reflection (above 500 cm^{-1}). Nujol mull, transmission (below 500 cm^{-1}).

Source: Jaquet and Haeuseler (2008).

Wavenumbers (IR, cm^{-1}): 3526, 3427, 2924s (broad), 2397, 2326, 1626, 1223, 1097, 908, 746s, 623s, 608s, 576s, 553s, 492, 424, 380, 353, 320s, 277, 226, 214, 171, 145sh, 108, 99, 42w.

Note: The wavenumbers were partly determined by us based on spectral curve analysis of the published spectrum. In the cited paper, Raman spectrum is given.

Wavenumbers (Raman, cm^{-1}): 805s, 756, 746s, 704, 671, 630s, 620s, 443, 413, 396, 356w, 345w, 329, 311.

I20 Sodium titanium iodate $Na_2Ti(IO_3)_6$

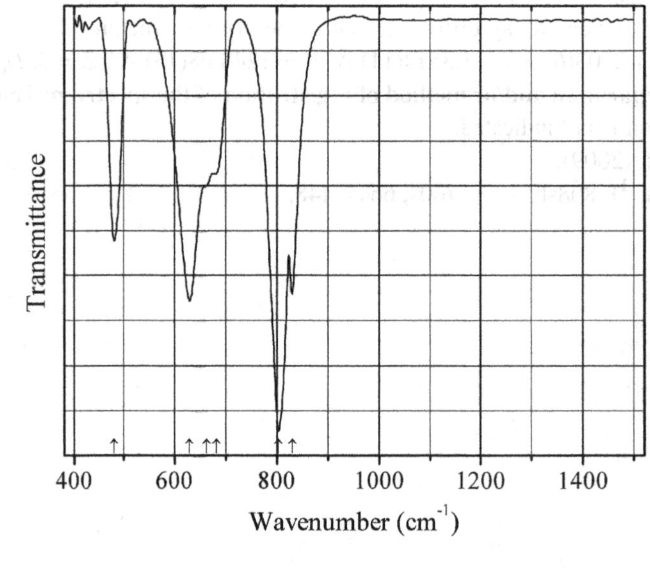

Origin: Synthetic.

Description: Colorless acicular crystals synthesized hydrothermally from Na_2CO_3, TiO_2, and HIO_3 at 230 °C for 4 days. Characterized by powder X-ray diffraction data. The crystal structure is solved. Hexagonal, space group $P6_3$, $a = 9.649(3)$, $c = 5.198(3)$ Å, $V = 419.1(3)$ Å^3, $Z = 1$. D_{calc}= 4.530 g/cm^3. The Ti^{4+} cation is disordered over two half-occupied sites

Kind of sample preparation and/or method of registration of the spectrum: Transmission. Kind of sample preparation is not indicated. Possibly, a procedure of baseline correction has been applied.

Source: Chang et al. (2009).

Wavenumbers (cm^{-1}): 830, 803s, 682sh, 663sh, 630, 481.

I21 Laurionite I-analogue Pb(OH)I

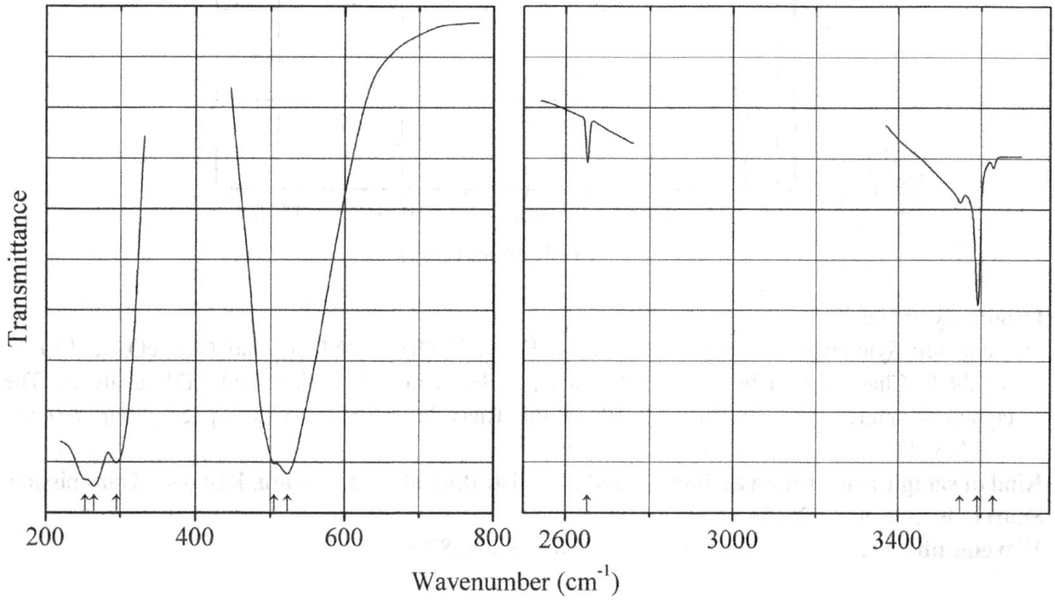

Origin: Synthetic.

Description: Characterized by powder X-ray diffraction data and thermoanalytical methods. Isostructural with laurionite.

Kind of sample preparation and/or method of registration of the spectrum: KBr or CsI disc, and Nujol or poly(chlortrifluorethene) mull. Transmission.

Source: Lutz et al. (1995).

Wavenumbers (IR, cm^{-1}): 3629, 3592, 3550, 523s, 505sh, 295, 265s, 253.

Note: In the cited paper, Raman spectrum is given.

Wavenumbers (Raman, cm^{-1}): 3496s, 318w, 244w, 172w, 104, 72, 55s, 50s

2.20 Xenates

Xe1 Double perovskite $KBa(XeNaO_6)$ $KBa(XeNaO_6)$

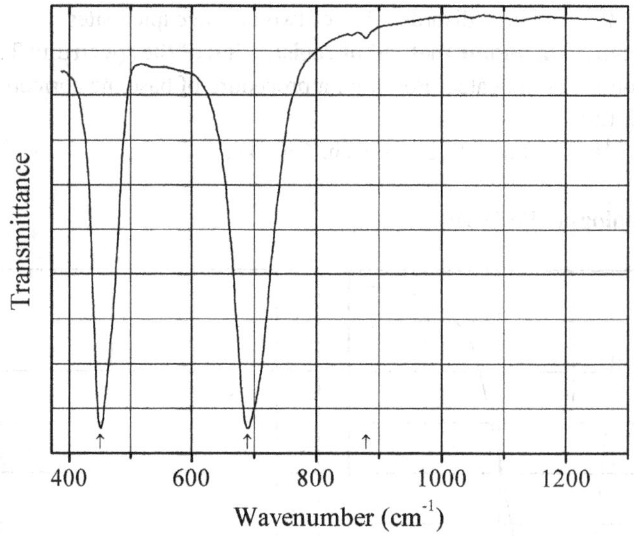

Origin: Synthetic.

Description: Synthesized hydrothermally from KOH, NaOH, $Ba(NO_3)_2$, and Na_4XeO_6 at 170 °C for 24 h. Characterized by TGA-DSC data, powder X-ray diffraction, and EDX analyses. The crystal structure is solved and refined by the Rietveld method. Cubic, space group *Fm*-3*m*, $a = 8.3188(2)$ Å, $V = 575.67(3)$ Å^3, $Z = 4$.

Kind of sample preparation and/or method of registration of the spectrum: KBr disc. Transmission.

Source: Britvin et al. (2015).

Wavenumbers (cm^{-1}): 1626w, 1434w, 1383w, 1392w, 879w, 689s, 451s.

Xe2 Double perovskite $KCa(XeNaO_6)$ $KCa(XeNaO_6)$

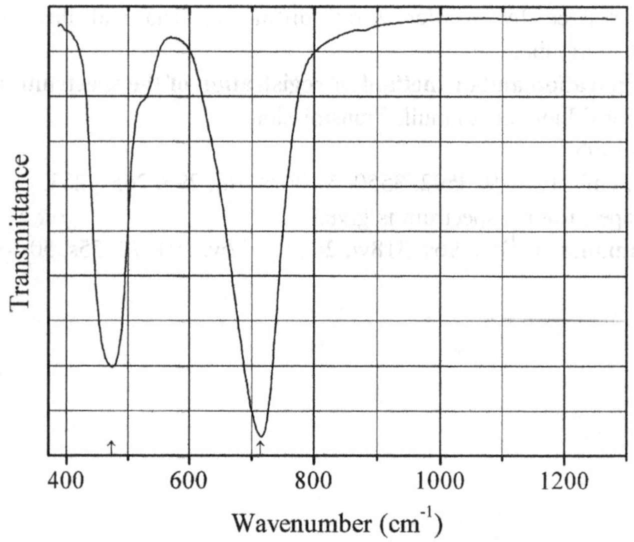

Origin: Synthetic.

Description: Synthesized hydrothermally from KOH, NaOH, $Ca(NO_3)_2{\cdot}4H_2O$, and Na_4XeO_6 at 170 °C for 24 h. Characterized by TGA-DSC data, powder X-ray diffraction, and EDX analyses. The crystal structure is solved and refined by the Rietveld method. Tetragonal, space group *I*4/*m*, $a = 5.7500(1)$, $c = 8.1558(2)$ Å, $V = 269.66(1)$ Å^3, $Z = 2$.

Kind of sample preparation and/or method of registration of the spectrum: KBr disc. Transmission.

Source: Britvin et al. (2015).

Wavenumbers (cm^{-1}): 715s, 474s.

Xe3 Double perovskite KSr(XeNaO$_6$) KSr(XeNaO$_6$)

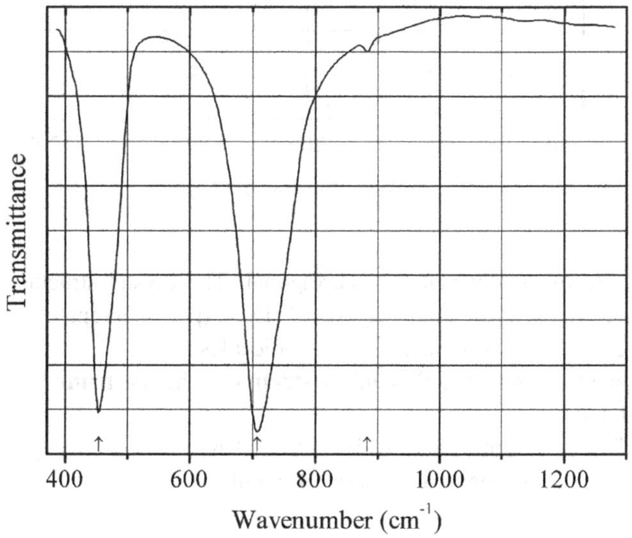

Origin: Synthetic.

Description: Synthesized hydrothermally from KOH, NaOH, $Sr(NO_3)_2$, and Na_4XeO_6 at 170 °C for 24 h. Characterized by TGA-DSC data, powder X-ray diffraction, and EDX analyses. The crystal structure is solved and refined by the Rietveld method. Cubic, space group *Fm*-3*m*, $a = 8.1920(1)$ Å, $V = 549.76(2)$ Å^3, $Z = 4$.

Kind of sample preparation and/or method of registration of the spectrum: KBr disc. Transmission.

Source: Britvin et al. (2015).

Wavenumbers (cm^{-1}): 1634w, 1444w, 1396w, 883w, 707s, 454s.

Xe4 Layered perovskite $K_4Xe_3O_{12}$ $K_4Xe_3O_{12}$

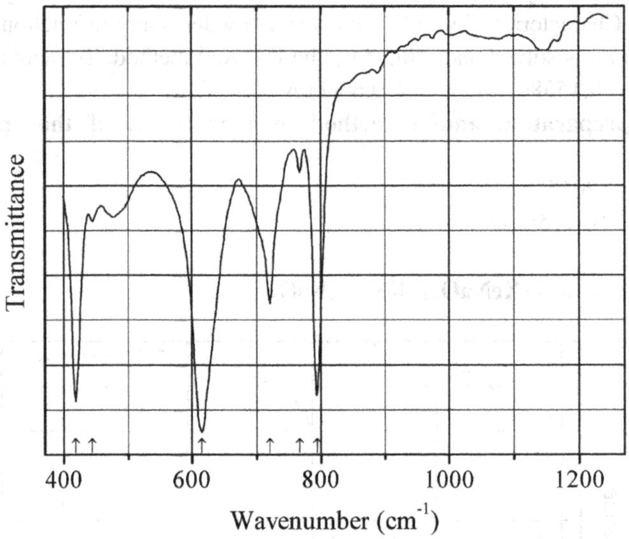

Origin: Synthetic.

Description: Aggregate of yellow platelets. Hexagonal. The crystal structure contains three-layer perovskite slabs composed of inner layers of $[XeO_6]^{4-}$ (perxenate) octahedra, which are sandwiched between the layers of neutral XeO_3 molecules.

Kind of sample preparation and/or method of registration of the spectrum: Nujol mull. Transmission.

Source: Britvin et al. (2016).

Wavenumbers (cm^{-1}): 794s, 768w, 721, 615s, 445w, 419s.

Note: The band at 721 cm^{-1} is due to absorption of Nujol.

2.21 Tungstates and W-Bearing Oxides

W18 Hydrotungstite $WO_2(OH)_2 \cdot H_2O$

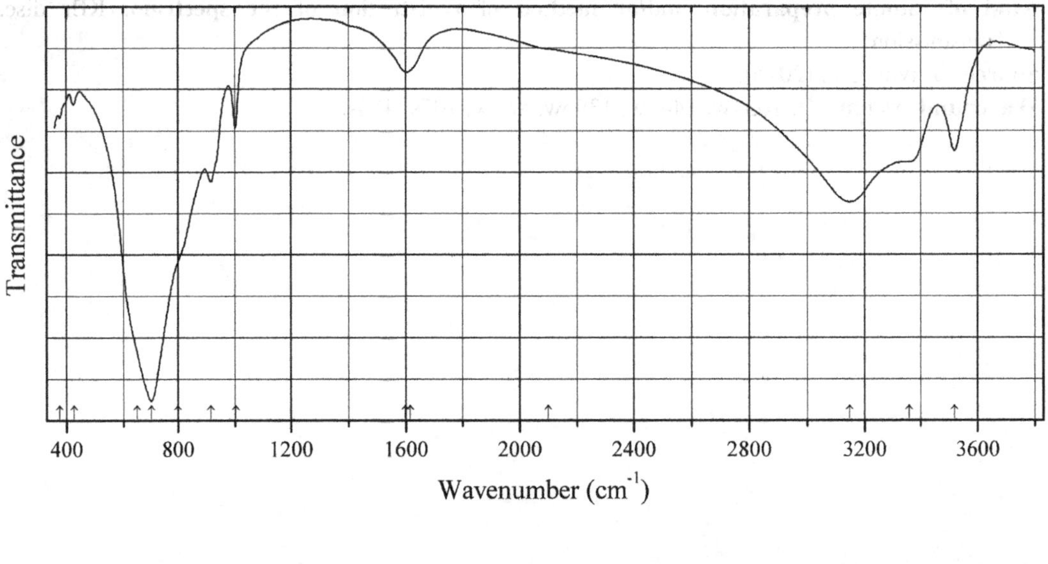

Origin: Alyaskitovoe Sn-W deposit, eastern Sakha Republic (Yakutia), Siberia, Russia.

Description: Yellow, massive. Investigated by I.V. Pekov. Characterized by qualitative electron microprobe analyses. The strongest lines of the powder X-ray diffraction pattern [*d*, Å (*I*, %)] are: 6.93 (100), 3.70 (60), 3.44 (30), 3.25 (40), 2.62 (70), 2.54 (50), 1.95 (70).

Kind of sample preparation and/or method of registration of the spectrum: KBr disc. Absorption.

Wavenumbers (cm^{-1}): 3518, 3359, 3147, 2100sh, 1615sh, 1599, 1002, 916, 800sh, 702s, 650sh, 426w, (375w).

Note: The spectrum was obtained by N.V. Chukanov.

W19 Ammonium paratungstate tetrahydrate $(NH_4)_{10}[H_2W_{12}O_{42}]·4H_2O$

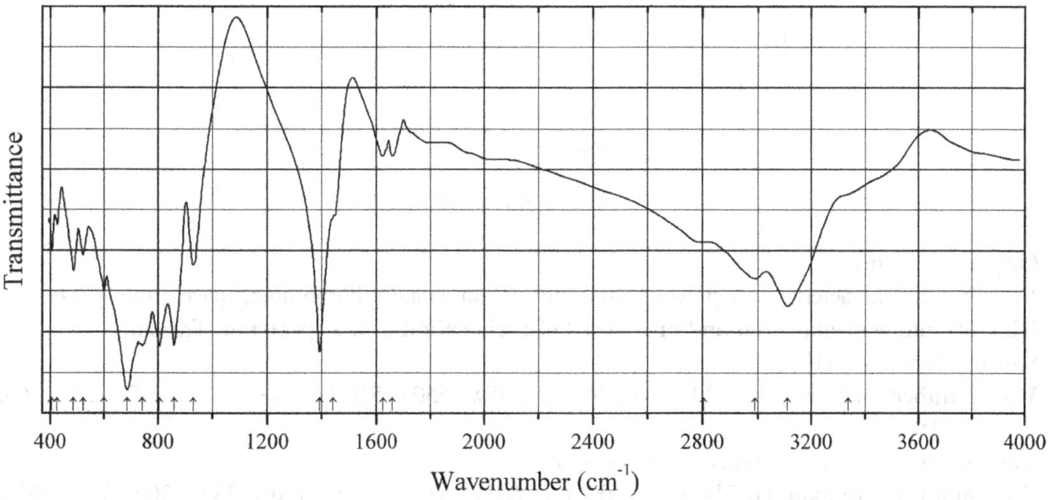

Origin: Synthetic.

Description: Reactant produced by H.C. Starck. Confirmed by elemental analysis. Characterized by powder X-ray diffraction data.

Kind of sample preparation and/or method of registration of the spectrum: KBr disc. Transmission.

Source: Szilágyi et al. (2004).

Wavenumbers (cm^{-1}): 3337sh, 3110, 2992, 2805sh, 1661w, 1624w, 1442sh, 1391s, 929, 859s, 805s, 741s, 687s, 601, 522, 486, 426, 406.

Note: The wavenumbers were determined by us based on spectral curve analysis of the published spectrum.

W20 Bismuth tungstate Bi_2WO_6

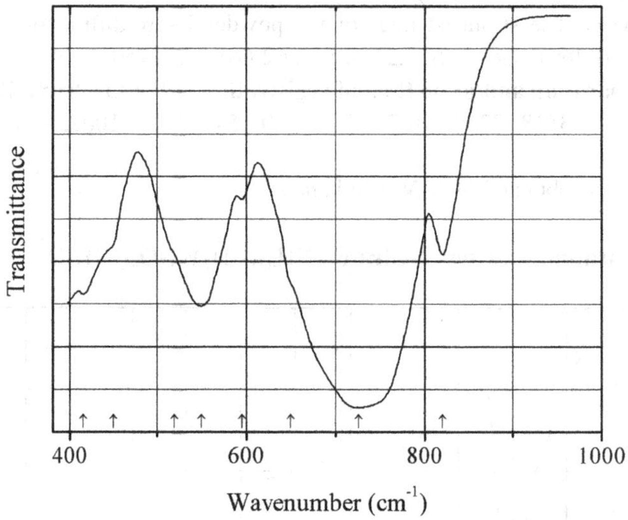

Origin: Synthetic.
Description: Characterized by powder X-ray diffraction data. Orthorhombic, space group *B*2*cb*.
Kind of sample preparation and/or method of registration of the spectrum: Transmission.
Source: Bode et al. (1973).
Wavenumbers (IR, cm^{-1}): 820, 725s, 650sh, 595w, 550s, 520sh, 450sh, 415, 345s, 290, 265, 245, 225.
Note: In the cited paper, Raman spectrum is given.
Wavenumbers (Raman, cm^{-1}): 820, 800s, 720, 605w, 525w, 460w, 420w, 335, 420w, 310s, 285s, 265, 225, 210.

W21 Cadmium tungstate $CdWO_4$

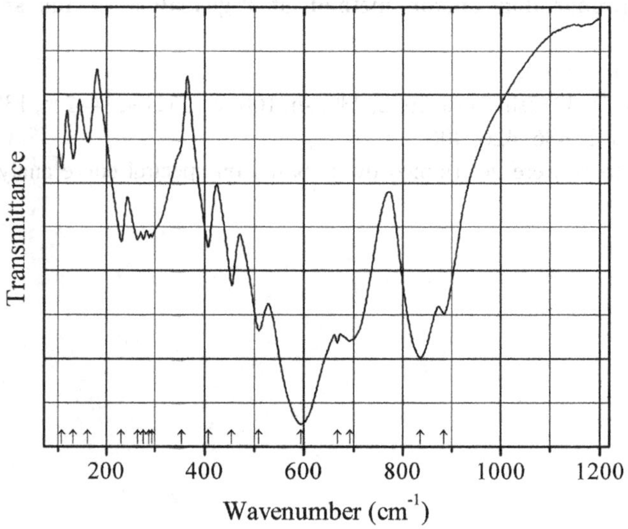

Origin: Synthetic.

Description: Prepared hydrothermally from corresponding oxides at 473 K for 48 h. Characterized by powder X-ray diffraction data and Ritveld crystal structure refinement. Isostructural with wolframite. Monoclinic, space group *P*2/*c*, $a = 5.026(1)$, $b = 5.078(1)$, $c = 5.867(1)$ Å, $\beta = 91.47(1)°$, $Z = 2$.

Kind of sample preparation and/or method of registration of the spectrum: KBr and polyethylene discs. Transmission.

Source: Daturi et al. (1997).

Wavenumbers (IR, cm^{-1}): 884, 835s, 693, 667, 595s, 510, 455, 408, 354sh, 294, 287, 276, 263, 230, 161w, 131w, 107w.

Note: The wavenumbers were partly determined by us based on spectral curve analysis of the published spectrum. In the cited paper, Raman spectrum is given.

Wavenumbers (Raman, cm^{-1}): 896s, 771, 706, 687, 547, 514, 387, 351, 307s, 269, 248w, 229, 177, 148, 133, 117, 99, 77.

W22 Cesium uranyl tungstate $Cs_4[(UO_2)_4(WO_5)(W_2O_8)O_2]$ $Cs_4[(UO_2)_4(WO_5)(W_2O_8)O_2]$

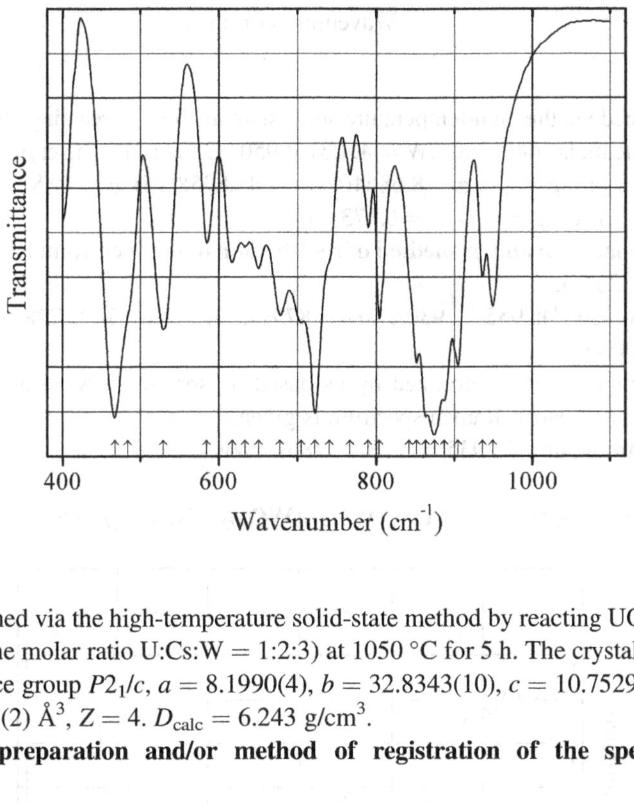

Origin: Synthetic.

Description: Obtained via the high-temperature solid-state method by reacting $UO_2(NO_3)_2$ with WO_3 and $CsNO_3$ (at the molar ratio U:Cs:W = 1:2:3) at 1050 °C for 5 h. The crystal structure is solved. Monoclinic, space group *P*2$_1$/*c*, $a = 8.1990(4)$, $b = 32.8343(10)$, $c = 10.7529(6)$ Å, $\beta = 117.594(4)°$, $V = 2565.5(2)$ Å^3, $Z = 4$. $D_{calc} = 6.243$ g/cm^3.

Kind of sample preparation and/or method of registration of the spectrum: KBr disc. Transmission.

Source: Xiao et al. (2015).

Wavenumbers (IR, cm^{-1}): 949, 936, 904, 887, 875s, 863s, 852, 842sh, 804, 789, 766w, 740sh, 722s, 705, 678, 651, 634, 618, 585, 530, 484sh, 467s.

Note: The wavenumbers were determined by us based on spectral curve analysis of the published spectrum. In the cited paper, Raman spectrum is given.

Wavenumbers (Raman, cm^{-1}): 952, 934, 782s, 765s, 548.

W23 Cesium uranyl tungstate $Cs_4[(UO_2)_7(WO_5)_3O_3]$ $Cs_4[(UO_2)_7(WO_5)_3O_3]$

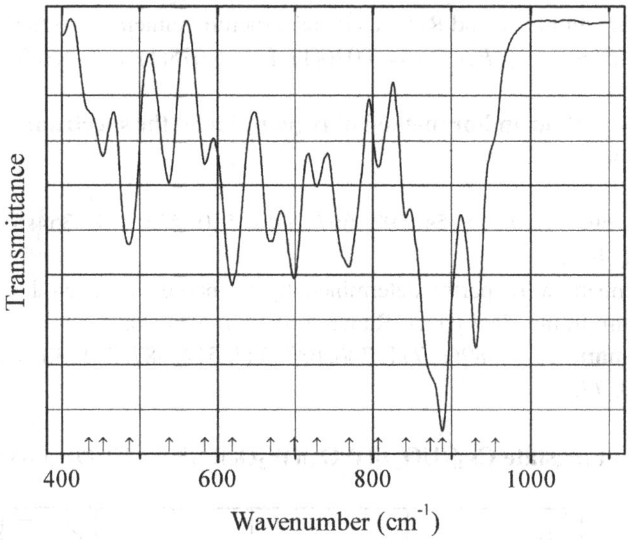

Origin: Synthetic.

Description: Obtained via the high-temperature solid-state method by reacting $UO_2(NO_3)_2$ with WO_3 and $CsNO_3$ (at the molar ratio U:Cs:W = 4:4:3) at 950 °C for 15 h. The crystal structure is solved. Monoclinic, space group $P2_1/c$, $a = 8.6864(4)$, $b = 41.8958(15)$, $c = 10.8213(7)$ Å, $\beta = 116.467(4)°$, $V = 3525.4(3)$ Å^3, $Z = 1$. $D_{calc} = 6.173$ g/cm^3.

Kind of sample preparation and/or method of registration of the spectrum: KBr disc. Transmission.

Source: Xiao et al. (2015).

Wavenumbers (IR, cm^{-1}): 955sh, 930s, 888s, 873sh, 842, 807, 769, 728, 699, 669, 619, 584, 538, 486, 453, 435sh.

Note: The wavenumbers were determined by us based on spectral curve analysis of the published spectrum. In the cited paper, Raman spectrum is given.

Wavenumbers (Raman, cm^{-1}): 935, 871, 828, 786s, 766s.

W24 Cesium uranyl tungstate $Cs_8(UO_2)_4(WO_4)_4(WO_5)_2$ $Cs_8(UO_2)_4(WO_4)_4(WO_5)_2$

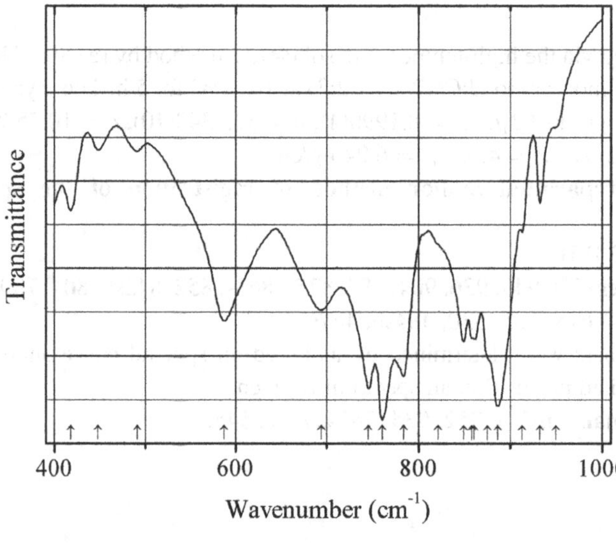

Origin: Synthetic.

Description: Synthesized from a mixture of $CsNO_3$, WO_3, and U_3O_8 taken in the ratio Cs:U:W = 4:2:3 by solid-state reaction in air at 650 °C for 1 week with intermediate grindings. Characterized by powder X-ray diffraction data. The crystal structure is solved. Monoclinic, space group $P2_1/n$, a = 11.2460(3), b = 13.8113(3), c = 25.7287(6) Å, β = 90.00°, V = 3996.23(17) Å^3, Z = 4. D_{meas} = 6.079(2) g/cm^3, D_{calc} = 6.087(2) g/cm^3. In the structure, the U atoms are in pentagonal bipyramid coordination, while W atoms are in two different environments, with tetrahedral and square pyramidal coordinations.

Kind of sample preparation and/or method of registration of the spectrum: KBr disc. Transmission.

Source: Yagoubi et al. (2007).

Wavenumbers (cm^{-1}): 950w, 932, 913sh, 886s, 875sh, 861, 858, 849, 821sh, 784s, 761s, 746s, 694, 587, 491w, 448w, 418.

W25 Lithium iron(III) tungstate wolframite-type $LiFe(WO_4)_2$

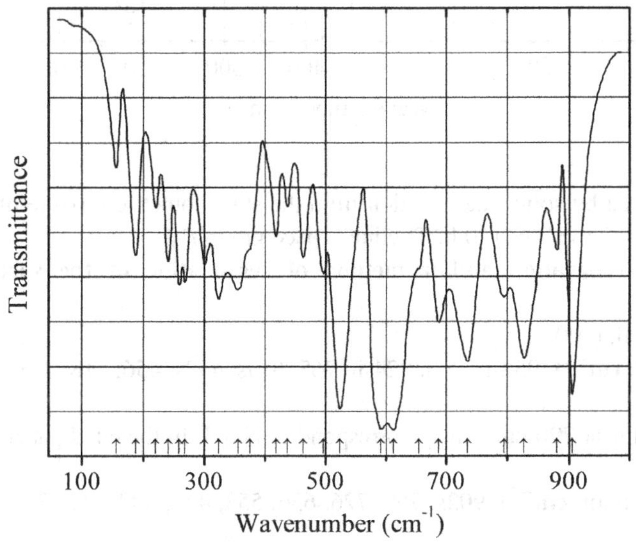

Origin: Synthetic.

Description: Synthesized from lithium carbonate, iron, and tungsten oxides using a ceramic technique.

Kind of sample preparation and/or method of registration of the spectrum: KBr disc and Nujol mull. Transmission.

Source: Fomichev and Kondratov (1994).

Wavenumbers (IR, cm^{-1}): 905s, 880, 827, 795, 734, 687, 655, 613s, 592s, 524s, 497, 463, 437, 418, 375sh, 355, 323, 300, 268, 258, 240, 220, 187, 155.

Note: In the cited paper, Raman spectrum is given.

Wavenumbers (Raman, cm^{-1}): 920s, 875, 786s, 772, 710, 655, 617, 539, 497, 463, 416, 381, 354, 321, 290, 268, 239, 211, 150, 103, 87.

W26 Lithium copper tungstate $Li_2Cu(WO_4)_2$

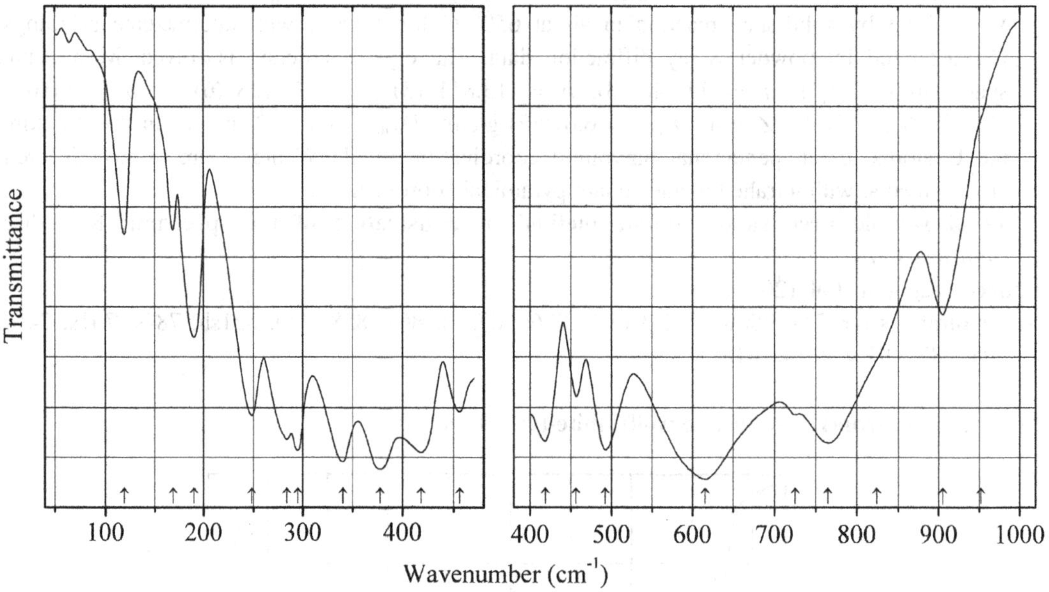

Origin: Synthetic.

Description: Prepared by solid-state reaction from the stoichiometric mixture of Li_2CO_3, WO_3, and CuO at 650–700 °C for 146–160 h. Triclinic, space group *P*-1, *Z* = 1.

Kind of sample preparation and/or method of registration of the spectrum: Nujol mull. Absorption.

Source: Mączka et al. (2002).

Wavenumbers (IR, cm^{-1}): 952sh, 905, 824sh, 765, 615s, 492s, 456, 418, 378s, 340, 295, 284, 248, 190, 169w, 119.

Note: A weak band near 720 cm^{-1} may correspond to Nujol. In the cited paper, Raman spectrum is given.

Wavenumbers (Raman, cm^{-1}): 902s, 769, 726, 636, 553, 470, 398, 363, 307, 256, 211, 174, 127, 112.

W27 Lithium nickel tungstate $Li_2Ni(WO_4)_2$

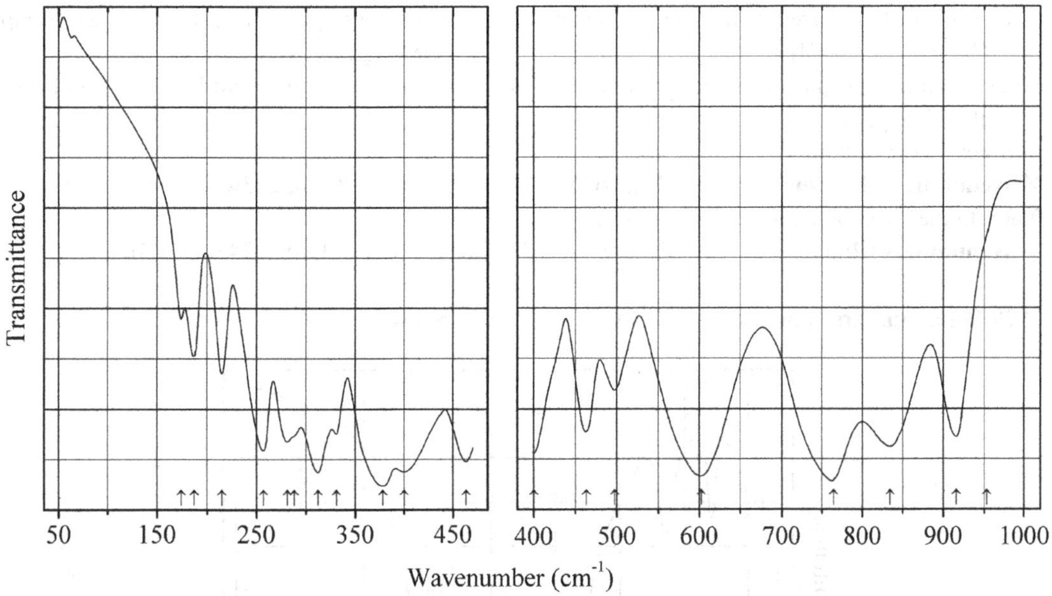

Origin: Synthetic.

Description: Prepared by solid-state reaction from the stoichiometric mixture of Li_2CO_3, WO_3, and CuO at 650–700 °C for 146–160 h. Triclinic, space group *P*-1, *Z* = 1. Structurally related to wolframite.

Kind of sample preparation and/or method of registration of the spectrum: Nujol mull. Absorption.

Source: Mączka et al. (2002).

Wavenumbers (IR, cm^{-1}): 953sh, 916, 834s, 764s, 602s, 497, 463, 400, 378s, 331, 312, 288sh, 281, 257, 215, 187w, 174w.

Note: In the cited paper, Raman spectrum is given.

Wavenumbers (Raman, cm^{-1}): 913s, 792, 754, 617w, 553w, 476w, 447w, 418, 387, 353, 311, 282, 266, 222w, 194, 143, 112.

W28 Potassium antimonate tungstate $KSbWO_6$

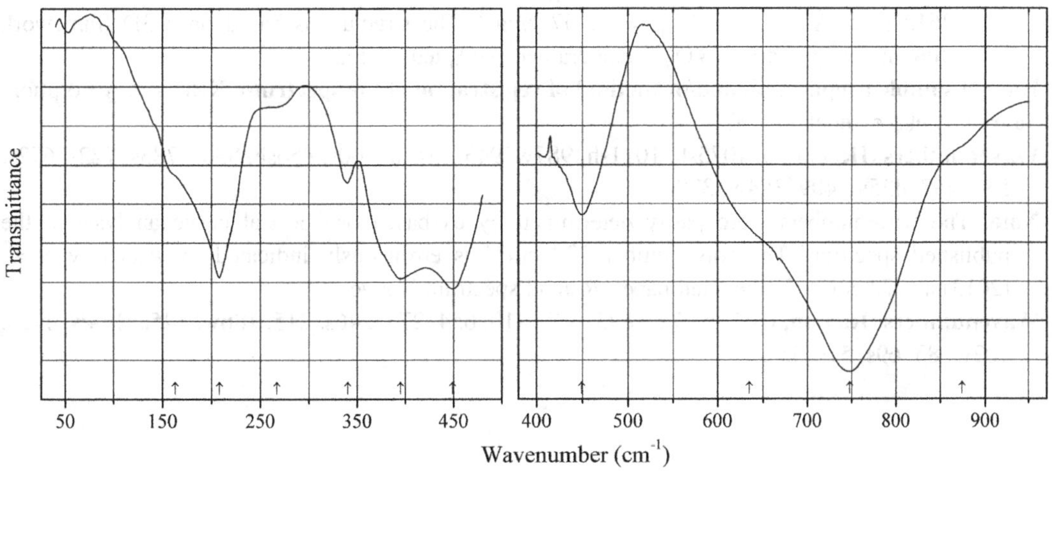

Origin: Synthetic.

Description: Prepared by the solid-state reaction between WO_3, Sb_2O_3, and KNO_3 at 973 K. The crystal structure solved by the Rietveld method is related to that of pyrochlore. Cubic, space group *Fd*3*m*, $a = 10.23671(7)$, $V = 1072.71(1)$ Å^3. $D_{calc} = 5.4886$ g/cm^3.

Kind of sample preparation and/or method of registration of the spectrum: KBr disc and Nujol mull. Absorption.

Source: Knyazev et al. (2010).

Wavenumbers (IR, cm^{-1}): 874sh, 748s, 635sh, 449, 395, 340w, 267sh, 208w, 163sh.

Note: In the cited paper, Raman spectrum is given.

Wavenumbers (Raman, cm^{-1}): 960, 909, 864, 721s, 654, 506s, 453, 382, 343, 239, 169s.

W29 Potassium arsenate tungstate K(AsW_2O_9) K(AsW_2O_9)

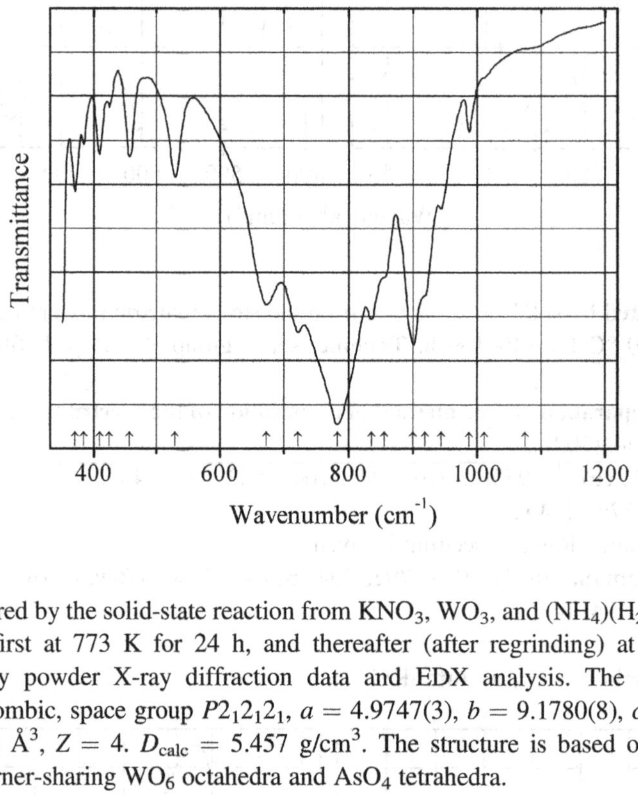

Origin: Synthetic.

Description: Prepared by the solid-state reaction from KNO_3, WO_3, and $(NH_4)(H_2AsO_4)$ in the molar ratio of 1:2:1, first at 773 K for 24 h, and thereafter (after regrinding) at 1033 K for 24 h. Characterized by powder X-ray diffraction data and EDX analysis. The crystal structure is solved. Orthorhombic, space group $P2_12_12_1$, $a = 4.9747(3)$, $b = 9.1780(8)$, $c = 16.6817(19)$ Å, $V = 761.65(12)$ Å^3, $Z = 4$. $D_{calc} = 5.457$ g/cm^3. The structure is based on a 3D framework consisting of corner-sharing WO_6 octahedra and AsO_4 tetrahedra.

Kind of sample preparation and/or method of registration of the spectrum: KBr disc. Absorption.

Source: Alekseev et al. (2013).

Wavenumbers (IR, cm^{-1}): 1075sh, 1011sh, 987w, 943, 918sh, 900s, 856sh, 836s, 783s, 722s, 672s, 530, 457, 425w, 409, 384w, 370.

Note: The wavenumbers were partly determined by us based on spectral curve analysis of the published spectrum. The wavenumber 722 cm^{-1} is erroneously indicated by Alekseev et al. (2013) as 772 cm^{-1}. In the cited paper, Raman spectrum is given.

Wavenumbers (Raman, cm^{-1}): 972s, 893, 805, 710, 654, 273, 246s, 215, 186w, 145s, 134w, 115s, 102w, 83, 69s, 51s, 35s.

W30 Potassium bismuth(III) tungstate $KBi(WO_4)_2$

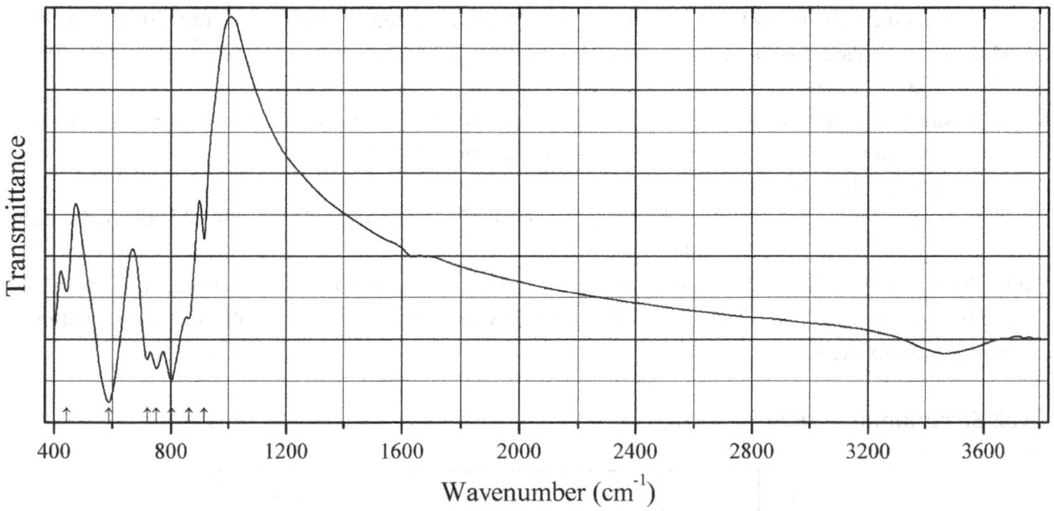

Origin: Synthetic.
Description: Crystals grown from the solution in $K_2W_2O_7$ melt. Monoclinic. $D_{meas} = 7.57$ g/cm^3, $D_{calc} = 7.51$ g/cm^3.
Kind of sample preparation and/or method of registration of the spectrum: Transmission. Kind of sample preparation is not indicated.
Source: Xie et al. (2007).
Wavenumbers (IR, cm^{-1}): 916, 864, 803s, 751s, 719s, 588s, (443), 410.
Note: In the cited paper, Raman spectrum is given.
Wavenumbers (Raman, cm^{-1}): 922w, 868s, 774, 753s, 709, 693, 637, 517, 430, 395sh, 389w, 363, 330, 302w, 287, 251w, 230w, 218.

W31 Potassium ytterbium tungstate $KYb(WO_4)_2$

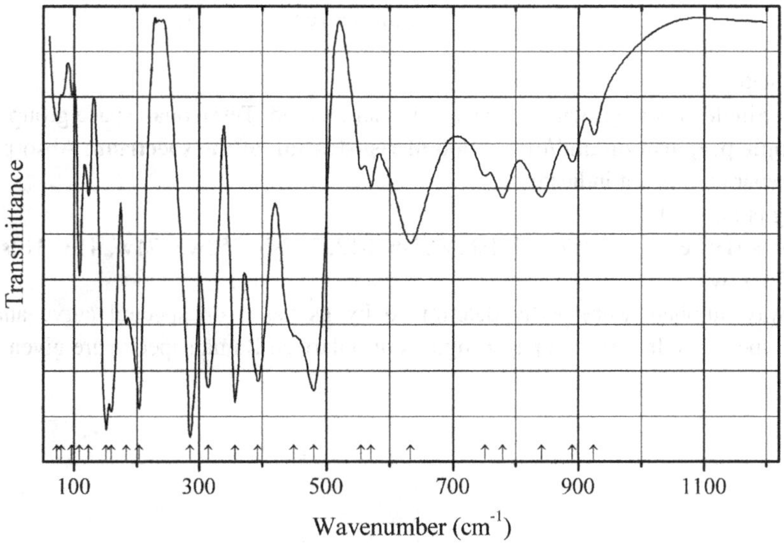

Origin: Synthetic.

Description: Crystal grown from the solution of K_2CO_3, WO_3, and Yb_2O_3 by the top-seeded solution growth method using $K_2W_2O_7$ as the solvent. Characterized by powder X-ray diffraction data. Monoclinic, space group *C2/c*, $a = 10.590(4)$, $b = 10.290(6)$, $c = 7.478(2)$ Å, $\beta = 130.70(2)°$, $V = 617.8(5)$ Å^3, $Z = 4$.

Kind of sample preparation and/or method of registration of the spectrum: KBr disc. Transmission. Conditions of far IR spectrum registration are not characterized.

Source: Zhao et al. (2008a).

Wavenumbers (cm^{-1}): 925, 891, 841, 779, 751, 633, 571, 555, 481s, 450sh, 393s, 356s, 314s, 285s, 204s, 183, 159s, 151s, 123, 108, 96w, 79sh, 72.

Note: The wavenumbers were partly determined by us based on spectral curve analysis of the published spectrum. In the cited paper, figures of Raman spectra with different scattering configurations are given.

W32 Strontium tungstate $SrWO_4$

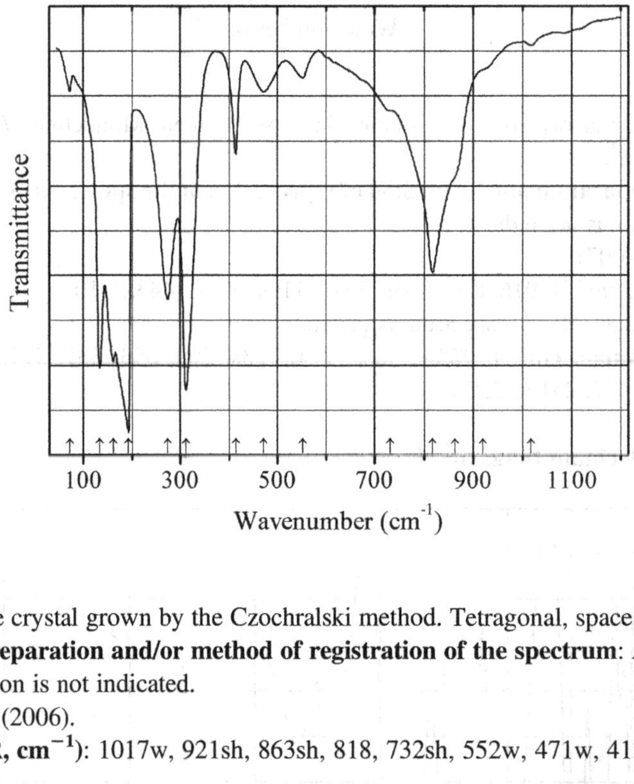

Origin: Synthetic.

Description: Single crystal grown by the Czochralski method. Tetragonal, space group $I4_1/a$, $Z = 2$.

Kind of sample preparation and/or method of registration of the spectrum: Absorption. Kind of sample preparation is not indicated.

Source: Ling et al. (2006).

Wavenumbers (IR, cm^{-1}): 1017w, 921sh, 863sh, 818, 732sh, 552w, 471w, 413, 312s, 274s, 193s, 162s, 135s, 73w.

Note: The wavenumbers were partly determined by us based on spectral curve analysis of the published spectrum. In the cited paper, figures of polarized Raman spectra are given.

W33 Huanzalaite $MgWO_4$

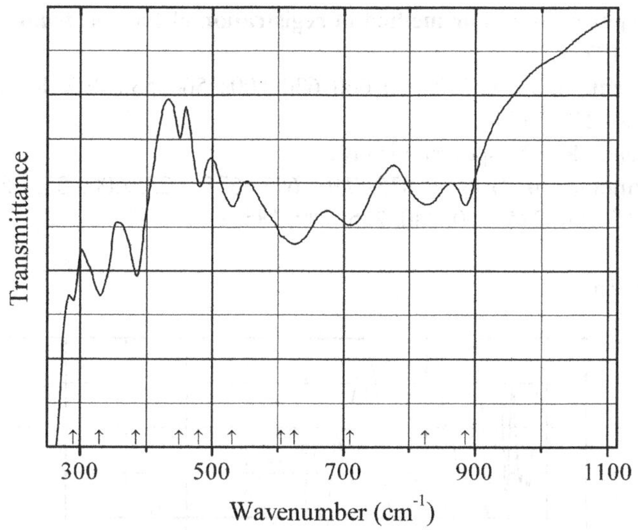

Origin: Synthetic.

Description: Prepared hydrothermally from sodium tungstate and magnesium nitrate at a temperature above 250 °C for 4 days. Characterized by powder X-ray diffraction data. Monoclinic, space group *P2/c*, $a = 4.687$, $b = 5.675$, $c = 4.928$ Å, $\beta = 90.71°$, $Z = 2$.

Kind of sample preparation and/or method of registration of the spectrum: KBr disc. Transmission.

Source: Günter and Amberg (1989).

Wavenumbers (cm^{-1}): 885, 825, 710, 625s, 605sh, 530, 480, 450w, 385s, 330s, 290.

W34 Yttrium tungstate Y_2WO_6

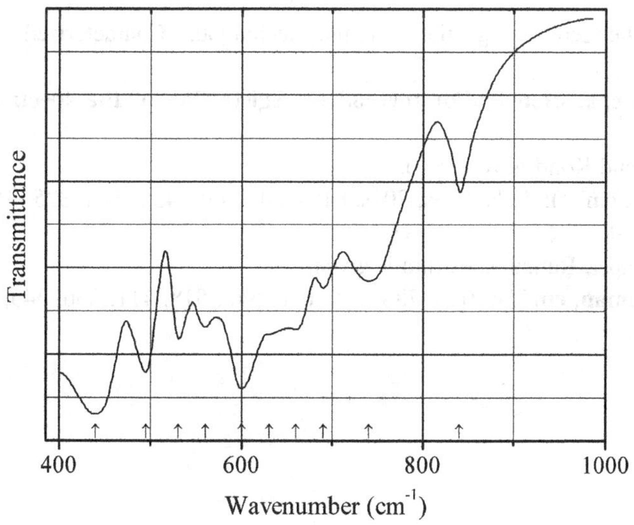

Origin: Synthetic.

Description: Characterized by powder X-ray diffraction data. Orthorhombic, space group *B*2*cb*.

Kind of sample preparation and/or method of registration of the spectrum: Transmission.

Source: Bode et al. (1973).

Wavenumbers (IR, cm^{-1}): 840w, 740, 690, 660, 630, 600s, 560, 530, 495, 440s, 390, 350, 335, 310, 290, 270, 255, 240, 230, 215.

Note: In the cited paper, Raman spectrum is given.

Wavenumbers (Raman, cm^{-1}): 935, 835, 710, 695, 675, 625, 600, 555, 525, 505, 450, 430, 400, 370, 345, 315, 290, 275, 260, 240, 225, 200, 185, 145.

W35 Sanmartinite $ZnWO_4$

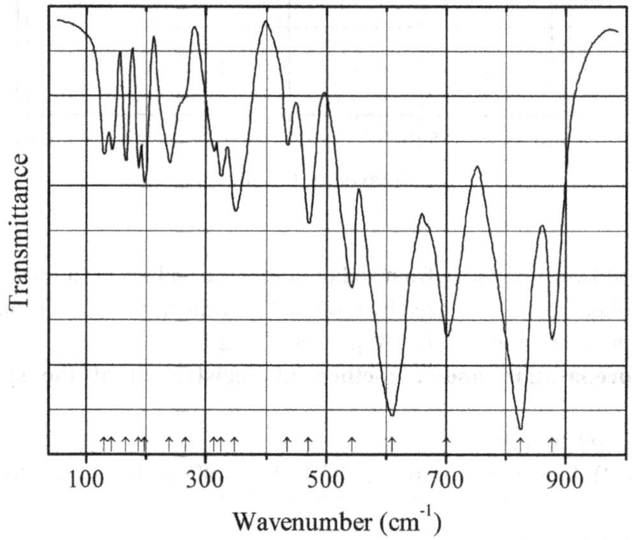

Origin: Synthetic.

Description: Synthesized using the ceramic technique. Characterized by powder X-ray diffraction data.

Kind of sample preparation and/or method of registration of the spectrum: Nujol mull (?). Transmission.

Source: Fomichev and Kondratov (1994).

Wavenumbers (IR, cm^{-1}): 877s, 825s, 701s, 610s, 542, 470, 435, 348, 325, 314, 267sh, 240, 197, 187, 166, 143, 130.

Note: In the cited paper, Raman spectrum is given.

Wavenumbers (Raman, cm^{-1}): 910s, 788, 710, 676, 548, 518, 411, 356, 345, 316, 276, 199, 168, 150, 127, 94s, 59.

Re1 Uranyl perrhenate hydrate $(UO_2)_2(ReO_4)_4 \cdot 3H_2O$

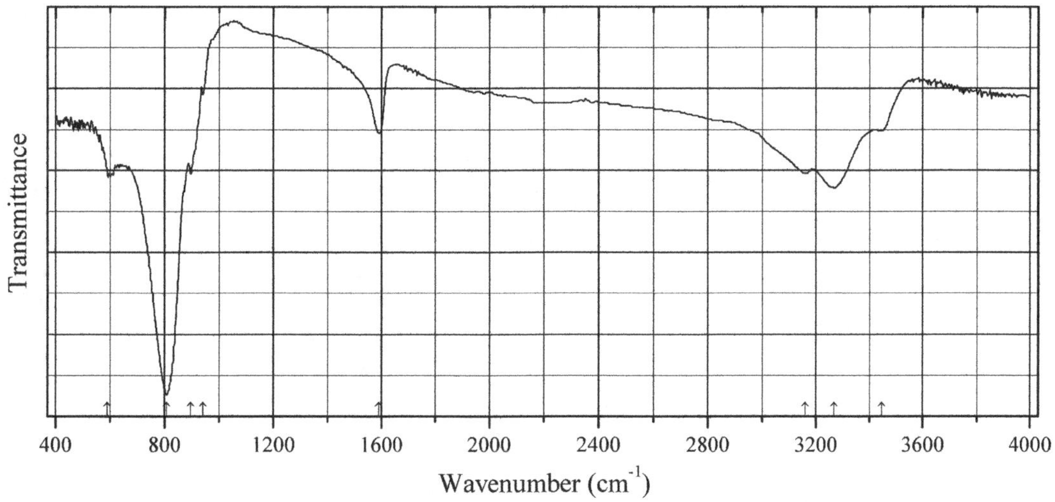

Origin: Synthetic.

Description: Crystals obtained by evaporation of a solution of equimolar amounts of UO_3 and Re_2O_7. Characterized by qualitative energy dispersive analysis. The crystal structure is solved. Triclinic, space group *P*-1, $a = 5.2771(7)$, $b = 13.100(2)$, $c = 15.476(2)$ Å, $\alpha = 107.180(2)°$, $\beta = 99.131(3)°$, $\gamma = 94.114(2)°$, $V = 1001.12$ Å^3, $Z = 2$. $D_{calc} = 5.291$ g/cm^3. The structure contains complex chains of uranyl pentagonal bipyramids bridging perrhenate groups via common vertices.

Kind of sample preparation and/or method of registration of the spectrum: Attenuated total reflection of a powdered sample.

Source: Karimova and Burns (2007).

Wavenumbers (cm^{-1}): 3446w, 3267, 3159, 1590, 941w, 897, 807s, 590.

Note: The wavenumbers were partly determined by us based on spectral curve analysis of the published spectrum.

9783030268053V01